ASTRONOMY AND ASTROPHYSICS ABSTRACTS

A Publication of the Astronomisches Rechen-Institut Heidelberg
Member of the Abstracting Board of the International
Council of Scientific Unions

Volume 6
Literature 1971, Part 2

Edited by
S. Böhme · W. Fricke · U. Güntzel-Lingner
F. Henn · D. Krahn · U. Scheffer · G. Zech

Springer-Verlag Berlin Heidelberg GmbH 1972

Astronomisches Rechen-Institut
Heidelberg
Director: Prof. Dr. W. Fricke

Astronomy and Astrophysics Abstracts
Editor-in-Chief: F. Henn

Astronomy and Astrophysics Abstracts
is prepared under the auspices
of the International Astronomical Union

ISBN 978-3-662-12280-8   ISBN 978-3-662-12278-5 (eBook)
DOI 10.1007/978-3-662-12278-5

© by Springer-Verlag Berlin Heidelberg 1972
Originally published by Springer-Verlag Berlin Heidelberg New York in 1972
Softcover reprint of the hardcover 1st edition 1972

Library of Congress Catalog Card Number 72-104650.

Preface

Astronomy and Astrophysics Abstracts, which has appeared in semi-annual volumes since 1969, is devoted to the recording, summarizing and indexing of astronomical publications throughout the world. It is prepared under the auspices of the International Astronomical Union (according to a resolution adopted at the 14th General Assembly in 1970).

Astronomy and Astrophysics Abstracts aims to present a comprehensive documentation of literature in all fields of astronomy and astrophysics. Every effort will be made to ensure that the average time interval between the date of receipt of the original literature and publication of the abstracts will not exceed eight months. This time interval is near to that achieved by monthly abstracting journals, compared to which our system of accumulating abstracts for about six months offers the advantage of greater convenience for the user.

Volume 6 contains literature published in 1971 and received before March 15, 1972; some older literature which was received late and which is not recorded in earlier volumes is also included.

The authors of papers who have sent us abstracts on request have effectively contributed to the success of our service. We should like to express our gratitude to them. We acknowledge with thanks contributions to this volume by Dr. J. Bouska, who surveyed journals and publications in the Czech language and supplied us with abstracts in English, and by the Commonwealth Scientific and Industrial Research Organization (C.S.I.R.O.), Sydney, for providing titles and abstracts of papers on radio astronomy.

We also extend our warmest thanks to Mrs Monika Betz, Miss Helga Ballmann, and Mrs Karola Gudé, who typed the text of this volume on IBM 72 Composers and compiled the pages from abstract slips in a perfect form for offset reproduction, to Miss Gisela Nollert, for punching all the material for author index and subject index, and to Mr Hartmut Jahreiß and Mr Heiner Schwan for helping with the layout of the text.

Heidelberg, April 1972

Siegfried Böhme
Walter Fricke
Ulrich Güntzel-Lingner
Frieda Henn
Dietlinde Krahn
Ute Scheffer
Gert Zech

Contents

Planetary System

Stars

Interstellar Matter, Gaseous Nebulae, Planetary Nebulae

Radio Sources, Quasars, Pulsars, X Ray-, Gamma Ray-Sources, Cosmic Radiation

Stellar Systems

Introduction

Astronomical bibliographies

Astronomy and Astrophysics Abstracts begins documentation and abstracting as from the year 1969. For information on astronomical literature before this date consultation of one of the following bibliographies is suggested:
(1) J. J. de Lalande, Bibliographie Astronomique, Paris 1803 (this work covers the time from 480 B. C. to the year 1803, VIII + 966 pages).
(2) J. C. Houzeau, A. Lancaster, Bibliographie générale de l'astronomie, Volume I (in two parts), Bruxelles 1882, 1887, Volume II, Bruxelles 1889. The complete title of Volume II is "Bibliographie générale de l'astronomie ou catalogue méthodique des ouvrages, des mémoires et des observations astronomiques, publiés depuis l'origine de l'imprimerie jusqu'en 1880". A new edition of these volumes was prepared by D. W. Dewhirst in 1964.
(3) Bibliography of Astronomy, 1881 - 1898. The literature of this period was recorded on standard slips by the Observatoire Royal de Belgique. From the material (some 52.000 items) a microfilm version was produced by University Microfilms Limited, Tylers Green, High Wycombe, Buckinghamshire, England, in 1970.
(4) Astronomischer Jahresbericht, 1899 gegründet von Walter Wislicenus, herausgegeben vom Astronomischen Rechen-Institut in Heidelberg (formerly in Berlin), Verlag W. de Gruyter, Berlin. For the period from 1899 to 1968 sixty-eight volumes were published, each of which, in general, covers the literature of one year.
(5) Bulletin Signalétique – Section Astronomie, Astrophysique, Physique du Globe. Published by Centre de Documentation du Centre National de la Recherche Scientifique, Paris. This publication is a continuation of "Bibliographie Mensuelle de l'Astronomie" founded in 1933 by the Société Astronomique de France. The publication is continued.
(6) Referativnyj Zhurnal. Founded in 1953 and published by Vsesoyuznyj Institut Nauchnoj i Tekhnicheskoj Informatsii, Akademiya Nauk, Moskva. The publication is continued.

Concept of Astronomy and Astrophysics Abstracts

This abstracting service aims to present a comprehensive documentation of the literature in all fields of astronomy and astrophysics. It appears in semi-annual volumes, two of which cover the literature of a calendar year. The half-yearly period of issue is regarded as an optimal period of time for summarizing papers into subject categories and for the presentation of abstracts as quickly as possible after the publication of the original literature. The time limits at which the documentation begins and ends for a volume are not sharply defined, except in the sense that all literature will be covered which was received by the editors within these limits.
Vol. 6 is devoted to the recording, summarizing and indexing of astronomical publications of the year 1971 received from August 15, 1971 to March 15, 1972; it also records a number of papers issued before 1971 but received within the given period of time.

The main characteristics of the concept of Astronomy and Astrophysics Abstracts may be summarized briefly.

(1) Titles of papers are given in the language of their authors whenever possible. If they are not in English but supplied with English translations they will be given in English. Abstracts are presented in English, French or German. Titles of papers in Russian are given in English.
(2) Authors' abstracts are used whenever possible. As a rule, popular articles were not abstracted; however their titles are usually given with the notation "Popular article".
(3) As a rule, each paper has been classified into one of 108 numbered subject categories and allocated a serial number within the category. In this way each item is numbered by six figures, the first three of which indicate the number of the category. Three further figures indicate the serial number within the category, which was allocated in the order of the receipt of the abstract. Reference to an abstract in Volume 1 is indicated by "01" before the number of the category; for example, 01.074.028, denotes Volume 1, category 074, abstract 028. Vol. 2 is indicated by "02", etc., Vol. 6 by "06".
A paper may have been classified into more than one category. Then its abstract has been allocated a number in one of the categories involved, and in the other category (or categories) the paper has been indicated by the title and a reference to the abstract number.
Papers whose authors are not named were treated like those with authors' names, with one exception: reports from correspondents of journals whose names were unknown were not numbered.

(4) There are categories which suggest the presentation of the material in subject groups. For instance, a subject group may be formed by all information received on the same solar eclipse, comet, nova, etc. The unsorted presentation of such material in a subject category would be inconvenient for the user, even if the individual comet, etc. were included in the subject index.

The following subject categories are subdivided into subject groups:
008 Observatories, Institutes. The publications of observatories and astronomical institutes are listed in alphabetical order of the towns of the institutions, each town forming a numbered subject group. For each publication a reference to an abstract number is made.
010 Societies, Associations, Organizations. The publications of each one form a subject group. The groups are presented in alphabetical order.
079 Solar eclipses. All publications related to one solar eclipse form a subject group.
103 Comets: Listed Objects. All publications related to the same comet form a numbered group.
124 Novae. All publications related to one nova form a subject group.
125 Supernovae. All publications related to one supernova form a subject group.

(5) Border fields of astronomy and astrophysics have been taken into account by presenting titles of papers occasionally without abstracts. The selection of papers for inclusion has been made according to the degree of relevance to astronomical research.

Introduction

Transliteration of the Russian alphabet

The transliteration of the Russian alphabet in use in Astronomy and Astrophysics Abstracts is presented here.

А	а	a	Р	р	r	
Б	б	b	С	с	s	
В	в	v	Т	т	t	
Г	г	g	У	у	u	
Д	д	d	Ф	ф	f	
Е	е	e	Х	х	kh	
Ё	ё	e	Ц	ц	ts	
Ж	ж	zh	Ч	ч	ch	
З	з	z	Ш	ш	sh	
И	и	i	Щ	щ	shch	
Й	й	j	Ъ	ъ	''	
К	к	k	Ы	ы	y	
Л	л	l	Ь	ь	'	
М	м	m	Э	э	eh	
Н	н	n	Ю	ю	yu	
О	о	o	Я	я	ya	
П	п	p				

This transliteration was recommended by the Abstracting Board of the International Council of Scientific Unions in 1969. It is essentially the same as the transliteration proposed by the Academy of Sciences, Moscow, and used by the Referativnyj Zhurnal (see Referativnyj Zhurnal, 51. Astronomiya, 1969 No. 1). It may be noted that the letters can be read and printed by usual data processing machines.
In the literature however the names of Russian authors can be found transliterated in different ways. We present the names in the form in which they are given in the references cited.

Sources of information

The majority of sources of information for this volume are given in section **001 Periodicals** and in section **008 Observatories, Institutes.** The term "periodical" has been used in its widest sense for publications in a sequence of undetermined duration, even if the intervals of appearance are not regular. Section 001 records 288 periodicals with their full titles and with abbreviations which are in use in Astronomy and Astrophysics Abstracts. It may be noted that the titles of the periodicals are given in their original languages, and that Russian titles have been transliterated applying the transliteration given above. Section 008 records 150 periodicals; these are publication series of observatories and astronomical institutes which have not been included in section 001. The abbreviations of the titles of the periodicals have been given so that in most cases they permit recognition of the full title without recourse to the key in section 001. The steadily growing number of periodicals makes it necessary to use more extensive abbreviations and to abandon the use of very condensed ones.
Other abstracting journals have been consulted in order to examine the degree of completeness of our service. Occasionally, in particular in Physics Abstracts, Referativnyj Zhurnal, and Bulletin Signalétique abstracts of papers were found which had not come to our attention. In such cases Astronomy and Astrophysics Abstracts cites these papers, but also gives reference to the abstracting service which acted as the source.

Classification into a scheme of subject categories

The subdivision of astronomy and its border fields into subject categories is facilitated by the fact that the astronomical objects appear to be particularly well suited for the formation of categories. Sun, moon, earth, planets, comets, and meteorites, the various kinds of stars, galaxies, radio sources, quasars, and pulsars etc. suggest natural subdivisions. It may be assumed that such subdivisions can be maintained for long periods of time. Experience shows, however, that progress in research may imply changes in the classification scheme, in particular, in fields where the expansion of knowledge is explosive.

A few explanatory remarks may be in order on some of the subject categories. Section 002 includes short news notes whose titles and authors are given, but the authors of the notes have not been included in the author index. In section 003 books on astronomy and astrophysics and its border fields are listed which came to our notice from August 1971 to March 1972. References to book reviews are given if the review appeared quickly.
For completeness of documentation, personal notes (section 006) and obituaries (section 007) are listed. In section 012 (Proceedings of Colloquia, Congresses, Meetings, and Symposia) the proceedings etc. are listed with titles and editors. The individual papers are classified into their corresponding subject categories.

Author index and subject index

The subject category and the serial number forming six figures for each abstract have been used as a means of reference in the author index and the subject index. These references are more precise than page references. They offer considerable advantages in indexing by means of data processing machines, and they are more convenient for the user.
The author index of this volume contains 7124 names. A complete reference comprises six figures, three for the subject category and three for the serial number within the category. In the case of more than one reference to abstracts in one category, the number of the category is given only once and not repeated in the immediately following references. The total number of papers (some do not give names of authors) recorded in this volume is about 6800.
We consider the subject index as only a first approximation to an optimal index covering all fields of astronomy and astrophysics and their border fields. Several iterative steps appear to be necessary until an index has been compiled for one of the subsequent volumes which may then serve as a kind of standard for the near future. The assigning of one or more key words to a paper is undoubtedly a difficult task. Some journals have started giving key words together with the titles of papers. These key words are chosen by the authors themselves and are in many cases identical with our designations of subject categories with no additional specification. In fact, in some cases it may be more useful to refer to a subject category as a whole than to an item number, in particular, if the total number of abstracts in a category is very small, and if more specific key words do not provide a proper description of the paper.
While each volume is scheduled to contain an author index and a subject index, the magnetic tapes containing the index information will be used to produce separate index volumes (authors and subjects) at intervals of a few years.
The text of the publication was typed on IBM 72 Composers in the editorial office, and it was given to the printer in a form ready for offset reproduction. The author index and the subject index were compiled and printed by means of electronic computer (Siemens 2002).

Abbreviations

AAS	American Astronomical Society	Geogr.	Geography, etc.
AAVSO	American Association of Variable Star Observers	Geophys.	Geophysics, etc.
		Ges.	Gesellschaft
Abh.	Abhandlungen	Glav.	Glavnyj (Main)
Abstr.	Abstract	Gos.	Gosudarstvennyj (State)
Abt.	Abteilung	HRD	Herzsprung-Russell diagram
Acad.	Academy, etc.	Hydrogr.	Hydrography, etc.
Accad.	Accademia	IAF	International Astronautical Federation
Adv.	Advances	IAU	International Astronomical Union
AG	Astronomische Gesellschaft	ICSU	International Council of Scientific Unions
AIAA	American Institute of Aeronautics and Astronautics	IEEE	Institute of Electrical and Electronics Engineers
AJB	Astronomischer Jahresbericht	Industr.	Industry, etc.
Akad.	Akademie	Inform.	Information
An.	Anales, etc.	Inst.	Institute, etc.
Ann.	Annals, etc.	Instn.	Institution
Arch.	Archiv, etc.	Ionosph.	Ionosphere, etc.
Ark.	Arkiv	Issled.	Issledovaniya (Research)
ASA	Astronomical Society of Australia	Ist.	Istituto
Asoc.	Asociación	Izv.	Izvestiya (News)
ASP	Astronomical Society of the Pacific	Jb.	Jahrbuch
Ass.	Association	JO	Journal des Observateurs
ASSA	Astronomical Society of Southern Africa	Journ.	Journal
Astrofis.	Astrofisica, etc.	Kl.	Klasse
Astrofiz.	Astrofizika, etc.	Lab.	Laboratory
Astron.	Astronomy, etc.	Mag.	Magazine
Astronaut.	Astronautics, etc.	Mat.	Matematica, etc.
Astrophys.	Astrophysics, etc.	Math.	Mathematics, etc.
ASV	Astronomical Society of Victoria	Mech.	Mechanics, etc.
ASWA	Astronomical Society of Western Australia	Med.	Mededelingen
Atmosph.	Atmosphere, etc.	Medd.	Meddelande, Meddelser
BA	Bulletin Astronomique	Mekhan.	Mekhanika, etc.
BAA	British Astronomical Association	Mém.	Mémoires
BAN	Bulletin of the Astronomical Institutes of the Netherlands	Mem.	Memoirs, Memorandum, etc.
		Meteorol.	Meteorology, etc.
Ber.	Berichte	MIT	Massachusetts Institute of Technology
BIH	Bureau International de l'Heure (Paris)	Mitt.	Mitteilungen
Bol.	Boletin	MVS Sonneberg	Mitteilungen über Veränderliche Sterne, Sonneberg
Boll.	Bolletino		
Bull.	Bulletin	Nachr.	Nachrichten
Byull.	Byulleten' (Bulletin)	NASA	National Aeronautics and Space Administration
Circ.	Circular		
Cl.	Classe	Nat.	Naturwissenschaftlich, etc.
Coll.	Collection	Naut.	Nautics, etc.
Commun.	Communication	NBS	National Bureau of Standards
Comun.	Comunicazioni	NRAO	National Radio Astronomy Observatory (Green Bank)
Contr.	Contributions, etc.		
COSPAR	Committee on Space Research	NRL	Naval Research Laboratory (Washington)
C.S.I.R.O.	Commonwealth Scientific Industrial Research Organization	Obs.	Observatory, etc.
		OSA	Optical Society of America
Dep.	Department	Oss.	Osservatorio, Osservazioni, etc.
Diss.	Dissertation	Ped.	Pedagogika, etc. (Pedagogics)
Div.	Division	Phil.	Philosophical
Dokl.	Doklady (Reports)	Phys.	Physics, etc.
ESO	European Southern Observatory	Planet.	Planetary
ESRO	European Space Research Organization	Priklad.	Prikladnoj (Applied)
Fis.	Fisica, etc.	Proc.	Proceedings
Fiz.	Fizika, etc.	Progr.	Progress, etc.
Fys.	Fysica, etc.	Pubbl.	Pubblicazioni
Géod.	Géodésie, etc.	Publ.	Publications
Geod.	Geodesy, etc.	Rap.	Raportoj
Geofis.	Geofisica, etc.	RAS	Royal Astronomical Society
Geofiz.	Geofizika, etc.	RAS Canada	Royal Astronomical Society of Canada
Geofys.	Geofysik, etc.	Rech.	Recherches
Geol.	Geology, etc.	Rend.	Rendiconti

Abbreviations

Rep.	Report		Techn.	Technics, etc.
Repr.	Reprint		Tekhn.	Tekhnika, etc.
Res.	Research		Teor.	Teoreticheskij
Rev.	Review, etc.		Terr.	Terrestrial, etc.
Ric.	Ricerche		TH	Technische Hochschule
Roy.	Royal, etc.		Theor.	Theoretical
SAF	Société Astronomique de France		Tidssk.	Tidsskrift
SAI	Società Astronomica Italiana		Trans.	Transactions
SAO	Smithsonian Astrophysical Observatory		Trudy	Trudy (Publications)
SAS	Société Astronomique de Suisse		Tsentr.	Tsentral'nyj (Central)
Sci.	Science, etc.		Tsirk.	Tsirkulyar (Circular)
Sect.	Section		TU	Technical University
Ser.	Series, etc.		Uch. Zap.	Uchenye Zapiski (Treatise)
S. I. R.	Service International Rapide des Latitudes		Univ.	University, etc.
Sitz.-Ber.	Sitzungsberichte		URSI	Union Radio Scientifique Internationale
Soc.	Society		Verh.	Verhandlungen
Soobshch.	Soobshcheniya (Communications)		Veröff.	Veröffentlichungen
Sternw.	Sternwarte		Wet.	Wetenschappen
Stud. Cerc.	Studii şi Cercetari		Wiss.	Wissenschaften, etc.
Supl.	Suplemento		Zeitschr.	Zeitschrift
Suppl.	Supplement		ZfA	Zeitschrift für Astrophysik
SuW	Sterne und Weltraum		Zhurn.	Zhurnal (Journal)

Periodicals, Proceedings, Books, Activities

001 Periodicals

AAS Photo-Bull.
AAS (American Astronomical Society) Photo-Bulletin. Published by Eastman Kodak Company, Rochester, N.Y., for the Working Group on Photographic Materials.

Abh. Deutsch. Akad. Wiss. Berlin
Abhandlungen der Deutschen Akademie der Wissenschaften zu Berlin. Klasse für Mathematik, Physik und Technik. Publisher: Akademie-Verlag, Berlin.

Acad. Roy. Belgique, Bull. Cl. Sci.
Académie Royale de Belgique, Bulletin de la Classe des Sciences (Koninklijke Academie van België, Mededelingen van de Klasse der Wetenschappen). 5e Série. Palais des Académies, Bruxelles.

Acta Astron.
Acta Astronomica. Publisher: Komitet Astronomii, Polskiej Akademii Nauk, Warszawa - Kraków.

Acta Phys. Austriaca
Acta Physica Austriaca. Publisher: Springer-Verlag, Wien.

Acta Univ. Carolinae Math. Phys.
Acta Universitatis Carolinae, Mathematica et Physica. Administrace: Matematicko-fyzikálni fakulta University Karlovy, Praha.

Actas Acad. Nacional Cienc. Lima
Actas de la Academia Nacional de Ciencias Exactas, Fisicas y Naturales de Lima. Lima - Peru.

Adv. Astron. Astrophys.
Advances in Astronomy and Astrophysics. Publisher: Academic Press, New York – London.

AIAA Journ.
AIAA Journal. A Publication of the American Institute of Aeronautics and Astronautics devoted to Aerospace Research and Development. Published by the American Institute of Aeronautics and Astronautics, Easton, Pa.

Am. Scient.
American Scientist. Society of Sigma Xi, New Haven, Conn.

Ann. d'Astrophys.
Annales d'Astrophysique. Revue internationale bimestrielle publiée par le Centre National de la Recherche Scientifique et éditée par son Service d'Astrophysique, Paris. After Vol. 31 replaced by "Astronomy and Astrophysics".

Ann. Françaises Chronométrie Micromécanique
Annales Françaises de Chronométrie et de Micromécanique, publication annuelle de l'Observatoire de Besançon, du Centre Technique de l'Industrie Horlogère et de la Société Française de Chronométrie et de Micromécanique. Rédaction et administration: Observatoire de Besançon. Publiées avec le concours du Centre National de la Recherche Scientifique et des organismes corporatifs.

Ann. Géophys.
Annales de Géophysique. Revue Internationale trimestrielle, publiée par le Centre National de la Recherche Scientifique, Paris.

Ann. Obs. Astron. Météorol. Toulouse
Annales de l'Observatoire Astronomique et Météorologique de Toulouse. Publisher: Gauthier-Villars, Paris.

Ann. Physics
Annals of Physics. Publisher: Academic Press Inc., New York, N.Y.

Ann. Physik
Annalen der Physik. 7. Folge. Publisher: Johann Ambrosius Barth, Leipzig.

Ann. Physique
Annales de Physique. Publisher: Masson et Cie., Paris.

Ann. Soc. Sci. Bruxelles
Annales de la Société Scientifique de Bruxelles. Série I: Sciences Mathématiques, Astronomiques et Physiques. Published by Institut de Physique, Heverlé-Louvain.

Annual Rep. Astron. Inst. Greece
Annual Reports of the Astronomical Institutes of Greece. Published by the Greek National Committee for Astronomy. Academy of Athens, Research Center for Astronomy and Applied Mathematics.

Annual Rev. Astron. Astrophys.
Annual Review of Astronomy and Astrophysics. Publisher: Annual Reviews Inc., Palo Alto, California.

Ann. Univ.-Sternw. Wien
Annalen der Universitäts-Sternwarte Wien. In Kommission bei Ferd. Dümmlers Verlag, Bonn.

Anzeiger. Österreich. Akad. Wiss. Math.-Nat. Kl.
Anzeiger. Österreichische Akademie der Wissenschaften. Mathematisch-Naturwissenschaftliche Klasse. Publisher: Springer-Verlag, Wien.

Applied Optics
Applied Optics. A monthly publication of the Optical Society of America. Published for the Optical Society of America by the American Institute of Physics, Easton, Pa.

Arch. Sci. Genève
Archives des Sciences, éditées par la Société de Physique et d'Histoire Naturelle de Genève. Publisher: Imprimerie Kundig, Genève. Subscription address: Librairie Payot, Genève.

Ark. Astron.
Arkiv för Astronomi. Utgivet av Kungliga Svenska Vetens-

kapsakademien, Stockholm. Printed by Almqvist & Wiksell, Stockholm.

Ark. Geofys.
Arkiv för Geofysik. Kungliga Svenska Vetenskapsakademien, Stockholm.Printed by Almqvist & Wiksell, Stockholm.

Artificial Satellites
Artificial Satellites. Publication of Polish Scientific Institutions. Polish Academy of Sciences, National Committee of Geophysics and Geodesy, National Committee for Space Research, Warsaw. Publishing Office: Palac Kultury i Nauki, Warszawa.

Asoc. Argentina Astron. Bol.
Asociación Argentina de Astronomía. Boletin. Editor: Instituto Argentino de Radioastronomía, Provincia de Buenos Aires, Argentina. Printer: Talleres Gráficos "Renovación", La Plata, República Argentina.

Astrofizika
Astrofizika. Izdatel'stvo Akademii Nauk Armyanskoj SSR, Erevan. [An English translation is published in "Astrophysics".]

Astrofiz. Issled. Izv. Spets. Astrofiz. Obs.
Astrofizicheskie Issledovaniya. Izvestiya Spetsial'noj Astrofizicheskoj Observatorii. Akademiya Nauk SSSR. Publishers: Izdatel'stvo "Nauka", Leningradskoe Otdelenie, Leningrad.

Astron. Astrophys.
Astronomy and Astrophysics. A European Journal. Published by Springer-Verlag, Berlin – Heidelberg– New York.

Astron. Astrophys. Suppl. Ser.
Astronomy and Astrophysics. Supplement Series. A European Journal. Published by the Astronomical Institute Lausanne and Geneva Observatory, Switzerland, on behalf of the Board of Directors.

Astronaut. Acta
Astronautica Acta. An Archive Journal of the International Academy of Astronautics. Published by Pergamon Press, New York – Oxford.

Astronaut. Aeronaut.
Astronautics & Aeronautics. A Publication of the American Institute of Aeronautics and Astronautics. Published monthly by the American Institute of Aeronautics and Astronautics, Easton, Pennsylvania.

Astron. in der Schule
Astronomie in der Schule. Zeitschrift für die Hand des Astronomielehrers. Herausgegeben vom Verlag Volk und Wissen, Berlin. Redaktion: Sternwarte Bautzen.

Astron. Journ.
The Astronomical Journal. Published for the American Astronomical Society by the American Institute of Physics, New York, N. Y. Editorial Office: Department of Astronomy, Columbia University, New York, N. Y.

Astron. Nachr.
Astronomische Nachrichten. Publisher: Akademie-Verlag, Berlin.

Astron. Soc. Pacific Leaflet
Astronomical Society of the Pacific. Leaflet. Edited by the Astronomical Society of the Pacific, San Francisco, California.

Astron. Tidssk.
Astronomisk Tidsskrift. Edited by Astronomisk Selskab, København; Norsk Astronomisk Selskap, Oslo; Svenska Astronomiska Sällskapet, Stockholm. Printed by John Griegs Boktrykkeri, Bergen.

Astron. Tsirk.
Astronomicheskij Tsirkulyar, izdavaemyj Byuro Astronomicheskikh Soobshchenij Akademii Nauk SSSR. Tipografiya Astrosoveta AN SSSR, Moskva.

Astron. Vestn.
Astronomicheskij Vestnik. Publishers: Izdatel'stvo "Nauka", Moskva.

Astron. Zhurn. Akad. Nauk SSSR
Astronomicheskij Zhurnal. Akademiya Nauk SSSR. Publishers: Izdatel'stvo "Nauka", Moskva. [An English translation is published in "Soviet Astronomy AJ"].

Astrophysics
Astrophysics. The Faraday Press cover-to-cover translation of Astrofizika. The Faraday Press, Inc., New York, N. Y.

Astrophys. Journ.
The Astrophysical Journal. Published in collaboration with the American Astronomical Society by the University of Chicago Press, Chicago, Illinois.

Astrophys. Journ. Suppl. Ser.
The Astrophysical Journal. Supplement Series. Published in collaboration with the American Astronomical Society by the University of Chicago Press, Chicago, Illinois.

Astrophys. Letters
Astrophysical Letters. An International *EXPRESS* Journal. Published monthly by Gordon and Breach Science Publishers Ltd., New York – London – Paris.

Astrophys. Norvegica
Astrophysica Norvegica. Edited by The Institute of Theoretical Astrophysics, University of Oslo (Det Norske Videnskaps-Akademi i Oslo). Universitets-forlaget, Oslo.

Astrophys. Space Sci.
Astrophysics and Space Science. An International Journal of Cosmic Physics. Published by D. Reidel Publishing Company, Dordrecht – Holland.

Atti Accad. Nazionale Lincei. Mem.
Atti della Accademia Nazionale dei Lincei. Serie Ottava. Memorie. Classe di Scienze fisiche, matematiche e naturali. Sezione I: Matematica, Meccanica, Astronomia, Geodesia e Geofisica. Published by Accademia Nazionale dei Lincei, Roma.

Atti Accad. Nazionale Lincei. Rend.
Atti della Accademia Nazionale dei Lincei. Serie Ottava. Rendiconti. Classe di Scienze fisiche, matematiche e naturali. Published by Accademia Nazionale dei Lincei, Roma.

Australian Journ. Phys.
Australian Journal of Physics. Published by the Commonwealth Scientific and Industrial Research Organization, East Melbourne, Victoria.

Australian Journ. Phys. Astrophys. Suppl.
Australian Journal of Physics, Astrophysical Supplement.
Published by Commonwealth Scientific and Industrial
Research Organization, East Melbourne, Victoria.

BAV Rundbrief
BAV Rundbrief. Mitteilungsblatt der Berliner Arbeitsgemeinschaft für Veränderliche Sterne. Editor: BAV Berliner Arbeitsgemeinschaft für Veränderliche Sterne eV.,
Berlin.

Bol. Inst. Mat., Astron., Fis. Univ. Nacional Córdoba
Boletin del Instituto de Matematica, Astronomia y
Fisica, Universidad Nacional de Córdoba (R. A.). Direc-
ción General de Publicaciones, Córdoba (Argentina).

Bol. Liga Latinoamericana Astron.
Boletin de la Liga Latinoamericana de Astronomia. Publi-
cado por la Asociacion Argentina Amigos de la Astronomia, Buenos Aires, Argentina.

Boll. Geod. Sci. Affini
Bolletino di Geodesia e Scienze Affini. Pubblicazione
dell'Istituto Geografico Militare, Firenze.

Boundary-Layer Meteorology
Boundary-Layer Meteorology. An International Journal
of Physical and Biological Processes in the Atmospheric
Boundary Layer. Published by D. Reidel Publishing Company, Dordrecht–Holland.

British Astron. Ass. Circ.
British Astronomical Association, Circular. Editorial
Office: 97 Hawkswood Drive, Hailsham, Sussex.

Bull. American Astron. Soc.
Bulletin of the American Astronomical Society. Published
for the American Astronomical Society by the American
Institute of Physics Inc., New York, N. Y.

Bull. Astron. (BA)
Bulletin Astronomique. 3e Série. Publié par le Centre
National de la Recherche Scientifique, Paris. After Vol.
3 (1968) replaced by "Astronomy and Astrophysics".

Bull. Astron. Inst. Czechoslovakia (BAC)
Bulletin of the Astronomical Institutes of Czechoslovakia.
Published under the auspices of the Czechoslovak Acade-
my of Sciences by Academia, Praha. Editor: Astronomic-
al Institutes of the Czechoslovak Academy of Sciences,
Praha.

Bull. Astron. Inst. Netherlands (BAN)
Bulletin of the Astronomical Institutes of the Nether-
lands. Publisher: North-Holland Publishing Company,
Amsterdam. After Vol. 20 replaced by "Astronomy and
Astrophysics".

Bull. Astron. Inst. Netherlands, Suppl. Ser.
Bulletin of the Astronomical Institutes of the Nether-
lands. Supplement Series. Published by the Astronomical
Institutes. Replaced by "Astronomy and Astrophysics",
Supplement Series.

Bull. Géod.
Bulletin Géodésique, being the Journal of the International Association of Geodesy. Nouvelle Série. Publié
par le Bureau Central de l'Association Internationale
de Géodésie, Paris.

Bull. Geograph. Survey Inst.

Bulletin of the Geographical Survey Institute. Published
by the Geographical Survey Institute, Ministry of Construction, Tokyo, Japan.

Bull. Hor.
Bulletin Horaire du Bureau International de l'Heure (BIH).
Rédaction: BIH, Observatoire de Paris.

Bull. Mesures Ionosph.
Bulletin de Mesures Ionosphériques. Publié par le Centre
National d'Etudes des Télécommunications, Issy-les-Mou-
lineaux.

Bull. Obs. Astron. Beograd
Bulletin de l'Observatoire Astronomique de Béograd.
Editor: Observatoire Astronomique de Béograd. Printed
by Naucna delo, Béograd.

Bull. Sci. Yougoslavie
Bulletin Scientifique. Conseil des Academies des Sciences
et des Arts de la RSF de Yougoslavie. Section A: Sciences
Naturelles, Techniques et Médicales. Redaction et Administration: Opatička ul. 18/II, Zagreb (Yougoslavie).

Bull. Signal.
Bulletin Signalétique. Section 120: Astronomie, Physique
spatiale, Géophysique. Centre de Documentation du
Centre Nationale de la Recherche Scientifique, Paris.

Bull. Soc. Roy. Sci. Liège
Bulletin de la Société Royale des Sciences de Liège.
L'Université, Liège.

Byull. Abastuman. Astrofiz. Obs.
Abastumanskaya Astrofizicheskaya Observatoriya, Gora
Kanobili. Byulleten'. Akademiya Nauk Gruzinskoj SSR.
Publishers: Izdatel'stvo "Metsniereba", Tbilisi.

Byull. Stantsij Optichesk. Nablyud. Iskusstv. Sputnikov Zemli
Byulleten' Stantsij Opticheskogo Nablyudeniya Iskusstvennykh Sputnikov Zemli. Published by Astronomicheskij Sovet Akademii Nauk SSSR, Moskva.
Beginning with number 60 (1971) the title of the publication changed in Nablyudeniya Iskusstvennykh Nebesnykh Tel.

Canadian Journ. Phys.
Canadian Journal of Physics. Published by the National
Research Council of Canada, Ottawa. Printed in Canada
by the University of Toronto Press, Toronto, Ont.

Celestial Mechanics
Celestial Mechanics. An International Journal of Space
Dynamics. Publishers: D. Reidel Publishing Company,
Dordrecht–Holland.

Ciel et Terre
Ciel et Terre. Bulletin de la Société Belge d'Astronomie,
de Météorologie et de Physique du Globe. Administration: Avenue Circulaire, 3, Bruxelles. Printed by Imprimerie R. Louis, Bruxelles.

Circ. d'Information
Circulaire d'Information. Union Astronomique Internationale. Commission des Etoiles Doubles. Address: Observatoire de Meudon, Meudon, France.

Coelum
Coelum. Periodico bimestrale per la Divulgazione dell'
Astronomia. Editor: Osservatorio Astronomico Univer-
sitario di Bologna.

Comments Astrophys. Space Phys.
Comments on Astrophysics and Space Physics. A Journal of Critical Discussion of the Current Literature. Publishers: Gordon and Breach Science Publishers, Inc., New York – London.

Comptes Rendus Acad. Bulg. Sci.
Comptes Rendus de l'Académie bulgare des Sciences. (Doklady Bolgarskoj Akademii Nauk). Sofia.

Comptes Rendus Acad. Sci. Paris
Comptes Rendus hebdomadaires des Séances de l'Académie des Sciences, publié avec le concours du Centre National de la Recherche Scientifique. Imprimerie: Gauthier-Villars, Paris.

Contr. Atmosph. Phys.
Contributions to Atmospheric Physics – Beiträge zur Physik der Atmosphäre. Publisher: Friedrich Vieweg & Sohn, Braunschweig.

Cosmic Electrodynamics
Cosmic Electrodynamics. An International Journal devoted to Geophysical and Astrophysical Plasmas. Printed in The Netherlands by D. Reidel Publishing Company, Dordrecht–Holland.

COSPAR Inform. Bull.
COSPAR. Information Bulletin. Address: COSPAR Secretariat, Paris.

Deutsche Geod. Kommission Bayer. Akad. Wiss.
Deutsche Geodätische Kommission bei der Bayerischen Akademie der Wissenschaften. Reihe A: Höhere Geodäsie; Reihe B: Angewandte Geodäsie; Reihe C: Dissertationen; Reihe D: Tafelwerke; Reihe E: Geschichte und Entwicklung der Geodäsie. Published by Verlag der Bayerischen Akademie der Wissenschaften, München.

Documentat. Observateurs
Documentation des Observateurs. Rédaction: Station d'Astrophysique de Forcalquier.

Documentat. Observateurs Circ.
Documentation des Observateurs. Circulaire. Rédaction: Station d'Astrophysique de Forcalquier.

Dokl. Akad. Nauk
Doklady Akademii Nauk SSSR. Seriya Matematika, Fizika. Publishers: Izdatel'stvo "Nauka", Moskva.

Dunsink Obs. Publ.
Dunsink Observatory Publications. The Observatory of the School of Cosmic Physics, Dublin Institute for Advanced Studies, Dublin.

Earth Extraterr. Sci.
Earth and Extraterrestrial Sciences. Published by Gordon and Breach Science Publishers, London.

Earth Planet. Sci. Letters
Earth and Planetary Science Letters. A Letter Journal devoted to the Development in Time of the Earth and Planetary System. Publisher: North-Holland Publishing Company, Amsterdam.

El Universo
El Universo. Organo de la Sociedad Astronomica de Mexico, Mexico, D. F.

Endeavour
Endeavour. A review of the progress of science, published in four languages by Imperial Chemical Industries Limited, London.

ESO Bull.
European Southern Observatory, Bulletin. Edited by European Southern Observatory. Office of the Director: Hamburg.

Fortschritte Phys.
Fortschritte der Physik. Publisher: Akademie-Verlag, Berlin.

Gaz. Astron. Mém.
Gazette Astronomique. Mémoires van het Sterrenkundig Genootschap van Antwerpen, (de la Société d'Astronomie d'Anvers), Antwerpen. Printer: «De Voorzorg», A. Van Leuvenhaege, Antwerpen.

Geochim. Cosmochim. Acta
Geochimica et Cosmochimica Acta. Journal of the Geochemical Society. Publishing House: Pergamon Press, Ltd., Oxford.

Geodezja Kartografia
Geodezja i Kartografia. Komitet Geodezji Polskiej Akademii Nauk. Publisher: Państwowe Wydawnictwo Naukowe, Warszawa.

Geomagn. Aeronom.
Geomagnetizm i Aehronomiya. Akademiya Nauk SSSR. Izdatel'stvo "Nauka", Moskva [An English translation is published in "Geomagnetism and Aeronomy".]

Geophys. Journ.
The Geophysical Journal of the Royal Astronomical Society. Published for the Royal Astronomical Society by Blackwell Scientific Publications, Oxford – Edinburgh.

Gerlands Beiträge Geophys.
Gerlands Beiträge zur Geophysik. Publisher: Akademische Verlagsgesellschaft Geest & Portig K.-G., Leipzig.

Glasnik Mat.
Glasnik Matematicki. Published by the Society of Mathematicians and Physicists of the S. R. of Croatia. Publisher: Drustvo Matematicara i Fizicara S. R. Hrvatske, Zagreb.

Helvetica Phys. Acta
Helvetica Physica Acta. Schweizerische Physikalische Gesellschaft. Publisher: E. Birkhäuser, Basel.

Hemel en Dampkring
Maandblad van de Nederlandse Vereniging voor Weer-en Sterrenkunde en van de Vereniging voor Sterrenkunde, Meteorologie, Geophysica en Aanverwante Wetenschappen in Belgie. Publisher: Wolters-Noordhoff N. V., Groningen.

IAU Circ.
International Astronomical Union, Circular. Central Bureau for Astronomical Telegrams, Smithsonian Astrophysical Observatory, Cambridge, Mass.

IBM Journ. Res. Development
IBM Journal of Research and Development. Published bimonthly by International Business Machines Corporation, Armonk, New York.

Icarus
Icarus. International Journal of Solar System Studies. Publisher: Academic Press, New York – London.

ICSU Bull.
ICSU Bulletin. International Council of Scientific Unions. Secretariat: 7, Via Cornelio Celso, Rome, Italy.

IEEE Spectrum
IEEE Spectrum. Published monthly by the Institute of Electrical and Electronics Engineers, Inc., New York, N. Y.

Inform. Bull. Southern Hemisphere
Information Bulletin of the Southern Hemisphere. Editorial Office: Observatorio Astronómico, La Plata, Argentina.

Inform. Bull. Variable Stars
Commission 27 of the I.A.U. Information Bulletin on Variable Stars. Konkoly Observatory, Budapest.

Infrared Physics
An International Research Journal. Publisher: Pergamon Press Ltd., Oxford – London – New York.

International Journ. Theor. Phys.
International Journal of Theoretical Physics. Publisher: Plenum Publishing Company, Donington House, London.

Irish Astron. Journ.
The Irish Astronomical Journal. A Quarterly Publication under the auspices of the Observatories of Armagh and Dunsink. Subscription address: Managing Editor, Irish Astronomical Journal, Armagh Observatory, Northern Ireland.

Izv. Akad. Nauk Armyan. SSR
Izvestiya Akademii Nauk Armyanskoj SSR. Fizika. Publisher: Izdatel'stvo AN Armyanskoj SSR, Erevan.

Izv. Glav. Astron. Obs. Pulkovo
Izvestiya Glavnoj Astronomicheskoj Observatorii v Pulkove. Akademiya Nauk SSSR. Izdanie Glavnoj astronomicheskoj observatorii v Pulkove, Leningrad.

Izv. Komissii Fiz. Planet
Izvestiya Komissii po Fizike Planet. Akademiya Nauk SSSR. Astronomicheskij Sovet. Moskva.

Izv. Krymskoj Astrofiz. Obs.
Izvestiya Krymskoj Astrofizicheskoj Observatorii. Akademiya Nauk SSR. Publishers: Izdatel'stvo "Nauka", Moskva.

Jenaer Rundschau (Jena Review)
Jenaer Rundschau (Jena Review). Publisher: VEB Verlag Technik, Berlin.

JETP Letters
JETP Letters. A translation of JETP Pis'ma v Redaktsiyu of the Academy of Sciences in the USSR. Published semimonthly by the American Institute of Physics, Lancaster, Pennsylvania.

Journ. Astronaut. Sci.
The Journal of the Astronautical Sciences. Published by the American Astronautical Society Inc., Baltimore, Md.

Journ. Astron. Soc. Victoria
The Journal of the Astronomical Society of Victoria.

Printed by D. Buscombe Printers, Glen Waverley, Victoria.

Journ. Astron. Soc. Western Australia
The Journal of the Astronomical Society of Western Australia. Edited by the Astronomical Society of Western Australia, Perth, W. A.

Journ. Atmosph. Terr. Phys.
Journal of Atmospheric and Terrestrial Physics. Publishers: Pergamon Press, Oxford – London – New York.

Journ. British Astron. Ass.
Journal of the British Astronomical Association. Subscription address: British Astronomical Association, Burlington House, Piccadilly, London.

Journ. British Interplanet. Soc.
Journal of the British Interplanetary Society. Printed in Great Britain by Unwin Brothers Ltd., The Gresham Press, Old Woking, Surrey, and published by The British Interplanetary Society, London.

Journ. Fluid Mechanics
Journal of Fluid Mechanics. Published by Cambridge University Press, London – New York.

Journ. Geophys. Res.
Journal of Geophysical Research. An International Scientific Publication. Published three times a month by the American Geophysical Union, Washington, D. C. First section: Space Physics; Second section: Physics and chemistry of the solid earth, planetology, geodesy; Third section: Oceans and atmospheres.

Journ. History Astron.
Journal for the History of Astronomy. Published by Macdonald and Co. (Publishers) Ltd., London. Printed in Great Britain by W. Heffer and Sons Ltd., Cambridge.

Journ. Inst. Navigation
Journal of the Institute of Navigation. Published quarterly by the Institute of Navigation, London.

Journ. Math. Phys.
Journal of Mathematical Physics. Published by the American Institute of Physics, New York, N. Y.

Journ. Observateurs (JO)
Journal des Observateurs. Publié avec le concours de l'Université d'Aix-Marseille par le Centre National de la Recherche Scientifique, Paris. After Vol. 51 replaced by "Astronomy and Astrophysics".

Journ. Optical Soc. America
Journal of the Optical Society of America. Publisher: American Institute of Physics, New York.

Journ. Phys. A. General Phys.
Journal of Physics A. General Physics. Europhysics Journal. Published by the Institute of Physics and the Physical Society, London, England, in association with the American Institute of Physics, New York.

Journ. Physique
Journal de Physique. Publication de la Société Française de Physique, Paris.

Journ. Plasma Phys.
Journal of Plasma Physics. Publishers: Cambridge University Press, London.

Journ. Proc. Roy. Soc. New South Wales
Journal and Proceedings of the Royal Society of New South Wales. Published by the Society, Science House, Sydney.

Journ. Quant. Spectrosc. Radiat. Transfer
Journal of Quantitative Spectroscopy & Radiative Transfer. Publisher: Pergamon Press, Oxford – New York.

Journ. Roy. Astron. Soc. Canada
The Journal of the Royal Astronomical Society of Canada, devoted to the advancement of astronomy and allied sciences. Printed by the University of Toronto Press, Toronto, Ontario, Canada.

Kometn. Tsirk. *Kiev*
Kometnyj Tsirkulyar. Gruppa po Issledovaniyu Komet Astrosoveta i Mezhduvedomstvennyj Geofizicheskij Komitet, Akademii Nauk SSSR. Kievskij Universitet im. T. G. Shevchenko.

Komety i Meteory
Komety i Meteory. Akademiya Nauk Tadzhikskoj SSR. Astronomicheskij Sovet Akademii Nauk SSSR. Publishers: Izdatel'stvo "Donish", Dushanbe.

Kosmich. Issled.
Kosmicheskie Issledovaniya. Akademiya Nauk SSSR. Publishers: Izdatel'stvo "Nauka", Moskva.

Kozmos
Kozmos. Popular Astronomical Journal of the Slovak Central Observatory in Hurbanovo. Publisher: Slovenská ústredná hvezdáren v Hurbanove.

L'Astronomie
L'Astronomie et Bulletin de la Société Astronomique de France. Revue mensuelle. Rédaction: Société Astronomique de France, Paris.

L'Universo
L'Universo. Rivista dell'Instituto Geografico Militare. Direzione, Redazione e Amministrazione: Istituto Geografico Militare, Firenze.

Magnitnye Polya Solnech. Pyaten
Magnitnye Polya Solnechnykh Pyaten. (Supplements to Solnechnye Dannye. Byulleten' (*Solar Data*)). Publishers: Izdatel'stvo "Nauka", Leningrad.

Math. Rev.
Mathematical Reviews. Published by the American Mathematical Society, Providence, R. I.

Mem. Fac. Sci. Kyoto Univ.
Memoirs of the Faculty of Science, Kyoto University. Series of Physics, Astrophysics, Geophysics, and Chemistry. Printed by Yamashiro Printing Publishing Co. Ltd., Kamigyo, Kyoto.

Mem. Japan Astron. Study Ass.
Memoirs of the Japan Astronomical Study Association. Izumi 59, Yugawara-machi, Kanagawa-ken, Japan.

Mem. Roy. Astron. Soc.
Memoirs of the Royal Astronomical Society. Published for the Royal Astronomical Society by Blackwell Scientific Publications, Oxford – Edinburgh.

Mem. Soc. Astron. Italiana
Memorie della Società Astronomica Italiana. Nuova Se-

rie. Pubblicate sotto gli auspici del Consiglio Nazionale dell Ricerche. Publisher: Tipografia Baccini & Chiappi, Firenze.

Messtechnik
Messtechnik (Zeitschrift für Instrumentenkunde). Publishers: Verlag Friedrich Vieweg & Sohn GmbH, Braunschweig.

Meteoritics
Meteoritics. The Journal of the Meteoritical Society. Published quarterly by The Meteoritical Society and Arizona State University Bureau of Publications. Editorial address: Center for Meteorite Studies, The Arizona State University, Tempe, Arizona.

Meteoritika
Akademiya Nauk SSSR. Komitet po Meteoritam. Publishers: Izdatel'stvo "Nauka", Moskva.

Mitt. Astron. Ges.
Mitteilungen der Astronomischen Gesellschaft, Hamburg. Printed by G. Braun, GmbH, Karlsruhe.

Monatsber. Deutsch. Akad. Wiss. Berlin
Monatsberichte der Deutschen Akademie der Wissenschaften zu Berlin. Mitteilungen aus Mathematik, Naturwissenschaft, Medizin und Technik. Publisher: Akademie-Verlag, Berlin.

Monthly Notes Astron. Soc. Southern Africa
Monthly Notes of the Royal Astronomical Society of Southern Africa. Published by the Astronomical Society of Southern Africa, Royal Observatory, Cape Province, South Africa.

Monthly Notes International Polar Motion Service
Monthly Notes of the International Polar Motion Service. Published by the Central Bureau, International Latitude Observatory of Mizusawa, Mizusawa-shi, Iwate-ken, Japan.

Monthly Notices Roy. Astron. Soc.
Monthly Notices of the Royal Astronomical Society. Published for the Royal Astronomical Society by Blackwell Scientific Publications, Oxford – Edinburgh.

Moon
The Moon. An International Journal of Lunar Studies. Publisher: D. Reidel Publishing Company, Dordrecht – Holland.

MVS Sonneberg
Mitteilungen über Veränderliche Sterne. Edited by Sternwarte Sonneberg (Zentralinstitut für Astrophysik, Bereich Sternphysik) der Deutschen Akademie der Wissenschaften zu Berlin.

Nablyud. Iskusstv. Nebesn. Tel
Nablyudeniya Iskusstvennykh Nebesnykh Tel. Published by Astronomicheskij Sovet Akademii Nauk SSSR, Moskva.

Nachr. Akad. Wiss. Göttingen
Nachrichten der Akademie der Wissenschaften in Göttingen. II. Mathematisch-Physikalische Klasse. Vandenhoeck & Ruprecht, Göttingen.

Nachr. Karten-, Vermessungswesen
Nachrichten aus dem Karten- und Vermessungswesen. Editor: Institut für Angewandte Geodäsie (Abt. II des

Deutschen Geodätischen Forschungsinstituts). Published by Verlag des Instituts für Angewandte Geodäsie, Frankfurt a. M.

Nature
Nature. Editorial and Publishing Offices: Macmillan Journals Limited, 4 Little Essex Street, London; 711 National Press Building, Washington, D. C.

Nature, Phys. Sci.
Nature, Physical Science. Editorial and Publishing Offices: Macmillan Journals Limited, London – Washington.

Naturwissenschaften
Die Naturwissenschaften. Publisher: Springer-Verlag, Berlin – Heidelberg – New York.

Nauchn. Informatsii
Nauchnye Informatsii. Astronomicheskij Sovet Akademii Nauk SSSR, Moskva.

Numerische Math.
Numerische Mathematik. Publisher: Springer-Verlag, Berlin – Heidelberg – New York.

Nuovo Cimento
Il Nuovo Cimento. Rivista Internazionale e Organo della Società Italiana di Fisica, Series A, B. Publisher: Nicola Zanichelli, Editore, Bologna.

Nuovo Cimento Lettere
Lettere al Nuovo Cimento, a Cura della Società Italiana di Fisica. Editrice Compositori, Bologna.

Nuovo Cimento Rivista
Rivista del Nuovo Cimento a cura della Società Italiana di Fisica. Editrice Compositori, Bologna.

Nuovo Cimento Suppl.
Supplemento al Nuovo Cimento. Publisher: Nicola Zanichelli, Editore, Bologna.

Observations Artificial Earth Satellites
Observations of Artificial Satellites of the Earth (Nablyudeniya Iskusstvennykh Sputnikov Zemli). Magyar Tudományos Akadémia Csillagvizsgáló Intézete, Budapest.

Observatory
The Observatory. A Review of Astronomy. Publishers: The Editors of "The Observatory", Royal Greenwich Observatory, Herstmonceaux Castle, Hailsham, Sussex, England.

Optik
Optik. Zeitschrift für das gesamte Gebiet der Licht- und Elektronenoptik. Publishers: Wissenschaftliche Verlagsgesellschaft mbH., Stuttgart.

Orion Schaffhausen
Orion. Zeitschrift der Schweizerischen Astronomischen Gesellschaft (SAG). Bulletin de la Société Astronomique de Suisse (SAS). Administration: Generalsekretariat der SAG, Schaffhausen.

Österreich. Zeitschr. Vermessungswesen
Österreichische Zeitschrift für Vermessungswesen. Editor and Publisher: Österreichischer Verein für Vermessungswesen, Wien.

Peremennye Zvezdy, Prilozhenie
Peremennye Zvezdy, Prilozhenie (The Variable Stars, Supplement). Astronomicheskij Sovet Akademii Nauk SSSR, Moskva.

Phil. Mag.
The Philosophical Magazine. A Journal of Theoretical, Experimental and Applied Physics. Eighth Series. Publisher: Taylor & Francis, Ltd., London.

Phil. Trans. Roy. Soc. London
Philosophical Transactions of the Royal Society of London. Series A, Mathematical and Physical Sciences. Published by the Royal Society, London.

Phys. Abstr.
Physics Abstracts. Science Abstracts, Series A. An INSPEC Publication, published by The Institution of Electrical Engineers, London.

Phys. Ber.
Physikalische Berichte. Herausgegeben von der Deutschen Physikalischen Gesellschaft e. V.und von der Deutschen Akademie der Wissenschaften zu Berlin. Friedrich Vieweg & Sohn, Braunschweig.

Phys. Blätter
Physikalische Blätter. Physik-Verlag, Mosbach/Baden.

Phys. Bull.
Physics Bulletin. Published by the Institute of Physics and the Physical Society, London, England.

Phys. Earth Planet. Interiors
Physics of the Earth and Planetary Interiors. A journal devoted to observational and experimental studies of the Earth and Planetary interiors and their theoretical interpretation by the physical sciences. Publisher: North-Holland Publishing Company, Amsterdam, Netherlands.

Phys. Fluids
The Physics of Fluids. Published by the American Institute of Physics, New York, N.Y.

Phys. Letters
Physics Letters. Volumes A and B. Publisher: North-Holland Publishing Company, Amsterdam.

Phys. Rev. A
Physical Review A, General Physics. Published for the American Physical Society by the American Institute of Physics, Lancaster, Pa., and New York, N.Y.

Phys. Rev. B
Physical Review B, Solid State. Published for the American Physical Society by the American Institute of Physics, Lancaster, Pa., and New York, N. Y.

Phys. Rev. C
Physical Review C, Nuclear Physics. Published for the American Physical Society by the American Institute of Physics, Lancaster, Pa., and New York, N.Y.

Phys. Rev. D
Physical Review D, Particles and Fields. Published for the American Physical Society by the American Institute of Physics, Lancaster, Pa., and New York, N.Y.

Phys. Rev. Letters
Physical Review Letters. Published weekly by The Amer-

ican Physical Society, New York, N. Y.

Phys. Today
Physics Today. Published by the American Institute of Physics, New York, N.Y.

Physica
Physica. Publishers: North-Holland Publishing Company, Amsterdam, The Netherlands, on request of the Foundation "Physica", Utrecht.

Planet. Space Sci.
Planetary and Space Science. Pergamon Press, Oxford – London – New York.

Plasma Physics
Plasma Physics. Publisher: Pergamon Press, Oxford, England.

Pokroky
Pokroky matematiky, fyziky a astronomie. Editor: Jednota čs. matematiků a fyziků. Publisher: Academia, Praha.

Postępy Astron.
Postępy Astronomii. Czasopismo Poświecone Upowszechnianiu Wiedzy Astronomicznej. Polskie Towarzystwo Astronomiczne, Warszawa. Printed in Poland by Pánstwowe Wydawnictwo Naukowe, Lódź.

Priroda
Priroda. Publishers: Izdatel'stvo "Nauka", Moskva.

Proc. Astron. Soc. Australia
Proceedings of the Astronomical Society of Australia. Published for the Society by Sydney University Press, Sydney.

Proc. Cambridge Phil. Soc.
Proceedings of the Cambridge Philosophical Society (Mathematical and Physical Sciences). Publishers: Cambridge University Press, London.

Proc. IEEE
Proceedings of the IEEE. Published monthly by the Institute of Electrical and Electronics Engineers, Inc., New York, N. Y.

Proc. Koninkl. Nederl. Akad. Wet.
Koninklijke Nederlandse Akademie van Wetenschappen. Proceedings. Series B, Physical Sciences. Publishers: North-Holland Publishing Company, Amsterdam.

Proc. National Acad. Sci. U.S.A.
Proceedings of the National Academy of Sciences of the United States of America. Published monthly by the National Academy of Sciences, Washington, D.C.

Proc. Roy. Irish Acad.
Proceedings of the Royal Irish Academy, Section A: Mathematical, Astronomical and Physical Science. Published by the Royal Irish Academy, Dublin.

Proc. Roy. Soc. London
Proceedings of the Royal Society of London. Series A: Mathematical and Physical Sciences. Published by the Royal Society, London.

Progr. Theor. Phys. Japan
Progress of Theoretical Physics. Published for the Research Institute for Fundamental Physics and the Physic-

al Society of Japan. Publication Office: Progress of Theoretical Physics, Yukawa Hall, Kyoto University, Kyoto, Japan.

Progr. Theor. Phys. Suppl.
Supplement of the Progress of Theoretical Physics. Published for the Research Institute for Fundamental Physics and The Physical Society of Japan. Publication Office: Progress of Theoretical Physics, Yukawa Hall, Kyoto University, Kyoto, Japan.

PTB Mitt.
PTB Mitteilungen. Amts- und Mitteilungsblatt der Physikalisch-Technischen Bundesanstalt, Braunschweig – Berlin.

Publ. Astron. Soc. Japan
Publications of the Astronomical Society of Japan. Published by the Astronomical Society of Japan. Office of the Society: Tokyo Astronomical Observatory, Mitaka, Tokyo. Agent: Maruzen Co. Ltd. (Export Department), Nihonbashi, Tokyo, Japan.

Publ. Astron. Soc. Pacific
Publications of the Astronomical Society of the Pacific. Published in Provo, Utah, by the Astronomical Society of the Pacific, San Francisco, California. Printed by Brigham Young University Press, Provo, Utah.

Publ. Roy. Obs. Edinburgh
Publications of the Royal Observatory, Edinburgh. Published by The Royal Observatory, Edinburgh, Scotland.

Publ. Tartu Astrofiz. Obs.
W. Struve nimelise Tartu Astrofüüsika Observatooriumi, Publikatsioonid. Eesti NSV Teaduste Akadeemia, Tartu.

Quarterly Journ. Roy. Astron. Soc.
Quarterly Journal of the Royal Astronomical Society. Published for the Royal Astronomical Society by Blackwell Scientific Publications, Oxford.

Radio Sci.
Radio Science. Published by the American Geophysical Union, Richmond, Virginia.

Referativ. Zhurn. 51. Astron.
Referativnyj Zhurnal. 51. Astronomiya. Vsesoyuznyj Institut Nachnoj i Tekhnicheskoj Informatsii. Moskva.

Referativ. Zhurn. 52. Geod. i Aehros"emka
Referativnyj Zhurnal. 52. Geodeziya i Aehros"emka. Vsesoyuznyj Institut Nauchnoj i Tekhnicheskoj Informatsii. Moskva.

Referativ. Zhurn. 62. Issled. kosm. prostranstva
Referativnyj Zhurnal. 62. Issledovanie Kosmicheskogo Prostranstva. Vsesoyuznyj Institut Nauchnoj i Tekhnicheskoj Informatsii. Moskva.

Rep. Progr. Phys.
Reports on Progress in Physics. Published by The Institute of Physics and the Physical Society, London.

Rev. Geophys. Space Phys.
Reviews of Geophysics and Space Physics (formerly Reviews of Geophysics). Published by the American Geophysical Union, Richmond, Virginia.

Revista Astron.
Revista Astronomica. Organo de la Asociación Argentina Amigos de la Astronomia, Buenos Aires.

Rev. Modern Phys.
Reviews of Modern Physics. Published for The American Physical Society by the American Institute of Physics, Lancaster, Pa., and New York, N.Y.

Rev. Sci. Instruments
Reviews of Scientific Instruments. Published by the American Institute of Physics, Lancaster, Pa., and New York, N.Y.

Rezul'taty Nablyud. Sovet. Iskusstv. Sputnikov Zemli
Rezul'taty Nablyudenij Sovetskikh Iskusstvennykh Sputnikov Zemli. Published by Astronomicheskij Sovet Akademii Nauk SSSR, Moskva.

Ric. Astron.
Ricerche Astronomiche. Specola Vaticana, Città del Vaticano.

Ric. Sci.
La Ricerca Scientifica. Serie Seconda. Rivista del Consiglio Nazionale delle Ricerche. Consiglio Nazionale delle Ricerche, Roma.

Ric. Spettrosc.
Ricerche Spettroscopiche. Laboratorio Astrofisico della Specola Vaticana. Specola Vaticana, Città del Vaticano.

Říše hvězd
Říše hvězd. Czechoslovak popular astronomical journal. Publisher: Orbis, Praha.

Roy. Astron. Soc. New Zealand Circ.
Royal Astronomical Society of New Zealand, Variable Star Section, Circular. Publication Office: Greerton, Tauranga, New Zealand.

Roy. Astron. Soc. New Zealand Variable Star Sect. Repr.
Royal Astronomical Society of New Zealand, Variable Star Section, Reprint. Publication Office: Greerton, Tauranga, New Zealand.

Rumanian Sci. Abstr.
Rumanian Scientific Abstracts. Natural Sciences. Publishers: The Scientific Documentation Centre of the Academy of the Socialist Republic of Romania, Bucureşti.

Sci. American
Scientific American. Published monthly by Scientific American, Inc., New York, N.Y.

Science
Science. American Association for the Advancement of Science, Washington, D.C.

Sci. Progr. Découverte
Science Progrès Découverte (formerly Science Progrès, La Nature). Revue publiée avec la participation du Palais de la Découverte. Published by Dunod, Editeur, Paris. Imprimerie Bayeusaine, Bayeux.

Sci. Rep. Tôhoku Univ.
The Science Reports of the Tôhuku University. First Series (Physics, Chemistry, Astronomy). Published by the Faculty of Science, Tôhoku University, Sendai, Japan.

Sitz.-Ber. Bayer. Akad. Wiss.
Bayerische Akademie der Wissenschaften. Mathematisch-Naturwissenschaftliche Klasse. Sitzungsberichte. Publisher: Verlag der Bayerischen Akademie der Wissenschaften, München.

Sitz.-Ber. Deutsch. Akad. Wiss. Berlin
Sitzungsberichte der Deutschen Akademie der Wissenschaften zu Berlin. Klasse für Mathematik, Physik und Technik. Publisher: Akademie-Verlag, Berlin.

Sitz.-Ber. Heidelberger Akad. Wiss.
Sitzungsberichte der Heidelberger Akademie der Wissenschaften. Mathematisch-Naturwissenschaftliche Klasse. Publisher: Springer-Verlag, Heidelberg.

Sitz.-Ber. Österreich. Akad. Wiss.
Sitzungsberichte. Österreichische Akademie der Wissenschaften. Mathematisch-Naturwissenschaftliche Klasse. Abteilung II: Mathematik, Astronomie, Meteorologie und Technik. Publisher: Springer-Verlag, Wien.

Sky Telescope
Sky and Telescope. Published by Sky Publishing Corporation, Cambridge, Mass.

Smithsonian Contr. Astrophys.
Smithsonian Contributions to Astrophysics. Smithsonian Institution Astrophysical Observatory, Cambridge, Mass. Printed by Smithsonian Institution Press, City of Washington. For sale by the Superintendent of Documents, U. S. Government Printing Office, Washington, D. C.

Smithsonian Year
Smithsonian Year. Annual Report of the Smithsonian Institution, including the financial report of the Executive Committee of the Boards of Regents. Published by the Smithsonian Institution, Washington, D.C.

Solar Physics
Solar Physics. A Journal for Solar Research and the Study of Solar Terrestrial Physics. Publishers: D. Reidel Publishing Company, Dordrecht—Holland.

Solnechnye Dannye Byull.
Solnechnye Dannye. Byulleten'. *(Solar Data).* Publishers: Izdatel'stvo "Nauka", Leningradskoe Otdelenie, Leningrad.

Soobshch. Byurakan. Obs.
Soobshcheniya Byurakanskoj Observatorii. Akademiya Nauk Armyanskoj SSR, Erevan.

Soobshch. Gos. Astron. Inst. Shternberg
Soobshcheniya Gosudarstvennogo Astronomicheskogo Instituta im P.K. Shternberga. Publishers: Izdatel'stvo Moskovskogo Universiteta, Moskva.

Southern Stars
Southern Stars. The Journal of the Royal Astronomical Society of New Zealand (Inc.). Address of the Society: P.O. Box 3181, Wellington C1, New Zealand.

Soviet Astron. AJ
Soviet Astronomy AJ. A translation of the Astronomical Journal of the Academy of Sciences of the USSR. Published by the American Institute of Physics, Inc., New York, N.Y.

Spaceflight
Spaceflight. A Publication of the British Interplanetary

Society. Printed by Eyre & Spottiswoode Limited at Grosvenor Press, Portsmouth, and published by the British Interplanetary Society, London.

Space Science Rev.
Space Science Reviews. Publishers: D. Reidel Publishing Company, Dordrecht—Holland.

Springer Tracts Modern Phys.
Springer Tracts in Modern Physics. (Ergebnisse der exakten Naturwissenschaften). Springer-Verlag, Berlin—Heidelberg—New York.

Sterne
Die Sterne. Zeitschrift für alle Gebiete der Himmelskunde. Johann Ambrosius Barth, Leipzig.

Sternenbote
Sternenbote. Monatsschrift für Österreichs Amateurastronomen. Publisher: Astronomisches Büro, Hermann Mucke, Wien.

Stockholms Obs. Ann.
Stockholms Observatoriums Annaler. Printed by Almquist & Wiksell, Stockholm.

Strolling Astronomer
The Strolling Astronomer. The Journal of The Association of Lunar and Planetary Observers, Publication Office: The Strolling Astronomer, Box 3 AZ, University Park, New Mexico.

Stud. Cerc. Astron.
Studii şi Cercetări de Astronomie. Editura Academiei Republicii Socialiste România. Editorial Office: Observatorul Astronomic, Bucureşti.

Stud. Geophys. Geod.
Studia geophysica et geodaetica. Published for the Geophysical Institute of the Czechoslovak Academy of Sciences by Academia, Praha.

Stud. Soc. Sci. Torunensis
Studia Societatis Scientiarum Torunensis, Toruń – Polonia. Sectio F (Astronomia).

Stud. Univ. Babeş-Bolyai
Studia Universitatis Babeş-Bolyai. Series Mathematica-Physica. Publishers: Intreprinderea Poligrafica, Cluj.

SuW
Sterne und Weltraum. Astronomische Monatsschrift. Publisher: Verlag Sterne und Weltraum Dr. Vehrenberg, Düsseldorf, Germany.

Tellus
Tellus, a bi-monthly Journal of Geophysics. Svenska Geofysiska Foreningen. Printed in Sweden by Almqvist & Wiksells Boktryckeri AB, Uppsala.

Trans. Astron. Obs. Yale Univ.
Transactions of the Astronomical Observatory of Yale University. Published by the Observatory, New Haven.

Trans. Roy. Soc. Canada
Transactions of the Royal Society of Canada. Published by the Royal Society of Canada, National Research Building, Ottawa.

Trudy Astrofiz. Inst. Alma-Ata
Trudy Astrofizicheskogo Instituta, Alma-Ata. Akademiya

Nauk Kazakhskoj SSR. Publishers: Izdatel'stvo "Nauka" Kazakhskoj SSR, Alma-Ata.

Trudy Glav. Astron. Obs. Pulkovo
Trudy Glavnoj Astronomicheskoj Observatorii v Pulkove. Akademiya Nauk SSSR. Izdanie Glavnoj astronomicheskoj observatorii v Pulkove, Leningrad.

Trudy Inst. Teor. Astron.,Leningrad
Trudy Instituta Teoreticheskoj Astronomii. Akademiya Nauk SSSR. Publishers: Izdatel'stvo "Nauka", Leningrad.

Trudy Tashkent. Astron. Obs.
Trudy Tashkentskoj Astronomicheskoj Observatorii. Akademiya Nauk Uzbekskoj SSR. Publishers: Izdatel'stvo "FAN" Uzbekskoj SSR, Tashkent.

Tsirk. Astron. Inst. Tashkent
Tsirkulyar Astronomicheskogo Instituta. Akademiya Nauk Uzbekskoj SSR. Izdatel'stvo "FAN" Uzbekskoj SSR, Tashkent.

Tsirk. Astron. Obs. L'vov
Tsirkulyar. Astronomicheskaya Observatoriya. L'vovskij Ordena Lenina Gosudarstvennyj Universitet imeni Ivana Franko. Publisher: Izdatel'stvo L'vovkogo Universiteta, L'vov.

Umschau
Umschau in Wissenschaft und Technik. Umschau-Verlag Frankfurt a. M.

Urania Barcelona
Urania. Revista de Astronomia y Ciencias Afines. Organo de la Sociedad Astronómica de España y América, Barcelona; Unión Nacional de Astronomia y Ciencias Afines, Madrid.

Urania Kraków
Urania. Miesiecznik Polskiego Towarzystwa Miłośników Astronomii, Kraków. Publisher: Krakowska Drukarnia Prasowa, Kraków.

Vasiona
Vasiona. Revue d'Astronomie et d'Astronautique. Bulletin de la Société Astronomique "R. Bosković", Beograd.

VdS Nachrichtenblatt
Nachrichtenblatt der Vereinigung der Sternfreunde e.V. After Vol. 18, No. 3 published in combination with "Sterne und Weltraum". Verlag Sterne und Weltraum Dr. Vehrenberg, Düsseldorf, Germany.

Veröff. Astron. Rechen-Inst. Heidelberg
Veröffentlichungen des Astronomischen Rechen-Instituts Heidelberg. Verlag G. Braun, Karlsruhe.

Veröff. Sternw. Sonneberg
Deutsche Akademie der Wissenschaften zu Berlin. Institut für Sternphysik. Veröffentlichungen der Sternwarte in Sonneberg. Publisher: Akademie-Verlag, Berlin.

Vesmír
Vesmír. Přírodovědecky časopis Čs. akadmie věd. Publisher: Academia, Praha.

Vestn. Khar'kov. Univ.
Vestnik Khar'kovskogo Universiteta. Seriya Astronomicheskaya. Publishers: Izdatel'stvo Khar'kovskogo Universiteta, Khar'kov.

Vestn. Kiev. Univ.
Vestnik Kievskogo Universiteta. Seriya Astronomii.
Publishers: Izdatel'stvo Kievskogo Universiteta, Kiev.

VJS Naturforsch. Ges. Zürich
Vierteljahresschrift der Naturforschenden Gesellschaft
in Zürich. Printer and Publisher: Leeman AG, Zürich.

Weltraumfahrt
Weltraumfahrt, Raketentechnik. Publisher: Umschau-
Verlag, Frankfurt a/Main.

Wiss. Zeitschr. Friedrich-Schiller Univ. Jena
Wissenschaftliche Zeitschrift der Friedrich-Schiller-Uni-
versität. Jena. Mathematisch-Naturwissenschaftliche
Reihe. Edited by the Rektor der Friedrich-Schiller-Uni-
versität Jena.

Wiss. Zeitschr. Humboldt-Univ. Berlin
Wissenschaftliche Zeitschrift der Humboldt-Universität
zu Berlin. Mathematisch-Naturwissenschaftliche Reihe.
Edited by the Rektor der Humboldt-Universität, Berlin.

Yamamoto Circ.
Yamamoto Circular. Published by the Yamamoto Observatory, Kamitanakami – Kiryutyo, Otu, Siga-ken, Japan.

Zeitschr. Angew. Physik
Zeitschrift für Angewandte Physik. Publisher: Springer-
Verlag, Berlin–Heidelberg–New York.

Zeitschr. Astrophys. (ZfA)

Zeitschrift für Astrophysik. Publisher: Springer-Verlag,
Berlin–Heidelberg–New York. After Vol. 69 (1968)
replaced by "Astronomy and Astrophysics".

Zeitschr. Geophys.
Zeitschrift für Geophysik. Publisher: Physica-Verlag,
Würzburg, Germany.

Zeitschr. Naturforschung
Zeitschrift für Naturforschung. Europhysics Journal.
Teil a: Astrophysik, Physik, Physikalische Chemie.
Published by Verlag der Zeitschrift für Naturforschung,
Tübingen, Germany.

Zeitschr. Physik
Zeitschrift für Physik. Publisher: Springer-Verlag, Berlin–
Heidelberg–New York.

Zemlya i Vselennaya
Zemlya i Vselennaya. Astronomiya, Geofizika, Issledovaniya Kosmicheskogo Prostranstva. Nauchno-Populyar-
nyj Zhurnal Akademii Nauk SSSR. Publishers: Izdatel'-
stvo "Nauka", Moskva.

Zentralblatt Math. Grenzgebiete
Zentralblatt für Mathematik und ihre Grenzgebiete. Publisher: Springer-Verlag, Berlin–Heidelberg–New York.

Zvaigžņota Debess
Latvijas PSR Zinātņu Akadēmijas Radioastrofizikas
Observatorijas Populārzinatnisks Gadalaiku Izdevums.
Izdevnieciba "Zinātne", Riga.

002 Bibliographical Publications

002.001 **News.**
Nature, Phys. Sci., Vol. 232, 113 - 116, 133 - 136, 157 - 160, 177 - 180 (1971). − (1) Magnetosphere: A trough wind; Astronomy: Future of Fred's place; Neutrino astronomy: Underwater detector. (2) Rapid variations of 3C 279 revealed as illusory; X-ray astronomy: Radio identifications; Supernovae: Shock waves in Vela. (3) Disks and gravitation; Close-up on X-ray stars. (4) Are cosmic rays still useful? ; Earth-moon system: Accretion of volatiles? ; Meteors and comets: Masses and ages.

002.002 **News and views.**
Nature, Vol. 232, 367 - 374, 440 - 448, 519 - 526 (1971). − (1) Galaxies: Circular polarization; Some pulsars may emit thermal X-rays. (2) The universe considered as hole; Exploding magnetosphere may cause Crab glitches; Accurate radio source positions from Cambridge. (3) Jupiter: Great Red Spot; Microwave molecular line astronomy.

002.003 **News notes.**
Sky Telescope, Vol. 42, 12 - 14, 84 - 85 (1971). (1) The Cetus radio arc; The remarkable variable star V 1057 Cygni; Lick Observatory open nights. (2) More interstellar compounds; Amino acids in meteorites; Mass of Pluto; Central Florida Museum opens enlarged planetarium.

002.004 **Science news.**
Priroda, No. 7.71, p. 102 - 110; No. 8.71, p. 98 - 106 (1971). In Russian.
(7) The magnetic moment of a neutrino; On the polarization of the radiation of pulsars; Does the Red Spot of Jupiter consist of solid helium? Organic matter of a meteorite. (8) Radio radiation of Uranus at a wavelength of 8.22 mm; Magnetic field near the pulsar in the Crab nebula; Formic acid in space; Origin of the radiation belts; New method of testing cosmological models.

002.005 **Bibliographical references to astrophysical papers published in the journal "Radiofizika", Vol. 13, Nos. 1 - 12 for the year 1970.**
Astron. Zhurn. Akad. Nauk SSSR, Vol. 48, 872 - 875 (1971). In Russian.

002.006 **Kurzberichte aus der Forschung.**
SuW, Vol. 10, 193 - 195, 231 - 233 (1971).
(1) Die Suche nach superschweren Elementen; 100µ-Beobachtungen des galaktischen Zentrums; Laufzeitverzögerung von Radiosignalenim Schwerefeld der Sonne; Radiostrahlung von Novae und Antares nachgewiesen; Grand-Tour-Vorbereitungen beginnen; Strahl-Folien-Spektroskopie;Technetium-Sterne.
(2) Eine weitere Galaxie mit nicht-kosmologischer Rotverschiebung ; Zweiter Röntgenpulsar entdeckt; Ist die Rotverschiebung als Entfernungsindikator für den Quasar 3C 279 ungeeignet? Entdeckung des galaktischen Zentrums bei 1 µm; Neues Radioteleskop in Indien; Helme der Apollo-Astronauten als Detektoren galaktischer Strahlung; ESRO-II in der Atmosphäre verglüht; Keine Radiostrahlung von Maffei 1; Wie groß sind „Schwarze Löcher"? (H.-M. Hahn).

002.007 **Nouvelles brèves.**
Ciel et Terre, Vol. 87, 464 - 467 (1971).
Observations d'une occultation rasante; 1970, une année cométaire exceptionnelle; L'influence du système planétaire sur les orbites cométaires presque paraboliques.

002.008 **News from science and other informations.**
Zemlya i Vselennaya, No. 4 (1971). In Russian.

New Soviet starts to Mars, p. 3; Lunokhod continues its work, p. 12; News on Mars (G. A. Lejkin), p. 13; Large eruption on the sun (V. F. Chistyakov), p. 40 - 41; Meteor trails − indicators of the physical state of the upper layers of the earth's atmosphere (A. M. Bakharev), p. 63.

002.009 **Bibliography of the astronomical scientific, educational, reference and popular science literature (1960 − 1969).** E. K. Kharadze, N. I. Bolkvadze.
Byull. Abastumansk. Astrofiz. Obs., No. 40, p. 221 - 296 (1971). In Georgian and Russian.

002.010 **Science news.**
Priroda, No. 9.71, p. 100 - 109 (1971). In Russian.
Discovery of a supernova, (A. D. Chuadze); "Black hole" in the Galaxy; News on a pulsar in the Crab nebula; Potassium granites on the moon (V. A. Baskina); Riddles of the earth's atmosphere.

002.011 **News and views.**
Nature, Vol. 233, 8 - 14, 161 - 166, 231 - 238 (1971). − (1) Astronomy: Universal distance; Model for Sco X-1; Infrared variations in Seyfert galaxies? ; Supermassive rotors and active galactic nuclei. (2) Satellite lines in the solar X-ray spectrum. (3) Limits on Seyfert galaxies; Selection in quasar redshift measurements; History of science: Kepler honoured; More about interstellar grains.

002.012 **News.**
Nature, Phys. Sci., Vol. 233, 1 - 4, 21 - 22, 65 - 68 (1971).
(1) Controlling radiation belts; Infrared Seyferts. (2) Spread of radon on the moon. (3) Selection and redshifts; Shocks and instabilities in the solar wind.

002.013 **Annotations on papers on geomagnetism and aeronomy, published in "News of higher educational establishments. Radiophysics", 1968, Vol. 11, Nos. 4 - 12.**
Geomagn. Aeronom., Vol. 11, 751 - 756 (1971). In Russian.

002.014 **Kurzberichte aus der Forschung.**
SuW, Vol. 10, 271 - 273 (1971).
Interstellares Formamid entdeckt; Helle G-Überriesen; Wassergewinnung auf dem Mond; Erfolgreiches Raketenexperiment der Landessternwarte Heidelberg; Röntgenstrahlung von Seyfert-Galaxien; Interstellarer Schwefel in Molekülen nachgewiesen; Identifikation zweier Radioquellen mit hellen Sternen; Solare Neutrinos.

002.015 **Annotation on papers on geomagnetism and aeronomy, published in "News of higher educational establishments, Radiophysics", 1969, Vol. 12, Nos. 1 - 4.**
Geomagn. Aeronom., Vol. 11, 938 - 940 (1971). In Russian.

002.016 **Nouvelles brèves.**
Ciel et Terre, Vol. 87, 558 - 562 (1971).
La distance du Pluton; La petite planète Amor; La rotation de Lydia; Orbites définitives de trois comètes; Nouvelles molécules interstellaires; Occultation de Beta Scorpion par Jupiter et Io.

002.017 **News notes.**
Sky Telescope, Vol. 42, 146, 205 - 206 (1971).
(1) OH molecules in other galaxies; Big southern Schmidt telescope; Supernova of A.D. 1181; Solar eclipse information for 1973. (2) Nova Cephei 1971; Infrared measures of Ganymede; Radio "photography" of the sky; Martian weather fore-

cast; Theta Coronae Borealis as a double star.

002.018   **Source materials for the recent history of astronomy and astrophysics: A checklist of manuscript collections in the United States.**
C. Weiner, J. N. Warnow.
Journ. History Astron., Vol. 2, 210 - 218 (1971).

002.019   **News.**
Nature, Phys. Sci., Vol. 233, 85 - 86, 105 - 108, 125 - 128, 145 - 148 (1971).
(1) Molecules: Spectroscopy at Dijon. (2) X-ray astronomy: GX3 + 1 pinned down. (3) Cosmic gamma-rays: No high-energy break. (4) Galaxies: Neighbourly BSOs.

002.020   **Science news.**
Priroda, No. 10.71, p. 96 - 105 (1971). In Russian.
Unique meteorite; The riddle of the lunar albedo; Microcraters on lunar rocks; Cosmic promethium; Have the stars an electric charge? Experiments in space; From the lunar soil; Unusual electrical phenomenon in the atmosphere.

002.021   **News and views.**
Nature, Vol. 233, 302 - 309, 370 - 378, 519 - 526, 585 - 592 (1971).
(1) Locating a quasar; Cosmic rays: Hobart conference; Atlantic islands and continental drift. (2) Anomalous redshifts explained? ; Cosmic X-rays not from galaxy clusters. (3) Similarities between pulsar models; Composition of lunar spinels. (4) Dilatons and gravity.

002.022   **News from science and other informations.**
Zemlya i Vselennaya, No. 5 (1971). In Russian.
Ten working days of Lunokhod 1, p. 24; Flight of Apollo 15, p. 30; How the density of the Martian atmosphere has been, p. 30; Conference of Czechoslovakian teachers in astronomy (*J. Siroký*), p. 36; Luna 18, p. 48; Elements of astronautics and astronomy for college students (*V. P. Karnitskaya*), p. 65; Amateur astronomers beyond the polar circle, p. 73.

002.023   **Kurzberichte aus der Forschung.**
SuW, Vol. 10, 303 - 304 (1971).
Zirkumstellare Hüllen in dem jungen Sternhaufen NGC 2264; Sternphotographie mit sehr hoher Winkelauflösung; Alter des Planetensystems 4.6 × 10⁹ Jahre? Die elliptische Polarisation in der optischen Astronomie; Erste diskrete extragalaktische Gammaquelle entdeckt; Interstellares SiO im Sagittarius; Mariner 9 Zwischenbericht; Neuer Flug von Stratoscope II; Apollo-17-Besatzung ausgewählt; Fehlschlag mit Luna 18.

002.024   **Science news.**
Priroda, No. 11.71, p. 103 - 112 (1971). In Russian.
Cosmic matter on earth; Investigation of the filamentary nebula in Cygnus (*Yu. P. Pskovskij*); Some problems of extragalactic astronomy; Apollo 15 on the moon; Fourth verification of the general theory of relativity; Composition of the earth's interior.

002.025   **Kurzberichte aus der Forschung.**
SuW, Vol. 10, 329 - 330 (1971).
Centaurus X-3 — ein weiterer Röntgenpulsar; Jupitermonde; Über den Bau der Pulsare.

002.026   **Centre de Données Stellaires. Inform. Bull. No. 2.**
J. Jung (Editor).
Printed at Observatoire de Strasbourg. 28 pp. (1971).
Contents: 1st supplement to the list of spectroscopic and photometric catalogues, *B. Hauck*, p. 1 - 4; Star catalogues available at the NASA – Goddard Space Flight Center, *J. Mead*, p. 5 - 6; A table of correspondences between Durch-

musterungen, *J. Jung, M. Bischoff*, p. 7; The General Catalogue of Stellar Identifications (CSI), *J. Jung, M. Bischoff*, p. 8 - 17; List of errors found in the Henry Draper Catalogue, *J. Jung, M. Bischoff*, p. 18 - 22; List of catalogues available at the Centre de Données Stellaires, *J. Jung*, p. 23 - 28.

002.027   **News notes.**
Sky Telescope, Vol. 42, 272 - 273, 279, 334 - 336 (1971).
(1) Is man changing earth's temperature? ; Green Bank's 300-foot radio telescope resurfaced; Nearest galaxy or richest globular? ; Promethium in S stars; Extending the range of a radar telescope. (2) New companion to M31? ; Study of a reflection nebula; CH Cygni: An eclipsing binary? ; Radio emission from X-ray sources; Statistics of quasars; Gerhard Herzberg honored; Minor planet news; Kepleriana.

002.028   **Forschung und Technik.**
Phys. Blätter, 27. Jahrgang, p. 326 - 329, 369 - 371, 519 - 522, 569 - 570 (1971).
(1) Niederfrequente Gravitationsstrahlung? Epsilon Aurigae ein Kollapsar? Nachweis der Existenz eines fossilen Isotops in Meteoriten; Promethium in der Photosphäre von HR 465. (2) Erste Beobachtung von Neutrinos aus dem Sonneninnern; Kugel aus dem Mondstaub des Fra Mauro Gebietes. (3) Magnetometerdaten vom Mond; Meßergebnisse von Venera 7; Ionenspektren nach Folienstripping. (4) Neuer Gesichtspunkt für die Suche nach Kollapsaren (Schwarzen Löchern).

002.029   **News.**
Nature, Phys. Sci., Vol. 234, 1 - 4, 21 - 22, 41 - 44, 61 - 64, 81 - 84, 101 - 102, 121 - 123, 141 - 142, 161 - 162 (1971).
(1) Scale invariance and gravity; Neutron stars: X-ray stars dissected; GX3 + 1: Search narrows; White dwarfs: Role of convection; Radio telescopes: Atmospheric limitations; Radio sources: Interacting galaxies; Local group: Nature of Maffei 2; Circinus XR-1: New X-ray pulsar. (2) Looking for flashes. (3) X-ray astronomy: Galactic sources; Galaxies: Radio nuclei and disks; Cosmic rays: Source of electrons. (4) Atmosphere: Artificial aurora. (5) Experimenting with the ionosphere; Infrared astronomy: Hot stars in Orion; Cygnus X-1: Trains of pulses; Centaurus A: Nuclear hot spot. (6) Models for the lunar interior. (7) Quasars: Luminosity function; Particles: Antimatter distribution. (8) Space: Tracking down noise; Magellanic Clouds: X-ray structure. (9) Antimatter and the γ-ray background spectrum.

002.030   **News and views.**
Nature, Vol. 234, 11 - 16, 68 - 74, 123 - 128, 172 - 178, 248 - 254, 380 - 386, 438 - 443, 505 - 510 (1971).
(1) Water on the moon and the Apollo programme; Origin of cosmic rays; Infrared and the black-body background. (2) Craters on the earth and moon; Pulsars and supernova remnants. (3) Plutonium-244 on the earth; Selenology: Astronauts in Britain; Variability of OJ 287. (4) Gravitational lenses and the structure of the universe. (5) Subsolidus reduction of lunar spinels; Evolutionary behaviour of quasars. (6) Black holes against neutron stars. (7) The earth's cores; Righting the applecart; GX3 + 1: Two candidates; A new magnetic reversal at 12,500 years? (8) Nonvelocity redshifts in galaxies? ; What colour is the sky on Venus? ; The earth's atmosphere and quasar redshifts.

002.031   **Science news.**
Priroda, No. 12.71, p. 79 - 88 (1971). In Russian.
End of the flight of Salyut; Development of the universe and origin of galaxies (*V. M. Tomozov*); Life on Mars? Is the gravitational constant constant? Electrons, ultrasonics and gravitational waves.

002.032 Nouvelles brèves.
Ciel et Terre, Vol. 87, 650 - 653 (1971).
L'occultation de Mars du 16 Mai 1971; Une nouvelle petite planète exceptionnelle: 1971 FA; La masse de Pluton; La comète périodique Vaisala (1942 II); La comète Toba (1971a); Redécouverte de comètes périodiques; Un nouveau catalogue d'orbites d'étoiles doubles.

002.033 Chronicle.
Urania Kraków, Vol. 42, 212 - 215, 253 - 257, 298 - 308 (1971). In Polish.
(1) Investigation of lunar samples (12) *(A. Marks);* The flight of Salyut and Soyus 10 *(A. Marks);* About one proposal of a flight to Jupiter *(M. Pańków).* (2) The first manned observatory in space *(A. Marks);* Investigations of lunar samples (13) *(A. Marks);* Source of X-radiation in Centaurus *(B. Kuchowicz);* Intensive source of X-radiation in Centaurus *(B. Kuchowicz);* Molecular ions in the sun's photosphere *(B. Kuchowicz);* Neutral hydrogen revolves in distant galaxies *(B. Kuchowicz).* (3) X rays and cosmic rays from pulsars *(B. Kuchowicz);* A second X ray pulsar *(B. Kuchowicz);* A galaxy with the largest mass *(B. Kuchowicz);* A map of the galactic center *(B. Kuchowicz);* Quasars in clusters of galaxies *(B. Kuchowicz);* Quasar PKS 2251 + 11, object Ton 256 and the relationship between quasars and galaxies *(B. Kuchowicz);* Gigantic magnetic fields on white dwarfs *(B. Kuchowicz);* Against the hierarchical model of universe *(B. Kuchowicz);* Measurement of distances by use of formaldehyde lines *(B. Kuchowicz);* On the possibility of life on Mars and Venus *(B. Kuchowicz);* A new hypothesis on the red shade of the surface of Mars *(I. Janos);* Phobos does not fall *(A. Marks);* Table of Jovian satellites *(M. Pańków);* Attempts at direct investigation of cometary material *(M. Pańków).*

002.034 Mitteilungen aus Wissenschaft und Literatur.
Sterne, 47. Jahrgang, p. 150 - 151, 213 - 216 (1971).
(1) Die Rentabilität großer Teleskope *(F. Schmeidler);* Neues Mond-Kolloquium in Houston *(J. Classen).* (2) Der Kern der Milchstrasse *(F. Schmeidler);* Ein neuer Katalog planetarischer Nebel *(F. Schmeidler);* Entdeckungen und neue Beobachtungen interstellarer Moleküle *(H. Lambrecht);* Gesicherter Nachweis solarer Neutrinos *(K.-H. Schmidt).*

002.035 Annual index of authors, to Astrophysical Journal, volumes 163 - 170, parts 1 and 2 and to the Supplement Series, volume 22, pages 319 - 484, and volume 23.
Astrophys. Journ., Vol. 170, No. 3, Part 3, 51 pp. (1971).

002.036 Bibliography of Soviet solar system space research in 1969.
Icarus, Vol. 15, 140 - 146 (1971).
The bibliography on Soviet solar system space research for the calendar year 1969 is taken from the report of the 13th Plenary Session of the Committee for Space Research (COSPAR), Leningrad; edited by Yu. I. Yefremov. The subject matter is grouped under the following topics: Investigations of the planets of the solar system; Lunar research; Meteors and meteorites; Investigations of interplanetary space; Physics of sun–earth couplings, Solar winds; and comets.

002.037 News.
Phys. Today, Vol. 24, No. 1, p. 19 - 22, 91 - 94; No. 2, p. 17 - 20, 61 - 64; No. 3, p. 17 - 20, 69 - 72; No. 4, p. 17 - 20, 65 - 68; No. 5, p. 17 - 20, 61 - 64; No. 6, p. 17 - 20, 61 - 64 (1971).
(1) Polarized light from white-dwarf star; ASTRA project monitors atmospheric pollution. (2) Amino acids in both moon and meteorite; European groups join forces to design 3.6-meter telescope. (3) Is the Gum nebula the ghost of an exploding supernova? ; Lifetime of heliumlike metastable ions measured; NSF astronomy reorganized into five separate programs; Plan

for economical Venus exploration proposed; New faces at the Astrophysical Journal. (4) New telescope at Palomar. (5) New coordination between universal and atomic time; Mauna Kea Observatory has 88-inch telescope running. (6) Pulsating X-ray source might be neutron star or black hole; Laser earth-strain gauge to search for gravity waves.

002.038 News.
Phys. Today, Vol. 24, No. 7, p. 13 - 16, 57 - 60; No. 8, p. 17 - 20, 69 - 72; No. 9, p. 17 - 20, 61 - 64; No. 10, p. 17 - 20, 69 - 72; No. 12, p. 17 - 20, 69 - 72 (1971). – (1) Quasars are relatively local, except those that are not. (2) Bok and Salpeter elected to Astronomical Society posts. (3) Astronomical Society is founded in Canada. (4) Arecibo Observatory is restructured and renamed; AURA to admit PhD-granting institutions to membership. (5) Toro makes a threesome with earth and moon; Theory of early universe explains particle production; Radio interferometer at Stanford undergoes tests.

002.039 Astronomy and Astrophysics Abstracts. Vol. 5, Literature 1971, Part I.
S. Böhme, W. Fricke, U. Güntzel-Lingner, F. Henn, D. Krahn, G. Zech (Editors).
Published for the Astronomisches Rechen-Institut, Heidelberg by Springer-Verlag, Berlin – Heidelberg – New York. 10 + 505 pp. Price DM 72.00 [Subcription price per volume DM 57.60] (1971).

002.040 Annotations on papers on geomagnetism and aeronomy published in "News of higher educational establishments. Radiophysics", 1969, Vol. 12, Nos. 5, 6, 8 - 10; Vol. 13, Nos. 1 - 3, 5.
Geomagn. Aeronom., Vol. 11, 1130 - 1134 (1971). In Russian.

002.041 News from science and other informations.
Zemlya i Vselennaya, 1971, No. 6. In Russian.
Start of Vertical 2, p. 7; Observations of a comet from an artificial earth satellite, p. 7; With one carrier rocket – eight artificial earth satellites, p. 7; First photographs of Mars during the great opposition in 1970 *(I. K. Koval'),* p. 14; Is Pluto rich in iron? p. 14; How did Cassini's division arise? p. 14; Can "black holes" be discovered? p. 14 - 15; One more nearby galaxy *(N. N. Chugaj),* p. 15; Preliminary scientific results of the flight of Apollo 14 *(D. Yu. Gol'dovskij),* p. 27 - 28; Astronomical observations by prehistoric man? p. 33; New minor planet, p. 65; Solar activity and radio radiation of Jupiter, p. 65; Traces of promethium in a star, p. 65; Quasars and clusters of galaxies *(B. V. Komberg),* p. 66; Possible excitation source of cosmic infrared radiation, p. 66.

002.042 Bibliography of far infrared spectroscopy.
D. Bloor.
Infrared Phys., Vol. 10, 1 - 55 (1970).
Contains references to spectroscopic studies within the range 50–2000 microns (0.05–2 mm) wavelength. Contains 1756 references plus 63 in the addendum covering the period 1923 to 1969. – *RAB*

002.043 Hydrographische Bibliographie.
(Ozeanographie, Erdmagnetismus, Nautik).
Separate prints from Deutsche Hydrographische Zeitschr. [edited by Deutsches Hydrographisches Institut, Hamburg], Jahrgang 23 (1970), 101 pp. (1971).

002.044 Rassegna delle riviste e notizie brevi. P. Maffei.
Coelum, Vol. 39, 139 - 144, 184 - 191, 228 - 232 (1971).

002.045 Noticiero astronomico.
Rev. Astron., Vol. 42, No. 174, p. 39 - 41 (1971).

002.046 Science and the citizen.
Sci. American, Vol. 225, No. 1, p. 42 - 45; No. 2, p. 44 - 47; No. 3, p. 74 - 75, 76, 80, 84; No. 4, p. 40 - 42, 44; No. 5, p. 46 - 50 (1971). − (1) Venera on Venus; (2) Woher der Gegenschein?; (3) Layered moon; Universal radicals, (4) Pioneers to Jupiter; (5) The export of lunar glass.

002.047 Bibliography.
Z. Kopal, M. Moutsoulas, J. W. Salisbury (Editors). The Moon, Vol. 3, 90 - 156, 239 - 262, 352 - 362 (1971). Current critical bibliography of the entire field of lunar studies.

003 Books (Astronomy and Astrophysics)

003.001 Annual Review of Astronomy and Astrophysics.
L. Goldberg, D. Layzer, J. G. Phillips (Editors). Annual Reviews Inc., Palo Alto, California. 11 + 394 pp. Price $ 10.50 (1971). − The individual papers are included in their corresponding subject categories − see 033.003, 063.004, 065.019, 065.020, 076.005, 097.008, 107.002, 113.010, 117.004, 131.016, 133.004, 155.008, 158.006.

003.002 Catalogue of orbital elements, masses and luminosities of close binaries.   M. A. Svechnikov.
Ministerstvo vysshego i srednego spetsial'nogo obrazovaniya RSFSR, Ural'skij gosudarstvennyj universitet imeni A. M. Gor'kogo, Uchenye zapiski, No. 88, Seriya astronomicheskaya, vyp. 5, Sverdlovsk. 179 pp. Price 1 Rbl. 60 Kop. (1969). In Russian.

003.003 Questions of astrophysics.
V. I. Voroshilov (Editor).
Respublikanskij Mezhvedomstvennyj Sbornik. Ser. Astrometriya i Astrofizika, Vyp. (No.) 12, Akademiya Nauk Ukrainskoj SSR, Glav. Astron. Obs. Izdatel'stvo "Naukova Dumka", Kiev. 88 pp. Price 70 Kop. (1971). In Russian. − The papers included are abstracted in their corresponding subject categories − see 022.026, 022.027, 031.007, 064.002, 071.00ᵍ 102.003, 113.015, 114.026 - 114.028, 122.025 - 122.027, 124.101.

003.004 Carbon stars.   Z. Alksne, J. Ikaunieks.
Academy of Sciences of the Latvian SSR, Radioastrophysical Observatory. Publishing House "Zinātne", Riga. 257 pp. Price 99 Kop. = Trans. Riga Obs., Vol. 13 (1971). In Russian. − Review in Referativ. Zhurn. 51. Astron., 2.51. 486 (1972).

003.005 Long period variable stars.
J. Ikaunieks, (Editor: B. V. Kukarkin).
Academy of Sciences of the Latvian SSR, Radioastrophysical Observatory. Publishing House "Zinātne", Riga. 135 pp. Price 60 Kop. = Trans. Riga Obs., Vol. 12 (1971). In Russian. Review in Referativ. Zhurn. 51. Astron., 11.51.605 (1971).

003.006 Close binary systems with spherical components.
A. M. Shul'berg.
Akademiya Nauk SSSR, Vsesoyuznoe Astronomo-Geodezi-

cheskoe Obshchestvo. Izdatel'stvo "Nauka", Moskva. 246 pp. Price 89 Kop. (1971).   In Russian.

003.007 Physical characteristics of the giant planets.
Handbook−review.
A. N. Aksenov, Z. N. Grigor'eva, V. F. Kartashov, N. V. Priboeva, L. P. Sorokina, V. G. Tejfel', G. A. Kharitonova. V. G. Tejfel' (Editor).
Akademiya Nauk Kazakhskoj SSR, Astrofizicheskij Institut. Izdatel'stvo "Nauka" Kazakhskoj SSR, Alma-Ata. 175 pp. Price 1 Rbl. 44 Kop. (1971).   In Russian.

003.008 Instationary stars and methods of their investigation.
Methods of investigation of variable stars.
V. B. Nikonov (Editor).
Izdatel'stvo "Nauka"; Glavnaya Redaktsiya Fiziko-Matematicheskoj Literatury, Moskva. 334 pp. Price 1 Rbl. 72 Kop. (1971).   In Russian. − The papers included are abstracted in their corresponding subject categories − see 113.018, 113.019, 120.002 - 120.008.

003.009 English-Russian astronomical dictionary (about
20000 terms).   A. A. Mikhailov (Editor).
Compiled by O. A. Melnikov, A. A. Nemiro, Z. I. Kadla, V. M. Pererva.
"Soviet Encyclopaedia" Publishing House, Moscow. 504 pp. Price 1 Rbl. 32 Kop. (1971).

003.010 Tables K 11: Two-star fix without use of altitude
difference method. Vol. V N (Lat. 40° − 49° 30′ N).
S. M. Kotlarić.
Edited by Hidrografskog Instituta Jugoslavenske Ratne Mornarice, Split. 29 + 367 pp. (1971). In Serbo-Croatian.

003.011 Astrophysics and general relativity, Vol. 2.
Brandeis University Summer Institute in Theoretical Physics, 1968.   M. Chrétien, S. Deser, J. Goldstein (Editors) Gordon and Breach Science Publishers, New York − London − Paris. 11 + 383 pp. Price DM 90.50 (1971). − The individual papers are included in their corresponding subject categories − see Abstracts 065.072, 143.032, 151.023, 162.034, 162.035.

003.012 New qualitative methods of celestial mechanics.
E. A. Grebenikov, Yu. A. Ryabov.

Izdatel'stvo "Nauka". Glavnaya Redaktsiya Fiziko-Matematicheskoj Literatury, Moskva, 444 pp. Price 2 Rbl. 2 Kop. (1971). In Russian.

003.013 **Encyclopedia of physics.** (Handbuch der Physik). S. Flügge (Chief editor). Vol. 49/3: Geophysics III (Part III). K. Rawer (Editor).
Springer-Verlag, Berlin–Heidelberg–New York. 7 + 537 pp. Price DM 150.00 (1971). – The individual contributions to this volume are included in their corresponding subject categories – see Abstracts 034.081 - 034.083, 084.272 - 084.277.

003.014 **Physics of the moon and planets.** I. K. Koval' (Editor).
Respublikanskij Mezhvedomstvennyj Sbornik. Ser. Astrometriya i Astrofizika, No. 14. Akademiya Nauk Ukrainskoj SSR, Glavnaya Astronomicheskaya Observatoriya AN USSR. Izdatel'stvo "Naukova Dumka", Kiev. 76 pp. Price 70 Kop. (1971). In Russian. – The papers included are abstracted in their corresponding subject categories – see 094.194, 097.058 - 097.060, 099.048, 099.049, 100.009.

003.015 **Investigation of the ionosphere and meteors.** I. B. Bijbosunov (Editor).
Akademiya Nauk Kirgizskoj SSR. Institut Fiziki i Matematiki. Izdatel'stvo "Ilim", Frunze. 105 pp. Price 36 Kop. (1971). In Russian. – The individual papers are included in their corresponding subject categories – see 082.072, 104.066 - 104.075.

003.016 **Annuaire 1972** du Bureau des Longitudes. Encyclopédie Physique et Spatiales. Gauthier-Villars, Editeur, Paris. 14 + 683 + A18 + B8 + C10 + D115 pp. (1971). Part 1: Ephémérides astronomiques; Part 2: Environment terrestre; Part 3: Systèmes d'étoiles; Part 4: Physiques des solides et des fluides; Part 5: Démographie de la France; Part 6: Prédictions astronomiques pour l'année 1973; Part 7: Notices; Part 8: Tables analytiques.

003.017 **XIIIth international congress on the history of science. Section No. 6. History of physics and astronomy.**
"Nauka", Moskva. 145 pp. (1971). Contributions in Russian, French, English and German. – Review in Referativ. Zhurn. 51. Astron., 12. 51.15 (1971). – Moscow, 1971 August 18 - 24.

003.018 **Transactions of the international seminar on the problem: "Generation of cosmic rays on the sun".**
Akademiya Nauk SSSR, Ministerstvo vyssh. i sredn. spets. obrazovaniya SSSR, Moskovskij universitet. Nauchno-issledovatel'skij institut yadernoj fiziki. Moskva. 468 pp. (1971). In Russian. – Leningrad, 8 - 12 Dec. 1970.

003.019 **Methodics of teaching physics, astronomy and technical sciences at colleges and high schools.**
Papers presented at the XIth zonal conference of lecturers of pedagogical colleges of the Ural, Siberia and Far East on the methodics of science, 1968, Sept. 9 - 14. (Sci. papers – Uch. zap.) Perm State Pedagogical College, Vol. 72. Perm'. 493 pp. Price 3 Rbl. 50 Kop. (1969). In Russian.

003.020 **Astronomy. Educational textbook for students of geography and natural science faculties at pedagogical colleges.** B. A. Volynskij.
"Prosveshchenie", Moskva. 208 pp. Price 71 Kop. (1971). In Russian. – Review in Referativ. Zhurn. 51. Astron., 12.51.33 (1971).

003.021 **Handbook for the amateur astronomer.** P. G. Kulikovskij.

4th revised and enlarged edition.
"Nauka", Moskva. 632 pp. Price 2 Rbl. 37 Kop. (1971). In Russian. – Corrections in Astron. Tsirk., No. 667, p. 8 (1971).

003.022 **Collection of tasks in general astronomy.** Nikolov, N. Stefanov, R. M. Rusev.
"Nauka i izkustvo", Sofia. 203 pp. Price 0.84 Lv. (1971). In Bulgarian.

003.023 **Theory of relativity and gravitation.**
Transactions of the meeting of the section of physics on the theory of relativity and gravitation, 22 - 23 February 1971.
Mosk. obshchestvo ispyt. prirody. Sekts. M. 104 pp. Price 30 Kop. (1971). In Russian.

003.024 **Physik und Kosmologie.** Stand und Zukunftsaspekte naturwissenschaftlicher Forschung in Deutschland. Forschung und Information, Vol. 7, with a preface by R. Kurzrock.
Colloquium Verlag, Berlin. 148 pp. Price DM 14.80 (1971). – The individual contributions within the subject scope of Astronomy and Astrophysics Abstracts are included in their corresponding categories – see Abstracts 021.005, 051.033, 065.104, 094.237, 094.238, 141.213, 143.051, 157.009, 158.107, 162.050.

003.025 **Selected exercises in galactic astronomy.** I. Atanasijević.
Astrophysics and Space Science Library, Vol. 26. D. Reidel Publishing Company, Dordrecht–Holland. 12 + 144 pp. Price Dfl. 35.00 (1971). – Contents: (1) Determination of the position of the galactic equator; (2) The apparent distribution and the colors of the globular clusters; (3) Determination of the vertex of the Hyades cluster; (4) Solar motion and velocity distribution of a group of stars; (5) The distribution of the residual velocities: A numerical method; (6) The asymmetry of stellar motions; (7) The theory of galactic rotation; (8) The determination of the galactic orbit of a star.

003.026 **Highlights of Astronomy, Volume 2,** as presented at the XIVth general assembly of the I.A.U. 1970. C. de Jager (Editor).
D. Reidel Publishing Company, Dordrecht–Holland. 12 + 793 pp. Price Dfl. 140.00, $ 42.00 respectively (1971).

003.027 **Handbook of elemental abundances in meteorites.** B. Mason (Editor).
Gordon and Breach Science Publishers, New York – Paris – London. Series on Extraterrestial Chemistry, Vol. 1. 9 + 555 pp. Price £ 14.60 (1971). – Contents: Introduction, *B. Mason;* Hydrogen, *I. R. Kaplan;* The inert gases, *D. Heymann;* Lithium, *W. Nichiporuk;* Beryllium, *P. R. Buseck;* Boron, *P. A. Baedecker;* Carbon, *G. P. Vdovykin, C. B. Moore;* Nitrogen, *C. B. Moore,* Oxygen, *W. D. Ehmann;* Fluorine, *G. W. Reed, Jr.;* Sodium, *G. G. Goles;* Magnesium, *B. Mason;* Aluminium, *B. Mason;* Silicon, *C. B. Moore;* Phosphorus, *C. B. Moore;* Sulfur, *C. B. Moore;* Chlorine, *G. W. Reed, Jr.;* Potassium, *G. G. Goles;* Calcium, *B. Mason;* Scandium, *G. G. Goles;* Titanium, *B. Mason;* Vanadium, *W. Nichiporuk;* Chromium, *G. G. Goles;* Manganese, *G. G. Goles;* Iron, *B. Mason;* Cobalt, *C. B. Moore;* Nickel, *C. B. Moore;* Copper, *G. G. Goles;* Zinc, *C. B. Moore;* Gallium, *P. A. Baedecker, J. T. Wasson;* Germanium, *P. A. Baedecker, J. T. Wasson;* Arsenic, *M. E. Lipschutz;* Selenium, *I. Z. Pelly, M. E. Lipschutz;* Bromine, *G. W. Reed, Jr.;* Rubidium, *G. G. Goles;* Strontium, *K. Gopalan, G. W. Wetherill;* Yttrium, *B. Mason;* Zirconium and hafnium, *W. D. Ehmann, T. V. Rebagay;* Niobium, *P. R. Buseck;* Molybdenum, *M. E. Lipschutz;* Ruthenium, *W. Nichiporuk;* Rhodium, *W. Nichiporuk;* Palladium, *W. Nichiporuk;* Silver, *P. R. Buseck;* Cadmium,

P. R. Buseck; Indium, *P. A. Baedecker;* Tin, *P. R. Buseck;* Antimony, *W. D. Ehmann;* Tellurium, *I. Z. Pelly, M. E. Lipschutz;* Iodine, *G. W. Reed, Jr.;* Caesium, *G. G. Goles;* Barium, *C. C. Schnetzler;* Rare earths, *J. A. Philpotts, C. C. Schnetzler;* Tantalum, *W. D. Ehmann;* Tungsten, *W. D. Ehmann;* Rhenium, *J. W. Morgan;* Osmium, *J. W. Morgan;* Iridium, *P. A. Baedecker,* Platimun, *W. D. Ehmann;* Gold, *W. D. Ehmann;* Mercury, *G. W. Reed, Jr.;* Thallium, *M. E. Lipschutz;* Lead, *V. M. Oversby;* Bismuth, *M. E. Lipschutz;* Thorium, *J. W. Morgan;* Uranium, *J. W. Morgan;* Index of meteorites.

003.028   Determination of the azimuth of a terrestrial object with the sun.   V. G. Vasil'ev.
"Nedra", Moskva. 70 pp. Price 23 Kop. (1971). In Russian. – Review in Referativ. Zhurn. 52. Geod. Aehros"emka, 12.52.106 (1971).

003.029   Model of a closed magnetosphere.
Issled. po geomagnetizmu, aehron. i fiz. Solntsa, vyp. (No.) 14. Irkutsk. 376 pp. Price 50 Kop. (1970). In Russian.

003.030   Moon rocks and minerals. Scientific results of the study of the Apollo 11 lunar samples with preliminary data on Apollo 12 samples.   A. A. Levinson, S. R. Taylor.
Pergamon Press, New York – Toronto – Oxford – Sydney – Braunschweig. 14 + 222 pp. Price $ 11.50 (1971). – Contents: (1) Introduction; (2) The rocks and soils; (3) The minerals; (4) Chemistry of samples brought by Apollo 11 and 12; (5) Bioscience and organic matter; (6) Petrology: Experimental studies and origin of the lavas; (7) Age of the lunar rocks, isotope studies, cosmic ray and solar wind effects; (8) Physical properties; (9) Origin of the moon.

003.031   Physics and chemistry of the earth, Vol. 8.
L. H. Ahrens, F. Press, S. K. Runcorn, H. C. Urey (Editors).
Pergamon Press, Oxford – New York – Toronto – Sydney – Braunschweig. 8 + 333 pp. Price DM 138.75 (1971).

003.032   Selected works of H. Poincaré. I. New methods of celestial mechanics.
"Nauka", Moskva. 772 pp. Price 3 Rbl. 22 Kop. (1971). In Russian. – Review in Referativ. Zhurn. 62. Issled. kosm. prostranstva, 1.62.229; 51. Astron., 2.51.108 (1972).

003.033   Practical work on stellar astronomy.
P. G. Kulikovskij (Editor).
"Nauka", Moskva. 187 pp. Price 53 Kop. (1971). In Russian. Review in Referativ. Zhurn. 51. Astron., 5.51.33 (1972).

003.034   Universal gravitation.   A. F. Bogorodskij.
Akademiya Nauk Ukrainskoj SSR, Glavnaya Astronomicheskaya Observatoriya. Izdatel'stvo "Naukova Dumka", Kiev. 352 pp. Price 2 Rbl. 52 Kop. (1971). In Russian. Review in Referativ. Zhurn. 51. Astron., 1.51.863 (1972).

003.035   The Year Book of the International Council of Scientific Unions 1971.
Edited by ICSU Secretariat, Rome, Italy. 241 pp. (1971). – Included is a list of the Commissions of the IAU.

003.036   Time in science and philosophy. An international study of some current problems.
J. Zeman (Editor).
Elsevier Publishing Company, Amsterdam – London – New York. 305 pp. Price Dfl. 55.00 (1971).

003.037   Instationary stars and methods of their investigation. Eclipsing variables.
V. P. Tsesevich (Editor).

Izdatel'stvo "Nauka"; Glavnaya Redaktsiya Fiziko-Matematicheskoj Literatury, Moskva. 352 pp. Price 1 Rbl. 74 Kop. (1971). In Russian. – The individual papers are included in their corresponding subject categories – see Abstracts 121.079 - 121.087.

003.038   Histoire du principe de relativité.
M.-A. Tonnelat.
Flammarion, Éditeur, Paris, 561 pp. Price F 58.00 (1971).

003.039   Astronomical investigations.
E. P. Fedorov (Editor).
Respublikanskij Mezhvedomstvennyj Sbornik. Astrometriya i Astrofizika, No. 13. Akademiya Nauk Ukrainskoj SSR, Glavnaya Astronomicheskaya Observatoriya. Izdatel'stvo "Naukova Dumka", Kiev. 124 pp. Price 1 Rbl. 3 Kop. (1971). In Russian. – The papers included are abstracted in their corresponding subject categories – see 032.033, 034.117, 041.038 - 041.042, 045.022 - 045.024, 093.026, 094.303 - 094.306, 098.037, 103.133, 124.103.

003.040   Problems of cosmic physics. Vypusk 6.
S. K. Vsekhsvyatskij (Editor).
Mezhvedomstvennyj Nauchnyj Sbornik. Izdatel'stvo Kievskogo Universiteta, Kiev. 220 pp. Price 1 Rbl. 17 Kop. (1971). In Russian. – The individual papers are included in their corresponding subject categories – see 022.132 - 022.134, 074.087, 077.052, 077.053, 084.305, 091.033, 102.018 - 102.020, 103.133, 103.135, 104.105, 105.160, 106.032, 107.016, 122.142, 124.100, 160.017.

003.041   Základy nebeské mechaniky.   P. Andrle.
Academia, nakladatelstvi Československé akademie věd, Praha. 308 pp. Price Kčs. 30.00 (1971).

003.042   Practical work in elementary astronomy.
M. Minnaert.
Translated from the English edition. "Mir", Moskva. 240 pp. Price 1 Rbl. 30 Kop. (1971). In Russian.

003.043   Theories of equilibrium figures of a rotating homogeneous fluid mass.   Y. Hagihara.
Scientific and Technical Information Office, National Aeronautics and Space Administration, Washington, D.C. NASA SP-186. [For sale by the Superintendent of Documents, U.S. Government Printing Office, Washington, D.C.], 8 + 168 pp. Price $ 1.75 (1970). – Contents Chapter I. General properties of equilibrium figures; Chapter II. Ellipsoidal figures of equilibrium; Chapter III. Lamé functions; Chapter IV. Theory of Poincaré; Chapter V. Theory of Jeans; Chapter VI. Theory of Liapounov; Chapter VII. Theory of Lichtenstein; Appendix A. Poincaré's tidal theory.

003.044   Wave generation and shaping.   L. Strauss.
McGraw-Hill Book Company, New York. Second edition. 775 pp. (1970).

003.045   Handbook on celestial mechanics and astrodynamics.   V. K. Abalakin, E. P. Aksenov, E. A. Grebenikov, Yu. A. Ryabov.
"Nauka", Moskva. 584 pp. Price 2 Rbl. 2 Kop. (1971). In Russian. – Review in Referativ. Zhurn. 62. Issled. kosm. prostranstva, 10.62.324; 51. Astron., 11.51.126 (1971).

003.046   Stars, galaxies, and metagalaxy.   T. A. Agekian.
Izdatelstvo Nauka, Moscow. Second edition. 324 pp. (1970). – Review in Astrophys. Space Sci., Vol. 12, 258 - 259; 1971 (Z. Kopal).

003.047   The sun and planets.   S. M. Ajvazyan.
"Ajastan", Erevan. 164 pp. Price 1 Rbl. 3 Kop.

(1971). In Russian. — Review in Referativ. Zhurn. 51. Astron., 8.51.60 (1971).

003.048 Atome, Mensch und Universum. H. Alfvén.
Suhrkamp Verlag, Frankfurt. 100 pp. Price DM 10.00 (1971). — Review in SuW, Vol. 10, 313; 1971 (*T. Schmidt*).

003.049 Atoms, stars, and nebulae. L. H. Aller.
Harvard University Press, Cambridge, Mass. Second edition. 12 + 351 pp. Price $ 11.95 (1971). — Reviews in Sky Telescope, Vol. 42, 167, 226 - 228; 1971 (*L. Motz*).

003.050 Investigation of the outer space 1969. Cosmic automatic instruments for the investigation of the moon and the circumlunar space (1958–1968). A. V. Baevskij.
Itogi nauki VINITI AN SSSR, Moskva. 194 pp. Price 1 Rbl. 40 Kop. (1971). In Russian.

003.051 Celestial navigation for yachtsmen. M. Blewitt.
Edward Stanford Ltd., London. 5th edition. 66 pp. Price £ 1.00 (1971). — Review in Journ. Inst. Navigation, Vol. 24, 576 - 577; 1971 (*M. Richey*).

003.052 Geodesy. G. Bomford.
Clarendon Press, Oxford. Third edition. 10 + 731 pp. Price £ 10.00 (1971). — Review in Geophys. Journ. Roy. Astron. Soc., Vol. 24, 210; 1971 (*A. H. Cook*).

003.053 Astronomical use of television-type image sensors. V. R. Boscarino (Editor).
National Aeronautics and Space Administration, Washington, D.C. NASA SP-256, 217 pp. [Available from National Technical Information Service, Springfield, Va.], Price $ 3.00 (1971). — Review in Sky Telescope, Vol. 42, 231 - 232 (1971).

003.054 Noctilucent clouds. V. A. Bronshtehn, N. I. Grishin.
"Nauka", Moskva. 360 pp. Price 1 Rbl. 84 Kop. (1970). In Russian. — Review in Priroda, No. 11.71, p. 115 - 116; 1971 (*N. V. Vasil'ev*).

003.055 Problems in the foundations of physics. M. Bunge (Editor).
Springer-Verlag, Berlin–Heidelberg–New York. 6 + 162 pp. Price DM 54.00 (1971).

003.056 Interference of electromagnetic waves. A. H. Cook.
International Series of Monographs on Physics. Clarendon Press, Oxford; Oxford University Press, London. 8 + 253 pp. Price £ 5.00 (1971). — Review in Nature, Vol. 233, 431; 1971 (*J. Dyson*).

003.057 The conceptual foundations of contemporary relativity theory. J. Cowperthwaite Graves.
M.I.T. Press, Cambridge, Mass., 12 + 362 pp. Price $ 15.00 (1971).

003.058 The view from space. Photographic exploration of the planets. M. E. Davies, B. C. Murray.
Columbia University Press, New York. 12 + 166 pp. Price $ 14.95 (1971).

003.059 Early Greek astronomy to Aristotle. D. R. Dicks.
Thames and Hudson, London. 272 pp. Price £ 2.50 (1971). — Review in Journ. British Astron. Ass., Vol. 82, 70 - 71; 1971 (*C. A. Ronan*).

003.060 Kenntnis vom erdnahen Raum im Wandel eines Jahrhunderts. W. Dieminger.
Johann Ambrosius Barth, Leipzig. 31 pp. Price DM 6.00 (1971).

003.061 Dem roten Planeten auf der Spur. Mars und das Sonnensystem. G. Doebel.
Verlag M. DuMont Schauberg, Köln. 240 pp. Price DM 19.80 (1971). — Reviews in Orion Schaffhausen, 29. Jahrgang, p. 122 - 123; 1971 (*N. Hasler-Gloor*); SuW, Vol. 10, 251; 1971 (*G. D. Roth*).

003.062 Evolution of the earth. R. H. Dott, R. L. Batten.
McGraw Hill, Book Company, Maidenhead. 576 pp. Price £ 6.50 (1971).

003.063 Das Weltall im Bild. Photographischer Himmelsatlas.
Edited by A. Eisenhuth under collaboration of H. Haffner. Verlag Styria, Graz–Wien–Köln. Second edition. 28 pp. + 205 illustrations. Price DM 39.00 (1971). — Review in SuW, Vol. 10, 252; 1971 (*G. Klare*).

003.064 Methods of astrodynamics. P. R. Escobal.
Translated from the English edition. "Mir", Moskva. 341 pp. Price 2 Rbl. 14 Kop. (1971). In Russian. — Review in Referativ. Zhurn. 62. Issled. kosm. prostranstva, 8.62.267 (1971).

003.065 Die Werke von Jakob Bernoulli. Band 1. J. O. Fleckenstein (Editor).
Birkhäuser Verlag, Basel. 12 + 541 pp. Price sfr. 76.00 (1969). Review in Journ. History Astron., Vol. 2, 207; 1971 (*A. E. Shapiro*).

003.066 Das Weltall. Eine moderne Kosmogonie. C. Friedemann.
Urania Verlag, Leipzig–Jena–Berlin. Second revised edition. 224 pp. Price DM 6.80 (1971).

003.067 My world line: An informal autobiography. G. Gamow.
Viking Press, New York. 178 pp. Price $ 5.95 (1970). — Review in Physics Today, Vol. 24, No. 3, p. 51 - 52; 1971 (*R. A. Alpher, R. Herman*).

003.068 Introduction to orbital mechanics. F. T. Geyling, H. R. Westerman.
Addison-Wesley, Reading, Mass., 349 pp. Price $ 15.00 (1971).

003.069 Boundaries of the universe. J. S. Glasby.
Georg Allen & Unwin, London. 296 pp. Price £ 4.50 (1971). — Review in Journ. British Astron. Ass., Vol. 82, 71 - 72; 1971 (*P. Moore*).

003.070 Atlas du ciel de l'astronome amateur. D. Godillon.
Editions Doin, Paris. 413 pp. Price F 78.00 (1971). — Reviews in Orion, 29. Jahrgang, p. 121 - 122; 1971 (*E. Antonini*); Sky Telescope, Vol. 42, 167 - 168 (1971).

003.071 Al-Bitruji: On the principles of astronomy.
An edition of the Arabic and Hebrew versions with translation, analysis and an Arabic-Hebrew-English glossary. B. R. Goldstein.
Yale University Press, New Haven. Two volumes, 24 + 610 pp. Price $ 35.00 (1971).

003.072 Introduction to space science. R. C. Haymes.
Wiley & Sons, New York. 556 pp. Price $ 14.95 (1971). — Review in Sky Telescope, Vol. 42, 232 (1971).

003.073 **Basic Astronomy.** H. Haysham.
Thomas Reed Publications Ltd., London. 234 pp.
Price £ 3.25 (1971). – Review in Sky Telescope, Vol. 42, 301 (1971).

003.074 **Doppelsterne.** W. D. Heintz.
Wilhelm Goldmann Verlag, München. 186 pp. Price DM 24.00 (1971). – Review in Sky Telescope, Vol. 42, 301 (1971).

003.075 **The radio universe.** J. S. Hey.
Pergamon International Popular Science Series. Pergamon Press, Oxford–New York. 8 + 248 pp. Price £ 2.50, $ 6.75 respectively (1971). – Reviews in Journ. British Astron. Ass., Vol. 81, 496; 1971 (*R. J. J. Langton*); Orion, 29. Jahrgang, p. 163; 1971 (*H. Rohr*); Observatory, Vol. 91, p. 229 - 230; 1971 (*N. J. B. A. Branson*).

003.076 **Atomspektren.** W. R. Hindmarsh.
Wissenschaftliche Taschenbücher. Akademie-Verlag, Berlin. 424 pp. Price DM 25.00 (1971).

003.077 **The Tompion clocks at Greenwich and the dead beat escapement.**
D. Howse, with an appendix by B. Hutchinson.
National Maritime Museum, Greenwich. 40 pp. Price 30 p. (1971). – Review in Journ. British Astron. Ass., Vol. 82, 72 - 73; 1971 (*C. A. Ronan*).

003.078 **The riddle of the individual and his universe.** J. Hyman.
Fundamental Research Press, Berkeley, Calif., 6 + 128 pp. Price $ 8.50 (1970). – Review in Space Sci. Rev., Vol. 12, 532; 1971 (*C. de Jager*).

003.079 **Red giants and white dwarfs. Man's descent from the stars.** R. Jastrow.
Harper and Row, New York. Second edition. 16 + 192 pp. Price $ 6.95 (1971). – Review in Sky Telescope, Vol. 42, 167 (1971).

003.080 **The Harvard College Observatory. The first four directorships, 1839–1919.**
B. Z. Jones, L. G. Boyd.
Belknap (Harvard University Press), Cambridge, Mass. 16 + 496 pp. Price $ 15.00 (1971). – Reviews in Sky Telescope, Vol. 42, 233, 366 - 369; 1971 (*D. Hoffleit*).

003.081 **Advances in heat transfer, Vol. 7.**
T. F. Irvine, Jr., J. P. Hartnett (Editors).
Academic Press, New York. 406 pp. Price £ 10.25 (1971). – Review in Journ. British Interplanet. Soc., Vol. 24, 622 (1971).

003.082 **Geometrie und Wirklichkeit.** B. Kanitschneider.
Erfahrung und Denken, Vol. 36. Duncker & Humblot, Berlin. 389 pp. Price DM 76.00 (1971).

003.083 **The astrological history of Masha 'Allah.**
E. S. Kennedy, D. Pingree.
Harvard University Press, Cambridge, Mass. 14 + 206 pp. Price $ 10.00 (1971).

003.084 **Epoch of great astronomical discoveries (XIIIth - XIXth centuries).**
V. G. Kharalampiev, N. S. Nikolov.
"Nauka i izkustvo", Sofia. 368 pp. Price 1.95 Lv. (1970). In Bulgarian.

003.085 **A new photographic atlas of the moon.** Z. Kopal.
Taplinger Publishing Co., New York. 8 + 312 pp. Price $ 20.00 (1971). – Review in Sky Telescope, Vol. 42,

299 - 300; 1971 (*J. Ashbrook*).

003.086 **Widening horizons: Man's quest to understand the structure of the universe.** Z. Kopal.
Taplinger Publishing Company Inc., New York. 176 pp. Price $ 6.95 (1971). – Review in Physics Today, Vol. 24, No. 10, p. 56 - 57; 1971 (*F. M. Branley*).

003.087 **Elements of astronautics in courses of physics and astronomy. (Handbook for teachers at colleges).**
I. V. Kosheurov.
Translated from the Russian edition. "Nar. prosv.", Sofia. 144 pp. Price 0.50 Lv. (1970). In Bulgarian.

003.088 **History of the earth. An introduction to historical geology.** B. Kummel.
W. H. Freeman and Company, San Francisco. Second edition. 707 pp. Price £ 5.20 (1971). – Reviews in Geophys. Journ. Roy. Astron. Soc., Vol. 24, 208 - 209; 1971 (*A. J. Smith*); Journ. British Interplanet. Soc., Vol. 24, 560 (1971).

003.089 **Electromagnetic exploration of the moon.** W. I. Linlor (Editor).
Mono Book Corp., Baltimore. 240 pp. Price £ 20.00 (1970). Reviews in Phys. Earth Planet. Interiors, Vol. 4, 429 - 430; 1971 (*D. C. Tozer*); Space Sci. Rev., Vol. 12, 528; 1971 (*G. H. Pettengill*).

003.090 **Electromagnetic fields and waves.** P. Lorrain, D. R. Corson.
W. H. Freeman and Company, San Francisco. Second edition. 706 pp. Price £ 7.00, £ 4.00 respectively (1970). – Review in Journ. British Interplanet. Soc., Vol. 24, 560 (1971).

003.091 **Dark nebulae, globules, and protostars.** B. T. Lynds (Editor).
The University of Arizona Press, Tucson, Arizona. 8 + 150 pp. Price $ 7.50 (1971). – Reviews in Journ. Roy. Astron. Soc. Canada, Vol. 65, 306; 1971 (*S. van den Bergh*); Sky Telescope, Vol. 42, 232 (1971).

003.092 **Time and the space-traveller.** L. Marder.
Allen and Unwin, London. 216 pp. Price £ 3.25 (1971). – Review in Spaceflight, Vol. 13, 396 - 397; 1971 (*D. F. Lawden*).

003.093 **Course of general astrophysics. Text-book for students of astronomy.** D. Ya. Martynov.
"Nauka", Moskva. Second edition. 616 pp. Price 1 Rbl. 77 Kop, (1971). In Russian. – Review in Referativ. Zhurn. 51. Astron., 11.51.36 (1971).

003.094 **Pioneering in outer space.** H. Messel, S. T. Butler (Editors).
Shakespeare Head Press, Sydney. 505 pp. Price 8.40 Australian dollars (1970). – Review in Spaceflight, Vol. 13, 395; 1971 (*J. Sved*).

003.095 **Astronomy.** D. H. Menzel.
Thames & Hudson, London. 320 pp. Price £ 6.30 (1971). – Review in Journ. British Astron. Ass., Vol. 81, 495; 1971 (*P. Moore*).

003.096 **Guide des étoiles et des planètes.**
D. H. Menzel. Translated from English by M. Egger, F. Egger, with a preface by P. Couderc.
Editions Delachaux et Niestlé, Neuchâtel–Paris. 405 pp. (1971). – Reviews in L'Astronomie, 85ᵉ année, p. 358; 1971 (*B. Morando*), Orion, 29. Jahrgang, p. 121; 1971 (*E. Antonini*).

003.097 **Les trois étapes de la cosmologie.**

J. Merleau-Ponty, B. Morando.
Collection "Science Nouvelle". Éditions Robert Laffont, Paris. 316 pp. Price F 27.90(1971).—Reviews in L'Astronomie, 85e année, p. 438 (1971); Sky Telescope, Vol. 42, 301 (1971).

003.098 **Exploring the planets.** I. Nicolson (Editor).
P. Hamlyn, London. 159 pp. Price 30 p. (1970). — Review in Spaceflight, Vol. 13, 273; 1971 (*P. R. Skidmore*).

003.099 **Tycho Brahe: A biography.** W. Norlind.
Gleerup, Lund, Sweden. 422 pp. Price 60 kr. (1970). In Swedish. — Review in Journ. History Astron., Vol. 2, 205 - 207; 1971 (*V. E. Thoren*).

003.100 **The optical aurora.** A. Omholt.
Physics and Chemistry in Space, Vol. 4. Springer-Verlag, Berlin—Heidelberg—New York. 13 + 198 pp. Price DM 58.00 (1971).

003.101 **Lens aberration data.** J. M. Palmer.
Adam Hilger, London; American Elsevier, New York. 11 + 118 pp. Price £ 5.00, $ 17.00 respectively (1971). — Review in Observatory, Vol. 91, 167; 1971 (*C. F. W. Harmer*).

003.102 **The infinite worlds of Giordano Bruno.** A. Mann Paterson.
C. C. Thomas, Springfield, Ill. 12 + 227 pp. (1970). — Review in Journ. History Astron., Vol. 2, 204 - 205; 1971 (*W. H. Donahue*).

003.103 **Raumstationen. Laboratorien im All.** J. von Puttkamer.
Verlag Chemie, Weinheim (Germany). 216 pp. Price DM 29.80 (1971).

003.104 **Geology.** W. C. Putman, revised by A. B. Bassett.
Oxford University Press, New York. Second edition. 496 pp. Price £ 4.25 (1971). — Review in Journ. British Interplanet. Soc., Vol. 24, 744 (1971).

003.105 **Goals and means in the conquest of space.** R. G. Perel'man.
Translation of the Russian edition.
National Aeronautics and Space Administration. NASA TT F-595 [Available from Superintendent of Documents, U. S. Government Printing Office, Washington, D. C.], 6 + 178 pp. Price $ 3.00 (1971). — Review in Spaceflight, Vol. 13, 473 - 474; 1971 (*A. V. Cleaver*).

003.106 **Le soleil, la terre et la radio.** J. A. Ratcliffe.
Translated from English by C. N. Martin.
Collection L'Univers des Connaissances, éditions Hachette, Paris. 250 pp. (1970). — Review in L'Astronomie, 85e année, p. 438 (1971).

003.107 **The nature of light: An historical survey.** V. Ronchi, translated by V. Barocas.
Harvard University Press, Cambridge, Mass. 12 + 288 pp. Price $ 16.00 (1970). — Review in Journ. Optical Soc. America, Vol. 61, 1717; 1971 (*E. S. Barr*).

003.108 **Advances in geophysics. Vol. 15.** H. E. Landsberg, J. van Mieghem (Editors).
Academic Press, New York. 10 + 332 pp. Price $ 18.00 (1971).

003.109 **Interplanetary flight and communication. Vol. 1, No. 1: Dreams, legends and early fantasies.** N. A. Rynin.
Translated from the Russian edition (Leningrad, 1928). Israel Program for Scientific Translations, Jerusalem. [Available from National Technical Information Service, Springfield, Va.], 4 +

114 pp. Price $ 3.00 (1970).

003.110 **The solar spectrum from lambda 7498 to lambda 12016. A table of measures and identifications.**
J. W. Swenson, W. S. Benedict, L. Delbouille, G. Roland.
Mém. Soc. Roy. Sci., Liège, Special Vol. 5, 450 pp. Price B.fr. 600 (1970). — Review in Solar Physics, Vol. 22, 240; 1972 (*C. de Jager*).

003.111 **States, waves, and photons: A modern introduction to light.** J. W. Simmons, M. J. Guttmann.
Addison-Wesley Publishing Co., Inc., Reading, Mass. 279 pp. Price $ 9.50 (1970). — Review in IEEE Spectrum, Vol. 8, No. 9, p. 101; 1971 (*W. A. Miller*).

003.112 **Lectures on celestial mechanics.**
C. L. Siegel, J. K. Moser. Translation by C. L. Kalme. Revised and enlarged translation of "Vorlesungen über Himmelsmechanik" by C. L. Siegel (1956). Grundlehren der mathematischen Wissenschaften, Vol. 187. Springer-Verlag, Berlin—Heidelberg—New York. 10 + 290 pp. Price DM 78.00 (1971).

003.113 **Physics of solar continuum radio bursts.** A. Krüger.
Akademie-Verlag, Berlin. 164 pp. Price DM 39.00 (1971).

003.114 **Introduction into the theory of cosmic navigation.** V. S. Shebshaevich.
Sovetskoe radio, Moskva. 296 pp. Price 1 Rbl. 28 Kop. (1971). In Russian. — Review in Referativ. Zhurn. 62. Issled. kosm. prostranstva, 10.62.472 (1971).

003.115 **Modern cosmology.** D. W. Sciama.
Cambridge University Press, London. 8 + 212 pp. Price £ 3.60, $ 8.95 respectively (1971). — Reviews in Nature, Vol. 234, 54; 1971 (*J. V. Narlikar*); Sky Telescope, Vol. 42, 373 (1971).

003.116 **The physical foundations of general relativity.** D. W. Sciama.
Translated from the English edition by V. A. Ugarov. "Mir", Moscow. (1971). In Russian.

003.117 **Planeten. Geschwister der Erde.** W. Sandner.
Verlag Chemie, Weinheim (West Germany). 224 pp. Price DM 38.00 (1971).

003.118 **J. C. Poggendorff, Biographisch-Literarisches Handwörterbuch der exakten Naturwissenschaften.**
Vol. VIIb, Part 4, 1st number. H. Salié (Editor).
Published by Akademie-Verlag, Berlin. 160 pp. Price DM 24.00 (1971).

003.119 **Unser astronomisches Weltbild heute.** K. Schütte.
Herder-Bücherei, Freiburg im Breisgau (West Germany). 351 pp. Price DM 7.90 (1971). — Review in Sky Telescope, Vol. 42, 373 (1971).

003.120 **Kursbuch für das Sonnensystem. Weltraumfahrt bis zum Jahr 2000.** B. Stanek.
Hallwag Verlag, Bern - Stuttgart. 167 pp. Price DM 22.00 (1971).

003.121 **The earth sciences.** A. N. Strahler.
Harper and Row Publishers, New York - London. 824 pp. Price $ 15.95 (1971). — Review in Sky Telescope, Vol. 42, 372 (1971).

003.122 **Continental drift.** A study of the earth's moving

surface.   D. H. Tarling, M. P. Tarling.
G. Bell & Sons, Ltd., London; Doubleday & Co., New York.
112 pp. Price £ 1.50 (1971). – Reviews in Naturwissenschaften, 58. Jahrgang, p. 464; 1971 (*F. L. Boschke*); Sci.American, Vol. 225, No. 6, p. 111 - 112 (1971).

003.123   **The haven-finding art, a history of navigation from Odysseus to Captain Cook.**   E. G. R. Taylor.
American Elsevier, New York. 310 pp. Price $ 9.50 (1971).
Review in Sky Telescope, Vol. 42, 372 (1971).

003.124   **Shock waves in collisionless plasmas.**
D. A. Tidman, N. A. Krall.
Wiley-Interscience, New York; John Wiley & Sons Ltd., Chichester. 12 + 176 pp. Price $ 10.50, £ 4.90 respectively (1971). – Review in Journ. British Interplanet. Soc., Vol. 24, 624 (1971).

003.125   **Gravitationstheorie und Äquivalenzprinzip.**
H.-J. Treder.
Akademie-Verlag, Berlin. Mathematische Lehrbücher und Monographien, II. Abteilung. 8 + 122 pp. Price DM 22.00 (1971).

003.126   **An introduction to meteorological optics.**
R. A. R. Tricker.
American Elsevier Publishing Co., New York. 285 pp. Price $ 11.00 (1971). – Review in Publ. Astron. Soc. Pacific, Vol. 83, 509 (1971).

003.127   **Space methods of terrestrial studies.**
B. V. Vinogradov, K. Ya. Kondratyev.
Gidromet Izdat., Leningrad. 188 pp. Price 1 Rbl. 44 Kop. (1971). – Review in Applied Optics, Vol. 10, A12 (1971).

003.128   **Pulsars.** (Collected papers).
V. V. Vitkevich (Editor). Translated from the English edition.
"Mir" Moskva. 270 pp. Pric 1 Rbl. 36 Kop. (1971).
In Russian.

003.129   **Conquista de la luna.**   J. J. Wahl (Editor).
Ediciones Emeuve, Barcelona. 407 pp. (1970).
Review in Spaceflight, Vol. 13, 396; 1971 (*H. T. Mills*).

003.130   **Brockhaus ABC Astronomie.**   A. Weigert, H. Zimmermann. Revised by H. Zimmermann.
VEB F. A. Brockhaus-Verlag, Leipzig. Third edition. 453 pp.
Price M. 12.70 (1971). – Review in Astron. in der Schule,
8. Jahrgang, p. 93 - 94; 1971 (*R. Kollar*).

003.131   **Astronomy.**   R. Whittingham.
Hubbard Press, Northbrook, Ill.,48 pp. Price $ 2.95 (1971). – Review in Sky Telescope, Vol. 42, 168 (1971).

003.132   **Einführung in die Vermessungstechnik.**   H. Wittke.
Ferd.–Dümmlers–Verlag, Bonn. 520 pp. Price
DM 78.00 (1971). – Review in Österreich. Zeitschr. Vermessungswesen, 59. Jahrgang, p. 155 - 156; 1971 (*F. Ackerl*).

003.133   **Meteorites and the origin of planets.**   J. A. Wood.
Translated from the English edition.
"Mir", Moskva. 176 pp. Price 59 Kop. (1971). In Russian.
Reviews in Priroda, No. 11.71, p. 120 (1971); Referativ.
Zhurn. 51. Astron. 11.51.307 (1971).

003.134   **Analysis of Apollo 10 photography and visual observations.**
NASA Manned Spacecraft Center. National Aeronautics and Space Administration, NASA SP-232, [available from Superintendent of Documents, U. S. Government Printing Office,
Washington, D. C.], 226 pp. Price $4.25 (1971). – Review in Sky Telescope, Vol. 42, 233 (1971).

003.135   **Apollo 11 mission report, 1971.**
National Aeronautics and Space Administration,
NASA SP-238, [available from the National Technical Information Service, Springfield, Va.], 217 pp. Price $3.00 (1971). – Review in Sky Telescope, Vol. 42, 168 (1971).

003.136   **Apollo 14: Preliminary science report.**
NASA Manned Spacecraft Center. National Aeronautics and Space Administration, NASA SP-272, [available from Superintendent of Documents, U. S. Government Printing Office, Washington, D. C.], 309 pp. Price $3.00 (1971).
Review in Sky Telescope, Vol. 42, 233 (1971).

003.137   **The Pioneer mission to Jupiter, 1971.**
National Aeronautics and Space Administration,
NASA SP-268, [available from Superintendent of Documents, U. S. Government Printing Office, Washington, D. C.], 46 pp.
Price 30 cents (1971). – Review in Sky Telescope, Vol. 42, 301 (1971).

003.138   **Lunar photographs from Apollo 8, 10, and 11.**
R. G. Musgrove (Editor).
National Aeronautics and Space Administration, Washington, D. C. [Available from Superintendent of Documents, U. S. Government Printing Office, Washington, D. C.], 119 pp.
Price $4.00 (1971). – Review in Sky Telescope, Vol. 42, 231 (1971).

003.139   **Geschichte der Naturwissenschaften und der Technik im 19. Jahrhundert.**
VDI-Verlag, Düsseldorf. 247 pp. Price DM 32.00 (1971). –
Review in SuW, Vol. 10, 313; 1971 (*G. D. Roth*).

003.140   **New technique in astronomy.** (Materials of the conference of the Commission for instrument construction of the Astronomical Council of the USSR Academy of Sciences, Pulkovo, 27 - 30 Nov. 1967. Vypusk 3).
"Nauka", Leningrad. 237 pp. Price 1 Rbl. 50 Kop. (1970).
In Russian.

003.141   **Concepts of space: The history of theories of space.**
M. Jammer.
Harvard Univ. Press, Cambridge, Mass. Second edition. 221 pp.
Price $ 5.50 (1969). – Review in Physics Today, Vol. 24, No. 5, p. 55; 1971 (*A. Harvey*).

003.142   **Apollo 14: Science at Fra Mauro.**   W. Froehlich.
National Aeronautics and Space Administration.
EP-91, [available from Superintendent of Documents, U. S. Government Printing Office, Washington, D. C.], 48 pp. Price $ 1.25 (1971). – Review in Sky Telescope, Vol. 42, 233 (1971).

003.143   **Kepler als Forscher.**   H. C. Freiesleben.
Wissenschaftliche Buchgesellschaft, Darmstadt.
82 pp. (1970).

003.144   **Il cielo, luci e ombre nell'universo.**   G. Cecchini.
UTET, Torino, Two volumes. 1149 pp. Price
L. 24000 (1969). – Review in L'Universo, Anno 51, 1211;
1971 (*O. Mannelli*).

003.145   **Science year – The world book science annual 1972**
Field Enterprises Educational Corporation, Chicago, Ill.
444 pp. Price $ 7.95 (1971). – Review in Sky Telescope, Vol. 42, 372 (1971).

003.146 **XXIVth Herzen readings. Intercollege conference. Theoretical physics and astronomy. Paper abstracts.**
Leningradskij gosudarstvennyj pedagogicheskij institut im. A.I. Herzena, Leningrad. 95 pp. Price 42 Kop. (1971). In Russian.

003.147 **An original theory or new hypothesis of the universe, 1750.** T. Wright of Durham.
A facsimile Reprint together with the first publication of 'A theory of the Universe', (1734). Introduction and transcription by M. A. Hoskin.
Macdonald, London; American Elsevier Publishing Company, New York. 38 + 178 pp. Price $ 32.00 (1971). — Reviews in Journ. History Astron., Vol. 2, 208 - 209; 1971 (*G. J. Whitrow*); L'Astronomie, 85ᵉ année, p. 438 (1971); Sky Telescope, Vol. 42, 297 - 299; 1971 (*J. W. Streeter).*

003.148 **Electronic and ionic impact phenomena. Vol. 3: Slow collisions of heavy particles.**
H. S. W. Massey, E. H. S. Burhop, H. B. Gilbody (Editors).
Clarendon Press, Oxford. Second edition. 20 + 850 pp. Price $ 12.00 (1971). — Review in Planet. Space Sci., Vol. 19, 1197; 1971 (*D. R. Bates*).

003.149 **Das große Projekt.** Raumfahrt und Apollo-Programm: Die wissenschaftlichen Erkenntnisse und der praktische Nutzen für die Menschheit.
Edited by Carl Zeiss in cooperation with United States Information Service. Karl Weinbrenner & Söhne, Stuttgart. 150 pp. Price DM 29.50 (1971). — Review in SuW, Vol. 10, 341 - 342; 1971 (*U. Hehlgans*).

003.150 **Optical telescope technology.**
Office of Space Science and Applications. National Aeronautics and Space Administration. NASA SP-233, [available from Superintendent of Documents U.S. Government Printing Office, Washington, D.C.], 783 pp. Price $ 6.25 (1970). — Review in Sky Telescope, Vol. 41, 107 (1971).

003.151 **Flares of red dwarf stars.** R. E. Gershberg.
Izdatel'stvo Nauka, Moskva. 168 pp. Price 75 Kop. (1970). English translation by D. J. Mullan. Available from the Secretary of Armagh Observatory, Armagh, N. Ireland. Price $ 2.50, £ 1.00 respectively. — Review in Inform. Bull. Variable Stars, (IAU Commission 27), Konkoly Obs., Budapest, No. 591 (1971).

003.152 **Radio activity in the earth's history.**
"Nauka", Moskva. (1970). In Russian. — Review in Zemlya i Vselennaya, No. 3, p. 93 (1971).

003.153 **Transaction of the USSR conference on the problem "Astrophysical phenomena and the radio-carbon", Tbilisi, 1969, November 25 - 27.**
Tbilisi. universitet, Tbilisi. 148 pp. Price 70 Kop. (1970). In Russian.

003.154 **Unser Platz im Weltall. Entwicklung des astronomischen Denkens.** P. Moore.
Translated from English into German. Umschau Verlag, Frankfurt. 124 pp. Price DM 9.80 (1971). — Review in SuW, Vol. 10, 342; 1971 (*G. D. Roth*).

003.155 **Clerks and craftsmen in China and the west. Lectures and addresses on the history of science and technology.**
J. Needham, Wang Luig, Lu Gwei-Djen, Ho Ping-Yü.
Cambridge University Press, Cambridge. 20 + 470 pp. Price £ 7.50 (1970). — Review in Journ. History Astron., Vol. 2, 120 - 121; 1971 (*G. L. E. Turner*).

003.156 **Three Copernican treatises. The Commentariolus of Copernicus, The Letter against Werner, The Narratio prima of Rheticus.**
Third edition, revised with a biography of Copernicus and Copernicus bibliographies, 1939 - 1958 and 1959 - 1970. Translated with introduction and notes by E. Rosen.
Octagon, New York, 18 + 426 pp. Price $ 20.00 (1971).

003.157 **John Wilins, 1614 - 1672: An intellectual biography.** B. J. Shapiro.
University of California Press, Berkeley, Calif. 333 pp. Price £ 4.55 (1970). — Review in Journ. History Astronomy, Vol. 2, 46 - 48; 1971 (*A. J. Turner*).

003.158 **L'Astronomie à la portée de tous.** M. Oliveau.
Editions maritimes et d'outre-mer. 78 pp. (1971). Review in L'Astronomie, 85ᵉ année, p. 438 (1971).

003.159 **Storia dell'astronomia da Talete a Keplero.**
Editore Feltrinelli, Milano. 398 pp. Price L. 4.000. Review in Coelum, Vol. 39, 200; 1971 (*A Retti*).

003.160 **Early scientific books in Schaffer Library, Union College.** W. Somers, B. Hindle.
Union College, Schenectady, New York. 70 pp. (1971). Review in Journ. History Astron., Vol. 2, 130 (1971).

003.161 **Proceedings of the symposium on turbulence of fluids and plasmas. 18th international symposium, April 1968.** J. Fox (Editor).
John Wiley & Sons, Ltd., Chichester. 511 pp. Price £ 8.00 (1970). — Review in Journ. British Interplanet. Soc., Vol. 24, 127 (1971).

003.162 **Space research in the United Kingdom 1968 - 1969.**
The Royal Society and The Science Research Council, London. 13 + 140 pp. Price 13s. 6d., $ 1.75 respectively (1970). — Review in Planet. Space Sci., Vol. 19, 1198 1971 (*G. G. Shepherd*).

003.163 **Astronomy.** R. H. Baker, L. W. Fredrick.
Van Nostrand Reinhold, New York. 9th edition. 14 + 632 pp. (1971).

003.164 **The nature of the universe.** C. Kilmister.
Thames and Hudson, London. 216 pp. Price £ 2.25 (1971).

003.165 **Guide to the planets.** P. Moore.
Lutterworth, London. Revised edition. 224 pp. Price £ 2.50 (1971).

003.166 **Fundamentals of astrodynamics.**
R. R. Bate, D. D. Mueller, J. E. White.
Dover Publications Inc., New York; Constable and Co., Ltd., London. 12 + 455 pp. Price £ 2.25 net. (1971).

003.167 **Geology and physics of the moon: A study of some fundamental problems.** G. Fielder (Editor).
Elsevier Publishing Company, Amsterdam — London — New York. 8 + 159 pp. Price Dfl. 75.00, $ 22.00 respectively (1971) (1971).

003.168 **The astronomy of Birr Castle.** P. Moore.
Mitchell Beazley, London. 12 + 81 pp. Price £ 1.75 (1971).

004 History of Astronomy, Chronology

004.001 **Nuremberg's astronomical heritage.** E. G. Forbes.
Journ. British Astron. Ass., Vol. 81, 391 - 393
(1971).

004.002 **Die außereuropäischen Sternwarten vor 100 Jahren.**
J. Classen.
Sterne, 47. Jahrgang, p. 147 - 150 (1971).

004.003 **Die "Harmonia Macrocosmica" des Andreas
Cellarius.** F. Lombard.
Orion, 29. Jahrgang, p. 67 - 68 (1971).

004.004 **Zum Gaußschen Kryptogramm von 1812.**
K.-R. Biermann.
Monatsber. Deutsch. Akad. Wiss. Berlin, Vol. 13, 152 - 157
(1971).

004.005 **Observatories and angle-measuring devices in Georgia of the Xth – XIIIth centuries.**
G. G. Georgobiani.
Byull. Abastumansk. Astrofiz. Obs., No. 40, p. 207 - 220
(1971). In Georgian.
The existence and the chronology of the observatories
in Tbilisi and Kutaisi in Medieval Georgia are considered. The
angle-measuring devices are being studied, informations on
which are found in chronicles, manuscripts and on coins.

004.006 **A minor anniversary for 1971.** C. M. Botley.
Journ. British Astron. Ass., Vol. 81, 469 (1971).

004.007 **Cassini and the Paris Observatory.** M. Friedjung.
Journ. British Astron. Ass., Vol. 81, 479 - 480
(1971).

004.008 **Glaciation and the stones of Stonehenge.**
G. A. Kellaway.
Nature, Vol. 233, 30 - 35 (1971).
More geological evidence can now be brought to bear on
the problem of the origin of Stonehenge. Spreading ice sheets
rather than human activity could have brought the rocks to
Salisbury Plain.

004.009 **The astronomical significance of the large Carnac
menhirs.** A. Thom, A. S. Thom.
Journ. History Astron., Vol. 2, 147 - 160 (1971).

004.010 **The Milky Way before Galileo.** S. L. Jaki.
Journ. History Astron., Vol. 2, 161 - 167 (1971).

004.011 **Apianus's Astronomicum Caesareum and its Leipzig
facsimile.** O. Gingerich.
Journ. History Astron., Vol. 2, 168 - 177 (1971).

004.012 **The curved and the straight: Cometary theory from
Kepler to Hevelius.** J. A. Ruffner.
Journ. History Astron., Vol. 2, 178 - 194 (1971).

004.013 **Gauss and the discovery of Ceres.**
E. G. Forbes.
Journ. History Astron., Vol. 2, 195 - 199 (1971).

004.014 **Reconstructing the planetary motions of the Eudoxean system.** D. Hargreave.
Scripta math., Vol. 28, 335 - 345 (1970).
Interest in the homocentric spheres revived after Schiaparelli in 1875 showed how the heavenly machinery of Eudoxos
worked and inhowfar it could imitate the apparent motions
of the planets.

004.015 **Colours of lunar eclipses according to Indian tradition.** W. Petri.
Veröff. Forsch.-Inst. Deutsch. Museums Gesch. der Naturw.
Techn., Ser. A, No. 68, 8 pp. (1970). – Abstr. in Zentralblatt
Math. Grenzgebiete, Vol. 211, 301 (1971).

004.016 **Zur Kinematik der Planetenbewegung in Copernicus'
Commentariolus.** W. S. Contro, J. V. Feitzinger,
R. Hartmann, F. W. Ihloff, H.-G. Märtl, F. Rex, M. Schramm,
H. Zehe.
Arch. History Exact. Sci., Vol. 6, 360 - 371 (1970).
Owing to the success of Kepler's astronomy, important
aspects of the kinematic models of Copernicus have become
neglected or misunderstood. In response to this situation, the
article aims to interpret the kinematic models of the planetary
motions in longitude set out by Copernicus in the 'Commentariolus'.

004.017 **Greenwich register of source material on the history
of astronomy.** D. Howse.
Quarterly Journ. Roy. Astron. Soc., Vol. 12, 335 (1971).

004.018 **Some West European scientific museums.**
J. C. Albergotti.
Astron. Soc. Pacific, Leaflet No. 505, 8 pp. (1971).

004.019 **Het begin van de radiosterrekunde: Jansky en Reber.** F. P. Israel.
Hemel en Dampkring, Vol. 69, 279 - 283 (1971).

004.020 **Zodiaken och myterna. En kulturhistorisk studie, II.**
H. Laurell.
Astron. Tidsskr., Årg. 4, 97 - 122 (1971).

004.021 **Beknopt historisch overzicht van bijdragen van amateurastronomen aan de astronomie.**
W. Schmidt Sr.
Hemel en Dampkring, Vol. 69, 305 - 317 (1971).

004.022 **Stonehenge.** U. Losacco.
L'Universo, Anno 51, p. 1213 - 1224 (1971).
In Italian.

004.023 **Die Astrologie des Johannes Kepler.** G. D. Roth.
SuW, Vol. 10, 320 - 322 (1971).

004.024 **Johannes Kepler and the Rudolphine Tables.**
O. Gingerich.
Sky Telescope, Vol. 42, 328 - 333 (1971).

004.025 **The "first" Newtonian.** R. A. Wells.
Sky Telescope, Vol. 42, 342 - 344 (1971).

004.026 **Christopher Scheiner's observations of an object
near Jupiter.** J. Ashbrook.
Sky Telescope, Vol. 42, 344 - 345 (1971).

004.027 **Die Geschichte der Keplerschen Gesetze.**
U. Hoyer.
Phys. Blätter, 27. Jahrgang, p. 542 - 548 (1971).

004.028 **The rational basis of Kepler's laws.** E. G. Forbes.
Journ. British Astron. Ass., Vol. 82, 33 - 37 (1971).

004.029 **'New astronomy' by J. Kepler.** (Introduction into

this work).
Translated from the German edition by L. A. Vajnshtejn.
Priroda, No. 12.71, p. 54 - 58 (1971). In Russian.

004.030 Kepler and modern physics.
Yu. A. Danilov, Ya. A. Smorodinsky.
Priroda, No. 12.71, p. 59 - 63 (1971). In Russian.

004.031 Aus der Geschichte der Rudolfinischen Tafeln.
Sternenbote, 14. Jahrgang, p. 170 - 174 (1971).
Reproduced of W. Gerlach, M. List, "Johannes Kepler — Do-
kumente zu Lebenszeit und Lebenswerk".

004.032 Die instrumentelle Einrichtung einer Sternwarte
vor 100 Jahren. J. Classen.
Sterne, 47. Jahrgang, p. 210 - 212 (1971).

004.033 The quest in natural sciences. Methodological prob-
lems. V. Ambartsumyan, V. Kazyutinskij.
Nauka i zhizn', 1971, No. 6, p. 26 - 32. In Russian. — Abstr.
in Referativ. Zhurn. 51. Astron., 12.51.1 (1971).

004.034 Basic material on the history of physical and mathe-
matical sciences (XIIIth international congress on
the history of science, Moskva, 18 - 24 Aug., 1971).
"Nauka", Moskva. 62 pp. (1971). In Russian. — Review in
Referativ. Zhurn. 51. Astron., 12.51.16 (1971).

004.035 Kepler und die Gravitationstheorie. H.-J. Treder.
Sterne, 47. Jahrgang, p. 239 - 241 (1971).

004.036 Die Astronomenfamilie Kirch. P. Aufgebauer.
Sterne, 47. Jahrgang, p. 241 - 247 (1971).

004.037 Ergänzung zu den Beiträgen über die Sternwarten
vor hundert Jahren von J. Classen [Sterne, 47. Jahr-
gang, p. 29 - 31, 85 - 89 (1971)].** D. B. Herrmann.
Sterne, 47. Jahrgang, p. 213 (1971).

004.038 Notas de la visita que miembros de nuestra Sociedad
hicimos el domingo 26 de julio de 1971, a las grutas
de Xochicalco, con el objeto de observar el paso del sol por la
proximidad del zenit del lugar.
F. J. Reyna, A. González Solís.
El Universo, Vol. 25, 119 - 121 (1971).

004.039 La naturaleza de la luz. G. M. Mallén Fullerton.
El Universo, Vol. 25, 135 - 137 (1971).

004.040 The beginnings of photometry.
W. E. K. Middleton.
Applied Optics, Vol. 10, 2592 - 2594 (1971).
A brief account of the beginnings of photometry, with
emphasis on some of the early attempts to quantify light that
preceded the actual invention of useful photometers, is presented.

004.041 Kepler und seine Bedeutung für die Weltraumfahrt.
K. Schütte.
Astronautik, Hermann-Oberth-Ges., 1971, No. 3/4, 3 pp.
(1971).

004.042 Two lunar texts of the Achaemenid period from
Babylon. A. Aaboe, A. Sachs.
Centaurus, Vol. 14, 1 - 22 (1969). — Abstr. in Zentralblatt
Math. Grenzgebiete, Vol. 213, 3 (1971).

004.043 The origin of "System B" of Babylonian astronomy.

O. Neugebauer.
Centaurus, Vol. 12, 209 - 214 (1968). — Abstr. in Zentralblatt
Math. Grenzgebiete, Vol. 213, 3 (1971).

004.044 Three Copernican tables. O. Neugebauer.
Centaurus, Vol. 12, 97 - 106 (1968). — Abstr. in
Zentralblatt Math. Grenzgebiete, Vol. 213, 3 - 4 (1971).

004.045 Prophatius Judaeus and the medieval astronomical
tables. R. I. Harper.
Isis, Vol. 62, 61 - 68 (1971).

004.046 Lunar and solar velocities and the length of luna-
tion intervals in Babylonian astronomy.
A. Aaboe.
Mat.-fys. Medd., Danske Vid. Selsk., Vol. 38, No. 6, 27 pp.
(1971).
The fragments of late-Babylonian cuneiform texts published here extend our evidence of the elegant and consistent
manner in which account was made of the influence of lunar
and solar anomalies upon the variable time intervals between
syzygies of the same kind of sun and moon. Several new functions appear for the first time, most notably two associated
with the lengths of six-month intervals. ACT No. 55 is repub-
lished as an appendix. It turns out to be a lunar ephemeris in
which the 223 months of the Saros are broken up into twenty
intervals: eighteen 12-month intervals, one six-month, and one
one-month interval.

004.047 Die arabische Herkunft von zwei Sternverzeich-
nissen in cod. Vat. gr. 1056. P. Kunitzsch.
Zeitschr. der Deutschen Morgenländischen Ges., Vol. 120,
281 - 287 (1970).

004.048 James Cook and the transit of Venus.
W. H. Robertson.
Proc. Roy. Soc. New South Wales, Vol. 103, Part 1, p. 5 - 9
= Sydney Obs. Papers No. 63 (1970).

004.049 Conceptions of the universe.
F. M. Branley, with paintings by H. K. Wimmer.
Natural History, Vol. 79, No. 10, p. 30 - 35 = Contr. American Museum—Hayden Planetarium, Ser. 1, Publ. 33 (1970).
Ancient man devised elaborate explanations for what he
observed in the sky — but no more elaborate than those modern man happily accepts.

004.050 Några svenska kometframställningar under 1570
till 1750-talen. U. R. Johansson.
Astron. Tidsskr., Årg. 4, p. 145 - 166 (1971).

004.051 From Hipparchus to Kepler. J. M. Mohr.
Říše hvězd, Vol. 52, 201 - 207, 225 - 230 (1971).
In Czech.

004.052 The color of Altair. T. Condos, G. Reaves.
Publ. Astron. Soc. Pacific, Vol. 83, 834 - 835 (1971).
A discrepancy was found between the ancient and mod-
ern colors of the bright star Altair (α Aquilae).

004.053 Die Bedeutung der Fortschritte auf den Gebieten
der Optik und der Phototechnik für die Astronomie
im 19. Jahrhundert. B. Sticker.
Technikgeschichte in Einzeldarstellungen, [VDI-Verlag, Düssel-
dorf], No. 19, p. 103 - 136 (1971).

004.054 Variable star observing in America before AAVSO.
J. Ashbrook.
AAVSO Abstr., October 1971, p. 1 - 3.

005 Biography

005.001 **John Herschel zum 100. Todestag.**
G. Buttmann.
SuW, Vol. 10, 228 - 230 (1971).

005.002 **A country curate.** V. Barocas.
Quarterly Journ., Roy. Astron. Soc., Vol. 12,
179 - 182 (1971).

005.003 **Personal profile. Patrick Moore – Amateur 'Moon–Man'.** P. Moore.
Spaceflight, Vol. 13, 181 - 182 (1971).

005.004 **J. G. Hagen and his cosmic clouds.**
J. Ashbrook.
Sky Telescope, Vol. 42, 215 - 216 (1971).

005.005 **Nikolaus Kopernikus.** A. Oberstatter.
SuW, Vol. 10, 295 - 298 (1971).

005.006 **Marian Kowalski.** D. Ya. Martynov.
Observatory, Vol. 91, 227 - 228 (1971). – Letter.

005.007 **Johannes Kepler. Bild und synchronoptische Tafel über Leben und Zeitgeschichte.**
Phys. Blätter, 27. Jahrgang, p. 529 - 530 (1971). – Zusammengestellt aus Gerlach-List 'Johannes Kepler' von M. List, München.

005.008 **J. Kepler, Mathematicus. Zur vierhundertsten Wiederkehr seines Geburtstages.** W. Gerlach.
Phys. Blätter, 27. Jahrgang, p. 531 - 541 (1971).

005.009 **Johannes Kepler's unusual life.** S. P. Kapitsa.
Priroda, No. 12.71, p. 52 - 53 (1971). In Russian.

005.010 **Kepler the man.** C. A. Ronan.
Journ. British Astron. Ass., Vol. 82, 25 - 32 (1971).

005.011 **Nicholas Copernicus (3), (4), (5).**
S. R. Brzostkiewicz.
Urania Kraków, Vol. 42, 194 - 200, 232 - 239, 291 - 298 (1971). In Polish.

005.012 **Johannes Kepler - aus seinem Leben und Schaffen.**
W. Büttner.
Astron. in der Schule, 8. Jahrgang, p. 122 - 125 (1971).

005.013 **400. Wiederkehr des Geburtstages von Johannes Kepler.** H. Lambrecht.
Sterne, 47. Jahrgang, p. 225 - 226 (1971).

005.014 **Johannes Kepler, Leben, Mensch und Werk.**
W. Gerlach.
Sterne, 47. Jahrgang, p. 226 - 238 (1971).

005.015 **Johannes Kepler und Tübingen.** K. Walter.
Kleine Tübinger Schriften, Heft 7, 20 pp. (1971).
Lecture to the opening of the Kepler-exhibition in Tübingen, 1971 July 9.

005.016 **Sir John Frederick William Herschel (1792 - 1871).**
E. S. Barr.
Applied Optics, Vol. 10, 2494 (1971).

005.017 **Leben und Wirken von Johannes Kepler, des Begründers der »neuen Astronomie« zum 400. Geburtstag.** J. Hoppe.
Jenaer Rundschau, (Jena Review), 16. Jahrgang, p. 263 - 268 (1971).

005.018 **Ernst Abbe als Astronom.**
D. Wattenberg.
Jenaer Rundschau, (Jena Review), 16. Jahrgang, p. 269 - 272 (1971).

005.019 **Johannes Kepler.** A. M. Mikisha.
Zemlya i Vselennaya, No. 6, p. 34 - 43 (1971).
In Russian.

005.020 **Theodor Brorsen (1819 - 1895).**
Říše hvězd, Vol. 52, 174 - 177 (1971). In Czech.

005.021 **Johannes Kepler.** Z. Horský.
Pokroky, Vol. 16, 281 - 285 (1971). In Czech.

005.022 **Johannes Kepler.** Z. Horský.
Vesmír, Vol. 50, 341 - 343 (1971). In Czech.

005.023 **August Seydler (1849 - 1891).** M. Matyáš.
Pokroky, Vol. 16, 289 - 292 (1971). In Czech.

005.024 **Die Astrologie des Johannes Kepler.**
F. Hammer.
Sudhoffs Archiv, [Franz Steiner Verlag GmbH, Wiesbaden], Vol. 55, No. 2, p. 113 - 135 (1971).

005.025 **The scientific work of Georges Lemaître.**
P. A. M. Dirac.
Commentarii Pontificia Acad. Sci., Vol. 2, No. 11, p. 1 - 18 (1969).

005.026 **Sergejs Blažko (1870–1956).** I. Daube.
Zvaigžņotā debess, 1970./71. gada ziema, p. 38 - 40.

005.027 **D. Maksutovs, 1896–1964.** I. Daube.
Zvaigžņotā debess, 1971. gada pavasaris, p. 47 - 49.

005.028 **Ceļa uz tāliedarbības atklāšanu.** I. Rabinovičs.
Zvaigžņotā debess, 1971. gada pavasaris, p. 52 - 59.

005.029 **Koperniks un katoļu prelāti.** J. Veselovskis.
Zvaigžņotā debess, 1971. gada vasara, p. 39 - 40.

006 Personal Notes

J. N. Bahcall received the Helen B. Warner Price of the American Astronomical Society.
Physics Today, Vol. 24, No. 3, p. 75 (1971).

B. J. Bok received the 9th annual medal of the "Association pour le Développement International de l' Observatoire de Nice".
Bull. d' Information, Ass. Développement International Obs. Nice, No. 8, p. 87 - 89 (1971).

B. J. Bok received la neuvième médaille de l'A.D.I.O.N.
L'Astronomie, 85ᵉ année, p. 400 - 402 (1971).

E. Buchar, 70th birthday. B. Šternberk.
Říše hvězd, Vol. 52, 151 - 152 (1971). In Czech.

E. Buchar, 70th birthday.
Studia, Vol. 15, 205 - 209 (1971). In Czech.

E. Buchar, 70th birthday. J. Kabeláč.
Vesmír, Vol. 50, 250 (1971). In Czech.

E. M. Burbidge, Director of the Royal Greenwich Observatory.
Observatory, Vol. 91, 235 (1971).

M. Burbidge, Director of Royal Greenwich Observatory.
Physics Today, Vol. 24, No. 12, p. 63 (1971).

S. Chandrasekhar received the Henry Draper Medal.
Physics Today, Vol. 24, No. 7, p. 53 (1971).

A. W. J. Cousins received the Jackson-Gwilt Medal.
B. Lovell.
Quarterly Journ., Roy. Astron. Soc., Vol. 12, 139 (1971).

I. Curea, 70th birthday.
C. Drâmbă.
Stud. Cerc. Astron., Vol. 16, 119 - 122 (1971). In Roumanian.

R. H. Dicke received the National Medal of Science.
Physics Today, Vol. 24, No. 5, p. 71 (1971).

F. D. Drake, director of Arecibo Observatory, Puerto Rico.
Sky Telescope, Vol. 42, 14 (1971).

W. A. Fowler, 60th birthday.
Nature, Vol. 232, 607 (1971).

L. Goldberg, director of Kitt Peak National Observatory.
Sky Telescope, Vol. 42, 14 (1971).

H. Haupt, director of Univ.-Sternw. Graz and Sonnenobs. Kanzelhöhe.
Astron. Nachr., Vol. 292, 282 (1971).

O. Heckmann, 70th birthday.
SuW, Vol. 10, 183 (1971).

W. Heisenberg, 70. birthday.
P. Jordan.
Phys. Blätter, 27. Jahrgang, p. 559 - 562 (1971).

E. I. Kazimirchak-Polonskaya was awarded the F. A. Bredikhin prize. G. A. Chebotarev.
Byull. Inst. Teoret. Astron., *Leningrad*, Vol. 12, 639 - 640 (1971). In Russian.

K. Kellerman received the Helen B. Warner Prize of the American Astronomical Society.
Physics Today, Vol. 24, No. 12, p. 63 (1971).

D. G. King-Hele received the Charles Chree Medal.
Physics Today, Vol. 24, No. 8, p. 77 (1971).

D. G. King-Hele received the Eddington Medal.
B. Lovell.
Quarterly Journ., Roy. Astron. Soc., Vol. 12, 138 (1971).

E. L. Krinov received the Leonard Medal of the Meteoritical Society.
Meteoritics, Vol. 6, 329 - 330 (1971).

A. Lallemand received the 8th annual medal of the "Association pour le Développement International de l' Observatoire de Nice" (A.D.I.O.N.).
Bull. d' Information, Ass. Développement International Obs. Nice, No. 7, p. 35 - 37 (1970).

J. M. Mohr, 70th birthday. V. Vanýsek.
Říše hvězd, Vol. 52, 215 - 216 (1971). In Czech.

J. M. Mohr, 70th birthday. V. Vanýsek.
Vesmír, Vol. 50, 346 (1971). In Czech.

Eh. R. Mustel', received the Lenin order.
Zemlya i Vselennaya, No. 6, p. 13 (1971). In Russian.

F. Press received the Gold Medal of the Royal Astronomical Society. B. Lovell.
Quarterly Journ., Roy. Astron. Soc.,Vol. 12, 133 - 134 (1971).

I. Rabinovičs, 60th birthday. N. Cimahoviča.
Zvaigžņotā debess, 1971. gada vasara, p. 57 - 60.

E. Roemer received the Apthorp Gould Prize.
Physics Today, Vol. 24, No. 7, p. 53 (1971).

A. R. Sandage received the National Medal of Science.
Physics Today, Vol. 24, No. 5, p. 71 (1971).

M. Schwarzschild received the Janssen medal 1970 of the Société Astronomique de France. J. Rösch.
L'Astronomie, 85ᵉ année, p. 277 (1971).

R. Woolley received the Gold Medal of the Royal Astronomical Society. B. Lovell.
Quarterly Journ., Roy. Astron. Soc., Vol. 12, 135 - 137 (1971).

007 Obituaries

R. Bacher, 1909 - 1971 July 9.
Monthly Notes Astron. Soc. Southern Africa, Vol. 30, 101 (1971).

N. P. Barabashov, 1894 - 1971, April 29.
V. I. Ezerskij, V. A. Fedorets, A. T. Chekirda.
Astron. Tsirk., No. 663, p. 6 - 8 (1971). In Russian.

N. P. Barabashov, 1894 March 29 - 1971 April 29.
Z. Kopal.
The Moon, Vol. 3, 265 (1971).

C.-P. De Bièvre, 1888 - 1971.   H. Michel.
Ciel et Terre, Vol. 87, 385 - 386 (1971).

B. Boss died 1970 October 17.
Physics Today, Vol. 24, No. 1, p. 89 (1971).

P. D. Cameron died 1971 July 19.
Southern Stars, Vol. 24, 75 (1971).

S. Chapman, 1888 January 29 - 1970 June 16.
H. Friedman.
Astronaut. Acta, Vol. 16, 299 (1971).

S. Chapman died 1970 June 16.
V. C. A. Ferraro.
Bull. London math. Soc., Vol. 3, 221 - 250 (1971).

P. Clemente, died 1971 June.
Bull. Géod., Nouvelle Sér., Année 1971, No. 102, p. 348 (1971).

I. Depmans, 1885–1970.
I. Rabinovičs, J. Gaiduks.
Zvaigžņotā debess, 1971. gada pavasaris, p. 49 - 51.

J. Dick, died 1971 March 22.
Astron. Nachr., Vol. 292, 282 (1971).

J. Dick, died March 22, 1971.
H. Pauscher.
Sterne, 47. Jahrgang, p. 201 - 204 (1971).

W. J. Eckert died 1971 August 24.
Physics Today, Vol. 24, No. 11, p. 73 (1971).

W. J. Eckert, died 1971 August 24.
Science, Vol. 173, 1115 (1971).

W. J. Eckert died 1971 Aug. 24.
J. Ashbrook.
Sky Telescope, Vol. 42, 207 (1971).

A. Ehmert, 1910 - 1971.   G. Pfotzer.
Astronaut. Acta, Vol. 16, 299 (1971).

F. J. Hargreaves, 1891 February 10 - 1970 September 4.   E. J. Hysom.
Journ. British Astron. Ass., Vol. 82, 43 - 44 (1971).

F. J. Hargreaves, 1891 February 10 – 1970 September 4.   W. H. Steavenson.
Quarterly Journ. Roy. Astron. Soc., Vol. 12, 336 - 337 (1971).

W. A. Heiskanen, died 1971 October.
Bull. Géod., Nouvelle Sér., Année 1971, No. 102, p. 348 (1971).

J. Ikaunieks died 1970 April 28.
I. Daube, N. Cimahoviča, L. Vāczemnieks, R. Saveļjeva.
Zvaigžņotā debess, 1970./71. gada ziema, p. 40 - 48.

A. Kahrstedt, died 1971 January 11.
Astron. Nachr., Vol. 292, 282 (1971).

H. Knox-Shaw, 1885 Oct. 12 - 1970 April 11.
A. D. Thackeray.
Quarterly Journ., Roy. Astron. Soc., Vol. 12, 197 - 201 (1971).

C. Lombardi, died 1971 June 30.
Mem. Soc. Astron. Italiana, Nuova Ser., Vol. 42, 263 (1971).

D. F. Martyn, 1906 June 27 - 1970 March 5.
J. H. Carven.
Astronaut. Acta, Vol. 16, 300 (1971).

M. G. J. Minnaert, 1893 February 12 - 1970 October 26.   J.-C. Pecker.
Icarus, Vol. 15, 147 - 148 (1971).

M. G. J. Minnaert, 1893 February 12 – 1970 October 26.   C. de Jager.
Quarterly Journ. Roy. Astron. Soc., Vol. 12, 338 - 341 (1971).

A. de Moraes, 1916 - 1970 December 11.
Inform. Bull. Southern Hemisphere, No. 18, p. 30 (1971).

W. R. Van Nattan, died 1971 October 10th.
Sky Telescope, Vol. 42, 353 - 354 (1971).

Đ. Nikolić, died on August 9, 1971.
N. Janković.
Vasiona, Vol. 19, 53 - 54 (1971). In Serbo-Croatian.

C. F. Powell died 1969 August 10.
Proc. 11th international conference on cosmic rays, Budapest 1969, (see 012.025), p. 1 (1970).

J. Sadil died 1971 January 19.   R. Valach.
Vesmír, Vol. 50, 251 (1971).   In Czech.

V. M. Slipher, 1875 - 1969.   J. S. Hall.
Year Book American Phil. Soc. 1970, p. 161 - 166.

A. N. Spitz, died April 14, 1971.
Strolling Astronomer, Vol. 23, 56 - 57 (1971).

V. M. Tabachnik, 1928 January 6 - 1971 October 6.
A. M. Shul'berg.
Astron. Tsirk., No. 661, p. 7 - 8 (1971). In Russian.

Y. Väisälä, died 1971 August.
Bull. Géod., Nouvelle Sér., Année 1971, No. 102, p. 348 (1971).

Y. Väisälä, died on July 21, 1971.
Sky Telescope, Vol. 42, 273 (1971).

S. V. Voroshilova-Romanskaya, 1886 - 1969 Nov. 26.
Izv. Glav. Astron. Obs. v Pulkove, No. 187, p. 3 (1971).
In Russian.

C. B. Watts, 1889 - 1971 July 17.
Sky Telescope, Vol. 42, 131 (1971).

I. Weinberg died 1971 September.
Monthly Notices Astron. Soc. Southern Africa, Vol. 30, 152 (1971).

A. Yramategui died 1970 January 28.
E. A. King, Jr.
Meteoritics, Vol. 6, 331 (1971).

J. Zähringer, died 1970 July 22.
Geochim. Cosmochim. Acta, Vol. 35, 861 (1971).

E. Zinner, 1886 February 2 – 1970 August 30.
D. Wattenberg.
Astron. Nachr., Vol. 293, 79 - 80 (1971).

008 Observatories, Institutes

Reports, communications and publications of observatories and astronomical institutes are recorded in this section; included are numbered series of reprints. Whenever possible, the numbers of the abstracts referring to the publications are given. Observatories and institutes are listed in alphabetical order of their towns. In some cases observatory publications do not give the name of the town; the following list which gives names and towns of some institutions may serve as an aid in such cases.

Aarne Karjalainen Observatory	Oulu, Finland
Algonquin Radio Observatory	Lake Traverse, Ontario, Canada
Allegheny Observatory	Pittsburgh, Pennsylvania
Arthur J. Dyer Observatory	Nashville, Tennessee
Astronomical Latitude Station, Polish Academy of Sciences	Borowiec, Poland
Bosscha Observatory	Lembang, Indonesia
Boyden Observatory	Bloemfontein, South Africa
Bureau International de l'Heure	Paris, France
Cajigal Observatory	Caracas, Venezuela
California Institute of Technology	Pasadena, California
Cape of Good Hope	Cape Town, South Africa
Carter Observatory	Wellington, New Zealand
Catalina Station	Tucson, Arizona
Cavendish Laboratory	Cambridge, England
Ceskoslovenská Akademie Ved Astronomický Ustav	Praha, Czechoslovakia
Chamberlin Observatory, University of Denver	Denver, Colorado
Commonwealth Observatory	Canberra, Australia
Coralitos Observatory	Las Cruces, New Mexico
David Dunlap Observatory, University of Toronto	Richmond Hill, Ontario
Dearborn Observatory	Evanston, Illinois
Department of Astronomy and Observatory, Univ. California	Los Angeles, California
Department of Astronomy, University of Texas	Austin, Texas
Division Radiophysics, C.S.I.R.O. University Grounds	Sydney, N.S.W., Australia
Dominion Astrophysical Observatory	Victoria, British Columbia
Dominion Observatory	Ottawa, Ontario
Dominion Radio Astrophysical Observatory	Penticton, British Columbia
Dudley Observatory	Albany, New York
Dunsink Observatory	Dublin, Ireland
Engelhardt Observatory	Kazan, R.S.F.S.R.
European Southern Observatory	Hamburg, Federal German Republic
Five College Observatories	Amherst, Massachusetts
Florida State University Radio Observatory	Tallahassee, Florida
Flower and Cook Observatories, University of Pennsylvania	Philadelphia, Pennsylvania
Fraunhofer Institut	Freiburg, Federal German Republic
Georgetown Observatory	Washington, D.C.
Goddard Space Flight Center	Greenbelt, Maryland
Goethe Link Observatory, University of Indiana	Bloomington, Indiana
Hale Observatories	Pasadena, California
Harvard College Observatory	Cambridge, Massachusetts
Harvard Radio Astronomy Station	Cambridge, Massachusetts

Heinrich-Hertz-Institut	Berlin, Germany
High Altitude Observatory, University of Colorado	Boulder, Colorado
Institute for Astronomy, University of Hawaii	Honolulu, Hawaii
Institute for Theoretical Astronomy (Institut Teoreticheskoj Astronomii)	Leningrad, R.S.F.S.R.
Institute of Theoretical Astrophysics, Blindern	Oslo, Norway
Inter-American Observatory	Cerro-Tololo, (La Serena), Chile
International Latitude Observatory	Mizusawa, Japan
Joint Institute for Laboratory Astrophysics (JILA)	Boulder, Colorado
Kandilli Observatory	Istanbul, Turkey
Kapteyn Astronomical Laboratory	Groningen, Netherlands
Karl-Schwarzschild-Observatorium	Tautenburg, German Democratic Republic
Kenneth Mees Observatory	Rochester, New York
Kwasan Observatory	Kyoto, Japan
Lamont-Hussey Observatory	Bloemfontein, South Africa
Leander McCormick Observatory University of Virginia	Charlottesville, Virginia
Lee Observatory	Beirut, Lebanon
Leopold-Figl-Observatorium	Wien, Austria
Leuschner Observatory	Berkeley, California
Lick Observatory	Santa Cruz, (Mount Hamilton), California
Lindheimer Astronomical Research Center	Evanston, Illinois
Lockheed Solar Observatory	Saugus, California
Lohrmann-Institut für Geodätische Astronomie	Dresden, German Democratic Republic
Louisiana State University Observatory	Baton Rouge, Louisiana
Lowell Observatory	Flagstaff, Arizona
Lunar and Planetary Laboratory	Tucson, Arizona
Max-Planck-Institut für Astronomie	Heidelberg, Federal German Republic
Max-Planck-Institut für Phyik und Astrophysik	München, Federal German Republic
Max-Planck-Institut für Radioastronomie	Bonn, Federal German Republic
McDonald Observatory	Fort Davis, Texas
McMath Hulbert Observatory	Pontiac, Michigan
Molonglo Radio Observatory, University of Sydney	Sydney, New South Wales
Mount Cuba Observatory	Wilmington, Delaware
Mount Palomar Observatory	Pasadena, California
Mount Wilson Observatory	Pasadena, California
Mullard Radio Astronomy Observatory	Cambridge, England
Narrabri Observatory, University of Sydney	Sydney, New South Wales
National Bureau of Standards	Washington, D. C.
National Observatory, USA	Kitt Peak, Arizona
National Radio Astronomy Observatory	Charlottesville, Virginia Green Bank, West Virginia Tucson, Arizona
New Mexico State University Observatory	Las Cruces, New Mexico

Nizamiah Observatory — Hyderabad, India
Nuffield Radio Astronomy
 Laboratories, Jodrell Bank
 University of Manchester — Manchester, England
Observatoire Royal de Belgique — Uccle, Belgium
Observatorio de Cartuja — Granada, Spain
Observatorio del Ebro — Tortosa, Spain
Observatorio Fabra — Barcelona, Spain
Observatory, University of
 Michigan — Ann Arbor, Michigan
Ohio State University
 Radio Observatory — Columbus, Ohio
Ole Roemer-Observatoriet — Aarhus, Denmark
Perkins Observatory, Ohio State
 and Wesleyan Universities — Delaware, Ohio
Purple Mountain Observatory — Nanking, China
Radcliffe Observatory — Pretoria, South Africa
Remeis-Sternwarte — Bamberg,
 Federal German Republic
Republic Observatory — Johannesburg, South Africa
Rosemary Hill Observatory — Gainesville, Florida
Royal Radar Establishment,
 Radio Astronomy Division — Malvern, England
Sagamore Hill Radio Observatory Bedford, Massachusetts
Saint-Michel, l'Observatoire — Haute Provence, France
San Fernando Observatory — El Segundo, California

Smithsonian Astrophysical
 Observatory — Cambridge, Massachusetts
Specola Astronomica Vaticana — Castel Gandolfo, Italy
Specola di Padova — Asiago, Italy
Sproul Observatory — Swarthmore, Pennsylvania
Sternberg Observatory — Moscow, R.S.F.S.R.
Steward Observatory,
 University of Arizona — Tucson, Arizona
United States Naval Observatory Washington, D.C.
University of Florida,
 Radio Observatory — Gainesville, Florida
University of Illinois Observatory Urbana, Illinois
University of Michigan
 Observatories — Ann Arbor, Michigan
University of South Florida
 Observatory — Tampa, Florida
Uttar Pradesh State Observatory Naini Tal, India
Van Vleck Observatory — Middletown, Connecticut
Warner and Swasey Observatory Cleveland, Ohio
Washburn Observatory — Madison, Wisconsin
West Melton Observatory — Christchurch, New Zealand
Yale University Observatory — New Haven, Connecticut
Yerkes Observatory — Williams Bay, Wisconsin
Zentralinstitut für Astrophysik,
 Sternwarte Babelsberg, (Fach-
 bereich Kosmische Physik) — Potsdam-Babelsberg, German
 Democratic Republic

008.001 Abastumani

Abastumani Astrophysical Observatory at Mount Kanobili. V. V. Krivoshchekov.
AN GruzSSR. Abastumansk. astrofiz. observ. "Metsniereba", Tbilisi. 65 pp. Price 35 Kop. (1971). In Russian. – Review in Referativ. Zhurn. 51. Astron., 12.51.98 (1971).

Chronicle.
Byull. Abastumansk. Astrofiz. Obs., No. 40, p. 297 - 298 (1971). In Russian.

Abastumanskaya Astrofizicheskaya Observatoriya, Gora Kanobili, Byulleten', Akademiya Nauk Gruzinskoj SSR, No. 40 (N. L. Magalashvili, J. I. Kumsishvili, 06.121.022; A. Sh. Khatisov, 06.124.003; M. V. Dolidze, 06.114.040; N. B. Kalandadze, L. N. Kolesnik, V. I. Kuznetsov, 06.155.016; Ts. S. Khetsuariani, E. I. Tetruashvili, R. I. Kiladze, G. N. Salukvadze, A. Sh. Khatisov, 06.074.016; Ts. S. Khetsuriani, E. I. Tetruashvili, 06.073.030; A. S. Tskhovrebadze, 06.072.014; R. M. Dzigvashvili, 06.151.006; G. A. Malasidze, 06.151.007; Yu. S. Efimov, A. G. Thotochava, 06.120.009; A. Sh. Khatisov, 06.032.003; G. G. Georgobiani, 06.004.005; E. K. Kharadze, N. I. Bolkvadze, 06.002.009).

008.002 Alma-Ata

Akademiya Nauk Kazakhskoj SSR. **Trudy Astrofizicheskogo Instituta,** *Alma-Ata,* Vol. 16 (K. G. Dzhakusheva, 06.132.046; Ju. I. Glushkov, E. S. Eroshevich, 06.132.047; L. N. Kondratjeva, 06.155.041; Sh. N. Sabitov, 06.032.029; V. S. Matjagin, 06.031.054; I. D. Kupo, 06.114.124; Z. N. Chumak, 06.064.057; K. K. Kalchaev, 06.122.139; D. A. Rozkovsky, 06.131.135; I. L. Genkin, 06.151.070; I. L. Genkin, 06.151.071; I. L. Genkin, 06.061.039; Z. Kh. Kurmakaev, 06.066.083; E. K. Denisjuk, 06.158.120; Sh. N. Sabitov, 06.066.084; D. A. Rozkovsky, 06.132.048), 17 (D. A. Rozhkovsky, 06.155.042; K. Kalchayev, 06.117.028; I. D. Kupo, 06.031.055; V. S. Matyagin, 06.031.056; V. M. Tereshchenko,

A. V. Kharitonov, 06.114.125; E. S. Yeroshevich, 06.132.049; L. M. Genkina, 06.158.121; N. N. Pavlova, 06.158.122; I. L. Genkin, 06.151.072; I. L. Genkin, 06.151.073; I. L. Genkin, 06.061.040; V. A. Semenenya, 06.066.085; M. Abdildin, A. Junusov, O. Sakibayev, 06.066.086; V. F. Bogdanov, P. N. Boiko, A. V. Kharitonov, 06.034.116; P. N. Boiko, M. I. Musorin, 06.032.030).

008.003 Auckland

Auckland Observatory.
Inform. Bull. Southern Hemisphere, No. 18, p. 21 (1971). Current research report.

008.004 Austin

The Astronomy Department, The McDonald Observatory, and the Radio Astronomy Observatory (UTRAO) of the University of Texas at Austin. H. J. Smith.
Bull. American Astron. Soc., Vol. 3, 418 - 432 (1971).

008.005 Bamberg

Veröffentlichungen der Remeis-Sternwarte Bamberg, Astronomisches Institut der Universität Erlangen–Nürnberg, Vol. 8, Nos. 93 (R. Knigge, 06.123.064), 94 (D. Friedrich, E. Schöffel, 05.123.026), 95 (D. Friedrich, E. Schöffel, 06.123.065), 96 (W. Strohmeier, 05.123.030), 97 (R. Bloomer, 06.123.022), 98 (R. Bloomer, 06.123.023).

008.006 Baton Rouge

Contributions of the Louisiana State University Observatory, Nos. 43 (A. G. D. Philip, J. S. Drilling,

04.114.127), 44 (J. S. Drilling, A. G. D. Philip, 04.114.128), 45 (A. U. Landolt, 05.122.032), 46 (H. E. Bond, D. J. MacConnell, 05.113.015), 47 (D. J. MacConnell, R. L. Frye, W. P. Bidelman, H. E. Bond, 05.114.053) 48 (P. D. Lee, C. L. Perry, 05.153.035), 50 (H. E. Bond, 06.141.023).

008.007 Beirut

Lee Observatory, American University of Beirut, Lebanon. Monthly Bulletin, Astronomical Section, 1971 April - October (F. Bruin, H. Hourani, N. G. Bustati, 06.075. 030).

008.008 Belo Horizonte

Instituto de Ciencias Exatas da Universidade Federal de Minas Gerais. (Institute of Exact Sciences of the Federal University of Minas Gerais). Inform. Bull. Southern Hemisphere, No. 18, p. 14 (1971). Current research report.

008.009 Beograd

Rapport sur l'activité de l'Observatoire Astronomique de Belgrade en 1970. P. M. Djurković. Bull. Obs. Astron. Beograd, Vol. 28, (No. 124), 195 - 198 (1970).
Bulletin de l'Observatoire Astronomique de Beograd, Vol. 28, No. 123 (A. Kubičela, J. Arsenijević, 06.122.108; J. Arsenijević, 06.122.109; M. Djokić, 06.045.008; A. S. Kharin, 06.032.023; G. Teleki, 06.032.024; D. Olević, 06.098.022), No. 124 (G. Teleki, 06.082.081; G. M. Popović, 06.118.007; D. Olević, 06.118.008; D. J. Zulević, 06.118.009; P. M. Djurković, G. M. Popović, D. J. Zulević, D. M. Olević, 06.118.010; D. Djurović, 06.044.010; D. Djurović, 06.044.011; D. Djurović, 06.045.009; D. Šaletić, S. Sadžakov, 06.032.025; V. Erceg, 06.118.011; G. M. Popović, 06.118.012; G. M. Popović, T. D. Angelov, 06.115.009; V. Milovanović, R. Grujić, M. Djokić, 06.041.033; M. Jovanović, D. Djurović, D. Vesić, M. Lončarević, D. Mandić, 06.044.012; M. B. Protitch, 06.098.023; M. Simić, 06.096.016; M. Simić, 06.096.017; M. B. Protitch, 06.096.018; V. Milovanović, 06.045.010; A. Kubičela, 06.071.054; P. M. Djurković, 06.008.000).

008.010 Berlin

Heinrich-Hertz-Institut. Solare Beobachtungsergebnisse. Deutsche Akademie der Wissenschaften zu Berlin, Zentralinstitut für Solar-Terrestrische Physik, Berlin-Adlershof. HHI Solar Data, Vol. 22, 1971 June — October (E. A. Lauter, A. Böhme, F. W. Jäger, F. Fürstenberg, H. Künzel, D. Scholz, S. Böhm, 06.075.019).

Heinrich-Hertz-Institut. Supplement Series of Solar Data. Deutsche Akademie der Wissenschaften zu Berlin, Zentralinstitut für Solar-Terrestrische Physik, Berlin-Adlershof. HHI Suppl. Ser. Solar Data, Vol. 2, No. 5 (A. Krüger, F. Fürstenberg, K. H. Fricke, 06.077.043).

Veröffentlichungen der Wilhelm-Foerster-Sternwarte Berlin, Nos. 21 (H.-B. Brenske, 06.009.023), 22—25 (C. Kowalec, 04.099.029), 26 (H. Haug, 06.079.101), 27 (W. Meyer, 06.031.057).

008.011 Bochum

Sternwarte der Stadt Bochum, Separate print (H.-U. Keller, 06.162.076).

008.012 Bonn

Zur Einweihung des 100-m-Radioteleskops am 12. Mai 1971. O. Hachenberg. SuW, Vol. 10, 185 - 188 (1971).

Mitteilungen der Astronomischen Institute Bonn, Nos. 124 (M. Roemer, 06.082.052), 125 (M. Roemer, 06.082.056), 126 (C. Wulf-Mathies, 06.082.058), 128 (V. P. Bhatnagar, 06.083.004), 130 (D. Ellinger, G. V. Schultz, 06.034.114), 135 (M. Grewing, M. Walmsley, 05.141.019), 136 (O. Hachenberg, W. Priester, H. Schmidt, 05.008.022), 137 (M. Grewing, H. Heintzmann, 05.141.121), 138 (H. Volland, H. G. Mayr, 06.082.131).

Veröffentlichungen der Astronomischen Institute Bonn, Nos. 82 (W. Seggewiss, 06.113.035), 83 (W. Seggewiss, 06.113.036), 84 (H. van Schewick, 06.153.020).

008.013 Borowiec

Polish Academy of Sciences, Astronomical Latitude Station, Circular, No. 118 (06.044.022).

008.014 Brno

Contributions of the Observatory and Planetarium in Brno, Nos. 10 (Z. Pokorný, 06.091.041), 11 (M. Druckmüller, 06.103.101; Z. Okáč, 06.103.101; M. Druckmüller, 06.103.101), 12 (J. Šilhán, O. Obůrka, 06.121.095; V. Znojil, 06.122.160).

008.015 Bucarest

Observations solaires, Académie de la République Socialiste de Roumanie, Observatoire de Bucarest, Secteur Solaire. Rotations 1556 - 1569 (C. Popovici, E. Țifrea, V. Dinulescu, A. Dimitriu, S. Nicolescu, G. Mariș, 06.075.031).

008.016 Buenos Aires

Facultad de Ciencias Exactas y Naturales, Universidad Nacional de Buenos Aires. (School of Exact and Natural Sciences, National University of Buenos Aires). Inform. Bull. Southern Hemisphere, No. 18, p. 7 (1971). Current research report.

Instituto de Astronomía y Física del Espacio. (Institute of Space Astronomy and Physics). Inform. Bull. Southern Hemisphere, No. 18, p. 7 - 8 (1971). Current research report.

008.017 Byurakan

Byurakan Astrophysical Observatory, Preprint
No. 3 (V. A. Ambartsumian, 06.122.042).

008.018 Cambridge, Engl.

University of Cambridge. Report of the Observatories Syndicate for the year ending 1971 September 30.
R. O. Redman.
Separate print Cambridge, Engl., 8 pp. (1971).

008.019 Cambridge, Mass.

Smithsonian Contributions to Astrophysics. Smithsonian Institution Astrophysical Observatory, (United States Government Printing Office, Washington), Nos. 12 (B.-A. Lindblad, 06.104.076; B.-A. Lindblad, 06.104.077), 13 (C. H. Payne-Gaposchkin, 06.159.010).

Smithsonian Institution. Astrophysical Observatory. Research in Space Science. SAO Special Reports, Nos. 309 (R. L. Kurucz, 06.064.042), 321 (D. W. Latham, 06.114.092), 323 (H. E. Mitler, 06.071.048), 334 (J. E. Grindlay, 06.142.075), 335 (J. A. Hoffman, 06.034.001), 336 (R. E. McCrosky, A. Posen, G. Schwartz, C.-Y. Shao, 05.105.081).

008.020 Canberra

Mount Stromlo and Siding Spring Observatories. Research School of Physical Sciences, the Australian National University. − Report for the year ending 1970 December 31. O. J. Eggen.
Quarterly Journ. Roy. Astron. Soc., Vol. 12, 305 - 319 (1971).

008.021 Cape Town

Royal Observatory at the Cape of Good Hope.
Inform. Bull. Southern Hemisphere, No. 18, p. 26 (1971). Current research report.

Royal Greenwich Observatory and Royal Observatory, Cape of Good Hope. − Report for the year ending 1970 December 31.   R. Woolley, G. A. Harding.
Quarterly Journ. Roy. Astron. Soc., Vol. 12, 277 - 294 (1971).

Department of Astronomy, University of Cape Town.
Inform. Bull. Southern Hemisphere, No. 18, p. 23 (1971). Current research report.

National Institute for Telecommunications Research, Radio Astronomy Division.   G. C. Nicolson.
Inform. Bull. Southern Hemisphere, No. 18, p. 23 - 24 (1971). Current research report.

008.022 Carloforte

International Astronomical Station of Carloforte,

Cagliari. E. Proverbio.
Mem. Soc. Astron. Italiana, Nuova Ser., Vol. 42, 571 - 577 (1971). − Annual Report for 1970.

008.023 Castel Gandolfo

Ricerche Astronomiche. Specola Vaticana, Città del Vaticano, Vol. 8, No. 10 (W. J. Miller, 06.123.029).

Ricerche Spettroscopiche, Laboratorio Astrofisico della Specola Vaticana, Vol. 3, No. 7 (J. Junkes, 06.031.030).

Specola Vaticana, *Castel Gandolfo,* Comunicazione, Nos. 50 (P. J. Treanor, 04.034.066); 51 (D. J. K. O'Connell, 06.032.018); 52 (O. van de Vyver, 05.094.107).

008.024 Catania

The old Mount Etna Observatory and the new.
G. Godoli
Sky Telescope, Vol. 42, 20 - 21 (1971).

Osservatorio Astrofisico di Catania, Pubblicazione, No. 146 (06.075.020).

008.025 Cerro Tololo

Cerro Tololo-Inter-American Observatory, Contributions, Nos. 104 (W. A. Hiltner, D. E. Mook, 04.142.007), 113 (J. A. Graham, 04.121.013), 114 (G. A. Welch, 04.158.047), 115 (K. Y. Chen, 05.121.013), 116 (P. C. Keenan, 04.114.074), 117 (W. E. Kunkel, 04.122.137), 118 (H. E. Bond, 04.121.034), 119 (R. E. Schild, 04.153.024), 120 (J. S. Drilling, A. G. D. Philip, 04.114.127), 121 (J. S. Drilling, A. G. D. Philip, 04.114.128), 122 (A. G. D. Philip, 04.113.019), 123 (R. M. Humphreys, 04.122.094), 124 (W. Kunkel, P. Osmer, M. Smith, A. Hoag, D. Schroeder, W. A. Hiltner, H. Bradt, S. Rappaport, H. W. Schnopper, 04.142.077), 125 (W. E. Kunkel, V. M. Blanco, 04.142.028).

008.026 Cincinnati

Minor Planet Circulars, (MPC), Nos. 3165 - 3292 (P. Herget, 06.098.040).

008.027 Cleveland

Publications of the Warner and Swasey Observatory, Case Western Reserve University, Vol. 1, No. 1 (C. B. Stephenson, N. Sanduleak, 06.041.023).

008.028 Coimbra

O 'presente' e um presumível 'futuro' do Observatório Astronómico da Universidade de Coimbra. Opúsculo integrado no contexto das comemorações do bicentenário (1972) da criação do Observatório.   A. Simões da Silva. Comun. Obs. Astron. Univ. Coimbra, No. 9, 17 pp. (1971).

Comunicações do Observatório Astronómico da Universidade de Coimbra, Nos. 8 (M. Moreirinhas Pinheiro, 06.079.101), 9 (A. Simões da Silva, 06.008.000), 10 (A. Simões da Silva, M. Coelho Balça, 06.118.023).

R. S. Osherov, 06.031.032), 58 (L. S. Marochnik, A. A. Suchkov, 06.151.025; L. S. Marochnik, A. A. Suchkov, 06.151. 026; L. S. Marochnik, A. A. Suchkov, 06.151.027; L. S. Marochnik, A. A. Suchkov, 06.151.028).

008.029   Córdoba

The first century of the Observatorio Astronómico de Córdoba.   J. L. Sérsic.
Inform. Bull. Southern Hemisphere, No. 18, p. 1 - 6 (1971).

Observatorio Astronómico. (Astronomical Observatory, National University of Córdoba).
Inform. Bull. Southern Hemisphere, No. 18, p. 8 - 9 (1971). Current research report.

The first century of Cordoba Observatory.
J. L. Sersic.
Sky Telescope, Vol. 42, 347 - 350 (1971).

Observatorio Astronómico *(Universidad Nacional de Córdoba, Argentina),* Tirada Aparte, Nos. 168 (J. L. Sérsic, 03.160.009), 177 (R. F. Sisteró, C. R. Fourcade, 03.154.007), 180 (G.J. Carranza, 06.159.025), 182 (J. L. Sérsic, 06.158. 123), 183 (J. L. Sérsic, 04.158.081), 189 (J. Landi Dessy, J. R. Laborde, 06.159.019).

008.030   Cracow

Cracow Observatory, Reprints Nos. 85 (H. Brancewicz, 04.065.116), 86 (Z. Dworak, 04.082.146), 87 (R. Szafraniec, 05.121.019), 88 (W. Z. Wiśniewski, 06.122.050), 89 (J. M. Kreiner, 06.121.027).

008.031   Delaware

Contributions from the Perkins Observatory, Ohio State – Ohio Wesleyan Universities, Series I, Nos. 119 (A. E. Greene, R. F. Wing, 05.114.006), 120 (M. Plavec, 04.117.025), 121 (P. A. Ianna, 04.153.019), 122 (G. H. Newsom, 05.022.055), 123 (E. Capriotti, 05.133.030), 124 (J. W. Warner, R. F. Wing, 06.113.008).

Contributions from the Perkins Observatory. The Ohio State University and Ohio Wesleyan University, Series II, Nos. 23 (A. Slettebak, R. K. Brundage, 05.114.072), 24 (P. C. Keenan, 05.114.081), 25 (J. H. Baumert, 05.113.043), 26 (N. M. White, 05.113.044), 27 (R. F. Wing, 05.113.042).

008.032   Dushanbe

Byulleten' Instituta Astrofiziki, Akademiya Nauk Tadzhikskoj SSR, Nos. 55 (R. Sh. Bibarsov, 06.104.053; R. Sh. Bibarsov, 06.104.054; R. Sh. Bibarsov, 06.104.055; R. Sh. Bibarsov, 06.104.056; R. P. Tshebotaryov, V. N. Sidorin, G. A. Polushkin, R. Sh. Bibarsov, Sh. O. Isamutdinov, V. M. Kolmakov, 06.104.057; R. P. Tshebotaryov, V. N. Sidorin, 06.104.058; R. P. Tshebotaryov, Sh. O. Isamutdinov, 06.104. 059; V. M. Kolmakov, 06.104.060; U. Shodiev, 06.104.061), 56 (O. P. Vasiljanovskaja, Ń. N. Kiselev, T. K. Kiseleva, 06.122.085; S. G. Pomagaev, 06.151.024), 57 (R. Sh. Bibarsov, 06.104.062; R. Sh. Bibarsov, 06.104.063; O. P. Vasiljanovskaja, G. E. Erleksova, 06.122.090; T. G. Nikulina, 06.122.091;

008.033   Edinburgh

Royal Observatory, Edingburgh. Report for the year ending 1970 December 31.   H. A. Brück.
Quarterly Journ., Roy. Astron. Soc., Vol. 12, 183 - 191 (1971).

Communications from the Royal Observatory, Edinburgh, Nos. 97 (J. W. Campbell, 05.034.072), 99 (J. W. Campbell, 05.034.073), 100 (J. W. Campbell, 05.113.041), 101 (V. C. Reddish, 05.032.010), 102 (W. McD. Napier, 05.117.030), 103 (G. E. Bromage, 05.114.045), 104 (V. C. Reddish, C. Sloan, 05.115.013), 105 (P. W. J. L. Brand, 06.031.035), 106 (W. McD. Napier, 05.117.020), 107 (T. J. Lee, 06.022.081), 109 (G. C. Sudbury, 06.114.005), 110 (V. C. Reddish, 05.131.129), 111 (W. McD. Napier, 06.093.006), 112 (R. D. Wolstencroft, K. Nandy, 06.131.023), 113 (K. Nandy, W. M. Napier, G. I. Thompson, 06.131.022), 116 (P. W. J. L. Brand, 06.113.059), 118 (K. Nandy, N. C. Wickramasinghe, 06.132.021).

Publications of the Royal Observatory Edinburgh, Vol. 7, Nos. 5 (M. T. Brück, 06.132.034), 6 (K. Nandy, F. Smriglio, 06.114.073).

008.034   Flagstaff

Lovell Observatory Bulletin, *Flagstaff, Arizona,* Nos. 156 = Vol. 7, No. 19 (M. J. Price, J. S. Hall, P. B. Boyce, R. Albrecht, 06.099.074), 157 = Vol. 7, No. 20 (C. F. Capen, L. J. Martin, 06.097.093).

008.035   Fort Davis

The Astronomy Department, The McDonald Observatory, and the Radio Astronomy Observatory (UTRAO) of the University of Texas at Austin.   H. J. Smith.
Bull. American Astron. Soc., Vol. 3, 418 - 432 (1971).

008.036   Frankfurt

Veröffentlichungen des Astronomischen Instituts der Universität Frankfurt (Main), Nos.32 (R. Hartmann, 06.072.085), 33 (W. Gleissberg, 05.072.022), 34 (W. Gleissberg, T. Damboldt, 05.072.035), 35 (W. Gleissberg, 05.072.066).

008.037   Freiburg

Fraunhofer Institut, Map of the Sun.
1971 July 1 – December 31 (06.075.016).

Mitteilungen aus dem Fraunhofer Institut, *Freiburg,* Nos. 101 (F.-L. Deubner, 05.071.025), 102 (W. Mattig, 06.072.005), 104 (J. P. Mehltretter, 06.071.008), 106 (R. Göhring, 06.080.016), 106a (F.-L. Deubner, R. Göhring, 06.072.025), 107 (K. O. Kiepenheuer, 06.073.096).

008.038 Genève

Publications de l'Observatoire de Genève, Série A, Fasc. 78 (G. Goy, 06.113.057; F. Rufener, A. Maeder, 06.113.058; C. Jaschek, 06.119.015).

008.039 Graz

Mitteilungen der Universitätssternwarte Graz, No. 9 (H. J. Schober, 06.033.029).

008.040 Green Bank

National Radio Astronomy Observatory, *Green Bank,* Reprints, Series A, Nos. 202 (G. L. Verschuur, 05.131.038), 203 (D. F. Dickinson, C. A. Gottlieb, 05.131.037), 204 (M. A. Gordon, T. B. Williams, 05.132.015), 205 (G. K. Miley, C. M. Wade, 05.141.093), 206 (G. L. Verschuur, 05.131.083), 207 (D. Buhl, L. E. Snyder, 06.131.088), 208 (B. E. Turner, 05.133.021), 209 (K. B. Jefferts, A. A. Penzias, R. W. Wilson, M. Kutner, P. Thaddeus, 05.131.063), 210 (G. L. Verschuur, G. R. Knapp, 05.131.106) 211 (K. I. Kellermann, I. I. K. Pauliny-Toth, 05.141.120), 212 (H. J. Wendker, 05.131.100), 213 (J. F. C. Wardle, 05.141.122), 214 (J. Edrich, 06.033.025), 215 (J. Pfleiderer, 06.141.007), 216 (T. Nakano, 06.131.133), 217 (W. J. Wilson, A. H. Barrett, 05.114.085), 218 (K. I. Kellermann, 06.033.009).

National Radio Astronomy Observatory, *Green Bank,* Reprints, Series B, Nos. 233 (W. J. Wilson, A. H. Barrett, J. M. Moran, 03.114.073), 234 (K. I. Kellermann, I. I. K. Pauliny-Toth, 05.114.065), 235 (K. J. Johnston, S. H. Knowles, W. T. Sullivan III, J. M. Moran, B. F. Burke, K. Y. Lo, D. C. Papa, G. D. Papadopoulos, P. R. Schwartz, C. A. Knight, I. I. Shapiro, W. J. Welch, 05.131.081), 236 (W. J. Wilson, 05.114.064), 237 (V. Herrero, R. M. Hjellming, C. M. Wade, 05.124.105), 238 (W. R. Burns, M. S. Roberts, 05.158.079), 239 (M. C. H. Wright, 05.158.084), 240 (D. Buhl, 05.094.086), 241 (T. Nakano, 04.107.001), 242 (R. H. Rubin, B. E. Turner, 05.131.077), 243 (M. A. Gordon, 05.131.111), 244 (R. L. Brown, 05.142.031), 245 (C. M. Wade, H. Gent, R. L. Adgie, J. H. Crowther, 04.141.117), 246 (J. H. Taylor, G. R. Huguenin, 06.141.022), 247 (M. A. Gordon, D. C. Wallace, 06.131.010), 248 (D. S. De Young, 06.141.004), 249 (M. H. Andrews, R. M. Hjellming, E. Churchwell, 06.131.011), 250 (R. A. Sramek, 06.066.004), 251 (P. D. Jackson, F. J. Kerr, 06.131.014), 252 (R. W. Wilson, A. A. Penzias, K. B. Jefferts, M. Kutner, P. Thaddeus, 06.131.004), 253 (R. N. Manchester, 06.141.005), 254 (R. M. Hjellming, C. M. Wade, 06.142.011), 255 (R. M. Hjellming, 05.066.074), 256 (J. W. Findlay, 06.033.003), 257 (A. A. Penzias, P. M. Solomon, R. W. Wilson, K. B. Jefferts, 06.131.030), 258 (M. A. Gordon, S. T. Gottesman, 06.131.050), 259 (R. N. Manchester, 05.141.156), 260 (R. N. Manchester, 05.141.170), 261 (P. M. Solomon, K. B. Jefferts, A. A. Penzias, R. W. Wilson, 06.131.052), 262 (K. B. Jefferts, A. A. Penzias, R. W. Wilson, P. M. Solomon, 06.131.053), 263 (R. M. Hjellming, C. M. Wade, 06.141.074), 264 (M. Kutner, P. Thaddeus, 06.132.016), 265 (G. K. Miley, 05.141.128), 266 (K. I. Kellermann, D. L. Jauncey, M. H. Cohen, B. B. Shaffer, B. G. Clark, J. Broderick, B. Rönnäng, O. E. H. Rydbeck, L. Matveyenko, I. Moiseyev, V. V. Vitkevitch, B. F. C. Cooper, R. Batchelor, 06.141.151), 267 (R. N. Manchester, 06.141.191), 268 (H. M. Johnson, 06.132.002), 269 (J. M. Sutton, D. H. Staelin, R. M. Price, 05.141.152), 270 (K. I. Kellermann, 06.141.079).

008.041 Greenwich

Royal Greenwich Observatory and Royal Observatory, Cape of Good Hope. — Report for the year ending 1970 December 31. R. Woolley, G. A. Harding. Quarterly Journ. Roy. Astron. Soc., Vol. 12, 277 - 294 (1971).

Museums: Telescope transported. Nature, Vol. 233, 515 (1971).

Royal Observatory Bulletins, (Joint Publications of the Royal Greenwich Observatory, Herstmonceux; Royal Observatory, Cape of Good Hope), Nos. 153 (R.v.d.R. Woolley, 06.084.287), 160 (D. V. Thomas, R. E. Wallis, 06.041.021), 161 (S. V. M. Clube, Z. Aslan, T. W. Russo, E. D. Clements, 06.122.059), 162 (C. A. Murray, R. H. Tucker, E. D. Clements, 06.141.026), 163 (D. H. P. Jones, 06.122.011), 164 (G. A. Harding, M. P. Candy, 06.034.004), 165 (G. A. Harding, F. Fahim, C. M. Haslam, 06.155.027), 166 (R. Woolley, S. B. Pocock, E. A. Epps, R. Flin, 06.115.007), 169 (J. B. Alexander, B. S. Carter, 06.113.046).

008.042 Groningen

Nederlandse Vereniging voor Weer- en Sterrenkunde. Observations of Variable Stars. Report (Kapteyn Astronomical Laboratory, Groningen–Netherlands), No. 20 (L. Plaut, H. Feijth, 06.123.031).

008.043 Hamburg

Deutsches Hydrographisches Institut, Hamburg. **Astronomische Zeit- und Breitenbestimmungen, Empfangszeiten von Zeitsignalen,** 1971 January - September (06.044.020).

008.044 Hannover

Technische Universität Hannover. Astronomische Station des Instituts für Theoretische Geodäsie, No. 7 (R. Liese, 06.072.084).

008.045 Heidelberg

Astronomy and Astrophysics Abstracts, Vol. 5 (S. Böhme, W. Fricke, U. Güntzel-Lingner, F. Henn, D. Krahn, G. Zech, 06.002.039).

Astronomisches Rechen-Institut in Heidelberg, Mitteilungen, Serie A, Nos. 41 (W. Fricke, 05.041.027), 42 (P. Brosche, J. Sündermann, 05.081.047), 43 (W. Gliese, 05.126.018), 44 (J. Schubart, 04.098.013), 45 (R. Wielen, 04.151.014), 46 (W. Lohmann, 05.155.036), 47 (R. Wielen, 05.151.048), 48 (W. Gliese, 05.115.020), 49 (P. Brosche, 05.158.118), 50 (R. E. Laubscher, 05.097.071), 51 (H. Scholl, 05.099.079), 52 (W. Fricke, 06.043.003), 53 (J. Schubart, 06.043.012), 54 (H. Scholl, 06.099.050), 55 (R. Wielen, 06.151.033).

Astronomisches Rechen-Institut in Heidelberg, Mitteilungen, Serie B, Nos. 25 (P. Brosche, 05.158.098), 26 (W. Fricke, 05.043.005), 27 (R. Wielen, 05.153.040), 28 (R. E. Laubscher, 06.097.002), 29 (W. Lohmann, 06.153.003).

008.046 Helsinki

Reprints from the Astrophysics Laboratory, University of Helsinki, Nos. 34 (I. V. Tuominen, AJB 68, 106.20), 35 (O. Vilhu, AJB 68, 104.180), 36 (I. V. Tuominen, O. Vilhu, 04.064.027), 37 (I. V. Tuominen, O. Vilhu, 04.116.001), 38 (J. Tuominen, 04.080.040), 39 (O. Vilhu, I. V. Tuominen, 05.064.040).

008.047 Hyderabad

Nizamiah Observatory Reprint, Nos. 45 (A. Peraiah, AJB 66, 112.44), 46 (K. D. Abhyankar, 06.064.054), 47 (K. D. Abhyankar, M. B. K. Sarma, AJB 66, 113.01), 49 (K. S. Sastry, S. M. Alladin, 03.151.050).

The Osmania University. Contributions from the Nizamiah Observatory, Hyderabad. Contribution Nos. 3 (M. B. K. Sarma, 06.112.016), 4 (S. Aravamudan, 06.041.037).

008.048 Izmir

Scientific Reports of the Faculty of Science, Ege University, Izmir, Nos. 118—Astronomy No. 11 (Ş. Bozkurt, 06.158.106), 120—Astron. No. 12 (A. Kizilirmak, 06.121.067), 122—Astron. No. 13 (M. Ü. Akyol, 06.082.082).

008.049 Jena

Mitteilungen der Universitäts-Sternwarte zu Jena, Nos. 97 (W. Wenzel, J. Dorschner, C. Friedemann, 06.121.007), 101 (J. Dorschner, 06.132.008), 102 (S. Marx, 06.131.018), 103 (S. Marx, 06.131.019), 104 (K.-H. Schmidt, 06.131.036), 105 (C. Friedemann, 06.131.037), 106 (J. Dorschner, 06.131.042), 107 (K.-H. Schmidt, 06.131.043), 108 (J. Dorschner, 06.131.044), 109 (H. Zimmermann, 06.131.045), 110 (K.-H. Schmidt, 06.132.019).

008.050 Johannesburg

Republic Observatory. J. Hers.
Inform. Bull. Southern Hemisphere, No. 18, p. 24 - 26 (1971). Report for year ending 31 December 1970.

Republic Observatory, Johannesburg. – Report for the year ending 1970 December 31. J. Hers.
Quarterly Journ. Roy. Astron. Soc., Vol. 12, 274 - 276 (1971).

South African Council for Scientific and Industrial Research. Republic Observatory Johannesburg. Circulars. Vol. 8, No. 131 (J. A. Bruwer, M. Klerk, 06.098.021; J. A. Bruwer, M. Klerk, 06.103.101, 06.103.122; J. L. Newburg, 06.118.004; G. F. G. Knipe, 06.121.060, G. F. G. Knipe, 06.121.061; G. F. G. Knipe, 06.123.042; G. F. G. Knipe, 06.121.062).

008.051 Kiel

Sonderdrucke der Sternwarte Kiel, Nos. 170 (V. Weidemann, 05.126.025), 171 (I. Bues, 05.126.029), 172 (H. Holweger, 05.071.029), 173 (V. Weidemann, 05.011.028), 174 (V. Weidemann, 06.065.126), 175 (D. Reimers, 06.074.005), 176 (D. Richter, 06.126.002), 177 (A. Unsöld, 05.007.000), 178 (A. Unsöld, 06.071.060), 179 (V. Weidemann, 06.142.064).

008.052 Kitt Peak

Kitt Peak National Observatory, Contributions, Nos. 515 (K. I. Hudson, H.-Y. Chiu, S. P. Maran, F. E. Stuart, P. R. Vokac, 05.122.053), 534 (R. A. Wells, 05.097.051), 558 (W. Livingston, J. Harvey, 06.034.123).

008.053 Kodaikanal

Annual report of the Kodaikanal Observatory for the year 1968. M. K. V. Bappu.
Printed by the Manager Government of India Press Coimbatore and published by the Manager of Publications, Delhi. 12 pp. (1970).

Kodaikanal Observatory, Bulletins, Series B, Nos. 192 (M. K. V. Bappu, 06.075.026), 196 - 197 (M. K. V. Bappu, 06.075.037).

Kodaikanal Observatory, Bulletins, Series C, No. 194 (M. K. V. Bappu, 06.075.029).

Kodaikanal Observatory, Bulletin, No. 177 (M. K. V. Bappu, 06.075.028).

Kodaikanal Observatory, Reprints, Nos. 38 (U. V. Gopala Rao, 04.077.021), 39 (C. V. Sastry, 05.077.018), 48 (M. K. V. Bappu, K. R. Sivaraman, 05.073.034).

008.054 La Plata

Observatorio Astronómico. (Astronomical Observatory, National University of La Plata).
Inform. Bull. Southern Hemisphere, No. 18, p. 9 - 11 (1971). Current research report.

Observatorio Astronómico de la Universidad Nacional de La Plata. Serie Astronómica, Vol. 37 (C. Jaschek, L. Ferrer, M. Jaschek, 06.114.129).

Observatorio Astronómico de la Universidad Nacional de La Plata, Serie Especial, No. 24 (S. J. Slaucitajs, 06.046.038).

Separata Astronómica, Observatório Astronómico — La Plata — Argentina, Nos. 105 (V. N. de Monteagudo, J. Sahade, 04.114.118), 106 (J. Sahade, J. Albano, 04.121.046), 107 (H. Levato, S. Malaroda, 04.116.006), 108 (A. Feinstein, H. G. Marraco, 05.153.030), 109 (L. Ferrer, C. Jaschek, 05.113.051).

008.055 Lembang

The Bosscha Observatory in Indonesia.
S. M. Larson.
Sky Telescope, Vol. 42, 70 - 72 (1971).

008.056 **Leningrad**

Jubilee of the Institute for Theoretical Astronomy of the U.S.S.R. Academy of Sciences (Chronicle).
Byull. Inst. Teoret. Astron., *Leningrad,* Vol. 12, 755 - 756 (1971). In Russian and English.

Main stages of the history of the Institute for Theoretical Astronomy of the U.S.S.R. Academy of Sciences (1919 - 1969).   G. A. Chebotarev.
Byull. Inst. Teoret. Astron., *Leningrad,* Vol. 12, 758 - 766 (1971). In Russian.

Bibliography on the history and activities of the Institute for Theoretical Astronomy of the U.S.S.R. Academy of Sciences for 50 years (1919 - 1969).   M. V. Lapteva.
Byull. Inst. Teoret. Astron., *Leningrad,* Vol. 12, 767 - 771 (1971). In Russian.

Chronicle.
Byull. Inst. Teoret. Astron., *Leningrad,* Vol. 12, 942 - 945 (1971). In Russian.

Byulleten' Instituta Teoreticheskoj Astronomii, Akademiya Nauk SSSR, Vol. 12, Nos. 8 (G. A. Chebotarev, 06.006.000; N. S. Samojlova-Yakhontova, 06.098.001; G. A. Chebotarev, M. J. Shmakova, 06.098.002; L. S. Evdokimova, 06.053.001; O. A. Kalinin, A. M. Finkelshtein, 06.162.001; G. R. Kastel, 06.103.100; A. B. Onegina, E. M. Sereda, 06.097.001; L. I. Chernykh, N. S. Chernykh, 06.041.001; L. I. Chernykh, 06.098.003), 9 (06.008.000; G. A. Chebotarev, 06.008.000; M. V. Lapteva, 06.008.000; V. K. Abalakin, 06.047.015; Yu. V. Batrakov, 06.054.014; V. A. Ivakin, 06.021.002; G. A. Chebotarev, 06.042.033; E. I. Kazimirchak-Polonskaya, 06.102.011; Yu. V. Batrakov, 06.054.015), 10 (I. D. Zongolovich, 06.046.036; M. S. Volkov, 06.042.060; I. V. Galibina, 06.104.104; V. G. Degtjarjov, 06.042.061; B. Popovic, 06.042.062; T. K. Schinkarik, 06.042.063, V. I. Voronenko, G. K. Gorel, F. F. Kalichevich, R. T. Fedorova, 06.098.038; H. K. Raudsaar, 06.103.007; N. S. Chernykh, 06.103.008; L. I. Chernykh, 06.098.039; 06.008.000), Vol. 13, No. 1 (A. Deprit, J. Henrard, A. Rom, 06.094.282; V. K. Abalakin, 06.094.283; V. K. Abalakin, 06.094.284; O. N. Barteneva, 06.103.131; O. N. Barteneva, 06.103.131; E. M. Nezhinsky, 06.102.013; V. V. Terentyev, 06.042.052; A. M. Fominov, 06.072.081; S. G. Braunfeld, Z. N. Grigoryeva, E. K. Denisyuk, E. S. Yeroshevich, V. F. Kartashov, L. N. Kondratyeva, V. S. Matyagin, L. P. Sorokina, L. A. Usoltseva, A. A. Schipenstein, 06.098.032; A. Sh. Khatisov, 06.098.033; T. A. Guseva, 06.098.034; H. H. Raudsaar, 06.098.035; L. I. Chernykh, 06.098.036).

Ephemerides of minor planets for 1972,
(G. A. Chebotarev, 06.098.006).

Trudy Astronomicheskoj Observatorii, (Transactions of the Astronomical Observatory), *Leningrad,* Vol. 28 (A. K. Kolesov, 06.063.012; V. M. Loskutov, 06.063.013; V. A. Dombrovsky, T. A. Polyakova, V. A. Jakovleva, 06.122.045; V. G. Derevjanko, 06.158.027; G. V. Khozov, 06.031.015; G. P. Apushkinsky, V. V. Vitkovsky, 06.141.072; T. E. Derviz, 06.114.043; M. K. Babadzhanianz, L. D. Parfinenko, 06.034.010; V. A. Antonov, 06.151.009; T. A. Agekian, S. P. Yakimov, 06.151.010; K. V. Kholshevnikov, E. I. Timoshkova, 06.042.013; R. A. Lyakh, 06.042.014; R. P. Eremenko, E. N. Poljakhova, 06.054.011; A. A. Nemirov, 06.041.009; V. A. Fomin, 06.041.010; V. N. Lvov, 06.041.011; A. A. Nikitin, 06.061.012).

008.057 **London**

University of London Observatory. – Report for the year ending 1970 December 31.   C. W. Allen.
Quarterly Journ., Roy. Astron. Soc., Vol. 12, 192 - 196 (1971).

008.058 **Louvain**

Station météorologique automatique.
G. Schayes, L. Zandarin, with a preface by O. Godart.
Inst. d'Astron. Géophys. Georges Lemaître, Louvain, Contr. No. 10, 25 pp. (1971).

008.059 **Lund**

Meddelande från Lunds Observatorium, Series I, No. 250 (B. A. Lindblad, 06.104.042).

008.060 **L'vov**

Tsirkulyar. Astronomicheskaya Observatoriya, L'vov, Nos. 45 (I. A. Klimishin, A. F. Novak, 06.122.074; V. V. Golovatyj, 06.131.082; M. B. Girnyak, 06.123.026; M. Yu. Skul'skij, E. B. Vovchik, 06.121.050; Yu. V. Fridel', 06.122.075; P. A. Olijnyk, M. M. Koval'chuk, I. S. Laba, 06.074.051), 46 (I. A. Klimishin, A. F. Novak, 06.064.055; I. A. Klimishin, B. M. Gura, 06.064.056; V. V. Golovatyj, 06.134.008; V. V. Golovatyj, 06.132.045; I. V. Shpychka, V. V. Golovatyj, M. B. Girnyak, 06.082.132; E. B. Vovchik, 06.123.061, O. S. Yatsyk, 06.123.062).

008.061 **Madrid**

Memoria de las actividades del Seminario de Astronomía y Geodesía de la Facultad de Ciencias de la Universidad de Madrid en 1969.   J. M. Torroja.
Urania Barcelona, Año 55, Nos. 271–272, p. 240 - 250 (1970).

Memoria de las actividades del Seminario de Astronomía y Geodesia de la Universidad Complutense de Madrid en 1970.   J. M. Torroja.
Urania Barcelona, Año 56, No. 273, p. 122 - 129 (1971).

Boletín Astronómico del Observatorio de Madrid.
Instituto Geografico y Catastral, Sección 2ª, Astronomía, Vol. 7, No. 6 (J. Pensado, 06.075.018).

Universidad de Madrid – Facultad de Ciencias.
Seminario de Astronomia y Geodesia, Publicación, Nos. 63 (J. M. Torroja, 05.009.005), 64 (M. J. Sevilla, 06.031.010), 65 (J. M. Torroja, 06.008.061), 66 (M. J. Sevilla, 06.046.008), 67 (M. E. Rego, 06.114.039), 68 (M. J. Fernandez-Figuero, 06.114.094).

008.062 **Manchester**

Astronomical Contributions from the University of Manchester, Series II, Jodrell Bank Reprints, Nos. 417 (B. J. Rickett, 04.141.012), 422 (S. T. Gottesman, G. de Jager, 04.158.085), 425 (P. C. Gregory, 06.033.014), 426 (W. H. McCutcheon, R. D. Davies, 04.158.035), 427 (J. G. Davies,

M. I. Large, 04.141.009), 429 (P. Thomasson, J. G. Davies, 04.133.015), 430 (B. G. Clark, G. K. Miley, 02.141.100), 432 (J. G. Davies, M. I. Large, A. C. Pickwick, 04.141.212), 434 (C. G. T. Haslam, C. J. Salter, 05.141.047), 435 (B. Lovell, 05.008.079), 436 (W. Donaldson, H. Smith, 05.141.038), 437 (W. N. Charman, J. V. Jelley, J. H. Fruin, E. R. Hodgson, P. F. Scott, J. R. Shakeshaft, G. A. Baird, T. Delaney, B. G. Lawless, R. W. P. Drever, W. P. S. Meikle, R. A. Porter, R. E. Spencer, 04.066.010), 438 (D. A. Graham, 05.141.232), 439 (W. Donaldson, G. K. Miley, H. P. Palmer, 05.141.099), 440 (J. R. Baker, F. G. Smith, 05.157.005). 449 (R. E. Schönhardt, 05.141.154; M. I. Large, 05.141.165; A. G. Lyne, 05.141.166; D. A. Graham, 05.141.169; R. G. Conway, 05.134.023; F. G. Smith, 05.141.182).

008.063 Milano

Osservazioni meteorologiche eseguite nell'Osservatorio Astronomico di Brera in Milano durante l'anno 1970. F. Zagar.
Pubbl. Oss. Astron. Milano, 29 pp. (1971).

008.064 Minneapolis

University of Minnesota, Minneapolis, Minnesota, Separate prints (W. J. Luyten, 06.112.017).

008.065 Mizusawa

Annual Report of the International Polar Motion Service 1969, (S. Yumi, 06.045.011).

Monthly Notes of the International Polar Motion Service, 1971 Nos. 5 - 10 (06.045.020).

008.066 Mons

Report for the years 1969, 1970, 1971.
L. Houziaux.
Centre Univ. Mons, Fac. Sci., Dép. d' Astrophys., Commun. No. 19, 12 pp. (1971).

Centre Universitaire de l'Etat à Mons. Faculté des Sciences. Département d'Astrophysique. Communications, Nos. 19 (L. Houziaux, 06.008.066), 20 (E. Blondelot, 06.114.126), 21 (E. Blondelot, 06.114.127), 22 (M. Duruy, 06.123.063).

008.067 Moskva

Soobshcheniya Gosudarstvennogo Astronomicheskogo Instituta im. P. K. Shternberga. Izdatel'stvo Moskovskogo Universiteta, Nos. 160 (S. N. Vashkov'yak, 06.097.025; V. P. Dolgachev, 06.052.011; V. P. Girichev, 06.042.024), 170 (V. V. Nesterov, 06.045.003; L. V. Rykhlova, 06.045.004; L. M. Khommik, 06.041.012; D. N. Ponomarev, 06.041.013; I. M. Kalinina, 06.112.007; A. P. Gulyaev, V. A. Korobova, 06.041.014; V. G. Fedosenko, 06.041.015), 171 (D. K. Karimova, E. D. Pavlovskaya, 06.041.016; T. A. Uranova, 06.113.023; T. A. Uranova, 06.152.005; O. D. Dokuchaeva, 06.036.004; O. D. Dokuchaeva, 06.036.005), 172 (N. M. Artiukhina,

06.153.027; G. A. Starikova, 06.118.028; V. M. Luyutyj, 06.034.110), 173 (O. G. Badalyan, 06.071.067; G. F. Sitnik, A. I. Kylystov, 06.034.111; M. G. Larionov, A. A. Kapustkin, 06.033.022; M. V. Popov, 06.033.023; L. P. Grishchuk, V. I. Ulin, 06.066.072), 175 (G. N. Duboshin, A. I. Rybakov, E. P. Kalinina, P. N. Kholopov, 06.151.067).

008.068 München

Mitteilungen aus dem Institut für Astron. und Physik. Geodäsie der Technischen Hochschule München, No. 80 (M. Schneider, 06.052.026).

008.069 Naini Tal

Uttar Pradesh State Observatory, *Naini Tal*, Reprints, Nos. 36 (P. P. Saxena, 06.082.151), 38 (C. D. Kandpal, J. B. Srivastava, 04.121.031), 39 (J. B. Srivastava, C. D. Kandpal, 04.121.032), 50 (T. R. Bhatt, S. D. Sinvhaj, 05.122.095).

008.070 New Haven

Yale University Observatory, New Haven, Connecticut. P. Demarque.
Bull. American Astron. Soc., Vol. 3, 432 - 435 (1971).

Transactions of the Astronomical Observatory of Yale University, Vol. 31, (P. K. Lü, D. Hoffleit, 06.041.030).

008.071 New York

Contributions of American Museum—Hayden Planetarium, Series 1, Publication 31 (F. M. Branley, 04.014.007), 33 (F. M. Branley, 06.004.049).

008.072 Nice

Rapport d' activité de l' Observatoire de Nice pour 1969. J. C. Pecker.
Bull. d' Information, Ass. Développement International Obs. Nice, No. 7, p. 39 - 63, 77 - 79 (1970).

Rapport sur la mise en place du centre de calcul de l' Observatoire de Nice dans le cadre de la création du centre international d' astrophysique de l' Observatoire de Nice. J.-P. Scheidecker.
Bull. d' Information, Ass. Développement International Obs. Nice, No. 7, p. 65 - 68 (1970).

L' Observatoire et l'Université de Nice. R. Dars.
Bull. d' Information, Ass. Développement International Obs. Nice, No. 8, p. 17 - 19 (1971).

Rapport d' activité de l' Observatoire de Nice pour 1970. P. Delache.
Bull. d' Information, Ass. Développement International Obs. Nice, No. 8, p. 43 - 75, 103 - 106 (1971).

Les statuts de l' Observatoire de Nice. M. Henon.
Bull. d' Information, Ass. Développement International Obs. Nice, No. 8, p. 77 - 85 (1971).

Les constructions et réalisations nouvelles de l' Observatoire de Nice.   J. Marchal.
Bull. d' Information, Ass. Développement International Obs. Nice, No. 8, p. 87 - 89 (1971).

008.073 Nikolayev

150th anniversary of the Nikolayev Observatory.
N. S. Kalikhevich.
Astron. Zhurn. Akad. Nauk SSSR, Vol. 48, 1101 (1971). In Russian. English translation in Soviet Astron. AJ, Vol. 15, No. 5.

008.074 Northfield

Publications of the Goodsell Observatory, No. 16 (R. B. Carr, 06.121.076).

008.075 Ottawa

Contributions of the National Research Council of Canada, Ottawa, Canada, Nos. 11957 (M. B. Bell, E. R. Seaquist, L. D. Braun, 06.141.009), 11983 (V. Gaizauskas, L. W. Avery, F. D. Manning, 05.079.103), 12093 (L. A. Higgs, 06.133.003), 12102 (J. M.MacLeod, B. H. Andrew, W. J. Medd, E. T. Olsen, 06.141.115), 12286 (B. H. Andrew, G. A. Harvey, W. J. Medd, 06.141.174), 12296 (L. A. Higgs, 06.157.007).

Contributions from the Earth Physics Branch, Department of Energy, Mines and Resources, Ottawa, Ontario, Canada, Nos. 317 (S. Chapman, J. C. Gupta, S. R. C. Mallin, 06.084.311), 323 (S. Chapman, J. C. Gupta, 06.084.310), 353 (M. R. Dence, J. A. V. Douglas, A. G. Plant, R. J. Traill, 06.094.329), 363 (M. R. Dence, 05.105.047).

Publications of the Earth Physics Branch, Department of Energy, Mines and Resources, Ottawa, Canada, Vol. 40, No. 9 (G. A. Brown, 06.084.313); Vol. 41, Nos. 6 (D. R. Auld, I. W. Fetterley, 06.084.314), 8 (R. W. Tanner, 06.032. 038), 10 (E. I. Loomer, G. Jansen van Beek, 06.084.315), 11 (C. M. Carmichael, T. R. Hartz, 06.084.316).

008.076 Oxford

Department of Astrophysics, University of Oxford. Report for the year ending 1970 December 31.
D. E. Blackwell.
Quarterly Journ. Roy. Astron. Soc., Vol. 12, 328 - 333 (1971).

008.077 Paris

Bureau International de l'Heure, Circulaires B/C, Nos. 184 - 188 (06.045.030).

Bureau International de l'Heure, Circulaires, D57 - D61 (06.044.032).

008.078 Pasadena

Hale Observatories, operated by Carnegie Institution of Washington and California Institute of Technology, Pasadena, California. Annual report of the director, 1970 - 1971.
H. W. Babcock.
Reprinted from Carnegie Institution, Washington, Year Book, Vol. 70, 389 - 454 (1971).

008.079 Pereyra Irarola

Instituto Argentino de Radioastronomía. (Argentine Radioastronomy Institute).
Inform. Bull. Southern Hemisphere, No. 18, p. 12 - 13 (1971). Current research report.

008.080 Perth

Perth Observatory, Western Australia. Communications, No. 2 (I. Nikoloff, M. P. Candy, 06.103.006).

008.081 Pittsburgh

Publications of the Allegheny Observatory of the University of Pittsburgh, Vol. 8, No. 7 (W. R. Beardsley, M. Matthew, E. M. Erskine, T. L. Gandet, Jr., E. N. Hubbell, M. Jacobson, G. Jones, K. Kobus, S. Levy, P. Lowrey, D. Namisnak, 06.112.012), Vol. 11, No. 1 (W. R. Beardsley, J. K. de Jonge, D. J. Haring, J. R. Hansen, 06.112.013).

008.082 Praha

Académie Tchécoslovaque des Sciences, Institut Astronomique, **Station de l'Heure à Prague**, Série 5, Nos. 13 - 16 (L. Webrová, V. Ptáček, 06.044.019).

008.083 Pretoria

Radcliffe Observatory, Pretoria. – Report for the year ending 1971 March 31.   A. D. Thackeray.
Quarterly Journ. Roy. Astron. Soc., Vol. 12, 320 - 327 (1971).

Communications from the Radcliffe Observatory, Pretoria, Nos. 108 (A. D. Thackeray, 04.119.002), 109 (A. J. Wesselink, 05.159.008), 110 (E. N. Walker, 05.119.005), 112 (A. D. Thackeray, 06.121.021), 113 (R. M. Catchpole, M. W. Feast, 06.114.044).

Radcliffe Observatory, *Pretoria*, **Reprints**, Nos. 83 (A. D. Thackeray, J. B. Alexander, P. W. Hill, 04.122.152), 84 (R. M. Catchpole, M. W. Feast, 04.132.012), 85 (M. W. Feast, 04.131.118), 86 (T. L. Evans, 04.155.027), 87 (M. W. Feast, 03.122.096), 88 (T. L. Evans, 04.122.099), 89 (R. M. Catchpole, M. W. Feast, 05.122.106), 90 (T. L. Evans, J. W. Menzies, 05.154.016), 91 (J. B. Alexander, A. D. Thackeray, 05.122.105), 92 (P. J. Andrews, 06.159.023), 93 (P. J. Andrews, T. L. Evans, 06.159.024), 94 (M. W. Feast, 05.114. 101), 95 (A. D. Thackeray, 06.159.014), 96 (A. D. Thackeray, 06.159.013), 97 (D. Crampton, A. D. Thackeray, 05.132.029), 98 (T. L. Evans, 05.122.107), 99 (G. Cayrel de Strobel,

06.114.140), 100 (T. L. Evans, 06.159.022), 101 (A. D. Thackeray, 06.007.000), 102 (T. L. Evans, 06.122.036),103 (T. L. Evans, R. S. Stobie, 06.122.037), 104 (A. D. Thackeray, 06.008.083).

008.084 Pulkovo

Trudy Glavnoj Astronomicheskoj Observatorii v Pulkove (R. S. Gnevysheva, 06.075.001).

008.085 Richmond Hill

Communications from the David Dunlap Observatory, University of Toronto, Richmond Hill, Ontario, Canada, Nos. 270 (S. van den Bergh, 04.158.102), 271 (S. van den Bergh, R. Heeringa, 04.153.022), 272 (R. F. Garrison, 04.152. 006), 273 (J. V. Wall, T. Y. Chu, J. L. Yen, 03.157.005), 274 (D. L. DuPuy, 04.158.074), 275 (S. van den Bergh, 05.158. 032), 276 (S. van den Bergh, 04.158.070), 277 (T. G. Barnes, N. R. Evans, 04.124.101), 278 (J. D. Fernie, 04.004.036), 279 (B. H. Andrew, S. van den Bergh, E. K. Conklin, J. D. Kraus, 05.141.081), 280 (J. E. Winzer, J. A. Roberts, 05.157.011), 281 (S. van den Bergh, 05.160.002), 282 (H. B. Sawyer-Hogg, 06.122.141), 283 (J. D. Fernie, J. P. Hagen, Jr., G. L. Hagen, L. McClure, 05.080.015), 284 (S. van den Bergh, 05.125.010), 285 (J. R. Percy, 05.122.125), 286 (S. van den Bergh, 05.113.018), 287 (W. W. Morgan, W. A. Hiltner, R. F. Garrison, 05.153.006), 288 (D. W. Marks, M. J. Clement, 05.065. 099), 289 (S. van den Bergh, 05.125.013), 290 (R. Racine, 05.154:007), 291 (S. van den Bergh, 05.158.078), 292 (M. Peimbert, S. van den Bergh, 06.141.021), 293 (W. E. Harris, M. J. Clement, 06.065.013), 294 (S. van den Bergh, 06.125. 003), 295 (S. van den Bergh, 05.125.028), 296 (R. Racine, 06.153.009), 297 (J. D. Fernie, J. O. Hube, 06.122.046), 298 (J. Kormendy, S. P. S. Anand, 06.154.005), 299 (S. van den Bergh, 05.158.067), 300 (J. R. Percy, 06.122.054).

Publications of the David Dunlap Observatory, University of Toronto, *Richmond Hill, Ontario, Canada,* Vol. 3, Nos. 2 (C. M. Coutts, H. Sawyer Hogg, 06.122.092), 3 (C. M. Coutts, 06.122.093), 4 (W. L. Gorza, J. F. Heard, 06.121.056).

008.086 Riga

Transactions of the Observatory, Vol. 12 (J. Ikaunieks, 06.003.005), 13 (Z. Alksne, J. Ikaunieks, 06.003.004).

008.087 Rio de Janeiro

Observatório Naciónal. (National Observatory). Inform. Bull. Southern Hemisphere, No. 18, p. 14 (1971). Current research report.

008.088 Rochester

C. E. Kenneth Mees Observatory, University of Rochester, Rochester, N. Y., **Reprints,** Nos. 26 (C. Sturch, H. L. Helfer, 05.113.036), 27 (L. G. Taff, U. DeAngelis, 06.160.006), 28 (P. Murdin, 06.096.010).

008.089 Roma

·**Monthly Bulletin.** Osservatorio Astronomico di Roma, Nos. 163 - 166 (M. Cimino, M. Torelli, A. Cacciani, V. Croce, R. Flamini, U. Bartolini, 06.075.017).

Osservatorio Astronomico di Roma, Monte Mario – Monte Porzio – Stazione Astrofisica sul Gran Sasso. **Contributi scientifici,** Serie III, Nos. 101 (V. Castellani, P. Giannone, L. Gratton, N. Panagia, 06.065.114), 102 (P. Giannone, 06.065.115), 103 (P. Giannone, 06.065.117), 104 (V. Castellani, P. Giannone, A. Renzini, 04.154.012), 105 (G. Agnelli, M. Cimino, M. Cutolo, M. Puglisi, 06.083.057), 106 (G. A. De Biase, 04.034.089), 107 (V. Castellani, P. Giannone, A. Renzini, 05.065.030), 108 (V. Castellani, P. Giannone, A. Renzini, 05.065.092), 109 (V. Castellani, P. Giannone, A. Renzini, 05.065.113), 110 (K. Nandy, F. Smriglio, 03.114.117), 111 (K. Nandy, F. Smriglio, L. Rossi, 05.131.120), 112 (V. Groce, G. A. De Biase, M. F. Toniolo, 06.021.008), 113 (G. A. De Biase, G. De Gregorio, 05.034.026), 114 (N. Virgopia, M. S. Vardya, 05.114.068), 115 (G. A. De Biase, F. Sacchetti, D. Trevese, 06.034.038), 116 (V. Castellani, P. Giannone, A. Renzini, 06.065.036), 117 (G. B. Baratta, G. Caprioli, 06.034.008), 118 (A. Cacciani, M. Cimino, M. Fofi, 06. 034.034).

Photographic Journal of the Sun, Osservatorio Astronomico di Roma, Nos. 45 - 52 (M. Cimino, 06.075.015).

008.090 San Fernando

Instituto y Observatorio de Marina, San Fernando, Separate print (M. López Palacios, 06.158.124).

008.091 San Miguel

Observatorio Nacional de Física Cósmica. (National Observatory of Cosmic Physics). Inform. Bull. Southern Hemisphere, No. 18, p. 13 (1971). Current research report.

La investigación solar en el Observatorio Nacional de Fisica Cosmica,San Miguel. M. E. Machado. Rev. Astron., Vol. 42, No. 174, p. 20 - 22 (1971).

Observatorio Nacional de Física Cósmica, San Miguel, Argentina. Boletín Meteorológico mensual, Vol. 24 (10) - (12), Vol. 25 (1) - (2), 1970 October - 1971 February (06.075.038).

008.092 Santa Cruz

The University of California. Contributions from the Lick Observatory, Santa Cruz, California, Nos. 266 (M. F. Walker, 06.034.087), 271 (J. S. Miller, 02.133.011), 273 (P. Bodenheimer, 05.065.078), 282 (G. H. Herbig, 02.132.010), 285 (E. J. Wampler, AJB 68, 145.132), 292 (M. F. Walker, 02.122.047), 295 (R. R. Zappala, 02.113.010), 296 (R. P. Kraft, W. Krzemiński, G. S. Mumford, 02.122.099), 298 (P. S. Conti, 03.114.029), 299 (G. H. Herbig, 02.114.107), 300 (P. S. Conti, 02.119.014), 301 (K. S. Anderson, R. P. Kraft, 02.158.065), 302 (G. H. Herbig, 05.065.067), 304 (E. P. J. van den Heuvel, 02.153.038; G. F. Gahm, R. R. Zappala, 02.131.131), 305 (E. P. J. van den Heuvel, 04.121.014), 307

(P. Bodenheimer, J. P. Ostriker, 04.065.064), 308 (T. D. Kinman, 02.141.138), 309 (P. S. Conti, 03.119.014), 310 (L. H. Aller, M. F. Walker, 04.132.018), 311 (G. F. Gahm, 03.122.078), 313 (J. D. Scargle, E. A. Harlan, 03.134.005), 314 (M. F. Walker, 04.082.043), 315 (J. McNall, L. Robinson, E. J. Wampler, 04.034.044), 316 (J. D. Scargle, L. J. Caroff, P. D. Noerdlinger, 04.141.032), 317 (G. H. Herbig, 04.122.095), 318 (J. S. Miller, 04.114.022), 319 (P. S. Conti, 05.065.083), 320 (E. J. Wampler, 05.158.040), 321 (J. Nelson, R. Hills, D. Cudaback, J. Wampler, 04.141.125), 322 (G. H. Herbig, R. R. Zappala, 04.114.076), 323 (P. S. Conti, 04.114.045), 324 (T. Barker, L. D. Baumgart, D. Butler, K. M. Cudworth, E. Kemper, R. P. Kraft, J. Lorre, N. K. Rao, G. H. Reagan, D. R. Soderblom, 05.122.039), 325 (P. Bodenheimer, 05.065.124).

University of California. **Lick Observatory Bulletin,** Nos. 602 (S. Vasilevskis, A. R. Klemola, E. A. Harlan, AJB 68, 85.38), 603 (E. P. J. van den Heuvel, 02.117.026), 605 (E. A. Harlan, 02.114.011), 606 (J. Faller, I. Winer, W. Carrion, T. S. Johnson, P. Spadin, L. Robinson, E. J. Wampler, D. Wieber, 02.094.067), 607 (E. A. Harlan, D. C. Taylor, 03.114.105), 608 (W. L. Burke, 06.022.076), 609 (B. F. Jones, W. F. van Altena, 04.153.013), 610 (R. R. Zappala, 05.034.075).

The University of California, Santa Cruz. Publications of the Lick Observatory, Vol. 22, Part 2 (A. R. Klemola, S. Vasilevskis, C. D. Shane, C. A. Wirtanen, 06.112.005), Part 3 (A. R. Klemola, S. Vasilevskis, 06.155.013).

008.093 Santiago

Departamento de Astronomía, Universidad de Chile. (Astronomy Department, University of Chile). A) National Astronomical Observatory, Cerro Calán, B) Cerro El Roble Astronomical Station, C) Maipú Radioastronomical Observatory. Inform. Bull. Southern Hemisphere, No. 18, p. 18 - 20 (1971). Current research report.

008.094 São José dos Campos

Observatório Astronómico do Instituto Tecnologico de Aeronautica. (I. T. A. Astronomical Observatory). Inform. Bull. Southern Hemisphere, No. 18, p. 16 - 17 (1971). Current research report.

008.095 São Paulo

Centro de Radioastronomía e Astrofísica da Universidade Mackenzie (C.R.A.A.M.). (Center of Radio Astronomy and Astrophysics, Mackenzie University). Inform. Bull. Southern Hemisphere, No. 18, p. 15 - 16 (1971). Current research report.

Instituto Astronômico e Astrofísico da Universidade de São Paulo. (Astronomical and Geophysical Institute). Inform. Bull. Southern Hemisphere, No. 18, p. 16 (1971). Current research report.

008.096 Sendai

Sendai Astronomiaj Raportoj, Nos. 116 (Y. Shibata, 05.065.094), 117 (M. Kubo, 06.162.051), 118 (K. Arai, 06.143.053), 119 (T. Aikawa, 06.065.107), 121 (E. Kobayashi, M. Takeuti, 06.064.051).

008.097 Shemakha

Soobshcheniya Shemakhinskoj Astrofizicheskoj, Observatorii, Akademiya Nauk Azerbajdzhanskoj SSR, vyp. (No.) 5 (S. M. Azimov, 06.114.041; E. A. Trebenikov, G. F. Sultanov, V. A. Cheprasov, 06.098.007; O. Kh. Gusejnov, 06.065.047; D. M. Kuli-Zade, 06.071.020; Z. F. Seidov, 06.065.048; G. T. Arazov, 06.052.007; R. Sh. Yakh'yaev, 06.022.033).

008.098 Sonneberg

Mitteilungen über Veränderliche Sterne, *Sonneberg,* Vol. 5, No. 10 (L. Meinunger, 06.123.006; 06.123.007), Vol. 6, No. 1 (H.-E. Fröhlich, S. Rößiger, 06.131.084, E. Scheller, 06.122.086, P. Ahnert, 06.123.034, I. Meinunger, 06.123.035, H. Geßner, 06.123.036, W. Wenzel, 06.123.037, E. Splittgerber, 06.123.038).

008.099 St. Andrews

Astronomy in the University of St. Andrews. D. W. N. Stibbs. Bull. d'Information, Ass. Développment International Obs. Nice, No. 7, p. 5 - 10 (1970).

008.100 Sydney

Sydney Observatory. – Report for the year ending 1970 December 31.   H. Wood. Quarterly Journ. Roy. Astron. Soc., Vol. 12, 334 (1971).

Astrographic Catalogue 1900.0, Vol. 53 (H. Wood, 06.041.025).

Sydney Observatory Papers, No. 63 (W. H. Robertson, 06.004.048).

Division of Radiophysics, C.S.I.R.O., Sydney, (Epping, New South Wales, Australia), Separate prints (J. V. Wall, 06.141.245; A. J. Shimmins, 06.141.246; E. K. Bigg, 06.082.144; E. K. Bigg, R. T. Meade, 06.082.145; S. Twomey, 06.082.146; D. K. Milne, 06.125.002; O. B. Slee, E. R. Hill, 06.141.015; J. V. Wall, A. J. Shimmins, J. K. Merkelijn, 06.141.017; J. L. Caswell, G. A. Dulk, W. M. Goss, V. Radhakrishnan, A. J. Green, 05.141.101; D. N. Cooper, 06.033.027; D. N. Cooper, 06.033.028; J. W. Brooks, J. D. Murray, V. Radhakrishnan, 05.114.071; O. B. Slee, P. S. Mulhall, 05.141.092; B. J. Robinson, J. L. Caswell, H. R. Dickel, 05.131.094; A. J. Shimmins, J. G. Bolton, B. A. Peterson, J. V. Wall, 05.141.111; J. P. Wild, 06.073.107; D. K. Milne, 06.125.025; M. W. Sinclair, F. J. Kerr, 06.157.008; R. J. Taylor, J. Warner, N. E. Bacon, 06.082.147; E. K. Bigg, Z. Kviz, W. J. Thompson, 06.105.162; B. J. Robinson, J. L. Caswell, W. M. Goss, 06.131.069; J. B. Whiteoak, F. F. Gardner,

05.141.139; J. G. Bolton, J. V. Wall, A. J. Shimmins, 06.141.242; F. F. Gardner, J. B. Whiteoak, 06.141.243; J. B. Whiteoak, F. F. Gardner, 06.141.244).

008.101 Tartu

Eesti NSV Taeduste Akadeemia (Akademiya Nauk Estonskoj SSR), **W. Struve nimeliese Tartu Astrofüüsika Observatooriumi, Publikatsioonid** (Publikatsii Tartuskoj Astrofizicheskoj Observatorii imeni V. Struve) Köide (Tom) 39 (T. Viik, 06.063.015; T. Viik, 06.063.016; A. Kruusmaa, 06.064.007; I. Kuusik, 06.064.008; V. D. Malyuto, 06.114. 045; H. Albo, 06.121.026; M. Ruusalepp, L. Luud, 06.124. 103; L. Luud, M. Ruusalepp, T. Kuusk, 06.122.047; T. Kuusk, M. Ruusalepp, L. Luud, 06.122.048; T. Nugis, L. Luud, 06.114.046; E.-M. Maasik, 06.034.011; K. Eerme, 06.082. 031; H. Eelsalu, 06.031.016; H. Eelsalu, 06.112.006; E. Saar, 06.162.018; E. Saar, I. Saar, 06.162.019; E. Saar, 06.162.020; E. Saar, 06.162.021; E. Saar, 06.162.022; A. Kruusmaa, 06.064.009; V. Malyuto, 06.064.010; H. Albo, 06.122.049; L. Luud, Ü. Ibrus, I. Kolka, 06.114.047; T. Kipper, 06.064. 011; U. Veismann, 06.034.012; U. Veismann, 06.034.013; V. Riives, 06.102.005).

Tartu Astronoomia Observatorium, Teated, Nos. 30 (A. Kipper, 06.141.069; E. Saar, 06.162.016), 31 (H. Eelsalu, 06.013.010; H. Eelsalu, 06.013.011), 32 (L. Luud, M. Ilmas, 06.064.033; M. Ilmas, 06.064.034; M. Ilmas, 06.064. 035; L. Luud, M. Ilmas, 06.064.036), 33 (T. Viik, 06.064. 037), 34 (T. Feklistova, 06.114.078).

008.102 Tautenburg

Visit to Tautenburg. H. G. Miles.
Journ. British Astron. Ass., Vol. 81, 394 - 395 (1971).

Zentralinstitut für Astrophysik. Mitteilungen des Karl - Schwarzschild - Observatoriums Tautenburg der Deutschen Akademie der Wissenschaften zu Berlin, Nos. 53 (R. Ziener, 03.034.057), 54 (F. Börngen, 04.098.017), 55 (W. Högner, N. Richter, 05.031.020), 56 (W. Högner, 06.036.001), 57 (R. Ziener, 06.113.012).

008.103 Teide (Tenerife)

Astronomi på ferierejsen – Erfaringer fra Tenerife. P. Darnell.
Astron. Tidsskr., Årg. 4, 135 - 136 (1971).

El nuevo Observatório del Teide, su organización y puesta en marcha. J. M. Torroja, F. Sánchez.
Urania Barcelona, Año 55, Nos. 271–272, p. 85 - 96 (1970).

El nuevo Observatório del Teide. La Sección de alta atmósfera y medio interplanetario. F. Sánchez.
Urania Barcelona, Año 55, Nos. 271–272, p. 111 - 120 (1970).

El nuevo Observatório del Teide. La Sección de física solar. J.Casanovas.
Urania Barcelona, Año 55, Nos. 271–272, p. 121 - 126 (1970).

Presencia esporadica de polvo Sahariano en la atmosfera de la isla de Tenerife. F. Sánchez.
Urania Barcelona, Año 55, Nos. 271 – 272, p. 195 - 207 =

Obs. Astron. Teide, Tenerife, Publ. No. 11 (1970).

El Observatório del Teide en 1969. J. M. Torroja.
Urania Barcelona, Año 55, Nos. 271–272, p. 241 - 244 (1970).

El Observatorio de Teide en 1970. J. M. Torroja.
Urania Barcelona, Año 56, No. 273, p. 129 - 136 (1971).

Observatorio Astronomico del Teide, Tenerife "Islas Canarias". Publicación Nos. 6 (F. Sánchez, 06.082.029), 7 (J. M. Torroja, F. Sánchez, 06.008.103), 8 (F. Sánchez, 06.008.103), 9 (J. Casanovas, 06.008.103), 10 (J. M. Torroja, 06.008.103), 11 (F. Sánchez, 06.008.103).

008.104 Thessaloniki

Université de Thessaloniki, Annuaire de l'Institut Météorologique et Climatologique, 30 - 33 (G. C. Livadas, 06.082.150).

008.105 Tokyo

Annals of the Tokyo Astronomical Observatory, University of Tokyo, Second Series, Vol. 12, No. 4 (S. Iijima, Y. Niimi, 06.041.007; S. Isobe, 06.131.048; S. Isobe, 06.131.049), Vol. 13, No. 1 (H. Yasuda, I. Kamijo, 06.041. 026; H. Yasuda, 06.041.027).

Contributions from the Department of Astronomy, University of Tokyo, Nos. 128 (E. Scalise, Jr., 04.077.060), 129 (Y. Osaki, 04.122.096), 130 (H. Yoshimura, S. Kato, 05.080.026), 131 (W. Unno, 04.064.042), 132 (T. Tsuji, 06.122.006), 133 (H. Maehara, 06.122.007), 134 (M. Yuasa, 06.042.006), 135 (O. Kaburaki, Y. Uchida, 06.062.002), 136 (I. Masaki, 06.063.002), 137 (T. Sasao, 06.151.002), 138 (K. Utsumi, Y. Yamashita, 06.114.016).

Data Report of Hydrographic Observations. Series of Astronomy and Geodesy, Maritime Safety Agency, Tokyo, Japan, No. 6 (06.096.009).

Time and Latitude Bulletins, Tokyo Astronomical Observatory, Vol. 45, Nos. 3 - 8 (06.044.021).

Tokyo Astronomical Bulletin, Tokyo Astronomical Observatory, Second Series, Nos. 208 (H. Yasuda, R. Fukaya, H. Hara, H. Ishii, T. Ina, 06.041.003), 209 (M. Kitamura, A. Yamasaki, 06.121.015), 210 (S. Isobe, 06.114.029), 211 (K. Saito, 06.121.016), 212 (S. Iijima, S. Fujii, H. Kobayashi, F. Ohtsuka, 06.041.029), 213 (K. Nagasawa, 06.104.078), 214 (H. Kosai, 06.124.008), 215 (K. Osawa, K. Ichimura, K. Tomita, 06.142.065).

Tokyo Astronomical Observatory, Reprints, Nos. 391 (T. Takakura, K. Ohki, N. Shibuya, M. Fujii, M. Matsuoka, S. Miyamoto, J. Nishimura, M. Oda, Y. Ogawara, S. Ota, 05.076.014), 393 (S. Isobe, 06.152.002), 394 (O. Kaburaki, Y. Uchida, 06.062.002), 395 (H. Yoshimura, K. Tanaka, M. Shimizu, E. Hiei, 06.071.002), 396 (K. Nariai, 06.117. 013), 397 (S. Kikuchi, 06.125.103), 398 (T. Takakura, 06.073.015), 399 (T. Takakura, 06.072.035), 405 (K. Nishi, Z. Suemoto, 06.031.028), 406 (Y. Kozai, 04.042.077), 407 (T. Hirayama, 06.073.024), 408 (H. Tanabe, K. Mori, 05.034.035).

008.106 Tonantzintla

Boletin de los Observatorios Tonantzintla y Tacu-
baya, Vol. 6, Nos. 36 (S. Torres-Peimbert, 06.153.002;
S. Torres-Peimbert, H. Spinrad, 06.065.021; M. Peimbert,
S. Torres-Peimbert, 06.133.005; M. Peimbert, 06.133.006;
G. A. Gurzadyan, 06.122.012; E. F. Schmitter, E. Recillas-
Cruz, 06.131.017; A. G. D. Philip, L. J. Relyea, 06.114.023;
E. E. Mendoza V., 06.152.003; E. E. Mendoza V., 06.113.011;
E. E. Mendoza V., 06.082.020), 37 (M. Peimbert, 06.158.119;
S. Torres-Peimbert, M. Peimbert, 06.131.134; S. Torres-Peim-
bert, 06.065.128; P. Pişmiş, 06.122.136; E. E. Mendoza V.,
06.122.137; E. E. Mendoza V., 06.113.060; B. Iriarte Erro,
06.122.138).

008.107 Toruń

Bulletin of the Astronomical Observatory of
N. Copernicus University in Toruń, No. 48 (S. Grudzińska,
04.124.103; J. Krempeć, 04.124.101; R. Głębocki, J. Strobel,
04.114.120; R. Głębocki, A. Stawikowski, 05.080.027;
L. D. G. Young, R. A. J. Schorn, E. S. Barker, A. Woszczyk,
06.093.009).

008.108 Treviso

Osservatorio Privato Specola 'Ariel', Treviso, Pubblicazione, Nos. 51 (G. Romano, 04.123.011), 52 (G. Romano,
G. Di Țullio Vanzani, 04.122.125), 53 (G. Romano, M. Perissi-
notto, 05.123.019), 54 (G. Romano, 05.122.062).

008.109 Trieste

Pubblicazione Osservatorio Astronomico di Trieste,
-Nos. 424 (L. Rusconi, G. Sedmak, 05.031.041), 427 (L. Rusconi, G. Sedmak, 05.031.002), 428 (R. Faraggiana,
05.114.054), 429 (P. Zlobec, 06.075.033), 430 (B. Cester,
05.121.044), 431 (A. Abrami, 06.075.034), 432 (P. Santin,
05.077.022), 433 (R. Faraggiana, M. Hack, 06.122.058), 434
(06.075.035).

008.110 Turku

Astronomia-Optika Institucio, Universitato de
Turku, Informo, No. 36 (Y. Väisälä, 06.031.067).

008.111 Uccle

Observatoire Royal de Belgique (Koninklijke Sterrenwacht van België), Communications(Mededelingen), Série
A, Nos. 14 (Série Géophysique, No. 103 – R. J. Dejaffe, P. J.
Melchior, 06.045.001), 16 (N. Grevesse, A. J. Sauval,
05.071.022), 17 (N. Grevesse, A. J. Sauval, 06.071.018).

Observatoire Royal de Belgique (Koninklijke Sterrenwacht van België), Communications (Mededelingen), Série
B, Nos. 32 (C. Delys, R. Gonze, 06.075.003), 36 (C. Delys, R.
Gonze, 06.077.018), 50 (C. Gonze-Delys, R. Gonze,
06.077.041), 57 (G. Evrard, C. Gonze, A. Koeckelenbergh,
06.075.002), 58 (M. Bonatz, P. Melchior, 06.081.050), 59 (A.
Koeckelenbergh, 05.008.135), 60 (P. Cugnon, 06.131.104),

62 (P. Melchior, 06.043.008).

008.112 Victoria

Dominion Astrophysical Observatory, Victoria, B. C.
J. B. Hutchings.
Journ. Roy. Astron. Soc. Canada, Vol. 65, 246 (1971).

Dominion Astrophysical Observatory Victoria, B.C.
Report for the year 1970 April 1 to 1971 March 31.
K. O. Wright.
Quarterly Journ. Roy. Astron. Soc., Vol. 12, 295 - 304
(1971).

Contributions from the Dominion Astrophysical
Observatory, Victoria, B. C., Nos. 137 (G. Hill, J. B.
Hutchings, 04.121.025), 142 (A. H. Batten, J. M. Fletcher,
05.118.016), 143 (C. D. Scarfe, 05.118.018),146 (J. B.
Hutchings, 04.064.002),147 (A. H. Batten, 04.117.010), 148
(A. H. Batten, J. M. Fletcher, F. R. West, 05.118.029), 149
(D. Crampton, J. Grygar, L. Kohoutek, R. Viotti, 03.133.022),
151 (G. Hill, J. V. Barnes, 05.153.012), 152 (J. B. Hutchings,
05.064.030), 153 (J. B. Hutchings, G. Hill, 05.121.046), 154
(J. B. Hutchings, G. Hill, 05.117.029), 155 (J. Grygar, J.
Smolínski, J. B. Hutchings, 05.124.100), 156 (J. B. Hutchings,
05.031.040), 157 (J. B. Hutchings, 05.031.033), 161 (J. A.
Pearce, G. Hill, 06.119.010), 164 (G. Hill, J. V. Barnes, J. B.
Hutchings, J. A. Pearce, 06.121.025), 170 (J. E. Penfold,
06.122.089).

008.113 Vilnius

Latvijas PSR Zinātņu Akadēmijas Radioastrofizikas
Observatorijā. I. Daube.
Zvaigžņotā debess, 1971. gada pavasaris, p. 64 - 65.

Astronomijos Observatorijos Biuletenis (Bulletin of
the Vilnius Astronomical Observatory), No. 32 (G. Kakaras,
06.113.054; G. Kavaliauskaité, V. Straižys, A. Ažusienis,
06.113.055; R. Bartkus, 06.113.056; R. Bartkus, 06.131.132).

008.114 Warsaw

Warsaw University Observatory and Astronomical
Institute, Polish Academy of Sciences, Reprints Nos. 307
(W. Krzemiński, J. Smak, 05.124.107), 308 (S. Grzędzielski,
05.064.046), 309 (M. A. Abramowicz, 05.065.038), 310
(J. P. Lasota, 05.131.108), 311 (M. Bielicki, 05.092.012),
312 (K. Ziołkowski, 05.098.010), 313 (B. Paczyński,
06.065.049), 314 (W. Dziembowski, 06.065.050), 315
(A. Kruszewski, 06.158.031), 316 (M. Karpowicz, 06.160.003),
317 (B. Paczyński, 06.133.028), 318 (I. Semeniuk, A. Kruszewski, 06.141.215), 319 (M. A. Abramowicz, 06.065.109),
320 (J. Smak, 06.119.014), 321 (J. Juchniewicz, 06.022.096),
322 (S. M. Ruciński, 06.117.027).

Politechnika Warszawska, Obserwatorium Astrono-
miczno-Geodezyjne w Józefosławiu, (Warsaw Technical University, Astronomic-Geodetical Observatory at Józefosław),
Latitude Circular, Nos. 37 - 38 (L. Pieczyński, 06.045.021).

Publikacje Działu Geodezji Wyższej i Astronomii,
Geodezyjnej Zg. PAN, No. 17 (T. Chojnicki, 04.081.051).

008.115 **Washington**

United States Naval Observatory, *Washington, Circular*, Nos. 131 (J. S. Duncombe, 06.079.103), 132 (L. B. Weston, 06.047.018), 133 (R. S. Harrington, M. Miranian, 06.072.053), 134 (P. M. Janiczek, P. K. Seidelmann, 06.041.028), 135 (J. S. Duncombe, 06.079.104).

U. S. Naval Observatory, *Washington, D. C.,* Reprints, No. 96 (K. Aa. Strand, 01.008.132/I).

008.116 **Waterloo**

Contributions of the University of Waterloo Observatory, No. 8 (G. A. Bakos, 06.114.048).

008.117 **Wellington**

Astronomical phenomena for 1971, public programme and general information.   W. J. H. Fisher.
Astron. Bull. Carter Obs. Wellington, New Zealand, No. 74, 22 pp. (1970).

Report of the Carter Observatory Board for the year ended 1971, March 31.   R. P. Gough.
Astron. Bull. Carter Obs. Wellington, New Zealand, No. 75, p. 1 - 2 (1971).

Carter Observatory Report for the year ended 1971, March 31.   W. J. H. Fisher.
Astron. Bull. Carter Obs. Wellington, New Zealand, No. 75, p. 3 - 6 (1971).

Carter Observatory.
Inform. Bull. Southern Hemisphere, No. 18, p. 21 - 22 (1971).
Current research report.

008.118 **Wien**

Astronomische Mitteilungen Wien, Nos. 7 (J. Hopmann, 06.118.030), 8 (W. Tscharnuter, 06.151.069).

008.119 **Zelenchukskaya**

Soobshcheniya Spetsial'noj Astrofizicheskoj Observatorii, Akademiya Nauk SSSR, No. 2 (E. L. Chentsov, L. I. Snezhko, 04.064.098).

008.120 **Zürich**

Tätigkeitsbericht der eidgenössischen Sternwarte Zürich für das Jahr 1970.   M. Waldmeier.
Zürich, 7 pp. (1971).

Astronomische Mitteilungen der Eidgenössischen Sternwarte Zürich, Nos. 299 (M. Waldmeier, 06.074.069), 300 (M. Waldmeier, 06.074.070), 301 (M. Waldmeier, 06.074.071), 302 (M. Waldmeier, 06.072.054), 303 (M. Waldmeier, 06.075.022), 304 (M. Waldmeier, 06.072.063), 305 (M. Waldmeier, 06.034.097), 307 (M. Waldmeier, 06.074.072), 308 (M. Waldmeier, S. E. Weber, 06.074.073), 309 (W. Stanek, 06.073.084).

Quarterly Bulletin on Solar Activity (Zürich), Nos. 171 - 172 (M. Waldmeier, R. Howard, R. Michard, G. Olivieri, M. Bernot, 06.075.036).

Publikationen der Eidgenössischen Sternwarte Zürich, Band 13, Heft 5 (M. Waldmeier, 06.075.021).

009 Notes on Observatories, Planetaria, and Exhibitions

009.001 **Observatory-planetarium at Kutztown, Pennsylvania.**
A. Kiasat, C. R. Chambliss.
Sky Telescope, Vol. 42, 76 - 77 (1971).

009.002 **Wat betekent dit planetarium? Wat wil men ermee, hier?** K. Cuijpers.
Hemel en Dampkring, Vol. 69, 213 - 215 (1971).

009.003 **Radiophysical Observatory of the Academy of Sciences of the Latvian SSR.** A. E. Balklavs.
LatvPSR Zinātņu Akad. vēstis, Izv. AN LatvSSR, 1971, No. 3, p. 69 - 79. In Russian. – Abstr. in Referativ. Zhurn. 51. Astron., 8.51.131 (1971).

009.004 **Rapporto sull'attività del « Gruppo di Astrofisica ».**
A. Masani.
Mem. Soc. Astron. Italiana, Nuova Ser., Vol. 42, 249 - 253 (1971).

009.005 **The Mullard Space Science Laboratory.**
K. J. H. Phillips.
Spaceflight, Vol. 13, 97 - 99 (1971).

009.006 **West Berlin observatory and planetarium.**
B. Wedel.
Sky Telescope, Vol. 42, 152 - 154 (1971).

009.007 **The Wollongong University College Observatory.**
C. A. M. Gray, R. W. Upfold.
Journ. British Astron. Ass., Vol. 81, 463 - 466 (1971).

009.008 **Die Sternwarte der astronomischen Arbeitsgemeinschaft des Volksbildungswerkes Burghausen/Obb.**
E. Schmidt.
SuW, Vol. 10, 284 (1971).

009.009 **Die größte Sternwarte auf der Südhalbkugel.**
H. Gollnow.
Umschau, 71. Jahrgang, p. 820 - 821 (1971). – Concerning the joint Mount Stromlo and Siding Spring Observatories.

009.010 **The first universal space exhibition in Moscow.**
B. M. Tsirkov.
Zemlya i Vselennaya, No. 5, p. 60 - 65 (1971). In Russian.

009.011 **The Ussurijskaya solar station.**
V. F. Chistyakov.
AN SSSR. Dal'nevostochnyj nauchnyj tsentr, Vladivostok. 45 pp. Price 20 Kop. (1971). In Russian. – Review in Referativ. Zhurn. 51. Astron., 11.51.19 (1971).

009.012 **North Mountain Observatory well and truly opened.**
T. Tothill.
Journ. Roy. Astron. Soc. Canada, Vol. 65, L25 - L26 (1971).

009.013 **McLaughlin Planetarium, Toronto, Ontario.**
H. Creighton.
Journ. Roy. Astron. Soc. Canada, Vol. 65, 299 - 300 (1971).

009.014 **Four amateur observatories.**
F. Morefield, J. Soder, C. Sherrod, J. E. Dalton.
Sky Telescope, Vol. 42, 280 - 282 (1971).

009.015 **Aus der Arbeit der Volkssternwarten: Tätigkeitsbericht der Fachgruppe Astronomie des Deutschen Kulturbundes in Dresden für die Jahre 1968 bis 1971.**
H.-J. Blasberg.
Sterne, 47. Jahrgang, p. 216 - 219 (1971).

009.016 **Das Astronomische Zentrum 'B. H. Bürgel' in Potsdam.** A. Zenkert.
Sterne, 47. Jahrgang, p. 219 - 221 (1971).

009.017 **Baikal–sol–Baikal.** S. Ostroúmov.
El Universo, Vol. 25, 122 - 123 (1971).

009.018 **Le CERGA. (Le centre d' études et de recherches géodynamiques et astronomiques).** J. Levy.
Bull. d' Information, Ass. Développement International Obs. Nice, No. 8, p. 11 - 15 (1971).

009.019 **Osservatorio Astronomico Nazionale: Rapporto N. 1.**
Printed by Tipografia Antoniana, Padova. 55 pp. (1971).
Included are a short history of the project OAN by G. Righini; A preliminary report (January 1959); The first proposal to the ministry of Pubblica Istruzione (1961). The intended development of the project OAN.

009.020 **Der Weg zum Planetarium des Raumzeitalters.**
H. Letsch.
Jenaer Rundschau, (Jena Review), 16. Jahrgang, p. 273 - 277 (1971).

009.021 **Von den Volks- und Schulsternwarten der Deutschen Demokratischen Republik.** H. Wolf.
Jenaer Rundschau, (Jena Review), 16. Jahrgang, p. 282 - 286 (1971).

009.022 **Die Jena-Instrumente der Archenhold-Sternwarte in Berlin-Treptow.** D. Wattenberg.
Jenaer Rundschau, (Jena Review), 16. Jahrgang, p. 287 - 293 (1971).

009.023 **Die Wilhelm-Foerster-Sternwarte und die Astronomie in Berlin.** H.-B. Brenske.
Veröff. Wilhelm-Foerster-Sternw. Berlin, No. 21, 25 pp. (1970).

009.024 **Exposition "Kepler and Prague".** O. Hlad.
Říše hvězd, Vol. 52, 148 - 149 (1971). In Czech.

009.025 **Astronomija Latvijas PSR Zinātņu Akadēmijā 25 gados.**
Zvaigžņotā debess, 1971. gada vasara, p. 1 - 22.

Solar station to close.
Sky Telescope, Vol. 42, 345 (1971).
The solar observing station near Climax, Colorado, will close permanently in July, 1972.

010 Societies, Associations, Organizations

010.001 American Association of Variable Star Observers (AAVSO)

Personalities in the AAVSO. R. N. Mayall.
AAVSO Abstr., October 1971, p. 3 - 5.

Photoelectric photometry in the AAVSO.
J. J. Ruiz.
AAVSO Abstr., October 1971, p. 5 - 7.

History of the Solar Division. D. W. Rosebrugh.
AAVSO Abstr., October 1971, p. 7 - 8.

The AAVSO: Perspectives. W. M. Lowder.
AAVSO Abstr., October 1971, p. 8 - 9.

AAVSO report 29: A preview. B. L. Welther.
AAVSO Abstr., October 1971, p. 16 - 17.

American Association of Variable Star Observers. Abstracts of papers presented at Cambridge meeting, 16 October 1971.
AAVSO Abstr., October 1971, 19 pp.

Variable star observers meet in Massachusetts.
J. Ashbrook.
Sky Telescope, Vol. 42, 352 - 353 (1971).

010.002 American Astronomical Society (AAS)

The 135th meeting of the American Astronomical Society, held 24 through 27 August 1971 at Amherst, Massachusetts.
Bull. American Astron. Soc., Vol. 3, (No. 3, Part I), 357 - 412 (1971).

American astronomers report.
Sky Telescope, Vol. 42, 10 - 12 (1971).
Some highlights of the 134th meeting of the American Astronomical Society at Baton Rouge, Louisiana, March 29th to April 1st, 1971: Disk photography of a Jovian satellite; Very high-resolution photography with large reflectors.

010.003 Association of Lunar and Planetary Observers (ALPO)

Announcements.
Strolling Astronomer, Vol. 23, 75 - 76, 108 - 110 (1971).

The 1971 Astronomical League – A.L.P.O. convention at Memphis. J. L. Benton, Jr.
Strolling Astronomer, Vol. 23, 105 - 108 (1971).

010.004 Astronomical Society of Australia (ASA)

The fifth annual general meeting: Report of the Council, May 1971.
Proc. Astron. Soc. Australia, Vol. 2, 66 - 67 (1971).

010.005 Astronomical Society of Czechoslovakia

No publication received.

010.006 Astronomical Society of the Pacific (ASP)

Activities of the Society.
Publ. Astron. Soc. Pacific, Vol. 83, 694 - 696 (1971).

Activities of the Society: Minutes of the meeting of the directors, 7 May 1971.
Publ. Astron. Soc. Pacific, Vol. 83, 512 - 514 (1971).

Minutes of the 82nd annual meeting of the Astronomical Society of the Pacific, 7 May 1971.
Publ. Astron. Soc. Pacific, Vol. 83, 514 - 518 (1971).

Annual report of the treasurer for the year ending 31 December 1970.
Publ. Astron. Soc. Pacific, Vol. 83, 519 - 520 (1971).

010.007 Astronomical Society of Southern Africa (ASSA)

Notices.
Monthly Notes Astron. Soc. Southern Africa, Vol. 30, 83, 95, 115, 127, 151 (1971).

Proceedings of the annual general meeting, 1971.
Monthly Notes Astron. Soc. Southern Africa, Vol. 30, 101 - 108 (1971). – Included are: Report of Council for 1970–71, T. W. Russo, W. C. Bennett; Section reports: Occultation Section, A. G. F. Morrisby; Comet & Meteor Section, J. C. Bennett; Variable Star Section, R. P. de Kock.

Centre reports (Natal, Bloemfontein, Cape) for 1970–71.
Monthly Notes Astron. Soc. Southern Africa, Vol. 30, 117 - 120 (1971).

010.008 Astronomical Society of Victoria (ASV)

No publication received.

010.009 Astronomical Society of Western Australia (ASWA)

Report of proceedings – annual general meeting, 1971 July 12.
Journ. Astron. Soc. Western Australia, Vol. 28, July, p. 1 - 4 (1971).

Annual Junior Section report 1970/71.
Journ. Astron. Soc. Western Australia, Vol. 28, July, p. 4 - 5 (1971).

Reports of proceedings – 227th – 231st general meeting.
Journ. Astron. Soc. Western Australia, Vol. 29 - 33, August - December (1971).

Junior Section report.
Journ. Astron. Soc. Western Australia, Vol. 30, September, p. 6 (1971).

010.010 Astronomische Gesellschaft (AG)

Astronomische Gesellschaft. F. Egger.
Orion, 29. Jahrgang, p. 88 - 89 (1971).

010.011 **Astronomisk Selskab Kobenhavn**

No publication received.

010.012 **British Astronomical Association (BAA)**

Notices.
Journ. British Astron. Ass., Vol. 81, 338 - 339, 448; Vol. 82, 2 - 5 (1971).

Meetings of the Association.
Journ. British Astron. Ass., Vol. 81, 340 - 345; Vol. 82, 13 - 15 (1971).

Report of exhibits at the meeting of 1971 May 26.
A. C. Curtis.
Journ. British Astron. Ass., Vol. 81, 346 - 349 (1971).

Report of the council on work during the session 1970 July 1 to 1971 June 30 to be presented to members of the Association at the annual general meeting, 1971 October 27.
Journ. British Astron. Ass., Vol. 81, 422 - 448 (1971). — Included are Sections' reports.

Lunar Section.   P. A. Ringsdore.
Journ. British Astron. Ass., Vol. 81, 470 - 474 (1971).

Provincial meeting of the Association.
Journ. British Astron. Ass., Vol. 82, 6 - 9 (1971).

The annual general meeting of the Association.
Journ. British Astron. Ass., Vol. 82, 10 - 12 (1971).

The quatercentenary of Johannes Kepler.
Introduction to the commemorative meeting of the Association on 1971 June 30, by the director of the Historical Section.
Journ. British Astron. Ass., Vol. 82, 24 (1971).

Solar Section.   W. M. Baxter.
Journ. British Astron. Ass., Vol. 82, 49 (1971).

Variable Star Section.
Journ. British Astron. Ass., Vol. 82, 49 (1971).

010.013 **British Interplanetary Society (BIS)**

Society news.
Spaceflight, Vol. 13, 115 - 117, 151 - 153, 155, 271 - 272, 393 - 394 (1971).

Society meetings.
Spaceflight, Vol. 13, 435 - 440 (1971).

Report on 26th annual general meeting.
Spaceflight, Vol. 13, 389 - 390 (1971).

010.014 **Committee on Space Research (COSPAR)**

Brief report of the XIVth meeting of COSPAR.
Astronaut. Acta, Vol. 16, 381 - 384 (1971).

Committee on Space Research: 14 meeting of COSPAR and 12 International Space Science Symposium.
Z. Niemirowicz.
ICSU Bull., No. 24, p. 3 - 7 (1971).

Saules pētījumi turpinas.   N. Cimahoviča.
Zvaigžņotā debess, 1970./71. gada ziema, p. 49 - 61.

010.015 **European Space Research Organization (ESRO)**

No publication received.

010.016 **International Astronautical Federation (IAF)**

Report on the XXIst International Astronautical Congress, Constance, FRG, 5 – 10 October 1970.
Astronaut. Acta, Vol. 16, 307 - 315 (1971).

The 21st International Astronautical Congress.
Spaceflight, Vol. 13, 265 - 268 (1971). – Constance, 1970 Oct. 4 - 10.

010.017 **International Astronomical Union (IAU)**

Starptautiskās Astronomu Savienības 14. kongress.
A. Alksnis.
Zvaigžņotā debess, 1971. gada pavasaris, p. 1 - 14.

010.018 **Meteoritical Society**

Abstracts of papers, presented at the 34th annual meeting of the Meteoritical Society, August 20–28, 1971, Tübingen, Germany.
Meteoritics, Vol. 6, 246 - 327 (1971).

010.019 **Nederlandse Vereniging voor Weer- en Sterrenkunde**

Verenigingsnieuws.
Hemel en Dampkring, Vol. 69, 187 (1971).

Jongerenwerkgroep.
Hemel en Dampkring, Vol. 69, 196 - 197, 236 - 240, 297 - 304, 334 - 339 (1971).

010.020 **Polskie Towarzystwo Astronomiczne (PTA)**

No publication received.

010.021 **Polskie Towarzystwo Miłośników Astronomii**

PTMA Chronicle.
Urania Kraków, Vol. 42, 258 - 261, 311 - 314 (1971).

Some remarks about the activity of PTMA.
K. Rudnicki.
Urania Kraków, Vol. 42, 226 - 232 (1971).   In Polish.

Quingentenary anniversary of the Polskie Towarzystwo Miłośników Astronomii.   T. Grzesło, J. Rolewicz.
Separate print, Polskie Towarzystwo Miłośników Astronomii, Kraków. 23 pp. (1971). In Polish. – Historical note.

010.022 **Royal Astronomical Society (RAS)**

Meetings of the Society.
Observatory, Vol. 91, 133 - 138, 169 - 176, 176 - 180 (1971).

Meetings of the Society.
Quarterly Journ. Roy. Astron. Soc., Vol. 12, 87 - 97, 211 - 213 (1971).

Royal Astronomical Society meeting on storm effects in the ionosphere and magnetosphere, 1970 November 20. P. C. Kendall, H. Rishbeth.
Quarterly Journ., Roy. Astron. Soc., Vol. 12, 169 - 178 (1971).

Annual general meeting 1971 March 12.
B. Lovell.
Quarterly Journ. Roy. Astron. Soc., Vol. 12, 207 - 210 (1971)

Report of the council to the one hundred and fifty-first annual general meeting of the Society.
Quarterly Journ. Roy. Astron. Soc., Vol. 12, 214 - 220 (1971). – This report refers to the calendar year 1970.

Report of the honorary auditors for the year 1970.
C. Jordan, R. S. Peckover.
Quarterly Journ. Roy. Astron. Soc., Vol. 12, 221 (9171).

Treasurer's account for the year ending 1970 December 31.
Quarterly Journ. Roy. Astron. Soc., Vol. 12, 222 - 230 (1971).

Progress and present state of the Society.
Quarterly Journ. Roy. Astron. Soc., Vol. 12, 231 (1971).

010.023 Royal Astronomical Society of Canada (RAS Canada)

The meeting of the National Research Council associate Committee on Astronomy at Victoria, May 13–15, 1971. G. A. H. Walker.
Journ. Roy. Astron. Soc. Canada, Vol. 65, 169 - 183 (1971).

010.024 Royal Astronomical Society of New Zealand (RAS New Zealand)

Royal Astronomical Society of New Zealand: Variable Star Section.
Inform. Bull. Southern Hemisphere, No. 18, p. 21 (1971). Current research report.

Report of Variable Star Section. F. M. Bateson.
Southern Stars, Vol. 24, 12 - 17 = Roy. Astron. Soc. New Zealand, Repr. No. 23 (1971).

Royal Astronomical Society of New Zealand, Reprint No. 24 (F. M. Bateson, 05.122.130).

010.025 Schweizerische Astronomische Gesellschaft (SAG)

Aus der SAG und den Sektionen.
Orion Schaffhausen, 29. Jahrgang, p. 94, 125 - 129, 160 - 162, 199 - 200 (1971).

Bericht über die Tagung der SAG anlässlich der Generalversammlung vom 5./6. Juni 1971 in Burgdorf.
K. Büchler.
Orion, 29. Jahrgang, p. 158 - 159 (1971).

010.026 Sociedad Astronómica de México

Actividades de la Sociedad.

El Universo, Vol. 25, 57 - 61, 145 - 146 (1971).

Programa de actividades para el observatorio.
F. Diego Q.
El Universo, Vol. 25, 63 - 65 (1971).

010.027 Società Astronomica Italiana (SAI)

Verbale di scrutinio delle schede di votazione per la elezione del consiglio direttivo della Società Astronomica Italiana per il biennio Ottobre 1971 – Settembre 1973.
Mem. Soc. Astron. Italiana, Nuova Ser., Vol. 42, 657 - 658 (1971).

010.028 Société Astronomique de France (SAF)

Les séances de la Société. B. Clouet.
L'Astronomie, 85ᵉ année, p. 311 - 314, 359 - 360, 384 - 387 (1971).

Assemblée générale statutaire du lundi 14 juin 1971.
B. Clouet.
L'Astronomie, 85ᵉ année, p. 379 - 384 (1971).

Prix et médailles décernés par la Société.
L'Astronomie, 85ᵉ année, p. 388 - 390 (1971).

La vie de la Société Astronomique de France.
L. Tartois.
L'Astronomie, 85ᵉ année, p. 391 - 399 (1971).

010.029 Société Astronomique "R. Bosković"

No publication received.

010.030 Société Chronométrique de France

No publication received.

010.031 Société Belge d'Astronomie, de Méteorologie et de Physique du Globe

Séances mensuelles.
Ciel et Terre, Vol. 87, 563 - 572 (1971).

Assemblée générale statutaire du 20 mars 1971.
M. Ducuroir.
Ciel et Terre, Vol. 87, 573 - 579 (1971).

Rapport du Président du Comité de rédaction.
P. Paquet.
Ciel et Terre, Vol. 87, 580 (1971).

010.032 Svenska Astronomiska Sällskapet

Svenska Astronomiska Sällskapet; Astronomiska Sällskapet Tycho Brahe; Göteborgs Astronomiska Klubb.
Styrelsens berättelse för år 1970.
E. Holmberg, P. O. Lindblad, G. Darsenius, T. Elvius, B. Höglund, B.-A. Lindblad.
Separate print, Svenska Astronomiska Sällskapet, Stockholm. 8 pp. (1971).

010.033 **VAGO** (Astronomical-Geodetical Society of the USSR)

VAGB 5. kongress. A. Alksnis.
Zvaigžņotā debess, 1971. gada vasara, p. 47 - 52.

010.034 **Vereniging voor Sterrenkunde, België**

No publication received.

010.035 **Argentine Astronomical Association**

No publication received.

010.036 **Asociacion Argentina Amigos de la Astronomia**

Noticias de la Asociación.
Rev. Astron., Vol. 42, No. 174, p. 34 - 38 (1971).

010.037 **Founding meeting of the Canadian Astronomical Society, La Société Astronomique du Canada, at Victoria, May 15, 1971.** G. A. H. Walker.

Journ. Roy. Astron. Soc., Canada, Vol. 65, 170 - 171 (1971).

010.038 **Fünfzig Jahre Olbers-Gesellschaft Bremen.**
W. Stein.
Sterne, 47. Jahrgang, p. 204 - 209 (1971).

010.039 **Institut de France. Académie des Sciences. Annuaire pour 1972.**
Gauthier–Villars, Paris. 255 pp. (1971).
Cet annuaire a pour but d'exposer l'état de l'Académie des sciences dans le présent et dans le passé. Il contient en outre des indications sur les concours de ses prix et sur ses fondations.

010.040 **Rapport d' activité de l' A.D.I.O.N.**
Bull. d' Information, Ass. Développement International Obs. Nice, No. 7, p. 25 - 33, No. 8, p. 27 - 37 (1970/71). – Report 1969, 1970.

010.041 **Festrede am Jubiläumstag der Olbers-Gesellschaft.**
S. L. Jaki.
Separate print Olbers Ges., Bremen. 15 pp. (1971).

011 Reports on Colloquia, Congresses, Meetings, Symposia, and Expeditions

011.001 Supermassive objects in astrophysics.
V. Trimple.
Nature, Vol. 232, 607 - 611 (1971).
Report on a symposium held at the Institute of Theoretical Astronomy, Cambridge, Engl., on July 19 to 21, 1971 in honour of the sixtieth birthday of W. A. Fowler. Discussions on the following topics are presented: Black and white holes; Supermassive stars and disks; Observations; Nucleosynthesis and reaction rates.

011.002 The asteroid conference in Tucson.
M. S. Matthews.
Sky Telescope, Vol. 42, 22 - 24 (1971). — 1971 March 8 - 10.

011.003 Scientific session of the Division of General Physics and Astronomy of the USSR Academy of Sciences (September 30 - October 1, 1970).
Uspekhi fiz. nauk, Vol. 103, 769 - 776 (1971). In Russian.
Abstr. in Referativ. Zhurn. 51. Astron., 8.51.26 (1971).

011.004 Program of the USSR conference on the physics of pulsars (Moscow, FIAN, 17 - 19 December 1969).
Izv. vyssh. uchebn. zavedenij. Radiofizika, Vol. 13, 1900 - 1901 (1970). In Russian.

011.005 VIth Inter-American seminar on cosmic rays, La Paz, July 19 - 24, 1970. N. A. Dobrotin.
Vestn. AN SSSR, 1971, No. 1, p. 117 - 118. In Russian.
Abstr. in Referativ. Zhurn. 51. Astron., 8.51.35 (1971).

011.006 Report from a conference on eclipsing variables, (Kraków, December 11 - 12, 1970). J. M. Kreiner.
Postępy Astron., Vol. 19, 275 - 276 (1971). In Polish.

011.007 Kolloquium über relativistische Astrophysik.
Umschau, 71. Jahrgang, p. 718 (1971). — Observatory "Hoher List", Eifel, 1971 June 18 - 20.

011.008 The 4th European Space Conference.
A. V. Cleaver.
Spaceflight, Vol. 13, 108 - 110 (1971). — Brussels, 1970 Nov. 4 - 5.

011.009 The 4th Eurospace US-European conference.
C. R. Turner.
Spaceflight, Vol. 13, 193 - 196 (1971). — Venice, 1970 Sept. 21 - 25.

011.010 International moon conference. P. Goodwin.
Spaceflight, Vol. 13, 282 - 283 (1971). — Newcastle-upon-Tyne, 1971 March 22 - 26.

011.011 Physics of cosmic rays. Conference in Moscow, 26 Oct. - 2 Nov. 1970. G. B. Zhdanov.
Vestn. AN SSSR, 1971, No. 3, p. 109 - 111. In Russian.
Abstr. in Referativ. Zhurn. 51. Astron., 9.51.7 (1971).

011.012 Investigation of solar cosmic rays. Symposium in Leningrad, 8 - 12 December 1970.
L. I. Miroshnichenko.
Vestn. AN SSSR, 1971, No. 3, p. 111 - 113. In Russian.
Abstr. in Referativ. Zhurn. 51. Astron., 9.51.9 (1971).

011.013 16th annual conference of the South African Institute of Physics. P. A. T. Wild.
Monthly Notes Astron. Soc. Southern Africa, Vol. 30, 98 - 99 (1971). — Cape Town, 1971 July 14 - 16.

011.014 Residential week-end course on observational astronomy at King Alfred's training college, Winchester, 1971 April 23 - 25. P. Richards—Jones.
Journ. British Astron. Ass., Vol. 81, 493 - 494 (1971).

011.015 Stellafane holds Porter centennial.
D. Milon.
Sky Telescope, Vol. 42, 208 - 211 (1971).

011.016 The fourth Arizona conference on planetary atmospheres: Motions of planetary atmospheres.
P. J. Gierasch.
Earth Extraterr. Sci., Vol. 1, 171 - 184 (1970).

011.017 Solar proton events.
Journ. British Interplanet. Soc., Vol. 24, 685 - 686 (1971). — Boston, 1971 June 16 - 18.

011.018 The tenth planet.
Journ. British Interplanet. Soc., Vol. 24, 686 - 687 (1971). — Seattle, 1971 June 19.

011.019 The fifteenth Herstmonceux Conference.
Observatory, Vol. 91, 180 - 203 (1971). — 1971 April 5 and 6. — The 1971 Herstmonceux Conference was devoted chiefly to discussion of the abundances of the chemical elements.

011.020 Amatörastronomkongressen i Stockholm.
M. Lundblad.
Astron. Tidsskr., Årg. 4, 92 (1971).

011.021 Second USSR conference of young amateur astronomers. B. G. Pshenichner.
Zemlya i Vselennaya, No. 5, p. 74 - 77 (1971). In Russian.

011.022 Preliminary program of the Kiev conference on comets (1971, October 25 - 29, Plenary session of the Commission for Comets and Meteors of the Astronomical Council).
Kometn. Tsirk., *Kiev,* No. 120 (1971). In Russian.

011.023 Program for the 6th conference on comets (Kiev, 1971, Nov. 1 - 5).
Kometn. Tsirk., *Kiev,* No. 123 (1971). In Russian.

011.024 Soviet-American conference urges search for other worlds.
Science, Vol. 174, 130 - 131 (1971).

011.025 Astronomy from a space platform.
G. W. Morgenthaler.
Science, Vol. 174, 324 - 325 (1971). — Philadelphia, 1971 December 27 - 28.

011.026 400th anniversary of Johannes Kepler's birth.
R. J. Seeger.
Science, Vol. 174, 325 (1971). — Philadelphia, 1971 December 27.

011.027 Scientific session of the Department of General Physics and Astronomy of the USSR Academy of Sciences with the Department of Nuclear Physics, 23 - 24

Dec. 1970.
Uspekhi fiz. nauk, Vol. 104, 323 (1971). In Russian. − Abstr. in Referativ. Zhurn. 51. Astron., 11.51.17 (1971).

011.028 **Adunarea generală a Comitetului Naţional Român de Astronomie (CNRA).** C. Cristescu.
Stud. Cerc. Astron., Vol. 16, 223 - 224 (1971).

011.029 **Colocviul „Studiul fizic al micilor planete"**
C. Cristescu.
Stud. Cerc. Astron., Vol. 16, 224 - 226 (1971).

011.030 **Scottish Astronomical Societies' conference.**
D. B. Taylor.
Journ. British Astron. Ass., Vol. 82, 45 (1971). − Dundee, 1971 Sept. 25.

011.031 **Internationale Konferenz für Erziehung in Astronomie und Geschichte der modernen Astronomie,**
New York, 30. August bis 1. September 1971. H. Rohr.
Orion Schaffhausen, 29. Jahrgang, p. 196 - 197 (1971).

011.032 **Plasma in the laboratory and in the universe.**
J. Mergentalor.
Urania Kraków, Vol. 42, 308 - 310 (1971). In Polish.

011.033 **Annual meeting of the USSR Academy of Sciences, 3 - 4 March, 1971.**
Vestn. AN SSSR, 1971, No. 5, p. 5 - 64. In Russian. − Abstr. in Referativ. Zhurn. 51. Astron., 12.51.19 (1971).

011.034 **Scientific session of the Division of General Physics and Astronomy of the USSR Academy of Sciences (January 21, 1971).**
Uspekhi fiz. nauk, Vol. 104, 669 - 672 (1971). In Russian. Abstr. in Referativ. Zhurn. 51. Astron., 12.51.20 (1971).

011.035 **1971 Lunar Science Conference.** N. W. Hinners.
Icarus, Vol. 15, 135 - 139 (1971). − Houston, Texas, 1971 January 11 - 14.

011.036 **The 1971 meeting of the AAS Division for Planetary Sciences.** D. Morrison.
Icarus, Vol. 15, 343 - 348 (1971).

011.037 **I.A.U. colloquium (No. 21) on variable stars in globular clusters and in related systems.**
M. W. Feast.
Inform. Bull. Variable Stars, (I.A.U. Commission 27), Konkoly Obs., *Budapest,* No. 595 (1971). − Toronto, Canada, 1972 Aug. 29 - 31.

011.038 **International conference on the problem of "Communication with Extraterrestrial Intelligence (CETI)".**
Astron. Tsirk., No. 653, p. 1 - 5 (1971). In Russian.

011.039 **NATO Advanced Study Institute on lunar studies, Patras, Greece, September 14–25, 1971.**
The Moon, Vol. 3, 364 - 365 (1971).

011.040 **The Lunar Science Institute conference on lunar geophysics, Houston, Texas, October 18–21, 1971.**
The Moon, Vol. 3, 365 - 367 (1971).

011.041 **XVème assemblée générale de l'Association Internationale de Géodésie Moscou (URSS),** with a presidental address by A. Marussi; and a report by the general secretary J. J. Levallois.
Bull., Géod., Nouvelle Sér., Année 1971, No. 102, p. 349 - 439. 1971 August 1 - 14.

011.042 **Colloque de l' Union Astronomique Internationale Nice, septembre 1969: Sur la cohésion entre les différents procédés de mesure des étoiles doubles.**
P. Couteau.
Bull. d' Information, Ass. Développement International Obs. Nice, No. 7, p. 11 - 14 (1970).

011.043 **Scientific session of the general physics and astronomy division and the nuclear physics division of the USSR Academy of Sciences (February 17–18, 1971).**
Uspekhi fiz. nauk, Vol. 104, 672 - 680 (1971). In Russian. − Abstr. in Referativ. Zhurn. 51. Astron., 1.51.15 (1972).

011.044 **Scientific session of the general physics and astronomy division of the USSR Academy of Sciences (March 5, 1971).**
Uspekhi fiz. nauk, Vol. 104, 681 - 687 (1971). In Russian. − Abstr. in Referativ. Zhurn. 51. Astron., 1.51.16 (1972).

011.045 **Asteroids and planetesimals.** W. K. Hartmann.
Icarus, Vol. 15, 349 - 351 (1971). − IAU Colloquium No. 12, Tucson, Arizona, 1971 March 8–20.

011.046 **The relationship of interstellar molecules to the origin of life.** M. W. Werner.
Icarus, Vol. 15, 352 - 355 (1971). − A review of the conference held at the NASA Ames Research Center, February 25–26, 1971.

011.047 **Symposium über die Beobachtung veränderlicher Sterne durch Amateure in Darmstadt.**
H. Zipprich.
BAV Rundbrief, 20. Jahrgang, p. 22 - 23 (1971).

011.048 **Il quinto colloquio sulle stelle variabili.**
F. Ciatti.
Mem. Soc. Astron. Italiana, Nuova Ser., Vol. 42, 641 - 646 (1971). − Bamberg, Germany, 1971 August 31 – September 3.

011.049 **Le riunioni del CESRA ad Utrecht e Trieste.**
F. Chiuderi Drago, G. Tofani.
Mem. Soc. Astron. Italiana, Nuova Ser., Vol. 42, 647 - 648 (1971).

011.050 **Czechoslovak conference on the stellar astronomy.**
J. Grygar.
Pokroky, Vol. 16, 273 (1971). In Czech.

011.051 **Third Czechoslovak conference on the teaching of astronomy.** J. Široký.
Pokroky, Vol. 16, 272 (1971). In Czech.

011.052 **Symposium on the rotation of the earth, Morioka (Japan), 1971, 9–15 May.** E. P. Fedorov.
Vestn. AN SSSR, 1971, No. 10, p. 79 - 81. In Russian. Abstr. in Referativ. Zhurn. 51. Astron., 2.51.17; 52. Geod. i Aehros"emka, 2.52.107 (1971).

011.053 **Međunarodni simposijum posvećen 400-godišnjici rođenja Johanesa Keplera.** (International symposium on the occasion of the 400th anniversary of the birthday of Johannes Kepler). D. Trifunović.
Vasiona, Vol. 19, 59 - 62 (1971).

011.054 **Kometu pētnieku apspriede.** M. Dīriķis.
Zvaigžņotā debess, 1970./71. gada ziema, p. 61 - 66.

011.055 **Dienas kārtība − cefeīdas.** G. Carevskis.
Zvaigžņotā debess, 1970./71. gada ziema, p. 66 - 68.

011.056 **Pre-main-sequence stellar evolution. A colloquium held at Liège, June 29 − 30, July 1, 1969.**
A. Boury.
Earth Extraterr. Sci., Vol. 1, 155 - 170 (1970).

012 Proceedings of Colloquia, Congresses, Meetings, and Symposia

012.001 **The Menzel Symposium on Solar Physics, Atomic Spectra, and Gaseous Nebulae,** in honor of the contributions made by Donald H. Menzel. Proceedings of a symposium held at the Harvard College Observatory, Cambridge, Massachusetts April 8 - 9, 1971.
K. B. Gebbie (Editor).
National Bureau of Standards Special Publication 353, 213 pp.
[For sale by the Superintendent of Documents, U.S. Government Printing Office, Washington, D.C.] Price $ 1.75 (1971).
The individual papers are included in their corresponding subject categories − see Abstracts 022.043 - 022.046, 032.005, 032.006, 061.014, 071.021, 073.034, 074.028, 132.023, 133.012 - 133.014.

012.002 **Study week on 'Nuclei of galaxies',** Città del Vaticano, 1970 April 13 − 18, with a preface by
D. J. K. O'Connell.
Pontificiae Academiae Scientiarum, Città del Vaticano, Scripta Varia, Vol. 35. North-Holland Publishing Company, Amsterdam − London; American Elsevier Publishing Company, Inc., New York. 49 + 800 pp. Price hfl. 134.00 (1971). − The individual papers are included in their corresponding subject categories − see Abstracts 066.021, 141.077 - 141.086, 142.024, 151.017, 155.017, 158.035 - 158.050, 160.008.

012.003 **Solar magnetic fields.**
International Astronomical Union, Symposium No. 43, held at the Collège de France, Paris, France, August 31 to September 4, 1970. R. Howard (Editor).
D. Reidel Publishing Company, Dordrecht − Holland. 16 + 782 pp. Price Dfl. 140.00 (1971). − The individual papers are included in their corresponding subject categories − see Abstracts 033.007, 034.024 - 034.035, 062.013, 062.014, 071.028 - 071.041, 072.021 - 072.041, 073.039 - 073.050, 074.031 - 074.041, 077.023, 077.024, 080.011 - 080.027.

012.004 **Space Research XI.** Proceedings of open meetings of Working Groups of the Thirteenth Plenary Meeting of COSPAR, Leningrad, USSR, 20 - 29 May 1970, and of the Symposium on remote sounding of the atmosphere (jointly sponsored by COSPAR, WMO and IAMAP/IUGG), Leningrad, USSR, 22, 25 and 26 May 1970. Vol. 1, 2.
K. Ya. Kondratyev, M. J. Rycroft, C. Sagan (Editors).
Akademie-Verlag, Berlin. 34 + 1414 pp. Price DM 200.00 (1971). − The individual papers within the subject scope of Astronomy and Astrophysics Abstracts are included in their corresponding categories − see 013.009, 032.011, 074.049,

076.012 - 076.014, 078.016 - 078.020, 080.039, 080.040, 081.026 - 081.030, 082.050 - 082.058, 083.046 - 083.048, 084.024, 084.268, 084.269, 085.008, 091.018, 093.016 - 093.018, 094.178 - 094.191, 097.048 - 097.051, 098.017, 099.043, 100.008, 104.042 - 104.048, 105.052 - 105.060, 106.018 - 106.021, 141.137, 142.049 - 142.051, 143.027 - 143.031, 155.030.

012.005 **New techniques in space astronomy.**
International Astronomical Union, Symposium No. 41, held in Munich, Germany, August 10 - 14, 1970.
F. Labuhn, R. Lüst (Editors).
D. Reidel Publishing Company, Dordrecht−Holland. 15 + 419 pp. Price hfl. 80.00 (1971). − The individual papers are included in their corresponding subject categories − see Abstracts 031.021 - 031.028, 032.012, 032.013, 034.044 - 034.077, 051.020 - 051.029, 061.020 - 061.022, 076.015 - 076.017, 077.028, 142.052 - 142.054.

012.006 **Proceedings of IAU Colloquium No. 9, 'The IAU System of Astronomical Constants', Heidelberg, 12−14 August 1970.**
Introduction by B. Emerson, G. A. Wilkins.
Celestial Mechanics, Vol. 4, (No. 2), 128 - 280 (1971). − The papers presented are included in their corresponding subject categories − see Abstracts 043.003 - 043.015, 081.032, 099.050.

012.007 **Giornate di studio sulla cronologia dell'universo,** Padova, 6 - 7 novembre 1970.
Introduction by N. Dallaporta.
Mem. Soc. Astron. Italiana, Nuova Ser., Vol. 42, (N. 3), 266 - 418 (1971). − The individual contributions are included in their corresponding subject categories − see Abstracts 065.090 - 065.092, 080.044, 107.011, 141.182, 141.183, 158.096, 162.040, 162.041.

012.008 **Theory of the stellar atmospheres. (Théorie des atmosphères stellaires).**
D. Mihalas, B. Pagel, P. Souffrin.
Ier Cours avance de la Société Suisse d'Astronomie et d'Astrophysique, Saas-Fee. Edited by Observatoire de Genève, Sauverny, Suisse. 10 + 312 pp. Price Swiss f. 24.00 (1971).
The individual contributions are included in their corresponding subject categories − see Abstracts 064.046 - 064.048.

012.009 **A discussion on solar studies with special reference** to space observations. Arranged by the British National Committee on Space Research under the leadership of H. Massey, C. W. Allen, A. H. Gabriel, B. E. J. Pagel, R. Wilson. Phil. Trans. Roy. Soc. London, Ser. A, Math., Phys. Sci., Vol. 270, No. 1202, 195 pp. (1971). – The individual papers are included in their corresponding subject categories – see Abstracts 064.049, 064.050, 071.059 - 071.063, 073.093 - 073.097, 076.030 - 076.034, 078.031, 078.032, 114.120.

012.010 **The Magellanic Clouds.** A European Southern Observatory presentation: Principal prospects, current observational and theoretical approaches, and prospects for future research, based on the symposium on the Magellanic Clouds held in Santiago de Chile, March 1969, on the occasion of the dedication of the European Southern Observatory. A. B. Muller (Editor).
Astrophysics and Space Science Library, Vol. 23.
D. Reidel Publishing Company, Dordrecht–Holland. 12 + 189 pp. Price Dfl. 44.00 (1971). – The individual papers are included in their corresponding subject categories – see Abstracts 065.105, 158.108, 159.013 - 159.035.

012.011 **Papers presented at the fifth annual general meeting of the Astronomical Society of Australia, held at the University of Sydney on 19, 20 and 21 May 1971.**
Proc. Astron. Soc. Australia, Vol. 2, No. 1, 68 pp. (1971). The individual papers are included in their corresponding subject categories – see Abstracts 015.014, 022.063 - 022.064, 031.020, 032.010, 034.043, 063.027 - 063.028, 065.066 - 065.069, 071.044, 072.043, 074.046, 077.025 - 077.027, 080.030 - 080.037, 114.062 - 114.063, 117.018, 122.069, 125.017, 126.007, 131.078, 141.135.

012.012 **The Cowling symposium on magnetic problems in astronomy.**
G. A. J. Ferris, F. A. Goldsworthy (Editors).
Quarterly Journ. Roy. Astron. Soc., Vol. 12, (No. 4), 347 - 452, with an introduction by F. A. Goldsworthy, p. 347 (1971). – The individual papers are included in their corresponding subject categories – see 061.030, 061.031, 062.032, 062.033, 074.078, 080.057, 102.012, 116.010, 131.113.

012.013 **Special meeting on direct exploration of the moon.**
Highlights of Astronomy, Vol. 2,(see 003.026), 123 - 188 (1971). – The invited papers presented at this session are included in their corresponding subject categories – see Abstracts 053.026, 094.242 - 094.244.

012.014 **The origin of the earth and planets.**
Joint discussion during the XIVth general assembly of the IAU, Brighton 1970.
B. M. Middlehurst (Editor).
Highlights of Astronomy, Vol. 2, (see 003.026), 191 - 243 (1971). – The individual contributions are included in their corresponding subject categories – see Abstracts 091.027, 091.028, 107.014.

012.015 **Helium in the universe.** Joint discussion during the XIVth general assembly of the IAU, Brighton 1970.
J. S. Mathis (Editor), with opening remarks by D. E. Osterbrock.
Highlights of Astronomy, Vol. 2, (see 003.026), 245 - 331 (1971. – The individual contributions are included in their corresponding subject categories – see Abstracts 061.032, 061.033, 064.052, 065.110 - 065.112, 132.039, 162.054.

012.016 **Interstellar molecules.** Joint discussion during the XIVth general assembly of the IAU, Brighton 1970.
D. McNally (Editor), with introductory remarks by G. H. Herbig.
Highlights of Astronomy, Vol. 2, (see 003.026), 333 - 462,

with a Panel discussion on "Excitation mechanisms", p. 445 - 459 (1971). – The individual contributions are included in their corresponding subject categories – see Abstracts 064.053, 131.114 - 131.128.

012.017 **Atomic data of importance for ultraviolet and X-ray astronomy.** Joint discussion during the XIVth general assembly of the IAU, Brighton 1970.
C. Jordan (Editor), with an introduction by R. Wilson.
Highlights of Astronomy, Vol. 2, (see 003.026), 463 - 583 (1971). – The individual contributions are included in their corresponding subject categories – see Abstracts 021.006, 022.097 - 022.104, 061.034, 062.034, 071.064, 074.079, 076.035 - 076.038, 114.121.

012.018 **Photoelectric observations of stellar occultations.**
Joint discussion during the XIVth general assembly of the IAU, Brighton 1970.   T. J. Deeming (Editor).
Highlights of Astronomy, Vol. 2, (see 003.026), 585 - 722 (1971). – The individual contributions are included in their corresponding subject categories – see Abstracts 096.022 - 096.039, 115.011, 115.012.

012.019 **Pulsars, cosmic rays, and background radiation.**
Joint discussion during the XIVth general assembly of the IAU, Brighton 1970.   M. J. Rees (Editor).
Highlights of Astronomy, Vol. 2, (see 003.026), 723 - 767 (1971). – The individual contributions are included in their corresponding subject categories – see Abstracts 065.102, 066.069, 141.220 - 141.222, 143.057.

012.020 **The absolute magnitudes of the RR Lyrae stars.**
Joint meeting of Commissions 24, 27, 30, 33 and 37 during the XIVth general assembly of the IAU, Brighton 1970.   W. S. Fitch (Editor).
Highlights of Astronomy, Vol. 2, (see 003.026), 769 - 793 (1971). –The individual contributions are included in their corresponding subject categories – see Abstracts 122.123 - 122.126, 122.151.

012.021 **The radiating atmosphere.** Proceedings of a symposium organized by the Summer Advanced Study Institute, held at Queen's University, Kingston, Ontario, August 3–14, 1970.
B. M. McCormac (Editor),with an Institute review by A. V. Jones and Conclusions by B. M. McCormac, J. N. Bradbury, J. E. Evans.
Astrophysics and Space Science Library, Vol. 24.
D. Reidel Publishing Company, Dordrecht–Holland.9+455 pp. Price Dfl. 100.00 (1971). – The individual papers are included in their corresponding subject categories – see Abstracts 082.098 - 082.107, 083.065, 084.033 - 084.057, 084.295.

012.022 **General relativity and cosmology.** Proceedings of the International School of Physics «Enrico Fermi».
Course 47, Italian Physical Society, Varenna on Lake Como, Italy, 30th June - 12th July 1969. R. K. Sachs (Editor).
Academic Press, New York – London. 12 + 387 pp. Price $ 21.00 (1971). – The individual contributions are included in their corresponding subject categories – See Abstracts 065.123, 066.077 - 066.081, 141.227, 151.068, 158.118, 162.056 - 162.061.

012.023 **Physics of high energy density.** Proceedings of the International School of Physics «Enrico Fermi».
Course 48, Italian Physical Society, Varenna on Lake Como, Italy, 14th–26th July 1969.
P. Caldirola, H. Knoepfel (Editors), with an introduction by P. Caldirola.
Academic Press, New York – London. 14 + 418 pp. Price $ 23.00 (1971). – The individual contributions within the

subject scope of Astronomy and Astrophysics Abstracts are included in their corresponding categories — see Abstracts 022.114, 062.037, 062.038, 125.030.

012.024 **Mesospheric models and related experiments.** Proceedings of the fourth ESRIN-ESLAB symposium, held in Frascati, Italy, 6—10 July, 1970.
G. Fiocco (Editor).
Astrophys. Space Sci. Library, Vol. 25.
D. Reidel Publishing Company, Dordrecht—Holland. 8 + 298 pp. Price Dfl. 75.00 (1971). — The individual contributions within the subject scope of Astronomy and Astrophysics Abstracts are included in their corresponding categories — see Abstracts 082.113 - 082.130, 083.068 - 083.072.

012.025 **Proceedings of the 11th international conference on cosmic rays.** Invited papers and rapporteur talks. Budapest, 25 August—4 September 1969.
G. Bozóki, É. Gombosi, Á. Sebestyén, A. Somogyi.
Published by the Central Research Institute for Physics, Budapest, Hungary. 4 + 612 pp. Price DM 44.00 (1970). — The individual papers within the subject scope of Astronomy and Astrophysics Abstracts are included in their corresponding categories — see Abstracts 061.037, 065.124, 078.036 - 078.038, 084.411, 106.031, 141.228, 143.063 - 143.068.

012.026 **UMC symposium on geophysical theory and computers held at Stockholm in August 1970.** Upper Mantle Project Scientific Report No. 35.
Geophys. Journ. Roy. Astron. Soc., Vol. 25, Nos. 1—3, 305 pp. (1971). — The individual contributions within the subject scope of Astronomy and Astrophysics Abstracts are included in their corresponding categories — see Abstracts 045.017, 062.040, 062.041.

012.027 **Proceedings of the 11th international conference on cosmic rays. Volume 3. High energy interactions extensive air showers.** Budapest, August 25 — September 4, 1969. A. Somogyi (Editor).
Publishing House of the Hungarian Academy of Sciences (Akadémiai Kiadó), Budapest, Hungary. Acta Phys. Acad. Sci. Hungaricae, Vol. 29, Suppl. 3, 767 pp. Price $ 42.00 (1970). — The individual papers within the subject scope of Astronomy and Astrophysics Abstracts are included in the corresponding categories — see Abstracts 143.069 - 143.077.

012.028 **Proceedings of the 11th international conference on cosmic rays. Volume 4. Muons and neutrinos, techniques.** Budapest, August 25 - September 4, 1969.
A. Somogyi (Editor).
Publishing House of the Hungarian Academy of Sciences (Akadémiai Kiadó), Budapest, Hungary. Acta Phys. Acad. Sci. Hungaricae, Vol. 29, Suppl. 4, 607 pp. Price $ 28.00 (1970). — The individual papers within the subject scope of Astronomy and Astrophysics Abstracts are included in their corresponding categories — see Abstracts 065.125, 080.058, 080.059, 143.078 - 143.084.

012.029 **5th consultation on solar physics and hydromagnetics, Potsdam, 30. 9. — 4. 10. 1968**, with a preface by F. W. Jäger.
Geod. Geophys. Veröff., Nationalkomitee für Geodäsie und Geophysik, Deutsche Demokratische Republik, Deutsche Akad. Wiss. Berlin, Ser. 2, No. 13, 149 pp. (1969). — The individual papers are included in their corresponding subject categories — see Abstracts 062.050, 062.051, 071.068, 071.069, 072.087 - 072.092, 073.102 - 073.104, 074.089 - 074.091, 077.054, 080.060 - 080.062.

012.030 **Proceedings of the 11th international conference on cosmic rays. Vol. 2. Solar cosmic rays, modula-**tion of galactic radiation, magnetospheric and atmospheric effects. Budapest, August 25 - September 4, 1969.
A Somogyi (Editor).
Publishing House of the Hungarian Academy of Sciences (Akadémiai Kiadó), Budapest, Hungary. Acta Phys. Acad. Sci. Hungaricae, Vol. 29, Suppl. 2, 767 pp. Price $ 42.00 (1970). The individual papers within the subject scope of Astronomy and Astrophysics Abstracts are included in their corresponding subject categories — see Abstracts 05.073.082, 05.074.089, 05.074.090, 05.078.036 - 05.078.038, 05.078.040 - 05.078. 057, 05.106.037, 05.106.038, 05.143.072 - 05.143.080, 05.143.144 - 05.143.151; 06.076.039, 06.078.046 - 06.078. 059, 06.084.307, 06.084.308, 06.143.092 - 06.143.127.

012.031 **Conference on photographic astrometric technique.** H. Eichhorn (Editor).
Prepared by Dep. Astron. Univ. South Florida, Tampa, Florida, for National Aeronautics and Space Administration, Washington, D.C. NASA Contractor Report, NASA CR-1825, 8 + 267 pp. [For sale by the National Technical Information Service, Springfield, Virginia. Price $ 3.00] (1971). — Tampa, Florida, 1968 February 18—20. — The individual contributions are included in their corresponding subject categories — see Abstracts 031.068 - 031.081, 034.124 - 034.126, 041.046 - 041.054, 054.029, 054.030.

012.032 **Beam-foil spectroscopy.** Proceedings of the second international conference on beam-foil spectroscopy, Lysekil, Sweden, 7 — 12 June 1970.
I. Martinson, J. Bromander, H. G. Berry (Editors).
North-Holland Publishing Company, Amsterdam. 7 + 371 pp. Price $ 25.00 (1970). — Reprinted from: Nuclear Instruments & Methods — Vol. 90 (1970). — Review in Journ. Optical Soc. America, Vol. 61, 1719; 1971 (*J. A. R. Samson*). — The individual contributions within the subject scope of Astronomy and Astrophysics Abstracts are included in their corresponding categories — see Abstracts 022.136 - 022.154, 031.082, 034.127, 034.128, 062.053, 062.054, 071.070.

012.033 **Proceedings of the Sixth Winter School on Space Physics, Part I.** (Trudy Shestoi Vsesoyuznoi ezhegodnoi zimnei shkoly po kosmofizike). (Apatity, 18 March — 1 April 1969).
S. N. Vernov, G. E. Kocharov (Editors).
Academy of Sciences of the USSR, Scientific Council on Cosmic Rays. Kola Branch im. S.M. Kirov, Polar Geophysical Institute. Translated from Russian.
Israel Program for Scientific Translations, Jerusalem. 6 + 332 pp. Price: Part I, II, $ 28.00 (1971). — The individual contributions within the subject scope of Astronomy and Astrophysics Abstracts are included in their corresponding categories — see Abstracts 061.057 - 061.059, 062.074 - 062.076, 064.064, 065.155 - 065.157, 066.147, 074.102, 074.103, 078.061, 080.063, 080.064, 084.063, 084.318 - 084.320, 084.415 - 084.419, 085.013, 104.116, 106.038 - 106.041, 116.013, 142.097 - 142.099, 143.133 - 143.136, 158.140, 160.019, 162.097 - 162.100.

012.034 **Proceedings of the Sixth Winter School on Space Physics, Part II.** (Trudy Shestoi Vsesoyuznoi ezhegodnoi zimnei shkoly po kosmofizike). (Apatity, 18 March — 1 April 1969).
S. N. Vernov, G. E. Kocharov (Editors).
Academy of Sciences of the USSR, Scientific Council on Cosmic Rays. Kola Branch im. S.M. Kirov, Polar Geophysical Institute. Translated from Russian.
Israel Program for Scientific Translations, Jerusalem. 5 + 192 pp. Price: Part I, II, $ 28.00 (1971). — The individual contributions within the subject scope of Astronomy and Astrophysics Abstracts are included in their corresponding categories — see Abstracts 072.098, 078.062, 084.321, 094.327,

102.027 - 102.030, 103.119, 103.135, 105.170, 143.137 - 143.149.

012.035   International Union of Geodesy and Geophysics.
Report on the symposium on coastal geodesy, held in Munich 20th – 24th July 1970.   R. Sigl (Editor).
Institute for Astronomical and Physical Geodesy, Technical Univ., München. [Printed by Inst. Angewandte Geodäsie, Frankfurt, a.M.], 644 pp. (1970).

012.036   Report of the meeting of the Committee of European Solar Radio Astronomers (CESRA) 18 and 19 February 1971.
Edited by Committee of European Solar Radio Astronomers (CESRA), 77 pp. (1971). – The individual contributions are included in their corresponding subject categories – see Abstracts 077.060 - 077.078.

012.037   Institution of Electrical Engineers conference on earth station technology, 14 - 16 October 1970,
organised by the Electronics Division of the Institution of Electrical Engineers in association with the Institute of Electrical and Electronics Engineers, the Institution of Electronic and Radio Engineers.
Institution of Electrical Engineers, London. IEE Conference Publ. No. 72, 446 pp. (1971). – This report containing 70 papers which cover the whole field of space communication, equipment, aerials, amplifying and drive systems. – *BMT*

012.038   Space radiocommunications and techniques.
(Articles prepared on the occasion of the opening of the Second World Administrative Radio Conference for Space Telecommunications).
Telecommun. Journ., (*Switzerland*), Vol. 38, 227 - 429 (1971).
This special issue summarizes the papers prepared for the Second World Administrative Radio Conference for Space Telecommunications. It includes a review paper on the "Progress of radio astronomy" and on "Space radio astronomy". *BMT*

013   Reports on Astronomy in Various Countries and Particular Fields, International Cooperation

013.001   Astronomy in New Zealand.   G. A. Eiby.
Sky Telescope, Vol. 42, 18 - 20 (1971).

013.002   Which problems of physics and astrophysics seem to be now most important and interesting?
V. Ginzburg.
Nauka i zhizn', 1971, No. 2, p. 9 - 17.   In Russian.

013.003   Cooperation of the astronomical institutes of the Polish Academy of Sciences with foreign institutes.
J. Juchniewicz.
Postępy Astron., Vol. 19, 257 (1971).   In Polish.

013.004   Astronomy in the mid 20th century.
D. Ya. Martynov.
Zemlya i Vselennaya, No. 4, p. 20 - 27 (1971).   In Russian.

013.005   Methodology and progress of modern natural sciences.   V. A. Ambartsumyan, V. V. Kazyutinskij.
Vestn. AN SSSR, 1971, No. 3, p. 28 - 39.   In Russian.
Abstr. in Referativ. Zhurn. 51. Astron., 9.51.1 (1971).

013.006   The early history of astronomical activity in the Canadian public service.   R. M. Stewart.
Journ. Roy. Astron. Soc. Canada, Vol. 65, 206 - 216 (1971).
This article is adapted with only slight changes from the introductory sections of a lengthy report entitled "Activities of the Astronomical Branch, Department of the Interior" prepared in 1930.

013.007   Progrès récents de l'astronomie.
J. Kovalevsky.
L'Astronomie, 85e année, p. 365 - 376 (1971).

013.008   Über die wissenschaftliche Arbeit tschechoslowakischer Sternwarten.   O. Oburka.
Orion, 29. Jahrgang, p. 145 - 147 (1971).

013.009   Ultraviolet spectrophotometric measurements with the Orbiting Astronomical Observatory.
A. D. Code.
Space Research XI, (see 012.004), 1339 - 1344 (1971).

013.010   Stellar-statistical problems. I. On relations between astronomical disciplines and their system-theoretical description.   H. Eelsalu.
Tartu Astron. Obs. Teated, No. 31, p. 3 - 14 (1971).

013.011   Stellar-statistical problems. II. General considerations about the descriptive background from a stellar-statistical viewpoint.   H. Eelsalu.
Tartu Astron. Obs. Teated, No. 31, p. 15 - 20 (1971).

013.012   Nonprofessional astronomy in Hawaii.
G. W. Bunton.
Publ. Astron. Soc. Pacific, Vol. 83, 603 (1971). – Abstr. Astron. Soc. Pacific.

013.013   Astronomy of the future.   E. H. Thompson.
Observatory, Vol. 91, 224 - 225 (1971). – Letter.

013.014   Report on the co-operation of the Academies of Sciences of socialist countries in the field of scientific research through artificial satellite observations from November 1968 to December 1969.
A. G. Massevitch, S. K. Tatevjan.
Byull. Stantsij Optichesk. Nablyud. Iskusstv. Sputnikov Zemli, No. 57, p. 7 - 13 (1971). In Russian.

013.015 Some recent results and potential advances in radio astronomy. H. K. Bourne.
U. K. Sci. Mission (North America) U.K.S.M. Rep. No. 71/16, 6 pp. (1971). – *JCR*

013.016 Progress in radio astronomy. J. P. Hagen.
Telecommun. Journ., (*Switherland*), Vol. 38, 354 -

360 (1971). – *BMT*

013.017 Kosmische Physik mit neuen Akzenten. G. Skuridin.
Ideen Exakten Wissens, [Deutsche Verlags-Anstalt, Stuttgart], 9. 71, p. 575 - 581 (1971).

014 Teaching in Astronomy

014.001 Astronomical arrangement for a school. T. T. Gavrilyuk.
Zemlya i Vselennaya, No. 4, p. 64 - 66 (1971). In Russian.

014.002 School astronomical societies. I. R. Gordon.
Southern Stars, Vol. 24, 42 - 44 (1971).

014.003 Where is astronomical education headed? S. A. Africk.
Sky Telescope, Vol. 42, 277 - 278 (1971).

014.004 Pulsare und ihre Darstellung im Unterricht. K.-H. Schmidt, H. Albert.
Astron. in der Schule, 8. Jahrgang, p. 75 - 81 (1971).

014.005 Fachlich-methodische Bemerkungen zur Unterrichtseinheit 1. 1. "Einführung in das Unterrichtsfach Astronomie". H. Eckert.
Astron. in der Schule, 8. Jahrgang, p. 81 - 84 (1971).

014.006 Schülerbeobachtungen zum Stoffgebiet Astrophysik. G. Leistner.
Astron. in der Schule, 8. Jahrgang, p. 87 - 90 (1971).

014.007 Forschungsvorhaben zur Entwicklung der Methodik des Astronomieunterrichts. J. Stier.
Astron. in der Schule, 8. Jahrgang, p. 103 - 104 (1971).

014.008 Obligatorische Schülerbeobachtungen im Astronomieunterricht. H. Albert.
Astron. in der Schule, 8. Jahrgang, p. 105 - 110 (1971).

014.009 Hinweise zur Unterrichtseinheit 1.4.4. Erörterung eines aktuellen Beispiels der Erforschung des Mondes oder eines Planeten. H. J. Nitschmann.
Astron. in der Schule, 8. Jahrgang, p. 111 - 113 (1971).

014.010 Komplexe Planung der Stoffeinheit 2: Astrophysik und Stellarastronomie. K. Lindner.
Astron. in der Schule, 8. Jahrgang, p. 130 - 133 (1971).

014.011 Verwendungsmöglichkeiten des Himmelglobus im Astronomieunterricht. J. Schön, H. Albert.
Astron. in der Schule, 8. Jahrgang, p. 135 - 139 (1971).

014.012 Student's grade exercise works in astronomy. E. A. Koryakina.
(Uch. zap.) Perm. gos. ped. in-t, Vol. 72, (see 003.019), 428 - 431 (1969). In Russian. – Abstr. in Referativ. Zhurn. 51. Astron. 12.51.32 (1971).

014.013 On teaching of the fundamentals of spherical astronomy at high schools. E. I. Kovyazin.
(Uch. zap.) Perm. gos. ped. in-t, Vol. 72, 411 - 413 (1969). In Russian. – Abstr. in Referativ. Zhurn. 51. Astron., 1.51.29 (1972).

014.014 On equinoxes, solstices and related questions. A. A. Kaverin.
(Uch. zap.) Perm. gos. ped. in-t, Vol. 72, 401 - 405 (1969). In Russian. – Abstr. in Referativ. Zhurn. 51. Astron., 1.51.30 (1972).

014.015 Auxiliary illustrative astronomical devices made of organic glass with use of an internal illumination effect. E. G. Demidovich, A. A. Medvedev.
(Uch. zap.) Perm. gos. ped. in-t, Vol. 72, 406 - 410 (1969). In Russian. – Abstr. in Referativ. Zhurn. 51. Astron., 1.51.31 (1972).

014.016 Astronomy in the standard schedules of universities. B. A. Vorontsov-Vel'yaminov, V. V. Arsent'ev.
Zemlya i Vselennaya, No. 6, p. 57 - 58 (1971). In Russian.

014.017 Slobodne aktivnosti mladih astronoma. (Astronomy in school). M. Obradović.
Vasiona, Vol. 19, 74 - 76 (1971).

015 Miscellanea

015.001 **Extraterrestrial civilizations – problems and views.**
B. N. Panovkin.
Priroda, No. 7.71, p. 56 - 61 (1971). In Russian.

015.002 **Long-wavelength ultraviolet photoproduction of amino acids on the primitive earth.**
C. Sagan, B. N. Khare.
Science, Vol. 173, 417 - 420 (1971).

015.003 **Un nuovo cosmoscopio.** S. Zavatti.
Coelum, Vol. 39, 134 - 138 (1971).

015.004 **Cours d'astronomie: 1. Les propriétés de la lumière, les instruments astronomiques.** M. Dumont.
L'Astronomie, 85ᵉ année, p. 296 - 310 (1971).

015.005 **Ideas concerning the human factor in the development of modern science.** D. H. Levy.
Journ. Roy. Astron. Soc. Canada, Vol. 65, 145 - 151 (1971).

015.006 **Cours d'astronomie de la S.A.F.: 2. La terre, le temps et la sphère céleste.** G. Oudenot.
L'Astronomie, 85ᵉ année, p. 325 - 337 (1971).

015.007 **Astronomical notebook.** J. S. Griffith.
Spaceflight, Vol. 13, 111 - 115, 146 - 147, 152, 263 - 264, 300 - 304 (1971).

015.008 **Landscapes in space.** D. A. Hardy.
Journ. British Astron. Ass., Vol. 81, 460 - 462 (1971).

015.009 **Astronomical notebook.** J. S. Griffith.
Spaceflight, Vol. 13, 391 - 392 (1971).

015.010 **Interactions between astronomy and physics.**
G. Contopoulos.
Tekhn. khron., 1971, No. 3, p. 155 - 158. In Greek. – Abstr. in Referativ. Zhurn. 51. Astron., 10.51.3 (1971).

015.011 **Appropriation of the moon: Some perspectives of legal regulation.** Eh. G. Vasilevskaya.
Sov. gosudarstvo, 1971, No. 4, p. 92 - 99. In Russian. – Abstr. in Referativ. Zhurn. 62. Issled. kosm. prostranstva, 10.62.12 (1971).

015.012 **A formulation for the number of communicative civilizations in the Galaxy.** J. G. Kreifeldt.
Icarus, Vol. 14, 419 - 430 (1971).
    This paper presents the derivation of a dynamic formulation for the number of communicative civilizations in the Galaxy. The formulation is more complete than previous ones in that it allows the effects of different star generation rates, civilization lifetime, and development time distributions to be investigated.

015.013 **Catalogue of reproductions of astronomical photographs available from the Royal Astronomical**

Society.
Roy. Astron. Soc., London. 48 pp. (1971). – The catalogue is included in Quarterly Journ. Roy. Astron. Soc., Vol. 12, No. 3 (1971). – Available from The Librarian, Royal Astronomical Society, Burlington House, Piccadilly, London WIV 0NL, England.

015.014 **Extraterrestrial civilizations.**
J. C. Ribes, F. Biraud.
Proc. Astron. Soc. Australia, Vol. 2, 11 - 13 (1971).

015.015 **La emigración a otros planetas, una perspectiva planteada en el porvenir de la humanidad.** A. Bastidas.
Bol. Acad. Cie. Fis., Mat., Nat., Venezuela, Vol. 30, No. 89, p. 93 - 97 (1971).

015.016 **The moon in popular reliefs and in the humour of science fiction.** C. Naselli.
L'Universo, Anno 51, p. 1271 - 1288 (1971). In Italian.

015.017 **Zeitmarken im Alpenraum.** G. Innerebner.
SuW, Vol. 10, 323 - 326 (1971).

015.018 **Auch das Weltall hat Geschichte. Die Naturwissenschaftler entdecken die historische Perspektive.**
E. Verhülsdonk.
SuW, Vol. 10, 327 - 328 (1971).

015.019 **Más allá de la vía sábana.** S. de la Macorra.
El Universo, Vol. 25, 98 - 99 (1971).

015.020 **Astronomical notebook.** J. S. Griffith.
Spaceflight, Vol. 13, 471 - 472 (1971).

015.021 **An identified flying object.**
A. D. Thackeray, M. D. Overbeck.
Monthly Notices Astron. Soc. Southern Africa, Vol. 30, 155 (1971). – Letter.

015.022 **Johannes Kepler on stamps.** V. A. Orlov.
Zemlya i Vselennaya, No. 6, p. 72 - 73 (1971). In Russian.

015.023 **Astronomy and medicine.** R. Danić.
Vasiona, Vol. 19, 55 - 59 (1971). In Serbo-Croatian.

015.024 **On the infinity of the universe.** K. Mike.
Vasiona, Vol. 19, 63 - 65 (1971). In Serbo-Croatian.

015.025 **Jaunākie atklājumi un eksperimenti.**
I. Daube, Z. Alksne, A. Alksnis, A. Spektors, G. Ozoliņš.
Zvaigžņotā debess, 1971. gada pavasaris, p. 19 - 32.

015.026 **Diskussionsbemerkungen zur exobiologischen Hypothese.** P. Jordan
Abh. Akad. Wiss. Literatur Mainz, Math.-Nat. Kl. 1971, No. 1, p. 1 - 28.

Applied Mathematics, Physics

021 Mathematics, Computing, Machine Programs

021.001 **Lamé functions of the first kind generated by computer.** H. G. Walter.
Celestial Mechanics, Vol. 4, 15 - 30 (1971).
Proceeding from Lamé's differential equation the four classes of Lamé functions of the first kind are generated by computer with the aid of formula manipulation techniques. For this purpose algebraic expressions for the coefficients in the Lamé polynomials are constructed by virtue of recurrence formulae and presented in tabular as well as machine readable form for further processing.

021.002 **Computing equipment at the Institute for Theoretical Astronomy.** V. A. Ivakin.
Byull. Inst. Teoret. Astron., *Leningrad,* Vol. 12, 785 - 791 (1971). In Russian.

021.003 **Telcom-versatile real time instrument control system.** D. E. Trumbo.
Publ. Astron. Soc. Pacific, Vol. 83, 608 (1971). − Abstr. Astron. Soc. Pacific.

021.004 **The Lowell Observatory data system.** R. Albrecht, P. Boyce, J. Chastain.
Publ. Astron. Soc. Pacific, Vol. 83, 683 - 686 (1971).
Two computerized data-acquisition systems have been designed and built at the Lowell Observatory. In two programs involving rapid scanning we have realized a six-fold gain in the efficiency of data taking. Further gains are anticipated as more sophisticated programs of data analysis become available at the telescope.

021.005 **Problemlösung in der Physik mit Hilfe elektronischer Rechenmaschinen.** A. Schlüter.
Physik und Kosmologie, (see 003.024), p. 9 - 16 (1971).

021.006 **Computer programs for calculating atomic data for ions.** W. Eissner.
Highlights of Astronomy, Vol. 2, (see 012.017), 509 - 511 (1971).

021.007 **A new-concise computer treatment of X-ray microprobe data after Bence and Albee.**
A. S. Doan, Jr., R. L. Schmadebeck.
Meteoritics, Vol. 6, 260 - 261 (1971). − Abstract.

021.008 **Collegamento veloce in tempo reale con un elaboratore UNIVAC 1108.**
V. Groce, G. A. De Biase, M. F. Toniolo.
Calcolo, ['Marves', Roma], Vol. 5, Suppl. 1, 386 - 393 (1968)= Oss. Astron. Roma, Contr. Sci., Ser. 3, No. 112 (1970).

021.009 **An astronomical computer program package for the Onsala Space Observatory.**
P. G. Landgren, B. O. Rönnäng.
Res. Lab. Electronics, Chalmers Univ. Technology, Gothenburg, Sweden, Res. Report No. 102, 38 pp. = Preprint Onsala Space Obs., Onsala (1971).
This report describes the development, and practical usage of a series of FORTRAN programs for astronomy. The programs are: (1) Doppler velocity calculation program; (2) Program to calculate the source precession; (3) Program to calculate the nutation and aberration; (4) Julian day number calculator; (5) Program to calculate the sidereal time; (6) Program to determine the parameters of a very long baseline interferometer; (7) A pre-VLBI program to calculate fringe rate, delay and spatial frequency components for a very long baseline interferometry experiment.

021.010 **Datamaskinen − ett nytt astronomiskt instrument II.** Y. Ekedahl.
Astron. Tidssk., Årg. 4, p. 167 - 192 (1971).

Automated analysis of astronomical spectra.
See Abstr. 031.017.

Experiments with digital processing of stellar spectrograms. See Abstr. 031.037.

Data processing for the Westerbork Synthesis Radio Telescope. See Abstr. 033.016.

On the theory, techniques, and data processing of very long baseline interferometry. See Abstr. 033.021.

Automated algebraic manipulation in celestial mechanics. See Abstr. 042.010.

Method for calculation of perturbations for the satellite problem of three bodies with arbitrary inclinations and small eccentricities. See Abstr. 042.052.

ATLAS: A computer program for calculating model stellar atmospheres. See Abstr. 064.042.

General relativity and the application of algebraic manipulative systems. See Abstr. 066.014.

A data acquisition system with on-line computer.
See Abstr. 096.028.

Computer instrumentation for the Jupiter occultation expedition. See Abstr. 099.027.

OH spectral line measurements of radiation from the galactic H II regions W 3 and W 49 at 1665 MHz. See Abstr. 131.138.

022 Physical Papers Related to Astronomy and Astrophysics

022.001 **Numerical calculations of atomic structure constants. I. Angular parameters.** F. Bely.
Astron. Astrophys., Vol. 13, 336 - 344 (1971).

A description is given of a numerical method of obtaining the expression of the H matrix of complex including spin-orbit interaction and configuration mixing. The matrix elements are expressed in terms of Slater and spin-orbit integrals. As an example configurations are studied with special emphasis on the case of Fe XIII. The program can be used for atoms having up to nine electrons outside closed shells.

022.002 **Matrix method for the Liapunov stability analysis of cyclic discrete mechanical systems.**
P. W. Likins, R. E. Roberson.
Celestial Mechanics, Vol. 3, 491 - 507 (1971).

A matrix formalism is developed for the purpose of facilitating the Liapunov stability analysis of discrete, holonomic, mechanical systems with cyclic coordinates and with the Hamiltonian free of explicit time dependence. Matrix expressions are developed for the kinetic energy, the Routhian, the Hamiltonian, and the quadratic approximation of the dynamic potential energy, with cyclic coordinates, cyclic-coordinate velocities, and cyclic-coordinate generalized momenta not explicitly involved in the last of these functions. The final result is an expression for the quadratic approximation of the dynamic potential energy that is calculated much more readily than by scalar analysis. From the condition for positive-definiteness of this function, Liapunov stability conditions are available. The method is applied to a dual-spin satellite to illustrate the procedure.

022.003 **Effects of collisional excitation on the intensities of the 5876 Å and 4471 Å lines of neutral helium.**
D. P. Cox, E. Daltabuit.
Astrophys. Journ., Vol. 167, 257 - 259 (1971).

Collisional excitation is shown to make small but non-negligible contributions to the lower members of the $(n^3D \rightarrow 2^3P)$ series lines in neutral helium.

022.004 **Statistical mechanics of light elements at high pressure. I. Theory and results for metallic hydrogen with simple screening.** W. B. Hubbard, W. L. Slattery.
Astrophys. Journ., Vol. 168, 131 - 139 (1971).

A numerical algorithm for calculating the thermodynamic state variables of light elements at multi-megabar pressures and a wide range of temperatures is presented. A number of results are presented for liquid metallic hydrogen at densities of 2.679 and 2679 g cm^{-3} and temperatures ranging from 313000° to 4170°K.

022.005 **Spin-change scattering of C II and O I by atomic hydrogen.**
S. Wofsy, R. H. G. Reid, A. Dalgarno.
Astrophys. Journ., Vol. 169, 161 - 167 (1971).

Cross-sections for spin-change scattering are calculated for collisions of positive carbon ions C$^+$ with atomic hydrogen H and of neutral oxygen atoms O with H. An elastic-scattering model is used. The associated cooling efficiencies in the interstellar medium due to fine-structure transitions of C$^+$ and O are comparable at temperatures above 250°K, but C$^+$-H cooling dominates at low temperatures.

022.006 **Photoionization cross-sections for atoms and ions of sulfur.** R. D. Chapman, R. J. W. Henry.
Astrophys. Journ., Vol. 168, 169 - 171 (1971).

We calculate photoionization cross-sections for all levels belonging to the configuration $1s^2 2s^2 2p^6 3s^2 3p^q$ of atoms and ions of sulfur. The results are fitted to an interpolation formula which gives the cross-sections as a function of wavelength from threshold to $\sim$ 100 Å.

022.007 **The ratio of the Franck-Condon factors $q(0, 0)/q(0, 1)$ of the infrared atmospheric band system of oxygen.** K. H. Becker, W. Groth, U. Schurath.
Planet. Space Sci., Vol. 19, 1009 - 1010 (1971). – Research note.

022.008 **New method for treating systems containing elementary composite particles (nonrelativistic theory).**
R. H. Stolt, W. E. Brittin.
Phys. Rev. Letters, Vol. 27, 616 - 619 (1971).

022.009 **On the degree of circular polarization of synchrotron radiation.** D. B. Melrose.
Astrophys. Space Sci., Vol. 12, 172 - 192 (1971).

Formulas describing synchrotron radiation are extended to include the effect of the presence of an ambient medium and the effect of reabsorption and Faraday rotation on the degree of circular polarization. The transfer equation including the effects of the polarization is discussed in detail.

022.010 **Arc study of the oxygen Schumann-Runge system.** K. L. Wray, S. S. Fried.
Journ. Quant. Spectrosc. Radiat. Transfer, Vol. 11, 1171 - 1180 (1971).

From the photographic spectra it has been demonstrated that the radiation is a rotational/vibrational electronic spectrum and that the radiation obtained for arc heated air is similar in detailed structure to that for arc heated O$_2$-noble gas mixtures. This rotational line spectrum has been identified as the O$_2$ (Schumann-Runge) system as excited by a variety of sources.

022.011 **Perturbation analysis and constants for the red system of the cyanide radical.**
T. Fay, I. Marenin, W. van Citters.
Journ. Quant. Spectrosc. Radiat. Transfer, Vol. 11, 1203 - 1214 = Publ. Geothe Link Obs., Indiana Univ., *Bloomington*, No. 119 (1971).

A χ^2 test has been used to obtain improved molecular constants for the red system of the C^{12}N^{14} radical. The new constants predict wavenumbers that agree with the observed valued to 0.1 Å for rotational levels up to N''= 80 for the lower vibrational bands and up to N''= 50 for the higher bands. Wavenumbers for the (2,0) and (3,1) vibration-bands of the C^{13}N^{14} radical have also been computed with these molecular constants.

022.012 **Intensités absolues et forces d'oscillateur de quelques raies des bandes de vibration-rotation 1–0 et 2–1 du radical OH.** J. d'Incan, C. Effantin, F. Roux.
Journ. Quant. Spectrosc. Radiat. Transfer, Vol. 11, 1215 - 1224 (1971).

022.013 **Transition probabilities of some ArI and ArII spectral lines.** D. van Houwelingen, A. A. Kruithof.
Journ. Quant. Spectrosc. Radiat. Transfer, Vol. 11, 1235 - 1243 (1971).

The transition probabilities have been calculated from the temperatures and pressures found for the five spectral lines of the argon atom and the four lines of singly ionized argon, for which the power radiated per unit volume was measured. The results are given in tables together with literature values. The estimated random errors are also given as far as they are known.

022.014 **Calculation of the partition function for $^{14}N_2{}^{16}O$.**
L. D. Gray Young.
Journ. Quant. Spectrosc. Radiat. Transfer, Vol. 11, 1265 - 1270 (1971).

Internal and vibrational partition functions are tabulated for $^{14}N_2{}^{16}O$ for the temperature range $200-350°$K at $10°$K intervals. Rotational partition functions are also computed for the five lowest vibrational states.

022.015 **Non-adiabatic effects in Van der Waals broadening.**
D. N. Stacey, J. Cooper.
Journ. Quant. Spectrosc. Radiat. Transfer, Vol. 11, 1271 - 1274 (1971).

022.016 **Recomputation of the absorption strengths of the methane $3\,\nu_3$ J-manifolds at 9050 cm^{-1}.**
J. T. Bergstralh, J. S. Margolis.
Journ. Quant. Spectrosc. Radiat. Transfer, Vol. 11, 1285 - 1287 (1971).

Absorption intensities of the R-branch J-manifolds ($0 \leq J \leq 7$) of the $3\,\nu_3$ methane band have been computed using half-widths measured for the individual components of the J-manifolds. The effects of small pressure shifts of the lines are included. The results yield a value for the rotational temperature in agreement with the laboratory temperature and are very close to the results obtained assuming no J dependence of the half-widths.

022.017 **Spectral measurements of nitrogen continuum radiation behind incident shocks at speeds up to 13 km/sec.**
D. L. Ciffone, J. G. Borucki.
Journ. Quant. Spectrosc. Radiat. Transfer, Vol. 11, 1291 - 1310 (1971).

Spectral measurements of nitrogen continuum radiation behind incident shocks, at pressures of $0.2-0.8$ atm and $8000-13000°$K, have been made. The results are compared with both theory and other experiments.

022.018 **Theory of pressure-induced vibrational and rotational absorption of diatomic molecules at high temperatures.**
R. W. Patch.
Journ. Quant. Spectrosc. Radiat. Transfer, Vol. 11, 1311 - 1330 (1971).

A derivation is given for the integrated absorption coefficient of pressure-induced pure rotational and vibrational transitions in binary collisions of homonuclear diatomic molecules of the same chemical species. The previously neglected effects of excited vibrational states, mechanical anharmonicity, and vibration-rotation interaction are taken into account to obtain more accurate absorption coefficients at high temperatures, which are needed for radiation transfer calculations for late-type stars and entry into certain planetary atmospheres.

022.019 **Absorption coefficients for hydrogen–II. Calculated pressure-induced H_2-H_2 vibrational absorption in the fundamental region.**
R. W. Patch.
Journ. Quant. Spectrosc. Radiat. Transfer, Vol. 11, 1331 - 1353 (1971).

The coefficient for pressure-induced vibrational absorption in H_2-H_2 collisions was calculated for temperatures from 298 to $7000°$K and wave numbers between 100 and 40,000 cm^{-1} for local thermodynamic equilibrium. The model included electronic configuration interaction, mechanical anharmonicity, vibration-rotation interaction, excited vibrational states, and more realistic intermolecular potential and line shapes than previously used. An approximate formula for the absorption coefficient is given for rapid calculation.

022.020 **Franck-Condon factors and r-centroids for halogen molecules–I. The $B^3\Pi(0_u{}^+)-X^1\Sigma_g{}^+$ system of $^{35}Cl_2$.**
J. A. Coxon.
Journ. Quant. Spectrosc. Radiat. Transfer, Vol. 11, 1355 - 1364 (1971).

This report is the first of a series of papers concerned with Franck-Condon factors for halogen molecules, for which no previous data are available. The results of r-centroid calculations are also presented.

022.021 **Numerical evaluation of the redistribution function $R_{II-A}(x,x')$ and of the associated scattering integral.**
T. F. Adams, D. G. Hummer, G. B. Rybicki.
Journ. Quant. Spectrosc. Radiat. Transfer, Vol. 11, 1365 - 1376 (1971).

Methods are presented for generating numerical values for this redistribution function. Procedures using natural cubic spline representations are given for approximating the integrals involving this function that appear in line transfer calculations.

022.022 **Widths of optically thick lines.**
R. J. Exton.
Journ. Quant. Spectrosc. Radiat. Transfer, Vol. 11, 1377 - 1383 (1971).

022.023 **Synthetic spectral methods and the interpretation of shock-excited molecular spectra.**
R. W. Nicholls.
Journ. Roy. Astron. Soc. Canada, Vol. 65, 181 (1971).
Abstr. RAS Canada.

022.024 **Recent improvements in the beam-foil technique for assigning unidentified atomic spectrum lines.**
E. H. Pinnington.
Journ. Roy. Astron. Soc. Canada, Vol. 65, 181 (1971).
Abstr. RAS Canada.

022.025 **Faraday rotation measurements of absorption line parameters.**
D. Camm, F. L. Curzon.
Journ. Roy. Astron. Soc. Canada, Vol. 65, 181 - 182 (1971).
Abstr. RAS Canada.

022.026 **On the accuracy of the ratio of two electrical signals measured with EhPP-09.**
A. I. Shameka.
Astrometriya i Astrofiz., *Kiev*, Vyp. (No.) 12, (see 003.003), p. 70 - 73 (1971). In Russian.

022.027 **Formation of excited particles in collisions of He$^+$ ions with CO molecules.**
G. N. Polyakova, V. F. Erko, A. V. Sats, Ya. M. Fogel.
Astrometriya i Astrofiz., *Kiev*, Vyp. (No.) 12, (see 003.003), p. 79 - 83 (1971). In Russian.

Effective cross-sections of the formation of excited particles $CO^+(A^2\pi)$, He, C^+ and O^+ in collisions of He$^+$ ions (energy 0.16–30 keV) with CO molecules were measured.

022.028 **Zur Modifikation der Hamiltonschen Kräftefunktion durch Legendresche Transformationen.**
H.-J. Treder.
Monatsber. Deutsch. Akad. Wiss. Berlin, Vol. 13, 42 - 46 (1971).

022.029 **Radiative recombination coefficients for complex ions.**
C. B. Tarter.
Astrophys. Journ., Vol. 168, 313 - 316 (1971).

Radiative recombination coefficients are calculated for complex ions of astrophysical interest. A least-squares procedure is used to obtain three-term parametric expressions which agree with the computed values to better than 3 percent over the temperature ranges of interest.

022.030 The rate of the $^{12}C + {}^{16}O$ reaction.
C. J. Hansen, C. S. Zaidins.
Astrophys. Journ., Vol. 168, 317 - 318 (1971).

The rate of the $^{12}C + {}^{16}O$ reaction has been calculated to within a factor of 2 from new experimental data. It is shown that its contribution to energy generation is negligible compared with $^{12}C + {}^{12}C$ under normal circumstances, though it may have some effects on nucleosynthesis.

022.031 Errata: 'Band strengths in forbidden transitions: The Cameron bands of CO' [Journ. Quant. Spectrosc.
Radiat. Transfer, Vol. 10, 1321 - 1328 (1970)].
A. R. Fairbairn.
Journ. Quant. Spectrosc. Radiat. Transfer, Vol. 11, 1289 (1971).

022.032 On the normalization of Hönl-London factors.
A. Schadee.
Astron. Astrophys., Vol. 14, 401 - 404 (1971).

A new normalization of Hönl-London factors, which govern the intensity distribution within a molecular band, is presented. Problems posed by previous normalizations are solved.

022.033 On possibilities of determining complex form factors from an experiment with polarized particles.
R. Sh. Yakh'yaev.
Soobshch. Shemakhinsk. Astrofiz. Obs., vyp. (No.) 5, p. 90 - 97 (1971). In Russian.

022.034 Excitation of N_2 and N_2^+ systems by electrons – I. Absolute transition probabilities.
D. E. Shemansky, A. L. Broadfoot.
Journ. Quant. Spectrosc. Radiat. Transfer, Vol. 11, 1385 - 1400 (1971).

Quantitative optical measurements of the N_2 $1P$, $2P$ and N_2^+ $1N$ and Meinel systems, excited by electrons, have allowed measurements of transition probabilities, excitation cross-sections, and afterglow effects. The experiment and observations are described in detail in this article. Absolute transition probabilities have been derived for the N_2 $1P$ and the $N_2^+ M$ systems.

022.035 Excitation of N_2 and N_2^+ systems by electrons – II. Excitation cross sections and N_2 $1PG$ low pressure afterglow.   D. E. Shemansky, A. L. Broadfoot.
Journ. Quant. Spectrosc. Radiat. Transfer, Vol. 11, 1401 - 1439 (1971).

The analyses of recent observations of the systems are presented which we believe may indicate the sources of most of the discrepancies among the previously published work. The results of the present observations differ substantially with the earlier articles on the N_2 $1P$ and $N_2^+ M$ systems. The origin of the disagreement appears to be due to instrumental calibration only to a relatively minor degree. According to our interpretation the disagreement is partly due to the fact that the comparatively long lifetimes of these two systems have been ignored and partly because the N_2 $1PG$ band is affected by a low-pressure afterglow.

022.036 The computation of photoionization cross sections by means of the scaled Thomas-Fermi potential.
H. Kähler.
Journ. Quant. Spectrosc. Radiat. Transfer, Vol. 11, 1521 - 1535 (1971).

Photoionization cross sections of lighter neutral elements (Na, K, Mg, C, N, O) have been calculated by means of the scaled Thomas-Fermi potential and compared with results obtained by other methods. In contrast to the quantum-defect

method, the present method is applicable and yields reasonable results also for ultraviolet transitions, that is, for electron energies up to several Rydbergs and for low-lying bound states with large quantum defect.

022.037 Rotational line width of methane.
G. Yamamoto, M. Hirono.
Journ. Quant. Spectrosc. Radiat. Transfer, Vol. 11, 1537 - 1545 (1971).

022.038 An impact theory for Doppler and pressure broadening – I. General theory.
E. W. Smith, J. Cooper, W. R. Chappell, T. Dillon.
Journ. Quant. Spectrosc. Radiat. Transfer, Vol. 11, 1547 - 1565 (1971).

A quantum-mechanical impact theory for the combined effects of Doppler and pressure broadening is developed from quantum radiation theory. The results are compared with other semiclassical theories and certain simplifying approximations relevant to cases of experimental and theoretical interest are discussed.

022.039 An impact theory for Doppler and pressure broadening – II. Atomic and molecular systems.
E. W. Smith, J. Cooper, W. R. Chappell, T. Dillon.
Journ. Quant. Spectrosc. Radiat. Transfer, Vol. 11, 1567 - 1576 (1971).

A semiclassical theory for Doppler and pressure broadening in neutral gases is derived as a limiting case of a more general quantum mechanical theory. This theory is compared with other semiclassical theories and methods of calculation are discussed.

022.040 Absolute oscillator strengths for some resonance multiplets of Ca I, II, Mg I, II, B I, and Al I.
W. H. Smith, H. S. Liszt.
Journ. Optical Soc. America, Vol. 61, 938 - 941 (1971).

Radiative-lifetime studies of the more prominent resonance transitions of Ca I, II, Mg I, II, B I, and Al I between 1800 and 4250 Å are presented. The results for the lowest member of the resonance series are in good agreement with most previous-literature values obtained with Hanle-effect or beam-foil techniques. Results for the higher members of the resonance series for Mg I and Ca I are compared with previous results. These data are discussed and oscillator strengths are given.

022.041 Energy levels and transition probabilities for highly ionized atoms in the B I isoelectronic sequence.
L. J. Shamey.
Journ. Optical Soc. America, Vol. 61, 942 - 946 (1971).

Hartree-Fock calculations have been performed for Ne VI, Al IX, Si X, S XII, Ar XIV, and Fe XXII.

022.042 Radiative lifetime of the $A^1 \Pi$ state of CO.
J. G. Chervenak, R. A. Anderson.
Journ. Optical Soc. America, Vol. 61, 952 - 954 (1971).

The lifetime of the $A^1 \Pi$ state of CO has been measured by using a pulsed invertron-excitation source and a delayed-coincidence measurement technique. The fourth positive system was examined to measure this lifetime. It is of major importance in many absorption processes because it originates on the first excited state and terminates on the ground state of CO, and because of the presence of CO in the atmospheres of the earth and stellar bodies.

022.043 Experimental studies of atomic spectra and transition probabilities.   W. R. S. Garton.
National Bureau Standards Special Publ. 353, (see 012.001), p. 3 - 19 (1971).

022.044 Towards an elementary theory of the periodic table. D. Layzer.
National Bureau Standards Special Publ. 353, (see 012.001), p. 20 - 36 (1971).

022.045 Many-body calculations of energies and transition probabilities. H. P. Kelly.
National Bureau Standards Special Publ. 353, (see 012.001), p. 37 - 46 (1971).

022.046 Radioactive transitions in the helium isoelectronic sequence. A. Dalgarno.
National Bureau Standards Special Publ. 353, (see 012.001), p. 47 - 57 (1971).

022.047 Electron broadening of spectral lines of Si II and Si III ions. E. A. Yukov.
Astron. Zhurn. Akad. Nauk SSSR, Vol. 48, 1094 - 1096 (1971). In Russian. English translation in Soviet Astron. AJ, Vol. 15, No. 5.

Values of line widths, shifts, quadratic Stark-effect constants and oscillator strengths are presented for the lines of Si II and Si III ions.

022.048 A curve of growth determination of the f-values for the fourth positive system of CO and the Lyman-Birge-Hopfield system of N_2. M. J. Pilling, A. M. Bass, W. Braun.
Journ. Quant. Spectrosc. Radiat. Transfer, Vol. 11, 1593 - 1604 (1971).

The curve of growth method has been employed to determine f-values for the fourth positive system of CO and the magnetic dipole and electric quadrupole components of the Lyman-Birge-Hopfield system of N_2.

022.049 Self-broadened half-width measurements in the CO fundamental. J. E. Lowder.
Journ. Quant. Spectrosc. Radiat. Transfer, Vol. 11, 1647 - 1657 (1971).

A method for measuring rotational line half-widths for the Lorentzian line shape and a slit function described by three Gaussian curves has been applied to lines in the CO fundamental. This technique uses apparent line-center transmission measurements to determine half-widths of pressure-broadened absorption lines. The half-widths of 12 lines in the R-branch of the CO fundamental were measured and found to compare well with two previous measurements.

022.050 Cross sections for ionization of ions by electron impact. D. N. Tripathi, D. K. Rai.
Journ. Quant. Spectrosc. Radiat. Transfer, Vol. 11, 1665 - 1673 (1971).

Cross sections for ionization of ions by electron impact have been calculated using various classical and empirical methods. Empirical modifications have been suggested in few known empirical formulae to improve the results. Calculated cross sections have been compared with other theoretical calculations and observed cross sections. The dependence of the ionization cross sections on ionic charge is also discussed.

022.051 Radiation from a theta-pinch with oscillatory density. T. S. Green.
Journ. Quant. Spectrosc. Radiat. Transfer, Vol. 11, 1691 - 1698 (1971).

Studies of radiation from theta-pinches have provided a powerful method for investigation of excitation rates of ionic levels, and for identification of spectral lines observed in solar spectra. In the present paper, we consider the possibility that oscillation of the electron density in the pinch discharge producing an oscillation in line intensities can facilitate the interpretation of the observations. One finds that the amplitude of the oscillation of the line intensity, relative to its unperturbed value, depends sensitively on the excitation mechanism and the de-population mode.

022.052 Collision-broadened half-widths and shapes of methane lines. P. Varanasi.
Journ. Quant. Spectrosc. Radiat. Transfer, Vol. 11, 1711 - 1724 (1971).

Determination of methane abundance in the terrestrial and Jovian atmospheres requires line-shape parameters in methane bands in air-broadening and hydrogen- (and helium-) broadening, respectively. High-resolution measurements are reported on a few rotational lines in the ν_3-fundamental and the $2\nu_3$-overtone of methane. Half-width data are presented in self-broadening and in broadening by H_2, He, N_2, O_2 and air. Comparison with available data on $3\nu_3$-band shows, within experimental error, that the line-widths are the same for ν_3 and its overtones in all cases of broadening.

022.053 Gauss-Laguerre calculation of free-free Gaunt factors. B. H. Armstrong.
Journ. Quant. Spectrosc. Radiat. Transfer, Vol. 11, 1731 - 1734 (1971).

An analysis is made of some earlier calculations by Grant of the Maxwell-averaged free-free Gaunt factor in the Born approximation by means of Gauss-Laguerre integration.

022.054 The pressure broadening of radio recombination lines. M. Brocklehurst, S. Leeman.
Astrophys. Letters, Vol. 9, 35 - 36 (1971).

The pressure broadening of radio recombination lines has been calculated in the impact approximation. Results obtained from accurate collision cross-sections, and the theory of Baranger, are in fairly close agreement with the results obtained by Griem, who used classical path approximations.

022.055 The bending of the synchrotron spectrum at high energies. J. Jaffe, A. Treves.
Astrophys. Letters, Vol. 9, 39 - 41 (1971).

The synchrotron emission of ultrarelativistic electrons crossing a region of high magnetic field is considered. It is shown in a quite simple way that under certain conditions the resultant photon spectrum has a break in the X- or gamma-ray band. This result may be related to observations in the high energy range of the Crab nebula pulsar.

022.056 Accurately measured and calculated ground-term combinations of Ar II. L. Minnhagen.
Journ. Optical Soc. America, Vol. 61, 1257 - 1262 (1971).

022.057 $4p^3 - 4p^2 5s$ transitions in Nb IX. M. Said-Uz-Zafar Chaghtai.
Journ. Optical Soc. America, Vol. 61, 1264 (1971).

022.058 Transition probabilities for Ar I. C. J. Chen.
Journ. Optical Soc. America, Vol. 61, 1267 - 1268 (1971).

022.059 Errors in neutral iron oscillator strengths. R. A. Bell, W. L. Upson, II.
Astrophys. Letters, Vol. 9, 109 - 112 (1971).

It is suggested that the errors in the Fe I oscillator strengths of Corliss and co-workers are produced by errors in their photographic intensity scale. The suggestion that the errors are caused by errors in the adopted arc temperature is disproved by the solar curve of growth.

022.060 Bremsstrahlung – a new instrument for research. S. P. Kapitsa.
Priroda, No. 10.71, p. 22 - 27 (1971). In Russian.

022.061 Terrestrial and extraterrestrial limits on the photon mass. A. S. Goldhaber, M. M. Nieto.
Rev. Modern Phys., Vol. 43, 277 - 296 (1971).

We give a review of methods used to set a limit on the mass μ of the photon. Direct tests for frequency dependence of the speed of light are discussed, along with more sensitive techniques which test Coulomb's law and its analog in magnetostatics. The link between dynamic and static implications of finite μ is deduced from a set of postulates that make Proca's equations the unique generalization of Maxwell's. We note one hallowed postulate, that of energy conservation, which may be tested severely using pulsar signals. We present the merits of the old methods and of possible new experiments, and discuss other physical implications of finite μ. The best results from past experiments are (a) terrestrial measurements of c at different frequencies $\mu \lesssim 2 \times 10^{-43} g \equiv 7 \times 10^{-6}$ cm^{-1} $\equiv 10^{-10}$ eV; (b) measurements of radio dispersion in pulsar signals (whistler effect) $\mu \lesssim 10^{-44} g \equiv 3 \times 10^{-7}$ cm$^{-1} \equiv 6 \times 10^{-12}$ eV. Observations of the galactic magnetic field could improve the limit dramatically.

022.062 Inelastic collisions of fast charged particles with atoms and molecules – The Bethe theory revisited.
M. Inokuti.
Rev. Modern Phys., Vol. 43, 297 - 347 (1971).

The current understanding is summarized from a unified point of view, which Bethe initiated four decades ago and which enables one to put a variety of theoretical and experimental data into a coherent picture. Properties of the generalized oscillator strength, which plays the central role in the theory, are treated in detail. The integrated cross section for inelastic scattering and related quantities at the high-velocity limit also are discussed.

022.063 Line interactions in saturated masers.
A. G. Bromley.
Proc. Astron. Soc. Australia, Vol. 2, 34 - 36 (1971).

022.064 Determination of interatomic potentials.
R. J. Dyne.
Proc. Astron. Soc. Australia, Vol. 2, 38 - 40 (1971).

022.065 Absorption by vibrationally excited molecular oxygen in the Schumann-Runge continuum.
A. C. Allison, A. Dalgarno, N. W. Pasachoff.
Planet. Space Sci., Vol. 19, 1463 - 1473 (1971).

022.066 Quantum treatment of electron emission in a strong electromagnetic wave. S. A. Bonometto.
Astron. Astrophys., Vol. 15, 193 - 199 (1971).

The emission by electrons in a strong electromagnetic wave is studied by using quantum mechanics. The applicability of quantum results to astrophysical situations is discussed, outlining the possibility that strong electromagnetic waves, emitted by pulsars according to several models, may need quantum treatment. Energy losses by electrons are studied in the quantum regime and the possible importance of electron pair production in the wave is stressed.

022.067 Transition probabilities in intermediate-coupling and configuration mixing for Fe XVII. M. Loulergue.
Astron. Astrophys., Vol. 15, 216 - 220 (1971).

Transition probabilities between the seven configurations $2p^6$, $2p^5 \, 3l$, $2s \, 2p^6 \, 3l$ ($l = 0, 1, 2$) of Fe XVII are computed in intermediate coupling, including the full configuration mixing within a given complex as suggested by Layzer (1959).

022.068 Beam-foil lifetimes in neutral chromium.
C. L. Cocke, B. Curnutte, J. H. Brand.
Astron. Astrophys., Vol. 15, 299 - 303 (1971).

Beams of chromium from 50 to 100 keV were excited in passage through thin carbon foils, and the resulting downstream decay of the ensuing radiation was used to determine the lifetimes of the emitting states.

022.069 Remarks on the theory of fireballs and on the relativistic "non local theories". G. Wataghin.
Atti Accad. Nazionale Lincei, Rend. Sci. fis., mat., nat., Ser. 8, Vol. 49, 387 - 388 (1970).

022.070 Satellite bands of the γ'-system of titanium oxide.
J. G. Phillips.
Astrophys. Journ., Vol. 169, 185 - 189 (1971).

The $\lambda\lambda$ 6148 and 6174 bands in the spectrum of TiO are shown to be satellite bands of the (0, 0) band of the γ'-system.

022.071 On the oscillator strengths of the $3s \, ^2S-4p \, ^2P$ and $3d \, ^2D-4p \, ^2P$ transitions in sodiumlike ions.
H.-J. Kunze, R. U. Datla.
Astrophys. Journ., Vol. 169, 425 - 427 (1971).

The oscillator strength for the $3d-4p$ transition has been derived from the measured intensity ratio of the $3d-4p$ and $3s-4p$ transitions emitted by Ar VIII ions in a theta-pinch plasma. Theoretical values along the isoelectronic sequence of sodium have been computed by using a method of A. Burgess.

022.072 Laboratory measurement of the 6-centimeter formaldehyde transitions.
K. D. Tucker, G. R. Tomasevich, P. Thaddeus.
Astrophys. Journ., Vol. 169, 429 - 440 (1971).

The $1_{10} - 1_{11}$ rotational transition of $H_2 \, ^{12}C^{16}O$ in its ground and first two excited vibrational states, and that of $H_2 \, ^{13}C^{16}O$, $H_2 \, ^{12}C^{18}O$, and $HD^{12}C^{16}O$ in their vibrational ground states, all of which lie at a wavelength of about 6 cm, have been studied with a beam-maser spectrometer. Line widths in the range 1–3 kHz have been attained. All significant features of the observed spectra have been successfully interpreted in terms of the current theory of molecular hyperfine structure, and the appropriate hyperfine coupling constants have been determined.

022.073 Analogue of the primary solar reaction.
J. E. Brolley.
Astrophys. Journ., Vol. 169, 443 - 444 (1971).

Cross-sections for $n + n \rightarrow D + e^- + \bar{\nu}$ have been calculated and are compared with those of $p + p \rightarrow D + e^+ + \nu$.

022.074 A laboratory determination of the frequency of the 10-centimeter radio line of CH by optical measurements. K. M. Baird, H. Bredohl.
Astrophys. Journ., (Letters), Vol. 169, L83 - L86 (1971).

A new method of photographic Fabry-Perot interferometry has been used to measure the Λ doubling of the $J = 1/2$ level in the electronic ground state of CH. Our value of 3374 ± 15 MHz agrees with Douglas and Elliott's extrapolated value but not with Miller Goss's direct measurements.

022.075 Electronic recombination coefficient of molecular helium ions. A. W. Johnson, J. B. Gerardo.
Phys. Rev. Letters, Vol. 27, 835 - 838 (1971).

022.076 Runaway solutions: Remarks on the asymptotic theory of radiation damping. W. L. Burke.
Phys. Rev. A, General Phys., Vol. 2, 1501 - 1505 = Lick Obs. Bull., No. 608 (1970).

022.077 Theory of Stark broadening – I. Soluble scalar model as a test. U. Frisch, A. Brissaud.
Journ. Quant. Spectrosc. Radiat. Transfer, Vol. 11, 1753 - 1766 (1971).

The purpose of the present paper is to test, on an exactly

soluble model, the validity of standard approximations used in the theory of Stark broadening of atomic lines.

022.078   Theory of Stark broadening – II. Exact line profile with model microfield.   A. Brissaud, U. Frisch.
Journ. Quant. Spectrosc. Radiat. Transfer, Vol. 11, 1767 - 1783 (1971).

A semi-classical theory of Stark broadening is presented which differs basically from current theories. It is shown that the correct profiles are obtained in both the high- and low-density limit. Results are presented for hydrogen and helium lines.

022.079   Theoretical oscillator strengths for the boron iso-electronic sequence.   R. P. McEachran, M. Cohen.
Journ. Quant. Spectrosc. Radiat. Transfer, Vol. 11, 1819 - 1826 (1971).

The frozen core version of the Hartree-Fock approximation has been employed to calculate a large number of s-p and p-d oscillator strengths for BI, CII, NIII and OIV. With the exception of transitions originating in the ground $2p$-state (which require a multi-configuration treatment), the present results are generally in good agreement with experimental, as well as with other theoretical, determinations.

022.080   Fluorescence of the γ, ϵ and δ systems of nitric oxide; polarization and use of calculated intensities for spectrometer calibration.   H. M. Poland, H. P. Broida.
Journ. Quant. Spectrosc. Radiat. Transfer, Vol. 11, 1863 - 1876 (1971).

Positive and negative polarizations have been calculated for various rotational lines of diatomic molecules. As a check on these predictions, the polarization of individual rotational lines of NO fluorescense excited by the 2144 Å line of Cd$^+$ have been measured. Although measured polarization was less than calculated, the signs of the polarization of individual lines agreed with theory.

022.081   The condensation and evaporation of hydrogen on liquid-helium-cooled surfaces.   T. J. Lee.
Commun. Roy. Obs. Edinburgh, No. 107, 5 pp. (1971). – Reprinted from Proceedings of the Third International Cryogenic Engineering Conference, Berlin, 25 - 27 May 1970, p. 388 - 392 (1971).

The condensation of hydrogen has been studied in the temperature region of 2–4 K to support the development of high-performance hydrogen cryopumps. These experiments indicate departures from the expected vapour pressure – temperature dependence in the region of 3 K, that of greatest astronomical interest. The observed anomalies appear to depend on the nature of the substrate and on the thermal radiation flux to the condensed layer. The present programme of experiments has been undertaken to investigate these anomalies and thus provide reliable data on the saturated vapour pressure of hydrogen under conditions met in interstellar space.

022.082   Suprathermal proton bremsstrahlung by the Weizsäcker-Williams method.   F. C. Jones.
Astrophys. Journ., Vol. 169, 503 - 506 (1971).

The Weizsäcker-Williams method for calculating the bremsstrahlung cross-section utilizes the rest frame of the electron as the natural frame in which to carry out the calculation in terms of Compton scattering the virtual photons of the proton electromagnetic field. In the present paper the cross-section for suprathermal proton bremsstrahlung is calculated, and the results are shown to differ considerably from those of a previous paper in which the conventional and much more difficult method was used. In particular we conclude that the suprathermal proton bremsstrahlung process cannot explain the 2–6-MeV γ-rays reported by Vette *et al.*

022.083   A new look at the laboratory microwave spectrum of cyanoacetylene.   D. R. Johnson, F. Lovas.
Astrophys. Journ., Vol. 169, 617 - 619 (1971).

Laboratory measurements have been made on several rotational transitions in cyanoacetylene with potential application to radio astronomy. Observations of $J = 1 \leftarrow 0$ and $J = 2 \leftarrow 1$ rotational transitions are reported for the isotopic species $H^{12}C^{12}C^{12}C^{14}N$, $H^{13}C^{12}C^{12}C^{14}N$, and $H^{12}C^{12}C^{12}C^{15}N$. Measurement errors and resolution of hyperfine structure are discussed in some detail.

022.084   Collision broadening by neutral hydrogen.   K. A. Brueckner.
Astrophys. Journ., Vol. 169, 621 - 632 (1971).

The theory of collision broadening by neutral hydrogen is summarized and given in simplified form. The exact expression is evaluated, avoiding the multipole expansion but using scaled hydrogenic functions, and is applied to the lines of Fe I. The predictions for the line width are considerably larger than the dipole-dipole estimate and are in reasonable agreement with the average broadening observed experimentally.

022.085   Erratum: "Effects of collisional excitation on the intensities of the 5876 Å and 4471 Å lines of neutral helium." [Astrophys. Journ., Vol. 167, 257 - 259 (1971)].   D. P. Cox, E. Daltabuit.
Astrophys. Journ., Vol. 169, 635 (1971).

022.086   Rotational and vibrational temperatures of BaO from a barium release at 170 km, and the synthetic spectrum of BaO in the region 4700 Å to 15,500 Å.   V. Degen, N. Brown, G. J. Romick.
Planet. Space Sci., Vol. 19, 1625 - 1636 (1971).

022.087   The absorption of electrons in atomic oxygen.   A. Dalgarno, G. Lejeune.
Planet. Space Sci., Vol. 19, 1653 - 1667 (1971).

A detailed description is given of the energy degradation of energetic electrons absorbed in a weakly ionized gas of atomic oxygen that takes account of the discreteness of the energy losses through excitation. The results are briefly applied to atmospheric phenomena.

022.088   Precise laboratory determination of rotational transition frequencies in cyanoacetylene.   R. L. de Zafra.
Astrophys. Journ., Vol. 170, 165 - 168 (1971).

The quadrupole coupling constant and rotational constants B_0 and D_J for the ground vibrational state of cyanoacetylene (HC_3N) have been measured with high precision in the laboratory. Resulting values allow a more precise determination of interstellar HC_3N velocities, and simplify the identification of additional interstellar rotational transitions.

022.089   Semiempirical cross-sections and rates for excitation and for ionization of hydrogenic ions by electron impact.   D. H. Sampson, L. B. Golden.
Astrophys. Journ., Vol. 170, 171 - 180 (1971).

Semiempirical electron-impact cross-sections and rates are given for hydrogenic ions of any Z both for ionization and for excitation of the type $n \rightarrow n'$, where n and n' are the initial and final principal quantum numbers.

022.090   Electron-impact cross-sections and rates for $nl \rightarrow n'l'$ transitions in hydrogenic ions and hydrogen.   L. B. Golden, D. H. Sampson.
Astrophys. Journ., Vol. 170, 181 - 190 (1971).

Calculations of Coulomb-Born cross-sections are made for the $2l \rightarrow 3l'$ transitions for $Z = 2$, for the $1s \rightarrow 3l'$ transitions for $Z = 2$ and $Z = \infty$, and for the $1s \rightarrow 2l'$ transitions at threshold for $Z = 2, 3, 5$, and ∞.

022.091 Photoionization cross-sections for helium-like positive ions. R. T. Brown.
Astrophys. Journ., Vol. 170, 387 - 391 (1971).

The photoionization cross-section for a two-electron atomic system has been calculated with a two-parameter variational bound-state wave function and a Coulomb continuum wave function. The relatively simple closed-form expression obtained gives numerical results for He I and Li II that lie within or very near the envelope of curves resulting from much more elaborate calculations.

022.092 Transforms of the hypergeometric function that allow for more rapid convergence in calculations of free-free Gaunt factors. J. T. O'Brien.
Astrophys. Journ., Vol. 170, 613 - 615 (1971).

The expressions used by Karzas and Latter for calculating the free-free Gaunt factors are transformed into expressions which converge more rapidly; in some cases with several hundred fewer terms. The recursion formulae are also simpler.

022.093 Selected tables of atomic spectra. A. Atomic energy levels-second edition; B. Multiplet tables: N IV, N V, N VI, N VII. C. E. Moore.
National Standard Reference Data System, NSRDS–NBS 3, Section 4, 46 pp. (1971). [For sale by the Superintendent of Documents, U.S. Government Printing Office, Washington, D.C., Price 55 cents].

The present publication is the fourth Section of a series being prepared in response to the persistent need for a current revision of two sets of tables containing data on atomic spectra as derived from analyses of optical spectra. As in the previous Sections, Part A contains the atomic energy levels and Part B the multiplet tables. Four spectra of nitrogen, N IV, N V, N VI and N VII, are included. The form of presentation is described in detail in the text to Section 1.

022.094 Isoelectronic wavelength calculations for argon spectra. M. D. Williams.
Solar Physics, Vol. 21, 38 - 39 (1971). – Research note.

022.095 On the accuracy of Hartree-Fock theoretical wavelengths in the XUV. J. P. Connerade.
Solar Physics, Vol. 21, 40 - 41 (1971). – Research note.

022.096 Hard X-ray spectrum produced by thermal bremsstrahlung. J. Juchniewicz.
Acta Astron., Vol. 21, 479 - 485 (1971).

This paper points out the importance of relativistic corrections for the hard X-ray spectrum of a weakly relativistic plasma.

022.097 Atomic data of importance for ultra-violet and X-ray astronomy: A review of theory.
M. J. Seaton.
Highlights of Astronomy, Vol. 2, (see 012.017), 503 - 508 (1971).

022.098 Collision strengths for electron excitation of coronal ions. D. R. Flower.
Highlights of Astronomy, Vol. 2, (see 012.017), 512 - 517 (1971).

022.099 Relativistic contributions to energies of highly ionized atoms. R. Snyder.
Highlights of Astronomy, Vol. 2, (see 012.017), 544 - 548 (1971).

022.100 Relativistic corrections to atomic energy levels.
M. Jones.
Highlights of Astronomy, Vol. 2, (see 012.017), 549 - 554 (1971).

022.101 Magnetic multipole transition probabilities.
R. H. Garstang.
Highlights of Astronomy, Vol. 2, (see 012.017), 555 - 560 (1971).

022.102 Pressure broadening of UV lines.
H. van Regemorter.
Highlights of Astronomy, Vol. 2, (see 012.017), 561 - 565 (1971).

022.103 New results on electron broadening of some UV lines of N II, C II/IV and Si II/III/IV.
S. Sahal-Brechot, E. Segre.
Highlights of Astronomy, Vol. 2, (see 012.017), 566 - 574 (1971).

022.104 Atomic data of importance for ultraviolet and X-ray astronomy: Summary of the joint discussion.
A. G. Hearn.
Highlights of Astronomy, Vol. 2, (see 012.017), 580 - 583 (1971).

022.105 On the determination of properties of a synchrotron radiation source from its spectrum. V. N. Sazonov.
Astron. Zhurn. Akad. Nauk SSSR, Vol. 48, 1190 - 1194 (1971). In Russian. English translation in Soviet Astron. AJ, Vol. 15, No. 6.

It is shown that in the case of any distribution of the magnetic field in the source in value and directions and in the case of any distribution of radiating electrons in energy, the spectral index of the synchrotron radiation of the transparent source (i.e. source without absorption) cannot be changed with a too rapid frequency change.

022.106 Transition radiation as a secondary standard source in the VUV. W. Böhm, D. Labs.
Applied Optics, Vol. 10, 2021 - 2023 (1971).

The optical radiation caused by electron bombardment of metallic surfaces was tested for its use as a secondary standard source in the VUV. Aluminium of high purity was found to be a suitable target material.

022.107 Kalibrierung von Strahlungsquellen im Vakuum-Ultraviolett durch Anschluß an die Synchrotronstrahlung des Deutschen-Elektronen-Synchrotrons (DESY).
W. Böhm, D. Labs, D. Lemke, E. Pitz.
Bundesministerium für Bildung und Wissenschaft, Forschungsber. BMwF–FB W 69-09, [Available from Zentralstelle für Luftfahrtdokumentation und -information (ZLDI), Deutsche Forschungs- und Versuchsanstalt für Luft- und Raumfahrt, München. Price DM 22.47], 107 pp. (1969).

By means of the synchrotron radiation spectrographs or light sources — especially for astrophysical experiments — can be calibrated with an accuracy of ± 4 % in the vacuum-ultraviolet. In the spectral range from 1650 Å to 2700 Å gasdischarge lamps have been calibrated with this accuracy. Furthermore the transition radiation as a possible secondary standard light source is discussed.

022.108 Present state of the analysis of Nd I and Nd II.
J. Blaise, J. F. Wyart, R. Hoekstra, P. J. G. Kruiver.
Journ. Optical Soc. America, Vol. 61, 1335 - 1342 (1971).

022.109 Configurations $4f^{N-1}6s^26p$ in neutral gadolinium, dysprosium, erbium, and ytterbium. N. Spector.
Journ. Optical Soc. America, Vol. 61, 1350 - 1354 (1971).

022.110 Energies of the electronic configurations of the singly, doubly, and triply ionized lanthanides and actinides. L. Brewer.
Journ. Optical Soc. America, Vol. 61, 1666 - 1682 (1971).

022.111 **Energy differences between two spectroscopic systems in neutral, singly ionized, and doubly ionized lanthanide atoms.** W. C. Martin.
Journ. Optical Soc. America, Vol. 61, 1682 - 1686 (1971).

022.112 **Energy levels and mean lives of Cl II—Cl VII.** S. Bashkin, I. Martinson.
Journ. Optical Soc. America, Vol. 61, 1686 - 1692 (1971).

022.113 **One-electron spectrum of doubly ionized lutetium (Lu III) and nuclear magnetic dipole moment of ^{175}Lu.** V. Kaufman, J. Sugar.
Journ. Optical Soc. America, Vol. 61, 1693 - 1698 (1971).

022.114 **Acceleration of projectiles to hypervelocities.** J. G. Linhart.
Physics of high energy density. Course 48, Italian Phys. Soc., 1969, (see 012.023), p. 151 - 167 (1971).

022.115 **Zur Verbreiterung von Heliumlinien und der Wasserstofflinie Hβ durch Stark-Effekt der Mikrofelder.** H. J. Kusch.
Zeitschr. Naturforschung, Vol. 26a, 1970 - 1972 (1971).
Stark broadening of the neutral helium lines λ 5015 Å and λ 3889 Å was compared with the width of Hβ in a quasistationary plasma produced in helium-hydrogen mixtures of various composition. Electron densities derived from the broadening of Hβ differ from those determined from helium line profiles by a factor about 1.7.

022.116 **Atomic transition probabilities of the halogens.** R. D. Bengtson, M. H. Miller, D. W. Koopman, T. D. Wilkerson.
Phys. Rev. A, General Phys., Third Ser., Vol. 3, 16 - 24 (1971).
Absolute transition probabilities in the visible and near-infrared spectra of F I, Cl I, Cl II, and Br I have been measured using a gas-driven shock tube. Measured transition probabilities are compared with theoretical calculations and with other experimental values.

022.117 **Collision cross sections for the excitation of the Schumann-Runge dissociation continuum in molecular oxygen by 20—110-keV protons.** J. T. Park, F. D. Schowengerdt, D. R. Schoonover.
Phys. Rev. A, General Phys., Third Ser., Vol. 3, 679 - 684 (1971).
Absolute cross sections for the excitation of the Schumann-Runge dissociation continuum of molecular oxygen have been obtained from inelastic energy-loss spectra induced by 20—110-keV protons incident on gaseous targets of molecular oxygen. Apparent differential energy-loss cross sections, ionization cross sections, and total inelastic cross sections are also obtained from the energy-loss spectra.

022.118 **Stark broadening of H_β, H_γ, and H_δ: A comparison of theory and experiment.** R. A. Hill, J. B. Gerardo, P. C. Kepple.
Phys. Rev. A, General Phys., Third Ser., Vol. 3, 855 - 862 (1971).
A precision comparison is made between the Stark-broadened profiles of H_β, H_γ, and H_δ measured by Hill and Gerardo and the theoretical profiles calculated by Kepple and Griem. In addition, the effects of inelastic collisions between perturbing electrons and the radiating atom are investigated.

022.119 **Theory of relativistic magnetic dipole transitions: Lifetime of the metastable 2 ^3S state of the heliumlike ions.** G. W. F. Drake.
Phys. Rev. A, General Phys., Third Ser., Vol. 3, 908 - 915 (1971).
It has recently been established that the radiative lifetime of the metastable 2 ^3S state of helium and the heliumlike ions is determined by single-photon magnetic dipole (M1) transitions to the ground state, rather than the two-photon process proposed by Breit and Teller. The theory of $nl-n'l M$1 transitions with $n \neq n'$ is developed in the Pauli approximation and extended to two-electron systems. The results are compared with recent solar coronal observations by Gabriel and Jordan, and with a measurement of the 2 ^3S state lifetime in Ar XVII by Schmieder and Marrus.

022.120 **Measurements of lowest-S-state lifetimes of gallium, indium, and thallium.** M. Norton, A. Gallagher.
Phys. Rev. A, General Phys., Third Ser., Vol. 3, 915 - 927 (1971).
The lifetimes of the gallium 5 $^2S_{1/2}$ state, the indium 6 $^2S_{1/2}$ state, and the thallium 7 $^2S_{1/2}$ state were measured using the zero-field level-crossing (Hanle-effect) technique.

022.121 **Lifetimes of metastable CO and N_2 molecules.** W. L. Borst, E. C. Zipf.
Phys. Rev. A, General Phys., Third Ser., Vol. 3, 979 - 989 (1971).

022.122 **Stark and Zeeman effects on neutral-helium lines.** C. Deutsch, H. W. Drawin, L. Herman.
Phys. Rev. A, General Phys., Third Ser., Vol. 3, 1879 - 1890 (1971).
The effect of a static electromagnetic field on the $2P-nS$ and $2S-nP$ He I isolated lines is systematically investigated. Exact expressions are given for the shift of the allowed component, and the behavior of the corresponding transition probability under static perturbation is considered in detail. The usual hydrogenic approximations for the Stark constants are discussed with reference to the most recent values for the oscillator strengths available.

022.123 **Broadening of the sodium D lines by atomic hydrogen. An analysis in terms of the NaH molecular potentials.** E. L. Lewis, L. F. McNamara, H. H. Michels.
Phys. Rev. A, General Phys., Third Ser., Vol. 3, 1939 - 1948 (1971).
The interatomic forces between sodium and hydrogen atoms which are responsible for the broadening of the sodium D lines are discussed in terms of calculated interatomic potential curves for the NaH molecule. The importance of including overlap interactions and of considering both upper and lower states of the transitions is emphasized.

022.124 **Derivation of the blackbody radiation spectrum by classical statistical mechanics.** O. Theimer.
Phys. Rev. D, Particles and Fields, Third Ser., Vol. 4, 1597 - 1601 (1971).

022.125 **Collapsed nuclei.** A. R. Bodmer.
Phys. Rev. D, Particles and Fields, Third Ser., Vol. 4, 1601 - 1606 (1971).
We discuss the observational consistency, possible properties, and detection of collapsed nuclei C_A. The properties of C_A are discussed using composite baryon and quark models; small charges and hypercharges and, especially, neutral C_A are possible.

022.126 **Relativistic and realistic classical mechanics of two interacting point particles.** C. Fronsdal.
Phys. Rev. D, Particles and Fields, Third Ser., Vol. 4, 1689 - 1706 (1971).
Classical mechanics of two point particles interacting at a distance is given a Lorentz - covariant formulation without introducing unphysical degrees of freedom such as usually ac-

company the two-time formalism. The theory is then quantized and compared with quantum field theory to allow the determination of realistic potentials. Exact solutions are obtained for an inverse distance potential; classical orbits as well as quantum energy levels are determined.

022.127 Photon splitting and photon dispersion in a strong magnetic field.   S. L. Adler.
Ann. Physics, Vol. 67, 599 - 647 (1971).

We determine the refractive indices for photon propagation, and the absorption coefficient and polarization selection rules for photon splitting, in a strong constant magnetic field.

022.128 Atoms in superstrong magnetic fields.
R. G. Newton.
Phys. Rev. D, Particles and Fields, Third Ser., Vol. 3, 626 - 627 (1971).

It is pointed out that because of possibly large radiative corrections, the mass and magnetic moment of an electron in the presence of magnetic fields above 10^{13} G, which may be present in pulsars , are unknown.

022.129 Spectrum of high-energy electrons undergoing Klein-Nishina losses.   G. R. Blumenthal.
Phys. Rev. D, Particles and Fields, Third Ser., Vol. 3, 2308 - 2311 (1971).

022.130 Attempt to determine the elastic proton-nucleon cross section at 83 GeV.
E. R. Goza, E. G. Stafford.
Phys. Rev. D, Particles and Fields, Third Ser., Vol. 3, 2577 - 2581 (1971).

022.131 Balmer–α excitation by electron-impact excitation on atomic hydrogen.   H. Mahan, S. J. Smith.
Journ. Optical Soc. America, Vol. 61, 1587 - 1588 (1971).
Abstr. Optical Soc. America.

022.132 The interaction of hydrogen protons and atoms with molecules of H_2 and N_2.   V. V. Afrosimov, G. A. Leyko, Yu. A. Mamajev, M. N. Panov, N. V. Fedorenko.
Problems of cosmic physics. Vyp. (No.) 6, (see 003.040), p. 129 - 137 (1971). In Russian.

022.133 Energy distribution of electrons ejected in atomic collisions.
G. N. Ogurtsov, I. P. Flaks, S. V. Avakyan.
Problems of cosmic physics. Vyp. (No.) 6, (see 003.040), p. 138 - 142 (1971). In Russian.

022.134 Exothermal capture with ionization in collisions of He^{2+} ions.
Z. Z. Latypov, I. P. Flaks, A. A. Shaporenko.
Problems of cosmic physics. Vyp. (No.) 6, (see 003.040), p. 143 - 147 (1971). In Russian.

022.135 Multi-mode wave coupling in inhomogeneous anisotropic media – with some magneto – ionic applications.   O. E. H. Rydbeck.
Res. Lab. Electronics, Chalmers Univ. Technology, Gothenburg, Sweden, Res. Report No. 88, 5 + 287 pp. (1968).

This report deals with coupled waves, or coupled modes in arbitrary inhomogeneous n-wave systems. The partial waves, which make up the total wave field (of some physical kind) are expressed either in terms of so called inhomogeneous wave numbers, or of regular homogeneous ones (with corresponding phase integrals). A great number of coupled mode systems are analyzed, compared, and discussed in considerable detail, with applications for example to magneto-ionic, and to active (amplifying), inhomogeneous media. This report represents a major effort to relate all coupled wave phenomena to each other,

irrespective of their particular physical background; in short to present a unified theory of multi-mode couplings in inhomogeneous media.

022.136 Atomic spectra. References to analyses of atomic spectra; a critical compilation as of June 1970.
B. Edlén.
Beam-foil spectroscopy. Conference 1970, (see 012.032), p. 1 - 2 (1970).

022.137 Beam-foil spectroscopy as of June 1970.
S. Bashkin.
Beam-foil spectroscopy. Conference 1970, (see 012.032), p. 3 - 13 (1970).

022.138 Line identification problems in beam-foil spectroscopy.   M. Dufay.
Beam-foil spectroscopy. Conference 1970, (see 012.032). p. 15 - 23 (1970).

022.139 Atomic transition probabilities. A survey of our present knowkedge and future needs.   W. L. Wiese.
Beam-foil spectroscopy. Conference 1970, (see 012.032), p. 25 - 33 (1970).

022.140 Lifetimes and transition probabilities for some Fe II levels by the beam-foil method.
P. L. Smith, W. Whaling, D. L. Mickey.
Beam-foil spectroscopy. Conference 1970, (see 012.032), p. 47 - 50 (1970).

022.141 Mean life measurements of 4p and 4p$'$ levels in argon II using an image tube spectrograph.
G. E. Assousa, L. Brown, W. K. Ford, Jr.
Beam-foil spectroscopy. Conference 1970, (see 012.032), p. 51 - 54 (1970).

022.142 Mean life measurements in Na, Mg, Al, Si and K.
J. Bromander, H. G. Berry, R. Buchta.
Beam-foil spectroscopy. Conference 1970, (see 012.032), p. 55 - 58 (1970).

022.143 Relative intensities and transition probabilities of some $2s^2S_{1/2} - 2p^2P_{3/2}, _{1/2}$ transitions in the lithium isoelectronic sequence.
L. Barrette, E. J. Knystautas, B. Neveu, R. Drouin.
Beam-foil spectroscopy. Conference 1970, (see 012.032), p. 59 - 61 (1970).

022.144 A comparison of Ne II lifetimes obtained by several techniques.
J. H. Brand, C. L. Cocke, B. Curnutte, C. Swenson.
Beam-foil spectroscopy. Conference 1970, (see 012.032), p. 63 - 70 (1970).

022.145 Radiative mean lives of some electronic states in beam-foil excited atomic and ionic carbon.
D. J. Pegg, E. L. Chupp, L. W. Dotchin.
Beam-foil spectroscopy. Conference 1970 (see 012.032), p. 71 - 75 (1970).

022.146 Beam-foil spectroscopy at 1 MeV/nucleon energy: Preliminary results.
M. Dufay, A. Denis, J. Desesquelles.
Beam-foil spectroscopy. Conference 1970, (see 012.032), p. 85 - 91 (1970).

022.147 Radiative-lifetime measurements for ions of nitrogen and oxygen.   E. H. Pinnington.
Beam-foil spectroscopy. Conference 1970, (see 012.032), p. 93 - 102 (1970).

022.148 **New results of lifetime measurements in C, N. O, Ne and Ar ions.** P. Ceyzeriat, A. Denis, J. Desesquelles, M. Druetta, M. C. Poulizac.
Beam-foil spectroscopy. Conference 1970, (see 012.032), p. 103 - 108 (1970).

022.149 **A review of theoretical developments in atomic *f*-values.** A. W. Weiss.
Beam-foil spectroscopy. Conference 1970, (see 012.032), p. 121 - 131 (1970).

022.150 **Transition probabilities: New theory vs recent experimental results.**
C. Nicolaides, O. Sinanoğlu.
Beam-foil spectroscopy. Conference 1970, (see 012.032), p. 133 - 136 (1970).

022.151 **Theoretical transition probabilities and lifetimes in noble gas spectra.**
M. Aymar, S. Feneuille, M. Klapisch.
Beam-foil spectroscopy. Conference 1970, (see 012.032), p. 137 - 143 (1970).

022.152 **Transition probabilities and radiative lifetimes for Ne II.** B. F. J. Luyken.
Beam-foil spectroscopy. Conference 1970, (see 012.032), p. 145 - 147 (1970).

022.153 **Relativistic Hartree-Fock-Slater oscillator strengths for Tl.** C. P. Bhalla.
Beam-foil spectroscopy. Conference 1970, (see 012.032), p. 149 - 155 (1970).

022.154 **Absolute transition probabilities in Fe I.**
W. Whaling, M. Martinez-Garcia, D. L. Mickey, G. M. Lawrence.
Beam-foil spectroscopy. Conference 1970, (see 012.032), p. 363 - 368 (1970).

022.155 **H⁻ as a member of the He iso-electronic sequence.**
W. van Rensbergen.
Diss. Rijksuniv. Utrecht, 75 pp. (1970).

022.156 **Experimental oscillator strengths and the solar abundances of iron and nickel.**
T. Garz, H. Heise, H. Holweger, J. Richter.
Second Conference on atomic spectroscopy. Hannover, Germany, 1970, 4 pp. – See Phys. Abstr., Vol. 74, No. 46309 (1971).

022.157 **Microwave spectrum, vibration-rotation interaction, and ring puckering vibration in silacyclobutane and silacyclobutane-1,1-D$_2$.** W. C. Pringle, Jr.
Journ. Chem. Phys., Vol. 54, 4979 - 4988 (1971).
Laboratory measurements of lines of complex molecules containing C, H and Si. Line frequencies range from 8 to 55 GHz. – *RAB*

022.158 **Electron resonance of vibrationally excited OH radicals.**
K. P. Lee, W. G. Tam, R. Larouche, G. A. Woonton.
Canadian Journ. Phys., Vol. 49, 2207 - 2210 (1971).

022.159 **Stark effect and hyperfine structure of HCN measured with an electric resonance maser spectrometer.** H. E. Radford, C. V. Kurtz.
Journ. Res. National Bureau of Standards, Section A, Vol. 74A, 791 - 799 (1970).
The 449 MHz L-type doubling spectrum of HCN has been measured to high precision. The electric dipole moment of the V$_2$ = 1 state of HCN is found to be 2.94 Debye. – *RXM*

022.160 **Microwave spectrum, structure, dipole moment and quadrupole coupling constants of cis- and trans-nitrous acids.**
A. P. Cox, A. W. Brittain, D. J. Finnigan.
Trans. Faraday Soc., Vol. 67, 2179 - 2194 (1971).
Calculated and measured spectral constants are listed.

Observations of a high-altitude barium cloud.
Sky Telescope, Vol. 42, 382 - 386 (1971).

Instruments and Astronomical Techniques

031 Optics, Methods of Observation and Reduction

031.001 **Permanent control of the stigmatic point in the focal plane of a Cassegrain telescope.** J. Rösch.
Astron. Astrophys., Vol. 14, 143 - 153 (1971). In French.

A solution is proposed for adjusting the stigmatic point of the field onto the mechanical center of the tail-part of the telescope by an optical control which does not require pointing on a star, and thus can be performed at any time. It is based on the fact that two conditions have to be fulfilled, a) that the first focus of the hyperboloid coincides with the focus of the paraboloid, and b) that the point of the focal plane where the image is observed is on the revolution axis of the hyperboloid.

031.002 **Further development and properties of the spectral analysis by least-squares.** P. Vaníček.
Astrophys. Space Sci., Vol. 12, 10 - 33 (1971).

The concept of spectral analysis using least-squares is further developed to remove any undesired influence on the spectrum. The influence of such a 'systematic noise' can be eliminated without the necessity of knowing the magnitudes of the noise constituents. The technique can be used for irregularly spaced as well as equidistantly spaced data.

031.003 **Schwierige Objekte für Astrophotographen.**
H. Vehrenberg.
SuW, Vol. 10, 200 (1971).

031.004 **Surface quality estimates of large astronomical image details by the thread and slit method.**
I. Ya. Bubis.
Optiko-mekh. prom-st', 1971, No. 1, p. 7 - 117. In Russian.
Abstr. in Referativ. Zhurn. 51. Astron., 8.51.104 (1971).

031.005 **Problems involved in designing a system for minimizing stresses in astronomical mirrors.**
L. L. Voskresenskij.
Izv. vyssh. uchebn. zavedenij. Geod. i aehrofotos"emka, 1970, No. 2, p. 132 - 139. In Russian. – Abstr. in Referativ. Zhurn. 51. Astron., 8.51.106 (1971).

031.006 **Korrektoren zu Teleskop-Systemen.**
E. Wiedemann.
Orion, 29. Jahrgang, p. 83 - 85 (1971).

031.007 **On the influence of the background on the photographic star image.**
V. N. Sincheskul, B. F. Sincheskul.
Astrometriya i Astrofiz., *Kiev*, Vyp. (No.) 12, (see 003.003), p. 44 - 50 (1971). In Russian.

031.008 **On the transportation of high speed astronomical films in dry ice.** O. D. Dokuchaeva.
Astron. Tsirk., No. 613, p. 6 - 8 (1971). In Russian.

031.009 **Artificial star.** M. I. Malyshev.
Astron. Tsirk., No. 624, p. 5 - 7 (1971). In Russian.

031.010 **Reducción automática de posiciones de estrellas.**
M . J. Sevilla.
Urania Barcelona, Año 55, Nos. 271–272, p. 129 - 151 (1970).

031.011 **A radio astronomical method of measuring distances between continents and of clock comparisons.**
V. S. Troitskij.
Izv. vyssh.uchebn. zavedenij.Geod. i aehrofotos"emka, 1970, No. 1, p. 112 - 120. In Russian. – Abstr. in Referativ. Zhurn. 51. Astron., 9.51.141(1971).

031.012 **Geometrische Verbesserung des ausschliesslich auf Reflexion basierenden Schmidt-Teleskops.**
L. Epstein.
Orion, 29. Jahrgang, p. 141 - 142 (1971).

031.013 **The high-resolution filtration of signals of complicated form.** S. S. Nenjukov.
Solnechnye Dannye 1971 Byull., No. 6, p. 105 - 110 (1971). In Russian.

Conditions of registration of signals of complicated form with best resolution are determined. The influence of fluctuation noises was taken into account. The properties of a smoothing RC-filter are considered. The possibility of constructing a high-resolution adaptive RC-filter is shown.

031.014 **Stonyhurst disks for planets.** A. C. Gilmore.
Southern Stars, Vol. 24, 44 - 46 (1971).

031.015 **Technical problems of infrared observations.**
G. V. Khozov.
Trudy Astron. Obs., *Leningrad,* Vol. 28 (= Uchenye Zapiski Leningr. Un-ta, No. 359 = Seriya Matem. Nauk, vyp. (No.) 47), p. 39 - 46 (1971). In Russian.

Technical problems of observations in the infrared region are considered. Methods of registration and their accuracy are discussed.

031.016 **On the generalization of the theory of faint images in astrophotography.** H. Eelsalu.
Publ. Tartu Astrofiz. Obs., Vol. 39, 157 - 162 (1971). In Russian.

031.017 **Automated analysis of astronomical spectra.**
R. B. Hutchison.
Astron. Journ., Vol. 76, 711 - 718 (1971).

A description is given of a computer program which automates the analysis of high-resolution, infrared astronomical spectra. Procedures for the detection of spectral features, and for the determination of accurate line frequencies, line depths, and equivalent widths are presented. Line profile analysis, identification, and other specialized operations are discussed.

031.018 **Optik für Astro-Amateure.** E. Wiedemann.
Orion, 29. Jahrgang, p. 153 - 157 (1971).

031.019 **Impulstaellingsteknik anvendt i fotoelektrisk fotometri.** R. F. Nielsen.
Astron. Tidsskr., Årg. 4, 49 - 55 (1971).

031.020 **Pairs of spherical mirrors as field correctors for paraboloid mirrors.** N. J. Rumsey.
Proc. Astron. Soc. Australia, Vol. 2. 22 - 23 (1971).

031.021 **Holographically made zone plates for use in X-ray-telescopes.** D. Rudolph, G. Schmahl.
IAU Symposium No. 41, (see 012.005), p. 205 - 206 (1971).

031.022 **Construction of apodised zone plates for solar X-ray image formation.**
J. H. Dijkstra, W. de Graaff, L. J. Lantwaard.
IAU Symposium No. 41, (see 012.005), p. 207 - 210 (1971).

031.023 **Review of methods of intensity calibration in the spectral range 10–4000 Å.** R. W. P. McWhirter.
IAU Symposium No. 41, (see 012.005), p. 369 - 385 (1971).

031.024 **Absolute UV calibration of rocket photometers used to up-date the OAO calibration.** A. Gaide.
IAU Symposium No. 41, (see 012.005), p. 386 - 389 (1971).

031.025 **The wall-stabilized hydrogen arc as a radiation standard in the vacuum UV.** W. L. Wiese.
IAU Symposium No. 41, (see 012.005), p. 390 (1971).
Abstract.

031.026 **The transition-radiation as a light-source in the VUV.** W. Böhm, D. Labs.
IAU Symposium No. 41, (see 012.005), p. 391 (1971).
Abstract.

031.027 **Use of synchrotron radiation from an electron storage ring as an absolute standard of radiant flux for wavelengths from 1100 to 3000 Å.** E. T. Fairchild.
IAU Symposium No. 41, (see 012.005), p. 392 (1971).
Abstract.

031.028 **Attempts to observe the absolute intensity and the centre-to-limb variations of the sun in the vacuum ultraviolet region.** K. Nishi, Z. Suemoto.
IAU Symposium No. 41, (see 012.005), p. 393 - 397 (1971).

031.029 **Electronic optical astronomy: Philosophy and practice.** E. W. Dennison.
Science, Vol. 174, 240 - 244 (1971).

031.030 **Über das geometrische Spektrum gekreuzter Halbprismenpaare. I. Geometrisch und optisch gleichartige Halbprismen in nicht-gekreuzter Lage.** J. Junkes.
Ric. Spettrosc., Lab. Astrofis. Specola Vaticana, *Castel Gandolfo*, Vol. 3, (No. 7), 265 - 378 (1970).
Crossed prisms, especially when of equal geometric and optical qualities, give a dispersion which can be continuously varied, and therefore can be used as predispersers of variable dispersion for grating spectrographs. A comprehensive description of the spectra produced by such arrangements of crossed prisms is still needed. The present work is a contribution to the study of these spectra, and deals with the dispersion of the geometric spectrum of pairs of half-prisms of the Herschel-Risley and the Young-Thollon type, the elements of which are of equal geometric and optical characteristics. Part I handles the case when these units are used in the normal position.

031.031 **Sur une méthode d'exploitation des images d'étoiles doubles obtenues au moyen de la caméra électronique.** J. Rösch.
Comptes Rendus Acad. Sci. Paris, Sér. B, Vol. 273, 876 - 879 (1971).
Le but de la présente note est de montrer que la méthode n'exige pas l'emploi de lumière cohérente, et de préciser de façon concrète les conditions dans lesquelles elle peut être mise en œuvre efficacement.

031.032 **Investigation of three objectives "Industar-52".**
R. S. Osherov.
Byull. Inst. Astrofiz., *Dushanbe,* No. 57, p. 36 - 42 (1970).
In Russian.

031.033 **Ein Verfahren zur automatischen Auswertung von Sterndurchgangsbeobachtungen mit einem Passageninstrument.** S. Heitz, H. Walter.
Nachr. Karten- und Vermessungswesen, Ser. I, No. 50, p. 15 - 29 (1971).
This report describes a procedure for the automatic recording and evaluation of transit observations for the determination of astronomical longitudes and azimuths as developed and used by the Institut für Angewandte Geodäsie, Frankfurt a. M.

031.034 **A method of computing periods of cyclic phenomena.** I. Jurkevich.
Astrophys. Space Sci., Vol. 13, 154 - 167 (1971).
Use of an analysis of expected mean square deviations to search for periodicities in an observational data sample is described. The statistic for testing the null hypothesis of non-periodicity is derived from a partitioning of the total sum of squared deviations from the mean. The method is illustrated by numerical examples.

031.035 **The optical effects of a double calcite plate on a converging beam of light.** P. W. J. L. Brand.
Optica Acta, Vol. 18, 403 - 413 = Commun. Roy. Obs. Edinburgh, No. 105 (1971).
This work was undertaken to analyse possible systematic and random errors in a method of magnitude calibration for star field photographs taken with a Schmidt telescope. The aberrations and variation in transmission of a beam of light converging to a focus and passing through plates of calcite have been investigated, in order to estimate possible photometric errors due to these effects when a crossed calcite plate is used to calibrate photographic plates.

031.036 **Das atmosphärische Spektrum und seine Beseitigung.** B. Wedel.
SuW, Vol. 10, 339 - 341 (1971).

031.037 **Experiments with digital processing of stellar spectrograms.** W. K. Bonsack.
Publ. Astron. Soc. Pacific, Vol. 83, 602 (1971). – Abstr. Astron. Soc. Pacific.

031.038 **Technologie eines dünnen 32 cm–Teleskopspiegels.** H. Koberger.
Sternenbote, 14. Jahrgang, p. 158 - 163 (1971).

031.039 **How to built an amateur telescope.** L. Newelski.
Urania Kraków, Vol. 42, 249 - 253 (1971).
In Polish.

031.040 **Coherent processing and depth of focus of annular aperture imagery.** J. T. McCrickerd.
Applied Optics, Vol. 10, 2226 - 2230 (1971).
Image from high aspect ratio annulus are spatially filtered in a coherent optical processor for equalization of the modulation transfer function. The imagery is relatively insensitive to spherical aberation, field curvature, and longitudinal color; and the depth of focus is substantially greater than that of conventional imagery with comparable resolution.

031.041 **Jones's matrix representation of optical instruments. I: Beam splitters.** A. L. Fymat.
Applied Optics, Vol. 10, 2499 - 2505 (1971).
A general method is provided for constructing Jones's reflection and transmission matrices of any beam splitter. Derivations are presented for the various known configurations.

The method uses Abelès's matrices and pays special consideration to the different expressions of Jones's matrices relative to the various beams in an interferometric arrangement. The reversibility of the beam splitter in its action on the amplitude or phase, or both, of an incident light is studied.

031.042 **Application of the equidensity method to photographic photometry of star images.**
S. B. Vladimirov.
Astron. Tsirk., No. 651, p. 5 - 7 (1971). In Russian.

031.043 **1971 - 1972: Guide to scientific instruments.**
Compiled by B. J. Sheffer (Assistant Editor).
Science, Vol. 174A, No. 4010A, 184 + 130A pp. (1971).

031.044 **Introduction of corrections for irregularities of the figure of pivots into results of astronomical observations.** V. A. Sidorov.
Geod. i kartografiya, 1971, No. 6, p. 30 - 32. In Russian.
Abstr. in Referativ. Zhurn. 52. Geod. Aehros"emka, 12.52.104 (1971).

031.045 **Objektivprismenaufnahmen mit prismatischer Schmidt-Platte.** S. Marx, W. Pfau, N. Richter.
Jenaer Rundschau, (Jena Review), 16. Jahrgang, p. 294 - 298 (1971).

031.046 **Device and method of registration of star transit moments in the case of multiple contacts.**
K. Šteins, M. Ogriņš.
Latv. ordena trud. krasn. znameni gos. univ. im. P. Stuchki, Uch. zap., Vol. 148, vyp. (No.) 6, p. 3 - 11 (1971). In Russian.
A new principle is described for deducing the value of the mean moment of a periodic signal in the case of multiple contacts.

031.047 **The influence of lateral movements of stellar image on the precision of transit moments.** R. Kalniņa.
Latv. ordena trud. krasn. znameni gos. univ. im. P. Stuchki, Uch. zap., Vol. 148, vyp. (No.) 6, p. 12 - 28 (1971). In Russian.

031.048 **Formula for the standard deviation of star transit moments depending upon movement of images.**
K. Šteins.
Latv. ordena trud. krasn. znameni gos. univ. im. P. Stuchki, Uch. zap., Vol. 148, vyp. (No.) 6, p. 29 - 38 (1971). In Russian.

031.049 **The influence of brightness scintillation on transit moments of a star.** R. Kalniņa.
Latv. ordena trud. krasn. znameni gos. univ. im. P. Stuchki, Uch. zap., Vol. 148, vyp. (No.) 6, p. 39 - 47 (1971). In Russian.

031.050 **On the ideal equipment for registration of star transit moments.**
K. Šteins, R. Kalniņa, P. Rozenbergs, O. Judrups.
Latv. ordena trud. krasn. znameni gos. univ. im. P. Stuchki, Uch. zap., Vol. 148, vyp. (No.) 6, p. 48 - 59 (1971). In Russian.

031.051 **On the calculation of the standard deviation of star transit moments.** R. Kalniņa.
Latv. ordena trud. krasn. znameni gos. univ. im. P. Stuchki, Uch. zap., Vol. 148, vyp. (No.) 6, p. 60 - 66 (1971). In Russian.

031.052 **Photoelectric registration of star transits and spectra of observed stars.** L. Roze.
Latv. ordena trud. krasn. znameni gos. univ. im. P. Stuchki,

Uch. zap., Vol. 148, vyp. (No.) 6, p. 67 - 72 (1971). In Russian.
A small dependence of the lag of photoelectric registration of a star transit on the spectrum has been stated discussing the observations of the Latvian State University time service during the years 1968 - 1970.

031.053 **Automation and observations with a photoelectric zenith tube.** M. Ābele.
Latv. ordena trud. krasn. znameni gos. univ. im. P. Stuchki, Uch. zap., Vol. 148, vyp. (No.) 6, p. 73 - 86 (1971). In Russian.
The automatic driving system of the photoelectric zenith tube is discussed. Observations are made automatically without presence of an observer.

031.054 **A program for the computer BESM-3M for the treatment of photographic observations of extended objects.** V. S. Matjagin.
Trudy Astrofiz. Inst., *Alma-Ata,* Vol. 16, 38 - 64 (1971). In Russian.

031.055 **Identification of spectral lines by a computer.**
I. D. Kupo.
Trudy Astrofiz. Inst., *Alma-Ata,* Vol. 17, 26 - 32 (1971). In Russian.
A program for the identification of spectral lines by a computer has been prepared. The program allows to determine the wave-lengths of lines and equivalent widths for photometric measurements carried out with an intensity microphotometer as well as with a usual microphotometer.

031.056 **Comparison of different empirical formulae of characteristic curves.** V. S. Matyagin.
Trudy Astrofiz. Inst., *Alma-Ata,* Vol. 17, 33 - 39 (1971). In Russian.
Results of the application of different empirical formulas for the characteristic curves of extended cosmical objects are given.

031.057 **Beobachtungsobjekte für kleine und mittlere Fernrohre.** W. Meyer.
Veröff. Wilhelm-Foerster-Sternw. Berlin, No. 27, 38 pp. (1971).

031.058 **Determining the angular diameter of a small luminous coherent disk through the diffracted amplitude on axis of imaging lens.** G. Dufresne, A. Boivin.
Journ. Optical Soc. America, Vol. 61, 1579 (1971). — Abstr. Optical Soc. America.

031.059 **Double-option technique for testing large astronomical mirrors.** J. E. Simmons, I. Ghozeil.
Journ. Optical Soc. America, Vol. 61, 1586 (1971). — Abstr. Optical Soc. America.

031.060 **Fabrication and test of a lightweight 1.2-m-diameter $f/2$ parabola.**
W. H. Augustyn, Jr., R. R. Rigby, J. Vrabel.
Journ. Optical Soc. America, Vol. 61, 1586 (1971). — Abstr. Optical Soc. America.

031.061 **Wave-aberration evaluation and tolerancing of optimized extreme-ultraviolet glancing-incidence telescopes.** J. D. Mangus.
Journ. Optical Soc. America, Vol. 61, 1586 (1971). — Abstr. Optical Soc. America.

031.062 **Phase modulation in far infrared (submillimetre-wave) interferometers. I—Mathematical formulation.**

J. Chamberlain.
Infrared Phys., Vol. 11, 25 - 55 (1971).

Details of the theory of phase modulation are given for two-beam interferometers and compared with the more common amplitude modulation. The application to Fourier spectrometry is discussed in detail and the forms of the transmission and refraction spectra are given. The effects of noise are considered.

031.063 **Phase modulation in far infrared (submillimetre-wave) interferometers II—Fourier spectrometry and Terametrology.** J. Chamberlain, H. A. Gebbie.
Infrared Phys., Vol. 11, 57 - 73 (1971).

The advantages of phase modulation in astronomical spectrometry have been revealed in the work of Connes et al. in the near infrared; our purpose is to demonstrate more generally the benefits expected in a range of far infrared applications.

031.064 **On the determination of the instrumental parameters in differential observations of stars by computers.** L. L. Vagushchenko, V. A. Sinyaev.
Astron. Tsirk., No. 654, p. 2 - 4 (1971). In Russian.

031.065 **Increasing of signal-to-noise ratio by digital filtering in spectrophotometry.**
N. I. Grachev, A. V. Soloviev.
Astron. Tsirk., No. 659, p. 1 - 3 (1971). In Russian.

031.066 **L'immagine di coma al primo fuoco del telescopio OAN di 3.5 m.** C. Barbieri.
Mem. Soc. Astron. Italiana, Nuova Ser., Vol. 42, 635 - 637 (1971).

031.067 **Recherches portant sur des mètres à bouts.**
Y. Väisälä.
Ann. Acad. Sci. Fennicae, Ser. A, VI. Phys. 368, 6 pp. = Astron.-Optika Inst. Univ. Turku, Informo No. 36 (1971).

031.068 **The origin and goals of the automated stellar proper motion survey.** W. J. Luyten.
Conference on photographic astrometric technique, Tampa 1968, (see 012.031), p. 1 - 3 (1971).

031.069 **The automation of the stellar proper motion survey.**
J. S. Newcomb.
Conference on photographic astrometric technique, Tampa 1968, (see 012.031), p. 3 - 13 (1971).

031.070 **Computer controlled precision digitizers of optical image data extraction.** R. C. Strand.
Conference on photographic astrometric technique, Tampa 1968 (see 012.031), p. 15 - 23 (1971).

031.071 **The extraction of accurate coordinates of images on photographic plates by means of a scanning type measuring machine.** B. E. Ross.
Conference on photographic astrometric technique, Tampa 1968, (see 012.031), p. 55 - 59 (1971).

031.072 **Improvements in Ross type astrometric objectives.**
J. G. Baker.
Conference on photographic astrometric technique, Tampa 1968, (see 012.031), p. 61 - 84 (1971).

031.073 **Comparison of photogrammetric and astrometric data reduction results for the Wild BC-4 camera.**
D. H. Hornbarger, I. I. Mueller.
Conference on photographic astrometric technique, Tampa 1968, (see 012.031), p. 85 - 97 (1971).

031.074 **Automatic plate mensuration and reduction at**

Aeronautical Chart and Information Center.
T. O. Seppelin.
Conference on photographic astrometric technique, Tampa 1968, (see 012.031), p. 109 - 117 (1971).

031.075 **The reduction of the National Aeronautics and Space Administration plates at the New Mexico State University.** W. H. Haas.
Conference on photographic astrometric technique, Tampa 1968, (see 012.031), p. 119 - 122 (1971)

031.076 **Observations and reduction techniques of the U.S. Coast and Geodetic Survey.** H. H. Schmid.
Conference on photographic astrometric technique, Tampa 1968, (see 012.031), p. 133 - 139 (1971).

031.077 **An F1 Schmidt satellite camera and the methods of plate measurement and reduction.** J. Hewitt.
Conference on photographic astrometric technique, Tampa 1968, (see 012.031), p. 141 - 154 (1971).

031.078 **Plate measurement techniques and reduction methods used by the West German satellite observers, and resulting consequences for the observation.** H. Deker.
Conference on photographic astrometric technique, Tampa 1968, (see 012.031), p. 155 - 160 (1971).

031.079 **Photographic astrometry and overlap reduction techniques.** W. D. Googe, C. F. Lukac.
Conference on photographic astrometric technique, Tampa 1968, (see 012.031), p. 209 - 218 (1971).

031.080 **The overlap methods.**
P. Lacroute, with an appendix on "Complements to theoretical studies on overlap methods", by P. Lacroute, A. Valbousquet.
Conference on photographic astrometric technique, Tampa 1968, (see 012.031), p. 219 - 238 (1971).

031.081 **Inversion of very large matrices encountered in large scale problems of photogrammetry and photographic astrometry.** D. C. Brown.
Conference on photographic astrometric technique, Tampa 1968, (see 012.031), p. 249 - 263 (1971).

031.082 **Optical design for beam-foil experiments.**
G. S. Bakken, J. A. Jordan, Jr.
Beam-foil spectroscopy. Conference 1970, (see 012.032), p. 181 - 185 (1970).

031.083 **Planetary imaging and topographic mapping by radar interferometry.** A. E. E. Rogers, S. H. Zisk.
IEEE International Convention Digest 1971, (IEEE, New York), p. 110 - 111. — See Phys. Abstr., Vol. 74, No. 46353 (1971).

031.084 **Image error estimation for X-ray telescope.**
H. Wolter.
Optica Acta (GB), Vol. 18, 425 - 429 (1971). In German.

031.085 **The infrared universe.** A. Tucker.
Electronics Australia, Vol. 33, No. 5, p. 12 - 13 (1971).

The article describes the aspects of stars that can be seen from the earth's surface using infrared techniques. It also describes briefly some of these methods, and the telescopes used.

031.086 **Testing concave telescope mirrors.** F. L. Leroux.
Journ. Astron. Soc. Western Australia, Vol. 33, December, p. 6 - 11 (1971).

031.087 Ein Redundanzreduktionsverfahren für die Regi-
strierung der solaren Radioburststrahlung.
G. Zimmermann.
Zeitschr. Angew. Phys., Vol. 30, 370 - 376 (1971).

The usefulness of a flying spot digitizer for the
measurement of the coordinates of star images on photo-
graphic plates.   See Abstr. 034.124.

Experiments with the digital reduction of stellar
spectrograms.   See Abstr. 114.079.

032 Astronomical Instruments

032.001 Le futur grand télescope soviétique.
L'Astronomie, 85ᵉ année, p. 345 - 346 (1971).
Short note on a big telescope of Zelenchukskaya (Northern
Caucasus).

032.002 Home-made photoheliograph.   Yu. A. Grishin.
Zemlya i Vselennaya, No. 4, p. 66 - 67 (1971).
In Russian.

032.003 An investigation of the possibility to use the Abas-
tumani Observatory 70-cm meniscus telescope for
astrometric works.   A. Sh. Khatisov.
Byull. Abastumansk. Astrofiz. Obs., No. 40, p. 185 - 206
(1971). In Russian.

032.004 A working model of a 70-inch telescope.
M. Kaufman.
Sky Telescope, Vol. 42, 170 - 173 (1971).

032.005 Solar instrumentation (Part I).   J. W. Evans.
National Bureau Standards Special Publ. 353,
(see 012.001), p. 61 - 70 (1971).

032.006 Solar instrumentation (Part II).   R. B. Dunn.
National Bureau Standards Special Publ. 353, (see
012.001), p. 71 - 83 (1971).

032.007 Das 1,2-m-Teleskop für das Max-Planck-Institut für
Astronomie.   R. Schlegelmilch.
SuW, Vol. 10, 268 - 270 (1971).

032.008 Analysis of telescope costs.   L. Mertz.
Bull. American Astron. Soc., Vol. 3, 386 (1971).
Abstr. AAS.

032.009 Figure of merit definition for centrally obscured
orbiting large stellar telescopes.
W. N. Peters, A. B. Wissinger, B. M. Boyce.
Bull. American Astron. Soc., Vol. 3, 386 - 387 (1971).
Abstr. AAS.

032.010 Progress on the 150-inch Anglo-Australian telescope.
H. C. Minnett.
Proc. Astron. Soc. Australia, Vol. 2, 2 - 6 (1971).

032.011 Utilization of Zvenigorod's large satellite camera for
tracking deep space probes.   A. Losinsky.
Space Research XI, (see 012.004), 113 (1971).

032.012 A satellite experiment to measure the intensity and
the energy spectrum of gamma rays from solar
flares in the range 50–500 MeV.
G. F. Bignami, C. J. Bland, O. Citterio, A. J. Dean, P. Inzani.

IAU Symposium No. 41, (see 012.005), p. 44 (1971).
Abstract.

032.013 Glancing incidence optics for X-ray and ultraviolet
astronomy.
J. H. Underwood, W. M. Neupert, R. B. Hoover.
IAU Symposium No. 41, (see 012.005), p. 192 - 204 (1971).

032.014 Test: Heidenhain-Spiegelfernrohr 150/750/3400.
G. D. Roth.
SuW, Vol. 10, 311 - 312 (1971).

032.015 El mecanismo de relojeria del telescopio cancela.
L. Hordij.
Rev. Astron., Vol. 42, No. 174, p. 23 - 26 (1971).

032.016 De 1.52 meter-spiegeltelescoop van de Universiteits-
Sterrenwacht van Wenen.   A. Greve.
Hemel en Dampkring, Vol. 69, 320 - 324 (1971).

032.017 Telescope drives and guidance by stepping motors.
D. Clarke.
Observatory, Vol. 91, 215 - 217 (1971).
Stepping motors have now been on the market for several years but, as far as the writer is aware, their capabilities have
not been explored for use in providing telescope drive systems.
A simple pulsed drive was designed to operate the Grubb Parsons 510 mm reflector at Glasgow and has now been in use for
two years. The arrangement has been very reliable and indicates that the system has a potential allowing for various degrees of elaboration.

032.018 Astronomical researches carried out with the Zeiss
instruments of the Vatican Observatory.
D. J. K. O'Connell.
Specola Vaticana, Comun. No. 51, 9 pp. (1971). – Original
English text of an article published in the "Jena Review"
(Zeiss-Jena), 15. Jahrgang, p. 326 - 329 (1970).

032.019 An Australian 12¹/₂-inch Buchroeder relay telescope.
A. E. Coombs.
Sky Telescope, Vol. 42, 302 - 308 (1971).

032.020 Neue Beobachtungsstation für Satelliten in Berlin.
R. Lukas.
Orion Schaffhausen, 29. Jahrgang, p. 177 - 179 (1971).

032.021 Stellaraufnahmen mit 25 cm Newton-Teleskop.
K. Rihm.
Orion Schaffhausen, 29. Jahrgang, p. 179 - 181 (1971).

032.022 Bau einer Sternwarte mit Polyester-Kuppel.
K. Oechslin.

Orion Schaffhausen, 29. Jahrgang, p. 188 - 190 (1971).

032.023 **Investigation of the adjustment of Belgrade Vertical Circle tube.** A. S. Kharin.
Bull. Obs. Astron. Beograd, Vol. 28, (No. 123), 23 - 30 (1970).

032.024 **Belgrade large vertical circle examination by auto-collimational method.** G. Teleki.
Bull. Obs. Astron. Beograd, Vol. 28, (No. 123), 31 - 51 (1970).

By the autocollimational method, the stability of the tube's optical and mechanical unity of the Belgrade large vertical circle has been examined under various conditions. These examinations indicated that the general characteristics of the tube are changeable, in the function of the temperature factors chiefly, and that the same may have a great influence on the observational values.

032.025 **Corrections à courte période du cercle méridien et les méthodes de leur détermination.**
D. Šaletić, S. Sadžakov.
Bull. Obs. Astron. Beograd, Vol. 28, (No. 124), 117 - 136 (1970).

Du février au juillet 1968 nous avons déterminé les corrections du cercle méridien Askania N° 88077 (190/2580) de l'Observatoire de Beograd à chaque demi-degré. Nous avons obtenu 240 mesures indépendantes desquelles on peut former des équations de corrections à courte période. Notre résultats sont donnés dans une table.

032.026 **Ondrejov 2 m telescope.**
J. Grygar, P. Koubský.
Bull. d' Information, Ass. Développement International Obs. Nice, No. 8, p. 5 - 10 (1971).

Since August 1967 the telescope has been in use by the Astronomical Institute of the Czechoslovak Academy of Sciences at Ondrejov Observatory.

032.027 **Design and operation of the NASA 91.5-cm airborne telescope.**
R. M. Cameron, M. Bader, R. E. Mobley.
Applied Optics, Vol. 10, 2011 - 2015 (1971).

A 91.5-cm aperture telescope is being built for ir and submillimeter observations altitudes of 12 km to 14 km aboard a Star Lifter (Lockheed C-141 A) aircraft. The main optics will be totally reflecting, and aerodynamic boundary layer control will permit open-port operation (no material window). The elevation will be adjustable in flight between 35° and 75°.

032.028 **Das Jena-Astrolab: ein neues Feldinstrument zur Ortsbestimmung und Erkundung.**
G. Hemmleb, H. U. Sandig.
Jenaer Rundschau, (Jena Review), 16. Jahrgang, p. 302 - 304 (1971).

032.029 **An experience of using a rapid Schmidt camera.**
Sh. N. Sabitov.
Trudy Astrofiz. Inst., *Alma-Ata*, Vol. 16, 32 - 37 (1971). In Russian.

Some features and optical properties of a Schmidt camera with aperture 1 : 1 are described. The high effectivity of this instrument for photographing weak nebulae is shown.

032.030 **Triaxial instrument for photoelectric observations of artificial earth satellites during their entry into the shadow of the earth.** P. N. Boiko, M. I. Musorin.
Trudy Astrofiz. Inst., *Alma-Ata*, Vol. 17, 101 - 104 (1971). In Russian.

An instrument for photoelectric observations of artificial earth satellites is described. Observations carried out by means of this instrument enable the observer to estimate the ozone and atmospheric aerosol distribution relative to height.

032.031 **Self-made refractor.** A. K. Klejn.
Zemlya i Vselennaya, No. 6, p. 60 - 61 (1971). In Russian.

032.032 **Universal refractor.** A. N. Pod"yapol'skij.
Zemlya i Vselennaya, No. 6, p. 62 - 65 (1971). In Russian.

032.033 **Magnitude equation for the 400-mm astrograph.** A. B. Onegina, N. I. Solyanik.
Astrometriya i Astrofiz., *Kiev*, No. 13, (see 003.039), p. 81 - 83 (1971). In Russian.

032.034 **Extending the stellar field of view of Ritchey-Chrétien telescopes.** S. Rosin, M. Amon.
Journ. Optical Soc. America, Vol. 61, 1581 (1971). – Abstr. Optical Soc. America.

032.035 **Adaptation of the Schupmann medial telescope to a large-scale astronomical optical system.**
J. J. Villa.
Journ. Optical Soc. America, Vol. 61, 1581 (1971). – Abstr. Optical Soc. America.

032.036 **New instruments at the Skalnaté Pleso Observatory.** J. Sýkora, P. Bendík.
Říše hvězd, Vol. 52, 232 - 233 (1971). In Slovak.

032.037 **First observations with the 400-mm reflector at Mt. Maidanak.** M. I. Tertitzky.
Astron. Tsirk., No. 666, p. 7 - 8 (1971). In Russian.

032.038 **Problems in the development of a mirror transit telescope at Ottawa.** R. W. Tanner.
Publ. Earth Phys. Branch, Dep. of Energy, Mines and Resources, Ottawa, Canada, Vol. 41, (No. 8), 137 - 143 (1971).

In order to assist other observatories working on improvements in meridian circle techniques, some of the difficulties encountered at the Dominion Observatory in its mirror transit circle program are described. Some satisfactory aspects of the design, applicable perhaps to other instruments, are noted. An outline is given of the circumstances leading to abandonment of the project.

032.039 **Airborne infrared astronomy.** H. H. G. Aumann.
Thesis, Rice Univ., Houston, Texas. [Available from Univ. Microfilms, Ann Arbor, Mich., U.S.A. Order No. 70–23476], 120 pp. (1970).

An airborne infrared telescope has been developed that allows observations of astronomical objects beyond 25 microns, a region of the spectrum where the earth's atmosphere is opaque. Flux measurements of the infrared nebula in the Orion, the galactic center, and NGC 1068 are reported and analyzed.

032.040 **The design of fore-optics for analytical optical equipment used in astronomy.** J. F. Grainger.
Journ. Phys. E. Sci. Instruments, Vol. 4, 713 - 718 (1971). See Phys. Abstr., Vol. 74, No. 74748 (1971).

032.041 **Large telescope developments and the world trend.** M. Furuhata.
Astron. Herald, (*Japan*), Vol. 64, 38 - 39 (1971). In Japanese. See Phys. Abstr., Vol. 74, No. 77598 (1971).

032.042 **Large telescope design.** Y. Yamashita.
Astron. Herald, (*Japan*), Vol. 64, 326 - 329 (1971). In Japanese.

Reviews large telescopes under construction or under project in the world.

032.043   **The development of the optical telescope.**
R. Lincoln.
Journ. Astron. Soc. Western Australia, Vol. 31, October,
p. 2 - 8 (1971).

1971 - 1972: Guide to scientific instruments.
See Abstr. 031.043.

Maksutov spectrograph cameras.
See Abstr. 034.002.

033   Radio Telescopes and Equipment

033.001   **Microwave receivers for molecular line radio astronomy.**   D. Buhl, L. E. Snyder.
Nature, Phys. Sci. Vol. 232, 161 - 163 (1971).
  The development of electronic receivers, resulting in the
discovery of interstellar molecular lines, has encouraged further
instrumental improvements. In this article some aspects of the
NRAO spectral line receivers are discussed.

033.002   **Stanford's high-resolution radio interferometer.**
R. N. Bracewell, R. S. Colvin, K. M. Price,
A. R. Thompson.
Sky Telescope, Vol. 42, 4 - 9 (1971).

033.003   **Filled-aperture antennas for radio astronomy.**
J. W. Findlay.
Annual Rev. Astron. Astrophys., Vol. 9, (see 003.001),
271 - 292 (1971).
  This review is restricted to recent developments in the
theory and practice of filled-aperture antennas.

033.004   **The new supersynthesis radiotelescope for HI
spectroscopy at the Dominion Radio Astrophysical
Observatory.**
R. S. Roger, C. H. Costain. J. D. Lacey, T. L. Landecker.
Journ. Roy. Astron. Soc. Canada, Vol. 65, 179 (1971).
Abstr. RAS Canada.

033.005   **Precision 4.57 m antenna at the University of
British Columbia for millimetre wavelength astrono-
my.**   W. L. H. Shuter.
Journ. Roy. Astron. Soc. Canada, Vol. 65, 179 (1971).
Abstr. RAS Canada.

033.006   **Atmospheric limitations to the angular resolution
of aperture synthesis radio telescopes.**
R. Hinder, M. Ryle.
Monthly Notices Roy. Astron. Soc., Vol. 154, 229 - 253
(1971).
  The aim of the paper is to assemble the data available on
the phase variations imposed on a wave traversing the atmosphere, and to assess the effect they will have on limiting the
performance of very large radio telescopes of the synthesis
type.

033.007   **Reduction of the parasitical signal of circular polarization on an antenna of variable profile with the
help of a grating.**
N. A. Esepkina, V. Y. Petrunkin, N. S. Soboleva, G. M. Timofeeva, A. V. Reiner.
IAU Symposium No. 43, (see 012.003), p. 91 - 93 (1971).

033.008   **Application of the method of statistical estimates
in solving the problem of brightness reconstruction**

in radio astronomy.   L. G. Sodin.
Izv. vyssh. uchebn. zavedenij. Radiofizika, Vol. 14, 739 -
747 (1971). In Russian. − Abstr. in Referativ. Zhurn. 51.
Astron., 10.51.199 (1971).

033.009   **Joint Soviet-American radio interferometry.**
K. I. Kellermann.
Sky Telescope, Vol. 42, 132 - 133 (1971).

033.010   **A precision sidereal telescope drive based on a
solar time crystal clock.**
R. W. P. Drever, J. H. Fruin, J. V. Jelley.
Observatory, Vol. 91, 203 - 205 (1971).
  A requirement arose for a sidereal drive for a small
microwave radiometer on an equatorial mount, the equipment to run unattended to a precision of about 2 minutes
in hour angle over a 7-day period. The installation already
included a solar-time crystal clock based on a 10 MHz oscillator. The principle of the electronic system described here
is derived from the type of vernier timing system described
by Horowitz for generating precise periodicities in work on
pulsars.

033.011   **The influence of the "confusion" effect on the
sensitivity of large radio telescopes in observing
discrete radio sources.**   Yu. P. Ilyasov.
Izv. vyssh. uchebn. zavedenij. Radiofizika, Vol. 14, 536 - 546
(1971). In Russian. − Abstr. in Referativ. Zhurn. 51. Astron.,
11.51.115 (1971).

033.012   **Unsteerable radio telescope of the millimeter wavelength range.**
V. N. Glazman, A. G. Kislyakov, I. V. Mosalov.
Izv. vyssh. uchebn. zavedenij. Radiofizika, Vol. 14, 663 - 672
(1971). In Russian. − Abstr. in Referativ. Zhurn. 51. Astron.,
11.51.116 (1971).

033.013   **Polarization characteristics of radio telescope an-
tennae.**   N. A. Esepkina.
Izv. vyssh. uchebn. zavedenij. Radiofizika, Vol. 14, 673 - 679
(1971). In Russian. − Abstr. in Referativ. Zhurn. 51. Astron.,
11.51.117 (1971).

033.014   **Impedance measurements of a loop antenna in the
topside ionosphere.**   P. C. Gregory.
Planet. Space Sci., Vol. 18, 1357 - 1365 = Astron. Contr. Univ.
Manchester, Ser. 2, Jodrell Bank Repr., No. 425 (1970).

033.015   **A computer-run radio telescope spectrograph.**
A. Winnberg.
Sky Telescope, Vol. 42, 274 - 276 (1971).

033.016   **Data processing for the Westerbork Synthesis Radio**

Telescope. W. N. Brouw.
Edited by Sterrewacht Leiden. 111 pp. (1971). – Dissertation Univ. Leiden.

The paper consists of two main parts. In the first part a review is given of the theory underlying aperture synthesis methods in general, and the influence of imperfections in the individual components of the system is described. The second part of the paper describes the computer programs developed for the reduction of the Westerbork data. The last chapter describes some test observations made to establish the precision of the instrument. It turns out that without any special effort relative positions of radio sources can be determined with an accuracy of about $0''.1$, while absolute positions with an accuracy of $0''.5$ are easily attained. The relative errors in intensity for intense sources is of the order of 0.3 %.

033.017 Phase stability of two independently locked local oscillators at 22.235 GHz. G. D. Papadopoulos.
Proc. IEEE, Vol. 59, 1620 - 1622 (1971).

A technique for constructing phase-stable K-band radiometers suitable for very long baseline interferometric measurements is described. The local oscillators of two such radiometers were locked to separate rubidium standards and compared for phase stability.

033.018 Use of the Culgoora radioheliograph for the generation of diffraction patterns.
K. V. Sheridan, D. J. McLean.
Applied Optics, Vol. 10, 2427 - 2435 (1971).

A series of diffraction patterns produced with the electronic imaging system of the Culgoora radioheliograph is presented. These patterns serve both to illustrate the quality of this type of imaging system and to demonstrate the advantage, in special circumstances, of constructing a radio-frequency simulation of an optical imaging device.

033.019 Photon-noise-limited laser transducer for gravitational antenna.
G. E. Moss, L. R. Miller, R. L. Forward.
Applied Optics, Vol. 10, 2495 - 2498 (1971).

We have constructed and tested a long, wideband, laser-linked gravitational radiation antenna. Photon-noise-limited performance was achieved using 80 μW from a single mode Spectra-Physics 119 laser in a modified Michelson interferometer on a vibration isolation table in a quiet room.

033.020 23 cm receiver for Cracow 15 m radiotelescope.
J. Machalski.
Postępy Astron., Vol. 19, 325 - 334 (1971). In Polish.

This paper describes the construction and parameters of the apparatus for 23 cm, which has been developed for the Cracow 15 m radiotelescope.

033.021 On the theory, techniques, and data processing of very long baseline interferometry. B. O. Rönnäng.
Res. Lab. Electronics, Chalmers Univ. Technology, Gothenburg, Sweden, Res. Report No. 105, 7 + 83 pp. = Preprint Onsala Space Obs., Onsala (1971).

The following report gives the basic theoretical background for interferometric observations at radio wavelengths, and describes the receiver system for very long baseline interferometry and the computer programs which process the recorded data to get the fringe amplitude and fringe phase versus frequency. We derive the formulae giving the baseline parameters, the geometrical time delay and fringe rate for given station coordinates and radio source coordinates. The positions of some of the VLBI stations are tabulated and parameters for nine important interferometer baselines are listed.

033.022 Digital recording of radio astronomical data on a broad magnetic tape.
M. G. Larionov, A. A. Kapustkin.
Soobshch. Gos. Astron. Inst. Shternberga, No. 173, p. 29 - 34 (1971). In Russian.

A digital recording system using as storage device a magnetic tape of a 16-track tape-recorder is described. Recording of the observational data together with siderial minutes is made in a form of 11-digit parallel code.

033.023 The BESM-4 computer input system for processing of radio astronomical data recorded on a ShKhR-16 tape recorder. I. A. Avdakushin, V. V. Golubnichi, E. I. Gurevich, A. A. Kapustkin, M. G. Larionov, A. S. Nikanorov, M. V. Popov.
Soobshch. Gos. Astron. Inst. Shternberga, No. 173, p. 35 - 42 (1971). In Russian.

033.024 Errors with combined Rayleigh-Gaussian statistics: Application to antenna pointing.
D. C. Hogg, S. O. Rice.
Proc. IEEE, Vol. 59, 1715 - 1717 (1971).

033.025 A parametric amplifier for 46 GHz. J. Edrich.
Proc. IEEE, Vol. 59, 1125 - 1126 = National Radio Astron. Obs., *Green Bank,* Repr. Ser. A, No. 214 (1971).

A degenerate parametric amplifier for 46 GHz is described. It uses a Schottky barrier varactor packaged in a modified sharpless wafer. It will be used in spectral line receivers for radio astronomy.

033.026 Rutile traveling wave masers for the frequency range 1300–3400 MHz and their application in the Onsala 84 foot radio telescope to galactic spectral line emission studies. O. E. H. Rydbeck, E. L. Kollberg.
Res. Lab. Electronics, Chalmers Univ. Technology, Gothenburg, Sweden, Res. Report No. 89, 1 + 33 pp. (1968).

This report describes the development, and practical application of a series of extremely compact high gain traveling wave masers using chromium doped rutile as active material. These masers, which are electronically tunable through bands of 200 MHz, or more, are especially built for microwave emission studies, for example of anomalous galactic OH 18 cm radiation, on the 84 foot, equatorially mounted Cassegrainian radio telescope at the Onsala Space Research Observatory, Sweden.

033.027 Complex propagation coefficients and the step discontinuity in corrugated cylindrical waveguide.
D. N. Cooper.
Electronic Letters, Vol. 7, No. 5/6, p. 135 - 136 = Separate print Division Radiophys., C.S.I.R.O., Sydney, Australia (1971).

033.028 Orthogonality relationship for class of waveguide with anisotropic walls. D. N. Cooper.
Electronic Letters, Vol. 7, No. 5/6, p. 137 = Separate print Division Radiophys., C.S.I.R.O., Sydney, Australia (1971).

033.029 Prinzip des Phasensynchronfilters.
H. J. Schober.
Internationale Elektronische Rundschau, Jahrgang 25, No. 7, 3 pp. = Mitt. Univ.-Sternw. Graz, No. 9 (1971).

033.030 A compound interferometer for solar radio observation. A. Tsuchiya.
Journ. Inst. Electrical Commun. Engineers (*Japan*), Vol. 54, 349 - 359 (1971). In Japanese.

033.031 Multi-level correlation spectrometer for radioastronomy. F. K. Bowers.
IEEE International Convention Digest 1971, (IEEE, New

York), p. 156 - 157. – See Phys. Abstr., Vol. 74, No. 50133 (1971).

033.032 Potentials of masers for millimeter wave radio astronomy. S. Yngvesson.
IEEE International Convention Digest 1971, (IEEE, New York), p. 160 - 161. – See Phys. Abstr.,Vol. 74, No. 50134 (1971).

033.033 Potentials of parametric amplifiers for millimeter wave radio astronomy. J. Edrich.
IEEE International Convention Digest 1971, (IEEE, New York), p. 162 - 163. – See Phys. Abstr., Vol. 74, No. 50135 (1971).

033.034 Potentials of Josephson junctions for millimeter-wave radio astronomy. R. Y. Chiao.
IEEE International Convention Digest 1971, (IEEE, New York), p. 164. – See Phys. Abstr., Vol. 74, No. 50136 (1971).

033.035 Cal predicts performance of NASA's proposed orbiting radio telescope.
Research Trends, Vol. 19, 20 - 22 (1971).
Prediction of performance of NASA's proposed 100 me tre diameter orbiting radio telescope. Operating frequency 15–150 MHz. – *JWB*

033.036 Interferometer phase measurements: Comparison with theory. D. K. Barton.
IEEE Trans. Antennas Propagation, Vol. AP-19, 566 - 569 (1971).
Recent measured results for long baselines are compared with theoretical predictions of interferometer phase fluctuation. The measured values, from the Green Bank Observatory in West Virginia, are somewhat smaller than predicted from refractivity spectra data gathered earlier in Colorado. The effects of different lengths and observation times are predicted quite well.

033.037 High frequency part of a dm wave spectrograph for the solar radioastronomy.
E. Hiss, G. Zimmermann.
Zeitschr. Angew. Phys., Vol. 31, 205 - 209 (1971). In German.
The electronic design of a RF-multichannel spectrograph for radioastronomic measurements in the dm-range is described. In particular, characteristics and temperature behavior of this instrument are discussed.

033.038 Optical synthetic aperture analogues of two radio interferometers.
D. E. Yansen, G. O. Reynolds, D. J. Cronin.
Optica Acta (*GB*), Vol. 18, 167 - 180 (1971). – See Phys. Abstr., Vol. 74, No. 46356 (1971).

033.039 Atomic clocks in long baseline radioastronomical interferometry. C. Audoin, P. Grivet.
Revue Phys. Appliquée, Vol. 6, 247 - 254 (1971). In French.
Recalls the requirements of very long baseline interferometry, concerning the independent local oscillators associated with each antenna. The authors describe the different atomic frequency standards used in this application; they compare their respective frequency stability, and give the method to specify it.

033.040 Image correction in high-resolution radio interferometer. M. Ishiguro.
Proc. Res. Inst. Atmosph. Nagoya Univ., (*Japan*), Vol. 18, 73 - 88 (1971). – See Phys. Abstr., Vol. 74, No. 66674 (1971).

033.041 A giant ear turned towards the cosmos: Inauguration of the Effelsberg radiotelescope.
Telecommun. Journ., (*Switzerland*), Vol. 38, 626 - 627 (1971).

033.042 Experimental laser system for monitoring deformations in large radio reflectors.
G. A. Clark, R. H. Slater.
Proc. Inst. Electrical Engineers, (*GB*), Vol. 118, 1562 - 1568 (1971). – See Phys. Abstr., Vol. 75, No. 2009 (1972).

033.043 Listening to the universe. E. J. Blum.
Sci. Progr. Découverte, No. 3435, p. 4 - 14 (1971).
In French. – See Phys. Abstr., Vol. 75, No. 7574 (1972).

033.044 KU-band interferometry. G. D. Papadopoulos.
Mass. Inst. Technology, Res. Lab. Electronics, Techn. Rep., No. 481, 105 pp. (1970). – Thesis, Mass. Inst. Technology, Pasadena.

033.045 Arbitrarily shaped dual-reflector antennas. C. Yeh.
California Inst. Technology, Jet Propulsion Lab., Techn. Rep. 32–1503, 9 pp. (1971).
An analysis, using geometric optics, of a dual reflector antenna system with two arbitrarily shaped reflectors is given.

033.046 Investigation of the effects of precipitation on parabolic antennas employing linear orthogonal polarisation at 11 GHz. B. G. Evans, A. J. Fryatt, J. Read, P. T. Thompson, D. J. W. Turner.
Electronics Letters, Vol. 7, 375 - 377 (1971).
Results of artificial paraboloid antenna-wetting experiments, and their effects on cross-polarization measurements are given. – *BMT*

033.047 The prediction of polar diagrams of large Cassegrain antennas. T. Pratt, B. Claydon.
Marconi Rev., Vol. 34, 1 - 26 (1971).
The paper includes theoretical and measured near-field and far–field radiation patterns of a 10-ft diameter Cassegrain reflector. – *BMT*

033.048 Parametric amplifiers designed for receiving systems. C. S. Aitchison.
Electronic Engineering, Vol. 43, 56 - 59 (1971).
The designs of three complex parametric amplifiers for use in radio astronomy are summarized. – *JWB*

033.049 Near-field radiation characteristics of corrugated horns.
P. J. B. Clarricoats, A. D. Olver, P. K. Saha.
Electronics Letters, Vol. 7, 446 - 448 (1971).
The amplitude and phase patterns of corrugated feed horns are predicted in the near-field region by means of a spherical mode-expansion method which has been previously applied only in the far field. – *MWS*

033.050 A 20-GHz integrated balanced mixer. T. Araki, M. Hirayama.
IEEE Trans. Microwave Theory and Techn., Vol. MTT-19, 638 - 643 (1971).
Practical aspects of mixer and mount are discussed. – *ACM*

033.051 Propagation and radiation characteristics of low-permittivity dielectric cones.
P. J. B. Clarricoats, C. E. R. C. Salema.
Electronics Letters, Vol. 7, 483 - 485 (1971).
Theoretical radiation patterns are predicted. Their predicted similarity to those obtained for corrugated conical horns is supported by experimental results obtained at 9 GHz. *ACM*

033.052 Mutual interference between linear crosspolarised radio channels at 11 GHz.
P. A. Watson, F. Goodall, S. Ghobrial.
Electronics Letters, Vol. 7, 374 - 375 (1971). – *BMT*

033.053 4-GHz integrated-circuit mixer.
M. Katoh, Y. Akaiwa.
IEEE Trans. Microwave Theory and Techn., Vol. MTT-19, 634 - 637 (1971).
Conversion loss for microwave diode mixers is calculated. The relations between conversion loss and parameters are clarified. A 4 GHz integrated-circuit low-noise mixer is developed. – *ACM*

033.054 New technique for beam steering with fixed parabolic reflectors. A. W. Rudge, M. J. Withers.
Proc. Instn. Electr. Engineers, (*GB*), Vol. 118, 857 - 863 (1971). – *JWB*

033.055 Parametric amplification of millimeter and submillimeter waves: Results, potentials and limitations.
J. Edrich.
IEEE International Convention Digest 1970, (IEEE, New York), p. 104 - 110.
A discussion of mm wave parametric amplifiers using Schottky barrier diodes is presented. – *DNC*

033.056 Optimization criterion for illuminating circular antenna apertures. R. Holland.
IEEE Trans. Antennas Propagation, Vol. AP-19, 436 - 443 (1971).
After defining the ideal antenna radiation pattern as one in which power is spread uniformly over a cone, the author develops a theory for optimizing illumination functions based on the minimum antenna gain within such a cone. The technique gives a more reliable estimation of system limitations than the more usual rms techniques. – *DNC*

033.057 Synthesis of the fields of a transverse feed for a spherical reflector. L. J. Ricardi.
IEEE Trans. Antennas Propagation, Vol. AP-19, 310 - 320 (1971).
The author presents a technique for synthesizing the fields on a spherical cap in the focal region of a spherical reflector that will produce a specified reflected field at the reflector surface. – *DNC*

033.058 Discriminator design uses acoustic delay devices.
A. D. Robertson, G. W. Hurley, J. R. Yeager.
Microwaves, (*USA*), Vol. 10, No. 7, p. 27 - 30 (1971).
A discussion of microwave acoustic delay devices for use as the long arm in "interferometer" type frequency sensitive circuits in phase discriminators. – *DNC*

033.059 Helical feeds at millimetre wavelengths.
P. R. Cowles, E. A. Parker.
Electronics Letters, Vol. 7, 513 - 515 (1971).
The design of helical feeds for the 26 to 40 GHz frequency range is considered. – *BMT*

033.060 A continuously variable dielectric phase shifter.
W. T. Joines.
IEEE Trans. Microwave Theory Techn., Vol. MTT-19, 729 - 732 (1971).
A stripline phase shifter is described and its operation over the 1 to 2 GHz band are examined. – *JWB*

033.061 Radiation from a plane spiral antenna array.
K. K. Dey, P. Khastgir.
Proc. Instn. Radio Electronics Engineers Australia, Vol 32, 349 - 351 (1971). – *ACM*

033.062 Radiation from a continuous spiral antenna array.
K. K. Dey, P. Khastgir.
Proc. Instn. Radio Electronics Engineers Australia, Vol. 32, 351 - 353 (1971). – *ACM*

033.063 Measurement of sky noise temperature at 16 GHz and 35 GHz. Y. Otsu.
Journ. Radio Res. Lab., (*Japan*), Vol. 18, 87 - 111 (1971).
Detailed data, calculations and results are given from an experiment to measure sky noise temperature and to estimate antenna loss factor which causes apparent increase in temperature measurement. Relationship between rainfall rate and rise in temperature due to rain and cloud is described. Some aspects of temperature scintillation due to cloud are also discussed. – *ACM*

033.064 A study of the feasibility of applying capacitive displacement-measuring techniques to open-mesh grid structures. R. Deloach.
National Aeronautics and Space Administration, NASA Techn. Note, TN-D-6341, 42 pp. (1971).
The report discusses the use of a noncontacting capacitive displacement-measuring transducer to measure displacements in open-mesh grid structures. This technique has application to the measurement of radio telescope deflections. – *BMT*

033.065 Lenses improve horn antennas.
L. C. Gunderson, G. T. Holmes.
Microwaves, (*U.S.A.*), Vol. 10, No. 8, p. 33 - 35 (1971).
A description and measured patterns of a conical horn using a microwave lens across the aperture. – *BMT*

033.066 Synthesis of broadband microwave transistor amplifiers. R. S. Tucker.
Electronics Letters, Vol. 7, 455 - 456 (1971).
The design of broadband microwave transistor amplifiers is discussed in terms of direct synthesis of distributed commensurate matching networks. – *MWS*

033.067 A scaled hybrid integrated multiplier from 10 to 30 GHz. M. V. Schneider, W. W. Snell, Jr.
Bell System Techn. Journ., Vol. 50, 1933 - 1942 (1971). – *MWS*

033.068 Dual-gate gallium-arsenide microwave field-effect transistor.
J. A. Turner, A. J. Waller, E. Kelly, D. Parker.
Electronics Letters, Vol. 7, 661 - 662 (1971). – *ACM*

033.069 Propagation and radiation behaviour of corrugated feeds. 1. Corrugated-waveguide feed.
P. J. B. Clarricoats, P. K. Saha.
Proc. Instn. Electr. Engineers, (*GB*), Vol. 118, 1167 - 1176 (1971).
An investigation of the propagation and radiation characteristics of circular corrugated waveguides is described. –*JWB*

033.070 Propagation and radiation behiour of corrugated feeds. 2. Corrugated-conical-horn feed.
P. J. B. Clarricoats, P. K. Saha.
Proc. Instn. Electr. Engineers, (*GB*), Vol. 118, 1177 - 1186 (1971).
A theoretical and experimental investigation of the radiation behaviour of conical corrugated horns is described. – *JWB*

033.071 A microwave solar spectrograph. N. Fourikis.
Proc. Instn. Radio Electronics Engineers Australia, Vol. 32, 361 - 366 (1971).
The instrument is essentially a broad-band superhet total-power radiometer whose local oscillator is swept at the rate of

8 Hz. In order to obtain the dynamic spectra of burst radiation from the sun, the spectrum is divided into two ranges (2—4 GHz and 4—8 GHz) which are recorded on film on a time-sharing basis. — *ACM*

033.072 Efficiency of a stepped reflector when fed from an off-axis source. G. Poulton.
Electronics Letters, Vol. 7, 666 - 667 (1971).
A correlation technique has been used to find the efficiency of a stepped reflector, given the focal-region field and the aperture field of the feed. The scanning properties of the antenna predicted by this theory agree well with existing experimental data. — *ACM*

033.073 Radiation patterns of paraboloid with log-periodic dipole feed.
P. A. McInnes, E. W. Munro, A. J. T. Whitaker.
Electronics Letters, Vol. 7, 669 - 671 (1971).
The effect of axial displacement of phase centre on the secondary performance of a parabolic reflector is described, with particular reference to a log-periodic dipole-array feed. Computed and experimental results are presented. — *ACM*

033.074 Design and performance of series-mode frequency multipliers using the step-recovery diode.
S. A. Boctor, D. J. Roulston.
International Journ. Electronics, Vol. 31, 333 - 351 (1971).
A theoretical treatment with some experimental results for multipliers by 2 and 3 when the input frequency is 333 and 350 MHz respectively. — *NF*

033.075 An approximate solution for the radar echo pulse response for planetary radar altimeters.
W. T. Bundick.
National Aeronautics and Space Administration, NASA Techn. Note, TN—D—6434, 27 pp. (1971).
An approximate solution for the radar echo pulse response has been developed to compute the expected return pulse amplitude and shape for planetary radar altimeters. The present solution is more accurate at low altitudes and longer pulse widths than previous solutions. — *JWB*

033.076 Polarised interferometric spectrometry for the

millimetre and submillimetre spectrum.
D. H. Martin, E. Puplett.
Infrared Phys., Vol. 10, 105 - 109 (1970).

033.077 Thermal distortion of a paraboloidal shell.
R. J. Platt, Jr., M. D. Rhodes.
National Aeronautics and Space Administration, NASA Techn. Note, TN—D—6471, 24 pp. (1971). — *JWB*

033.078 Noise measure of metal-semi-conductor-metal Schottky-barrier microwave diodes.
H. A. Haus, H. Statz, R. A. Pucel.
Electronics Letters, Vol. 7, 667 - 669 (1971).
Impedance and noise measure of M.S.M. microwave diodes are calculated. — *ACM*

033.079 Parametric amplifiers for a radio-astronomy interferometer. R. Davies, R. E. Pearson.
Philips Techn. Rev., Vol. 32, No. 1, p. 20 - 31 (1971).
Article describes a cooled parametric amplifier system for radio interferometer at Defford, England. Operates at 2.695 GHz. — *JS*

033.080 Nonlinearity and noise in the avalanche transit-time oscillator. K. Mouthaan, H. P. M. Rijpert.
Philips Res. Rep., Vol. 26, 391 - 413 (1971).
Theory and experimental verification of noise behaviour of 6 GHz avalanche transit-time oscillator. — *JS*

033.081 Theoretical performance of prime-focus paraboloids using cylindrical hybrid-mode feeds. B. M. Thomas.
Proc. Instn. Electr. Engineers, (*GB*), Vol. 118, 1539 - 1549 (1971).
An expression is derived for the radiation pattern of a hybrid-mode field distributed over a circular aperture. Synthesising a desired paraboloid-aperture distribution by superposition of the beams radiating from a hybrid-mode feed at the prime focus is discussed. — *JS*

Jodrell Bank in Wales.
Nature, Vol. 233, 80 - 81 (1971).

New results and techniques in space radio astronomy. See Abstr. 051.029.

034 Astronomical Accessories

034.001 A gas-Čerenkov telescope experiment to observe cosmic gamma rays.   J. A. Hoffman.
SAO, *Cambridge, Mass.* Special Rep. No. 335, 13 + 155 pp., with an appendix by H. Helmken, J. A. Hoffman (1971).

The gas-Čerenkov balloon-borne cosmic gamma ray detector described here has a variable energy threshold and is sensitive down to 15 MeV with a half-angle resolution of 5° to 7°. Gamma rays are detected by a coincidence between light pulses in a scintillator at the top of the detector and pulses of Čerenkov light collected by a mirror at the bottom of the detector. Gamma rays convert or Compton scatter in the scintillator, and the resulting electrons emit Čerenkov radiation as they move through the 2 m gas column between the scintillator and the mirror. The detector was flown in a balloon from Palestine, Texas on 29 September, 1970 and observed M87, the Crab nebula, and the atmospheric gamma ray background.

034.002 Maksutov spectrograph cameras.   C. G. Wynne.
Monthly Notices Roy. Astron. Soc., Vol. 153, 261 - 277 (1971).

Designs are given, and performance described, for a simple $f/2$ Maksutov camera, an $f/2.2$ Maksutov-Cassegrain and an $f/1.4$ semi-solid Maksutov-Cassegrain camera, the two latter designed for use with image tubes. In each case, the advantages of these systems are discussed, as compared with the corresponding Schmidt constructions. An appendix deals with the designing of Maksutov systems.

034.003 On focusing, guiding and pointing of a siderostat.   J. Zicha.
Bull. Astron. Inst. Czechoslovakia, Vol. 22, 200 - 210, 218a - 218b (1971).

The instrument for automatic guiding of a siderostat pointed to a star and for occasional check of focus is described. The prototype of the instrument was manufactured and its parameters were tested. Next a method of pointing a siderostat is presented. Finally, it is suggested that the described principle of automatic guiding can be used in a coudé spectrograph.

034.004 The Isaac Newton Telescope Cassegrain spectrograph: Description and results of radial-velocity observations of stars near δ^2 Lyrae.
G. A. Harding, M. P. Candy.
Roy. Obs. Bull., Greenwich—Cape, No. 164, p. 251 - 256 (1971).

The spectrograph is mounted on a turntable fixed to the rear plate of the mirror cell, the main optical axes within the spectrograph being at right angles to the optical axis of the main telescope. The spectrograph will have three Schmidt cameras of varying focal lengths and a variety of gratings which can be combined to give a large range of dispersions varying from 12.5 Å/mm to about 360 Å/mm. New radial-velocity observations for 18 stars within about 15′ of δ^2 Lyrae are presented.

034.005 Das Protuberanzenfernrohr als Hochleistungsinstrument II, III.   G. Nemec.
SuW, Vol. 10, 197 - 199, 234 - 238 (1971).

034.006 The use of echelle gratings in single-pass spectrometers.
A. D. Petford, D. E. Blackwell, B. S. Collins, P. A. Ibbetson, E. A. Mallia, G. Smith, D. Emerson.

Solar Physics, Vol. 19, 264 - 269 (1971).

Tests of the performance of replicas of the latest echelle gratings are reported.

034.007 A complete Stokes-meter.   A. Cacciani, M. Fofi.
Solar Physics, Vol. 19, 270 - 276 (1971).

A new polarimeter is described which allows the simultaneous determination of the four Stokes parameters analysing the electric signal both in frequency and phase. Magnetographic applications in solar physics and improvements as compared to previous magnetographs are suggested.

034.008 Performance of a modified Askania iris photometer.   G. B. Baratta, G. Caprioli.
Mem. Soc. Astron. Italiana, Nuova Ser., Vol. 42, 245 - 247 (1971). – Letter.

034.009 The control system for the four-camera diffractional spectrograph of the Pulkovo Observatory.
L. M. Kotljar, G. I. Kalinina.
Solnechnye Dannye 1971 Byull., No. 5, p. 94 - 101 (1971). In Russian.

034.010 The possibilities of superorthicons in astrophysics.   M. K. Babadzhanianz, L. D. Parfinenko.
Trudy Astron. Obs., *Leningrad,* Vol. 28 (= Uchenye Zapiski Leningr. Un-ta, No. 359 = Seriya Matem. Nauk, vyp. (No.) 47), p. 57 - 63 (1971).   In Russian.

A method of the comparative study of TV and photographic methods of image registration are described.

034.011 An electrometer amplifier for photoelectric photometers.   E.-M. Maasik.
Publ. Tartu Astrofiz. Obs., Vol. 39, 137 - 140 (1971). In Russian.

A simple two-stage direct-coupled preamplifier circuit for the photomultiplier tube designed and used in Tartu is described.

034.012 On the problems of designing high-precision photoelectric stellar photometers.   U. Veismann.
Publ. Tartu Astrofiz. Obs., Vol. 39, 334 - 346 (1971). In Russian.

Problems of designing high-precision (0.1% on 1 sec) and high-stability stellar photometers are discussed. The technological possibilities of improving the elements of the system telescope—photometer are analysed. The statistical methods of dumping of atmospheric noise are described.

034.013 Remote indication of telescope coordinates by closed circuit television.   U. Veismann.
Publ. Tartu Astrofiz. Obs., Vol. 39, 347 - 350 (1971). In Russian.

A two-camera closed circuit television system for transmitting the image of telescope setting circles into the control board is described.

034.014 Photoelectron collection efficiency in photomultipliers.   A. T. Young, R. E. Schild.
Applied Optics, Vol. 10, 1668 - 1672 (1971).

Direct measurement of the photoelectron collection efficiency in the EMI 9558 photomultiplier yields a value of 75 ± 10%. The authors of this paper believe the substantially lower values recently reported to be spurious and discuss their reasons for this.

034.015 Use of photomultiplier tubes for photon counting.
A. T. Young.
Applied Optics, Vol. 10, 1681 - 1683 (1971).

034.016 Das Protuberanzenfernrohr als Hochleistungsinstrument IV.   G. Nemec.
SuW, Vol. 10, 276 - 278 (1971).

034.017 Apparatus for investigating the streams of neutron
albedo in the near-earth cosmic space. I.
I. A. Antonova, E. V. Korolko, V. I. Lazarev, B. V. Mariin,
Yu. A. Samonenko, A. V. Smirnov.
Geomagn. Aeronom., Vol. 11, 883 - 887 (1971).
In Russian.

034.018 Neutron spectrograph of cosmic rays using regis-
tering channels with different τ.
Y. L. Blokh, L. I. Dorman, N. S. Kaminer, I. N. Kapustin.
Geomagn. Aeronom., Vol. 11, 891 - 892 (1971).
In Russian. − Brief information.

034.019 Motion picture photography through the telescope.
R. J. Wood.
Journ. Roy. Astron. Soc. Canada, Vol. 65, 239 - 244 (1971).

034.020 Coudé spectral line scanner.   D. F. Gray.
Bull. American Astron. Soc., Vol. 3, 387 (1971).
Abstr. AAS.

034.021 A radial-velocity photometer.
T. R. Dennis, W. Liller.
Bull. American Astron. Soc., Vol. 3, 387 (1971). − Abstr.
AAS.

034.022 Efficiency of commercial electron-sensitive plates
for astronomical electronography.
P. Griboval, D. Griboval, M. Marin, J. Martinez.
Bull. American Astron. Soc., Vol. 3, 387 (1971). − Abstr.
AAS.

034.023 A photon noise limited film reducing microphoto-
meter.   I. L. Kofsky, C. S. Miller.
Bull. American Astron. Soc., Vol. 3, 387 - 388 (1971).
Abstr. AAS.

034.024 The Culgoora magnetograph.
J. V. Ramsay, R. G. Giovanelli, H. R. Gillett.
IAU Symposium No. 43, (see 012.003), p. 24 - 29 (1971).

034.025 A complete Stokes vector polarimeter.
F. Q. Orrall.
IAU Symposium No. 43, (see 012.003), p. 30 - 36 (1971).

034.026 Digital videomagnetograms in real time.
T. J. Janssens, N. K. Baker.
IAU Symposium No. 43, (see 012.003), p. 44 - 50 (1971).

034.027 The Kitt Peak magnetograph. IV: 40-channel probe
and the detection of weak photospheric fields.
W. Livingston, J. Harvey.
IAU Symposium No. 43, (see 012.003), p. 51 - 61 (1971).

034.028 A pressure scanning Fabry-Perot magnetometer.
T. D. Fay, A. A. Wyller.
IAU Symposium No. 43, (see 012.003), p. 62 - 64 (1971).

034.029 Sacremento Peak magnetograph.   R. B. Dunn.
IAU Symposium No. 43, (see 012.003), p. 65 - 70
(1971).

034.030 Systematic errors of the Crimean vector magneto-

graph.   V. A. Kotov.
IAU Symposium No. 43, (see 012.003), p. 71 - 75 (1971).

034.031 Analog video magnetograms in real time.
R. C. Smithson, R. B. Leighton.
IAU Symposium No. 43, (see 012.003), p. 76 - 83 (1971).

034.032 A new completely digitized filter magnetograph.
G. E. Brueckner.
IAU Symposium No. 43, (see 012.003), p. 84 - 88 (1971).

034.033 Difficulties in the simultaneous measurement of all
Stokes parameters.   E. Wiehr.
IAU Symposium No. 43, (see 012.003), p. 89 - 90 (1971).

034.034 A short report on the magnetic beam absorption
filter research at the Rome Astronomical Observatory.   A. Cacciani, M. Cimino, M. Fofi.
IAU Symposium No. 43, (see 012.003), p. 94 - 98 (1971).

034.035 Spectra-spectroheliograph observations.
A. M. Title, J. P. Andelin, Jr.
IAU Symposium No. 43, (see 012.003), p. 298 - 309 (1971).

034.036 A device for photographic reading of the declination
micrometer drum.
A. M. Stafeev, L. L. Vagushchenko.
Vestn. Kiev. Un-ta, Ser. Astron., No. 12, p. 90 - 92 (1970).
In Russian.

034.037 Extended-field large-aperture interferometer-spec-
trometer for airglow surveys.
A. M. Despain, D. J. Baker, A. J. Steed, T. Tohmatsu.
Applied Optics, Vol. 10, 1870 - 1876 (1971).
    The design of a field-of-view-widened interferometer for
airglow survey work is discussed, and some preliminary airglow results are presented.

034.038 Some thoughts on the information obtained by a
high resolution grating spectrometer.
G. A. De Biase, F. Sacchetti, D. Trevese.
Applied Optics, Vol. 10, 1885 - 1891 (1971).
    The possibility of obtaining a transfer function of a
diffraction grating affected by blank and ruling errors is examined and a simple experimental procedure for an approximate
determination of aberration function is described. Using this
aberration function, we have obtained a wavelength-dependent
transfer function in order to perform the best restoration of
the grating output.

034.039 2.5-km low-temperature multiple-reflection cell.
D. Horn, G. C. Pimentel.
Applied Optics, Vol. 10, 1892 - 1898 (1971).
    We describe the design and the operation of a multiple
reflection cell continuously controllable in temperature within a range of 120 K to 300 K and with a maximum path
length of 2540 m.

034.040 Continuous discharge line source for the extreme
ultraviolet.   F. Paresce, S. Kumar, C. S. Bowyer.
Applied Optics, Vol. 10, 1904 - 1908 (1971).
    A continuous gaseous discharge source suitable for use
between 100 Å and 1000 Å is described. The operating characteristics of this source and its spectral output with various
gases are presented.

034.041 Microphotometer for photometry of meteors.
A. S. Benyuch, V. V. Benyuch.
Vestn. Kiev. Un-ta, Ser. Astron., No. 13, p. 86 - 90 (1971).
In Russian.

034.042 Echellegitteret og dets anvendelse i astronomien.
P. E. Rathcke, P. E. Nissen.
Astron. Tidsskr., Årg. 4, 83 - 88 (1971).

034.043 An échelle grating for the 74-inch coudé.
H. R. Butcher.
Proc. Astron. Soc. Australia, Vol. 2, 21 - 22 (1971).

034.044 Techniques for gamma rays. C. E. Fichtel.
IAU Symposium No. 41, (see 012.005), p. 14 - 36 (1971).

034.045 Gamma-ray spectrometry of galactic sources in the energy range 0.2–3.0 MeV.
A. Bui-Van, G. Vedrenne, P. Mandrou.
IAU Symposium No. 41, (see 012.005), p. 45 - 57 (1971).

034.046 An experiment to measure the direction and energy spectrum of extraterrestrial gamma rays in the energy range 1–10 MeV from balloon altitudes.
A. J. Dean, A. Bellomo, P. Coffaro, M. Fatta, G. Gerardi, F. Madonia, A. Russo, L. Scarsi.
IAU Symposium No. 41, (see 012.005), p. 63 - 72 (1971).

034.047 Investigation of the power of resolution of a spark chamber for gamma-ray astronomy.
H. A. Mayer-Hasselwander, K. Pinkau, K. H. Schenkl, W. Voges, H. J. Schneider.
IAU Symposium No. 41, (see 012.005), p. 73 - 74 (1971).

034.048 A large-area gas-Čerenkov detector for high-energy gamma-ray astronomy.
J. Delvaille, K. Greisen, D. Koch, B. McBreen, G. Fazio, D. Hearn, H. Helmken.
IAU Symposium No. 41, (see 012.005), p. 75 - 76 (1971).

034.049 Gas-Čerenkov detector for 10 to 100 MeV gamma rays. H. Helmken, J. Hoffman.
IAU Symposium No. 41, (see 012.005), p. 77 - 78 (1971).

034.050 Chambre à étincelles optique pour la recherche de sources de rayons gamma.
J. Vasseur, J. Paul, B. Parlier, J. P. Leray, M. Forichon, B. Agrinier, G. Boella, L. Maraschi, A. Treves, R. Buccheri, L. Scarsi.
IAU Symposium No. 41, (see 012.005), p. 79 - 86 (1971).

034.051 A Bragg spectrometer for stellar X-ray astronomy.
H. Kestenbaum, J. R. P. Angel, R. Novick.
IAU Symposium No. 41, (see 012.005), p. 137 - 144 = Columbia Astrophys. Lab., Columbia Univ., New York, Contr. No. 34 (1971).

034.052 A focusing collector for long wavelength X-ray astronomy.
D. J. Yentis, J. R. P. Angel, D. Mitchell, R. Novick, P. vanden Bout.
IAU Symposium No. 41, (see 012.005), p. 145 - 158 = Columbia Astrophys. Lab., Columbia Univ., New York, Contr. No. 35 (1971).

034.053 A large area Thomson-scattering stellar X-ray polarimeter. R. Novick, R. S. Wolff.
IAU Symposium No. 41, (see 012.005), p. 159 - 164 = Columbia Astrophys. Lab., Columbia Univ., New York, Contr. No. 31 (1971).

034.054 X-ray spectrometry of galactic sources in the energy range 30–200 keV.
A. Bui-Van, G. Vedrenne, Y. Cezac, A. Bouigue, L. Sabaud.
IAU Symposium No. 41, (see 012.005), p. 168 - 179 (1971).

034.055 A new optical system for solar soft X-ray spectrophotometry. J. H. Dijkstra, W. Werner.
IAU Symposium No. 41, (see 012.005), p. 180 (1971). Abstract.

034.056 A rocket-borne X-ray spectrometer/monochromator system for mapping the solar corona.
L. W. Acton, R. C. Catura, J. L. Culhane, A. J. Meyerott.
IAU Symposium No. 41, (see 012.005), p. 181 (1971). Abstract.

034.057 Rocket prototype of an X-ray optical system for surveying and locating cosmic X-ray sources.
P. C. Fisher, L. W. Acton, R. C. Catura, P. Kirkpatrick, A. J. Meyerott, D. T. Roethig.
IAU Symposium No. 41, (see 012.005), p. 182 (1971). Abstract.

034.058 A rocket payload using focusing X-ray optics for the observation of soft cosmic X-rays.
P. Gorenstein, B. Harris, H. Gursky, R. Giacconi.
IAU Symposium No. 41, (see 012.005), p. 183 (1971). Abstract.

034.059 A balloon-borne X-ray telescope.
M. Fujii, M. Matsuoka, S. Miyamoto, J. Nishimura, M. Oda, Y. Ogawara, S. Ohta.
IAU Symposium No. 41, (see 012.005), p. 184 (1971). Abstract.

034.060 An automatic stabilized detection system for measuring soft celestial X-rays.
A. J. F. den Boggende, H. F. van Beek, A. C. Brinkman, H. T. J. A. Lafleur.
IAU Symposium No. 41, (see 012.005), p. 211 - 212 (1971).

034.061 A gas proportional chamber for use in cosmic X-ray research. P. Serlemitsos.
IAU Symposium No. 41, (see 012.005), p. 213 (1971). Abstract.

034.062 The use of silicium solid-state detector in solar X-ray measurements. B. Valníček.
IAU Symposium No. 41, (see 012.005), p. 214 (1971). Abstract.

034.063 The use of echelle gratings in ultraviolet space astronomy.
B. C. Boland, W. M. Burton, B. B. Jones, N. K. Reay.
IAU Symposium No. 41, (see 012.005), p. 254 - 261 (1971).

034.064 High resolution interference spectroscopy applied to astronomical investigations (2000 to 3000 Å).
B. Bates.
IAU Symposium No. 41, (see 012.005), p. 262 (1971). Abstract.

034.065 High resolution balloon-borne spectrograph for the near solar ultraviolet. P. Lemaire.
IAU Symposium No. 41, (see 012.005), p. 263 - 270 (1971).

034.066 The use of a Michelson interferometer on a coarsely stabilized spacecraft to obtain high resolution (0.1 Å) over 1300–3300 Å. D. D. Clark.
IAU Symposium No. 41, (see 012.005), p. 302 - 303 (1971).

034.067 Spectroheliographs for the ultraviolet.
Y. Öhman.
IAU Symposium No. 41, (see 012.005), p. 313 - 315 (1971).

034.068 Reflectors and polarizers for the vacuum ultraviolet.

B. Feuerbacher, B. Fitton.
IAU Symposium No. 41, (see 012.005), p. 316 (1971).
Abstract.

034.069 **A tunable birefringent filter for the UV region.**
K. Fredga, J. A. Högbom.
IAU Symposium No. 41, (see 012.005), p. 317 - 324 (1971).

034.070 **Fabry-Pérot interferometers as narrow band optical filters.** A. M. Title.
IAU Symposium No. 41, (see 012.005), p. 325 - 332 (1971).

034.071 **Instrumentation of the NRL solar eclipse rocket, 7 March 1970.** G. E. Brückner.
IAU Symposium No. 41, (see 012.005), p. 348 (1971).
Abstract.

034.072 **Instrumentation for high-resolution stellar UV-spectrophotometry.**
C. de Jager, A. Hammerschlag, W. Werner.
IAU Symposium No. 41, (see 012.005), p. 349 - 351 (1971).

034.073 **A spectrograph for cometary UV observations.**
L. Haser.
IAU Symposium No. 41, (see 012.005), p. 355 (1971).
Abstract.

034.074 **Housing of photoelectric equipment in rocket experiments.** H. Hessberg.
IAU Symposium No. 41, (see 012.005), p. 357 - 359 (1971).

034.075 **A SEC vidicon system for satellite applications.**
G. E. Brückner.
IAU Symposium No. 41, (see 012.005), p. 360 (1971).
Abstract.

034.076 **Rocket experiment with electronic camera for studying the metallic discontinuities in the ultraviolet spectrum of 'A' stars.** M. Combes.
IAU Symposium No. 41, (see 012.005), p. 361 - 362 (1971).

034.077 **Integral image tube-optical systems for the far UV.**
G. Chincarini.
IAU Symposium No. 41, (see 012.005), p. 363 - 365 (1971).

034.078 **Das Protuberanzenfernrohr als Hochleistungsinstrument V.** G. Nemec.
SuW, Vol. 10, 305 - 307 (1971).

034.079 **A versatile birefringent filter.**
K. Fredga, J. A. Högbom.
Solar Physics, Vol. 20, 204 - 227 (1971).
We discuss the properties of the Šolc filter and point out some new possibilities for filters based upon this general design.

034.080 **The lag of the narrow-band amplifier of a phase photoelectric installation.** M. I. Malyshev.
Astron. Tsirk., No. 640, p. 4 - 7 (1971). In Russian.

034.081 **Three-component airborne magnetometers.**
P. H. Serson, K. Whitham.
Encyclopedia of physics, Vol. 49/3, (see 003.013), 384 - 394 (1971).

034.082 **Aeromagnetic surveying with the fluxgate magnetometer.** J. R. Balsley.
Encyclopedia of physics, Vol. 49/3, (see 003.013), 395 - 421 (1971).

034.083 **Geophysical applications of high resolution magne-**

tometers. P. J. Hood.
Encyclopedia of physics, Vol. 49/3, (see 003.013), 422 - 460 (1971).

034.084 **Analysis of the calibration process of the channel of the longitudinal component of a magnetic field in a magnetograph with one photoelectric multiplier.**
O. V. Nikonov, E. S. Nikonova.
Solnechnye Dannye 1971 Byull., No. 7, p. 95 - 104 (1971).
In Russian.

034.085 **An echelle spectrograph for astronomical use.**
D. J. Schroeder, C. M. Anderson.
Publ. Astron. Soc. Pacific, Vol. 83, 438 - 446 (1971).
A Cassegrain spectrograph which uses a 73.5 groove/mm echelle grating as its main dispersing element has been designed and constructed at the University of Wisconsin. The optical design and, in particular, the effects of various aberrations are discussed in detail. The mechanical design is discussed in general terms and preliminary observations of various astronomical objects are given.

034.086 **On the neutrality of terrestrial cloud extinction.**
R. K. Honeycutt.
Publ. Astron. Soc. Pacific, Vol. 83, 502 - 503 = Publ. Goethe Link Obs., Indiana Univ., *Bloomington,* No. 127 (1971).
A rapid-scan spectrometer has recently been put into operation at the Goethe Link Observatory (Honeycutt 1971). The rapid-scan rate avoids the low-frequency noise components in the night-sky transparency and allows the possibility of making spectral scans through thin clouds.

034.087 **Performance of the spectracon in astronomical spectroscopy.** M. F. Walker.
Advances Electronics Electron. Phys., Vol. 28, 773 - 781 = Contr. Lick Obs., No. 266 (1970).
The observations show that despite the optical inefficiency caused by working at the coudé focus, the combination of the spectracon and Bowen camera provides a substantial gain over the results obtainable photographically with the prime-focus spectrograph of the 120-in. reflector.

034.088 **Investigation of the instrument KIM-3.**
P. P. Pavlenko.
Vestn. Khar'kov. Univ., No. 65, (Ser. Astron., No. 6), p. 63 - 83 (1971). In Russian.

034.089 **Spectromètre pour l'infrarouge lointain élargissement par l'argon de la première raie de rotation de HF gazeux à 40 cm^{-1}.** G. Bachet, R. Coulon.
Journ. Quant. Spectrosc. Radiat. Transfer, Vol. 11, 1827 - 1837 (1971).
A laboratory spectrometer, using small gratings (68.6 × 68.6 mm) and built for studies on far-infrared spectra of gases, is described. Spectra of water vapour and ammonia are presented, to show the performances of this instrument.

034.090 **Das Protuberanzenfernrohr als Hochleistungsinstrument VI.** G. Nemec.
SuW, Vol. 10, 330 - 331 (1971).

034.091 **Two amateurs set up an observatory-telescope cooperative.** A. E. Morton.
Sky Telescope, Vol. 42, 374 - 377 (1971).

034.092 **An amateur-built chain drive for a 12$^1/_2$-inch reflector.** C. Nash.
Sky Telescope, Vol. 42, 378 - 380 (1971).

034.093 **Glass filaments for eyepiece cross-wires.**
E. G. Moore.

Journ. British Astron. Ass., Vol. 82, 42 (1971).

034.094 **Der Dispersionskompensator.** C. Albrecht.
Orion Schaffhausen, 29. Jahrgang, p. 191 - 192
(1971).

034.095 **Ein Blinkkomparator für Amateure.** C. Albrecht.
Orion Schaffhausen, 29. Jahrgang, p. 192 - 194
(1971).

034.096 **Contribution to the background rate of a satellite
X-ray detector by spallation products in a caesium
iodide crystal.** C. S. Dyer, G. E. Morfill.
Astrophys. Space Sci., Vol. 14, 243 - 258 (1971).
The energy spectra observed in a CsI crystal in the
20 keV−2 MeV range, due to the decay of radioactive isotopes
produced in the crystal by bombardment with 155 MeV protons, are presented as a function of time after irradiation. The
Rudstam formula is used to predict the spallation that would
occur in such a crystal on board a satellite due to cosmic rays
and passages through the South Atlantic Anomaly.

034.097 **A tunable Hα-filter and its work.** M. Waldmeier.
Astron. Mitt. Sternw. Zürich, No. 305, 14 pp.
(1971).
A description is given of an Hα-filter, whose passband of
0.25 Å can be shifted up to ± 16 Å by mechanical-optical
means. As the exposures are short, pictures at different wavelengths can be obtained almost simultaneously. Therefore the
filter allows the study of three-dimensional motions in active
regions. Examples of limb prominences and surges on the disc
are given.

034.098 **Errata: 'A versatile birefringent filter'** [Solar Physics,
Vol. 20, 204 - 227 (1971)].
K. Fredga, J. A. Högbom.
Solar Physics, Vol. 21, 249 (1971).

034.099 **An analysis of errors in the channel of the longitu-
dinal component of the magnetic field for a magne-
tograph with one photomultiplier.**
O. V. Nikonov, E. S. Nikonova.
Solnechnye Dannye 1971 Byull., No. 8, p. 100 - 108 (1971).
In Russian.

034.100 **The main optical characteristics of two monochro-
matic filters of the firm "Halle".**
Z. B. Korobova.
Solnechnye Dannye 1971 Byull., No. 9, p. 87 - 92 (1971).
In Russian.
The results of investigations of the Halle monochromatic
filters for the Hα and K CaII lines are discussed. Recommenda-
tions for the work with these filters are given.

034.101 **Lasers: Applications in physics research.**
J. A. Armstrong.
Physics Today, Vol. 24, No. 3, p. 34 - 39 (1971).
Examples for applications are: Time and distance meas-
urements; Moon ranging; Light scattering.

034.102 **Electronic devices for infrared astronomical observations.** T. Maihara.
Mem. Fac. Sci., Kyoto, Univ., Ser. Phys., Astrophys., Geophys., Chemistry, Vol. 33, 251 - 256 (1971).
A circuit system of electronics is presented along with its
characteristics which has been designed and used for the purpose of the infrared photometry.

034.103 **New multichannel optical correlator.**
G. J. M. Aitken, L. Wang.
Applied Optics, Vol. 10, 2475 - 2481 (1971).

This paper describes a multichannel optical correlator
that uses two spatial dimensions to achieve multichannel capa-
bility and operates in real time. Given I signals, $r_i(t)$, and J
signals, $s_j(t)$, it can produce the $I \times J$ cross products, $r_i(t)s_j(t)$,
averaged over some interval. An experimental correlator and
its performance are described along with applications.

034.104 **Standards for photometry.** G. A. W. Rutgers.
Applied Optics, Vol. 10, 2595 - 2599 (1971).
A survey is given of the standards used in photometry
and spectroradiometry. The candela and lumen standards are
discussed in Sec. II. Section III deals with standards for spectroradiometry − the tungsten strip lamp, the anode of the carbon arc, and the high pressure xenon arc − and also with
some irradiance standards. A comparison of photometric and
radiometric standards is given in Sec. IV. It is concluded that
with the present standards an accuracy of 1% is hardly obtainable.

034.105 **Preliminary results from data reduction schemes for
an X-ray microprobe equipped with a nondispersive
detector and a multichannel analyser.**
A. S. Doan, Jr., P. A. Comella.
Meteoritics, Vol. 6, 260 (1971). − Abstract.

034.106 **The employment of the synchrotron radiation for
the energetic calibration of an astronomical apparatus.** G. A. Gurzadyan, J. B. Ohanesyan.
Astron. Zhurn. Akad. Nauk SSSR, Vol. 48, 1289 - 1300
(1971). In Russian. English translation in Soviet Astron. AJ,
Vol. 15, No. 6.
The paper describes a special installation attached to the
electronic circular accelerator for the extraction of synchrotron radiation. The latter is used as a shortwave and X-ray
source for the energetic calibration of an astronomical apparatus designed for functioning in the outer space. The above
installation has proved singularly efficient in operations in the
region of X-radiation.

034.107 **Helios-Zodiakallicht-Photometer. Optischer und
mechanischer Aufbau.** C. Leinert, E. Pitz.
Bundesministerium für Bildung und Wissenschaft, Forschungsber. BMBW−FB W 70-09, [Available from Zentralstelle für
Luftfahrtdokumentation und -information (ZLDI), Deutsche
Forschungs- und Versuchsanstalt für Luft- und Raumfahrt,
München. Price DM 15.55], 74 pp. (1970).
The optical and mechanical concept of the HELIOS Zodiacal Light Photometers is described. By a study of the single
components it is shown, that a sufficient straylight reduction
is possible by adequate fitting of the photometers into the
satellite, by the construction of the baffle systems and by the
selection of the optical elements.

034.108 **Untersuchungen an Komponenten eines Infrarot-
Photometers für das Ballonteleskop THISBE.**
W. Hofmann, D. Lemke, C. Thum.
Bundesministerium für Bildung und Wissenschaft, Forschungs-
ber. BMBW−FB W 71-31, [Available from Zentralstelle für
Luftfahrtdokumentation und -information (ZLDI), Deutsche
Forschungs- und Versuchsanstalt für Luft- und Raumfahrt,
München. Price DM 10.50], 50 pp. (1971).
The main components of a surface photometer for the
measurement of the night sky radiation in the infrared region
with the balloon borne telescope THISBE were examined. Especially suited PbS-detectors were chosen, an electronics able
for balloon flights was developed and the design of the photometer was fixed after climate-tests. The planned observations
are discussed briefly.

034.109 **Electronic imaging devices in astronomy.**
G. R. Carruthers.

Astrophys. Space Sci., Vol. 14, 332 - 377 (1971).

Presented are the basic physical principles of electronic imaging devices, and descriptions of the types of devices currently in use or under development, classified into two categories: those which produce a final image on film, as does conventional photography, and those which produce a television signal output. Criteria for device selection are discussed, and some typical examples of the application of these devices in astronomy and related areas are given.

034.110 Automatic electrophotometer with photon counting. V. M. Lyutyi.
Soobshch. Gos. Astron. Inst. Shternberga, No. 172, p. 30 - 41 (1971) In Russian.

An automatic pulse counting stellar electrophotometer is described. The photometer can be used both with one program (automatically) and with different programs as a remote controlled device. The photomultiplier together with filters gives the colour system quite close to UBV of Johnson and Morgan.

034.111 Calculation of the instrument function of a spectrometer on the basis of telluric O_2 lines.
G. F. Sitnik, A. I. Khlystov.
Soobshch. Gos. Astron. Inst. Shternberga, No. 173, p. 19 - 28 (1971). In Russian.

A method of calculating a spectrometer's instrument function using the telluric O_2 lines is proposed. The profiles of O_2 lines are calculated theoretically using the Voigt function for the absorption coefficient, the model of the earth's atmosphere being inhomogeneous and non-isothermic.

034.112 Telescopio automático para fotometría y polarimetría de la luz zodiacal y otras fuentes difusas.
C. Sánchez-Magro, F. Sánchez.
Urania Barcelona, Año 56, No. 273, p. 10 - 23 (1971).

034.113 Handsteuerung zur Automatischen Kamera für Astrogeodäsie aus Jena. M. Steinbach.
Jenaer Rundschau, (Jena Review), 16. Jahrgang, p. 299 - 301 (1971).

034.114 Erprobung eines Bildverstärkers für die Aufnahme schwacher astronomischer Objekte.
D. Ellinger, G. V. Schultz.
Messtechnik 4/71, p. 92 - 98 = Mitt. Astron. Inst. Bonn, No. 130 (1971).

A four-stage image intensifier was tested with regard to its utilization in the astronomical detection technique. Exposure time gain, threshold sensitivity, resolving power, and extent of density fluctuations are given.

034.115 Autocollimation device for determining the lateral flexure of the tube of astronomical universal instruments. V. G. L'vov.
Geod. i kartografiya, 1971, No. 7, p. 11 - 13. In Russian. – Abstr. in Referativ. Zhurn. 52. Geod. Aehros"emka, 1.52.96 (1972).

034.116 Spectrometer for absolute spectrophotometry of stars.
V. F. Bogdanov, P. N. Boiko, A. V. Kharitonov.
Trudy Astrofiz. Inst., Alma-Ata, Vol. 17, 94 - 100 (1971). In Russian.

A spectrometer for absolute spectrometry of stars is described. Seya-Namioka's scheme is used. The demands to the design of the instrument, the influence of inaccuracies in the construction of separate details and the methods of compensation of these inaccuracies are considered.

034.117 Investigation of the electromechanical system for leading of the wire along the right ascension at the Odessa Astronomical Observatory.
L. L. Vagushchenko, A. M. Stafeev.
Astrometriya i Astrofiz., Kiev, No. 13, (see 003.039), p. 84 - 88 (1971). In Russian.

034.118 International comparison of radiometers: Techniques and results. M. P. Thekaekara, M. W. Wilson.
Journ. Optical Soc. America, Vol. 61, 1555 - 1556 (1971). Abstr. Optical Soc. America.

034.119 Mathematical simulation of solar concentrators.
G. L. Schrenk.
Journ. Optical Soc. America, Vol. 61, 1566 (1971). – Abstr. Optical Soc. America.

034.120 Diffraction-limited $f/3$ focal reducer for a 1.22-m $f/6.5$ Newtonian telescope. R. A. Buchroeder.
Journ. Optical Soc. America, Vol. 61, 1567 (1971). – Abstr. Optical Soc. America.

034.121 Static infrared attitude sensor for spacecraft.
E. J. Fjarlie, T. Doyle.
Journ. Optical Soc. America, Vol. 61, 1583 (1971). – Abstr. Optical Soc. America.

034.122 The heat-pipe oven, a new instrument in spectroscopy. C. R. Vidal.
Journ. Optical Soc. America, Vol. 61, 1588 (1971). – Abstr. Optical Soc. America.

034.123 The Kitt Peak magnetograph. I. Principles of the instrument. W. Livingston, J. Harvey.
Kitt Peak National Obs., Contr. No. 558, 29 pp. (1971).

034.124 The usefulness of a flying spot digitizer for the measurement of the coordinates of star images on photographic plates. H. Eichhorn, G. Gatewood.
Conference on photographic astrometric technique, Tampa 1968, (see 012.031), p. 25 - 32 (1971).

034.125 The Lick-Gaertner automatic measuring system.
S. Vasilevskis, W. A. Popov.
Conference on photographic astrometric technique, Tampa 1968, (see 012.031), p. 33 - 48 (1971).

034.126 A semi-automatic measuring machine.
K. A. Strand.
Conference on photographic astrometric technique, Tampa 1968, (see 012.031), p. 49 - 54 (1971).

034.127 Introduction to a discussion on spectrometers and detectors. L. Heroux.
Beam-foil spectroscopy. Conference 1970, (see 012.032), p. 173 - 180 (1970).

034.128 A computer controlled beam-foil spectroscopy system with an application to argon.
L. Bridwell, L. M. Beyer, W. E. Maddox, R. C. Etherton.
Beam-foil spectroscopy. Conference 1970, (see 012.032), p. 187 - 196 (1970).

034.129 Performance of an automated computerized plate scanner. W. J. Luyten.
Proc. National Acad. Sci. U.S.A., Vol. 68, 513 - 516 (1971).

A brief summary is given of the initial performance of an automated computerized plate-measuring machine, and it is seen that the machine is working entirely satisfactorily.

034.130 A thin film silver detector for the direct measurement of atomic oxygen in the upper atmosphere.

R. J. Thomas.
Thesis, Utah State Univ., Logan. [Available from Univ. Microfilms, Ann Arbor, Mich., U.S.A. Order No. 70–26993], 155 pp. (1970). – See Phys. Abstr., Vol. 74, No. 54052(1971).

034.131 **A hydrogen flame ionization detector for Martian/ lunar life detection experiments.**
D. P. Lucero, P. H. Smith, R. D. Johnson.
ISA (*Instrument Society America*) Trans., Vol. 10, 58 - 66 (1971). – See Phys. Abstr., Vol. 74, No. 70627 (1971).

034.132 **Hydrostatic bearings for the world's great telescopes.**
B. V. Barlow.
Contemporary Phys., Vol.12,419 - 436 (1971). – See Phys. Abstr., Vol. 74, No. 74745 (1971).

034.133 **Stellar interferometer.** P. L. Kebabian.
Quarterly Progr. Report, (*USA*), No. 101, p. 1 - 11 (1971).

034.134 **Observations of the Apollo-12 liquid oxygen cloud with a new type of spectrograph.**
L. J. Lantwaard, H. van de Stadt.
Journ. Phys. E. Sci. Instruments, Vol. 4, 879 - 881 (1971).

034.135 **Astronomical infrared spectroscopy with a Connes-type interferometer. I. Instrumental.**
R. Beer, R. H. Norton, C. H. Seaman.
Rev. Sci. Instruments, Vol. 42,1393 - 1403 (1971).
The construction and operation of a Connes-type Fourier spectrometer are discussed in an astronomical context with a sample of the type of spectra obtained with the system.

034.136 **Construction and testing of large-area X-ray collimators.**
D. J. Stewart, W. M. Glencross, D. H. Brabban.
Journ. Phys. E. Sci. Instruments, Vol. 4, 966 - 968 (1971). See Phys. Abstr., Vol. 75, No. 5030 (1972).

034.137 **Systems design of telescope arrays for intensity interferometry of astronomical sources.** V. Herrero.
Thesis, Univ. Wisconsin, Madison. [Available from Univ. Microfilms, Ann Arbor, Mich., U.S.A. Order No. 70–20847], 165 pp. (1970).
A new type of optical intensity interferometer has been designed. An array of telescopes, with photoelectric detectors, single bit analog to digital conversion, and digital data transmission and correlation techniques is suggested for performing intensity interferometry of stellar sources with angular sizes larger than 50 microseconds of arc.

034.138 **Evaluation of an LN_2-temperature solid state spectrometer for rocket borne experiments.**
G. A. MacGregor, I. Turiel, R. Bettenhausen, A. Iantuono.
Rev. Sci. Instruments, Vol. 42, 35 - 39 (1971).

034.139 **Balloon flight instrumentation for solar cell measurements.**
F. W. Sarles, Jr., W. C. Haase, P. F. McKenzie.
Rev. Sci. Instruments, Vol. 42, 346 - 353 (1971).

034.140 **Elektronenoptische Bildverstärker in der Astronomie.**
H. Schmidt, J.-D. Schumann.
Jahrb. Landesamt Forsch. Nordrhein-Westfalen 1970, p. 323 - 354 (1971).

034.141 **Telephoto lenses for X-ray telescopes and neutron cameras.** H. Wolter.
Zeitschr. Angew. Phys., Vol. 31, No. 3, p. 152 - 155 (1971). In German.

Astronomical use of television-type image sensors.
See Abstr. 003.053.

Holographically made zone plates for use in X-ray-telescopes. See Abstr. 031.021.

Construction of apodised zone plates for solar X-ray image formation. See Abstr. 031.022.

1971 - 1972: Guide to scientific instruments.
See Abstr. 031.043.

Computer controlled precision digitizers of optical image data extraction. See Abstr. 031.070.

Solar instrumentation (Part I).
See Abstr. 032.005.

Solar instrumentation (Part II).
See Abstr. 032.006.

Survey on new techniques for X-ray astronomy.
See Abstr. 051.020.

High resolution interferometric studies of the solar magnesium II doublet spectral region. See Abstr. 071.061.

An inexpensive pulse counter for photometry of occultations. See Abstr. 096.008.

Accuracy of radial velocities determined with a fiber optic electrostatic image tube. See Abstr. 112.063.

A study of the wavelength dependence of interstellar polarization using a scanning spectropolarimeter.
See Abstr. 131.023.

Experiment to measure hard solar and celestial X-rays from the fifth Orbiting Solar Observatory.
See Abstr. 142.054.

035   Clocks and Frequency Standards

035.001   **Two sundials of Canadian design.**
M. M. Thomson.
Journ. Roy. Astron. Soc. Canada, Vol. 65, 159 - 163 (1971).

035.002   **Time-keeping.**   Eh. I. Bauman.
Zemlya i Vselennaya, No. 4, p. 37 - 40 (1971).
In Russian.

035.003   **Synchronous storage of a signal during observations with a photoelectric transit instrument.**
A. D. Egorov.
Vestn. Khar'kov. Univ., No. 65, (Ser. Astron., No. 6), p. 84 - 88 (1971). In Russian.

035.004   **Diurnal frequency variation and refraction index.**
D. Latham.
Nature, Phys. Sci., Vol. 234, 157 - 158 (1971).
   A diurnal variation in the frequency difference between atomic clocks at two different locations is shown to be due to a diurnal variation in the index of refraction acting on the radio-frequency signal used to compare the clocks.

035.005   **On the comparison of time etalons using meteor trains.**
B. S. Dudnik, B. L. Kashcheev, A. N. Smirnov.
Vestn. Khar'kov. politekhn. in-ta, 1971, No. 54, p. 29 - 34.
In Russian. − Abstr. in Referativ. Zhurn. 51. Astron., 2.51.131 (1972).

036   Photographic Auxiliaries

036.001   **Zur Optimierung astronomischer Photogramme durch das FAH-Verfahren.**   W. Högner.
Sterne, 47. Jahrgang, p. 136 - 147 (1971).

036.002   **Relative detective quantum efficiency measurements of some emulsions when used with a new developer for astronomical plates.**   S. Jeffers.
Journ. Roy. Astron. Soc. Canada, Vol. 65, 176 - 177 (1971).
Abstr. RAS Canada.

036.003   **Response of type IIIa-J Kodak spectroscopic plates to baking in various controlled atmospheres.**
A. G. Smith, H. W. Schrader, W. W. Richardson.
Applied Optics, Vol. 10, 1597 - 1599 (1971).
   A new emulsion with excellent signal-to-noise characteristics has been hypersensitized to speeds practicable for astronomical applications by baking in a controlled nitrogen atmosphere. Important gains are achieved in speed and fog suppression relative to the usual procedure of baking in air.

036.004   **Comparison of the speed and fog density of the new astronomical photographic films made by the scientific institute Techphotoproject (Kazan) and the photographic materials of other firms.**   O. D. Dokuchaeva.
Soobshch. Gos. Astron. Inst. Shternberga, No. 171, p. 23 - 33 (1971). In Russian.
   The new astronomical films A-500 Y and possibly A-600 Y were found to have a speed equal to the speed of the best astronomical photomaterials. The panchromatic films, especially A-700 Y, have higher speed than photomaterials made by firms Kodak (London) and ORWO (DDR).

036.005   **An attempt of using dry ice for short-time storage of astronomical photographic films of high speed.**
O. D. Dokuchaeva.
Soobshch. Gos. Astron. Inst. Shternberga, No. 171, p. 34 - 46 (1971). In Russian.

036.006   **Relative detective quantum efficiency measurements of four astronomical emulsions when used with a new developer.**   S. Jeffers.
Astron. Astrophys., Vol. 15, 221 - 223 (1971).
   Relative detective quantum efficiency measurements have been made on four emulsions (IIa−0, 103a−0, IIa−D, IIIa−J) when used with their recommended developers and with MWP 2 developer. In both cases, the densities giving optimum detective quantum efficiency have been determined. A significant improvement in detective quantum efficiency is found for 103a−0 and IIIa−J emulsions.

036.007   **Automatic plate holder drive for astronomical photography.**   G. Jeansaume.
Electron. Microelectron. Industr., (*France*), No. 147, p. 53 - 54 (1971). In French. − See Phys. Abstr., Vol. 75, No. 2006 (1972).

Positional Astronomy. Celestial Mechanics

041 Positional Astronomy, Star Catalogues and Atlases

041.001 **Observations of Pluto, minor planets 10 Hygiea and 433 Eros, Saturn's satellites VII, VIII, IX made at the Crimean Astrophysical Observatory.**
L. I. Chernykh, N. S. Chernykh.
Byull. Inst. Teoret. Astron., *Leningrad*, Vol. 12, 739 - 741 (1971). In Russian.

041.002 **Erratum: "Orientation of the FK4 catalogue from meridian observations of the moon". [Astron. Journ.,** Vol. 75, 851 - 856 (1970)]. B. L. Klock, D. K. Scott.
Astron. Journ., Vol. 76, 581 (1971).

041.003 **Meridian observations of major planets and some minor planets 1963 - 1967.** H. Yasuda, R. Fukaya, H. Hara, H. Ishii, T. Ina.
Tokyo Astron. Bull., Second Ser., No. 208, p. 2435 - 2449 (1971).

This paper contains positions of major planets (Mercury, Venus, Mars, Jupiter, Saturn, Uranus, and Neptune) and some principal minor planets (Ceres, Pallas, Juno, and Vesta) observed with the eight-inch meridian circle during the period 1963−1967.

041.004 **An attempt of calculating an ephemeris for physical observations of the sun from the surfaces of planets.**
V. K. Abalakin, N. N. Petrova, T. A. Polozhentseva, B. M. Rubashev.
Solnechnye Dannye 1971 Byull., No. 4, p. 111 - 114 (1971). In Russian.

041.005 **Corrections to the circumpolar zone of AGK 3.**
V. V. Telnyuk-Adamchuk.
Astron. Tsirk., No. 618, p. 1 - 3 (1971). In Russian.

041.006 **Proposal for a new "Southern Durchmusterung".**
N. J. Rumsey.
Southern Stars, Vol. 24, 59 - 60 (1971). − Letter.

041.007 **PZT star systems as compared with the SAO catalog.** S. Iijima, Y. Niimi.
Ann. Tokyo Astron. Obs., Second Ser., Vol. 12, 241 - 262 (1971).

Values of star places in PZT star systems are compared with those in the Smithsonian Astrophysical Observatory Star Catalog for nine of PZT observatories, Tokyo, Mizusawa, Mount Stromlo, Washington, Richmond, Ottawa, Neuchâtel, Hamburg, and Greenwich. Results of comparison in right ascension, declination, and the proper motions, all in the sense of PZT minus SAO, are expressed by the following form as $K + A_1 \sin \alpha + B_1 \cos \alpha + A_2 \sin 2\alpha + B_2 \cos 2\alpha$. Periodic terms in this expression are considered to be almost ascribable to the scattering character of the SAO itself. However, some parts of them must be due to PZT star systems. Discussions are given also on this problem.

041.008 **Catalogue of right ascensions of 645 FKSZ stars in the FK4 system.** L. L. Vagushchenko.
Astron. Tsirk., No. 635, p. 5 - 7 (1971). In Russian.

041.009 **Determination of absolute right ascensions from observations of star pairs symmetric relative to the zenith.** A. A. Nemiro.
Trudy Astron. Obs., *Leningrad,* Vol. 28 (= Uchenye Zapiski Leningr. Un-ta, No. 359 = Seriya Matem. Nauk, vyp. (No.) 47), p. 136 - 143 (1971). In Russian.

A method is proposed for the determination of absolute right ascensions of stars using observations at two observatories of star pairs symmetric relative to their zeniths. It is shown that the results of the observations will be free from the influence of pivot errors and azimuth of the transit instrument.

041.010 **On the use of lunar observations for the improvement of the zero-points of star catalogues and of the elements of the lunar orbit.** V. A. Fomin.
Trudy Astron. Obs., *Leningrad,* Vol. 28 (= Uchenye Zapiski Leningr. Un-ta, No. 359 = Seriya Matem. Nauk, vyp. (No.) 47), p. 144 - 154 (1971). In Russian.

The equations of condition, connecting the differences between the observed and the tabular co-ordinates of the moon with the equinox and the equator errors of the star catalogue, ΔA and ΔB, respectively, the correction to the obliquity of the ecliptic and the corrections to the elements of the lunar orbit are derived. Some inaccuracies of the method proposed by D. P. Duma (1964) for the reduction of short series of lunar observations are pointed out and corrected.

041.011 **On the possibility of determining the clock correction with an astronomical universal instrument by observations of stars symmetric with regard to the zenith.** V. N. Lvov.
Trudy Astron. Obs., *Leningrad,* Vol. 28 (= Uchenye Zapiski Leningr. Un-ta, No. 359 = Seriya Matem. Nauk, vyp. (No.) 47), p. 154 - 162 (1971). In Russian.

The accuracy of the method is found to be sufficient for first-class astronomical determinations.

041.012 **Some questions of the total calculation of the instrument system from differential determinations of right ascensions.** L. M. Khommik.
Soobshch. Gos. Astron. Inst. Shternberga, No. 170, p. 14 - 19 (1971). In Russian.

041.013 **Derivation of a declination system of stars of the Moscow photographic zenith tube.**
D. N. Ponomarev.
Soobshch. Gos. Astron. Inst. Shternberga, No. 170, p. 20 - 32 (1971). In Russian.

041.014 **Representation of a reference system by the Moscow catalogue of faint stars (zone 30° - 45°).**
A. P. Gulyaev, V. A. Korobova.
Soobshch. Gos. Astron. Inst. Shternberga, No. 170, p. 37 - 40 (1971). In Russian.

041.015 **Investigation of the systematic errors on the plates of the large astrograph of the Moscow Observatory.**
V. G. Fedosenko.
Soobshch. Gos. Astron. Inst. Shternberga, No. 170, p. 41 - 54 (1971). In Russian.

041.016 **The weights of 45 star catalogues.**
D. K. Karimova, E. D. Pavlovskaya.
Soobshch. Gos. Astron. Inst. Shternberga, No. 171, p. 3 - 9 (1971). In Russian.

041.017 **Empirical measures of the effect of coma on photographic astrometric measurements.**
K. W. Kamper, Jr.
Bull. American Astron. Soc., Vol. 3, 372 (1971). – Abstr. AAS.

041.018 **Comparison of the circumpolar zones of seven catalogues with AGK 2 – AGK 3.**
V. V. Telnjuk-Adamchuk.
Vestn. Kiev. Un-ta, Ser. Astron., No. 12, p. 68 - 72 (1970). In Russian.

The comparison of seven catalogues with AGK 2 – AGK 3 shows that the proper motions of AGK 2 – AGK 3 in the circumpolar zone contain essential systematic errors.

041.019 **Catalogue of the declinations of FKSZ stars in the system of FK 4.** A. Ja. Gregul.
Vestn. Kiev. Un-ta, Ser. Astron., No. 12, p. 73 - 89 (1970). In Russian.

041.020 **Observations of major planets.** N. D. Kovalenko.
Vestn. Kiev. Un-ta, Ser. Astron., No. 12, p. 93 - 95 (1970). In Russian.

041.021 **Results obtained with a Danjon astrolabe at Herstmonceux. II. Analysis and discussion: Herstmonceux Astrolabe Catalogue.** D. V. Thomas, R. E. Wallis.
Roy. Obs. Bull., Greenwich–Cape, No. 160, p. 113 - 171 (1971).

This Bulletin contains an analysis and assessment of the Herstmonceux astrolabe observations, details of which have previously been published. The Herstmonceux Astrolabe Catalogue contains values of $\Delta\alpha$ or $\Delta\delta$ or both for each of the 253 stars observed. Possible sources of systematic errors have been examined. A comparison has been made of the Herstmonceux Astrolabe Catalogue with N30, FK4 and some selected observational catalogues with a similar mean epoch of observation. Estimates of the systematic accuracy of each catalogue have been obtained from consideration of the internal precision of the catalogues and the scatter of the differences between them.

041.022 **On the astrometric use of the prime focus of the Isaac Newton Telescope with the Wynne field corrector.** C. A. Murray.
Monthly Notices Roy. Astron. Soc., Vol. 154, 429 - 444 (1971).

The results of measurement of 23 plates taken on the Pleiades with the Isaac Newton Telescope through the Wynne prime focus field corrector are discussed. Coefficients of radial distortion terms, including cubic and fifth powers of the radius are derived using the star positions in the catalogue of Eichhorn *et al.* as a standard reference frame. It is shown that the astrometric accuracy of the system is as good as that generally achieved with conventional long focus refractors.

041.023 **Luminous stars in the Southern Milky Way.**
C. B. Stephenson, N. Sanduleak.
Publ. Warner and Swasey Obs., Vol. 1, No. 1, 100 + 100 pp. (1971).

This publication represents an extension to the entire Southern Milky Way of the objective-prism survey for intrinsically luminous stars in the Northern Milky Way that was carried out a few years ago jointly by the Hamburg and Warner and Swasey observatories. These northern surveys were published as a series of six monographs titled "Luminous

Stars in the Northern Milky Way". Unfortunately, Vols. V and VI, which should have had a boundary in common in the sky, did not, a gap of 5° in declination lying between them; in the present volume we close that gap. We also publish herewith identification charts, or references to published charts, for all of the 5132 stars of the catalogue.

041.024 **Amelioration of declinations for 238 FKSZ stars from the –20° to +24° declination zone.**
M. Tudor, E. Toma.
Stud. Cerc. Astron., Vol. 16, 191 - 195 (1971). In Roumanian.

041.025 **Astrographic Catalogue 1900.0. Sydney Section: Dec. –51° to –65°,** from photographs taken at the Sydney Observatory, New South Wales, Australia. Vol. 53: Explanation. H. Wood.
Published by V. C. N. Blight, Government Printer, New South Wales, Sydney, Australia. 64 pp. (1971). – Contents: History; Personnel; The telescope; The measurement of the plates; The reference stars; Calculation of standard co-ordinates; Calculation of equatorial co-ordinates; Computation of plate constants; Accuracy; Magnitudes; Particulars of plates; Errata.

041.026 **Northern PZT stars observation program.**
H. Yasuda, I. Kamijo.
Ann. Tokyo Astron. Obs., Second Ser., Vol. 13, 1 - 79 (1971).

The outline of northern PZT stars observation program is explained. Moreover the final accuracy of these observations is estimated from the experiences in AGK3R observations.

041.027 **Status of PZT catalogs in current use.** H. Yasuda.
Ann. Tokyo Astron. Obs., Second Ser., Vol. 13, 80 - 92 (1971).

From the comparison of the common stars in AGK3R and PZT catalogs, the systematic errors in PZT catalogs in current use are shown. Some PZT catalogs have larger errors. The systematic errors are reflected in the determination of polar motion.

041.028 **Normalized observations of Venus 1901 - 1949.**
P. M. Janiczek, P. K. Seidelmann.
United States Naval Obs., *Washington,* Circ. No. 134, 75 pp. (1971).

The reconstructed observations described and tabulated in this circular were derived from residuals that were systematically reduced and employed by R. L. Duncombe in a discussion of the motion of Venus (Astron. Papers prepared for the use of the American Ephemeris and Nautical Almanac, Vol. 16, Part 1, 1958). These data are available in machine readable form and may be obtained through inquiry to the Superintendent, U. S. Naval Obs., Washington, D. C.

041.029 **Improvement of mean places for supplementary stars in the Tokyo PZT star list.**
S. Iijima, S. Fujii, H. Kobayashi, F. Ohtsuka.
Tokyo Astron. Bull., Second Ser., No. 212, p. 2491 - 2504 (1971).

29 stars taken from the Smithsonian Astrophysical Observatory Star Catalog were supplemented to the Tokyo PZT star list in 1969. Errors of the adopted mean places are calculated here by use of their observational data obtained during the past 19 months. The final star list including such supplemented stars with the mean places improved by these results has been used since the beginning of 1971. The star list is tabulated.

041.030 **Preliminary catalogue of the positions and proper motions of stars between declinations –70° and –90°, reduced to the equinox of 1950 without applying proper motions.** P. K. Lü, with an introduction by

D. Hoffleit, P. K. Lü.
Trans. Astron. Obs. Yale Univ., *New Haven,* Vol. 31, 37 + 274 pp. (1971). – The catalogue lists 18702 stars.

041.031 Is the "astrometric" plate reduction method rigorous?   H. Eichhorn.
Astron. Nachr., Vol. 293, 127 - 130 (1971).

It is shown that there is no intrinsic difference between the so-called "astrometric" and "photogrammetric" methods of reducing star plates, and that either, when applied by competent investigators, will yield equivalent results.

041.032 Meridiankreis - Beobachtungen von Planeten 1950 bis 1955.   P. Labitzke.
Astron. Nachr., Vol. 293, 131 - 135 (1971).

The positions observed at the meridian circle of the observatory Munich in the years 1950–1955 and their deviations from the ephemerides are given for the planets Mars, Jupiter, Saturn, Uranus, Neptun, Ceres, Pallas, Juno, Vesta.

041.033 Observations à la lunette zénithale (de 110 mm) du service de latitude de l'Observatoire de Beograd en 1969.   V. Milovanović, R. Grujić, M. Djokić.
Bull. Obs. Astron. Beograd, Vol. 28, (No. 124), 159 - 163 (1970).

Les observations et les réductions furent prépondéramment suivant le programme et le procédé décrits au Bull. Obs. Beograd, Vol. 24, (No. 3 - 4), 19 (1959). Les déclinaisons sont corrigées par les nouvelles valeurs de corrections des déclinaisons des sous-groupes.

041.034 An overlap reduction of the measurements used to produce the Cape Photographic Catalogs for −52° to −56° and −56° to −60°.
C. F. Lukac, J. K. Murphy, W. D. Googe, H. K. Eichhorn.
TOPOCOM, Technical Report, [U.S. Army Topographic Command, Washington], No. 1-2, 6 + 185 pp. (1970).

The plate overlap technique applied here incorporates all the information on adjoining plates into one solution rather than treat each plate as a separate entity. The Cape Photographic Zone Catalogs for −52° to −56° and −56° to −60° were selected to test the overlap method because of the availability of the measurements, as well as the need to improve the position of the southern stars. The newly derived positions indicate an improvement in accuracy when compared with the positions previously derived by the classical method.

041.035 El catálogo Messier.   A. D. S. Vázquez A.
El Universo, Vol. 25, 94 - 97 (1971).

041.036 Catálogo Messier.   A. D. S. V. Aguirre.
El Universo, Vol. 25, 111 - 114 (1971).

041.037 Stars with large proper motions in the astrographic zones +32° and +33°. (Final list).   S. Aravamudan.
Nizamiah Obs., Osmania Univ., *Hyderabad,* Contr. No. 4, 33 pp. (1969).

This contribution gives the relative proper motions of 995 stars in these regions from R.A. 8h to 12h together with the corrections for converting them into absolute motions.

041.038 General principles of studying differences in positions and proper motions of stars as a random field.
Ya. S. Yatskiv.
Astrometriya i Astrofiz., *Kiev,* No. 13, (see 003.039), p. 3 - 13 (1971). In Russian.

The methods applied at present for comparing star catalogues have a number of drawbacks. In this connection calculation methods of the random field theory are suggested to be used, the main advantage of which is objectivity (independence of the person carrying out the analysis) and homogeneity

of the results obtained. Some of the problems considered are illustrated by an example of comparing the Goloseevo catalogue of latitude stars with the declinations of the GC.

041.039 On the relative orientation of the fundamental catalogue coordinate systems.   N. T. Mironov.
Astrometriya i Astrofiz., *Kiev,* No. 13, (see 003.039), p. 13 - 18 (1971). In Russian.

To concretize the statement on the noncoincidence of equator and vernal equinox of different catalogues, the angles between axes and planes of the catalogue coordinate systems are proposed to be determined. When comparing catalogues, it would be reasonable to exclude systematic differences due to a relative rotation of the coordinate systems of the catalogues under comparison. The presented considerations are illustrated by comparing the catalogues FK4 and GC, FK4 and N30.

041.040 On taking into account the wire thickness when reducing observations of planets to the centre of the disk.   I. I. Bozhko, A. S. Kharin.
Astrometriya i Astrofiz., *Kiev,* No. 13, (see 003.039), p. 89 - 91 (1971). In Russian.

041.041 Processing of meridian observations by an electronic computer.   L. L. Vagushchenko, V. A. Sinyaev.
Astrometriya i Astrofiz., *Kiev,* No. 13, (see 003.039), p. 92 - 100 (1971). In Russian.

041.042 On determination of star declinations from observations in upper and lower culminations.
A. K. Korol.
Astrometriya i Astrofiz., *Kiev,* No. 13, (see 003.039), p. 100 - 105 (1971). In Russian.

Various procedures are discussed to determine weights when estimating declinations from the observations in upper and lower culminations. It is shown that in case of small number of observations the determination of individual weights by means of empiric dispersions give rough results. Average weights are preferable to be computed for declination zones.

041.043 International Information Bureau on Astronomical Ephemerides.
Bull. B.I.I.E.A. (I.A.U.–COSPAR), Paris. Information Cards, Nos. 1 - 29, with an introduction by B. Morando (1971).

During its 14th general assembly, the International Astronomical Union has established an International Information Bureau on Astronomical Ephemerides (B.I.I.E.A., see Proceedings of the 14th general assembly of the I.A.U., p. 84). The purpose of this Information Bureau is to provide information to the international scientific community on the availability of astronomical ephemerides, catalogues of star positions and lists of reduced positional observations for use in astronomical and space research. The Information Bureau is located at the Bureau des Longitudes in Paris, under the management of B. Morando. The Information collected by the Bureau shall be distributed free of charge to the institutions that request it. It will be given on information cards.

041.044 Radio location, minor planets and astrometry.   V. N. Bojko.
Vestn. Leningr. un-ta, 1971, No. 13, p. 144 - 150. In Russian. Abstr. in Referativ. Zhurn. 51. Astron., 2.51.130 (1972).

041.045 Results of differential observations of right ascensions with Sucharev's horizontal meridian circle in Pulkovo.   G. I. Pinigin.
Astron. Tsirk., No. 667, p. 4 - 6 (1971). In Russian.

041.046 Astrometric investigations at the Vienna Observatory and astrometric plans for the 60″ reflector of the L. Figl Observatory.   J. Meurers.

Conference on photographic astrometric technique, Tampa 1968, (see 012.031), p. 99 - 108 (1971).

041.047 **The AGK3, a basis for a general (northern) reference catalogue of positions and proper motions.**
W. Dieckvoss.
Conference on photographic astrometric technique, Tampa 1968, (see 012.031), p. 161 - 167 (1971).

041.048 **An analysis of the AGK3 comparison star positions.**
P. Herget.
Conference on photographic astrometric technique, Tampa 1968, (see 012.031), p. 169 - 171 (1971).

041.049 **Recent progress on the Yale Zone Catalogues.**
D. Hoffleit.
Conference on photographic astrometric technique, Tampa 1968, (see 012.031), p. 173 - 179 (1971).

041.050 **Comparison of the SAO and AGK3R star catalogues.** F. P. Scott, C. A. Smith, Jr.
Conference on photographic astrometric technique, Tampa 1968, (see 012.031), p. 181 - 190 (1971).

041.051 **Plans for a standard region for long focus astrographs.** G. D. Gatewood, L. W. Fredrick.
Conference on photographic astrometric technique, Tampa 1968, (see 012.031), p. 191 - 195 (1971).

041.052 **Cape photographic catalogues.** R. H. Stoy.
Conference on photographic astrometric technique, Tampa 1968, (see 012.031), p. 197 - 199 (1971).

041.053 **The improvement of star positions by photographic methods.** S. V. M. Clube.
Conference on photographic astrometric technique, Tampa 1968, (see 012.031), p. 199 - 207 (1971).

041.054 **The behavior of magnitude dependent systematic errors.** H. Eichhorn.
Conference on photographic astrometric technique, Tampa 1968, (see 012.031), p. 241 - 247 (1971).

041.055 **FK4 and fundamental reference system.** A. Sinzi.
Astron. Herald, (*Japan*), Vol. 64, 96 - 99 (1971).
In Japanese. — See Phys. Abstr., Vol. 74, No. 77452 (1971).

041.056 **What is the fundamental coordinate system?**
S. Aoki.
Astron. Herald, (*Japan*), Vol. 64, 100 - 102 (1971). In Japanese. — See Phys. Abstr., Vol. 74, No. 77593 (1971).

Centre de Données Stellaires. Inform. Bull. No. 2.
See Abstr. 002.026.

An investigation of the possibility to use the Abastumani Observatory 70-cm meniscus telescope for astrometric works. See Abstr. 032.003.

Catálogo de nebulosas extragalácticas de la zona −5°/−25° de declinación, seleccionadas para la determinación de un sistema absoluto de movimientos propios estelares. II — Coordenadas rectilíneas de nebulosas, estrellas de referencia y estrellas de control. See Abstr. 158.124.

042 Celestial Mechanics

042.001 The possibility of capture in the restricted problem of three bodies and formation of bridges between galaxies.   S. Yabushita.
Monthly Notices, Roy. Astron. Soc., Vol. 153, 97 - 109 (1971).

An investigation is made of the possibility of capture process where a planet of negligible mass, initially in Keplerian orbit about a star, is perturbed by a passing star with hyperbolic velocity at great distance and becomes a planet of the star escaping to infinity with it. It is shown that this process occurs under a fairly wide range of initial conditions. If a galaxy is represented by a heavy nucleus and a disk part with negligible mass compared with the nucleus, the process of capture makes it possible for a passing galaxy to extract a fraction of mass originally in the disk.

042.002 Redundant variables in celestial mechanics.   R. Broucke, H. Lass, M. Ananda.
Astron. Astrophys., Vol. 13, 390 - 398 (1971).

The theory of redundant variables is used in the Lagrangian and the Lagrangian equations of motion of the perturbed two-body problem with 3 degrees of freedom. Several properties of the osculating plane of motion are given and in particular a strong emphasis is made on separating the in-plane and out-of-plane effects. Several new Lagrangians and new sets of equations of motions are given.

042.003 Erratum: "A recursive von Zeipel algorithm for the ideal resonance problem". [Astron. Journ., Vol. 76, 157 - 166 (1971)].   B. Garfinkel, A. Jupp, C. Williams.
Astron. Journ., Vol. 76, 581 (1971).

042.004 Existence of quasi-periodic solutions to the three-body problem.   B. B. Lieberman.
Celestial Mechanics, Vol. 3, 408 - 426 (1971).

A rigorous proof is given for the existence of quasi-periodic solutions with only two degrees of freedom to a planar three-body problem. The solution corresponds physically to the small bodies moving on different, nearly elliptical orbits about a large mass located at a focus. The perihelia of the two orbits are locked in such a way that the difference of the two perihelia has mean value zero.

042.005 Periodic collision orbits in the elliptic restricted three-body problem.   R. Broucke.
Celestial Mechanics, Vol. 3, 461 - 477 (1971).

This article considers the two-dimensional elliptic restricted three-body problem, and in particular some of its aspects related to regularization and periodic collision orbits. The mechanism of regularization with Birkhoff coordinates and with the energy differential equation is described. Then the initial conditions for collision orbits are established. The theory is illustrated with the description of a new family of symmetric periodic collision orbits. In the high eccentricity ranges, some relations with the triple collision problem are pointed out.

042.006 The comparison of Hori's perturbation theory and von Zeipel's theory.   M. Yuasa.
Publ. Astron. Soc. Japan, Vol. 23, 399 - 403 (1971).

The equivalence of Hori's (1966) perturbation theory and von Zeipel's (1916) theory is established up to the third order approximations in perturbed elliptic motions.

042.007 On the stability of the triangular Lagrangian solutions for the restricted three-body problem in the three-dimensional circular case.   A. P. Markeev.
Astron. Zhurn. Akad. Nauk SSSR, Vol. 48, 862 - 868 (1971). In Russian. English translation in Soviet Astron. AJ, Vol. 15, No. 4.

Stability of the triangular Lagrangian solutions for the restricted three-body problem in the three-dimensional circular case is discussed. The stability has been proved for most of initial conditions at all values of μ from the stability area in the first order approximation with the exception of those two, for which instability for the plane case was previously proved.

042.008 Apples in a spacecraft.   H. Alfvén.
Science, Vol. 173, 522 - 525 (1971).

Some consequences of Newtonian mechanics, previously overlooked, results in a new understanding of the behavior of small bodies in the solar system. Collisions between such bodies lead not to a scattering of these bodies over an increasing volume but instead to a contraction resulting in a "jet stream", with application to meteor streams and streams of asteroids. It is possible that comets are formed by bunching in such streams.

042.009 A restricted three-body problem in a resisting medium.   G. Horedt.
Astron. Astrophys., Vol. 14, 223 - 225 (1971).

For two circularly moving bodies whose masses are in the ratio 1 : 999, the transition of a particle from the domain of attraction of the large mass to that of the small mass is studied by backward integration in a resisting medium of constant density, where the resistance is proportional to the square of the particle's velocity.

042.010 Automated algebraic manipulation in celestial mechanics.   W. H. Jefferys.
Commun. Ass. Computing Machinery, Vol. 14, 538 - 541 (1971).

In this paper we consider some of the applications of automated algebraic manipulation which have been made in celestial mechanics. Particular attention is paid to the use of Poisson series, and a typical problem in perturbation theory is described. The requirements of processors for use in celestial mechanics are considered and compared with those for general manipulation packages. Some future directions for research using these systems are briefly outlined.

042.011 Non-classical problems of celestial mechanics.   V. V. Radzievskij.
Zemlya i Vselennaya, No. 4, p. 14 - 19 (1971). In Russian.

042.012 On the derivatives in the gravitational field of an axisymmetric planet.   A. N. Kovalenko.
Vestn. Leningr. un-ta, 1971, No. 1, p. 123 - 134. In Russian. Abstr. in Referativ. Zhurn. 51. Astron., 9.51.107; 62. Issled. kosm. prostranstva, 9.62.324 (1971).

042.013 Analytical theory of satellite motion in a non-central gravitational field.
K. V. Kholshevnikov, E. I. Timoshkova.
Trudy Astron. Obs., *Leningrad*, Vol. 28 (= Uchenye Zapiski Leningr. Un-ta, No. 359 = Seriya Matem. Nauk, vyp. (No.) 47), p. 97 - 118 (1971). In Russian.

An analytical theory is developed of satellite motion in the gravitational field of a rotating non-spherical planet. The orbit of the problem of two fixed centers is used as intermediate. It enables one to take into account the second zonal harmonic of the potential. The first order perturbations caused by all zonal and tesseral harmonics are obtained in the non-resonance case. A general method of integration of the equations in the resonance case is proposed.

042.014 Development of the disturbing function in the case that the mutual inclination of the orbits is near $\pi/2$. R. A. Lyakh.
Trudy Astron. Obs., *Leningrad,* Vol. 28 (= Uchenye Zapiski Leningr. Un-ta, No. 359 = Seriya Matem. Nauk, vyp. (No.) 47), p. 119 - 125 (1971). In Russian.
The development of the disturbing function is derived with the cosine of the mutual inclination used as a small parameter. The development is given for the case of circular orbits; it may be generalized for the case of elliptic motions.

042.015 On the practical stability of the triangular points in the restricted three-body problem. J. Olszewski.
Celestial Mechanics, Vol. 4, 3 - 14 (1971).
This paper contains a proposal of a new way of treating astrodynamical stability problems. A definition of a practical stability and a direct method of its examination are presented. The method has been applied to the triangular points problem for variety of μ and e values in the case of the linearized equation system as well as in the general one. The results are shown in a form which facilitates the comparison with results published by other authors.

042.016 Periodic, consecutive-collision orbits in the restricted problem for $\mu \neq 1/2$. E. M. Standish, Jr.
Celestial Mechanics, Vol. 4, 31 - 43 (1971).
Sequences of periodic, consecutive-collision orbits in the restricted problem of three bodies have been found for cases where the two primaries are not of equal mass. Some of the orbits are shown graphically.

042.017 Sufficient conditions for escape in the three-body problem. E. I. Standish, Jr.
Celestial Mechanics, Vol. 4, 44 - 48 (1971).
Sufficient conditions are given for the escape of a member of a three-body system. The set of conditions are similar to those given previously by Tevzadze (1962). The new set compares favorably in most cases with that of Tevzadze.

042.018 On the integrability cases of the equation of motion for a satellite in an axially symmetric gravitational field. A. Ghaffari.
Celestial Mechanics, Vol. 4, 49 - 53 (1971).
Two integrability cases of the second-order differential equation which defines the projection (on a plane $z =$ const.) of the motion of a satellite around an axially symmetric body have been investigated. Analytical and physical properties are expressed.

042.019 Stability of and motion about L_4 at three-to-one commensurability. K. T. Alfriend.
Celestial Mechanics, Vol. 4, 60 - 77 (1971).
The stability of L_4 and the motion about L_4 in the restricted problem of three bodies is investigated when there is three-to-one commensurability between the long and short periods of motion, that is, when the mass ratio μ has the value $\mu = \mu_3 = 0.013516...$. The two time scale method is used (1) to show that L_4 is an unstable equilibrium point when $\mu = \mu_3$, (2) to determine for what initial conditions periodic orbits occur when $\mu \approx \mu_3$, (3) to determine the stability of the periodic orbits, and (4) to investigate the boundedness of the motions about L_4 when $\mu \approx \mu_3$.

042.020 Attitude stability of a symmetric satellite at the equilibrium points in the restricted three-body problem. T. R. Kane, E. L. Marsh.
Celestial Mechanics, Vol. 4, 78 - 90 (1971).
A problem of attitude motion of the smallest body for the restricted three-body problem is analyzed. Axial symmetry is assumed for the body, and attention is focused on the case in which the symmetry axis is normal to the orbit plane.

042.021 Periodic orbits near L_4 for mass ratios near the critical mass ratio of Routh. K. R. Meyer, D. S. Schmidt.
Celestial Mechanics, Vol. 4, 99 - 109 (1971).
The Hamiltonian for orbits near L_4 and mass ratios near μ_1 is brought into a normal form. A theorem shows that two coefficients in this expansion predict the behavior of the periodic orbits.

042.022 Solution of the N-body problem with recurrent power series. R. Broucke.
Celestial Mechanics, Vol. 4, 110 - 115 (1971).
A recurrent power series solution is given for the classical N-body problem. The application to numerical integration is also pointed out.

042.023 Classification of the motions of three bodies in a plane. V. Szebehely.
Celestial Mechanics, Vol. 4, 116 - 118 (1971).

042.024 The parabolic case in the generalized problem of two fixed centres. V. P. Girichev.
Soobshch. Gos. Astron. Inst. Shternberga, No. 160, p. 35 - 47 (1971). In Russian.

042.025 The systems of canonical variables of Adel Soudan. J. Meffroy.
Bull. American Astron. Soc., Vol. 3, 386 (1971). – Abstr. AAS.

042.026 The equations of perturbed motion of a satellite. E. I. Timoshkova.
Astron. Zhurn. Akad. Nauk SSSR, Vol. 48, 1061 - 1066 (1971). In Russian. English translation in Soviet Astron. AJ, Vol. 15, No. 5.
The satellite motion in the gravitational field of a non-spherical rotating planet is considered. The solution of the problem of two fixed centres is used as undisturbed intermediate orbit. The equations allow to find the perturbations of the elements with an accuracy up to 10^{-9}.

042.027 On the theory of the figures of liquid planets. V. N. Zharkov, V. P. Trubitsyn.
Izv. AN SSSR. Fiz. Zemli, 1971, No. 5, p. 3 - 10. In Russian. Abstr. in Referativ. Zhurn. 51. Astron., 10.51.103 (1971).

042.028 Kepler's laws and the Mars orbiters. J. R. Millburn.
Spaceflight, Vol. 13, 422 - 426 (1971).

042.029 Theory of general perturbations for non-canonical systems. G.-i. Hori.
Publ. Astron. Soc. Japan, Vol. 23, 567 - 587 (1971).
The theory of general perturbations originally developed for the solution of canonical systems is generalized so as to be applicable to non-canonical systems. Perturbed Kepler motion is discussed as an application of the general theory. Another application to the non-linear oscillations is shown by the solution of van der Pol's equation. Finally the present theory is compared with the original theory applied to the generalized canonical system.

042.030 On the real singularities of the N-body problem. H. J. Sperling.
Journ. reine angew. Math., Vol. 245, 15 - 40 (1970).
We study the singularities of the solution of the classical N-body problem under some restrictions. The major goals of

this paper are a proof of von Zeipel's conjecture and the extension of the results about the collision of all bodies at one point and instant to the case of several simultaneous collisions at different points.

042.031 **A new class of periodic solutions in the restricted three body problem.** K. R. Meyer, J. I. Palmore.
Journ. differential Equations, Vol. 8, 264 - 276 (1970).

By a classical theorem of Liapunov there are two families of periodic orbits in the restricted three body problem which eminate from the Lagrange triangular libration points for a certain range of values of the mass ratio parameter. For certain values of the mass ratio parameter these two families are near resonance and additional families of periodic orbits of longer period orbits are found. In some cases these additional families are found to terminate on the two classical families of Liapunov. The new families of periodic orbits are found by using a modified version of Birkhoff's fixed point theorem.

042.032 **Computerized series solution of relativistic equations of motion.** R. Broucke.
Astrophys. Space Sci., Vol. 12, 366 - 377 (1971).

A method of solution of the equations of planetary motion is described. It consists of the use of numerical general perturbations in orbital elements and in rectangular coordinates. The solution is expanded in Fourier series in the mean anomaly with the aid of harmonic analysis and computerized series manipulation techniques. A detailed application to the relativistic motion of the planet Mercury is described both for Schwarzschild and isotropic coordinates.

042.033 **Some topical problems of celestial mechanics.**
G. A. Chebotarev.
Byull. Inst. Teoret. Astron., *Leningrad,* Vol. 12, 792 - 795 (1971). In Russian.

042.034 **The Roche problem in an eccentric orbit.**
A. Nduka.
Astrophys. Journ., Vol. 170, 131 - 142 (1971).

The equations governing the evolution of a homogeneous satellite describing an eccentric Keplerian orbit about a rigid spherical central mass are formulated. A consistent formulation of these equations requires that we allow internal motions of uniform vorticity in the satellite as well as the possibility of varying orientation of its ellipsoidal figure. The equations have been solved numerically, and it appears from the examination of two different cases that if during the description of the Kepler orbit the satellite transgresses the classical Roche limit, its figure begins to change dramatically toward a disklike shape.

042.035 **The effect of gravitational and aerodynamic perturbations on the rotational motion around the mass centre of an asymmetrical solid body.** Yu. A. Pupyshev.
Vestn. Leningr. un-ta, 1971, No. 7, p. 129 - 134. In Russian.
Abstr. in Referativ. Zhurn. 51. Astron., 12.51.119; 62. Issled. kosm. prostranstva, 12.62.425 (1971).

042.036 **Power series representation of partial derivatives required in orbit determination.** A. F. Schanzle.
Celestial Mechanics, Vol. 4, 287 - 294 (1971).

A program to integrate the equations of motion by series in powers of the time step can be easily modified to furnish the elements of the matrizant in power series of the time step. In particular, if the series representing the motion are obtained recursively, differentiation of the recurrence relations will provide immediately a recursive scheme for computing the coefficient in the power series for the elements of the matrizant.

042.037 **A note on independent variables for restricted three-body problems.** T. A. Heppenheimer.
Celestial Mechanics, Vol. 4, 326 - 328 (1971).

It is found, that the true anomaly of the motion of the primaries if the only independent variable, so that the equations of motion show invariancy in form from the circular case.

042.038 **A problem of orbital dynamics, which is separable in KS-variables.** U. Kirchgraber.
Celestial Mechanics, Vol. 4, 340 - 347 (1971).

The motion of a rocket (with constant thrust) in the gravitational field of a mass point is studied. It is shown that, for the three dimensional case, the KS-variables produce the separability. A closed solution is found by using elliptic functions.

042.039 **The influence of the great inequality on the secular disturbing function of the planetary system.**
P. Musen.
Celestial Mechanics, Vol. 4, 378 - 396 (1971).

This paper discusses the influence of the great inequality, between Jupiter and Saturn, on the secular (long-period) disturbing function of the principal planets. There are several modern methods of eliminating periodic terms from the Hamiltonian and deriving a purely secular disturbing function: For example the Krylov-Bogolubov method is suggested for eliminating periodic terms, if it is desired to include the secular perturbations of the fifth and higher order in the heliocentric elements.

042.040 **Lie transforms and the Hamiltonization of non-Hamiltonian systems.** A. A. Kamel.
Celestial Mechanics, Vol. 4, 397 - 405 (1971).

To develop the perturbation solution of the non-Hamiltonian system of differential equations $\dot{y} = g(y, t; \epsilon)$, it is sufficient to obtain the perturbation solution of a Hamiltonian system represented by the Hamiltonian $K = Y \cdot g(y, t; \epsilon)$ which is linear in the adjoint vector Y. This Hamiltonization allows the direct use of the perturbation methods already established for Hamiltonian systems.

042.041 **Étude topologique générale des équations différentielles conservatives (indépendantes du temps ou périodiques). Application à la mécanique céleste.**
C. Marchal.
Celestial Mechanics, Vol. 4, 406 - 422 (1971).

The topological study of the conservative differential equations (either independant of the time or periodical) leads to 4 main types of trajectories: The 'open trajectories'; coming from infinity and going to infinity. The 'limited trajectories'; they always come again into any neighbourhood of any past position. The 'oscillating trajectories', and the 'abnormal trajectories'. All of these types may be encountered in the three-body problem of celestial mechanics.

042.042 **On the generalized restricted problem of three bodies.** G. N. Duboshin.
Celestial Mechanics, Vol. 4, 423 - 441 (1971).

The particular case of the complete generalized three-body problem (Duboshin, 1969, 1970) where one of the body-points does not exert influence on the other two is analyzed. It is not supposed that generally the third axiom of mechanics (action = reaction) takes place. Then we determine conditions for some particular solutions.

042.043 **Characteristic exponents at L_4 and L_5 in the elliptic restricted problem of three bodies.**
G. E. O. Giacaglia.
Celestial Mechanics, Vol. 4, 468 - 489 (1971).

We study the linear stability of the triangular points in the elliptic restricted problem by determining the characteristic exponents with a convergent method of iteration which in essence was introduced by Cesari (1940). We obtain the general term of such exponents as a power series in the eccentricity ϵ of the primaries, valid for sufficiently small ϵ and at all values of μ except one in the interval of stability of the circular problem.

042.044   **Expansion of the planetary disturbing function.**
R. Broucke, G. Smith.
Celestial Mechanics, Vol. 4, 490 - 499 (1971).

Some methods are described for the expansion of the disturbing function in planetary theory. One method uses the classical binomial expansion theorem or a successive approximation process derived from it. Another method is a direct application of the Laplace series expansions.

042.045   **Numerical experiments on the N-body problem.**
S. J. Aarseth.
Astrophys. Space Sci., Vol. 14, 20 - 34 (1971).

The different types of numerical methods available for integrating the equations of motion of N-body systems are discussed. N-body computations have been performed for a whole range of initial conditions and the general results are summarized. Some important aspects of cluster evolution are clarified and the qualitative behaviour of small stellar systems is now quite well understood.

042.046   **A multi-particle regularisation technique.**
D. C. Heggie.
Astrophys. Space Sci., Vol. 14, 35 - 39 (1971).

Certain features of a practical regularisation method, in which the potential or kinetic energy is used as a time-regularising function, are described.

042.047   **The use of integrals in numerical integrations of the N-body problem.**   P. E. Nacozy.
Astrophys. Space Sci., Vol. 14, 40 - 51 (1971).

The method presented here yields solutions of a higher accuracy while using less time of calculation than conventional procedures of numerical integration that do not use the integrals directly. The results of an application of the method to numerical integrations of a gravitational system of 25-bodies are given.

042.048   **Direct integration methods of the N-body problem.**
S. J. Aarseth.
Astrophys. Space Sci., Vol. 14, 118 - 132 (1971).

A fourth-order polynomial method for the integration of N-body systems is described in detail together with the computational algorithm. A discussion is given of the Kustaanheimo-Stiefel regularization procedure which is used to integrate dominant two-body encounters as well as close binaries.

042.049   **Treatment of close approaches in the numerical integration of the gravitational problem of N-bodies.**
D. G. Bettis, V. Szebehely.
Astrophys. Space Sci., Vol. 14, 133 - 150 (1971).

This paper discusses transformations to eliminate singularities and describes in considerable detail the numerical approaches to more accurate and faster integration. The basic ideas of smoothing and regularization are explained and applications are given.

042.050   **On some generalisations of the problems of two and three bodies.**   K. Ziołkowski.
Postępy Astron., Vol. 19, 345 - 350 (1971).   In Polish.

042.051   **New expansions of the perturbation function.**
K. Ziołkowski.
Postępy Astron., Vol. 19, 351 – 354 (1971).   In Polish.

042.052   **Method for calculation of perturbations for the satellite problem of three bodies with arbitrary inclinations and small eccentricities.**   V. V. Terentyev.
Byull. Inst. Teoret. Astron., *Leningrad*, Vol. 13, 36 - 48 (1971). In Russian.

A method for calculation of perturbations of elements of a satellite due to the attraction of a third body has been developed for small and equal to zero eccentricities and arbitrary inclinations. Corresponding ALGOL - 60 procedures have been applied to Jupiter X. The validity of the new method has been confirmed by comparison with the results of the existing theories.

042.053   **On a system of kinematic variables in celestial mechanics.**   L. A. Meleshkov.
Trudy pyatykh chtenij, posvyashch. razrabotke nauch. nasledija i razvitiyu idej K. Eh. Tsiolkovskogo, 1970. Sekts. "Mekh. kosmich. poleta". Moskva,1971, p. 161 - 172. In Russian.
Abstr. in Referativ. Zhurn. 62. Issled. kosm. prostranstva, 12.62.420 (1971).

042.054   **Teoria de perturbaciones para sistemas no canónicos.**
M. C. Pinilla
Urania Barcelona, Año 56, No. 273, p. 51 - 60 (1971).

Mersman has recently presented Hori's method from a more general point of view; here this theory is extended to the case of non-canonical systems of differential equations. Introduction of adequate differential operators permits to present the results in nearly the same form as in the Hamiltonian case.

042.055   **On the determination of minimal distance between orbits.**   L. Lautsenieks.
Latv. ordena trud. krasn. znameni gos. univ. im. P. Stuchki, Uch. zap., Vol. 148, vyp. (No.) 6, p. 87 - 90 (1971). In Russian.

042.056   **Two-body problem with slowly decreasing mass.**
F. Verhulst, W. Eckhaus.
International Journ. non-linear Mech.,Vol. 5, 617 - 624 (1970).

Evolution of the orbital elements in the two-body problem with slowly decreasing mass is described by a non-linear, non-autonomous system of differential equations with appropriate initial values. It is shown that the system of differential equations can be replaced by an integral equation and that three domains of the phase plane can be distinguished where the behaviour of the solutions is entirely different.

042.057   **On the elliptical class of trajectories in the problem of two fixed centres.**   I. S. Kozlov.
Trudy pyatykh chtenij, posvyashch. razrabotke nauch. nasledija i razvitiyu idej K. Eh. Tsiolkovskogo, 1970. Sekts. "Mekh. kosmich. poleta". Moskva, 1971, p. 93 - 113. In Russian.
Abstr. in Referativ. Zhurn. 51. Astron., 1.51.113 (1972).

042.058   **On the ellipticity of slowly rotating configurations.**
Z. F. Seidov.
Izv. AN AzSSR. Ser. Fiz.-tekhn. i mat. n., 1970, No. 6, p. 92 - 96. In Russian. – Abstr. in Referativ. Zhurn. 51. Astron., 1.51.127 (1972).

042.059   **Undisturbed eccentric anomaly difference as the independent variable in the perturbation differential equations.**   T. Godal, T. V. Johansen, E. L. Liipola.
Astronaut. Acta, Vol. 16, 259 - 264 (1971).

The osculating two-body orbit is described by departure position and velocity vectors restricted to vary so that the time derivative of the difference θ of the eccentric anomalies of the instantaneous position and departure position becomes equal to the time derivative of the eccentric anomaly E of the instantaneous undisturbed motion. The exact variational equations for both these vectors, the angular momentum vector and the semi major axis then all become simpler with θ as independent variable than in the classical case with the departure position at the pericenter and with E as the independent variable.

042.060   **The equilibrium figures in post-Newtonian approximation of general relativity theory.**   M. S. Volkov.
Byull. Inst. Teoret. Astron., *Leningrad,* Vol. 12, 866 - 869 (1971). In Russian.
    Krefetz derived an integro-differential equation defining the equilibrium figures of axisymmetrical celestial bodies in post-Newtonian approximation. In the paper presented this equation has been solved by Lichtenstein's method.

042.061   **The stability of circular satellite orbits and the estimate of the deviations.**   V. G. Degtjarjov.
Byull. Inst. Teoret. Astron., *Leningrad,* Vol. 12, 882 - 889 (1971). In Russian.
    For dynamical systems with quadratic Lyapunov function the deviations from stable motion are estimated. For example the circular orbits of an axisymmetrical field of gravitation are investigated. For a central field the result derived coincides with the theory of minor perturbations.

042.062   **Les équations des perturbations des petites planètes et comètes pour toutes les excentricités.**
B. Popović.
Byull. Inst. Teoret. Astron., *Leningrad,* Vol. 12, 890 - 898 (1971). In Russian.

042.063   **On the stability of a class of regular motions of a satellite.**   T. K. Schinkarik.
Byull. Inst. Teoret. Astron., *Leningrad,* Vol. 12, 899 - 909 (1971). In Russian.

042.064   **On a system of kinematic variables in celestial mechanics.**   L. A. Meleshkov.
Trudy pyatykh chtenij, posvyashch. razrabotke nauch. naslediya i razvitiyu idej K. Eh. Tsiolkovskogo, 1970. Sekts. "Mekh. kosmich. poleta". Moskva, 1971, p. 161 - 172. In Russian. Abstr. in Referativ. Zhurn. 51. Astron., 2.51.102 (1972).

042.065   **Conditionally periodic motions of a particle under attraction of an axisymmetric planet.**
A. I. Sazhin.

Vestn. Leningr. un-ta, 1971, No. 13, p. 166 - 172. In Russian. Abstr. in Referativ. Zhurn. 51. Astron., 2.51.103; 62. Issled. kosm. prostranstva, 2.62.212 (1972).

042.066   **Some periodic orbits in the elliptic restricted problem of three bodies.**   P. J. Shelus.
Thesis, Univ. Virginia, Charlottesville. [Available from Univ. Microfilms, Ann Arbor, Mich., U.S.A. Order No. 70–26546], 120 pp. (1970).
    This paper gives a preliminary report of the numerical exploration of the effects of the ellipticity of the orbits of the primaries on the linear stability characteristics of certain periodic orbits in the circular case.

042.067   **Periodic orbits in the rectilinear restricted three-body problem.**   R. Broucke.
Journ. Mécanique, Vol. 10, 449 - 465 (1971).
    A detailed study has been made of the two-dimensional motion of a massless particle under the gravitational action of two primaries which are moving in oscillatory Keplerian motion on a straight line. Some numerical results are given, relating to two families of periodic orbits and thirteen isolated periodic orbits.

042.068   **Isolated fixed points in celestial mechanics.**
R. B. Barrar.
Journ. Math. Analysis Applications, Vol. 36, 506 - 517 (1971).
– See Phys. Abstr., Vol. 75, No. 10814 (1972).

New qualitative methods of celestial mechanics.
See Abstr. 003.012.

Handbook on celestial mechanics and astrodynamics.   See Abstr. 003.045.

Lectures on celestial mechanics.
See Abstr. 003.112.

Matrix method for the Liapunov stability analysis of cyclic discrete mechanical systems.
See Abstr. 022.002.

On the perturbations of the five outer planets by the four inner ones.   See Abstr. 091.001.

Recent developments of integrating the gravitational problem of *n*-bodies.   See Abstr. 151.038.

043 Astronomical Constants

043.001 On determination of correction to precession from stellar proper motions.
S. Vasilevskis, A. R. Klemola.
Astron. Journ., Vol. 76, 508 - 512 = Lick Obs. Bull., No. 618 (1971).

Recent values for correction to centennial precession, with their mean errors, derived by Fricke (1967) from fundamental proper motions are: $\Delta p_1 = +1\!''\!.10 \pm 0\!''\!.15$ and $\Delta\lambda + \Delta e = +1\!''\!.20 \pm 0\!''\!.16$, as a consequence of the directly determined $\Delta n = +0\!''\!.44 \pm 0\!''\!.06$ and $\Delta k = -0\!''\!.19 \pm 0\!''\!.07$. Proper motions with reference to galaxies, derived at Lick and compared with those in AGK3, yield $\Delta n = +0\!''\!.31 \pm 0\!''\!.07$ and $\Delta k = -0\!''\!.78 \pm 0\!''\!.09$. The disagreement between both sets, particularly for Δk, can be explained either by systematic errors in the Lick proper motions or by local star streamings that vitiate the correction to precession derived from the fundamental proper motions.

043.002 Nearly diurnal nutation from time measurements.
S. Débarbat.
Astron. Astrophys., Vol. 14, 306 - 310 (1971). In French.

The results are given in order to compare the value deduced from Danjon astrolabe observations (Paris Observatory, 1956.5−1970.0) with theoretical values obtained from earth models.

043.003 Determinations of precession. W. Fricke.
Celestial Mechanics, Vol. 4, 150 - 162 (1971). − Invited review paper presented at IAU Colloquium No. 9 (see 012.006).

043.004 On the correction to precession from proper motions referred to galaxies.
S. Vasilevskis, A. R. Klemola.
Celestial Mechanics, Vol. 4, 163 - 170 (1971). − Presented at IAU Colloquium No. 9 (see 012.006).

043.005 The excess secular change in the obliquity of the ecliptic and its relation to the internal motion of the earth. S. Aoki, C. Kakuta.
Celestial Mechanics, Vol. 4, 171 - 181 (1971). − Presented at IAU Colloquium No. 9 (see 012.006).

043.006 Notes on equinox motion and corrections to precession. T. C. Van Flandern.
Celestial Mechanics, Vol. 4, 182 - 185 (1971). − Presented at IAU Colloquium No. 9 (see 012.006).

043.007 Observed and theoretical values of the nutations.
R. O. Vicente.
Celestial Mechanics, Vol. 4, 186 - 189 (1971). − Presented at IAU Colloquium No. 9 (see 012.006).

043.008 Precession-nutations and tidal potential.
P. Melchior.
Celestial Mechanics, Vol. 4, 190 - 212 (1971). − Presented at IAU Colloquium No. 9 (see 012.006).

043.009 Détermination des masses des planètes.
J. Kovalevsky.
Celestial Mechanics, Vol. 4, 213 - 223 (1971). − Invited review paper presented at IAU Colloquium No. 9 (see 012.006).

043.010 A determination of the masses of the five outer planets.
R. L. Duncombe, W. J. Klepczynski, P. K. Seidelmann.
Celestial Mechanics, Vol. 4, 224 - 232 (1971). − Presented at IAU Colloquium No. 9 (see 012.006).

043.011 Simultaneous solution for the masses of the principal planets from analysis of optical, radar, and radio tracking data. J. H. Lieske, W. G. Melbourne, D. A. O'Handley, D. B. Holdridge, D. E. Johnson, W. S. Sinclair.
Celestial Mechanics, Vol. 4, 233 - 245 (1971). − Presented at IAU Colloquium No. 9 (see 012.006).

043.012 The planetary masses and the orbits of the first four minor planets. J. Schubart.
Celestial Mechanics, Vol. 4, 246 - 249 (1971). − Presented at IAU Colloquium No. 9 (see 012.006).

043.013 The masses of the principal planets.
W. J. Klepczynski, P. K. Seidelmann, R. L. Duncombe.
Celestial Mechanics, Vol. 4, 253 - 272 (1971). − Presented at IAU Colloquium No. 9 (see 012.006).

043.014 Corrections to the lunisolar precession and the motion of the equinox from proper motions of cepheids. A. M. Sinzi.
Celestial Mechanics, Vol. 4, 273 - 276 (1971). − Presented at IAU Colloquium No. 9 (see 012.006).

043.015 Sur l'effet d'une correction de la constante de la précession. A. Bec.
Celestial Mechanics, Vol. 4, 277 - 278 (1971). − Presented at IAU Colloquium No. 9 (see 012.006).

043.016 Finding a better value for G. J. W. Beams.
Physics Today, Vol. 24, No. 5, p. 34 - 40 (1971).

We know the gravitational constant only to within about half a percent. With this elegant method, we hope for a precision of at least one part in ten thousand.

043.017 The system of planetary masses.
M. E. Ash, I. I. Shapiro, W. B. Smith.
Science, Vol. 174, 551 - 556 (1971).

New results show that Pluto's mass cannot be determined reliably from existing data.

043.018 Determination of the earth's and moon's masses based on observations of the motion of the automatic interplanetary stations Venus 4, Venus 5, Venus 6 and Venus 7. E. L. Akim, V. A. Stepaniants, Z. P. Vlasova.
Dokl. Akad. Nauk SSSR, Ser. Mat. Fiz., Vol. 201, 1303 - 1306 (1971). In Russian.

044 Time, Rotation of the Earth

044.001 Enlarged conditions in the selection of star pairs for time determination by Zinger's method.
V. G. Medvedev.
Izv. vyssh. uchebn. zavedenij. Geod. i aehrofotos"emka, 1970, No. 1, p. 121 - 129. In Russian. – Abstr. in Referativ. Zhurn. 51. Astron., 9.51.157; 52. Geod. Aehros"emka, 9.52.100 (1971).

044.002 Analysis of observations with transit instruments.
G. P. Pilnik.
Astron. Zhurn. Akad. Nauk SSSR, Vol. 48, 1067 - 1078 (1971). In Russian. English translation in Soviet Astron. AJ, Vol. 15, No. 5.
The origin of seasonal waves in astronomical time determinations is considered. It is shown that it is possible to take into account the errors in the coordinates of stars at definite moments of time. In the observations of 1962 these errors caused waves with periods of 12.0, 6.0, 4.0, 3.0, 2.4 and 2.0 months. This should be taken into account in the study of the short-period variations of the rotation of the earth. The results of a study of the stability of a system of transit instruments from year to year are presented.

044.003 Turning back of the Siberian rivers and its influence on the earth's rotation. Yu. V. Batrakov.
Astron. Zhurn. Akad. Nauk SSSR, Vol. 48, 1079 - 1084 (1971). In Russian. English translation in Soviet Astron. AJ, Vol. 15, No. 5.
The changes in the earth's rotation and the motion of the poles due to the planned turning back of the flow of the great Siberian rivers have been estimated. The effects expected have been shown to be small compared with those observed at present.

044.004 R. T. A. Innes and the variable rotation of the earth, with a note on some recent developments in the definition of time. J. Hers.
Monthly Notes Astron. Soc. Southern Africa, Vol. 30, 129 - 134 (1971).

044.005 Time reckoning in Russian marine and the chronology of events during the First Russian Antarctic Expedition. Ya. P. Koblents.
Inform. byull. Sov. antarkt. ehkspeditsii, 1970, No. 80, p. 5 - 23. In Russian. – Abstr. in Referativ. Zhurn. 51. Astron., 11.51.7 (1971).

044.006 Propagation of very long radio waves and irregularities in the earth's rotation. A. G. Fleer.
Trudy Sib. NII metrol., 1971, vyp. (No.) 11, p. 23 - 52. In Russian. – Abstr. in Referativ. Zhurn. 51. Astron., 11.51.176 (1971).

044.007 The correlation between the short-period irregularities of the earth's rotation and the motion of the instantaneous pole. A. G. Fleer.
Trudy Sib. NII metrol., 1971, vyp. (No.) 11, p. 53 - 58. In Russian. – Abstr. in Referativ. Zhurn. 51. Astron., 11.51.177 (1971).

044.008 Canada adopts atomic time. M. M. Thomson.
Journ. Roy. Astron. Soc. Canada, Vol. 65, 302 - 303 (1971).

044.009 Die Zeit. H.-U. Keller.
Orion Schaffhausen, 29. Jahrgang, p. 171 - 175 (1971).

044.010 La précision des divers types des instruments utilisés aux services de l'heure. D. Djurović.
Bull. Obs. Astron. Beograd, Vol. 28, (No. 124), 91 - 97 (1970).

044.011 Les irrégularités saisonières de la rotation de la terre en 1968 et les systèmes de l'heure. D. Djurović.
Bull. Obs. Astron. Beograd, Vol. 28, (No. 124), 99 - 107 (1970).

044.012 Détermination astronomique de l'heure en 1969.
M. Jovanović, D. Djurović, D. Vesić, M. Lončarević, D. Mandić.
Bull. Obs. Astron. Beograd, Vol. 28, (No. 124), 165 - 170 (1970).
Ce nouveau programme fut composé par D. Djurović et V. Radogostić-Sekulović. Il contient 297 étoiles.

044.013 The irregularity of the earth's rotation from data of astronomical observations for 1968.0 – 1971.0.
N. S. Sidorenkov.
Astron. Zhurn. Akad. Nauk SSSR, Vol. 48, 1305 - 1307 (1971). In Russian. English translation in Soviet Astron. AJ, Vol. 15, No. 6.

044.014 The rotation of the earth.
D. E. Smylie, L. Mansinha.
Sci. American, Vol. 225, No. 6, p. 80 - 88 (1971).

044.015 Was ist eine Zeitskala? G. Becker.
PTB Mitt., 81. Jahrgang, p. 405 - 411 (1971).
The following subjects are discussed: Does a theoretical time scale exist? Distinction between time concept and time scale; problems of the practical realization of time scales; the problem of the validity of time scales; definition of the origin and of the scale unit; definition of a time scale; what is a time scale composed of? Comparison with temperature scales; distinction between time units and scale units; is the mean solar time a time? – Besides these fundamental problems proposals and remarks are given with respect to the definition of the "International Atomic Time".

044.016 Übergang auf die neue UTC-Zeitskala zum 1. 1. 1972.
G. Becker.
PTB Mitt., 81. Jahrgang, p. 475 (1971).

044.017 Determination of Ephemeris Time by meridian observations of the moon with a visual transit instrument of the time service. B. A. Baranov.
Trudy VNII fiz.-tekhn. i radiotekhn. izmerenij, 1970, vyp. (No.) 3 (33), 265 - 273. In Russian. – Abstr. in Referativ. Zhurn. 51. Astron., 1.51.151; 52. Geod. Aehros"emka, 1.52.101 (1972).

044.018 On the correlation of variations of the phase propagation velocity of very long radio waves with the earth's motion around the mass centre. A. G. Fleer.
Uspekhi fiz. nauk, Vol. 104, 332 - 334 (1971). In Russian. Abstr. in Referativ. Zhurn. 51. Astron., 1.51.152 (1972).

044.019 Détermination astronomique de l'heure et heures demi-définitives de réception des signaux horaires.
L. Webrová, V. Ptáček.
Acad. Tchécoslov. Sci., Inst. Astron., Station de l'Heure, Prague, Sér. 5, Nos. 13 - 16 (1971). – 1971 January – August.

044.020 Astronomische Zeit- und Breitenbestimmungen, Empfangszeiten von Zeitsignalen.

Edited by Deutsches Hydrographisches Institut, Hamburg. 1971 January - September (1971).

044.021 **Time and Latitude Bulletins, Tokyo Astronomical Observatory.**
Tokyo Astron. Obs. Mitaka, Tokyo, Japan. Vol. 45, Nos. 3 - 8, p. 17 - 54 (1971). — 1971 March — August: Coordinates of instantaneous pole and corrections for UT2; Times of emission of radio time signals on UT2; Times of arrival of GBR and Iwôjima Loran-C signals; Astronomical observations made with the PZT.

044.022 **Time and Latitude Service.**
Polish Acad. Sci., Astron. Latitude Station, Borowiec, Circ. No. 118 (1971). — 1971 April - June.

044.023 **Time and rotation of the earth.** V. Ptáček.
Vesmír, Vol. 50, 295 - 297 (1971). In Czech.

044.024 **Corrections to Czechoslovak time signals.** V. Ptáček.
Říše hvězd, Vol. 52, 155, 181, 198, 223, 237 (1971). 1971 May - Sept.

044.025 **The atomic clock and the definition of the second.** E. Muff.
Schweiz. Archiv, Vol. 37, No. 3, p. 67 - 71 (1971). In German.
The relation of proper time of a standard clock against time dilatation and gravitational potential is established. In the special case of an observer on the surface of the earth this relation is analyzed. The theory is applied to a clock of a satellite of the earth and to the relativistic effect of a displacement of the clock on its rate, and finally a paradox is explained by means of the relativistic synchronization.

044.026 **Atomic standards of time.** L. Essen.
Alta Frequenza, (*Italy*), Vol. 40, 679 - 683 (1971). See Phys. Abstr. Vol. 74, No. 74813 (1971).

044.027 **Characteristics of a scale of time.** B. Guinot.
Alta Frequenza, (*Italy*), Vol. 40, 684 - 686 (1971). See Phys. Abstr., Vol. 74, No. 74814 (1971).

044.028 **Time and frequency: A bibliography of NBS literature published July 1955-Dec. 1970.** B. E. Blair.
National Bureau of Standards, Washington, D.C. Report Special Publ. 350, 50 pp. (1971).

044.029 **The physical basis of atomic frequency standards.** A. S. Risley.
National Bureau of Standards, Washington, D. C., Report TN 399, 54 pp. (1971).
A discussion ot the phyical basis of atomic frequency standards is given. These principles are related to the conditions under which an atom can be used as the working substance of a stable and accurate frequency standard. Three examples of atomic frequency standards, the hydrogen maser, the cesium beam, and the rubidium gas cell, are then discussed in terms of these principles and conditions.

044.030 **UTC time scale to change in 1972.**
Hewlett-Packard Journ., (*USA*), Vol. 23, No. 2, p. 16 (1971).

044.031 **Atomic time scales.** A. G. Mungall.
Metrologia, (*Germany*), Vol. 7, 146 - 153 (1971).
Techniques of construction of atomic time scales are discussed and experiments at the National Research Council using a group of commercial cesium standards and the NRC 2.1 m laboratory cesium standard, Cs III are described.

044.032 **Universal time and coordinates of the pole; Emission time of time signals; Coordinated time; Informations.**
Bureau International de l'Heure, Paris, Circ. D57 - D61 (1971).

Les coordonnées du pôle instantané en 1968 déterminées de mesures de l'heure. See Abstr. 045.009.

Sur le traitement des latitudes et heures observées pour en tirer les coordonnées du pôle et les irregularités de la rotation de la terre d'après le procédé du BIH et d'après celui de M. D. Djurović. See Abstr. 045.010.

Einiges über den Zweck und über die Methoden astronomisch-geodätischer Positionsbestimmungen. See Abstr. 046.001.

The elasticity theory of dislocations in real earth models and changes in the rotation of the earth. See Abstr. 081.012.

Comments on paper by D. E. Smylie and L. Mansinha: 'The elasticity theory of dislocations in real earth models and changes in the rotation of the earth'. See Abstr. 081.013.

Reply to comments on 'The elasticity theory of dislocations in real earth models and changes in the rotation of the earth'. See Abstr. 081.014.

045 Latitude Determination, Polar Motion

045.001 On periodicities found in investigating the closing errors in all the International Latitude Service Stations. R. J. Dejaiffe, P. J. Melchior.
Astron. Astrophys., Vol. 14, 468 - 472 (1971).

General analysis of the closing error in each northern International Latitude Station is presented on the base of observations made during 50 years (1899 - 1949). A long period is found around 24 to 29 months; other periods appear about 14 - 18, 9 and/or 8 months.

045.002 Latitude determination by measurements of azimuth differences of stars. V. N. Baranov.
Izv. vyssh. uchebn. zavedenij. Geod. i aehrofotos"emka, 1969, No. 6, p. 63 - 71. In Russian. – Abstr. in Referativ. Zhurn. 51. Astron., 9.51.155; 52. Geod. Aehros"emka, 9.52.98 (1971)

045.003 On density functions of the distributions of mean latitudes during night. V. V. Nesterov.
Soobshch. Gos. Astron. Inst. Shternberga, No. 170, p. 3 - 5 (1971). In Russian.

045.004 Analysis of Chandler's component in the polar motion during 119 years. L. V. Rykhlova.
Soobshch. Gos. Astron. Inst. Shternberga, No. 170, p. 6 - 13 (1971). In Russian.

045.005 Determination of the earth's pole of rotation from laser range observations to satellites.
K. Lambeck.
Bull. Géod., Nouvelle Sér., Année 1971, No. 101, p. 263 - 281 (1971).

The methods of using earth satellites for determining the motion of the earth's axis of rotation and of the earth's principal axis of maximum inertia are discussed. Some simple formulae are also presented for evaluating the influence of various error sources in the orbital calculations on the pole coordinates and these offer some explanations of the frequencies found in the spectrum of the pole coordinates obtained by Anderle and Beuglass (1970).

045.006 Refined geodetic results based on Doppler satellite observations. R. J. Anderle.
U. S. Naval Weapons Lab., Dahlgren, Virginia, NWL Techn. Report No. TR-2889, 3 + 14 + A3 + B18 pp. (1971).

Locations of Doppler satellite observing stations have been recomputed to determine a set which is more self-consistent and more consistent with the CIO pole. The motion of the earth's pole computed on the basis of these coordinates agrees with the astronomical determinations to about one meter, while the coordinates of individual stations determined on the basis of five days of observations of one satellite show a consistency of about two meters. Solutions for the positions of a station based on two satellite passes observed within a few hours were found to be accurate to about ten meters.

045.007 Analytical solution of some problems when observing two stars by the equal-altitude method.
A. I. Piotrovskaya.
Trudy Novosib. in-ta inzh. geod., aehrofotos "emki i kartogr., 1971, Vol. 24, p. 27 - 32. In Russian. – Abstr. in Referativ. Zhurn. 51. Astron., 12.51.148; 52. Geod. Aehros"emka, 12.52.103 (1971).

045.008 Analyse de l'influence de variation d'inclinaison des axes du tube zénithal (à Belgrade) sur la valeur de la latitude. M. Djokić.

Bull. Obs. Astron. Beograd, Vol. 28, (No. 123), 15 - 22 (1970).

045.009 Les coordonnées du pôle instantané en 1968 déterminées de mesures de l'heure. D. Djurović.
Bull. Obs. Astron. Beograd, Vol. 28, (No.124), 109 - 115 (1970).

045.010 Sur le traitement des latitudes et heures observées pour en tirer les coordonnées du pôle et les irregularités de la rotation de la terre d'après le procédé du BIH et d'après celui de M. D. Djurović. V. Milovanović.
Bull. Obs. Astron. Beograd, Vol. 28, (No. 124), 181 - 185 (1970).

045.011 Annual report of the International Polar Motion Service for the year 1969. S. Yumi.
Published for the International Council of Scientific Unions by Central Bureau of the International Polar Motion Service, Mizusawa. 4 + 171 pp. (1971).

This volume is a continuation of earlier reports. It contains the results of latitude observations made in 1969 in collaboration of 54 stations and observatories all over the globe.

045.012 Alcuni dati supplementari sulle variazioni della latitudine a Carloforte in funzione dell'angolo orario e della declinazione lunare. T. Nicolini.
Separate print Istituto Univ. Navale, Cattedra Astron. Generale e Sferica, Napoli, 4 pp. (1969).

The numerical evaluations reported here, refer to the residual variations of the latitude of Carloforte in function of the lunar hour angle and declination at the times of observation, for the years 1935 - 1940 and 1946 - 1948.

045.013 Rapporto sui lavori del Consiglio scientifico dell' IPMS (International Polar Motion Service).
E. Fichera.
Separate print Istituto Univ. Navale, Astron. Generale e Sferica, Napoli, 8 pp. (1971). – Presented at XIVth general assembly of the IAU, Brighton, August 1970.

045.014 Studio del catalogo SIL di Melchior – Dejaiffe dalle osservazioni 1935.0 - 1947.0. E. Fichera.
Separate print Istituto Univ. Navale, Astron. Generale e Sferica, Napoli, 115 pp. (1971). – Contribution to the global revision of the calculations of the International Latitude Service, (Working Group - UAI - Com. 19).

045.015 On the motion of the earth's poles.
A. A. Mikhailov.
Astron. Zhurn. Akad. Nauk SSSR, Vol. 48, 1301 - 1304 (1971). In Russian. English translation in Soviet Astron. AJ, Vol. 15, No. 6.

The recently published rectangular coordinates of the earth's poles referred to the CIO were used for a determination of the three main components of the polar motion. The mean secular motion during 1901–1968 was found to be 102 mm per year in the direction of 76°W longitude, but was far from uniform. The mean annual motion was very close to an ellipse with the semiaxes $0\overset{''}{.}098$ and $0\overset{''}{.}076$, the major axis directed towards 13°W. The Chandler wobble was very changeable, the mean period being close to 7/6 of a year, the mean amplitude for three intervals of 21 years varying from $0\overset{''}{.}06$ to $0\overset{''}{.}19$, a smaller amplitude corresponding to a shorter period and vice versa.

045.016 Über die Empfindlichkeit astronomisch-geodätischer Lotabweichungen gegenüber Dichte-Anoma-

lien des Untergrundes, nebst einer Anwendung auf die Bestimmung der Tiefenstrukturen im Nördlinger Ries.
A. Tuğluoğlu.
Dissertation, Inst. Theoretische Geodäsie der Rheinischen Friedrich-Wilhelms-Univ., Bonn. 99 pp. (1971).

045.017 **The excitation of the Chandler wobble by earthquakes.** F. A. Dahlen.
Geophys. Journ. Roy. Astron. Soc., Vol. 25, Nos. 1–3, (see 012.026), p. 157 - 206 (1971).

The computed polar shifts produced by the 1960 Chilean and 1964 Alaskan earthquakes showed a complete lack of agreement with the polar shifts inferred from the astronomical data by Smylie & Mansinha (1968). Changes in the earth's inertia tensor produced by large earthquakes are sufficiently large that it is extremely plausible that seismic activity is sufficient to maintain the earth's Chandler wobble.

045.018 **Ein Versuch zur Bestimmung der absoluten Lotabweichung aus visuellen Satellitenbeobachtungen.**
G. Gerstbach.
Österreich. Zeitschr. Vermessungswesen, 59. Jahrgang, p. 139 - 148 (1971).

045.019 **Distributions of vertical deflections in Western Germany and their group theoretical structure.**
E. Grafarend.
Mitt. Inst. Theor. Geod. Univ. Bonn, No. 1, 41 pp. (1971). In German.

045.020 **Monthly Notes of the International Polar Motion Service.**
IPMS Monthly Notes, International Latitude Obs. Mizusawa (Japan). 1971 Nos. 5 - 10 (1971). — Announces the values of latitudes observed at the collaborating stations during 1971 May - October.

045.021 **Results of the determination of latitude in Józefosław.** L. Pieczyński.
Latitude Circ., Warsaw Techn. Univ., Astron.-Geod. Obs. Józefosław, Nos. 37 - 38 (1971). — 1971 April - October.

045.022 **Study of the free diurnal nutation of the earth based on latitude observations in Pulkovo 1904–1941.** A. I. Emets, Ya. S. Yatskiv.
Astrometriya i Astrofiz., *Kiev*, No. 13, (see 003.039), p. 61 - 75 (1971). In Russian.

The power spectrum was studied of the latitude variation in Pulkovo from 1904 to 1941. The variations with periods 219, 208, 194 mean days and almost the same amplitudes $0''.004$ were found in the frequency region near the frequency of free diurnal nutation. A hypothesis was considered on the nutation amplitude modulation and a conclusion was drawn on the quasi-harmonic character of the power spectrum of the Pulkovo latitude variation in the frequency region near the frequency of free diurnal nutation.

045.023 **On comparison of results of separate determinations of the free diurnal nutation of the earth.**
Ya. S. Yatskiv.
Astrometriya i Astrofiz., *Kiev*, No. 13, (see 003.039), p. 75 - 81 (1971). In Russian.

A comparison of parameter estimates for the free diurnal nutation of the earth is considered. The parameters are based on latitude and time observations at different observatories. A report of basic determinations of this nutation parameters is presented.

045.024 **Latitude variations in Poltava from observations with the Bamberg zenith telescope from 1949 to 1961 in a new declination system.** E. I. Obrezkova.

Astrometriya i Astrofiz., *Kiev*, No. 13, (see 003.039), p. 114 - 117 (1971). In Russian.

When processing by the chain method a 12-year series of latitude observations with the Bamberg zenith telescope, a new declination system was obtained. Data are presented on latitude variations in Poltava from 1949 to 1961 calculated in this system.

045.025 **Treatment of latitude observations by adjustment to a standard.** N. P. Godisov.
Astron. Tsirk., No. 654, p. 4 - 7 (1971). In Russian.

045.026 **Analysis of the Chandler period of polar coordinates calculated with Orlov's method.**
E. Proverbio, F. Carta, F. Mazzoleni.
Mem. Soc. Astron. Italiana, Nuova Ser., Vol. 42, 497 - 515 (1971).

The coordinates of the pole related to the mean pole of the epoch in the period 1900-1962 for each ILS station were calculated using the method proposed by Orlov (1955). Successively the values of the amplitudes and periods of the polar motion were calculated. The comparison of the data seems to confirm a correlation between the amplitudes and periods of Chandler motion; that is the validity of the time-variable model suggested by Melchior (1949) and other authors.

045.027 **Instrumental constants for latitude reduction at the ILS Carloforte Stations.**
E. Proverbio, S. Uras.
Mem. Soc. Astron. Italiana, Nuova Ser., Vol. 42, 547 - 554 (1971).

The results of observations carried out during 1970 – 1971 for determining and studying the micrometer constants of the *VZT* of the international latitude Station of Carloforte are given.

045.028 **Wobble excitation by earthquakes in real earth models.** D. E. Smylie, L. Mansinha.
Nature, Vol. 232, 621 - 622 (1971).

The most important point discussed in some recent articles is the question of the efficacy of the mass displacements associated with major earthquakes in exciting the observed polar motion. We compute that the Chile earthquake of 1960 shifted the pole 1.03 centiseconds of arc towards 269°E, and that the Alaska earthquake of 1964 shifted the pole 1.20 centiseconds of arc toward 51°E. These provide equal and opposite contributions to Chandler wobble excitation.

045.029 **Latitude observation data processing by digital computer.** E. Proverbio, S. Uras.
Boll. Geod. Sci. Affini, Anno 30, p. 263 - 293 (1971).

The authors explain the procedure used in latitude calculation by the International Latitude Service based on the Talcott method. The formulae and the constants used in the calculation of declinations and zenith distances observed are given with an accuracy of $0''.001$. The latitude reduction program is then analyzed by means of the IBM 1130/32 computer of the University of Cagliari and an example of latitude calculation is discussed.

045.030 **Coordonnées du pôle instantané rapportées à l'origine conventionnelle internationale et corrections de longitude TU 1 – TU 0, à 0h TU.**
Bureau International de l'Heure, (BIH), Paris, Circ. B/C, Nos. 184 - 188 (1971). — Valeurs interpolées et extrapolées.

The correlation between the short-period irregularities of the earth's rotation and the motion of the instantaneous pole. See Abstr. 044.007.

Astronomische Zeit- und Breitenbestimmungen, Empfangszeiten von Zeitsignalen. See Abstr. 044.020.

Time and Latitude Bulletins, Tokyo Astronomical Observatory. See Abstr. 044.021.

Time and Latitude Service. See Abstr. 044.022.

The prime geocentric meridian, longitudes and the universal time. See Abstr. 046.029.

046 Geodetic Astronomy, Navigation

046.001 **Einiges über den Zweck und über die Methoden astronomisch-geodätischer Positionsbestimmungen.** H. Müller.
Orion, 29. Jahrgang, p. 99 - 108 (1971).

046.002 **Possibility of using orbital stations for cartographic and geodetic purposes.** G. Birardi.
L'Universo, Anno 51, p. 770 - 776 (1971). In Italian.

046.003 **Entwicklung und Perspektive des modernen geodätischen Gerätebaues.** H. Peschel.
Jenaer Rundschau, (Jena Rev.), 16. Jahrgang, p. 159 - 161 (1971).

046.004 **Perspektiven der geodätisch-astronomischen Arbeiten mit Fundamentalinstrumenten.** H. Kautzleben.
Jenaer Rundschau, (Jena Rev.), 16. Jahrgang, p. 193 - 195 (1971).

046.005 **Zur geodätischen Vermessung der Erde mit Satelliten.** K. Arnold.
Jenaer Rundschau, (Jena Rev.), 16. Jahrgang, p. 196 - 199 (1971).

046.006 **A satellite navigation system for general aviation and marine use.** T. M. B. Wright.
Journ. British Interplanet. Soc., Vol. 24, 591 - 602 (1971).
The aims of this paper are to establish the conditions under which a satellite system for marine and aviation use is economically viable, the features it must incorporate to make it operationally attractive, and the basis for the technical solution.

046.007 **The longitude of Herstmonceux.** N. P. J. O'Hora.
Observatory, Vol. 91, 155 - 159 (1971).

046.008 **Los cálculos de estación en triangulación espacial.** M. J. Sevilla.
Urania Barcelona, Año 55, Nos. 271–272, p. 153 - 193 (1970).

046.009 **Use of an alidade level for azimuth determinations.** V. G. L'vov.
Geod. i kartografiya, 1971, No. 4, p. 21 - 25. In Russian.
Abstr. in Referativ. Zhurn. 52. Geod. Aehros"emka, 9.52.101 (1971).

046.010 **On the absolute determination of the deflections of the plumb-line.** F. Andersson.
Bull. Géod., Nouvelle Sér., Année 1971, No. 101, p. 335 - 340 (1971).

046.011 **Polar navigation – A new transverse Mercator technique.** G. C. Dyer.
Journ. Inst. Navigation, Vol. 24, 484 - 495 (1971).

046.012 **Simplified methods of position fixing using earth satellites.** T. M. B. Wright.
Journ. Inst. Navigation, Vol. 24, 496 - 511 (1971).

046.013 **Beitrag zur Objektivierung von Meridiandurchgangsbeobachtungen.** S. Wächter.
Arbeiten aus dem Vermessungs- und Kartenwesen der Deutschen Demokratischen Republik. Vol. 17. Geodätischer Dienst – Zentrale Leitstelle für Information und Dokumentation für Vermessungs- und Kartenwesen, Leipzig. 87 pp. (1969). – Diss. Techn. Univ. Dresden.

046.014 **Midnatsolens varighet som breddebestemmelse.** K. Petersen.
Astron. Tidsskr., Årg. 4, 56 - 66 (1971).

046.015 **The weight in azimuth determinations by Polaris observations.** G. Folloni.
Boll. Geod. Sci. Affini, Anno 30, p. 223 - 229 (1971). In Italian.

046.016 **Compensation des coordonnées d'une station laser et des positions du satellite Geos–B, à l'aide des observations simultanées optiques et laser.** I. Șerban.
Stud. Cerc. Astron., Vol. 16, 151 - 161 (1971).
On présente un exemple pratique de détermination de la position d'une station laser par une méthode de géodesie géométrique spatiale en utilisant des mesures de directions et distances. On compense les coordonnées de la station et les positions du satellite.

046.017 **Détermination des coordonnées d'une station laser à l'aide des observations laser et photographiques d'un satellite artificiel.** I. Predeanu.
Stud. Cerc. Astron., Vol. 16, 163 - 168 (1971).
On utilise une méthode géométrique pour déterminer les coordonnées d'une station laser, quand on donne des observations faites simultanément à cette station et à deux autres stations dont les coordonnées sont connues (l'une optique et l'autre laser): 62 mesures simultanées du satellite Geos B.

046.018 **Ableitung differentieller Beziehungen in Vektoren- bzw. Matrizenschreibweise in der astronomisch geodätischen Ortsbestimmung.** H. Lichtenegger.
Österreich. Zeitschr. Vermessungswesen, 59. Jahrgang, p. 106 - 113 (1971).

046.019 **The mean latitude of the Engelhardt Observatory**

for 20 years.   A. M. Zulliev.
Astron. Tsirk., No. 646, p. 6 - 8 (1971). In Russian.

046.020   **The use of L band in a satellite system for aiding air navigation.**   G. Quaglione, E. Vitali.
Journ. British Interplanet. Soc., Vol. 24, 707 - 727 (1971).
  The paper discusses the main technical and operational characteristics of a satellite navigational aid system, particular attention being given to voice communications.

046.021   **Geodätische Astronomie.**   R. Sigl, E. Wolf.
Deutsche Geod. Kommission Bayer. Akad. Wiss., München, Ser. B, No. 187, p. 30 - 46 (1971).

046.022   **Numerical investigations on the strength of figure in a European satellite network using fictitious satellite positions in heights of 1200 kms and 4000 kms.**
W. Ehrnsperger, M. Näbauer, H. Seifers, H. Wolf.
Deutsche Geod. Komission Bayer. Akad. Wiss., München, Ser. B, No. 188, p. 5 - 22 (1971). In German.

046.023   **A combined gravimetric-astrogeodetic method for telluroid and vertical deflection analysis.**
E. Grafarend.
Deutsche Geod. Kommission Bayer. Akad. Wiss., München, Ser. B, No. 188, p. 23 - 36 (1971). – Presented to the IUGG-Congress, Moscow 1971.

046.024   **On the ill-conditioning of the normals in the camera calibration.**   M. Näbauer.
Deutsche Geod. Kommission Bayer. Akad. Wiss., München, Ser. B, No. 188, p. 51 - 64 (1971). In English and German.

046.025   **Connection of satellite triangulation stations by three dimensional traverses.**   K. Ramsayer.
Deutsche Geod. Kommission Bayer. Akad. Wiss., München, Ser. B, No. 188, p. 65 - 78 (1971).

046.026   **Über die Anwendung des Verfahrens zur Integration der Doppler-Frequenzverschiebung von Signalen künstlicher, erdnaher Satelliten für die geodätische Ortsbestimmung.**   A. Wallenhauer.
Deutsche Geod. Kommission Bayer. Akad. Wiss., München, Ser. C, No. 161, 54 pp. (1971). – Thesis Techn. Univ. Braunschweig.

046.027   **Bericht über die Entwicklung des Sonderforschungsbereiches 78 Satellitengeodäsie und die Arbeiten 1970 der Gruppe A "Dynamische Methode" und Bericht über die Arbeiten 1970 der Gruppe B "Geometrische Methode".**
R. Sigl, M. Kneißl.
Veröff. Bayer. Kommission für Internationale Erdmessung, Bayer. Akad. Wiss., München, Astron.-Geod. Arbeiten, No. 28, 49 pp. (1971).

046.028   **Determination of the astronomical azimuth of a terrestrial object from observations of transits of stars through the vertical of the object.**   A. V. Gozhij.
Dopovidi AN URSR, B, No. 7, p. 616 - 618, 666. In Ukrainian. – Abstr. in Referativ. Zhurn. 52. Geod. Aehros"emka, 12.52.108 (1971).

046.029   **The prime geocentric meridian, longitudes and the universal time.**   I. D. Zhongolovich.
Astron. Zhurn. Akad. Nauk SSSR, Vol. 48, 1308 - 1313 (1971). In Russian. English translation in Soviet Astron. AJ, Vol. 15, No. 6.
  In the present paper a definition has been given for the location of a system of terrestrial rectangular coordinate axes in the earth, and the dependence of this location upon the earth's polar motion is shown. The formulae for the polar mo-

tion influence on the results of some astronomical and geodetic observations have been derived. The justification of a new notion of the prime geocentric meridian is given. Several statements for discussion are made aiming at possible rationalization of some terms and methods now in use at the IPMS and BIH.

046.030   **On a possible increase of the accuracy of gravimetric results in geodesy.**   V. V. Brovar.
Astron. Zhurn. Akad. Nauk SSSR, Vol. 48, 1327 - 1332 (1971). In Russian. English translation in Soviet Astron. AJ, Vol. 15, No. 6.
  At present quasi-geoidal heights are known to be found with errors smaller than ± 10 m. This fact enables pure gravity anomalies to be obtained with a high degree of certainty. Thus the errors of disturbing gravity potential and those of vertical deflections obtained from pure anomalies are decreased nearly by half.

046.031   **Die Bestimmung des Richtungsvektors Riga-Sofia aus Beobachtungen des Satelliten "Echo 2".**
K. Arnold, V. Hristov, J. Klětnieks, N. Georgiev, K. Lapuška, D. Schoeps, J. Balodis, Z. Darakčiev, K.-H. Marek, H. Montag.
Deutsche Akad. Wiss. Berlin, Forschungsbereich Kosmische Phys., Veröff. Zentralinst. Phys. der Erde, *Potsdam*, No. 8, 60 pp. (1971).
  Simultaneous observations of the satellite "Echo 2" were made in Riga and Sofia during the period 1967 to 1968. From these observations the azimuth of the vector connecting these two stations was computed by a method published formerly. Using 13 pairs of simultaneous observations a standard error in azimuth of ± 0."28 was obtained.

046.032   **Analyse der Beobachtungsergebnisse der astronomisch-geodätischen Längenbestimmung Borowiec - Dresden - Potsdam aus dem Jahre 1966.**   J. Höpfner.
Deutsche Akad. Wiss. Berlin, Forschungsbereich Kosmische Phys., Veröff. Zentralinst. Phys. der Erde, *Potsdam*, No. 11, 123 pp. (1971).

046.033   **Untersuchungen zur Plattenreduktion von Sternaufnahmen.**   M. Mimus.
Deutsche Geod. Kommission, Bayer. Akad. Wiss., München, Ser. C, No. 169, 3 + 84 pp. (1971). – Dissertation Techn. Univ. Berlin.

046.034   **Automation der Isoliniendarstellung mit Hilfe des Wiener- und des Kalman-Filters.**   K. R. Koch, S. Lauer.
Mitt. Inst. Theor. Geod. Univ. Bonn, No. 2, 14 pp. (1971).

046.035   **Determination of the true azimuth by means of a meridian finder.**   P. I. Baran, V. Ya. Bantash.
Inzh. geodeziya. Mezhved. resp. nauch. sb., 1971, vyp. (No.) 9, p. 60 - 69. In Russian. – Abstr. in Referativ. Zhurn. 52. Geod. Aehros"emka, 1.52.95 (1972).

046.036   **On the determination of a terrestrial chord by means of laser observations of AES.**
I. D. Zhongolovich.
Byull. Inst. Teoret. Astron., *Leningrad*, Vol. 12, 851 - 865 (1971). In Russian.
  The present paper deals with a method of determining the length of the chord connecting two points on the terrestrial surface. The method is based on synchronous photographic observations of AES with simultaneous determinations of the AES topocentric distances. The direction of the chord has been supposed to be determined with sufficient accuracy from a series of special observations.

046.037   **The arctic-antarctic geodetic traverse project.**

I. D. Zhongolovitch.
Byull. Stantsij Optichesk. Nablyud. Iskusstv. Sputnikov Zemli, No. 57, p. 14 - 20 (1971). In Russian.

046.038 Longitud geografica del Observatorio Astronómico de la Universidad Nacional de La Plata.
S. J. Slaucitajs.
Obs. Astron. Univ. Nacional La Plata, Ser. Especial, No. 24, 21 pp. (1971).

046.039 Precision distance measurement based on the velocity of light.   K. D. Froome.
Sci. Progr., (*GB*), Vol. 59, 199 - 223 (1971).
Since the war, geodetic surveying has relied increasingly upon direct distance measurement in preference to triangulation related to perhaps a single calibrated base-line per country. This article describes the problems involved with precise

measurement through the earth's atmosphere and describes the basic forms of instrument which have been developed for this purpose.

Tables K 11: Two-star fix without use of altitude difference method. Vol. V N (Lat. 40° −49° 30′ N).
See Abstr. 003.010.

Geodesy.   See Abstr. 003.052.

Ein Verfahren zur automatischen Auswertung von Sterndurchgangsbeobachtungen mit einem Passageninstrument.
See Abstr. 031.033.

Geodetic applications of grazing occultations.
See Abstr. 096.024.

047 Ephemerides, Almanacs, Calendars

047.001   The Air Almanac 1972, January − April.
Her Majesty's Stationery Office, London; United States Naval Observatory, Washington. 244 + A82 + F4 pp. Price £ 1.75 (1971).

047.002   Nautisches Jahrbuch für das Jahr 1972.
Edited by Deutsche Demokratische Republik, Seehydrographischer Dienst, Rostock. 22nd year. 29 + 367 pp. (1971).

047.003   Nautisches Jahrbuch oder Ephemeriden und Tafeln für das Jahr 1972, zur Bestimmung der Zeit, Länge und Breite zur See nach astronomischen Beobachtungen.
Edited by "Deutsches Hydrographisches Institut", Hamburg. 121. Jahrgang, 4 + 39 + 367 + 30 pp. (1971).

047.004   The Star Almanac for Land Surveyors for the Year 1972.
Prepared by *H. M. Nautical Almanac Office,* published by Order of *The Science Research Council.* Her Majesty's Stationary Office, London. 14 + 70 pp. Price 40p. (1971).

047.005   Anuario del Observatorio Astronomico Nacional 1971.
Published by Observatorio Astronomico Nacional, Facultad de Ciencias, Universidad Nacional de Colombia, Bogotá. 112pp. (1971).

047.006   Annuaire de l'Observatoire Royal de Belgique [Jaarboek van de Koninklijke Sterrenwacht van België] 1972.
Imprimerie Hayez, Bruxelles. 139e année (jaargang). 215 pp. (1971).

047.007   Ephémérides Nautiques pour l'an 1972. Ouvrage publié par le Bureau des Longitudes spécialement à l'usage des marins.
Gauthier-Villars, Editeur, Paris. 479 pp. (1971).

047.008   1972 Nautical Almanac. Pub. No. 681.
Published by Hydrographic Office of Japan, Tokyo. 4 + 467 pp. (1971).

047.009   Astronomical phenomena for the year 1974.
Issued by the Nautical Almanac Office, United States Naval Observatory.
U. S. Government Printing Office, Wasnington, D.C. 66 pp. Price 55 cents (1970).

047.010   The Indian Ephemeris and Nautical Almanac for the Year 1972.
Office of preparation: Nautical Almanac Unit, Regional Meteorological Centre, Alipore, Calcutta. Printed by the Manager, Government of India Press, Calcutta. 22 + 469 pp. Price Rs. 17.00, 39s. 8d., $ 6 12 cents, respectively (1971).

047.011   Utrekning føreåt av oppgang og nedgang ved naturleg horisont.   O. Befring.
Astron. Tidsskr., Årg. 4, 89 - 92 (1971).

047.012   Ephémérides Astronomiques pour 1972.
Publiées par la Société Astronomique de France.
L'Astronomie, 86e année, Suppl. pour Janvier 1972. 65 pp. Price F 8.00 (1971).

047.013   1972 Abridged Nautical Almanac. Pub. No. 683.
Published by Hydrographic Office of Japan, Tokyo. 3 + 236 + 7 pp. (1971).

047.014   Anuarul Observatorului din Bucureşti − 1972.
Editura Academiei Republicii Socialiste România. 245 pp. Price Lei 25 (1971).

047.015   A sketch of the history of the "Annuaire Astronomique de l'URSS".   V. K. Abalakin.
Byull. Inst. Teoret. Astron., *Leningrad*, Vol. 12, 772 - 776 (1971). In Russian.

047.016   **Efemérides Astronómicas 1972.**
Published by Instituto y Observatorio de Marina,
San Fernando (Cádiz). Printed in Spain by Imprenta del Observatorio de Marina, San Fernando. Vol. 181, VIII + 597 pp.
Price 200 pesetas (1971).

047.017   **Nautički Godišnjak 1972.**
Published by Hidrografski Institut Jugoslavenske
Ratne Mornarice, Split. H I–N–31, Godina 39, 11 + 213 +
68 pp. Price 30.– Din. (1971).

047.018   **Sunlight, moonlight, and twilight for Antarctica,
1972 - 1974.**   L. B. Weston.
United States Naval Obs., *Washington,* Circ. No. 132, 19 pp.
(1971).
The graphs in this Circular give data concerning the
rising and setting of the sun and moon and the duration of
twilight for high southern latitudes.

047.019   **Sterregids 1972.**
Compiled by J. Meeus with a contribution by T.
de Groot.
Edited by the "Nederlandse Vereniging voor Weer- en Sterrenkunde" and the "Vereniging voor Sterrenkunde in België".
Printed by Wolters-Noordhoff nv, Groningen. 97 pp. (1971).

047.020   **Himmelskalender 1972.**
Ein astronomisches Jahrbuch für Österreich.
H. Mucke, K. Mayrhofer (Editors).
Verlag, Astronomisches Büro, H. Mucke, Wien. 83 pp. Price
öS 25.00 (1971).

047.021   **Der Sternenhimmel 1972.**
Kleines astronomisches Jahrbuch für Sternfreunde.
R. A. Naef (Editor).
Verlag Sauerländer, Aarau. 32. Jahrgang, 190 pp. Price Fr.
15.00 (1971).

047.022   **Almanacco Astronomico della Rivista Coelum per
l'anno 1972.**
Coelum Suppl., Vol. 39, Fasc. 11 - 12 [Osservatorio Astronomico Universitario, Bologna], 28 + 40 pp. Price L. 2000
(1971).

047.023   **Hvězdářská ročenka 1972.**
Compiled by J. Bouška, V. Guth, B. Onderlička, J.
Ruprecht.
Ročník 48. Academia nakladatelství Československé akademie
věd, Praha. 219 pp. Price Kčs. 14.00 (1971).

047.024   **Anuario del Observatorio Astronómico de Madrid
para 1972.**
Published by Instituto Geográfico y Catastral, Madrid. 384 pp.
Price 100 pesetas (1971).

047.025   **The Nautical Almanac for the Year 1973.**
Issued by Her Majesty's Nautical Almanac Office,
London; and Nautical Almanac Office United States Naval
Observatory, Washington. Printed and published by Her Majesty's Stationery Office, London. 276 + 35 pp. Price £ 1.50
(1971).

047.026   **Almanaque Nautico y Aeronautico para el año 1972.**
Observatorio Naval. Republica Argentina. Armada
Argentina, Servicio de Hidrografia Naval, Buenos Aires. 384 pp.
Price $ 7,00 (1971).

047.027   **Suplemento al Almanaque Nautico y Aeronautico
para el año 1972. Sol, Planetas y Estrellas.**
Observatorio Naval, Republica Argentina. Armada Argentina,
Servicio de Hidrografia Naval, Buenos Aires. 8 + 133 pp. Price
$ 2,50 (1971).

047.028   **Anuário do Observatório de S. Paulo para 1972.**
Published by Instituto Astronômico e Geofísico,
Universidade de São Paulo, São Paulo, Brasil. 11 + 114 + 183 +
2 pp. (1971).

047.029   **Astronomical calendar of the Observatory in Sofia
for the year 1972.**
N. Bonev (Editor).
Izdatelstvo na B'lgarskata Akademiya na Naukite, Sofiya.
88 pp. Price 0.80 Lv. (1971). In Bulgarian.

047.030   **Astronomical Handbook for Southern Africa 1972.**
Edited by the Astronomical Society of Southern
Africa, Cape Town. 42 pp. (1971).

047.031   **Rocznik Astronomiczny na Rok 1972.**
Prepared under the supervision of J. Radecki.
Instytut Geodezji i Kartografii, Państwowe przedsiębiorstwo
wydawnictw kartograficznych, Warszawa. Vol. 27, 135 pp.
Price zł 68.– (1971).

047.032   **Visibility of the planets, 1972.**   L. P. Lee.
Southern Stars, Vol. 24, 76 - 77 (1971).

047.033   **Astronomical ephemeris for the year 1972.**
R. Danić, A. Kubičela.
Vasiona, Vol. 19, 79 - 92 (1971). In Serbo-Croatian.

047.034   **Fourier series representation of the position of the
sun.**   J. W. Spencer.
Search, Vol. 2, No. 5, p. 172 (1971).
Explicit Fourier series as function of day number are
given for solar declination, equation of time and solar parallax.

047.035   **Astronomiskais Kalendārs 1972. Gadam.**
J. Bikše, I. Daube, M. Dīriķis, I. Rabinovics (Editors).
Latvijas PSR Zinātņu Akadēmija, Radioastrofizikas Observatorija Vissavienibas Astronomijas un Ģeodēzijas Biedrības
Latvijas Nodaļa. Izdevniecība «Zinātne», Rīgā, 177 pp. Price
29 Kop. (1971). In Latvian.

Annuaire 1972 du Bureau des Longitudes.
See Abstr. 003.016.

**International Information Bureau on Astronomical
Ephemerides.**   See Abstr. 041.043.

Space Research

051 Extraterrestrial Research, Spaceflight Related to Astronomy

051.001 **On the use of night luminous layer effect for purposes of automatic navigation and orientation of manned spaceships.** K. J. Kondratiev, N. F. Romanteev, S. I. Smoktii, E. V. Khrunov.
Dokl. Akad. Nauk SSSR, Ser. Mat. Fiz., Vol. 199, 1278 - 1281 (1971). In Russian.

051.002 **New astronomical results obtained by use of space vehicles (based on reports of the International Astronautical Congress, October 4 - 10, 1970).** O. Wołczek.
Postępy Astron., Vol. 19, 269 - 274 (1971). In Polish.

051.003 **From the history of astronautics.** V. P. Glushko.
Zemlya i Vselennaya, No. 4, p. 4 - 12 (1971). In Russian.

051.004 **Infra-red interstellar communication.** A. T. Lawton.
Spaceflight, Vol. 13, 83 - 85 (1971).

051.005 **Space report.**
Spaceflight, Vol. 13, 89 - 93, 122 - 137, 170 - 171, 174, 203 - 209, 255 - 256, 258, 284 - 286, 352 - 354, 356 (1971).
(1) Lunokhod 1. (2) Experiments for HEAO (High Energy Astronomy Observatory); Skylab X-ray telescope; Lunokhod experiments; Extraterrestrial amino acids; Einstein on test; Rover trainer; Lunar water process; Apollo 13: Revised data; First moon tyre; Lunar magnetic field; Water on the moon; France launches sixth satellite (PEOLE); Grand tour invitations. (3) ATS-F experiments. (4) Life on Mars?; Lunokhod still active; Fossil nebula; Vertical 1 probes solar corona. (5) Docking with Salyut; Apollo micro-organism experiment. (6) New X-ray star. (7) New Mariner Mars plan; Robots to Mars.

051.006 **The concept of a general-purpose laboratory in space.** E. Stuhlinger, J. Downey.
AIAA Paper, No. 814, 6 pp. (1971).

051.007 **1970: A Soviet space year.** K. L. Plummer.
Spaceflight, Vol. 13, 175 - 179 (1971).

051.008 **The protection of astronauts against solar flares.** P. M. Molton.
Spaceflight, Vol. 13, 220 - 224 (1971).

051.009 **Startalk — The problems of interstellar communication.** A. T. Lawton.
Spaceflight, Vol. 13, 241 - 244 (1971).

051.010 **Orbital bases. Space Station Situation Report — 1: The North American Rockwell proposal.** D. Baker.
Spaceflight, Vol. 13, 318 - 334 (1971).

051.011 **Space Station Situation Report — 2. The McDonnell-Douglas proposal.** D. Baker.
Spaceflight, Vol. 13, 344 - 351 (1971).

051.012 **Barium exhaust spectrum during the expansion of** combustion products through a nozzle into vacuum.
C. Batalli-Cosmovici, K.-W. Michel.
Zeitschr. Naturforschung, Vol. 26a, 1147 - 1155 (1971).

051.013 **Raumtransporter, die nächste Raketengeneration.** H. W. Köhler.
SuW, Vol. 10, 278 - 280 (1971).

051.014 **Scattered and reflected light intensities above the atmosphere.** B. C. Thompson, M. B. Wells.
Applied Optics, Vol. 10, 1539 - 1549 (1971).
We describe a calculational method developed to predict the reflected intensities at a spacecraft position for monochromatic sunlight. The computation uses as input a library of the intensities reflected from different plane-slab atmospheres that was generated by use of the LITE-II Monte Carlo procedure.

051.015 **Astronautique 1970.** J. Meeus.
Ciel et Terre, Vol. 87, 473 - 514 (1971).

051.016 **A small astronomical satellite for obtaining stellar spectra.** A. B. Underhill.
Bull. American Astron. Soc., Vol. 3, 400 (1971). – Abstr. AAS.

051.017 **Advances in aeronautics during recent years and its importance for the cognition of the universe.** V. G. Fesenkov.
Vestn. AN KazSSR, 1971, No. 4, p. 3 - 10. In Russian.

051.018 **Some peculiarities of the forms of craters produced by high-speed particles in a semi-infinite obstacle.** L. V. Leont'ev, A. V. Tarasov, I. A. Tereshkin.
Kosmich. Issled., Vol. 9, 796 - 798 (1971). In Russian. Brief information.

051.019 **Interstellar flights.** L. M. Gindilis.
Zemlya i Vselennaya, No. 5, p. 42 - 48 (1971). In Russian.

051.020 **Survey on new techniques for X-ray astronomy.** R. Giacconi.
IAU Symposium No. 41, (see 012.005), p. 104 - 133 (1971).

051.021 **Instrumentation for the measurement of high energy phenomena on NASA spacecraft.** A. G. Opp, N. G. Roman.
IAU Symposium No. 41, (see 012.005), p. 134 (1971). Abstract.

051.022 **Results of astronomical studies in the far UV region.** V. G. Kurt.
IAU Symposium No. 41, (see 012.005), p. 219 - 232 (1971).

051.023 **Optical systems for UV space researches.** G. Courtès.
IAU Symposium No. 41, (see 012.005), p. 273 - 301 (1971).

051.024 **A rocket payload using Cassegrain-echelle optics**

with image intensification for high resolution ultraviolet stellar spectroscopy.
W. M. Burton, N. K. Reay, D. B. Shenton, R. Wilson.
IAU Symposium No. 41, (see 012.005), p. 304 - 312 (1971).

051.025 **The OAO-B telescope.** A. Boggess III.
IAU Symposium No. 41, (see 012.005), p. 333
(1971). – Abstract.

051.026 **The Princeton experiment on the Orbiting Astronomical Observatory.** E. B. Jenkins.
IAU Symposium No. 41, (see 012.005), p. 334 - 335 (1971).

051.027 **Real time control of the observing program of an Orbiting Solar Observatory.**
E. M. Reeves, M. C. E. Huber, G. L. Withbroe, R. W. Noyes.
IAU Symposium No. 41, (see 012.005), p. 336 - 347 (1971).

051.028 **Solutions of some high potential problems in rocket experiments.** H. Hessberg, J. Niekerke.
IAU Symposium No. 41, (see 012.005), p. 356 (1971). Abstract.

051.029 **New results and techniques in space radio astronomy.** J. K. Alexander.
IAU Symposium No. 41, (see 012.005), p. 401 - 418 (1971).

051.030 **Twenty-five years of rocket and satellite astronomy.** H. Friedman.
Nature, Vol. 234, 181 - 183 (1971).
On October 10, 1946, a V-2 rocket, instrumented by the Naval Research Laboratory, was launched from White Sands, New Mexico, and carried an ultraviolet spectrograph above the ozone layer to photograph the solar ultraviolet spectrum. After twenty-five years of rocket astronomy, it is therefore appropriate to trace some of the history of astronomy from space platforms.

051.031 **Programme der wissenschaftlichen Weltraumforschung.** W. Regula.
Umschau, 71. Jahrgang, p. 867 - 872 = Weltraumfahrt, 22. Jahrgang, p. 133 - 138 (1971).

051.032 **Raumstationen – wozu?** K. Schwenzfeger.
Umschau, 71. Jahrgang, p. 910 - 914 = Weltraumfahrt, 22. Jahrgang, p. 176 - 180 (1971).

051.033 **Weltraumforschung und Kosmologie.**
R. Lüst.
Physik und Kosmologie, (see 003.024), p. 120 - 128 (1971).

051.034 **Space report.**
Spaceflight, Vol. 13, 376 - 381, 385, 426 - 427, 458 - 465 (1971).
(1) Apollo 16 landing site. (2) Hadley base. (3) The Fra Mauro samples. (4) Moon found serene despite monthly quakes; Origin of life; Gravity–assist to Mercury; Salyut – Apollo docking; Space clock; The Mark VA radio telescope; Observations of Luna 18; X-ray mapper for OSO; Seeking extra-terrestrial civilizations.

051.035 **Special relativity and interstellar flights.**
V. V. Dobronravov.
Trudy pyatykh chtenij, posvyashch. razrabotke nauch. nasledyia i razvitiyu idej K. Eh. Tsiolkovskogo, 1970. Sekts. "Mekh. kosmich. poleta". Moskva, 1971, p. 3 - 16. In Russian.
Abstr. in Referativ. Zhurn. 51. Astron., 1.51.875 (1972).

051.036 **On investigations according to the SPIN program.**
V. M. Grigorevsky.
Byull. Stantsij Optichesk. Nablyud. Iskusstv. Sputnikov Zemli, No. 57, p. 43 - 46 (1971). In Russian.

051.037 **Data Catalog of Satellite Experiments.**
NSSDC 71-20, National Space Science Data Center, NASA, Goddard Space Flight Center, Greenbelt, Md. 16 ۱ 525 pp. (1971).
The purposes of the Data Catalog of Satellite Experiments are to announce the availability of experimental space science data, to describe these data, and to inform potential data users of the services provided by the National Space Science Data Center (NSSDC). This edition of the Catalog is the first cumulative edition published since January 1969 and supersedes all previous editions.

051.038 **Sasniegumi kosmosa apgūšanā.**
P. Petrovs, J. Lipskis, J. Timuks, J. Francmanis.
Zvaigžņotā debess, 1970./71. gada ziema, p. 24 - 37.

051.039 **ASV astronomu novērojumi kosmosā (Orbitālās astronomiskās observatorijas).**
J. Timuks, J. Francmanis.
Zvaigžņotā debess, 1971. gada pavasaris, p. 39 - 42.

051.040 **Space radio astronomy.** R. G. Stone.
Telecommun. Journ., (*Switzerland*), Vol. 38, 361 - 365 (1971). – *BMT*

051.041 **Astronautica.**
Coelum, Vol. 39, 145 - 150, 192 - 196, 233 - 234 (1971).

051.042 **Noticiero astronautico.**
Rev. Astron., Vol. 42, No. 174, p. 42 - 44 (1971).

Zehn Jahre bemannte Weltraumfahrt.
Weltraumfahrt, 22. Jahrgang, p. 106 - 107 (1971).

Ultraviolet spectrophotometric measurements with the Orbiting Astronomical Observatory. See Abstr. 013.009.

052 Astrodynamics and Navigation of Space Vehicles

052.001 **Nonlinear longitudinal dynamics of an orbital lifting vehicle.** N. X. Vinh, A. J. Dobrzelecki.
Celestial Mechanics, Vol. 3, 427 - 460 (1971).
This paper presents an analytical study of the longitudinal dynamics of a thrusting, lifting, orbital vehicle in an nearly circular orbit. The translational motion is composed of a nonlinear oscillation, or phugoid, and a spiral mode which results in either decay or dilatation of the orbit depending on the perturbed initial conditions. Elements of the orbit such as radial distance, velocity, and flight path angle were obtained explicitly as functions of time. The behavior of the variations of these elements is correctly predicted. Explicit expressions for period and damping of the angle-of-attack mode were derived.

052.002 **The orbit computation system for the IRIS satellite.** I. M. Wales, H. G. Walter.
Journ. British Interplanet. Soc., Vol. 24, 497 - 509 (1971).
The various stages of the orbit computation system for the IRIS satellite are described without detailed theoretical formulations. The Differential Correction Orbit Program is the main link in the system and its eight principal sections are outlined.

052.003 **Perturbations of the elements of an intermediate orbit by air resistance.** B. N. Noskov.
Vestn. Mosk. un-ta. Fiz., astron., Vol. 12, No. 1, p. 26 - 32 (1971). In Russian. – Abstr. in Referativ. Zhurn. 51. Astron., 8.51.149; 62. Issled. kosm. prostranstva, 8.62.257 (1971).

052.004 **On the effects of gravitational absorption on orbits of artificial earth satellites.** F. Bocchio.
Mem. Soc. Astron. Italiana, Nuova Ser., Vol. 42, 131 - 143 (1971).
Assuming the earth as a shielding body, the influence of the absorption of the gravitational force of the moon on the osculating elements of artificial earth satellites is computed.

052.005 **On the double-point optimum transition.** V. A. Kuz'minykh.
Kosmich. Issled., Vol. 9, 610 - 613 (1971). In Russian.
Brief information.

052.006 **Trans–stellar navigation.** J. Strong.
Spaceflight, Vol. 13, 252 - 255 (1971).

052.007 **On the representation of the coordinates of an artificial satellite of a spheroidal planet by means of series.** G. T. Arazov.
Soobshch. Shemakhinsk. Astrofiz. Obs., vyp. (No.) 5, p. 70 - 89 (1971). In Russian.

052.008 **A recurrence relation for inclination functions.** R. H. Gooding.
Celestial Mechanics, Vol. 4, 91 - 98 (1971).
When the terms of the series expansion for the gravitational potential of the earth are expressed in terms of the orbital elements of an arbitrary earth satellite, the orbital inclination appears in each term as the argument of a function of inclination only. The present paper gives a recurrence relation for a general normalized inclination function with three parameters.

052.009 **A note on 'The main problem of satellite theory for small eccentricities, by A. Deprit and A. Rom, 1970'** [Celestial Mechanics, Vol. 2, 166 - 206 (1970)].

K. Aksnes.
Celestial Mechanics, Vol. 4, 119 - 121 (1971).

052.010 **On the rotation of the orbital plane during ballistic re-entry.** R. Arho.
Planet. Space Sci., Vol. 19, 1215 - 1224 (1971).
Consequences of atmospheric rotation at re-entry of a spacecraft during the final phase are drawn. A method to determine Ω (longitude of imaginary ascending node) and i (inclination of orbital plane to equator) as functions of altitude is outlined.

052.011 **Long-period parallactic terms in the motion of distant artificial earth satellites.** V. P. Dolgachev.
Soobshch. Gos. Astron. Inst. Shternberga, No. 160, p. 25 - 34 (1971). In Russian.

052.012 **Genauigkeit der Raumnavigation.** E. A. Steinhoff.
Umschau, 71. Jahrgang, p. 786 (1971).

052.013 **Analytical studies of luni-solar effects on the motion of artificial satellites.** R. L. Felsentreger, J. P. Murphy.
National Aeronautics and Space Administration, NASA-TM-X-65485 (1971). – Abstr. in The Moon, Vol. 3, 239 (1971).

052.014 **Lunisolar perturbations of the motion of artificial satellites.** D. Fisher.
National Aeronautics and Space Administration, NASA-TM-X-65476 (1971). – Abstr. in The Moon, Vol. 3, 239 (1971).

052.015 **Field tests using dead reckoning and monoptic video for remote lunar surface navigation.** W. C. Mastin, P. H. Broussard, Jr.
National Aeronautics and Space Administration, NASA-TM-X-64467 (1971). – Abstr. in The Moon, Vol. 3, 261 (1971).

052.016 **The most probable characteristics of the initial elements of nearly circular orbits.** V. G. Degtyarev.
Kosmich. Issled., Vol. 9, 781 - 784 (1971). In Russian.
Brief information.

052.017 **Liapunov stability of spinning satellites with long flexible appendages.** P. C. Hughes, J. C. Fung.
Celestial Mechanics, Vol. 4, 295 - 308 (1971).
The equations of motion are derived for the case of a spinning satellite which has a central rigid body, and long flexible appendages which are nominally in the spin plane. The 'major-axis theorem' is found to be necessary, but not sufficient for this case, and appropriate sufficiency criteria are derived from a Liapunov function.

052.018 **Tests and comparisons of gravity models.** J. G. Marsh, B. C. Douglas.
Celestial Mechanics, Vol. 4, 309 - 325 (1971).
Optical observations of the GEOS satellites were used to obtain orbital solutions with different sets of geopotential coefficients. The solutions were compared before and after modification to high order terms (necessary because of resonance) and then analyzed by comparing subsequent observations with predicted trajectories. The most important source of error in orbit determination and prediction for the GEOS satellites is the effect of resonance.

052.019 **Lageabweichung einer drallstabilisierten Sonnenson-de infolge des Solardrucks.** H. W. Zipse.
Celestial Mechanics, Vol. 4, 329 - 339 (1971).

Based on the derived relations it is possible to determine the gross attitude deviations of spin stabilized axisymmetric probes, which move along elliptic solar orbits. It is supposed for the derivation, that these deviations are so small, that their reaction on the disturbing momentum is negligible.

052.020 **Motion of a space station. I.**
G. E. O. Giacaglia, W. H. Jefferys.
Celestial Mechanics, Vol. 4, 442 - 467 (1971).

In this first part of the work we develop the equations of motion of a triaxial space station in orbit around the oblate earth. A first order solution of the problem is presented and the method of complete integration of the system is outlined up to second order of approximation.

052.021 **Die Berücksichtigung der Ellipsoidgestalt der Erde in der Flugnavigation.** L. Kiefer.
Deutsche Geod. Kommission, Bayer. Akad. Wiss., München, Ser. C, No. 166, 3 + 115 pp. (1971). – Dissertation Univ. Stuttgart.

052.022 **Integral invariants of some problems of space flight mechanics.** Yu. P. Surkov.
Trudy pyatykh chtenij, posvyashch. razrabotke nauch. naslediya i razvitiyu idej K. Eh. Tsiolkovskogo, 1970. Sekts."Mekh. kosmich. poleta". Moskva, 1971, p. 173 - 179. In Russian.
Abstr. in Referativ. Zhurn. 62. Issled. kosm. prostranstva, 12.62.421 (1971).

052.023 **A second- and higher-order perturbation analysis of two-body trajectories.** T. T. Soong, N. A. Paul.
AIAA Journ., Vol. 9, 589 - 593 (1971).

An approach to the development of a second- and higher-order perturbation theory for two-body trajectories is presented. It is shown that the higher order analysis can be developed in a systematic manner in terms of series solutions to the two-body problem as functions of the time variable. Simple algebraic recursive formulas for the determination of the series coefficients are derived and the radii of convergence for these series solutions are determined. Their accuracy and the rate of convergence are investigated in a number of numerical cases.

052.024 **Integral invariants in some mechanical problems of spaceflight.** Yu. P. Surkov.
Trudy pyatykh chtenij, posvyashch. razrabotke nauch. naslediya i razvitiyu idej K. Eh. Tsiolkovskogo, 1970. Sekts. "Mekh. kosmich. poleta". Moskva, 1971, p. 173 - 179. In Russian.
Abstr. in Referativ. Zhurn. 51. Astron., 1.51.115 (1972).

052.025 **Plane vibrations of a solid body on an elliptic orbit.**
I. I. Romanyuk.
Vestn. Leningr. un-ta, 1971, No. 13, p. 126 - 134. In Russian.
Abstr. in Referativ. Zhurn. 51. Astron., 1.51.128; 62. Issled. kosm. prostranstva, 1.62.241 (1972).

052.026 **Theorie der Satellitenbahnen III.** M. Schneider.
Bundesministerium für Bildung und Wissenschaft, Forschungsber. BMBW–FB W 71-36, [available from Zentralstelle für Luftfahrtdokumentation und -information (ZLDI), Deutsche Forschungs- und Versuchsanstalt für Luft- und Raumfahrt, München. Price DM 35.70], 170 S. = Mitt. Inst. Astron. Physik. Geod. Techn. Univ. München, No. 80 (1971).

Lagrange's planetary equations are solved for a satellite moving in a noncentral gravitational field of the primary.

052.027 **Minimum ballistic factor missile shapes.**
S. C. Jain, V. B. Tawakley.
Astronaut. Acta, Vol. 16, 277 - 279 (1971).

Power law axisymmetric body profiles having minimum ballistic factor have been determined in the two situations when (a) surface area and diameter are prescribed and (b) surface area and length are prescribed. A general Newton-Busemann drag law has been employed and the results compared with those obtained by using Newtonian drag law.

052.028 **The non-optimality of Lawden's spiral.**
D. J. Bell.
Astronaut. Acta, Vol. 16, 317 - 324 (1971).

The intermediate-thrust arcs are candidates for minimum-fuel transfer orbits in space navigation. Lawden's spiral is such an arc when the time of flight is unspecified. Several methods have been used in the last five years to show that this spiral is non-optimal. One method is the examination of the second variation using specially constructed control variations. A similar method is used in the present paper.

052.029 **Perturbations in the motion of an artificial satellite caused by Callisto and Jupiter's non-sphericity.**
N. B. Batueva.
Vestn. Leningr. un-ta, 1971, No. 13, p. 139 - 143. In Russian.
Abstr. in Referativ. Zhurn. 51. Astron., 1.51.122; 62. Issled. kosm. prostranstva, 1.62.236 (1972).

052.030 **Absolute optimum and two-impulse optimal interplanetary transfers with applications to earth-Mars and earth-Venus.** J. P. Gravier.
Thesis, Univ. Colorado, Boulder. [Available from Univ. Microfilms, Ann Arbor, Mich., U.S.A. Order No. 70–23714], 203 pp. (1970).

Concerns optimal transfers and rendezvous between two elliptical and inclined planetary orbits taking account of the attraction of the planets. This analysis is concerned with the exact numerical solution for the optimal transfer from the standpoint of the characteristic velocity between two real planets.

052.031 **On the problem of quadratic programming in questions of optimum planning of orbit measurements.** L. Yu. Belousov.
Kosm. Issled., Vol. 9, 813 - 822 (1971). In Russian.

052.032 **Optimum planning of measurements in the case of some measurable parameters.** L. Yu. Belousov.
Kosm. Issled., Vol. 9, 823 - 830 (1971). In Russian.

052.033 **Potential accuracy of observing dynamical systems and problems of spaceflight.** B. A. Reznikov.
Kosm. Issled., Vol. 9, 831 - 842 (1971). In Russian.

052.034 **Analysis of iteration processes in problems of treating measurements in cosmic collectives.**
L. M. Romanov.
Kosm. Issled., Vol. 9, 843 - 849 (1971). In Russian.

A problem of orbital dynamics, which is separable in KS-variables. See Abstr. 042.038.

053 Lunar and Planetary Probes and Satellites

053.001 **Theory of the motion of artificial lunar satellites.**
L. S. Evdokimova.
Byull. Inst. Teoret. Astron., *Leningrad,* Vol. 12, 685 - 713 (1971). In Russian.
The motion of artificial lunar satellites in the gravitational field of the non-spherical moon, the earth and the sun is investigated. The short and intermediate periodic perturbations are derived by von Zeipel's method. The system of the differential equations for the secular and the long periodic perturbations is obtained.

053.002 **The lunar roving vehicle.** R. N. Watts, Jr.
Sky Telescope, Vol. 42, 14 - 15 (1971).

053.003 **The exploration of Mars.** R. N. Watts, Jr.
Sky Telescope, Vol. 42, 15 (1971).

053.004 **Sonnensonde Helios bis 0,25 AE Sonnenabstand.**
H. Porsche.
Umschau, 71. Jahrgang, p. 674 - 675 (1971).

053.005 **Ruimtevaartuigen onderzoeken de planeten.**
C. Titulaer, T. de Vries.
Hemel en Dampkring, Vol. 69, 199 - 206 (1971).

053.006 **De Amerikaanse bemande marsvluchten.**
T. de Vries.
Hemel en Dampkring, Vol. 69, 207 - 210 (1971).

053.007 **Kunstmatige objecten op de maan.** J. Meeus.
Hemel en Dampkring, Vol. 69, 183 - 184 (1971).

053.008 **The manned exploration of Mars.**
D. Dooling, Jr.
Spaceflight, Vol. 13, 198 - 202 (1971).

053.009 **Apollo 14: A visit to Fra Mauro, 1, 2.**
D. Baker.
Spaceflight, Vol. 13, 164 - 169, 210 - 212 (1971).

053.010 **Orbiters to Mars.** D. Baker.
Spaceflight, Vol. 13, 187 - 192 (1971).

053.011 **Lunar roving vehicle: Design report.** D. Baker.
Spaceflight, Vol. 13, 234 - 240 (1971).

053.012 **Objects in heliocentric orbit − 2.**
Compiled by G. Falworth.
Spaceflight, Vol. 13, 298 - 299 (1971).

053.013 **Apollo 15 television from the moon.**
R. N. Watts, Jr.
Sky Telescope, Vol. 42, 136 - 138 (1971).

053.014 **Apollo Laser Ranging Retro-Reflector experiment (SO78). Final report.** C. O. Alley.
National Aeronautics and Space Administration, NASA-CR-114951 (1971). − Abstr. in The Moon, Vol. 3, 259 - 260 (1971).

053.015 **Apollo 14 Laser Ranging Retro-Reflector experiment: Design certification review report.**
National Aeronautics and Space Administration, NASA-CR-114900 (1971). − Abstr. in The Moon, Vol. 3, 260 (1971).

053.016 **Lunar traverse missions.**
R. G. Brereton, J. D. Burke, R. B. Coryell, L. D. Jaffe.
Jet Propulsion Lab. Quarterly Techn. Rev., Vol. 1, 125 - 137 (1971). − Abstr. in The Moon, Vol. 3, 260 (1971).

053.017 **From Apollo 11 to the space stations of the future.**
H. Cohen.
National Aeronautics and Space Administration, NASA-TM-X-67072 (1971). − Abstr. in The Moon, Vol. 3, 260 (1971).

053.018 **The 1971 Mariner Mars spacecraft.**
A. A. J. Hooke.
Spaceflight, Vol. 13, 408 - 421 (1971).

053.019 **Sowjetische Roboter auf dem Mond.**
A. D. Podolski.
Umschau, 71. Jahrgang, p. 905 - 909 = Weltraumfahrt, 22. Jahrgang, 171 - 175 (1971).

053.020 **Le programme Mariner 1971 et ses répercussions sur la connaissance de la planète Mars.**
J. Dragesco.
L'Astronomie, 85e année, p. 439 - 446 (1971).

053.021 **Mariner 9 approaches Mars.** R. N. Watts, Jr.
Sky Telescope, Vol. 42, 270 (1971).

053.022 **Lunokhod and Salyut end their missions.**
R. N. Watts, Jr.
Sky Telescope, Vol. 42, 346 (1971).

053.023 **Some Soviet space notes.** R. N. Watts, Jr.
Sky Telescope, Vol. 42, 279 (1971).

053.024 **Das wissenschaftliche Programm von "Apollo 15".**
H. Mucke.
Sternenbote, 14. Jahrgang, p. 126 - 130 (1971).

053.025 **Marssonden unterwegs.** L. Fritsch.
Sternenbote, 14. Jahrgang, p. 138 - 142, 146 - 147 (1971).

053.026 **The Apollo missions.** L. R. Scherer.
Highlights in Astronomy, Vol. 2, (see 012.013), 125 - 141 (1971).

053.027 **Second generation of Soviet automatic lunar landers.**
G. N. Nikolaev, M. K. Rozhdestvenskij, V. I. Shkirina.
Vestn. AN SSSR, 1971, No. 6, p. 15 - 32. In Russian. − Abstr. in Referativ. Zhurn. 62. Issled. kosm. prostranstva, 12.62.253 (1971).

053.028 **A rolling laboratory on the moon − Lunokhod 1.**
AN SSSR. "Nauka", Moskva. 128 pp. Price 1 Rbl. 26 Kop. (1971). In Russian.

053.029 **Start to a planet − at what time?**
V. I. Levantovskij.
Zemlya i Vselennaya, No. 6, p. 22 - 26 (1971). In Russian.

053.030 **Rolling laboratory on the moon − Lunokhod 1.**
Yu. N. Lipskij, V. V. Shevchenko.
Zemlya i Vselennaya, No. 6, p. 76 - 78 (1971).

053.031 **Post-occultation reception of lunar ship Endeavour radio transmission.** W. W. Salisbury, D. L. Fernald

Nature, Vol. 234, 95 (1971).

053.032 **Early manned exploration of the planets.**
R. R. Titus.
Journ. Spacecraft and Rockets, Vol. 8, 517 - 522 (1971).
Outlines a means of performing early manned flyby and stopover missions to Mars and Venus with chemical propulsion and the Saturn V booster.

053.033 **Investigation of radio signals of the Soviet interplanetary station Venera 7 during the landing on the surface of Venus by means of a computer.**
Yu. N. Aleksandrov, O. N. Rzhiga, A. M. Shakhovskoj.
Kosm. Issled., Vol. 9, 904 - 911 (1971). In Russian.

053.034 **Laser location of the light reflector attached to Lunokhod 1.** Yu. L. Kokurin, V. V. Kurbasov, V. F. Lobanov, A. N. Sukhanovskij, N. S. Chernykh.
Kosm. Issled., Vol. 9, 912 - 919 (1971). In Russian.

053.035 **Apollo 15 – fourth landing of men on the moon.**
D. Kneževié.
Vasiona, Vol. 19, 65 - 66 (1971). In Serbo-Croatian.

053.036 **Venus 7.** W. Alexejew.
Ideen exakten Wissens, [Deutsche Verlags-Anstalt, Stuttgart], 8.71, p. 509 - 516 (1971).

Space science: Probing Mars.
Nature, Vol. 233, 82 - 83 (1971).

Hard way to Jupiter.
Nature, Vol. 234, 246 (1971).

Where to go after Mars.
Nature, Vol. 234, 246 - 247 (1971).

Imaging of Mercury and Venus from a flyby.
See Abstr. 092.004.

Lunochod 1. See Abstr. 094.031.

054 Artificial Earth Satellites

054.001 **The first laboratory in space.**
R. N. Watts, Jr.
Sky Telescope, Vol. 42, 83, 85 (1971).

054.002 **Optical characteristics of artificial satellites.**
G. A. McCue, J. G. Williams, J. M. Morford.
Planet. Space Sci., Vol. 19, 851 - 868 (1971).
This paper presents a comprehensive study of the observed optical characteristics of several hundred satellites. The primary data for the study were 22,000 optical observations gathered by volunteer observers. These observations were reduced to standard observing conditions and subjected to numerous statistical tests. The observational results were compared with theoretical satellite brightness predicted by equations derived by the authors.

054.003 **On the prediction of trajectories of artificial earth satellites.** Yu. I. Paraev.
Trudy Sib. fiz.-tekhn. in-ta pri Tomsk. un-te, 1970, vyp. (No.) 51, p. 36 - 45. In Russian. – Abstr. in Referativ. Zhurn. 51. Astron., 8.51.157 (1971).

054.004 **Measurement of the light pressure acting on Echo 1.**
L. A. Vasil'ev, L. P. Nazarova.
Kosmich. Issled., Vol. 9, 587 - 591 (1971). In Russian.

054.005 **Sur un modèle mathématique pour l'étude des effets de la pression de radiation solaire sur le mouvement des satellites artificiels.** S. Ferraz-Mello.
Comptes Rendus Acad. Sci. Paris, Sér. A, Vol. 273, 197 - 200 (1971).
La principale caractéristique de ce modèle est que les paramètres géométriques liés à l'ombre de la Terre ne sont pas calculés d'après l'orbite réelle mais d'après une orbite moyenne. Le modèle conduit à la non-existence de perturbations séculaires dans les éléments métriques. Les effets angulaires de nature séculaire sont très petits comme le montrent quelques résultats numériques.

054.006 **Intercosmos.** K. E. Speranskij.
Zemlya i Vselennaya, No. 4, p. 28 - 32 (1971). In Russian.

054.007 **Skylab.** D. Baker.
Spaceflight, Vol. 13, 335 - 337 (1971).

054.008 **Salyut.** K. W. Gatland.
Spaceflight, Vol. 13, 338 - 340 (1971).

054.009 **Optimum stabilization of plane oscillations of a satellite in a Keplerian orbit.**
I. F. Vereshchagin, V. V. Malanin.
Avtomatika i telemekhanika, 1971, No. 2, p. 173 - 176. In Russian. – Abstr. in Referativ. Zhurn. 62. Issled. kosm. prostranstva, 9.62.413 (1971).

054.010 **The ill-fated Soyuz 11.** R. N. Watts, Jr.
Sky Telescope, Vol. 42, 138 (1971).

054.011 **Determination of the borders of shadow parts of artificial satellite orbits in case of conical earth's shadow.** R. P. Eremenko, E. N. Poljakhova.
Trudy Astron. Obs., *Leningrad*, Vol. 28 (= Uchenye Zapiski Leningr. Un-ta, No. 359 = Seriya Matem. Nauk, vyp. (No.) 47), p. 125 - 135 (1971). In Russian.
Two forms of the canonical shadow equation are investigated for the conical earth's shadow model. For both cases the formulas are found for the coefficients of the shadow equation.

054.012 **French weather satellite.**
R. N. Watts, Jr.

Sky Telescope, Vol. 42, 212 (1971).

054.013   The astrometric series for equatorial coordinates and the generalized method of dependences in astrography for artificial satellites of the earth.   V. Tomelleri.
Boll. Geod. Sci. Affini, Anno 30, p. 119 - 155 (1971).
In Italian.
   Expressions of the serial remainders, after the first order terms of standard coordinates, in the series for the celestial equatorial coordinates, are given for a possible use of the method of dependences in a wider than usual region even without diminishing the simplicity of the method.

054.014   Contributions of the Institute for Theoretical Astronomy to the theory of the motion of satellites and to satellite geodesy.   Yu. V. Batrakov.
Byull. Inst. Teoret. Astron., *Leningrad,* Vol. 12, 777 - 784 (1971). In Russian.

054.015   Perturbations of orbital elements of an earth satellite from arbitrary zonal harmonics.
Yu. V. Batrakov.
Byull. Inst. Teoret. Astron., *Leningrad,* Vol. 12, 813 - 847 (1971). In Russian.
   Two groups of formulae have been obtained for both the secular and the long-periodic perturbations of Keplerian elements of an earth satellite from arbitrary zonal harmonics of the earth's gravitational potential. The control showed perfect identity between these formulae and those of Brouwer.

054.016   On the determination of the difference between the draconitic and sidereal orbital periods of the artificial satellites.   T. Oproiu.
Stud. Cerc. Astron., Vol. 16, 215 - 219 (1971).

054.017   Seventh OSO in orbit.   R. N. Watts, Jr.
Sky Telescope, Vol. 42, 271, 279 (1971).

054.018   OSO 7 begins observations.   R. N. Watts, Jr.
Sky Telescope, Vol. 42, 346 (1971).

054.019   Approximate graphical method for computing ephemerides of an artificial earth satellite with a circular polar orbit.   V. I. Evseev.
(Uch. zap.) Perm. gos. ped. int, Vol. 72, (see 003.019), 393 - 397 (1969). In Russian. – Abstr. in Referativ. Zhurn. 51. Astron., 12.51.127; 62. Issled. kosm. prostranstva, 12.62.455 (1971).

054.020   Satellite digest.   G. Falworth.
   Spaceflight, Vol. 13, 94 - 96, 138 - 139, 172 - 174, 213 - 215, 257 - 258, 287 - 289, 355 - 356, 386 - 388, 428 - 431, 456 - 457, 480 (1971). – Monthly listing of all known artificial satellites and spacecraft – 1970 September – 1971 June.

054.021   Seguimiento Doppler de satélites artificiales.   A. Rius.
Urania Barcelona, Año 56, No. 273, p. 24 - 38 (1971).

054.022   Circular orbit determination of AES by the method of minimization.   L. Lautsenieks.
Latv. ordena trud. krasn. znameni gos. univ. im. P. Stuchki, Uch. zap., Vol. 148, vyp. (No.) 6, p. 98 - 107 (1971). In Russian.

054.023   Evolution of the orbits of light artificial satellites under the influence of perturbing effects of solar radiation pressure and the earth's oblateness.
E. N. Polyakhova.
Vestn. Leningr. un-ta, 1971, No. 13, p. 159 - 165. In Russian.

Abstr. in Referativ. Zhurn. 51. Astron., 1.51.123; 62. Issled. kosm. prostranstva, 1.62.234 (1972).

054.024   On satellite deceleration after a solar flash.
   M. J. Tovadrovs, B. B. Bagkhos, A. S. Asaad, V. M. Grigorevsky.
Byull. Stantsij Optichesk. Nablyud. Iskusstv. Sputnikov Zemli, No. 57, p. 47 - 53 (1971). In Russian.

054.025   Man-made objects in space (Vyp. (No.) 4). (1966, Jan. 1 - 1967, Dec. 31).   G. A. Lejkin (Editor).
Byull. Stantsij Optichesk. Nablyud. Iskusstv. Sputnikov Zemli, No. 58, 109 pp. (1971). In Russian.

054.026   Man-made objects in space (Vyp. (No.) 5). (1968, Jan. 1 - Dec. 31).   G. A. Lejkin (Editor).
Byull. Stantsij Optichesk. Nablyud. Iskusstv. Sputnikov Zemli, No. 59, 54 pp. (1971). In Russian.

054.027   Man-made objects in space (Vyp. (No.) 6). (1969, Jan. 1 - Dec. 31).   G. A. Lejkin (Editor).
Nablyud. Iskusstv. Nebesn. Tel, No. 61, 74 pp. (1971). In Russian.

054.028   The astrometric series for equatorial coordinates and the generalized method of dependences in astrography for artificial satellites of the earth.   V. Tomelleri.
Boll. Geod. Sci. Affini, Anno 30, p. 295 - 330 (1971).
   Expression of the serial remainders, after the first order terms of standard coordinates, in the series for the celestial equatorial coordinates, are given for a possible use of the method of dependences in a wider than usual region even without diminishing the simplicity of the method. Elaborations of formulae need an algebraic introduction about two new groups of polynomes and some new analogies of combinatorial analysis.

054.029   Comments on the accuracy of Baker-Nunn observations.   K. Lambeck.
Conference on photographic astrometric technique, Tampa 1968, (see 012.031), p. 123 - 131 (1971).

054.030   Appendix: Comparison of measuring and reducing techniques of satellite observations and their accuracy.
Conference on photographic astrometric technique, Tampa 1968, (see 012.031), p. 265 - 267 (1971).

054.031   Table of artificial satellites launched from 1957 to 1970.
Telecommun. Journ. (*Switzerland*), Vol. 38, No. 5, Appendix 1, p. 1 - 143 (1971). – See Phys. Abstr., Vol. 74, No. 53732 (1971).

054.032   Analytic estimates of the accuracy of determining the parameters of the motion of artificial earth satellites from position measurements of stars relative to a probe started from a satellite.
L. F. Porfir'ev, V. V. Smirnov, Yu. A. Sereda.
Kosm. Issled., Vol. 9, 850 - 858 (1971). In Russian.

054.033   Sad end during the landing of space ship Soyuz 11.   D. Knežević.
Vasiona, Vol. 19, 54 - 55 (1971). In Serbo-Croatian.

054.034   Kunstmanen.   J. Meeus.
   Hemel en Dampkring, Vol. 69, 215 - 217 (1971). – 1971 January – April.

054.035   Künstliche Erdsatelliten und Raumsonden: Situationsbericht.
Weltraumfahrt, 22. Jahrgang, p. 120 - 121 (1971). – 1971

March 1 – June 30.

Sojus 11 – Salut. Erste Raumstation erfolgreich.
Weltraumfahrt, 22. Jahrgang, p. 101 - 105 (1971).

An F1 Schmidt satellite camera and the methods of plate measurement and reduction.   See Abstr. 031.077.

Plate measurement techniques and reduction methods used by the West German satellite observers, and resulting consequences for the observation.   See Abstr. 031.078.

Correlation between the solar activity indices used in the study of motion of artificial earth satellites. See Abstr. 072.081.

055   Observations of Earth Satellites, Lunar and Planetary Probes

055.001   **Observations of artificial earth satellites in Uzhgorod.**
M. V. Bratijchuk.
Zemlya i Vselennaya, No. 4, p. 48 - 51 (1971). In Russian.

055.002   **Telescopic observations of lunar missions.**
J. O. Cappellari, Jr., W. I. McLaughlin.
Spaceflight, Vol. 13, 363 - 367, 369 (1971).

055.003   **Optical tracking of spacecraft.**
V. J. Slabinski.
Sky Telescope, Vol. 42, 202 - 204 (1971).

055.004   **Le choix du degré de polynôme pour l'approximation d'une section de l'orbite d'un satellite.**
J. Łatka.
Geod. Kartografia, Vol. 20, 255 - 258 (1971). In Polish.
    L'auteur s'occupe du critérium permettant de définir le degré du polynôme d'approximation pour une section de l'orbite d'un satellite photographié. Le cas traité correspond à l'orbite circulaire.

055.005   **Satellites artificiels: Observations de périodes photométriques, 1968 - 1971.**   J. Meeus.
Ciel et Terre, Vol. 87, 606 - 618 (1971).

055.006   **The program for precise photographic artificial satellite observations.**
E. P. Aksionov, S. K. Tatevjan.
Byull. Stantsij Optichesk. Nablyud. Iskusstv. Sputnikov Zemli, No. 57, p. 20 - 22 (1971). In Russian.

055.007   **Optical observations of distant space bodies.**
P. P. Dobronravin, V. M. Mozherin, V. K. Prokofjev, N. S. Chernykh.
Byull. Stantsij Optichesk. Nablyud. Iskusstv. Sputnikov Zemli, No. 57, p. 22 - 24 (1971). In Russian.

055.008   **On possibilities of using the AFU-75 camera for photographic observations of satellites.**
K. K. Lapushka.
Byull. Stantsij Optichesk. Nablyud. Iskusstv. Sputnikov Zemli, No. 57, p. 25 - 27 (1971). In Russian.

055.009   **On standardization of astrometric reduction of satellitograms (satellite photographs).**
N. P. Erpylev.
Byull. Stantsij Optichesk. Nablyud. Iskusstv. Sputnikov Zemli, No. 57, p. 27 - 30 (1971). In Russian.

055.010   **Observations of active artificial satellites with AFU-75 cameras in the mode of weak satellites photography.**
J. K. Ballodis, K. K. Lapushka, L. K. Lautzenieks.
Byull. Stantsij Optichesk. Nablyud. Iskusstv. Sputnikov Zemli, No. 57, p. 30 - 33 (1971). In Russian.

055.011   **Results of photographic observations of artificial earth satellites.**
Nablyud. Iskusstv. Nebesn. Tel, No. 60, 34 pp. (1971). In Russian.

Theoretical Astrophysics

061 General Theoretical Problems of Astrophysics, Gravitational Instability, Neutrino Astronomy, X Ray- and Gamma Ray-Astronomy, Frequency and Origin of Elements etc.

061.001 **Kinematic-dynamo theory. III. The effect of turbulent diffusivity in the dynamo equations.**
I. Lerche.
Astrophys. Journ., Vol. 168, 115 - 121 (1971).
The kinematic-dynamo equations, developed previously for a constant diffusivity, are further generalized to include the effect of turbulent diffusivity. This generalization is necessary since current ideas concerning galactic, and solar, dynamo action indicate that turbulent diffusivity is larger than "classical" diffusivity by about three orders of magnitude. Some simple cases are worked through to illustrate the physical importance of turbulent diffusivity on dynamo action.

061.002 **Kinematic-dynamo theory. IV. Dynamo action in non-rotating spheres with isotropic turbulence.**
I. Lerche.
Astrophys. Journ., Vol. 168, 123 - 129 (1971).
Application of the kinematic-dynamo equations to the problem of dynamo action in a nonrotating conducting sphere is discussed. The point of the paper is to establish by direct computation that large-scale dynamo action is directly associated with small scale isotropic velocity turbulence in an object of finite size.

061.003 **Accretion of cosmic dust.** B. A. Trubnikov.
Priroda, No. 8.71, p. 76 - 77 (1971). In Russian.

061.004 **On electron acceleration in an alternating magnetic field under astrophysical conditions.**
M. F. Bakhareva, V. N. Lomonossov, B. A. Tverskoi.
Astron. Zhurn. Akad. Nauk SSSR, Vol. 48, 697 - 709 (1971). In Russian. English translation in Soviet Astron. AJ, Vol. 15, No. 4.
For the mechanism of particle acceleration in an alternating magnetic field the solution of the stationary problem of the energy spectrum of accelerated electrons is given. It is shown that synchrotron X-radiation of the Crab nebula and the pulsar, placed in the centre of the nebula, could be generated by the given mechanism.

061.005 **The emission of magneto-sonic waves by planets and binaries moving in circular orbits.**
V. P. Dokuchaev.
Astron. Zhurn. Akad. Nauk SSSR, Vol. 48, 726 - 729 (1971). In Russian. English translation in Soviet Astron. AJ, Vol. 15, No. 4.
The emission of magneto-sonic waves in the motion of planets and stars in circular orbits in cosmic space is investigated. Density perturbations of the interstellar gas in the wave zone of sources are determined. It is shown that the emission spectrum consists of frequencies multiple to the frequency of rotation of sources along the orbit. The power of emission by planets and binaries is estimated.

061.006 **Iodine-129 in terrestrial ores.**

B. Srinivasan, E. C. Alexander, Jr., O. K. Manuel.
Science, Vol. 173, 327 - 328 (1971).
Xenon extracted from natural iodyrite (silver iodide) from Broken Hill, New South Wales, Australia, contains excess xenon-129 from the in situ decay of naturally occurring iodine-129 and excess xenon-128 from neutron capture on iodine-127.

061.007 **The URCA process at high temperatures and densities.** R. W. McFee, D. W. Schlitt.
Astrophys. Space Sci., Vol. 12, 4 - 9 (1971).
The URCA neutrino loss rate from a hot stellar environment is investigated. Rates are calculated for some typical odd mass isobar pairs and for the even mass isobar fifty-six for temperatures between 5×10^8 K to 5×10^{10} K.

061.008 **Nonlinear thermal convection with free boundaries.**
J. O. Murphy.
Australian Journ. Phys., Vol. 24, 587 - 592 (1971).
The equations governing thermal convection in a horizontal fluid layer heated from below have been solved by numerical integration when free boundary conditions apply. The computed forms for the vertical velocity, the mean temperature, and the temperature fluctuation are given for several values of the Rayleigh number in the range 900–2×10^4. For larger values of R comparison is made with the results predicted by an asymptotic theory.

061.009 **Density fluctuations driven by Alfvén waves.**
J. V. Hollweg.
Journ. Geophys. Res., Vol. 76, 5155 - 5161 (1971).
The equations for a linearly polarized Alfvén wave, propagating parallel to the direction of the average magnetic field in a perfectly conducting fluid, are solved to second order in the wave quantities for cases where the fluid obeys single adiabatic or double adiabatic equations of state. To this order, we find no change in the wave magnetic field or transverse wave velocity, but longitudinal wave velocity and density fluctuations appear, driven by gradients in the wave magnetic-field pressure.

061.010 **Modern state of X-ray astronomy.**
I. S. Shklovsky.
Priroda, No. 9.71, p. 24 - 28 (1971). In Russian.

061.011 **Observational γ-astronomy.** W. L. Kraushaar.
Priroda, No. 9.71, p. 89 (1971). In Russian.

061.012 **Review of some results of theoretical spectroscopy of complex atoms in connection with some new problems of astrophysics.** A. A. Nikitin.
Trudy Astron. Obs., *Leningrad*, Vol. 28 (= Uchenye Zapiski Leningr. Un-ta, No. 359 = Seriya Matem. Nauk, vyp. (No.) 47), p. 163 - 172 (1971). In Russian.
Recent theoretical results on the spectra of complex

atoms are reviewed from the point of view of an astrophysicist. Special attention is given to the results found in Vilnius by A. P. Jucys and his group.

061.013 **Kinematic-dynamo action under incompressible, isotropic velocity turbulence.**
I. Lerche, B.-C. Low.
Astrophys. Journ., Vol. 168, 503 - 508 (1971).
We demonstrate that incompressible, isotropic velocity turbulence on its own in an infinite medium leads to regenerative dynamo action.

061.014 **Optical line spectrum.**
S. J. Czyzak, T. K. Krueger.
National Bureau Standards Special Publ. 353, (see 012.001), p. 151 - 160 (1971).

061.015 **Cross-section for ^{10}Be production of high energy fragmentation of oxygen.**
G. M. Raisbeck, F. Yiou.
Nature, Phys. Sci., Vol. 233, 73 - 74 (1971).
It is pointed out that an indirect estimation for the cross section ^{16}O(p-)^{10}Be at high energies, deduced from activities in meteorite samples, requires an unstated and unsupportable assumption.

061.016 **Finite amplitude convection in a compressible medium.** R. Van der Borght.
Publ. Astron. Soc. Japan, Vol. 23, 539 - 551 (1971).
A system of ordinary non-linear differential equations, describing finite amplitude convection in a compressible medium, is derived using a variational principle first developed by Prigogine and Glansdorff (1964, 1965). It is shown that these equations reduce to well known results in the incompressible non-linear case and in the compressible linear one.

061.017 **Towards a theory of jet streams.** J. Trulsen.
Astrophys. Space Sci., Vol. 12, 329 - 348 (1971).
A kinetic equation for a jet stream consisting of identical, partially inelastic grains in neighbouring orbits around a central gravitating body is derived and given a preliminary discussion.

061.018 **Neutrino bremsstrahlung from a degenerate electron gas.** S. K. Saha.
Astrophys. Space Sci., Vol. 12, 493 - 500 (1971).
The neutrino bremsstrahlung process is studied according to the photon-neutrino weak coupling theory by considering the electrons as relativistic and degenerate and by adding some lattice effects at high density. The neutrino energy loss rate due to the process is then compared with that obtained according to the current-current coupling theory. It is concluded that the process is important only in cases of high dense stars when the temperature is below 10^7K.

061.019 **Detection of plutonium-244 in nature.**
D. C. Hoffman, F. O. Lawrence, J. L. Mewherter, F. M. Rourke.
Nature, Vol. 234, 132 - 134 (1971).
Mass spectrometric measurements of plutonium isolated from Precambrian bastnasite confirm the presence of ^{244}Pu in nature. Although the existence of ^{244}Pu as an extinct radioactivity has been postulated to explain the xenon isotope ratios observed in meteorites this is the first indication of its present existence in nature.

061.020 **The present state of gamma-ray astronomy.**
G. Clark.
IAU Symposium No. 41, (see 012.005), p. 3 - 13 (1971).

061.021 **Spectral analysis of gamma rays with the COS-B**

satellite. The caravane collaboration.
H. C. van de Hulst, A. Scheepmaker, B. N. Swanenburg, H. A. Mayer-Hasselwander, E. Pfeffermann, K. Pinkau, H. Rothermel, H. Schneider, W. Voges, J. Labeyrie, P. Keirle, J. Paul, G. Bellomo, G. Bignami, G. Boella, L. Scarsi, G. W. Hutchinson, A. J. Pearce, D. Ramsden, R. D. Wills, P. J. Wright.
IAU Symposium No. 41, (see 012.005), p. 37 - 43 (1971).

061.022 **Gamma-ray spectrometry in the energy range 0.5−5 MeV.**
F. Albernhe, C. Doulade, I. M. Martin, R. Talon, G. Vedrenne.
IAU Symposium No. 41, (see 012.005), p. 58 - 62 (1971).
In French.

061.023 **"Bootstrap" equation of state for cold ultradense matter.** J. C. Wheeler.
Astrophys. Journ., Vol. 169, 105 - 111 (1971).
An equation of state for zero-temperature ultradense ($\gg$ nuclear density) matter is derived, based on the "bootstrap" concept of the hadron mass spectrum.

061.024 **Determinación de la presencia de compuestos quimicos en atmósfcras estelares y planetarias, cometas y espacio interestelar.** F. P. Huberman.
Rev. Astron., Vol. 42, No. 174, p. 5 - 12 (1971).

061.025 **On the non-linear absorption coefficient of electromagnetic radiation under astrophysical conditions.**
S. A. Kaplan, G. M. Khaplanov, Yu. G. Khronopulo.
Astrofizika, Vol. 7, 501 - 505 (1971). In Russian. − English translation in Astrophysics, Vol. 7, No. 3.
The influence of strong electromagnetic radiation on the absorption coefficient in a low frequency spectral line of an atom with three levels is considered.

061.026 **Turbulent velocity waves and kinematic dynamo activity.** I. Lerche.
Astrophys. Space Sci., Vol. 13, 137 - 147 (1971).
It is demonstrated that a turbulent distribution of small amplitude velocity waves gives rise to kinematic dynamo activity on its own in an infinite medium. In view of the fact that most astrophysical objects apparently contain turbulent wave motions, the present calculation is indicative of the extent to which turbulent dynamo activity may be physically important in such objects.

061.027 **On the oscillations of a magnetic polytrope.**
S. K. Trehan, D. F. Billings.
Astrophys. Journ., Vol. 169, 567 - 584 (1971).
The second-order-tensor virial formulation for gaseous configurations with a prevalent magnetic field is here applied to determine the effect of a weak poloidal magnetic field on the fundamental modes of oscillations of a polytrope. The theory of polytropes with a poloidal magnetic field is reviewed and simplified. It is found that of the nine modes of oscillation, three are neutral, four are nonradial, and two are the coupled radial and the nonradial modes. The results of the effect of a weak magnetic field ($h \ll 1$) on the various modes of oscillation are tabulated for polytropic indices $n = 1, 1.5, 2, 3, 3.5,$ and 4.

061.028 **Nonlinear motions of rotating gaseous ellipsoids.**
M. Fujimoto.
Astrophys. Journ., Vol. 170, 143 - 152 (1971).
We apply methods considered by Rossner (1967) and Fujimoto (1968) to a numerical study of the nonlinear behavior of rotating ellipsoids on the basis of a somewhat special but sufficiently reasonable formula for the viscosity. We present also numerical studies of the dynamical evolution and gravitational collapse of gaseous ellipsoids in free precession.

061.029 Equilibrium configurations (4), (5). T. Kwast.
Urania Kraków, Vol. 42, 244 - 249, 287 - 291
(1971). In Polish.

061.030 Magnetic fields in astronomy. T. G. Cowling.
Quarterly Journ. Roy. Astron. Soc., Vol. 12, No. 4,
(see 012.012), 348 - 351 (1971).

061.031 The dynamo problem. N. O. Weiss.
Quarterly Journ. Roy. Astron. Soc., Vol. 12, No. 4,
(see 012.012), 432 - 446 (1971).

061.032 Helium in the universe: Introductory talk.
R. J. Tayler.
Highlights of Astronomy, Vol. 2, (see 012.015), 248 - 253
(1971).

061.033 Helium in the universe: Final summary.
G. Burbidge.
Highlights of Astronomy, Vol. 2, (see 012.015), 328 - 331
(1971).

**061.034 The relevance to astrophysics of the results of re-
cent experiments with colliding charged-particle
beams.** K. T. Dolder.
Highlights of Astronomy, Vol. 2, (see 012.017), 527 - 536
(1971).

061.035 Axially symmetric explosion in a spheroid.
S. Sakashita.
Astrophys. Space Sci., Vol. 14, 431 - 437 (1971).
The method of Laumbach and Probstein is applied to a
point explosion in a spheroid with exponential density distribution. It is shown that the shock wave propagates strongly
along the direction of symmetry axis and the envelope of the
shock front elongates to the same direction. The rate of elongation of the shock envelope increases with the eccentricity of
the spheroid and finally the blowout of the shock wave along
the polar axis occurs when the eccentricity exceeds some critical value.

**061.036 The ordinary URCA process in the presence of posi-
trons and large magnetic fields.** R. W. McFee.
Astrophys. Space Sci., Vol. 14, 446 - 453 (1971).
The effect of positron capture on the ordinary URCA
neutrino luminosity in a zero magnetic field is investigated for
several values of the degeneracy parameter and the range of
temperatures $5 \times 10^8 K - 5 \times 10^{10} K$. The rate for this process is
then compared with those in large magnetic fields. The results
indicate that positron capture reduces the effect of large magnetic fields on this process at high temperatures.

061.037 Natural neutrinos. G. Marx.
Proc. 11th international conference on cosmic rays,
Budapest 1969, (see 012.025), p. 591 - 612 (1970).

061.038 On Schwarzschild's stability criterion.
D. Lortz.
Zeitschr. Naturforschung, Vol. 26a, 1992 - 1994 (1971).
The stability of a gas in an external gravitational field ϕ is
investigated for arbitrary initial conditions.

061.039 On the bases of gravithermodynamics.
I. L. Genkin.
Trudy Astrofiz. Inst., *Alma-Ata*, Vol. 16, 111 - 115 (1971).
In Russian.
The principles of constructing thermodynamics of homo-
geneous gravitating mediums are considered. Expressions for
free energy, entropy and specific heats are found.

061.040 On gravitational instability. I. L. Genkin.

Trudy Astrofiz. Inst., *Alma-Ata*, Vol. 17, 82 - 85
(1971). In Russian.
The existence of two types of gravitational instability is
shown. The first type includes the study of the evolution of
two different density distributions of matter which have equal
initial velocities. The second type includes the study of the
evolution of density fluctuations in the medium, which are
connected with the velocity fluctuations by the continuity
equation.

**061.041 Non-linear theory of gravitational instability in the
expanding universe. II.** K. Tomita.
Progr. Theor. Phys., Japan, Vol. 45, 1747 - 1762, with an
errata list to part I (1967), p. 1762 (1971).
The general relativistic non-linear theory of gravitational
instability is reformulated in order to get a hydrodynamical
system of equations which are applicable at any early stage of
cosmic expansion. On this basis, the second-order density
perturbations associated with the first-order rotational and
gravitational waves are derived, and the amplitudes of the
density perturbations are shown to become comparable with
the square of the amplitudes of the first-order waves. Moreover the second-order density perturbations corresponding to
the first-order ones are derived and examined.

061.042 Electron-nucleus neutrino bremsstrahlung.
P. Cazzola, G. De Zotti, A. Saggion.
Phys. Rev. D, Particles and Fields, Third Ser., Vol. 3, 1722 -
1727 (1971).
The neutrino energy rate arising from bremsstrahlung
reactions of the type $nucleus + e^- \rightarrow nucleus + e^- + \nu + \bar{\nu}$ is
computed in a relativistic context at various densities and
temperatures of matter.

061.043 Ultraviolet stellar astronomy.
M. Grün, P. Koubský.
Vesmír, Vol. 50, 264 - 268 (1971). In Czech.

061.044 New prospect for gamma-ray-line astronomy.
D. D. Clayton.
Nature, Vol. 234, 291 - 292 (1971).
It is suggested that detectable amounts of ^{60}Fe exist in
nearby explosive remnants. ^{60}Fe has a half-life $\tau_{1/2} = 3 \times 10^5$ yr,
and the subsequent decay of its ^{60}Co daughter is accompanied
by a two-gamma-ray cascade with energies $E_\gamma = 1.17$ MeV and
1.33 MeV. This pair of equal-intensity lines is the signature
that can identify ^{60}Fe in interstellar space.

**061.045 The role of a quasi-static field in multiple scattering
of electromagnetic waves.** Yu. N. Barabanenkov.
Izv. vyssh. uchebn. zavedenij. Radiofizika, Vol. 14, 1292 -
1294 (1971). In Russian. – Abstr. in Referativ. Zhurn. 51.
Astron., 2.51.216 (1972).

**061.046 Radiation from magnetic and electric dipoles of a
dielectric sphere.** M. S. Kovner, G. A. Lupanov.
Izv. vyssh. uchebn. zavedenij. Radiofizika, Vol. 14, 1294 -
1296 (1971). In Russian. – Abstr. in Referativ. Zhurn. 51.
Astron., 2.51.217 (1972).

061.047 Advances in X-ray astronomy. I. S. Shklovskij.
Vestn. AN SSSR, 1971, No. 8, p. 81 - 84. In Russian. – Abstr. in Referativ. Zhurn. 51. Astron., 2.51.520
(1972).

061.048 Magnetic fields in astrophysics. E. N. Parker.
Journ. Applied Phys. (*USA*), Vol. 42, 1464 - 1467
(1971). – See Phys. Abstr. Vol. 74, No. 46112 (1971).

061.049 Infra-red astronomical background radiation.
M. Harwit.

Nuovo Cimento Rivista, Vol. 2, 253 - 277 (1970).

In this review article the author discusses measurement techniques, the interplanetary background, the galactic background including stellar and interstellar components, the intergalactic background and the relationship between cosmic rays and background radiation.

061.050 **Astrophysical significance of the dissipation of turbulence in a dense baryon fluid.**   P. B. Jones.
Proc. Roy. Soc. London, Ser. A, Vol. 323, 111 - 125 (1971).
– See Phys. Abstr., Vol. 74, No. 46130 (1971).

061.051 **Thermal instability and Jeans criterion.**
J. Manfroid.
Bull. Soc. Roy. Sci. Liège, Vol. 40, p. 24 - 36 (1971). In French.

The author investigates the influence of thermal effects on gravitational instability and derives sufficient and necessary criteria for instability from a linear analysis.

061.052 **On a class of nonlinear differential equations of astrophysics.**   R. V. Ramnath.
Journ. Math. Analysis Applications, Vol. 35, 27 - 47 (1971).

Investigation is made of asymptotic approximations to the solutions of a class of nonlinear differential equations of astrophysics, occurring in the study of polytropic gas spheres. The general form is that of Emden's equation of index n. More general equations with variable coefficients are also considered, representing changes in the polytropic gas law as a function of the radius.

061.053 **Proposed experimental test of neutrino sea as a subquantic medium.**   H. C. Dudley.
Nuovo Cimento B, Ser. 11, Vol. 4B, 68 - 72 (1971).

Astrophysical findings coupled with known properties of a generalized neutrino-antineutrino flux suggest that there exists a finite medium which may be capable of transmitting electromagnetic radiation. The authors suggest a critical experimental test using high-velocity emitters with flows parallel to the earth's galactic motion.

061.054 **New complexities for the cosmic chemist.**
A. H. Barrett.
Comments Atomic Molecular Phys., Vol. 2, No. 6, p. 165 - 170 (1971).

Discusses the recent discoveries of hitherto undetected molecules in cosmic space.

061.055 **Cosmic X-ray astronomy.**   D. J. Adams.
Contemporary Phys., Vol. 12, 471 - 493 (1971).

The results obtained from observations of the celestial X-radiation over the past nine years are reviewed. Emphasis is placed on the study of the discrete sources, particularly those which have been identified with astronomical objects

known from optical and radio studies. The isotropic background X-radiation is dealt with briefly, and the theories which account for X-ray production on the astronomical scale are mentioned where applicable.

061.056 **Classical distributions of charged dust.** A. K. Datta.
Indian Journ. Phys., Vol. 44, 278 - 282 (1971).

The paper considers the equations of classical hydrodynamics and electromagnetism for a distribution of charged dust. Some general theorems and formulae are obtained.

061.057 **Geochemical features of the composition of the universe and the origin of atomic nuclei.**
V. V. Cherdyntsev.
Proc. Sixth Winter School on Space Physics, Part I. Apatity 1969, (see 012.033), p. 102 - 107 (1971).

061.058 **The properties of nuclei far from the beta-stability strip.**   E. E. Berlovich.
Proc. Sixth Winter School on Space Physics, Part I. Apatity 1969, (see 012.033), p. 108 (1971). – Abstract.

061.059 **Application of the methods of measurement of small and ultrasmall quantities of various isotopes to astrophysical research.**   V. O. Naidenov.
Proc. Sixth Winter School on Space Physics, Part I. Apatity 1969, (see 012.033), p. 310 - 318 (1971).

061.060 **Physics and astrophysics.**   E. Schatzman.
CERN Rep. No. 70-31, 110 pp. (1970).

061.061 **Conductive transfer in relativistic medium.**
E. Suhonen.
Ann. Acad. Sci. Fennicae A. VI. (Phys.), No. 356, 22 pp. (1971).

061.062 **Phénomènes de fluorescence dans les astres.**
P. Swings.
Mém. Soc. Sci. Liège, 6. Ser., Vol.1, 131 - 141 (1971).

Theories of equilibrium figures of a rotating homogeneous fluid mass.   See Abstr. 003.043.

The solar light-element abundances and primeval helium.   See Abstr. 071.048.

Neutrino astronomy: Probing the sun's interior.
See Abstr. 080.010.

More solar models and neutrino fluxes.
See Abstr. 080.049.

μ Cassiopeiae and the primordial helium abundance – A critique.   See Abstr. 114.020.

062 Magneto-Hydrodynamics, Plasma

062.001 The structure of transverse hydromagnetic shocks in regions of low ionization. D. J. Mullan.
Monthly Notices, Roy. Astron. Soc., Vol. 153, 145 - 170 (1971).

The compression of a magnetic field by a shock in a predominantly neutral gas is discussed. By solving the fluid equations, a steady-state structure is determined for a one-dimensional shock front propagating through a partially ionized gas in a direction perpendicular to the field lines. Temperatures and velocities of ions, electrons, and atoms are calculated as a function of the spatial coordinate moving with the shock frame. It is suggested that ion-atom velocity differences within shocks lead to efficient removal of the magnetic field from a contracting protostar, thereby permitting fragmentation into masses as small as one solar mass. Application of the results to a model which has been proposed for solar flares, and to conditions in laboratory experiments are noted.

062.002 Magnetohydrodynamic wave-mode coupling. Quantum field-theoretical approach to weakly nonlinear case with application to solar coronal heating.
O. Kaburaki, Y. Uchida.
Publ. Astron. Soc. Japan, Vol. 23, 405 - 423 (1971).

The problem of non-linear coupling of magnetohydrodynamic wave modes is discussed by the use of a quantum field-theoretical approach with the Hamiltonian formulation. By defining the interaction Hamiltonian for a weakly non-linear coupling by the next lowest order terms, we can treat the problem in terms of the level population equation by using the transition probabilities calculated from the matrix element of this interaction Hamiltonian in quantum theory. An application to solar coronal heating is given as an example of problems of astrophysical interest.

062.003 Bremsstrahlung power density: Long-range interactions. A. Oppenheim.
Phys. Rev. Letters, Vol. 27, 3 - 5 (1971).

The power density of bremsstrahlung and the recombination rate are found for a fully ionized, nonrelativistic hydrogen plasma dominated by long-range collisions. The results show that the bremsstrahlung energy radiated per unit volume per second is greater than previously predicted.

062.004 Ion-acoustic instability of a two-temperature, collisional, fully ionized plasma.
T. D. Rognlien, S. A. Self.
Phys. Rev. Letters, Vol. 27, 792 - 795 (1971).

From a perturbation analysis of the fluid equations for a homogeneous, unmagnetized plasma it is shown that long-wavelength ion waves are unstable when $T_e \gtrsim T_i$. The additional destabilizing effect of a current is also investigated.

062.005 The influence of spatial dispersion on the coefficient of spontaneous emission of cyclotron radiation in a non-equilibrium plasma. H. Jahn.
Astrophys. Space Sci., Vol. 12, 34 - 39 (1971).

The coefficient of spontaneous emission of cyclotron radiation propagating in the direction of an external magnetic field in non-equilibrium plasma is calculated including the effect of spatial dispersion on the emission process. The resulting part of the emission coefficient may become important in the vicinity of the cyclotron frequency.

062.006 Laser absorption measurement of electron temperature in an argon plasma. M. F. Weisbach.
Journ. Quant. Spectrosc. Radiat. Transfer, Vol. 11, 1225 - 1234 (1971).

062.007 Diffusion acceleration rates of charged turbulent plasma particles in a magnetic field.
S. A. Kaplan, V. N. Tsytovich, A. S. Chikhachev.
Izv. vyssh. uchebn. zavedenij. Radiofizika, Vol. 14, 204 - 216 (1971). In Russian. – Abstr. in Referativ. Zhurn. 51. Astron., 8.51.209 (1971).

062.008 Instabilitäten und kollektive Energiedissipation in Plasmastoßwellen. M. Keilhacker, K.-H. Steuer.
Naturwissenschaften, 58. Jahrgang, p. 377 - 383 (1971).

The paper deals with plasma shock waves in which the dissipation of energy is accomplished by collective plasma interaction rather than by binary particle collisions. The collisionless dissipation is shown to be due to microinstabilities, which in low Mach number shocks cause an anomalously high plasma resistivity accompanied by strong electron heating, and in high Mach number shocks a collisionless viscosity resulting in ion heating. The results have a bearing on problems in astrophysics.

062.009 Zur Elektrodynamik turbulent bewegter leitender Medien unter Berücksichtigung des Halleffektes.
G. Helmis.
Cosmic Electrodynamics, Vol. 2, 197 - 210 (1971).

The mean value of $v \times B$ in a turbulently moving electrically conducting medium is calculated taking into account the influence of the Hall effect. The turbulence is assumed to be homogeneous, isotropic, but not symmetric with respect to reflection. General expressions are developed for small turbulence velocities; exactly valid formulas are derived for special correlation tensors and any value of conductivity and the Hall parameter.

062.010 Asymptotic solutions in nonlinear convection with free boundaries. R. Van der Borght.
Australian Journ. Phys., Vol. 24, 579 - 585 (1971).

An investigation is made of the problem of nonlinear convection in an incompressible fluid at high Rayleigh number, using an asymptotic theory. An estimate is derived for the preferred horizontal wave number.

062.011 Propagation of hydromagnetic waves in a stochastic magnetic field. G. C. Valley.
Astrophys. Journ., Vol. 168, 251 - 264 (1971).

The propagation of coherent small-amplitude hydromagnetic waves through a medium in which the magnetic field has a large-scale ordered component and a stationary random component is investigated. The important new result obtained for hydromagnetic-wave propagation in a magnetic field which has a random component is that in many cases a coherent wave damps into random noise. The results and discussion of these various cases are given. Application of these results shows that coherent hydromagnetic waves propagate at most 1 a.u. in the solar wind and 1 kpc in the Galaxy before damping into random noise.

062.012 On magnetosonic disturbances caused by expansion of an ideal conducting sphere in a cold plasma.
L. P. Gorbatchev, Yu. N. Savchenko.
Geomagn. Aeronom., Vol. 11, 898 - 900 (1971). In Russian. Brief information.

062.013 Evolution of turbulent magnetic fields – approach to a steady state. S. Nagarajan.
IAU Symposium No. 43, (see 012.003), p. 487 - 504 (1971).

062.014 Observation of solar flare type processes in the laboratory.
W. H. Bostick, V. Nardi, L. Grunberger, W. Prior.
IAU Symposium No. 43, (see 012.003), p. 512 - 525 (1971).

062.015 Electron energy equation for an atomic radiating plasma. S. S. R. Murty.
Journ. Quant. Spectrosc. Radiat. Transfer, Vol. 11, 1681 - 1690 (1971).

Energy transfer between electrons and heavy particles by elastic collisions has been adequately treated in the literature and the aim of this work is to bring out explicitly the effect of radiation in formulating an electron-energy equation valid for all optical thicknesses of the plasma. Detailed consideration is given to an optically thin plasma and a derivation of the Kramers-Unsöld approximation for hydrogenic gases is presented.

062.016 Compton scattering of plasma and magnetohydrodynamic waves on relativistic electrons as a source of radio emission from metagalactic objects.
G. G. Getmantsev.
Izv. vyssh. uchebn. zavedenij. Radiofizika, Vol. 14, 659 - 662 (1971). In Russian. – Abstr. in Referativ. Zhurn. 51. Astron., 10.51.165 (1971).

062.017 The dependence of the number of the extreme resolved line of Lyman series on electron density.
L. N. Kurochka, L. B. Maslennikova.
Vestn. Kiev. Un-ta, Ser. Astron., No. 12, p. 25 - 30 (1970). In Russian.

The dependence of the number of the last resolved line of Lyman series on electron density and on velocity of hydrogen atoms in a plasma is obtained. The paper gives the dependence of the number of the extreme resolved line on electron density at different values of velocities of atoms for the joint action of Stark and Doppler effects.

062.018 The relativistic quasilinear theory of particle acceleration by hydromagnetic turbulence.
R. M. Kulsrud, A. Ferrari.
Astrophys. Space Sci., Vol. 12, 302 - 318 (1971).

The quasilinear theory of acceleration of relativistic particles by hydromagnetic turbulence is treated in the adiabatic limit of small gyration radius. The theory is based on the relativistic Vlasov equation; however, a given pitch-angle scattering rate by microturbulence is postulated and is added to this equation.

062.019 Maximum temperatures for radiation from plasma waves. D. F. Smith, P. A. Sturrock.
Astrophys. Space Sci., Vol. 12, 411 - 414 (1971).

We derive upper limits to the radiation temperatures for emission near the fundamental and second harmonic of the electron plasma frequency in terms of the effective temperature for plasma waves.

062.020 Measurement of electron-impact broadening of ionized berillium and barium lines in an electric shock tube plasma.
M. Platiša, J. Purić, N. Konjević, J. Labat.
Astron. Astrophys., Vol. 15, 325 - 328 (1971).

The aim of this paper is to provide more data on the Stark broadening of lithium like ions which can be compared with existing theory. The plasma source was an electromagnetically driven shock tube. The electron density was determined by laser interferometry and the electron temperature from relative intensities of A II spectral lines.

062.021 Plasma, der vierte Zustand der Materie.
W. H. Kegel.

SuW, Vol. 10, 298 - 302 (1971).

062.022 Modelle kraftfreier Magnetfelder. R. Wagner.
Zeitschr. Naturforschung, Vol. 26a, 1753 - 1762 (1971).

A method for computing force-free magnetic fields of known anomality $\alpha \neq$ const is described. The formulae derived can be used to decide if to a given geometry a force-free field does exist. Existing fields can be computed immediately. The results are illustrated by examples.

062.023 Strong electromagnetic waves in overdense plasmas.
C. Max, F. Perkins.
Phys. Rev. Letters, Vol. 27, 1342 - 1345 (1971).

062.024 Nonlinear Landau damping of Alfvén waves.
J. V. Hollweg.
Phys. Rev. Letters, Vol. 27, 1349 - 1352 (1971).

062.025 Rotational discontinuities in an anisotropic plasma.
P. D. Hudson.
Planet. Space Sci., Vol. 19, 1693 -1699 (1971).

A general theory of rotational discontinuities is developed and the changes in the components of the plasma pressure, $p_{\parallel}$ and $p_{\perp}$, and in the magnetic induction, B, are found. Some special solutions are analysed and the identification of rotational discontinuities in the solar wind is discussed.

062.026 Computer simulation of plasmas. J. M. Dawson.
Astrophys. Space Sci., Vol. 13, 446 - 467 (1971). – Review paper.

The work that has been carried out at Princeton and at the Naval Research Laboratory is summarized.

062.027 Enhancement of relaxation processes by collective effects. R. M. Kulsrud.
Astrophys. Space Sci., Vol. 13, 468 - 477 (1971).

The various collective processes in plasma physics, with an emphasis on their possible relevance to stellar dynamics are reviewed.

062.028 The dynamics of a toroidal magnetic ring.
C. G. Lilliequist, M. D. Altschuler, Y. Nakagawa.
Solar Physics, Vol. 20, 348 - 361 (1971).

Solving the nonlinear partial differential equations of magnetohydrodynamics numerically, we examine (1) the time development of a purely toroidal magnetic field (a magnetic ring) and (2) the interaction of a magnetic ring with a poloidal magnetic field. Axisymmetry and incompressibility are assumed. Parameters are chosen to correspond to photospheric conditions. We conjecture that toroidal magnetic fields may be involved in the bright rings of sunspots or in the dynamics of spicules.

062.029 Energy and momentum exchange in transverse plasma waves. J. V. Hollweg, H. J. Völk.
Journ. Geophys. Res., Vol. 76, 7527 - 7541 (1971).

We calculate, by a perturbation analysis, the energy and momentum changes both of a single particle and of a distribution of particles moving in a transverse electromagnetic wave propagating parallel to the direction of the average magnetic field.

062.030 Plasma radiation from collisionless MHD shock waves. I. Shock-region analysis. D. F. Smith.
Astrophys. Journ., Vol. 170, 559 - 571 (1971).

The structures of perpendicular collisionless magnetohydrodynamic shock waves in which the magnetic pressure is much larger than the particle pressure and charge neutrality applies are examined as potential sources of electron plasma waves. It is shown that significant excitation of such waves cannot occur in shocks with a laminar solution, due to their extreme velocity inhomogeneity. Turbulent shocks are characterized by a highly nonthermal energy density in ion-acoustic waves which effectively suppress any instability of electron plasma waves and preferentially heat the electrons.

062.031 **Plasma physics applied to cosmology.** H. Alfvén.
Physics Today, Vol. 24, No. 2, p. 28 - 31, 33 (1971).

062.032 **Magnetohydrodynamic stability problems in astrophysics.** R. J. Tayler.
Quarterly Journ. Roy. Astron. Soc., Vol. 12, No. 4, (see 012.012), 352 - 362 (1971).

In laboratory studies of plasmas confined by a magnetic field, it is found that most configurations in which the plasma and magnetic pressures are comparable are unstable, and that the instability growth times are not much greater than the dynamical timescale of the system. Similar instabilities probably occur in many astrophysical situations. Two such problems, the interaction of the cosmic rays and the galactic magnetic field and the structure of radio galaxies are discussed briefly.

062.033 **Magnetohydrodynamics of rotating fluids: A summary of some recent work.** R. Hide.
Quarterly Journ. Roy. Astron. Soc., Vol. 12, No. 4, (see 012.012), 380 - 383 (1971).

062.034 **Results obtained from observations of laboratory plasmas.** W. Lochte-Holtgreven.
Highlights of Astronomy, Vol. 2, (see 012.017), 537 - 543 (1971).

062.035 **Measurement of hydromagnetic waves as a method of estimating a photon's rest mass.** M. A. Gintsburg.
Dokl. Akad. Nauk SSSR, Ser. Mat. Fiz., Vol. 201, 817 - 819 (1971). In Russian.

062.036 **Reverse deflection and contraction of a plasma beam moving along curved magnetic field lines.**
L. Lindberg, L. Kristoferson.
Cosmic Electrodynamics, Vol. 2, 305 - 308 (1971).

A plasma beam which originally moves in a longitudinal magnetic field and enters a region of curved magnetic field is investigated experimentally. It is found that the beam contracts perpendicular to the $v \times B$-direction and the whole beam becomes deflected in the direction opposite to that in which the magnetic field is curved.

062.037 **Cumulation processes. Self-similar solutions in gas dynamics.** J. P. Somon.
Physics of high energy density. Course 48, Italian Phys. Soc., 1969, (see 012.023), p. 189 - 216 (1971).

062.038 **The physics of strong shock waves in gases.**
R. A. Gross.
Physics of high energy density. Course 48, Italian Phys. Soc., 1969, (see 012.023), p. 245 - 277 (1971).

062.039 **On the motion of a charged particle in an axially symmetric magnetic field.** V. L. Bharadwaj.
Zeitschr. Naturforschung, Vol. 26a, 2068 - 2070 (1971).

The object of this note is to study the motion of a charged particle entering the magnetic field due to a steady current inside a plasma column.

062.040 **Simple wave motion in magnetoelasticity.**
J. Bazer, F. Karal.
Geophys. Journ. Roy. Astron. Soc., Vol. 25, Nos. 1–3, (see 012.026), p. 127 - 156 (1971).

Simple one-dimensional wave motion in an infinitely

conducting, electrically neutral, elastic medium in the presence of a magnetic field is studied. As in magnetogasdynamics, there are slow, fast, and Alfvén-like simple wave solutions. These waves afford a mechanism for generating intense magnetic fields and together with the corresponding magnetoelastic shocks they enable one to solve one-dimensional propagation problems with sufficiently simple initial conditions.

062.041 **Geometrical magnetoelasticity.**
J. Bazer.
Geophys. Journ. Roy. Astron. Soc., Vol. 25, Nos. 1–3, (see 012.026), p. 207 - 238 (1971).

An account is given of propagation phenomena in an electrically neutral, infinitely conducting, inhomogeneous elastic medium in the presence of a magnetic field.

062.042 **Chocs et ondes rotatoires de la magnétohydrodynamique relativiste.** I. Lukačević.
Ann. Inst. Henri Poincaré, Nouvelle Sér., Sect. A, Vol. 14, 219 - 248 (1971).

Such types of shock waves are investigated where the magnetic field strength remains unchanged in magnitude. Starting from the basic equations of relativistic magnetohydrodynamics for an ideal fluid the fundamental shock relations are constructed and the possible types of shocks are classified.

062.043 **Profiles of the He I 4471.5- and 4922-Å lines in a dc arc.** J. W. Birkeland, M. E. Bacon, W. G. Braun.
Phys. Rev. A, General Phys., Third Ser., Vol. 3, 354 - 358 (1971).

062.044 **Collisional ionization rates for lithium- and beryllium-like ions.** H.-J. Kunze.
Phys. Rev. A, General Phys., Third Ser., Vol. 3, 937 - 942 (1971).

Collisional ionization rates for C IV, N V, and O VI as well as for O V and Ne VII are deduced from the time history of spectral lines emitted by these ions in a hot plasma.

062.045 **The Langevin equation of weak turbulence.**
K. Elsässer, P. Gräff.
Ann. Physics, Vol. 68, 305 - 336 (1971).

The hierarchy equations describing weakly interacting waves in a fluid are solved by the method of characteristic functionals, combined with the time asymptotic method of Bogoliubov and Mitropolski. The result to lowest nontrivial order allows one to characterize the stochastic state of the fluid in close analogy to the Brownian motion of a test particle.

062.046 **Thomson scattering in a strong magnetic field.**
V. Canuto, J. Lodenquai, M. Ruderman.
Phys. Rev. D, Particles and Fields, Third Ser., Vol. 3, 2303 - 2308 (1971).

The effect of a strong magnetic field on neutron stars or white dwarfs is calculated for Thomson scattering in a fully ionized collisionless plasma.

062.047 **Compton Fokker-Planck equation for hot plasmas.**
G. Cooper.
Phys. Rev. D, Particles and Fields, Third Ser., Vol. 3, 2312 - 2316 (1971).

The Fokker-Planck equation for Compton scattering in a plasma is developed without recourse to a non-relativistic expansion.

062.048 **Some general relations in relativistic magnetohydrodynamics.** P. Yodzis.
Phys. Rev. D, Particles and Fields, Third Ser., Vol. 3, 2941 - 2945 (1971).

A method is outlined for obtaining general relations governing the behavior of magnetofluids in general relativity. Several such relations are obtained for the case of infinite conductivity, and their possible relevance to galactic cosmogony, gravitational collapse, and pulsar theory is briefly discussed.

062.049 Final stages of evolution of a magnetoid and observations. L. M. Ozernoy.
Astron. Tsirk., No. 661, p. 4 - 6 (1971). In Russian.

062.050 Calculations of sunlike dynamos for alternating fields. M. Steenbeck, F. Krause.
Geod. Geophys. Veröff., Ser. 2, No. 13, (see 012.029), p. 129 (1969). – The full text of the present paper is published in Astron. Nachr., Vol. 291 (1969). – See 02.062.008.

062.051 On some electromagnetic phenomena in electrically conducting turbulently moving matter, especially in the presence of Coriolis forces. K.-H. Rädler.
Geod. Geophys. Veröff., Ser. 2, No. 13, (see 012.029), p. 131 - 135 (1969).

062.052 Plasma motion in a strong increasing dipole magnetic field. B. V. Somov, S. I. Syrovatskij.
Zhurn. ehksperim. i teor. tiz., Vol. 61, 621 - 628 (1971). In Russian. – Abstr. in Referativ. Zhurn. 51. Astron., 2.51.211 (1972).

062.053 Plasma light sources. A. H. Gabriel.
Beam-foil spectroscopy. Conference 1970, (see 012.032), p. 157 - 161 (1970).

062.054 Observation and identification of highly stripped iron transitions produced in the plasma focus.
J. P. Connerade, N. J. Peacock, R. J. Speer.
Beam-foil spectroscopy. Conference 1970, (see 012.032), p. 163 - 166 (1970).

062.055 The structure of hydromagnetic shocks in regions of very low ionization. D. J. Mullan.
Thesis, Univ. Maryland, Catonsville. [Available from Univ. Microfilms, Ann Arbor, Mich., U.S.A. Order No. 70–16026], 173 pp. (1969).
Gas flows in the interstellar medium in all likelihood contain shocks. Detailed models of several shocks are computed to determine the heating and drifting of ions relative to the atoms resulting from the momentum transfer.

062.056 On the dispersion of electromagnetic waves in interstellar space. A. R. Lee.
Phys. Letters A, Vol. 36a, 283 - 284 (1971).
The author draws attention to the fact that the dispersion effect of electromagnetic waves in a plasma is indistinguishable in form from that due to assigning a non-zero rest-mass to the photon.

062.057 Plasma physics, space research, and the origin of the solar system. H. Alfvén.
Cesk. Casopis Fis., Ser. A, Vol. 21, 502 - 510 (1971). In Czech. – See Phys. Abstr., Vol. 75, No. 7359 (1972).

062.058 The interaction of homogeneous wave turbulence and a magnetohydrodynamic tangential discontinuity. C. K. W. Tam.
Journ. Plasma Phys., Vol. 5, 265 - 274 (1971).
The interaction of homogeneous wave turbulence and a magnetohydrodynamic tangential discontinuity is studied. A magnetohydrodynamic description is used which is believed to be adequate for plasma problems in interplanetary space. On applying the present theory to the problem of interaction between the turbulent waves in the magnetosheath and the magnetopause, it is found that the turbulent shear stress pro-

duced is too weak to produce any large-scale internal magnetospheric convection as was previously contemplated.

062.059 The hydromagnetic Kelvin-Helmholtz problem in a Hall plasma. J. F. McKenzie.
Journ. Plasma Phys., Vol. 5, 275 - 288 (1971).
The hydromagnetic analogue of the Kelvin-Helmholtz problem is extended to include the effects of the Hall term. The special case of a hot, unmagnetized fluid on one side of the interface and a cold, magnetized fluid on the other is studied in some detail.

062.060 Relativistic kinetic theory of the large-amplitude transverse Alfvén wave.
A. Barnes, G. C. J. Suffolk.
Journ. Plasma Phys., Vol. 5, 315 - 329 (1971).
It is shown that the relativistic Vlasov-Maxwell equations admit a solution very much like the transverse Alfvén wave of magnetohydrodynamic theory. The dispersion relation yields a criterion for the firehose instability which turns out to be the same as that derived from linearized theory.

062.061 The gravitational instability of a rotating magneto-plasma in the guiding centre approximation.
A. D. Lunn.
Journ. Plasma Phys., Vol. 5, 365 - 373 (1971).
Jeans's criterion for instability is found to be changed only for the case of perturbations perpendicular to the magnetic field and rotation parallel to it, when rotation and finite Larmor radius effects tend to stabilize the plasma.

062.062 Analysis of electromagnetic instabilities parallel to the magnetic field. W. Pilipp, H. J. Völk.
Journ. Plasma Phys., Vol. 6, 1 - 17 (1971).
Transverse waves and instabilities propagating along the magnetic field in a homogeneous plasma are discussed analytically and numerically for frequencies of the order of the ion cyclotron frequency and below. Extensive discussions of the energy losses and gains of the ions and electrons are given for all unstable modes.

062.063 Ideally conducting magnetostatic equilibria and associated time dependent, resistive flows.
J. C. Stevenson.
Journ. Plasma Phys., Vol. 6, 125 - 136 (1971).
Several types of two-dimensional solutions for the equations of magnetohydrodynamics are described. For all these solutions the magnetic field contains at least one hyperbolic neutral point. Two new magnetostatic equilibria are introduced for the ideally conducting case. The magnetic field associated with one of these is used to construct an exact time-dependent solution of the MHD equations where the fluid is necessarily at rest.

062.064 Electrostatic shielding of a test charge in a non-neutral plasma. R. C. Davidson.
Journ. Plasma Phys., Vol. 6, 229 - 235 (1971).
The electrostatic shielding of a test electron embedded in a magnetically confined pure electron gas column is investigated.

062.065 Stability of an anisotropic plasma jet.
M. R. Raghavachar.
Journ. Plasma Phys., Vol. 6, 237 - 248 (1971).
The stability of an anisotropic plasma jet has been investigated using equations of Chew, Goldberger & Low (1965) in both plane and cylindrical geometries.

062.066 Cyclotron radiation in hot magnetoplasmas.
J. Trulsen.
Journ. Plasma Phys., Vol. 6, 367 - 400 (1971).

The effects of thermal motions on the cyclotron radiation from test particles gyrating in a homogeneous magnetoplasma are studied. These effects take care of all singularities that exist in the theory of cyclotron radiation in cold magnetoplasma, e.g. the divergence in energy loss for small particle energies.

062.067   **Existence conditions for collisionless hydromagnetic shock waves along the magnetic field.**
Y. Kato, M. Tajiri, T. Taniuti.
Journ. Plasma Phys., Vol. 6, 467 - 493 (1971).

This paper is concerned with existence conditions for steady hydromagnetic shock waves propagating in a collisionless plasma along an applied magnetic field. The electrostatic waves are excluded. The conditions are based on the requirement that solutions of the Vlasov-Maxwell equations deviate from a uniform state ahead of a wave.

062.068   **Stability of anisotropic plasmas to almost-perpendicular magnetosonic waves.**
R. W. Landau, S. Cuperman.
Journ. Plasma Phys., Vol. 6, 495 - 512 (1971).

The stability of anisotropic plasmas to the magnetosonic (or right-hand compressional Alfvén) wave, near the ion cyclotron frequency, propagating almost perpendicular to the magnetic field, is investigated.

062.069   **Higher branches of the dispersion relation for perpendicular magnetosonic waves in relativistic anisotropic plasmas.**   S. Cuperman, N. Metzler.
Journ. Plasma Phys., Vol. 6, 541 - 545 (1971).

Dispersion relations are calculated and solved for frequencies near the first two harmonics of perpendicular magnetosonic waves in relativistic anisotropic plasmas.

062.070   **Electron pitch-angle diffusion driven by oblique whistler-mode turbulence.**
L. R. Lyons, R. M. Thorne, C. F. Kennel.
Journ. Plasma Phys., Vol. 6, 589 - 606 (1971).

A general description of cyclotron harmonic resonant pitch-angle scattering is presented. Quasi-linear diffusion coefficients are prescribed in terms of the wave normal distribution of plasma wave energy. Numerical computations are performed for the specific case of relativistic electrons interacting with a band of low frequency whistler-mode turbulence. A parametric treatment of the wave energy distribution permits normalized diffusion coefficients to be presented graphically solely as a function of the electron pitch-angle.

062.071   **Exact nonlinear evolution of Alfvén modes in the guiding-center model.**   B. Patterson.
Phys. Fluids, Vol. 14, 1127 - 1136 (1971).

Alfvén waves parallel to a uniform magnetic field in a uniform but anisotropic plasma are studied in the Chew–Goldberger–Low and guiding-center models. Exact solutions are found in the stable and unstable cases.

062.072   **Arbitrary amplitude magnetoacoustic waves under gravity: An exact solution.**   Y. T. Chiu.
Phys. Fluids, Vol. 14, 1717 - 1724 (1971).

Exact Riemann wave solutions to the magnetohydrodynamic equations for a compressible medium under the influence of a gravitational field are obtained. The resultant magnetoacoustic waves cannot be classified as slow and fast modes, and their propagation characteristics are found to be quite different from those in a uniform medium. In particular, the phenomenon of large amplitude wave ducting is examined. Certain wave properties correspond exceedingly well with the observed fine structure of waves propagating away from solar flares.

062.073   **Signal dispersion and Faraday rotation in hot plasmas.**
J. Skilling.
Phys. Fluids, Vol. 14, 2523 - 2531 (1971).

The signal dispersion and Faraday rotation of high-frequency radiation passing through a hot, possibly relativistic plasma are investigated. Corrections are found to the formulas given by cold-plasma theory. Results for the damping are compared with synchrotron self-absorption formulas.

062.074   **Space-physical investigations in the laboratory.**
I. M. Podgornyi.
Proc. Sixth Winter School on Space Physics, Part I. Apatity 1969, (see 012.033), p. 240 - 248 (1971).

062.075   **Self-similar flow in rarefied plasma.**
M. A. Gintsburg.
Proc. Sixth Winter School on Space Physics, Part I. Apatity 1969, (see 012.033), p. 323 - 324 (1971).

062.076   **Particle acceleration in a plasma in outer space.**
M. A. Gintsburg.
Proc. Sixth Winter School on Space Physics, Part I. Apatity 1969, (see 012.033), p. 325 (1971). – Abstract.

Radiative recombination coefficients for complex ions.   See Abstr. 022.029.

On the correct form of the equation of radiative transfer in a nonuniform magnetoactive plasma.
See Abstr. 063.008.

Radiation from a high-temperature, low-density plasma: The X-ray spectrum of the solar corona.
See Abstr. 074.017.

Collisionless stellar dynamics.   See Abstr. 151.039.

Stability properties for encounterless self-gravitational stellar gas and plasma.   See Abstr. 151.044.

The generation of magnetic fields in astrophysical bodies. VII. The internal small-scale fields.
See Abstr. 156.002.

The generation of magnetic fields in astrophysical bodies. VIII. Dynamical considerations.
See Abstr. 156.003.

063 Radiative Transfer

063.001 Non-linear coupling between thermal conduction and radiative transfer.   H. Frisch.
Astron. Astrophys., Vol. 13, 359 - 366 (1971).

The problem of coupled thermal conduction and radiative transfer is solved numerically for a plane slab, of finite thickness, heated by thermal conduction, the temperatures at the boundaries and the radiation entering the slab being given. In the optically thick case a simple physical interpretation of the solution can be given in terms of a region of pure radiative equilibrium and transition regions where absorption (but not emission) of radiative energy is negligible.

063.002 Energy-momentum tensor for radiation and radiative viscosity.   I. Masaki.
Publ. Astron. Soc. Japan, Vol. 23, 425 - 431 (1971).

The energy-momentum tensor for radiation in an optically thick matter is calculated for the case in which the matter interacts with photons by the process of Thomson scattering as well as by the processes of absorption and emission of photons.

063.003 Radiative transfer with Rayleigh scattering. II. Finite atmosphere.   H. Domke.
Astron. Zhurn. Akad. Nauk SSSR, Vol. 48, 777 - 789 (1971). In Russian. English translation in Soviet Astron. AJ, Vol. 15, No. 4.

The equation of transfer of polarized light in a conservative Rayleigh scattering plane-parallel atmosphere of finite optical thickness with arbitrary distribution of primary sources is reduced to scalar integral equations with displacement kernels. The four constants which appear in these equations are expressed in terms of the emergent radiation. As an example the solution is found of the problem of diffuse reflection and transmission.

063.004 The formation of spectral lines.
D. G. Hummer, G. Rybicki.
Annual Rev. Astron. Astrophys., Vol. 9, (see 003.001), 237 - 270 (1971).

This review is intended to provide an introduction to the physics and phenomenology of the radiation field in spectral lines. Our attention is confined to problems involving the transfer of radiation in which optical-depth effects are, or can be, significant and in which it cannot be assumed a priori that atomic levels are populated according to the laws of Saha and Boltzmann.

063.005 Pade approximants and the X and Y functions of radiative transfer theory.
H. Cohen, K. Cureton.
Journ. Quant. Spectrosc. Radiat. Transfer, Vol. 11, 1279 - 1283 (1971).

We demonstrate that the [1, 1] Pade approximant to the X and Y functions is a reasonable alternative to the exact solutions.

063.006 On the calculation of the point-source radiation transfer in a scattering medium.
B. M. Golubitskij, M. V. Tantashev.
Izv. vyssh. uchebn. zavedenij. Radiofizika, Vol. 14, 325 - 327 (1971). In Russian. – Abstr. in Referativ. Zhurn. 51. Astron., 8.51.194 (1971).

063.007 Radiative transfer in spherically symmetric systems— II. The non-conservative case and linearly polarized radiation.   J. P. Cassinelli, D. G. Hummer.
Monthly Notices Roy. Astron. Soc., Vol. 154, 9 - 21 (1971).

The method for the solution of transfer problems in spherically symmetric systems developed recently by Hummer and Rybicki is here generalized to the non-conservative case. This procedure, which depends on the iterative determination of the Eddington factor $f = K/J$, handles in a natural way the outward peaking of the radiation field which occurs in extended atmospheres. To illustrate the present extension of this method, solutions are obtained for the problem of scattering of linearly polarized radiation by an extended electron-scattering atmosphere.

063.008 On the correct form of the equation of radiative transfer in a nonuniform magnetoactive plasma.
V. V. Zheleznyakov.
Astrophys. Journ., Vol. 168, 281 - 282 (1971).

The criticism by Enome and Walsh of Zheleznyakov's equation of radiative transfer in a nonuniform magnetoactive plasma is shown to be incorrect. The discrepancy between the form of the equation given by Bekefi and that given by Zheleznyakov is a consequence of the ambiguous definition of specific intensity: Bekefi used specific intensity per unit solid angle along the group velocity vector while Zheleznyakov used specific intensity per unit solid angle along the wave vector.

063.009 A new computational method for the X and Y functions of radiative transfer.   J. K. Shultis.
Astron. Astrophys., Vol. 14, 463 - 467 (1971).

The X and Y functions of radiative transfer are shown to satisfy a pair of Fredholm integral equations. The equations are derived by considering a particular problem of diffuse reflection and transmission whose emergent intensities are closely related to these functions. Application of the singular eigenfunction expansion technique of neutron transport theory to this particular problem yields the Fredholm equations for both nonconservative and conservative atmospheres.

063.010 The spectrum of steady state turbulent convection.   F. Winterberg.
Zeitschr. Naturforschung, Vol. 26 a, 1140 - 1146 (1971).

Based on Heisenberg's statistical theory of turbulence, a model for steady state turbulent convection is herein proposed, and on the basis of this model, equations for the energy spectrum for steady state turbulent convection are derived. The energy spectrum has a substantial deviation from the Kolmogoroff law, as a result of the buoyancy force acting on the rising and falling eddies. The presented theory may be applicable to convection in planetary and stellar atmospheres wherein the radiative heat transport is small.

063.011 Non-grey radiative heat transfer in conservative plane-parallel media with reflecting boundaries.
R. J. Reith, Jr., C. E. Siewert, M. N. Özişik.
Journ. Quant. Spectrosc. Radiat. Transfer, Vol. 11, 1441 - 1462 (1971).

In the present paper, we report the required extension of the analysis of finite-slab problems, defined in terms of the picket-fence model, to include the effects of specularly and diffusely reflecting boundaries. We cast our analysis in the familiar H-function notation of Chandrasekhar, and we illustrate that computations relevant to the considered model can be made with a suitably high degree of precision. Further, we present the results of a rather extensive parameter survey and investigate several approximations to the 'exact' solution.

063.012 H-functions for some scattering indicatrixes with different values of the asymmetry factor.

A. K. Kolesov.
Trudy Astron. Obs., *Leningrad,* Vol. 28 (= Uchenye Zapiski Leningr. Un-ta, No. 359 = Seriya Matem. Nauk, vyp. (No.) 47), p. 3 - 14 (1971). In Russian.

The functions H (η) and p (η, ζ) are calculated for the scattering indicatrix of Henyey-Greenstein with different values of asymmetry factor and of particle albedo. The results are given in tables.

063.013 Light scattering by spherical particles with refraction index 1.38. V. M. Loskutov.
Trudy Astron. Obs., *Leningrad,* Vol. 28 (= Uchenye Zapiski Leningr. Un-ta, No. 359 = Seriya Matem. Nauk, vyp (No.) 47), p. 14 - 24 (1971). In Russian.

Functions describing the scattered intensity and polarization are computed for particles with refraction index 1.38 and gamma-distribution of particle sizes.

063.014 The effect of recoil in resonance-line scattering. T. F. Adams.
Astrophys. Journ., Vol. 168, 575 - 577 (1971).

A discussion of recoil from a microscopic point of view and direct numerical calculations show that recoil can be neglected for resonance-line scattering which involves Doppler redistribution. Further arguments show that recoil can be neglected in problems of physical interest even when the damping wings are important.

063.015 Solution of the equation of transfer for a spherical atmosphere by the method of Grant and Hunt. T. Viik.
Publ. Tartu Astrofiz. Obs., Vol. 39, 3 - 33 (1971). In Russian.

The method of solving the equation of transfer elaborated by Grant and Hunt (1968) is extended to the case of a spherical shell atmosphere by approximating the derivative of intensity with respect to μ with series of Legendre polynomials.

063.016 Solution of the equation of transfer by the method of regional averaging. T. Viik.
Publ. Tartu Astrofiz. Obs., Vol. 39, 34 - 43 (1971). In Russian.

The scattering of radiation in a spherical shell atmosphere with optical thickness τ_1 is considered. The equation of transfer is solved by the method of regional averaging, assuming that the opacity varies inversely as the n-th power of the geometric radius of the layer.

063.017 Multiple scattering calculations for technology. P. S. Mudgett, L. W. Richards.
Applied Optics, Vol. 10, 1485 - 1502 (1971).

A many-flux (discrete ordinate) radiative transfer calculation procedure is described with the goal of making the mathematics easy to learn and use. The major approximation is the neglect of polarization. Emission within the scattering medium is not included, and the formulas are restricted to a scattering medium bounded by parallel planes.

063.018 On the theory of nonstationary radiation transfer for anisotropic scattering. V. P. Grinin.
Astrofizika, Vol. 7, 203 - 209 (1971). In Russian. – English translation in Astrophysics, Vol. 7, No. 2.

The nonstationary radiation transfer in a homogeneous atmosphere is considered. It is assumed that the mean lifetime of photons inside the medium is determined by the time interval between two subsequent scatterings. A modification of Sobolev's probability method is used in the case of anisotropic scattering to obtain the intensity of the emergent radiation. A similarity relation is given connecting the transparency and reflection coefficients for arbitrary x in the interval from 0 to 1 with the same coefficients corresponding to $x = \frac{1}{2}$ (isotropic scattering).

063.019 Spherically symmetric boundary-value problems of radiative transfer. M. Kanal.
Bull. American Astron. Soc., Vol. 3, 378 (1971). – Abstr. AAS.

063.020 Line source function for a two-level atom. S. Ueno.
Bull. American Astron. Soc., Vol. 3, 380 (1971). – Abstr. AAS.

063.021 Diffuse reflection and transmission of light by a semi-infinite atmosphere according to the four-term scattering indicatrix. A. K. Kolesov, O. I. Smokty.
Astron. Zhurn. Akad. Nauk SSSR, Vol. 48, 1013 - 1022 (1971). In Russian. English translation in Soviet Astron. AJ, Vol. 15, No. 5.

The strict theory of anisotropic light scattering developed by Sobolev (1969) is used for the solution of the problem of diffuse reflection and transmission of light by a semi-infinite atmosphere. The case of the four-term scattering indicatrix is considered. Exact formulae for the reflection and transmission coefficients are derived.

063.022 The scattering of line radiation—I. A generalized redistribution function. J. D. Argyros, D. Mugglestone.
Journ. Quant. Spectrosc. Radiat. Transfer, Vol. 11, 1621 - 1632 (1971).

The theory of the frequency redistribution of radiation during scattering is re-investigated with the aim of producing a generalized frequency redistribution function. The assumption, commonly made, that the atom performing the scattering of the radiation has a constant velocity for the duration of the scattering event is not made in the present investigation. As a result a formal theory is developed which does indeed lead to a generalized redistribution function. Two limiting cases are discussed.

063.023 The scattering of line radiation—II. Velocity-non-correlated scattering. J. D. Argyros, D. Mugglestone.
Journ. Quant. Spectrosc. Radiat. Transfer, Vol. 11, 1633 - 1646 (1971).

An investigation is carried out into the effects of using different atomic models of the radiation scattering event with velocity-noncorrelated scattering (a physical process introduced in a previous paper). Special emphasis is placed on those models of the scattering event which lead to complete frequency redistribution in the radiation viewed by an observer at a large distance from the atmosphere being investigated.

063.024 Numerical solution of the transfer equation for polarized continuum radiation. S. Dumont.
Journ. Quant. Spectrosc. Radiat. Transfer, Vol. 11, 1675 - 1680 (1971).

We show how to generalize Feautrier's numerical method to solve the transfer equation of polarized radiation in the case where axial symmetry of the radiation is not assumed.

063.025 The derivatives of Dawson's function. R. Barakat.
Journ. Quant. Spectrosc. Radiat. Transfer, Vol. 11, 1729 - 1730 (1971).

Higher-order derivatives of Dawson's function are given in terms of Hermite and other polynomials.

063.026 Rayleigh scattering by obliquely oriented uniform needles. G. A. Shah.
Nature, Phys. Sci., Vol. 232, 184 - 185 (1971).

This communication considers the Rayleigh scattering of electromagnetic waves by thin cylindrical particles with circumference to wavelength ratio, $\kappa \ll 1$. The relevant approximation for the electric and magnetic dipole as well as quadrupole Mie coefficients have been derived for an arbitrary angle of incidence. A sample calculation shows that the uniform needle approximation for oblique cylinder can give reasonably accurate results in the range $0 < \kappa \lesssim 0.1$ for moderate values of the complex dielectric constant.

063.027 Infinite fine structure coupling in radiative transfer problems. L. F. McNamara.
Proc. Astron. Soc. Australia, Vol. 2, 41 - 42 (1971).

063.028 A fast method for the determination of emergent intensities in radiative transfer theory.
C. J. Cannon.
Proc. Astron. Soc. Australia, Vol. 2, 42 - 43 (1971).

063.029 A note on finite-difference schemes for the surface and planetary boundary layers.
P. A. Taylor, Y. Delage.
Boundary-Layer Meteorology, Vol. 2, 108 - 121 (1971).

A scheme using an expanding grid, based on the form chosen for mixing length or eddy viscosity, is proposed which gives good results with or without a surface layer in the case of a neutrally stratified atmosphere.

063.030 The spectrum of radiation scattered by relativistic electrons. W. H. Kegel.
Astron. Astrophys., Vol. 15, 306 - 310 (1971).

The spectral distribution of radiation arising from nonlinear Thomson scattering of a strong electromagnetic wave by an ensemble of ultrarelativistic electrons is discussed. The derived spectrum is compared with the spectrum which is emitted by an ensemble of particles with the same energy distribution in the case of synchrotron radiation, because of the similarity of the two radiation mechanisms. The cases of finite and infinite radiative zone are considered.

063.031 Line transfer in the presence of two-dimensional velocity gradients. C. J. Cannon, D. E. Rees.
Astrophys. Journ., Vol. 169, 157 - 163 (1971).

Feautrier's (1964) method is generalized to solve the equation of radiative transfer for a spectral line in an atmosphere exhibiting horizontal fluctuations in all physical parameters together with multidimensional velocity gradients. Model two-dimensional velocity problems are discussed.

063.032 Errata: 'Radiative transfer in an ionized medium at high temperature'. [Astrophys. Journ., Vol. 166, 301 - 309 (1971)]. E. D. Loh, G. P. Garmire.
Astrophys. Journ., Vol. 169, 447 (1971).

063.033 Induced Compton scattering by relativistic particles.
D. B. Melrose.

Astrophys. Space Sci., Vol. 13, 56 - 69 (1971).

Equations describing the evolution of isotropic distributions of unpolarized electromagnetic waves and of scattering particles of any energy due to spontaneous and induced Compton scattering are derived in the semi-classical approximation for unshielded particles. It is shown that induced Compton scattering of high frequency waves by relativistic electrons in synchrotron sources is a negligible effect contrary to the conclusions of Oster (1968) and of Kaplan and Tsytovich (1969).

063.034 New approximate forms for the H-function for isotropic scattering. S. Karanjai, M. Sen.
Astrophys. Space Sci., Vol. 13, 267 - 272 (1971).

Three new approximate forms for the H-function for isotropic scattering have been developed. Each of the forms involve three unknown functions of albedo ω.

063.035 On the solution of the transfer equation system for a non-scattering medium with a homogeneous magnetic field. J. M. Katz.
Solar Physics, Vol. 20, 362 - 364 (1971).

This paper describes a method, which gives the exact solution of the boundary problem for a transfer equation system of any order with an arbitrary absorption matrix.

063.036 Radiative transfer in non-steady state media.
I. N. Minin.
Vestn. Leningr. un-ta, 1971, No. 7, p. 122 - 128. In Russian.
Abstr. in Referativ. Zhurn. 51. Astron., 12.51.202 (1971).

063.037 On combined operational method for transfer problems in homogeneous, spherical media.
T. H. Kho, K. K. Sen.
Astrophys. Space Sci., Vol. 14, 223 - 242 (1971).

In this paper, the combined operational method developed by Busbridge (1961) in connection with the radiative transfer problems in plane-parallel atmospheres has been extended to similar problems in isotropic scattering, homogeneous spherical media. The relevant auxiliary equation has been formulated, the scattering function defined and the integro-differential equation for such function deduced.

063.038 Milne's problem for polarized radiation with non-conservative Rayleigh scattering law.
M. G. Kuzmina.
Dokl. Akad. Nauk SSSR, Ser. Mat. Fiz., Vol. 201, 809 - 812 (1971). In Russian.

063.039 On the statistical theory of radiative transfer in a non-homogeneous scattering medium (plane stratified heterogeneity). Yu. N. Barabanenkov.
Izv. vyssh. uchebn. zavedenij. Radiofizika, Vol. 14, 1290 - 1292 (1971). In Russian. — Abstr. in Referativ. Zhurn. 51. Astron., 2.51.456 (1972).

064 Stellar Atmospheres , Stellar Envelopes

064.001 A non-LTE picket-fence model in radiative equilibrium. D. Mihalas, W. R. Luebke.
Monthly Notices, Roy. Astron. Soc., Vol. 153, 229 - 239 (1971).

Solutions for a non-LTE picket-fence model are obtained for several values of the thermal coupling parameter ϵ and line-strength parameter r. The effects of the lines upon the temperature at the boundary and at great depth (backwarming effect) are evaluated. The results based on this somewhat idealized model permit certain generalizations to be drawn which largely verify previous intuitive conclusions, and complement current numerical work on more realistic models.

064.002 Excitation and ionization of helium in chromospheric flares. K. V. Alikayeva.
Astrometriya i Astrofiz., *Kiev*, Vyp. (No.) 12, (see 003.003), p. 50 - 63 (1971). In Russian.

A calculation of excitation and ionization of helium was made for conditions appropriate for chromospheric flares. The radiation field of spectral lines was taken into account for different optical depths in the D_3 line.

064.003 The importance of circulation in radiative stellar envelopes. R. C. Smith.
Monthly Notices Roy. Astron. Soc., Vol. 153, P33 - P35 (1971).

It is shown that a recently proposed rotation law cannot be relevant to many observed early-type stars.

064.004 Convection and metal abundance. E. Böhm-Vitense.
Astron. Astrophys., Vol. 14, 390 - 395 (1971).

The influence of the metal abundance Z on the structure of the convection zone is studied. The velocities in the convective zones decrease and the depths of the convection zones increase with decreasing metal abundance. For lower metal content the transition from radiative to convective atmospheres therefore occurs at higher effective temperatures.

064.005 A transformation of the stellar wind equations. P. H. Roberts.
Astrophys. Letters, Vol. 9, 79 - 80 (1971).

A simple transformation is demonstrated which reduces the number of parameters in the stellar wind equations from two to one.

064.006 The temperature control bracket. K. B. Gebbie, R. N. Thomas.
Astrophys. Journ., Vol. 168, 461 - 479 (1971).

The factors determining the temperature distribution in a stellar atmosphere are divided into transfer effects and population effects. As a measure of the latter, we introduce the temperature control bracket [TCB], which, in radiative equilibrium, describes the control of T_e by the quantity and spectral distribution of the radiation field. Algebraic expressions for the [TCB] are given in terms of the microscopic rate processes for a pure hydrogen atmosphere. A caricatured computation is presented to demonstrate the influence of the various physical effects on the distribution of T_e in radiative equilibrium.

064.007 Approximate analytical solution for the equation of mechanical equilibrium. A. Kruusmaa.
Publ. Tartu Astrofiz. Obs., Vol. 39, 44 - 57 (1971). In Russian.

An attempt is made to construct an analytical expression for the dependence electronic pressure versus temperature in early type stellar atmospheres.

064.008 A dispersion formula for broadening of hydrogen lines for early-type stars. I. Kuusik.
Publ. Tartu Astrofiz. Obs., Vol. 39, 58 - 68 (1971). In Russian.

A dispersion formula for hydrogen lines as a result of natural, Doppler and Stark broadening mechanisms is given. The characteristic widths of these mechanisms for Hγ are evaluated for three atmospheres.

064.009 The mean absorption coefficient in the calculations of model stellar atmospheres. A. Kruusmaa.
Publ. Tartu Astrofiz. Obs., Vol. 39, 273 - 281 (1971). In Russian.

Six non gray model stellar atmospheres for parameters lg T_e = 4.4, 4.5, 4.6, lg g = 4.0 and He : H = 0.0512 with Planck and Rosseland mean absorption coefficients are calculated.

064.010 Quantitative analysis of the atmospheres of two F stars. V. Malyuto.
Publ. Tartu Astrofiz. Obs., Vol. 39, 282 - 300 (1971). In Russian.

The curve-of-growth technique has been used to find atmospheric parameters and abundances of some elements in the atmospheres of the δ Scuti-type star β Cas and the metal-rich star 10 UMa.

064.011 Molecular abundances in a solar composition gaseous mixture. T. Kipper.
Publ. Tartu Astrofiz. Obs., Vol. 39, 321 - 333 (1971).

064.012 The transfer of linearly polarized radiation in extended atmospheres. J. P. Cassinelli, D. G. Hummer.
Bull. American Astron. Soc., Vol. 3, 378 (1971). – Abstr. AAS.

064.013 Poincaré-Lighthill coordinate perturbation temperature correction technique in spherical geometry. R. D. Chapman.
Bull. American Astron. Soc., Vol. 3, 378 (1971). – Abstr. AAS.

064.014 Excitation of C III in Wolf-Rayet envelopes. J. I. Castor, H. Nussbaumer.
Bull. American Astron. Soc., Vol. 3, 378 - 379 (1971). Abstr. AAS.

064.015 Comments on the synthesized spectrum of the CN red system. I. Marenin, A. Greene.
Bull. American Astron. Soc., Vol. 3, 379 (1971). – Abstr. AAS.

064.016 The violet opacity in S, C-S and N stars and circumstellar silicon carbide grains. D. P. Gilra, A. D. Code.
Bull. American Astron. Soc., Vol. 3, 379 (1971). – Abstr. AAS.

064.017 Molecular opacitiy in cool carbon stars. D. Alexander, J. Collins, T. Fay, H. R. Johnson.
Bull. American Astron. Soc., Vol. 3, 380 (1971). – Abstr. AAS.

064.018 Microturbulence in main sequence A and F stars. G. Elste, J. Ionson.
Bull. American Astron. Soc., Vol. 3, 380 - 381 (1971). Abstr. AAS.

064.019 Envelope relaxation oscillations in luminous red giants. W. K. Rose, R. L. Smith.
Bull. American Astron. Soc., Vol. 3, 402 (1971). — Abstr. AAS.

064.020 Calculations of shock waves in an RR Lyrae atmosphere. S. J. Hill.
Bull. American Astron. Soc., Vol. 3, 402 (1971). — Abstr. AAS.

064.021 Determination of atmospheric parameters for G and K giants by means of photoelectric indices.
L. Hansen, P. Kjaergaard.
Astron. Astrophys., Vol. 15, 123 - 160 (1971).

The narrow-band photometry published by Dickow et al. (1970) has been used for calibration of atmospheric parameters for G and K giants. It is shown that the dwarfs can be distinguished from the giants, and that stars with indices affected by interstellar reddening or duplicity can be eliminated. $R - I$ and $[Fe/H]$ are assumed to represent effective temperature and chemical composition, while a residual Δn from a relation between the cyanogen index n, $R - I$, and $[Fe/H]$ is assumed to measure surface gravity. Calibrations giving $R - I$ and $M_V(K)$ as functions of the observed indices are derived for giants in the intervals $0.40 < R - I < 0.65$ (G8–K3) and $0.65 < R - I < 0.88$ (K4–K5).

064.022 Spontaneous fission of heavy transuranium elements in the surface layers of the peculiar A star HR 465 explains promethium abundance. B. Kuchowicz.
Nature, Vol. 232, 551 - 552 (1971).

The possibility of a presence of transuranium elements in the surfaces of the Ap stars was raised by Kuchowicz in 1970 when observational results obtained by Jaschek and Malaroda indicated a presence of uranium lines in the Ap star 73 Dra. In the meantime, Aller and Cowley found that the extremely unstable element promethium is present in HR 465. Arguments are given in favour of an indirect origin of this element from some previously occurring r-process in which there were produced superheavy transuranium elements (with atomic numbers in the range up to 114); their main decay mode is spontaneous fission in which Pm may constitute up to 5 per cent of the decay products. This mechanism provides a steady supply of Pm to the stellar atmosphere, with a time scale of the order of the half-lives of superheavy elements.

064.023 Meridional circulation in rotating stellar atmospheres—II. The effect of variable opacity.
D. Brand, R. C. Smith.
Monthly Notices Roy. Astron. Soc., Vol. 154, 293 - 300 (1971).

In a previous paper the circulation currents in a uniformly rotating stellar atmosphere with constant opacity were discussed. The present paper investigates the effect of removing the restriction of constant opacity. It is found that the circulation pattern is qualitatively the same as for constant opacity.

064.024 Remarks on the H⁻ equilibrium in stellar atmospheres. F. Praderie.
Astrophys. Letters, Vol. 9, 27 - 31 (1971).

The importance of the charge-exchange process $H^- + H^+ \rightarrow H(1s) + H(ns, p, d)$ is compared to that of the associative detachment process. It is shown that (1) the first process is dominant in main sequence A stars, and in lower effective temperature stars with gravity smaller than the main sequence stars; (2) in the sun, the charge-exchange process is of minor importance at every depth.

064.025 Decay of radiative magnetohydrodynamic shock waves in the atmosphere of B stars.
M. S. Bhatnagar, K. P. Kulshrestha, J. N. Tandon.

Astrophys. Letters, Vol. 9, 135 - 137 (1971).

Differential equations describing the decay of a hydromagnetic shock through a stellar atmosphere are derived. It is shown that an isothermal shock decays more rapidly than a non-magnetic shock. Comparison of the theoretical velocity curve with the radial-velocity observations of BW Vulpeculae, a β Cephei star is made.

064.026 Effect of molecular line absorptions on stellar opacities. T. Tsuji.
Publ. Astron. Soc. Japan, Vol. 23, 553 - 565 (1971).

Some properties of stellar opacities at relatively low temperatures are studied. The effects of several opacity sources such as H_2O (vibration-rotation and pure rotation bands), TiO (electronic bands), H_2 (collision-induced absorption), etc. are discussed. The applicability as well as the limitation of some approximate methods to evaluate the Rosseland mean opacity are examined. For diatomic molecules the Elsasser band model is applied to evaluate mean absorptions. Some approximate values of the Rosseland and the Planck mean opacities are given.

U64.027 Erratum: "On thermal-convective instability in a stellar atmosphere" [Publ. Astron. Soc. Japan, Vol. 23, 181 - 184 (1971)]. P. K. Bhatia.
Publ. Astron. Soc. Japan, Vol. 23, 597 (1971).

064.028 Cellular convection in model stellar envelopes. D. J. Mullan.
Monthly Notices Roy. Astron. Soc., Vol. 154, 467 - 489 (1971).

Model stellar envelopes have been constructed using a theory of cellular convection due to Öpik, in which turbulent heat exchange between rising and falling gas is allowed for by introducing an experimentally determined coefficient. Radiative heat losses are also included. Results for solar models and for main sequence models are presented. A comparison with other authors is given.

064.029 Curves of growth and line profiles for neutral helium lines in early type stars.
B. J. O'Mara, R. W. Simpson.
Monthly Notices Roy. Astron. Soc., Vol. 154, 505 (1971). — Summary. — The full text of this paper appears in Mem. of the Roy. Astron. Soc., Vol. 75, Part 2.

064.030 Radio emission from stellar coronas. L. Oster.
Astrophys. Journ., Vol. 169, 57 - 61 (1971).

Recent radio observations by Wade and Hjellming are used to derive a plausible base density for a hot ($10^6\,°K$) corona of M supergiants. It is argued that the observed radio emission might correspond to the equivalent of solar type IV microwave bursts, and it is pointed out that this interpretation could be checked with relatively little observational labor.

064.031 Models for the envelopes of Be stars. III. Pole-on stars. J. M. Marlborough, J.-R. Roy.
Astrophys. Journ., Vol. 169, 327 - 331 (1971).

Hα line profiles are presented for model envelopes for which the observer's line of sight is parallel to the rotation axis of the star. A comparison is made of these line profiles with the predicted Hα line profiles for the same model envelope when the observer's line of sight is perpendicular to the rotation axis.

064.032 Radiation pressure and accelerating stellar winds. J. M. Marlborough.
Astrophys. Journ., Vol. 169, 441 - 442 (1971).

It is shown that insofar as the hydrodynamics is concerned the conclusions of Marlborough and Roy concerning the effect of radiation pressure on stellar winds are completely consistent with the mechanism proposed by Lucy and Solomon regarding mass loss in hot stars.

064.033 Emission lines in stellar spectra. I. The formulation of the problem. L. Luud, M. Ilmas.
Tartu Astron. Obs. Teated, No. 32, p. 3 - 12 (1971). In Russian.
The equations of the statistical equilibrium of the level population with five free parameters are written in terms of Sobolev's theory of moving envelopes taking into account radiative and electron impact processes.

064.034 Emission lines in stellar spectra. II. Calculation of some transition probabilities for hydrogen.
M. Ilmas.
Tartu Astron. Obs. Teated, No. 32, p. 13 - 46 (1971). In Russian.
The probabilities of radiative ionization and induced recombinations are calculated and presented in tables. Using Gryzinski's theory the probabilities of ionization and excitation by electron collisions are calculated and presented in tables.

064.035 Emission lines in stellar spectra. III. Calculations of the hydrogen spectrum. M. Ilmas.
Tartu Astron. Obs. Teated, No. 32, p. 47 - 110 (1971). In Russian.
Using an iteration procedure the equations of the statistical equilibrium of the level population given previously are solved for the hydrogen atom.

064.036 Emission lines in stellar spectra. IV. Comparison of the observed and calculated intensities of hydrogen lines. L. Luud, M. Ilmas.
Tartu Astron. Obs. Teated, No. 32, p. 111 - 124 (1971). In Russian.
The observed and computed decrements of hydrogen lines are compared. It is found that in the case of P Cyg a good agreement exists. The agreement for Be stars is poorer. The agreement for symbiotic stars is satisfactory enough only for Z And.

064.037 Method of regional averaging in radiative transfer.
T. Viik.
Izv. Akad. Nauk Ehstonskoj SSR, Vol. 20, (Fiz., Mat., 1971, No. 3), 285 - 288 = Tartu Astron. Obs. Teated, No. 33.

064.038 Highly excited hydrogen lines in stellar spectra. I.
S. Barcza.
Astrophys. Space Sci., Vol. 13, 36 - 47 (1971).
The solutions of the hydrogenic Schrödinger equation are given with two boundary conditions imposed on the wave function, for distances of the order of magnitude of one hundred times the Bohr radius from the central nucleus. Thus the shifts and splitting of the $H\gamma - H_{26}$ lines are given which arise from the non-vacant environment. The gas density is derived from the number of the visible Balmer lines.

064.039 On the relativistic ejection of a particle by radiation pressure. P. D. Noerdlinger.
Astrophys. Space Sci., Vol. 13, 70 - 73 = Contr. Lick Obs., No. 335 (1971).
The Poynting-Robertson formulae are used to evaluate the motion and terminal velocity of a particle ejected from rest in a powerful radiation field due to a point source, generalizing the results of Chandrasekhar (1934) to the relativistic case.

064.040 Continuous absorption by neon ions.
M. R. C. McDowell.
Observatory, Vol. 91, 217 - 220 (1971).

In order to obtain reliable data on k_ν, the continuous absorption coefficient, accurate values of the elastic-scattering phase-shifts over a range of energies are needed. Since neon has a relatively high cosmic abundance it is of interest to examine its free-free absorption.

064.041 Curves of growth and line profiles for neutral helium lines in early type stars.
B. J. O'Mara, R. W. Simpson.
Mem. Roy. Astron. Soc., Vol. 75, 51 - 84 (1971).
Curves of growth are presented for those He I lines which are considered most useful in a helium abundance analysis. Line profiles are also presented for the three strongest lines — 4471, 4026 and 4388 Å. The results are given for line blanketed model atmospheres with θ_e in the range 0.176 to 0.4 and $\log g = 3.5, 4.0$. For the weak lines the curves are tabulated for the two values of microturbulent velocity — $\xi = 5$, 10 km s^{-1}. The value $\xi = 5$ km s^{-1} is used for the strong lines. Electron scattering effects, the L.T.E. assumption, and the profile of the line absorption coefficient are discussed in some detail.

064.042 ATLAS: A computer program for calculating model stellar atmospheres. R. L. Kurucz.
SAO, *Cambridge, Mass.,* Special Report, No. 309, 15 + 291 pp. (1970).
The computer program ATLAS calculates model stellar atmospheres in radiative and convective equilibrium for the complete range of stellar temperatures. The approximations used limit the program to plane-parallel, horizontally homogeneous, steady-state, nonmoving atmospheres with energy and abundances constant with depth. The program has been written to allow detailed statistical equilibrium calculations, but only hydrogen continua and H$^-$ are coded at present. Most of the published continuous opacities and hydrogen lines have been included, and provision is made for adding others easily. There is also provision for treating line opacity as line-absorption distribution functions.

064.043 On the theory of stellar winds.
B. R. Durney, P. H. Roberts.
Astrophys. Journ., Vol. 170, 319 - 323 (1971).
It has recently been shown by Roberts that solutions of the stellar wind equations depend essentially on a single parameter $K = \frac{1}{2} A \epsilon_\infty^{1/2}$, where ϵ_∞ is the (dimensionless) residual energy per particle at infinity and A is a nondimensional constant proportional to the reciprocal of the mass flux. This transformation makes it a comparatively simple matter to examine solutions for a wide variety of A and ϵ_∞. The calculations reported below are for the range $75 < K < 2000$, which covers many cases of astrophysical interest.

064.044 Turbulence velocities in the atmosphere of Alpha Orionis. R. B. Hutchison.
Astrophys. Journ., Vol. 170, 551 - 555 (1971).
The curve of line-width correlation has been applied to OH lines in the spectrum of α Ori, and the characteristic turbulence and thermal velocities in the atmosphere of this star have been determined: $\xi_m = 9.9 \pm 2.0$ km s^{-1} (microturbulence); $\xi_M \lesssim 3$ km s^{-1} (macroturbulence); $\xi_{th} \simeq 1.8$ km s^{-1} (thermal). Implications of these results are discussed.

064.045 Thermalization lengths and mean numbers of scattering for line photons.
R. G. Athay, A. Skumanich.
Astrophys. Journ., Vol. 170, 605 - 611 (1971).
A simple algebraic method is given for the derivation of approximate thermalization lengths, mean number of scatterings for line photons, and scaling laws for the maximum value of the line-source function in finite atmospheres. The method utilizes only the shape of the line absorption coefficient.

064.046 **Theoretical analysis of stellar spectra.** D. Mihalas.
Theory of the stellar atmospheres, Saas-Fee 1971,
(see 012.008), 1 - 156 (1971).

064.047 **Determination of stellar abundances.** B. Pagel.
Theory of the stellar atmospheres, Saas-Fee 1971,
(see 012.008), 157 - 237 (1971).

064.048 **Convection.** P. Souffrin.
Theory of the stellar atmospheres, Saas-Fee 1971,
(see 012.008), 238 - 312 (1971).

064.049 **The production of solar and stellar chromospheres
and coronae.** C. de Jager.
Phil. Trans. Roy. Soc. London A, Vol. 270, No. 1202, (see
012.009), 175 - 182 (1971).

064.050 **Observations of stellar chromospheres.** H. Zirin.
Phil. Trans. Roy. Soc. London A, Vol. 270, No.
1202, (see 012.009), 183 - 188 (1971).

064.051 **The atmosphere of Delta Cephei. I. A coarse analysis.** E. Kobayashi, M. Takeuti.
Sci. Rep. Tôhoku Univ., First Ser., Vol. 54, 33 - 58 = Sendai
Astron. Rap., No. 121 (1971).
High dispersion spectra at twelve phases of the population I cepheid δ Cephei are obtained in the blue region, and
are analysed by the differential curve of growth method relative to the sun. It is found that the variation in atmospheric
parameters is rather mild. The iron abundance relative to hydrogen in δ Cephei may be about four times larger than that
of the sun.

064.052 **Abundance of helium in stellar atmospheres.**
R. Cayrel.
Highlights of Astronomy, Vol. 2, (see 012.015), 254 - 268
(1971).

064.053 **Interstellar molecules from cool stars.**
N. C. Wickramasinghe.
Highlights of Astronomy, Vol. 2, (see 012.016), 438 - 444
(1971).

064.054 **Centre-limb variation of line profile in a moving
atmosphere.** K. D. Abhyankar.
Nizamiah Obs., Osmania Univ., *Hyderabad,* Repr. No. 46,
12 pp. (1967). – Reprinted from 'Modern Astrophysics',
[Gauthier-Villars, Paris], 1967, p. 199 - 210.

064.055 **On the afterglow of a shock wave coming to a star's
surface.** I. A. Klimishin, A. F. Novak.
Tsirk. L'vov. Astron. Obs., No. 46, p. 3 - 7 (1971). In Russian.

064.056 **Calculation of the degree of ionization behind a
shock wave front moving in a stellar atmosphere.**
I. A. Klimishin, B. M. Gura.
Tsirk. L'vov. Astron. Obs., No. 46, p. 8 - 15 (1971). In Russian.

064.057 **A spectrophotometric investigation of the star α^2
CVn. II.** Z. N. Chumak.
Trudy Astrofiz. Inst., *Alma-Ata,* Vol. 16, 70 - 76 (1971).
In Russian.
The velocity of microturbulence in the atmosphere,
ionization and excitation temperatures and the damping para-
meter have been derived. The dependence of some of these
quantities on the excitation potential has been investigated.
The abundances of several elements have been determined.

064.058 **Cellular convection in stellar envelopes.**
D. J. Mullan.
Irish Astron. Journ., Vol. 9, 310 - 315 (1970).

Cellular convection in stars has been treated by a theory
due to Öpik (1950). The solar convection zone is found to be
only 10^4 km deep, reaching a temperature of 9.5×10^4 deg K.
These numbers are not sensitive to the free parameters of the
theory. Quantitative agreement with solar and stellar observations has been found in more than a dozen observational tests
in some of which current theories have failed. This note summarizes the principal results which are in agreement with observations.

064.059 **Determinazione di parametri fisici di un'atmosfera
stellare: Un metodo.**
A. Preite-Martinez, A. Natta.
Mem. Soc. Astron. Italiana, Nuova Ser., Vol. 42, 561 - 569
(1971).
A method is proposed for the determination of the physical parameters (T_e, g, abundances, etc.) which characterize
the atmospheric structure of a star. This method is based on
the use of model atmosphere calculations.

064.060 **Thermal instability and the convective stability of
stellar chromospheres.** R. J. Defouw.
Thesis, California Inst. Technology, Pasadena. [Available from
Univ. Microfilms, Ann Arbor, Mich., U.S.A. Order No.
70–18033], 107 pp. (1970).
It is shown that the Schwarzschild criterion does not apply to chromospheres because it ignores the possibility of
thermal instability.

064.061 **Hydrogen-deficient model atmospheres for degenerate stars.** R. J. Doyle.
Thesis, Univ. Virginia, Charlottesville. [Available from Univ.
Microfilms, Ann Arbor, Mich., U.S.A. Order No. 71–6622],
134 pp. (1970).
A preliminary model atmosphere analysis was applied to
two types of degenerate stars that have cool hydrogen-deficient atmospheres.

064.062 **Model atmospheres and strong line profiles for late-
type dwarfs.** J. L. Hershey.
Thesis, Univ. Virginia, Charlottesville. [Available from Univ.
Microfilms, Ann Arbor, Mich., U.S.A. Order No. 70–8074],
164 pp. (1969).
Twenty non-grey model atmospheres have been computed in detail for stars on the lower main sequence. The models
form a loose grid ranging from 5200° to 3200° in effective
temperature.

064.063 **Radiative transfer effects on thermal-convective instability in hydromagnetics.** P. K. Bhatia.
Bull. Soc. Roy. Sci. Liège, Vol. 40, 137 - 141 (1971).
The effects of radiative transfer on the convective instability of a thermally unstable stellar atmosphere have been included in the presence of a uniform rotation and a uniform
magnetic field. The criterion for monotonic instability is
found to be the same as in the absence of these effects.

064.064 **Quasiequilibrium structures in the ionized atmospheres of stars and planets.** V. M. Fadeev.
Proc. Sixth Winter School on Space Physics, Part I. Apatity
1969, (see 012.033), p. 256 - 263 (1971).

064.065 **The effect of radiative equilibrium on the photo-
spheric angular velocity.** B. R. Durney.
National Center Atmosph. Res., Boulder, Colorado. NCAR MS
No. 71–175a, 15 pp. (1971). – To be published in Astrophys.
Journ.

The formation of spectral lines.
See Abstr. 063.004.

The spectrum of steady state turbulent convection. See Abstr. 063.010.

Consequences of Strömgren's theorem for radiative envelope stars.   See Abstr. 065.029.

Nuclear processes associated with peculiar A-type stars.   See Abstr. 065.095.

The solar chromosphere and the general structure of a stellar atmosphere.   See Abstr. 073.034.

Non-radial oscillations and energy transport in rotating solar (stellar) wind.   See Abstr. 074.076.

Neutral-helium line strengths. IV. Fourteen "normal" stars of population I.   See Abstr. 114.006.

Neutral-helium line strengths. V. The weak-helium-line stars of population I.   See Abstr. 114.007.

Neutral-helium line strengths. VI. The variations of the helium spectrum variable α Centauri. See Abstr. 114.008.

Determinación de las abundancias de los elementos en la atmósfera de la estrella de alta velocidad 31 Aql. See Abstr. 114.039.

Helium abundance determinations in main sequence B stars.   See Abstr. 114.063.

Opacity probability distribution functions for electronic systems of CN and C_2 molecules including their stellar isotopic forms.   See Abstr. 114.068.

Spectrophotometric studies of non-stable stars. II. On the spectrum of RW Aurigae in the region 3080–6100 Å. See Abstr. 114.081.

On the motion of matter in the envelope of P Cygni.   See Abstr. 114.082.

Nuclear and non-nuclear processes in the production of peculiar A stars.   See Abstr. 114.084.

Quantitative Analyse des Spektrums des Metalllinien-sterns 63 Tauri.   See Abstr. 114.086.

Curve-of-growth analysis of the spectrum of Procyon.   See Abstr. 114.087.

Abundances of the elements in Sirius and Merak. See Abstr. 114.092.

N III and C III emission in Of stars. See Abstr. 114.103.

The chromosphere of Arcturus. See Abstr. 114.107.

On the presence of He^3 in the photosphere of Rho Leonis.   See Abstr. 114.109.

A comment on the interpretation of the broad component of N III λλ4634–4640 emission in Of stars. See Abstr. 114.118.

The photometric variability of Ap stars. See Abstr. 116.011.

Rotationally extended stellar envelopes – III. The Be component of VV Cephei.   See Abstr. 121.057.

Atmospheric structure, mass loss, and chemical composition in R Andromedae and R Cygni. See Abstr. 122.006.

Atmospheres of pulsating stars. See Abstr. 122.014.

A study on differential radial velocities in the spectra of Chi Cygni and Omicron Ceti.   See Abstr. 122.061.

A note on models for the envelopes of long-period variable stars.   See Abstr. 122.099.

Considerations on the origin of the Platt particles. See Abstr. 131.041.

Polarimetry of red and infrared stars at 1 to 4 microns.   See Abstr. 131.086.

065 Stellar Structure, Stellar Evolution, Stellar Nucleosynthesis

065.001 Rapid differential rotation in main sequence stars.
J. J. Monaghan, N. C. Smart.
Monthly Notices, Roy. Astron. Soc., Vol. 153, 195 - 204 (1971).

The structure of massive stars in rapid differential rotation is investigated using a perturbation method. Only the spherically symmetric terms are retained, but the equation of energy transport, with the neglect of circulation, is treated exactly. The results are in qualitative agreement with those obtained by Mark and show that rapid differential rotation can lead to a very large reduction in the luminosity of a star. The circulation time scale is estimated using Eddington's method as corrected by Öpik. For those stars with the largest angular momentum the circulation time is comparable to the typical evolutionary time scale.

065.002 A reexamination of the post-Newtonian Maclaurin spheroids. J. M. Bardeen.
Astrophys. Journ., Vol. 167, 425 - 446 (1971).

The post-Newtonian corrections to Maclaurin spheroids are calculated in such a way that the Newtonian and relativistic configurations being compared have the same energy density, rest mass, and angular momentum. The singularity found by Chandrasekhar is interpreted as a splitting of the Maclaurin sequence into a spherelike sequence and a disklike to ringlike sequence. The post-Newtonian corrections are calculated to several quantities of physical interest, including the binding energy.

065.003 The post-Newtonian effects of general relativity on the equilibrium of uniformly rotating bodies. V. The deformed figures of the Maclaurin spheroids.
S. Chandrasekhar.
Astrophys. Journ., Vol. 167, 447 - 453 (1971).

The theory of the (deformed) post-Newtonian Maclaurin spheroid developed in an earlier paper is specialized suitably to make a comparison with the Newtonian spheroid having the same angular momentum and baryon number.

065.004 The post-Newtonian effects of general relativity on the equilibrium of uniformly rotating bodies. VI. The deformed figures of the Jacobi ellipsoids.
S. Chandrasekhar.
Astrophys. Journ., Vol. 167, 455 - 463 (1971).

The theory of the (deformed) post-Newtonian Jacobi ellipsoid developed in an earlier paper is specialized suitably to make a comparison with the Newtonian ellipsoid having the same angular momentum and baryon number.

065.005 Thermonuclear detonations in evolved stellar cores. J.-R. Buchler, J. C. Wheeler, Z. Barkat.
Astrophys. Journ., Vol. 167, 465 - 478 (1971).

An investigation has been made of the local effects of the propagation of a Chapman-Jouguet detonation (CJD) fueled by ^{12}C or ^{16}O burning under conditions of high electron degeneracy, the type of CJD of most immediate astrophysical interest. The temperature, density, and pressure behind the detonation front are tabulated as a function of the initial density and the energy released in the front. The astrophysical significance of the results is discussed.

065.006 The effect of rapid rotation on radiation from stars. III. Strong helium I lines.
J. Hardorp, M. Scholz.
Astron. Astrophys., Vol. 13, 353 - 358 (1971).

For main sequence models of rapidly rotating stars of 3.55 to 8.1 solar masses (identical with those of papers I and II of this series), line profiles of He I λ4471 and λ4026 are computed in the LTE approach. The conclusions of paper II are not changed by the new results — in particular, the broadest-lined stars can be understood as rotating near maximum speed. The new computations may aid detection of rapid rotators viewed under intermediate angles. The consequences of an error in the paper by Roxburgh *et al.* (1965) on the results of papers I, II and III are briefly discussed.

065.007 Neutron star models and pulsars.
G. Chanmugam, M. Gabriel.
Astron. Astrophys., Vol. 13, 374 - 379 (1971).

Neutron star models are constructed with the Nemeth-Sprung equation of state. Their dynamical stability is also discussed. It is suggested that the Crab pulsar has a central density $> 5 \times 10^{14}$ gm cm^{-3}.

065.008 The double valued nature of the hydrogen main sequence. A. S. Grossman, J. W. Opoien.
Astron. Astrophys., Vol. 13, 487 - 488 (1971).

A calculation has been made for a 0.1 $M_{\odot}$ star at the composition $X = 0.68$, $Y = 0.29$ and an l/H_p value of 1 which shows that there are two possible, stable, hydrogen main sequence configurations for the above parameters.

065.009 Magnetic accretion processes in peculiar A stars.
O. Havnes, P. S. Conti.
Astron. Astrophys., Vol. 14, 1 - 11 (1971).

A new hypothesis is proposed to account for the abundance anomalies and surface distributions of the various elements in the magnetic peculiar A stars. It is suggested that these stars have selectively captured atoms from the interstellar medium by its interaction with the rotating stellar magnetic field.

065.010 Stars with helium envelopes.
P. Biermann, R. Kippenhahn.
Astron. Astrophys., Vol. 14, 32 - 42 (1971).

Models are computed for stars with helium envelopes and carbon-oxygen cores, covering a wide range of total mass and of core mass. Three types of models occur, characterized by a) non-degenerate isothermal cores, b) degenerate cores and c) carbon burning cores. There is a group of models which for the same mass and the same chemical profile can belong either to type b) or type c) contradicting the Vogt-Russell theorem. The models are compared with observed helium and R CrB stars indicating their interior structures and therefore their possible evolutionary history.

065.011 Frictional heating in neutron stars.
G. Greenstein.
Nature, Phys. Sci., Vol. 232, 117 - 119 (1971).

If neutron star interiors are superfluid, heat is dissipated within the star as its rotation is slowed. This process sets a lower limit to the temperature such a star can attain: The limit is such that thermal X-ray emission from certain pulsars may be detectable.

065.012 Thermal stability of the helium-burning shell in stars of 15 solar masses. T. R. Dennis.
Astrophys. Journ., Vol. 167, 311 - 319 (1971).

The thermal stability of the helium-burning shell in a model for a red supergiant of 15 $M_{\odot}$ has been investigated. On the basis of a linear stability analysis the model is shown to be thermally stable by a large margin. A semiquantitative physical analysis shows that the primary stabilizing factor is the presence of substantial radiation pressure in the helium-burning shell.

065.013 Differential rotation in stars on the upper main sequence. II. W. E. Harris, M. J. Clement.
Astrophys. Journ., Vol. 167, 321 - 325 (1971).

Equilibrium models for slowly rotating stars of 16, 28, and 47 $M_\odot$ are presented. The interior distribution of angular velocity is uniquely determined by the requirement that the azimuthal force near the surface vanish, and that the steady state be free of meridian circulation. It is shown that radiation pressure brings the models closer to rigid rotation.

065.014 Evolution of stars containing ^{3}He.
R. K. Ulrich.
Astrophys. Journ., Vol. 168, 57 - 70 (1971).

This paper discusses the details of the pre-main-sequence evolution of stars containing an initial mass fraction of ^{3}He, X_3, less than or equal to 0.005. It is suggested that the apparent youth of the high-mass stars is due to ^{3}He burning. An estimate of the age of NGC 2264 and the stars near the Orion nebula from evolution of the upper main sequence and the gravitational ages of low-mass stars then yields an initial ^{3}He abundance which is much larger than the ^{3}He abundance in gas-rich meteorites.

065.015 Equilibrium and pulsational properties of the lower helium main sequence.
C. J. Hansen, W. H. Spangenberg.
Astrophys. Journ., Vol. 168, 71 - 78 (1971).

Results are presented for pure-helium models in complete equilibrium within the mass range $0.3 \leq M/M_\odot \leq 1.35$. The models are double valued with respect to mass whereas the central density is monotonic. Hydrogen-rich envelopes are added to some models for realism. Some discussion is included to argue against the higher-density models representing reasonable stellar objects.

065.016 A method for suppression of the thermal instability in helium-shell-burning stars. A. V. Sweigart.
Astrophys. Journ., Vol. 168, 79 - 97 (1971).

The principal goal of the present paper is to show that suppressing the helium shell flashes by ignoring the release of gravitational energy does give results in encouragingly close agreement with the mean evolution expected from detailed relaxation-cycle calculations. The present computations will be confined to a 1 $M_\odot$ population II star with the following composition parameters: $X = 0.90$, $Y = 0.099$, and $Z = 0.001$. As a secondary goal we will investigate some of the features of the second red-giant phase of this star.

065.017 Properties of low-density neutron-star matter.
J.-R. Buchler, Z. Barkat.
Phys. Rev. Letters, Vol. 27, 48 - 51 (1971).

A nuclear Thomas-Fermi model is used to determine the ground state of matter at subnuclear densities allowing for inhomogeneities on a nuclear scale ("clusters").

065.018 Astrophysical importance of the reaction $C^{12} + O^{16}$.
S. E. Woosley, W. D. Arnett, D. D. Clayton.
Phys. Rev. Letters, Vol. 27, 213 - 216 (1971).

The nuclear reactions between C^{12} and O^{16} are shown to be vitally important during explosive oxygen burning because they regulate the number of α particles produced per Si^{28} nucleus. The value of the $C^{12} + O^{16}$ total reaction cross section will help determine the nature of the explosions that have produced the elements.

065.019 Convection in stars. I. Basic Boussinesq convection.
E. A. Spiegel.

Annual Rev. Astron. Astrophys., Vol. 9, (see 003.001), 323 - 352 (1971).

Part I of this review is devoted to the problem of convective transfer in the laboratory situation. It should be stressed, however, that the equations and approximations used are the same as those now used in stellar structure calculations with a common goal—to predict the march of temperature through the convective fluid. In Part II the special problems of stellar convection such as large density variation, overshooting, rotation, and radiative transfer will be considered in the context of pure convection theory.

065.020 Recent developments in the theory of degenerate dwarfs. J. P. Ostriker.
Annual Rev. Astron. Astrophys., Vol. 9, (see 003.001), 353 - 366 (1971).

First considered are the modifications to the structure and mass–radius relation of spherical stars caused by the contributions of the ions, general relativity, etc. Then non-spherical stars – rotating and magnetic – are discussed with the surprising result noted that inclusion of non-zero angular momentum can drastically alter the concept of a "mass limit". Finally the problem of mechanical stability is surveyed for both spherical and non-spherical stars.

065.021 Ages and masses for nine "super-metal-rich" field stars. S. Torres-Peimbert, H. Spinrad.
Bol. Obs. Tonantzintla y Tacubaya, No. 36, Vol. 6, 15 - 19 (1971).

A comparison with theoretical evolutionary models was made for nine field K giants of known supermetallicity and distance, yielding the approximate age and mass of each star.

065.022 Magnetic neutron stars. I. Static field configurations.
G. Dautcourt, K. Fritze.
Astron. Nachr., Vol. 292, 211 - 219 (1971).

The geometry of fossil magnetic fields frozen in the highly conductive neutron star matter in a non-rotating (or weakly rotating) star is studied. Numerical results are obtained for a poloidal and a toroidal dipole field.

065.023 Sternentwicklung auf dem Horizontalast und Zustandsgrößen von RR-Lyrae-Sternen in Kugelhaufen verschiedenen Alters. W. Thänert.
Astron. Nachr., Vol. 292, 251 - 262 (1971).

The behaviour of the masses, the luminosities, the radii, and the colours of the RR Lyrae stars situated in 7 globular clusters is investigated in dependence upon the clusters' age. The ages were determined according to a method by Sandage; the determination of the masses is based on relations derived by Baker.

065.024 Second order rotational perturbation of non-radial oscillations of a star. P. Smeyers, J. Denis.
Astron. Astrophys., Vol. 14, 311 - 318 (1971).

The perturbation method is applied for the effects of a uniform rotation on non-radial oscillations of a star up to the second order in the angular velocity. The distortion of the equilibrium configuration is taken into account. Numerical results obtained for a rotating homogeneous and compressible spheroid agree with values already known from other methods. The stabilizing effect of a slow uniform rotation on convection is corroborated.

065.025 Neutron star models.
G. Börner, K. Sato.
Astrophys. Space Sci., Vol. 12, 40 - 46 (1971).

Neutron star models are calculated using an equation of state discussed in an earlier paper. A maximum mass for a neutron star of 1.74 solar masses is found. The central density of this star is 3.3×10^{15} g/cm^3. The lightest stars have masses of 0.02 (resp. 0.03) solar masses with central densities 2.2×10^{14} g/cm^3 (resp. 1.9×10^{14}).

065.026 Rapidly rotating polytropes in the post-Newtonian approximation to general relativity.
G. G. Fahlman, S. P. S. Anand.
Astrophys. Space Sci., Vol. 12, 58 - 82 (1971).
The study of uniformly rotating polytropes with axial symmetry is extended to include all rotational terms of order Ω^4, where Ω is the angular velocity, consistently within the first post-Newtonian approximation to general relativity. The equilibrium structure is determined by treating the effects of rotation and post-Newtonian gravitation as independent perturbations on the classical polytropic structure.

065.027 Rapidly rotating supermassive stars in the first post-Newtonian approximation to general relativity.
G. G. Fahlman, S. P. S. Anand.
Astrophys. Space Sci., Vol. 12, 83 - 97 (1971).
The structure and stability of rapidly rotating uniformly rotating supermassive stars is investigated using the full post-Newtonian equations of hydrodynamics. The standard model of a supermassive star, a polytrope of index three, is adopted. All rotation terms up to and including those of order Ω^4, where Ω is the angular velocity, are retained. The dynamical stability of the model is treated by using the binding energy approach. It is concluded that the uniformly rotating supermassive star does not provide a suitable base for a model of a QSO.

065.028 On the electric charge of stars. V. F. Shvartsman.
Zhurn. ehksperim. i teor. fiz., Vol. 60, 881 - 884 (1971). In Russian. − Abstr. in Referativ. Zhurn. 51. Astron., 8.51.398 (1971).

065.029 Consequences of Strömgren's theorem for radiative envelope stars. R. Gallino, E. Lovera.
Mem. Soc. Astron. Italiana, Nuova Ser., Vol. 42, 123 - 129 (1971).
Le théorème de Strömgren, qui lie la luminosité d'une étoile en equilibre radiatif au rapport entre la pression de radiation et la pression totale, est applicable même dans le cas où seulement l'enveloppe de l'étoile est radiative et la structure du noyau quelquonque. Il est utile pour déterminer directement la structure des étoiles massives avec coefficient d'opacité constant et pour en discuter la stabilité.

065.030 Pulsations of massive main-sequence stars.
D. D. Clayton.
Comments Astrophys. Space Phys., Vol. 3, 127 - 131 (1971).
Massive main-sequence stars derive their internal power from the CNO cycle, which has a steep temperature dependence − roughly T^{16} at 25×10^6 °K. Thus compression and heating releases power faster, causing expansion and cooling, with the possibility of self-sustained pulsations. For common upper-main-sequence stars, however, there is sufficient dissipation in the regions of Kramers opacity and the amplitude of the pulsational modes is sufficiently small at the center where the driving energy must come from that most stars cannot maintain a pulsation. Only if the stellar mass exceeds $M > M_c \approx 60-100$ M$_\odot$ will the dissipative effects be sufficiently small and radiation pressure sufficiently dominant to allow the pulsational amplitude to grow.

065.031 Points noirs dans la théorie de l'évolution stellaire.
M. Schwarzschild. Translated from English into French by J. Rösch.
L'Astronomie, 85e année, p. 277 - 291 (1971).

065.032 Dissipative processes in neutron-star crusts and the production of black-body X-ray sources.

R. N. Henriksen, P. A. Feldman, W. Y. Chau.
Journ. Roy. Astron. Soc. Canada, Vol. 65, 178 (1971).
Abstr. RAS Canada.

065.033 On selective absorption as a possible accelerating mechanism in hot stars. I. F. Malov.
Astron. Tsirk., No. 616, p. 4 - 6 (1971). In Russian.

065.034 From what do the stars originate?
Yu. N. Efremov.
Zemlya i Vselennaya, No. 4, p. 54 - 63 (1971). In Russian.

065.035 Ultradense matter. S. Frautschi, J. N. Bahcall, G. Steigman, J. C. Wheeler,
Comments Astrophys. Space Phys., Vol. 3, 121 - 126 (1971).
Since the early work of Oppenheimer and Volkoff (1939), various estimates have been made of the equation of state for a cold neutron star and the limiting mass above which the star is unstable. The equation of state depends on the varieties of particles and on the interactions among them. Recent estimates place the limiting mass anywhere between 0.37 M$_\odot$ and about 3 M$_\odot$.

065.036 Opacità radiativa e calcolo di modelli stellari.
V. Castellani, P. Giannone, A. Renzini.
Mem. Soc. Astron. Italiana, Nuova Ser., Vol. 42, 221 - 227 (1971). − Letter.

065.037 Quantitative results of stellar evolution and pulsation theories. K. Fricke, R. S. Stobie, P. A. Strittmatter.
Monthly Notices Roy. Astron. Soc., Vol. 154, 23 - 46 (1971).
The discrepancy between the masses of cepheid variables deduced from evolution theory and pulsation theory is examined. It is shown that plausible systematic errors in interior opacity, distance calibration and the $T_{eff}-B-V$ relation can cause substantial errors in the derived masses. The effects of these systematic errors in other astrophysical contexts (e.g. solar neutrinos, Hyades mass–luminosity relation) are examined and the reliability of previously derived quantities discussed. The extreme sensitivity to input physical data of the evolutionary tracks in the post red giant phase is emphasized. In particular the length of the second crossing track is shown to depend strongly on the hydrogen abundance profile in the neighbourhood of the shell source.

065.038 Stars without circulation. The effect of slight differential rotation in the convective core of upper main sequence stars. J. J. Monaghan.
Monthly Notices Roy. Astron. Soc., Vol. 154, 47 - 57 (1971).
The effect of relaxing the usual, but poorly justified assumption, that convective regions in stars rotate rigidly, is examined by considering rotating stars with no circulation. Upper main sequence stars are examined in detail and the differential rotation in the radiative envelope is found to depend sensitively on the differential rotation in the convective core. Some models are constructed which have zero angular velocity at the equator but they are found to be unstable near the edge of the star.

065.039 The site of the helium flash. P. R. Owen.
Monthly Notices Roy. Astron. Soc., Vol. 154, 59 - 78 (1971).
It is not well established theoretically that helium flash stars evolve smoothly into horizontal branch stars. One of the crucial points in this evolution is the site of the flash and it is the purpose of this paper to show how nitrogen burning can have a crucial effect on this. Only the core has been investigated. Since nearly all the carbon, nitrogen and oxygen in the core will have been converted into nitrogen by hydrogen burning it is assumed that the core initially consists only of

helium and nitrogen.

065.040 The oscillations and the stability of rotating masses with toroidal magnetic fields. II.
R. K. Kochhar, S. K. Trehan.
Astrophys. Journ., Vol. 168, 265 - 280 (1971).

The oscillations and the stability of rotating gaseous masses with prevalent toroidal magnetic fields are examined by using the second-order tensor virial equations. It is shown that in the presence of rotation (solid body and differential) and toroidal magnetic field the point of bifurcation, where the Jacobi ellipsoids branch off from the Maclaurin spheroids, is unaffected by the presence of the magnetic field and differential rotation. The presence of the magnetic field increases the critical value of the eccentricity, where the Maclaurin spheroids become unstable, above the value $e = 0.9529$ that obtains in the absence of a magnetic field. In the equipartition state it is the radial mode which becomes unstable for $\gamma \lesssim {}^8/_5$ whereas for $\gamma > {}^8/_5$ it is the Kelvin mode which becomes unstable. The solid-body rotation has a stabilizing influence on both these modes.

065.041 The Schönberg-Chandrasekhar limit and rotation.
A. Maeder.
Astron. Astrophys., Vol. 14, 351 - 358 (1971).

The effects of axial rotation on the Schönberg-Chandrasekhar (S-C) limit are studied by means of a first-order analytic method. The Virial theorem expressed for an isothermal rotating configuration enables the discussion of the effects of rotation on the stability of the stellar core and particularly on the maximum pressure which it can withstand. The expression of the S-C limit in presence of rotation is given.

065.042 Mass formula and properties of matter at subnuclear densities.   F. Tondeur.
Astron. Astrophys., Vol. 14, 451 - 462 (1971). In French.

We study how the properties of superdense matter depend on the mass formula. We calculate these properties using different sets of parameters for the mass formula; we also include supplementary terms in the Bethe-Weizsäcker formula.

065.043 Multiplicity of solutions for the equilibrium models of a helium-burning population I star.
M. Kozłowski.
Astrophys. Letters, Vol. 9, 65 - 67 (1971).

A nonevolutionary sequence of static models for a helium-burning star of $10\,M_\odot$ is constructed. Along this sequence the Vogt-Russell theorem is not valid. For a certain range of helium core masses it is possible to construct three different models of a star with a given total mass and a given chemical structure. These models possess different parameters at the surface but identical parameters at the centre. It is suggested that the erratic behaviour of the evolutionary tracks computed for models burning helium in the core may be partly due to the existence of multiple solutions of stellar structure equations.

065.044 Evidence for products of rapid neutron capture on the surfaces of peculiar A stars.
B. Kuchowicz.
Astrophys. Letters, Vol. 9, 85 - 89 (1971).

Observations indicate the presence of the unstable element promethium, as well as of certain products of rapid neutron capture in peculiar A stars. An analysis of these features indicates the possible presence of the heaviest elements in the surfaces of these stars.

065.045 Magnetic capture in the course of accretion. I.
P. R. Amnuél', O. Kh. Gusejnov.
Izv. AN AzSSR. Ser. fiz.-tekhn. i mat. n., 1970, No. 5, p. 74 -

82. In Russian. – Abstr. in Referativ. Zhurn. 51. Astron., 9.51.403 (1971).

065.046 Two probable paths of WR stars' evolution.
A. A. Gusejn-zade.
Izv. AN AzSSR. Ser. fiz.-tekhn. i mat. n., 1970, No. 5, p. 83 - 86. In Russian. – Abstr. in Referativ. Zhurn. 51. Astron., 9.51.415 (1971).

065.047 Neutronization of matter at ultra-high densities.
O. Kh. Gusejnov.
Soobshch. Shemakhinsk. Astrofiz. Obs., vyp. (No.) 5, p. 19 - 38 (1971). In Russian.

065.048 Polytropes with phase transition. II. Polytrope $n = 1$.   Z. F. Seidov.
Soobshch. Shemakhinsk. Astrofiz. Obs., vyp. (No.) 5, p. 58 - 69 (1971). In Russian.

065.049 Evolution of single stars. V. Carbon ignition in population I stars.   B. Paczyński.
Acta Astron., Vol. 21, 271 - 288 (1971).

The results of model computations are presented for population I stars of 3, 5, 7, 10, and 15 $M_\odot$ in the evolutionary phases from helium exhaustion in the centre up to carbon ignition in the centre.

065.050 Nonradial oscillations of evolved stars. I. Quasi-adiabatic approximation.   W. Dziembowski.
Acta Astron., Vol. 21, 289 - 306 (1971).

Linear oscillations of stars consisting of an extended envelope and a highly condensed core are investigated. For such stars, even high frequency modes become essentially internal gravity waves of short wavelength in the deep stellar interior. A suitable method of solution of fourth order equation for adiabatic oscillations is described. Numerical solution for $l = 0$ (radial pulsation) and $l = 2$ modes have been obtained for a cepheid model.

065.051 The internal characteristics and cooling time of hot barion stars.
G. S. Hajian, Yu. L. Vartanian.
Astrofizika, Vol. 7, 237 - 258 (1971). In Russian. – English translation in Astrophysics, Vol. 7, No. 2.

The internal characteristics and cooling time are considered for hot barion stars with masses 0.64 $M_\odot$ and 1.55 $M_\odot$. All calculations are made on the basis of general relativity. The results of calculations are shown in figures and tables.

065.052 On the rotation of configurations with homogeneous distribution of matter in general relativity.
G. G. Arutyunian, D. M. Sedrakian.
Astrofizika, Vol. 7, 259 - 270 (1971). In Russian. – English translation in Astrophysics, Vol. 7, No. 2.

The problem of uniform rotation of a model consisting of incompressible liquid is solved by taking into account the second order terms of the angular velocity. The most important integral parameters of configurations, as well as the characteristics of internal structure are calculated.

065.053 Dynamics of protostars in a forming cluster.
T. Arny, P. Weissman.
Bull. American Astron. Soc., Vol. 3, 367 (1971). – Abstr. AAS.

065.054 Accretion of matter by condensed objects.
F. C. Michel.
Bull. American Astron. Soc., Vol. 3, 393 (1971). – Abstr. AAS.

065.055 Longitudinal plasma wave contributions to transport processes in stellar media.

J. E. Littleton, H. M. van Horn, H. L. Helfer.
Bull. American Astron. Soc., Vol. 3, 394 (1971). – Abstr. AAS.

065.056   **An imitation of semiconvection.**
          B. M. Schlesinger.
Bull. American Astron. Soc., Vol. 3, 394 (1971). – Abstr. AAS.

065.057   **The evolution of a 20 M$_\odot$ star.**
          S. R. Sreenivasan, K. E. Ziebarth.
Bull. American Astron. Soc., Vol. 3, 394 (1971). – Abstr. AAS.

065.058   **Post-main sequence evolution of low mass old disk stars.**   B. M. Tinsley.
Bull. American Astron. Soc., Vol. 3, 394 - 395 (1971). Abstr. AAS.

065.059   **Do neutron stars have an ocean surface?**
          B. Carter, H. Quintana.
Nature, Vol. 232, 391 - 392 (1971).
          It is argued that the equation of state of cold neutron star matter may contain a large first order gas-liquid type phase discontinuity at subnuclear densities, which would imply the existence of a major density discontinuity between the solid crust of the star and the neutron superfluid below.

065.060   **The reduced width of the 7.115 MeV level in ^{16}O.**
          N. V. Vidal, G. Shaviv, B. Z. Kozlovsky.
Astrophys. Letters, Vol. 9, 131 - 133 (1971).
          Recent measurements of two parameters in the triple-α and the subsequent reaction warrant a new calculation of the abundance curve. An astrophysical low limit for the reduced width of the 7.115 MeV level in ^{16}O permits a fair narrowing of its range.

065.061   **Non-radial oscillations and the Beta Canis Majoris phenomenon.**   Y. Osaki.
Publ. Astron. Soc. Japan, Vol. 23, 485 - 502 (1971).
          Ledoux's (1951) theory of non-radial oscillations for the β Canis Majoris phenomenon is examined. Results for line profiles and radial velocity curves are compared with observations. It is concluded that purely non-radial oscillations can explain most of the fundamental properties of the β Canis Majoris stars.

065.062   **The limiting mass of a neutron star.**
          Y. C. Leung, C. G. Wang.
Nature, Phys. Sci., Vol. 233, 99 - 100 (1971).
          A brief description of the derivation of an equation of state for dense baryon matter in the density range between 10^{13} to 10^{18} gm/cm^3 employing methods and results of nuclear physics as well as high energy physics is presented. The limiting mass of a neutron star computed from this equation of state is put at half of a solar mass.

065.063   **Structure and evolution of supermassive rotating magnetic polytropes.**   L. M. Ozernoy, V. V. Usov.
Astrophys. Space Sci., Vol. 12, 267 - 301 (1971). In Russian.
          The structure of rotating magnetic polytropes is considered in Roche approximation. Investigation of the influence of poloidal as well as toroidal magnetic fields on the conditions of the beginning of matter outflow due to rotational instability is carried out. Both maximum possible energy output and duration of the quasi-statical evolution phase up to the appearance of hydrodynamic instability due to the effects of general relativity are calculated for supermassive magnetic polytropes of index three with uniform of differential rotation. The 'radius-mass' relation is obtained for supermassive differentially-rotating magnetic polytropes referring to the

longest part of the quasi-statistical evolution stage; some consequences are pointed out, including the 'period-luminosity' relation. The evolution of the considered models of supermassive rotating magnetic polytropes with different character of rotation and different geometry of a magnetic field is discussed.

065.064   **Neutrino emission, population II stars and big-bang cosmology.**   P. R. Chaudhuri.
Astrophys. Space Sci., Vol. 12, 325 - 328 (1971).
          The spectroscopic determination of the helium abundance in the oldest stars of the Galaxy is supported by the theory of stellar evolution when neutrino emission is considered according to the photon-neutrino coupling theory, if it is assumed that the population II stars started their life with a low surface helium content.

065.065   **The effect of a poloidal magnetic field on the radial and non-radial oscillations of a gaseous mass.**
          G. G. Fahlman.
Astrophys. Space Sci., Vol. 12, 424 - 455 (1971).
          The oscillations of a polytrope with infinite electric conductivity containing a weak internal poloidal magnetic field which is continuous with an external dipole field are examined with the aid of a variational equation. The corrections to the fundamental characteristic frequencies of the radial and non-radial l = 2 pulsation modes are calculated. The magnetic field removes a degeneracy which occurs between these two modes and the resulting frequency splitting is evaluated. The relevance of the results to the known magnetic stars is briefly discussed.

065.066   **Core helium burning evolution at 15 M$_\odot$.**
          J. W. Robertson.
Proc. Astron. Soc. Australia, Vol. 2, 23 - 24 (1971).
          Using the initial abundances (X, Y, Z) = (0.602, 0.354, 0.044), 15 M$_\odot$ models have been evolved from the initial main sequence to core helium exhaustion for the two hypotheses: composition gradient ignored (case A) and included (case B) in the criterion for convection.

065.067   **Core helium burning evolution at 5 M$_\odot$.**
          J. W. Robertson.
Proc. Astron. Soc. Australia, Vol. 2, 24 - 25 (1971).
          5 M$_\odot$ models have been evolved from the initial main sequence through the bluest point of the 'cepheid loops' traversed during core helium burning, for the three (X, Y, Z) compositions (0.602, 0.354, 0.044), (0.708, 0.272, 0.020) and (0.617, 0.363, 0.020), to determine the effects of Y and Z on the loops. The results of these calculations will be described in detail in a forthcoming paper.

065.068   **Population I helium burning red giants. Observations.**
          R. D. Cannon.
Proc. Astron. Soc. Australia, Vol. 2, 25 - 26 (1971).

065.069   **Population I helium burning red giants. Theory.**
          D. J. Faulkner.
Proc. Astron. Soc. Australia, Vol. 2, 26 - 27 (1971).
          In the preceding paper, Cannon has outlined the observational evidence for the existence of a distinct concentration of stars near the base of the red giant branch in intermediate-age galactic clusters, which he tentatively identifies with the core helium burning phase of evolution occurring after the helium flash. This paper reports preliminary results of evolutionary calculations to test this identification.

065.070   **Secular stability. II. An analytic discussion of the onset of complex roots.**
          M. L. Aizenman, J. Perdang.
Astron. Astrophys., Vol. 15, 200 - 205 (1971).

We show that near the main sequence it is possible to formulate the secular problem in such a way that the occurrence of complex roots can be expressed in terms of known physical parameters.

065.071   **On global dynamical stability of rotating stars.**
K. J. Fricke, R. C. Smith.
Astron. Astrophys., Vol. 15, 329 - 331 (1971).

The local criteria for axisymmetric dynamical stability of rotating stars are shown to be globally valid by the use of a variational principle. These criteria are necessary and sufficient so long as the perturbation of the gravitational potential can be neglected.

065.072   **Observational data on the extremes of stellar evolution.**   J. L. Greenstein.
Astrophysics and general relativity. Brandeis Univ. Summer Inst. 1968, Vol. 2, (see 003.011), 5 - 35 (1971).

065.073   **Red giant evolution to the helium flash of a super-metal-rich star.**   P. Demarque, J. N. Heasley.
Monthly Notices Roy. Astron. Soc., Vol. 155, 85 - 94 (1971).

An evolutionary sequence is presented for a star of 1.19 $M_\odot$, with composition parameters $(X, Z) = (0.69, 0.06)$. The evolution was followed from the main sequence to the onset of convection in the core during the helium flash.

065.074   **A pseudo-polytrope particular approach to stars.**
O. Kh. Gusejnov, F. K. Kasumov, Z. F. Seidov.
Tsirkulyar Shemakhin. Astrofiz. observ., 1971, No. 3 (9), p. 4 - 5. In Russian. – Abstr. in Referativ. Zhurn. 51. Astron., 11.51.556 (1971).

065.075   **An analytic approach to the evolution of degenerate carbon cores of stars.**   W. D. Arnett.
Astrophys. Journ., Vol. 169, 113 - 118 (1971).

An analytic procedure for calculation of the evolution of degenerate massive stellar cores is described and compared with detailed numerical results. This procedure provides a simple computational method for investigating such objects as well as a simple physical picture for understanding them.

065.076   **The stability of supermassive disks against fragmentation.**   W. J. Quirk, C. F. McKee.
Astrophys. Journ., Vol. 169, 119 - 124 (1971).

Applying Newtonian stability criteria for both uniform and differential rotation, we find that rotating, relativistic, supermassive (mass $> 10^5\ M_\odot$) disks with a height W greater than one-twentieth of the radius R are stable against fragmentation. The disk must be in a state of strong differential rotation to be stable against assuming an ellipsoidal configuration.

065.077   **On the use of linear interpolation in opacity tables.**
R. T. Rood.
Astrophys. Journ., Vol. 169, 191 - 193 (1971).

Insofar as they influence the results of stellar-interior calculations, errors resulting from the use of linear interpolation in opacity tables are not negligible. They may be large enough to negate many of the recent improvements in opacity tables.

065.078   **Importance of the internal velocity field in star formation.**   T. Arny.
Astrophys. Journ., Vol. 169, 289 - 292 (1971).

The assumption that opacity controls the mass of fragments in a collapsing cloud is shown to lead to results not in agreement with observation. It is shown that internal motions can set the mass of fragments, and that fragmentation at either constant angular-momentum density or kinetic-energy density reproduces the observed mass function. If the internal velocity field does not influence fragmentation, it is difficult

to understand why there are any massive stars formed.

065.079   **A radiative-transfer model of a cepheid.**
J. E. Bendt, C. G. Davis, Jr.
Astrophys. Journ., Vol. 169, 333 - 342 (1971).

A model due to Christy is established and studied in the gray diffusion approximation. This is then used in the study of transfer and frequency-dependent effects. The comparison with the UBV spectrally resolved photoelectric observations of XX Cen is made, and the conclusions are summarized.

065.080   **The photodisintegration rate of ^{24}Mg.**
R. G. Couch, K. C. Shane.
Astrophys. Journ., Vol. 169, 413 - 419 (1971).

The photodisintegration rate of ^{24}Mg is investigated in the context of the silicon-burning process in stars.

065.081   **Stellar rates for the ^{28}Si$(\alpha, \gamma)^{32}$S and ^{16}O$(\alpha, \gamma)^{20}$Ne reactions.**   J. W. Toevs, W. A. Fowler, C. A. Barnes, P. B. Lyons.
Astrophys. Journ., Vol. 169, 421 - 424 (1971).

The stellar interaction rate for the ^{28}Si$(\alpha, \gamma)^{32}$S reaction has been calculated from new cross-sections for this reaction for temperatures from $(0.3-5) \times 10^9$ °K. The astrophysical implications of new data for the ^{16}O$(\alpha, \gamma)^{20}$Ne reaction at low energy are also discussed.

065.082   **Algunos aspectos de la evolución estelar.**
J. C. Muzzio.
Rev. Astron., Vol. 42, No. 174, p. 13 - 17 (1971).

065.083   **Matter in superstrong magnetic fields: The surface of a neutron star.**   M. Ruderman.
Phys. Rev. Letters, Vol. 27, 1306 - 1308 (1971).

In huge magnetic fields ($B \gtrsim 10^{12}$G) matter forms a tightly bound, dense ($\gtrsim 10^4$g cm^{-3}) solid with properties of a one-dimensional metal and a work function of the order of a keV. Electron field emission from the sharp surface of a pulsar is much easier than ion emission; it is estimated to be cut off when the stellar rotation period exceeds several seconds.

065.084   **On nucleosynthesis of elements in s-process.**
V. S. Shorin, V. M. Gribunin, V. N. Kononov, I. I. Sidorova.
Astrofizika, Vol. 7, 489 - 500 (1971). In Russian. – English translation in Astrophysics, Vol. 7, No. 3.

The stellar nucleosynthesis s-process, which is considered as a step-by-step "slow" neutron capture by iron nuclei, has been modelled by an analog computer. The obtained results explain well the observed correlation between the abundance of nuclei in the solar system and the cross-section of neutron radiative capture, if one supposes, that the distribution function of neutron exposures has a special form. The role of isomeric states of nuclei, excited in the neutron capture process, is discussed.

065.085   **Structure and evolution of supermassive rotating magnetic polytropes.**
L. M. Ozernoy, V. V. Usov.
Astrophys. Space Sci., Vol. 13, 3 - 35 (1971).

The structure of rotating magnetic polytropes is considered in Roche approximation. Investigation of the influence of poloidal as well as toroidal magnetic fields on the conditions of the beginning of matteroutflow due to rotational instability is carried out. The influence of the turbulent convection and twisting of magnetic force lines on the time of smoothing of differential rotation is considered. The evolution of the considered models of supermassive rotating magnetic polytropes with different character of rotation and different geometry of a magnetic field is discussed.

065.086 **Axisymmetric multipole magnetic fields in polytropic stars.** T. C. Chiam, J. J. Monaghan.
Monthly Notices Roy. Astron. Soc., Vol. 155, 153 - 167 (1971).

The structure of axisymmetric, multipole magnetic fields of the poloidal type in polytropic stars is studied. The magnetic field is defined in terms of a stream function Ψ patterned after the stream function for a current ring. We rederive in a simple way the result that the current involves an arbitrary function of Ψ. By choosing this arbitrary function to be linear in Ψ and involving two arbitrary constants, various families of solutions are obtained by a perturbation method. Fields similar to those recently found to fit the observations are found and the structure of the field determined.

065.087 **Rotating stars with very large magnetic fields.** J. J. Monaghan, K. W. Robson.
Monthly Notices Roy. Astron. Soc., Vol. 155, 231 - 247 (1971).

Uniformly rotating main sequence stars with strong dipole magnetic fields are examined using a perturbation method. The spherically symmetric part of the perturbing forces is taken into account essentially exactly, while the non-spherical terms are treated as a perturbation. The form of the internal magnetic field, the perturbations to the pressure, temperature, luminosity and the surface are computed.

065.088 **Cen XR-3: A neutron star younger than the Crab?** R. N. Henriksen, P. A. Feldman, W. Y. Chau.
Nature, Vol. 234, 450 - 453 (1971).

A model for the pulsed X-ray source Cen XR-3 has been constructed, based on studies of wobbling, young, evolving neutron stars which are generating heat in their crusts by the conversion of rotational energy through precession-induced cyclic strain. The observational features can be understood if the observed period is interpreted as the free nutation period. We predict the actual rotational period to be ~ 8 ms and the slow-down rate $\sim 8 \times 10^{-13}$ s s^{-1}.

065.089 **Element production in simple helium burning.** C. J. Hansen.
Astrophys. Journ., Vol. 169, 585 - 588 (1971).

The final abundances of the α-nuclei (^{12}C, ^{16}O, ^{20}Ne, etc.) resulting from helium burning at constant temperature and density have been computed by using recently available reaction rates. For one density (10^5 g cm^{-3}) the calculation is extended up to a temperature high enough to demonstrate the emergence of the e-process.

065.090 **Problemi di evoluzione e nucleosintesi.** A. Masani.
Mem. Soc. Astron. Italiana, Nuova Ser., Vol. 42, 293 - 330 (1971).

We discuss the main theories about the formation of elements and we point out those aspects which have validity as well as the more uncertain ones.

065.091 **L'età delle stelle di popolazione I.** G. Barbaro.
Mem. Soc. Astron. Italiana, Nuova Ser., Vol. 42, 331 - 347 (1971).

The methods, for age determination of population I stars and in particular for open clusters, that are based on stellar evolutionary theory are analyzed. The influence of uncertainties connected with theory (opacity, chemical composition), with transformation relations between theoretical and experimental variables and with experimental data are evaluated. Some considerations about the kinematic age determination of associations are put forward.

065.092 **La determinazione dell'età delle stelle di popolazione II.** A. Renzini.

Mem. Soc. Astron. Italiana, Nuova Ser., Vol. 42, 349 - 362 (1971).

The problem of the age determination for population II stars is briefly discussed. The problem of fitting observed C-M diagrams of globular clusters with theoretical time constant loci is discussed. Then the classical method of age determination by means of the main-sequence turn-off luminosity is described. Some observational properties of population II stars in extragalactic objects such as dwarf galaxies in the local group are reported.

065.093 **A prediction of the rate of the ^{12}C(α, γ) ^{16}O reaction in helium burning.** W. D. Arnett.
Astrophys. Journ., (Letters), Vol. 170, L43 - L45 (1971).

A relatively simple use of current astrophysical theory relates the observed abundance ratio ^{12}C/^{16}O to the rate of the reaction ^{12}C(α, γ) ^{16}O during core helium burning in massive stars. This quantitative prediction may soon be compared with experimental measurements of ^{12}C(α, γ) ^{16}O at low energy which are now under way.

065.094 **Erratum: 'Frictional heating in neutron stars'.** [Nature, Phys. Sci., Vol. 232, 117 - 119 (1971)]. G. Greenstein.
Nature, Phys. Sci., Vol. 234, 180 (1971). − See Abstr. 065.011.

065.095 **Nuclear processes associated with peculiar A-type stars.** A. G. W. Cameron.
Publ. Astron. Soc. Pacific, Vol. 83, 585 - 591 (1971). − Invited symposium paper presented at the Hawaii meeting of the Astronomical Society of the Pacific, 22–25 June 1971.

065.096 **Evolution of a 0.6 $M_\odot$ white dwarf.** S. C. Vila.
Astrophys. Journ., Vol. 170, 153 - 156 (1971).

The evolution of a 0.6 $M_\odot$ white-dwarf star composed of an oxygen core and a helium envelope has been calculated. The calculations have taken into account the outer convection zone and the effect of the ion lattice on the specific heats. It is found that the star can become an invisible black dwarf in an interval of the order of 7×10^9 years, which is less than the present estimates of the age of the universe.

065.097 **The evolution of super-helium-rich 2 $M_\odot$ stars.** J. W. Liebert, C. Allen, F. Schweizer, J. Tarter.
Publ. Astron. Soc. Pacific, Vol. 83, 626 - 632 (1971).

Evolutionary calculations are presented for three 2 $M_\odot$ configurations, two having super-helium-rich compositions, $(X, Y, Z) = (0.30, 0.69, 0.01)$ and $(0.30, 0.695, 0.005)$, and one a moderately helium-rich composition $(0.60, 0.39, 0.01)$. The $Z = 0.005$ configuration is evolved past core helium exhaustion, the others only past core helium ignition. The structural evolution of these models is discussed with regard to general ideas on post-main-sequence evolution. Observable features of a hypothetical super-helium-rich cluster are also indicated.

065.098 **The ground state of matter at high densities: Equation of state and stellar models.**
G. Baym, C. Pethick, P. Sutherland.
Astrophys. Journ., Vol. 170, 299 - 317 (1971).

The equation of state of zero-temperature matter in complete nuclear equilibrium is given for mass densities below 5×10^{14} g cm^{-3}. We redetermine, taking into account the effect of the Coulomb lattice and using more recent nuclear mass extrapolations, the sequence of equilibrium nuclides present at mass densities between 10^4 and 4.3×10^{11} g cm^{-3}, the point of neutron drip; and calculate the equation of state here. We calculate zero-temperature white-dwarf and neutron-star models for the equation of state. The maximum stable white-dwarf mass, the Chandrasekhar limit, is $1.00 M_\odot$. The lightest stable neutron star has a mass 0.0925 $M_\odot$ and a central density 1.55 × 10^14 g cm^{-3}, neutron stars between 0.0925 and 0.11$M_\odot$ are entirely solid. Moments of inertia, surface deformations, and mass quadrupole moments are calculated for slowly rotating neutron stars.

065.099 5 $M_\odot$ evolution for population I compositions.
J. W. Robertson.
Astrophys. Journ., Vol. 170, 353 - 362 (1971).

Stellar models of 5 $M_\odot$ have been evolved from the initial main sequence through the bluest point of the loops during core helium burning for the three (X, Y, Z) compositions: (0.602, 0.354, 0.044), (0.708, 0.272, 0.02), and (0.617, 0.363, 0.02). Convective overshooting of the helium-burning core and helium-carbon-oxygen semiconvection are included in the evolution. The loops become longer and brighter as Y is increased or Z is decreased. With the inclusion of core convective overshooting, the loops become longer and more time is spent near the blue end of the loops.

065.100 Evolution of low-mass stars. IV. Effects of multi-level atomic partition functions for the ideal-gas region.   H. C. Graboske, Jr., A. S. Grossman.
Astrophys. Journ., Vol. 170, 363 - 370 (1971).

Several improvements have been introduced in the theory of equilibrium thermodynamic properties for densities and temperatures appropriate to evolution of low-mass stars. Effects of these modifications have been investigated for stars in the mass range 0.03−0.2 $M_\odot$. In the ideal-gas region, inclusion of the excited states in the atomic partition functions cause the early pre-main-sequence and deuterium-main-sequence models to shift toward lower T_e and L, an effect which increases as stellar mass increases. Modification of the Coulomb interaction free energy causes the late pre-main-sequence and hydrogen-main-sequence models to shift toward lower T_e and L, an effect which increases with decreasing stellar mass.

065.101 Properties of hadron matter. II. Dense baryon matter and neutron stars.   Y. C. Leung, C. G. Wang.
Astrophys. Journ., Vol. 170, 499 - 521 (1971).

We have provided certain details of a nuclear-matter computation, based on the Brueckner-Bethe-Goldstone theory of nuclear reaction, which leads to an equation of state for matter in the density region 10^13−5 × 10^14 g cm^{-3}. We also explore the possibilities that at very high baryon densities (>10^17 g cm^{-3}) or for very short baryon separations (<0.1 fm), the net baryon-baryon interaction may be negligible so that the results of dynamical models, like the statistical bootstrap model and the dual-resonance model, may be applicable to the study of dense baryon matter. Several plausible equations of state are constructed, and their effect on the limiting mass of the neutron star is examined.

065.102 Surface composition and cooling histories of neutron stars.   A. G. W. Cameron.
Highlights of Astronomy, Vol. 2, (see 012.019), 731 - 736 (1971).

065.103 Neutron stars and black holes.   S. Sofia.
Nature, Phys. Sci., Vol. 234, 155 - 157 (1971).

It is shown that rotational instability is probably the mechanism which detaches the envelope of an object undergoing gravitational collapse. In the case in which a high luminosity core (neutron star) forms, radiation pressure on the electron gas causes a violent expansion of the detached ring, and this can be identified with the supernova phenomenon. If the collapse continues towards the black hole stage, the ring cannot be expelled, and remains near the collapsar. It is also shown that the determining feature separating proto-neutron stars from proto-black holes is the angular momentum.

065.104 Probleme der Sternentstehung.
H. Elsässer.
Physik und Kosmologie, (see 003.024), p. 70 - 79 (1971).

065.105 Evolution of massive stars.   R. Kippenhahn.
The Magellanic Clouds. Astrophys. Space Sci. Library, Vol. 23, (see 012.010), 144 - 155 (1971).

065.106 Erratum: 'On Schwarzschild's theory of stellar rotation' [Sci. Rep. Tôhoku Univ., First Ser., Vol. 53, 21 - 29 (1970)].   T. Aikawa.
Sci. Rep. Tôhoku Univ., First Ser., Vol. 53, 158 (1970).

065.107 Perturbation theory of rotating polytropes: Series-expansion method.   T. Aikawa.
Sci. Rep. Tôhoku Univ., First Ser., Vol. 54, 13 - 27 = Sendai Astron. Rap., No. 119 (1971).

The second order effect of the series-expansion method for uniformly rotating polytropes is investigated. By using the numerical results, the validity of the method is discussed.

065.108 On Gabriel's mechanism of formation of semi-convective zone in massive main-sequence stars.
K. Mimura, K. Suda.
Sci. Rep. Tôhoku Univ., First Ser., Vol. 54, 28 - 32 (1971).

Gabriel's mechanism proposed to support the Schwarzschild and Härm's evolutionary stellar models is reconsidered. In contrast to his result, the time scale of inward motion of convection zone is estimated as larger by order of at least 10^7 than the local Kelvin time scale of the region passed by convection zone. From this point of view, it is difficult to accept Gabriel's mechanism.

065.109 Theory of level surfaces inside relativistic, rotating stars. I. Poincaré's limit.   M. A. Abramowicz.
Acta Astron., Vol. 21, 449 - 454 (1971).

Poincaré had proved that the angular velocity of rigid rotation of any stationary star has a limit. One can show that there is a similar, but different limit in the more general case of differential rotation. It has been shown that analogical conditions exist in general relativity. Violation of these conditions implies a continuous increase of the rate of expansion of a star.

065.110 The abundance of helium in stellar interiors.
J. Faulkner.
Highlights of Astronomy, Vol. 2, (see 012.015), 269 - 287 = Contr. Lick Obs., No. 333 (1971).

065.111 Production of helium by stellar evolution.
R. Kippenhahn.
Highlights of Astronomy, Vol. 2, (see 012.015), 296 - 300 (1971).

065.112 Production of helium in massive objects.
R. V. Wagoner.
Highlights of Astronomy, Vol. 2, (see 012.015), 301 - 317 (1971).

065.113 Rotating neutron stars.
V. V. Papoian, D. M. Sedrakian, E. V. Chubarian.
Astron. Zhurn. Akad. Nauk SSSR, Vol. 48, 1195 - 1200 (1971). In Russian. English translation in Soviet Astron. AJ, Vol. 15, No. 6.

The structure and integral parameters of uniformly rotating equilibrium models of neutron stars are determined in ω^4 approximation in Newton's theory of gravitation. It is shown that the results obtained differ slightly from those obtained in ω^2 approximation.

065.114 Il problema dell'elio.
V. Castellani, P. Giannone, L. Gratton, N. Panagia.

Giornate di Studio sull'Elio, Soc. Astron. Italiana, [Tipografia Baccini & Chiappi, Firenze], 1970, p. 3 - 10 = Oss. Astron. Roma, Contr. Sci., Ser. 3, No. 101 (1970).

065.115 **L'elio nell'evoluzione stellare.** P. Giannone.
Giornate di Studio sull'Elio, Soc. Astron. Italiana, [Tipografia Baccini & Chiappi, Firenze], 1970, p. 69 - 88 = Oss. Astron. Roma, Contr. Sci., Ser. 3, No. 102 (1970).

065.116 **On the dependence of Q metallicity index from basic parameters of stellar evolution.**
F. A. D'Antona.
Astrophys. Space Sci., Vol. 14, 314 - 316 (1971).
The possible dependence of van den Bergh's Q parameter on the mass distribution function in globular clusters is investigated. The resulting independence within the observational errors assures that Q is available as index of the chemical composition and age of the clusters.

065.117 **Evoluzione stellare.** P. Giannone.
Atti XIII Riunione Soc. Astron. Italiana 1970, p. 113 - 123 = Oss. Astron. Roma, Contr. Sci., Ser. 3, No. 103 (1970).

065.118 **Relativistic evolution of $10^3 M_\odot$ star.**
A. Kovetz, G. Shaviv.
Astrophys. Space Sci., Vol. 14, 378 - 388 (1971).
A fully relativistic evolution of $10^3 M_\odot$ is described for initial composition of $Y = Z = 0$. Our results show that (a) a great part of the star is in radiative equilibrium, (b) the maximal red-shift for main-sequence $10^3 M_\odot$ stars is significantly less than for isentropic models, and (c) a very low amount of CNO elements is formed at any stage before hydrogen is completely consumed and hence such stars cannot be the progenitors of population II stars.

065.119 **Explosive carbon-burning nucleosynthesis.**
C. J. Hansen.
Astrophys. Space Sci., Vol. 14, 389 - 395 (1971).
Results are presented which show the final abundances of the major nuclei produced in explosive, but partial, carbon-burning taking place in a rapidly expanding medium. The effects of energy deposition due to burning and expansive cooling are taken into account explicitly. The results are compatible with those obtained by other authors but show the sensitivity of the final abundances to the assumed initial conditions. A conjecture is made as to the physical site of the burning.

065.120 **Implosions and emplosions in supermassive objects.**
G. Shaviv, G. E. Tauber.
Astrophys. Space Sci., Vol. 14, 396 - 398 (1971).
It is shown that for small velocities the time dependent evolution of massive stars may be described by a series of static models. The fate of the dynamic system is analyzed by the behaviour of the static models. The method is illustrated for general relativistic polytropes.

065.121 **The evolution of hydrogen-helium stars.**
D. Ezer, A. G. W. Cameron.
Astrophys. Space Sci., Vol. 14, 399 - 421 (1971).
The structure and evolution of hydrogen-helium stars of 5, 10, 20, 30, 100, and 200 $M_\odot$ have been followed through the pre-main sequence phases of evolution and through hydrogen-burning on the main sequence.

065.122 **Contribution to the helium content of the Galaxy from pulsationally unstable stars evolving inhomogeneously.** Y. Tanaka, S. Sakashita.
Progr. Theor. Phys., Japan, Vol. 46, 1627 - 1628 (1971).
Letter.

065.123 **General relativistic neutron and hyperon star models.**
H. Heintzmann.
General relativity and cosmology. Course 47, Italian Phys. Soc., 1969, (see 012.022), p. 359 - 361 (1971).

065.124 **Nucleosynthesis in stars.** D. D. Clayton.
Proc. 11th international conference on cosmic rays, Budapest 1969, (see 012.025), p. 21 - 39 (1970).

065.125 **On a possibility of observing neutrinos from collapsing stars and neutrino oscillations.**
G. V. Domogatsky, G. T. Zatsepin.
Proc. 11th international conference on cosmic rays, Vol. 4, (see 012.028), 361 - 365 (1970).

065.126 **Vom Ende der Sterne.** V. Weidemann.
Bild der Wissenschaft, [Deutsche Verlags-Anstalt, Stuttgart], Vol. 7, 670 - 681 = Sonderdruck Sternw. Kiel, No. 174 (1971).

065.127 **On the possible evolutionary meaning of nitric and carbon sequences of Wolf-Rayet stars.**
A. A. Gusejn-zade.
Izv. AN AzSSR. Ser. fiz.-tekhn. i mat. n., 1970. No. 6, p. 97 - 100. In Russian. – Abstr. in Referativ. Zhurn. 51. Astron., 1.51.565 (1972).

065.128 **Evolution of stellar models with high metal content.**
S. Torres-Peimbert.
Bol. Obs. Tonantzintla y Tacubaya, No. 37, Vol. 6, 113 - 130 (1971).
Evolutionary sequences for stellar configurations between 1 and 1.45 $M_\odot$ with chemical abundances varying from $Z = 0.023$ to 0.10 and from $X = 0.60$ to 0.68 were computed using the Berkeley Stellar Evolution program. The effect of high metal abundances of the structure and evolution of models of 1.25 $M_\odot$ is studied in detail.

065.129 **Thermonuclear origin of rare neutron-rich isotopes.**
W. M. Howard, W. D. Arnett, D. D. Clayton, S. E. Woosley.
Phys. Rev. Letters, Vol. 27, 1607 - 1610 (1971).
Many rare neutron-rich isotopes in the range $16 < Z \lesssim 34$ can be synthesized from seed nuclei exposed to explosive carbon burning. This process, which involves no new astrophysical parameters, can solve most of the outstanding problems in the thermonuclear synthesis of elements in the range $Z \lesssim 34$.

065.130 **Equation of state of neutron star matter and neutron star models.**
S. Ikeuchi, S. Nagata, T. Mizutani, K. Nakazawa.
Progr. Theor. Phys., Japan, Vol. 46, 95 - 113 (1971).
The energy of neutron star matter is calculated by using the Brueckner theory for three different-type nuclear potentials in the range of densities $5 \times 10^{13} \lesssim \rho \lesssim 1.14 \times 10^{15}$ (g cm^{-3}). In this calculation, contributions from higher partial waves, three-body clusters and mixture of protons are included. Using the results, equations of state are determined and extended by several methods to density regions higher than 1.14×10^{15} g cm^{-3}. Neutron star models are constructed based on the derived equations of state. These models show that the maximum mass of the stable neutron star lies between 1 $M_\odot$ and 3 $M_\odot$. The $^3P_2$-pairing superfluidity is investigated and the energy gap of half a MeV has been obtained. Some physical effects by this superfluidity are discussed in relation to the pulsar and the thermal history.

065.131 **Superfluid state in neutron star matter. II. Properties of anisotropic energy gap of $^3P_2$ pairing.**
T. Takatsuka, R. Tamagaki.

Progr. Theor. Phys., Japan, Vol. 46, 114 - 134 (1971).

By applying the formulation developed in I and using several semiphenomenological two-nucleon potentials, properties of the $^3P_2$ pairing originating from spin-orbit forces are investigated at high density ($\rho \gtrsim 2 \times 10^{14}$ gcm^{-3}) in neutron star matter.

065.132 **Contribution of cosmic ray nuclei to metallic elements in population II stars.** S. Hayakawa.
Progr. Theor. Phys., Japan, Vol. 46, 994 - 995 (1971).

065.133 **Shock-wave propagation in stellar interiors.**
N. Virgopia.
Mem. Soc. Astron. Italiana, Nuova Ser., Vol. 42, 481 - 495 (1971).

The problem of isothermal shock waves propagation in the interior of spherically symmetric stellar models of 12 $M_\odot$ and 0.5 $M_\odot$ has been studied. The possibility of mass ejection when the particle speed exceeds the escape velocity from the surface of the stars is also discussed.

065.134 **Evolution of the stars.** J. Grygar.
Vesmír, Vol. 50, 323 - 325 (1971). In Czech.

065.135 **Calculation of the ^{12}C+α capture cross section at stellar energies.** F. C. Barker.
Australian Journ. Phys., Vol. 24, 777 - 792 (1971).

The ^{12}C(α, γ)^{16}O cross section is calculated at stellar energies, using R-matrix parameters obtained by fitting consistently the ^{12}C+α scattering phase shifts and the α-spectrum from ^{16}N β-decay. This limits the ^{12}C+α channel radius to the range 5−7 fm. The S-factor at E_a = 400 keV is calculated to lie in the range 0.05−0.33 MeV b.

065.136 **On the stability of super-massive stars.**
I. Nedyalkov.
Godishn. Vissh. tekhn. uchebni zaved. Fiz., Vol. 4, sb. 1, p. 117 - 122 (1967, 1970). In Russian. − Abstr. in Referativ. Zhurn. 51. Astron., 2.51.740 (1972).

065.137 **Stability of nonradial oscillations of cold nonrotating neutron stars. I. General method.**
L. Battiston, P. Cazzola, L. Lucaroni.
Nuovo Cimento B, Ser. 11, Vol. 3B, 295 - 318 (1971).

It is assumed that the star is made of a perfect fluid. A suitable frame of reference is chosen such that the Einstein's linearized equations for the oscillation turn out to be fairly simple. The four independent solutions of these equations can be interpreted as amplitudes describing ingoing and outgoing sound and gravitational waves coupled together. Approximate solutions are constructed.

065.138 **Comic rays, neutron stars and pulsars.**
P. R. Chaudhuri.
Journ. Phys. A. General Phys., Vol. 4, 508 - 516 (1971).

The cooling behaviour of neutron stars, taking into account several models proposed by Tsuruta and Cameron, has been studied on the basis of neutrino emission according to the photon-neutrino weak coupling theory.

065.139 **On the origin of magnetic fields in white dwarfs and meson stars.** R. F. O'Connell, K. M. Roussel.
Nuovo Cimento Lettere, Ser. 2, Vol. 2, 55 - 57 (1971).

065.140 **Fusion chain reactions in stellar cores.**
B. J. Smernoff.
Thesis, Brandeis Univ., Waltham, Mass. [Available from Univ. Microfilms, Ann Arbor, Mich., U.S.A. Order No. 70−24659], 80 pp. (1970).

The role played by fusion chain reactions in the advanced evolution of stars is studied using a simple method to estimate the multiplication factor for such chain reactions. The Lindhard dielectric formulation of the stopping power problem is used to calculate the energy loss of a test charge to a relativistic degenerate electron gas. The stopping powers of an ion gas and an ion lattice are also calculated, and the collective energy loss to the lattice is expressed as a general function of phonon parameters.

065.141 **Pulsational characteristics of a low mass cepheid model including convection effects.** E. M. Jones.
Thesis, Univ. Wisconsin, Madison. [Available from Univ. Microfilms, Ann Arbor, Mich., U.S.A. Order No. 70−13918], 47 pp. (1970).

The model parameters are: mass 0.93 $M_\odot$, M_{bol}=−1.7, and log T_e=3.75. Convection is treated by a time-dependent mixing-length type theory where the eddy velocity is assumed to exponentially approach the equilibrium value computed in Böhm−Vitense's formulation. The dynamical evolution of the model is computed by a code based on the method of Cox, Brownlee, and Eilers.

065.142 **Star formation.** M. V. Penston.
Contemporary Phys., Vol. 12, 379 - 394 (1971).

A brief review of the interstellar medium and its instabilities is given with particular emphasis to recent results on its 'two-phase' nature. The existing models of the collapse of gas cloud to form stars are discussed. Finally some aspects of the observation of young stars are considered.

065.143 **Stellar stability and the upper end of the main sequence.** K. E. Ziebarth.
Thesis, Univ. Colorado, Boulder. [Available from Univ. Microfilms, Ann Arbor, Mich., U.S.A. Order No. 70−23769], 139 pp. (1970).

The purposes of this thesis are to find the upper limit of stable masses as a function of composition and to investigate the mass loss hypothesis.

065.144 **The hydrodynamics of a helium shell flash in a star of one solar mass.** R. E. Zimmermann.
Thesis, Univ. California, Los Angeles. [Available from Univ. Microfilms, Ann Arbor, Mich., U.S.A. Order No. 71−10655], 123 pp. (1970).

The evolution of a metal-poor star of one solar mass has been calculated from the main sequence to the helium flash under the assumption that the plasma neutrino and photoneutrino processes act to cool the center during the giant phases. Some estimates are made as to the probable hydrodynamic developments in the envelope of the star.

065.145 **Neutron star models and nuclear forces.**
W. Hillebrandt, S. Kistler.
Zeitschr. Physik, Vol. 246, 60 - 70 (1971).

Equations of state of cold neutron matter are calculated by the method of unitary transformations for a hard-core and a soft-core potential. Equilibrium configurations are constructed in the Newtonian and the general relativistic theory of gravitation.

065.146 **Strong shock with radiation near the surface of a star.** P. L. Sachdev, S. Ashraf.
Phys. Fluids, Vol. 14, 2107 - 2110 (1971).

The propagation of a shock wave originating in a stellar interior, is considered when it approaches the surface of the star and assumes a self-similar character, 'forgetting' its initial conditions. The adiabatic and isothermal flows behind such a shock are compared.

065.147 **Magnetohydrodynamics in neutron stars.**
I. R. Rostron.
Thesis, Brigham Young Univ., Provo, Utah. [Available from

Univ. Microfilms, Ann Arbor,Mich.,U. S. A. Order No. 71 -
8840], 119 pp. (1970).

A theoretical study of some of the properties of neutron
stars is made. The principal interest is in magnetohydrodynamic behavior in the presence of gravitational potentials and large
magnetic fields.

065.148 **Acceleration and radiation around oblique rotators.**
A. Cavaliere.
Nuovo Cimento B, Ser. 2, Vol. 5 B, 110 - 118 (1971).

Particles accelerated by the e. m. fields existing near a
rotating, magnetized neutron star are widely held to be the
source of activity of NP 0531, and of the whole Crab nebula
at high frequencies at least. The author derives the main
features of the radiation from these particles and discusses a)
the production of flashes in IR to X-ray bands in the upper
magnetosphere, b) the sweeping of the high-energy electrons
beyond $r \approx c/\Omega$ and c) the nebular radiation as resulting from
the wave field configuration.

065.149 **Origin of the magnetic fields in early-type peculiar**
stars.   S. Kato.
Astron. Herald, (*Japan*), Vol. 64, 67 - 69 (1971). In Japanese.

There are three explanations of the origin of the stellar
magnetic field: 1. fossil theory; 2. dynamo mechanism; 3. processes other than that of dynamo. The author points out the
possibility of 1 and 3.

065.150 **Magnetic field decay in condensed objects. The in-**
fluence of space-time curvature.
M. Grewing, H. Heintzmann.
Zeitschr. Physik, Vol. 247, 223 - 226 (1971).

A generalization of Lamb's formula for the decay of a
magnetic field due to ohmic dissipation is given which takes
into account the influence of space-time curvature. It is found
that for condensed objects such as heavy neutron stars the decay time for the magnetic field is substantially lengthened.

065.151 **Neutron star matter.**
G. Baym, H. A. Bethe, C. J. Pethick.
Nuclear Phys. A, Vol. A175, 225 - 271 (1971).

The authors determine the constitution of the ground
state of matter and its equation of state in the regime from
4.3×10^{11} g/cm^3 up to densities $\approx 5 \times 10^{14}$ g/cm^3. They describe the energy of nuclei in the free neutron regime by a
compressible liquid-drop model designed to take into account
three important features.

065.152 **Thermal stability of the helium-burning shell in stars**
of 15 solar masses.   T. R. Dennis.
Thesis, Princeton Univ., N.J. [Available from Univ. Microfilms,
Ann Arbor, Mich., U.S.A. Order No. 71–14369], 34 pp.
(1970). – See Phys. Abstr., Vol. 75, No. 10867 (1972).

065.153 **Neutron stars: The equation of state of matter at**
very high densities.   J. W. Craft.
Thesis, Univ. California, Los Angeles. [Available from Univ.
Microfilms, Ann Arbor, Mich., U.S.A. Order No. 71–13989],
279 pp. (1970).

The equation of state of matter at very high densities is
calculated for a number of two-nucleon interactions. The nu-
cleon-nucleon interaction is described by a number of phenomenological two-body potentials designed to fit various low-energy nuclear physics data, such as scattering data and nuclear
matter binding properties. The equations of stellar structure
were then solved for these equations of state for various values
of central density. The gravitational mass and star radius, along
with the angular momentum for a slowly-rotating star, were
calculated and compared for the several models. These results
were compared with those obtained by using the Tsuruta-Ca-
meron model.

065.154 **Stellar evolution.**
R. Gallino, A. Masani, G. Silvestro.
Nuovo Cimento Rivista, Ser. 2, Vol. 1, 55 - 78 (1971).

The authors discuss semiconvection, the evolution of
massive red supergiants, red giants and supergiants of $M \lesssim 10$
$M_\odot$, the He flash in the core for stars with $M \lesssim 4 M_\odot$, the horizontal branch of globular clusters, the He-shell flash, mass loss
in the giant phase and the last evolutionary phases.

065.155 **Ejection of gas from stars in the late stages of evolu-**
tion.   G. S. Bisnovatyi-Kogan.
Proc. Sixth Winter School on Space Physics, Part I. Apatity
1969, (see 012.033), p. 49 - 53 (1971).

065.156 **Low-frequency oscillations of a magnetic rotating**
neutron star.   Yu. V. Vandakurov.
Proc. Sixth Winter School on Space Physics, Part I. Apatity
1969, (see 012.033), p. 54 - 61 (1971).

065.157 **Computation of the opacity of stars with allowance**
for line absorption.
A. F. Nikiforov, B. V. Uvarov.
Proc. Sixth Winter School on Space Physics, Part I. Apatity
1969, (see 012.033), p. 73 - 76 (1971).

065.158 **Calculation of stellar structure. II. Determination of**
the helium abundance of the sun by the theoretical
prediction abundance of the sun by the theoretical prediction
of line and continuum radiation from solar-model photo-
spheres.   C. A. Rouse.
Progr. High Temperature Phys. Chem., Vol. 4, 139 - 191
(1971).

065.159 **On stellar activity cycles.**
B. R. Durney, J. O. Stenflo.
National Center Atmosph. Res., Boulder, Colorado, NCAR MS
No. 71–206, 12 pp. (1971).

The relation between the average magnetic field B, the
angular velocity Ω, and the period P of stellar activity cycles
is studied.

The rate of the ^{12}C + ^{16}O reaction.
See Abstr. 022.030.

Thomson scattering in a strong magnetic field.
See Abstr. 062.046.

The collapse of rotating stars.
See Abstr. 066.010.

On the pressure inside a rotating star.
See Abstr. 066.013.

Possible existence of particles of imaginary mass,
energy and momentum.   See Abstr. 066.027.

Relativistic stars, black holes and gravitational waves
(including an in-depth review of the theory of rotating, relativistic stars).   See Abstr. 066.078.

Comments on the instability strip for halo popula-
tion variables.   See Abstr. 115.002.

Surface characteristics of the magnetic stars.
See Abstr. 116.007.

Theoretical aspects of magnetic stars.
See Abstr. 116.008.

Magnetic stars with an external non-linear force-free
field.   See Abstr. 116.012.

Evolution in close binary systems.
See Abstr. 117.022.

Short-period variables. VIII. Evolution and pulsation of δ Scuti stars.   See Abstr. 122.003.

Observational aspects of cepheid evolution.
See Abstr. 122.155.

The effect of beta processes on the dynamic evolution of carbon-detonation supernovae.
See Abstr. 125.011.

Supernovae and neutron stars.
See Abstr. 125.031.

Molecules in dense clouds and protostars.
See Abstr. 131.117.

On the evolution of Strömgren spheres.
See Abstr. 131.132.

Evolution of single stars. VI. Model nuclei of planetary nebulae.   See Abstr. 133.028.

Evolutionary origin of the magnetic field on pulsars and its relation to other types of stars.
See Abstr. 141.019.

Evolution of a stabilized oblique rotator: Behavior over short and long time scales.   See Abstr. 141.060.

Faraday rotation and signal dispersion: The geometrical optics approximation, an exact solution, and first order smoothing theory.   See Abstr. 141.162.

Pulsare und Neutronensterne.
See Abstr. 141.213.

Neutron starquakes and pulsar speedup.
See Abstr. 141.231.

X-ray sources and final stages of stellar evolution.
See Abstr. 142.064.

Formation stellaire dans l'association Sco OB_1.
See Abstr. 152.004.

The blue stars above the turn-off in M67: Horizontal branch or blue stragglers?   See Abstr. 153.029.

Globular-cluster stars: Results of theoretical evolution and pulsation studies compared with the observations.
See Abstr. 154.021.

066 Relativistic Astrophysics (without Cosmology), Background Radiation, Gravitation Theory

066.001 Relativistic disks. I. Uniform rotation.
J. M. Bardeen, R. V. Wagoner.
Astrophys. Journ., Vol. 167, 359 - 423 (1971).
The structure and gravitational field of uniformly rotating, infinitesimally thin disks is calculated numerically within the framework of general relativity. The methods of calculation are described in some detail, since they can easily be extended to differentially rotating disks. Most results are quite accurate, even in the extreme relativistic limit in which the redshift from the center of the disk to infinity becomes infinite. An investigation of the local instability to fragmentation indicates that a large amount of differential rotation and/or thickness is necessary to stabilize the disks. The behavior of rotating disks when their gravitational fields become strong is very different from the behaviour of spherical or slowly rotating configurations, and offers intriguing possibilities in astrophysics if the stability problem can be solved.

066.002 Ehrenfest's paradox. D. H. Weinstein.
Nature, Vol. 232, 548 (1971).
The purpose of this communication is to point out what one might expect to observe as a disk rotates and to develop the magnitude in a practical case.

066.003 A measurement of the gravitational deflection of radio waves by the sun. J. M. Hill.
Monthly Notices Roy. Astron. Soc., Vol. 153, 7P - 11P (1971).
The deflection of radio waves in the sun's gravitational field has been measured by using the Cambridge One-Mile telescope to observe the radio source 3C 279 before and after its occultation by the sun on 1970 October 8. The observed deflection was 1.07 ± 0.17 times that predicted by general relativity.

066.004 A measurement of the gravitational deflection of microwave radiation near the sun, 1970 October.
R. A. Sramek.
Astrophys. Journ., (*Letters*), Vol. 167, L55 - L60 (1971).
The position of the radio source 3C 279 was monitored during an occultation by the sun using a 2.7-km-baseline radio interferometer. Observations were made simultaneously at wavelengths of 11.1 and 3.7 cm. The two-wavelength observations gave a measurement of the refraction in the solar corona independent of the gravitational deflection. After correcting for this refraction, the gravitational bending was found to be $1''.57 \pm 0''.08$ at the limb of the sun. This is 0.90 ± 0.05 of the deflection predicted by Einstein's general theory of relativity.

066.005 Search for microwave pulses associated with gravitational radiation. R. B. Partridge.
Phys. Rev. Letters, Vol. 26, 912 - 915 (1971).
A microwave radiometer has been used to search for pulses of radio waves from the direction of the galactic center. The results were compared with data from Weber's gravitational-wave experiment. No strong evidence was found associating microwave pulses with pulses of gravitational radiation.

066.006 Cosmic background radiation at λ = 3.3 mm.
M. F. Millea, M. McColl, R. J. Pedersen, F. L. Vernon, Jr.
Phys. Rev. Letters, Vol. 26, 919 - 922 (1971).
Ground based measurements of the cosmic background radiation at λ = 3.3 mm were made at two high-altitude sites. A weighted average of the two measurements corresponds to

a blackbody temperature of 2.61 ± 0.25 K.

066.007 Criterion for the instability of a uniformly rotating configuration in general relativity.
S. Chandrasekhar, J. L. Friedman.
Phys. Rev. Letters, Vol. 26, 1047 - 1050 (1971).
Uniformly rotating configurations in general relativity are considered, and a condition is obtained that they can be quasistatically deformed without violating any of the requirements for equilibrium. This condition extends, into the domain of the rotating stars, the criterion for the onset of dynamical instability (via a neutral mode of oscillation) that occurs by radial pulsations in nonrotating stars.

066.008 Final states of gravitational collapse.
R. M. Wald.
Phys. Rev. Letters, Vol. 26, 1653 - 1655 (1971).
We examine all of the black-hole geometries which can be analytically developed in terms of a parameter from the Schwarzschild geometry. If general (nonspherical) gravitational collapse produces black holes and if analytic variation of the initial conditions of gravitational collapse causes analytic variation of the final space-time geometry of the black holes produced by the collapse, the generic final state of gravitational collapse is a Kerr-Newman black hole, fully specified by its mass, angular momentum, and charge.

066.009 Electromagnetic test fields around a Kerr-metric black hole. J. R. Ipser.
Phys. Rev. Letters, Vol. 27, 529 - 531 (1971).
Weak electromagnetic perturbations (test fields) in the exterior of a Kerr-metric black hole are studied. The proof of the nonexistence of time-independent, nonaxisymmetric test fields actually holds for perturbations associated with any physical (e.g., gravitational) field.

066.010 The collapse of rotating stars.
O. H. Guseinov. F. K. Kasumov.
Astron. Zhurn. Akad. Nauk SSSR, Vol. 48, 722 - 725 (1971). In Russian. English translation in Soviet Astron. AJ, Vol. 15, No. 4.
The problem of mass loss is considered for a rotating star during collapse.

066.011 A decisive test for the general relativity.
A. Blokland.
Astrophys. Space Sci., Vol. 12, 219 - 242 (1971).
A very accurate evaluation of the Schwarzschild constants is now possible from continuous observation of the spin axes of coupled gyroscopes, because the angular velocities of these axes can be more than 100000 times greater than the better known precession velocity of a single gyro in orbit around the earth.

066.012 On the gravitational potential tensor and the equations of motion in relativistic mechanics.
I. G. Fikhtengol'ts.
Zhurn. ehksperim. i teor. fiz., Vol. 60, 1206 - 1210 (1971). In Russian. – Abstr. in Referativ. Zhurn. 51. Astron., 8.51.685 (1971).

066.013 On the pressure inside a rotating star.
M. A. Abramowicz.
Postępy Astron., Vol. 19, 247 - 253 (1971). In Polish.

The pressure decreases outward inside any rotating, relativistic star.

066.014 **General relativity and the application of algebraic manipulative systems.**
D. Barton, J. P. Fitch.
Commun. Ass. Computing Machinery, Vol. 14, 542 - 547 (1971).
The paper describes some applications of symbolic algebra systems to problems of general relativity including the derivation of the field equations, the Petrov classification of a metric, and the solution of the field equations in the presence of matter in a simple case. Attention is drawn to the strictly algebraic difficulties encountered in this work.

066.015 **Possible interpretation of Weber's experiments.**
I. I. Kalinnikov, S. M. Kolesnikov.
Astron. Tsirk., No. 619, p. 7 - 8 (1971).   In Russian.

066.016 **General relativity and the orbit of Icarus.**
I. I. Shapiro, W. B. Smith, M. E. Ash, S. Herrick.
Astron. Journ., Vol. 76, 588 - 606 (1971).
Results from 413 photographic observations of the minor planet (1566) Icarus, including 342 from the 1968 close approach, were analyzed to improve the determination of Icarus' orbit and to test its consistency with the theory of general relativity. Introducing a parameter λ which would assume the value unity were general relativity correct and zero were Newtonian theory valid, we find $\lambda = 0.95 \pm 0.08$. Corresponding sensitivity studies indicate that, on the basis of Icarus data alone, the real uncertainty in λ is about 0.2. The Icarus data, as we had predicted previously, are of marginal use for the determination of the solar quadrupole moment and cannot be used to improve estimates of other astronomical constants such as the mass of Mercury.

066.017 **Some comments on 'The phenomenon of time dilation'-2.**   W. A. Elliott.
Spaceflight, Vol. 13, 86 - 87, 120 (1971).

066.018 **Interaction of gravitational radiation with an inviscid fluid in simple motion.**   F. P. Esposito.
Astrophys. Journ., Vol. 168, 495 - 502 (1971).
The equations which govern the response of an inviscid fluid to an incident gravitational wave are derived, and solved in the case of a uniform, incompressible fluid in uniform motion. The interaction is found to be purely reactive; the fluid does not extract energy from the gravitational wave.

066.019 **Interaction of vortex and potential motions in relativistic hydrodynamics. IV.**
A. D. Chernin, E. D. Eidelman.
Astrofizika, Vol. 7, 314 - 316 (1971). In Russian. – English translation in Astrophysics, Vol. 7, No. 2.
The interaction of motions in an ultrarelativistic fluid is considered on the basis of a new solution of the relativistic hydrodynamic equations.

066.020 **On the hypothesis of the diffuse background of gravitational radiation in the universe.**
N. R. Sibgatullin.
Dokl. Akad. Nauk SSSR, Ser. Mat. Fiz., Vol. 200, 308 - 310 (1971). In Russian.

066.021 **Two diffuse background radiation fields.**
P. Morrison.
Nuclei of galaxies. Conference 1970, (see 012.002), p. 699 - 706 (1971).

066.022 **Background microwave radiation at wavelengths shorter than the 2.7°K peak.**
J. E. Kapitzky, E. R. Harrison.
Bull. American Astron. Soc., Vol. 3, 391 - 392 (1971).
Abstr. AAS.

066.023 **Observable effects of primordial gravitational waves.**   M. J. Rees.
Bull. American Astron. Soc., Vol. 3, 392 (1971). – Abstr. AAS.

066.024 **Measurement of the gravitational deflection of microwave radiation near the sun.**
R. A. Sramek.
Bull. American Astron. Soc., Vol. 3, 416 (1971). – Abstr. AAS.

066.025 **Orbital periods in the restricted problem of two bodies in the general theory of relativity when the cosmological constant is taken into account.**
Z. Kh. Kurmakaev.
Astron. Zhurn. Akad. Nauk SSSR, Vol. 48, 1056 - 1060 (1971). In Russian. English translation in Soviet Astron. AJ, Vol. 15, No. 5.
Anomalistic and sidereal periods in coordinate and proper time are determined approximately. A formula for the displacement of the pericentre, when the cosmological constant is taken into account, is defined more precisely.

066.026 **Lense-Thirring type gravitational forces between disks and cylinders.**
R. F. O'Connell, S. N. Rasband.
Nature, Phys. Sci., Vol. 232, 193 - 195 (1971).
The magnitude of non-Newtonian gravitational forces are calculated within the context of linearized General Relativity and compared with each other as well as the Newtonian attraction. It is concluded that such forces are currently not measurable.

066.027 **Possible existence of particles of imaginary mass, energy and momentum.**   R. Fox.
Nature, Phys. Sci., Vol. 232, 129 - 130 (1971).
Particles with imaginary energy, momentum, and mass, called dybbuks, are investigated. Identifying the particle velocity with the ratio of momentum to energy make dybbuks timelike. As an application of the ideas presented, in which the dybbuk density determines the magnitude of coupling constants, the average gravitational constant, G, is evaluated in dense collapsing stars. It is found that G decreases as the fourth power of the radius implying that black holes can not form.

066.028 **On the adiabatic expansion of a relativistic gas.**
P. T. Landsberg, W. C. Saslaw, A. J. Haggett.
Monthly Notices Roy. Astron. Soc., Vol. 154, 7P - 8P (1971).
It is shown that contrary to some recent statements an ideal relativistic gas can expand adiabatically and quasi-statically adiabatically.

066.029 **The motion of a probe particle in the gravitational field of a flattened ellipsoid of rotation.**
V. S. Brezhnev, N. I. Maksyukov.
Vestn. Mosk. un-ta Fiz., astron., Vol. 12, 147 - 151 (1971). In Russian. – Abstr. in Referativ. Zhurn. 51. Astron., 10.51.683 (1971).

066.030 **Black holes as agents of magnetic fields.**
R. M. Wald.
Nature, Vol. 233, 52 - 53 (1971).
There is considerable theoretical evidence for believing that all black holes are of the Kerr-Newman type. The electromagnetic field (and all other properties) of such black holes are uniquely fixed by their mass, angular momentum,

and charge. We obtain here a simple upper limit on the amount of charge a black hole could reasonably be expected to have and show that this implies a very small magnetic field.

066.031 Interaction of gravitational and electromagnetic fields or another effect? C. Ferencz, G. Tarcsai.
Nature, Vol. 233, 404 - 406 (1971).

It has been demonstrated that an electromagnetic wave propagating in moving inhomogeneous media is undergoing a frequency variation. This theoretically derived effect can be demonstrated in independent measurements: in known terrestrial troposcatter measurements; in the frequency measurements of radio sources when occultated by the sun; in the extreme red-shift values observed at the solar limb. In this manner probably other anomalous frequency shifts will be explained also.

066.032 Sources and detection of gravitational radiation of high frequency. L. Halpern.
Nature, Phys. Sci., Vol. 233, 18 (1971).

The surprisingly large flux of cosmic gravitational radiation found by Weber in the kilocycle range and the theoretical development on black holes suggests the search for comparable intensities at higher frequencies. A method is suggested by which such a radiation can be detected.

066.033 Old and new approaches in the study of gravitational absorption. F. Bocchio.
Geophys. Journ. Roy. Astron. Soc. Vol. 24, 101 - 102 (1971).

The purpose of the letter is to summarize an attempt to evaluate the absorptional effect assuming the moon as the source, the earth as the screen and an artificial satellite as the probe. It appears that the computed perturbations are very small and beyond the possibility of detection.

066.034 Dilaton and possible non-Newtonian gravity. Y. Fujii.
Nature, Phys. Sci., Vol. 234, 5 - 7 (1971).

A model is proposed which allows a dilaton to show up in a possible non-Newtonian part of the gravitational force. By examining the available observational facts it can be shown that the force-range of the additional force, if it exists, will be either between 10 m and 1 km or smaller than ~ 1 cm.

066.035 Universal gravitation. A. F. Bogorodsky.
Vestn. Kiev. Un-ta, Ser. Astron., No. 13, p. 3 - 15 (1971). In Russian.

The progress of ideas on the physical nature of gravitation from the classical Newtonian law of universal attraction to contemporary Einstein's relativity theory is briefly considered. Some remarks on recent difficulties of gravitation theory are pointed out.

066.036 On a homogeneous gravitational field in relativity. A. F. Bogorodsky.
Vestn. Kiev. Un-ta, Ser. Astron., No. 13, p. 16 - 21 (1971). In Russian.

A solution of Einstein's field equations is obtained. The solution can be regarded as a simple relativistic expression of a homogeneous gravitational field.

066.037 Some solutions of Einstein's field equations in semi-reducible spaces. F. E. Khlistun.
Vestn. Kiev. Un-ta, Ser. Astron., No. 13, p. 22 - 30 (1971). In Russian.

Metrics of the type $ds^2 = e_1 \omega^2 dx_1^2 + e_2 \psi^2 dx_2^2 + \varphi^2 [e_3 dx_3^2 + e_4 U^2 dx_4^2]$ are considered. Some solutions of the field equations are obtained.

066.038 Gravitationswellen. P. Jakober.

Orion, 29. Jahrgang, p. 135 - 138 (1971).

066.039 La précession d'un gyroscope en relativité générale. J. Madore.
Comptes Rendus Acad. Sci. Paris, Sér. A, Vol. 273, 782 - 784 (1971).

Une dérivation simplifiée est donnée de la formule de Schiff de précession d'un gyroscope en relativité générale.

066.040 General relativity and gravitational collapse. R. U. Sexl.
Acta Phys. Austriaca Suppl., Vol. 7, 308 - 354 (1970). – Review article on the theory of general relativity.

066.041 On the detection of black holes. C. Leibovitz, D. P. Hube.
Astron. Astrophys., Vol. 15, 251 - 255 (1971).

It is shown that in principle the gravitational lens effect may lead to significant light variations when a collapsed object such as a black hole passes between the observer and a normal star. Light curves characteristic of such an event are computed, and the possibility of observing such an event is discussed.

066.042 Checking the equivalence principle. V. B. Braginsky, V. I. Panov.
Priroda, No. 11.71, p. 43 - 49 (1971). In Russian.

066.043 Theoretical frameworks for testing relativistic gravity. III. Conservation laws, Lorentz invariance, and values of the PPN parameters. C. M. Will.
Astrophys. Journ., Vol. 169, 125 - 140 (1971).

The Parametrized Post-Newtonian (PPN) formalism is used to prove that metric theories which have post-Newtonian integral conservation laws for energy, momentum, angular momentum, and center-of-mass motion must be of a particular form, i. e., their PPN parameter values must obey certain constraints. It is also shown that the post-Newtonian metric of any theory of gravity is invariant under a post-Galilean transformation if and only if certain four PPN parameters satisfy a set of constraints. For theories which do possess conservation laws, the transformation properties of the conserved integral quantities under post-Galilean transformations are determined.

066.044 Relativistic gravity in the solar system. II. Anisotropy in the Newtonian gravitational constant. C. M. Will.
Astrophys. Journ., Vol. 169, 141 - 155 (1971).

The Parametrized Post-Newtonian formalism is used to show that some theories of gravity predict an anisotropy in the Newtonian gravitational constant G, as measured locally by means of Cavendish experiments. Two such theories are Whitehead's theory and a theory of gravity (devised in this paper) which predicts different flat-space propagation speeds for gravity and for light.

066.045 Measurement of the far-infrared background radiation in the night sky. A. Blair, J. G. Beery, F. Edeskuty, R. D. Hiebert, J. P. Shipley, K. D. Williamson, Jr.
Phys. Rev. Letters, Vol. 27, 1154 - 1157 (1971).

A rocket-borne radiometer measurement of background radiation in the spectral range from 6 to 0.08 mm has yielded a flux which corresponds to an equivalent blackbody temperature of $3.1 ^{+0.5}_{-2.0}$ K.

066.046 New theory of gravitation. H. Yilmaz.
Phys. Rev. Letters, Vol. 27, 1399 - 1402 (1971).

A locally Lorentz-invariant curved-space theory of gravitation where the local field is a massless, spin-2 field φ^ν_μ of Pauli-Fierz type is presented. The static central body problem reduces exactly to author's 1958 theory in a special case so that as in that theory the three crucial tests are satisfied.

066.047 Gravitational radiation from a particle falling radially into a Schwarzschild black hole.
M. Davis, R. Ruffini, W. H. Press, R. H. Price.
Phys. Rev. Letters, Vol. 27, 1466 - 1469 (1971).
We have computed the spectrum and energy of gravitational radiation from a "point test particle" of mass m falling radially into a Schwarzschild black hole of mass $M \gg m$. The total energy radiated is about $0.0104 mc^2 (m/M)$, 4 to 6 times larger than previous estimates.

066.048 Massenverlust durch Gravitationsstrahlung?
D. W. Sciama.
Umschau, 71. Jahrgang, p. 944 - 945 (1971).

066.049 A tensor-tensor theory of gravitation. C. Firmani.
Astrophys. Space Sci., Vol. 13, 128 - 136 (1971).
One approach to a tensor-tensor theory of gravitation is proposed as an attempt to represent Mach's principle in a consistent way. In this theory the geometrodynamic properties of general relativity are still valid, but not necessarily its field equations. Einstein's equations may be obtained as a special case.

066.050 Guided gravitational waves.
W. B. Campbell, T. A. Morgan.
Nature, Phys. Sci., Vol. 234, 143 - 145 (1971).
By analogy with certain solutions of Maxwell's equations it can be shown that guided gravitational waves are a possibility.

066.051 Rotating relativistic ring. W. H. McCrea.
Nature, Vol. 234, 399 - 401 (1971).
Consideration of a rotating ring sheds light upon the problem of the rotating disk recently discussed in *Nature* by several authors (Atwater, Suzuki, Marsh, Noonen, Weinstein) and shows the disk problem to be more difficult than it appears.

066.052 Lichtablenkung im Gravitationsfeld der Sonne und Laufzeitverzögerung von Radarsignalen.
H. Dehnen, H. Hönl.
Naturwissenschaften, 58. Jahrgang, p. 619 (1971).

066.053 Radiation from particles falling into black-holes.
D. K. Ross.
Publ. Astron. Soc. Pacific, Vol. 83, 633 - 637 (1971).
The electromagnetic radiation emitted when a charged particle falls into a neutral Schwarzschild black-hole is calculated. The results for the particle falling with zero angular momentum and for the particle orbiting the black-hole are calculated in turn and compared with the corresponding results for gravitational radiation. For the orbit case we have calculated the time required for both a charged and an uncharged particle to spiral into the black-hole. We find that the gravitational radiation case is probably not of astrophysical importance.

066.054 Gravitational radiation. B. Kuchowicz.
Urania Kraków, Vol. 42, 208 - 211, 240 - 244 (1971). In Polish.

066.055 Tetrads, anholonomic coordinates, and space-time geometry. K. M. Gatha, R. C. Dutt.
Australian Journ. Phys., Vol. 24, 631 - 652 (1971).
Taking the tetrad vectors as the fundamental gravitational variables in Riemannian space-time and using anholonomic Minkowskian coordinates, a more fundamental tetrad-dependent tensor $\Lambda_{(\mu\nu)\omega}$ of gravitation is obtained instead of the metric-dependent curvature tensor of general relativity. $\Lambda_{(\mu\nu)\omega}$ gives a local dynamical structure to gravitation and plays an important part in cosmology.

066.056 A search for spectral features in the submillimeter background radiation.
J. C. Mather, M. W. Werner, P. L. Richards.
Astrophys. Journ., (*Letters*), Vol. 170, L59 - L65 (1971).
We have made mountaintop observations at 1 percent spectral resolution of atmospheric and sky emission in the frequency region $\nu = 6 - 14$ cm^{-1}. No emission features were found which can be related to the diffuse, isotropic flux $F_0 = 1.3 \times 10^{-9}$ W cm^{-2} sterad^{-1} reported from rocket and balloon experiments. We can thus set a lower limit to the spectral width of any feature responsible for F_0. Since we exclude the possibility that the radiation lies in a single narrow line such as might arise from a maser process in the upper atmosphere, our observations seem to require an extraterrestrial origin for F_0.

066.057 The gravitational field of a bounded source in general relativity. S. Persides.
Astrophys. Journ., Vol. 170, 479 - 498 (1971).
In this paper the gravitational field of a bounded and isolated material source is expressed explicitly in terms of the density, pressure, and other characteristics of the source in a scheme of successive approximations. Particular emphasis is given to the radiation zone and to the relation of the gravitational waves to the source. In the first part, the approximation method is established for a general source. In the second part, the method is applied for a bounded source of perfect fluid. Finally, the field in the radiation zone is related to the source by calculating explicitly the first nonzero term of the news function and the rate at which energy is radiated in the form of gravitational waves.

066.058 The stability of precessing elliptical orbits in a Schwarzschild field. R. O. Hansen.
Astrophys. Journ., Vol. 170, 557 - 558 (1971).
It is shown that orbits in a Schwarzschild metric of central mass M for which $l \leq (3 + e)2GM/c^2$, where e is the eccentricity and l is the semi-latus rectum of the elliptical orbit in a suitably defined rotating frame, are unstable.

066.059 Long wave trains of gravitational waves from a vibrating black hole. W. H. Press.
Astrophys. Journ., (*Letters*), Vol. 170, L105 - L108 (1971).
The vibrations of a black hole of mass M, perturbed from spherical symmetry, have been studied numerically. Initial perturbations of high spherical-harmonic index ($l \gg 1$) which contain Fourier components of long wavelength ($2\pi M \gtrsim \lambda \gg 2\pi M/l$) produce long-lasting vibrations. The vibrational energy is radiated away gradually in a long, nearly sinusoidal wave train of gravitational radiation with angular frequency $\omega \approx (27)^{-1/2} l/M$.

066.060 Introducing the black hole.
R. Ruffini, J. A. Wheeler.
Physics Today, Vol. 24, No. 1, p. 30 - 36, 39, 41 (1971).
According to present cosmology, certain stars end their careers in a total gravitational collapse that transcends the ordinary laws of physics.

066.061 Thermodynamic instability of a system of gravitating fermions. P. Hertel, W. Thirring.
Quanten und Felder, (Physikalische und philosophische Betrachtungen zum 70. Geburtstag von Werner Heisenberg, H. P. Dürr (Editor). [Friedrich Vieweg & Sohn, Braunschweig]), p. 309 - 324 (1971).

066.062 Zur Interpretation der Trederschen Tetradentheorie

der Gravitation. U. Kasper.
Monatsber. Deutsch. Akad. Wiss. Berlin, Vol. 13, 177 - 184 (1971).

066.063 **Zur Struktur der Feldgleichungen des Gravitationsfeldes in der Trederschen Tetradentheorie.**
U. Kasper.
Monatsber. Deutsch. Akad. Wiss. Berlin, Vol. 13, 184 - 188 (1971).

066.064 **Erhaltungssätze und Identitäten in der Tetraden-Theorie.** E. Kreisel.
Monatsber. Deutsch. Akad. Wiss. Berlin, Vol. 13, 302 - 310 (1971).

066.065 **Kreuzexperimente in der allgemeinen Relativitätstheorie und die experimentelle Bestimmung des Riemannschen Krümmungstensors.** H.-J. Treder.
Monatsber. Deutsch. Akad. Wiss. Berlin, Vol. 13, 310 - 317 (1971).

066.066 **A resolution of the clock paradox.** M. Sachs.
Physics Today, Vol. 24, No. 9, p. 23 - 29 (1971).

066.067 **Hair tonic for black holes.**
F. C. Michel.
Comments Astrophys. Space Phys., Vol. 3, 163 - 167 (1971).
    Some comments to a number of recently published articles describing the properties of black holes are presented.

066.068 **Particle creation by gravitational fields in collapse and singularity.** Ya. B. Zeldovich.
Comments Astrophys. Space Phys., Vol. 3, 179 - 184 (1971).
    A discussion on the theory of particle creation in strong and rapidly changing gravitational fields is presented. Some applications of this process are discussed.

066.069 **The cosmic background radiation: Some recent developments.** M. J. Rees.
Highlights of Astronomy, Vol. 2, (see 012.019), 757 - 767 (1971).

066.070 **Field approach to gravitation and its significance in astrophysics.** V. Majerník.
Astrophys. Space Sci., Vol. 14, 265 - 285 (1971).
    A field modification of classical gravitational theory which is analogous to the classical electrodynamics is proposed. Within its framework it is possible to account for some types of behaviour of matter occurring under certain extreme physical conditions. Especially, the energy release in quasars and pulsars may be calculated, under some plausible physical assumptions, to obtain values comparable with the observable ones. Several astrophysical effects (e.g. the occurrence of nonthermal radiation in pulsars and quasars, etc.) find reasonable explanations within this field approach to gravitation.

066.071 **Exact cosmological solutions in Brans and Dicke's scalar-tensor theory, I.** H. Dehnen, O. Obregón.
Astrophys. Space Sci., Vol. 14, 454 - 459 (1971).
    The exact solution is sought for the cosmological equations of Brans and Dicke's scalar-tensor theory when a power law exists between the gravitational constant and the radius of curvature of the universe.

066.072 **Some properties of the general solution of the gravitational equations for a dust-like medium.**
L. P. Grishchuk, V. I. Ulin.
Soobshch. Gos. Astron. Inst. Shternberga, No. 173, p. 43 - 53 (1971). In Russian.
    Analytical evidence is given of the statement which was made in previous works by one of the authors (Grishchuk,

1966, 1967).

066.073 **On the increase in entropy in an expanding universe.**
K. Sakai.
Progr. Theor. Phys., Japan, Vol. 46, 1292 - 1293 (1971).
Letter.

066.074 **Quantum theory of gravitation vs. classical theory.
– Fourth-order potential –.** Y. Iwasaki.
Progr. Theor. Phys., Japan, Vol. 46, 1587 - 1609 (1971).
    The perihelion-motion of Mercury depends on the fourth-order potential in quantum field theory; it is a "Lamb shift". In spite of the unrenormalizability of the theory, we have extracted a finite and physically meaningful quantity, a fourth-order potential, from fourth-order graphs. We have also discussed briefly renormalization of the Newtonian potential in the fourth-order perturbation. The Hamiltonian obtained is the same as the classical one and so it cannot explain the Dicke-Goldenberg experiment. We have calculated fourth-order potential also in Q.E.D.

066.075 **Quantum theory of gravity and the perihelion motion of Mercury.** K. Hiida, M. Kikugawa.
Progr. Theor. Phys., Japan, Vol. 46, 1610 - 1622 (1971).
    In the quantum theory of gravity the potential between two celestial bodies is calculated up to order $(v/c)^2$ by treating the celestial bodies as assemblies of nucleons. It is necessary to calculate three-body potentials in order to get the potential proportional to G^2, G being the gravitational constant. A method is discussed for determining the retarded potential uniquely. The potential obtained coincides with the classical one given by Einstein, Infeld and Hoffman. The perihelion motion of Mercury is also discussed.

066.076 **Some properties of a collapsing fluid sphere.**
A. Banerjee.
Progr. Theor. Phys., Japan, Vol. 46, 1625 - 1627 (1971).
Letter.

066.077 **General relativity and kinetic theory.** J. Ehlers.
General relativity and cosmology. Course 47, Italian Phys. Soc., 1969, (see 012.022), p. 1 - 70 (1971).

066.078 **Relativistic stars, black holes and gravitational waves (including an in-depth review of the theory of rotating, relativistic stars).** K. S. Thorne.
General relativity and cosmology. Course 47, Italian Phys. Soc 1969,(see 012.022), p. 237 - 283 (1971).

066.079 **Detection of gravitational waves.** B. Bertotti.
General relativity and cosmology. Course 47, Italian Phys. Soc., 1969, (see 012.022), p. 347 - 355 (1971).

066.080 **Cosmological density fluctuations during hadron stage.** W. Kundt.
General relativity and cosmology. Course 47, Italian Phys. Soc., 1969,(see 012.022), p. 365 - 372 (1971).

066.081 **Creation of particles by gravitational fields.**
H. Urbantke.
General relativity and cosmology. Course 47, Italian Phys. Soc., 1969, (see 012.022), p. 383 - 387 (1971).

066.082 **A new solution of Einstein's equations with cosmological term.** G. E. Gorelik.
Vestn. Mosk. un-ta. Fiz. astron., Vol. 12, 477 - 479 (1971). In Russian. – Abstr. in Referativ. Zhurn. 51. Astron., 1.51.843 (1972).

066.083 **On siderial periods in the gravitational field which is characterized by Schwarzschild's interior solu-**

tion.   Z. Kh. Kurmakaev.
Trudy Astrofiz. Inst., *Alma-Ata*, Vol. 16, 116 - 118 (1971).
In Russian.

066.084   Motions of fast particles in a spherically symmetric gravitational field.   Sh. N. Sabitov.
Trudy Astrofiz. Inst., *Alma-Ata*, Vol. 16, 123 - 127 (1971).
In Russian.

Motions of fast particles in the gravitation field of the sun without taking into account electromagnetic and other forces and rotation of the sun are discussed. The solutions of the equations in first and second approximation and the relativistic energy integral are given.

066.085   On the impossibility of changing the gravitational constant.   V. A. Semenenya.
Trudy Astrofiz. Inst., *Alma-Ata*, Vol. 17, 86 - 90 (1971). In Russian.

This paper contains arguments which show the impossibility of changing the gravitational constant in reciprocal proportion to time. In particular, the reduction of the gravitational constant is at variance with the age of the stars, with the age of the solar system, with the problem of the origin and development of life on the earth and with the three-dimensional character of the real physical space.

066.086   On proper rotation in Einstein's gravitational theory.   M. Abdildin, A. Junusov, O. Sakibayev.
Trudy Astrofiz. Inst., *Alma-Ata*, Vol. 17, 91 - 93 (1971). In Russian.

It is shown that in the general relativity theory the rotation of one body causes the rotation of another one. The equations of the rotational motion of a small object moving on an elliptical orbit around a massive rotating body are integrated.

066.087   Quantum theory of gravitation and the mass of the electron.   G. Rosen.
Phys. Rev. D, Particles and Fields, Third Ser., Vol. 4, 275 - 277 (1971).

Gravity-modified quantum electrodynamics formulated by Salam and collaborators is applied to the electron self-energy problem in all orders of perturbation theory.

066.088   Effect of gravitational light deflection on the proposed gyroscope test of the Lense-Thirring effect.
R. F. O'Connell, G. L. Surmelian.
Phys. Rev. D, Particles and Fields, Third Ser., Vol. 4, 286 - 288 (1971).

066.089   Motion of particles in Einstein's relativistic field theory. I. Introduction and general theory.
C. R. Johnson.
Phys. Rev. D, Particles and Fields, Third Ser., Vol. 4, 295 - 317 (1971).

066.090   Motion of particles in Einstein's relativistic field theory. II. Application of general theory.
C. R. Johnson.
Phys. Rev. D, Particles and Fields, Third Ser., Vol. 4, 318 - 339 (1971).

066.091   Aging of an electromagnetic wave group.
R. S. Hornbostel, C. J. Marcinkowski.
Phys. Rev. D, Particles and Fields, Third Ser., Vol. 4, 931 - 946 (1971).

066.092   Hydrostatic equilibrium and gravitational collapse of relativistic charged fluid balls.
J. D. Bekenstein.

Phys. Rev. D, Particles and Fields, Third Ser., Vol. 4, 2185 - 2190 (1971).

066.093   Theory of the detection of short bursts of gravitational radiation.   G. W. Gibbons, S. W. Hawking.
Phys. Rev. D, Particles and Fields, Third Ser., Vol. 4, 2191 - 2197 (1971).

It is argued that the short bursts of gravitational radiation which Weber reports most probably arise from the gravitational collapse of a body of stellar mass or the capture of one collapsed object by another. We shall analyze the response of a gravitational-wave detector to such a burst.

066.094   Theory of the angular dependence of a gravitational radiation detector.   A. Mehra.
Phys. Rev. D, Particles and Fields, Third Ser., Vol. 4, 2566 - 2569 (1971).

The directivity pattern for a gravitational radiation detector is obtained. A system of three orthogonal detectors is considered and relationships between their responses and the direction of the source are studied. The results are used for analyzing the variation in detector response with the earth's rotation.

066.095   A theory of gravity in the framework of the Lorentz covariant and second quantized formalism.
S. Sato.
Progr. Theor. Phys., Japan, Vol. 46, 282 - 296 (1971).

On the basis of the remark that the gravitational interaction really acts with the binding energy, the emphasis is laid on the necessity of treating the gravitational interaction by the method of particle physics irrespective of the general relativity. Accordingly, the interaction of a massless particle with spin 2 (graviton) is investigated by using the method which has been applied to the reformulation of quantum electrodynamics by the present author. The rotation of the perihelion is calculated by taking into account higher order corrections to the gravitational interaction, and a value which is about 8% smaller than the result of general relativity is obtained. This is exactly the value expected by Dicke and Goldenberg from their measurement of the solar oblateness. As regards the behaviour of the photon in the gravitational field, our theory reveals some features different from general relativity.

066.096   A role of the uncertainty principle in general relativity and the limiting size of collapsing Fermion spheres.   N. Hokkyo.
Progr. Theor. Phys., Japan, Vol. 46, 984 - 989 (1971).

An attempt is made to modify the Schwarzschild metric by the uncertainty principle in space regions of the linear size of the order of the Planck length, and the role of the modified metric in avoiding the unlimited gravitational collapse of superdense Fermion spheres is examined.

066.097   Electrodynamics of direct interparticle action. II. Relativistic treatment of radiative processes.
F. Hoyle, J. V. Narlikar.
Ann. Physics, Vol. 62, 44 - 97 (1971).

This paper is a sequel to an earlier paper that described nonrelativistic quantum electrodynamics in terms of the time symmetric theory of direct interparticle action. The restriction to a nonrelativistic treatment is removed in the present paper. The path integral approach to quantum mechanics is extended to include relativistic particles with spin as a preliminary to achieving this end.

066.098   On classical scalar field theories and the relativistic Kepler problem.   C. M. Andersen, H. C. von Baeyer.
Ann. Physics, Vol. 62, 120 - 134 (1971).

Two versions of classical relativistic field theory corresponding to massless scalar exchange are compared. It is found that they lead to identical field equations but different equations of motion. The motion of one particle bound by the field of another, infinitely massive particle, i.e., the Kepler problem, is examined. In one case the path is an ellipse precessing through an angle equal to minus one-sixth the value predicted by general relativity. In the other case the path is an ellipse which does not precess.

066.099 Spin-two massless-particle radiation as a gravitational effect.   R. H. Good, Jr.
Ann. Physics, Vol. 62, 590 - 601 (1971).
The theory for a spin-two massless-particle field interacting with sources is developed, in parallel with Maxwell's theory for photons interacting with charges.

066.100 The complete Schwarzschild solution.
N. Rosen.
Ann. Physics, Vol. 63, 127 - 133 (1971).
Some remarks are made about the complete Schwarzschild solution, consisting of the interior and exterior solutions, joined together.

066.101 Linearized gravitation theory in macroscopic media.
P. Szekeres.
Ann. Physics, Vol. 64, 599 - 630 (1971).
The refraction of gravitational waves is discussed by developing a macroscopic theory of gravitation along the lines of classical electromagnetism. A model of a medium whose molecules are harmonic oscillators is discussed and constitutive equations are derived. Gravitational waves are demonstrated to slow down in such a medium.

066.102 A conjecture regarding quantum fluctuation of gravitation and elementary particles as excitons in a turbulent gravitational field.   K. E. Woehler.
Ann. Physics, Vol. 64, 631 - 646 (1971).
Based on Wheeler's conjecture that the quantum fluctuations of the metric create a multiple connected foam-like structure of the vacuum with a structure constant of $\approx 10^{-33}$ cm and large virtual energy densities $\approx 10^{115}$ erg/cm^3 and that elementary particles are exciton-like weak coherent perturbations in the violent vacuum physics, a model theory is constructed in which real turbulent fluctuations are superimposed on the average metric with the fluctuating metric satisfying the free space Einstein equations.

066.103 Inertial and gravitational mass in the Brans-Dicke theory.   H. C. Ohanian.
Ann. Physics, Vol. 67, 648 - 661 (1971).
We investigate the equality of inertial and gravitational mass in the Brans-Dicke theory of gravitation in both the classical and quantum case. We derive a general expression for the ratio of inertial to gravitational mass for any classical static or quasistatic system.

066.104 Gravitational-scalar field coupling II.
J. H. Higbie.
Ann. Physics, Vol. 68, 521 - 540 (1971).
Particle-like solutions are sought for the system: gravitational field-electromagnetic field-massless scalar field-explicit source for all three fields. The scalar field is taken to be (1) the ordinary minimally coupled field, (2) the (nonminimally coupled) conformally invariant field, or (3) the Brans-Dicke field.

066.105 Comment on the spin precession of the Schiff satellite in the Brans-Dicke theory.
R. E. Morganstern.
Phys. Rev. D, Particles and Fields, Third Ser., Vol. 3, 616 - 617 (1971).

066.106 Near-field approximation for strong gravitational fields.   T. A. Morgan.
Phys. Rev. D, Particles and Fields, Third Ser., Vol. 3, 800 - 810 (1971).
In the first part of this paper we generalize the work of Morgan and Bondi to quasistatic systems without axial symmetry. The major result is the derivation of a near-field expansion for nonrotating, axially symmetric systems, to arbitrary order in L/λ, where L characterizes the size of the system and λ is a characteristic wavelength.

066.107 Relativistic hydrodynamics in one dimension.
M. H. Johnson, C. F. McKee.
Phys. Rev. D, Particles and Fields, Third Ser., Vol. 3, 858 - 863 (1971).
Hydrodynamic equations for one-dimensional motion, of interest in supernova explosions, are integrated in the relativistic limit. A simple solution is found for free expansion into a vacuum. The propagation of a shock into a medium of decreasing density is determined, and the solution for the subsequent flow behind the shock is also obtained.

066.108 Relativistic Kepler problem.   C. Fronsdal.
Phys. Rev. D, Particles and Fields, Third Ser., Vol. 3, 1299 - 1302 (1971).
Relativistic quantum field theory is used as a starting point to construct a classical, completely relativistic theory of planetary orbits.

066.109 Equivalence principle for massive bodies. IV. Planetary bodies and modified Eötvös-type experiments.
K. Nordtvedt.
Phys. Rev. D, Particles and Fields, Third Ser., Vol. 3, 1683 - 1689 (1971).
Employing a model of massive bodies as made up of an equilibrium assembly of particles interacting with each other via both gravitational and nongravitational forces, the gravitational-to-inertial-mass ratio (M_g/M_i) is calculated for the massive body.

066.110 Scale invariance of the second kind and the Brans-Dicke scalar-tensor theory.   J. L. Anderson.
Phys. Rev. D, Particles and Fields, Third Ser., Vol. 3, 1689 - 1691 (1971).

066.111 Electromagnetic field and wave propagation in gravitation.   Tse Chin Mo, C. H. Papas.
Phys. Rev. D, Particles and Fields, Third Ser., Vol. 3, 1708 - 1712 (1971).
From the physical three-vector Maxwell equations for an electromagnetic (E.M.) field in static gravitation, we examine the artifice of replacing the gravitation by an equivalent medium and we find modified Debye potentials for an E.M. wave in a simple, angularly homogeneous, material medium in a Schwarzschild gravitational field.

066.112 Theory of gravitation.   N. Rosen.
Phys. Rev. D, Particles and Fields, Third Ser., Vol. 3, 2317 - 2319 (1971).
The question arises whether one can set up a theory of gravitation which, like general relativity, is based on the equivalence principle, but which however does not accept the principle of covariance. This theory leads to the same results as general relativity in the three crucial tests. The formalism can be modified to take into account the solar oblateness observed by Dicke and Goldenberg.

066.113 Scattering of electromagnetic and gravitational waves by a static gravitational field: Comparison between the classical (general-relativistic) and quantum field-theoretic results.   P. J. Westervelt.

Phys. Rev. D, Particles and Fields, Third Ser., Vol. 3, 2319 - 2324 (1971).

The classical general-relativistic cross sections for the scattering of either an electromagnetic wave or a gravitational wave by a scalar particle are calculated and found to agree with the results of the quantized linearized field theory.

066.114 Brans-Dicke theory under a transformation of units and the three tests. R. E. Morganstern.
Phys. Rev. D, Particles and Fields, Third Ser., Vol. 3, 2946 - 2950 (1971).

066.115 Focusing of gravitational radiation by the galactic core. J. K. Lawrence.
Phys. Rev. D, Particles and Fields, Third Ser., Vol. 3, 3239 - 3240 (1971).

The possibility is considered that the gravitational radiation observed by Weber has an extragalactic origin and is focused by the galactic core acting as a gravitational lens. While sufficient intensification is possible, too few sources are correctly located for the effect to be important.

066.116 Rotverschiebungsverluste und Entwicklungseffekte bei kosmischer Hintergrundsstrahlung metagalaktischer Herkunft. G. Dautcourt.
Monatsber. Deutsche Akad. Wiss. Berlin, Vol. 13, 417 - 421 (1971).

The intensity of metagalactic background radiation arising from physical processes in intergalactic matter or resulting from the superimposed radiation of discrete sources depends upon the "strengths" of evolution effects: In the case of "power law radiation" above a critical strength which depends upon the spectral index there is a strong dependence on the cut-off redshift, where the radiation production commences.

066.117 Finite rest masses of wave quanta in material media. K. D. Cole.
Australian Journ. Phys., Vol. 24, 871 - 880 (1971).

The equivalence of a dispersion relationship and Einstein's mass-energy relationship leads to the specification of a particle in a vacuum which is equivalent to a "photon" in a medium. Applying dynamical equations to the equivalent particle in the case of a radiofrequency photon in a plasma around a star, a new gravitational redshift formula is deduced which reduces to the well-known expression in the appropriate limit. A new form of bending of photon trajectories in a gravitational field is also described.

066.118 On the connection between gravitational and electromagnetic fields. P. Penchev.
Godishn. Vissh. tekhn. uchebni zaved. Fiz., Vol. 4, sb. 1, p. 105 - 116 (1967, 1970). In Bulgarian. – Abstr. in Referativ. Zhurn. 51. Astron., 2.51.756 (1972).

066.119 On the absorption of gravitation. A. Karastoyanov.
Godishn. Vissh. tekhn. uchebni zaved. Fiz., Vol. 5, sb. 2, p. 19 - 23 (1968, 1970). In Bulgarian. – Abstr. in Referativ. Zhurn. 51. Astron., 2.51.757 (1972).

066.120 Investigation of the possibility to measure horizontal gravitational forces with the help of torsional weights. A. Karastoyanov.
Godishn. Vissh. tekhn. uchebni zaved. Fiz., Vol. 4, sb. 1, p. 29 - 32 (1967, 1970). In Bulgarian. – Abstr. in Referativ. Zhurn. 51. Astron., 2.51.769 (1972).

066.121 An approximate solution of the vacuum static case of spherical symmetry in Brans-Dicke theory. M. N. Mahanta, D. R. K. Reddy.
Journ. Math. Phys., Vol. 12, 929 - 932 (1971).

Starting from the usual variational principle, the gravitational field equations for the vacuum static case of spherical symmetry are obtained in Brans-Dicke's scalar-tensor theory. An approximate solution to the second order is presented. It is observed that the results of gravitational redshift, deflection of light, and the rotation of the perihelion of Mercury are in agreement with the earlier results obtained by Brans, Dicke, and Heckmann. But the method, being simpler, can be used in solving other problems in the theory.

066.122 On light tracks in the presence of a massive body. R. Burman.
Current Sci. (*India*), Vol. 40, No. 3, p. 61 - 62 (1971).

The null geodesics of the space-time metric external to a static spherically symmetric body are treated for the case of a non-zero cosmological constant.

066.123 Cosmological constant and fundamental length. J. L. Anderson, D. Finkelstein.
American Journ. Phys., Vol. 39, 901 - 904 (1971). – See Phys. Abstr., Vol. 74, No. 54181 (1971).

066.124 Particle gravitation theories of the Hoyle-Narlikar type. C. B. G. McIntosh.
Journ. Phys. A. General Phys., Vol. 4, 491 - 500 (1971).

Two possible generalizations of the action of the Hoyle-Narlikar theory are suggested and the two resultant particle theories developed.

066.125 Gravitational radiation experiments. J. Weber.
35. Phys. Meeting, Hannover (*Germany*), 1970 [B. G. Teubner, Stuttgart], p. 191 - 222 (1970).

An account is given of recent experimental research on gravitational radiation together with some discussion of the instrumentation.

066.126 A new experiment in general relativity. F. Occhionero.
Nuovo Cimento Lettere, Ser. 2, Vol. 2, 155 - 158 (1971).

The author suggests an artificial gravitational experiment based on the effect of linear displacement of the centre of mass of the gyro away from the geodesic followed by the whole satellite.

066.127 Sensitivity of Weber's antenna for the short pulses of gravitational radiation. V. N. Rudenko.
Phys. Letters A, Vol. 35A, 409 - 410 (1971).

Sensitivity of a gravitational mass-quadrupole detector is calculated if it registers according to the rate of change in the output signal. A limit is given for the connection coefficient of a piezo-transducer with detector.

066.128 Disc-cylinder Argonne-Maryland gravitational radiation experiments. J. Weber.
Nuovo Cimento B, Ser. 11, Vol. 4B, 197 - 204 (1971).

066.129 Cosmological solutions to a linear theory of gravitation. H. O. Girotti, D. Wisnivesky.
Nuovo Cimento B, Ser. 11, Vol. 4B, 205 - 216 (1971).

Two different cosmological solutions for a homogeneous isotropic universe are analysed using the linear theory of gravitation presented earlier. It is shown that the solutions are similar to the ones obtained from Einstein equations and that they agree with the observed properties of spacetime at large.

066.130 A simple relativistic theory of gravitation. C. J. Coleman.
Journ. Phys. A. General Phys., Vol. 4, 611 - 616 (1971).

It is shown that a particular theory with a Lorentz-invariant scalar potential ϕ and a simple gravitational metric involving ϕ leads to the same results for the three Einstein tests as general relativity.

066.131 Stellar aberration and apparent rotation: a direct link. D. C. Ferguson.
American Journ. Phys., Vol. 39, 1089 - 1090 (1971).

066.132 Gravitational red-shift in the theory of the generalized gravitational potential. N.-H. Cherry.
Nuovo Cimento Lettere, Ser. 2, Vol. 2, 619 - 620 (1971).

066.133 The conversion of gravitational into electromagnetic waves. D. Boccaletti, F. Occhionero.
Nuovo Cimento Lettere, Ser. 2, Vol. 2, 549 - 554 (1971).
The authors study the conversion of gravitational into electromagnetic waves in a static magnetic field in a vacuum.

066.134 Observable effects of torsion in space-time.
V. N. Ponomariev.
Bull. Acad. Polonaise Sci., Ser. Sci. Math., Astron., Phys., Vol. 19, 545 - 550 (1971). See Phys. Abstr., Vol. 74, No. 74841 (1971).

066.135 Radar evidence that the velocity of light in space is not c. B. G. Wallace.
Spectroscopy Letters, (USA), Vol. 4, No. 3–4, p. 79 - 84 (1971). – See Phys. Abstr., Vol. 74, No. 74845 (1971).

066.136 Gravitational wave astronomy: An interim survey.
P. B. Fellgett, D. W. Sciama.
Radio and Electronic Engineer, (GB), Vol. 42, 391 - 397 (1971). – See Phys. Abstr., Vol. 74, No. 77596 (1971).

066.137 On crucial tests of general relativity.
Tan Tung Arjun.
Chin. Journ. Phys., (Taiwan), Vol. 9, No. 1, p. 29 - 30 (1971). See Phys. Abstr.,Vol. 74, No. 81262 (1971).

066.138 Weber's antenna and the radiation sources.
J. Madore, A. Papapetrou.
Ann. Inst. Henri Poincaré, Ser. A, Vol. 14, 139 - 151 (1971).
Contains a discussion of some questions concerning the relation of events detected by Weber's antenna to the sources of the gravitational radiation.

066.139 General relativistic fluid spheres. IV. Differential equations for non-charged spheres of perfect fluid.
B. Kuchowicz.
Acta Phys. Polonica B, Vol. B 2, 657 - 667 (1971).
Assumptions underlying the search for new exact solutions of Einstein's field equations for space filled with matter are examined. Three assumptions are retained: spherical symmetry, macroscopic neutrality of matter, and the energy-momentum tensor of a perfect fluid. Schwarzschild canonical coordinates have been used in the previous investigations of this series, now other coordinate systems are introduced, and differential equations relating the metric tensor components and their derivatives are given.

066.140 General relativity in the large. R. Geroch.
General Relativity and Gravitation, Vol. 2, 61 - 74 (1971).

066.141 Energy release from colliding black holes.
D. K. Ross.
Nuovo Cimento B, Ser. 11, Vol. 6B, 83 - 87 (1971).
Using general principles, the amount of mass-energy which can be extracted from two extreme Kerr rotating black holes when they collide is found to be at most 21% of their total initial mass-energy.

066.142 Vacuum, matter and the universe. V. Lapchinskiy.

Soviet Sci. Rev., (GB), Vol. 2, 328 - 335 (1971).
See Phys. Abstr., Vol. 75, No. 4768 (1972).

066.143 Radiation and energy in an asymptotically Friedmann space-time. E. N. Glass.
Journ. Math. Phys., (USA), Vol. 12, 1222 - 1230 (1971).
A bounded gravitating system in an asymptotically hyperbolic Friedmann space is examined. It is shown that a bounded gravitating system in a cosmological space can be understood in much the same manner as a similar physical system in an asymptotically flat space-time.

066.144 Generalized field equations and the Brans-Dicke theory. B. O. J. Tupper.
Nuovo Cimento B, Ser. 11, Vol. 6B, 105 - 110 (1971).
Generalized field equations of general relativity formed from a set of tetrad vectors are written down for a scalar-dependent tetrad. The resulting equations are identical with the field equations of a particular case of the Brans-Dicke theory.

066.145 Event horizons and gravitational collapse.
W. Israel.
General Relativity and Gravitation, Vol. 2, 53 - 59 (1971).
Reviews the present status of the black hole conjecture and discusses how it can be understood in physical terms.

066.146 Focusing of gravitational radiation by interior gravitational fields. J. K. Lawrence.
Nuovo Cimento B, Ser. 11, Vol. 6B, 225 - 235 (1971).
The deflection of rays of gravitational radiation passing through the interior of a diffuse, static, spherically symmetric distribution of matter is examined along with the focusing of the radiation by this process.

066.147 Experimental verification of the gravitation theory in cosmology. A. M. Finkel'shtein.
Proc. Sixth Winter School on Space Physics, Part I. Apatity 1969, (see 012.033), p. 44 - 48 (1971).

066.148 Two relativistic matter distributions of radial symmetry. B. Kuchowicz.
Indian Journ. Pure Applied Math., Vol. 2, 297 - 300 (1971).
Two new solutions of Einstein's field equations are obtained, one in canonical Schwarzschild coordinates and the other in isotropic coordinates. The metric tensor components and matter density are expressed in terms of simple elementary functions and their derivatives.

Theory of relativity and gravitation.
See Abstr. 003.023.

Supermassive objects in astrophysics.
See Abstr. 011.001.

Collapsed nuclei. See Abstr. 022.125.

Computerized series solution of relativistic equations of motion. See Abstr. 042.032.

Some general relations in relativistic magnetohydrodynamics. See Abstr. 062.048.

Rapidly rotating polytropes in the post-Newtonian approximation to general relativity. See Abstr. 065.026.

Rapidly rotating supermassive stars in the first post-Newtonian approximation to general relativity.
See Abstr. 065.027.

Neutron stars and black holes.
See Abstr. 065.103.

Theory of level surfaces inside relativistic, rotating stars. I. Poincaré's limit. See Abstr. 065.109.

Relativistic evolution of $10^3 M_\odot$ star. See Abstr. 065.118.

General relativistic neutron and hyperon star models. See Abstr. 065.123.

On a possibility of observing neutrinos from collapsing stars and neutrino oscillations. See Abstr. 065.125.

On the Freundlich red shift. See Abstr. 071.052.

Point sources of neutrino radiation. See Abstr. 080.058.

Determination of astrodynamic constants and a test of the general relativistic time delay with S-band range and Doppler data from Mariners 6 and 7. See Abstr. 094.189.

Evidence for black holes in binary star systems. See Abstr. 117.012.

Black holes and binary stars. See Abstr. 117.025.

Further evidence for collapsed objects in binary star systems.   See Abstr. 117.026.

Further evidence for a black hole in β Lyrae. See Abstr. 121.089.

Rotating white dwarfs in general relativity. See Abstr. 126.013.

Quasi-stellar objects and gravitational lenses. See Abstr. 141.143.

Quasi-stellar objects: Their importance for cosmology and general relativity.   See Abstr. 141.227.

What flux limits can be set for X-ray pulses accompanying Weber's pulses?   See Abstr. 142.067.

Optical appearance of a collisionless gas of stars surrounding a black hole.   See Abstr. 151.014.

Formation and evolution of bright black holes. See Abstr. 158.046.

The effect of the Einstein light deflection on observed properties of clusters of galaxies. See Abstr. 160.010.

On the nature of mass. See Abstr. 162.015.

Expansion anisotropy and the spectrum of the cosmic background radiation.   See Abstr. 162.043.

The hadron barrier in cosmology and gravitational collapse.   See Abstr. 162.047.

On the removal of initial singularity in a big-bang universe in terms of a renormalized theory of gravitation. I. Examination of the present status and a new approach. See Abstr. 162.069.

On the removal of initial singularity in a big-bang universe in terms of a renormalized theory of gravitation. II. Criteria for obtaining a physically reasonable model. See Abstr. 162.070.

Sun

071 Solar Photosphere, Spectrum

071.001 Photometric analysis of solar granulation corrected for the blurring-effect. M. Lévy.
Astron. Astrophys., Vol. 14, 15 - 23 (1971). In French.

We have statistically analyzed ten microphotometer scans from solar granulation photographs at 5300 Å, taken during the partial eclipse of May 20, 1966 by Rösch and Hugon at the Pic-du-Midi observatory. By the computation of the apparent profile of the moon we have studied both atmospheric effect and instrumental influence determined by the same spread function φ. This determination allows us to correct the spatial power spectrum of the brightness fluctuations of the continuum from the blurring of the image; we are also able to correct the root mean square of the brightness fluctuations; this result enables us to determine the corrected root mean square of the temperature fluctuations. Finally, we compare all our results to Edmonds and Schwarzschild's calculations (Stratoscope flight of September 24, 1959).

071.002 Photospheric mass motions associated with a flare. H. Yoshimura, K. Tanaka, M. Shimizu, E. Hiei.
Publ. Astron. Soc. Japan, Vol. 23, 443 - 448 (1971).

Curved absorption lines were observed in a flare region. They are interpreted by large-scale mass motions with a dimension of 10^4 km in the photosphere and low chromosphere.

071.003 True central intensities of Fraunhofer lines. J. W. Brault, C. D. Slaughter, A. K. Pierce, R. S. Aikens.
Solar Physics, Vol. 18, 366 - 378 (1971).

From observations made at Kitt Peak the true central intensities of 40 Fraunhofer lines distributed between $\lambda\lambda$ 3083–7699, for the center of the solar disk, are derived and given in tabular form. The paper includes a description of the optical system − telescope and spectrograph. The methods of observation and reduction of the data are discussed.

071.004 Measurements of the oscillatory and slowly-varying components of the solar velocity field.
N. R. Sheeley, Jr., A. Bhatnagar.
Solar Physics, Vol. 18, 379 - 384 (1971).

Spectroheliograms with high spatial resolution are presented to illustrate the decomposition of the solar velocity field into its oscillatory and slowly-varying components. An analysis of data obtained in the lines FeI λ 5434 and FeII λ 4924 yield essentially the same principal results. It is concluded that spectroheliograms of the slowly-varying component represent the velocity field of the photospheric granulation.

071.005 Bright-dark asymmetry in solar granulation. M. I. Parvey, S. Musman.
Solar Physics, Vol. 18, 385 - 390 = Publ. Goethe Link Obs., Indiana Univ., *Bloomington*, No. 120 (1971).

Although a positive print of solar granulation gives the impression of bright irregular areas on a dark background, this impression is highly subjective and depends upon the nature of the photographic process. We developed an objective method for comparing bright and dark features and applied it to 40000 elements from a granulation photograph.

We found that dark features were fewer in number, larger, and had larger perimeter-to-area ratios than the bright features. Our results are consistent with the subjective impression that granulation is composed of bright features separated by dark lanes.

071.006 The intensity distribution of the D_3 helium line near the solar limb. R. A. Gulyaev.
Solar Physics, Vol. 18, 410 - 416 (1971).

Cinematographic observations of the slitless flash spectrum near the D_3 helium line were performed in Yurgamysh at the total solar eclipse of September 22, 1968. The intensity distribution of the D_3 line was obtained with a height resolution of 44 km within the height interval between −3400 and +1700 km above the limb. The absorption line D_3 on the disk near the limb was discovered.

071.007 Erratum: 'Studies of granular velocities. II. Statistical analysis of two high-resolution spectrograms'.
[Solar Physics, Vol. 16, 253 - 271 (1971)].
J. P. Mehltretter.
Solar Physics, Vol. 18, 510 (1971).

071.008 On the rms intensity fluctuation of solar granulation. J. P. Mehltretter.
Solar Physics, Vol. 19, 32 - 39 = Mitt. Fraunhofer Inst. *Freiburg*, No. 104 (1971).

Using a selected high definition granulation photograph obtained with a 40 cm aperture telescope from the ground, a new determination of the rms intensity fluctuation is attempted.

071.009 Radiation damping and the Goldberg-Unno method. V. I. Troyan.
Astrometriya i Astrofiz., *Kiev*, Vyp. (No.) 12, (see 003.003), p. 64 - 70 (1971). In Russian.

An effect of radiation damping on the results of investigation of turbulence in the solar photosphere with the Goldberg-Unno method is discussed.

071.010 On observations of solar granulation. M. B. Kerimbekov.
Solnechnye Dannye 1971 Byull., No. 4, p. 96 - 99 (1971). In Russian.

A new method of reduction of the photographs of granulation is discussed. Supposed that compression of the gas medium gives rise to the waves on the solar surface, the characteristic sizes of sound waves were calculated.

071.011 Determination of the turbulent velocities in the undisturbed photosphere by Goldberg's method.
O. G. Badalyan.
Astron. Tsirk., No. 622, p. 3 - 5 (1971). In Russian.

071.012 On the abundance of chlorine in the sun. D. L. Lambert, E. A. Mallia, J. Brault.
Solar Physics, Vol. 19, 289 - 296 (1971).

A low-noise photoelectric scan which includes the predicted position of the ClI transition $4s^4 P_{5/2} - 4p^4 D^0{}_{7/2}$ provides inconclusive evidence for the presence of the line in the solar photospheric spectrum. An upper limit $\log N(Cl) \leqslant 5.5$ is derived. A new derivation of the chlorine abundance for the Orion nebula is presented: $\log N(Cl) \approx 5.8$. It is suggested that a cosmic abundance $\log N(Cl) = 5.5$ to 5.8 be adopted.

071.013 A power spectrum analysis of granular intensity fluctuations and velocities.   H. Reiling.
Solar Physics, Vol. 19, 297 - 313 (1971).
An exceptionally highly resolved spectrogram has been used to derive rms values for both the intensity fluctuations in the continuum and the granular velocities in the Ba II line $\lambda 5853.69$ by means of the power spectrum analysis.

071.014 Interferometric measurements of 142 solar wavelengths.   J. E. O'Brien.
Solar Physics, Vol. 19, 314 - 322 (1971).
The wavelengths of 142 solar absorption lines, in light taken from the center of the solar disk, and in the wavelength region 4675 Å to 6275 Å, have been determined by interferometric comparison with the standard wavelengths of Hg^{198}. The wavelengths of 68 solar iron lines are compared with hollow cathode wavelengths, and differences between the resulting wavelength shift and that predicted by relativity theory calculated.

071.015 Absolute intensity calibrations of solar K-line profiles.   J. M. Pasachoff.
Solar Physics, Vol. 19, 323 - 329 (1971).
Individual K-line profiles from elements of fine structure on the surface of the sun are calibrated absolutely. The continuum calibrations of Labs and Neckel and of Houtgast and Namba are considered, and the average K-profile is scaled to that of White and Suemoto. The ranges of intensities across a high-resolution spectrogram are tabulated for various parts of the line profile.

071.016 Isotopes of magnesium in the solar atmosphere.   R. Boyer, J. C. Henoux, P. Sotirovski.
Solar Physics, Vol. 19, 330 - 337 (1971).
We have analysed MgH $A^2 \Pi - X^2 \Sigma$ (0.0), (1.1), (2.2), (0.1) and (1.2) absorption bands in a sunspot spectrum. By two different methods, which are almost independent of the estimated value of the correction for stray light, we have determined the solar isotopic ratios of magnesium.

071.017 Solar brightness measurement between 12 and 24 microns.   J. P. Baluteau.
Astron. Astrophys., Vol. 14, 428 - 436 (1971).
The solar spectrum between 12 and 24 microns is obtained with a scanning grating spectrometer. A continuous decrease of the solar brightness temperature is measured between 12 and 24 μ. Temperatures as low as 4500°K above 20μ are obtained. Our measurements are not consistent with the predictions of the B.C.A model (Gingerich and de Jager, 1968) but rather show a good agreement with the model elaborated by Léna (1970).

071.018 A search for CH⁺ in the solar photosperic spectrum.   N. Grevesse, A. J. Sauval.
Astron. Astrophys., Vol. 14, 477 - 480 (1971).
We present the results of a search for lines of CH⁺ in the solar photospheric spectrum. We conclude that if CH⁺ is present, the equivalent widths of the most intense lines cannot exceed 2 mÅ. An empirical solar f_{00}-value is derived and compared with a laboratory value and other available astrophysical values. Some possible reasons are given for explaining the large discrepancy between laboratory and astrophysical f_{00}-values.

071.019 A new calibration of Rowland intensities in the Fraunhofer lines.   V. S. Shkljarnik.

Solnechnye Dannye 1971 Byull., No. 5, p. 83 - 85 (1971). In Russian.

071.020 Investigation of the physical conditions in the solar photosphere by the method of the curves of growth.
D. M. Kuli-Zade.
Soobshch. Shemakhinsk. Astrofiz. Obs., vyp. (No.) 5, p. 39 - 57 (1971). In Russian.

071.021 The chemical composition of the photosphere and the corona.   G. L. Withbroe.
National Bureau Standards Special Publ. 353, (see 012.001), p. 127 - 148 (1971).

071.022 New magnetic flux.   E. N. Frazier.
Bull. American Astron. Soc., Vol. 3, 376 (1971). Abstr. AAS.

071.023 Granulation patterns and solar oblateness.   R. S. Kandel, S. L. Keil.
Bull. American Astron. Soc., Vol. 3, 376 - 377 (1971). Abstr. AAS.

071.024 Solar photospheric temperature profile.   A. J. Skalafuris.
Bull. American Astron. Soc., Vol. 3, 377 (1971). – Abstr. AAS.

071.025 The solar abundance of silicon from forbidden Si I lines.   N. Grevesse, J. P. Swings.
Bull. American Astron. Soc., Vol. 3, 377 (1971). – Abstr. AAS.

071.026 Remarks on the solar abundance of iron.   C. R. Cowley.
Bull. American Astron. Soc., Vol. 3, 377 - 378 (1971). Abstr. AAS.

071.027 Solar infrared limb profile measurements.   N. J. Johnson.
Bull. American Astron. Soc., Vol. 3, 385 (1971). – Abstr. AAS.

071.028 On the fine structure of the magnetic field in the undisturbed photosphere.
V. M. Grigoryev, G. V. Kuklin.
IAU Symposium No. 43, (see 012.003), p. 252 - 259 (1971).

071.029 Supergranulation at the center of the disk.   E. N. Frazier.
IAU Symposium No. 43, (see 012.003), p. 260 - 267 (1971).

071.030 Magnetographic and spectrographic observations of weakly active regions.   C. J. Durrant.
IAU Symposium No. 43, (see 012.003), p. 268 - 273 (1971).

071.031 The magnetic and velocity fields and brightness in the solar atmosphere.   S. I. Gopasyuk, T. T. Tsap.
IAU Symposium No. 43, (see 012.003), p. 274 - 278 (1971).

071.032 Observations and interpretation of supergranule velocity and magnetic fields.   S. Musman.
IAU Symposium No. 43, (see 012.003), p. 289 - 292 (1971).

071.033 The time dependence of magnetic, velocity, and intensity fields in the solar atmosphere.
N. R. Sheeley, Jr.
IAU Symposium No. 43, (see 012.003), p. 310 - 315 (1971).

071.034 Hα structures and small-scale magnetic field configurations.   S. F. Smith.

IAU Symposium No. 43, (see 012.003), p. 323 - 328 (1971).

071.035 Magnetic field spectroheliograms from the San Fernando Observatory. D. Vrabec.
IAU Symposium No. 43, (see 012.003), p. 329 - 339 (1971).

071.036 Five-minute oscillations in the solar magnetic field. A. S. Tanenbaum, J. M. Wilcox, R. Howard.
IAU Symposium No. 43, (see 012.003), p. 348 - 355 (1971).

071.037 An attempt to associate observed photospheric motions with the magnetic field structure and flare occurrence in an active region.
M. J. Martres, I. Soru-Escaut, J. Rayrole.
IAU Symposium No. 43, (see 012.003), p. 435 - 442 (1971).

071.038 The possibility of magnetic field origin in fine structure elements of solar features.
M. Kopecký, G. V. Kuklin.
IAU Symposium No. 43, (see 012.003), p. 534 - 541 (1971).

071.039 Anisotropy of electric conductivity and dissipation of magnetic fields. M. Kopecký, V. Kopecký.
IAU Symposium No. 43, (see 012.003), p. 542 - 544 (1971).

071.040 Time evolution of the large-scale solar magnetic field.
M. D. Altschuler, G. Newkirk, Jr., D. E. Trotter, R. Howard.
IAU Symposium No. 43, (see 012.003), p. 588 - 594 (1971).

071.041 Sector structure of the solar magnetic field.
J. M. Wilcox.
IAU Symposium No. 43, (see 012.003), p. 744 - 753 (1971).

071.042 Determination of line profiles of the solar spectrum using a double-pass spectrophotometer with digital recording on punched cards.
V. N. Karpinsky, E. K. Kokhan, N. I. Pechinskaja.
Astron. Zhurn. Akad. Nauk SSSR, Vol. 48, 1004 - 1012 (1971). In Russian. English translation in Soviet Astron. AJ, Vol. 15, No. 5.
Systematic investigations of the Fraunhofer line profiles begun in summer 1969 using the new double-pass spectrophotometer with digital recording on punched cards are presented. The influence of ghosts and common scattered light has been eliminated completely. The observed profiles of the Fe 5307.4 Å and Ni 5435.9 Å lines are given in a table.

071.043 Magnesium hydride in the sun.
D. L. Lambert, E. A. Mallia, A. D. Petford.
Monthly Notices Roy. Astron. Soc., Vol. 154, 265 - 278 (1971).
Low noise photoelectric scans of the MgH $A^2\Pi - X^2\Sigma$ (0,0) band in the spectrum of the solar disk are examined and accurate equivalent widths obtained. Analysis of these widths with a recent photospheric model atmosphere provides an estimate for the band oscillator strength: $f_{0,0} = 0.055$. Estimates of the isotopic abundance ratios have been made from the photospheric and penumbral MgH lines and from strong Mg I lines which have measureable isotopic wavelength shifts. The sunspot analyses and the significance of the contribution of the bright umbral dots to the observed spectrum are discussed.

071.044 Broadening of the solar sodium D-lines by atomic hydrogen. L. F. McNamara.
Proc. Astron. Soc. Australia, Vol. 2, 40 - 41 (1971).

071.045 Using CN λ 3883 spectroheliograms to map weak photospheric magnetic fields. N. R. Sheeley, Jr.
Solar Physics, Vol. 20, 19 - 25 (1971).

By photographically averaging time sequences of high-resolution CN λ 3883 spectroheliograms, the noise level due to the rapidly fluctuating intensity of the solar background has been reduced significantly. A comparison between these enhanced spectroheliograms and a photoelectric magnetogram suggests that the brightness-magnetic field correlation extends to much weaker field strengths and fainter faculae than can be detected on a single, high quality CN λ 3883 spectroheliogram.

071.046 Test of a solar streamline analysis on terrestrial wind data. H. J. E. Fischer.
Solar Physics, Vol. 20, 26 - 30 (1971).
Terrestrial wind data are used to test a technique for obtaining solar streamline patterns from line-of-sight velocity measurements. The method of analysis and its limitations are discussed. The major features of the reconstructed terrestrial streamline pattern agree satisfactorily with those of the actual wind field.

071.047 Choice of a reference object in the problem of measuring the luminescence in the contours of Fraunhofer lines. V. S. Tsvetkova.
Vestn. Khar'kov. Univ., No. 65, (Ser. Astron., No. 6), p. 37 - 52 (1971). In Russian.

071.048 The solar light-element abundances and primeval helium. H. E. Mitler.
SAO, *Cambridge, Mass.*, Special Report, No. 323, 5 + 15 pp. (1970).
From a comparison of recent determinations of photospheric element abundances with solar cosmic-ray measurements, the abundances of helium, carbon, nitrogen, oxygen, neon, magnesium, silicon, and sulphur in the solar surface are determined within a probable error of 12%. The abundance of iron is also determined this way, but the resulting value may be in error by a factor of 2. The abundance of helium is found to be lower than the usually accepted 10% (i.e., Y = 0.28); it is 5.6% (i.e., $Y_\odot = 0.188 \pm 0.020$). Using a simple model for the exchange of material between stars and the interstellar matter, we find that a reasonable value for the initial helium fraction is then $Y_0 \cong 0.133$, considerably smaller than the values generally obtained from hot big-bang models.

071.049 Interference in solar oscillations.
R. J. Reif, S. Musman.
Solar Physics, Vol. 20, 257 - 263 (1971).
We have analyzed magnetograph observations of the 5-min oscillations. We find that most of the oscillatory power is concentrated in space and frequency. Interference effects where these concentrations overlap can explain some of the variations in amplitude of the oscillation.

071.050 The use of the Goldberg-Unno method for the investigation of small-scale photosphere motions.
E. Gurtovenko, V. Troyan.
Solar Physics, Vol. 20, 264 - 274 (1971).
The small-scale motion field is due to the microturbulence and 'unresolved' macroturbulence (convection and wave motions). The microturbulence affects the absorption line coefficient, while the macroturbulence influences the line profiles. The Goldberg-Unno method is analyzed. It does not depend on the photospheric model and allows one to obtain in principle the microturbulent velocity $\xi_t(\tau)$ by measuring various parts of line profiles of the multiplet. The method has a differential character in which the various kind of errors affect the results only to a small degree. The method of accounting for the damping effect is considered and an attempt to analyse the microturbulent velocities by the Goldberg-Unno method under deviation from LTE is made.

071.051 A comparison of the intensity variations of the CN photospheric and K line chromospheric network with time. S. Y. Liu, N. R. Sheeley, Jr. Solar Physics, Vol. 20, 282 - 285 (1971).

As part of a program to study the behavior of various K-line spectral features both as a function of spatial position and time, time-lapse sequences of spectroheliograms have been obtained simultaneously in the bandhead of CN at λ3883 and in the violet emission peak (K_{2v}) of Ca II λ3934.

071.052 On the Freundlich red shift. A. G. Gasanalizade. Solar Physics, Vol. 20, 507 - 512 (1971).

The Freundlich red shift of wavelengths in the solar spectrum is discussed. The cross-section for scattering in the solar atmosphere appears to be $\langle\sigma_\odot\rangle = 1.58 \times 10^{-18}$ cm². A new expression for Freundlich's red shift formula is obtained. Some numerical examples are given and some interesting aspects of Freundlich's parameters are discussed.

071.053 Observation de la raie d'émission de C I dans le spectre solaire à 1993,6 Å. D. Samain. Comptes Rendus Acad. Sci. Paris, Sér. B, Vol. 273, 1133 - 1136 (1971).

Après avoir décrit l'appareillage, et présenté les mesures de la raie 1993 du carbone neutre dans le spectre solaire, les caractéristiques de ces mesures sont décrites: fort renforcement de la raie au bord solaire; extrême sensibilité à l'activité.

071.054 On the supergranulation intensity field in integrated light. A. Kubičela. Bull. Obs. Astron. Beograd, Vol. 28, (No. 124), 187 - 193 (1970).

A very high contrast picture of the photosphere at λ = 5200 ± 500 Å has been obtained, where the granular intensity field has been averaged and the supergranulation pattern to some extent revealed.

071.055 Ultraviolet solar spectrum recorded by echelle spectrograph (1970 to 1800 Å). H. C. McAllister. Solar Physics, Vol. 21, 27 - 37 (1971).

In the region from 1946.5 Å to 1963.5 Å 79 absorption features are measured and 33 identified. Most of the identified stronger lines are due to FeI. A significant feature of the solar spectrum in this region coincides with the raie ultime of SeI.

071.056 Observation on the detailed correspondence of magnetic and Hα features. S. A. Schoolman. Solar Physics, Vol. 21, 57 - 59 (1971). – Research note.

071.057 An intensity distribution of bright points observed on a CN spectroheliogram. B. Gillespie. Solar Physics, Vol. 21, 93 - 95 (1971).

This note is a report of photometry of bright photospheric network 'points' observed on a spectroheliogram taken under optimal seeing conditions.

071.058 Photometry of the sodium resonance doublet in the flash spectrum during the solar eclipse on September 22, 1968. R. A. Gulyaev. Solnechnye Dannye 1971 Byull., No. 10, p. 67 - 71 (1971). In Russian.

The slitless flash spectrograms obtained during the solar eclipse on September 22, 1968 were reduced. The distributions of the integrated intensities of the spectral lines were derived: for the sodium D-lines within the height interval from −300 to +1000 km, and for the NiI λ 5892.88 line within the height interval from −250 to +100 km above the solar limb.

071.059 A review of models of the solar photosphere and low chromosphere: The temperature–height profile. D. L. Lambert. Phil. Trans. Roy. Soc. London A, Vol. 270, No. 1202, (see 012.009), 3 - 21 (1971).

071.060 Abundance of iron in the photosphere. A. Unsöld. Phil. Trans. Roy. Soc. London A, Vol. 270, No. 1202, (see 012.009), 23 - 28 (1971).

071.061 High resolution interferometric studies of the solar magnesium II doublet spectral region. B. Bates, D. J. Bradley, D. A. McBride, C. D. McKeith, N. E. McKeith, W. M. Burton, H. J. B. Paxton, D. B. Shenton, R. Wilson. Phil. Trans. Roy. Soc. London A, Vol. 270, No. 1202, (see 012.009), 47 - 53 (1971).

071.062 A measurement of the brightness temperature of the sun in the range 65 to 180 cm^{-1}. T. A. Clark, G. R. Courts, R. E. Jennings. Phil. Trans. Roy. Soc. London A, Vol. 270, No. 1202, (see 012.009), 55 - 58 (1971).

071.063 Helium-like ion forbidden line emission, and solar active regions. F. F. Freeman, A. H. Gabriel, B. B. Jones, C. Jordan. Phil. Trans. Roy. Soc. London A, Vol. 270, No. 1202, (see 012.009), 127 - 133 (1971).

071.064 The relative intensities of lines from BeI-like ions in the solar spectrum. C. Jordan. Highlights of Astronomy, Vol. 2, (see 012.017), 519 - 526 (1971).

071.065 The determination of the turbulent velocity in the photosphere of the sun from profiles of CN molecule lines. G. A. Porfirjeva. Astron. Zhurn. Akad. Nauk SSSR, Vol. 48, 1227 - 1231 (1971). In Russian. English translation in Soviet Astron. AJ, Vol. 15, No. 6.

From profiles of four lines, belonging to the system of vibrational bands with $\Delta v = -1$ (electron transition $B^2\Sigma^+ - X^2\Sigma^+$) of the CN molecule, velocities of turbulent motions in the photosphere of the sun are determined. The measurements are carried out at six distances from the centre of the solar disk when $\sin\vartheta$ = 0.0, 0.7, 0.85, 0.9, 0.95, 0.98. In the first approximation the turbulent velocity does not depend on the value of $\sin\vartheta$ and is equal to 3.5 ± 0.12 km/sec.

071.066 Radiation temperature of CO molecules on the sun. A. P. Sarychev. Astron. Zhurn. Akad. Nauk SSSR, Vol. 48, 1232 - 1236 (1971). In Russian. English translation in Soviet Astron. AJ, Vol. 15, No. 6.

It is shown that under conditions of the solar atmosphere the source function for lines of the vibrating-rotating spectrum of CO coincides with Planck's function. The method of determination of the mean value T_{rad} from measurements of the line depth at various distances from the centre of the disk is considered.

071.067 Turbulent velocities and formation depths of a number of lines in the solar atmosphere. O. G. Badalyan. Soobshch. Gos. Astron. Inst. Shternberga, No. 173, p. 3 - 18 (1971). In Russian.

The formation depths of a number of lines were calculated by the method of weighting functions, Bilderberg's model of the solar atmosphere was used. It is shown that the values of the optical depths adopted in Unno's paper (1959) and the range of τ_{5000} while passing from the center of line towards the wing are overestimated. The turbulent velocities were determined using 16 pairs of lines with especially large formation depths in the solar photosphere by Goldberg's method using the Utrecht Atlas of Solar Spectrum.

071.068 Preliminary results of analysis of the weak magnetic field structure in the undisturbed photosphere.
V. M. Grigorjev, G. V. Kuklin.
Geod. Geophys. Veröff., Ser. 2, No. 13, (see 012.029), p. 109 - 111 (1969).

071.069 On the continuous solar emergent radiation.
C. Popovici.
Geod. Geophys. Veröff., Ser. 2, No. 13, (see 012.029), p. 147 - 149 (1969).

071.070 Solar abundance determination.
O. Engvold, Ö. Hauge.
Beam-foil spectroscopy. Conference 1970, (see 012.032), p. 351 - 362 (1970).

071.071 Isotopic abundance of magnesium in the sun.
C. K. Kumar.
Thesis, Univ. Michigan, Ann Arbor. [Available from Univ. Microfilms, Ann Arbor, Mich., U.S.A. Order No. 70−21706], 75 pp. (1970).

The isotopic abundance of magnesium in the sun has been determined by an analysis of the profiles of the lines of the $A^2\Pi-X^2\Sigma^+(\Delta\nu=0$ sequence) bands of MgH in the 5100−5160 Å region of the sunspot spectrum.

071.072 Effects of rotation on solar convection: Cyclones in the solar atmosphere. D. J. Mullan.
Irish Astron. Joun., Vol. 10, 12 - 24 (1971).

In the present paper we suggest that the large magnetic regularities are not convection cells involved in transporting heat vertically, but instead are somewhat analogous to terrestrial cyclones. If solar cyclones are confined within one pressure scale height of the upper surface of the convection zone, then the horizontal scale corresponding to $D/d \sim 8 \times 10^3$ agrees within a factor of 3 with that of the large magnetic regularities. Cyclone growth times are found to be about 50 days. 1967 Gilman has shown that in order to maintain the solar jetstream (the equatorial acceleration) against dissipating magnetic field stresses, the equator must be hotter than the pole such that $(T_e\text{-}T_p)/T_p \approx 7 \times 10^{-4}$. Here we investigate the modifications to Öpik's formulae of cellular convection in the presence of rotation in order to estimate $T_e\text{-}T_p$ theoretically. The result is $(T_e\text{-}T_p)/T_p = 3 \times 10^{-4}$.

071.073 Iron in the sun and the stars. R. H. Garstang.
Journ. Astron. Soc. Western Australia, Vol. 29, August, p. 2 - 3 (1971).

The solar spectrum from lambda 7498 to lambda 12016. A table of measures and identification.
See Abstr. 003.110.

Collegamento veloce in tempo reale con un elaboratore UNIVAC 1108. See Abstr. 021.008.

Beam-foil lifetimes in neutral chromium.
See Abstr. 022.068.

Isoelectronic wavelength calculations for argon spectra. See Abstr. 022.094.

Experimental oscillator strengths and the solar abundances of iron and nickel. See Abstr. 022.156.

Automated analysis of astronomical spectra.
See Abstr. 031.017.

The Kitt Peak magnetograph. IV: 40-channel probe and the detection of weak photospheric fields.
See Abstr. 034.027.

Spectra-spectroheliograph observations.
See Abstr. 034.035.

On the solution of the transfer equation system for a non-scattering medium with a homogeneous magnetic field.
See Abstr. 063.035.

The relation between electric conductivity in faculae and in the Bilderberg model of the photosphere.
See Abstr. 072.001.

A note on the effect of scattered light on the facula-to-photosphere contrast. See Abstr. 072.003.

The distribution of electric conductivity and its gradients in the photospheric layers of an active region.
See Abstr. 072.060.

Multi-channel magnetograph observations. III. Faculae. See Abstr. 072.064.

Solar magnetic fields in association with flares.
See Abstr. 073.042.

On arising of magnetic ropes in the convective zone. II. The Parker mechanism of arising.
See Abstr. 073.092.

The X-ray corona and the photospheric magnetic field. See Abstr. 074.032.

A new solar atlas from echelle rocket-ultraviolet spectra. See Abstr. 076.010.

On the classification of some highly ionized iron and nickel lines in the 200−400 Å region of the solar spectrum. See Abstr. 076.023.

A high resolution solar ultraviolet spectrum between 200 and 220 nm. See Abstr. 076.030.

Observations of the extreme ultraviolet solar spectrum. See Abstr. 076.031.

Solar oblateness and the abundance of lithium in the sun. See Abstr. 080.002.

Hydromagnetic structure at the boundary of a supergranule. See Abstr. 080.037.

Solar magnetic fields—small scale.
See Abstr. 080.046.

Submillimetre-wave solar observations using a double-output Michelson interferometer. See Abstr. 082.139.

072 Sunspots, Faculae, Solar Activity

072.001 **The relation between electric conductivity in faculae and in the Bilderberg model of the photosphere.**
M. Kopecký, E. Soytürk.
Bull. Astron. Inst. Czechoslovakia, Vol. 22, 154 - 156 (1971).

The value of the electric conductivity at different optical and geometrical depths in the Bilderberg model of the photosphere and low chromosphere and in the model of faculae is given. In the case of inhomogeneous electric conductivity in the active regions of the sun relatively complicated systems of electric currents must exist in the active regions.

072.002 **Dependence of molecular dissociation equilibrium on magnetic field strength in sunspots.**
V. P. Gaur, M. C. Pande, B. M. Tripathi, G. C. Joshi.
Bull. Astron. Inst. Czechoslovakia, Vol. 22, 157 - 160 (1971).

It is shown that the total number of CO, CN, C_2, OH, NH and CH molecules varies significantly with magnetic field strength in sunspots. The equivalent widths of two first overtone lines of CO belonging to 0−2 vibration-rotation band as a function of magnetic field strength are also given. It appears that up to 2500 gauss the equivalent widths change linearly with magnetic field strength.

072.003 **A note on the effect of scattered light on the facula-to-photosphere contrast.**
M. C. Pande, B. M. Tripathi, V. P. Gaur.
Bull. Astron. Inst. Czechoslovakia, Vol. 22, 161 - 162 (1971).

The measured facula-to-photosphere contrast in the continuum radiation corrected for the scattered light should yield a hotter average facula model than the models derived without such a correction.

072.004 **Complexes of activity of the solar cycle and very large scale convection.** H. Yoshimura.
Solar Physics, Vol. 18, 417 - 433 (1971).

The magnetic field pattern associated with large scale convective motions, which are much larger than the supergranules and have been conceived as a source of maintenance of the solar differential rotation, is calculated in the framework of a slowly and differentially rotating thin spherical shell, including the effects of thermal conductivity and viscosity. The approximation of Boussinesq are used and the initial state of the magnetic field is assumed to be purely toroidal.

072.005 **Observations of stray-light and sunspot intensities during the Mercury transit of 1970 May 9.**
W. Mattig.
Solar Physics, Vol. 18, 434 - 442 = Mitt. Fraunhofer Inst., No. 102 (1971).

In order to test the usual method for correcting sunspot intensity measurements for stray light, we have measured, during the Mercury transit of 1970 May 9, the intensities of Mercury, a sunspot umbra, and the aureole.

072.006 **Sunspot intensity observations during the 9 May 1970 Mercury transit.** P. Maltby, L. Staveland.
Solar Physics, Vol. 18, 443 - 449 (1971).

The intensity of a sunspot was measured in eight wavelength regions during the Mercury transit of 9 May 1970. The observations have been corrected for scattered light in the earth's atmosphere as well as in the instrument using two different methods plus a combination of these. One method consists of using Mercury as a calibration spot. In the second method the corrections for scattered light are determined from solar limb observations.

072.007 **The molecular spectrum of sunspot umbrae.**
P. Sotirovski.
Astron. Astrophys., Vol. 14, 319 - 336 (1971).

By means of spectrographic analysis we have studied the molecular spectra in three sunspot umbrae. Our first step was to try to identify all the molecular lines in the spectral range $\lambda\lambda 4900 - 6450$ Å using the method of coincidences. Fourteen molecules have been studied; three of them (CN, MgH and TiO) were extensively analysed and provided reliable information about the rotational temperatures. We obtain an empirical curve of growth from the MgH lines that allowed us to estimate the Mg isotopic ratios. Our last step was to compute the equivalent widths of molecular lines using Hénoux's sunspot model and to compare the calculations with the observations.

072.008 **A series of related active regions during January 14 – June 1, 1969.** S. W. Prata.
Solar Physics, Vol. 19, 92 - 109 (1971).

The history of a series of active regions is traced for the first half of 1969. At least four active regions successively occupied positions near Carrington longitude 270. In some cases the emergence of new regions in the midst of old regions was observed. All the larger flares observed in the studied region were associated with inverted polarity. In some cases the fields emerged inverted; in other cases the inverted polarity appeared to result from the interaction of emerging flux with fields already present at the surface.

072.009 **Physics of sunspots. Part IV.** J. Jakimiec.
Postępy Astron., Vol. 19, 199 - 213 (1971).
In Polish.

The problem of constructing complete, hydromagnetic models of the photospheric layers of a spot is considered, when all relevant observational data are taken into account. Much attention is devoted to the question of determining the distribution of the magnetic forces in the spot.

072.010 **The magnetic field asymmetry in a sunspot.**
V. F. Tshistyakov, V. A. Golubjev.
Astron. Tsirk., No. 621, p. 4 - 6 (1971). In Russian.

072.011 **Sunspot magnetic field observations measured from lines of neutral and ionized elements.**
G. V. Lyamova.
Astron. Tsirk., No. 622, p. 5 - 8 (1971). In Russian.

072.012 **Two-dimensional observations of the velocity field in and around sunspots.**
N. R. Sheeley, Jr., A. Bhatnagar.
Solar Physics, Vol. 19, 338 - 346 (1971).

Doppler spectroheliograms of sunspots and their surroundings have been obtained with a spatial resolution approaching one second of arc and a time resolution of 20 s per frame. Observations of 5 sunspots, located 18°, 45°, 56°, 60° and 72° from the disk center respectively, showed considerable long-lived fine structure and, in particular, indicated that the spatially-averaged horizontal flow was outward for roughly 10000 km beyond the outer boundary of the penumbra.

072.013 **Planetary influences on the large-scale distribution of solar activity.** P. Ambrož.
Solar Physics, Vol. 19, 480 - 482 (1971).

The aim of the present short note is to check the reality of existence of a spatial relationship between the solar surface area being influenced by the planetary tidal force and the distribution of calcium flocculi.

072.014 **On photospheric faculae contrast in blue light.**
A. S. Tskhovrebadze.
Byull. Abastumansk. Astrofiz. Obs., No. 40, p. 97 - 100 (1971). In Russian.
The contrast value increases from the center towards the limb without passing an extremal point.

072.015 **On the forms of the 11-year solar cycle curves as analyzed by a generalized method of principal components.** A. B. Wertlieb, G. V. Kuklin.
Solnechnye Dannye 1971 Byull., No. 5, p. 76 - 82 (1971). In Russian.
A generalization of the principal component method permits to study regularities in the forms of the 11-year cycle curves. The typical form and typical deviations from it are described.

072.016 **A study of short-period variations in the series of solar activity.** K. A. Kandaurova.
Solnechnye Dannye 1971 Byull., No. 5, p. 107 - 114 (1971). In Russian.
The existence of short periods in solar activity is confirmed. The duration of their existence and the instability of their occurrence have been determined. The frequency of these periods is apparently multiple to the principle 11-year cycle.

072.017 **Variations of latitude distribution in the index of total sunspot group area during several 11-year cycles.** G. V. Kuklin.
Solnechnye Dannye 1971 Byull., No. 6, p. 89 - 92 (1971). In Russian.
The method of principal components is used to study variations of latitude distribution of the P index during the cycles 12 - 18.

072.018 **On the influence of variations of the magnetic field of a sunspot on the emission of S-component sources of solar radio emission.** Sh. B. Akhmedov.
Solnechnye Dannye 1971 Byull., No. 6, p. 111 - 115 (1971). In Russian.
The influence of the sunspot magnetic field at photospheric level on the emission of the S-component sources of solar radio emission in cm-wavelength range is considered. The distribution of electron temperature with height in the region of formation of the S-component sources is determined as well as the gradient.

072.019 **The stability of sunspot magnetic fields and the origin of solar flares.** K. Sakurai.
Planet. Space Sci., Vol. 19, 1289 - 1296 (1971).
The steady motion in sunspot magnetic regions is considered for both current-free and force-free configurations. The sufficient condition for stability is obtained in the presence of both external current-free and force-free magnetic fields and a steady motion. It is shown that the pattern of such steady motion is most important in triggering an instability of sunspot magnetic fields, both for the current-free and force-free configuration. The onset of a solar flare seems to be associated with an instability connected to the steady motion within the sunspot magnetic regions.

072.020 **The connection of solar activity indices with cosmic ray intensity.**
E. V. Kolomeetz, Yu. A. Shakhova.
Geomagn. Aeronom., Vol. 11, 892 - 895 (1971). In Russian. Brief information.

072.021 **Paschen-Back effect of the lithium resonance**

doublet in sunspots. P. Maltby.
IAU Symposium No. 43, (see 012.003), p. 141 - 147 (1971).

072.022 **The crossover and magneto-optical effects in sunspot spectra.** V. M. Grigoryev, J. M. Katz.
IAU Symposium No. 43, (see 012.003), p. 148 (1971).

072.023 **On magnetic fields in sunspots and active regions.**
E. H. Schröter.
IAU Symposium No. 43, (see 012.003), p. 167 - 180 (1971).

072.024 **Magnetic field and turbulence in sunspots.**
J. Rayrole.
IAU Symposium No. 43, (see 012.003), p. 181 - 189 (1971).

072.025 **Photoelectric measurements of sunspot magnetic fields.** F.-L. Deubner, R. Göhring.
IAU Symposium No. 43, (see 012.003), p. 190 - 191 (1971).

072.026 **Some remarks on the statics and dynamics of magnetic field structure development in active regions.**
V. Bumba, J. Suda.
IAU Symposium No. 43, (see 012.003), p. 201 - 211 (1971).

072.027 **On the structure of magnetic field and electric currents of a unipolar sunspot.** V. A. Kotov.
IAU Symposium No. 43, (see 012.003), p. 212 - 219 (1971).

072.028 **Magnetic field strengths derived from various lines in the umbral spectrum.** C. Zwaan, J. Buurman.
IAU Symposium No. 43, (see 012.003), p. 220 - 222 (1971).

072.029 **Observations of the two-level structure of sunspot magnetic fields.** H. I. Abdussamatov.
IAU Symposium No. 43, (see 012.003), p. 231 - 234 (1971).

072.030 **On the circular polarization in active regions.**
E. Wiehr.
IAU Symposium No. 43, (see 012.003), p. 235 - 236 (1971).

072.031 **Evolution of the magnetic field configuration in an active region.** D. L. Schatz.
IAU Symposium No. 43, (see 012.003), p. 243 - 248 (1971).

072.032 **Line profiles in sunspot umbrae and penumbrae by atomic beam spectroscopy.** F. Roddier.
IAU Symposium No. 43, (see 012.003), p. 249 - 251 (1971).

072.033 **Vertical velocities associated with plage region magnetic fields.** R. G. Giovanelli, J. V. Ramsay.
IAU Symposium No. 43, (see 012.003), p. 293 - 297 (1971).

072.034 **On the reality of magnetic fine structure.**
C. Sawyer.
IAU Symposium No. 43, (see 012.003), p. 316 - 322 (1971).

072.035 **Sunspot magnetic fields and high energy electrons in flares.** T. Takakura.
IAU Symposium No. 43, (see 012.003), p. 390 - 396 (1971).

072.036 **Observations of magnetic field changes in active regions.**
K. L. Harvey, W. C. Livingston, J. W. Harvey, C. D. Slaughter.
IAU Symposium No. 43, (see 012.003), p. 422 - 427 (1971).

072.037 **Theories of small-scale magnetic fields.**
P. A. Sweet.
IAU Symposium No. 43, (see 012.003), p. 457 - 474 (1971).

072.038 **Sunspot magnetic fields and umbral dots.**
P. R. Wilson.

IAU Symposium No. 43, (see 012.003), p. 475 - 479 (1971).

072.039 Statistical model of small scale discrete structure of magnetoplasma in active regions of the sun.
E. I. Mogilevsky.
IAU Symposium No. 43, (see 012.003), p. 480 - 486 (1971).

072.040 Distribution of the magnetic force in the surface layers of sunspots. J. Jakimiec.
IAU Symposium No. 43, (see 012.003), p. 505 - 511 (1971).

072.041 A numerical study of the solar cycle.
Y. Nakagawa.
IAU Symposium No. 43, (see 012.003), p. 725 - 736 (1971).

072.042 The contours of Balmer lines in the spectrum of a sunspot. A. P. Stefanov.
Vestn. Kiev. Un-ta, Ser. Astron., Vol. 12, p. 17 - 24 (1970). In Russian.

A method of computing the line and continuous absorption coefficients for Balmer lines in a sunspot spectrum and the computation of the contours of Hα–Hγ for the Michard and Mattig models are given. These contours are compared with the observed ones.

072.043 Theory of the solar 22-year cycle.
J. H. Piddington.
Proc. Astron. Soc. Australia, Vol. 2, 7 - 10 (1971).

The objects of the present paper are, first, to indicate briefly why the cyclonic reversal mechanism, and hence the dynamo theory, is not applicable to the solar cycle, second, to propose a deeply penetrating magnetic field in place of the shallow poloidal field of the dynamo theory, and third, to introduce a theory of the solar cycle based on this truly 'general' field.

072.044 Beobachtung der Sonne. M. Reble.
SuW, Vol. 10, 307 (1971).

072.045 On the structure of solar faculae.
V. A. Krat, M. N. Stojanova.
Solar Physics, Vol. 20, 57 - 58 (1971). – Research note.

072.046 The topology of force-free magnetic field near bipolar sunspots. M. A. Raadu, Y. Nakagawa.
Solar Physics, Vol. 20, 64 - 77 (1971).

The topological character of a new type of solution representing a force-free magnetic field near bipolar sunspots is examined. It is shown that some of the observed topological features of chromospheric fibrils and filaments in Hα can be interpreted in terms of the configuration of the magnetic lines of force of the present solution.

072.047 Prominent Zeeman lines in sunspot spectra and their temperature sensitivity. A. Wittmann.
Solar Physics, Vol. 20, 78 - 80 (1971). – Research note.

072.048 The splitting of the π-component of the Zeeman triplet in a sunspot.
V. A. Golubev, V. F. Tshistjakov.
Astron. Tsirk., No. 638, p. 5 - 7 (1971). In Russian.

072.049 On the splitting of the π-component in a sunspot from different absorption lines. A. V. Baranov.
Astron. Tsirk., No. 641, p. 1 - 3 (1971). In Russian.

072.050 Investigations on the program "Rapid variations of solar magnetic fields", II. Rapid variations of the magnetic fields in two sunspot groups.
G. F. Vjalshin, H. Künzel.
Solnechnye Dannye 1971 Byull., No. 7, p. 70 - 74 (1971).

In Russian.

The results of investigations of rapid variations of maximum magnetic intensities in the main sunspots of the groups 207 and 209 (in the "Solar Data") during June 6 - 15, 1969 are given. Real variations of the intensities with an amplitude, on the average, of 500 Oe in 4 hours have been detected, which confirms the results previously obtained.

072.051 Investigations on the program "Rapid variations of solar magnetic fields", IV. Motions of sunspot umbrae. L. Dezsö, G. Gyertyános, B. Kálmán, Á. Kovács.
Solnechnye Dannye 1971 Byull., No. 7, p. 77 - 88 (1971). In Russian.

On the basis of a series of photoheliograms, obtained at the Heliophysical Observatory in Debrecen, accurate heliographic coordinates of several umbrae were determined and in many cases their areas have been measured also. In figures and tables various umbral motions and the variation in areas of some umbrae are shown.

072.052 Precise determinations of the magnetic vector in sunspots. M. G. Adam.
Monthly Notices Roy. Astron. Soc., Vol. 155, 169 - 183 (1971).

For the determination of magnetic field configurations Treanor's method, involving micrometer position measurements, is considerably more sensitive and more powerful than the usual methods depending upon line intensity measurements, but its full potential accuracy has not hitherto been achieved. Several modifications have now been made in the techniques of Treanor's method in order to obtain greater precision. The revised procedure has been used for magnetic field determinations across a medium sized spot and the consistency of the results for Fe I lines indicates that the direction of the magnetic vector is now obtained with a mean precision of 2–3°. The spot magnetic field, when the observations have this degree of certainty, is free from the anomalous effects suggested by earlier work.

072.053 Sunspot areas, 1907 - 1970.
R. S. Harrington, M. Miranian.
United States Naval Obs., *Washington*, Circ. No. 133, 18 pp. (1971).

Since 1907 the U. S. Naval Observatory has been taking daily photographs of the sun for the purpose of determining sunspot positions, areas, counts, and types. This present Circular summarizes the program and its results through 1970 and compares the area measurements with the more familiar Zurich mean daily sunspot numbers.

072.054 The asymmetry of solar activity in the years 1959– 1969. M. Waldmeier.
Solar Physics, Vol. 20, 332 - 344 = Astron. Mitt. Sternw. Zürich, No. 302 (1971).

In the decade 1959–1969 the solar activity showed a very strong asymmetry on the two hemispheres. On the northern hemisphere spots, faculae and prominences were more numerous and the white light corona was brighter than on the southern hemisphere. Over the more active hemisphere the corona was denser and hotter. Between density N_e and temperature T holds the relation: $N_e = 10^{-10} T^3$. The real asymmetry was strengthened by a phase difference of the two hemispheres. This phase shift is subject to a long period that contains 8 eleven-year cycles.

072.055 A feature of the secularly smoothed maxima of sunspot frequency. R. Henkel.
Solar Physics, Vol. 20, 345 - 347 (1971). – Research note.

072.056 On magneto-optical effects in sunspots.
A. Wittmann.

Solar Physics, Vol. 20, 365 - 368 (1971).

Magneto-optical effects on the circular polarization within the line Fe I $\lambda 6302.5$ are investigated. Quantitative results on the V-reversal near the line centre are given for homogeneous magnetic fields.

072.057 **Further study of H_2O lines in the umbral spectrum in the region of 0.93 μ.**
E. A. Mallia, D. E. Blackwell, A. D. Petford.
Solar Physics, Vol. 20, 369 - 371 (1971). – Research note.

072.058 **Über schnelle Änderungen der magnetischen Feldstärke von Sonnenflecken.** H. Künzel.
Astron. Nachr., Vol. 293, 105 - 110 (1971).

The results from hourly measurements of the magnetic field strengths of several main spots in three selected spot groups within the years 1968/69 are given. Besides a general tendency in the daily course of the field strength there are rapid variations up to 500 Gauss within a few hours. The development of the spot groups is described, and informations on the flare-activity are given.

072.059 **Driftende Sonnenflecke.** W. Schulze.
Sterne, 47. Jahrgang, p. 188 - 198 (1971).

072.060 **The distribution of electric conductivity and its gradients in the photospheric layers of an active region.**
M. Kopecký.
Bull. Astron. Inst. Czechoslovakia, Vol. 22, 343 - 346 (1971).

An approximate general model of the distribution of electric conductivity and of the gradients of the electric conductivity in the photospheric layers of an active region of the sun, namely in the vertical and the horizontal direction, is presented. At different locations and different geometrical depth the values of electric conductivity ($8 \times 10^9 - 3 \times 10^{12}$), the values of the gradients ($\Delta\sigma$ per 1 cm) of the electric conductivity ($0 - 10^6$) and the directions of these gradients vary. This means that relatively complicated systems of electric currents must exist in the active regions.

072.061 **Relation of surges to spotgroups and to flares.**
J. Kleczek, H. Kleczková, J. Kvíčala.
Bull. Astron. Inst. Czechoslovakia, Vol. 22, 347 - 351 (1971).

Nearly three thousand surges observed between July 1, 1957 and December 31, 1963 have been investigated. Their association with spotgroups and with flares on the disc studied in this paper corroborates earlier findings. The larger the spotgroup, the larger the probability of surge occurrence. There are, however, surges occurring far from spotgroups. There are also surges not associated with a flare. They are different phenomena even if they are often closely associated.

072.062 **The ratio of penumbral and umbral areas of sunspots in the 11-year solar activity cycle.**
A. Antalová.
Bull. Astron. Inst. Czechoslovakia, Vol. 22, 352 - 370 (1971).

The sun-spot observations published in the Greenwich Observations in the years 1916–1960 were the data used in investigating the ratio of the penumbral and umbral areas q in the years of the maximum of the 11-year solar cycle. The results are reported in detail.

072.063 **An objective calibration of the scale of sunspot-numbers.** M. Waldmeier.
Astron. Mitt. Sternw. Zürich, No. 304, 10 pp. (1971).

From the definition of the sunspot-relative-numbers it follows that this index is subject to a more or less degree to the quality of seeing and the personal method of counting spots. Therefore the observations have to be reduced to the common standard scale by applying an individual reduction factor. The Zürich standard scale has never been calibrated in an objective way. Nevertheless there are no indications of a change of that scale. There is a close correlation between the daily values of the sunspot-numbers R and the solar-radio-emission F at 10.7 cm wavelength. This correlation is even closer for the monthly and yearly means. As the intensity of the radio-emission can be measured in an objective way, it may yield a possibility to calibrate the scale of the sunspot-relative-numbers.

072.064 **Multi-channel magnetograph observations. III. Faculae.** E. N. Frazier.
Solar Physics, Vol. 21, 42 - 53 (1971).

Simultaneous observations of photospheric magnetic fields, CaII K emission, the 'photospheric network' and continuum faculae show that these four quantities are correlated in a complicated manner. Measurements of the photospheric network and the continuum faculae over a wide range of μ result in families of limb contrast curves. These curves indicate that the dependence on H is as important as the dependence on μ. They also indicate that the magnetic field has a preferred inclination of about 50°.

072.065 **The influence of the sunspot model on the Li-abundance.** G. Stellmacher, E. Wiehr.
Solar Physics, Vol. 21, 96 - 100 (1971).

The dependence of the Li-abundance on the equivalent width of the Li-resonance doublet at λ 6708 Å is calculated for different umbral models. The choice of the model strongly influences the deduced Li-abundance.

072.066 **Facular models and the sunspot energy deficit.**
P. R. Wilson.
Solar Physics, Vol. 21, 101 - 112 (1971).

The problem of the energy deficit in a sunspot is shown to be critically related to the depth of a given sunspot model. Recent facular models are discussed and a new model is derived from recent data using a two-dimensional radiative transfer analysis.

072.067 **Observation of filamentary structure in sunspot umbrae.** D. Papathanasoglou.
Solar Physics, Vol. 21, 113 - 115 (1971). – Research note.

During the transit of Mercury on 9 May 1970, some photographic observations of this phenomenon were carried out.

072.068 **The probable behaviour of sunspot cycle 21.**
W. Gleissberg.
Solar Physics, Vol. 21, 240 - 245 (1971).

After an explanation of the method of forecasting based upon the 80-yr sunspot cycle, reasons are given for assuming that the maximum of the present 80-yr cycle now has passed. Starting from this assumption some predictions about cycle 21 are made.

072.069 **A new representation of the 80-year cycle in sunspot frequency.** R. Hartmann.
Solar Physics, Vol. 21, 246 - 248 (1971).

The aim of the present note is to deduce the 80-year cycle from the yearly means $\bar{R}$ of the sunspot relative numbers immediately without using any kind of averaging or smoothing.

072.070 **Active longitudes of areas of spot groups in the 20-year cycle.** D. Mariş.
Solnechnye Dannye 1971 Byull., No. 8, p. 86 - 89 (1971).
In Russian.

Active longitudes of total areas of sunspots during 1964–1969 have been detected by the isoline method. They are shown to coincide practically with the active longitudes of the 18th and 19th solar cycles.

072.071 **On the nature of solar activity.** G. Ya. Vassilyeva, D. A. Kuznetsov, A. A. Shpitalnaya.
Solnechnye Dannye 1971 Byull., No. 8, p. 96 - 100 (1971). In Russian.

The regularities in the positions of Jupiter during the epochs of extreme solar activity for 33 cycles are considered. The cycles are supposed to be stimulated by the galactic magnetic field being disturbed by the planets in their orbital motion.

072.072 **On the configuration of magnetic fields of sunspots.** V. N. Milovanov.
Solnechnye Dannye 1971 Byull., No. 8, p. 108 - 111 (1971). In Russian.

The sunspots were classified according to the configuration of the magnetic field in relation to the depth. Three forms of magnetic tubes have been detected.

072.073 **Orientation of light bridges in sunspots.** G. S. Minasjants, S. O. Obashev, T. M. Minasjants.
Solnechnye Dannye 1971 Byull., No. 9, p. 72 - 74 (1971). In Russian.

The orientation of light bridges in sunspots with respect to the solar equator is investigated. It is shown that the bridges are situated in the north-south direction. This is an independent argument in favour of the statement that light bridges in most cases are facular chains penetrating into the sunspot umbra.

072.074 **Non-orthogonal expansions of curves of 11-year cycles.** A. B. Wertlieb, G. V. Kuklin.
Solnechnye Dannye 1971 Byull., No. 9, p. 79 - 87 (1971). In Russian.

072.075 **On fluctuations of Wolf numbers during the 20th solar cycle.** Y. I. Vitinsky.
Solnechnye Dannye 1971 Byull., No. 9, p. 106 - 108 (1971). In Russian.

Values of the fluctuation index for 1966–1970 and the lists of its positive and negative strong fluctuations during the rising branch of the 20-th solar cycle for Wolf numbers are given.

072.076 **Variable electrical conductivity in sunspot structure.** Y. Chen.
Thesis, Univ. Rochester, New York. [Available from Univ. Microfilms, Ann Arbor, Mich., U.S.A. Order No. 70–18050], 146 pp. (1970).

072.077 **Erratum: 'A study of short-period variations in the series of solar activity '** [Solnechnye Dannye 1971 Byull., No. 5, p. 107 - 114 (1971)]. K. A. Kandaurova.
Solnechnye Dannye 1971 Byull., No. 9, p. 113 (1971). In Russian.

072.078 **Dependence of faculae magnetic fields on height.** H. I. Abdussamatov, M. N. Stoyanova.
Solnechnye Dannye 1971 Byull., No. 10, p. 72 - 76 (1971). In Russian.

The value of the magnetic field intensity in some facula elements, measured simultaneously on two lines is shown to be constant in practice. Relative displacements of maxima of magnetic field intensities are observed at two levels.

072.079 **Analysis and calculation of magnetic fields of sunspots in the solar atmosphere (dipole approximation). I. The magnetic field of unipolar spots.** O. S. Korolev.
Solnechnye Dannye 1971 Byull., No. 10, p. 76 - 85 (1971). In Russian.

The analysis of the magnetic field of unipolar spots in the solar atmosphere was made by means of the normalized formulas of a dipole. It is shown that the magnetic field of small spots becomes localized in lower layers of the atmosphere, while that of large sunspots is due to the field in upper layers of the corona.

072.080 **On the problem of 11-year cycle pairs.** A. B. Wertlieb, G. V. Kuklin.
Solnechnye Dannye 1971 Byull., No. 10, p. 86 - 93 (1971). In Russian.

A significant difference between forms of odd and even cycle curves is shown. The study of coupling between cycles in pairs using the results of expansions of individual cycles and non-orthogonal expansions of cycle pairs in odd-even and even-odd combinations gives evidence of more close connections between cycles in pairs of an even-odd type. The nature of couplings between such cycle pairs seems to vary within a quasi-period of 170 - 180 years.

072.081 **Correlation between the solar activity indices used in the study of motion of artificial earth satellites.** A. M. Fominov.
Byull. Inst. Teoret. Astron., *Leningrad*, Vol. 13, 49 - 58 (1971). In Russian.

Correlations between the 10.7-cm solar flux, the visible solar spot area and the geomagnetic planetary index have been investigated for the 27-day intervals over the period 1957 – 1967.

072.082 **Low-frequency oscillatory convection in the strong magnetic field.** Y. D. Zhugzhda.
Cosmic Electrodynamics, Vol. 2, 253 - 266, 267 - 279 (1971). In Russian and English.

The conditions which can give rise to low-frequency convective oscillations in polytropic atmospheres in the presence of arbitrarily strong magnetic fields are studied. Two limiting cases are examined – quasi-adiabatic and quasi-isothermal oscillations. The mechanism of the growth of the type of instability considered is discussed as is the possibility of low frequency convection in sunspots.

072.083 **Relación entre el peso estadístico de manchas y la actividad solar.** F. H. Cabello.
Urania Barcelona, Año 56, No. 273, p. 61 - 64 (1971).

072.084 **Hinweise auf Zusammenhänge zwischen den Planeten und den Sonnenfleckenperioden.** R. Liese.
Techn. Univ. Hannover. Astron. Station Inst. Theor. Geod., No. 7, 48 pp. (1971).

072.085 **Die Verwendbarkeit von Monatsmitteln der Sonnenfleckenrelativzahlen zur Untersuchung kurzdauernder Schwankungen der Fleckenhäufigkeit.** R. Hartmann.
Veröff. Astron. Inst. Univ. Frankfurt, No. 32, 2 + 11 pp. (1971).

The fluctuations of the sunspot activity led to the discovery of the 11-year and the 80-year cycle. Until now one did not succeed in finding certain rules concerning the short-period variations. In this publication the temporary variations of sunspot activity are investigated using the monthly means of the daily sunpot relative numbers.

072.086 **"Free" pulsations of the contrast of sunspots in the chromosphere.** E. P. Surkov.
Astron. Tsirk., No. 657, p. 6 - 8 (1971). In Russian.

072.087 **A set of hydrostatic sunspot models.** A. Stankiewicz.
Geod. Geophys. Veröff., Ser. 2, No. 13, (see 012.029), p. 7 - 17 (1969).

072.088 **The statistical relation between the model para-**

meter $\Delta \Theta$ and magnetic field strength in sunspots.
A. Spodenkiewicz, A. Stankiewicz.
Geod. Geophys. Veröff., Ser. 2, No. 13, (see 012.029), p. 19 -
21 (1969).

072.089 Determination of the azimuthal electric currents in
the surface layers of sunspots.   J. Jakimiec.
Geod. Geophys. Veröff., Ser. 2, No. 13, (see 012.029), p. 23
(1969). – Abstract.

072.090 An attempt to construct the model of a facula.
J. Krawiecka.
Geod. Geophys. Veröff., Ser. 2, No. 13, (see 012.029), p. 25
(1969). – Abstract.

072.091 A magnetohydrodynamic model for the solar cycle.
I. K. Csada.
Geod. Geophys. Veröff., Ser. 2, No. 13, (see 012.029), p.
137 - 139 (1969).

072.092 On a fine structure of the latitude–time diagram.
M. Kopecky, G. V. Kuklin.
Geod. Geophys. Veröff., Ser. 2, No. 13, (see 012.029), p.
141 - 145 (1969).

072.093 Number of sunspot groups and their lifetimes in
the 19th cycle.   M. Kopecký.
Říše hvězd, Vol. 52, 146 - 148 (1971). In Czech.

072.094 New data on the tidal origin of solar activity dis-
turbances.   I. V. Maksimov, N. P. Smirnov.
Trudy Arkt. i antarkt. NII, 1971, No. 302, p. 26 - 38.
In Russian.

072.095 Solar activity and positions of the planets.
V. Shuvalov.
Nauka i zhizn', 1971, No. 10, p. 63 - 64. In Russian.

072.096 The molecular spectrum of sunspots.   J. C. Webber.
Thesis, California Inst. Technology, Pasadena.
[Available from Univ. Microfilms, Ann Arbor, Mich., U.S.A.
Order No. 70–18050], 80 pp. (1970).
Analysis led to estimates of the effective rotational temperatures of MgH, CaH and TiO. A new model was derived
from the molecular lines measured here, and shown to differ
widely from previous models. The usefulness of photographic
spectra for this purpose is seriously questioned, and suggestions are made for new observations.

072.097 On the spectral study of sunspot number and solar
radio flux between 600 and 9400 MHz.
D. R. K. Rao.
Proc. Indian Acad. Sci., Section A, Vol. 73, 223 - 231 (1971).
The relationship between the solar activity as represented
by sunspot number and the slowly varying component of solar
radio emission at frequencies from 600 to 9400 MHz, has
been studied for high solar activity (1957-58), declining phase
of moderate solar activity (1961-62) and ascending phase of
moderate solar activity (1965-66).

072.098 Analysis of hidden periodicities and the possibility
of statistical prediction.   O. B. Vasil'ev.
Proc. Sixth Winter School on Space Physics, Part II. Apatity
1969, (see 012.034), p. 121 - 128 (1971).

072.099 Die Fackelfeuer der Sonne.   H. Zirin.
Bild der Wissenschaft, [Deutsche Verlags-Anstalt,
Stuttgart], Vol. 8, 883 - 893 (1971).

The dynamics of a toroidal magnetic ring.
See Abstr. 062.028.

Isotopes of magnesium in the solar atmosphere.
See Abstr. 071.016.

The possibility of magnetic field origin in fine structure elements of solar features.   See Abstr. 071.038.

Anisotropy of electric conductivity and dissipation
of magnetic fields.   See Abstr. 071.039.

Measurement of the transversal magnetic field in
the chromosphere above a sunspot.
See Abstr. 073.009.

The position regularities of flares related to the
field maximum in sunspot groups.   See Abstr. 073.045.

Soft solar X-rays and solar acitivty. VI: Optical
identification of activity associated with X-ray background
fluctuations.   See Abstr. 076.029.

On the relations between development of active
regions on the sun and their succeeding events in cosmic ray
intensity.   See Abstr. 078.024.

Observations of the polar magnetic fields.
See Abstr. 080.022.

Dynamics of large-scale magnetic fields.
See Abstr. 080.023.

About the connection between the magnetic field
and movements in active regions of the sun.
See Abstr. 080.061.

Solar cycle control of $N_m(E)$.
See Abstr. 083.012.

Solar activity and planetary luminosity.
See Abstr. 091.005.

The interplanetary hydrogen cone and its solar
cycle variations.   See Abstr. 106.005.

The yearly variation of the cosmic ray intensity
from 1963 - 1967, its relation to the observed variation of
solar acitivity with heliolatitude and time.
See Abstr. 143.106.

On the role of the heliolatitudes of sunspots in the
11-year galactic cosmic ray modulation.
See Abstr. 143.114.

Solar activity and dimension of modulation region.
See Abstr. 143.115.

The 11-year intensity modulation of galactic cosmic
rays and the heliographic distribution of sunspots.
See Abstr. 143.143.

073 Solar Chromosphere, Flares, Prominences

073.001 The Harvard-Smithsonian reference atmosphere.
O. Gingerich, R. W. Noyes, W. Kalkofen, Y. Cuny.
Solar Physics, Vol. 18, 347 - 365 (1971).

We present a model of the solar atmosphere in the optical depth range from $\tau_{5000} = 10^{-8}$ to 25. It combines an improved model of the photosphere that incorporates recent EUV observations with a new model of the quiet lower chromosphere. The latter is based on OSO 4 observations of the Lyman continuum, on infrared observations, and on eclipse electron densities.

073.002 Hydrogen ionization and n=2 population for model spicules and prominences. A. Poland, A. Skumanich, R. G. Athay, E. Tandberg-Hanssen.
Solar Physics, Vol. 18, 391 - 402 (1971).

Using slab model atmospheres that are irradiated from both sides by photospheric, chromospheric, and coronal radiation fields we have determined the ionization and excitation equilibrium for hydrogen. We also present ionization curves for 6000 K, 7500 K, and 10000 K models ranging in total hydrogen density from $1 \times 10^{10} /cm^3$ to $3 \times 10^{12} /cm^3$. Using these curves it is possible to obtain the total hydrogen density from the n=2 population density in prominences and spicules.

073.003 Motions of Hα-spicules along the solar limb.
G. M. Nikolsky, A. G. Platova.
Solar Physics, Vol.18, 403 - 409 (1971).

A series of Hα-spectrograms obtained with the 21 in. Lyot coronagraph has been examined. Measurements of relative distances between spicules (nearly 50 features) embrace the time interval of about 21 min (38 pictures). It is found out that spicules oscillate along the limb with a characteristic time interval (period) about 1 min, characteristic amplitude of 1 arc sec and velocities about 10–15 km/s. The oscillations show no correlation for distant spicules.

073.004 Solar prominences and their magnetic fields—I.
E. Tandberg-Hanssen.
Sky Telescope, Vol. 42, 72 - 75 (1971).

073.005 Enhanced emission of iron nuclei in solar flares.
P. B. Price, I. Hutcheon, R. Cowsik, D. J. Barber.
Phys. Rev. Letters, Vol. 26, 916 - 919 (1971).

Etched tracks in an Apollo-12 spacecraft window and a Surveyor-3 camera-lens filter give the interplanetary Fe energy spectra from ~1 to ~30 MeV/nucleon during 1967–1969. The strongly energy-dependent Fe/He ratio suggests that heavy nuclei are preferentially emitted from accelerating regions because of their low ionization state and high magnetic rigidity. The Fe fluxes give rise to extremely high track densities that we have observed in lunar soil.

073.006 Particle acceleration in solar flares.
L. I. Miroshnichenko.
Priroda, No. 7.71, p. 37 - 41 (1971). In Russian.

073.007 On the mechanism of the glow of a flocculus in lines H, K and the infrared triplet of Ca⁺.
P. N. Polupan.
Astron. Zhurn. Akad. Nauk SSSR, Vol. 48, 730 - 737 (1971). In Russian. English translation in Soviet Astron. AJ, Vol. 15, No. 4.

It is shown that the principal mechanisms of glow of a flocculus in lines H, K and the infrared triplet of Ca⁺ are resonance scattering of the sun's radiation and electronic collision.

073.008 Ionization and excitation of helium in solar prominences. R. I. Kostik.
Astron. Zhurn. Akad. Nauk SSSR, Vol. 48, 738 - 746 (1971). In Russian. English translation in Soviet Astron. AJ, Vol. 15, No. 4.

The equations of stationarity for the five helium energy levels are derived and solved simultaneously with the transfer equations for the prominence with plane–parallel geometry. The calculation shows, the observed populations of triplet atomic energy levels can be interpreted by recombination mechanism of excitation with electron temperature $T_e = 5000° - 10000°$ and electron density $n_e = 10^{11} - 10^{12}$.

073.009 Measurement of the transversal magnetic field in the chromosphere above a sunspot.
V. A. Kotov.
Astron. Zhurn. Akad. Nauk SSSR, Vol. 48, 869 - 871 (1971). In Russian. English translation in Soviet Astron. AJ, Vol. 15, No. 4. – Short note.

073.010 High-resolution photography of the solar chromosphere. IX: Limb observations of high spectral purity. R. E. Loughhead, E. J. Tappere.
Solar Physics, Vol. 19, 44 - 51 (1971).

High-quality photographs of the solar limb and neighbouring regions of the disk at various wavelengths in the Hα line have been obtained through a tunable 1/8 Å filter used in tandem with a 1 Å Halle filter to eliminate parasitic light. The new observations throw further light on (a) the reappearance of the photospheric limb in the wings of Hα, and (b) the dark band lying immediately above the photospheric limb described by Loughhead (1969) and Nikolsky (1970).

073.011 Oscillations of visible chromosphere boundary and regularity in position of spicule groups along the limb. J. V. Platov, N. S. Shilova.
Solar Physics, Vol. 19, 52 - 58 (1971).

A series of Hα-filtergrams, covering about 42 min, was studied in order to obtain quantitative data on the oscillations of the visible chromosphere boundary and on the regularity in position of spicule groups along the limb.

073.012 Morphological relationships in the chromospheric Hα fine structure. P. Foukal.
Solar Physics, Vol. 19, 59 - 71 (1971).

A continuous relationship is proposed between the basic elements of the dark fine structure of the quiet and active chromosphere. A progression from chromospheric bushes to fibrils, then to chromospheric threads and active region filaments, and finally to diffuse quiescent filaments, is described.

073.013 On the topology of filaments and chromospheric fibrils near sunspots. Y. Nakagawa, M. A. Raadu, D. E. Billings, D. McNamara.
Solar Physics, Vol. 19, 72 - 85 (1971).

The similarity between the spiral topology of chromospheric fibrils and filaments observed in Hα near sunspots and the configuration of an axisymmetric force-free magnetic field is examined. It is suggested that some of the observed features could be interpreted in terms of the configuration of lines of force of an axisymmetric force-free chromospheric magnetic field. Implications of the results of analysis to the possible interpretations of other observed topological features near a sunspot are discussed.

073.014 **Thermal effects in the formation of loop prominences.** D. W. Goldsmith.
Solar Physics, Vol. 19, 86 - 91 (1971).

Loop prominences that appear after some solar flares may owe their form and duration to the behavior of the cooling function of the material (Cox and Tucker, 1969) and to the magnetic field configuration. Computer simulation of a model event shows that small temperature contrasts in a medium of several million degrees temperature may be enhanced as the medium cools by line and recombination radiation, on a time scale of several minutes. The loop prominence systems could be the result of this process repeated through successively higher levels of the solar atmosphere.

073.015 **Acceleration of electrons and solar flares due to quasi-static electric field.** T. Takakura.
Solar Physics, Vol. 19, 186 - 201 (1971).

The storage of flare energy, efficient acceleration of electrons and the trigger of the flares are suggested to be attributed to a quasi-static electric field caused by a gas motion near the photosphere without satisfying the frozen condition. The primary cause of the onset of flares would be the acceleration of electrons due to the electric field above a critical strength.

073.016 **On the flux of neutrons from flares.**
Z. Švestka.
Solar Physics, Vol. 19, 202 - 206 (1971).

Under the assumption that white-light flares are caused by energetic particles penetrating into the photosphere, the known number of protons needed for the white-light emission is used to obtain an estimate of the production of neutrons occurring at the same time.

073.017 **On the flux of gamma rays from solar flares.**
L. D. De Feiter.
Solar Physics, Vol. 19, 207 - 208 (1971). – Research note.

073.018 **Kurzer Bericht über Sonnenprotuberanzen 1970.**
E. Moser.
Orion, 29. Jahrgang, p. 85 - 88 (1971).

073.019 **Variability of the bright lines profiles in the spectra of quiet prominences.**
G. Y. Smolkov, V. S. Bashkirtsev.
Solnechnye Dannye 1971 Byull., No. 4, p. 77 - 86 (1971). In Russian.

The character and extent of variability of the most bright lines at the transition from one point of the prominence to the other as well as from one quiet prominence to the other were considered.

073.020 **Analysis of the weak chromospheric flare spectrum.**
E. Kurochka, L. Kurochka, T. Stembcovskaya.
Solnechnye Dannye 1971 Byull., No. 4, p. 87 - 95 (1971). In Russian.

Some parameters of the emission of the H, He and CaII lines profiles were obtained on the basis of spectrophotometric investigations of the limb flare spectrum. Optical depth, the total number of hydrogen atoms per 1 cm² of the flare in the different quantum states, the electron temperature and other characteristics of the matter state in the flare at some distances from the limb profiles of Lyman, Paschen and Bracket series have been calculated. The obtained results are analysed.

073.021 **On flare associations as a manifestation of "sympathy" of the flares within twenty-four hours.**

V. V. Kassinsky.
Solnechnye Dannye 1971 Byull., No. 4, p. 104 - 110 (1971). In Russian.

073.022 **Electron densities derived from line intensity ratios: Beryllium isoelectronic sequence.**
R. H. Munro, A. K. Dupree, G. L. Withbroe.
Solar Physics, Vol. 19, 347 - 355 (1971).

A direct method for determining electron densities from emission line intensities of ions in the beryllium isoelectronic sequence is described and then applied to the analysis of extreme ultraviolet CIII and O V spectra from both quiet and active areas in the solar transition region.

073.023 **Solar rotation: Direct evidence from prominences for a westward wind.** W. Livingston.
Solar Physics, Vol. 19, 379 - 383 (1971).

Hα and K-line spectra of quiescent prominences, taken with the slit placed normal to the limb, commonly reveal a gas streaming (5–50 km/s) that is peculiar to the upper edge of these objects. On the average this streaming is uni-directional and consistent with a hypothetical east–west wind.

073.024 **The abundance of helium in prominences and in the chromosphere.** T. Hirayama.
Solar Physics, Vol. 19, 384 - 400 (1971).

The abundance of helium relative to hydrogen is spectroscopically determined in prominences and in the chromosphere by using 1952, 1958, 1962 and 1966 eclipse data. A new chematic model of the chromosphere is presented where spicules have no hot region of emitting neutral helium lines. Here it is suggested that the kinetic temperature of spicules, 6000 ~ 8000 K, would be primarily determined by the radiation temperature of the corona and the transition region beyond the Lyman continuum of hydrogen which happens to be around those temperatures.

073.025 **Energy balance in cool quiescent prominences.**
A. Poland, U. Anzer.
Solar Physics, Vol. 19, 401 - 413 (1971).

The energy balance for cool quiescent prominences is examined using a 6000 km, 6000 K isothermal slab model prominence with a density gradient dictated by a modified Kippenhahn-Schlüter model. We show that for a reasonable combination of the temperature and the length scale for various temperature changes we can establish an energy balance in quiescent prominences between the gain through thermal conduction and the net radiative loss. Therefore we can construct a model of a quiescent prominence for which a steady state exists, once a prominence is formed.

073.026 **The helium radiation in prominences.**
N. A. Yakovkin, M. Yu. Zeldina.
Solar Physics, Vol. 19, 414 - 430 (1971).

The present paper shows that at temperatures of 7000 K strong 4686 Å (He⁺) emission in loop prominences is associated with the process of charge-exchange of helium ions caused by streams of particles. Because of the efficiency of charge-exchange collisions the emission is provided by low-density streams and having low velocities of the order of some hundred km/s. Under the action of these streams the prominences practically are not heated in consequence of high radiation losses.

073.027 **On the distribution of material as a function of temperature in the post-flare loop system of 12 August 1970.** R. R. Fisher.
Solar Physics, Vol. 19, 440 - 450 (1971).

Observations of the post-flare loop system formed after the east limb proton flare of 12 August 1970 include (a) sets of filtergrams from which photographic subtractions have been constructed and (b) spectra from which a distribution of electron density as a function of temperature for three coronal regions are derived. The filtergrams show no indications of radial velocities in excess of 80 km/s. The spectra indicate an increase in density at the tops of the loops with most of the material at a relatively cool temperature. The distribution functions obtained for areas just above and just below loops indicate a lower electron density and the presence of material at high temperatures.

073.028 **Mass motions in a flare spray.** M. K. McCabe.
Solar Physics, Vol. 19, 451 - 462 (1971).

The morphology of the March 12, 1969 flare prominence is described and the results of position and velocity measurements made on the fragments of ejected material are shown and the optical features are related to the radio emission and to the structure of the coronal magnetic field.

073.029 **Magnetic fields, bremsstrahlung and synchrotron emission in the flare of 24 October 1969.**
H. Zirin, G. Pruss, J. Vorpahl.
Solar Physics, Vol. 19, 463 - 471 (1971).

An impulsive flare October 24, 1969 produced two bursts with virtually identical time profiles of 8800 MHz emission and X-rays above 48 keV. The two spikes of hard X-rays correspond in time to the times of sharp brightening and expansion in the $H\alpha$ flare. A model of the flare based on $H\alpha$ observations at Big Bear shows that the density of electrons with energy above 10 keV is 5×10^7 if the field density is 10^{11}.

073.030 **On the observation of the solar limb flare on November 4, 1968.** Ts. S. Khetsuriani, E. I. Tetruashvili.
Byull. Abastumansk. Astrofiz. Obs., No. 40, p. 93 - 96 (1971).
In Russian.

073.031 **Errata: 'On the existence of solar-flare plasma of $T \geq 10^8$ °K' [Astrophys. Journ., Vol. 164, 365 - 368 (1971)].** S. Kahler.
Astrophys. Journ., Vol. 168, 319 - 320 (1971).

073.032 **Solar prominence and their magnetic fields – II.**
E. Tandberg-Hanssen.
Sky Telescope, Vol. 42, 142 - 145 (1971).

073.033 **On the "transparency" of prominences in the UV emission with $\lambda \leq 504$ Å.** N. N. Morozhenko.
Solnechnye Dannye 1971 Byull., No. 6, p. 67 - 72 (1971).
In Russian.

Quiet prominences are concluded to have fibrous structure, which facilitates penetration of the outer UV emission into them with $\lambda \leq 504$ Å. The rate of reducing this emission caused by prominences with various optical depths is shown in a figure.

073.034 **The solar chromosphere and the general structure of a stellar atmosphere.**
R. N. Thomas, K. B. Gebbie.
National Bureau Standards Special Publ. 353, (see 012.001), p. 84 - 111 (1971).

073.035 **Solar magnetic movie.**
N. K. Baker, T. J. Janssens.
Bull. American Astron. Soc., Vol. 3, 376 (1971). – Abstr. AAS.

073.036 **The time development of chromospheric flares.**
R. S. Kandel, M. D. Papagiannis, F. M. Strauss.
Bull. American Astron. Soc., Vol. 3, 376 (1971). – Abstr. AAS.

073.037 **Solar flares in the extreme ultraviolet.**
A. T. Wood, Jr.
Bull. American Astron. Soc., Vol. 3, 384 (1971). – Abstr. AAS.

073.038 **Thermal effects in the formation of loop prominences.** D. W. Goldsmith.
Bull. American Astron. Soc., Vol. 3, 416 (1971). – Abstr. AAS.

073.039 **Observations of magnetic fields in quiescent prominences.** E. Tandberg-Hanssen.
IAU Symposium No. 43, (see 012.003), p. 192 - 200 (1971).

073.040 **Application of the chromospheric magnetograph to active regions.** H. Zirin.
IAU Symposium No. 43, (see 012.003), p. 237 - 242 (1971).

073.041 **Magnetic fields measured with the 10830 Å HeI line.** J. Harvey, D. Hall.
IAU Symposium No. 43, (see 012.003), p. 279 - 288 (1971).

073.042 **Solar magnetic fields in association with flares.**
R. Michard.
IAU Symposium No. 43, (see 012.003), p. 359 - 366 (1971).

073.043 **Magnetic fields associated with solar flares.**
E. B. Mayfield.
IAU Symposium No. 43, (see 012.003), p. 376 - 389 (1971).

073.044 **Electric currents connected with the proton flares of 7 July and 2 September, 1966.** A. B. Severny.
IAU Symposium No. 43, (see 012.003), p. 417 - 421 (1971).

073.045 **The position regularities of flares related to the field maximum in sunspot groups.** V. V. Kasinsky.
IAU Symposium No. 43, (see 012.003), p. 432 - 434 (1971).

073.046 **Volume characteristics of magnetic-channel flares.**
L. Křivský.
IAU Symposium No. 43, (see 012.003), p. 443 - 449 (1971).

073.047 **The relation between dashes and flares. (Physical nature of the dash phenomena).**
D. A. Kuznetsov, A. A. Shpitalnaya.
IAU Symposium No. 43, (see 012.003), p. 450 - 453 (1971).

073.048 **Active regions at millimeter wavelengths and the measurement of magnetic fields.** M. R. Kundu.
IAU Symposium No. 43, (see 012.003), p. 642 - 651 (1971).

073.049 **On the orientation of magnetic fields in quiescent prominences.** U. Anzer, E. Tandberg-Hanssen.
IAU Symposium No. 43, (see 012.003), p. 656 - 662 (1971).

073.050 **Magnetic fields in polar prominences.**
G. Y. Smolkov.
IAU Symposium No. 43, (see 012.003), p. 710 - 713 (1971).

073.051 **Spectrophotometry of solar prominences.**
A. N. Sergeeva, N. A. Yakovkin.
Vestn. Kiev. Un-ta, Ser. Astron., No. 12, p. 3 - 10 (1970).
In Russian.

Photometric results of three bright prominence spectra are reduced. The tables contain central intensities in units of the continuous spectrum of the disk centre, complete and Doppler half-widths, equivalent widths and the numbers of atoms at the upper level along the line of sight. Kinetic temperatures, turbulent velocities are given. Complete number of the hydrogen atoms and $L\alpha$ temperatures are determined by the weak metal lines observed in prominences.

073.052 **On the mechanism of glow of a flocculus in the hydrogen lines.** P. N. Polupan.
Vestn. Kiev. Un-ta, Ser. Astron., No. 12, p. 11 - 16 (1970). In Russian.
A comparison of the calculated intensities of lines of Balmer series with the observed ones in the flocculus on June 15, 1967 is given.

073.053 **Extreme ultraviolet radiation from solar flares.** V. B. Bhatia, J. N. Tandon.
Astrophys. Letters, Vol. 9, 13 - 15 (1971).
To explain the similarity between the time structure of the extreme ultraviolet flashes from solar flares and the hard X-rays it is suggested that the extreme ultraviolet radiation is produced by the synchrotron emission of the same electrons which are responsible for hard X-rays. The flux of extreme ultraviolet radiation is calculated and compared with observations.

073.054 **Approximate determination of the source function in solar prominences.**
N. A. Yakovkin, M. Y. Zeldina.
Vestn. Kiev. Un-ta, Ser. Astron., No. 13, p. 31 - 47 (1971). In Russian.
The mean number of scattering acts of a quantum in dependence on the law of distribution of energy sources in a flat layer and a sphere of optical thickness $\tau_0 = 100$ is calculated. The connection of the scattering number with the self-absorption factor is also considered. It is shown that the physical parameters derived from the observations, characterizing the diffusion of radiation in a sphere and a flat layer, depend to a great extent on the assumed model. An approximate solution of the integral equation of the diffusion of radiation for a flat layer of any optical thickness and a sphere is derived.

073.055 **Excitation and ionization of hydrogen in a flocculus.** P. N. Polupan.
Vestn. Kiev. Un-ta, Ser. Astron., No. 13, p. 48 - 56 (1971). In Russian.
Electron temperature, electron density, efficient thickness, concentration of hydrogen atoms on the levels, and other parameters of a flocculus are determined.

073.056 **Observations of solar flares at Ondřejov Observatory during the year 1970.** F. Hřebík, J. Kvíčala, L. Křivský, J. Olmr.
Bull. Astron. Inst. Czechoslovakia, Vol. 22, 305 - 320 (1971).
This paper is a continuation of twelve preceding reports (which contained chromospheric flares observed at Ondřejov Observatory from 1948 up to the end of 1969) and covers the period of 1970. The data, about 208 flares associated with 9400, 808, 536 and 260 MHz events, and 27 KHz SEA observed and recorded during 1970 are summarized in tables. The corresponding curves of the Hα line-width changes are plotted in figures. The net times of observations of the chromosphere made by the spectrohelioscope and the net times of measurements of the solar flux at 536 MHz are also given.

073.057 **Prominences.** S. B. Pikel'ner.
Zemlya i Vselennaya, No. 5, p. 13 - 19 (1971). In Russian.

073.058 **Spectral investigation of chromospheric fine structure.** U. Grossmann-Doerth, M. von Uexküll.
Solar Physics, Vol. 20, 31 - 46 = Mitt. Fraunhofer Inst., *Freiburg*, No. 105 (1971).
Hα spectra and effectively simultaneous filtergrams were taken at the Fraunhofer Observatory on Capri with the 35 cm domeless Coudé. The spatial resolution of the 19 best spectra selected for analysis was estimated to be 1–2 arc sec. The comparison of several hundred Hα line profiles emitted by typical chromospheric structure elements with theoretical prediction yielded strong evidence to suggest that the chromosphere consists of two parts: a lower, rather uniform layer at rest superposed by 'clouds' (condensations of great spatial variability). For most image points the line-of-sight velocity, optical thickness, source function and Doppler broadening of these clouds could be determined.

073.059 **A study of the fine structure of the solar chromosphere at the limb.** C. E. Alissandrakis, C. J. Macris.
Solar Physics, Vol. 20, 47 - 56 (1971).
We have measured the dimensions, distances from the inner limb and the lifetime of bright mottles at the limb. Spicule lifetimes have been measured too. The problem of the 'dark band', lying just above the inner limb, as well as the relations between bright mottles and spicules are discussed.

073.060 **Radio observations of filaments during the eclipses of September 11, 1969 and March 7, 1970.**
M. Simon, B.-A. Wickström.
Solar Physics, Vol. 20, 122 - 129 (1971).
By high resolution observations at 3 mm wavelength during the eclipses of September 11, 1969 and March 7, 1970 we have observed two filaments on the solar disc in absorption. We analyze in detail the 3.3 mm eclipse data and 9.5 mm (non-eclipse) data we have pertaining to the large filament observed on September 11, 1969.

073.061 **A note on the acceleration phase of high-energy particles in the solar flare on 7 July, 1966.**
K. Sakurai.
Solar Physics, Vol. 20, 147 - 149 (1971). – Research note.

073.062 **Magnetic energy conversion processes and solar flares.** B. Coppi, A. Friedland.
Trudy Mezhdunar. seminara 1970 po probl. "Generatsiya kosmich. luchej na Solntse", p. 320 - 378. Moskva (1971). Abstr. in Referativ. Zhurn. 51. Astron., 11.51.513 (1971).

073.063 **Oscillatory motions in a solar prominence of March 26, 1964.** A. A. Shpitalnaya, E. Ţifrea.
Stud. Cerc. Astron., Vol. 16, 131 - 139 (1971).

073.064 **Eruptive phenomena during the period November 14–16, 1970 in the F type sunspot group and their geophysical correlations.** G. Mariş, S. Nicolescu.
Stud. Cerc. Astron., Vol. 16, 207 - 214 (1971). In Roumanian.
We analyse the Hα filtergrams of the eruptive phenomena in a large sunspot group during the period November 14–16, 1970, associated with radio, X-ray, solar protons and ionospheric events. The correlation in time between various examined aspects of the solar active regions seems to indicate that the sequence for the beginning of the different phenomena is in general: optical flares, X-ray events, radio events.

073.065 **Filament solar de mare latitudine, 3–11 aprilie 1971.** E. Ţifrea, S. Nicolescu.
Stud. Cerc. Astron., Vol. 16, 221 (1971). – Note.

073.066 **On non-thermal motions of matter in chromospheric spicules.** E. V. Kononovich.
Astron. Tsirk., No. 638, p. 1 - 2 (1971). In Russian.

073.067 **The center-limb variation of the intensity of radiation in the cores of the Hα- and K Ca II lines for the quiet chromosphere and flocculi.** I. F. Nikulin.
Astron. Tsirk., No. 638, p. 3 - 5 (1971). In Russian.

073.068 On the structure of the active region in the K₂ Ca II line of the sun. T. I. Redyuk.
Astron. Tsirk., No. 641, p. 3 - 5 (1971). In Russian.

073.069 On temporal intensity variations of the K₂ Ca II line in an active region of the sun. T. I. Redyuk.
Astron. Tsirk., No. 641, p. 5 - 7 (1971). In Russian.

073.070 Fine structure of motions in the chromosphere over a sunspot. M. Mamadazimov.
Solnechnye Dannye 1971 Byull., No. 7, p. 88 - 95 (1971). In Russian.

The character of motions at two chromospheric levels over a sunspot of the group 239 (1969) was investigated.

073.071 Processes of magnetic-energy conversion and solar flares. B. Coppi, A. B. Friedland.
Astrophys. Journ., Vol. 169, 379 - 404 (1971).

A model is presented for the onset and development of solar flares which relies on the transformation of magnetic energy to kinetic energy in a plasma. A two-dimensional configuration in which plasma counterflows cross magnetic fields of opposite polarities is considered. The flow velocities are assumed to be both subsonic and sub-Alfvénic, and no shockwave processes are considered. These velocities are taken sufficiently large to generate a relatively strong electric field and to set up a sequence of instabilities, the first of which is explosive in the region of zero magnetic field.

073.072 Efficiency of solar flares.
M. I. Pudovkin, A. D. Chertkov.
Dokl. Akad. Nauk SSSR. Ser. Mat. Fiz., Vol. 201, 75 - 77 (1971). In Russian.

073.073 A measurement of the non-thermal velocity in the low chromosphere. R. C. Canfield.
Solar Physics, Vol. 20, 275 - 281 (1971).

I have determined horizontally averaged non-thermal velocities from Jensen and Orrall's (1963) observations of Doppler widths of weak rare-earth emission lines in the wings of H and K. Combining these results with previous rare-earth line results, I conclude that this velocity in the low chromosphere (300–600 km) is 2.0 ± 0.2 km/s, and changes little with height.

073.074 Nature of the fine structure of the middle chromosphere. S. B. Pikel'ner.
Solar Physics, Vol. 20, 286 - 294 (1971).

Fine dark Hα filaments 'fibrils' form at the limb, apparently in most of the middle chromosphere corresponding to an altitude between 1500–2000 km and 4000 km. Their temperature is about 18000 K and the density about 5×10^9 cm^{-3}. The gas in the fibrils is ionized by electronic collisions and by the external ultraviolet radiation. Their calculated optical thickness in Hα is about 1. The fibrils in active regions are wider and show more contrast. The emission of the fibrils at the limb is explained by the scattering of the solar radiation.

073.075 Observation in the wing of the Hα line and identification of the spicular structure near the solar limb.
S. Koutchmy, C. Macris.
Solar Physics, Vol. 20, 295 - 297 (1971).

Composite photographs were constructed from a positive picture of the spicular structure and from the adjacent disk structure reproduced negatively. This method seems to be very suitable for the study of the morphology of the spicular structures.

073.076 Hα fine structure and the chromospheric field.
P. Foukal.
Solar Physics, Vol. 20, 298 - 309 (1971).

The physical characteristics of the Hα structures previously defined as fibrils and threads are studied. The interpretation of the fibrils as ends of flux tubes is useful in tracing the behavior of the transverse field component over the solar surface. The observed properties of fibrils and threads are consistent with the hypothesis that they are produced by a shock wave mechanism similar to that advanced by Parker to explain spicules.

073.077 A note on chromospheric fine structure at active region polarity boundaries. S. W. Prata.
Solar Physics, Vol. 20, 310 - 316 (1971).

High resolution Hα filtergrams from Big Bear Solar Observatory reveal that some filamentary features in active regions have fine structure and hence magnetic field transverse to the gross structure and the zero longitudinal field line. These features are distinct from the usual active region filament, in which fine structure, magnetic field and filament are all parallel to the zero longitudinal field line. The latter occur on boundaries between regions of weaker fields while the former occur at boundaries between regions of stronger field.

073.078 High dispersion spectroscopic study of quiescent prominences. O. Engvold, W. Livingston.
Solar Physics, Vol. 20, 375 - 388 (1971).

The utility of very high dispersion spectra (5–11 mm/Å) for the study of line profile and velocity structure in quiescent prominences is demonstrated by observations, taken with the spectrographic slit positioned normal to the limb in Hα λ6563, He D₃ λ5876, and Ca⁺K λ3933 Å. The emission profiles of both Hα and the K line often show a central reversal (absorption). Emission structures in the K-line can be complex with details as narrow as 0.04 Å.

073.079 Extreme-ultraviolet observations of a surge.
R. P. Kirshner, R. W. Noyes.
Solar Physics, Vol. 20, 428 - 437 (1971).

A flare surge at the limb was observed in C III 977 Å by the Harvard OSO 6 spectroheliometer. The kinematic behavior of the surge is the same in C III and in Hα. The amount of C III emission is consistent with a model in which the C III ions occupy sheaths with thickness ~ 100 km surrounding the cooler Hα-emitting threads. The mass of the material containing C III ions is about 10^{-2} times that emitting Hα.

073.080 On the relative shifts of Hα in the spectrum of the chromosphere of the sun. I. E. D. Khilov.
Vestn. Leningr. un-ta, 1971, No. 7, p. 142 - 148. In Russian.
Abstr. in Referativ. Zhurn. 51. Astron., 12.51.410 (1971).

073.081 Iron-line emission during solar flares.
G. A. Doschek, J. F. Meekins, R. W. Kreplin, T. A. Chubb, H. Friedman.
Astrophys. Journ., Vol. 170, 573 - 586 (1971).

Iron-line emission ~1.9 Å recorded by NRL Bragg crystal spectrometers on OSO-6 during intense solar soft X-ray flares is discussed. Individual emission features are resolved in second order, and variability in emission from Fe XXV relative to inner-shell emission from lower ionization stages is considered. A feature at 1.932 Å is attributed to radiation from Fe II in the photosphere.

073.082 An upper limit on the hardness of the nonthermal electron spectra produced during the flash phase of solar flares. S. R. Kane.
Astrophys. Journ., Vol. 170, 587 - 591 (1971).

The observations of impulsive solar-flare X-rays ≳ 10 keV made with the OGO-5 satellite have been analyzed in order to study the variation of the nonthermal electron spectrum from one flare to another. The X-ray spectrum at the maxima of

129 impulsive X-ray bursts is represented by $KE^{-\gamma}$ photons $cm^{-2} s^{-1} keV^{-1}$, and the frequency of occurrence of bursts with different values of γ is studied.

073.083   Eruptive prominence of 3 May 1971.
M. Rybanský.
Bull. Astron. Inst. Czechoslovakia, Vol. 22, 380 - 381, 382b - 382d (1971).

The prominence reached a height of 540000 km. The maximum velocity observed was 200 km s^{-1}.

073.084   Stationäre Filaments, ihre koronale Umgebung sowie ihre Beziehung zum solaren Magnetfeld.
W. Stanek.
Astron. Mitt. Sternw. Zürich, No. 309, 87 pp. (1971). – Dissertation ETH Zürich.

In order to make full use of the intensity contours over prominences, the intensity scale used at Arosa had to be calibrated by comparison with the measurements in absolute units of Pic du Midi Observatory. Model calculations for the distribution of the electrons in the surroundings of prominences up to 2 solar radii were made, based on photographs of solar streamers in integral light. The "streets of prominences" follow the boundaries between areas of opposite magnetic polarity. The situation of these areas is dependent on the solar cycle.

073.085   A comparison between MgII and CaII spectroheliograms.   K. Fredga.
Solar Physics, Vol. 21, 2, 60 - 81 (1971).

A detailed photometric comparison between a MgII K filterheliogram and a nearly simultaneous CaII K spectroheliogram shows a close correspondence in both location and intensity of the bright features on the sun. We also estimate theoretically at which heights in the solar atmosphere the radiation recorded in these heliograms originate. We arrive an average height of 1700 - 1900 km above the photosphere.

073.086   The effect of two-dimensional macroscopic velocity fields on models of the lower solar chromosphere.
C. J. Cannon.
Solar Physics, Vol. 21, 82 - 92 (1971).

It is found that the large scale fluctuation data can be explained by models of the lower solar chromosphere in which the inhomogeneous effects arise only from horizontal, two-dimensional macroscopic velocity fields. It is also shown, however, that the corresponding small scale fluctuation data cannot be explained in a similar manner.

073.087   Chromospheric absorbing features promising the appearance and the development of an active center.
M.-J. Martres, I. Soru-Escaut.
Solar Physics, Vol. 21, 137 - 145 (1971).

We describe short lived chromospheric dark features with strong velocity fields and we show their correlation with the birth and the further development of an active center. It is shown that radial velocities precede the modifications of magnetic fields. An attempt to compare these chromospheric velocities and photospheric ones is made.

073.088   Solar flares in the extreme ultraviolet.   L. A. Hall.
Solar Physics, Vol. 21, 167 - 175 (1971).

Measurements of flare-related impulsive enhancements in solar emission lines in the extreme ultraviolet, observed from the satellite OSO-III, are reported. The maximum enhancements of radiation from ions in the chromosphere-corona transition region precede the Hα maximum by an average of 2 min, and occur in the same period of time as the hard component of solar X-rays and the impulsive microwave bursts.

073.089   The transfer of Lyman continuum radiation in chromospheric flares.
R. S. Kandel, M. D. Papagiannis, F. M. Strauss.
Solar Physics, Vol. 21, 176 - 187 (1971).

We study the time evolution of a layer of the middle or lower chromosphere being heated by a stream of energetic particles during a solar flare. The region, which is not in LTE, is allowed to cool by the transfer of Lyman continuum radiation, with collisional as well as radiative processes being considered. The resulting time dependence of the electron density and the effective thickness of the layer are in good agreement with values derived from observations.

073.090   The sources of helium excitation in loop prominences on July 9 and 11, 1966.
N. N. Morozhenko.
Solnechnye Dannye 1971 Byull., No. 8, p. 81 - 85 (1971).   In Russian.

The excitation rate of metastable helium in loop prominences on July 9 and 11, 1966 is estimated. A comparison with quiet prominences gives evidence of the following: 1) In the loop prominences under investigation the metastable helium is 4−7 times more excited than in the quiet ones; 2) An increased UV-radiation of the chromosphere and corona, peculiar to active regions of the sun, can be the source of additional excitation.

073.091   On the magnetic field in emission knots of flares.
H. I. Abdussamatov.
Solnechnye Dannye 1971 Byull., No. 9, p. 67 - 72 (1971). In Russian.

The method of obtaining "pure" emission profiles of spectral lines for the whole flare, projected on the solar disc, is described. The first attempt is made to measure the magnetic field in the emission knots of small flares. The value of the magnetic field intensity in the investigated emission knots varies from 50 to 300 oerst.

073.092   On arising of magnetic ropes in the convective zone. II. The Parker mechanism of arising.
A. A. Soloviev.
Solnechnye Dannye 1971 Byull., No. 10, p. 93 - 98 (1971). In Russian.

It is shown that magnetic ropes arising in the body of the sun due to "magnetic buoyancy" have the shape of a pointed loop, whose edge is directed to the photosphere.

073.093   The manifold structure of the chromosphere and corona.   H. Zirin.
Phil. Trans. Roy. Soc. London A, Vol. 270, No. 1202, (see 012.009), 77 - 80 (1971).

073.094   The structure of the chromosphere−corona transition region from limb and disk intensities.
W. M. Burton, C. Jordan, A. Ridgeley, R. Wilson.
Phil. Trans. Roy. Soc. London A, Vol. 270, No. 1202, (see 012.009), 81 - 98 (1971).

073.095   Measurements of electron temperature in the solar chromosphere and corona.   L. Heroux, M. Cohen.
Phil. Trans. Roy. Soc. London A, Vol. 270, No. 1202, (see 012.009), 99 - 107 (1971).

073.096   The role and necessity of optical space observations in solar physics.   K. O. Kiepenheuer.
Phil. Trans. Roy. Soc. London A, Vol. 270, No. 1202, (see 012.009), 109 - 116 = Mitt. Fraunhofer Inst., *Freiburg*, No. 107 (1971).

073.097   Extreme ultraviolet emission during flares.
W. M. Glencross.

Phil. Trans. Roy. Soc. London A, Vol. 270, No. 1202, (see 012.009), 117 - 125 (1971).

073.098 On the model of initial (premaximum) phase of a chromospheric flare.
R. E. Guseinov, V. S. Imshennik, V. V. Paleichik.
Astron. Zhurn. Akad. Nauk SSSR, Vol. 48, 1217 - 1226 (1971). In Russian. English translation in Soviet Astron. AJ, Vol. 15, No. 6.
   A model of the initial (premaximum) phase of a chromospheric flare in terms of the strong explosion theory is discussed. The results of solution of the problem with an account of all significant dissipative processes are given.

073.099 The glow of the line He II λ 4686 in quiescent prominences.   N. N. Morozhenko.
Astron. Zhurn. Akad. Nauk SSSR, Vol. 48, 1237 - 1243 (1971). In Russian. English translation in Soviet Astron. AJ, Vol. 15, No. 6.
   The glow of the line He II λ 4686 in bright and faint prominences is considered.

073.100 The nature of the fine structure of the chromosphere.   S. B. Pikel'ner.
Astron. Zhurn. Akad. Nauk SSSR, Vol. 48, 1212 - 1216 (1971). In Russian. English translation in Soviet Astron. AJ, Vol. 15, No. 6.

073.101 Magnetic fields of active regions and flares.
V. F. Chistjakov.
Astron. Tsirk., No. 657, p. 4 - 6 (1971). In Russian.

073.102 Some characteristics of the upper parts of large flares.   V. G. Banin.
Geod. Geophys. Veröff., Ser. 2, No. 13, (see 012.029), p. 49 - 54 (1969).

073.103 The region of limb flares and active prominences of July 6, 1968.
B. Rompolt, T. Kozar, E. Szumiejko.
Geod. Geophys. Veröff., Ser. 2, No. 13, (see 012.029), p. 55 - 79 (1969).

073.104 On the magnetic field of active regions.
M. M. Molodensky.
Geod. Geophys. Veröff., Ser. 2, No. 13, (see 012.029), p. 124 - 127 (1969).

073.105 Il doppio bordo della cromosfera solare.
G. Godoli, O. Morgante, M. L. Sturiale.
Mem. Soc. Astron. Italiana, Nuova Ser., Vol. 42, 579 - 583 (1971).
   The solar chromospheric double limb cannot be due to scattered light. According to recent research the double limb is due to an instrumental effect. This effect can be utilized for the measurement of the height of the solar chromosphere.

073.106 La zona di transizione tra la cromosfera e la corona solare.   G. Poletto.
Mem. Soc. Astron. Italiana, Nuova Ser., Vol. 42, 585 - 616 (1971).
   Theoretical and empirical models of the transition region between the solar chromosphere and the corona are reviewed. The problem of abundance determination from *UV* data is also illustrated.

073.107 Solar explosions.   J. P. Wild.
   Search, Vol. 2, No. 7, p. 229 - 233 = Separate print Division Radiophys., C.S.I.R.O., Sydney, Australia (1971).

073.108 Heating of the chromospheric network by hydro-

magnetic waves.   R. W. Milkey.
Thesis, Indiana Univ., Bloomington. [Available from Univ. Microfilms, Ann Arbor, Mich., U.S.A. Order No. 70–19096], 82 pp. (1970).
   Looks at the propagation and dissipation of hydromagnetic waves for the purpose of demonstrating the effect of concentrated magnetic fields on the height distribution of energy deposition in the solar atmosphere, with the goal of illustrating a mechanism for the origin of the calcium network.

073.109 Velocity fields in magnetically disturbed regions of the Hα chromosphere.   P. H. Roberts, Jr.
Thesis, California Inst. Technology, Pasadena. [Available from Univ. Microfilms, Ann Arbor, Mich., U.S.A. Order No. 70–14849], 148 pp. (1970).
   Doppler movies taken in Hα show a long lived flow region connecting spots of opposite magnetic polarity in newly developing sunspot regions. A physical model for the arches is proposed, whereby they consist of material trapped by magnetic lines of force which are emerging from the photosphere in the forming sunspot region. The study of velocity features in the Hα chromosphere of the quiet sun has been extended to active regions.

073.110 Longitude distribution of solar flares.
   P. C. W. Fung, P. A. Sturrock, P. Switzer, G. van Hoven.
Stanford Univ. Inst. Plasma Res., SUIPR Rep., No. 393, 19 pp. (1971). – *RTS*

073.111 Spectral characteristics of impulsive solar flare X-rays > 10 keV.   S. R. Kane, K. A. Anderson.
California Space Sci. Lab., Univ. California, Berkeley, April 1970, 32 pp. – *NRL*

073.112 The emission and propagation of ~40 keV solar flare electrons. 2. The electron emission structure of large active regions.   R. P. Lin.
California Space Sci. Lab., Univ. California, Berkeley, June 1970, 33 pp. – *NRL*

A tunable Hα-filter and its work.
See Abstr. 034.097.

Observation of solar flare type processes in the laboratory.   See Abstr. 062.014.

The dynamics of a toroidal magnetic ring.
See Abstr. 062.028.

Arbitrary amplitude magnetoacoustic waves under gravity: An exact solution.   See Abstr. 062.072.

Remarks on the H⁻ equilibrium in stellar atmospheres.   See Abstr. 064.024.

The production of solar and stellar chromospheres and coronae.   See Abstr. 064.049.

Photospheric mass motions associated with a flare.
See Abstr. 071.002.

Solar brightness measurement between 12 and 24 microns.   See Abstr. 071.017.

Supergranulation at the center of the disk.
See Abstr. 071.029.

Five-minute oscillations in the solar magnetic field.
See Abstr. 071.036.

An attempt to associate observed photospheric motions with the magnetic field structure and flare occurrence in an active region.   See Abstr. 071.037.

A comparison of the intensity variations of the CN photospheric and K line chromospheric network with time. See Abstr. 071.051.

Photometry of the sodium resonance doublet in the flash spectrum during the solar eclipse on September 22, 1968. See Abstr. 071.058.

A review of models of the solar photosphere and low chromosphere: The temperature—height profile. See Abstr. 071.059.

A measurement of the brightness temperature of the sun in the range 65 to 180 cm^{-1}. See Abstr. 071.062.

Complexes of activity of the solar cycle and very large scale convection.   See Abstr. 072.004.

A series of related active regions during January 14 – June 1, 1969.   See Abstr. 072.008.

The stability of sunspot magnetic fields and the origin of solar flares.   See Abstr. 072.019.

Sunspot magnetic fields and high energy electrons in flares.   See Abstr. 072.035.

Theories of small-scale magnetic fields. See Abstr. 072.037.

The topology of force-free magnetic field near bipolar sunspots.   See Abstr. 072.046.

Relation of surges to spotgroups and to flares. See Abstr. 072.061.

Electron temperature and emission measure variations during solar X-ray flares.   See Abstr. 074.075.

Soft solar X-rays and solar activity. V: Relation of the course of soft X-ray fluctuations to the course of solar activity, 9 March, 1967 – 18 May, 1968. See Abstr. 076.007.

Extreme ultraviolet flashes of solar flares observed via sudden frequency deviations: Experimental results. See Abstr. 076.020.

The time behavior of temperature and emission measure in X-ray flares.   See Abstr. 076.026.

Anisotropy of solar hard X-radiation during flares. See Abstr. 076.027.

Solar flare X-ray spectra.   See Abstr. 076.034.

On the polarization and anisotropy of solar X-radiation during flares.   See Abstr. 076.038.

On the relations of radio emission flux with flare activity.   See Abstr. 077.010.

The anomalous distribution in heliocentric longitude of solar injected cosmic radiation.   See Abstr. 078.002.

The solar longitude dependence of proton event delay time.   See Abstr. 078.005.

Effect of the solar boundary condition on flare-particle propagation.   See Abstr. 078.010.

Propagation of low energy protons associated with the 24 January 1969 solar flare.   See Abstr. 078.021.

Inhomogeneities in the solar atmosphere from the Ca II infra-red lines.   See Abstr. 080.041.

Solar magnetic fields—large scale. See Abstr. 080.047.

A study of two flares on 8 July 1968 in the light of their ionospheric effects.   See Abstr. 083.054.

Orientation of interplanetary shock waves (sound measurements) and position of chromospheric flares. See Abstr. 106.013.

Effects of active solar regions on the galactic cosmic ray intensity.   See Abstr. 143.044.

074 Solar Corona, Solar Wind

074.001 Solar wind acceleration caused by the gradient of Alfvén wave pressure. G. Alazraki, G. Couturier.
Astron. Astrophys., Vol. 13, 380 - 389 (1971).
To investigate the possibility of an additional acceleration of the solar wind due to the presence of Alfvén waves propagating outward from the sun, we study a stationary, spherically symmetric model, we assume a radial wind flow and also a radial field. The variations of the temperature are described by a polytrope law. The variations of the solar wind velocity with the radial distance are deduced for different values of the coronal parameters: temperature, magnetic field and energy density in the form of Alfvén waves.

074.002 Nature and origin of directional discontinuities in the solar wind. L. F. Burlaga.
Journ. Geophys. Res., Vol. 76, 4360 - 4365 (1971).
We determine the ratio of rotational discontinuities to tangential discontinuities in the set of directional discontinuities obtained by Burlaga from Ness's Pioneer 6 magnetic field data for the period December 18 to 25, 1965, and discuss some hypotheses concerning the origin of these discontinuities, a problem that has not been studied experimentally until now.

074.003 Photometry of the outer solar corona from lunar-based observations. J. D. Bohlin.
Solar Physics, Vol. 18, 450 - 457 (1971).
Two-dimensional isophotes of the extreme solar corona ($r_{max} \sim 45\ R_\odot$) have been derived from integrated vidicon pictures taken from the moon's surface by the unmanned probes Surveyors 6 and 7. The resulting structure of the outer corona is compared to ground-based observations of the innermost corona ($1.125 \leqslant r/R_\odot \leqslant 2.0$) made by the High Altitude Observatory K-coronameter. The possible existence of a streamer seen by Surveyor 7 is analyzed over the region $15 \leqslant r \leqslant 22.5\ R_\odot$.

074.004 Radiative emittance and temperature structure of the coronal plasma.
E. Ya. Vilkovisky, S. O. Obashev.
Astron. Zhurn. Akad. Nauk SSSR, Vol. 48, 747 - 751 (1971).
In Russian. English translation in Soviet Astron. AJ, Vol. 15, No. 4.
A conclusion on the existence of three regions of thermal stability in the solar corona is made as a result of the analysis of the cooling rate in the heat-balance equation for the coronal plasma.

074.005 On the solar transition layer and solar active regions.
D. Reimers.
Astron. Astrophys., Vol. 14, 198 - 209 (1971).
An empirical minimum equator model of the solar transition layer and inner corona is constructed using observations of the solar radio and ultraviolet spectra and of the white and the monochromatic corona. The sudden temperature rise in the transition layer determined from ultraviolet spectra is confirmed by observations of the solar radio spectrum. Our one parameter model of solar active regions (Reimers, 1971) is confirmed by the extreme-ultraviolet spectroheliograms from OSO IV and by radio spectra of active regions as well as by observations of coronal rays and condensations.

074.006 Angular momentum and short-term directional fluctuations of the solar wind from Vela 3 data.
J. Hardorp.
Astron. Astrophys., Vol. 14, 210 - 214 (1971).
From recently published lists of detailed solar wind observations by the Vela 3 satellites it is shown that the sun's loss of angular momentum can be derived from time averages of solar wind density, velocity, and direction without danger of statistical bias. Directional fluctuations of the solar wind on a time scale of 3 hours or shorter show a dispersion of 1° only, in contrast to the 3° dispersion of slow variations.

074.007 Solar coronal streamers observed at 169 MHz with the Nançay east—west radioheliograph.
F. Axisa, Y. Avignon, M. J. Martres, M. Pick, P. Simon.
Solar Physics, Vol. 19, 110 - 127 (1971).
The slowly varying component of the sun at 169 MHz for the period August 1962 to March 1966 has been studied, using daily data from the east—west radioheliograph at Nançay. By relating these radio data to the optical features visible on the solar disc (calcium plages and filaments), this thermal radio-emission can be satisfactorily interpreted as originating in both helmets and active streamers. Finally, the observations are used to derive a morphological model for streamers in the medium corona, and several implications of such a model are briefly discussed.

074.008 Observational data on the solar wind.
M. Sroczyńska.
Postępy Astron., Vol. 19, 215 - 223 (1971). In Polish.
A short review of the observational data on the solar wind (near the orbit of the earth) is given.

074.009 Stoß- oder Strahlungsanregung in der Sonnenkorona? M. Waldmeier.
Naturwissenschaften, 58. Jahrgang, p. 413 - 414 (1971).

074.010 On a model of interaction between solar wind and interstellar medium.
V. B. Baranov, K. V. Krasnobaev.
Kosmich. Issled., Vol. 9, 620 - 622 (1971). In Russian.
Brief information.

074.011 Current sheet magnetic model for the solar corona. K. H. Schatten.
Cosmic Electrodynamics, Vol. 2, 232 - 245 (1971).
A new magnetic model is developed and compared with previous models and the observed solar corona. An attempt is made to more accurately compute the three dimensional currents flowing in the solar corona. A comparison with the axisymmetric, isothermal MHD solution of Pneuman and Kopp (1970) suggests that the model is able to simulate to high accuracy an isothermal corona. A comparison of the model with the May 30, 1965 solar eclipse and the November 12, 1966 solar eclipse shows the model is capable of computing many features including the polar plume orientations as well as radial and non-radial streamers in the solar corona.

074.012 Temperature variations of the coronal region near a prominence. Ts. Chultem.
Astron. Tsirk., No. 616, p. 1 - 3 (1971). In Russian.

074.013 Interstellar matter and the location of the shock front. H. J. Fahr.
Planet. Space Sci., Vol. 19, 1121 - 1129 (1971).
Interstellar hydrogen penetrating into the heliosphere undergoes charge exchange processes with the solar wind protons and ionization processes by the solar EUV radiation. This results in an extraction of momentum from the solar wind plasma. Changes of the geometry and the location of the shock front due to this interaction are studied in detail

and it is shown that the distance of the magnetic shock front from the sun decreases from 200 to 80 AU for an increase of the interstellar hydrogen density from 0.1 to 1.0 cm^{-3}.

074.014 On the abundance of calcium in the solar corona.
R. R. Fisher.
Solar Physics, Vol. 19, 431 - 435 (1971).
Measured values for the total intensity of the continuum and the ratio of integrated intensities $I(\lambda 5694)/I(\lambda 5446)$ are used to estimate the fraction of electrons along the line of sight contributing to the excitation of Ca XV. This estimate of electron density along with an estimate of the dimension of the emitting region are used to find a value of the abundance of Ca in the solar corona. The estimated abundance is log $N_{Ca}/N_H = -4.35$.

074.015 Monochromatic observations of a coronal loop.
R. R. Fisher.
Solar Physics, Vol. 19, 436 - 439 (1971).
An isophotal map of a small coronal loop, obtained from a coronagraph observation through a solid Fabry-Perot interferometer, is used to estimate the variation of emission per unit volume and the pressure gradient at the top and sides of the loop. The magnitude of the magnetic field necessary to maintain the estimated pressure gradients is found to be $|\vec{H}^2| = 30 \, G^2$.

074.016 Solar corona at the total eclipse of September 22, 1968. Ts. S. Khetsuriani, E. I. Tetruashvili, R. I. Kiladze, G. N. Salukvadze, A. Sh. Khatisov.
Byull. Abastumansk. Astrofiz. Obs., No. 40, p. 55 - 92 (1971). In Russian.
Isophotes of the corona have been obtained by means of the equidensity method. The structure of the corona has been studied with the prints obtained through a radial neutral filter. The values of polarization degrees are higher in the coronal rays than in gaps.

074.017 Radiation from a high-temperature, low-density plasma: The X-ray spectrum of the solar corona.
W. H. Tucker, M. Koren.
Astrophys. Journ., Vol. 168, 283 - 311 (1971).
The results of calculations of the 0.5−70 Å X-ray spectrum of a high-temperature, low-density plasma are presented. The temperature range is $6 \times 10^5 \, °-10^8 °K$, and the elemental abundances characteristic of the solar corona have been assumed. We have considered the processes of line emission following electron collisional excitation, radiation resulting from recombination, bremsstrahlung, and two-photon decay following the excitation of the metastable $2S$ state in hydrogenic and helium-like ions.

074.018 Variations of solar-wind plasma properties: Vela observations of a possible heliographic latitude-dependence.
A. J. Hundhausen, S. J. Bame, M. D. Montgomery.
Journ. Geophys. Res., Vol. 76, 5145 - 5154 (1971).
Twenty-seven-day averages of the solar-wind density and flow speed, observed by Vela 3 and 4 spacecraft between July 1965 and July 1968, are found to vary with the heliographic latitude of observation. The variations can be reasonably interpreted in terms of a latitude-dependence in the structure of high-speed solar-wind streams related to solar activity.

074.019 Nonthermal electrons and high-frequency waves in the upstream solar wind. 1. Observations.
F. L. Scarf, R. W. Fredricks, L. A. Frank, M. Neugebauer.
Journ. Geophys. Res., Vol. 76, 5162 - 5171 (1971).
We use OGO 5 interplanetary particle and wave observations from March 11 and 12, 1968, to demonstrate that oscillations near the characteristic upper hybrid and electron plas-

ma frequencies are produced when nonthermal electrons $(E_e \gtrsim 700-800$ ev) flow upstream. The results are discussed in terms of resonant interactions, with the streaming particle speed set equal to the wave phase speed.

074.020 Correction: 'Effect of interstellar neutral hydrogen on the termination of the solar wind' [Journ. Geophys. Res., Vol. 75, 6892 - 6898 (1970)]. C. L. Semar.
Journ. Geophys. Res., Vol. 76, 5374 - 5375 (1971).

074.021 The low brightness of the solar corona on September 22, 1968.
N. I. Dzjubenko, V. I. Ivanchuk, G. A. Rubo, K. I. Churjumov.
Solnechnye Dannye 1971 Byull., No. 6, p. 73 - 79 (1971). In Russian.
The results of an absolute photometry of the solar corona on Sept. 22, 1968 are given. A comparison of photometric details of the corona with the active and quiet areas on the solar surface has been made.

074.022 Photometry of the solar corona on March 7, 1970.
R. A. Gulyaev.
Solnechnye Dannye 1971 Byull., No. 6, p. 80 - 83 (1971). In Russian.
Isodensitometric analyses of a solar corona negative, obtained during the solar eclipse on March 7, 1970 in Mexico, were performed.

074.023 Shock-free deceleration of the solar wind?
M. Wallis.
Nature, Phys. Sci., Vol. 233, 23 - 25 (1971).
Ionizing interactions can cause shock-free trans-sonic deceleration both according to the hydrodynamic description of the solar wind interaction with interstellar hydrogen and because of the presence of supra-thermal ions ahead of comets and non-magnetic planets.

074.024 A connection of velocity and magnetic field variations in the solar wind with the variations of the K_p-index. N. V. Mikerina, K. G. Ivanov.
Geomagn. Aeronom., Vol. 11, 729 - 730 (1971). In Russian. Brief information.

074.025 Alfvénic wave pressures and the solar wind.
J. W. Belcher.
Astrophys. Journ., Vol. 168, 509 - 524 (1971).
Using a one-fluid polytrope model of the solar wind, we investigate the detailed behavior of outwardly propagating coronal Alfvén waves and their effects on the dynamics of the expanding solar corona as they propagate and are convected into interplanetary space. The waves exert an effective pressure on the solar wind, analogous to a radiation pressure, and inclusion of the energy fluxes of Alfvénic waves can result in significant changes in the large-scale streaming properties of the solar wind.

074.026 Observation of the angular-momentum flux carried by the solar wind. A. J. Lazarus, B. E. Goldstein.
Astrophys. Journ., Vol. 168, 571 - 574 (1971).
Using data from the Mariner 5 spacecraft, we have measured the angular-momentum flux carried by the solar wind and find it in agreement with earlier estimates based on observations of the deflection of comet tails.

074.027 Émission thermique de la couronne calme sur 169 MHz au cours du cycle solaire.
Y. Avignon, P. Lantos.
Comptes Rendus Acad. Sci. Paris, Sér. B, Vol. 273, 684 - 685 (1971).
L'étude des enregistrements solaires obtenus à l'aide du

radio-héliographe de Nançay a montré que le flux du soleil calme à 169 MHz varie proportionnellement aux variations de diamètre équatorial observées par ailleurs et que, de ce fait, la température de brillance et la température électronique de la couronne sont sensiblement constantes au cours du cycle solaire.

074.028 The corona. J. B. Zirker.
National Bureau Standards Special Publ. 353, (see 012.001), p. 112 - 126 (1971).

074.029 On spherical-symmetrical flowing of plasma from the sun.
I. I. Alexeev, V. P. Shabansky, A. R. Shister.
Geomagn. Aeronom., Vol. 11, 761 - 764 (1971).
In Russian.

074.030 Rotational breaks in the solar wind.
K. G. Ivanov.
Geomagn. Aeronom., Vol. 11, 765 - 770 (1971).
In Russian.

074.031 On coronal instability and moving radio features associated with a flare spray. G. Daigne.
IAU Symposium No. 43, (see 012.003), p. 367 - 375 (1971).

074.032 The X-ray corona and the photospheric magnetic field.
A. S. Krieger, G. S. Vaiana, L. P. van Speybroeck.
IAU Symposium No. 43, (see 012.003), p. 397 - 412 (1971).

074.033 Magnetic fields in the lower corona associated with the expanding limb burst on March 30th 1969 inferred from the microwave high-resolution observations.
S. Enomé, H. Tanaka.
IAU Symposium No. 43, (see 012.003), p. 413 - 416 (1971).

074.034 Interaction of coronal material with magnetic fields.
G. W. Pneuman, R. A. Kopp.
IAU Symposium No. 43, (see 012.003), p. 526 - 533 (1971).

074.035 Coronal magnetic fields above active regions.
D. M. Rust, J.-R. Roy.
IAU Symposium No. 43, (see 012.003), p. 569 - 579 (1971).

074.036 Experimental study of the orientation of magnetic fields in the corona. P. Charvin.
IAU Symposium No. 43, (see 012.003), p. 580 - 587 (1971).

074.037 The magnetic field structure in the active solar corona. K. H. Schatten.
IAU Symposium No. 43, (see 012.003), p. 595 - 608 (1971).

074.038 Optical and radio observations of large scale magnetic fields on the sun.
G. Daigne, M. F. Lantos-Jarry, M. Pick.
IAU Symposium No. 43, (see 012.003), p. 609 - 615 (1971).

074.039 Radio-astronomical evidence for magneto-hydrodynamical pulsations in the corona.
H. Rosenberg.
IAU Symposium No. 43, (see 012.003), p. 652 - 655 (1971).

074.040 Observations of the coronal network.
G. W. Simon, R. W. Noyes.
IAU Symposium No. 43, (see 012.003), p. 663 - 666 (1971).

074.041 Preliminary results of spectroscopic determination of the coronal rotation.
V. E. Stepanov, N. F. Tjagun.
IAU Symposium No. 43, (see 012.003), p. 667 - 671 (1971).

074.042 On temperature and velocity differences of solar wind components. Yu. B. Ponomarenko.
Astron. Zhurn. Akad. Nauk SSSR, Vol. 48, 976 - 984 (1971).
In Russian. English translation in Soviet Astron. AJ, Vol. 15, No. 5.

A model of the solar wind consisting of electrons, protons and α-particles is considered. It is found that molecular weight and relative velocity differences decrease while temperature differences increase with distance from the sun. Near the corona base the temperatures of the components are approximately equal, while the velocity of α-particles is considerably less than those of protons and electrons. At large distances the velocities of the components are approximately equal while the temperature of protons is considerably less than the temperatures of electrons and α-particles.

074.043 On structure, absolute photometry and polarization of the corona according to observations of the solar eclipse of September 22, 1968.
Ts. S. Khetsuriani, E. I. Tetruashvili, R. I. Kiladze, G. N. Salukvadze, A. Sh. Khatisov.
Astron. Zhurn. Akad. Nauk SSSR, Vol. 48, 985 - 997 (1971).
In Russian. English translation in Soviet Astron. AJ, Vol. 15, No. 5.

074.044 On the neutron theory of corona heating.
C.-C. Cheng.
Nature, Phys. Sci., Vol. 233, 112 - 113 (1971).
It is pointed out that if the neutron theory of corona heating proposed by Fowler and Hashemi (Nature, Vol. 230, 518, 1971) is true then the flux of the neutron capture gamma ray line at 2.23 MeV at the earth is too high to be consistent with the observational data.

074.045 The K corona and electron density.
M. Rybanský.
Bull. Astron. Inst. Czechoslovakia, Vol. 22, 321 - 324 (1971).
A method is described for the determination of the intensity of the K corona from photographs of the corona spectrum taken without the eclipse. Further, a simple method is submitted to determine the electron density from the K corona intensity in coronal condensations.

074.046 A coronal disturbance observed simultaneously with a white-light coronameter and the 80 MHz radioheliograph. R. T. Hansen, C. J. Garcia, R. J.-M. Grognard, K. V. Sheridan.
Proc. Astron. Soc. Australia, Vol. 2, 57 - 60 (1971).
Observations on 1970 August 11 and 12 at Mauna Loa (Hawaii) of the white-light corona revealed a rare solar event in the form of a short-lived coronal brightening off the eastern limb of the sun. A complex radio event in the same region was observed simultaneously with the 80 MHz radioheliograph at Culgoora (Australia). Associations between the optical and radio events will be described and the physical implications briefly discussed.

074.047 Recent developments in theory of solar wind.
E. N. Parker.
Rev. Geophys. Space Phys., Vol. 9, 825 - 835 (1971).
This paper reviews some of the recent theoretical work and outstanding problems in understanding the quiet solar wind that have been posed by the observations.

074.048 Nonthermal electrons and high-frequency waves in the upstream solar wind. 2. Analysis and interpretation. R. W. Fredricks, F. L. Scarf, L. A. Frank.
Journ. Geophys. Res., Vol. 76, 6691 - 6699 (1971).
A streaming instability caused by suprathermal electrons of bulk energy ~ 1 kev and random widths from ~60 ev to 4 kev is investigated as a model to explain the OGO 5 observa-

tions of electric-field fluctuations with frequencies at or near the local electron plasma frequency upstream from the earth's bow shock.

074.049 **Radar explorations of the sun and the origin of the solar wind in active regions.** I. M. Gordon.
Space Research XI, (see 012.004), 1201 - 1203 (1971).

074.050 **Gyro-resonance absorption of plasma waves in the corona and the fine structure of solar radio bursts.**
V. V. Zheleznyakov, E. Ya. Zlotnik.
Solar Physics, Vol. 20, 85 - 94 (1971).

We consider the important part played by plasma wave absorption at the higher harmonics of the electron gyrofrequency (under conditions of the so-called double plasma resonance) in the solar corona. An explanation is proposed for the fine structure of type III bursts on the basis of this effect.

074.051 **Photometric investigation of the solar corona on September 22, 1968.**
P. A. Olijnyk, M. M. Koval'chuk, I. S. Laba.
Tsirk. L'vov. Astron. Obs., No. 45, p. 37 - 46 (1971). In Russian.

074.052 **Depolarization of the solar corona on the 22nd of September, 1968 near prominences.**
Yu. N. Lipskij, Yu. P. Pskovskij, A. V. Bugaevskij.
Astron. Tsirk., No. 642, p. 1 - 2 (1971). In Russian.

074.053 **The radial electric field in the solar wind.**
S. Cuperman, A. Harten.
Astrophys. Journ., Vol. 169, 165 - 169 (1971).

. The radial electrostatic field in the solar wind, which is responsible for the equality of the electron and positively charged ion fluxes observed at about 1 a. u. as well as for the quasi-charge neutrality of the sun, is calculated. This is achieved by using expressions for particle density, bulk velocity, and electron and proton temperatures obtained from a two-fluid model with modified transport coefficients. The modifications of transport coefficients are purely phenomenological and are determined by the requirement that the predicted gross features of the solar wind match the available observations.

074.054 **Conversion of magnetic-field energy into kinetic energy in the solar wind.** Y. C. Whang.
Astrophys. Journ., Vol. 169, 369 - 378 (1971).

The conversion of magnetic-field energy into kinetic energy in the solar wind is identified as a source of energy which increases the velocity of the solar wind.

074.055 **Interaction of a collisionless shock wave and a stream of particles in the upper solar corona.**
C. Lacombe, B. Møller Pedersen.
Astron. Astrophys., Vol. 15, 406 - 418 (1971).

Decametre radio observations are reported which point to the interaction between type III bursts and a slowly ascending perturbation in the upper solar corona. Radiation enhancements of type III bursts are observed at the crossing with the perturbation. An interpretation is proposed in which this perturbation is assumed to be a collisionless shock wave with $T_e > T_i$, exciting a turbulence of ion acoustic waves.

074.056 **The temperature anisotropy and adiabatic cooling of the protons in the solar wind.**
M. Eyni, A.S. Kaufman.
Planet. Space Sci., Vol. 19, 1609 - 1614 (1971).

A simple theoretical expression for the mean kinetic temperature of the protons in a steady state as a function of heliocentric distance is derived. For an assumed base temperature of $5 \times 10^5 K$ at a distance of 0.05 AU, the calculated tempera-

ture at a distance of 1 AU is in the range $(2-4) \times 10^4 K$ for an average anisotropy factor of 3: this range of temperatures is close to the observed average value under so-called 'quiet' conditions. Measurement of the anisotropy factor at different heliocentric distances is required to test the basis of the model.

074.057 **A simple formula for the total dielectronic recombination coefficient.** M. Landini, B. C. Fossi.
Solar Physics, Vol. 20, 322 - 331 (1971).

Two simple relationships for the total dielectronic recombination coefficient are developed. The first is for isoelectronic sequences of H, He, Ne, K−Ni and the second for Li−F, Na−A and Cu−Kr. Comparison with the extended computations of Jordan and Elwert is made.

074.058 **Coronal magnetic field of the sun on 7 January 1969.** D. E. Trotter, G. Newkirk, Jr.
Solar Physics, Vol. 20, 372 - 374 (1971). − Research note.

074.059 **Coronagraphic observations of an enhanced coronal region: I, Fe XII and Ni XV emission line data.**
R. Fisher, T. Pope.
Solar Physics, Vol. 20, 389 - 399 (1971).

Nine coronal emission lines representing five stages of Fe ionization and one stage of Ni were observed in an enhanced coronal region. The data from these observations are presented along with a density model of the enhanced region obtained from the Fe XIII and Ni XV emission line ratios as a function of position angle. The electron densities obtained from Fe XIII lines range from $N_e = 10^8$ to 10^9 cm^{-3}, and are slightly lower for Ni XV line data. Estimates of the variation of temperature over the enhanced region are inferred from the observed line intensities.

074.060 **Correlation of solar wind velocity with λ5303 coronal intensity.** P. N. Pathak.
Solar Physics, Vol. 20, 462 - 473 (1971). − Presented at IUCSTP Symposium on Solar-Terrestrial Physics, Leningrad, May 1970.

Using solar wind velocity data obtained by Mariner-2 and IMP-1 spacecraft, an attempt has been made to study its correlation with λ5303 coronal intensity. It is shown that the long-lasting regions of enhanced λ5303 intensity in the solar corona are well correlated with recurrent streams of solar wind having high velocity. The time-lag between the central meridian passage (CMP) of the coronal features and the detection of the solar wind streams at the spacecraft is found to be smaller than that implied by a radial solar wind.

074.061 **Kinetic models of the solar wind.**
J. Lemaire, M. Scherer.
Journ. Geophys. Res., Vol. 76, 7479 - 7490 (1971).

A new kinetic model of the quiet solar wind is presented and compared with earlier exospheric, semikinetic, and hydrodynamical models. The bulk velocity, the density, the average electron and proton temperatures, and the energy flux, which are observed at 1 AU for quiet solar-wind conditions, are well represented by such a kinetic model. The average electron temperature is nearly independent of the bulk velocity, whereas a positive correlation between the average proton temperature and the bulk velocity is found. Consequently it is suggested that in the interplanetary medium $(r > 6Rs)$ no external heating mechanism is needed to explain the observed quiet solar-wind properties.

074.062 **Collisionless solar wind. 2. Variable electron temperature.** J. V. Hollweg.
Journ. Geophys. Res., Vol. 76, 7491 - 7502 (1971).

We consider a two-component 'model' for the solar wind, in which the protons become collisionless beyond $r_0 \gtrsim 10\,R_S$, where they are already highly supersonic. The proton temperatures are found from the double adiabatic equation of state. The electrons are highly subsonic, and their temperature profile is prescribed ad hoc. The principal results are reported and discussed.

074.063 Higher moment equations and the distribution function of the solar-wind plasma. Y. C. Whang.
Journ. Geophys. Res., Vol. 76, 7503 - 7507 (1971).
The higher moment equations for a collisionless fully ionized plasma are studied in this paper. We consider that the plasma is composed of protons and electrons. The new equations can be used to study the thermal anisotropy and the heat flux of the solar-wind proton.

074.064 Temperature profile of solar winds. T. Yeh.
Journ. Geophys. Res., Vol. 76, 7508 - 7515 (1971).
The temperature profile of the solar wind can be calculated from the energy equation by assuming that the velocity profile is known. When the logarithmic expansion rate of the solar wind is small, the heat-flow equation can be integrated analytically. If the coronal temperature is sufficiently high, the energy equation has a critical solution in which the temperature vanishes at infinity. A quiet-time temperature profile is calculated by using the observed data at the orbit of the earth.

074.065 Reverse and forward slow shocks in the solar wind.
L. F. Burlaga, J. K. Chao.
Journ. Geophys. Res., Vol. 76, 7516 - 7521 (1971).
Probable reverse and forward slow shocks were found in plasma and magnetic-field data from Pioneer 8. The shocks were oblique and weak (Mach ~ 1.2). Numerous nonshocklike discontinuities were found with $B_1 \neq B_2$ and $n_1 \neq n_2$. It was observed that discontinuities with $B_1 \neq B_2$, $n_1 = n_2$ seldom if ever occur.

074.066 On magnetic fields of the solar wind according to data on the diffusion coefficient of cosmic rays.
A. N. Charakhch'yan, T. N. Charakhch'yan.
"Trudy Mezhdunar. seminara po probl. "Generatsiya kosmich. luchej na Solntse", 1970". Moskva, 1971, p. 211 - 230. In Russian. – Abstr. in Referativ. Zhurn. 51. Astron., 12.51.658 (1971).

074.067 Erratum: "Radiation from a high-temperature, low-density plasma: The X-ray spectrum of the solar corona" [Astrophys. Journ., Vol. 168, 283 - 311 (1971)].
W. H. Tucker, M. Koren.
Astrophys. Journ., Vol. 170, 621 (1971).

074.068 Preliminary analysis of observations of the coronal condensation of 10 September 1970.
M. Rybanský.
Bull. Astron. Inst. Czechoslovakia, Vol. 22, 376 - 380, 382a (1971).
The paper describes the observations and photometry concerning part of the data pertaining to the coronal condensation of 10 September 1970. Fe XIV—5303 Å, Ca XIII—4086 Å, Ca XV 5446 Å and 5694 Å emission lines and the K corona were observed. The electron density in the centre of the condensation was determined as 1.7×10^{10} cm^{-3}.

074.069 Form und Struktur der Korona bei der Sonnenfinsternis vom 7. März 1970. M. Waldmeier.
Astron. Mitt. Sternw. Zürich, No. 299, 23 pp. (1971).
The corona of March 7, 1970 is of maximum-type. The shape of the corona is in close connection with the latitude distribution of the prominences. The maximum-type corona is observed when prominences appear at all latitudes. The maximum-type corona is characterized by almost circular isophotes. The ellipticity of the isophotes amounts, even in the innermost corona, to 0.07 only and decreases outward to 0.01.

The streamers stand out all around the sun's limb. At the "magnetic" equator as well as at the "magnetic" poles the streamers are radial, in the other regions they are inclined polward.

074.070 Beobachtungen der Sonnenkorona in den Jahren 1969 und 1970. M. Waldmeier.
Astron. Mitt. Sternw. Zürich, No. 300, 21 pp. (1971).

074.071 The solar corona in the eleven-year cycle.
M. Waldmeier.
Physics of the Solar Corona. Symposium Athens 1970, p. 130-139 = Astron. Mitt. Sternw. Zürich, No. 301 (1971).
The corona changes its shape, structure and brightness according to the phase of the 11-yr cycle. The following discussion is based on coronal variations from 1958 (maximum of cycle No. 19) through 1969 (maximum of cycle No. 20). The material used has been gathered at the Astrophysical Observatory Arosa (monochromatic corona) and on nine total eclipses of the sun (white light corona).

074.072 Monochromatische Grenzisophoten der Korona 1960–1971. M. Waldmeier.
Astron. Mitt. Sternw. Zürich, No. 307, 15 pp. (1971).

074.073 Die Helligkeitsverteilung in der Korona bei der Sonnenfinsternis vom 7. März 1970.
M. Waldmeier, S. E. Weber.
Astron. Mitt. Sternw. Zürich, No. 308, 16 pp. (1971).
A method is described, by which – using pictures of the corona that were taken with the same camera, with the same exposure time, on the same film, but with different apertures of the objective – relative intensities of the corona's brightness can be deduced. In tables the intensities are presented for distances up to r = 3.0 and for intervals of 5° in heliographic latitude.

074.074 A method for the evaluation of the brightness distribution in the solar corona. W. Stanek.
Solar Physics, Vol. 21, 121 - 129 (1971).
A method is given to calculate the relative brightnesses along a solar radius from a set of photographs of the solar corona taken at eclipses. This set must be taken with different exposures, the ratio of two successive exposures, however, being constant. From the photometry of the density the relative brightnesses can be derived.

074.075 Electron temperature and emission measure variations during solar X-ray flares. D. M. Horan.
Solar Physics, Vol. 21, 188 - 197 (1971).
X-ray emission from seventeen X-ray flares was analyzed to obtain electron temperatures and emission measures associated with the source region in the solar corona. In all X-ray flares studied the peak temperature chronologically preceded the peak X-ray flux and the peak flux never came after the peak emission measure.

074.076 Non-radial oscillations and energy transport in rotating solar (stellar) wind. S. Grzędzielski.
Solar Physics, Vol. 21, 225 - 236 (1971).
It seems that the gravity-shear waves may represent a means of wave energy and angular momentum transport in the solar wind that is little affected by shock wave dissipation. From the observational point of view the velocity pattern, stationary in the co-moving frame, that corresponds to properly oriented gravity shear waves may be identified with the well known fast-slow streams frequently occurring in the interplanetary medium.

074.077 On the radial structure of coronal rays.
A. T. Nesmjanovich.

Solnechnye Dannye 1971 Byull., No. 9, p. 74 - 79 (1971). In Russian.

An analysis of the large-scale structure of the solar corona is made from the results of many solar eclipses. It is shown that rectilinear rays occur more frequently at mean and low heliographic latitudes and most of them incline to the solar equator. The main characteristics of rectilinear rays and active formations, connected with them, are given.

074.078 **The solar wind.**   T. G. Cowling.
Quarterly Journ. Roy. Astron. Soc., Vol. 12, No. 4, (see 012.012), 447 - 452 (1971).

074.079 **The electron excitation rate for the green coronal line at 5303 Å.**   D. Petrini.
Highlights of Astronomy, Vol. 2, (see 012.017), 518 (1971). Abstract.

074.080 **A review of theoretical work on the effects of solar wind transport on energetic solar particles.**
W. I. Axford.
Trudy Mezhdunar. seminara po probl. "Generatsiya kosmich. luchej na Solntse",1970. Moskva, 1971, p. 152 - 183. — Abstr. in Referativ. Zhurn. 62. Issled. kosm. prostranstva, 12.62.345 (1971).

074.081 **The relative role of factors limiting the solar wind.**
L. I. Dorman.
Geomagn. Aeronom., Vol. 11, 945 - 948 (1971). In Russian.

074.082 **On disturbances of the parameters of the solar wind plasma.**   L. P. Smirnova, V. P. Shabanski.
Geomagn. Aeronom., Vol. 11, 1073 - 1075 (1971). In Russian. Brief information.

074.083 **Average solar wind proton properties from Pioneers 6 and 7.**   J. D. Mihalov, J. H. Wolfe.
Cosmic Electrodynamics, Vol. 2, 326 - 339 (1971).

Distributions and average values for the proton speed, azimuthal and polar flow directions, and the proton temperature and density in the solar wind are presented. The data are obtained from the Ames plasma probes on Pioneers 6 and 7.

074.084 **Hydromagnetic waves and discontinuities in the solar wind.**   L. F. Burlaga.
Space Sci. Rev., Vol. 12, 600 - 657 (1971).

074.085 **Electron energy flux in the solar wind.**
K. W. Ogilvie, J. D. Scudder, M. Sugiura.
Journ. Geophys. Res., Vol. 76, 8165 - 8173 (1971).

We describe electron observations in the solar wind that cover the energy range 10 ev to 9.9 kev, that is, from energies characteristic of the plasma electrons to energies just below those normally observed in solar-electron events. We show that transient effects can be important in determining the electron energy flux in the solar wind.

074.086 **Ratio of specific heats in the solar-wind plasma flow through the earth's bow shock.**   W.-W. Shen.
Journ. Geophys. Res., Vol. 76, 8181 - 8188 (1971).

The ratio of specific heats in the solar-wind plasma flow can be significantly different in the upstream and downstream regions of the earth's bow shock. On the basis of one-fluid theory it is found that jump ratios of fluid parameters are sensitively dependent on the ratio of the specific heats only in the downstream region. The appropriate ratio of specific heats of a thermodynamic analog to the reducible problems in the magnetohydrodynamic flow is investigated.

074.087 **The structure of the solar corona on September 22, 1968 and its connection with photospheric and**

chromospheric phenomena.   S. K. Vsekhsvyatsky, N. I. Dzyubenko, V. I. Ivanchuk, G. A. Rubo.
Problems of cosmic physics. Vyp. (No.) 6, (see 003.040), p. 3 - 18 (1971). In Russian.

074.088 **Screw-like structure of a coronal ray.**
A. T. Nesmjanovich, N. I. Dzyubenko, Yu. A. Chomenko, O. S. Popov.
Astron. Tsirk., No. 657, p. 1 - 3 (1971). In Russian.

074.089 **The radio investigation of the solar corona.**
S. Gorgolewski.
Geod. Geophys. Veröff., Ser. 2, No. 13, (see 012.029), p. 81 - 85 (1969).

074.090 **Coronal scattering at 32.5 MHz near sunspot minimum and maximum.**   B. Krygier, S. Gorgolewski.
Geod. Geophys. Veröff., Ser. 2, No. 13, (see 012.029), p. 87 - 89 (1969).

074.091 **Observation of radio-wave refraction in a large coronal irregularity.**   B. Krygier.
Geod. Geophys. Veröff., Ser. 2, No. 13, (see 012.029), p. 91 - 94 (1969).

074.092 **Comments on neutrons as an energy source for the solar corona.**   E. L. Chupp.
Nature, Phys. Sci., Vol. 232, 152 (1971).

A proposal has been made by Fowler and Hashemi (Nature, Vol. 230, 513, 1971) that neutrons produced in thermonuclear reactions just under the photosphere supply the energy for heating the solar corona through the ionization loss of the charged decay fragments. It is shown in this note that the 2.2 MeV neutron-proton capture gamma ray from those neutrons which do not leave the sun would be easily detectable at the earth. Since no steady gamma ray flux is seen at the earth, the observations cast considerable doubt on the model.

074.093 **Helium-like line emission from coronal features.**
L. W. Acton, R. C. Catura, A. J. Meyerott, J. L. Culhane.
Nature, Phys. Sci., Vol. 233, 75 - 77 (1971).

We present the preliminary results of a rocket experiment designed to study the O VII and Ne IX line emission from discrete coronal features. These results indicate that electron densities in normal, non-flare, coronal features are below the low density limits (6×10^9 and 1×10^{11} cm^{-3} respectively) given by Freeman et al. for these particular ions.

074.094 **Anomalous dispersion relation and instability in the solar wind plasma with thermal anisotropy.**
S. Watanabe.
Rep. Ionosph. Space Res. (*Japan*), Vol. 24, 298 - 312 (1970).

Wave-particle interactions in the solar wind plasma having a thermal anisotropy and a high-β ratio are investigated in detail for magnetosonic waves propagated parallel to a magnetic field. The results from the numerical and graphical analyses of the dispersion equation are compared with the results given by the FLR approximation.

074.095 **Laboratory experiment on the solar wind interaction with planetary bodies.**
H. Kubo, T. Itoh, A. Yamori.
Rep. Ionosph. Space Res. (*Japan*), Vol. 24, 313 - 321 (1970).

The electric field generated by the interaction of solar wind with planetary bodies is discussed with a laboratory simulation. The mechanisms generating the electric and the magnetic field by the solar wind interaction are also discussed, and they would respectively be due to charge separation and pressure gradient drift.

074.096 Coronal electron temperatures associated with solar flares. D. M. Horan.
Thesis, Catholic Univ. America, Washington, D.C. [Available from Univ. Microfilms, Ann. Arbor, Mich., U.S.A. Order No. 70–22133], 133 pp. (1970).

X-ray emission from seventeen X-ray flares was analyzed to obtain electron temperatures and emission measures associated with the source region in the solar corona. The bases for the analysis, and the time histories of the seventeen individual events studied are presented.

074.097 Energy source for the solar corona. J. Hashemi-Tafreshi.
Thesis, Univ. Oklahoma, Norman. [Available from Univ. Microfilms, Ann Arbor, Mich., U.S.A. Order No. 70–23009], 71 pp. (1970).

It is proposed that heat is transported to the corona through the agency of neutrons originating in the photospheric layer of the sun. A steady state model for the corona has been developed. A flux of neutrons is assumed to reach the base of the corona.

074.098 Hydromagnetic studies of the solar wind. M. S. Gussenhoven.
Thesis, Boston College, Chestnut Hill, Mass. [Available from Univ. Microfilms, Ann Arbor, Mich., U.S.A. Order No. 70–24601], 179 pp. (1970).

Of the eight non-linear, hydromagnetic equations describing a one fluid model of the solar wind which is inviscid, isothermal, infinitely conducting, time-independent and independent of azimuth, six are solved explicitly.

074.099 Preliminary observations of interplanetary scintillation at 69.3 MHz. T. Watanabe, H. Washimi, T. Kakinuma, M. Kojima, K. Maruyama, Y. Ishida.
Proc. Res. Inst. Atmosph. Nagoya Univ., (*Japan*), Vol. 18, 59 - 71 (1971).

Observations of interplanetary scintillation of radio sources for studying the solar wind have been carried out at two stations, Toyokawa and Fujigane, since June 1970.

074.100 Determination of diffusion losses of solar ^{4}He wind in the moon dust. L. Schultz, U. Frick, P. Signer.
Helvetica Phys. Acta, Vol. 44, 614 - 617 (1971). In German.

The isotope composition of solar wind previously investigated was supplemented by diffusion study of He and Ne in the moon dust and in meteorites.

074.101 Hydromagnetic waves generated at the bow shock by tangential discontinuities in the solar wind. R. K. Jaggi, R. A. Wolf.
Phys. Fluids, Vol. 14, 648 - 657 (1971).

The problem of interaction of an inhomogeneity frozen into the solar wind with the bow shock has been investigated. These frozen-in tangential discontinuities produce ripples that move along the shock, ripples that can cause magnetosonic waves in the downstream medium. There are five linearly independent types of frozen-in inhomogeneities.

074.102 A mechanism of acceleration of the solar wind plasma. M. V. Konyukov.
Proc. Sixth Winter School on Space Physics, Part I. Apatity 1969, (see 012.033), p. 254 - 255 (1971).

074.103 Radar astronomy of the sun. I. M. Gordon.
Proc. Sixth Winter School on Space Physics, Part I. Apatity 1969, (see 012.033), p. 264 - 268 (1971).

NASA satellites to study sun's corona.
IEEE Spectrum, Vol. 8, No. 10, p. 91 (1971).

Transition probabilities in intermediate-coupling and configuration mixing for Fe XVII. See Abstr. 022.067.

Collision strengths for electron excitation of coronal ions. See Abstr. 022.098.

A rocket-borne X-ray spectrometer/monochromator system for mapping the solar corona. See Abstr. 034.056.

Density fluctuations driven by Alfvén waves. See Abstr. 061.009.

Magnetohydrodynamic wave-mode coupling. Quantum field-theoretical approach to weakly non-linear case with application to solar coronal heating. See Abstr. 062.002.

Rotational discontinuities in an anisotropic plasma. See Abstr. 062.025.

Analysis of electromagnetic instabilities parallel to the magnetic field. See Abstr. 062.062.

The production of solar and stellar chromospheres and coronae. See Abstr. 064.049.

The chemical composition of the photosphere and the corona. See Abstr. 071.021.

Theories of small-scale magnetic fields. See Abstr. 072.037.

The asymmetry of solar activity in the years 1959–1969. See Abstr. 072.054.

Electron densities derived from line intensity ratios: Beryllium isoelectronic sequence. See Abstr. 073.022.

Stationäre Filamente, ihre koronale Umgebung sowie ihre Beziehung zum solaren Magnetfeld. See Abstr. 073.084.

The manifold structure of the chromosphere and corona. See Abstr. 073.093.

The structure of the chromosphere—corona transition region from limb and disk intensities. See Abstr. 073.094.

Measurements of electron temperature in the solar chromosphere and corona. See Abstr. 073.095.

La zona di transizione tra la cromosfera e la corona solare. See Abstr. 073.106.

On the identification of the λ 417 line in the solar extreme ultraviolet spectrum. See Abstr. 076.008.

80 MHz radioheliograph evidence on moving type IV bursts and coronal magnetic fields. See Abstr. 077.024.

A U-type solar radio burst originating in the outer corona. See Abstr. 077.030.

A moving Type IV radio burst and its relation to the coronal magnetic field. See Abstr. 077.040.

Solar radioastronomy and coronal magnetohydrodynamics. See Abstr. 077.060.

Radioastronomy and coronal microstructures. See Abstr. 077.076.

Physical processes in the corona, related to radio emission. See Abstr. 077.077.

27-day variations of solar activity and cosmic ray intensity and the determination of the effective size of the asymmetric solar wind. See Abstr. 078.046.

Cosmic ray currents in stationary streams of solar plasma. See Abstr. 078.050.

A statistical model of geophysical processes with asymmetry and excess of the density probability function. See Abstr. 083.032.

Conditions for magnetic interaction of asteroids with the solar wind. See Abstr. 098.010.

Decameter radio radiation of Jupiter as an indicator of high-speed streams and shock waves in the solar wind. See Abstr. 099.016.

Comets and their interaction with the solar wind. See Abstr. 102.012.

Interplanetary hydrogen and helium from cosmic dust and the solar wind. See Abstr. 106.001.

Theoretical constraints on the microscale fluctuations in the interplanetary medium. See Abstr. 106.026.

Angular distribution of radio waves scattered by the interplanetary medium. See Abstr. 106.033.

The effect of solar wind velocity inhomogeneities on the structure of the interplanetary field. See Abstr. 106.041.

Interaction between interstellar helium and the solar wind. See Abstr. 131.081.

The modulation of galactic protons by the solar wind: A Monte Carlo approach. See Abstr. 143.003.

Non-linear interaction of galactic cosmic rays with interplanetary magnetic fields and a possible geometry of the solar wind. See Abstr. 143.017.

Correlation of galactic cosmic-ray intensity with $\lambda 5303$ coronal intensity. See Abstr. 143.048.

Cosmic-ray decreases and the occurrence of solar flares. See Abstr. 143.091.

075 Solar Patrol

075.001 **Catalogue of solar activity for the year 1968.**
R. S. Gnevysheva.
Trudy Glav. Astron. Obs. Pulkovo, 160 pp. Price 1 Rbl.
61 Kop. (1971). In Russian.

075.002 **Activité solaire en 1969.**
G. Evrard, C. Gonze, A. Koeckelenbergh.
Ciel et Terre, Vol. 87, 413 - 434 (1971).

075.003 **Activité radioélectrique solaire en 1966.**
C. Delys, R. Gonze.
Ciel et Terre, Vol. 87, 435 - 443 (1971).

075.004 **Definitive Sonnenflecken-Relativzahlen für 1970.**
R. A. Naef.
Orion, 29. Jahrgang, p. 82 (1971).

075.005 **Datos relativos a la actividad solar y geomagnética en 1967.**
Urania Barcelona, Año 55, Nos. 271–272, p. 152 (1970).

075.006 **Datos relativos a la actividad solar y geomagnética en 1968.**
Urania Barcelona, Año 55, Nos. 271–272, p. 194 (1970).

075.007 **Sunspots in 1970.** P. S. Laurie.
Quarterly Journ. Roy. Astron. Soc., Vol. 12, 244
(1971). – Progress report.

075.008 **L'osservazione sistematica del sole a Catania.**
G. Godoli, V. Sciuto, M. L. Sturiale, R. A. Zappalà.
Coelum, Vol. 39, 213 - 227 (1971).

075.009 **Soliagttagelser i 1970.** G. Persson.
Astron. Tidsskr., Årg. 4, 136 - 137 (1971).

075.010 **Solar activity and geomagnetic storms 1970.**
P. S. Laurie, K. Dyson.
Observatory, Vol. 91, 233 - 235 (1971).

075.011 **Nombres relatifs de Wolf pour l'année 1970.**
M. Waldmeier.
L'Astronomie, 85^{e}année, p. 411 (1971).

075.012 **Die Sonnentätigkeit im zweiten Halbjahr 1970.**
R. Müller.
Sterne, 47. Jahrgang, p. 199 - 200 (1971).

075.013 **Definitive Sonnenflecken-Relativzahlen für 1970.**
M. Waldmeier.
Sterne, 47. Jahrgang, p. 200 (1971).

075.014 **L'activité solaire.** M.-J. Martres.
L'Astronomie, 85e année, p. 318 - 320, 362 - 363,
402 - 403, 449 - 450 (1971). – Rotations 1569 - 1574.

075.015 **Daily Hα chromosphere pictures, daily K$_{232}$ chromosphere pictures, daily white light photosphere pictures.** M. Cimino (Editor).
Photographic Journal of the Sun, Oss. Astron. Roma, Nos.
45 - 52 (1971). – 1971 April 3 – 1971 November 6. – Rotations 1573 - 1580.

075.016 **Map of the Sun.**
Edited by Fraunhofer Institut, Freiburg. – 1971
July 1 – December 31.

075.017 **Solar phenomena.** M. Cimino, M. Torelli, A. Cacciani, V. Croce, R. Flamini, U. Bartolini.
Oss. Astron. Roma, Monthly Bull. Nos. 163 - 166 (1971).
1971 July – October: Daily total areas of sunspot-groups; Heliographic position, classification and area of sunspot-groups; Longitudinal sunspot magnetic fields; Hours of K-line cinematographic patrol; Hours of Hα cinematographic patrol; S.C.N.A. and S.E.A.; Explanation.

075.018 **Actividad solar en 1969.**
J. Pensado.
Bol. Astron. Obs. Madrid, Vol. 7, No. 6, 128 pp. (1971). –
I. Números relativos de Wolf; II. Estadísticas de manchas y superficie de las mismas; III. Fáculas cromosféricas brillantes;
IV. Filamentos de hidrógeno; V. Protuberancias.

075.019 **Solare Beobachtungsergebnisse (Solar Data).**
E. A. Lauter, A. Böhme, F. W. Jäger, F. Fürstenberg, H. Künzel, D. Scholz, S. Böhm.
Heinrich-Hertz-Inst., Zentralinstitut für solar-terrestrische Physik, Deutsche Akad. Wiss. Berlin, HHI Solar Data, Vol. 22,
June – October (1971). – Solar radio emission; Sunspot magnetic data.

075.020 **Solar observations made at Catania Astrophysical Observatory during 1970.**
Oss. Astrofis. Catania, Pubbl. No. 146, 176 pp. (1971).
This bulletin includes all the data deduced from the solar observations made during 1970 at Catania Astrophysical Observatory: Sunspots; Hα and K faculae; Hα flares; Hα quiescent prominences; K quiescent prominences; Hα active prominences on disc and at limb; Hα patrol hours.

075.021 **Heliographic maps of the photosphere for the year 1970.** M. Waldmeier.
Publ. Sternw. Zürich, Vol. 13, (No. 5), 123 - 151 (1971).
The present publication gives heliographic maps of the photosphere and evolution tables of sunspot-groups for the year 1970. Maps and tables are based on daily drawings of spots and faculae using a projected solar image with a diameter of 25 cm. Such drawings are carried out at the Swiss Federal Observatory Zurich and at its two branch stations, the Astrophysical Observatory Arosa and the Specola Solare Locarno-Monti. At all three stations refractors of 225 cm focal length and 15 cm aperture are used for these observations.

075.022 **Die Sonnenaktivität im Jahre 1970.**
M. Waldmeier.
Vierteljahresschrift Naturforsch. Ges. Zürich, Jahrgang 116,
p. 253 - 271 = Astron Mitt. Sternw. Zürich, No. 303 (1971).
The present paper gives the frequency numbers of sunspots, photospheric faculae and prominences as well as the intensity of the coronal line 5303 Å and of the solar radio emission at the wavelength of 10.7 cm, all characterizing the solar activity in the year 1970.

075.023 **Grafikoni izlaza i zalaza sunca i mjeseca 1972.**
Edited by Hidrografski Institut Jugoslavenske Ratne
Mornarice, Split. HI-N-32, 29 pp. (1971).

075.024 **Daily maps of the sun and geophysical graphs.**
Solnechnye Dannye 1971 Byull., No. 4, p. 1 - 76;
No. 5, p. 1 - 75; No. 6, p. 1 - 66; No. 7, p. 1 - 69; No. 8, p. 1 - 80; No. 9, p. 1 - 66; No. 10, p. 1 - 66 (1971). In Russian.

075.025 **Magnetic fields of sunspots.**
Prilozheniya k Byulletenyu "Solnechnye Dannye",
1971, Nos. 4 - 10. In Russian.

075.026 **Summary of solar observations, January - June, 1966.** M. K. V. Bappu.
Kodaikanal Obs. Bull., Ser. B, No. 192, 21 pp. (1970).

075.027 **Datos relativos a la actividad solar y geomagnética en 1969.**
Urania Barcelona, Año 56, No. 273, p. 50 (1971).

075.028 **Summary of prominence and calcium flocculus observations, magnetic data, ionospheric data, for the second half of 1969.** M. K. V. Bappu.
Kodaikanal Obs. Bull., No. 177, 341 pp. (1970).

075.029 **Summary of magnetic and ionospheric observations, January - June 1966.** M. K. V. Bappu.
Kodaikanal Obs. Bull., Ser. C, No. 194, 303 pp. (1971).

075.030 **Solar photospheric observations.**
F. Bruin, H. Hourani, N. G. Bustati.
Lee Obs., American Univ. Beirut, Monthly Bull., Astron. Section, 1971 April - October (1971).
Sunspot relative numbers; Heliographic mean position and classification of the sunspot groups; Number of facular zones.

075.031 **Observations solaires.**
C. Popovici, E. Țifrea, V. Dinulescu, A. Dimitriu, S. Nicolescu, G. Mariș.
Obs. Bucarest, Section d'Astrophys., Secteur Solaire, Acad. République Socialiste Roumanie. 57 pp. Price Lei 3.00 (1971). Les rotations 1556 - 1569 (25 décembre 1969 - 10 janvier 1971).

075.032 **For solar patrol – Intercosmos 4.**
B. I. Valniček, I. P. Tindo, B. Stark.
Zemlya i Vselennaya, No. 6, p. 16 - 21 (1971). In Russian.

075.033 **Osservazioni solari (ottiche e radio) N° 20 – IV° trimestre 1970.** P. Zlobec.
Pubbl. Oss. Astron. Trieste, No. 429, 14 pp. (1971).

075.034 **Osservazioni solari (ottiche e radio) N° 21 – I° trimestre 1971.** A. Abrami.
Pubbl. Oss. Astron. Trieste, No. 431, 22 pp. (1971).

075.035 **Osservazioni solari N° 22 – II° trimestre 1971.**
Pubbl. Oss. Astron. Trieste, No. 434, 18 pp. (1971).

075.036 **Sunspots** (sunspot relative numbers and sunspot-areas); **Synoptic charts of solar magnetic fields** (Mount Wilson Observatory); **Eruptions chromosphériques brillantes; Intensité de la couronne solaire; Solar radio emission.**
M. Waldmeier, R. Howard, R. Michard, G. Olivieri, M. Bernot.
Quarterly Bull. Solar Activity (published by Eidgen. Sternw. Zürich), Nos. 171 - 172, p. 241 - 433 (1971). – Observations of the co-operating observatories for 1970 July - December are given.

075.037 **Summary of solar observations, January - December 1967.** M. K. V. Bappu.

Kodaikanal Obs. Bull., Ser. B, Nos. 196 - 197, p. B41 - B63, B65 - B84 (1970).

075.038 **Boletín Meteorológico mensual.**
Edited by Obs. Nacional de Física Cósmica, San Miguel, Argentina. Vol. 24 (10) - (12), 1970 October - December, Vol. 25 (1) - (2), 1971 January - February.

075.039 **Solar photosphere charts.** L. Schmied.
Říše hvězd, Vol. 52, 198 (1971).
Rotations Nos. 1570 - 1571.

075.040 **Geomagnetic and solar data.**
J. V. Lincoln (Editor).
Journ. Geophys. Res., Vol. 76, 4709, 5377 - 5379, 6212, 7020, 7825, 8457 (1971). – 1971 March – August.

075.041 **Provisional sunspot numbers.**
Yamamoto Circ., Nos. 1739 - 1742, 1744, 1745 (1971). In Japanese. – 1971 June – November.

075.042 **Catalogue of data on solar-terrestrial physics in World Data Center A Subcenters:** Solar and interplanetary phenomena, ionospheric phenomena, flare-associated events, geomagnetic phenomena, aurora, cosmic rays, airglow.
World Data Center A, Upper Atmosph. Geophys. Rep. UAG-15, 263 pp. (1971).

075.043 **Fenomeni solari.**
F. Mazzucconi, S. Delli Santi, M. L. Sturiale, A. Abrami.
Coelum, Vol. 39, 154 - 159, 202 - 207, 236 - 244 (1971). – 1971 March – August.

075.044 **Osservatorio Magnetico dell'Aquila. Bolletino magnetico.**
Coelum, Vol. 39, 160, 208, 245 (1971). – 1971 March – August.

075.045 **Centro Universitario Fenomeni fluttuanti–Firenze. Test P.**
Coelum, Vol. 39, 161, 209, 246 (1971). – 1971 March – August.

075.046 **Solar and solar system activity.**
R. J. J. Langton, J. R. Smith.
Journ. British Astron. Ass., Vol. 81, 405 - 407, 481 - 483; Vol. 82, 65 - 67 (1971). – 1971 March – August.

075.047 **Sunspot numbers.**
Sky Telescope, Vol. 42, 51, 121, 180, 245, 315, 389 (1971). – 1971 May – October.

075.048 **Zonnevlekkengetallen.**
Hemel en Dampkring, Vol. 69, 186, 260 (1971). – 1971 May – August.

075.049 **Indices of geomagnetic activity.**
Journ. Atmosph. Terr. Phys., Vol. 33, 1289, 1497, 1657, 1801, 1977 (1971). – 1971 April – August.

076   Solar UV, X Rays, Gamma Radiation

076.001   **Monitoring of the Lyman alpha emission line of the sun during the year 1969.**
J. E. Blamont, A. Vidal Madjar.
Journ. Geophys. Res., Vol. 76, 4311 - 4324 (1971).

The shape of the Lyman alpha line emitted by the whole solar disk has been measured with an experiment on board the OSO 5 satellite. These measurements showed that the shape of the line changes considerably with solar activity, owing to variations in intensity at the center of the line, which are much greater than on the blue wing or averaged over the whole line.

076.002   **A comparison of solar EUV intensities and K-coronameter measurements.**   G. L. Withbroe.
Solar Physics, Vol. 18, 458 - 473 (1971).

Characteristics of the emission observed above the solar limb in four EUV lines, Si XII λ 499, Mg X λ 625, Ne VIII λ 770, and O VI λ 1032 are discussed. The mean temperature of the corona derived from the ratios of the intensities of Si XII λ 499 and Mg X λ 625 is 1.8 million K. The EUV data are compared with K-coronameter measurements in order to yield new estimates of the abundances of Si, Mg, Ne and O relative to hydrogen.

076.003   **Satellite lines in the solar X-ray spectrum.**
W. M. Neupert.
Solar Physics, Vol. 18, 474 - 488 (1971).

Observations of solar X-ray line emission using crystal spectrometers during a large chromospheric flare have provided a list of wavelengths with a precision of 0.003 Å in first order of diffraction and correspondingly better in higher orders. In addition to the resonance, intersystem $(1 \, ^1S_0 - 2 \, ^3P_1)$ and forbidden $(1 \, ^1S_0 - 2 \, ^3S_1)$ transitions of ions of the He I isoelectronic sequence, we have recorded satellite lines arising from ions in the Li I, Be I and B I isoelectronic sequences. Apparent decreases in the ratio of forbidden and intersystem line intensities of Mg XI and Si XIII during the flare are used to derive electron densities. A search for satellite lines on the long-wavelength side of the Lyman-alpha line of H I-like ions has yielded no positive identifications.

076.004   **The deduction of energy spectra of non-thermal electrons in flares from the observed dynamic spectra of hard X-ray bursts.**   J. C. Brown.
Solar Physics, Vol. 18, 489 - 502 (1971).

The derivation of dynamic spectra of high energy electrons in flares from high resolution hard X-ray observations and the bearing of this analysis on different models of flare conditions are considered.

076.005   **Ultraviolet studies of the solar atmosphere.**
R. W. Noyes.
Annual Rev. Astron. Astrophys., Vol. 9, (see 003.001), 209 - 236 (1971).

Although we will discuss below the evidence for inhomogeneous structure and some of its implications, the main emphasis of this review will be on the mean structure of the solar atmosphere, in both quiet and active regions.

076.006   **Time-resolved rocket observations on the vacuum ultra-violet solar spectrum during the March 1970 solar eclipse.**
R. W. Nicholls, F. J. Morgan, C. Y. Yang.
Journ. Roy. Astron. Soc. Canada, Vol. 65, 173 (1971).
Abstr. RAS Canada.

076.007   **Soft solar X-rays and solar activity. V: Relation of the course of soft X-ray fluctuations to the course**
of solar activity, 9 March, 1967 — 18 May, 1968.
R. G. Teske.
Solar Physics, Vol. 19, 356 - 378 (1971).

Soft solar X-rays in the wavelength interval 8–12 Å were observed from OSO III. The totality of the observations that were made between 9 March, 1967, and 18 May, 1968, is summarized graphically and compared to the course of solar activity as observed at other wavelengths, with particular emphasis upon visible activity.

076.008   **On the identification of the λ 417 line in the solar extreme ultraviolet spectrum.**
D. R. Flower, C. Jordan.
Astron. Astrophys., Vol. 14, 473 - 476 (1971).

Collision strengths for transitions between terms of the first four configurations of Fe XV have been calculated in the distorted wave approximation. The computed ratio of the intensity of the transition at 417 Å to that of the transition at 284 Å agrees within a factor of two with the most recent observations by Hall and Hinteregger (1970) of the intensity ratio of the solar lines at 417 Å and 284 Å. The possibility of other contributors to the observed line at 417 Å is discussed.

076.009   **The observation of nonthermal solar X-radiation in the energy range $3 < E < 10$ keV.**
S. W. Kahler, R. W. Kreplin.
Astrophys. Journ., Vol. 168, 531 - 541 (1971).

The low-energy (3–10 keV) X-ray spectra observed during solar impulsive bursts of $E < 10$ keV X-rays reported by Kane and Anderson are analyzed. In two of these bursts we can separate the total low-energy X-ray emission into thermal and nonthermal components. The inferred nonthermal electron spectrum is discussed in relation to acceleration by electric fields. The electron spectrum allows a determination of the minimum value of the ratio of electric field strength to electron density.

076.010   **A new solar atlas from echelle rocket-ultraviolet spectra.**   E. F. Milone, W. P. Schneider.
Bull. American Astron. Soc., Vol. 3, 377 (1971). – Abstr. AAS.

076.011   **New observations of satellite lines in the solar X-ray spectrum.**   J. H. Parkinson.
Nature, Phys. Sci., Vol. 233, 44 - 45 (1971).

New observations of several satellite lines, of the form $1s^2.nl$ - $1s2p.nl$ close to the helium-like resonance lines of several elements are reported in the solar X-ray spectrum. The $n = 2$ lines are shown to be formed by the process of dielectronic recombination. It is suggested that higher members of the series appear in the wings of the resonance line.

076.012   **Variations in soft solar X-rays and in the upper atmosphere temperature.**   V. E. Chertoprud.
Space Research XI, (see 012.004), 929 - 933 (1971).

076.013   **Gradual rise and fall X-ray bursts observed aboard OSO-4.**
K. J. H. Phillips, J. L. Culhane, P. W. Sanford.
Space Research XI, (see 012.004), 1351 - 1354 (1971).

076.014   **Non-flare solar X-ray emission shorter than 4 Å.**
V. D. Ivanov, V. Letfus, S. L. Mandel'stam, I. P. Tindo.
Space Research XI, (see 012.004), 1355 - 1358 (1971).

076.015 Studies of the solar X-ray spectrum as a function of position on the disk.
D. H. Brabban, W. M. Glencross, J. R. H. Herring.
IAU Symposium No. 41, (see 012.005), p. 135 - 136 (1971).

076.016 Survey of new solar results. R. Tousey.
IAU Symposium No. 41, (see 012.005), 233 - 250 (1971).

076.017 New ultraviolet solar flux measurements at 2000 Å using a balloon borne instrument.
M. Ackerman, D. Frimout, R. Pastiels.
IAU Symposium No. 41, (see 012.005), p. 251 - 253 (1971).

076.018 Lα measurements during the solar eclipse of 12 November 1966. L. H. Weeks, L. G. Smith.
Solar Physics, Vol. 20, 59 - 63 (1971).
Rocket measurements of the solar Lα (1216 Å) flux during various phases of the total eclipse of 12 November, 1966 confirms that the radiation emanates from the chromosphere. A significant difference between two portions of the photosphere limb is attributed to enhanced emission in the region of a Ca K plage.

076.019 EUV and soft X-ray images of the sun on March 11th, 1971. H. Bräuninger, H. J. Einighammer, J. V. Feitzinger, H. H. Fink, D. H. Höhn, H. Koops, G. Krämer, U. Mayer, G. Möllenstedt, M. Mozer.
Solar Physics, Vol. 20, 81 - 84 = Mitt. Astron. Inst. Univ. Tübingen, No. 125 (1971). – Research note.

076.020 Extreme ultraviolet flashes of solar flares observed via sudden frequency deviations: Experimental results. R. F. Donnelly.
Solar Physics, Vol. 20, 188 - 203 (1971).
Properties of solar-flare EUV flashes measured via a type of ionospheric event, called a sudden frequency deviation (SFD), are presented. SFD's are sensitive to bursts of radiation in the 1–1030 Å wavelength range.

076.021 Solar X-ray observations from Vela satellites.
P. E. Fehlau.
Solar Physics, Vol. 20, 228 - 229 (1971). – Research note.

076.022 High-resolution extreme-ultraviolet solar spectrum recorded with a diffraction-filter spectrograph.
W. Schweizer, G. Schmidtke.
Astrophys. Journ., *(Letters).* Vol. 169, L27 - L29 (1971).
A diffraction filter for suppression of long-wavelength stray light was flown April 1971 in a grazing-incidence grating spectrograph. Although efficiency of the instrument was low, approximately 50 emission lines of the solar extreme-ultraviolet spectrum between 50 and 210 Å were recorded with a resolution of 0.05 Å.

076.023 On the classification of some highly ionized iron and nickel lines in the 200–400 Å region of the solar spectrum. K. G. Widing, G. D. Sandlin, R. D. Cowan.
Astrophys. Journ., Vol. 169, 405 - 411 (1971).
A number of previously unclassified multiplets of Fe XI-XIV are identified in the extreme-ultraviolet spectrum of the sun. The iron lines account for most of the previously unidentified strong lines between 330 and 370 Å. Other solar identifications include the resonance lines of Ni XVII and Ni XVIII, and one $3p-3d$ multiplet of Fe XIII. Oscillator strengths calculated in intermediate coupling for resonance transitions in Fe X–Fe XIV are tabulated.

076.024 Gamma ray and neutron emissions from the sun.
E. L. Chupp.
Space Sci. Rev., Vol. 12, 486 - 525 (1971).

Evidence for acceleration of charged particles in the solar atmosphere is reviewed with specific reference to production of gamma rays and neutrons at the sun. Fluxes of these components at the earth, based on theoretical assumptions are also reviewed and estimates and conditions for obtaining observable fluxes from Syrovatskii's dynamic dissipation model are considered. Knowledge about the sun, to be derived from such observations, is discussed. Finally, a brief review of the present status of experimental observations and suggestions for new experimental approaches are given.

076.025 Rocket observations of the ultraviolet solar spectrum during the total eclipse of 1970 March 7.
A. H. Gabriel, W. R. S. Garton, L. Goldberg, T. J. L. Jones, C. Jordan, F. J. Morgan, R. W. Nicholls, W. J. Parkinson, H. J. B. Paxton, E. M. Reeves, C. B. Shenton, R. J. Speer, R. Wilson.
Astrophys. Journ., Vol. 169, 595 - 614 (1971).
A sequence of thirty-five ultraviolet photographic spectra of the sun has been obtained in the wavelength region 850–2190 Å, as a function of time during the eclipse. These cover the range from before second contact until midtotality, with a spatial resolution of the order 2 arc sec. A general description of the experiment and data is given. Twenty-five new coronal lines have been seen, the majority of which have been identified as new forbidden transitions. The Lα corona is observed out to over 1.5 $R_\odot$, and a quantitative interpretation is presented.

076.026 The time behavior of temperature and emission measure in X-ray flares. R. W. Milkey, N. K. Blocker, W. H. Chambers, P. E. Fehlau, J. C. Fuller, W. E. Kunz.
Solar Physics, Vol. 20, 400 - 412 (1971).
X-ray observations from Vela-5 spacecraft of five flares occurring in November and December 1969 were reduced to temperatures and emission measures as a function of time. This reduction was done assuming a thermal spectrum including free-free and free-bound emission. A phenomenological model is proposed to explain the nature of the time behavior of the temperature and emission measure.

076.027 Anisotropy of solar hard X-radiation during flares.
G. Elwert, E. Haug.
Solar Physics, Vol. 20, 413 - 421 (1971).
The angular distribution of solar flare associated hard X-rays ($\gtrsim 10$ keV) is calculated on the assumption that they originate as bremsstrahlung emission of energetic electrons with a power law spectrum. Supposing the electrons to move in a fixed direction, the X-radiation is considerably anisotropic, especially at high photon energies. Taking into account a magnetic field, the anisotropy decreases with increasing pitch angles of the electrons. The anisotropic angular distribution of solar X-radiation seems to be connected with the centre-to-limb variation of hard X-ray bursts and with the correlation of shortwave fadeouts and geomagnetic crochets to Hα flares.

076.028 Discussion of paper 'On the polarization and anisotropy of solar X-radiation during flares', by G. Elwert and E. Haug. [Solar Physics, Vol. 15, 234 - 248 (1970)].
S. W. Kahler, G. A. Doschek, J. F. Meekins, D. M. Horan.
Solar Physics, Vol. 20, 422 - 424, with a reply by G. Elwert, E. Haug, p. 425 - 427 (1971). – Research note.

076.029 Soft solar X-rays and solar activity. VI: Optical identification of activity associated with X-ray background fluctuations. R. G. Teske.
Solar Physics, Vol. 21, 146 - 156 (1971).
Minor Hα activity, consisting of small brightenings and small, surgelike spikes, was observed to take place above an active center at the solar limb in good time-association with small fluctuations in the soft X-ray background flux, suggesting that even small dynamical events seen optically are associ-

ated with coronal heating.

076.030 **A high resolution solar ultraviolet spectrum between 200 and 220 nm.** B. C. Boland, B. B. Jones, R. Wilson, S. F. T. Engstrom, G. Noci.
Phil. Trans. Roy. Soc. London A, Vol. 270, No. 1202, (see 012.009), 29 - 46 (1971).

076.031 **Observations of the extreme ultraviolet solar spectrum.** R. Tousey.
Phil. Trans. Roy. Soc. London A, Vol. 270, No. 1202, (see 012.009), 59 - 70 (1971).

076.032 **Interpretation of extreme ultraviolet emissions from the sun.** C. W. Allen.
Phil. Trans. Roy. Soc. London A, Vol. 270, No. 1202, (see 012.009), 71 - 75 (1971).

076.033 **New results of X-ray flare studies.** S. Mandelstam.
Phil. Trans. Roy. Soc. London A, Vol. 270, No. 1202, (see 012.009), 135 - 142 (1971).

076.034 **Solar flare X-ray spectra.** W. M. Neupert.
Phil. Trans. Roy. Soc. London A, Vol. 270, No. 1202, (see 012.009), 143 - 155 (1971).

076.035 **The interpretation of XUV solar radiation.** L. Goldberg.
Highlights in Astronomy, Vol. 2, (see 012.017), 476 - 485 (1971).

076.036 **Some problems relating to solar line identification.** A. H. Gabriel.
Highlights in Astronomy, Vol. 2, (see 012.017), 486 - 494 (1971).

076.037 **Opacity sources in the UV spectrum of the sun.** R. M. Bonnet, D. Sacotte.
Highlights in Astronomy, Vol. 2, (see 012.017), 495 - 502 (1971).

076.038 **On the polarization and anisotropy of solar X-radiation during flares.** G. Elwert, E. Haug.
Highlights of Astronomy, Vol. 2, (see 012.017), 575 - 579 (1971).

076.039 **Solar gamma ray burst observed on 27 Sept. 1968.** Y. Hirasima, K. Okudaira, T. Yamagami.
Proc. 11th international conference on cosmic rays, Vol. 2, (see 012.030), 683 - 688 (1970).

076.040 **X-ray and radio emissions from the sun.** S. N. Ghosh.
Indian Journ. Pure and Applied Phys., Vol. 8, 801 - 805 (1970).
 Assuming a Maxwell distribution of electron velocities in the solar corona, the X-ray fluxes at the earth for temperatures 7×10^5, 1×10^6, 1×10^7 and 1×10^8 °K are calculated for free-free transitions. The fluxes are compared with those for free-bound transitions and line emissions obtained by G. Elwert (1948, 1954) and K. Kawabata (1960). The emissions of radio waves by free-free transitions for different conditions of the sun are also calculated.

076.041 **Studies on solar X-ray flares recorded by SOLRAD-9 on Explorer 37 satellite.** M. K. Das Gupta, S. K. Sarkar.

Indian Journ. Pure and Applied Phys., Vol. 9, 249 - 252 (1971). – See Phys. Abstr., Vol. 74, No. 59823 (1971).

076.042 **Solar soft X-radiation.** R. J. Thomas.
 Thesis, Univ. Michigan, Ann Arbor. [Available from Univ. Microfilms, Ann Arbor, Mich., U.S.A. Order No. 70–21802], 206 pp. (1970).
 Data from the University of Michigan's ion chamber photometer aboard the earth-orbiting satellite OSO-III are used to investigate relationships between solar soft X-radiation and phenomena observed at other wavelengths.

076.043 **Solar electron temperatures and X-ray flare activity.** D. M. Horan, R. W. Kreplin.
AIAA Journ., Vol. 9, 1634 - 1636 (1971).

 On the accuracy of Hartree-Fock theoretical wavelengths in the XUV. See Abstr. 022.095.

 Ultraviolet solar spectrum recorded by echelle spectrograph (1970 to 1800 Å). See Abstr. 071.055.

 High resolution interferometric studies of the solar magnesium II doublet spectral region. See Abstr. 071.061.

 Helium-like ion forbidden line emission, and solar active regions. See Abstr. 071.063.

 On the flux of gamma rays from solar flares. See Abstr. 073.017.

 Solar flares in the extreme ultraviolet. See Abstr. 073.037.

 Extreme ultraviolet radiation from solar flares. See Abstr. 073.053.

 Iron-line emission during solar flares. See Abstr. 073.081.

 An upper limit on the hardness of the nonthermal electron spectra produced during the flash phase of solar flares. See Abstr. 073.082.

 Solar flares in the extreme ultraviolet. See Abstr. 073.088.

 The transfer of Lyman continuum radiation in chromospheric flares. See Abstr. 073.089.

 Electron temperature and emission measure variations during solar X-ray flares. See Abstr. 074.075.

 Observations of solar bursts at microwave and extreme ultraviolet wavelengths. See Abstr. 077.051.

 Topics in solar cosmic ray and X-ray production. See Abstr. 078.041.

 Solar X-ray control of E and sporadic-E layers of the ionosphere. See Abstr. 083.076.

 Experiment to measure hard solar and celestial X-rays from the fifth Orbiting Solar Observatory. See Abstr. 142.054.

 Low-energy gamma rays (0.1–10 MeV) in the atmosphere and in outer space. See Abstr. 142.098.

077 Solar Radio Radiation

077.001 Inversion in sense of circular polarization with respect to frequency for solar microwave bursts.
E. Scalise, Jr., D. Basu, P. Marques dos Santos.
Astron. Astrophys., Vol. 13, 471 - 477 (1971).

Inspection of circular polarization data of solar bursts at 9400, 3750, 2000 and 1000 MHz over 1957 - 69 revealed that the polarization sense of a burst may reverse with respect to frequency once (single inversion) or twice or thrice (multiple inversion) between 9400 and 1000 MHz. Multiple inversion events, very few in number, may be associated with double sources both producing radio bursts simultaneously.

077.002 Wave propagation in the wave plasma and the spectrum of the solar radio bursts.   L. Mollwo.
Solar Physics, Vol. 19, 128 - 148 (1971).

The frequency bands of noise storms and type I-bursts as well as of type IVdm-bursts are shown to be in accordance with the consequences of a recently proposed mechanism (Mollwo, 1970). Interpretations are supposed of some features of type III- and of type IVmA-bursts. The magnetic field strength over active regions in two corona levels is deduced, too. The discussion leads to a conception of the corona parameters in the level of type $IV\mu$-bursts.

077.003 Results of a new absolute calibration of the solar flux density at 2980 MHz.   G. K. Schmidt.
Solar Physics, Vol. 19, 149 - 151 (1971).

The solar flux density was measured absolutely at the frequency 2980 MHz and the results were compared with those of observers, working in the same frequency range. We have found, that our values are integrating well into a smooth spectral curve, whereas it seems that even the corrected values of Ottawa 2800 MHz are 4% too high.

077.004 Solar microwave bursts as indicators of the occurrence of solar proton emission.   D. L. Croom.
Solar Physics, Vol. 19, 152 - 170 (1971).

A study has been made of the relation of 19 GHz (λ = 1.58 cm) solar radio bursts to solar proton emission, with particular reference to the usefulness of relatively long duration bursts with intensities exceeding 50% of the quiet sun flux as indicators of the occurrence of proton events during the four years from 1966–69. The complete microwave spectra of the proton events have also been studied, and have been used to extend the results obtained at 19 GHz to other frequencies, particularly in the 5–20 GHz band.

077.005 Forecasting the intensity of solar proton events from the time characteristics of solar microwave bursts.   D. L. Croom.
Solar Physics, Vol. 19, 171 - 185 (1971).

A survey of the main characteristics of solar microwave bursts in relation to their usefulness for indicating the intensity of associated solar proton emissions suggests that time parameters give much better results than intensity or spectrum parameters. In particular, best results are obtained by using the effective, or mean, burst duration. Using this parameter solar proton warnings and intensity estimates can be made with observations at only one frequency, preferably in the range 5–20 GHz.

077.006 Active solar radio regions at metric frequencies and the interplanetary sector structures.
K. Sakurai, R. G. Stone.
Solar Physics, Vol. 19, 247 - 256 (1971).

The possible relation between type I noise active regions and the polarity distribution of the interplanetary magnetic field is examined for the period from 13 March to 21 August, 1968 (Solar Rotation numbers 1842 - 1847) by using data from ground-based and satellite observations. The position of the source of the sector boundaries and its relation to the type I radio regions are investigated by taking into account the mean bulk velocity of solar winds as observed by space probes. A model of the large-scale structure of type I radio regions and their relation to the sector structure of the magnetic field as observed in the interplanetary space is briefly discussed.

077.007 Spectra of the different types of solar microwave bursts.   M. K. Das Gupta, S. K. Sarkar.
Journ. Roy. Astron. Soc. Canada, Vol. 65, 152 - 158 (1971).

Spectra of the different types of solar microwave bursts have been examined separately using simultaneous bursts recorded at 1415, 2695, 4995, 8800, 15,400 and 35,000 MHz by the Sagamore Hill Radio Observatory during the period December 1967 to March 1970. The variation of peak flux and of energy (average flux × duration) with frequency for the individual bursts under different morphological types have been examined.

077.008 A procedure for observing the solar five minute oscillation in two dimensions.
J. B. Rice, V. Gaizauskas.
Journ. Roy. Astron. Soc. Canada, Vol. 65, 174 (1971).
Abstr. RAS Canada.

077.009 Calibrations for the solar patrol at Algonquin Radio Observatory and Dominion Radio Astrophysical Observatory.   A. E. Covington.
Journ. Roy. Astron. Soc. Canada, Vol. 65, 174 (1971).
Abstr. RAS Canada.

077.010 On the relations of radio emission flux with flare activity.   V. N. Ikhsanova.
Solnechnye Dannye 1971 Byull., No. 4, p. 100 - 104 (1971). In Russian.

From observations of the decrease of spectral density of the radio emission flux in the centimeter wavelength range an attempt of short-termed (1–2 days) forecasting of chromospheric flares is made.

077.011 Observations of sporadic solar radio emission spectra in the region of meter and decimeter waves.
Yu. B. Vedenejev, N. M. Prytkov, K. A. Shmelev.
Astron. Tsirk., No. 615, p. 1 - 2 (1971).   In Russian.

077.012 3-m measurements of positions of local radio emission sources on the sun using a two-element interferometer.   Yu. B. Vedenejev, N. M. Prytkov, A. K. Chandaev, K. A. Shmelev.
Astron. Tsirk., No. 615, p. 2 - 4 (1971). In Russian.

077.013 Faraday rotation dispersion and the distribution of polarization characteristics of type III bursts.
A. D. Fokker.
Solar Physics, Vol. 19, 472 - 479 (1971).

Some results on the distribution of polarization characteristics of type III bursts obtained by Chin, Lusignan and Fung (1971) are analysed in terms of the effects of Faraday rotation dispersion on fully polarized signals with small axial ratios of the polarization ellipse.

077.014 Correlation between aurorae and solar radio bursts at 185 MHz.   E. Doylerush.

Journ. British Astron. Ass., Vol. 81, 449 - 453 (1971).

077.015 Mapping of solar polarization at 7.8 GHz.
D. W. Richards, R. M. Straka.
Nature, Phys. Sci., Vol. 233, 92 - 94 (1971).

Two-dimensional maps of circularly polarized emission
from the sun at 7.8 GHz, with a resolution of 4.4 arc minutes,
show polarization of local sources of up to 15% and variations
as small as 2% in flux density and 0.2% in polarization, as
functions of time. Both flux density and polarization in
McMath plage 11002 showed substantial increases thirty
minutes or more before a flare on October 24, 1970. It is
suggested that flares may be triggered when both of these
quantities surpass critical values.

**077.016 Observations of the solar oscillatory component at
a wavelength of 3 millimeters.**
M. Simon, F. I. Shimabukuro.
Astrophys. Journ., Vol. 168, 525 - 529 (1971).

The solar oscillatory component has been observed at a
wavelength of 3 mm by its modulation of the solar free-free
emission. The observations were carried out separately at 3.3
and 3.5 mm with two different radio telescopes to rule out
possible instrumental and atmospheric effects. Power spectra
of the data show that the strongest spectral component is
at 5.5×10^{-3} Hz (period 180 s).

**077.017 An attempt to interpret the agents generating the
bursts of the solar radio emission as strong hydrodynamical shock waves.** I. Garczyńska.
Acta Astron., Vol. 21, 395 - 414 (1971).

An attempt was made to connect the data on chromospheric flares, on radio bursts of different frequencies, and on
geomagnetic storms with sudden commencements. It is
assumed that all these effects are caused by strong shock
waves, produced during chromospheric flares. Three models
of such processes, the Sedov model, the Kompaneec model
and our own model, are discussed. For the last model numerical calculations were performed and the electron concentration above the active areas was found. The possibility of
predicting geomagnetic storms with an accuracy of about
3 hours by means of our model and using radio bursts data
is pointed out.

077.018 Activité radioélectrique solaire en 1967.
C. Delys, R. Gonze.
Ciel et Terre, Vol. 87, 534 - 544 (1971).

**077.019 Fine structure in meter wavelength solar radio type
III burst events.** G. L. Tarnstrom, K. W. Philip.
Bull. American Astron. Soc., Vol. 3, 374 - 375 (1971).
Abstr. AAS.

**077.020 Parallel drifting bands in meter wavelength solar
radio emissions.** K. W. Philip, G. L. Tarnstrom.
Bull. American Astron. Soc., Vol. 3, 375 (1971). — Abstr.
AAS.

077.021 Solar polarization mapping at 7.8 GHz.
D. W. Richards, R. M. Straka.
Bull. American Astron. Soc., Vol. 3, 375 (1971). — Abstr.
AAS.

**077.022 Polarized solar radio emission at millimeter wave-
lengths in active regions.** E. B. Mayfield,
S. Edelson, F. I. Shimabukuro.
Bull. American Astron. Soc., Vol. 3, 375 - 376 (1971).
Abstr. AAS.

**077.023 The magnetic fields and the polarization of radio
emission in the active center of October 1968.**

N. Erushev, A. B. Severny, T. Tsap.
IAU Symposium No. 43, (see 012.003), p. 428 - 431 (1971).

**077.024 80 MHz radioheliograph evidence on moving type
IV bursts and coronal magnetic fields.**
S. F. Smerd, G. A. Dulk.
IAU Symposium No. 43, (see 012.003), p. 616 - 641 (1971).

**077.025 Possible causes of line splitting in drift pair solar
bursts.** D. B. Melrose, W. Sy.
Proc. Astron. Soc. Australia, Vol. 2, 56 - 57 (1971).

In this paper possible causes of line splitting in emission
near the local plasma frequency are considered in connection
with drift pair solar radio bursts. The basic model envisaged
for the bursts involves a bunch of electrons streaming through
the solar corona at several times the thermal velocity of
electrons.

**077.026 Multiple magnetic-loop structure of a type IV solar
radio outburst.** R. T. Stewart, K. V. Sheridan.
Proc. Astron. Soc. Australia, Vol. 2, 60 - 62 (1971).

In this paper we discuss 80 MHz heliograph observations
of the multiple source structure and polarization of a type IV
solar radio outburst on 1970 November 16. At times during
the event six sources were present. Three of these were highly
circularly polarized in a L. H. sense and two in a R. H. sense.
The sixth source was extended and had oppositely polarized
edges. From the source behaviour we conclude that the radio
emission came from two expanding and one stationary magnetic arch.

**077.027 Evolution of a jet-like structure in the late phase of
a complex solar outburst.**
A. C. Riddle, K. V. Sheridan.
Proc. Astron. Soc. Australia, Vol. 2, 62 - 65 (1971).

A new feature in the form of a jet formed by close juxta-
position of a number of highly polarized, separately resolved
80 MHz sources was observed as the late phase of a very complex outburst on 1971 January 25. We present here a source
model which, we think, can explain the observed source properties. The early phase, also complex and involving numerous
moving sources, will be described first.

**077.028 Space radio-astronomy of solar bursts at all fre-
quencies: The STEREO project.**
J. L. Steinberg, C. Caroubalos.
IAU Symposium No. 41, (see 012.005), p. 419 (1971).
Abstract. — The full text is published in Astronomy and As-
trophysics, Vol. 9, 329 - 338 (1970).

**077.029 A pulsating regime of stream instability and the
origin of 'rain' type radio bursts.** V. V. Zaitsev.
Solar Physics, Vol. 20, 95 - 105 (1971).

An interpretation is suggested for the 'rain' type radio
bursts on the basis of a pulsating regime of the stream instability. This regime may occur in ejected stabilized ion streams
trapped in the region in which the magnetic field of the corona has a typical bipolar structure.

**077.030 A U-type solar radio burst originating in the outer
corona.** R. G. Stone, J. Fainberg.
Solar Physics, Vol. 20, 106 - 111 (1971).

The observation of a U-type solar radio burst with a
reversing frequency of approximately 0.7 MHz suggests the
presence of a magnetic bottle extending out to about 35 $R_\odot$.
A possible model of this loop structure is developed from the
data. The occurrence of low-frequency U-bursts seems to be
extremely rare although magnetic bottles may develop frequently during solar maximum.

077.031 Spectral features of large type IV bursts and inter-

relation to solar-terrestrial phenomena.
S. T. Akinyan, E. I. Mogilevsky, A. Böhme, A. Krüger.
Solar Physics, Vol. 20, 112 - 121 (1971).

During the present cycle of solar activity about 25 spectra of type IV bursts have been analysed derived from single frequency measurements of the HHI. These spectra in addition to polarization measurements give the opportunity of a judgement of different type IV burst components with respect to the effective emission mechanism. Basing on these data and supported by the results of former collections of type IV burst spectra on the one side, and by recent measurements of the X-ray emission as observed by the SOLRAD satellites on the other, a physical discussion of distinct components and their association with high-energy phenomena of solar-terrestrial physics will be tempted.

077.032 **On the polarization of solar microwave bursts observed at 17 GHz.** W. Wassenberg.
Solar Physics, Vol. 20, 130 - 135 (1971).

The polarization distribution of 17 GHz bursts is studied observed within a period of 1 yr after maximum solar activity. The fine structure of the polarization curve of complex bursts is shown and two possible interpretations of the observed inversion of the polarization at 17 GHz during a complex event are given.

077.033 **19 GHz (1.58 cm) solar radio bursts in the period July 1967 to December 1969.**
D. L. Croom, R. J. Powell.
Solar Physics, Vol. 20, 136 - 146 (1971).

The results of 2 1/2 yr (July 1967 – December 1969) monitoring of solar radio bursts at 19 GHz ($\lambda = 1.58$ cm) at the Radio and Space Research Station, Slough, are presented. Fifteen bursts with peak flux increases exceeding 1000×10^{-22} Wm^{-2} Hz^{-1} were observed during this period.

077.034 **Statistical investigations of the time-dependent fine structure of solar bursts.** G. Zimmermann.
Astron. Astrophys., Vol. 15, 433 - 438 (1971).

As there exists no theoretical interpretation of the time-dependent fine structure of solar radio bursts, the observed data must be analyzed statistically to give objective results. The purpose of the present work is to provide data for the proper design of a multichannel radiospectrograph for dm wavelengths, and a data reduction and recording system. Methods for the statistical investigation of fine structure have been tested. They have been applied to paper records of single-channel measurements at 240, 400 to 460, and 1420 MHz made in 1966 and 1967. The preliminary results are presented.

077.035 **Investigations on the program "Rapid variations of solar magnetic fields", III. Some peculiarities of solar radio emission during June 6–17, 1969.**
A. A. Gnezdilov, O. I. Judin.
Solnechnye Dannye 1971 Byull., No. 7, p. 74 - 77 (1971).
In Russian.

The observational data on fluctuations of the radio intensity and fluctuations of circularly polarized component at $\lambda = 3$ cm (Institute of Radio-Physics, Gorky) and activity at $\lambda = 1.5$ m (IZMIRAN, Moscow) have been compared with measurements of the magnetic fields in sunspot groups 207, 208 and 209 during the period from 6 to 17 June 1969.

077.036 **Regular pulses from the sun and a possible clue to the origin of solar cosmic rays.**
D. J. McLean, K. V. Sheridan, R. T. Stewart, J. P. Wild.
Nature, Vol. 234, 140 - 142 (1971).

Observations of extremely regular pulsations in metre-wave solar bursts suggest a model for their generation: the shock front responsible for associated type II emission accelerates electrons in a tube of strong magnetic field. Oscillations of this tube modulate the intensity of synchrotron emission by these electrons, causing the regular pulses. Solar cosmic rays observed after some of these bursts may be explained by the same model.

077.037 **A comparative study of the different aspects of impulsive and gradual rise and fall types of solar microwave bursts.** M. K. Das Gupta, S. K. Sarkar.
Journ. Roy. Astron. Soc. Canada, Vol. 65, 284 - 294 (1971).

Impulsive and GRF (gradual rise and fall) types of solar microwave bursts recorded at 606, 1415, 2695, 4995, 8800 and 15400 MHz by the Sagamore Hill Solar Radio Observatory for nearly a 4-year period covering the maximum phase of the current solar cycle have been examined in relation to their different characteristic features. Peak flux spectra of the simultaneous bursts have also been examined. The bursts have been examined in relation to optical and X-ray solar flares.

077.038 **Center-to-limb variation of the peak flux spectra of type IV radio bursts associated with solar proton flares.** K. Sakurai.
Planet. Space Sci., Vol. 19, 1685 - 1691 (1971).

Type IV radio bursts with wide band from microwave to metric-wave frequency are generally associated with solar proton flares. We have analyzed the center-to-limb variation of the peak flux spectra of type IV radio bursts associated with solar proton flares by using the observational data in 1954 through 1967. In this analysis, we show that the U-shaped peak flux spectra are typical for most solar proton flares and are generally obtained independent of the position of flares. The center-to-limb variation has been shown to be due solely to the sharp decrease of the peak flux at metric frequencies with increase of the angular distance from the central meridian of the solar disk.

077.039 **Wechselwirkung von Ionen- und Elektronenwellen mit intensitätsgleicher Emission bei der einfachen und doppelten Plasmafrequenz.** D. Beermann.
Zeitschr. Naturforschung, Vol. 26a, 1909 - 1918 (1971).

Nonlinear interaction of longitudinal plasma waves in two-fluid plasma exciting the solar type II burst emission is investigated. The interaction of ion and electron waves yields radiation at the fundamental plasma frequency, while emission at twice the plasma frequency is caused by self-interaction of the electron waves. Equal flux intensity results in both the radiation bands if the variation of electron density associated with the two wave modes is of the same order of magnitude.

077.040 **A moving Type IV radio burst and its relation to the coronal magnetic field.**
G. A. Dulk, M. D. Altschuler.
Solar Physics, Vol. 20, 438 - 447 (1971).

A moving Type IV burst, observed with the Culgoora radioheliograph on 1970 April 29, moved out to about 3 $R_\odot$ and attained high circular polarization before fading. The appearance of the moving Type IV source suggests an isolated, self-contained, synchrotron emitting plasmoid. To explain the observed source structure and high unipolar polarization, we suggest that a ring of electric current was ejected from the low corona and guided by coronal magnetic field lines; the radio emission was synchrotron radiation generated by mildly-relativistic electrons trapped in the poloidal magnetic field of the ring current.

077.041 **Activité radioélectrique solaire en 1968.**
C. Gonze-Delys, R. Gonze.
Ciel et Terre, Vol. 87, 619 - 635 (1971).

077.042 **The temperature of solar active regions on a wavelength of 2 cm.** A. Tlamicha.

Bull. Astron. Inst. Czechoslovakia, Vol. 22, 371 - 375 (1971).

The measurements of the temperature of the active regions on the sun on a wavelength of 2 cm by using high resolution radio telescope at the N.R.A.O., Green Bank, antenna 43 m in diameter, are presented. The variation of the temperature of the active regions is between 10000°K and 24000°K.

077.043 Catalogue of solar type IV radio events, 1956–1969.
A. Krüger, F. Fürstenberg, K. H. Fricke.
Heinrich-Hertz-Inst., Zentralinstitut für solar-terrestrische Physik, Deutsche Akad. Wiss. Berlin, HHI Suppl. Ser. Solar Data, Vol. 2, 187 - 240 (1971).

The catalogue lists a number of 346 major world-wide observed type IV radio bursts occurring since the beginning of the 19th solar cycle (1954) up to the maximum of the 20th solar cycle (1969). The compilation is based partially on the Švestka-Olmr-catalogue which has been completed by a few new events and other characteristics, and has been extended for the period after 1963. There is included an atlas of spectral diagrams of selected events.

077.044 Solar radio emission at 1.2 mm wavelength.
M. R. Kundu.
Solar Physics, Vol. 21, 130 - 136 (1971).

Some properties of solar active regions at 1.2 mm wavelength are discussed. Equatorial and polar brightness distributions of the quiet sun at 1.2 mm wavelength are also presented.

077.045 A comparison of type III solar radio burst theories using satellite radio observations and particle measurements. L. G. Evans, J. Fainberg, R. G. Stone.
Solar Physics, Vol. 21, 198 - 203 (1971).

The required electron density to excite a type III solar burst can be predicted from different theories, using the low frequency radio observations of the RAE-1 satellite. Electron flux measurements by satellite in the vicinity of 1 AU then give an independent means of comparing these predicted exciter electron densities to the measured density.

077.046 Some results of a study of fluctuations of the polarized solar radio emission at 3 cm.
M. S. Durasova, G. A. Lavrinov, A. K. Chandaev, O. I. Yudin.
Solnechnye Dannye 1971 Byull., No. 8, p. 90 - 96 (1971).
In Russian.

A preliminary analysis of fluctuations in the radiometer channels of intensity and polarization is made. Quasi-periodic fluctuations of the polarized component of radio emission have been detected at λ = 3 cm. They were found to occur rarer than those in non-polarized emission.

077.047 On mutual shifts of the minima of solar radio emission at various wavelengths during 1963 - 1965.
S. Zięba, A. Michalec.
Solnechnye Dannye 1971 Byull., No. 9, p. 101 - 106 (1971).
In Russian.

Radio observations of the sun at various wavelengths during 1963 – 1965 are discussed. The shifts of the minima can possibly be explained by synchrotron radiation emitted in the remnants of disappearing centres of activity.

077.048 On the frequency dependence of the degree of polarization of type III solar radio bursts.
V. V. Fomichev, I. M. Chertok.
Astron. Zhurn. Akad. Nauk SSSR, Vol. 48, 1244 - 1250 (1971). In Russian. English translation in Soviet Astron. AJ, Vol. 15, No. 6.

The analysis of many observational data shows that the mean degree of polarization of type III radio bursts is approximately constant close to 20% in a wide frequency range from 20 to 200 MHz. Possible interpretations of this effect are discussed.

077.049 On some peculiarities of groups of type III solar radio bursts.
A. K. Markeev, V. A. Styazhkin, I. M. Chertok.
Astron. Zhurn. Akad. Nauk SSSR, Vol. 48, 1251 - 1257 (1971). In Russian. English translation in Soviet Astron. AJ, Vol. 15, No. 6.

The structural peculiarities of type III events registered in the form of single bursts, short lifetime ($\tau < 1$ min) or long lifetime ($\tau > 1$ min) groups, are investigated by data of the observations with the IZMIRAN 45–90 MHz spectrograph in 1967 - 1969.

077.050 A study of the directivity effect of radio emission of two local sources on the sun at 0.8, 1.3, 3.2, 9.1, 21 and 43 cm. A. S. Grebinskij, A. N. Tsyganov.
Vestn. Leningr. un-ta, 1971, No. 13, p. 151 - 158. In Russian.
Abstr. in Referativ. Zhurn. 51. Astron., 1.51.436 (1972).

077.051 Observations of solar bursts at microwave and extreme ultraviolet wavelengths.
J. P. Castelli, D. W. Richards.
Journ. Geophys. Res., Vol. 76, 8409 - 8413 (1971).

During the several months in 1967 when solar emission was measured in the band 270 to 1310 Å by instrumentation aboard the OSO 3 satellite, numerous flare-associated EUV bursts were observed. These provided the first EUV burst data for direct comparison with microwave radio bursts. We point out the similarities between intensity-time structure of these EUV bursts and that of centimeter and decimeter radio bursts.

077.052 Investigations of the slowly varying component of solar radio emission. N. P. Stasjuk.
Problems of cosmic physics. Vyp. (No.) 6, (see 003.040), p. 45 - 62 (1971). In Russian.

077.053 Results of radio observations of the solar eclipse on September 22, 1968 at frequencies 204 and 600 MHz. A. T. Nesmyanovich, O. S. Popov, A. M. Sveeridov, V. V. Chmeel, V. A. Khaleemonenko, A. Ya. Yavlinsky.
Problems of cosmic physics. Vyp. (No.) 6, (see 003.040), p. 63 - 72 (1971). In Russian.

077.054 About the interpretation of the continuous radio burst emission of the sun. – Invited review paper.
A. Krüger.
Geod. Geophys. Veröff., Ser. 2, No. 13, (see 012.029), p. 27 - 47 (1969).

077.055 Interaction between solar burst and planetary atmosphere. M. Shimizu.
Kagaku (*Japan*), Vol. 40, 655 - 662 (1970). In Japanese.
Abstr. in Phys. Abstr., Vol. 74, No. 50095 (1971).

077.056 Analysis of occurrences of different types of solar microwave bursts at different frequencies.
M. K. Das Gupta, S. K. Sarkar.
Indian Journ. Pure and Applied Phys., Vol. 9, 187 - 188 (1971).

Occurrences of different types of solar microwave bursts recorded at five different frequencies over the period Nov. 1966 to Feb. 1970 at Sagamore Hill Solar Radio Observatory, USA, have been examined.

077.057 Improved radio mapping of the sun. M. Arisawa.
Proc. Res. Inst. Atmosph. Nagoya Univ., (*Japan*), Vol. 18, 89 - 101 (1971). – See Phys. Abstr., Vol. 74, No. 66645 (1971).

077.058 Lēni mainīgais Saules radiostarojums.
M. Kamenskis.
Zvaigžņotā debess, 1970./71. gada ziema, p. 1 - 10.

077.059 **Solar explosions.** J. P. Wild.
Search, Vol. 2, 229 - 233 (1971).
A review of selected aspects of solar radio astronomy.
Masson memorial lecture to 1971 ANZAAS congress, Brisbane,
25 May. – *DMCL*

077.060 **Solar radioastronomy and coronal magnetohydrodynamics.** M. Kuperus.
Rep. of the meeting of CESRA 1971, (see 012.036), p. 4 -
14 (1971).

077.061 **Solar radio astronomy from the optical solar
researcher's point of view.** Z. Švestka.
Rep. of the meeting of CESRA 1971, (see 012.036), p. 15 -
30 (1971).

077.062 **Observations of peculiar solar events at metric
waves.** A. Abrami.
Rep. of the meeting of CESRA 1971, (see 012.036), p. 31 -
38 (1971).

077.063 **Some preliminary results of an investigation into
the characteristics of chains of type I bursts between
320 and 160 MHz.** J. P. Loonen.
Rep. of the meeting of CESRA 1971, (see 012.036), p. 39
(1971).

077.064 **Peculiar absorption and emission microstructures in
a type IV solar radio outburst.** C. Slottje.
Rep. of the meeting of CESRA 1971, (see 012.036), p. 40
(1971).

077.065 **On stormbursts with an inclination in the time-
frequency plane.** A. D. Fokker.
Rep. of the meeting of CESRA 1971, (see 012.036), p. 41 -
42 (1971).

077.066 **Note on the occurrence of linear polarization in
type III bursts.** A. D. Fokker.
Rep. of the meeting of CESRA 1971, (see 012.036), p. 42 -
43 (1971).

077.067 **On periodic variations in the continuum level of
solar radio noise storms in the 200 MHz range.**
O. P. Sveen.
Rep. of the meeting of CESRA 1971, (see 012.036), p. 44 -
45 (1971).

077.068 **Properties of type III solar radio bursts.**
E. Lyngstad, Ø. Elgarøy.
Rep. of the meeting of CESRA 1971, (see 012.036), p. 46 -
47 (1971).

077.069 **Frequency structure in wide band radio bursts.**
Ø. Elgarøy.
Rep. of the meeting of CESRA 1971, (see 012.036), p. 48 -
49 (1971).

077.070 **Observations of the quiet-sun at decameter wavelength scattering effects on the brightness distributions.** M. Aubier.
Rep. of the meeting of CESRA 1971, (see 012.036), p. 50 -
51 (1971).

077.071 **Radio model of the quiet sun.** F. C. Drago.
Rep. of the meeting of CESRA 1971, (see 012.036),
p. 52 - 53 (1971).

077.072 **The slowly varying component.** M. Pick.
Rep. of the meeting of CESRA 1971, (see 012.036),
p. 54 (1971).

077.073 **Solar bursts at sub mm-wavelengths.** G. Feix.
Rep. of the meeting of CESRA 1971, (see 012.036),
p. 55 - 56 (1971).

077.074 **Report of the working discussion of group I.
February 19, 1971.** E. Schanda, G. Feix.
Rep. of the meeting of CESRA 1971, (see 012.036), p. 57 -
58 (1971).

077.075 **Solar joint project.** G. Feix.
Rep. of the meeting of CESRA 1971, (see 012.036),
p. 59 - 62 (1971).

077.076 **Radioastronomy and coronal microstructures.**
J. L. Steinberg.
Rep. of the meeting of CESRA 1971, (see 012.036), p. 63 -
66 (1971).

077.077 **Physical processes in the corona, related to radio
emission.** H. Rosenberg.
Rep. of the meeting of CESRA 1971, (see 012.036), p. 67 -
69 (1971).

077.078 **Report on the meeting of the working group III.**
A. Boischot.
Rep. of the meeting of CESRA 1971, (see 012.036), p. 70 -
71 (1971).

077.079 **3 millimeter observations of the 7 March 1970 total
solar eclipse.** P. N. Swanson.
Pennsylvania State Univ., Coll. Sci., Dep. Astron., Radio
Astron. Obs., Sci. Rep. No. 22, 41 pp. (1971).
A double peak in the radial distribution curve of the sun
at 3 mm wavelength has been observed. – *RTS*

077.080 **A catalog of distinct solar radio events for fixed
frequency observations made at the Pennsylvania
State University Radio Astronomy Observatory during the
period 1 July 1964 through 30 June 1970.**
J. P. Hagen, F. L. Wefer.
Pennsylvania State Univ., Coll. Sci., Dep. Astron., Radio
Astron. Obs., Sci. Rep. No. 23 (1971). – *NRL*

077.081 **Solar radio spike bursts.**
G. L. Tarnstrom, K. W. Philip.
Alaska, Univ. Geophys. Inst., Sci. Rep. UAG R–217, 186 pp.
(1971).
Spikes bursts, i.e. bursts less than 0.01 sec duration, have
been observed with a spectrograph in the 200 MHz region. It
appears from the results that there is a morphological continuity between spike bursts and short duration type III. – *NRL*

Physics of solar continuum radio bursts.
See Abstr. 003.113.

**On the influence of variations of the magnetic field
of a sunspot on the emission of S-component sources of
solar radio emission.** See Abstr. 072.018.

**On the spectral study of sunspot number and solar
radio flux between 600 and 9400 MHz.** See Abstr. 072.097.

**Radio observations of filaments during the eclipses
of September 11, 1969 and March 7, 1970.**
See Abstr. 073.060.

**On coronal instability and moving radio features
associated with a flare spray.** See Abstr. 074.031.

**Magnetic fields in the lower corona associated with
the expanding limb burst on March 30th 1969 inferred from**

the microwave high-resolution observations.
See Abstr. 074.033.

Optical and radio observations of large scale magnetic fields on the sun.   See Abstr. 074.038.

A coronal disturbance observed simultaneously with
a white-light coronameter and the 80 MHz Culgoora radioheliograph.   See Abstr. 074.046.

Gyro-resonance absorption of plasma waves in the

corona and the fine structure of solar radio bursts.
See Abstr. 074.050.

The radio investigation of the solar corona.
See Abstr. 074.089.

X-ray and radio emissions from the sun.
See Abstr. 076.040.

Interplanetary-particle associations with type III
solar bursts.   See Abstr. 078.014.

078 Solar Cosmic Radiation

078.001 **Statistical study of solar protons, alpha particles, and $Z \geqslant 3$ nuclei in 1967–1968.**
T. P. Armstrong, S. M. Krimigis.
Journ. Geophys. Res., Vol. 76, 4230 - 4244 (1971).
   Comprehensive measurements of protons, α particles, and $Z \geqslant 3$ nuclei of $E > 0.5$ Mev/nucleon obtained with the lunar-orbiting Explorer 35 spacecraft are reported. The data are discussed in the context of particle propagation and properties of the solar source region.

078.002 **The anomalous distribution in heliocentric longitude of solar injected cosmic radiation.**
E. P. Keath, R. P. Bukata, K. G. McCracken, U. R. Rao.
Solar Physics, Vol. 18, 503 - 509 (1971).
   Concurrent observations of the solar flare of March 12, 1969 by two spacecrafts separated in solar longitude by 38° show that the accessibility at 1 AU to cosmic ray particles is not a simple function of the relative solar longitude. The cosmic ray flux, degree of anisotropy, and rise time all indicate that the favored path for cosmic ray propagation in this event was some 40° to the east of the nominal Archimedes spiral line of force from the flare location.

078.003 **Anisotropy characteristics of low energy cosmic ray population of solar origin.** U. R. Rao,
K. G. McCracken, F. R. Allum, R. A. R. Palmeira, W. C. Bartley, I. Palmer.
Solar Physics, Vol. 19, 209 - 233 (1971).
   Comparing Explorer 34 observations on solar protons in the energy range 0.7–55 MeV with similar observations from other spacecrafts we investigate the energy dependence of the equilibrium anisotropy of the solar cosmic radiation during both the early and late decay phases of the flare event, and investigate a number of properties of the onset phase of the flare effect.

078.004 **The unusual anisotropic solar particle event of November 18, 1968.** S. P. Duggal, I. Guidi,
M. A. Pomerantz.
Solar Physics, Vol. 19, 234 - 246 (1971).
   We describe and analyze the neutron monitor observations during the ground level event of November 18, 1968, and deduce therefrom the characteristics of the propagation mechanism that prevailed throughout the entire duration of this event.

078.005 **The solar longitude dependence of proton event delay time.** E. Barouch, M. Gros, P. Masse.
Solar Physics, Vol. 19, 483 - 493 (1971).
   The relationship between heliographic longitude and the delay between flare occurrence and solar proton observation is studied using results obtained aboard HEOS A1 during 1969. The result obtained differs from previous findings. We ascribe this to the formation of a long-lived magnetic field configuration close to the sun associated with a particular group of active regions.

078.006 **Equatorial and precipitating solar protons in the magnetosphere. 1. Low-energy diurnal variations.**
L. J. Lanzerotti.
Journ. Geophys. Res., Vol. 76, 5235 - 5243 (1971).
   Recent experimental determinations of the total proton γ-ray production cross sections are used together with equatorial proton fluxes measured on ATS 1 to calculate the predicted diurnal variations in the precipitating low-energy solar proton-produced γ rays.

078.007 **Equatorial and precipitating solar protons in the magnetosphere. 2. Riometer observations.**
T. A. Potemra, L. J. Lanzerotti.
Journ. Geophys. Res., Vol. 76, 5244 - 5251 (1971).
   The equatorial solar protons measured on ATS 1 during the January 28, 1967, solar event are used together with the ionosphere model of Potemra et al. (1969) to calculate the expected daytime 30-MHz riometer absorption. The computed absorption values compare favorably with the riometer observations at Byrd, Antarctica, located on approximately the same L shell ($L = 7$) as ATS 1 ($L = 6.4$). Empirical linear relationships between the square root of the 2π-omnidirectional ATS 1 proton fluxes greater than some E_{min} and the daytime 30-MHz riometer absorptions are also studied.

078.008 **On the spectrum of solar particles in the initial period of the flare of February 23, 1956.**
L. I. Miroshnichenko.
Geomagn. Aeronom., Vol. 11, 705 - 707 (1971). In Russian.

078.009 **Low-energy (≥ 0.3 Mev) solar-particle observations at widely separated points (>0.1 AU) during 1967.**
S. M. Krimigis, E. C. Roelof, T. P. Armstrong, J. A. Van Allen.
Journ. Geophys. Res., Vol. 76, 5921 - 5946 (1971).
   Simultaneous observations of solar-particle events at low (<1 Mev) energies have been obtained by using intercalibrated solid-state detectors on Mariner 5, Explorer 35, and Mariner 4 during days 200 to 271, 1967. An examination of the intensity-time profiles, anisotropies intensity ratios of protons to α particles, ratios of α particles to $Z \geq 3$ (M-nuclei), and the proton energy spectrums from several events has led to the conclusion that, in view of the observations, multi-spacecraft studies should allow the development of more sophisticated models for the understanding of low-energy solar-particle events.

078.010 **Effect of the solar boundary condition on flare-particle propagation.** R. C. Englade.
Journ. Geophys. Res., Vol. 76, 6190 - 6192 (1971). – Letter.

078.011 **The increase of the cutt-off rigidity of solar cosmic rays in the period of PCA, connected with the commencement of a magnetic disturbance.**
V. M. Driatzky, A. V. Shirochkov.
Geomagn. Aeronom., Vol. 11, 895 - 898 (1971). In Russian. Brief information.

078.012 **The entry of solar protons over the polar caps.**
G. E. Morfill, J. J. Quenby.
Planet. Space Sci., Vol. 19, 1541 - 1577 (1971).
   Observations of solar protons at energies from 1 MeV to 360 MeV are examined in relation to the information that these particles give about the magnetosphere, magnetotail and magnetopause. Trajectory integrations in a realistic model of the geomagnetic field out to $25R_E$ and a tail field model fitted to observations from $15R_E$ to $80R_E$ are used to obtain a better understanding of the particle motion. We discuss separately the expected effects of anisotropies and gradients on the polar cap distribution in terms of possible magnetotail field models, in order to determine whether experimental data is able to distinguish various possibilities.

078.013 **Detailed directional and temporal properties of solar energetic particles associated with propagating interplanetary shock waves.** S. Singer, M. D. Montgomery.
Journ. Geophys. Res., Vol. 76, 6628 - 6642 (1971).

In the period May 1967 to May 1968, detailed Vela 4 observations of the properties of protons in the enhancements ('spikes') associated with propagating interplanetary shocks are reported. These results are combined with observations of the solar-wind plasma (for three of the events) and also compared with the predictions of the various models.

078.014 **Interplanetary-particle associations with type III solar bursts.** T. E. Graedel, L. J. Lanzerotti. Journ. Geophys. Res., Vol. 76, 6932 - 6938 (1971).

Interplanetary proton ($E \sim 0.6$ and $E \sim 1.2$ Mev) and electron ($E > 40$ kev; $E > 300$ kev) data from May 28 to November 30, 1967, are used to study the association of particle data with solar radio-burst data in the dekametric band from OGO 3 and with data reported in Solar Geophysical Data and the I. A. U. Quarterly Bulletin.

078.015 **Solar proton medium flux constancy over a million years.** A. K. Lavrukhina, G. K. Ustinova. Nature, Vol. 232, 462 - 463 (1971).

Depth distributions of cosmogenic radionuclides of different life times: Na^{22}, Fe^{55}, Al^{26}, Mn^{53} and Mn^{54}, produced by solar and galactic cosmic rays in lunar surface layer with chemical composition similar to that of B/17-33/ rock from Apollo-11 mission have been calculated. By the comparison with measured activities of these radionuclides in rock 10017 (Shedlovsky et al., Science, Vol. 167, 574, 1970) the constancy of medium solar proton flux over a million years is confirmed and its equality to present medium flux at about $31.1^{+5.6}_{-9.4}$ protons cm^{-2} sec^{-1} is determined.

078.016 **Operation PCA 69 – an investigation of the solar particle event of 2 November 1969.** J. C. Ulwick. Space Research XI, (see 012.004), 1181 - 1187 (1971).

078.017 **Solar proton structure across the polar cap and apparent anomalous latitude cut-off.** D. E. Page. Space Research XI, (see 012.004), 1189 - 1194 (1971).

078.018 **Time variations of the flux of electrons of energy between 7 MeV and 300 MeV.** C. J. Bland, C. Dilworth, D. Maccagni, E. G. Tanzi, J. P. Mercier, A. Raviart, L. Treguer. Space Research XI, (see 012.004), 1205 - 1211 (1971).

078.019 **Solar cosmic ray bursts with proton energies $E_p > 30$ MeV observed aboard Venus 6.** S. N. Vernov, E. A. Chuchkov, N. N. Kontor, G. P. Lyubimov, A. G. Nikolaev, N. V. Pereslegina. Space Research XI, (see 012.004), 1213 - 1228 (1971).

078.020 **Penetration of solar protons over the polar cap during the 25 February 1969 event.** F. Axisa, A. Bewick, A. C. Durney, J. Engelmann, R. Hynds, L. Koch, G. Morfill. Space Research XI, (see 012.004), 1229 - 1238 (1971).

078.021 **Propagation of low energy protons associated with the 24 January 1969 solar flare.** A. Balogh, P. C. Hedgecock, R. J. Hynds, J. Sear. Solar Physics, Vol. 20, 150 - 165 (1971).

This paper presents directional low energy proton measurements together with interplanetary magnetic field measurements. Propagation of 1 to 13 MeV solar protons is discussed in terms of the relative importance of field-aligned streaming compared to convection of the proton population in the solar wind. It is pointed out that the effect of adiabatic deceleration can be quite important.

078.022 **The propagation of solar cosmic-ray bursts.** C. K. Ng, L. J. Gleeson.

Solar Physics, Vol. 20, 166 - 185 (1971).

A model is presented in which we show analytically the three phases of anisotropy which occur during solar cosmic-ray events observed in the 7.5 MeV to 21 MeV kinetic-energy interval and reported by McCracken et al. (1971). The model is based on the cosmic-ray particles being convectively transported out from the sun, undergoing anisotropic diffusion along the interplanetary magnetic-field lines, and losing energy by adiabatic deceleration or by collision processes. The event is seen simply as a pulse moving outward from the sun after a cosmic-ray burst with a negative density-gradient in front of it and a positive gradient behind.

078.023 **27-day recurrence and long term modulation of cosmic ray intensity.** V. K. Balasubrahmanyan, D. Venkatesan. Solar Physics, Vol. 20, 186 - 187 (1971). – Research note.

078.024 **On the relations between development of active regions on the sun and their succeeding events in cosmic ray intensity.** N. P. Milovidova, V. P. Nefediev. Solnechnye Dannye 1971 Byull., No. 7, p. 109 - 114 (1971). In Russian.

For a determination of possible active regions on the sun, responsible for generation of a geoeffective agent, the situation in the active regions at different levels of the solar atmosphere is considered.

078.025 **The release of energetic particles from the sun.** G. M. Simnett. Solar Physics, Vol. 20, 448 - 461 (1971).

A model is developed to account for the release of solar cosmic rays from the sun. The solar atmosphere out to $3-5$ solar radii above the photosphere is permeated with magnetic field lines which trap low rigidity ($\lesssim 50$ MV) flare particles. Plasma heated by the flare process disturbs the trapping field, and not until the disturbance reaches $3-5$ solar radii can the low rigidity flare particles have access to interplanetary space, where they form corotating streams. Reference is made to satellite observations of solar electromagnetic radiation and charged particles.

078.026 **A search for solar neutrons near solar maximum, II.** P. Cortellessa, P. di Benedetto, C. Paizis. Solar Physics, Vol. 20, 474 - 490 (1971).

A plastic scintillator counter with anticoincidence screen has flown in four balloon flights at a floating altitude variable between 4 and 5 mb, during night to day, day-time and night-time flights. The analysis and comparison of the day-time and night-time parts of the flights has given for the continuous emission flux an upper limit of 5.5×10^{-3} $n/cm^2 s$ over the energy range from 10 MeV to 200 MeV.

078.027 **Additional electron flux detection after magnetic perturbations.** S. A. Volobuyev, A. M. Galper, V. V. Dmitrenko, V. G. Kirillov-Ugryumov, B. I. Luchkov, E. M. Shermanzon. Solar Physics, Vol. 20, 491 - 496 (1971).

An additional electron flux at an energy above 100 MeV was observed in the experiments carried out with high-altitude balloons flown at geomagnetic latitudes 46° and 49°, in the upper layers of the atmosphere, on the days following magnetic perturbations. Its intensity, equal to 6×10^{-2} cm^{-2} s^{-1} sr^{-1}, decreased over $20-30$ hours. The effect observed confirms the presence of high-energy electrons in the regions of the trapped radiation.

078.028 **On the propagation, acceleration and deceleration of cosmic rays in different regions of the interplanetary space.** L. I. Dorman. "Trudy Mezhdunar. seminara po probl. "Generatsiya kosmich.

luchej na Solntse", 1970". Moskva, 1971, p. 52 - 63. In Russian. – Abstr. in Referativ. Zhurn. 51. Astron., 12.51.660 (1971).

078.029 **Gradients and diffusion of cosmic rays in the interplanetary space.**
A. G. Zusmanovich, E. V. Kolomeets, Yu. A. Shakhova.
"Trudy Mezhdunar. seminara po probl. "Generatsiya kosmich. luchej na Solntse", 1970". Moskva, 1971, p. 255 - 268. In Russian. – Abstr. in Referativ. Zhurn. 51. Astron., 12.51.661 (1971).

078.030 **Low-energy proton increases associated with interplanetary shock waves.**
R. A. R. Palmeira, F. R. Allum, U. R. Rao.
Solar Physics, Vol. 21, 204 - 224 (1971).

Impulsive increases in the low energy proton flux observed by the Explorer 34 satellite, in very close time association with geomagnetic storm sudden commencements are described. It is shown that these events are of short duration (20–30 min) and occur only during the decay phase of a solar cosmic-ray flare event. The differential energy spectrum and the angular distribution of the direction of arrival of the particles are discussed.

078.031 **Solar particle events.**   Z. Švestka.
Phil. Trans. Roy. Soc. London A, Vol. 270, No. 1202, (see 012.009), 157 - 165 (1971).

078.032 **Solar cosmic rays.**   C. E. Fichtel.
Phil. Trans. Roy. Soc. London A, Vol. 270, No. 1202, (see 012.009), 167 - 174 (1971).

078.033 **Solar flare effects in lunar and meteoritic krypton and xenon.**
M. N. Rao, K. Gopalan, V. S. Venkatavaradan, L. L. Wilkening.
Meteoritics, Vol. 6, 303 (1971). – Abstract.

078.034 **The propagation of low-energy solar protons and electrons in the interplanetary medium.**
R. A. R. Palmeira, F. R. Allum, K. G. McCracken.
Trudy Mezhdunar. seminara po probl. "Generatsiya kosmich. luchej na Solntse",1970. Moskva, 1971, p. 75 - 151. – Abstr. in Referativ. Zhurn. 62. Issled. kosm. prostranstva, 12.62.344 (1971).

078.035 **On the integral multiplicities of generation for calculation of the spectrum of solar cosmic rays.**
L. I. Miroshnichenko.
Geomagn. Aeronom., Vol. 11, 1077 - 1080 (1971). In Russian. Brief information.

078.036 **Solar particles and related solar phenomena.**
K. A. Anderson.
Proc. 11th international conference on cosmic rays, Budapest 1969, (see 012.025), p. 315 - 339 (1970).

078.037 **Report on quasi-stationary modulation papers.**
E. N. Parker.
Proc. 11th international conference on cosmic rays, Budapest 1969, (see 012.025), p. 341 - 356 (1970).

078.038 **Modulation dynamics.**   A. Somogyi.
Proc. 11th international conference on cosmic rays, Budapest 1969, (see 012.025), p. 357 - 380 (1970).

078.039 **On the shock waves connected with energetic particle fluxes from the sun.**   S. I. Syrovatskij.
Trudy Mezhdunar. seminara po probl. "Generatsiya kosmich. luchej na Solntse", 1970. Moskva, 1971, p. 37 - 51. In Rus-

sian. – Abstr. in Referativ. Zhurn. 51. Astron., 1.51.418 (1972).

078.040 **On the character of fast particles propagation in the interplanetary space at different stages of increase of solar cosmic ray fluxes.**
S. N. Vernov, P. V. Vakulov, E. V. Gorchakov, V. G. Kurt, Yu. I. Logachëv, G. P. Lyubimov, G. A. Timofeev.
Trudy Mezhdunar. seminara po probl. "Generatsiya kosmich. luchej na Solntse", 1970. Moskva, 1971, p. 64 - 74. In Russian. – Abstr. in Referativ. Zhurn. 51. Astron., 1.51.419 (1972).

078.041 **Topics in solar cosmic ray and X-ray production.**
T. L. Cline.
Trudy Mezhdunar. seminara po probl. "Generatsiya kosmich. luchej na Solntse", 1970. Moskva, 1971, p. 390 - 436. Abstr. in Referativ. Zhurn. 51. Astron., 1.51.453 (1972).

078.042 **On the connection of solar cosmic rays with the solar and interplanetary magnetic fields.**
B. M. Vladimirskij, L. S. Levitskij, A. B. Severnyj.
Trudy Mezhdunar. seminara po probl. "Generatsiya kosmich. luchej na Solntse", 1970. Moskva, 1971, p. 312 - 319. In Russian. – Abstr. in Referativ. Zhurn. 51. Astron., 1.51.454 (1972).

078.043 **Generation of protons and electromagnetic radiation during solar flares.**
L. E. Gajnova, V. I. Ivanov, E. V. Kolomeets, V. A. Kobzev.
Trudy Mezhdunar. seminara po probl. "Generatsiya kosmich. luchej na Solntse, 1970. Moskva, 1971, p. 379 - 389. In Russian. – Abstr. in Referativ. Zhurn. 51. Astron., 1.51.455 (1972).

078.044 **North-south asymmetry of solar-particle fluxes in polar-cap regions.**   V. Domingo, D. E. Page.
Journ. Geophys. Res., Vol. 76, 8159 - 8164 (1971).

We present results obtained during the asymmetric solar particle event of January 24, 1969, and make a comparison with the event of November 2, 1967.

078.045 **Energy changes of cosmic rays in interplanetary space.**   L. J. Gleeson, C. K. Ng.
Journ. Geophys. Res., Vol. 76, 8434 - 8440 (1971). – Letter.

078.046 **27-day variations of solar activity and cosmic ray intensity and the determination of the effective size of the asymmetric solar wind.**
M. V. Alaniya, L. K. Shatashvili.
Proc. 11th international conference on cosmic rays, Vol. 2, (see 012.030), 199 - 202 (1970).

078.047 **27-day cosmic ray intensity variations in 1967.**
G. A. Bazilevskaya, V. P. Okhlopkov, T. N. Charakhchyan.
Proc. 11th international conference on cosmic rays, Vol. 2, (see 012.030), 203 - 207 (1970).

078.048 **On the 27-day recurrency in cosmic ray intensity and geomagnetic K_p-data.**
E. Dyring, H. Hauska, B. Kangedal.
Proc. 11th international conference on cosmic rays, Vol. 2, (see 012.030), 209 - 214 (1970).

078.049 **Cosmic ray intensity increases before Forbush effects.**
L. I. Dorman, N. S. Kaminer, T. V. Kebuladze.
Proc. 11th international conference on cosmic rays, Vol. 2, (see 012.030), 227 - 231 (1970).

078.050 Cosmic ray currents in stationary streams of solar plasma. G. F. Krymsky, A. M. Altukhov, A. I. Kuzmin, G. V. Skripin.
Proc. 11th international conference on cosmic rays, Vol. 2, (see 012.030), 249 - 251 (1970).

078.051 Modulation and heliocentric gradient of low energy cosmic rays near solar minimum, 1965.
S. R. Kane, J. R. Winckler.
Proc. 11th international conference on cosmic rays, Vol. 2, (see 012.030), 269 - 274 (1970).

078.052 Evidence for residual modulation left over from the 1954 - 1964 solar cycle. H. V. Neher.
Proc. 11th international conference on cosmic rays, Vol. 2, (see 012.030), 289 - 294 (1970).

078.053 Solar flare increases in cosmic ray intensity on November 18, 1968; February 25, 1969, and March 30, 1969.
P. Venkatarangan, D. Venkatesan, J. A. Van Allen.
Proc. 11th international conference on cosmic rays, Vol. 2, (see 012.030), 409 - 420 (1970).

078.054 Solar flare increases in cosmic ray intensity observed at Sulphur Mountain and Calgary between November 1968 and April 1969. T. Mathews, B. G. Wilson.
Proc. 11th international conference on cosmic rays, Vol. 2, (see 012.030), 433 - 438 (1970).

078.055 The entry of solar protons into the geomagnetic field during the event of June 9, 1968.
J. J. Quenby.
Proc. 11th international conference on cosmic rays, Vol. 2, (see 012.030), 475 - 480 (1970).

078.056 The effect of the asymmetric magnetosphere on the response of high-latitude neutron monitors to solar particle events. M. A. Shea, D. F. Smart.
Proc. 11th international conference on cosmic rays, Vol. 2, (see 012.030), 539 - 543 (1970).

078.057 Multiple neutron studies of unusual cosmic ray increases during 1967 - 1969.
M. Kodama, T. Suda, K. Ogura.
Proc. 11th international conference on cosmic rays, Vol. 2, (see 012.030), 571 - 576 (1970).

078.058 Production of neutrons and gamma-rays by nuclear interactions in solar flares. K. Ito, H. Okazoe.
Proc. 11th international conference on cosmic rays, Vol. 2, (see 012.030), 679 - 682 (1970).

078.059 The flux of fast neutrons in the atmosphere. 2. Solar flare effects.
E. S. Light, M. Merker, R. B. Mendell, S. A. Korff.
Proc. 11th international conference on cosmic rays, Vol. 2, (see 012.030), 745 - 751 (1970).

078.060 The solar particle event of March 1970 as observed over the polar cap and in the radiation belt with the satellite GRS-A/AZUR.
E. Achtermann, B. Häusler, D. Hovestadt, M. Scholer.
Zeitschr. Geophys., Vol. 37, 211 - 223 (1971).
Experimental results on energetic electrons, protons and alpha-particles are presented for the time of the March 1970 Solar Particle Events. The paper discusses (1) the time history of the solar particle event, (2) the proton to alpha ratio and its variation during the event, (3) the occurrence of delayed penetration of particles into the central polar cap region, (4) the apparent inward motion of outer belt electrons associated with the event.

078.061 Propagation of 1.5 MeV solar protons in the interplanetary space.
S. N. Vernov, E. V. Gorchakov, G. A. Timofeev.
Proc. Sixth Winter School on Space Physics, Part I. Apatity 1969, (see 012.033), p. 211 - 215 (1971).

078.062 Current concepts on the acceleration and propagation of solar cosmic rays. L. I. Miroshnichenko.
Proc. Sixth Winter School on Space Physics, Part II. Apatity 1969, (see 012.034), p. 33 - 37 (1971).

Transactions of the international seminar on the problem: "Generation of cosmic rays on the sun". See Abstr. 003.018.

Solar explosions. See Abstr. 073.107.

Gamma ray and neutron emissions from the sun. See Abstr. 076.024.

Solar microwave bursts as indicators of the occurrence of solar proton emission. See Abstr. 077.004.

Forecasting the intensity of solar proton events from the time characteristics of solar microwave bursts. See Abstr. 077.005.

Regular pulses from the sun and a possible clue to the origin of solar cosmic rays. See Abstr. 077.036.

Center-to-limb variation of the peak flux spectra of type IV radio bursts associated with solar proton flares. See Abstr. 077.038

Penetration of solar protons over the polar cap during the February 25, 1969, event. See Abstr. 084.201.

Asymmetric access of energetic solar protons to the earth's north and south polar caps. See Abstr. 084.202.

Solar particle cutoffs as observed at low altitudes. See Abstr. 084.203.

On the direct observation of cosmic-ray energy losses. See Abstr. 143.006.

The daily variation of cosmic-ray cutoff at the altitude of a geostationary satellite. See Abstr. 143.028.

The physics of cosmic X-ray, γ-ray, and particle sources. See Abstr. 143.032.

Radiation transfer in space physics. See Abstr. 143.136.

The diffusion coefficients of cosmic rays. See Abstr. 143.138.

079 Solar Eclipses

079.001 Periodiciteit in het optreden van eclipsen.
M. Drooger.
Hemel en Dampkring, Vol. 69, 169 - 173 (1971).

079.002 L'éclipse de soleil du 3 mai 1375 av. J.-C.
V. Maitre.
L'Astronomie, 85ᵉ année, p. 404 - 405 (1971).

079.003 On a graphical method for pre-computing circumstances of solar eclipses observable from the surface of the moon.   S. V. Drozdov.
(Uch. zap.) Perm. gos. ped. in-t, Vol. 72, 398 - 400 (1969).
In Russian. − Abstr. in Referativ. Zhurn. 51. Astron.,
1.51.140 (1972).

079.004 Interesting peculiarities on solar eclipses.
J. Meeus.
Zemlya i Vselennaya, No. 6, p. 29 - 33 (1971). In Russian.

079.100 Solar eclipse, 1970 March 7

Ionosphere-gravity wave interactions during the March 7, 1970, solar eclipse.   P. R. Arendt.
Journ. Geophys. Res., Vol. 76, 4695 - 4697 (1971). − Letter.

Atmospheric gravity waves induced by a solar eclipse, 2.   G. Chimonas, C. O. Hines.
Journ. Geophys. Res., Vol. 76, 7003 - 7005 (1971). − Letter.

Phase variation of a very accurate radio frequency signal due to the solar eclipse.   D. Sadeh.
Journ. Geophys. Res., Vol. 76, 8427 - 8429 (1971).
A delay in the time of arrival of time ticks during the solar eclipse on March 7, 1970, was observed. These time ticks are transmitted by the Loran C system from Cape Fear, North Carolina, and are received in Washington. The presence at 10 km altitude of 100 times more electrons per cm² than previously believed could explain all the experimental observations.

Recording the eclipse shadow bands on magnetic tape.   F. M. Susel.
Journ. Roy. Astron. Soc. Canada, Vol. 65, 273 - 276 (1971).
During the eclipse of March 7, 1970, the Lakeland Community College solar expedition devised several experiments to record the shadow bands that appeared both prior to and after totality.

1970 solar eclipse as "seen" by a torsion pendulum.
E. J. Saxl, M. Allen.
Phys. Rev. D, Particles and Fields, Third Ser., Vol. 3, 823 - 825 (1971).

The observation of the March 7, 1970 solar eclipse at 8.3-mm wavelength.   R. W. Haas.
Pennsylvania State Univ., Dep. Astron., Radio Astron. Obs., Sci. Rep. No. 21, 104 pp. (1971).
A detailed discussion of techniques and atmospheric and other corrections for 8.3-mm radiometer observations of the sun, with reference to the observation of the 1970 March 7 total eclipse at Miahuatlan, Mexico. Observations of lunar and solar disk temperatures and solar bursts are also briefly described. − NRL

Observations of the total eclipse on March 7, 1970

in Mexico.   G. F. Vjalshin, V. M. Sobolev, Y. S. Streletsky.
Solnechnye Dannye 1971 Byull., No. 10, p. 99 - 105 (1971).
In Russian.

Ergebnisse der Sonnenfinsternis-Beobachtung vom 7. März 1970.   M. Waldmeier.
Umschau, 71. Jahrgang, p. 668 - 670 (1971).

Form und Struktur der Korona bei der Sonnenfinsternis vom 7. März 1970.   See Abstr. 074.069.

Rocket observations of the ultraviolet solar spectrum during the total eclipse of 1970 March 7.
See Abstr. 076.025.

3 millimeter observations of the 7 March 1970 total solar eclipse.   See Abstr. 077.079.

079.101 Solar eclipse, 1971 February 25   ·

O eclipse parcial do sol de 25 de Fevereiro de 1971. Observações espectroheliográficas e determinação dos tempos de contacto.   M. Moreirinhas Pinheiro.
Comun. Obs. Astron. Univ. Coimbra, No. 8, 17 pp. (1971).

Osservazioni dell'eclisse parziale di sole del 25 febbraio 1971 a Pino Torinese.
Mem. Soc. Astron. Italiana, Nuova Ser., Vol. 42, 555 - 560 (1971).
From photographic observations during the partial eclipse the times of first and second contact and of maximum phase have been deduced and compared with the computed times.

Observations of the partial solar eclipse of 25 February 1971 at the Brno Observatory.
K. Jehlička, J. Medek, K. Raušal.
Říše hvězd, Vol. 52, 192 - 194 (1971). In Czech.

Observations of a partial solar eclipse at 9 mm wavelength.   A. M. Flett, P. R. Foster, P. Strachan, D. C. Thornton.
Solar Physics, Vol. 20, 317 - 321 (1971).
Observations at 9 mm wavelength of the partial solar eclipse of February 25th 1971 were made to investigate possible limb brightening of the sun. The results obtained show that less than 5 % of the solar disc power can be contained in any such brightening.

Rezultati promatranja djelomične pomrčine Sunca 25. II 1971. (Results of observations of the partial solar eclipse 25 February, 1971).   P. Krešimir.
Vasiona, Vol. 19, 68 - 71 (1971).

Einige Betrachtungen zur partiellen Sonnenfinsternis vom 25. Februar 1971.   H. Haug.
Veröff. Wilhelm-Foerster-Sternw. Berlin, No. 26, 7 pp. (1971).

Solar eclipse 1971, February 25.
Zemlya i Vselennaya, No. 4, p. 68 (1971). In Russian.

079.102 Solar eclipse, 1968 September 22

Observation of the solar eclipse on 22 September

1968 at frequencies of 214 and 950 MHz.
P. A. Kapustin, A. A. Petrovskij, V. V. Khrulev, I. P. Kuznetsova, A. M. Paseka.
Izv. vyssh. uchebn. zavedenij. Radiofizika, Vol. 14, 530 - 535 (1971). In Russian. – Abstr. in Referativ. Zhurn. 51. Astron., 10.51.349 (1971).

Some results of observations of the solar eclipse on September 22, 1968 in polarized emission at 3 cm.
M. S. Durasova, G. A. Lavrinov.
Solnechnye Dannye 1971 Byull., No. 5, p. 102 - 107 (1971). In Russian.

The results of observations of the solar eclipse on Sept. 22, 1968 with two radio telescopes at Zimenki station in the wavelengths of 3.2 cm and 3.06 cm are given. It is shown that the sizes of the sources of the polarized and non-polarized radio emission differ from each other negligibly.

Solar corona at the total eclipse of September 22, 1968. See Abstr. 074.016.

079.103 Solar eclipse, 1972 July 10

Site investigations for observations of the total solar eclipse of July 10, 1972.
A. T. Nesmjanovich, K. I. Churyumov.
Astron. Tsirk., No. 653, p. 5 - 8 (1971). In Russian.

Total solar eclipse of 10 July 1972.
J. S. Duncombe.
United States Naval Obs., *Washington,* Circ. No. 131, 51 pp. (1971).

This Circular contains the following information: The eclipse map from the American Ephemeris and Nautical Almanac; The path of the total phase; The central line, duration and width of path at 10,000 feet, 20,000 feet, 30,000 feet, and 40,000 feet; The central line, duration and width of path at 80 km, 100 km, 200 km, and 300 km; Local circumstances

for points on the central line; Local circumstances for geographic locations for both total and partial phase; Maps of portions of the path; Climatological data; Map of path.

079.104 Solar eclipse, 1973 June 30

An exploration near Agades and Timbuktu, an African diary.   J. M. Pasachoff.
Munger Africana Library Notes, No. 7, 52 pp. (1971). – This booklet may be a useful source of up-to-date information to help plan for observing the total eclipse of the sun on June 30, 1973, in Africa. It can be ordered for $ 2.00 from The Munger Africana Library, California Institute of Technology, Pasadena, Calif.

Total solar eclipse of 30 June 1973.
J. S. Duncombe.
United States Naval Obs., *Washington,* Circ. No. 135, 33 pp. (1971).

This Circular contains the following information: The eclipse map from the American Ephemeris and Nautical Almanac; The path of the total phase; The central line, duration and width of path at 10,000 feet, 20,000 feet, 30,000 feet, and 40,000 feet; The central line, duration and width of path at 80 km, 100 km, 200 km, and 300 km; Local circumstances for points on the central line; Local circumstances for geographic locations for both total and partial phase; Maps of portions of the path; Map of path.

079.105 Solar eclipse, 1961 February 15

Atmospheric effects of the total solar eclipse 1961.
B. Antić.
Archiv Meteor. Geophys. Biokl., Wien, Ser. B, Vol. 18, 197 - 201 (1970). – Abstr. in Bull. Sci. Yougoslavie, Sect. A, Vol. 16, 305 (1971).

080   Solar Figure, Internal Constitution, Rotation, Miscellanea

**080.001   Active cavity radiometric scale, international
pyrheliometric scale, and solar constant.**
R. C. Willson.
Journ. Geophys. Res., Vol. 76, 4325 - 4340 (1971).
    The active cavity radiometer type II, a new and accurate
standard detector, has been developed for the absolute measurement of optical radiant flux. In August 1968 two ACR's
measured the solar irradiance at an altitude of 25 km in a
balloon-flight experiment. The solar-constant value derived
from this measurement was H_0 = 137.0 mw/cm^2.

**080.002   Solar oblateness and the abundance of lithium in
the sun.**   R. H. Dicke.
Phys. Rev. Letters, Vol. 27, 210 - 213 (1971).
    The Goldreich-Schubert thermally driven turbulence to
the deep interiors of solar-type stars which exhibit rotational
slowing is precluded by spectroscopic observations showing
the presence of lithium and beryllium. These observations,
similar observations of the sun, the solar-wind torque, and the
solar oblateness are consistent with a model for which such a
turbulence is terminated at the surface of a rapidly rotating
core containing 95% of the star's mass.

**080.003   On the propagation of a spectrum of acoustic
waves in the solar atmosphere.**   P. Ulmschneider.
Astron. Astrophys., Vol. 14, 275 - 282 (1971).
    The propagation of acoustic waves, their transformation
into shock waves and their dissipation has been computed on
basis of the Harvard Smithsonian Reference Atmosphere
(HRA) for the sun. Acoustic frequency spectra of Stein (1968)
were used and the effect of radiative damping included. Good
agreement was found between the heating produced by these
waves and the computed radiative losses in the chromosphere.
Coronal heating proved more diffucult to explain.

**080.004   The solar constant. (A compilation of recent meas-
urements).**   D. Labs, H. Neckel.
Solar Physics, Vol. 19, 3 - 15 (1971).
    A detailed compilation of the most recent values of the
solar constant is given (13 values published from 1967 to
1970). The most probable value seems to be 1.95 cal cm^{-2}
min^{-1} or 1.36 kW m^{-2} with a formal rms error of ± 0.3%.
The corresponding effective temperature is 5770 K. Systematic errors of the order of ± 1%, but also a possible variability
of the same order cannot be excluded.

**080.005   The rotation of magnetic loop systems in the solar
atmosphere.**   G. W. Pneuman.
Solar Physics, Vol. 19, 16 - 31 (1971).
    The rotation of initially poloidal loop configurations of
this type, as influenced by differential rotation of the foot-
points, is investigated. The analysis is restricted to axially symmetric fields and it is assumed that the toroidal magnetic field
induced by differential rotation is small as compared to the
initial poloidal field. The most interesting physical situation is
that of flux tubes existing in one solar hemisphere only, one
end of the tube being fixed in the photosphere at a higher latitude than the other. Using the Newton and Nunn model for
the surface differential rotation rate, the angular velocity
distribution on two particularly simple types of closed magnetic loop systems is determined analytically. It is shown that
the angular velocity increases outward in the polar regions
but decreases outward near the equator – leading to a decrease
in differential rotation with height.

**080.006   A method for constructing streamlines for the sun's
large scale flow from Doppler velocities.**

P. A. Gilman.
Solar Physics, Vol. 19, 40 - 43 (1971).
    We show that, if the large scale departures from the mean
differential rotation, measured by Howard and Harvey, represent nearly horizontal flow, we may under certain assumptions
deduce a pattern of streamlines for these motions from the
Doppler line of sight velocities. The structure of the resulting
streamline pattern could be compared to large scale magnetic
patterns, and latitudinal transports of magnetic flux and momentum could be estimated.

080.007   On arising of magnetic ropes in the convective zone.
A. A. Solovjev.
Solnechnye Dannye 1971 Byull., No. 5, p. 86 - 93 (1971).
In Russian.

**080.008   On the structure of motions in the convective en-
velope of the sun.**   E. M. Drobyshevsky.
Solnechnye Dannye 1971 Byull., No. 6, p. 84 - 88 (1971).
In Russian.
    The results of research of laboratory scale laminar convec-
tion are supposed to be applicable to solar turbulent convection, if the criteria calculated with the use of turbulent transfer coefficients, coincide with the laminar convection criteria.
In this case rotation strongly affects the convection from a
depth of 10 - 20 thousands km and deeper.

**080.009   Irregularities in solar rotation and variation of its
radius.**   D. Nohonoj.
Solnechnye Dannye 1971 Byull., No. 6, p. 93 - 97 (1971).
    It is shown that irregularities in rotation of the sun and
variations of its radius are the consequence of the law of conservation of the angular momentum.

080.010   Neutrino astronomy: Probing the sun's interior.
G. L. Wick.
Science, Vol. 173, 1011 - 1012 (1971).

080.011   The measurement of solar magnetic fields.
J. M. Beckers.
IAU Symposium No. 43, (see 012.003), p. 3 - 23 (1971).

080.012   Measurements of magnetic fields.   M. Semel.
IAU Symposium No. 43, (see 012.003), p. 37 - 43
(1971).

**080.013   The interpretation of magnetograph results: The
formation of absorption lines in a magnetic field.**
J. O. Stenflo.
IAU Symposium No. 43, (see 012.003), p. 101 - 129 (1971).

**080.014   Coherence properties of polarized radiation in weak
magnetic fields.**   L. L. House.
IAU Symposium No. 43, (see 012.003), p. 130 - 140 (1971).

**080.015   The effect of collisions on spectral line formation
in solar magnetic regions.**   F. K. Lamb.
IAU Symposium No. 43, (see 012.003), p. 149 - 161 (1971).

080.016   Line formation in inhomogeneous magnetic fields.
R. Göhring.
IAU Symposium No. 43, (see 012.003), p. 162 - 164 (1971).

**080.017   The magnetic fields at different levels in the active
regions of the solar atmosphere.**   T. T. Tsap.
IAU Symposium No. 43, (see 012.003), p. 223 - 230 (1971).

080.018 On the time fluctuations of magnetic fields.
A. Severny.
IAU Symposium No. 43, (see 012.003), p. 340 - 347 (1971).

080.019 Large scale solar magnetic fields and their consequences. G. Newkirk, Jr.
IAU Symposium No. 43, (see 012.003), p. 547 - 568 (1971).

080.020 The polar fields and time fluctuations of the general
magnetic field of the sun. A. B. Severny.
IAU Symposium No. 43, (see 012.003), p. 675 - 695 (1971).

080.021 Opposite polarities in the development of some
regularities in the distribution of large-scale magnetic fields.
P. Ambrož, V. Bumba, R. Howard, J. Sýkora.
IAU Symposium No. 43, (see 012.003), p. 696 - 709 (1971).

080.022 Observations of the polar magnetic fields.
J. O. Stenflo.
IAU Symposium No. 43, (see 012.003), p. 714 - 724 (1971).

080.023 Dynamics of large-scale magnetic fields.
G. V. Kuklin.
IAU Symposium No. 43, (see 012.003), p. 737 - 743 (1971).

080.024 The sun as a magnetic rotator. J. Tuominen.
IAU Symposium No. 43, (see 012.003), p. 754
(1971).

080.025 Theories of large scale fields and the magnetic activity cycle. N. O. Weiss.
IAU Symposium No. 43, (see 012.003), p. 757 - 769 (1971).

080.026 Dynamo theory of the sun's general magnetic field
on the basis of a mean-field magnetohydrodynamics.
F. Krause, K.-H. Rädler.
IAU Symposium No. 43, (see 012.003), p. 770 - 779 (1971).

080.027 A dynamo model for the large scale fields.
I. K. Csada.
IAU Symposium No. 43, (see 012.003), p. 780 - 782 (1971).

080.028 Solar sector magnetism. J. M. Wilcox.
Comments Astrophys. Space Phys., Vol. 3, 133 -
139 (1971).
Solar sector magnetism is different from the classical
Babcock model in two fundamentals. (1) the solar sector
boundary on the average remains in approximately the north-
south direction and is not greatly sheared by the effects of
differential rotation, i.e., the solar sector pattern rotates in an
approximately rigidly rotating coordinate system, and (2) the
solar sectors have the same polarity on both sides of the equator. We emphasize again that it appears that both the Babcock
model and the sector pattern may coexist in solar magnetism.

080.029 Sun from the stratosphere. V. A. Krat.
Zemlya i Vselennaya, No. 5, p. 20 - 24 (1971).
In Russian.

080.030 Asymptotic methods in the theory of non-linear
convection. R. van der Borght.
Proc. Astron. Soc. Australia, Vol. 2, 45 - 46 (1971).

080.031 An improved Vitense model of the solar convection
zone. R. van der Borght.
Proc. Astron. Soc. Australia, Vol. 2, 46 - 47 (1971).

080.032 Non-linear convection with variable viscosity and
thermal conductivity.
R. van der Borght, B. E. Waters.

Proc. Astron. Soc. Australia, Vol. 2, 47 - 48 (1971).

080.033 Depth dependence on the various scale lengths in a
quasi-Vitense model of the solar convection zone.
B. E. Waters.
Proc. Astron. Soc. Australia, Vol. 2, 48 - 50 (1971).

080.034 Depth dependence of physical parameters associated
with convection in the solar atmosphere.
B. E. Waters.
Proc. Astron. Soc. Australia, Vol. 2, 50 - 51 (1971).

080.035 Non-linear convection with free boundaries at high
Rayleigh number. J. O. Murphy.
Proc. Astron. Soc. Australia, Vol. 2, 51 - 52 (1971).

080.036 Non-linear convection with hexagonal planform.
J. O. Murphy.
Proc. Astron. Soc. Australia, Vol. 2, 53 - 54 (1971).

080.037 Hydromagnetic structure at the boundary of a
supergranule. E. Ribes, W. Unno.
Proc. Astron. Soc. Australia, Vol. 2, 54 - 55 (1971).

080.038 Eddy viscosity and the solar rotation. V. P. Starr.
Nature, Vol. 233, 186 - 187 (1971).
The possibility is proposed that the interior of the sun
may be in a state of high rotation compared with the exterior
through the action of a negative eddy viscous process. As has
been established through much observational evidence, such
eddy negative viscous processes are active in other fluid
systems, especially in the earth's atmosphere.

080.039 Recent measurements of the solar radiation incident
on the atmosphere. A. J. Drummond.
Space Research XI, (see 012.004), 681 - 693 (1971).

080.040 The solar constant from data of balloon investigations in the USSR and the USA.
K. Ya. Kondratyev, G. A. Nikolsky, D. G. Murcray, J. J. Kos-
ters, P. R. Gast.
Space Research XI, (see 012.004), 695 - 703 (1971).

080.041 Inhomogeneities in the solar atmosphere from the
CaII infra-red lines. P. Mein.
Solar Physics, Vol. 20, 3 - 18 (1971).
From a time sequence of high dispersion spectra taken
by Evans, the solar fine structures are studied in the CaII infrared triplet. The Doppler shifts and the intensity fluctuations
in different points of the profiles are converted into fluctuations of the model atmosphere. A weighting function method is worked out in that purpose. The theoretical line profiles are computed in non LTE from a program written by
Dumont.

080.042 Natural modes of the acoustic-gravity type in the
solar atmosphere. J. F. McKenzie.
Astron. Astrophys., Vol. 15, 450 - 459 (1971).
We derive the dispersion equation for the natural modes
of a simple model of the solar atmosphere, which consists of
a slab of gas bounded above by a hot gas and below by a cool
gas, all under the action of gravity. By identifying the observed oscillations with these natural modes other observed
features of the oscillations, which are in apparent conflict with
each other, can be resolved.

080.043 The sun, C^{12}/C^{13} abundance ratio and neutrino
emission. P. Raychaudhuri.
Astrophys. Space Sci., Vol. 13, 231 - 233 (1971).
It is pointed out that the effect of neutrino emission
process according to the photon-neutrino coupling theory can

be taken to interpret the observed ratio of C^{12}/C^{13} abundance and other solar activities.

080.044 **Considerazioni sulla cronologia solare.**
G. Righini.
Mem. Soc. Astron. Italiana, Nuova Ser., Vol. 42, 273 - 277 (1971).
The solar models developed up to now are unable to explain completely the constitution and nature of our sun. The best approximation of the solar age is therefore to assume that the age of the earth and the age of the sun are equal inside the range of uncertitude of $\pm 0.03 \times 10^9$ years.

080.045 **Primary solar reaction dependence on deuteron structure.** J. E. Brolley.
Solar Physics, Vol. 20, 249 - 256 (1971).
The energy dependence of the process $p + p \rightarrow D + e^+ + \nu$ has been investigated as a function of the $\%D$ state. The pp wavefunction was generated from a Schrödinger equation utilizing a velocity-dependent potential, containing the effect of vacuum polarization, which was fitted to the most recent pp phases. The present likely uncertainty in $\%D$ state implies $1\,^1/_2\,\%$ uncertainty in the reaction rate. The present calculation of $\Lambda^2(0)$ is 4 % less than a previous result. New values of $S(0)$, incorporating recent ft values, have been obtained.

080.046 **Solar magnetic fields—small scale.** J. Harvey.
Publ. Astron. Soc. Pacific, Vol. 83, 539 - 549 (1971).
Invited symposium paper presented at the Hawaii meeting of the Astronomical Society of the Pacific, 22–25 June 1971. — Observations related to the small-scale structure of solar magnetic fields are reviewed with emphasis on recent photospheric measurements.

080.047 **Solar magnetic fields—large scale.** R. Howard.
Publ. Astron. Soc. Pacific, Vol. 83, 550 - 560 (1971).
Invited symposium paper presented at the Hawaii meeting of the Astronomical Society of the Pacific, 22–25 June 1971.

080.048 **Solar magnetic fields—extended.** J. M. Wilcox.
Publ. Astron. Soc. Pacific, Vol. 83, 561 - 570 (1971).
Invited symposium paper presented at the Hawaii meeting of the Astronomical Society of the Pacific, 22–25 June 1971.

080.049 **More solar models and neutrino fluxes.**
Z. Abraham, I. Iben, Jr.
Astrophys. Journ., Vol. 170, 157 - 163 (1971).
We present neutrino fluxes from a sequence of solar models that differ from one another in regard to opacity, equation of state, and nuclear cross-section factors. Using current estimates of the relevant input parameters, we obtain capture rates that range between 3 and 10 times the most recent result of the Davis ^{37}Cl neutrino-capture experiment. The contribution to a theoretical capture rate due to neutrinos from all reactions other than ^{8}B decay ranges from 0.5 to 1.5 times the latest observational result.

080.050 **The solar constant.** V. Barocas.
Journ. British Astron. Ass., Vol. 82, 16 - 23 (1971).

080.051 **Solar neutrinos. III. Composition and magnetic-field effects and related inferences.**
J. N. Bahcall, R. K. Ulrich.
Astrophys. Journ., Vol. 170, 593 - 603 (1971).
An improved solar model is shown to yield a calculated capture rate for the ^{37}Cl neutrino experiment that is a factor of 6 larger than the observed rate reported by Davis, Rogers, and Radeha. The ^{8}B branch must occur a factor of 10 less often than expected on the basis of standard solar models. The effects on the calculated capture rates of a variety of composition uncertainties are considered. The effects of a strong pri-

mordial magnetic field are also investigated. The results obtained by several investigators who used different stellar-evolution programs are compared, and it is concluded that the different programs give consistent, but observationally incorrect, results.

080.052 **Large-scale motions in the sun.** J. H. Piddington.
Solar Physics, Vol. 21, 4 - 20 (1971).
Large-scale solar motions comprise differential rotation, axially symmetric meridional motions, and possible asymmetric motions. These motions must be basic in any satisfactory theory of the changing pattern of solar magnetic fields and of the 22-yr cycle. The present paper provides a critical review of the large, and sometimes conflicting observational data, and includes some discussion of the possible origins of the motions.

080.053 **On the solar oblateness: The combined effect of a pole-equator difference in effective temperature and mechanical heating.** B. R. Durney, N. E. Werner.
Solar Physics, Vol. 21, 21 - 26 (1971).
With the help of a model atmosphere of the sun we evaluate the pole-equator difference in flux assuming the following type of pole-equator temperature difference ($\Delta T = T_e - T_p$): (a) $\Delta T \approx 2$K for $\tau > \tau_0 (\tau_0 \approx 0.05)$; (b) $\Delta T \approx 10$K for $\tau < \tau_0$. We compare the results of our calculations with Dicke and Goldenberg's observations.

080.054 **Use of a birefringent element to separate magnetic polarity.** H. E. Ramsey.
Solar Physics, Vol. 21, 54 - 56 (1971). — Research note.

080.055 **On D–C electrical conductivity in a partially ionized solar magnetoplasma.** A. A. Wyller, H. S. Yun.
Solar Physics, Vol. 21, 116 - 120 (1971).
A numerical application to the extreme cases of high and low degree of ionization of the binary collision theory of Burgers (1960) and the multiple collision theory of Shkarofsky (1960) shows very good agreement in the values of the magnetic tensor components for solar electrical conductivity. It is pointed out that the inclusion of ion motions in Burgers theory favors its use in the future evaluation of the solar thermal conductivity tensor.

080.056 **Magnetically trapped particles in the lower solar atmosphere.** A. O. Benz, T. Gold.
Solar Physics, Vol. 21, 157 - 166 (1971).
The lifetime of electrons with kinetic energies up to about 10 MeV was found to be completely determined by the motion of the mirror points, provided the gyro-synchrotron loss can be neglected. Some fluid and streaming instabilities were also considered. The altitude of injection of energetic particles was estimated.

080.057 **Time fluctuations of the general magnetic field of the sun.** A. B. Severny.
Quarterly Journ. Roy. Astron. Soc., Vol. 12, No. 4, (see 012.012), 363 - 379 (1971).

080.058 **Point sources of neutrino radiation.**
G. E. Kocharov, Yu. N. Starbunov.
Proc. 11th international conference on cosmic rays, Vol. 4, (see 012.028), 353 - 359 (1970).

080.059 **A progress report on the Brookhaven solar neutrino experiment.** R. Davis, Jr.
Proc. 11th international conference on cosmic rays, Vol. 4, (see 012.028), 371 - 374 (1970).

080.060 **Structure of solar magnetoplasma.**
E. I. Mogilevsky.

Geod. Geophys. Veröff., Ser. 2, No. 13, (see 012.029), p. 95 - 108 (1969).

080.061 About the connection between the magnetic field and movements in active regions of the sun.
E. I. Mogilevsky, B. D. Shelting.
Geod. Geophys. Veröff., Ser. 2, No. 13, (see 012.029), p. 113 - 122 (1969).

080.062 Influence of inhomogeneities of magnetic fields on line contours and magnetographic measurements.
J. Staude.
Geod. Geophys. Veröff., Ser. 2, No. 13, (see 012.029), p. 123 (1969). — The full text of the present paper is published in Solar Physics, Vol. 8 (1969). — See 02.080.002.

080.063 The neutrino emission of the sun. G. E. Kocharov.
Proc. Sixth Winter School on Space Physics, Part I.

Apatity 1969, (see 012.033), p. 158 - 176 (1971).

080.064 Differential rotation and nonsphericity of the sun's surface. S. S. Mandrykin, Yu. N. Starbunov.
Proc. Sixth Winter School on Space Physics, Part I. Apatity 1969, (see 012.033), p. 177 - 185 (1971).

Analogue of the primary solar reaction.
See Abstr. 022.073.

Granulation patterns and solar oblateness.
See Abstr. 071.023.

Solar rotation: Direct evidence form prominences for a westward wind. See Abstr. 073.023.

A rotating solar magnetic "dipole" observed from 1926 to 1968. See Abstr. 106.027.

Earth

081 Figure, Composition, and Gravity of the Earth

081.001 **Basic magmatism in Guiana and continental drift.**
A. Choudhuri, M. W. Milner.
Nature, Phys. Sci., Vol. 232, 154 - 155 (1971).

081.002 **A mechanism of gravitational differentiation of the earth's matter.** V. P. Miasnikov, O. G. Sorokhtin, G. B. Udintsev, S. A. Ushakov.
Dokl. Akad. Nauk SSSR, Ser. Mat. Fiz., Vol. 199, 86 - 89 (1971). In Russian.

081.003 **Earth's gravity field to the sixteenth degree and station coordinates from satellite and terrestrial data.** E. M. Gaposchkin, K. Lambeck.
Journ. Geophys. Res., Vol. 76, 4855 - 4883 (1971).
Geodetic parameters describing the earth's gravity field and the positions of satellite-tracking stations in a geocentric reference frame have been computed. These parameters were estimated by means of a combination of four different types of data: routine and simultaneous satellite observations, observations of deep-space probes, and measurements of terrestrial gravity.

081.004 **New solution for the anomalous gravity potential.**
H. Baussus-von Luetzow.
Journ. Geophys. Res., Vol. 76, 4884 - 4891 (1971).

081.005 **Potential expansion for a nonhomogeneous oblate spheroid.** S. L. Levie, Jr.
Journ. Geophys. Res., Vol. 76, 4897 - 4900 (1971).
An expression for the external potential of a nonhomogeneous oblate spheroidal mass distribution is provided in this paper. The solution begins with the known potential expansion for a homogeneous oblate spheroid and develops the expansion valid for an arbitrary number of oblate spheroidal layers of different densities surrounding an oblate spheroidal core.

081.006 **A procedure for combining satellite and gravimetric data for position and gravity field determinations.**
R. H. Rapp.
Journ. Geophys. Res., Vol. 76, 4986 - 4990 (1971).
A procedure is described in which geocentric station positions derived from satellite data can be combined with similar positions derived by using astronomic information, deflection of the vertical, and height anomaly information.

081.007 **Composition of the earth: Chondritic or achondritic?**
J. W. Larimer.
Geochim. Cosmochim. Acta, Vol. 35, 769 - 786 (1971).
The composition of chondrites and achondrites has long provided a useful framework in which to consider the bulk composition of the earth. In this report, a simple test is developed in an attempt to see which, if either, comes closest to matching the earth in trace element content. The elements considered include: the alkalies (Na, K and Rb), eight refractory elements (Ba, Ca, La, Sc, Sr, Th, U and Y) and eleven volatile elements (Ar, Bi, C, Cs, Hg, In, Kr, N, Pb, Tl and Xe) plus H_2O.

081.008 **Erdgezeitenregistrierungen in der Arktis.**

M. Bonatz.
Umschau, 71. Jahrgang, p. 675 (1971).

081.009 **The role of the gravitational field in geodesy.**
V. V. Brovar.
Izv. vyssh. uchebn. zavedenij. Geod. i aehrofotos"emka, 1970, No. 2, p. 66 - 72. In Russian. – Abstr. in Referativ. Zhurn. 52. Geod. Aehros"emka, 8.52.71 (1971).

081.010 **On the connection between spherical and ellipsoidal functions with regard to the earth's gravitational field.** L. A. Savrov, M. U. Sagitov.
Vestn. Mosk. un-ta. Fiz., astron., Vol. 12, 223 - 226 (1971). In Russian. – Abstr. in Referativ. Zhurn. 52. Geod. Aehros"-emka, 8.52.72 (1971).

081.011 **Pleistocene glaciation and the viscosity of the lower mantle.** R. J. O'Connell.
Geophys. Journ. Roy. Astron. Soc., Vol. 23, 299 - 327 (1971).

081.012 **The elasticity theory of dislocations in real earth models and changes in the rotation of the earth.**
D. E. Smylie, L. Mansinha.
Geophys. Journ. Roy. Astron. Soc., Vol. 23, 329 - 354 (1971).
We extend the general elasticity theory of dislocations to self-gravitating, radially inhomogeneous earth models with liquid cores, which we call real earth models and compute the inertia tensor changes for a number of earthquakes whose fault parameters have been obtained by other workers. Implications for the rotation of the earth are discussed.

081.013 **Comments on paper by D. E. Smylie and L. Mansinha: 'The elasticity theory of dislocations in real earth models and changes in the rotation of the earth'.**
F. A. Dahlen.
Geophys. Journ. Roy. Astron. Soc., Vol. 23, 355 - 358 (1971).

081.014 **Reply to comments on 'The elasticity theory of dislocations in real earth models and changes in the rotation of the earth'.** D. E. Smylie, L. Mansinha.
Geophys. Journ. Roy. Astron. Soc., Vol. 23, 359 - 360 (1971).

081.015 **The dynamical properties and internal constitutions of the earth, the moon and the planets.**
A. H. Cook.
Quarterly Journ., Roy. Astron. Soc., Vol. 12, 154 - 168 (1971).
The aim of this lecture is to show how information about the internal constitutions of the planets may be inferred from their dynamical properties coupled with knowledge of the behaviour of solids under very high pressures.

081.016 **Die Bedeutung der remanenten Magnetisierung von Gesteinen für die Theorie der Kontinentaldrift.**
H. Soffel.
Phys. Blätter, 27. Jahrgang, p. 417 - 423 (1971).

081.017 **A reply to Lyttleton: 'The effect of core formation on the radius of a planet' [Observatory, Vol. 90, 80 - 82 (1970)].** S. K. Runcorn.
Observatory, Vol. 91, 164 - 165 (1971).

081.018 **On Lyttleton's extensions of Ramsey's theory of the nature of the core of the earth.**
F. C. Frank.
Geophys. Journ. Roy. Astron. Soc., Vol. 23, 461 - 464 (1971).

081.019 **Isotopic composition of xenon from a Greenland anorthosite.** P. M. Jeffery.
Nature, Vol. 233, 260 - 261 (1971).

Following a study of the data resulting from the lunar α-scattering experiment of Turkevich et al., Olsen pointed out that the rocks in the Mare Tranquillitatus region were similar in oxide composition to certain terrestrial anorthosites. Samples of these perhaps very special rocks are used to study the isotopic compositions of the noble gases "trapped" within them. For six samples of Group III anorthosites from West Greenland the presence of xenon has been examined.

081.020 **Barycentric equilibrium tide theory.** I. Michelson.
Bull. American Astron. Soc., Vol. 3, 385 - 386 (1971). – Abstr. AAS.

081.021 **The free nonlinear boundary value problem of physical geodesy.** E. Grafarend, W. Niemeier.
Bull. Géod., Nouvelle Sér., Année 1971, No. 101, p. 243 - 262 (1971).

Within potential theory of Poisson-Laplace equation the boundary value problem of physical geodesy is classified as free and nonlinear. For solving this typical nonlinear boundary value problem four different types of nonlinear integral equations corresponding to singular density distributions within single and double layer are presented.

081.022 **Methods for the computation of geoid undulations from potential coefficients.** R. H. Rapp.
Bull. Géod., Nouvelle Sér., Année 1971, No. 101, p. 283 - 297 (1971).

081.023 **Density layer models for the geopotential.**
F. Morrison.
Bull. Géod., Nouvelle Sér., Année 1971, No. 101, p. 319 - 328 (1971).

Spherical harmonic representation of the geopotential suffers from convergence problems in both mathematical and numerical senses; it also is not suitable for use with the terrestrial gravity data, which is distributed very unevenly over the surface of the earth. Of the alternative mathematical models, the simple density layer offers the most mathematical and computational simplicity and avoids the problems associated with spherical harmonics. In practice low order spherical harmonic models have been combined with a density layer for a complete model of the geopotential.

081.024 **Viscosity of the earth's core.** R. Hide.
Nature, Phys. Sci., Vol. 233, 100 - 101 (1971).

Various lines of geophysical evidence, including the recently discovered correlation between global features of the earth's magnetic and gravitational fields, imply that shallow bumps are present on the core-mantle interface and that core motions interact strongly with the bumps. Tentative theoretical considerations of the magnetohydrodynamics of the interaction process yield a critical value of the effective kinematical viscosity (i. e. eddy plus molecular) above which strong interactions would not be expected to occur, showing that $\nu < 10^6$ cm² s⁻¹ for the core.

081.025 **The earth and the moon.** F. Press.
Quarterly Journ. Roy. Astron. Soc., Vol. 12, 232 - 243 (1971). – Harold Jeffreys Lecture, delivered at Burlington House on 12 March 1971.

081.026 **Application of space techniques to solid-earth and ocean physics.** C. A. Lundquist.
Space Research XI, (see 012.004), 439 - 456 (1971).

081.027 **Temporal variations of the geopotential derived from satellite observations.** Y. Kozai.
Space Research XI, (see 012.004), 469 - 477 (1971).

081.028 **New geodetic parameters for a standard earth.**
K. Lambeck, E. M. Gaposchkin.
Space Research XI, (see 012.004), 479 - 492 (1971).

081.029 **The determination of the fine structure of the earth's gravitational field.** L. P. Pellinen.
Space Research XI, (see 012.004), 493 - 498 (1971).

081.030 **Determination of the equations of condition for the zonal harmonics using the DIAL satellite.**
A. Cazenave, F. Forestier.
Space Research XI, (see 012.004), 521 - 524 (1971).

081.031 **Early eras in the earth's history.** M. V. Muratov.
Priroda, No. 11.71, p. 16 - 23 (1971). In Russian.

081.032 **The 1964 IAU system and the geodetic reference system 1967.** J. Kovalevsky.
Celestial Mechanics, Vol. 4, 279 - 280 (1971). – Presented at IAU Colloquium No. 9 (see 012.006).

081.033 **Interpretation of global gravity anomalies.**
J. W. Higbie, F. D. Stacey.
Nature, Phys. Sci., Vol. 234, 130 - 132 (1971).

081.034 **A second-order Markov gravity anomaly model.**
J. F. Kasper, Jr.
Journ. Geophys. Res., Vol. 76, 7844 - 7849 (1971).

A second-order Markov process is proposed as a statistical model for gravity anomalies in a local region. The method is applied to the analysis of errors in an inertial navigation system caused by unknown gravity phenomena.

081.035 **Air-borne gravimetry errors associated with geoidal undulations.** R. E. Wall.
Journ. Geophys. Res., Vol. 76, 7293 - 7295 (1971). – Brief report.

081.036 **Representation of the earth's gravitational potential.** J. P. Vinti.
Celestial Mechanics, Vol. 4, 348 - 367 (1971).

The paper represents the earth's gravitational potential V, outside a sphere bounding the earth, by means of its difference δV from the author's spheroidal potential. The difference δV is in turn represented as arising from a surface density σ on the sphere bounding the earth. Values of σ are computed from potential coefficients obtained from two sources, Rapp and the Smithsonian Astrophysical Observatory. In two appendices values and contour maps of σ are given.

081.037 **Berechnungsverfahren für die Bestimmung des Erdschwerepotentials aus Doppler-Beobachtungen mit Hilfe des Modells einer einfachen Schicht.** B. Witte.
Deutsche Geod. Kommission, Bayer. Akad. Wiss., München, Ser. C, No. 167, 2 + 45 pp. (1971).

081.038 **Series solutions of Molodensky's problem.**
H. Moritz.
Deutsche Geod. Kommission, Bayer. Akad. Wiss., München, Ser. A, No. 70, 4 + 92 pp. (1971).

081.039 **Time of formation of the earth's core.**
V. M. Oversby, A. E. Ringwood.
Nature, Vol. 234, 463 - 465 (1971).

The time interval between accretion of the earth and formation of the core is calculated as a function of the amount of lead in the core. The accretion age of the earth was taken as 4.55×10^9 yr. and model lead isotopic composition as that of East Pacific Rise tholeiite dredge samples. Experimental measurement of the distribution coefficient for Pb between silicate and various metal phases fixes the range of lead contents likely for the earth's core. A most probable upper limit of 10^8 yr. between accretion and core formation is found.

081.040 **Solidity of the inner core of the earth inferred from normal mode observations.** A. M. Dziewonski, F. Gilbert.
Nature, Vol. 234, 465 - 466 (1971).
Potential energy calculations show that eigenfrequencies of some of the recently observed normal modes are highly sensitive to the elastic parameters in the inner core of the earth. One group of these modes is particularly sensitive to the bulk modulus, the other to the shear modulus. Only models of the earth with a rigid inner core can satisfy both groups of the data.

081.041 **The low abundance of sulfur in the earth's core.** R. Brett.
Meteoritics, Vol. 6, 251 - 252 (1971). – Abstract.

081.042 **On a possibility of determining the earth's gravitational potential as referred to the geoid from gravimetric measurements.** I. F. Monin.
Dopovidi AN URSR, 1971, B, No. 7, p. 625 - 628, 666. In Ukrainian. – Abstr. in Referativ. Zhurn. 52. Geod. Aehros"-emka, 12.52.64 (1971).

081.043 **On the gravitational potential as referred to the ellipsoid in the theory of the figure of the earth.** I. F. Monin.
Dopovidi AN URSR, 1971, B, No. 8, p. 696 - 699. In Ukrainian. – Abstr. in Referativ. Zhurn. 52. Geod. Aehros"emka, 12.52.65 (1971).

081.044 **Method of determining the gravitational field of the earth from perturbations of resonant satellites.** K. P. Ivanovskaya.
Astron. Zhurn. Akad. Nauk SSSR, Vol. 48, 1314 - 1321 (1971). In Russian. English translation in Soviet Astron. AJ, Vol. 15, No. 6.
A numerical example shows the possibility of determining the coefficients of the earth's potential from longperiodic perturbations of satellites with commensurable mean motions. The errors of tesseral coefficients are estimated. It has been found that the sectorial coefficients and the tesseral ones with large second indices have the smallest errors. The needed accuracy on non-resonance perturbations has been also considered.

081.045 **Determinación de armónicos zonales pares del potencial terrestre.** C. Simó.
Urania Barcelona, Año 56, No. 273, p. 39 - 49 (1971).

081.046 **Planetología comparada.** A. J. Fillias.
Urania Barcelona, Año 56, p. 65 - 76 (1971).

081.047 **Origin of the earth.** B. Yu. Levin.
Zemlya i Vselennaya, No. 6, p. 8 - 13 (1971). In Russian.

081.048 **Theory and computation of amplitudes of terrestrial line spectra.**
A. Ben-Menahem, M. Israel, U. Levité.
Geophys. Journ. Roy. Astron. Soc., Vol. 25, 307 - 406 (1971).
Universal tables for spectral line amplitudes of surface displacements and strains of a gravitating radially-heterogeneous earth, have been prepared. Given a structural model and a dislocation source of arbitrary orientation and depth, spheroidal and toroidal line-amplitudes are calculable for $l \leqslant 100$, $n \leqslant 4$ everywhere on the earth's surface.

081.049 **Earth's gravity field represented by a simple-layer potential from Doppler tracking of satellites.**
K.-R. Koch, B. U. Witte.
Journ. Geophys. Res., Vol. 76, 8471 - 8479 (1971).

081.050 **Erdgezeitenregistrierungen in der Arktis – International Astro-Geo-Project Spitzbergen 1968/70.**
M. Bonatz, P. Melchior.
Zeitschr. Vermessungswesen, 96. Jahrgang, p. 305 - 309 = Obs. Roy. Belgique, Commun. Sér. B, No. 58 (1971).

081.051 **Station: Longyearbyen (Spitsbergen). (Astro-geo Project Spitsbergen 1968–1970). Mesures faites dans les trois composantes avec six pendules horizontaux VM et trois gravimètres Askania.**
M. Bonatz, P. Melchior, B. Ducarme.
Obs. Roy. Belgique, *Bruxelles,* Bull. d'Observations des Marées Terrestres, (Section Géodynamique), Vol. 4, fasc. 1, 23 + 110 pp. (1971).

081.052 **Marées terrestres.**
P. Melchior (Editor).
Bull. d'Informations, (Obs. Roy. Belgique, Bruxelles), Nos. 61 - 62, p. 3000 - 3237 (1971).

081.053 **Westward displacement of the lithosphere.**
R. C. Bostrom.
Nature, Vol. 234, 536 - 538 (1971).
The high value of the seismic propagation quality factor Q in the earth's lower mantle requires that the retarding torque is engendered almost wholly in the outer part including the seas. Evidence is adduced that the zone of low seismic velocities (70 to 400 kms from surface) is the site of decoupling, permitting the displacement west of the overlying lithosphere. A preliminary estimate of the rate is 5 cms yr^{-1} at the equator. Due to the variance in rate with latitude and its regional variance in mechanical characteristics, the lithosphere should divide into segments; these should migrate west at greatly different velocities; and they should rotate about moments of impedance to their displacement. The available energy is not less than 10^{26} erg yr^{-1}.

081.054 **Observaciones gravimetricas en la Provincia de Buenos Aires, (Año 1967).**
B. Gudoias, G. Font de Affolter, A. Mateo.
Obs. Astron. Univ. Nacional La Plata, Ser. Geod., Vol. 8, 2 + 41 pp. (1970).

The excess secular change in the obliquity of the ecliptic and its relation to the internal motion of the earth. See Abstr. 043.005.

Spectrophotometry of the environment from space. See Abstr. 082.060.

Internal friction in moon and earth rocks. See Abstr. 094.235.

Mariner photography of Mars and aerial photography of earth: Some analogies. See Abstr. 097.088.

The origin of the system earth – moon. See Abstr. 107.009.

082 The Earth's Atmosphere including Refraction, Scintillation, Extinction, Airglow, Site Testing

082.001 Sunlight photodetachment of O_3^-.
R. Byerly, Jr., E. C. Beaty.
Journ. Geophys. Res., Vol. 76, 4596 - 4601 (1971).

This paper reports a determination of an approximate photon-energy dependence of the cross section for photodetachment of O_3^- and a determination of the sunlight photodetachment rate of O_3^-.

082.002 Semiannual variation in the heterosphere: A reappraisal. L. G. Jacchia.
Journ. Geophys. Res., Vol. 76, 4602 - 4607 (1971).

082.003 Observations of the O I 1304-Å airglow from OGO 4.
R. R. Meier, D. K. Prinz.
Journ. Geophys. Res., Vol. 76, 4608 - 4620 (1971).

082.004 Molecular oxygen density measurements from 80 to 140 kilometers. P. J. Brannon, J. M. Hoffman.
Journ. Geophys. Res., Vol. 76, 4630 - 4633 (1971).

082.005 The seasonal variation of the atmospheric temperature near the mesopause level: Preliminary results.
F. Verniani, G. Cevolani.
Journ. Geophys. Res., Vol. 76, 5101 - 5103 (1971).

By using the mass and luminosity equations of meteor theory, a method of employing photographic meteor data yields values of the atmospheric density scale height. This method was applied to 413 precisely reduced meteors photographed by Super-Schmidt cameras.

082.006 Comments on the emission feature at 11.7 cm^{-1}.
H. A. Gebbie, R. A. Bohlander, D. G. Murcray.
Nature, Phys. Sci., Vol. 232, 156 (1971).

Some comments on the article of Beery et al. (Nature, Phys. Sci., Vol. 230, 36, 1971) attributing an emission feature observed at 11.7 cm^{-1} in the atmosphere to nitrous oxide are given.

082.007 Atmospheric total ozone increase during the 1960s.
W. D. Komhyr, E. W. Barrett, G. Slocum, H. K. Weickmann.
Nature, Vol. 232, 390 - 391 (1971).

082.008 On the dynamic unity of the earth's atmosphere.
E. S. Kazimirovskii, V. F. Loginov, G. I. Sukhomazova.
Dokl. Akad. Nauk SSSR, Ser. Mat. Fiz., Vol. 199, 76 - 77 (1971). In Russian.

082.009 On theoretical models of the structure and dynamics of the earth's thermosphere. M. N. Izakov.
Space Sci. Rev., Vol. 12, 261 - 298 (1971). – Review article.

082.010 The use of directional observations to give accurate orbital decay rates for studies of upper-air density.
B. R. May.
Planet. Space Sci., Vol. 19, 685 - 692 (1971).

It is shown how directional observations can be combined with supplementary orbital data to give accurate rates of decay of the orbital period of an artificial earth satellite.

082.011 The 2.53 mm molecular rotation line of atmospheric O_2. D. L. Croom.
Planet. Space Sci., Vol. 19, 777 - 789 (1971).

082.012 More comments on the validity of Jeans' escape rate.
R. T. Brinkmann.
Planet. Space Sci., Vol. 19, 791 - 794 (1971). – Research note.

082.013 Hydroxyl emission of the upper atmosphere–IV. Correlation with the molecular oxygen emission.
N. N. Shefov.
Planet. Space Sci., Vol. 19, 795 - 796 (1971). – Research note.

082.014 Number densities of atomic oxygen and molecular nitrogen in the thermosphere. R. H. G. Reid.
Planet. Space Sci., Vol. 19, 801 - 812 (1971).

The absorption by the earth's atmosphere of some solar ultraviolet lines was observed by instruments on the satellite OSO-IV. We calculate the absorption curves using a static diffusion model of the atmosphere (Jacchia, 1965) and determine the number densities of atomic oxygen and molecular nitrogen at 120 km by fitting to the observations.

082.015 Atomic oxygen concentration from the measurement of the [OI] 5577 Å emission of the airglow.
B. S. Dandekar, J. P. Turtle.
Planet. Space Sci., Vol. 19, 949 - 957 (1971).

The [OI] green line emission of the nightglow originates in two layers, a sharp one around 100 km and diffuse one above 150 km. The peak concentrations of atomic oxygen for the Chapman and the Barth mechanism are 7×10^{11} and 9×10^{10} cm^{-3} respectively.

082.016 Requirements for rocket auroral spectroscopy in the vacuum ultraviolet. J. R. Catchpoole.
Journ. Atmosph. Terr. Phys., Vol. 33, 1329 - 1336 (1971).

Measurements of the auroral UV spectrum are important for an improved understanding of the physical processes of the upper atmosphere. This paper considers certain aspects of this requirement dealing in particular with preliminary calculations relating atmospheric and experimental parameters.

082.017 Measurements in Jamaica of the optical transmission coefficient of the atmosphere for the wavelength 694.3 nm. M. T. Ottway, R. W. H. Wright, G. S. Kent.
Journ. Atmosph. Terr. Phys., Vol. 33, 1337 - 1343 (1971).

The importance of a knowledge of the behaviour of the atmospheric optical transmission coefficient T is discussed with particular reference to the use of laser radar methods for studying the atmosphere. The results obtained show that on clear cloudless nights Jamaica is a good site for laser radar studies.

082.018 San Marco II measurements of equatorial atmospheric density at heights between 200 and 300 km.
L. Broglio.
Journ. Atmosph. Terr. Phys., Vol. 33, 1473 - 1480 (1971).

The atmospheric density was continuously measured at heights between 200 and 400 km by using the San Marco II 'balance'. Density values of perturbed and quiet days are presented. The orbital decay deduced from the San Marco II density data agrees well with that obtained refined ephemeris.

082.019 The definition of the atmospheric transmissivity from data of sky brightness distribution.
V. I. Kushpil, L. F. Petrova.
Astron. Zhurn. Akad. Nauk SSSR, Vol. 48, 850 - 853 (1971). In Russian. English translation in Soviet Astron. AJ, Vol. 15,

No. 4.

A method of defining the atmospheric transmissivity from the data of sky brightness distribution is presented. The errors of the method are considered. The definition of optical thickness of the atmosphere resulting from the distribution of sky brightness taken from theoretical and experimental works is given.

082.020 Preliminary report on the astronomical site at San Pedro Martir, Baja California, Mexico.
E. E. Mendoza V.
Bol. Obs. Tonantzintla y Tacubaya, No. 36, Vol. 6, 95 - 96 (1971).

The present note is intended to give only a brief account of the main characteristics of astronomical interest at the San Pedro Martir Observatory.

082.021 Analisi preliminare sulla qualità dell'immagine solare alla cupola « Amici » dell'Osservatorio Astrofisico di Arcetri. F. Mazzucconi, A. Righini.
Mem. Soc. Astron. Italiana, Nuova Ser., Vol. 42, 229 - 238 (1971). − Letter.

082.022 Le condizioni del cielo notturno a Serra la Nave dal 1965 al 1970. C. Blanco.
Mem. Soc. Astron. Italiana, Nuova Ser., Vol. 42, 239 - 243 (1971). − Letter.

082.023 Polarisatie van haloverschijnselen.
G. P. Können.
Hemel en Dampkring, Vol. 69, 187 - 189 (1971).

082.024 Site-testing for a proposed 12 MHz radio telescope in Alberta. F. S. Chute, C. G. Englefield,
P. J. R. Harding, C. R. James, D. Routledge.
Journ. Roy. Astron. Soc. Canada, Vol. 65, 179 - 180 (1971). Abstr. RAS Canada.

082.025 The disturbed variations of the hydrogen emission.
N. M. Martsvaladze, L. M. Fishkova, N. N. Shefov.
Astron. Tsirk., No. 619, p. 5 - 6 (1971). In Russian.

082.026 Atmospheric emission of atomic hydrogen Hα from observations at Zvenigorod, Abastumani and Alma-Ata. V. I. Krassovsky.
Journ. Atmosph. Terr. Phys., Vol. 33, 1499 - 1514 (1971).

This paper deals with the study of the behaviour of the atmospheric Hα emission on the basis of the data accumulated during IGY and IQSY. The comparison has been made with the known data about the orthohelium and atomic hydrogen Lα emissions. The results obtained make it possible to assume that besides an atomic hydrogen geocorona there exist also atomic hydrogen clouds in the thermosphere.

082.027 Covariation in 6300 Å and 5577 Å emissions in nightglow. S. R. Pal.
Planet. Space Sci., Vol. 19, 1087 - 1093 (1971).

The general nocturnal variation of 6300 Å and 5577 Å emissions of the atomic oxygen is discussed on the basis of the nightglow observation during 1964−1967 at Mt. Abu, India. On some nights both the emissions showed intensity covariations and the short lived enhancements.

082.028 Upper atmospheric densities between 155 and 165 km by observations of AlO clouds.
M. Ackerman, P. Simon.
Planet. Space Sci., Vol. 19, 1193 - 1195 (1971). − Research note.

082.029 Estudio de las condiciones del Observatório del Teide para la observación astronomica. IV. Calidad

de las imágenes telescópicas. F. Sánchez.
Urania Barcelona, Año 55, Nos. 271−272, p. 97 - 110 (1970).

082.030 Seeing and scintillation. A. T. Young.
Sky Telescope, Vol. 42, 139 - 141, 150 (1971).

082.031 On the problem of noctilucent cloud observations by methods of space astrophysics. K. Eerme.
Publ. Tartu Astrofiz. Obs., Vol. 39, 141 - 156 (1971).
In Russian.

The paper deals with problems connected with noctilucent cloud observations from space vehicles. One of the first tasks for planning such observations is to find values for observation time for various orbital inclinations as a function of season and geographical latitude. The problem is solved in the first approximation for circular orbits for the most favourable cases of optical axis orientation.

082.032 Sounding rocket observation of helium 304- and 584-Å glow. T. Ogawa, T. Tohmatsu.
Journ. Geophys. Res., Vol. 76, 6136 - 6145 (1971).

The altitude distribution of the extreme ultraviolet emissions of neutral helium (He 584 Å) and singly ionized helium (He⁺ 304 Å) was measured by aluminum and bismuth thin-film photon counters aboard an exospheric sounding rocket flown from Uchinoura, Japan, on September 19, 1970, 2030 JST. Evidence for the presence of extraterrestrial helium glow is discussed.

082.033 OGO-4 observations of the Lyman-Birge-Hopfield emission in the day airglow.
D. K. Prinz, R. R. Meier.
Journ. Geophys. Res., Vol. 76, 6146 - 6158 (1971).

The results of measurements of the day airglow in the wavelength interval 1356−1600 Å from detectors aboard the polar-orbiting satellite OGO 4 are reported for the period August 1967 through July 1968.

082.034 The rotation rate of the upper atmosphere: A non-linear effect. Y. T. Chiu.
Planet. Space Sci., Vol. 19, 1201 - 1208 (1971).

In order to study the apparent 'super-rotation' of the earth's upper atmosphere, we examine a non-linear solution to the continuity, momentum conservation and state equations of gasdynamics for the steady state motion of a gravitationally stratified, cylindrical, inviscid atmosphere under the influence of a steady state, local time dependent temperature distribution.

082.035 Wachsende Sorge um sauberen Himmel.
G. Klare.
SuW, Vol. 10, 274 (1971).

082.036 Luftverhältnisse und allgemeine Wetterbedingungen für astronomische Beobachtungen im Ruhrgebiet.
G. Kreuz.
SuW, Vol. 10, 281 - 282 (1971).

082.037 Estimate of height and concentration of the atomic oxygen layer in the upper atmosphere of the earth.
G. S. Ivanov-Kholodny, V. V. Katyushina.
Geomagn. Aeronom., Vol. 11, 819 - 824 (1971).
In Russian.

082.038 The moment of inertia of the earth's atmosphere.
N. S. Sidorenkov, D. I. Stekhnovsky.
Astron. Zhurn. Akad. Nauk SSSR, Vol. 48, 1096 - 1098 (1971). In Russian. English translation in Soviet Astron. AJ, Vol. 15, No. 5.

Using recent data on fields of the atmospheric pressure and temperature at sea level, as well as data on the height of land, the moment of inertia of the atmosphere relative to the rotation axis of the earth is calculated by numerical integration.

082.039 Global temperature and density structure in the lower thermosphere. D. Rees.
Journ. British Interplanet. Soc., Vol. 24, 643 - 658 (1971).

Neutral temperature and molecular diffusion coefficient measurements at mid-latitude and in the auroral zone are used to derive global models of temperature, air density and pressure in the lower thermosphere.

082.040 Reaction rate of vibrationally excited hydroxyl with ozone.
R. N. Coltharp, S. D. Worley, A. E. Potter.
Applied Optics, Vol. 10, 1786 - 1789 (1971).

The reaction of vibrationally excited hydroxyl with ozone was studied in a fast-flow system by observing the effect of ozone concentration on the intensity of hydroxyl emission bands in the 1.4-μ to 2.2-μ wavelength region. Fourier transform spectroscopy was used to obtain spectra of the faint emissions from the excited hydroxyl. Rate constants for the ozone-hydroxyl reaction were calculated for vibrational levels $v = 2$ to $v = 9$.

082.041 Vibrationally excited nitric oxide in the upper atmosphere. T. C. Degges.
Applied Optics, Vol. 10, 1856 - 1860 (1971).

A brief summary is presented of the relative importance of the different physical mechanisms that determine the vibrational temperature of nitric oxide in the earth's atmosphere between 60 km and 160 km. Collisional, radiative, and chemiluminescent processes are considered and estimates of vibrational temperatures are made for extreme normal conditions and for IBC type I aurorae.

082.042 Internal energy balance and energy transfer in the lower thermosphere.
E. Bauer, R. Kummler, M. H. Bortner.
Applied Optics, Vol. 10, 1861 - 1869 (1971).

The present work is designed to provide an over-all model of the physical state of the earth's lower thermosphere. To do this, excitation of the different modes of motion – the kinetic, rotational, vibrational, and chemical energy of the heavy particles; electronic excitation and ionization; electron kinetic energy; and over-all macroscopic fluid mechanical motions – is considered.

082.043 Decrease in upper-atmosphere rotation rate at heights above 350 km. D. G. King-Hele.
Nature, Vol. 233, 325 - 326 (1971).

The average rate of rotation of the upper atmosphere can be evaluated from the small changes in the inclinations of satellite orbits to the equator, and results from about 30 orbits have shown that the rotation rate increases from about 1.1 rev/day at a height of about 200 km to about 1.4 rev/day at a height of 350 km. Recently several new values have been obtained, and these show that the rotation rate decreases above 350 km, to about 1.0 rev/day at 420 km and 0.7 rev/day at 500 km. The maximum rate is at a height between 300 and 350 km, near the height of maximum electron concentration in the F-layer of the ionosphere.

082.044 Night-sky background measurement in the 100 μm spectral region. A. G. Blair.
Nature, Phys. Sci., Vol. 234, 26 - 28 (1971).

A recent rocket-borne radiometer experiment has measured fluxes in two spectral ranges, 65 to 170 μm and 0.8 to 6 mm. The 65 to 170 μm results are presented in this communication and the longer wavelength results and some of the operating characteristics of the instrument have been de-

scribed previously (Phys. Rev. Letters, Vol. 27, 1154, 1971).

082.045 Meteoritic contamination of the upper atmosphere by the Quadrantid shower. F. Link, R. Robley.
Planet. Space Sci., Vol. 19, 1585 - 1587 (1971).

Continuing the studies of the optical effects of the cosmic dust in the upper atmosphere (Link and Weill, 1970) we have tried to detect the influence of Quadrantids upon the brightness of the twilight and the night sky.

082.046 Primordial oil slick. A. C. Lasaga, H. D. Holland, M. J. Dwyer.
Science, Vol. 174, 53 - 55 (1971).

Calculations and some preliminary experiments suggest that an early methane atmosphere would have been polymerized by solar ultraviolet radiation in geologically short periods of time. An oil slick 1 to 10 meters thick could have been produced in this way and might well have been of considerable importance in the development of life.

082.047 Photoelectron excitation of atomic-oxygen resonance radiation in the terrestrial airglow.
E. C. Zipf, E. J. Stone.
Journ. Geophys. Res., Vol. 76, 6865 - 6874 (1971).

Recent laboratory measurements of the absolute cross sections for the excitation of the O I (^3S) resonance state by electron impact on O and O_2 when combined with in situ measurements of the photoelectron energy distribution from 120 to 300 km show that photoelectron impact is the principal excitation mechanism for $\lambda 1304$-A resonance radiation in the dayglow; dissociative excitation of O_2 is found to play a minor role.

082.048 O_2 Herzberg I bands in the night airglow: Covariation with O I (5577). K. A. Dick, G. G. Sivjee.
Journ. Geophys. Res., Vol. 76, 6987 - 6989 (1971).

On the basis of measurements obtained during the NASA 1968 air-borne auroral expedition, the Herzberg I bands of O_2 and the 5577-A line of O I were found to covary in intensity under nonauroral conditions.

082.049 Search for MgO in the atmosphere.
T. Wentink, Jr., N. B. Brown.
Journ. Geophys. Res., Vol. 76, 7006 - 7007 (1971). – Letter.

082.050 On the nature of the circumterrestrial dust cloud.
V. G. Fessenkov.
Space Research XI, (see 012.004), 347 - 350 (1971).

082.051 The results of investigation of dust in the upper atmosphere by the twilight method. N. B. Divari.
Space Research XI, (see 012.004), 351 - 355 (1971).

082.052 Recent observational results on the neutral upper atmosphere. M. Roemer.
Space Research XI, (see 012.004), 761 - 778 = Mitt. Astron. Inst. Bonn, No. 124 (1971).

082.053 Gas flow-satellite surface interactions and deduced atmospheric density variations.
M. Ya. Marov, A. A. Pyarnpuu, G. I. Zmievskaya.
Space Research XI, (see 012.004), 935 - 939 (1971).

082.054 Determination of diurnal and semi-annual variations in the density of the upper atmosphere from observations of Cosmos satellite drag. M. I. Voiskovsky,
B. V. Kugaenko, V. M. Synitsyn, P. E. Elyasberg.
Space Research XI, (see 012.004), 953 - 957 (1971).

082.055 Variations of density in the upper atmosphere correlated with the instantaneous solar flux during

1967 - 1969. J. P. Lespés, J. L. Falin, M. Ill.
Space Research XI, (see 012.004), 959 - 963 (1971).

082.056 **Geomagnetic activity effect on atmospheric density in the 250 to 800 km altitude region.** M. Roemer.
Space Research XI, (see 012.004), 965 - 974 = Mitt. Astron. Inst. Bonn, No. 125 (1971).

082.057 **An "equivalent duration" to characterize atmospheric disturbances connected with geomagnetic storms.** I. Almár, E. Illés-Almár.
Space Research XI, (see 012.004), 975 - 979 (1971).

082.058 **Semi-annual variation of atmospheric densities near solar activity maximum.** C. Wulf-Mathies.
Space Research XI, (see 012.004), 981 - 986 = Mitt. Astron. Inst. Bonn, No. 126 (1971).

082.059 **Use of astronomical telescopes to measure aerosol pollution.** W. M. Porch, E. J. Mannery, R. J. Charlson, P. W. Hodge.
Nature, Vol. 233, 326 - 327 (1971).
This paper compares the results of atmospheric extinction measurements made at three astronomical observatories (Kitt Peak, Arizona, Rattlesnake Ridge, Washington, and Cerro Tololo, Chile) with atmospheric aerosol light scattering measurements made in other remote clean air environments. The closeness in the values of scattering coefficient determined by these various methods establishes the astronomical telescope as a potentially valuable tool in monitoring worldwide aerosol pollution.

082.060 **Spectrophotometry of the environment from space.** K. Ya. Kondratyev.
Naturwissenschaften, 58. Jahrgang, p. 529 - 541 (1971).
The visual observations of the earth carried out by the cosmonauts produced voluminous data on the colorimetry of the planet and the optical structure of the horizon, and made it possible to follow such phenomena as the aurora and noctilucent clouds. Still more interesting results were obtained from spectrophotometric measurements of the brightness of the earth's twilight aureole and of various natural formations.

082.061 **On the role of the structure of the wind in the development of optical heterogeneities in the atmosphere.**
V. G. Khetselius, V. F. Umarov, N. N. Kiliachkov.
Astron. Tsirk., No. 643, p. 1 - 2 (1971). In Russian.

082.062 **Results of annual seeing observations on Mt. Majdanak.** V. I. Kardopolov, V. E. Slutsky, V. G. Khetselius, V. S. Shevchenko.
Astron. Tsirk., No. 643, p. 2 - 4 (1971). In Russian.

082.063 **Site testing on Mt. Majdanak.** S. P. Yatsenko.
Astron. Tsirk., No. 643, p. 4 - 6 (1971). In Russian.

082.064 **On air temperature measurements at Mt. Sanglok.** G. V. Novikova.
Astron. Tsirk., No. 643, p. 6 - 8 (1971). In Russian.

082.065 **On variations of the nightglow near Hα.**
V. F. Zhidkov.
Astron. Tsirk., No. 644, p. 4 - 6 (1971). In Russian.

082.066 **Astroclimate in the area of the high-altitude expedition of the Sternberg Astronomical Institute (1957 - 1971).** V. M. Kovalenko, V. I. Bulavina, O. N. Kovalenko.
Astron. Tsirk., No. 645, p. 1 - 4 (1971). In Russian.

082.067 **On the connection between astronomical and meteorological data on astroclimate.**

V. M. Kovalenko, O. N. Kovalenko.
Astron. Tsirk., No. 645, p. 4 - 7 (1971). In Russian.

082.068 **A diffuse-photochemical study of the mesosphere and lower thermosphere and the associated conservation mechanisms.** B. G. Hunt.
Journ. Atmosph. Terr. Phys., Vol. 33, 1869 - 1892 (1971).
Results are presented for the photochemical composition of an oxygen-hydrogen atmosphere extending from 60 to 160 km, in which allowance was made for both molecular and eddy diffusion. Results are also given for the OH airglow emission and for the 5577 Å airglow of atomic oxygen.

082.069 **Astronomi og luftforurening.** P. E. Nissen.
Astron. Tidsskr., Årg. 4, 140 - 142 (1971).

082.070 **Determination of heights of atmospheric inhomogeneities from cinematographing the sun.**
U. I. Iljasov.
Solnechnye Dannye 1971 Byull., No. 7, p. 105 - 108 (1971). In Russian.
The optical inhomogeneities of lower layers of the earth's atmosphere have been investigated by means of cinematographing the variations of the solar limb.

082.071 **Polar star-trail observations of astronomical seeing in Arizona, Baja California, Chile, and Australia.**
M. F. Walker.
Publ. Astron. Soc. Pacific, Vol. 83, 401 - 422 = Lick Obs. Bull., No. 617 (1971).
The results of polar star-trail seeing observations at Kitt Peak and Flagstaff (Arizona), San Pedro Mártir (Baja California), Cerro Tololo (Chile), and Mount McKinlay (Australia), are presented. The purpose of these observations was twofold: (1) to determine the seeing at several existing observatories, or potential observatory sites, on the system established for the California Site Survey, and (2) to further investigate the conditions required for the best seeing.

082.072 **On the circulation of air masses in upper layers of the atmosphere during the IYQS.**
I. B. Bijbosunov, K. A. Karimov.
Investigation of the ionosphere and meteors, (see 003.015), p. 3 - 20 (1971). In Russian.

082.073 **Excitation mechanism of N I 5199 Å lines in the day airglow.** S. K. Gupta.
Astrophys. Space Sci., Vol. 13, 203 - 210 (1971).
Production rates of N (^2D) metastable atoms in the daytime atmosphere have been calculated for different possible processes, using positive ion composition and reaction rate data, available presently. Emission rates of the 5199 Å doublet of NI are calculated by separately and jointly considering the deactivation of N (^2D) atoms by electrons and molecular oxygen. From a comparison of the computed results with the observational data, an attempt has been made to identify the processes of importance to the production and deactivation of N (^2D). For further confirmation of the proposed mechanism the variation of integrated intensity of this radiation with solar activity has been obtained.

082.074 **Wave dynamics in the thermosphere: I: Tidal motion.** S. Kato.
Space Sci. Rev., Vol. 12, 421 - 445 (1971).
The atmospheric tides, their transmission and excitation in the thermosphere, are discussed in reviewing various investigations in this field. We shall restrict our discussion to the solar tide, in particular its first two components, i. e. the diurnal and semi-diurnal tides.

082.075 **Air density at heights near 150 km in 1970, from**

the orbit of Cosmos 316 (1969 - 108A).
D. G. King-Hele, D. M. C. Walker.
Planet. Space Sci., Vol. 19, 1637 - 1651 (1971).

Analysis of the orbit of Cosmos 316 has yielded 102 values of air density at heights near 150 km between 24 December 1969 and 28 August 1970. The observed variations in density justify three conclusions: (1) There are substantial increases in density at the times of geomagnetic disturbances. (2) There is usually little resemblance between variations in density and in the solar 10.7-cm radiation. (3) The most important long-term feature in the density variations is a semiannual variation, the maximum density (about 10 April 1970) being 1.3 times greater than the minimum density (about 13 January and 10 August 1970) for heights near 160 km.

082.076 Simultaneous measurements of the hydrogen airglow emissions of Lyman alpha, Lyman beta, and Balmer alpha. C. S. Weller, R. R. Meier, B. A. Tinsley.
Journ. Geophys. Res., Vol. 76, 7734 - 7744 (1971).

Lyman-α, 740- to 1050-Å, and Balmer-α airglow measurements made at 134° solar-zenith angle on October 13, 1969, are compared with resonance-scattering models of solar radiation. This paper compares the results with a variety of multiple scattering hydrogen models in an attempt not only to determine the atomic hydrogen concentration but also to resolve the discrepancy between the magnitudes of the theoretical and experimental Balmer-α emission rates.

082.077 Seasonal variation of the atomic-oxygen concentration in the lower thermosphere.
F. Barlier, D. Perret, C. Jaeck.
Journ. Geophys. Res., Vol. 76, 7797 - 7800 (1971).

The exospheric temperature can be deduced from incoherent-scatter measurements as well as from an analysis of the orbital decay of satellites. However, it appears that there are systematic differences that vary with season in the results obtained by the two methods. In this paper an interpretation of these differences is attempted by using static-diffusion models.

082.078 Diurnal variation of the atmospheric density at 400 kilometers.
D. Alcayde, P. Bauer, C. Jaeck, J. Falin.
Journ. Geophys. Res., Vol. 76, 7814 - 7816 (1971).

We report evidence that the detailed behavior of the neutral oxygen density throughout the day exhibits steeper changes in atmospheric density with time than would be predicted by the model, especially in the late afternoon.

082.079 Rocket measurement of OH in the mesosphere.
J. G. Anderson.
Journ. Geophys. Res., Vol. 76, 7820 - 7824 (1971). − Letter.

082.080 An analytic model for upper atmosphere densities based upon Jacchia's 1970 models.
C. E. Roberts, Jr.
Celestial Mechanics, Vol. 4, 368 - 377 (1971).

An analytic model is presented which gives upper atmospheric densities (between 90 and 125 km) as a function of the exospheric temperature and the altitude.

082.081 Need for an international agreement on astronomical refraction. G. Teleki.
Bull. Obs. Astron. Beograd, Vol. 28, (No. 124), 67 - 68 (1970).

082.082 The monochromatic extinction at Ege Univ. Observatory. M. Ü. Akyol.
Sci. Rep. Fac. Sci. Ege Univ., *Izmir*, No. 122 (Astron. No. 13), 9 pp. (1971).

Monochromatic extinction at Ege University Observatory was studied spectrophotometrically from a series of four night observations. The atmospheric pollution is considerable.

082.083 On the elimination of seeing effects from solar intensity measurements. A. Wittmann.
Solar Physics, Vol. 21, 237 - 239 (1971). − Research note.

082.084 Frequency spectrum of atmospheric oscillations of the solar limb. O. B. Vassiljev, U. I. Iljasov.
Solnechnye Dannye 1971 Byull., No. 9, p. 93 - 101 (1971). In Russian.

Frequency spectra of atmospheric oscillations of the solar limb at a wavelength from a unit to some hundreds Hertz are investigated. These spectra have been taken during photographing the sun in September 1968. The process under investigation is shown to be non-stationary.

082.085 A contribution to the knowledge of atmospheric turbidity and its influence upon solar radiation.
I. Penzar.
Dissertation Faculty of Natural Sciences and Mathematics, University Zagreb (1970). − Abstr. in Bull. Sci. Yougoslavie, Sect. A, Vol. 16, 272 - 273 (1971).

082.086 Reflection of light pulses from clouds.
G. N. Plass, G. W. Kattawar.
Applied Optics, Vol. 10, 2304 - 2310 (1971).

The reflection of light pulses from clouds is calculated by a Monte Carlo technique for all orders of multiple scattering. The returned flux is determined as a function of the photon path length for: (1) three different size distributions for the cloud particles (haze C, nimbostratus, ice crystal); (2) two different variations of the number density with height of the scattering centers within the cloud; (3) six different zenith angles of the source. The influence of detector half-width and of the atmosphere on the returned flux is studied.

082.087 Atmospheric emission measurements with a balloon-borne Michelson interferometer.
R. E. Jennings, A. F. M. Moorwood.
Applied Optics, Vol. 10, 2311 - 2318 (1971).

Astronomical measurements at mid- and far-infrared wavelengths can be made by using aircraft or balloons to carry telescopes above the strongly absorbing lower atmosphere. To measure the atmospheric background radiation present at these altitudes, a Michelson interferometer has been built and flown to a height of 38 km on a balloon. This paper describes the instrument and the methods of data reduction and calibration, and presents initial spectra in the range $100-1000$ cm^{-1} ($10-100\mu$) obtained to a mean resolution of 34 cm^{-1} during flight.

082.088 Visual observations and spectral investigations of the earth's atmosphere twilight aureole from the Soyuz-5 spacecraft. K. Ya. Kondratyev, B. V. Volynov, A. P. Galtsev, O. I. Smokty, E. V. Khrunov.
Applied Optics, Vol. 10, 2521 - 2533 (1971).

Results are given of the spectrophotometric measurements of the earth's atmosphere twilight aureole as first performed from the Soyuz-5 spacecraft. The analysis is made of the experimental findings depending on the wavelength, the perigee height of the sight ray above the earth's surface, the sunset angle, etc. The comparison is carried out of the vertical profiles for the twilight aureole monochromatic brightness with the results of the corresponding theoretical calculations for the aerosol Elterman model (1968).

082.089 Calculated circumsolar radiation as a function of aerosol type, field of view, wavelength, and optical depth. H. Grassl.
Applied Optics, Vol. 10, 2542 - 2543 (1971). − Letter.

082.090 Bidirectional reflectance of the moonlit earth.

W. B. Fowler, E. I. Reed, J. E. Blamont.
Applied Optics, Vol. 10, 2657 - 2660 (1971).

From OGO-4 airglow photometer data and computed lunar spectral irradiances at the subsatellite point, the highest radiance over clouds and lowest radiance over open ocean are examined near 3914 Å, 5577 Å, 5893 Å, 6225 Å, and 6300 Å in terms of bidirectional reflectance. The results are compared to and are consistent with mathematical models of the atmosphere developed by Plass and Kattawar, and with daytime measurements from OSO-3 by Neel, Griffin, and Millard.

082.091 **Emission of atmospheric helium 10830 Å in the predawn period.** T. I. Toroshelidze.
Astron. Tsirk., No. 652, p. 1 - 3 (1971). In Russian.

082.092 **Ozone production rates in an oxygen-hydrogen-nitrogen oxide atmosphere.** P. J. Crutzen.
Journ. Geophys. Res., Vol. 76, 7311 - 7327 (1971).

082.093 **Gas phase destruction of tropospheric ozone.**
L. A. Ripperton, F. M. Vukovich.
Journ. Geophys. Res., Vol. 76, 7328 - 7333 (1971).

082.094 **The xenon record of extinct radioactivities in the earth.** M. S. Boulos, O. K. Manuel.
Science, Vol. 174, 1334 - 1336 (1971).

Results of analyses of xenon from well gas rich in carbon dioxide place narrow limits on any age difference between the earth and the oldest meteorites. The occurrence of excess radiogenic xenon-129 in well gas also suggests that any quantitative degassing of existing solid materials to form the atmosphere must have been limited to a very early period of the earth's history, approximately the first 10^8 years. Alternatively, this observation is consistent with a model of the earth's continuous, but still incomplete, degassing since its time of formation.

082.095 **Investigations of the upper atmosphere and outer space made in the USSR in 1970.**
Dokl. COSPAR. 14th plenary meeting, Seattle, USA. "Nauka", Moskva. 144 pp. (1971). In Russian. – Abstr. in Referativ. Zhurn. 62. Issled. kosm. prostranstva, 12.62.54 (1971).

082.096 **Some results of photometric investigations of the daytime horizon of the earth made by Soyuz 4 and Soyuz 5.** A. B. Sandomirskij, G. V. Rozenberg, N. P. Al'tovskaya, T. A. Sushkevich.
Izv. AN SSSR. Fiz. atmosf. i okeana, Vol. 7, 590 - 598 (1971). In Russian. – Abstr. in Referativ. Zhurn. 62. Issled. kosm. prostranstva, 12.62.383 (1971).

082.097 **Temperature measurement of the upper atmosphere by the method of homodyne detection.**
L. A. Katasev, V. F. Chepura.
Geomagn. Aeronom., Vol. 11, 1118 - 1120 (1971). In Russian. Brief information.

082.098 **Airglow – Introduction and review.**
D. M. Hunten.
Astrophys. Space Sci. Library, Vol. 24, (see 012.021), 3 - 16 (1971).

082.099 **The dayglow.**
E. J. Llewellyn, W. F. J. Evans.
Astrophys. Space Sci. Library, Vol. 24, (see 012.021), 17 - 33 (1971).

082.100 **Dayglow nitrogen band systems.**
A. L. Broadfoot.
Astrophys. Space Sci. Library, Vol. 24, (see 012.021), 34 - 44 (1971).

082.101 **Some observations of the earth's ultraviolet dayglow.** J. B. Pearce.
Astrophys. Space Sci. Library, Vol. 24, (see 012.021), 45 - 48 (1971).

082.102 **Interpretation of airglow in terms of excitation mechanisms. D region nightglow.** R. L. Gattinger.
Astrophys. Space Sci. Library, Vol. 24, (see 012.021), 51 - 63 (1971).

082.103 **Interpretation of F region nightglow.** J. F. Noxon.
Astrophys. Space Sci. Library, Vol. 24, (see 012.021), 64 - 72 (1971).

082.104 **Fluctuations in the intensity and excitation temperature in the OH airglow (8–3) band.**
G. Visconti, F. Congeduti, G. Fiocco.
Astrophys. Space Sci. Library, Vol. 24, (see 012.021), 82 - 89 (1971).

082.105 **Dayglow and twilight excitation mechanisms for airglow.** H. N. Rundle.
Astrophys. Space Sci. Library, Vol. 24, (see 012.021), 90 - 104 (1971).

082.106 **Airglow observations and the interpretation of ionospheric disturbances associated with magnetic storms.** L. Thomas.
Astrophys. Space Sci. Library, Vol. 24, (see 012.021), 105 - 115 (1971).

082.107 **Atmospheric dynamics and airglow fluctuations.**
R. D. Sears.
Astrophys. Space Sci. Library, Vol. 24, (see 012.021), 116 - 122 (1971).

082.108 **Derivation of the photometric flux of daylight from filtered measurements of global (sun and sky) radiant energy.** A. J. Drummond, A. K. Ångström.
Applied Optics, Vol. 10, 2024 - 2030 (1971).

The study reported on here is a continuation of an earlier investigation by the same authors into the relationship between natural illumination and shortwave solar radiation. The results now given extend this work to parallel determination of the (more generally applicable) illumination of integral daylight on the basis of similarly filtered global (sun and sky) radiation. Included is an extension to recent solar radiation measurements obtained on high-altitude aircraft.

082.109 **Excitation of O_2 Herzberg bands in the nightglow.**
R. D. Srivastava, R. V. Shukla, A. N. Srivastava.
Astrophys. Space Sci., Vol. 14, 317 - 322 (1971).

The excitation mechanism for O_2 Herzberg bands as given by Young and Black (1966) is examined. The emission of Herzberg bands is found to arise from two layers centred at 80 and 100 km. The rate coefficients of a number of quenching reactions involving atmospheric gases are obtained theoretically.

082.110 **Phase variations in atmospheric optical propagation.**
S. F. Clifford, G. M. B. Bouricius, G. R. Ochs, M. H. Ackley.
Journ. Optical Soc. America, Vol. 61, 1279 - 1284 (1971).

We obtained excellent agreement between the observed phase structure function and Tatarski's theoretical curve.

082.111 **Temporal-frequency spectra for a spherical wave propagating through atmospheric turbulence.**
S. F. Clifford.
Journ. Optical Soc. America, Vol. 61, 1285 - 1292 (1971).

The spectra, calculated for spherical waves, reveal contributions at higher frequencies for amplitude scintillations, nearly identical phase results, and a phase-difference spectrum with no nulls, in contrast with the plane-wave results. Comparison with recent data is shown.

082.112 **Multiple-scattering model for light transmission through optically thick clouds.** H. M. Heggestad.
Journ. Optical Soc. America, Vol. 61, 1293 - 1300 (1971).

082.113 **Aeronomic reactions of hydrogen and ozone.**
M. Nicolet.
Astrophys. Space Sci. Library, Vol. 25, (see 012.024), 1 - 51 (1971).

082.114 **A meridional model of the oxygen-hydrogen atmosphere.** E. Hesstvedt.
Astrophys. Space Sci. Library, Vol. 25, (see 012.024), 52 - 64 (1971).

082.115 **Oxygen, hydrogen and nitrogen constituents in the mesosphere and ionization processes.** L. Thomas.
Astrophys. Space Sci. Library, Vol. 25, (see 012.024), 65 - 77 (1971).

082.116 **Photochemistry and the escape efficiency of terrestrial hydrogen.** R. T. Brinkmann.
Astrophys. Space Sci. Library, Vol. 25, (see 012.024), 89 - 102 (1971).

082.117 **Dynamical modelling of the stratosphere and mesosphere.** R. J. Murgatroyd.
Astrophys. Space Sci. Library, Vol. 15, (see 012.024), 104 - 121 (1971).

082.118 **Tides and gravity waves in the upper atmosphere.**
R. S. Lindzen.
Astrophys. Space Sci. Library, Vol. 25, (see 012.024), 122 - 130 (1971).

082.119 **The meteorological structure of the mesosphere including seasonal and latitudinal variations.**
J. S. Theon, W. S. Smith.
Astrophys. Space Sci. Library, Vol. 25, (see 012.024), 131 - 146 (1971).

082.120 **Nocturnal variations in the OH (8–3) band airglow.**
F. Congeduti, G. Fiocco, G. Visconti.
Astrophys. Space Sci. Library, Vol. 25, (see 012.024), 147 - 148 (1971).

082.121 **Ultraviolet solar radiation related to mesospheric processes.** M. Ackerman.
Astrophys. Space Sci. Library, Vol. 25, (see 012.024), 149 - 159 (1971).

082.122 **Penetration of solar radiation in the Schumann-Runge bands of molecular oxygen.** G. Kockarts.
Astrophys. Space Sci. Library, Vol. 25, (see 012.024), 160 - 176 (1971).

082.123 **Relaxation of the 2.7 μ and 4.3 μ bands of carbon dioxide.** A. P. Williams.
Astrophys. Space Sci. Library, Vol. 25, (see 012.024), 177 - 187 (1971).

082.124 **The photochemistry of ozone and singlet molecular oxygen in the atmosphere.** R. P. Wayne.
Astrophys. Space Sci. Library, Vol. 25, (see 012.024), 240 - 252 (1971).

082.125 **The airglow reaction $NO + O + (M) \rightarrow NO_2^* + (M)$ at low pressures.**
K. H. Becker, W. Groth, D. Thran.
Astrophys. Space Sci. Library, Vol. 25, (see 012.024), 261 - 265 (1971).

082.126 **On the origin of noctilucent clouds.**
G. Fiocco, G. Grams.
Astrophys. Space Sci. Library, Vol. 25, (see 012.024), 266 - 267 (1971).

082.127 **On the semiannual variation of the upper atmosphere.** V. P. Bhatnagar.
Astrophys. Space Sci. Library, Vol. 25, (see 012.024), 273 (1971).

082.128 **Rocket mass spectrometry in Japan.** N. Fugono.
Astrophys. Space Sci. Library, Vol. 25, (see 012.024), 284 - 285 (1971).

082.129 **Measurement of submillimeter-wave stratospheric emission spectra.**
J. Chamberlain, W. J. Burroughs, J. E. Harries.
Astrophys. Space Sci. Library, Vol. 25, (see 012.024), 286 - 291 (1971).

082.130 **Some effects of dimers of the water molecule in the atmosphere.**
W. J. Burroughs, J. E. Harries, J. Chamberlain, H. A. Gebbie.
Astrophys. Space Sci. Library, Vol. 25, (see 012.024), 292 - 298 (1971).

082.131 **A theory of the diurnal variations of the thermosphere.** H. Volland, H. G. Mayr.
Ann. Géophys., Vol. 26, 907 - 919 = Mitt. Astron. Inst. Bonn, No. 138 (1970).

Applying the concept of characteristic waves in a two dimensional model atmosphere it is possible to separate the contributions of the density and temperature variations due to various energy mechanisms. It is shown that the diurnal density variations within the thermosphere are excited by solar EUV heat input and by a tidal gravity wave.

082.132 **Nomogram for determining air masses during electro-photometric observations.**
I. V. Shpychka, V. V. Golovatyj, M. B. Girnyak.
Tsirk. L'vov. Astron. Obs., No. 46, p. 28 - 31 (1971).
In Russian.

082.133 **Statistical theory of light propagation in a turbulent medium.**
V. I. Klyatskin, V. I. Tatarskii.
Journ. Optical Soc. America, Vol. 61, 1551 - 1552 (1971).
Abstr. Optical Soc. America.

082.134 **Beam-wave correlation function in a turbulent atmosphere.** P. M. Livingston.
Journ. Optical Soc. America, Vol. 61, 1552 (1971). – Abstr. Optical Soc. America.

082.135 **Vacuum ultraviolet spectra of the late twilight airglow.** J. L. Buckley, H. W. Moos.
Journ. Geophys. Res., Vol. 76, 8378 - 8383 (1971).

Sounding rocket spectra of the late twilight (solar-zenith angle of 120°) ultraviolet airglow between 1260 and 1900 Å are reported. The only observed features are O I 1304 and 1356. When the instrument looked at an elevation of 17° above the western horizon, the brightnesses were 70 and 33 rayleighs, respectively.

082.136 **Odd nitrogen in the mesosphere.** D. F. Strobel.

Journ. Geophys. Res., Vol. 76, 8384 - 8393 (1971).

Comparison of previous theoretical work with observational data has suggested a downward NO flux through the mesopause on the order of $(1-5) \times 10^8$ cm^{-2} sec^{-1}. The importance of this flux as a source of odd nitrogen in the mesosphere and upper stratosphere is investigated.

082.137 Temperature coefficients for N (^2D) quenching by O_2 and N_2O.
T. G. Slanger, B. J. Wood, G. Black.
Journ. Geophys. Res., Vol. 76, 8430 - 8433 (1971).
Brief report.

082.138 Rate constants for the reactions of CO_2^+ + H and $H^+ + CO_2$. F. C. Fehsenfeld, E. E. Ferguson.
Journ. Geophys. Res., Vol. 76, 8453 - 8454 (1971).

Because of the potential interest in Mars and Venus aeronomy, we recently investigated the reaction of CO_2^+ ions with atomic hydrogen and H^+ ions with CO_2. The measurements were made in a flowing afterglow system.

082.139 Submillimetre-wave solar observations using a double-output Michelson interferometer.
W. J. Burroughs, J. Chamberlain.
Infrared Phys., Vol. 11, 1 - 23 (1971).

A double-output Michelson-type interferometer having Golay cell detectors has been used at an altitude of 3580 m to study, by means of Fourier spectrometry, the atmospheric absorption of solar radiation in the submillimetre-wave region. Spectra having 0.5, 0.25 and 0.125 cm^{-1} resolution were obtained. The 0.5 cm^{-1} results have been compared with spectra obtained with the same instrument in the laboratory using a mercury lamp source.

082.140 Some problems of atmospheric investigations.
M. Ill.
Byull. Stantsij Optichesk. Nablyud. Iskusstv. Sputnikov Zemli, No. 57, p. 34 (1971). In Russian.

082.141 Some aspects of the study of the earth's upper atmosphere from ground satellite observations.
V. E. Chertoprud.
Byull. Stantsij Optichesk. Nablyud. Iskusstv. Sputnikov Zemli, No. 57, p. 35 - 40 (1971). In Russian.

082.142 On the possibility of revealing irregular oscillations in the upper atmosphere. I. Almar.
Byull. Stantsij Optichesk. Nablyud. Iskusstv. Sputnikov Zemli, No. 57, p. 40 - 42 (1971). In Russian.

082.143 Certain possibilities to study atmospheric density variations. E. P. Aksionov.
Byull. Stantsij Optichesk. Nablyud. Iskusstv. Sputnikov Zemli, No. 57, p. 42 - 43 (1971). In Russian.

082.144 Stratospheric pollution and volcanic eruptions.
E. K. Bigg.
Weather, Vol. 26, No. 1, p. 13 - 18 = Separate print Division Radiophys., C.S.I.R.O., Sydney, Australia (1971).

082.145 High-level stratocumulus clouds.
E. K. Bigg, R. T. Meade.
Weather, Vol. 26, No. 2, p. 55 - 57 = Separate print Division Radiophys., C.S.I.R.O., Sydney, Australia (1971).

082.146 The composition of cloud nuclei. S. Twomey.
Journ. Atmosph. Sci., Vol. 28, 377 - 381 = Separate print Division Radiophys., C.S.I.R.O., Sydney, Australia (1971).

082.147 Scale length in atmospheric turbulence as measured

from an aircraft.
R. J. Taylor, J. Warner, N. E. Bacon.
Quarterly Journ. Roy. Meteorol. Soc., Vol. 96, 750 - 755= Separate print Division Radiophys., C.S.I.R.O., Sydney, Australia (1970).

082.148 Solar cycle variation of planetary exospheric temperatures. S. J. Bauer.
Nature, Phys. Sci., Vol. 232, 101 - 102 (1971).

This communication shows that a simple scaling law, based on an approximate solution of the heat balance equation, can be used to determine the variation of exospheric temperature T_∞ over a solar cycle. T_∞ is defined as the temperature of the isothermal region (on account of heat conduction) extending into the exosphere and so T_∞ also controls the escape of gases from planetary atmospheres. Earth and Venus seem to have exospheres which are stable even at solar maximum although they are subject to escape ($B < 15$), while the exosphere of Mars is stable only near solar minimum and escape should be excessive near solar maximum.

082.149 Rocket measurement of the far infrared sky background.
D. P. McNutt, K. Shivanandan, P. D. Feldman.
Nature, Phys. Sci., Vol. 234, 25 - 26 (1971).

In our experiment we have observed a high background signal in a band around 280 μm. Two high measurements were made soon after astronomical twilight whereas all of the low results were obtained near local midnight. It is possible that the radiation originates in the outer magnetosphere or in the interplanetary medium near the earth and is subject to diurnal variability. But it is also possible that the observed radiation is variable and dependent on the occurrence of geomagnetic disturbances.

082.150 Observations météorologiques de Thessaloniki 1962-1965.
Published by G. C. Livadas.
Univ. Thessaloniki, Annuaire Inst. Météorologique et Climatologique, 30 - 33, 30 + 30 + 30 + 30 pp. (1971).

082.151 Estimation of atomic nitrogen and nitric oxide in the night-time atmosphere in the altitude region 80-110 kms and their contribution to the night airglow continuum.
P. P. Saxena.
Ann. Géophys., Vol. 26, 771 - 776 = Uttar Pradesh State Obs., Repr. No. 36 (1970).

082.152 Radiometer studies of atmospheric attenuation of solar emission at 19 GHz. P. G. Davies.
Proc. Instn. Electr. Engineers, (GB), Vol. 118, 737 - 741 (1971). $- DNC$

Blotting out the stars.
Southern Stars, Vol. 24, 41 - 42 (1971).

Noctilucent clouds. See Abstr. 003.054.

On the neutrality of terrestrial cloud extinction.
See Abstr. 034.086.

Variations in soft solar X-rays and in the upper atmosphere temperature. See Abstr. 076.012.

Some parameters of the ionosphere and neutral atmosphere measured by the rocket MP-12 on March 30, 1968. See Abstr. 083.041.

The sunspot cycle 1968/70 in ionospheric absorption and stratospheric temperature. See Abstr. 083.050.

Infrared investigations of aurora and airglow.
See Abstr. 084.040.

Canadian national report on geomagnetism and
aeronomy.   See Abstr. 084.316.

Atmospheric oscillations—III.
See Abstr. 091.009.

Laser radar observations of dust from comet Bennett.
See Abstr. 103.101.

Characteristics of the interplanetary medium and
changes of the kinetic energy of the long waves in the troposphere.   See Abstr. 106.008.

083 Ionosphere

083.001 **Westward electric fields as the cause of nighttime enhancements in electron concentrations in mid-latitude *F* region.** C. G. Park.
Journ. Geophys. Res., Vol. 76, 4560 - 4568 (1971).

083.002 **Solar cycle variations in daytime ionospheric E_S parameters in the South Pacific Area.**
W. J. Baggaley.
Journ. Geophys. Res.,Vol. 76, 4674 - 4678 (1971).
Brief report.

083.003 **On small-scale electrostatic fields in the ionosphere.** B. N. Gershman, A. V. Samsonov.
Dokl. Akad. Nauk SSSR, Ser. Mat. Fiz., Vol. 199, 73 - 75 (1971). In Russian.

083.004 **The implications of the semi-annual variations in the F2-region for the lower thermosphere.**
V. P. Bhatnagar.
Planet. Space Sci., Vol. 19, 723 - 729 (1971).

083.005 **Improved accuracy in *in-situ* probe measurements of ionospheric n_e and T_e using techniques requiring only low bandwidth telemetry channels.** C. V. Goodall.
Planet. Space Sci., Vol. 19, 827 - 840 (1971).

083.006 **Time lags in the auroral zone ionosphere.**
J. S. Reid, J. Phillips.
Planet. Space Sci., Vol. 19, 959 - 969 (1971).

083.007 **Effective collision frequency of electrons in atmospheric gases.** Y. Itikawa.
Planet. Space Sci., Vol. 19, 993 - 1007 (1971).
Numerical values are obtained for the effective collision frequencies for a number of atmospheric gases, i.e. for N_2, O_2, H_2, CO_2, H_2O, O, H and He, over the range of the electron temperature from 100 to 5000°K.

083.008 **Comparison of scintillation, spread *F* and electrostatic probe observations of electron density irregularities.** P. L. Dyson.
Journ. Atmosph. Terr. Phys., Vol. 33, 1185 - 1192 (1971).

083.009 **An ionospheric *E*-region nighttime model.**
W. M. Chen, R. D. Harris.
Journ. Atmosph. Terr. Phys., Vol. 33, 1193 - 1207 (1971).

083.010 **The solar cycle constant for equatorial ionospheric absorption.** E. W. Mbipom.
Journ. Atmosph. Terr. Phys., Vol. 33, 1263 - 1265 (1971).
Short paper.

083.011 **Effect of sun–earth distance on *E*-region ionization.** L. M. Muggleton.
Journ. Atmosph. Terr. Phys., Vol. 33, 1299 - 1305 (1971).

083.012 **Solar cycle control of $N_m(E)$.**
L. M. Muggleton.
Journ. Atmosph. Terr. Phys., Vol. 33, 1307 - 1310 (1971).
The regression coefficients between *E*-region peak electron density, $N_m(E)$, and sunspot number, R, depend on the particular solar cycle investigated. It is found that a representative worldwide relationship for the 1949–59 solar cycle exists.

083.013 **On the corpuscular "hypothesis" of ionization of the ionospherical D-layer in mean latitudes.**

Yu. A. Bragin.
Kosmich. Issled., Vol. 9, 625 - 626 (1971). In Russian.
Brief information.

083.014 **Electric coupling between the magnetosphere and the ionosphere as a cause of polar magnetic disturbances and auroral break-up.** T. Oguti.
Cosmic Electrodynamics, Vol. 2, 164 - 183 (1971).
We stress the importance of the transient nature of the interaction between the ionosphere, the magnetosphere and the solar wind in order to investigate the origin of the polar disturbance phenomena.

083.015 **The numerical calculation of reflection and transmission coefficients for thin highly ionised layers including the effect of the earth's magnetic field.**
C. I. Chessell.
Journ. Atmosph. Terr. Phys., Vol. 33, 1515 - 1532 (1971).

083.016 **The semi-annual variation in the height of the F2-layer peak.**
D. Eccles, J. W. King, H. Kohl, R. J. Pratt.
Journ. Atmosph. Terr. Phys., Vol. 33, 1641 - 1646 (1971).
Short paper.

083.017 **Dynamical behavior of thermal protons in the mid-latitude ionosphere and magnetosphere.**
P. M. Banks, A. F. Nagy, W. I. Axford.
Planet. Space Sci., Vol. 19, 1053 - 1067 (1971).

083.018 **Study of the total electron content and slab-thickness of the ionosphere at Tortosa.**
E. Galdón.
Urania Barcelona, Año 55, Nos. 171–172, p. 208 - 228 (1970).

083.019 **Solar activity and total electron content correlation in the ionosphere at Tortosa.**
E. Galdón, L. F. Alberca.
Urania Barcelona, Año 55, Nos. 271–272, p. 229 - 240 (1970).

083.020 **Determination of electron content at a low-latitude station with off-zenith transverse point.**
S. Basu, Sunanda Basu.
Journ. Geophys. Res., Vol. 76, 5337 - 5343 (1971).
The importance of the choice of mean ionospheric height (h_M) in the determination of latitude variation of total electron content (TEC) from satellite transmissions at a low-latitude station is investigated. By considering the concept of magneto-ionic mode coupling in the vicinity of the transverse region, the Faraday rotation as a function of latitude for a frequency of 40 MHz has been theoretically computed for Calcutta (12° N geomagnetic) by using typical electron concentration profiles representing different ionospheric conditions. This result was used to determine the criterion for the proper choice of h_M consistent with least error in the determination of TEC over the entire latitude range of interest.

083.021 **Contribution à l'étude de l'absorption ionosphérique par riomètre.** J. Lavergnat.
Comptes Rendus Acad. Sci. Paris, Sér. B, Vol. 273, 577 - 580 (1971).
On propose une nouvelle méthode de calcul de l'absorption ionosphérique à partir des enregistrements obtenus par riomètre.

083.022 **Simultaneous sudden frequency deviations and sudden enhancements of the total electron content**

of the ionosphere.   M. D. Papagiannis, D. A. Matsoukas.
Nature, Phys. Sci., Vol. 233, 55 - 56 (1971).

083.023   On results of investigations of local and integral
electron concentration of the ionosphere by means
of coherent radio waves radiated from artificial earth satellites. II.
Ya. L. Alpert, L. N. Vitshas, V. I. Krayushkina, V. M.
Sinelnikov.
Geomagn. Aeronom., Vol. 11, 595 - 601 (1971).   In Russian.

083.024   On a conjugate of disturbances in the ionospheric
F2 region.
N. P. Benkova, G. V. Bukin, Chin Khong Tien.
Geomagn. Aeronom., Vol. 11, 602 - 607 (1971).   In Russian.

083.025   The definition of the absorption value in the lower
ionosphere by data on the density and spectrum
of the solar cosmic ray stream.
V. M. Driatzky, T. M. Krupitzkaya, A. V. Shirochkov.
Geomagn. Aeronom., Vol. 11, 716 - 718 (1971).   In Russian.
Brief information.

083.026   On a method of defining the parameters of the
ionospheric plasma.   G. I. Terina.
Geomagn. Aeronom., Vol. 11, 726 - 727 (1971).   In Russian.
Brief information.

083.027   Polar ionosphere: Measured ion density enhancements and soft electron precipitation.
F. A. Morse, H. H. Hilton, P. F. Mizera.
Journ. Geophys. Res., Vol. 76, 6099 - 6111 (1971).
    The subject of this paper is the relationship of ionospheric
positive ion densities and soft particle fluxes obtained during
the magnetic storm period August 13 to 18, 1968, over the
northern polar cap.

083.028   Dependence of the ionospheric absorption on lunar
phase.   A. S. M. Rao, B. R. Rao.
Journ. Geophys. Res., Vol. 76, 6208 - 6209 (1971). – Letter.

083.029   An empirical relationship between ionospheric
equivalent slab thickness and mean gradient of the
electron temperature in the F-region.
P. Amayenc, F. Bertin, J. Papet-Lepine.
Planet. Space Sci., Vol. 19, 1313 - 1317 (1971).

083.030   Lunar-solar tides in electron density at fixed
heights of the ionosphere.
R. P. Sharma, R. G. Rastogi.
Planet. Space Sci., Vol. 19, 1349 - 1357 (1971).
    The paper describes the annual average lunar semi-diurnal
variations in the electron density at fixed heights (at intervals
of 10 km) of the ionosphere over Huancayo and Puerto Rico
computed separately for different solar hours.

083.031   Ionizing radiation at heights between 200 and
300 km.
V. L. Maduev, B. M. Makhmudov, S. N. Kuznetzov.
Geomagn. Aeronom., Vol. 11, 771 - 775 (1971).
In Russian.

083.032   A statistical model of geophysical processes with
asymmetry and excess of the density probability
function.
I. S. Vsekhsvyatskaya, N. P. Sergeenko, L. A. Yudovitch.
Geomagn. Aeronom., Vol. 11, 785 - 789 (1971).
In Russian.

083.033   Absorption in the night lower ionosphere as an
indicator of ionization processes under normal

conditions and in disturbances. I.   G. Nestorov.
Geomagn. Aeronom., Vol. 11, 798 - 807 (1971).
In Russian.

083.034   Electric fields in the ionosphere during pulsed
aurorae.
V. K. Roldugin, B. N. Belenkaya, N. F. Malzeva.
Geomagn. Aeronom., Vol. 11, 813 - 818 (1971).
In Russian.

083.035   On motion and diffusion of the sporadic E-layer.
Yu. A. Ignatiev, Z. N. Krotova.
Geomagn. Aeronom., Vol. 11, 906 - 909 (1971). In Russian.
Brief information.

083.036   Interpretation of low-frequency impedance measure-
ments in the near-earth plasma.
Yu. M. Mikhailov.
Geomagn. Aeronom., Vol. 11, 916 - 919 (1971). In Russian.
Brief information.

083.037   Ambipolar diffusion along magnetic field lines in
the presence of an electric current.
M. Scholer, G. Haerendel.
Planet. Space Sci., Vol. 19, 915 - 927 (1971).

083.038   On the possible relationship between substorm
electron drift and cosmic noise absorption on the
morning side of the auroral zone.   M. A. Abdu.
Journ. Atmosph. Terr. Phys., Vol. 33, 1703 - 1710 (1971).
    Calculation of the energy loss in the atmosphere of drif-
ting electrons during auroral substorms show a harder energy
spectrum for the electrons reaching the morning side than
those near the midnight side of the auroral zone. The possibility of detecting the precipitation effect of these electrons on
the morning cosmic noise absorption events is examined.

083.039   Dynamical behaviour of the nighttime ionosphere
at Arecibo.   G. J. Nelson, L. L. Cogger.
Journ. Atmosph. Terr. Phys., Vol. 33, 1711 - 1726 (1971).

083.040   Ionospheric total electron content behaviour during
geomagnetic storms.   M. Mendillo.
Nature, Phys. Sci., Vol. 234, 23 - 24 (1971).
    This article describes a new study of the total content
storm effects and gives the first set of average storm patterns
for the mid-latitude ionosphere.

083.041   Some parameters of the ionosphere and neutral
atmosphere measured by the rocket MP-12 on
March 30, 1968.
N. M. Klyueva, A. A. Pokhunov, Yu. K. Chasovitin.
Kosmich. Issled., Vol. 9, 736 - 740 (1971). In Russian.

083.042   Composition of the earth's ionosphere at heights
between 220 and 4360 km from measurements of
the "vertical cosmic probe".
V. A. Ershova, L. D. Sivtseva, V. I. Yarin.
Kosmich. Issled., Vol. 9, 741 - 747 (1971). In Russian.

083.043   Profiles of electron density and electron tempera-
tures at heights between 80 and 170 km.
G. P. Komrakov, V. G. Khryukin, Yu. K. Chasovitin.
Kosmich. Issled., Vol. 9, 791 - 793 (1971). In Russian.
Brief information.

083.044   Ionospheric electron content at Ahmedabad (near
the crest of equatorial anomaly) by using beacon
satellite transmissions during half a solar cycle.
R. G. Rastogi, R. P. Sharma.
Planet. Space Sci., Vol. 19, 1505 - 1517 (1971).

083.045 Electron density enhancements in the E and F regions of the ionosphere during solar flares.
G. D. Thome, L. S. Wagner.
Journ. Geophys. Res., Vol. 76, 6883 - 6895 (1971).

Using the method of Thomson scatter, it has been possible to observe detailed ionospheric effects due to solar flares. Observations have been limited to four 2B flares, and two of these events have been examined intensively.

083.046 Progress in the study of interaction between neutral and ionized parts of the atmosphere.
E. A. Lauter, J. Taubenheim.
Space Research XI, (see 012.004), 1005 - 1012 (1971).

083.047 Models of lower ionosphere electron density profiles. A. P. Mitra, D. K. Chakrabarty.
Space Research XI, (see 012.004), 1013 - 1018 (1971).

083.048 Models for the effective recombination coefficients in the ionosphere. A. P. Mitra, P. Banerjee.
Space Research XI, (see 012.004), 1019 - 1025 (1971).

083.049 Striations in ionospheric ion clouds, 1.
H. J. Völk, G. Haerendel.
Journ. Geophys. Res., Vol. 76, 4541 - 4559 (1971).

083.050 The sunspot cycle 1958/70 in ionospheric absorption and stratospheric temperature.
H. Schwentek.
Journ. Atmosph. Terr. Phys., Vol. 33, 1839 - 1852 (1971).

It is shown that the effect of the sunspot cycle 1958/70 on ionospheric absorption in the D-region determined daily at Lindau at 2.6 MHz (method A3) and temperature in the stratosphere measured daily by radiosonde launchings at West Berlin is very pronounced in winter but small in summer.

083.051 The definition of collision frequencies and their relation to the electron conductivity of the ionosphere. K. Suchy, K. Rawer.
Journ. Atmosph. Terr. Phys., Vol. 33, 1853 - 1868 (1971).

083.052 Photoionization yields of O, O_2 and N_2 for high and low solar activity. K. Schlegel.
Journ. Atmosph. Terr. Phys., Vol. 33, 1923 - 1931 (1971).

Photoionization yields of O, O_2 and N_2 were calculated with the help of a Monte-Carlo simulation of a model ionosphere. The photoionization yields depending on altitude, photon wavelength and solar activity were expressed in a simple analytical form for easy use in calculations.

083.053 A radio scintillation method of estimating the small-scale structure in the ionosphere. C. L. Rufenach.
Journ. Atmosph. Terr. Phys., Vol. 33, 1941 - 1951 (1971).

The power spectra of the intensity fluctuations for the radio source Cygnus A were analyzed for 44 nights during the summer of 1970. When the scattering was weak, oscillations were observed in the power spectra. These oscillations are attributed to a Fresnel filtering effect.

083.054 A study of two flares on 8 July 1968 in the light of their ionospheric effects. P. R. Sengupta.
Journ. Atmosph. Terr. Phys., Vol. 33, 1953 - 1971 (1971).

The X-ray flare of 8 July 1968 is of considerable importance, both from the ionospheric and solar physics point of view. The associated optical flare was of importance 3B. In the present work we have studied in detail the observed ionospheric effects of the flare in order to derive a suitable physical model of flare-induced ionization below 100 km and hence to estimate the spectral characteristics of the ionizing radiation from the observed SID's.

083.055 The spectrum of electron content fluctuations in the ionosphere. J. E. Titheridge.
Planet. Space Sci., Vol. 19, 1593 - 1608 (1971).

Continuous records of the electron content of the ionosphere, from 1965 to 1970, are used to obtain power spectra covering periods from 30 sec to 2 yr at latitudes of 34° S and 42° S. The spectra consist basically of three sections: (1) A gravity wave region, from about 16 to 80 min, in which the r.m.s. amplitude per octave A_0 is reasonably constant at about 1×10^{15} electrons/m^2. (2) A storm region (12 hr−27 day) attributed largely to magnetic disturbances, in which A_0 is approximately constant at 20×10^{15} els/m^2. This is joined to the gravity wave section by a smooth increase of A_0. (3) Semiannual, annual and solar cycle changes of about the same amplitude as the storm region. This section is separated from the storm region by a slight decrease (about 40 per cent) in A_0.

083.056 A latitudinal survey of the daytime polar F layer.
C. P. Pike.
Journ. Geophys. Res., Vol. 76, 7745 - 7753 (1971).

083.057 Electromagnetic resonance phenomena and equipments to study the relation between solar activity and the magnetoplasma (ionosphere).
G. Agnelli, M. Cimino, M. Cutolo, M. Puglisi.
Advances Phys., Vol. 78, 217 - 268 = Oss. Astron. Roma, Contr. Sci., Ser. 3, No. 105 (1970).

An exact definition of the lower ionosphere is given. The reason for the non-linear behaviour of the plasma in this region is explained. The theories of Graffi and Marziani for a general non-linear equation of an electromagnetic field are given.Some non-linear effects as gyro-interaction, self-modulation, self-demodulation, detection effect, are considered. The concept of 'ionospheric detection effect' is given. The experiments carried out in Italy are briefly described, and the methods of measurement are explained.

083.058 On the proper radio radiation of the auroral ionosphere. A. I. Bozhkov, N. K. Osipov.
Geomagn. Aeronom., Vol. 11, 1021 - 1025 (1971). In Russian.

083.059 On the stability of a dynamic model of the F region of the ionosphere. N. A. Sidorov.
Geomagn. Aeronom., Vol. 11, 1084 - 1085 (1971). In Russian.
Brief information.

083.060 A common system of coordinates for description of the planetary distribution of the ionospheric parameters. O. V. Chernishev.
Geomagn. Aeronom., Vol. 11, 1086 - 1087 (1971). In Russian.
Brief information.

083.061 The role of meteors in the formation of E_s.
V. N. Donii.
Geomagn. Aeronom., Vol. 11, 1089 - 1090 (1971). In Russian.
Brief information.

083.062 The profiles N (h) in the D region of the ionosphere in a period of sudden ionospheric disturbance.
V. V. Belikovich, E. A. Benedictov, L. V. Grishkevich, V. A. Ivanov, G. P. Komrakov, V. V. Podmoskov, F. A. Flat.
Geomagn. Aeronom., Vol. 11, 1090 - 1092 (1971). In Russian.
Brief information.

083.063 Finding of the electron concentration profile in the lower ionosphere in high latitudes from riometric observations. Yu. P. Novikov.
Geomagn. Aeronom., Vol. 11, 1092 - 1095 (1971). In Russian.
Brief information.

083.064 Measurements of electron concentration and elec-

tron temperature in the ionosphere by sounding methods. Yu. B. Burakov, G. P. Komrakov, I. V. Popkov, V. G. Khriukin, Yu. K. Chasovitin.
Geomagn. Aeronom., Vol. 11, 1105 - 1108 (1971). In Russian.
Brief information.

083.065 Ionospheric effects of auroral electric fields.
J. C. G. Walker.
Astrophys. Space Sci. Library, Vol. 24, (see 012.021), 207 - 216 (1971).

083.066 Correction: 'A simplified model of the geomagnetic S_q current system and the electric fields within the ionosphere' [Cosmic Electrodynamics, Vol. 1, 428 - 459 (1971)]. H. Volland.
Cosmic Electrodynamics, Vol. 2, 375 (1971).

083.067 Seasonal and solar cycle variation of total electron content at temperate latitudes.
E. Galdón, L. F. Alberca.
Urania Barcelona, Año 56, No. 273, p. 110 - 121 (1971).
From data of the BE-B beacon satellite, values of the total electron content of the ionosphere (T.E.C.) at the three hours around midday have been obtained through the Faraday rotation method in three stations of different latitude: Observatorio del Ebro, Val-Joyeux and Lindau. The variation of TEC with the solar activity has been obtained. The similarity of the seasonal variation in the three stations is notorious. The results are discussed and compared with those of other authors.

083.068 The roles of water vapor and nitric oxide in determining electron densities in the D-region.
G. C. Reid.
Astrophys. Space Sci. Library, Vol. 25, (see 012.024), 198 - 209 (1971).

083.069 Discussion of the formation of major positive and negative ions up to the 50 km level.
V. A. Mohnen.
Astrophys. Space Sci. Library, Vol. 25, (see 012.024), 210 - 219 (1971).

083.070 The rate constants of some ion-molecule and ion-electron reactions, from in situ ionospheric meas- urements. A. Giraud.
Astrophys. Space Sci. Library, Vol. 25, (see 012.024), 268 - 272 (1971).

083.071 Rocket measurements of ion and electron densities in the D-region during sunrise.
A. Pedersen, J. A. Kane.
Astrophys. Space Sci. Library, Vol. 25, (see 012.024), 274 - 278 (1971).

083.072 Ion densities in the lower D-region measured with a parachuted probe. R. Jaeschke, A. Pedersen.
Astrophys. Space Sci. Library, Vol. 25, (see 012.024), 279 - 282 (1971).

083.073 Neutral composition variation above 400 kilometers during a magnetic storm.
D. R. Taeusch, G. R. Carignan, C. A. Reber.
Journ. Geophys. Res., Vol. 76, 8318 - 8325 (1971).

083.074 Probe and radar electron temperatures in an isotropic nonequilibrium plasma. W. R. Hoegy.
Journ. Geophys. Res., Vol. 76, 8333 - 8340 (1971).

083.075 Studies of positive-ion composition in the equatorial D-region ionosphere.
R. A. Goldberg, A. C. Aikin.
Journ. Geophys. Res., Vol. 76, 8352 - 8364 (1971).

083.076 Solar X-ray control of E and sporadic-E layers of the ionosphere. K. K. Vij, R. Hislop.
Journ. Geophys. Res., Vol. 76, 8441 - 8444 (1971). – Letter.

083.077 Distribution of ion density in the ionosphere at a height of 600 km in lower and mean latitudes.
V. N. Ponomarev.
Kosm. Issled., Vol. 9, 878 - 884 (1971). In Russian.

Impedance measurements of a loop antenna in the topside ionosphere. See Abstr. 033.014.

Airglow observations and the interpretation of ionospheric disturbances associated with magnetic storms.
See Abstr. 082.106.

084 Aurorae, Geomagnetic Field, Radiation Belts

Aurorae

084.001 **Unified theory of SAR arc formation at the plasmapause.** J. M. Cornwall, F. V. Coroniti, R. M. Thorne.
Journ. Geophys. Res., Vol. 76, 4428 - 4445 (1971).
We propose that stable auroral red arcs are generated at the plasmapause as a consequence of the turbulent dissipation of ring current energy. We postulate that the intense ion cyclotron wave turbulence generated by the ring current is an important SAR arc electron heat source.

084.002 **Rocket measurements of low-energy auroral electrons.** S. Rearwin.
Journ. Geophys. Res., Vol. 76, 4505 - 4517 (1971).

084.003 **Pulsating auroral patches exhibiting sudden intensity-dependent spatial expansion.**
M. W. J. Scourfield, N. R. Parsons.
Journ. Geophys. Res., Vol. 76, 4518 - 4524 (1971).

084.004 **Errata: 'Auroral spectroscopy'.** [Space Sci. Rev., Vol. 11, 776 - 826 (1971)]. A. V. Jones.
Space Sci. Rev., Vol. 12, 258 (1971).

084.005 **Observations on pulsating auroras.**
R. E. Miller, W. M. Zeitz.
Planet. Space Sci., Vol. 19, 693 - 696 (1971).
Optical auroral pulsations in $\lambda 3914$ N_2^+ and $\lambda 5577$ OI have been measured in the frequency range 0.01 to 1.0 Hz by a double photometer system during the NASA Airborne Auroral Expedition.

084.006 **Rocket measurements of the magnetic fields associated with visual aurorae.** K. Burrows, J. D. Stolarik, J. P. Heppner.
Planet. Space Sci., Vol. 19, 877 - 890 (1971).

084.007 **On the correlation between pulsating aurora and cosmic radio noise absorption.**
A. Brekke.
Planet. Space Sci., Vol. 19, 891 - 896 (1971).

084.008 **Excitation of stable auroral red arcs: Simultaneous HF radar, photometer and Alouette I observations.**
D. S. Lund, R. D. Hunsucker.
Journ. Atmosph. Terr. Phys., Vol. 33, 1177 - 1183 (1971).

084.009 **The E × B instability in auroral arcs.**
E. B. Hooper, Jr., J. C. G. Walker.
Journ. Atmosph. Terr. Phys., Vol. 33, 1441 - 1455 (1971).

084.010 **A Monte Carlo analysis of the passage of auroral X-rays through the atmosphere.**
G. R. Pilkington, C. D. Anger.
Planet. Space Sci., Vol. 19, 1069 - 1085 (1971).
The deep penetration of auroral X-rays into the atmosphere has been examined by means of a Monte Carlo calculation in order to determine the effects of atmospheric scattering and absorption on spectra observed at balloon altitudes.

084.011 **Erratum:'Lifetimes and heights of O(^1S) in pulsating aurorae.** [Planet. Space Sci., Vol. 19, 119 - 124 (1971)]. J. G. Moore.
Planet. Space Sci., Vol. 19, 1199 (1971).

084.012 **Correlation between electron and proton fluxes in postbreakup aurora.** A. D. Johnstone.
Journ. Geophys. Res., Vol. 76, 5259 - 5267 (1971).
Simultaneous measurements of the electron and proton spectrum between 2 and 30 kev in the recovery phase of a large auroral substorm are reported. Increases in the proton intensity are correlated with increases in the electron intensity.

084.013 **Relationship between magnetospheric electric fields and the motion of auroral forms.**
M. C. Kelley, J. A. Starr, F. S. Mozer.
Journ. Geophys. Res., Vol. 76, 5269 - 5277 (1971).
Twenty-two hours of balloon-measured electric fields obtained on five different nights are compared with simultaneous all-sky camera pictures of auroral forms for the purpose of interpreting the large-scale motions of visible auroras.

084.014 **The auroral X-ray radiation in the midnight sector in the period of the storm of March 8, 1970.**
J. Gasset, I. A. Zhulin, F. Chambou, Kh. D. Kanonidi, Yu. M. Kopylov, L. Saint-Marc, J.-P. Treilhou.
Geomagn. Aeronom., Vol. 11, 637 - 642 (1971). In Russian.

084.015 **Simultaneous observation of proton precipitation and auroral radar echoes.** T. Hagfors, R. G. Johnson, R. A. Power.
Journ. Geophys. Res., Vol. 76, 6093 - 6098 (1971).

084.016 **Simultaneous spectrographic triangulation of an SAR arc and a hydrogen arc.**
R. J. Hoch, C. R. Batishko, K. C. Clark.
Journ. Geophys. Res., Vol. 76, 6185 - 6189 (1971).
An "Stable Auroral Red" arc at about $L = 2.8$ and a hydrogen (H) arc between $L = 3.4$ and $L = 4.0$ were monitored from Richland and Banff throughout the night of March 23–24, 1969. The measurements show the SAR arc and H arc to be clearly separated by about 5° of latitude and apparently quite uncorrelated in behavior.

084.017 **Correction: 'Auroral phenomena driven by the magnetospheric plasma'.** [Journ. Geophys. Res., Vol. 76, 272 - 277 (1971)]. H. E. Taylor, F. W. Perkins.
Journ. Geophys. Res., Vol. 76, 6210 (1971).

084.018 **The influence of thermospheric winds on the auroral red-line profile of atomic oxygen.**
P. B. Hays, S. K. Atreya.
Planet. Space Sci., Vol. 19, 1225 - 1228 (1971).

084.019 **Electrodynamical effects of interaction of charged particle streams with the atmosphere. II.**
N. K. Osipov, V. G. Pivovarov.
Geomagn. Aeronom., Vol. 11, 845 - 852 (1971).
In Russian.

084.020 **The development of instability in the earth's auroral plasma and the escape of particles.**
I. N. Men'shutina, M. I. Pudovkin.
Kosmich. Issled., Vol. 9, 729 - 735 (1971). In Russian.

084.021 **New mechanism of auroral arcs.**
T. Ogawa, T. Sato.
Planet. Space Sci., Vol. 19, 1393 - 1412 (1971).
A new mechanism of quiet auroral arcs is presented in which the leading part of arc formation is played by the ionosphere. A model is numerically examined for a number of parameter sets to find that the ionospheric irregularity can

certainly result in auroral arcs. The calculated results are compared with recent experiments of auroral arcs. Deformation of an arc is also discussed in terms of the cross-field instability.

084.022 **Theoretical N_2 vibrational distribution in an aurora.**
R. W. Schunk, P. B. Hays.
Planet. Space Sci., Vol. 19, 1457 - 1461 (1971).
The N_2 vibrational distribution in an aurora is investigated. During the auroral bombardment the vibrational distribution is non-Boltzmann.

084.023 **Comments on paper by L. R. Megill et al., 'Oxygen atmospheric and infrared atmospheric bands in the aurora'** [Journ. Geophys. Res., Vol. 75, 4775 - 4785 (1970)].
W. F. J. Evans, E. J. Llewellyn.
Journ. Geophys. Res., Vol. 76, 7013 - 7017, with a reply by L. R. Megill, A. M. Despain, D. J. Baker, K. D. Baker, p. 7018 - 7019 (1971).

084.024 **Auroral luminosity profile in the far ultraviolet.**
W. Landensperger.
Space Research XI, (see 012.004), 1195 - 1197 (1971).

084.025 **Deactivation of $N_2 A\ ^3\Sigma_u{}^+$ molecules in the aurora.**
D. E. Shemansky, E. C. Zipf, T. M. Donahue.
Planet. Space Sci., Vol. 19, 1669 - 1683 (1971).
Recent rocket observations of the N_2 V–K (Vegard–Kaplan) system in the aurora have been reinterpreted using an atmospheric model based on mass spectrometer measurements in an aurora of similar intensity at the same time of year. Below 130 km we find that it is possible to account for the deactivation in bright auroras by invoking large nitric oxide concentrations, similar to those recently observed mass spectrometrically and using a rate constant of 8×10^{-11} cm^3 sec^{-1} for both the $\nu = 1$ levels.

084.026 **Calculated and observed features of stable auroral red arcs during three geomagnetic storms.**
R. G. Roble, R. B. Norton, J. A. Findlay, E. Marovich.
Journ. Geophys. Res., Vol. 76, 7648 - 7662 (1971).
Satellite electron temperature and density data are used to calculate the structure of several stable auroral red arcs (SAR arcs) according to the thermal conduction model of the arc. There is general agreement between the calculated and observed $\lambda 6300$ emission features for SAR arcs observed during the geomagnetic storm periods of October 29 to November 2, 1968, May 14–15, 1969, and March 8–9, 1970.

084.027 **$\lambda 5577$ [O I] and $\lambda 4278\ N_2{}^+$ emissions in a SAR arc.**
R. J. Hoch, L. L. Smith, K. C. Clark.
Journ. Geophys. Res., Vol. 76, 7663 - 7668 (1971).
Observations with scanning photometers at the Rattlesnake Mountain Observatory, Richland, Washington, show that at 1950–1953 PST the bright SAR arc of March 8–9, 1970, was definitely not exclusively emissions of atomic oxygen at $\lambda\lambda 6300$–6364 Å. In this intense SAR arc (3.5–4.0 kR at $\lambda 6300$) emissions at $\lambda 5577$ and $\lambda 4278$ were also recorded.

084.028 **Coordinated auroral-electron observations from a synchronous and a polar satellite.**
R. D. Sharp, D. L. Carr, R. G. Johnson, E. G. Shelley.
Journ. Geophys. Res., Vol. 76, 7669 - 7682 (1971).

084.029 **Latitude and time variations in precipitated electron energy inferred from measurements of auroral heights.** J. S. Boyd, A. E. Belon, G. J. Romick.
Journ. Geophys. Res., Vol. 76, 7694 - 7700 (1971).

084.030 **Rocket-based measurement of particle fluxes and currents in an auroral arc.**
R. R. Vondrak, H. R. Anderson, R. J. Spiger.

Journ. Geophys. Res., Vol. 76, 7701 - 7713 (1971).

084.031 **Rocket-based measurement of Birkeland currents related to an auroral arc and electrojet.**
R. J. Park, P. A. Cloutier.
Journ. Geophys. Res., Vol. 76, 7714 - 7733 (1971).

084.032 **Emissions of atomic oxygen at 5577 and 6300 Å during the aurora of July 9/10, 1970.**
Yu. L. Truttse, V. G. Sobolev.
Astron. Tsirk., No. 650, p. 4 - 6 (1971). In Russian.

084.033 **The relationship of the plasmasphere and the stable auroral red arcs in the magnetic storm of October 29 to November 7, 1968.**
C. R. Chappell, K. K. Harris, G. W. Sharp.
Astrophys. Space Sci. Library, Vol. 24, (see 012.021), 73 - 80 (1971).

084.034 **Recent observations of red arcs and mid-latitude aurora.** H. Lauche.
Astrophys. Space Sci. Library, Vol. 24, (see 012.021), 81 (1971).

084.035 **Auroral morphology.**
C. S. Deehr, A. Egeland, F. Søraas.
Astrophys. Space Sci. Library, Vol. 24, (see 012.021), 125 - 149 (1971).

084.036 **Auroral excitation and ionization intensity and temperature variations.** G. G. Shepherd.
Astrophys. Space Sci. Library, Vol. 24, (see 012.021), 150 - 159 (1971).

084.037 **Auroral conjugacy and time-dependent geometry of auroras.**
T. N. Davis, T. J. Hallinan, H. C. Stenbaek-Nielsen.
Astrophys. Space Sci. Library, Vol. 24, (see 012.021), 160 - 169 (1971).

084.038 **Equatorial modulation in a pulsating aurora.**
D. A. Bryant, G. M. Courtier, G. Bennett.
Astrophys. Space Sci. Library, Vol. 24, (see 012.021), 170 - 175 (1971).

084.039 **Infrared spectrum of aurora.**
A. V. Jones, R. L. Gattinger.
Astrophys. Space Sci. Library, Vol. 24, (see 012.021), 176 - 184 (1971).

084.040 **Infrared investigations of aurora and airglow.**
A. T. Stair, Jr., E. R. Huppi, B. P. Sandford, R. E. Murphy, R. R. O'Neil, A. M. Hart, R. J. Huppi, W. R. Pendleton.
Astrophys. Space Sci. Library, Vol. 24, (see 012.021), 185 - 204 (1971).

084.041 **Basic processes and cross sections in proton auroras.**
R. J. McNeal, J. H. Birely.
Astrophys. Space Sci. Library, Vol. 24, (see 012.021), 217 - 222 (1971).

084.042 **Particle precipitation patterns.** T. R. Hartz.
Astrophys. Space Sci. Library, Vol. 24, (see 012.021), 225 - 238 (1971).

084.043 **Low energy auroral particle measurements from polar satellites.** R. D. Sharp, R. G. Johnson.
Astrophys. Space Sci. Library, Vol. 24, (see 012.021), 239 - 254 (1971).

084.044 **Auroral precipitation patterns.**
R. H. Eather, S. B. Mende.
Astrophys. Space Sci. Library, Vol. 24, (see 012.021), 255 - 266 (1971).

084.045 **Direct observation of temporal and spatial structure in auroral electrons.**   D. S. Evans.
Astrophys. Space Sci. Library, Vol. 24, (see 012.021), 267 - 278 (1971).

084.046 **Preliminary report on the observation of low energy ions in the aurora.**
B. A. Whalen, J. R. Miller, I. B. McDiarmid.
Astrophys. Space Sci. Library, Vol. 24, (see 012.021), 279 - 287 (1971).

084.047 **X-rays associated with auroras.**   H. Trefall.
Astrophys. Space Sci. Library, Vol. 24, (see 012.021), 288 - 298 (1971).

084.048 **Survey of radio reflections from aurora.**
A. G. McNamara.
Astrophys. Space Sci. Library, Vol. 24, (see 012.021), 301 - 313 (1971).

084.049 **Optical and radio observations of aurora.**
P. Czechowsky, G. Lange-Hesse.
Astrophys. Space Sci. Library, Vol. 24, (see 012.021), 314 - 326 (1971).

084.050 **Progress in ELF and VLF emission studies at high latitudes.**   T. S. Jørgensen.
Astrophys. Space Sci. Library, Vol. 24, (see 012.021), 327 - 335 (1971).

084.051 **VLF radio emissions and the aurora.**
A. R. W. Hughes, T. R. Kaiser.
Astrophys. Space Sci. Library, Vol. 24, (see 012.021), 336 - 344 (1971).

084.052 **Rocket, ground and satellite measurements of ELF emissions during a PCA.**
J. Holtet, A. Egeland, N. C. Maynard.
Astrophys. Space Sci. Library, Vol. 24, (see 012.021), 345 - 354 (1971).

084.053 **The proton aurora and auroral substorm.**
L. E. Montbriand.
Astrophys. Space Sci. Library, Vol. 24, (see 012.021), 366 - 373 (1971).

084.054 **Auroral infrasonic substorms.**   C. R. Wilson.
Astrophys. Space Sci. Library, Vol. 24, (see 012.021), 374 - 383 (1971).

084.055 **Dynamics of auroral formation.**   A. Hasegawa.
Astrophys. Space Sci. Library, Vol. 24, (see 012.021), 384 - 393 (1971).

084.056 **Physical processes involving striation formation in barium clouds.**   H. Völk, G. Haerendel.
Astrophys. Space Sci. Library, Vol. 24, (see 012.021), 394 - 406 (1971).

084.057 **Field aligned continuity of Hall current electrojets and other consequences of density gradients in the auroral ionosphere.**
J. P. Heppner, J. D. Stolarik, E. M. Wescott.
Astrophys. Space Sci. Library, Vol. 24, (see 012.021), 407 - 426 (1971).

084.058 **Poleward expansion of the auroral oval and associated phenomena in the magnetotail during auroral substorms, 2.**
E. W. Hones, Jr., S.-I. Akasofu, S. J. Bame, S. Singer.
Journ. Geophys. Res., Vol. 76, 8241 - 8257 (1971).

084.059 **Measurements of highly collimated short-duration bursts of auroral electrons and comparison with existing auroral models.**   B. J. O'Brien, D. L. Reasoner.
Journ. Geophys. Res., Vol. 76, 8258 - 8278 (1971).

084.060 **Field-aligned particle currents near an auroral arc.**
L. W. Choy, R. L. Arnoldy, W. Potter, P. Kintner, L. J. Cahill, Jr.
Journ. Geophys. Res., Vol. 76, 8279 - 8298 (1971).

084.061 **Vibrational population of the $A^3\Sigma_u^+$ and $B^3\Pi_g$ states of N_2 in normal auroras.**
D. C. Cartwright, S. Trajmar, W. Williams.
Journ. Geophys. Res., Vol. 76, 8368 - 8377 (1971).

084.062 **Midday auroral observations in the oval, cusp region, and polar cap.**   G. J. Romick, N. B. Brown.
Journ. Geophys. Res., Vol. 76, 8420 - 8424 (1971).
Brief report.

084.063 **Penetration of auroral particles into the earth's atmosphere.**   L. L. Lazutin.
Proc. Sixth Winter School on Space Physics, Part I. Apatity 1969, (see 012.033), p. 269 - 286 (1971).

The optical aurora.   See Abstr. 003.100.

Correlation between aurorae and solar radio bursts at 185 MHz.   See Abstr. 077.014.

Equatorial and precipitating solar protons in the magnetosphere. 1. Low-energy diurnal variations.
See Abstr. 078.006.

Geomagnetic Field

084.201 **Penetration of solar protons over the polar cap during the February 25, 1969, event.**
J. Engelmann, R. J. Hynds, G. Morfill, F. Axisa, A. Bewick, A. C. Durney, L. Koch.
Journ. Geophys. Res., Vol. 76, 4245 - 4261 (1971).

During the February 25, 1969, solar proton event, particle measurements from the close-orbiting satellite Esro 2, and particle and magnetic field measurements outside the magnetosphere from the Heos A1 satellite are available. We discuss these observations and show that the auroral zone high-energy 'peaks' are caused by a different mechanism than that responsible for the low-energy enhancements referred to above. We also discuss the high-latitude increases and their possible cause in terms of coupling between polar cap lines of force and the interplanetary magnetic field.

084.202 **Asymmetric access of energetic solar protons to the earth's north and south polar caps.**
J. A. Van Allen, J. F. Fennell, N. F. Ness.
Journ. Geophys. Res., Vol. 76, 4262 - 4275 (1971).

During the energetic solar particle event that began on January 24, 1969, the ratio N/S of the intensity of protons $E_p > 0.3$ Mev over the earth's north (N) polar cap to that over its south (S) polar cap varied from a value greater than 20 to about 1, as observed with satellite Injun 5 in a low-altitude polar orbit. The interplanetary intensity of protons was measured simultaneously with similar detectors on Explorer 33 and 35. The results are much clearer than earlier ones by virtue of the extraordinarily strong anisotropy of interplanetary particle intensity early in this event and the comprehensiveness of the observations.

084.203 **Solar particle cutoffs as observed at low altitudes.**
W. L. Imhof, J. B. Reagan, E. E. Gaines.
Journ. Geophys. Res., Vol. 76, 4276 - 4290 (1971).

The purpose of this paper is to present a series of satellite-borne measurements on the low-latitude cutoff boundaries of solar particles for several solar events during the 1969–1970 period.

084.204 **Correlated observations of electrons and magnetic fields at the earth's bow shock.**
M. Neugebauer, C. T. Russell, J. V. Olson.
Journ. Geophys. Res., Vol. 76, 4366 - 4380 (1971).

084.205 **Magnetotail changes in relation to the solar wind magnetic field and magnetospheric substorms.**
M. P. Aubry, R. L. McPherron.
Journ. Geophys. Res., Vol. 76, 4381 - 4401 (1971).

In this paper we investigate whether the magnetotail changes occur only as a consequence of substorms or also as a direct consequence of changes in the solar wind parameters. Using data from several satellites (Ogo 5, ATS 1, Imp 4, Explorer 33 and 35) and 17 ground magnetic observatories, we conclude that the tail responds to both changes in the north-south orientation of the interplanetary field and substorm activity.

084.206 **Time variations of the magnetotail plasma sheet at 18 R_E determined from concurrent observations by a pair of Vela satellites.** E. W. Hones, Jr., J. R. Asbridge, S. J. Bame.
Journ. Geophys. Res., Vol. 76, 4402 - 4419 (1971).

084.207 **Relativistic electron precipitation during magnetic storm main phase.** R. M. Thorne, C. F. Kennel.
Journ. Geophys. Res., Vol. 76, 4446 - 4453 (1971).

084.208 **Geomagnetic dynamos in a stable core.**
E. Bullard, D. Gubbins.
Nature, Vol. 232, 548 - 549 (1971).

084.209 **Some processes in the magnetosphere.**
V. P. Shabansky.
Space Sci. Rev., Vol. 12, 299 - 418 (1971).

This paper is a continuation of the previous review (Shabansky, 1968). It considers the problems related to the processes of the interaction of charged particles with the geomagnetic field, and also contains the original results obtained by the author.

084.210 **Polar substorm and interplanetary magnetic field.**
S. Kokubun.
Planet. Space Sci., Vol. 19, 697 - 714 (1971).

In the present paper, relationships between the development of a polar magnetic substorm and the interplanetary magnetic field are examined in detail, using magnetic field data obtained with the IMP 3 satellite. One of the purposes of the analysis is to discuss the characteristics of geomagnetic variations before the onset of a strong auroral electrojet.

084.211 **Satellite observations of magnetospheric radio noise–I. Emissions between the plasma and upper hybrid resonance frequencies.** P. C. Gregory.
Planet. Space Sci., Vol. 19, 813 - 820 (1971).

084.212 **Comparison of geomagnetic storms and trough development at solar activity maximum and minimum.** D. D. Woodbridge.
Planet. Space Sci., Vol. 19, 821 - 826 (1971).

084.213 **A constant-pressure, tail-like extrusion of plasma into the confined magnetic field of a line dipole.**
R. C. Hewson-Browne, P. C. Kendall.
Planet. Space Sci., Vol. 19, 869 - 876 (1971).

084.214 **A comment on trajectory calculations in the Williams and Mead geomagnetic field model.**
G. E. Morfill.
Planet. Space Sci., Vol. 19, 1016 - 1018 (1971). – Research note.

084.215 **Solar and lunar hydromagnetic tides in the earth's magnetosphere.** H. Maeda.
Journ. Atmosph. Terr. Phys., Vol. 33, 1135 - 1146 (1971).

The purpose of this paper is to estimate in some detail the solar and lunar hydromagnetic tides in the magnetosphere, and to discuss their effects on the distribution of the plasma density.

084.216 **The influence of the plasma layer on the proper oscillations of the magnetosphere's tail.**
A. A. Nusinov.
Kosmich. Issled., Vol. 9, 615 - 617 (1971). In Russian. Brief information.

084.217 **Calculation of the finite conductivity of the magnetosphere's plasma.** V. M. Barsukov.
Kosmich. Issled., Vol. 9, 617 - 620 (1971). In Russian. Brief information.

084.218 **Separation of lunar daily geomagnetic variations into parts of oceanic and ionospheric origin in the Indian region.** N. S. Sastri, D. R. K. Rao.
Geophys. Journ. Roy. Astron. Soc., Vol. 23, 269 - 272 (1971).

084.219 **An empirical determination of aerodynamic factors for the magnetosphere.**
J. P. Schieldge, G. L. Siscoe.

Cosmic Electrodynamics, Vol. 2, 141 - 163 (1971).

The empirical drag coefficient, the north-south component of the lift due to the inclination of the dipole axis to the solar wind direction, and the empirical shape of the magnetosphere have been determined from satellite data associated with the five geomagnetically quiet days of each month for the first half of 1967. Twenty-eight days of Fourier analyzed ATS-1 magnetic field data and the averages of thirteen coincident days of Explorer 33 solar wind data were used in the study.

084.220 **Characteristics of storm-time geomagnetic daily variation.** R. P. Kane.
Journ. Atmosph. Terr. Phys., Vol. 33, 1585 - 1595 (1971).

084.221 **Variations in the high latitude proton trapping boundary associated with polar magnetic substorms.**
H. R. Lindalen, F. Søraas, K. Aarsnes, R. Amundsen.
Planet. Space Sci., Vol. 19, 1041 - 1048 (1971).

Protons in the energy range (100–200) keV in the noon-midnight meridian plane are studied from data obtained by the ESRO IA satellite. The results are summarized in three statements.

084.222 **Expansion of an ion cloud in the earth's magnetic field.** W. G. Pilipp.
Planet. Space Sci., Vol. 19, 1095 - 1119 (1971).

The expansion of an ion cloud perpendicular to a homogeneous magnetic field is especially considered with regard to the physical conditions of the HEOS cloud experiment carried out in March 1969 by the Institute of Extraterrestrial Physics. The results of the numerical calculations are given and the basic assumptions, as well as the results of the calculations, are discussed with respect to the HEOS cloud experiment.

084.223 **Locating the magnetospheric ring current.**
R. M. Thorne, C. F. Kennel.
Comments Astrophys. Space Phys., Vol. 3, 115 - 120 (1971).

It is the purpose of this note to point out that 10–100 keV protons, which dominate ring current energetics have two preferred regions of cyclotron instability, and consequently loss, which serve as stable trapping boundaries for ring current protons. The pertinent electromagnetic ion cyclotron instability therefore limits the ring current location, and, using the empirical notion that within the stable region the energy density does not exceed the geomagnetic energy density, also the ring current energy density, and therefore, the main phase depression.

084.224 **The uniform electrical conductivity model for the earth and the geomagnetic disturbance daily variation.** H. F. Petersons.
Australian Journ. Phys., Vol. 24, 593 - 600 (1971).

Chapman's uniform conducting core model has been investigated using S_q and SD results for the IGY. Values for the conductivity and core radius could not be found for which there was agreement between the S_q and SD results at the likely 95% confidence limit of S_q, and consequently either the uniform core model is inadequate for representing S_q and SD or the errors in the analysis of S_q and SD are larger than estimated statistically.

084.225 **On the equilibrium of the magnetopause current layer.** S.-Y. Su, B. U. Ö. Sonnerup.
Journ. Geophys. Res., Vol. 76, 5181 - 5188 (1971).

A numerical study was made of the confinement of a uniform magnetic field by a warm plasma with a net velocity parallel to the confined field. Parker's assumptions of a vanishing electric field and vanishing trapped-particle currents and pressures were adopted. For a given proton temperature equi-

libriums were found whenever the electron temperature exceeded a certain critical value. The properties of the equilibrium solutions are discussed. Finally, the equilibrium confinement with the incident streaming plasma tangential to the magnetopause but perpendicular to the confined field is treated.

084.226 **Plasma in the earth's polar magnetosphere.**
L. A. Frank.
Journ. Geophys. Res., Vol. 76, 5202 - 5219 (1971).

First observations of the plasmas in the dayside polar magnetosphere were obtained with the earth-satellite IMP 5 during July-August 1969. The results are interpreted in terms of a proposed magnetospheric model that identifies the geomagnetic field lines threading the polar cusp of plasmas found in the vicinity of the neutral point with the field lines in the distant plasma sheet via a convection of field lines from polar cusp to plasma sheet.

084.227 **Evidence for a normal geomagnetic field polarity event at 263 ± 5 m.y. BP within the Late Palaeozoic reversed interval.**
K. M. Creer, J. G. Mitchell, D. A. Valencio.
Nature, Phys. Sci., Vol. 233, 87 - 89 (1971).

Basalts, radiometrically dated at 263 ± 5 m.y., have stable remanent magnetization with normal polarity. This demonstrates the existence of a previously unrecognized event within the Late Palaeozoic interval of reversed geomagnetic polarity.

084.228 **On excitation of ionosonic waves in the magnetosphere.** V. M. Chmirev.
Geomagn. Aeronom., Vol. 11, 643 - 646 (1971). In Russian.

084.229 **On the physical nature of the 11-year cyclic variation of magnetic disturbances.** A. I. Ol.
Geomagn. Aeronom., Vol. 11, 647 - 650 (1971). In Russian.

084.230 **On annual variations of the geomagnetic field in the polar caps.** S. M. Mansurov, L. G. Mansurova.
Geomagn. Aeronom., Vol. 11, 660 - 664 (1971). In Russian.

084.231 **High-altitude pulsations of the geomagnetic field and their use to diagnose the parameters of the magnetosphere.**
O. M. Raspopov, S. A. Chernous, B. V. Kiselev.
Geomagn. Aeronom., Vol. 11, 669 - 673 (1971). In Russian.

084.232 **The morphology of the 60-year variations of the geomagnetic field in Europe.**
V. P. Golovkov, G. I. Kolomiitzeva.
Geomagn. Aeronom., Vol. 11, 674 - 678 (1971). In Russian.

084.233 **A numerical solution of the equations of the dynamo theory of terrestrial magnetism. I. The method.** E. P. Kropachev.
Geomagn. Aeronom., Vol. 11, 690 - 696 (1971) In Russian.

084.234 **The magnetic field of vertical currents in the magnetosphere.** N. M. Rudneva.
Geomagn. Aeronom., Vol. 11, 731 - 733 (1971). In Russian.
Brief information.

084.235 **Modelling of rapid particle penetration through neutral points of the magnetosphere.**
E. M. Dubinin, G. G. Managadze, I. M. Podgorny.
Geomagn. Aeronom., Vol. 11, 733 - 734 (1971). In Russian.
Brief information.

084.236 **Multiple crossings of the earth's bow shock at large geocentric distances.** B. Bavassano, F. Mariani, U. Villante, N. F. Ness.

Journ. Geophys. Res., Vol. 76, 5970 - 5977 (1971).

A series of very distant crossings of the earth's bow shock has been identified by the magnetic field experiment on board Pioneer 8 at geocentric distances between 120 and 200 R_E, during the period December 19 to 25, 1967. The experimental observations are presented and compared with theoretical predictions before the results are discussed.

084.237 Magnetospheric convection pattern inferred from magnetic disturbance and auroral motions.
T. N. Davis.
Journ. Geophys. Res., Vol. 76, 5978 - 5984 (1971).

084.238 Association of magnetotail phenomena with visible auroral features. S.-I. Akasofu, E. W. Hones, Jr., M. D. Montgomery, S. J. Bame, S. Singer.
Journ. Geophys. Res., Vol. 76, 5985 - 6003 (1971).

084.239 Plasma-sheet structure and the onset of magnetospheric substorms. T. W. Hill, A. J. Dessler.
Planet. Space Sci., Vol. 19, 1275 - 1288 (1971).

In this paper we have presented a self-consistent model that describes the role of the plasma sheet in causing the onset of the magnetospheric substorm. The source of the plasma sheet in this model is magnetosheath plasma entering the dayside magnetopause along the demarcation lines separating open and closed field lines.

084.240 Geomagnetic storm fields near a synchronous satellite. K. Kawasaki, S.-I. Akasofu.
Planet. Space Sci., Vol. 19, 1339 - 1347 (1971).

An apparent early recovery of the main phase of geomagnetic storms at the distance of the synchronous satellite is examined in terms of changing electric current distributions in the magnetosphere during magnetic storms.

084.241 A model of a magnetospheric substorm
V. B. Lyatsky, Yu. P. Maltzev.
Geomagn. Aeronom., Vol. 11, 839 - 844 (1971). In Russian.

084.242 AU- and AL-indices on international quiet days and variations of the geomagnetic field in high latitudes. R. G. Afonina, Ya. I. Fel'dshtein.
Geomagn. Aeronom., Vol. 11, 860 - 865 (1971). In Russian.

084.243 On form and character of the variation of energy spectra of anomalous magnetic and gravitational fields. V. N. Lugovenko, S. A. Serkerov.
Geomagn. Aeronom., Vol. 11, 871 - 874 (1971). In Russian.

084.244 Numerical solution of the equations of the dynamo theory of terrestrial magnetism. II.
E. P. Kropachev.
Geomagn. Aeronom., Vol. 11, 875 - 882 (1971). In Russian.

084.245 Geomagnetic effects of breaks and the models of the deformation of the magnetosphere.
N. V. Mikerina, K. G. Ivanov.
Geomagn. Aeronom., Vol. 11, 920 - 922 (1971). In Russian. Brief information.

084.246 On the possibility of generation of long-period oscillations at the boundaries of magnetospheric convective streams. Yu. P. Maltzev.
Geomagn. Aeronom., Vol. 11, 925 - 927 (1971). In Russian. Brief information.

084.247 Spherical analyses of the geomagnetic field by
angular data and the extrapolated value g_1^0. II.
S. I. Braginsky, N. V. Kulanin.
Geomagn. Aeronom., Vol. 11, 931 - 933 (1971). In Russian. Brief information.

084.248 Palaeomagnetism of Suriname dolerites.
J. Veldkamp, F. G. Mulder, J. D. A. Zijderveld.
Phys. Earth Planet. Interiors, Vol. 4, 370 - 380 (1971).

084.249 Projection of the meridional boundary of the magnetosphere for different orientation of the geomagnetic dipole.
A. E. Antonova, E. G. Eroshenko, Ya. I. Fel'dshtejn.
Kosmich. Issled., Vol. 9, 786 - 787 (1971). In Russian. Brief information.

084.250 The patterns and sources of high-latitude particle precipitation. G. A. Paulikas.
Rev. Geophys. Space Phys., Vol. 9, 659 - 701 (1971).

We survey, in this paper, the properties of the zones of particle precipitation at high latitudes. We consider the average properties of particle influx from the outer magnetosphere as determined by satellite and rocket observations, as well as ground-based techniques. The point of view of this paper is to try to associate source regions, and, if possible, precipitation mechanisms with each of the families of precipitating particles that collectively form the auroral zone. We consider the plasma sheet, the extraterrestrial ring current, the polar cusp, and the outer radiation belt as source regions of precipitating particles. Brief surveys of the precipitation of helium ions and of possible means of artificially inducing particle precipitation are also included.

084.251 Plasma instabilities in the magnetosphere.
A. Hasegawa.
Rev. Geophys. Space Phys., Vol. 9, 703 - 772 (1971).

An introductory review of theories of plasma instabilities in the magnetosphere is presented. Part A is a review of theories of plasma instabilities that are relevant to magnetospheric plasmas. Instabilities arising from velocity-distribution anisotropies, such as a pitch-angle anisotropy or the presence of beams, as well as instabilities from nonuniform distributions of plasmas and magnetic fields, are discussed. Part B is a summary of works related to actual plasma instabilities in the magnetosphere. In view of the observed plasma parameters, it is shown that the magnetosphere is rather stable against most macroscopic instabilities, and its dynamics are predominantly governed by microscopic instabilities.

084.252 Plasma-sheet ions at lunar distance preceding substorm onset. H. B. Garrett, T. W. Hill, M. A. Fenner.
Planet. Space Sci., Vol. 19, 1413 - 1418 (1971).

In this paper we report the measurement of enhanced ion fluxes observed at a distance of 60 R_E in the geomagnetic tail, preceding the onset of magnetospheric substorms. These ions are tentatively identified as plasma-sheet ions that have escaped from the near-earth region and are observed at the lunar surface as they flow down the tail away from the earth.

084.253 Observations of earth's bow shock for low Mach numbers. V. Formisano, P. C. Hedgecock, G. Moreno, J. Sear, D. Bollea.
Planet. Space Sci., Vol. 19, 1519 - 1531 (1971).

Four anomalous earth's bow shock locations observed during 1969 are studied, using plasma and magnetic field data from the European satellite HEOS-1. For low magnetosonic Mach numbers ($1 < M_{MS} \lesssim 3$), the magnetosheath magnetic field is characterized by the unusual absence of low frequency turbulence. This experimental result seems to indicate that under such conditions the bow shock is a laminar shock.

084.254 Average and unusual locations of the earth's magnetopause and bow shock. D. H. Fairfield.
Journ. Geophys. Res., Vol. 76, 6700 - 6716 (1971).

A best-fit ellipse and hyperbola have been calculated to represent several hundred magnetopause and bow-shock positions observed by six Imp spacecraft. The aim has been to establish an accurate two-dimensional representation of the average position and shape of the magnetopause and bow shock and to investigate the magnitude and the causes of the variations from the average.

084.255 Magnetopause structure during the magnetic storm of September 24, 1961. B. U. Ö. Sonnerup.
Journ. Geophys. Res., Vol. 76, 6717 - 6735 (1971).

It is the purpose of this report to present the magnetometer data from a set of Explorer-12 magnetopause penetrations during the inbound pass of September 24, 1961, when a moderate magnetic storm was in progress, and to propose an interpretation of these data in terms of the so-called open magnetosphere model (Dungey, 1961).

084.256 Magnetopause attitudes during OGO 5 crossings. B. G. Ledley.
Journ. Geophys. Res., Vol. 76, 6736 - 6742 (1971).

The attitude of the magnetopause current layer is determined for 31 OGO 5 crossings located at subsatellite local times between 0430 and 1020 and solar magnetospheric latitudes between $-8.5°$ and $+40°$.

084.257 OGO 5 observations of the polar cusp on November 1, 1968.
C. T. Russell, C. R. Chappell, M. D. Montgomery, M. Neugebauer, F. L. Scarf.
Journ. Geophys. Res., Vol. 76, 6743 - 6764 (1971).

During the large magnetic storm of November 1, 1968, the OGO 5 spacecraft encountered the apparent direct penetration of magnetosheath plasma into the dayside magnetosphere at magnetic latitudes as low as $43°$. Because this region of magnetosheath plasma occurred on magnetospheric field lines immediately adjacent to the zone of trapped energetic particles, it is interpreted to be the polar cusp.

084.258 Magnetospheric substorms on September 14, 1968.
E. W. Hones, Jr., R. H. Karas, L. J. Lanzerotti, S.-I. Akasofu.
Journ. Geophys. Res., Vol. 76, 6765 - 6780 (1971).

Several moderate magnetospheric substorms on September 14, 1968, were observed by two satellites (Vela 4 A and ATS 1), balloon-borne X-ray detectors, and extensive arrays of ground-based instruments in Alaska. Combined results from those observations and their interpretations are described in detail. Several interesting aspects of magnetospheric substorms are suggested by the analysis.

084.259 Storm sudden commencements and polar magnetic substorms.
K. Kawasaki, S.-I. Akasofu, F. Yasuhara, C.-I. Meng.
Journ. Geophys. Res., Vol. 76, 6781 - 6789 (1971).

The purpose of this paper is to show that a simple compression of the magnetosphere alone may sometimes trigger substorms.

084.260 Structured variations of the plasmapause: Evidence of a corotating plasma tail.
H. A. Taylor, Jr., J. M. Grebowsky, W. J. Walsh.
Journ. Geophys. Res., Vol. 76, 6806 - 6814 (1971).

We examine one of the more complete sets of high-resolution proton distributions analyzed to date, obtained from OGO 4 during September 1967. A time-dependent analysis of the plasmapause motion using a simple plasma convection model will provide evidence that a plasma tail may be formed by the tendency of plasma associated with the storm-distorted duskside plasmasphere bulge to corotate with the earth during a period of quieting magnetic activity.

084.261 Observations of striation formation in a barium ion cloud. N. W. Rosenberg.
Journ. Geophys. Res., Vol. 76, 6856 - 6864 (1971).

We have completed an analysis of the optical data from a release of a barium ion cloud in which the onset and development of striations have been more clearly recorded than in other releases studied by our laboratory or reported elsewhere. The results of the analysis are presented here for evaluation of instability characteristics.

084.262 Diurnal variation of the geomagnetic field. 1. The solar variation. J. S. Jackson.
Journ. Geophys. Res., Vol. 76, 6896 - 6908 (1971).

The dynamo theory of the magnetic variations of the earth's atmosphere is treated as a three-dimensional problem, considering the finite thickness of the ionospheric-current layer. The anisotropy of the conductivity of the ionosphere is also considered.

084.263 Diurnal variation of the geomagnetic field. 2. The lunar variation. J. S. Jackson.
Journ. Geophys. Res., Vol. 76, 6909 - 6914 (1971).

The dynamo theory of the lunar magnetic variations is developed as a three-dimensional problem. The finite thickness of the ionospheric-current layer and the anisotropy of the conductivity are considered. The currents responsible for the lunar magnetic variations are assumed to flow in the same layer of the ionosphere as those responsible for the solar magnetic variations, i. e., from 100 to 120 km above the earth's surface.

084.264 Estimate of the dependence of the magnetospheric electric field on the velocity of the solar wind.
M. Mendillo, M. D. Papagiannis.
Journ. Geophys. Res., Vol. 76, 6939 - 6943 (1971).

A model is developed that predicts that the magnetospheric dawn-dusk electric field varies as the square of the solar-wind speed V. This relation is used to compute the radius of the plasmapause for given V at 1800 LT, where the dawn-dusk electric field cancels the electric field due to the rotation of the earth.

084.265 Localized character of magnetotail magnetic fluctuations during polar magnetic substorms.
G. Rostoker, F. P. Camidge.
Journ. Geophys. Res., Vol. 76, 6944 - 6951 (1971).

The magnetic fluctuations observed in the magnetotail in conjunction with polar magnetic substorms are discussed, and evidence is presented that only a confined portion of the magnetotail is affected during any given phase of a substorm.

084.266 Impulsive energetic electron fluxes in the distant magnetotail associated with the onset of magnetospheric substorms.
S.-I. Akasofu, E. W. Hones, Jr., S. Singer.
Journ. Geophys. Res., Vol. 76, 6976 - 6979 (1971). – Letter.

084.267 Frequency tables of the geomagnetic index Kp, 1932 - 1970. E. J. Zawalick, A. L. Cage.
Journ. Geophys. Res., Vol. 76, 7009 - 7012 (1971). – Letter.

084.268 Acceleration of electrons and protons in the magnetosphere during magnetic storms.
S. N. Vernov, S. N. Kuznetsov, E. N. Sosnovets, L. V. Tverskaya, M. V. Teltsov, O. V. Khorosheva.
Space Research XI, (see 012.004), 1253 - 1258 (1971).

084.269 Low-energy (12–48 keV) electrons and protons in the magnetosphere at solar activity maximum.
D. R. Parsignault, R. C. Chase, L. Katz.
Space Research XI, (see 012.004), 1265 - 1275 (1971).

084.270 An experimental study of palaeosecular variation.
A. Brock.
Geophys. Journ. Roy. Astron. Soc., Vol. 24, 303 - 317 (1971).
A total of 83 selected palaeomagnetic results are used in a latitude analysis of palaeosecular variation to show that Cox's 1970 model for secular variation fits the data, provided the dipole wobble is between $1\,^1/_2\,°$ and $2°$.

084.271 Stürme in der Magnetosphäre.
G. Kremser, J. W. Münch.
Umschau, 71. Jahrgang, p. 888 - 891 = Weltraumfahrt, 22. Jahrgang, p. 154 - 157 (1971).

084.272 Morphology of magnetic disturbance.
T. Nagata, N. Fukushima.
Encyclopedia of physics, Vol. 49/3, (see 003.013), 5 - 130 (1971).

084.273 Theoretical aspects of the worldwide magnetic storm phenomenon.   E. N. Parker, V. C. A. Ferraro.
Encyclopedia of physics, Vol. 49/3, (see 003.013), 131 - 205 (1971).

084.274 Maßzahlen der erdmagnetischen Aktivität.
M. Siebert.
Encyclopedia of physics, Vol. 49/3, (see 003.013), 206 - 275 (1971).

084.275 Classical methods of geomagnetic observations.
V. Laursen, J. Olsen.
Encyclopedia of physics, Vol. 49/3, (see 003.013), 276 - 322 (1971).

084.276 Neuere Meßmethoden der Geomagnetik.
H. Schmidt, V. Auster.
Encyclopedia of physics, Vol. 49/3, (see 003.013), 323 - 383 (1971).

084.277 Phénomènes T.B.F. d'origine magnétosphérique.
R. Gendrin.
Encyclopedia of physics, Vol. 49/3, (see 003.013), 461 - 525 (1971).

084.278 Magnetic fluxes in the magnetosphere's tail at different phases of a substorm.   J. I. Feldstein.
Dokl. Akad. Nauk SSSR. Ser. Mat. Fiz., Vol. 201, 82 - 85 (1971). In Russian.

084.279 The radial diffusion of trapped particles induced by fluctuating magnetospheric fields.   M. Walt.
Space Sci. Rev., Vol. 12, 446 - 485 (1971).
Small fluctuations in magnetospheric electric and magnetic fields lead to random changes in the radial positions of trapped particles. The characteristics of this radial diffusion are described theoretically in terms of the statistical properties of the field fluctuations, in particular the power spectra of the various spatial components. This radial diffusion has a major influence on the structure of the radiation belts; and since the net flow of particles is inward at most positions of the magnetosphere, the process acts as a strong source of trapped particles.

084.280 Structure of the magnetopause.   D. M. Willis.
Rev. Geophys. Space Phys., Vol. 9, 953 - 985 (1971).
This paper reviews present understanding of the internal structure of the thin boundary layer, termed the magnetopause, that separates the distorted geomagnetic field in the magnetosphere from the flow of solar plasma in the magnetosheath. The fundamental theoretical concepts of the subject are introduced by considering the structure of the boundary layer that exists when a cold, unmagnetized stream of ions and electrons impinges normally on a vacuum magnetic field. This idealized model indicates how the earth's magnetic field is confined by the impact pressure of the solar wind, the geomagnetic field being terminated by induced shielding currents flowing in the magnetopause. The various idealizations and approximations in this model, such as the assumption of a cold solar plasma and the neglect of the interplanetary magnetic field, are examined critically.

084.281 Mathematical models of the open magnetosphere: Application to dayside auroras.
T. G. Forbes, T. W. Speiser.
Journ. Geophys. Res., Vol. 76, 7542 - 7551 (1971).
Two static mathematical models of the open or Dungey model of the magnetosphere are constructed. If the models presented here are applicable to the physical magnetosphere, then the magnetosphere is continuously open, and reconnection is maximized for a due south interplanetary magnetic field. The first model indicates that on the average the neutral point associated with magnetic merging will be located on the MCS in longitude at 1100 to 1000 hours geomagnetic local time, a result that may explain the 1100 to 1000 hours alignment of polar-cap auroras and the recently discovered polar cusps. The second model has been used to show that a small component of the interior magnetospheric field can be expected to depend on the polarity of the interplanetary field, especially when a southward component of the interplanetary field exists. (MCS = magnetopause current sheet).

084.282 Magnetospheric-field distortions observed by OGO 3 and 5.
M. Sugiura, B. G. Ledley, T. L. Skillman, J. P. Heppner.
Journ. Geophys. Res., Vol. 76, 7552 - 7565 (1971).

084.283 Geomagnetic storm particles in the high-latitude magnetotail.   S. J. Bame, E. W. Hones, Jr., S.-I. Akasofu, M. D. Montgomery, J. R. Asbridge.
Journ. Geophys. Res., Vol. 76, 7566 - 7583 (1971).

084.284 Magnetospheric substorms in the distant magnetotail observed by IMP 3.
C.-I. Meng, S.-I. Akasofu, E. W. Hones, Jr., K. Kawasaki.
Journ. Geophys. Res., Vol. 76, 7584 - 7594 (1971).

084.285 The dayside of the plasmasphere.
C. R. Chappell, K. K. Harris, G. W. Sharp.
Journ. Geophys. Res., Vol. 76, 7632 - 7647 (1971).

084.286 Pioneer 8 plasma-wave measurements at distant bow-shock crossings.   F. L. Scarf.
Journ. Geophys. Res., Vol. 76, 7769 - 7777 (1971). – Brief report.

084.287 Magnetic results 1962, 1963 and 1964 (Hartland).
R. v. d. R. Woolley.
Roy. Obs. Bull., Herstmonceux – Cape of Good Hope, No. 153, [London: Her Majesty's Stationery Office], p. D173 - D346. Price £ 1.50 (1971).
The present bulletin is the last in the series of Royal Observatory Bulletins, Series D. In future Hartland magnetic results will be published under the auspices of the Institute of Geological Sciences.

084.288 Magnetic results 1968, Eskdalemuir, Hartland and

Lerwick Observatories.
Natural Environment Research Council, Inst. Geol. Sci., Geomagnetic Bull. No. 2, [London: Her Majesty's Stationery Office], 5 + 161 pp. Price £ 2.50 (1970).

084.289 Current layers and acceleration of particles in space.
S. I. Syrovatskij.
Trudy Mezhdunar. seminara po probl."Generatsiya kosmich. luchej na Solntse",1970. Moskva, 1971, p. 15 - 36. In Russian. Abstr. in Referativ. Zhurn. 62. Issled. kosm. prostranstva, 12.62.351 (1971).

084.290 Space-time correlations at running of breaks on to the magnetosphere. K. G. Ivanov.
Geomagn. Aeronom., Vol. 11, 1026 - 1031 (1971). In Russian.

084.291 Rotational break in the solar wind and magnetic disturbance on the earth.
K. G. Ivanov, N. V. Mikerina.
Geomagn. Aeronom., Vol. 11, 1032 - 1037 (1971). In Russian.

084.292 Shock wave in the magnetosphere's tail as a reason of the active phase of a magnetospheric substorm.
V. B. Latski, Yu. P. Maltsev.
Geomagn. Aeronom., Vol. 11, 1038 - 1043 (1971). In Russian.

084.293 Magnetohydrodynamic waves and magnetic variations. Study of fluctuations at the time of geomagnetic storms. A. Best, G.-R. Leman.
Geomagn. Aeronom., Vol. 11, 1044 - 1052 (1971). In Russian.

084.294 Spectral characteristics of the earth's and the interplanetary magnetic field in the frequency band
$1 - 10^{-4}$ Hz. V. G. Dubrovskiy, S. A. Kramarenko.
Geomagn. Aeronom., Vol. 11, 1053 - 1060 (1971). In Russian.

084.295 Polar magnetic substorms. R. Boström.
Astrophys. Space Sci. Library, Vol. 24, (see 012.021), 357 - 365 (1971).

084.296 Anomalous pliocene paleomagnetic pole positions from Baja California.
D. W. Strangway, B. E. McMahon, T. R. Walker, E. E. Larson.
Earth Planet. Sci., Letters, Vol. 13, 161 - 166 (1971).
In a study of modern red sediments and the way in which they become magnetized, it has been possible to determine a pliocene paleomagnetic pole position from Baja California, Mexico.

084.297 Evidence that the Laschamp polarity event did not occur 13 300–30 400 years ago.
C. R. Denham, A. Cox.
Earth Planet. Sci. Letters, Vol. 13, 181 - 190 (1971).
No evidence for a geomagnetic reversal between 30 400 and 13 300 yr ago has been found in a paleomagnetic study of sediments from Mono Lake, California, with ages controlled by ^{14}C dating. If the Laschamp reversed event occurred during this time, its duration can have been no longer than 1700 yr.

084.298 On the stability of an electron beam injected along the geomagnetic field.
Ju. K. Alekhin, V. I. Karpman, D. D. Rjutov, R. Z. Sagdeev.
Cosmic Electrodynamics, Vol. 2, 280 - 291, 292 - 304 (1971). In English and Russian.
The dynamics of an electron beam injected into the magnetospheric plasma is investigated. The beam structure is considered theoretically and the dispersion in space of the beam is given. Instabilities of the beam are considered near the point of injection both for cold and warm cases. Kinetic instabilities of the beam are also investigated and the conditions of stability and instability are given for both high and low altitudes.

084.299 Deformation of the duskside plasmapause.
A. Nishida.
Cosmic Electrodynamics, Vol. 2, 340 - 349 (1971).
The sharp westward edge of the plasmapause bulge frequently detected by the whistler method is examined by using the geomagnetic data in the dusk sector as a measure of the magnetospheric convection activity in the bulge region. It is found that the detection of the above feature is invariably associated with the increase in the northward geomagnetic disturbance field on the duskside.

084.300 Interplanetary origin of electric fields in the magnetosphere. A. Nishida.
Cosmic Electrodynamics, Vol. 2, 350 - 374 (1971).
Correlation analysis between the activity of polar magnetic disturbances and the condition of the interplanetary magnetic field is summarized and the origin of the magnetospheric electric fields responsible for the disturbances is discussed.

084.301 Algunas características y períodos evidenciados en el W. de Europa para la variación secular de la dirección geomagnética. C. Gaibar-Puertas.
Urania Barcelona, Año 56, No. 273, p. 77 - 109 (1971).
A series of values of the geomagnetic declination in San Fernando (Spain) for the centuries XV to XIX has been deduced from several sources of indirect information. The distribution of these values is good enough to allow a determination of the main events of the secular variation, such as the transit of the agonic line, the maximum elongation E. and W., etc.

084.302 Physical structure of hydromagnetic disturbances in the inner magnetosheath.
R. L. Kaufmann, J.-T. Horng.
Journ. Geophys. Res., Vol. 76, 8189 - 8198 (1971).
The purpose of the present work is to study deviations from the average bahavior in the inner magnetosheath. Magnetometer data are plotted in a manner that helps us to visualize the magnetic-field distortions. The resulting picture will make it easier to study the other smaller-amplitude waves that are also seen in the magnetosheath and to interpret time variations of particle fluxes in this region.

084.303 Correlated changes between the daily variation of geomagnetic field and interplanetary plasma parameters. R. P. Kane.
Journ. Geophys. Res., Vol. 76, 8199 - 8210 (1971).

084.304 Adiabatic particle orbits in a magnetic null sheet.
B. U. Ö. Sonnerup.
Journ. Geophys. Res., Vol. 76, 8211 - 8222 (1971).

084.305 Catalogue of geomagnetic disturbance indices for the period 1873–1910 (Petersburg and Pavlovsk observations). I. D. Zosimovich, D. A. Andrienko.
Problems of cosmic physics. Vyp. (No.) 6, (see 003.040), p. 168 - 192 (1971). In Russian.

084.306 On the inflation of the geomagnetic field during a magnetic storm. V. C. A. Ferraro.
Geophys. Journ. Roy. Astron. Soc., Vol. 25, 419 - 423 (1971).
A more accurate discussion than in an earlier paper is given of the inflation of the geomagnetic field during the main phase of a magnetic storm. The radial velocity of the injected plasma is found to be of the order of 1 km s^{-1}.

084.307 Interaction of energetic particles with the neutral sheet of the geomagnetic tail.
I. I. Alekseev, A. P. Kropotkin.
Proc. 11th international conference on cosmic rays, Vol. 2, (see 012.030), 521 - 524 (1970).

084.308   A short lived increase in the charged particle flux at the equator during the magnetic storm of October 31, 1968.   R. K. Sood, C. A. Green, H. Elliot.
Proc. 11th international conference on cosmic rays, Vol. 2, (see 012.030), 657 - 662 (1970).

084.309   Conjugate conductivity asymmetry and the precipitation of magnetospheric electrons.
J. R. Catchpoole.
Australian Journ. Phys., Vol. 24, 881 - 887 (1971).
Using a simplified model of the atmosphere, a calculation is made of the wind-induced vertical redistribution of ions derived from the 100 km level. For various values of the parameters involved, an estimate is made of the consequent precipitation to ionospheric heights of previously trapped electrons of the magnetosphere.

084.310   The solar and lunar daily geomagnetic variation of declination at Trevandrum, 1853–1869.
S. Chapman, J. C. Gupta.
Pure Applied Geophys. (PAGEOPH), Vol. 87, 1971/IV, p. 1 - 9 = Contr. Earth Phys. Branch, Ottawa, Canada, No. 323 (1971).
The eye readings of easterly declination at Trevandrum observatory have been analysed to determine solar and lunar daily variations. Hourly readings during February 1853 to February 1865 are analysed by Chapman-Miller method. A brief discussion of the results is also given.

084.311   The sunspot cycle influence on the solar and lunar daily geomagnetic variations.
S. Chapman, J. C. Gupta, S. R. C. Malin.
Proc. Roy. Soc. London, Ser. A, Vol. 324, 1 - 15 = Contr. Earth Phys. Branch, Ottawa, Canada, No. 317 (1971).
The relative importance of the sunspot cycle influence on solar and lunar daily geomagnetic variations has been discussed for over a century. It is shown that the sunspot cycle influence on the lunar variations, measured by the Wolf ratio, is significant, but is only about one third of the sunspot cycle influence on the solar variations.

084.312   The response of the magnetosphere to sunspots as indicated by geomagnetic storminess levels.
E. J. Chernosky.
Radio Sci., Vol. 6, 579 - 582 (1971).
The level of geomagnetic activity that occurs with developing sunspots has been examined by using 38 years of daily geomagnetic Ap activity indices and sunspot number Rz. It is found that the Ap values between 70 and 160 are characteristically associated with sunspots that develop rapidly and reach a peak number about three days before the occurrence of these levels of Ap.

084.313   Record of observations at St. John's magnetic observatory 1968-69.   G. A. Brown.
Publ. Earth Phys. Branch, Dep. of Energy, Mines and Resources, Ottawa, Canada, Vol. 40, (No. 9), 535 - 613 (1971).

084.314   Record of observations at Victoria magnetic observatory 1969.   D. R. Auld, I. W. Fetterley.
Publ. Earth Phys. Branch, Dep. of Energy, Mines and Resources, Ottawa, Canada, Vol. 41, (No. 6), 77 - 133 (1971).

084.315   Magnetic substorms, December 5, 1968.
E. I. Loomer, G. Jansen van Beek.
Publ. Earth Phys. Branch, Dep. of Energy, Mines and Resources, Ottawa, Canada, Vol. 41, (No. 10), 179 - 198 (1971).

084.316   Canadian national report on geomagnetism and aeronomy.   C. M. Carmichael, T. R. Hartz.
Publ. Earth Phys. Branch, Dep. of Energy, Mines and Re-

sources, Ottawa, Canada, Vol. 41, (No. 11), 199 - 236 (1971).

084.317   On the relation between the cyclotron and Langmuir frequencies of electrons in the earth's magnetosphere.   K. G. Ivanov, A. D. Shevnin.
Kosm. Issled., Vol. 9, 873 - 877 (1971). In Russian.

084.318   Magnetospheric electric fields and the rotation of the earth.   B. E. Bryunelli.
Proc. Sixth Winter School on Space Physics, Part I. Apatity 1969, (see 012.033), p. 287 - 298 (1971).

084.319   Magnetic storm development in the earth's magnetosphere.
M. I. Pudovkin, S. I. Isaev, S. A. Zaitseva.
Proc. Sixth Winter School on Space Physics, Part I. Apatity 1969, (see 012.033), p. 299 - 309 (1971).

084.320   Interaction of high-energy particles with the neutral layer of the magnetosphere tail.
I. I. Alekseev, A. P. Kropotkin.
Proc. Sixth Winter School on Space Physics, Part I. Apatity 1969, (see 012.033), p. 328 - 329 (1971).

084.321   Comparative analysis of different geomagnetic-field models.   V. A. El'tekov, V. F. Nikitin.
Proc. Sixth Winter School on Space Physics, Part II. Apatity 1969, (see 012.034), p. 92 (1971). – Abstract.

Model of a closed magnetosphere.
See Abstr. 003.029.

Ratio of specific heats in the solar-wind plasma flow through the earth's bow shock. See Abstr. 074.086.

Equatorial and precipitating solar protons in the magnetosphere. 2. Riometer observations.
See Abstr. 078.007.

The entry of solar protons over the polar caps.
See Abstr. 078.012.

The entry of solar protons into the geomagnetic field during the event of June 9, 1968.
See Abstr. 078.055.

Electric coupling between the magnetosphere and the ionosphere as a cause of polar magnetic disturbances and auroral break-up.   See Abstr. 083.014.

Dynamical behavior of thermal protons in the mid-latitude ionosphere and magnetosphere.
See Abstr. 083.017.

A statistical model of geophysical processes with asymmetry and excess of the density probability function.
See Abstr. 083.032.

Comet brightness variations and outer space conditions.   See Abstr. 102.002.

Dynamic processes in comets and their analogy with the dynamics of the earth's magnetosphere.
See Abstr. 102.030.

Further studies of the effect of the earth's magnetic field on meteor trains.   See Abstr. 104.051.

The interplanetary medium and its interaction with the earth's magnetosphere.   See Abstr. 106.003.

Signature in the interplanetary medium for substorms. See Abstr. 106.010.

Relationship of interplanetary parameters and occurrence of magnetospheric substorms. See Abstr. 106.017.

Reconnecting of force lines of the interplanetary

magnetic field and the geomagnetic tail.
See Abstr. 106.028.

The geomagnetic cut-off at Ft. Churchill and the primary cosmic ray electron spectrum from 10 MeV to 12 GeV in 1968. See Abstr. 143.124.

Radiation Belts

084.401 Harnessing the energy in the radiation belts.
N. Brice.
Journ. Geophys. Res., Vol. 76, 4698 - 4701 (1971). – Letter.

084.402 Composition of fast charged particles in the inner radiation belt. E. V. Gorchakov, M. V. Ternovskaya.
Kosmich. Issled., Vol. 9, 622 - 624 (1971). In Russian.
Brief information.

084.403 Nonadiabatic behavior of radiation-belt particles.
H. E. Taylor, R. J. Hastie.
Cosmic Electrodynamics, Vol. 2, 211 - 223 (1971).
This paper is concerned with calculating the rigidity (momentum per unit charge) at which the nonadiabatic changes become significant as a function of magnetic field line. This is first done without approximation by numerically integrating the equation of motion of the particle, both in a dipole field and in a more realistic magnetospheric field model, to obtain the magnitude of the jumps in μ ('magnetic moment') which are then compared with analytic estimates. Next heuristic arguments are used to obtain an empirical expression which serves as an upper bound for the adiabatic region. Finally, the results are compared with experimental measurements in the real magnetosphere.

084.404 Comments on 'Radial diffusion of outer-zone electrons'. M. Walt, L. L. Newkirk.
Journ. Geophys. Res., Vol. 76, 5368 - 5370, with a reply by L. J. Lanzerotti, C. G. Maclennan, M. Schulz, p. 5371 - 5373 (1971). – Letter.

084.405 A connection between some parameters of the outer radiation belt and the interplanetary medium.
S. N. Kuznetzov, V. G. Stolpovsky.
Geomagn. Aeronom., Vol. 11, 703 - 704 (1971). In Russian.
Brief information.

084.406 Flux profile of protons measured in 1967 at the equator between $L = 1.15$ and 1.21.
E. Achtermann, G. Paschmann, D. Hovestadt.
Journ. Geophys. Res., Vol. 76, 6004 - 6013 (1971).

084.407 Motion of energetic particles in a magnetospheric model including a convection electric field.
J. C. Kosik.
Planet.Space Sci., Vol. 19, 1209 - 1214 (1971).

084.408 Dynamics of the belt of DR-currents and mean latitude red arcs. S. A. Zaitzeva, M. I. Pudovkin, V. V. Dryakhlov, V. N. Diyachenko.
Geomagn. Aeronom., Vol. 11, 853 - 859 (1971). In Russian.

084.409 Equatorial proton and electron angular distributions in the loss cone and at large angles.
F. H. Bogott, F. S. Mozer.
Journ. Geophys. Res., Vol. 76, 6790 - 6805 (1971).
The present study of data from the Berkeley experiment on ATS 5 was motivated by a desire to make coordinated measurements of both proton and electron pitch-angle distributions at lower energies than those previously measured. Distributions near the loss cone around local midnight are examined, at quiet times and during substorm disturbances. Data are presented for 8 days in August and September 1969 when the detector sampled a wide range of angles. Both diurnal effects and some substorm variations are studied for these days.

084.410 Observations of low-energy (0.3- to 1.8-Mev) differential spectrums of trapped protons.
D. Venkatesan, S. M. Krimigis.
Journ. Geophys. Res.,Vol. 76, 7618 - 7631 (1971).

084.411 The earth's radiation belts. S. N. Vernov.
Proc. 11th international conference on cosmic rays, Budapest 1969, (see 012.025), p. 85 - 162 (1970).

084.412 Source and loss processes of protons of the inner radiation belt. T. A. Farley, M. Walt.
Journ. Geophys. Res., Vol. 76, 8223 - 8240 (1971).

084.413 Long-period variations of the spectra of protons in the inner radiation zone.
O. I. Savun, P. I. Shavrin.
Kosm. Issled., Vol. 9, 885 - 889 (1971). In Russian.

084.414 Investigation of the spectra of electrons in the inner radiation zone aboard Cosmos 219.
O. I. Savun, P. I. Shavrin.
Kosm. Issled., Vol. 9, 890 - 895 (1971). In Russian.

084.415 The trapping of charged particles by the magnetic field with allowance for ionization losses.
D. Kh. Morozov.
Proc. Sixth Winter School on Space Physics, Part I. Apatity 1969, (see 012.033), p. 249 - 252 (1971).

084.416 Plasma containment in adiabatic traps.
B. B. Kadomtsev.
Proc. Sixth Winter School on Space Physics, Part I. Apatity 1969, (see 012.033), p. 319 - 320 (1971).

084.417 Distribution of the outer belt protons in the μ, L space. Yu. I. Gubar', V. P. Shabanskii.
Proc. Sixth Winter School on Space Physics, Part I. Apatity 1969, (see 012.033), p. 321 - 322 (1971).

084.418 Comparison of the variations of high-energy elec-

trons and ultra-low-frequency radiation.
A. B. Malyshev, P. M. Svidskii, V. A. Burov, I. G. Kleimenova,
V. A. Troitskaya.
Proc. Sixth Winter School on Space Physics, Part I. Apatity
1969, (see 012.033), p. 326 (1971).

084.419 Application of electrical fields in interplanetary
    space as radiation shielding against protons and
electrons.   T. Ya. Ryabova, K. A. Trukhanov.
Proc. Sixth Winter School on Space Physics, Part I. Apatity
1969, (see 012.033), p. 331 - 332 (1971).

085   Solar-Terrestrial Relations

085.001 Relation of sunspot and earthquake activity.
    J. Gribbin.
Science, Vol. 173, 558 (1971).

085.002 Neues über solar-terrestrische Beziehungen.
    W. Carnuth.
SuW, Vol. 10, 196 (1971).

085.003 Relazioni tra fenomeni solari e terrestri in biologia.
    A. Giordano.
Coelum, Vol. 39, 125 - 133 (1971).

085.004 Verband tussen zonnevlekkenactiviteit en meteo-
    rologische grootheden?   H. J. Manie.
Hemel en Dampkring, Vol. 69, 189 - 192 (1971).

085.005 Geophysical coordinate transformations.
    C. T. Russell.
Cosmic Electrodynamics, Vol. 2, 184 - 196 (1971).
    The common coordinate systems in use in solar-terres-
trial relationships are described and then the transformation
matrices required to convert vectors in one system to another
are derived.

085.006 Solar activity and phenomena in the biosphere.
    M. N. Gnevyshev, K. F. Novikova.
Zemlya i Vselennaya, No. 4, p. 33 - 36 (1971). In Russian.

085.007 Inter-seasonal breaks of synoptic processes and
    geomagnetic activity.   M. G. Gubanova.
Solnechnye Dannye 1971 Byull., No. 6, p. 97 - 105 (1971).
In Russian.
    It is shown that the relations of geomagnetic activity and
the start points of synoptic seasons reveal themselves best of
all for cold seasons. The character of these relations involves
lengthening of winter seasons and shortening of intermediate
seasons during the years with heightened geomagnetic activity.

085.008 Corpuscular fluxes and mesospheric temperatures
    during periods of elevated solar activity.
G. A. Kokin, Yu. N. Moiseev, V. F. Tulinov, I. G. Shapiro.
Space Research XI, (see 012.004), 923 - 928 (1971).

085.009 On the relationship between solar activity and the
    barometric pressure field of the earth's northern
hemisphere.   I. V. Maximov, B. A. Sleptsov-Shevlevich.
Dokl. Akad. Nauk SSSR. Ser. Mat. Fiz., Vol. 201, 339 - 341
(1971). In Russian.

085.010 Some peculiarities of relations between seasonal
    repetitions of circulation types and solar activity.
M. G. Gubanova.
Solnechnye Dannye 1971 Byull., No. 9, p. 108 - 112 (1971).
In Russian.

085.011 Variations of the direct radiation intensity and solar
    activity.   V. F. Loginov, G. N. Martjanova.
Solnechnye Dannye 1971 Byull., No. 10, p. 106 - 108 (1971).
In Russian.
    The relation between the intensity of visible direct solar
radiation, the solar radio flux at λ = 10.7 cm and the index of
geomagnetic activity Ap for 1961−1963 is studied. It is shown
that variations of the direct radiation are connected apparently
with solar activity and the frequency of appearing meteors.

085.012 On a mechanism of solar-tropospheric relations.
    B. I. Sazonov.
Solnechnye Dannye 1971 Byull., No. 10, p. 108 - 111 (1971).
In Russian.

085.013 Localization of active solar regions responsible for
    the generation of the geoeffective agent.
V. P. Nefed'ev.
Proc. Sixth Winter School on Space Physics, Part I. Apatity
1969, (see 012.033), p. 330 (1971).

085.014 The influence of even and odd 11-year solar cycles
    on the rate of growth of oak-trees.   S. I. Kostin.
Solnechnye Dannye 1971 Byull., No. 8, p. 111 - 114 (1971).
In Russian.

    The magnetic structure of interplanetary space.
See Abstr. 106.031.

Planetary System

091 Physics of the Planetary System (Planetary Atmospheres, Figure, Interior, Magnetic Fields, Rotation, etc.)

091.001 On the perturbations of the five outer planets by the four inner ones. J. S. Griffith.
Celestial Mechanics, Vol. 3, 478 - 490 (1971).

The expressions given by Clemence were checked by the comparison of numerical values of their second differentials with the numerical values of the perturbing forces. Agreement was good in most cases, save that the use of the second differentials unduly magnified some terms rightly neglected by Clemence. In the appendix the work of Clemence is compared with that of Carpenter and the differences are found to be all smaller than the contributions from the secular variation of the eccentricity, which are so small that they were neglected by Clemence.

091.002 The greenhouse effect in a gray planetary atmosphere. An exact solution for the emergent radiation. D. W. N. Stibbs.
Astrophys. Journ., Vol. 168, 155 - 160 (1971).

The principles of invariance are applied to the generation and scattering of thermal radiation in a semi-infinite plane-parallel atmosphere with gray absorption and parallel insolation. The exact solution for the emergent thermal radiation is derived, and the conditions for limb brightening, limb darkening and the greenhouse effect are deduced.

091.003 Comments on the rate of evaporation of a non-Maxwellian atmosphere.
J. W. Chamberlain, G. R. Smith.
Planet. Space Sci., Vol. 19, 675 - 684 (1971).

In recent Monte-Carlo calculations by Chamberlain and Campbell (Astrophys. Journ., Vol. 149, 687 - 705, 1967), and by Chamberlain (Astrophys. Journ., Vol. 155, 711 - 714, 1969) on a light gas escaping from a planetary atmosphere, the exospheric density gradients were incorrect because of a conceptual error in the analysis. The calculated escape fluxes and the velocity distributions were correct, however. Here we explain the density error and show that the earlier method for computing escape fluxes was equivalent to the more rigorous technique used by Brinkmann.

091.004 Charged particle induced ionization rates in planetary atmospheres. J. Dubach, W. A. Barker.
Journ. Atmosph. Terr. Phys., Vol. 33, 1287 - 1288 (1971).

We have derived a non-relativistic formula for calculating the rate at which charged particles from a solar flare produce ion pairs in planetary atmospheres. Our method is simpler to use than the elaborate expression developed by Velinov.

091.005 Solar activity and planetary luminosity.
V. K. Balasubrahmanyan, D. Venkatesan.
Solar Physics, Vol. 19, 257 - 263 (1971).

The Chree superposition analysis of the luminosities of the planets Jupiter, Saturn, Uranus and Neptune indicates a correlation between solar activity and planetary luminosity. The variations of the solar constant in the visible range are considered to be too small to explain the observed changes in brightness. The interaction of solar extreme ultraviolet or solar wind particles with the atmospheres of these planets is probably responsible for the increased albedo during periods of high solar activity.

091.006 Solar-planetary relationships. D. Venkatesan.
Journ. Roy. Astron. Soc. Canada, Vol. 65, 174 - 175 (1971). – Abstr. RAS Canada.

091.007 The importance of methane to pressure-induced absorption in the atmospheres of the outer planets.
K. Fox, I. Ozier.
Journ. Roy. Astron. Soc. Canada, Vol. 65, 182 (1971).
Abstr. RAS Canada.

091.008 Identical frequency correlations of two irrotational force fields of a nearly spherical planet.
M. A. Khan.
Geophys. Journ. Roy. Astron. Soc., Vol. 23, 411 - 415 (1971).

Formulas for comparing the identical frequencies of two irrotational force fields of a nearly spherical planet are given. Such comparisons are extremely useful in several geophysical studies.

091.009 Atmospheric oscillations–III. M. N. Jones.
Planet. Space Sci., Vol. 19, 1359 - 1385 (1971).

The theory of the oscillations of a planetary atmosphere, developed in earlier papers, is extended to the case of a perturbation arising from Hall conductivity. The horizontal modes of semi-diurnal oscillation are computed for a uniform Hall region of a rotating atmosphere, and are interpreted in terms of an isothermal model of the E-region.

091.010 Ein zehnter Planet zwischen Merkur und Sonne? H. Heuseler.
SuW, Vol. 10, 275 (1971).

091.011 Occultations as planetary probes. R. T. Brinkmann.
Bull. American Astron. Soc., Vol. 3, 374 (1971). – Abstr. AAS.

091.012 A radiative transfer model for planetary atmospheres. R. D. Cess.
Journ. Quant. Spectrosc. Radiat. Transfer, Vol. 11, 1699 - 1710 (1971).

The method of matched asymptotic expansions is applied to the analysis of infrared radiative energy transfer within a planetary atmosphere. The goal is to illustrate, by means of a simple mathematical model, qualitative features of atmospheric thermal structure. The matched asymptotic expansion is employed in the upper portion of the atmosphere, and the analysis illustrates that the temperature distribution in the lower portion of the atmosphere is independent of both concentration of absorbing gas and rotational line structure of the vibrational-rotation bands.

091.013 Planetary temperatures at 9.55 mm wavelength. R. W. Hobbs, S. L. Knapp.
Icarus, Vol. 14, 204 - 209 (1971).

Measurements of Mercury, Mars, Venus, Jupiter, and Saturn have been made with the 85-ft reflector of the Naval Research Laboratory at 9.55 mm wavelength. The measurements were made in the interval August 1966 to April 1969. The gain of the antenna at the zenith is calibrated using several radio sources with easily extrapolated spectra; the change of gain with antenna position is calibrated using measurements of Jupiter. Average temperatures of $260 \pm 18°K$, $207 \pm 13°K$, $441 \pm 30°K$, $157 \pm 8°K$, and $126 \pm 6°K$ are found for Mercury, Mars, Venus, Jupiter, and Saturn, respectively.

091.014 The meaning of Russell's law. J. Veverka.
Icarus, Vol. 14, 284 - 285 (1971).

Using a two point Gaussian quadrature, it is shown that Russell's law is a direct consequence of the fact that in general a planet's brightness decreases rapidly with increasing phase angle.

091.015 A simple technique for calculating line profiles of inhomogeneous planetary atmospheres.
R. A. Stokes, W. C. DeMarcus.
Icarus, Vol. 14, 307 - 311 (1971).

The computation of synthetic spectra for planetary atmospheres in which multiple scattering is important usually requires lengthy numerical work. We obtain solutions to the appropriate equation of radiative transfer by use of a variational principle whose extremum is essentially the reflectivity. When considering light diffusely reflected from the atmospheres of the outer planets, the technique reduces the computational work enormously and is applicable to models in which the albedo varies strongly with optical depth as well.

091.016 Laboratory simulation of absorption spectra in cloudy atmospheres.
D. J. McCleese, J. S. Margolis, G. E. Hunt.
Nature, Phys. Sci., Vol. 233, 102 - 103 (1971).

For the first time measurements in the multiple scattering regime of the diffuse reflectivity as a function of single scattering albedo have been made in a geometry which may be simulated by a plane parallel atmosphere of large optical depth. A comparison between the measurements and a theoretical computation of the diffuse reflectivity is presented. The measurements are within 1% agreement with the theoretical calculations for two different sizes of scattering particles which are larger than and smaller than the wavelength of the incident light, corresponding to the Mie and Rayleigh regimes respectively.

091.017 Solar-wind-induced mass loss from magnetic field-free planets. F. C. Michel.
Planet. Space Sci., Vol. 19, 1580 - 1583 (1971).

The point of this note is to estimate how much atmospheric matter can be carried away by the solar wind. We will restrict our discussion to those planets whose atmosphere is not shielded by a strong magnetic field, as is the earth's, from direct interaction with the solar wind.

091.018 Some problems of atmospheric physics for terrestrial planets. A. M. Oboukhov, G. S. Golitsyn.
Space Research XI, (see 012.004), 121 - 128 (1971).

091.019 On the integral for backscattering from a randomly rough surface. A. K. Fung, H. L. Chan.
Proc. IEEE, Vol. 59, 1280 - 1281 (1971).

To study the angular behavior of the usual backscattering integral, a numerical example is given and the integral is evaluated without approximating the surface correlation function. Results obtained are compared with approximate analytic evaluations. It is shown that either one of the two possible approximations may be acceptable, depending upon the incident frequency.

091.020 Interpretation of the light curves of some non-atmospheric bodies in the solar system. K. Lumme.
Astrophys. Space Sci., Vol. 13, 219 - 230 (1971).

A model to explain the phase curves of some non-atmospheric bodies in the solar system has been developed. It is found that, for instance, phase curves of the moon, the Galilean satellites of Jupiter, some asteroids and Saturn's rings are in agreement with the given model.

091.021 On methods for determining the smoothness factor for planetary surfaces. N. P. Barabashov.
Vestn. Khar'kov. Univ., No. 65, (Ser. Astron., No. 6), p. 13 - 19 (1971). In Russian.

091.022 The photolysis of CO_2 at 1849 and 2139 Å.
D. C. Krezenski, R. Simonaitis, J. Heicklen.
Planet. Space Sci., Vol. 19, 1701 - 1703 (1971).

The principal atmospheric component of both Mars and Venus is CO_2. Therefore the photochemistry of this molecule is important in the atmospheres of these planets. We have now examined the photodissociation of CO_2 at wavelengths with insufficient energy ($\lambda > 1658$ Å) to produce $O(^1D)$ atoms and have found that at 25°C photodissociation occurs.

091.023 Die Planetenmonde und ihre Bewegung.
G. Schimek.
Sternenbote, 14. Jahrgang, p. 110 - 118 (1971).

091.024 Natural water on other planets (1).
W. F. Dörpholz, G. N. Katterfeld.
Urania Kraków, Vol. 42, 200 - 207 (1971). In Polish.

091.025 A semiempirical model for the mean transmission of a molecular band an application to the 10μ and 16μ bands of NH_3. L. Trafton.
Icarus, Vol. 15, 27 - 38 (1971).

A new method for constructing accurate semiempirical band models for the mean transmission is presented and applied to the 10μ and 16μ bands of NH_3. The assumptions and approximations that are commonly made are discussed and new ones proposed.

091.026 Particle formation during water-vapor photolysis.
I. D. Clark, J. F. Noxon.
Science, Vol. 174, 941 - 944 (1971).

Micrometer-sized particles have been produced by the photolysis of water vapor at wavelengths between 1500 and 1700 angstroms. Although the composition of the particles was not determined, the rate of particle production is consistent with the efficient conversion of water photolysis products to particulate matter. Such a process may be of importance in planetary atmospheres.

091.027 Internal constitution and thermal histories of the terrestrial planets. B. Levin (Invited speaker).
Highlights of Astronomy, Vol. 2, (see 012.014), 204 - 227 (1971).

091.028 Internal constitution of the giant planets.
W. De Marcus (Invited speaker).
Highlights of Astronomy, Vol. 2, (see 012.014), 228 - 238 (1971).

091.029 Composition and structure of planetary atmospheres.
D. M. Hunten.
Space Sci. Rev., Vol. 12, 539 - 599 (1971).

091.030 Geological evolution of planets. K. Beneš.
Vesmír, Vol. 50, 363 - 366 (1971). In Czech.

091.031 Satellites of the outer planets: Their physical and

chemical nature.   J. S. Lewis.
Icarus, Vol. 15, 174 - 185 (1971).

Steady-state thermal models for the icy satellites are constructed in which the energy released by radioactive decay in the interiors of the satellites is exactly balanced by the net radiative loss from their surfaces. It is shown that the Galilean satellites of Jupiter and the larger satellites of Saturn, Uranus, and Neptune very likely have extensively melted interiors, and most probably contain a core of hydrous silicates, an extensive mantle of ammonia-rich liquid water, and a relatively thin crust of ices. Consequences of this model relating to the Galilean satellites and the rings of Saturn are briefly described. The atmospheric compositions and densities of the large icy satellites and certain features of the retention of volatiles during accretion are discussed.

091.032   Photometric color indices of extrasolar planets.
          G. L. Matloff.
Icarus, Vol. 15, 341 - 342 (1971).

Color indices in U−B and B−V of planets resembling the earth, Mars, Jupiter, and Venus are calculated as a function of planet relative reflectivity and primary star color indices. earth-like planets will appear bluer than their primary stars, while planets resembling Mars, Jupiter, or Venus will be redder. In principle, if very large space telescopes are used, it is possible therefore to photometrically differentiate between earthlike planets and other known planet types.

091.033   The dust ring of the earth.   K. Kordylewski.
          Problems of cosmic physics. Vyp. (No.) 6, (see
003.040), p. 164 - 167 (1971). In Russian.

091.034   Rotation of planets as the result of direct growth.
          R. I. Kiladze.
Astron. vestn., Vol. 5, 159 - 166 (1971). In Russian. − Abstr. in Referativ. Zhurn. 51. Astron., 2.51.250 (1972).

091.035   The effects of gravity decrease on planetary orbits.
          E. J. Kelsey.
American Journ. Phys., Vol. 39, 795 - 797 (1971).

The effects of gravity decrease are discussed in connection with three models of planetary motion. These are (I) two bodies in a circular orbit, (II) two bodies in an elliptical orbit, and (III) many bodies in a solar system with interactions between the planets. The effect on Bode's law is discussed along with the possibilities of detecting gravity decrease in planetary motion.

091.036   Motions in planetary atmospheres: A review.
          R. Hide.
Meteorol. Magazine, (*GB*), Vol. 100, 268 - 276 (1971).

091.037   Checking up the agreement between the thermo-
          dynamical parameters of a planet's atmosphere
measured with an automatic station and data concerning its descent velocity.   G. S. Golitsyn, V. V. Kerzhanovich.
Kosm. Issled., Vol. 9, 896 - 903 (1971). In Russian.

091.038   Planetary atmospheres and spectroscopy.
          M. Shimizu.
Journ. Spectroscop. Soc. Japan, Vol. 20, No. 2, p. 75 - 84 (1971). In Japanese.

091.039   Model for rotating and nonuniform planetary exos-
          pheres.   R. E. Hartle.
Phys. Fluids, Vol. 14, 2592 - 2598 (1971).

The model neutral exosphere with a uniformly rotating exobase is generalized by allowing variations in exobase density and temperature which characterize the thermosphere just below the base. The corresponding velocity distribution function, satisfying the collisionless Boltzmann equation, is constructed and used to form a general expression for the velocity moments. Resulting density profiles of rotating exospheres with nonuniform densities and temperatures on the exobase are compared with corresponding nonrotating exospheres.

091.040   The international planetary patrol.   B. J. Harris.
          Journ. Astron. Soc. Western Australia, Vol. 30,
September, p. 2 - 5 (1971).

091.041   Observing planets.   Z. Pokorný.
          Contr. Obs. and Planetarium Brno, No. 10, 135 pp.
(1970). In Czech.

Some fundamental terms from the planetary topography are mentioned. Instructions for visual and photographic observations are given and special measurements of Venus, Mars, Jupiter and Saturn are described. The photoelectric photometry as well as the process of obtaining spectrograms of planets and the way of their calibration are briefly discussed.

Physical characteristics of the giant planets.
Handbook−review.   See Abstr. 003.007.

Physics of the moon and planets.
See Abstr. 003.014.

Meridian observations of major planets and some minor planets 1963 − 1967.   See Abstr. 041.003.

Meridiankreis-Beobachtungen von Planeten 1950 bis 1955.   See Abstr. 041.031.

Détermination des masses des planètes.
See Abstr. 043.009.

A determination of the masses of the five outer planets.   See Abstr. 043.010.

Simultaneous solution for the masses of the principal planets from analysis of optical, radar, and radio tracking data.   See Abstr. 043.011.

The masses of the principal planets.
See Abstr. 043.013.

Results of astronomical studies in the far UV region.
See Abstr. 051.022.

The spectrum of steady state turbulent convection.
See Abstr. 063.010.

Quasiequilibrium structures in the ionized atmospheres of stars and planets.   See Abstr. 064.064.

The dynamical properties and internal constitutions of the earth, the moon and the planets.
See Abstr. 081.015.

Solar cycle variation of planetary exospheric temperatures.   See Abstr. 082.148.

The reflection of circularly polarized ultra-short radio waves from the moon and planets.
See Abstr. 094.146.

092 Mercury

092.001 **Isophotometric measurements of Mercury.**
M. Dizer, M. N. Yazici.
Astrophys. Space Sci., Vol. 13, 106 - 109 (1971).
Isophotometric measurements have been made on films, taken during the transit of Mercury, on May 9, 1970. This note gives the isophotes of Mercury in front of the sun in white light and some results of the comparison between the intensity of the sunspot's umbra and Mercury in H-alpha and white light.

092.002 **Tagesbeobachtungen des Merkur.** F. Dorst.
SuW, Vol. 10, 336 (1971).

092.003 **Radar observations of Mercury.**
R. M. Goldstein.
Astron. Journ., Vol. 76, 1152 - 1154 (1971).
Radar scattering properties of the planet Mercury at $12^1/_2$-cm wavelength are presented. Data from two inferior conjunctions show that backscattering anomalies can be attributed to specific regions on the planetary surface. The rota-

tion period of Mercury is measured to better than 0.4%.

092.004 **Imaging of Mercury and Venus from a flyby.**
B. C. Murray, M. J. S. Belton, G. E. Danielson, M. E. Davies, G. P. Kuiper, B. T. O'Leary, V. E. Suomi, N. J. Trask.
Icarus, Vol. 15, 153 - 173 (1971).
This paper describes the results of study of an imaging experiment planned for the 1973 Mariner Venus/Mercury fly-by mission. Scientific objectives, mission constraints, analysis of alternative systems, and the rationale for final choice are presented.

Quantum theory of gravitation vs. classical theory. —Fourth-order potential —. See Abstr. 066.074.

Quantum theory of gravity and the perihelion motion of Mercury. See Abstr. 066.075.

A note on radiative transfer in lunar and Mercurian surfaces. See Abstr. 094.224.

093 Venus

093.001 **A non-gray greenhouse model and microwave absorption in the atmosphere of Venus.**
K. Iwasaki.
Publ. Astron. Soc. Japan, Vol. 23, 387 - 397 (1971).
With data from Venera 4, 5, and 6 and Mariner 5, we investigate the lower atmosphere of Venus. First, the greenhouse effect due to the atmosphere of Venus is evaluated with the use of a non-gray radiative model. Next, the microwave absorption in the atmosphere of Venus is calculated. Based on these models, pressure-induced absorption of CO_2 and water vapor is considered.

093.002 **A search for deuterium on Venus.**
L. Wallace, F. E. Stuart, R. H. Nagel, M. D. Larson.
Astrophys. Journ., *(Letters)*, Vol. 168, L29 - L31 (1971).
The interpretation of observations obtained from Mariner 5 has suggested that the bulk of the Venus Lα emission might be due to deuterium rather than hydrogen. The present observation, obtained at high enough spectral resolution to resolve the two components, indicates that the converse is true and, consequently, that the deuterium interpretation is not correct.

093.003 **Preliminary results of an investigation of the atmosphere of Venus by means of Venera 7.**
M. Ya. Marov, V. S. Avduevskij, M. K. Rozhdestvenskij, N. F. Borodin, V. V. Kerzhanovich.
Kosmich. Issled., Vol. 9, 570 - 579 (1971). In Russian.

093.004 **Sur les irrégularités de la rotation en 4 jours des taches nuageuses en forme de Y dans la haute atmosphère de Vénus, et leur corrélation avec la rotation lente du sol.** C. Boyer, P. Guérin.
Comptes Rendus Acad. Sci. Paris, Sér. B, Vol. 273, 154 - 157 (1971).

093.005 **Once more about anomalous water in the Venus clouds.** V. A. Bronshten.
Astron. Tsirk., No. 627, p. 7 - 8 (1971). In Russian.

093.006 **The ashen light on Venus.** W. McD. Napier.
Planet. Space Sci., Vol. 19, 1049 - 1051 (1971).
It is suggested that the glow visible over the nocturnal hemisphere of Venus is due to reflected earthlight and starlight.

093.007 **How did Venus lose its angular momentum?**
B. M. French, S. F. Singer.
Science, Vol. 173, 169 - 170 (1971).

093.008 **Absorption by Venus in the 3–4-micron region.**
R. Beer, R. H. Norton, J. V. Martonchik.
Astrophys. Journ., (*Letters*), Vol. 168, L121 - L124 (1971).
The continuum absorption by Venus in the $3-4-\mu$ region has been determined from a detailed comparison of high-resolution spectra of Venus and the sun. The resultant ratio shows a major depression centered near 2580 cm^{-1} and a less well-defined dip near 3050 cm^{-1}. The possibility that these may be due to bicarbonates in the upper clouds of Venus is discussed.

093.009 **High-dispersion spectroscopic observations of Venus during 1968 and 1969. I. The carbon dioxide bands at 7820 Å and 7883 Å.**
L. D. G. Young, R. A. J. Schorn, E. S. Barker, A. Woszczyk.
Acta Astron., Vol. 21, 329 - 363 (1971).
Twenty-two well exposed photographic plates of the spectra of the 7820 Å and 7883 Å CO_2 bands on Venus were obtained during 1968 and 1969. Assuming constant width for the CO_2 lines, we find rotational temperatures ranging from 229° K to 258° K. The average value of the rotational tem-

perature is 244 ± 3° K (s. d.).

093.010 Measurements of phase dependence of the Venus radio emission at 8.2 mm wavelength.
Yu. N. Vetukhnovskaya, A. D. Kuzmin, B. Ya. Losovsky.
Astron. Zhurn. Akad. Nauk SSSR, Vol. 48, 1033 - 1037 (1971). In Russian. English translation in Soviet Astron. AJ, Vol. 15, No. 5.

093.011 A polarimetric search for a Venus halo during the 1969 inferior conjunction.   J. Veverka.
Icarus, Vol. 14, 282 - 283 (1971).
Polarimetric observations of Venus during the 1969 inferior conjunction show no evidence of a halo effect due to hexagonal ice crystals. From this I conclude that the magnitude of the anomaly cannot exceed 0.2% in polarization. These results may not be incompatible with the photometric measurements of O'Leary which imply, on a simple model, a polarization anomaly of ⩾ 0.2%.

093.012 A nongrey calculation of the runaway greenhouse: Implications for Venus' past and present.
J. B. Pollack.
Icarus, Vol. 14, 295 - 306 (1971).
We obtain the dependence of the surface temperature of a planet covered with bodies of liquid water upon the amount of incident solar flux. To determine this relationship we carry out nongrey calculations of both the thermal flux emitted to space and the planetary albedo. Water vapor is the only source of infrared opacity in our models. For most of the model atmospheres with a 50% cloud cover, the solar flux incident upon Venus at the present time is sufficiently high to cause complete evaporation of any hypothetical oceans, thus assuring that any carbon dioxide outgassed into the atmosphere remains there. Our calculations indicate that four and a half billion years ago, when the solar luminosity was 30% lower than its present value, Venus may have had a very moderate surface temperature and bodies of liquid water on its surface.

093.013 On Marov's model of the atmosphere of Venus.
F. Link.
Icarus, Vol. 14, 382 - 387 (1971).
At the 1970 COSPAR meeting in Leningrad, Marov proposed a provisional model of the atmosphere of Venus, based mainly on the results obtained by Mariner V and by Veneras IV, V, and VI. This model is compared with optical phenomena observed from earth, such as the transit of Venus across the sun, and the occultation of Regulus in 1959.

093.014 Height variations of Venusian clouds.
G. E. Hunt, R. A. J. Schorn.
Nature, Phys. Sci., Vol. 233, 39 - 40 (1971).
There is now strong evidence that the visible Venus clouds consist of at least two layers. A dense cloud whose top is at ~ 0.2 atm and a diffuse stratospheric haze extending from this level to a pressure of ~ 5 mb. The observations made by Moroz at phase angles of 141 — 158°, refer to absorption lines formed in the haze and not in the tropospheric cloud as he has claimed. His observations cannot be interpreted in terms of variations of the structure of the Venusian clouds, as we have shown, thus disputing the claim made by Moroz.

093.015 Peculiarities of propagation of radio waves in the atmosphere of Venus and in interplanetary plasma from data of Venera 7.
O. I. Yakovlev, A. I. Efimov, T. S. Timofeeva.
Kosmich. Issled., Vol. 9, 748 - 753 (1971). In Russian.

093.016 Chemical composition of the Venus atmosphere.
A. P. Vinogradov, Yu. A. Surkov, B. M. Andreichikov, O. M. Kalinkina, I. M. Grechishcheva.

Space Research XI, (see 012.004), 129 - 140 (1971).

093.017 Lower atmosphere of Venus from radio astronomical and space measurements.
A. D. Kuzmin, A. P. Naumov, T. V. Smirnova, Yu. N. Vetukhnovskaya.
Space Research XI, (see 012.004), 141 - 145 (1971).

093.018 Radio wave fluctuations and refraction coefficient variations in the atmosphere of Venus.
M. A. Kolosov, O. I. Yakovlev, G. D. Yakovleva, A. I. Yefimov.
Space Research XI, (see 012.004), 147 - 153 (1971).

093.019 The atmosphere of Venus.   V. I. Moroz.
Uspekhi fiz. nauk, Vol. 104, 255 - 296 (1971). In Russian. – Abstr. in Referativ. Zhurn. 51. Astron., 11.51.312 (1971).

093.020 O I and H I emissions from the upper atmosphere of Venus.   H. W. Moos, G. J. Rottman.
Astrophys. Journ., (*Letters*), Vol. 169, L127 - L130 (1971).
Ultraviolet spectra of Venus were obtained from a rocket-borne telescope-spectrophotometer in 1971 January. The two strong features between 1190 and 1340 Å are O I 1304 and H I 1216 with brightnesses of 5.7 and 27 kilorayleighs, respectively.

093.021 Report on the elongation of Venus: 1971 January.
J. H. Robinson.
Journ. British Astron. Ass., Vol. 82, 50 - 53 (1971). – Report of the Mercury and Venus Section.

093.022 Flux and polarization calculations of the radiation reflected from the clouds of Venus.
G. W. Kattawar, G. N. Plass, C. N. Adams.
Astrophys. Journ., Vol. 170, 371 - 386 (1971).
The flux and polarization reflected from a spherically symmetric planetary atmosphere is calculated by the Monte Carlo technique. Spherical geometry is used throughout the calculation with no plane-parallel approximations. The scattering angle for the photons is chosen from an appropriate single-scattering function calculated from Mie theory. The photons are followed through all orders of multiple scattering. Models considered for the Venus atmosphere include several single-layered models with various particle size distributions and one multilayered model.

093.023 Comments on "The Venus spectrum: New evidence for ice".   R. A. Schorn, L. G. Young.
Icarus, Vol. 15, 103 - 109 (1971).
The authors refer to a recent article by Plummer (Icarus, Vol. 12, 233 - 237, 1970), who has attempted to show that high-altitude infrared spectra exhibit evidence for ice-crystal clouds on Venus. In this note it is pointed out that the airborne and ground-based spectra offer no convincing evidence for an ice-cloud composition.

093.024 Erratum: 'High-dispersion spectroscopic observations of Venus. IX. The carbon dioxide bands at 12,030 Å and 12,177 Å' [Icarus, Vol. 13, 74 - 81 (1970)].
L. D. G. Young.
Icarus, Vol. 15, 151 (1971).

093.025 Calculation of general circulation of the Venus atmosphere.   V. G. Turikov, D. V. Chalikov.
Izv. AN SSSR. Fiz. atmosf. i okeana, Vol. 7, 705 - 722 (1971). In Russian. – Abstr. in Referativ. Zhurn. 51. Astron., 1.51.274 (1972).

093.026 Results of daytime observations of Venus and Mars obtained with the Wanschaff vertical circle in 1969.

V. K. Tarady, A. S. Kharin.
Astrometriya i Astrofiz., *Kiev*, No. 13, (see 003.039), p. 105 - 108 (1971). In Russian.

093.027 Difficulties with thermal protons in the Venusian topside ionosphere. P. M. Banks.
Journ. Geophys. Res., Vol. 76, 8455 - 8456 (1971).

Thermal protons of planetary origin are rapidly destroyed in the Venusian ionosphere, and thus they cannot be the principal topside ions. He^+ is conjectural, whereas there might be a possibility that O^+ is present in significant density.

093.028 Venus: a measurement and interpretation of radiation near 1.35 centimeters wavelength and the microwave phase effect. D. M. Wrathall.
Thesis, Brighton Young Univ., Provo, Utah. [Availabel from Univ. Microfilms, Ann. Arbor, Mich., U.S.A. Order No. 70–22096], 248 pp. (1971). – See Phys. Abstr., Vol. 74, No. 50101 (1971).

093.029 The structure and circulation of the lower atmosphere of Venus. J. M. Diamante.
Thesis, New York Univ., N.Y. [Available from Univ. Microfilms, Ann Arbor, Mich., U.S.A. Order No. 70–19280], (1969).

Utilizing the available experimental data, an expression has been obtained for the microwave absorption coefficient in a planetary atmosphere composed principally of carbon dioxide, with some nitrogen and trace amounts of water vapor. This expression has been applied to a model of the atmosphere of Venus derived from the measurements of the Mariner 5 and Venera 4 spacecraft to determine the microwave brightness temperature and radar attenuation as functions of frequency.

093.030 Soft landing of Venera 7 on the Venus surface and preliminary results of investigations of the Venus atmosphere. V. S. Avduevsky, M. Ya. Marov, M. K. Rozhdest-vensky, N. F. Borodin, V. V. Kerzhanovich.
Journ. Atmosph. Sci., Vol. 28, 263 - 269 (1971).

Planetary imaging and topographic mapping by radar interferometry. See Abstr. 031.083.

Normalized observations of Venus 1901 - 1949. See Abstr. 041.028.

Imaging of Mercury and Venus from a flyby. See Abstr. 092.004.

The atmospheres of Mars and Venus. See Abstr. 097.008.

Atomic carbon in the atmospheres of Mars and Venus. See Abstr. 097.046.

Der Stand der Mars- und Venusforschung. See Abstr. 097.055.

Ultraviolet radiation in space and in the Venus atmosphere. See Abstr. 155.030.

094 Moon

094.001 **The moon's librations.** H. Jeffreys.
Monthly Notices, Roy. Astron. Soc., Vol. 153,
73 - 81 (1971).

Koziel's results on the librations are rediscussed. Serious departures from independence of the errors are found and allowed for. The results do not differ much from Koziel's values but the uncertainties are larger. The effect of elasticity is considered. It is shown that the elastic strain contributes nothing to the librations but would affect the perturbations of a satellite.

094.002 **Report on lunar nomenclature.**
The Working Group of Commission 17 of the IAU:
D. H. Menzel, M. Minnaert, B. Levin, A. Dollfus, B. Bell.
Space Sci. Rev., Vol. 12, 136 - 186 (1971). – Contents:
Names proposed for craters on the moon's far side, with identifying biographical data; Biographical data on persons proposed to be added to names already on the moon; Names proposed for lunar craters: Living astronauts and cosmonauts; Checked positions of craters named in 1961; Biographical data for names adopted in 1961; Biographical data for names adopted in 1964.

094.003 **Moon: Origin and evolution of multi-ring basins.**
W. K. Hartmann, C. A. Wood.
The Moon, Vol. 3, 3 - 78 (1971).

This paper summarizes current data and new observations on lunar basin systems. Parts 1—4 review earlier literature and give new crater-counts used to reconstruct basin histories. Parts 5—11 interpret the results in terms of origin and evolution of basins. Section 12 summarizes the results and compares them to those of other authors.

094.004 **Classification, scales sequence and nomenclature of lunar maps.** K. P. Florensky, A. A. Gurshtein,
V. I. Korablev, L. M. Bougaevsky, K. B. Shingareva.
The Moon, Vol. 3, 78 - 89 (1971).

094.005 **Some very thin lunar crescents.**
J. Ashbrook.
Sky Telescope, Vol. 42, 78 - 79 (1971).

094.006 **Some plans for Apollo 15.**
Sky Telescope, Vol. 42, 80 - 83 (1971).

094.007 **Preliminary data on the lunar soil brought to earth by automatic probe 'Luna 16'.**
A. P. Vinogradov.
Journ. British Interplanet. Soc., Vol. 24, 475 - 495 (1971).

Results are presented of the analysis of the Luna-16 soil sample in terms of physical properties, rock types, mineral composition and chemical composition. Comparisons are made with the results of the Apollo-11 and Apollo-12 analyses. The material from all three sites is surprisingly similar in petrological, mineralogical and chemical composition though certain details are different.

094.008 **On inferring elastic properties of the deep lunar interior.** G. H. A. Cole.
Planet. Space Sci., Vol. 19, 929 - 947 (1971).

The moon is represented as a spherical body with a crust enclosing an interior in hydrostatic equilibrium. The moment of inertia of the model is calculated, for the observed mass and radius values of the moon, and the model most likely to represent the moon is taken to be that one with the correct value of the moment of inertia. In the present paper we attempt to invert this procedure and obtain information about the constituent material from the observed value of the moment of inertia.

094.009 **Functional dependences of bistatic-radar frequency spectra and cross sections on surface scattering laws.**
G. L. Tyler, D. H. H. Ingalls.
Journ. Geophys. Res., Vol. 76, 4775 - 4785 (1971).

This paper presents several results connecting surface scattering laws with bistatic-radar echo signal characteristics. In particular, we model the case of signals transmitted from a spacecraft, scattered by the lunar surface and received on earth.

094.010 **Distribution and origin of helium, neon, and argon isotopes in Apollo 12 samples measured by in situ analysis with a laser-probe mass spectrometer.**
G. H. Megrue.
Journ. Geophys. Res., Vol. 76, 4956 - 4968 (1971).

Distributions of the relative isotopic abundances of helium, neon, and argon were measured in situ by laser-probe mass spectrometric analysis of Apollo samples 12070,67, 12018,24, and 12010,13. On a scale of 10^{-3} to 10^{-5} gram, the distributions of helium, neon, and argon isotopes in these samples are highly variable.

094.011 **On the origin of the moon. III. Some aspects of the dynamics of the circumterrestrial swarm.**
E. L. Ruskol.
Astron. Zhurn. Akad. Nauk SSSR, Vol. 48, 819 - 829 (1971).
In Russian. English translation in Soviet Astron. AJ, Vol. 15, No. 4.

The formation of the moon is considered in the vicinity of the earth as a process which has accompanied the accumulation of the earth itself. It is concluded that the general trend of the evolution of the swarm was the formation of a single or a few large bodies (protomoons) which comprised the main portion of the mass of the swarm and of a cloud of small collisional debris which played a great role in capturing new interplanetary particles.

094.012 **On the selection of the optimal method of astrometrical determinations on the lunar surface.**
A. A. Gurshtein.
Astron. Zhurn. Akad. Nauk SSSR, Vol. 48, 830 - 832 (1971).
In Russian. English translation in Soviet Astron. AJ, Vol. 15, No. 4.

In connection with the publication of a number of papers concerning a rational method of the determination of selenographic latitudes and longitudes directly from the lunar surface, limitations which can be introduced into such a method due to principal constructive complexity of the appropriate astrometrical instrument are considered. It is concluded that the most perspective one is the method of registration of moments of transit of a series of stars in different azimuths through a fixed almucantur.

094.013 **On the problem of astrometrical observations from the lunar surface by the method of equal zenith distances.** Sh. T. Khabibullin, A. N. Sanovich.
Astron. Zhurn. Akad. Nauk SSSR, Vol. 48, 833 - 842 (1971).
In Russian. English translation in Soviet Astron. AJ, Vol. 15, No. 4.

The problems of determining selenographic coordinates by means of astrometrical observations from the lunar surface by the method of equal zenith distances are discussed. Such observations give the possibility to define principal parameters of the moon's rotation: inclination of Cassini's equator to the ecliptic I and the parameter of the physical libration f.

094.014 Of time and the moon. G. W. Wetherill.
Science, Vol. 173, 383 - 392 (1971).
Dating of lunar materials reveals the early history of an observable planetary body.

094.015 Evidence for compounds hydrolyzable to amino acids in aqueous extracts of Apollo 11 and Apollo 12 lunar fines.
K. Harada, P. E. Hare, C. R. Windsor, S. W. Fox.
Science, Vol. 173, 433 - 435 (1971).

094.016 Flux of micrometeoroids: Lunar sample analyses compared with flux model. J. S. Dohnanyi.
Science, Vol. 173, 558 (1971).

094.017 Chemical composition of Apollo 14 soils 14163 and 14259. C. C. Schnetzler, D. F. Nava.
Earth Planet. Sci. Letters, Vol. 11, 345 - 350 (1971).
Major and minor element concentrations have been determined by atomic absorption spectrophotometry and isotope dilution in Apollo 14 soils 14163 and 14259. The two soils have quite similar compositions. These soils have higher Si, Al, K, rare earths, Ba, Li, Rb, Zr and Hf, and lower Fe, Ti, Mn and Cr concentrations than soils from the mare areas sampled by Apollo 11, Apollo 12 and Luna 16.

094.018 Determination of 36 elements in Apollo 14 bulk fines 14163 by activation analysis.
A. O. Brunfelt, K. S. Heier, E. Steinnes, B. Sundvoll.
Earth Planet. Sci. Letters, Vol. 11, 351 - 353 (1971).
36 elements were determined in Apollo 14 bulk fines 14163. A one gram bulk sample was split from the 4.960 g received of sample 14163,154, and analysed according to the method described by Brunfelt and Steinnes with some slight modifications. The analytical results are shown in a table.

094.019 Fluorine and other trace elements in lunar plagioclase concentrates.
G. W. Reed, S. Jovanovic, L. H. Fuchs.
Earth Planet. Sci. Letters, Vol. 11, 354 - 358 (1971).
The F contents of plagioclase concentrates from Apollo 11 fines and an anorthosite inclusion from an Apollo 12 breccia are of the order of 100 ppm. This is very much lower than the ~2500 ppm F found by Surveyor VII in soil of anorthositic composition at the crater Tycho.

094.020 Thermal studies of lunar material and its terrestrial analogues.
A. P. Dmitriev, E. A. Dukhovskoi, G. Ya. Novik, R. G. Petrochenkov.
Dokl. Akad. Nauk SSSR, Ser. Mat. Fiz., Vol. 199, 1036 - 1037 (1971). In Russian.

094.021 The device TOR-I for the investigation of engineering-physical properties of lunar ground.
M. P. Drozhzhina, V. V. Dymov, V. M. Krylov, V. V. Markachev, A. A. Silin, V. V. Shvarev.
Dokl. Akad. Nauk SSSR, Ser. Mat. Fiz., Vol. 199, 1038 - 1040 (1971). In Russian.

094.022 Investigation of frictional and abrasive properties of lunar material.
A. S. Buialo, V. I. Kvochka, V. V. Maltsev, V. N. Lavrenchik, B. P. Lobashev, E. A. Motovilov, M. I. Smorodinov, V. V. Shvarev.
Dokl. Akad. Nauk SSSR, Ser. Mat. Fiz., Vol. 199, 1268 - 1270 (1971). In Russian.

094.023 Electrical studies of lunar material and its terrestrial analogues by the Q-meter technique.
A. R. Golovkin, E. A. Dukhovskoi, G. Ya. Novik, R. G. Petrochenkov, V. V. Rzhevskii.
Dokl. Akad. Nauk SSSR, Ser. Mat. Fiz., Vol. 199, 1271 - 1273 (1971). In Russian.

094.024 Preliminary results of lunar regolith mechanical studies.
V. I. Druzhininskaia, V. V. Markachev, A. V. Semenov, Yu. A. Surkov, I. I. Cherkasov.
Dokl. Akad. Nauk SSSR, Ser. Mat. Fiz., Vol. 199, 1274 - 1277 (1971). In Russian.

094.025 Laser measurements of the earth–moon distance.
B. Kołaczek.
Postępy Astron., Vol. 19, 259 - 262 (1971). In Polish.

094.026 The mechanical and magnetic properties of the moon. S. K. Runcorn.
Journ. British Astron. Ass., Vol. 81, 350 - 364 (1971).

094.027 Extension of the chronological catalogue of reported lunar events: October 1967 – June 1971.
P. Moore.
Journ. British Astron. Ass., Vol. 81, 365 - 390 (1971).

094.028 First drilling experiment on the moon.
V. E. Grafov, V. P. Bulekov, D. D. Dryuchenko, B. V. Zakhar'ev, Eh. A. Motovilov, M. I. Smorodinov, Yu. N. Strelov, V. V. Shvarev.
Kosmich. Issled., Vol. 9, 580 - 586 (1971). In Russian.

094.029 Lunar induction and highly conducting subsurface bodies. G. Schubert, K. Schwartz.
Cosmic Electrodynamics, Vol. 2, 224 - 231 (1971).
Lunar induction in the presence of a relatively small subsurface body of high electrical conductivity is investigated. The body is assumed to be spherical, of infinite conductivity and buried beneath the surface of an otherwise nonconducting moon. The amplification of tangential magnetic field components at the lunar surface and directly above the subsurface body is determined.

094.030 The magnetism of the moon.
P. Dyal, C. W. Parkin.
Sci. American, Vol. 225, No. 2, p. 62 - 73 (1971).

094.031 Lunochod 1. L. Lebedew, S. Nikitin.
Bild der Wissenschaft, (DVA, Stuttgart), 8. Jahrgang, p. 780 - 789 (1971).

094.032 Equalization results of the positions of centers of selenodetic catalogues in the Mills-2 system.
V. A. Nikonov, T. P. Skobeleva.
Astron. Tsirk., No. 627, p. 1 - 3 (1971). In Russian.

094.033 Position of the center of the figure of the moon's visible side in the system of nine selenodetic catalogues. Yu. N. Lipskij, V. A. Nikonov.
Astron. Tsirk., No. 627, p. 3 - 5 (1971). In Russian.

094.034 Reduction of nine selenodetic catalogues to the center of the figure of the moon's visible side.
V. A. Nikonov.
Astron. Tsirk., No. 627, p. 5 - 7 (1971). In Russian.

094.035 Ferromagnetic resonance of lunar samples.
F.-D. Tsay, S. I. Chan, S. L. Manatt.
Geochim. Cosmochim. Acta, Vol. 35, 865 - 875 (1971).
Evidence is presented to support that the electron spin resonance spectra observed for a selection of Apollo 11 lunar samples (10087-10, 11; 10046-29, 30; 10062-26, 27; 10017-35, 36) arise from the ferromagnetic centers consisting

of metallic Fe.

094.036 Exotic armalcolite and the origin of Apollo 11 ilmenite basalts. A. T. Anderson, Jr.
Geochim. Cosmochim. Acta, Vol. 35, 969 - 973 (1971).
Armalcolite is probably partly exotic in lunar basalt 10022 as suggested by K_2O, SiO_2 rich devitrified glass included within it. Assimilation of armalcolite (and possibly ilmenite) and addition of a K_2O, SiO_2 rich liquid probably helped produce the high TiO_2 content and variable alkali contents of Apollo 11 ilmenite basalts.

094.037 Trace element chemistry of Apollo 14 lunar soil from Fra Mauro.
S. R. Taylor, P. Muir, M. Kaye.
Geochim. Cosmochim. Acta, Vol. 35, 975 - 981 (1971).
Analytical data are presented for Apollo 14 fines (< 1 mm) sample 14163,136 for 31 trace elements. Their abundances are reported in a table.

094.038 Fabric and compositions of olivine in three Apollo 12 igneous rocks. P. Butler, Jr.
Bull. Geol. Soc. America, Vol. 2, 512 - 513 (1970). – Abstr. in The Moon, Vol. 3, 91 - 92 (1971).

094.039 Grain size and modal analyses of lunar regolith material returned by Apollo 11 and 12.
J. C. Butler, E. A. King, Jr., M. F. Carman.
Bull. Geol. Soc. America, Vol. 2, 512 (1970). – Abstr. in The Moon, Vol. 3, 92 (1971).

094.040 Genesis of lunar fines.
M. B. Duke, G. A. Sellers, C. C. Woo, M. L. Bird.
Bull. Geol. Soc. America, Vol. 2, 543 (1970). – Abstr. in The Moon, Vol. 3, 96 (1971).

094.041 Petrology of rocks from the Apollo 11 and Apollo 12 landing sites. I. D. MacGregor.
Bull. Geol. Soc. America, Vol. 2, 612 - 613 (1970). – Abstr. in The Moon, Vol. 3, 99 (1971).

094.042 Composition and origin of glasses and lithic fragments in Apollo 11 samples.
M. Prinz, T. E. Bunch, K. Keil.
Bull. Geol. Soc. America, Vol. 2, 657 (1970). – Abstr. in The Moon, Vol. 3, 101 (1971).

094.043 Lunar petrology of silicate melt inclusions, Apollo 11 and 12, and terrestrial equivalents.
E. Roedder, P. W. Weiblen.
Bull. Geol. Soc. America, Vol. 2, 666 (1970). – Abstr. in The Moon, Vol. 3, 102 (1971).

094.044 On the origin of lunar rocks. A. P. Vinogradov.
Geochemistry International, Vol. 7, 1 - 11 (1970). Abstr. in The Moon, Vol. 3, 105 (1971).

094.045 Strontium-rubidium ages of lunar rocks.
G. J. Wasserburg, D. A. Papanastassiou.
Bull. Geol. Soc. America, Vol. 2, 715 - 716 (1970). – Abstr. in The Moon, Vol. 3, 106 (1971).

094.046 A peculiar lunar sample – Apollo 12013.
A. L. Albee, D. S. Burnett, A. A. Chodos, J. C. Huneke, D. A. Papanastassiou, F. A. Podosek, G. P. Russ II, F. Tera, G. J. Wasserburg.
Bull. Geol. Soc. America, Vol. 2, 480 (1970). – Abstr. in The Moon, Vol. 3, 110 (1971).

094.047 On the cryptic component of the Apollo 12 soil.
U. L. Bottino, C. C. Schnetzler, P. D. Fullagar,

J. A. Philpotts.
EOS, Trans. American Geophys. Union, Vol. 51, 772 (1970). Abstr. in The Moon, Vol. 3, 111 (1971).

094.048 Total carbon and nitrogen abundances and the chemical evolution of the lunar regolith.
C. B. Moore, J. W. Larimer, C. F. Lewis, F. M. Delles, R. C. Gooley.
Bull. Geol. Soc. America, Vol. 2, 628 (1970). – Abstr. in The Moon, Vol. 3, 130 (1971).

094.049 Apollo 12 clinopyroxenes: Subsolidus relations.
J. J. Papike, A. E. Bence, C. T. Prewitt, G. E. Brown.
Bull. Geol. Soc. America, Vol. 2, 644 (1970). – Abstr. in The Moon, Vol. 3, 133 (1971).

094.050 Trace element abundances in Apollo 12 samples.
C. C. Schnetzler, J. A. Philpotts.
Bull. Geol. Soc. America, Vol. 2, 676 (1970). – Abstr. in The Moon, Vol. 3, 137 (1971).

094.051 Uranium-thorium-lead isotope relations in lunar materials. L. T. Silver.
Bull. Geol. Soc. America, Vol. 2, 684 - 685 (1970). – Abstr. in The Moon, Vol. 3, 137 - 138 (1971).

094.052 Stable isotope geochemistry of lunar samples.
H. P. Taylor, Jr., S. Epstein.
Bull. Geol. Soc. America, Vol. 2, 700 - 701 (1970). – Abstr. in The Moon, Vol. 3, 140 (1971).

094.053 A comparison of some Apollo 11 and Apollo 12 samples. L. A. Taylor, G. Kullerud, W. F. Bryan.
Bull. Geol. Soc. America, Vol. 2, 701 (1970). – Abstr. in The Moon, Vol. 3, 140 (1971).

094.054 Abundances of 35 elements, 14 rare-earths included, in ten Apollo 12 rock and soil samples.
H. Wakita, R. A. Schmitt, P. Rey.
EOS, Trans. American Geophys. Union, Vol. 51, 772 (1970). Abstr. in The Moon, Vol. 3, 142 (1971).

094.055 Trapping of lunar atmosphere in the lunar surface.
R. H. Manka.
EOS, Trans. American Geophys. Union, Vol. 51, 772 (1970). Abstr. in The Moon, Vol. 3, 145 (1971).

094.056 Radiation history of the moon.
P. B. Price, D. J. Barber, D. O'Sullivan.
Bull. Geol. Soc. America, Vol. 2, 656 (1970). – Abstr. in The Moon, Vol. 3, 145 (1971).

094.057 A classification of impact craters.
F. Hörz, L. B. Ronca.
Modern Geol., Vol. 2, 65 - 69 (1971). – Abstr. in The Moon, Vol. 3, 146 (1971).

094.058 Geology of the Apollo 11 and Apollo 12 landing sites. G. A. Swann, G. G. Schaber, R. L. Sutton.
Bull. Geol. Soc. America, Vol. 2, 697 - 698 (1970). – Abstr. in The Moon, Vol. 3, 147 (1971).

094.059 Possible endogenetic craters associated with formations in the rim of Copernicus.
R. Greeley, D. E. Gault.
EOS, Trans. American Geophys. Union, Vol. 51, 773 (1970). Abstr. in The Moon, Vol. 3, 148 (1971).

094.060 Lunar mare ridges, rings and volcanic ring complexes. R. G. Strom.
Modern Geol., Vol. 2, 133 - 157 (1971). – See Phys. Abstr.,

Vol. 74, No. 53927 (1971).

094.061 **Preliminary examination of lunar samples from Apollo 14.** The Lunar Sample Preliminary Examination Team: D. H. Anderson, M. N. Bass, A. D. Bennett, D. D. Bogard, R. Brett, L. G. Bromwell, P. Butler, Jr., W. D. Carrier III, R. S. Clark, T. Cobleigh, M. B. Duke, P. W. Gast, E. K. Gibson, Jr., W. R. Hart, G. H. Heiken, W. C. Hirsch, F. Hörz, E. D. Jackson, P. H. Johnson, J. E. Keith, C. F. Lewis, J. F. Lindsay, J. R. Martin, W. C. Melson, E. D. Mitchell, C. B. Moore, D. A. Morrison, W. B. Nance, W. C. Phinney, A. M. Reid, M. A. Reynolds, K. A. Richardson, W. I. Ridley, E. Schonfeld, A. B. Shepard, R. L. Sutton, N. J. Trask, J. Warner, R. B. Wilkin, H. G. Wilshire, D. R. Wones,
Science, Vol. 173, 681 - 693 (1971).
A physical, chemical, mineralogical, and biological analysis of 43 kilograms of lunar rocks and fines is presented.

094.062 **Structural control of plains distributions, lunar southern highlands.**
R. S. Saunders.
Bull. Geol. Soc. America, Vol. 2, 674 (1970). – Abstr. in The Moon, Vol. 3, 149 (1971).

094.063 **Mechanical properties of lunar surface materials estimated from secondary impact craters.**
H. J. Moore.
EOS, Trans. American Geophys. Union, Vol. 51, 772 (1970). Abstr. in The Moon, Vol. 3, 152 (1971).

094.064 **Transient magnetic field measurements of the surface of the moon.** P. Dyal, C. W. Parkin, C. P. Sonett, D. S. Colburn.
EOS, Trans. American Geophys. Union, Vol. 51, 774 (1970). Abstr. in The Moon, Vol. 3, 154 - 155 (1971).

094.065 **Lunar soil adhesion due to electrostatic forces stabilized by solar radiation.** J. J. Grossman.
EOS, Trans. American Geophys. Union, Vol. 51, 772 - 773 (1970). – Abstr. in The Moon, Vol. 3. 155 (1971).

094.066 **Morphology and origin of lunar craters.**
R. S. Saunders, E. L. Haines, J. E. Conel.
Polarforschung, Vol. 7, 33 - 35 (1971). – Abstr. in The Moon, Vol. 3, 149 (1971).

094.067 **The Times atlas of the moon.**
H. A. G. Lewis (Editor).
Times Newspapers, London. (U.S. distributor: Quadrangle, Chicago). 37 + 111 pp. Price $ 25.00 (1969). – Review in Science, Vol. 173, 712 - 713; 1971 (*H. Masursky*).

094.068 **Geologic setting of the Apollo 14 samples.**
G. A. Swann, N. J. Trask, M. H. Hait, R. L. Sutton.
Science, Vol. 173, 716 - 719 (1971).

094.069 **Concerning the electrical conductivity of the moon.**
C. P. Sonett, J. D. Mihalov, N. F. Ness.
Journ. Geophys. Res., Vol. 76, 5172 - 5180 (1971).
The response of the moon to a large discontinuity in the interplanetary magnetic field with $|\Delta B| = 6\gamma$, observed by the lunar satellite Explorer 35, is examined on the basis of presently available theoretical models.

094.070 **The moon. Hypotheses and discoveries.**
V. Vladimirov, L. Pipko.
Priroda, *Sofia,* Vol. 20, No. 2, p. 58 - 60 (1971).
In Bulgarian.

094.071 **A dark haloed craters program.** K. J. Delano.

Strolling Astronomer, Vol. 23, 51 - 54 (1971).

094.072 **Eratosthenes from sunrise to sunset.**
H. D. Jamieson.
Strolling Astronomer, Vol. 23, 57 - 61 (1971).

094.073 **Lunar notes.**
C. L. Ricker, H. D. Jamieson, J. E. Westfall.
Strolling Astronomer, Vol. 23, 62 - 67 (1971).

094.074 **Formation of glass spheres on the moon.**
C. A. Cross.
Nature, Vol. 233, 185 - 186 (1971).
A remarkable morphological similarity has been discovered between lunar glass spheres and the glass shot which is formed during the manufacture of mineral wool.

094.075 **Lunar Apennine-Hadley region: Geological implications of earth-based radar and infrared measurements.** S. H. Zisk, M. H. Carr, H. Masursky, R. W. Shorthill, T. W. Thompson.
Science, Vol. 173, 808 - 812 (1971).
Recently completed high-resolution radar maps of the moon contain information on the decimeter-scale of the surface. When this information is combined with eclipse thermal-enhancement data and with high-resolution Lunar Orbiter photography, the surface morphology is revealed in some detail.

094. 076 **On some approximations in Brown's lunar theory.**
J. S. Griffith.
Celestial Mechanics, Vol. 4, 54 - 59 (1971).
A re-evaluation of Brown's lunar theory has long been awaited. While working on this problem a number of questions about the adequacy of Brown's approach have arisen, and some of these questions including that of the need for a literal solution of the main problem are discussed in this paper.

094.077 **An analysis of the distribution of boulders in the vicinity of small lunar craters.**
W. S. Cameron, G. J. Coyle.
The Moon, Vol. 3, 159 - 188 (1971).
Nine Orbiter 3 high-resolution photographs were examined at three sites for distributions of boulders around craters ≥ 110 m in diameter; three kinds of distributions were noted. Crater morphologies were also classified. We find that the examination of craters reveals that both the morphology (crater type) and the distributions of boulders indicate the existence of at least two distinct populations of primary craters. All the data used for boulder-distribution analysis are given in a table.

094.078 **The directional characteristics of lunar infrared radiation.**
R. U. Sexl, H. Sexl, H. Stremnitzer, D. G. Burkhard.
The Moon, Vol. 3, 189 - 213 (1971).
A theory of the directional characteristics of the lunar infrared radiation measured by Saari and Shorthill has been derived. This theory is in excellent agreement with experiment at all angles of observation and at all phase angles. The radiation law used to describe the angular dependence of the infrared radiation emitted by a flat element of the lunar surface is $0.85 \cos \theta + 0.22 \cos^2 \theta$, where θ is the angle between the surface normal and the direction of observation. This radiation law is subsequently modified by taking into account lunar surface roughness.

094.079 **On possible mechanisms of gas accumulation in the surface layer of the moon.**

A. M. Gutkin, M. S. Markov, T. M. Raitburd, M. V. Slonims-kaya.
The Moon, Vol. 3, 214 - 220 (1971).

094.080 **Objectives and requirements of unmanned rover exploration of the moon.**
D. B. Nash, J. E. Conel, F. P. Fanale.
The Moon, Vol. 3, 221 - 230 (1971).
    The scientific value of unmanned rovers for continued lunar exploration is considered in light of Apollo findings which suggest that the moon's surface is more heterogeneous than expected. Unmanned rovers are well-suited for low-cost, low-risk preliminary reconnaissance where measurement of a few definitive parameters over a wide area is more important than obtaining a wide array of detailed results at a given site.

094.081 **On the systems of selenographic coordinates, their determination and terminology.**
Sh. T. Habibullin.
The Moon, Vol. 3, 231 - 238 (1971).
    Attention is drawn to the absence in literature of the precise definitions of selenographic and celestial selenocentric coordinate systems. In certain cases inaccuracies in the formulation of the first Cassini law occur. This is due to the fact that the principal directions dealt with in the theory of lunar rotation are being constantly confused. A clear-cut definition of the principal coordinate systems concerned with the lunar rotation is given.

094.082 **Electrical conductivity and temperature of the lunar interior from magnetic transient-response measurements.** P. Dyal, C. W. Parkin.
Journ. Geophys. Res., Vol. 76, 5947 - 5969 (1971).
    The response of the moon to magnetic-field step transients in the solar wind has been investigated for over 100 events, by using simultaneous data from the Apollo-12 lunar surface magnetometer and the lunar-orbiting Explorer-35 magnetometer. It is concluded that the simplest model, which qualitatively explains all the general aspects of the dark-side transient-response data, is a spherically symmetric three-layer model having a thin outer crust of very low conductivity. Temperatures for the three layers are calculated for three possible lunar material compositions.

094.083 **Een foefje bij kleurenfotografie van de maan.**
J. Klinkspoor.
Hemel en Dampkring, Vol. 69, 266 - 267 (1971).

094.084 **Lunar gravity analysis from long-term effects.**
A. S. Liu, P. A. Laing.
Science, Vol. 173, 1017 - 1020 (1971).
    The global lunar gravity field was determined from a weighted least-squares analysis of the averaged classical element of the five Lunar Orbiters. The observed-minus-computed residuals have been reduced by a factor of 10 from a previously derived gravity field. The values of the second-degree zonal and sectorial harmonics are compatible with those derived from libration data.

094.085 **Expedition to Hadley-Apennine—1.**
D. Baker.
Spaceflight, Vol. 13, 358 - 362, 383 (1971).

094.086 **Apollo seismic experiments.** J. E. Davies.
Spaceflight, Vol. 13, 370 - 373 (1971).

094.087 **Apollo 14: A visit to Fra Mauro—3.**
D. Baker.
Spaceflight, Vol. 13, 373 - 376 (1971).

094.088 **Effects of physical librations of the moon on the**

orbital elements of a lunar satellite.
A. J. Ferrari, W. G. Heffron.
National Aeronautics and Space Administration, NASA-CR-117841 (1971). — Abstr. in The Moon, Vol. 3, 240 (1971).

094.089 **Simple mass distribution for the lunar potential.**
S. L. Levie.
National Aeronautics and Space Administration, NASA-CR-116507 (1971). — Abstr. in The Moon, Vol. 3, 241 (1971).

094.090 **Electrical conductivity and temperature of the lunar interior from magnetic transient response measurements.** P. Dyal, C. W. Parkin, C. P. Sonett, D. S. Colburn.
National Aeronautics and Space Administration, NASA-TM-X-62012 (1971). — Abstr. in The Moon, Vol. 3, 241 (1971).

094.091 **Examination of lunar scientific objectives and evaluation and development studies for possible post-Apollo period.** J. R. Booker, R. L. Kovach.
National Aeronautics and Space Administration, NASA-CR-116418 (1971). — Abstr. in The Moon, Vol. 3, 241 - 242 (1971).

094.092 **NASA lunar sample analysis program: A summary of phase analysis on Apollo 12 samples.**
K. Fredrikson, J. Nelen, C. A. Andersen, J. R. Hinthorne.
National Aeronautics and Space Administration, NASA-CR-114871 (1971). — Abstr. in The Moon, Vol. 3, 242 (1971).

094.093 **Microchemical, microphysical and adhesive properties of Apollo 11 and 12. Final report.**
J. J. Grossman, J. A. Ryan, N. R. Mukherjee, M. W. Wegner.
National Aeronautics and Space Administration, NASA-CR-114916 (1971). — Abstr. in The Moon, Vol. 3, 242 (1971).

094.094 **Thermoluminescence of lunar sample 10084 Apollo 11.** B. Hess, R. Huber, W. Herr.
Modern Geol., Vol. 2, 55 - 60 (1971). — Abstr. in The Moon, Vol. 3, 242 - 243 (1971).

094.095 **Luminescence petrography of the Apollo 12 rocks and comparative features in terrestrial rocks and meteorites.** R. F. Sippel.
National Aeronautics and Space Administration, NASA-CR-114842 (1971). — Abstr. in The Moon, Vol. 3, 243 (1971).

094.096 **Data on Apollo 13 and 12 samples. Speculations on petrologic differentiation. Final report.**
J. V. Smith, A. T. Anderson, Jr., R. C. Newton.
National Aeronautics and Space Administration, NASA-CR-114917 (1971). — Abstr. in The Moon, Vol. 3, 243 - 244 (1971).

094.097 **Oxygen isotope fractionation in Apollo 12 rocks and soils.** R. N. Clayton, N. Onuma, T. K. Mayeda.
National Aeronautics and Space Administration, NASA-CR-114925 (1971). — Abstr. in The Moon, Vol. 3, 245 (1971).

094.098 **Lunar sample analysis program. High sensitivity isotropic analysis: Apollo 11 and 12. Isotropic abundances of actinide elements in Apollo 12 samples.**
P. R. Fields, H. Diamond, D. H. Metta, C. M. Stevens, D. J. Rokop.
National Aeronautics and Space Administration, NASA-CR-114870 (1971). — Abstr. in The Moon, Vol. 3, 245 (1971).

094.099 **Research on stimulated exoelectron emission from lunar materials.**
R. B. Gammage, K. Becker.
Technical Progress Report, NASA-CR-117853. — Abstr. in The Moon, Vol. 3, 245 (1971).

094.100 **Application of the Mössbauer technique to the study of probably lunar and planetary surface material and to the study of returned lunar surface samples. Final report.**
C. L. Herzenberg, C. L. Riley, R. B. Moler.
National Aeronautics and Space Administration, NASA-CR-114887 (1971). − Abstr. in The Moon, Vol. 3, 246 (1971).

094.101 **Infrared vibrational spectroscopic studies of minerals from Apollo 11 and 12 lunar samples. Final report.**
C. Kara, Jr., P. A. Ester, J. J. Kovach.
National Aeronautics and Space Administration, NASA-CR-114890 (1971). − Abstr. in The Moon, Vol. 3, 246 - 247 (1971).

094.102 **Analyses of Apollo 11 and 12 rocks and soils by neutron activation.**
C. P. Kharkar, K. K. Turekian.
National Aeronautics and Space Administration, NASA-CR-114926 (1971). − Abstr. in The Moon, Vol. 3, 247 (1971).

094.103 **A search for carbon and its compounds in lunar samples from Mare Tranquillitatis.**
K. A. Kvenvolden, C. Ponnamperuma (Editors).
National Aeronautics and Space Administration, NASA-SP-257 (1971). − Abstr. in The Moon, Vol. 3, 247 (1971).

094.104 **Mineralogy and petrography of some Apollo 12 samples.** B. Mason.
National Aeronautics and Space Administration, NASA-CR-114840 (1971). − Abstr. in The Moon, Vol. 3, 247 - 248 (1971).

094.105 **Search for C^{15} and C^{30} alkanes in lunar soils. Final report.**
W. G. Meinschein, E. Cordes, V. J. Shiner, Jr.
National Aeronautics and Space Administration, NASA-CR-114919 (1971). − Abstr. in The Moon, Vol. 3, 248 (1971).

094.106 **Particle detector studies on lunar orbital missions.**
K. A. Anderson.
National Aeronautics and Space Administration, NASA-CR-114858 (1971). − Abstr. in The Moon, Vol. 3, 249 (1971).

094.107 **Investigation of lunar materials.**
G. M. Comstock, A. O. Evwaraye, R. L. Fleischer, M. R. Hart.
National Aeronautics and Space Administration, NASA-CR-114843 (1971). − Abstr. in The Moon, Vol. 3, 249 - 250 (1971).

094.108 **Surveyor 3 parts and materials returned from the moon by Apollo 12: Evaluation of lunar effects.**
National Aeronautics and Space Administration, NASA-CR-114855 (1971). − Abstr. in The Moon, Vol. 3, 250 (1971).

094.109 **A classification of meteorite impact craters.**
F. Hörz, L. B. Ronca.
National Aeronautics and Space Administration, NASA-CR-116890 (1971). − Abstr. in The Moon, Vol. 3, 252 (1971).

094.110 **Relative heights of photographic features of the moon.** Z. Kopal, M. Moutsoulas.
Air Force Cambridge Res. Lab., AFCRL-70-0487; AD-713679 (1971). − Abstr. in The Moon, Vol. 3, 252 (1971).

094.111 **Evolution of mare surface.** T. Gold.
National Aeronautics and Space Administration, NASA-CR-114845 (1971). − Abstr. in The Moon, Vol. 3, 253 (1971).

094.112 **Lunar Hadley rille: Considerations of its origin.**
R. Greeley.
National Aeronautics and Space Administration, NASA-TM-X-62011 (1971). − Abstr. in The Moon, Vol. 3, 253 (1971).

094.113 **Lava tubes and channels in the lunar Marius Hills.**
R. Greeley.
National Aeronautics and Space Administration, NASA-TM-X-62013 (1971). − Abstr. in The Moon, Vol. 3, 253 (1971).

094.114 **Physical properties of the Apollo 12 lunar fines.**
T. Gold, B. T. O'Leary, M. Campbell.
National Aeronautics and Space Administration, NASA-CR-114844 (1971). − Abstr. in The Moon, Vol. 3, 254 (1971).

094.115 **Physical properties of the Apollo 12 lunar fines.**
T. Gold, B. T. O'Leary, M. Campbell.
National Aeronautics and Space Administration, NASA-CR-114893 (1971). − Abstr. in The Moon, Vol. 3, 254 (1971).

094.116 **Micrometeorite craters and related features on lunar rock surfaces.**
F. Hörz, J. B. Hartung, D. E. Gault.
National Aeronautics and Space Administration, NASA-TM-X-66705 (1971). − Abstr. in The Moon, Vol. 3, 254 - 255 (1971).

094.117 **Permafrost.** J. S. Laurence, S. H. Ward.
National Aeronautics and Space Administration, NASA-CR-116879 (1971). − Abstr. in The Moon, Vol. 3, 255 (1971).

094.118 **Optical properties of Apollo 12 moon samples.**
F. Briggs, B. O'Leary.
National Aeronautics and Space Administration, NASA-CR-114894 (1971). − Abstr. in The Moon, Vol. 3, 255 - 256 (1971).

094.119 **Luminescence of Apollo 11 and 12 lunar sample analysis program.**
N. N. Greenman, H. G. Gross.
National Aeronautics and Space Administration, NASA-CR-114899 (1971). − Abstr. in The Moon, Vol. 3, 256 (1971).

094.120 **Investigation of the photometric response of the moon and the planets.** W. M. Sinton.
Air Force Cambridge Res. Lab., AFCRL-70-0435; AD-713688 (1971). − Abstr. in The Moon, Vol. 3, 258 (1971).

094.121 **Lunar sample analysis program; Magnetic properties of Apollo 12 lunar samples 12052 and 12065. Final report.** C. S. Gromme.
National Aeronautics and Space Administration, NASA-CR-114891 (1971). − Abstr. in The Moon, Vol. 3, 258 (1971).

094.122 **Neutron diffraction study of lunar materials.**
S. J. Pickart, H. A. Alperin.
National Aeronautics and Space Administration, NASA-CR-114853 (1971). − Abstr. in The Moon, Vol. 3, 258 - 259 (1971).

094.123 **Crater size and impact flash predicted for the S-4B stage lunar impact on Apollo 13 flight.**
D. W. Jex, G. Hintze.
National Aeronautics and Space Administration, NASA-TM-X-64517 (1971). − Abstr. in The Moon, Vol. 3, 261 (1971).

094.124 **On the estimate of precision of contemporary data on absolute and relative heights of details of the lunar surface.** A. A. Gurshtein, A. A. Konopikhin.
Astron. Zhurn. Akad. Nauk SSSR, Vol. 48, 1051 - 1055

(1971). In Russian. English translation in Soviet Astron. AJ, Vol. 15, No. 5.

A method of comparing absolute heights of details of the lunar surface from various catalogues having no common points by means of enlisting the data on relative heights from maps of the moon is proposed. The method gives a possibility to estimate also the precision of the used relative heights.

094.125 **Apollo 15 pictorial.**
V. D. R. Scott, J. B. Irwin, A. M. Worden.
Sky Telescope, Vol. 42, 192 - 201 (1971).

094.126 **Solar flare effects in lunar xenon.**
M. N. Rao, K. Gopalan, V. S. Venkatavaradan, L. Wilkening.
Nature, Phys. Sci., Vol. 233, 114 - 117 (1971).

Detailed analysis of the xenon isotopes data from Apollo 11 lunar samples shows that certain anomalies observed in xenon^{132} can be attributed to the low energy solar flare proton reactions on barium. The observed isotopic excesses reasonably compare with the theoretical estimates made using derived cross-sections from nuclear systematics and an average solar cosmic ray flux derived from lunar and meteoritic radioactivities.

094.127 **Selenography and selenodesy with Apollo whole-disk lunar photographs. I. Selenography.**
D. W. G. Arthur.
Icarus, Vol. 14, 388 - 418 (1971).

This paper presents a selenographic methodology for the far-encounter Apollo pictures of the moon, and gives results for two Apollo 8 photographs.

094.128 **Experimental studies on the formation of lunar surface features by fluidization.** S. A. Schumm.
Bull. Geol. Soc. America, Vol. 81, 2539 - 2552 (1970). – Abstr. in Icarus, Vol. 14, 447 (1971).

094.129 **Petrologic and mineralogic investigation of some crystalline rocks returned by the Apollo 14 mission.**   A. J. Gancarz, A. L. Albee, A. A. Chodos.
Earth Planet. Sci. Letters, Vol. 12, 1 - 18 (1971).

We describe the petrology and mineralogy of seven crystalline samples, on which Rb/Sr, ^{40}Ar-^{39}Ar, or cosmic ray exposure ages have been determined. These isotopic investigations indicate the existence of two rock groups characterized by different crystallization and cosmic ray exposure ages.

094.130 **^{40}Ar-^{39}Ar ages and cosmic ray exposure ages of Apollo 14 samples.**
G. Turner, J. C. Huneke, F. A. Podosek, G. J. Wasserburg.
Earth Planet. Sci. Letters, Vol. 12, 19 - 35 (1971).

We have used the ^{40}Ar-^{39}Ar dating technique on eight samples of Apollo 14 rocks, breccia fragments, and soil fragments. The exposure age data are summarized in a table.

094.131 **Rb-Sr ages of igneous rocks from the Apollo 14 mission and the age of the Fra Mauro formation.**
D. A. Papanastassiou, G. J. Wasserburg.
Earth Planet. Sci. Letters, Vol. 12, 36 - 48 (1971).

In this study, we report the results of analyses on several basaltic rocks, a breccia, and four soil samples returned from the Fra Mauro site by the Apollo 14 mission. In addition to these samples from Apollo 14, we analyzed sample 12004 from the Apollo 12 mission.

094.132 **Apollo 14: Nature and origin of rock types in soil from the Fra Mauro formation.**
Apollo Soil Survey: F. K. Aitken, D. H. Anderson, M. N. Bass, R. W. Brown, P. Butler, Jr., G. Heiken, P. Jakeš, A. M. Reid,

W. I. Ridley, H. Takeda, J. Warner, R. J. Williams.
Earth Planet. Sci. Letters, Vol. 12, 49 - 54 (1971).

Compositions of glasses in the Apollo 14 soil correspond to four types of Fra Mauro basalts, to mare basalts and soils, and, in minor amounts, to gabbroic anorthosite and potash granite. The Fra Mauro basalts can be related by simple low pressure crystal-liquid fractionation that implies a parent composition like that of Apollo 14 sample 14310.

094.133 **The age of the Fra Mauro formation: A radiometric older limit.**
W. Compston, M. J. Vernon, H. Berry, R. Rudowski.
Earth Planet. Sci. Letters, Vol. 12, 55 - 58 (1971).

The internal Rb-Sr age of a 0.17 g basaltic clast from an Apollo 14 breccia is 4.15 ± 0.10 by. This limits the Imbrium event to not older than 4.25 by on current interpretations of lunar stratigraphy.

094.134 **On the development of the crater population on the moon with time under meteoroid and solar wind bombardment.**   G. Neukum, H. Dietzel.
Earth Planet. Sci. Letters, Vol. 12, 59 - 66 (1971).

A theory is evaluated to describe the development of the lunar crater population with time under the bombardment by meteoroids and solar wind. The theory permits the calculation of absolute formation ages of the lunar surface as well as the particle flux, supposing the crater distributions on the moon have been measured. As an important result it includes a D^{-2} equilibrium crater distribution law (D = crater diameter), actually measured in Mare Tranquillitatis and Oceanus Procellarum.

094.135 **Mixing models and the recognition of end-member groups in Apollo 11 and 12 soils.** J. F. Lindsay.
Earth Planet. Sci. Letters, Vol. 12, 67 - 72 (1971).

Lunar soils returned from the Apollo 11 and 12 sites appear to consist of more than one source material. The paper attempts to define more rigorously the end members present in the lunar soil and to establish mixing models to allow some understanding of the provenance of the soil.

094.136 **Uranium-bearing minerals of lunar rock 12013.**
E. L. Haines, A. L. Albee, A. A. Chodos, G. J. Wasserburg.
Earth Planet. Sci. Letters, Vol. 12, 145 - 154 (1971).

The U distribution in rock 12013 was studied by fission track and elemental mapping techniques. Major U-bearing phases are whitlockite, apatite, zircon, and phase β, which is a Zr-Ti mineral rich in Fe, Nb, Y, REE and containing up to 3.6 % UO_2, 4.7 % ThO_2, and 4.2 % PbO.

094.137 **Search for exoelectrons in Apollo 12 materials.**
R. B. Gammage, K. Becker.
Earth Planet. Sci. Letters, Vol. 12, 155 - 158 (1971).

The search for exoelectron emission from Apollo 12 materials usually produced negative results, suggesting that the concentration of naturally occurring traps is very low. Weak exoelectron activity was found scattered inhomogeneously within a pigeonite basalt rock and in a deep core tube sample.

094.138 **Oxidation state of iron in plagioclase from lunar basalts.** S. S. Hafner, D. Virgo, D. Warburton.
Earth Planet. Sci. Letters, Vol. 12, 159 - 166 (1971).

The purpose of the present work was to determine the oxidation state of iron in the plagioclase from the coarse-grained basalts 10044 and 12021, using Mössbauer spectroscopy, and to investigate the location of iron in the crystal structure.

094.139 **The $(^{78}Kr/^{83}Kr)_{sp}$ – $(^{131}Xe/^{126}Xe)_{sp}$ correlation in**

Apollo 12 rocks.
H. Schwaller, P. Eberhardt, J. Geiss, H. Graf, N. Grögler.
Earth Planet. Sci. Letters, Vol. 12, 167 - 169 (1971).

The Apollo 12 rocks show, similar to the Apollo 11 rocks, a good correlation between the $^{78}Kr/^{83}Kr$ and the $^{131}Xe/^{126}Xe$ ratios of the spallation component. The correlation line is distinctly different for Apollo 11 and 12 rocks, reflecting the difference in the Sr/Zr abundance ratio.

094.140 Single domain grain distributions. I. A method for the determination of single domain grain distributions. A. Stephenson.
Phys. Earth Planet. Interiors, Vol. 4, 353 - 360 (1971).

094.141 Single domain grain distributions. II. The distribution of single domain iron grains in Apollo 11 lunar dust. A. Stephenson.
Phys. Earth Planet. Interiors, Vol. 4, 361 - 369 (1971).

A quantitative analysis of some of the magnetic properties of an Apollo 11 lunar dust sample is given. It is shown that the experimental results are in good agreement with those predicted from single domain theory described in the foregoing cited paper and can be explained on the basis of an assembly of single domain iron grains distributed such that the number of grains within a given volume range is inversely proportional to the square of the volume.

094.142 Orthopyroxene and orthopyroxene-bearing rock fragments rich in K, REE, and P in Apollo 14 soil sample 14163. L. H. Fuchs.
Earth Planet. Sci. Letters, Vol. 12, 170 - 174 (1971).

The purpose of this paper is threefold: (1) to present a variety of orthopyroxene compositions found in a small sample of the Apollo 14 soil; (2) to indicate that the K and REEP components of KREEP are not positively correlated in orthopyroxene-containing rock fragments from the soil; and (3) to present some X-ray powder results obtained on the rare-earth calcium phosphate mineral.

094.143 Expedition to Hadley-Apennine–2. D. Baker.
Spaceflight, Vol. 13, 431 - 435 (1971).

094.144 Laser location of the moon. Yu. L. Kokurin.
Priroda, No. 10.71, p. 42 - 46 (1971). In Russian.

094.145 Rb-Sr ages and elemental abundances of K, Rb, Sr, and Ba in samples from the Ocean of Storms.
V. R. Murthy, N. M. Evensen, B.-M. Jahn, M. R. Coscio, Jr.
Geochim. Cosmochim. Acta, Vol. 35, 1139 - 1153 (1971).

This paper deals with the Rb-Sr isotopic relationships, ages and the abundances of K, Rb, Sr and Ba in the crystalline rocks and fines collected by the Apollo 12 mission from the Ocean of Storms. The samples allocated for our study included 5 crystalline rocks, fines from the contingency sample and a sample of fines from the bottom of a 15 cm trench dug on the northwest rim of Head crater. In addition to the studies on Apollo 12 samples, we include here internal isochron data on an Apollo 11 sample.

094.146 The reflection of circularly polarized ultra-short radio waves from the moon and planets.
M. P. Dolukhanov, Yu. Eh. Udal'ev.
Kosmich. Issled., Vol. 9, 754 - 758 (1971). In Russian.

094.147 Density variations of the surface layer of the moon. N. N. Krupenio.
Kosmich. Issled., Vol. 9, 759 - 766 (1971). In Russian.

094.148 Mechanical properties of a lunar sample returned by Luna 16.

V. V. Gromov, A. D. Dmitriev, A. K. Leonovich, V. A. Lozhkin, P. S. Pavlov, A. V. Rybakov, V. V. Shvarev.
Kosmich. Issled., Vol. 9, 767 - 774 (1971). In Russian.

094.149 Determination of the mutual positions of the centres of selenodetic catalogues from hypsometric maps. Yu. N. Lipskij, V. A. Nikonov, T. P. Skobeleva.
Kosmich. Issled., Vol. 9, 775 - 780 (1971). In Russian.

094.150 How to use magnetic fields for fun and profit. C. P. Sonett, S. K. Runcorn.
Comments Astrophys. Space Phys., Vol. 3, 149 - 154 (1971).

For solving the question of the origin and the subsequent evolution of the moon, it would be important to have an understanding of how the remanent magnetism observed in the Apollo samples originated and at the same time to understand other manifestations of lunar magnetism obtained by the Explorer 35 lunar orbiter, the Apollo 12 Lunar Surface Magnetometer, and the Apollo 14 Lunar Portable Magnetometer. The results are discussed.

094.151 Skandinaviska namn på månformationer.
Å. Wallenquist.
Astron. Tidsskr., Årg. 4, 67 - 75 (1971).

094.152 Hvor langt er der til Månen? J. Baerentzen.
Astron. Tidsskr., Årg. 4, 76 - 82 (1971).

094.153 The lunar rocks. B. Mason.
Sci. American, Vol. 225, No. 4, p. 48 - 58 (1971).

Although it is too soon for an analysis of the rocks returned by Apollo 15, the material provided by earlier missions tells of a moon whose outer shell crystallized more than three billion years ago.

094.154 The new moon: A review. N. W. Hinners.
Rev. Geophys. Space Phys., Vol. 9, 447 - 522 (1971)

094.155 Lunar atmosphere. F. S. Johnson.
Rev. Geophys. Space Phys., Vol. 9, 813 - 823 (1971).

A rarified atmosphere should be present on the moon because of contributions from the solar wind, meteoric volatilization, and internal degassing. Of these three natural sources, definite predictions can be made only on the basis of solar wind input. The lunar atmosphere of solar origin is expected to consist mainly of neon. Daytime neon concentrations are expected to be near 6×10^4 cm^{-3}, and nighttime concentrations are expected to be near 1.5×10^6 cm^{-3}.

094.156 Ages of crystalline rocks from Fra Mauro.
L. Husain, J. F. Sutter, O. A. Schaeffer.
Science, Vol. 173, 1235 - 1236 (1971).

Crystallization ages for six rocks from Fra Mauro have been measured by the argon-40–argon-39 method. All six rocks give an age of $3.77 \pm 0.15 \times 10^9$ years, which is the same as for fragmental rocks from this site. It is concluded that the Imbrium event and the crystallization of a significant portion of the pre-Imbrian basalts were essentially contemporaneous.

094.157 Die Geochemie des Mondes. G. M. Brown.
Endeavour, No. 111, Vol. 30, 147 - 152 (1971).

094.158 A dunite-norite lunar microbreccia.
G. J. Taylor, U. B. Marvin.
Meteoritics, Vol. 6, 173 - 179 (1971).

094.159 Petrological character of the Luna 16 sample from Mare Fecunditatis. J. A. Wood, J. B. Reid, Jr., G. J. Taylor, U. B. Marvin.
Meteoritics, Vol. 6, 181 - 193 (1971).

094.160 **Specific heat of Palisades diabase at liquid helium temperatures and some comments on the vibrational spectrum of complex structures.**
J. A. Morrison, P. R. Norton.
Journ. Geophys. Res., Vol. 76, 4993 - 4996 (1971).

The specific heat of Palisades (quartz) diabase at liquid helium temperatures is found to be very similar to that of lunar rocks 10017 and 10046 and to be very much larger than the value corresponding to measured acoustic wave velocities.

094.161 **A surface-layer representation of the lunar gravitational field.** L. Wong, G. Buechler, W. Downs, W. Sjogren, P. Muller, P. Gottlieb.
Journ. Geophys. Res., Vol. 76, 6220 - 6236 (1971).

A surface-layer representation of the lunar gravitational field has been derived dynamically from the analysis of Doppler observations on both polar and equatorial lunar orbiters. The force model contained 600 discrete masses located on the mean lunar surface between the approximate boundaries of ±60° latitude and ±95° longitude. The derived major mascons were generally in agreement with a model based on polar orbits alone. A technique for combining the discrete mass gravitational field for the front side with a spherical harmonics expansion for the back side is described.

094.162 **Oblique electromagnetic reflection from layered lunar models based on data from Apollo 11 and Apollo 12.** G. R. Jiracek, S. H. Ward.
Journ. Geophys. Res., Vol. 76, 6237 - 6245 (1971).

In this report we present calculations that pertain to electromagnetic experiments conducted in the frequency range 10^4 to 10^{10} Hz. The results are considered directly applicable to the interpretation of certain experimental data. We have studied the effect of a gradual densification in the lunar regolith and of values of magnetic permeability other than the value of free space, because the Apollo missions have confirmed these occurrences. Results are obtained at various angles of incidence so that both astatic and bistatic soundings are treated.

094.163 **Seismic velocity models of the lunar near surface and their implications.** J. S. Watkins.
Journ. Geophys. Res., Vol. 76, 6246 - 6252 (1971).

Models of the lunar near-surface seismic velocity distribution are derived that represent hypothetical moons in which the near-surface character is due to (1) cold accretion of particles from space and (2) steady-state impact. Comparison of model travel times with observed travel times favors the cold-accretion model.

094.164 **Lunar fines and terrestrial rock powders: Relative surface areas and heats of adsorption.**
F. P. Fanale, D. B. Nash, W. A. Cannon.
Journ. Geophys. Res., Vol. 76, 6459 - 6461 (1971).

Surface area measurements by Kr adsorption (BET method) indicate that Apollo 11 lunar fines and ground terrestrial mafic rock powders have similar effective surface areas that are a factor of 10−100 higher than their geometrical or surficial surface areas.

094.165 **Spherulitic textures in glassy and crystalline rocks.**
G. Lofgren.
Journ. Geophys. Res., Vol. 76, 5635 - 5648 (1971).

Keith and Padden (1963) have developed a theory of spherulite growth that relates changes of spherulite morphology to changes of specific growth parameters. It is the purpose of this paper to adapt this theory, which was developed for organic polymers, to systems of geologic interest.

094.166 **Investigation of glass recovered from Apollo 12 sample 12057.** B. P. Glass.
Journ. Geophys. Res., Vol. 76, 5649 - 5657 (1971).

Glass particles from a 500-mg sample of <1-mm fines from lunar soil (sample 12057) have been studied in detail. On the basis of their physical and chemical properties, the glasses from sample 12057 can be divided into at least six groups. All the glasses were probably produced by meteorite impact.

094.167 **Lunar breccias.** D. S. McKay, D. A. Morrison.
Journ. Geophys. Res., Vol. 76, 5658 - 5669 (1971).

This paper reviews some of the available data on Apollo 11 and 12 breccias with particular emphasis on the origin of 'welded' breccias, in an attempt toward contributing to the understanding of these rocks.

094.168 **Lunar near-side tectonic patterns from Orbiter 4 photographs.** W. E. Elston, A. W. Laughlin, J. A. Brower.
Journ. Geophys. Res., Vol. 76, 5670 - 5674 (1971).

For this study, 14,000 plots were made of linear features on Lunar Orbiter 4 photographs, including near-side mare ridges, segments of straight rills, highland ridges, and walls of polygonal craters. This paper complements the analysis of nonrandomness in direction of overlap of lunar craters (Elston et al., 1971) and uses the same statistical techniques.

094.169 **Nonrandom distribution of lunar craters.** W. E. Elston, M. J. Aldrich, E. I. Smith, R. C. Rhodes.
Journ. Geophys. Res., Vol. 76, 5675 - 5682 (1971).

It has long been known that some lunar craters of all sizes and ages are aligned in nonrandom groups or chains. This paper deals with an objective evaluation of this phenomenon for craters >10 km in diameter. It complements three other contributions to this issue.

094.170 **Determination of origin of small lunar and terrestrial craters by depth diameter ratio.** E. I. Smith.
Journ. Geophys. Res., Vol. 76, 5683 - 5689 (1971).

This article introduces a technique for determining impact or volcanic origin of lunar craters less than 3.5 km in diameter by means of depth-diameter ratio.

094.171 **Evidence for lunar volcano-tectonic features.** W. E. Elston.
Journ. Geophys. Res., Vol. 76, 5690 - 5702 (1971).

This article tries to show that moon and earth have probably had a history of devolatilization and, hence, volcanism.

094.172 **Terraced depressions in lunar maria.** R. Holcomb.
Journ. Geophys. Res., Vol. 76, 5703 - 5711 (1971).

Some shallow, irregularly shaped mare depressions are bordered by terraces. Morphologic study suggests that some of these depressions are drained lava lakes. If so, mare materials in which they occur originated as lava. Some associated sinuous rills and craters may have had volcanic origins. The terraced depressions may yield clues to the development of the large multi-ring basins.

094.173 **Cauldron subsidence in lunar craters Ritter and Sabine.** R. A. De Hon.
Journ. Geophys. Res., Vol. 76, 5712 - 5718 (1971).

The best analogies for Ritter and Sabine are found in terrestrial caldera such as Valles, New Mexico. Ritter and Sabine exhibit a wide variety of internal features similar to those of cauldrons of subsidence, including subsidence along ring faults, postsubsidence volcanism controlled by ring fractures, and probable resurgence of magma with accompanying uplift of the caldera floor.

094.174 **Copernicus as a lunar caldera.** J. Green.

Journ. Geophys. Res., Vol. 76, 5719 - 5731 (1971).
All morphological features observed in and around the crater Copernicus are in agreement with a simple volcanic origin. The crater appears to be a caldera.

094.175 **Laboratory simulation of impact cratering with high explosives.**   V. R. Oberbeck.
Journ. Geophys. Res., Vol. 76, 5732 - 5749 (1971).

094.176 **Craters produced by missile impacts.**
H. J. Moore.
Journ. Geophys. Res., Vol. 76, 5750 - 5755 (1971).

094.177 **Micrometeorite craters on lunar rock surfaces.**
F. Hörz, J. B. Hartung, D. E. Gault.
Journ. Geophys. Res., Vol. 76, 5770 - 5798 (1971).

It is the object of this report to summarize detailed stereomicroscopic studies of nearly 5000 craters in the 0.1- to 3-mm size range observed on 7 whole rocks returned during Apollo 12. It is intended to evaluate the conditions of formation of these craters and to relate the observed crater populations to the flux of primary cosmic particles.

094.178 **Apollo lunar exploration.**
G. M. Low, L. R. Scherer.
Space Research XI, (see 012.004), 1 - 13 (1971).

094.179 **Lunar surface exploration.**   N. Armstrong.
Space Research XI, (see 012.004), 15 - 29 (1971).

094.180 **Results of recent manned and unmanned lunar exploration.**   L. D. Jaffe.
Space Research XI, (see 012.004), 31 - 49 (1971).

094.181 **The nature of the surface of the moon**   T. Gold.
Space Research XI, (see 012.004), 51 - 61 (1971).

094.182 **Lunar surface layer density from spacecraft radar measurements.**   N. N. Kroupenio.
Space Research XI, (see 012.004), 63 - 67 (1971).

094.183 **The bimodality of the microstructure of lunar ground.**
T. E. Shvidkovskaya, G. A. Leikin, V. A. Krasnopolsky.
Space Research XI, (see 012.004), 69 - 71 (1971).

094.184 **Structural-mechanical properties of lunar soils and their terrestrial analogues.**
I. I. Cherkasov, V. V. Shvarev, G. S. Steinberg.
Space Research XI, (see 012.004), 73 - 83 (1971).

094.185 **Electrical conductivity and the age of the moon.**
T. Nagata, T. Rikitake, M. Kono.
Space Research XI, (see 012.004), 85 - 88 (1971).

094.186 **The distribution of radioactive elements in the moon's interior.**   V. S. Troitsky.
Space Research XI, (see 012.004), 89 - 92 (1971).

094.187 **Global mapping of the moon.**
Y. N. Lipsky, Y. P. Pskovsky, J. F. Rodionova, V. V. Schevchenko, V. I. Chikmachev, L. I. Volchkova.
Space Research XI, (see 012.004), 93 - 95 (1971).

094.188 **Preliminary results of laser ranging to a reflector on the lunar surface.**   J. D. Mulholland, C. O. Alley, P. L. Bender, D. G. Currie, R. H. Dicke, J. E. Faller, W. M. Kaula, G. J. F. MacDonald, H. H. Plotkin, D. T. Wilkinson.
Space Research XI, (see 012.004), 97 - 104 (1971).

094.189 **Determination of astrodynamic constants and a test**

of the general relativistic time delay with S-band range and Doppler data from Mariners 6 and 7.
J. D. Anderson, P. B. Esposito, W. Martin, D. O. Muhleman.
Space Research XI, (see 012.004), 105 - 112 (1971).

094.190 **Lunar dust potential.**   J. W. Rhee.
Space Research XI, (see 012.004), 275 - 277 (1971).

094.191 **Lunar Explorer 35 and OGO 3: Dust particle measurements in selenocentric and cislunar space from**
1967 to 1969.   W. M. Alexander, C. W. Arthur, J. L. Bohn.
Space Research XI, (see 012.004), 279 - 285 (1971).

094.192 **Höhenmessungen auf dem Mond.**
F. Schmeidler.
SuW, Vol. 10, 292 - 294 (1971).

094.193 **An instrument for demonstrating the lunar soil.**
V. K. Baranov.
Optiko-mekh. prom-st', 1971, No. 3, p. 61 - 62. In Russian.
Abstr. in Referativ. Zhurn. 62. Issled. kosm. prostranstva, 11.62.54 (1971).

094.194 **On the possibility of obtaining automatically statistical morphological characteristics of lunar surface regions from their photographic image.**
L. A. Akimov, Yu. V. Kornienko.
Astrometriya i Astrofiz., *Kiev*, No. 14, (see 003.014), p. 63 - 71 (1971). In Russian.

The possibility is discussed of reproducing the relief of a lunar region from its photographic image and of obtaining statistical morphological characteristics from its relief. Practical ways of realizing this possibility are analyzed. These results, with few exceptions, are applicable to other planets, such as the earth, Mars, Mercury as well.

094.195 **Some questions of photogrammetry of Luna 9- and Luna 13-panoramas.**   V. V. Kiselev.
Izv. vyssh. uchebn. zavedenij. Geod. aehrofotos"emka, 1970, No. 1, p. 155 - 165.   In Russian. – Abstr. in Referativ. Zhurn. 62. Issled. kosm. prostranstva, 11.62.196 (1971).

094.196 **Craters and stones telling of the moon's history.**
A. A. Gurshtein, K. B. Shingareva, A. A. Konopikhin, V. P. Shashkina, A. A. Pronin, Z. V. Popova, V. D. Popovich, R. B. Zezin, A. T. Bazilevsky, P. A. Dubin, A. V. Ivanov, O. D. Rode, V. P. Polosukhin.
Priroda, No. 11.71, p. 2 - 15 (1971). In Russian.

094.197 **Chemismus und Entstehungsgeschichte der Mondlandschaften.**   H. Wänke.
Umschau, 71. Jahrgang, p. 873 - 878 = Weltraumfahrt, 22. Jahrgang, p. 139 - 144 (1971).

094.198 **Aufbau und Entwicklung des Mondes.**   R. Meißner.
Umschau, 71. Jahrgang, p. 879 - 886 = Weltraumfahrt, 22. Jahrgang, p. 145 - 152 (1971).

094.199 **On the origin of excess ^{131}Xe in lunar rocks.**
P. Eberhardt, J. Geiss, H. Graf, H. Schwaller.
Earth Planet. Sci. Letters, Vol. 12, 260 - 262, with a correction Vol. 13, 222 (1971).

A Ba-feldspar sample was irradiated in a reactor with epithermal (> 0.4 eV) and fast neutrons. A large amount of excess ^{131}Xe was found in the irradiated feldspar. A discussion of the possible neutron induced reactions on Ba leading to ^{131}Xe shows that only resonance capture can be of importance. This experiment confirms the conclusion that epithermal neutron capture by ^{130}Ba is responsible for the large and variable ^{131}Xe$_{sp}$ yield observed in lunar rocks.

094.200 Noble gas analysis of KREEP fragments in lunar soil 12033 and 12070. J. Funkhouser.
Earth Planet. Sci. Letters, Vol. 12, 263 - 272 (1971).

Solar wind noble gases of bulk Apollo 12 soil samples are related to the percentage KREEP component. The source material of KREEP had a prior solar wind irradiation, the products of which were largely outgassed during deposition. Nobel gas analysis of individual KREEP fragments from 12033 and 12070, coupled with pertinent chemical data, enable specific events that have occurred at the Apollo 12 landing site to be dated. The age of Surveyor Crater is estimated to be about 240 my, while the original KREEP stratum was deposited at least 2 billion years ago.

094.201 Li, Be and B abundances in fines from the Apollo 11, Apollo 12, Apollo 14 and Luna 16 missions.
O. Eugster.
Earth Planet. Sci. Letters, Vol. 12, 273 - 281 (1971).

This paper reports determinations of Li, Be and B abundances in the Apollo fine samples 10084, 12070, 14141 and 14259 and in sample L16–19 no. 118 belonging to the fraction 'Regolith C' of the core tube returned by the automatic probe Luna 16. The major purpose of the present study is to compare the Li, Be and B elemental abundances with the analogous abundance patterns for related terrestrial and meteoritic material.

094.202 Selective volatilization on the lunar surface: Evidence from Apollo 14 feldspar-phyric basalts.
G. M. Brown, A. Peckett.
Nature, Vol. 234, 262 - 266 (1971).

Analysis of Apollo 14 samples has revealed a basalt that could be parental to the highlands anorthosites.

094.203 Compositional variations in lunar spinels.
S. E. Haggerty.
Nature, Phys. Sci., Vol. 233, 156 - 160 (1971).

Compilation of 167 electron microprobe analyses of spinels from the Apollo 11, 12 and 14 sites show, (a) that the major substitutional parameters are (1) Fe + Ti for 2Cr, (2) Mg for Fe and (3) Cr for Al; (b) that solid solutions are present (1) between 0.75 $FeCr_2O_4$–0.25 $FeAl_2O_4$ and Fe_2TiO_4, and (2) between $MgAl_2O_4$ and $FeAl_2O_4$; (c) compositional discontinuities exist between and within these series. Subsolidus reduction of Apollo 14 chromian-ulvospinels decompose to $FeTiO_3$ + Fe and titanian-chromites comparable in composition to Apollo 11. Spinel stability relationships, considered in terms of $T°C$ and fO_2, and thermodynamic data for the multi-component spinel-prism show that the paragentic sequence (FeMg) Cr_2O_4, Fe(AlCr)$_2O_4$, Fe_2TiO_4 is probably related to increasing fO_2 with crystallization.

094.204 Organic analyses of selected areas of Surveyor III recovered on the Apollo 12 mission.
B. R. Simoneit, A. L. Burlingame.
Nature, Vol. 234, 210 - 211 (1971).

Sections of the mirror, and the middle and lower shrouds of Surveyor III were analyzed for organic contaminants by high resolution mass spectrometry of the solvent washings. The major contaminants found were hydrocarbons, silicones, and dioctyl phthalate, probably all adsorbed during preliminary quality examinations and from the bagging materials used after recovery. The exhaust products from the LM descent engine and possibly from the Surveyor III engines were evident in trace amounts. Some possible outgassing products from the spacecraft were also detected.

094.205 On reasons for the different brightness distribution on the disks of the moon and Mars during phases near zero. N. P. Barabashov.
Vestn. Khar'kov. Univ., No. 65, (Ser. Astron., No. 6), p. 3 - 13

(1971). In Russian.

094.206 Results of luminescence measurements of the lunar surface in the H and K Ca II lines.
V. S. Tsvetkova.
Vestn. Khar'kov. Univ., No. 65, (Ser. Astron., No. 6), p. 20 - 37 (1971). In Russian.

094.207 Mineralogy of Apollo 15415 'genesis rock': Source of anorthosite on moon. I. M. Steele, J. V. Smith.
Nature, Vol. 234, 138 - 140 (1971).

Petrographic, microprobe and X-ray analysis of Apollo rock 15415 shows an anorthosite with minor pyroxene. The plagioclase has a high anorthite and low minor element content suggesting an early product of crystal-liquid differentiation. The pyroxene analyses fall in two distinct groups indicating crystallization (or recrystallization) at low temperatures and at depth with respect to the mare basalts. Reverse compositional trends for coexisting olivine-plagioclase in some lunar rock fragments compared to terrestrial igneous intrusions suggest that lunar and terrestrial differentiation may not be analogous.

094.208 Subsolidus reduction of lunar spinels.
S. E. Haggerty.
Nature, Phys. Sci., Vol. 234, 113 - 117 (1971).

Ulvospinel-rich Mg-Al-chromites decompose at high temperatures and low fO_2 to ilmenite + iron + Mg Al enriched chromite. Spinel decomposition is accompanied by the breakdown of fayalite to iron + cristobalite. Coexisting ilmenite and chromite are stable, and calculated $T°C/fO_2$ curves show that close limits are set on the conditions of dissociation, which at $1000°C$ are 10^{-15} to 10^{-16} atms. Spinels in the breccias are more intensely reduced than those in the crystalline rocks, suggesting that reduction may be related to contact with iron vaporized during an impact event.

094.209 Lunar surface mechanical properties from Surveyor data. R. H. Jones.
Journ. Geophys. Res., Vol. 76, 7833 - 7843 (1971).

During the Surveyor program spacecraft were successfully landed at five widely separated lunar locations. Recent computer simulations of each landing have provided more comprehensive data on the mechanical properties of the lunar surface than have been obtained previously by this method of analysis. Results show that the variations in surface bearing pressure observed at the various lunar sites are probably due to surface slope effects and do not necessarily indicate differences in soil properties at these sites.

094.210 Discussion of paper by A. E. Ringwood, 'Petrogenesis of Apollo 11 basalts and implications for lunar origin'. [Journ. Geophys. Res., Vol. 75, 6453 - 6479 (1970)].
S. F. Singer.
Journ. Geophys. Res., Vol. 76, 8071 - 8074, with a reply by A. E. Ringwood, p. 8075 - 8076 (1971).

094.211 Development of the moon in the light of modern data. B. Yu. Levin.
Priroda, No. 12.71, p. 2 - 9 (1971). In Russian.

094.212 Why is the moon not of cast iron?
M. A. Korets, Z. L. Ponizovsky.
Priroda, No. 12.71, p. 101 (1971). In Russian.

094.213 Some stereoscopic studies of the moon.
R. J. Livesey.
Journ. British Astron. Ass., Vol. 82, 38 - 41 (1971).

094.214 The characteristics of lunar domes.
K. J. Delano.

Strolling Astronomer, Vol. 23, 85 - 90 (1971).

094.215 **Statistical analysis of lunar dome characteristics.**
J. E. Westfall.
Strolling Astronomer, Vol. 23, 90 - 91 (1971).

094.216 **Statistical analysis of lunar dome distribution.**
J. E. Westfall.
Strolling Astronomer, Vol. 23, 91 - 98 (1971).

094.217 **Some new methods and goals for the selected areas
program.** H. D. Jamieson, C. A. Vaucher.
Strolling Astronomer, Vol. 23, 102 - 105 (1971).

094.218 **Lunar notes.** C. L. Ricker.
Strolling Astronomer, Vol. 23, 105 (1971).

094.219 **At the foot of lunar Apennines.**
S. R. Brzostkiewicz.
Urania Kraków, Vol. 42, 277 - 287 (1971). In Polish.

094.220 **Lunar gravity estimate: Independent confirmation.**
W. L. Sjogren.
Journ. Geophys. Res., Vol. 76, 7021 - 7026 (1971).
Reduction of $2^{1}/_{2}$ days of Lunar Orbiter 4 radio tracking
data has provided an independent estimate of the low-degree
spherical harmonic coefficients in the lunar potential model.
The estimate is in good agreement with previous results and
confirms that the moon is essentially homogeneous.

094.221 **Discussion of paper by O.B. James and E.D. Jack-
son, 'Petrology of the Apollo 11 ilmenite basalts'**
[Journ. Geophys. Res., Vol. 75, 5793 - 5824 (1970)].
P. Butler, Jr.
Journ. Geophys. Res., Vol. 76, 7298 - 7300, with a reply by
E. D. Jackson, p. 7301 - 7303 (1971).

094.222 **Soviet—French experiment on the laser location of
the moon.** Yu. L. Kokurin, L. A. Vedeshin.
Vestn. AN SSSR, 1971, No. 6, p. 33 - 38. In Russian.

094.223 **Limited-interval definitions of the photometric
functions of lunar crater walls by photography from**
orbiting Apollo. R. L. Wildey.
Icarus, Vol. 15, 93 - 99 (1971).
By the use of only relative photometry (intraframe) it is
shown that the photometric functions of material reposed on
the inner walls of some of the younger lunar craters photographed on the far side of the moon from the Apollo 11 Command Module are not of a form which can be reduced to a
dependence on phase angle and brightness-longitude alone.
Some other dependence on the completely general degrees of
freedom described by phase angle, angle of incidence, and
angle of emergence seems to be required.

094.224 **A note on radiative transfer in lunar and Mercurian
surfaces.** R. D. Cess, J. Srinivasan.
Icarus, Vol. 15, 100 - 102 (1971).
A linearized counterpart to the analysis of Ulrichs and

094.225 **Geologic maps of early Apollo landing sites of set C.**
N. J. Trask.
National Aeronautics and Space Administration, NASA-CR-
116408 (1971). − Abstr. in The Moon, Vol. 3, 261 (1971).

094.226 **Apollo 12 lunar sample information.** J. Warner.
National Aeronautics and Space Administration,
NASA-TR-R-353 (1971). − Abstr. in The Moon, Vol. 3, 261
(1971).
Campbell (1969) is presented for the thermal emission from
lunar and Mercurian surfaces. It is shown by illustration that

the linearized analysis may easily be extended to incorporate
additional features of the transient cooling process.

094.227 **Apollo mission 14: Lunar photography indexes.**
Prepared under the direction of the Department of
Defense by the Aeronautical Chart and Information Center,
United States Air Force for the National Aeronautics and
Space Administration. Photography indexes compiled by
Manned Spacecraft Center, Mapping Sciences Branch. 3 sheets
(1971).

094.228 **Apollo 14 lunar photography (NSSDC ID No.
71-008A-01).**
Prepared by A. T. Anderson, M. A. Niksch.
NSSDC 71-16a, Part I, Data User's Note, National Space Science Data Center, Goddard Space Flight Center, National
Aeronautics and Space Administration, Greenbelt, Maryland,
5 + 25 pp. (1971).

094.229 **Apollo 14 photography: 70-mm, 35-mm, 16-mm,
and 5-in. frame index.**
Prepared by Mapping Sciences Branch, Manned Spacecraft
Center, National Aeronautics and Space Administration Hous-
ton, Texas. NSSDC preparation, directed by A. T. Anderson.
National Space Science Data Center, NSSDC 71-16b, NASA,
Goddard Space Flight Center, Greenbelt, Maryland. 5 + 139 pp.
(1971).

094.230 **Apollo 14 photographic catalog.**
Prepared by Mapping Sciences Branch, Manned
Spacecraft Center, National Aeronautics and Space Adminis-
tration Houston, Texas. NSSDC preparation, directed by A.
T. Anderson.
National Space Science Data Center, NSSDC 71-16c, NASA −
Goddard Space Flight Center, Greenbelt, Maryland. 5 + 375 pp
(1971).
This catalog contains proof prints of 70-mm and 5-in.
photographs taken during the Apollo 14 mission; uncontrolled
mosaics of surface panoramas and continuous orbital photo-
graphy are included.

094.231 **Measurement of sulphur concentrations and the
isotope ratios $^{33}S/^{32}S$, $^{34}S/^{32}S$ and $^{36}S/^{32}S$ in Apollo**
12 samples. H. G. Thode, C. E. Rees.
Earth Planet. Sci. Letters, Vol. 12, 434 - 438 (1971).
Sulphur concentrations in the five rock samples and one
fines sample from the Apollo 12 mission range from 550 ppm
to 903 ppm. $\delta^{34}S$ values for the rocks range from +0.37 $^{0}/_{00}$
to +0.68 $^{0}/_{00}$ while for the fines sample it is +8.70 $^{0}/_{00}$ (all with
respect to Canyon Diablo troilite). The corresponding values
obtained for $\delta^{33}S$ and $\delta^{36}S$ show that the isotope effects
involved are mass dependent and that the amounts of sulphur
present in the samples mask spallogenic contributions to ^{33}S
and ^{36}S.

094.232 **Moonquakes.**
G. Latham, M. Ewing, J. Dorman, D. Lammlein,
F. Press, N. Toksoz, G. Sutton, F. Duennebier, Y. Nakamura.
Science, Vol. 174, 687 - 692 (1971).
Over 100 events believed to be moonquakes have been
recorded by the two seismic stations installed on the lunar
surface during Apollo missions 12 and 14. With few exceptions, the moonquakes occur at monthly intervals near times
of perigee and apogee and show correlations with the longer-
term (7-month) lunar gravity variations. The repeating moonquakes are believed to occur at not less than 10 different locations. However, a single focal zone accounts for 80 percent of
the total seismic energy detected. The moonquakes appear to
be releasing internal strain of unknown origin, the release
being triggered by tidal stresses.

094.233 Gravity measured at the Apollo 14 landing site.
R. L. Nance.
Science, Vol. 174, 1022 - 1023 (1971).
The gravity at the Apollo 14 landing site has been determined from the accelerometer data that were telemetered from the lunar module. The values for the lunar gravity measured at the Apollo 11, 12, and 14 sites were reduced to a common elevation and were then compared between sites. The observed gravity was also used to compute the lunar radius at each landing site.

094.234 Cracking of lunar mare soil.
L. D. Jaffe.
Nature, Vol. 234, 402 - 403 (1971).
Pictures from Surveyor 3 and Apollos 11 and 12 give the impression that lunar surface in Oceanus Porcellarum and Mare Tranquillitatis soil tends to crack into thin flat "tiles" when disturbed, suggesting a thin rigid crust over a softer substrate. These pictures were taken at illumination angles high above the horizontal. Photographs of the same areas taken at low sun during Apollo 12 activity and stereopairs of them obtained from Apollos 11 and 12 show that the disturbed material consists of roughly equiaxed clods. The impression of flat "tiles" and crusting is an illusion.

094.235 Internal friction in moon and earth rocks.
W. P. Mason.
Nature, Vol. 234, 461 - 463 (1971).
Using a low frequency displacement component as well as the Granato-Lücke velocity damping type for dislocations, it is shown that the combination agrees with measurements of Q^{-1} for three rocks. Most all of the Q^{-1} is connected with grain boundaries. By applying high hydrostatic pressures, the Q^{-1} reduces to that for flawless rocks. This Q^{-1} is less for moon rocks since the gravitational stresses are not large enough to start dislocation mills whereas the stresses are larger for the outer mantle and higher dislocation densities result with an associated higher Q^{-1}.

094.236 Potential of the mascon-free moon. C. L. Goudas.
Nature, Phys. Sci., Vol. 234, 85 - 86 (1971).
The mascons were formed after the first eon of the life of the moon. Question, therefore, rises whether the moon was closer to a condition of hydrostatic equilibrium when the mascons were not present. The answer can be obtained by subtracting the effect of the mascons on the general force function of the moon. The subtraction is presented in this paper in which the mascons are treated as mass-points. The conclusion drawn from the results is that indeed the moon was closer to a condition of hydrostatic equilibrium before the formation of the mascons.

094.237 Altersbestimmungen an Mondproben.
J. Zähringer.
Physik und Kosmologie, (see 003.024), p. 56 - 62 (1971).

094.238 Mineralogische Zusammensetzung der Mondmaterie aus dem Meer der Ruhe. A. El Goresy.
Physik und Kosmologie, (see 003.024), p. 63 - 69 (1971).

094.239 Los FTLs. P. Corvan.
El Universo, Vol. 25, 100 - 102 (1971).

094.240 Posibles FTLs en el area de Aristarchus.
T. Moseley.
El Universo, Vol. 25, 102 - 103 (1971).

094.241 Posibles FTLs en el area de Aristarchus.
P. Moore.
El Universo, Vol. 25, 103 (1971).

094.242 Lunar igneous activity and differentiation.
G. Fielder.
Highlights of Astronomy, Vol. 2, (see 012.013), 142 - 154 (1971).

094.243 Seismology of the moon and implications on internal structure, origin and evolution.
M. Ewing, G. Latham, F. Press, G. Sutton, J. Dorman, Y. Nakamura, R. Meissner, F. Duennebier, R. Kovach.
Highlights of Astronomy, Vol. 2, (see 012.013), 155 - 172 (1971).

094.244 Induced and permanent magnetism on the moon: Structural and evolutionary implications.
C. P. Sonett, P. Dyal, D. S. Colburn, B. F. Smith, G. Schubert, K. Schwartz, J. D. Mihalov, C. W. Parkin.
Highlights of Astronomy, Vol. 2, (see 012.013), 173 - 188 (1971).

094.245 Selection of points for the development of a fundamental control system on the lunar surface.
A. A. Gurshtein, N. P. Slovokhotova.
The Moon, Vol. 3, 266 - 288 (1971).
The present paper, taking into consideration the pressing necessity for the creation of a unique lunar fundamental control system proposes a general principle for approach to the selection of features for the development of such system. While estimating the quality of the selected craters images we followed the requirements suggested by Franz (1901), which are valid up to the present time: (1) The diameter of craters as small as possible. (2) The contrast against the background as large as possible. (3) The shapes as clear as possible. (4) The form as near to a circle as possible. In an appendix a list of 192 craters for development of a first order fundamental selenodetic system is given.

094.246 Lava tubes and channels in the lunar Marius Hills.
R. Greeley.
The Moon, Vol. 3, 289 - 314 (1971).
This paper presents quantitative and qualitative geomorphic evidence for the existence of lava channels and partly collapsed lava tubes in the lunar Marius Hills region. Analogs are offered as terrestrial counterparts to the lunar structures described.

094.247 Simple mass distribution for the lunar potential.
S. L. Levie, Jr.
The Moon, Vol. 3, 315 - 325 (1971).
A set of twenty-one point masses gravitationally equivalent to the L1 lunar potential model is presented. By construction, the equivalence is valid only in a region of space 'sampled' by Apollo spacecraft. That region is taken to be a finite, torus-shaped shell. When used in place of the L1 model for Apollo 12 lunar orbit determination, the solution set gives spacecraft positions identical to within about 100 m.

094.248 A large scale surface pattern associated with the ejecta blanket and rays of Copernicus.
J. E. Guest, J. B. Murray.
The Moon, Vol. 3, 326 - 336 (1971).
Lunar Orbiter photographs of large, fresh rayed craters show that V-shaped features commonly occur on the outer part of the ejecta blanket and on the rays, and are associated with satellitic craters thought to be of secondary impact origin. A description of the form and distribution of the V-features surrounding Copernicus and a preliminary quantitative analysis are given.

094.249 Bearing strength of lunar soil. L. D. Jaffe.
The Moon, Vol. 3, 337 - 345 (1971).

Bearing capacities of lunar soil returned from Surveyor 3 vary from 0.02–0.04 N cm^{-2} at a bulk density of 1.15 g cm^{-3} to 30–100 N cm^{-2} at 1.9 g cm^{-3}. The relation between bulk density and logarithm of the bearing capacity is roughly linear. Preliminary comparison with bearing measurements made in-situ on the moon by remote-control techniques, prior to return of samples from the moon, suggests good agreement if the lunar material has a bulk density of about 1.6 g cm^{-3} at a depth of 2.5 cm.

094.250 Lunar surface temperatures from Apollo 12.
C. J. Cremers, R. C. Birkebak, J. E. White.
The Moon, Vol. 3, 346 - 351 (1971).
The diurnal variation of temperatures in the lunar surface layer is calculated using the measured properties of the Apollo 12 samples. The results are compared with similar calculations made using data from the Apollo 11 samples and with previous infrared temperature measurements. Comparisons are also made with prior calculations which used assumed properties. The maximum temperature at lunar noon is 389.4 K and the minimum temperature just before sunrise is 87.8 K.

094.251 "Goethite" on the moon and the possible existence of other oxidized and hydroxyl-bearing phases in lunar impactites.   S. O. Agrell, M. G. Bown, P. Gay, J. V. P. Long, J. D. C. McConnell.
Meteoritics, Vol. 6, 247 - 248 (1971). – Abstract.

094.252 Tritium in lunar material.   P. Bochsler, P. Eberhardt, J. Geiss, H. Loosli, H. Oeschger, M. Wahlen.
Meteoritics, Vol. 6, 251 (1971). – Abstract.

094.253 Grain size frequency distributions of lunar samples and estimation of lunar surface ages.
J. C. Butler, E. A. King, Jr., M. F. Carman.
Meteoritics, Vol. 6, 254 - 255 (1971). – Abstract.

094.254 Spallation noble gases in lunar rocks.
P. Eberhardt, J. Geiss, H. Graf, N. Grögler, M. Mörgeli, H. Schwaller, A. Stettler.
Meteoritics, Vol. 6, 263 (1971). – Abstract.

094.255 Trapped noble gases in lunar material.
P. Eberhardt, J. Geiss, H. Graf, N. Grögler, M. D. Mendia, M. Mörgeli, H. Schwaller, A. Stettler, U. Krähenbühl, H. R. von Gunten.
Meteoritics, Vol. 6, 264 (1971). – Abstract.

094.256 Mechanisms of lunar crater formation.   F. El-Baz.
Meteoritics, Vol. 6, 264 - 265 (1971). – Abstract.

094.257 The opaque mineralogy of Apollo 14 crystalline rocks.   A. El Goresy, P. Ramdohr, L. A. Taylor.
Meteoritics, Vol. 6, 266 (1971). – Abstract.

094.258 Evidence of meteorite impacts found in lunar soil and breccias of the Apollo landing sites.
W. v. Engelhardt, J. Arndt, W. F. Müller, D. Stöffler.
Meteoritics, Vol. 6, 267 (1971). – Abstract.

094.259 Li, Be and B abundances in fines from the Apollo 11, 12, 14 and Luna 16 missions.   O. Eugster.
Meteoritics, Vol. 6, 268 (1971). – Abstract.

094.260 On the origin of lunar albedo.
O. Eugster, M. Maurette.
Meteoritics, Vol. 6, 269 (1971). – Abstract.

094.261 Micron-sized impact craters on lunar samples and related simulation experiments.
H. Fechtig, A. Mehl, G. Neukum, E. Schneider.

Meteoritics, Vol. 6, 269 (1971). – Abstract.

094.262 Lunar chondrules.
K. Fredriksson, J. Nelen, A. Noonan.
Meteoritics, Vol. 6, 270 - 271 (1971). – Abstract.

094.263 Active and inert gases in Apollo 12 and 11 samples released by crushing at room temperature and by heating at low temperatures.
J. Funkhouser, E. Jessberger, O. Müller, J. Zähringer.
Meteoritics, Vol. 6, 271 - 272 (1971). – Abstract.

094.264 A dynamic mechanism for the formation of shatter cones.   P. J. S. Gash.
Meteoritics, Vol. 6, 273 (1971). – Abstract.

094.265 Impacts with oblique trajectories.   D. E. Gault.
Meteoritics, Vol. 6, 273 - 274 (1971). – Abstract.

094.266 Microcraters on lunar rock 12054.
J. B. Hartung, F. Horz, D. E. Gault.
Meteoritics, Vol. 6, 276 - 277 (1971). – Abstract.

094.267 On hydrogen, deuterium, helium, nitrogen and their solar wind components in lunar matter.
J. Hintenberger, S. Specht, H. Voshage.
Meteoritics, Vol. 6, 278 (1971). – Abstract.

094.268 Individual variations of cosmogenic and radiogenic rare gases in lunar soil components.   T. Kirsten.
Meteoritics, Vol. 6, 281 - 282 (1971). – Abstract.

094.269 Chondrules of lunar origin.
G. Kurat, K. Keil, M. Prinz, C. E. Nehru.
Meteoritics, Vol. 6, 285 - 286 (1971). – Abstract.

094.270 Geologic mapping of the moon, 1961–1971.
D. J. Milton.
Meteoritics, Vol. 6, 293 - 294 (1971). – Abstract.

094.271 Morphology of sprays from the moon and elsewhere.
G. Mueller.
Meteoritics, Vol. 6, 294 - 295 (1971). – Abstract.

094.272 A lunar feldspathic peridotite (12036) and its melt inclusions.   M. Prinz, K. Keil, G. Kurat, T. E. Bunch.
Meteoritics, Vol. 6, 301 - 302 (1971). – Abstract.

094.273 Petrology of Fra Mauro basalt 14310.
W. I. Ridley, R. J. Williams, H. Takeda, R. W. Brown, R. Brett.
Meteoritics, Vol. 6, 304 - 305 (1971). – Abstract.

094.274 Large scale cratering and cometary impacts.
D. J. Roddy.
Meteoritics, Vol. 6, 305 - 306 (1971). – Abstract.

094.275 Trace element studies of lunar samples.
C. C. Schnetzler, J. A. Philpotts.
Meteoritics, Vol. 6, 310 (1971). – Abstract.

094.276 Characteristic rim features of craters.   C. R. Seeger.
Meteoritics, Vol. 6, 313 - 314 (1971). – Abstract.

094.277 A dunite-norite lunar microbreccia.
G. J. Taylor, U. B. Marvin.
Meteoritics, Vol. 6, 320 (1971). – Abstract.

094.278 Composition and origin of lunar fines from Apollo 12 and 14.
H. Wänke, F. Wlotzka, A. Balacescu, F. Teschke.

Meteoritics, Vol. 6, 321 (1971). – Abstract.

094.279 **Summary of lunar stratigraphy and structure: A perspective for lunar sample analysis.**
D. E. Wilhelms, J. F. McCauley, E. C. T. Chao.
Meteoritics, Vol. 6, 324 - 326 (1971). – Abstract.

094.280 **Lunar metal.** F. Wlotzka, E. Jagoutz, H. Wänke.
Meteoritics, Vol. 6, 326 - 327 (1971). – Abstract.

094.281 **Petrological character of the Luna 16 sample from Mare Fecunditatis.**
J. A. Wood, J. B. Reid, G. J. Taylor, U. B. Marvin.
Meteoritics, Vol. 6, 327 (1971). – Abstract.

094.282 **Analytical lunar ephemeris. The mean motions.**
A. Deprit, J. Henrard, A. Rom.
Byull. Inst. Teoret. Astròn., *Leningrad*, Vol. 13, 1 - 12 (1971). In Russian.

Delaunay's program of solving in literal form the solar part of the lunar theory has been re-examined. The concept of a Delaunay operation has been replaced by that of a canonical transformation generated by the algorithm of Lie transforms. Delaunay's simplifications concerning the mass ratios have been removed. New elements have been introduced to make differentiation rules trivial and to keep apparent the d' Alembert characteristics of the formulas. All periodic terms to order 10 have been eliminated from the Hamiltonian of the main problem. The reduction of the Hamiltonian ended with the removal of the terms of long period. The reduced Hamiltonian has been differentiated to produce the mean motions in the longitude of the moon (to order 10), of its perigee and of its node (to order 8).

094.283 **On determining the topocentric distances to points of the lunar surface.** V. K. Abalakin.
Byull. Inst. Teoret. Astron., *Leningrad*, Vol. 13, 13 - 16 (1971). In Russian.

The present paper deals with the determination of geometric distance from a given point on the earth surface to a point with known selenographic coordinates on the surface of the moon. The relevant formulae have been derived for the geocentric rectangular frame of reference with the celestial equator as the fundamental plane.

094.284 **On computing the position of the selenographic zero-point in the geocentric frame of reference.**
V. K. Abalakin.
Byull. Inst. Teoret. Astron., *Leningrad*, Vol. 13, 17 - 20 (1971). In Russian.

The solution of the problem of determining the geocentric position of the zero-point of selenographic coordinates has been given in terms of rectangular coordinates.

094.285 **Gravitational anomalies and deviations of the plumb-line on the moon.** N. A. Chujkova.
Astron. Zhurn. Akad. Nauk SSSR, Vol. 48, 1322 - 1326 (1971). In Russian. English translation in Soviet Astron. AJ, Vol. 15, No. 6.

The map obtained by us of gravitational anomalies and deviations of the plumb-line of the moon, based on the Lorrell model (the model JPL-3), represents the general structure of the whole moon; the map, based on the Blackshear model, represents only the gravitational inhomogeneity of the equatorial belt on the front of the moon.

094.286 **On possible differences in the chemical composition of the earth and of the moon formed in the circumterrestrial swarm.** E. L. Ruskol.
Astron. Zhurn. Akad. Nauk SSSR, Vol. 48, 1336 - 1338 (1971). In Russian. English translation in Soviet Astron. AJ,

Vol. 15, No. 6.

The depletion of the lunar material in volatile elements and its relative enrichment in silicates as compared to the earth's material is discussed under the assumption of the origin of the moon in a circumterrestrial swarm.

094.287 **Expedition to Hadley-Apennine–3.** D. Baker.
Spaceflight, Vol. 13, 468 - 470 (1971).

094.288 **α-corundum from the lunar dust.**
B. Kleinmann, P. Ramdohr.
Earth Planet. Sci. Letters, Vol. 13, 19 - 22 (1971).

White grains with a bright luster were extracted from a heavy fraction ($D > 4.03 \, g/cm^3$) of the Apollo 11 fines. The diameters of these grains range from 40 to $100 \mu m$. They were identified as α-corundum by X-ray diffraction and by electron microprobe analysis. Scanning electron microscopy showed that these grains are composed of innumerable tiny, euhedral α-Al_2O_3 crystals with their typical habit.

094.289 **Lunar seismograms for LM and S-IVB impacts interpreted as modulation mirage,** E. Strick.
Earth Planet. Sci. Letters, Vol. 13, 23 - 31 (1971).
It is the general consensus of opinion of the scientific community that the unusual seismic properties observed from LM and S-IVB impacts in the Apollo experiments are not consistent with any of the proposed models of the lunar exterior. I submit that the reason a seeming incompatibility with the existing seismic theory occurs is due to an instrumental effect associated with nonlinear and mode of detection properties of the capacity seismometer.

094.290 **Neutron capture effects in lunar gadolinium and the irradiation histories of some lunar rocks.**
G. W. Lugmair, K. Marti.
Earth Planet. Sci. Letters, Vol. 13, 32 - 42 (1971).

The Gd isotopic composition in 19 lunar rock and soil samples from three Apollo sites is reported. Enrichments in $^{158}GdO/^{157}GdO$ due to neutron capture range up to 0.75%. A model is constructed which gives both average cosmic-ray irradiation depths and effective neutron exposure ages (T_n) for some rocks. Rock 14310 is the first lunar sample where Kr anomalies due to resonance neutron capture in Br are observed. A ^{81}Kr-Kr exposure age of 262 ± 7 my is calculated for this rock.

094.291 **Remanent magnetization of lunar samples.**
D. W. Strangway, G. W. Pearce, W. A. Gose, R. W. Timme.
Earth Planet. Sci. Letters, Vol. 13, 43 - 52 (1971).

The remanent magnetization of samples returned from the moon by the Apollo 11 and 12 missions consists, in most cases, of two distinct components: An unstable one and one of high stability, which is probably a thermoremanent magnetization due to cooling from above 800°C in the presence of a field of a few thousand gammas. Our data imply that the moon experienced a magnetic field that lasted at least from about 3.0 by to 3.8 by, which is the age range of Apollo 11 and 12 samples.

094.292 **Neutron capture on ^{149}Sm in lunar samples.**
G. P. Russ III, D. S. Burnett, R. E. Lingenfelter, G. J. Wasserburg.
Earth Planet. Sci. Letters, Vol. 13, 53 - 60 (1971).

High precision isotopic composition measurements of Sm have been carried out for two terrestrial and seven lunar samples from three Apollo sites. The lunar samples, selected to show a wide variation in cosmic ray exposure ages, have a wide range of enrichments in $^{150}Sm/^{154}Sm$ (up to 0.8%) and depletions in $^{149}Sm/^{154}Sm$ which are due to neutron capture. The ratio of the number of neutrons captured per atom by

^{149}Sm to ^{157}Gd is 0.9 and reflects a hardened lunar neutron spectrum.

094.293 Petrology of Apollo 11 sample 10071. A differentiated mini-igneous complex.
M. J. Drake, D. F. Weill.
Earth Planet. Sci. Letters, Vol. 13, 61 - 70 (1971).

Sample 10071, 33 is a thin section of Apollo 11 ferrobasalt showing an unusual dual texture: The thin section includes material with a distinct variolitic texture. The mineralogy and chemistry of the variolitic portion show it to be the product of rapid cooling of a liquid, intermediate between the typical Apollo 11 ferrobasalt and the associated Si and K-rich mesostasis.

094.294 Chemical composition of lunar anorthosites and their parent liquids. N. J. Hubbard, P. W. Gast, C. Meyer, L. E. Nyquist, C. Shih, H. Wiesmann.
Earth Planet. Sci. Letters, Vol. 13, 71 - 75 (1971).

Two types of lunar anorthosites, associated anorthositic materials and parent non-mare basalts exist on the moon. One type is low in alkalies and trace elements and includes 15415 and Apollo 11 anorthositic materials and derives from low K non-mare basalts. The second type is high in alkalies and trace elements and derived from the KREEP type of non-mare basalts.

094.295 Experimental petrology of Apollo 12 basalts: Part 1, sample 12009.
D. H. Green, N. G. Ware, W. O. Hibberson, A. Major.
Earth Planet. Sci. Letters, Vol. 13, 85 - 96 (1971).

The mineralogy of five examples of Apollo 12 basalts has been examined quantitatively using the electron microprobe for chemical analysis of olivine, pyroxenes, feldspar, spinels and ilmenite. The crystallization sequence and compositions of precipitated phases for each of these basalts have been determined experimentally at atmospheric pressure and at high pressure. The lunar sample, 12009, is a rapidly quenched basalt with microphenocrysts of olivine ($\sim$7%) and spinel in a cryptocrystalline matrix with many small microlites.

094.296 Age of an Apollo 15 mare basalt: Lunar crust and mantle evolution.
G. J. Wasserburg, D. A. Papanastassiou.
Earth Planet. Sci. Letters, Vol. 13, 97 - 104 (1971).

An internal Rb-Sr isochron for the large basalt boulder 15555 yields an age 3.32 ± 0.06 AE and an initial ^{88}Sr/^{86}Sr, $I = 0.69934 \mp 5$. These values may indicate that extensive lava flows occurred at $\sim$3.3 AE over widespread areas of the moon. The Sr composition of the anorthosite 15415 is as low as that of plagioclase extracted from the Apollo 11 low K rocks. We also report results on basaltic rock 14276 from Apollo 14 and summarize the Rb-Sr ages reported previously on Apollo 14 samples.

094.297 Lunar basalt genesis: The origin of the europium anomaly. A. L. Graham, A. E. Ringwood.
Earth Planet. Sci. Letters, Vol. 13, 105 - 115 (1971).

A peculiarity observed in the Apollo 11 and 12 basalts is the low relative abundance of the rare earth europium. The trace element abundances of lunar basalts are examined and the implications of the near-chondritic relative abundances of incompatible elements are discussed. Relations between partial melting processes in the earth's mantle and the lunar interior are discussed.

094.298 The luminescent and thermoluminescent properties of Apollo 12 lunar samples.
I. M. Blair, J. A. Edgington, R. A. Jahn.
Earth Planet. Sci. Letters, Vol. 13, 116 - 120 (1971).

We have studied the natural thermoluminescence (TL),

direct luminescence and induced TL of eight Apollo 12 lunar samples. The value of such studies in supplying information about both the past thermal and surface activity history of the lunar surface, and the possible causes of transient lunar phenomena has already been noted.

094.299 The geochemistry of the opaque minerals in Apollo 14 crystalline rocks.
A. El Goresy, P. Ramdohr, L. A. Taylor.
Earth Planet. Sci. Letters, Vol. 13, 121 - 129 (1971).

Of the many rocks collected during the Apollo 14 mission, only two are wholly basaltic in origin. Rock 14310, which has a chemical composition not unlike that of a high-alumina basalt, consists of euhedral plagioclase laths ($\sim$60%) with intersertal anhedral pyroxene, the opaques, and a late-stage mesostasis of fayalite, glass, tridymite, a phosphate mineral, and opaque minerals. No forsteritic olivine was encountered. Rock 14053 is more mafic than 14310, has an ophitic texture, and consists of 50% pyroxene, 40% plagioclase, and 10% of olivine, opaques, an unusual texture involving fayalite, as well as mesostasis.

094.300 Rb-Sr ages on density and size fractions of Apollo 11 fine soil sample 10084.
W. H. Pinson, Jr., P. M. Hurley.
Earth Planet. Sci. Letters, Vol. 13, 130 - 133 (1971).

Size and density fractions of Apollo 11 soil show that the materials $< 60\mu$m and < 3.0 g/cm^3 have ^{87}Rb/^{86}Sr and ^{87}Sr/^{86}Sr values that are closer to a 4.6 AE model isochron than coarser and more dense fractions. The results support the conclusion that the allochthonous part of the aggregate came in as impact ejecta from source materials which differentiated early in lunar history and were not subsequently modified in Rb/Sr ratio during a remelting or impact event.

094.301 Die Vermessung des Mondes.
M. Moutsoulas.
Jenaer Rundschau, (Jena Review), 16. Jahrgang, p. 278 - 281 (1971).

094.302 Radar studies of the moon. N. N. Krupenio.
"Nauka", Moskva. 172 pp. Price 67 Kop. (1971).
In Russian. – Review in Referativ. Zhurn. 51. Astron., 1.51.317; 62. Issled. kosm. prostranstva, 1.62.183 (1972).

094.303 Comparative analysis of modern selenodetic reference networks. V. S. Kislyuk.
Astrometriya i Astrofiz., *Kiev*, No. 13, (see 003.039), p. 19 - 29 (1971). In Russian.

Results are presented of comparison and analysis of lunar object location accuracy in the Schrutka-Rechtenstamm, Baldwin, AMS, ACIC, MAO AS Ukr. SSR, Mills and Arthur catalogues. Systematic and random differences of these catalogues are considered.

094.304 Attempt of constructing an independent system of basic points on the moon. V. S. Kislyuk.
Astrometriya i Astrofiz., *Kiev*, No. 13, (see 003.039), p. 30 - 43 (1971). In Russian.

A procedure is presented for compiling a positional catalogue of 50 points on the moon in a system whose scale and orientation are derived independently from those given by other authors. The catalogue is compiled on the basis of measuring 18 lunar plates taken by means of the 400-mm astrograph of the Main Astronomical Observatory of the Ukrainian Academy of Sciences.

094.305 Comparison of the charts compiled by Hayn, Weimer, Watts and Nefedyev for the lunar limb zone. L. N. Kizyun.
Astrometriya i Astrofiz., *Kiev*, No. 13, (see 003.039), p. 43 -

54 (1971). In Russian.

Corrections are calculated to the moon's orbital elements from meridian observations carried out at the United States Naval Observatory during 1956–1967 and discussed before and after application of corrections for limb profile irregularities from Hayn's, Weimer's, Watt's and Nefedyev's charts. All the current charts of the marginal zone have systematic errors which should be taken into consideration when reducing observations by these charts. Watt's chart is the most perfect with regard to random errors.

094.306 **Procedure for determination and reduction to a single system of absolute heights of the lunar marginal zone.** I. V. Gavrilov, A. S. Duma.
Astrometriya i Astrofiz., *Kiev*, No. 13, (see 003.039), p. 54 - 61 (1971). In Russian.

To determine reliably absolute heights of lunar points it is necessary to measure the moon's limb within a single system with selected basis points on the lunar disk. The article deals with a description and discussion of the procedure for reduction of such measurements and determining from them absolute heights of the marginal zone. An example of reduction of 18 lunar plates shows that the proposed procedure yields results in a single system.

094.307 **Thin highly conducting layer in the moon: Consistent interpretation of dayside and nightside electromagnetic responses.** G. Schubert, D. S. Colburn.
Journ. Geophys. Res., Vol. 76, 8174 - 8180 (1971).

The vacuum transient response of the moon to a time-varying spatially uniform magnetic field is determined for a lunar electrical conductivity model that was based on the harmonic analysis of Apollo 12 and Explorer 35 dayside magnetometer data. A model containing a conducting core and a highly conducting thin subsurface layer is presented, and its transient behavior is discussed.

094.308 **Differential thermal analysis and gas release studies of Apollo 11 samples.**
F. M. Wachi, D. E. Gilmartin, J. Oró, W. S. Updegrove.
Icarus, Vol. 15, 304 - 313 (1971).

The search for carbon-containing compounds was performed by direct pyrolysis of lunar fines in the mass spectrometer. Trace quantities of organic matter observed could be attributed to contaminants either from the exhaust of the lunar module or from terrestrial handling of the lunar fines.

094.309 **Possibility of a layered moon.**
Z. F. Daneš, D. R. McNeely.
Icarus, Vol. 15, 314 - 318 (1971).

Recent lunar data can be interpreted as indicating the moon has a hard "crust" a few km thick, an "asthenosphere" about a hundred km thick, and a high viscosity "lower mantle."

094.310 **Rima Goclenius II.** R. B. Baldwin.
Journ. Geophys. Res., Vol. 76, 8459 - 8465 (1971).

The prominent rille Rima Goclenius II, which crosses the lunar crater Goclenius nearly centrally, appears to have been formed as a secondary effect after the opening of a deep-seated fracture. The main fracture was formed with an opening of about 175 meters along much of its length. A secondary dip-slip fracture then appeared, and a wedge-shaped piece of ground slid downward along this secondary fault, joining the primary fracture about 1 km below the present mare filling of Goclenius. The width of the resulting rille is a linear function of the absolute elevation of the ground along each point of the rille.

094.311 **Determination of the parameters of a selenocentric reference system and the deflections of the vertical at the lunar surface.** M. Burša.

Studia, Vol. 15, 210 - 227 (1971).

094.312 **Distribution of radon-222 on the surface of the moon.** D. Heymann, A. Yaniv.
Nature, Phys. Sci., Vol. 233, 37 - 39 (1971).

094.313 **Model of early lunar differentiation.**
V. R. Murthy, N. M. Evenson, H. T. Hall.
Nature, Vol. 234, 267, 290 (1971).

094.314 **On the thermal properties of the moon.**
M. S. Krass.
Vestn. Mosk. un-ta. Geologiya, 1971, No. 4, p. 72 - 82. In Russian. – Abstr. in Referativ. Zhurn. 51. Astron., 2.51.281 (1972).

094.315 **Rules for depth distribution of cosmogenetic isotopes in lunar rocks.**
A. K. Lavrukhina, G. K. Ustinova.
Astron. vestn., Vol. 5, 144 - 149 (1971). In Russian. – Abstr. in Referativ. Zhurn. 51. Astron., 2.51.313; 62.Issled. kosm. prostranstva, 2.62.184 (1972).

094.316 **Statistical analysis of lunar crater chains.**
G. Fielder.
Separate print from "Geological problems in lunar and planetary research". Science and Technology, [American Astronaut. Soc., Tarzana, California], Vol. 25, 393 - 397 (1971).

Linear arrays of lunar craters were investigated by statistical techniques using photography of the A, B, and P cameras of the Ranger 8 and 9 probes.

094.317 **Lava flows and the origin of small craters in Mare Imbrium.** G. Fielder, J. Fielder.
Separate print from "Geology and physics of the moon" [Elsevier Publishing Company, Amsterdam], p. 15 - 26 (1971).

094.318 **Untersuchungen an den Mondproben.**
H. Wänke, F. Wlotzka.
Universitas, [Wissenschaftliche Verlagsanstalt, Stuttgart], Vol. 26, 845 - 852 (1971).

094.319 **Temperature distribution in lunar rilles.**
A. S. Adorjan.
Journ. Spacecraft and Rockets, Vol. 8, 669 - 674 (1971).

An analytical investigation on the temperature distribution in lunar rilles is presented. The governing equations of the heat balance in the rille are formulated in Fredholm's integral equations of the second kind. A numerical solution method is selected to solve the governing equations by using successive substitution in the form of Neumann's series. The convergence of these series is discussed and a method developed on the solution of the heat balance equations, numerical examples are given to illustrate the temperature distribution in rilles with different width to depth ratios at various solar elevation angles.

094.320 **Spectrophotometry of the moon, Mars, and Uranus.** R. H. Vainkin.
Thesis, Univ. California, Los Angeles. [Available from Univ. Microfilms, Ann Arbor, Mich., U.S.A. Order No. 71–3853], 204 pp. (1970).

Photoelectric measurements of the irradiance from the discs of Mars and Uranus and from local areas on the moon and Mars, have been made with the Mount Wilson 60 inch reflector and scanning spectrometer. The radiance factors and albedos of the objects were calculated with solar irradiance values.

094.321 **Lunar gravity measurement from Apollo 14 television.** W. Hooper.
American Journ. Phys., Vol. 39, 974 (1971).

094.322 Brief review of thermoluminescence studies in lunar samples.
H. P. Hoyt, Jr., M. Miyajima, R. M. Walker.
Modern. Geol., (*GB*), Vol. 2, 263 - 264 (1971). – See Phys. Abstr., Vol. 74, No. 74628 (1971).

094.323 Rotational velocity of the moon–earth system and the mean motion of the moon. Y. Hatanaka.
Astron. Herald, (*Japan*), Vol. 64, 40 - 44 (1971). In Japanese. See Phys. Abstr., Vol. 74, No. 77535 (1971).

094.324 Heat conduction calculation for a model of the surface of the moon. M.-K. Liu, F. A. Williams.
International Journ. Heat Mass Transfer, (*GB*), Vol. 14, 1843 - 1851 (1971).

Four layers, a dust layer, an anorthosite layer, a basalt layer and a dunite layer, have been used to model the surface of the moon 4.6 billion years ago. The objective of the analysis is to determine whether the solidification of the basalt could be delayed about 1 billion years to produce the observed age difference between the dust and the crystalline rocks found on the surface of the moon.

094.325 Magnetic properties of Apollo 12 lunar samples.
S. K. Runcorn, D. W. Collinson, W. O'Reilly, A. Stephenson, M. H. Battey, A. J. Manson, P. W. Readman.
Proc. Roy. Soc. London, Ser. A, Vol. 325, 157 - 174 (1971).

094.326 Lunar crater origin in the maria from analysis of Orbiter photographs. G. Fielder, R. J. Fryer, C. Titulaer, A. K. Herring, B. Wise.
Phil. Trans., Ser. A, No. 1215, Vol. 271, 361 - 409 (1971).

One third of a million counts on 73127 craters in 22 test areas of lunabase have been made with the aim of diagnosing the origin of the craters that are predominantly between 100 and 2000 m in diameter by the use of a statistical method that is capable of measuring both the chaining and clustering of craters. Independent measurements of clustering have been made using photometry and using equi-areal counts to determine the number density of craters.

094.327 Morphology of the moon and the analysis of the chemical composition of the lunar surface.
S. V. Viktorov.
Proc. Sixth Winter School on Space Physics, Part II. Apatity 1969, (see 012.034), p. 28 (1971). – Abstract.

094.328 The ages of the lunar seas. L. B. Ronca.
Proc. National Acad. Sci. USA, Vol. 68, 1188 - 1189 (1971).

Using four possible relationships between geomorphic index and age, the author concludes that the age of the youngest effusions is less than 3×10^9 years and the age of the oldest effusions is more than 4×10^9 years. The results of the analyses of the Russian Luna 16 samples, although preliminary, fit in this interpretation.

094.329 Mineralogy and petrology of some Apollo 12 samples.
M. R. Dence, J. A. V. Douglas, A. G. Plant, R. J. Traill.
Proc. Second Lunar Sci. Conference, [The M. I. T. Press, Cambridge, Mass.], Vol. 1, 285 - 299 = Contr. Earth Phys. Branch, Ottawa, Canada, No. 353 (1971).

Five crystalline rocks, one breccia, and a sample of coarse (1–2 mm) fines have been examined.

Physics of the moon and planets.
See Abstr. 003.014.

Moon rocks and minerals. Scientific results of the study of the Apollo 11 lunar samples with preliminary data on Apollo 12 samples. See Abstr. 003.030.

A new photographic atlas of the moon.
See Abstr. 003.085.

Electromagnetic exploration of the moon.
See Abstr. 003.089.

On the use of lunar observations for the improvement of the zero-points of star catalogues and of the elements of the lunar orbit. See Abstr. 041.010.

Apollo 14: A visit to Fra Mauro. 1, 2.
See Abstr. 053.009.

Apollo 15 television from the moon.
See Abstr. 053.013.

The Apollo missions. See Abstr. 053.026.

The dynamical properties and internal constitutions of the earth, the moon and the planets.
See Abstr. 081.015.

Isotopic composition of xenon from a Greenland anorthosite. See Abstr. 081.019.

The earth and the moon. See Abstr. 081.025.

Planetología comparada. See Abstr. 081.046.

The value of photoelectric occultation timings in lunar motion studies. See Abstr. 096.022.

The investigation of lunar limb structure by means of stellar occultations. See Abstr. 096.025.

Photoelectric measurements of lunar occultations. V. Observational results. See Abstr. 096.040.

Some modern aspects of cosmic mineralogy. I.
See Abstr. 105.006.

Crustal thickness and the forms of impact craters.
See Abstr. 105.050.

Brecciation processes in extraterrestrial matter.
See Abstr. 105.095.

Some limits on the micrometeoroid flux at 1 au imposed by lunar erosion rates. See Abstr. 105.101.

Metallic spherules – their formation on the earth and on the moon. See Abstr. 105.103.

A comparison between a Java tektite (J2) and lunar rock 12013. See Abstr. 105.140.

Tektites and the moon. See Abstr. 105.163.

Chemical links between stony meteorites and the lunar surface. See Abstr. 105.169.

An investigation of the lunar ejecta theory of the cause of the Gegenschein. See Abstr. 106.004.

The origin of the system earth – moon.
See Abstr. 107.009.

Lunar composition as a clue to the early history of the solar system. See Abstr. 107.015.

095   Lunar Eclipses

095.001   **Chart of isochrones of the lunar eclipse of 1971 August 6.**   S. M. Kozik.
Astron. Tsirk., No. 618, p. 3 - 5 (1971).   In Russian.

095.002   **Algunos comentarios sobre el eclipse total de luna del 10 de febrero de 1971.**   F. Diego Q.
El Universo, Vol. 25, 36 - 38 (1971).

095.003   **Eclipse total de luna.**   J. Rubí Garza.
El Universo, Vol. 25, 39 - 40 (1971).

095.004   **The total eclipse of the moon on August 6th.**
Sky Telescope, Vol. 42, 243 - 245 (1971).

095.005   **Amateur's photographs of August lunar eclipse.**   N. B. Dumas.
Monthly Notes Astron. Soc. Southern Africa, Vol. 30, 128 (1971).

095.006   **Total eclipse of the moon.**   M. M. Dagaev.
Priroda, No. 11.71, p. 15 (1971). In Russian.

095.007   **Total lunar eclipse on August 6, 1971.**   V. Lasarevski.
Astron. Tsirk., No. 645, p. 8 (1971). In Russian.

095.008   **Visual observation of contacts of lunar features during the lunar eclipse 6 August 1971.**
M. Dujnič.
Říše hvězd, Vol. 52, 238 (1971).   In Slovak.

095.009   **Observation of the total lunar eclipse. (6. 8. 1971).**   G. Branislav, M. Milan.
Vasiona, Vol. 19, 71 - 73 (1971). In Serbo-Croatian.

095.010   **Totalna pomrčina Mjeseca 10. II 1971.** (Total lunar eclipse 10 February, 1971).   P. Krešimir.
Vasiona, Vol. 19, 73 (1971).

Periodiciteit in het optreden van eclipsen.
See Abstr. 079.001.

096   Lunar Occultations

096.001   **Mars occultation roundup.**
Sky Telescope, Vol. 42, 48 - 51 (1971).

096.002   **Three grazing occultations.**
J. Hers, M. D. Overbeek.
Monthly Notes Astron. Soc. Southern Africa, Vol. 30, 85 - 92 (1971).

096.003   **Observations of occultations of stars by the moon at the Astronomical Observatory of the Kiev University.**   A. K. Osipov.
Vestn. Kiev. Un-ta, Ser. Astron., No. 12, p. 96 - 97 (1970). In Russian.

096.004   **Occultations rasantes en France, janvier—juin 1972.**   J. Meeus.
L'Astronomie, 85ᵉ année, p. 377 - 378 (1971).

096.005   **Occultation of ZC 3091 during the total eclipse of the moon on 1971 August 6.**   G. J. Muller.
Monthly Notes Astron. Soc. Southern Africa, Vol. 30, 120 - 122 (1971).

096.006   **Some notes on the timing of grazing occultations.**   K. J. Sterling, P. J. Adair.
Monthly Notes Astron. Soc. Southern Africa, Vol. 30, 123 (1971).

096.007   **Rakende sterbedekkingen, januari - juni 1972.**   J. Meeus.
Hemel en Dampkring, Vol. 69, 295 (1971).

096.008   **An inexpensive pulse counter for photometry of occultations.**   R. A. Berg.
Publ. Astron. Soc. Pacific, Vol. 83, 433 - 437 (1971).
RUFAS is a two-channel pulse counter for photoelectric occultation observations. The minimum pulse pair resolution is 80 nanoseconds, although under normal observing conditions the dead time is at least 200 nanoseconds. The counter can be used in other photon-counting applications; because of the large amount of data output, computer interfacing is desirable and possible.

096.009   **Occultation observation in 1970.**
Data Rep. Hydrographic Observations, Ser. Astron., Geod., *Tokyo*, (Pub. 691), No. 6, p. 1 - 17 (1971).
This is a continuation of the series of the report of occultation observations made by the Hydrographic Department and some cooperators in Japan, and contains the data for 1970.

096.010   **Effect upon occultations of lunar surface structure along the line of sight.**   P. Murdin.
Astrophys. Journ., Vol. 169, 615 - 616 (1971).
It is made clear why Fresnel diffraction at a straight edge of negligible thickness is an appropriate model for describing occultations of stars by the moon.

096.011   **L'occultation de Mars du 15 mai 1972.**   J. Meeus.
L'Astronomie, 85ᵉ année, p. 447 - 449 (1971).

096.012   **The reappearance of stars from occultation.**   R. L. Waterfield.
Journ. British Astron. Ass., Vol. 82, 46 - 48 (1971).

096.013   **Vorausberechnete Sternbedeckungen durch den**

Mond 1972.
Astron. Nachr., Vol. 293, 137 - 142 (1971).

096.014   Streifende Sternbedeckung am 30. September
          1971.   W. Jaschek.
Sternenbote, 14. Jahrgang, p. 131 - 133 (1971).

096.015   Streifende Sternbedeckung am 29. Dezember 1971.
          W. Jaschek.
Sternenbote, 14. Jahrgang, p. 176 - 177 (1971).

096.016   Observations des occultations faites à Beograd en
          1968, 1969 et 1970.   M. B. Protitch.
Bull. Obs. Astron. Beograd, Vol. 28, (No. 124), 173 - 175
(1970).

096.017   Réductions des occultations observées à Beograd
          en 1969 et 1970.   M. Simić.
Bull. Obs. Astron. Beograd, Vol. 28, (No. 124), 177 - 178
(1970).

096.018   Sur la correction ΔT = TE − TU, d'après les observa-
          tions des occultations à Beograd en 1969.
M. B. Protitch.
Bull. Obs. Astron. Beograd, Vol. 28, (No. 124), 179 - 180
(1970).

096.019   Ocultación parcial de Marte del 16 mayo de 1971.
          F. Diego Q.
El Universo, Vol. 25, 80 - 84 (1971).

096.020   Ocultación parcial de Marte por la luna.
          J. G. Hernández.
El Universo, Vol. 25, 85 - 86 (1971).

096.021   Occultations of stars by the moon observed at the
          Cracow Astronomical Observatory in the year 1970.
M. Winiarski.
Acta Astron., Vol. 21, 529 - 532 (1971).
    Results are given of 115 observations of occultations of
stars by the moon made at the old Cracow Observatory and
the new observatory "Fort Skała".

096.022   The value of photoelectric occultation timings in
          · lunar motion studies.   T. C. van Flandern.
Highlights of Astronomy, Vol. 2, (see 012.018), 587 - 588
(1971).

096.023   A comparative study of visual and photoelectric
          timing of occultations.   L. V. Morrison,
Highlights of Astronomy, Vol. 2, (see 012.018), 589 - 591
(1971).

096.024   Geodetic applications of grazing occultations.
          D. W. Dunham.
Highlights of Astronomy, Vol. 2, (see 012.018), 592 - 600
(1971).

096.025   The investigation of lunar limb structure by means
          of stellar occultations.   D. S. Evans.
Highlights of Astronomy, Vol. 2, (see 012.018), 601 - 606
(1971).

096.026   Optical and radio occultation analysis.
          C. Hazard.
Highlights of Astronomy, Vol. 2, (see 012.018), 607 - 621
(1971).

096.027   Seeing effects on occultation curves.   A. T. Young.
          Highlights of Astronomy, Vol. 2, (see 012.018),
622 - 623 (1971).

096.028   A data acquisition system with on-line computer.
          E. Høg.
Highlights of Astronomy, Vol. 2, (see 012.018), 624 - 625
(1971).

096.029   Lunar occultation theory and techniques.
          K. R. Lang.
Highlights of Astronomy, Vol. 2, (see 012.018), 626 - 635
(1971).

096.030   Photoelectric observations of occultations in Japan.
          A. M. Sinzi.
Highlights of Astronomy, Vol. 2, (see 012.018), 636 - 637
(1971).

096.031   A preliminary analysis of photoelectric occultation
          measurements.   E. Pansch, C. de Vegt.
Highlights of Astronomy, Vol. 2, (see 012.018), 638 - 645
(1971).

096.032   The effects of filters and colour on stellar occulta-
          tions and appropriate deconvolution procedures.
T. Krishnan.
Highlights of Astronomy, Vol. 2, (see 012.018), 646 - 661
(1971).

096.033   Remarks on the restoration of occultation observa-
          tions.   T. J. Deeming.
Highlights of Astronomy, Vol. 2, (see 012.018), 662 - 667
(1971).

096.034   Analysis of lunar occultation data.
          M. M. McCants, R. E. Nather.
Highlights of Astronomy, Vol. 2, (see 012.018), 668 - 674
(1971).

096.035   Photometric observations of the occultations of
          stars by the moon.   K. D. Rakos.
Highlights of Astronomy, Vol. 2, (see 012.018), 675 - 687
(1971).

096.036   Some recent observations of occultations by the
          moon.   N. M. White.
Highlights of Astronomy, Vol. 2, (see 012.018), 688 - 691
(1971).

096.037   Photoelectric occultation observations of Regulus
          and the Pleiades.   R. A. Berg.
Highlights of Astronomy, Vol. 2, (see 012.018), 700 - 707
(1971).

096.038   Occultation studies at the Dominion Astrophysical
          Observatory.   C. L. Morbey, J. B. Hutchings.
Highlights of Astronomy, Vol. 2, (see 012.018), 708 - 712
(1971).

096.039   Photoelectric observations of stellar occultations:
          Closing remarks.   D. S. Evans.
Highlights of Astronomy, Vol. 2, (see 012.018), 721 - 722
(1971).

096.040   Photoelectric measurements of lunar occultations.
          V. Observational results.   D. S. Evans.
Astron. Journ., Vol. 76, 1107 - 1116 (1971).
    Observational timings of 254 lunar occultations, almost
all disappearances, made between December 1968 and April
1971 with the McDonald 36-inch reflector, using equipment
and techniques previously described, are reported. For some
40% of the observations in which the observational noise was
sufficiently small and the diffraction pattern well enough defined, a lunar limb slope has been derived. The determination

of limb slopes is closely bound up with the use of occultation observations for astrometry of double stars. Seven cases of discovery of double stars are included in notes to the main table of results.

096.041 **Lunar occultations of two multiple systems.**
W. I. Beavers, J. J. Eitter.
Astron. Journ., Vol. 76, 1131 - 1132 (1971).

The photoelectric measurement of the lunar occultation of BD+ 27° 943 reveals that the object is a triple system. Similar simultaneous two-color measurements of the known binary BD+ 18° 2057 provide a measurement of projected separation and magnitude difference.

096.042 **Practical observability code for total occultations.**
L. Pazzi.

Monthly Notices Astron. Soc. Southern Africa, Vol. 30, 153 - 154 (1971).

096.043 **Grazing occultations, 1972.**
Southern Stars, Vol. 24, 78 - 79 (1971).

Angular diameters of the red giants 46 Leo and ϕ Aqr and parameters of some binary systems from occultation observations. See Abstr. 115.011.

Stellar diameters from occultations.
See Abstr. 115.014.

Lunar occultation observations of PKS 1514–24 (=AP Lib). See Abstr. 141.179.

097 Mars

097.001 Photographic observations of Deimos at the Main Astronomical Observatory of the Ukrainian Academy of Sciences in 1967. A. B. Onegina, E. M. Sereda. Byull. Inst. Teoret. Astron., *Leningrad*, Vol. 12, 732 - 738 (1971). In Russian.

097.002 The motion of Mars: 1751 - 1969. R. E. Laubscher. Astron. Astrophys., Vol. 13, 426 - 436 (1971).

Clemence's new theory for the motion of Mars, along with Newcomb's Tables of the Sun, are compared with all available meridian and radar ranging observations of Mars. It is shown that Clemence's theory is wholly adequate for representing the meridian observations of Mars, and definitive mean elements of the orbit of the planet are derived. The long-standing problem of accurately determining the mass of Venus from observations of Mars is resolved, and a new value for that mass presented. New determinations of the corrections to the obliquity of the ecliptic, the mean longitude of the earth, the equinox and equator point of the adopted star reference system, and their rates, are also presented.

097.003 Mars: Narrow-band photometry, from 0.3 to 2.5 microns, of surface regions during the 1969 apparition. T. B. McCord, J. A. Westphal. Astrophys. Journ., Vol. 168, 141 - 153 = Contr. 13, Planet. Astron. Lab. = Contr. 1911, Division Geolog. Sci. (1971).

Narrow-band spectrophotometric observations were made of areas on Mars 200 km in diameter and of the whole disk of Mars during the 1969 apparition. The spectral region from 0.3 to 2.5 μ was covered with a resolution of 0.02–0.07 μ. The spectral reflectivities of six Martian areas relative to the bright region Arabia were obtained near opposition. Syrtis Major and Arabia were also observed before opposition. Geometric albedos have been calculated for five areas between 0.3 and 1.1 μ and for two additional areas and the whole disk of Mars between 0.3 and 2.5 μ.

097.004 A Mariners' 1969 closeup map of Mars. C. A. Cross. Sky Telescope, Vol. 42, 16 - 17 (1971).

097.005 The ionosphere of Mars below 80 km altitude–II. Solar cosmic ray event. R. C. Whitten, I. G. Poppoff, J. S. Sims, W. A. Barker, P. T. McCormick, J. Dubach. Planet. Space Sci., Vol. 19, 971 - 979 (1971).

Models of the lower ionosphere of Mars during a solar proton event are developed using the ion chemistry discussed in a previous paper (Planet. Space Sci., Vol. 19, 243 - 250 , 1971). The computed electron number densities for two representative proton spectra appear to be large enough to allow one to easily determine the entire profiles of electron concentration by means of a bistatic radar occultation experiment.

097.006 The experiment of the determination of differences of heights on Mars on the basis of intensities of CO_2 bands λ 1.6μ. V. I. Moroz, N. A. Parfentjev, D. P. Cruikshank, L. V. Gromova. Astron. Zhurn. Akad. Nauk SSSR, Vol. 48, 790 - 794 (1971). In Russian. English translation in Soviet Astron. AJ, Vol. 15, No. 4.

Apparatus, method and first results of the determination of differences of heights on Mars from the relative intensity of CO_2 bands in various parts of the planet are described.

097.007 An optical model of the Martian surface in the visible region of the spectrum. A. V. Morozhenko, E. G. Yanovitsky. Astron. Zhurn. Akad. Nauk SSSR, Vol. 48, 795 - 809 (1971). In Russian. English translation in Soviet Astron. AJ, Vol. 15, No. 4.

It is assumed that the surface layer of Mars is composed of semitransparent particles and is consistent with the Hapke model. The reflection law for the Martian surface is written down for that case. Values of particles packing density in the surface layer, of optical parameters and scattering indicatrices of these particles are obtained using phase curves of Mars in the spectral region 0.359 – 1.064 μ and the disk distribution of brightness in the wavelength 0.626 μ. This gives the possibility to calculate the intensity of radiation reflected by the Martian surface at any angles of incidence and reflection. A comparison of the results of calculations with observations shows a satisfactory agreement.

097.008 The atmospheres of Mars and Venus. A. P. Ingersoll, C. B. Leovy. Annual Rev. Astron. Astrophys., Vol. 9, (see 003.001), 147 - 182 (1971).

In this review, we shall compare the compositions, thermal structures, and dynamics of the lower and upper atmospheres of Mars and Venus, and shall consider some of the problems of their evolution. The remainder of the Introduction consists of brief reviews of attempts to detect an atmosphere on Mercury, and of the abundances of volatile constituents of the earth's atmosphere and oceans.

097.009 Mariner ultraviolet spectrometer: Topography and polar cap. C. A. Barth, C. W. Hord. Science, Vol. 173, 197 - 201 (1971).

Ultraviolet measurements reveal the topography of Mars and show that ozone may be adsorbed on the polar cap.

097.010 Zur Wolkenstatistik des Planeten Mars. W. Schlosser. Astron. Nachr., Vol. 292, 205 - 206 (1971).

The distribution of the Martian clouds observed at Hamburg Observatory during the oppositions 1948, 1958, 1963, 1965 and 1967 is graphically displayed. The number of clouds per degree longitude shows a positive correlation with the 21°5 North radar-altitude.

097.011 Gedanken zur Mars-Nomenklatur. J. Blunck. SuW, Vol. 10, 214 - 216 (1971).

097.012 Contribution à l'étude de la physicochimie de Mars. A. Dauvillier. Comptes Rendus Acad. Sci. Paris, Sér. B, Vol. 273, 410 - 414 (1971).

Il existe une analogie entre les hautes atmosphères de Mars et de Vénus qui pourraient renfermer le même aérosol. L'accélération séculaire de Phobos est rapprochée du phénomène analogue présenté par une douzaine de comètes à courte période. L'assombrissement saisonnier est interprété par l'inversion du relief. Les calottes polaires ne sont pas constituées d'hydrate d'anhydride carbonique.

097.013 Mars. J. L. Perdrix. Journ. Astron. Soc. Victoria, Vol. 24, 50 - 64 (1971).

097.014 Der Mars, Topographie und Atmosphäre I. D. Horn.

Phys. Blätter, 27. Jahrgang, p. 406 - 416 (1971).

097.015 On the secular acceleration of Phobos.
V. A. Shor, N. I. Glebova, L. I. Sorokina.
Astron. Tsirk., No. 617, p. 3 - 6 (1971). In Russian.

097.016 Sunspot areas and Martian blue clearing.
N. N. Petrova.
Astron. Tsirk., No. 622, p. 1 - 2 (1971). In Russian.

097.017 On falling of meteor matter on the Martian surface.
E. N. Kramer, I. S. Shestaka.
Astron. Tsirk., No. 625, p. 3 - 5 (1971). In Russian.

097.018 Earth/Mars orbit demonstrations. J. R. Millburn.
Spaceflight, Vol. 13, 259 - 262 (1971).

097.019 Telescopic observations of Mars in 1971.
G. de Vaucouleurs.
Sky Telescope, Vol. 42, 134 - 135 (1971).

097.020 An amateur's map of Mars in 1971.
R. E. Stencel.
Sky Telescope, Vol. 42, 181 (1971).

097.021 Marte oposición 1971. A. M. de Velasco.
El Universo, Vol. 25, 41 - 43 (1971).

**097.022 Observing Mars III - The ALPO 1971 observing
program.** C. F. Capen.
Strolling Astronomer, Vol. 23, 41 - 44 (1971).

097.023 Mars 1969 - The north polar region - ALPO report II.
C. F. Capen, T. R. Cave.
Strolling Astronomer, Vol. 23, 67 - 75 (1971).

097.024 Cloud formations on Mars in the summer of 1969.
V. V. Prokofjeva, N. A. Uschakova, A. N. Abra-
menko, A. K. Dabachov.
Astron. Tsirk., No. 635, p. 7 - 8 (1971). In Russian.

**097.025 The influence of the sun on the motion of the
Martian satellites.** S. N. Vashkov'yak.
Soobshch. Gos. Astron. Inst. Shternberga, No. 160, p. 3 - 24
(1971). In Russian.

097.026 Exploration of Mars after Viking.
H. D. Greyber.
Bull. American Astron. Soc., Vol. 3, 385 (1971). – Abstr.
AAS.

097.027 Sounding Mars atmosphere by limb scanning.
J. C. Gille, P. J. Gierasch.
Bull. American Astron. Soc., Vol. 3, 415 (1971). – Abstr.
AAS.

**097.028 A spectroscopic determination of pressure in the
Martian atmosphere from CO_2 bands.**
V. I. Moroz, D. P. Cruikshank.
Astron. Zhurn. Akad. Nauk SSSR, Vol. 48, 1038 - 1045
(1971). In Russian. English translation in Soviet Astron. AJ,
Vol. 15, No. 5.

**097.029 Mariner 1969 infrared radiometer results: Tempera-
tures and thermal properties of the Martian surface.**
G. Neugebauer, G. Münch, H. Kieffer, S. C. Chase, Jr., E.
Miner.
Astron. Journ., Vol. 76, 719 - 728, 747 - 749 (1971).
In this paper, refined estimates of the temperatures cha-
racterizing the top layer of the Martian soil are presented. On
the basis of these final results the thermophysical properties

of the various Martian terrains are inferred and discussed in
relation to their visual characteristics. Special consideration
is given to the measurements obtained over the south polar
cap flyby of Mariner 7 to derive the temperature of the
frost deposit forming the cap.

**097.030 On the minimal resolution needed for the observa-
tion of Martian surface features from earth distance.** M. S. Bobrov.
Astron. vestn., Vol. 5, 117 - 118 (1971). In Russian. – Abstr.
in Referativ. Zhurn. 51. Astron., 10.51.227 (1971).

**097.031 Measurements of Mars radio emission at 8.22 mm
and evaluation of thermal and electrical properties
of its surface.**
A. D. Kuzmin, B. Ya. Losovsky, Yu. N. Vetukhnovskaya.
Icarus, Vol. 14, 192 - 195 (1971).
In order to improve our knowledge of Mars we performed
new and more accurate measurements of Martian radio emission at 8.22-mm wavelength. These results were used to evaluate thermal and electrical properties of the Martian surface.

**097.032 Mars and Jupiter: Radio emission at 2.3 mm and
8.15 mm.**
V. A. Efanov, I. G. Moiseev, A. G. Kislyakov, A. I. Naumov.
Icarus, Vol. 14, 198 - 203 (1971).
Mars observations were made during the 1969 opposition
using the 22 m radio telescope of the Crimean Astrophysical
Observatory of the Academy of Sciences of the USSR. The
brightness temperature of Mars was determined by comparing
the intensity of its radio emission with that of Jupiter, the
brightness temperature of which was taken to be 140 ± 20° K
at λ = 2.3 mm and 144 ± 23°K at λ = 8.15 mm. With this
assumption, the Mars radio temperatures were found to be
240 ± 30°K at 2.3 mm and 210 ± 30°K at 8.15 mm.

**097.033 Mars: Measurements of its brightness temperature
at 1.85 and 3.75 cm wavelength.** M. J. Klein.
Icarus, Vol. 14, 210 - 213 (1971).
New measurements of the microwave temperature of
Mars are reported. The brightness temperatures measured
during the planet's close approach in 1967 were 182° ± 15°K
(m.e.) at 1.85 cm, and 200° ± 11°K (m.e.) at 3.75 cm.

**097.034 Mars: A possible discrepancy between the radio
spectrum and elementary theory.** E. E. Epstein.
Icarus, Vol. 14, 214 - 221 (1971).
The available radio observations of Mars have been
assessed. The radio spectrum does not turn up at short wavelengths, as predicted by elementary theory, but appears to be
flat or perhaps slightly convex. Our ideas about the origin of
the radio spectrum of Mars may require re-examination if
future accurate and precise short millimeter-wave observations
confirm the present spectrum.

097.035 The microwave spectrum of Mars: An analysis.
C. Sagan, J. Veverka.
Icarus, Vol. 14, 222 - 234 (1971).
A weighted least squares fit to the best available data on
the Martian microwave spectrum indicates that the brightness
temperature decreases from long to short wavelengths, rather
than increasing as expected from the solution of the one-
dimensional equation of heat conduction. Reasonable assumptions on the ratio of electrical to thermal skin depths, on internal heat sources, on ferromagnetic materials, on radiative
conduction, on compaction with depth, and on surface roughness all fail in reproducing the deduced spectrum. A layer of
liquid water some tens of microns thick, on the average, localized in the top few millimeters of a Martian epilith with refractive index $\simeq$ 1.6 fits the microwave spectrum, and the infrared and radar data as well. The origin of such a layer of

liquid water and its possible exobiological significance are discussed.

097.036 **The effect of haze on the visibility of Martian surface features.** D. J. van Blerkom.
Icarus, Vol. 14, 235 - 244 = Contr. Five College Obs., Univ. Mass., Amherst, Mass.,No. 84 (1971).

It is possible that an optically thin aerosol layer exists in the Martian atmosphere. This investigation questions whether such a layer can be identified with the "blue haze" which obscures surface detail at wavelengths shorter than λ 4500. It is assumed that a Martian haze is not unlike a terrestrial one, and may be approximated by a Henyey-Greenstein function. Radiation transfer theory for single scattering is used to compute the reduction in contrast of surface features due to diffuse reflection and diffuse transmission of radiation from areas of different surface albedo (blurring). It is concluded that forward scattering hazes cannot cause the strong reduction of surface contrast observed even for optical thickness as high as $\tau = 0.3$.

097.037 **Mars: The spectral albedo (03.-2.5μ) of small bright and dark regions.**
T. B. McCord, J. H. Elias, J. A. Westphal.
Icarus, Vol. 14, 245 - 251 (1971).

A review of the available spectral geometric albedo measurements for Mars was presented earlier for the spectral region 0.3 to 1.1μ. A new observational study has greatly increased the store of data, especially for small Martian regions and for the infrared spectral region 1.0 to 2.5μ. Here we combine the new data with data both from the earlier review and, for the infrared spectral region, from the literature. We present a more complete picture of Martian spectral reflectivity properties than was available. This study should provide a more firm basis upon which models of Martian surface composition can be built.

097.038 **Carbon suboxide on Mars: A working hypothesis.**
T. A. Perls.
Icarus, Vol. 14, 252 - 264 (1971).

The evidence for and against the formation of carbon suboxide polymer in the Martian atmosphere is examined critically. Present data allow photolysis and radiolysis of carbon monoxide to form carbon suboxide gas which rapidly polymerizes. These processes are expected to be greatly enhanced by ionizing radiation from the sun and it is noted that the great 1956 yellow clouds occurred after an exceptional solar flare. Many Mars observations could be explained in terms of particles of carbon suboxide polymer in the atmosphere and on the ground. Preliminary tests show that monomer in equilibrium with the polymer is not expected to be detectable at Mars surface temperatures.

097.039 **A preliminary assessment of Martian wind regimes.**
P. Gierasch, C. Sagan.
Icarus, Vol. 14, 312 - 318 (1971).

Elevation differences of the order of an atmospheric scale height are now known to exist abundantly over the Martian surface. The time-independent and frictionless thermal wind equation is solved for radiative-convective atmospheres at the level of scaling analysis, to determine the influence of topography on the mean wind, for large horizontal length scales.

097.040 **Mapping the surface of Mars.** C. A. Cross.
Spaceflight, Vol. 13, 402 - 407 (1971).

097.041 **Extensive cloud activity on Mars.**
D. Milon, G. P. Kuiper, S. Larson, R. B. Minton,
P. B. Boyce, K. Czuia, C. F. Capen.
IAU Circ., No. 2358 (1971).

097.042 **Disappearance of Martian south polar cap.**
P. B. Boyce, K. Czuia.
IAU Circ., No. 2359 (1971).

097.043 **Mars.** C. F. Capen.
IAU Circ., No. 2364 (1971).

097.044 **Planet Mars.** I. K. Koval'.
Zemlya i Vselennaya, No. 5, p. 26 - 30 (1971).
In Russian.

097.045 **Mariner 6 and 7 ultraviolet spectrometer experiment: Analysis of hydrogen Lyman-alpha data.**
D. E. Anderson, Jr., C. W. Hord.
Journ. Geophys. Res., Vol. 76, 6666 - 6673 (1971).

Mariner 6 and 7 ultraviolet spectrometers that flew by Mars in 1969 observed the Lyman-α dayglow of atomic hydrogen. Data in the altitude range 200 to 24000 km are analyzed to determine the structure of the Martian exosphere.

097.046 **Atomic carbon in the atmospheres of Mars and Venus.** M. B. McElroy, J. C. McConnell.
Journ. Geophys. Res., Vol. 76, 6674 - 6690 (1971).

Photochemical processes involving atomic carbon in the upper atmospheres of Mars and Venus are critically examined. It is shown that carbon is produced mainly by electron- and photon-induced dissociation of CO and CO_2, with CO relatively more important for Mars. Detailed calculations are presented for the dayglow emissions from C at 1657 and 1561 Å and results are compared with the Mariner 6 and 7 results for Mars.

097.047 **Size classification of Mars simulation samples.**
W. G. Egan.
Journ. Geophys. Res., Vol. 76, 6213 - 6219 (1971).

A detailed physical analysis was made of a particulate sample of limonite (1.19- to 2.38-mm ASTM sieve range, Venango County, Pennsylvania) shown previously to produce the best optical and physical match to the Martian bright areas.

097.048 **Mariner 6 and 7 television results.** B. A. Smith.
Space Research XI, (see 012.004), 155 - 164 (1971).

097.049 **Summary of Mariner 6 and 7 radio occultation results on the atmosphere of Mars.**
A. J. Kliore, G. Fjeldbo, B. L. Seidel.
Space Research XI, (see 012.004), 165 - 175 (1971).

097.050 **Comparison of parameters of the Mars ionosphere according to Mariners 4, 6 and 7 measurements and to calculations of radiowave absorption.** A. N. Kazantsev.
Space Research XI, (see 012.004), 177 - 179 (1971).

097.051 **On Gamma-radiation of the atmosphere and surface of Mars.** Yu. A. Surkov, L. P. Moskaleva, A. N. Khalemsky, V. P. Kharyukova.
Space Research XI, (see 012.004), 181 - 190 (1971).

097.052 **A unified procedure for the detection of life on Mars.** R. Radmer, B. Kok.
Science, Vol. 174, 233 - 239 (1971).

A mass spectrometer can be used to perform a variety of remote biologically oriented experiments.

097.053 **Mars in 1965, 1967, 1969.** A. K. Suslov.
Izv. Vses. geogr. o-va, Vol. 103, 261 - 263 (1971).
In Russian. — Abstr. in Referativ. Zhurn. 51. Astron., 11.51.327 (1971).

097.054 **Ground-based TV observations of the atmospheric features on Mars during photographing aboard**

Mariner 6 and 7.
V. V. Prokofjeva, V. K. Prokofjev, N. A. Ushakova.
Astron. Tsirk., No. 636, p. 6 - 8 (1971). In Russian.

097.055 **Der Stand der Mars- und Venusforschung.**
H. Heuseler.
Umschau, 71. Jahrgang, p. 892 - 898 = Weltraumfahrt, 22.
Jahrgang, p. 158 - 164 (1971).

097.056 **Mars op de korrel genomen.**
W. F. Gielingh, G. J. M. Besamusca.
Hemel en Dampkring, Vol. 69, 300 - 301 (1971).

097.057 **South polar and equatorial differences in central peaked Martian craters.**
B. M. Cordell, R. E. Lingenfelter, G. Schubert.
Nature, Vol. 234, 335 - 337 (1971).
    Central peaks in Martian craters in the south polar region occur more frequently than their equatorial counterparts and the peak frequency statistics and crater morphologies imply that preferential peak production has taken place.

097.058 **On the violet clouds of Mars.**
N. B. Ibragimov, I. K. Koval.
Astrometriya i Astrofiz., *Kiev*, No. 14, (see 003.014), p. 3 - 8 (1971). In Russian.
    Results are given of photometric investigations of Martian violet clouds based on observations carried out during 1958 - 1961, 1965 and 1969.

097.059 **Results of laboratory photometric measurements of artificial planets.**   O. I. Bugaenko, I. K. Koval, V. D. Krugov, A. V. Morozhenko, L. F. Slutsky.
Astrometriya i Astrofiz., *Kiev*, No. 14, (see 003.014), p. 33 - 48 (1971). In Russian.
    Results are given of photoelectric measurements of brightness distribution and brightness-phase dependence for six spheres the surface of which have different colours and roughness. These results are applied to verify the validity of some empirical formulas used earlier for determining optical parameters of the Martian surface.

097.060 **On the glow of a spherical planetary atmosphere from the limb side.**   E. G. Yanovitskiy.
Astrometriya i Astrofiz., *Kiev*, No. 14, (see 003.014), p. 55 - 63 (1971). In Russian.
    An approximate formula is obtained for computing the diffuse radiation of a spherical planetary atmosphere from the limb side. The case is considered when the atmosphere adjoins a surface that reflects light isotropically. For the case of very small optical thickness of the atmosphere a simple analytical expression is found for computing the diffuse radiation intensity on the assumption that the scattering coefficient of the aerosol and gas components of the atmosphere changes exponentially with altitude. This formula is used for the calculation of the limb side radiation intensity of the Martian atmosphere.

097.061 **Marstekeningen 1971.**   H. Nieuwenhuis.
Hemel en Dampkring, Vol. 69, 330 - 331 (1971).

097.062 **Mars 1971.**   G. Nemec.
SuW, Vol. 10, 336 (1971).

097.063 **Some high-resolution photographs of Mars.**
S. M. Larson, R. B. Minton.
Sky Telescope, Vol. 42, 260 - 261 (1971).

097.064 **International planetary patrol results.**
L. Martin, W. A. Baum.
Sky Telescope, Vol. 42, 261 - 262 (1971).

097.065 **Telescopic observations of Mars in 1971—II.**
G. de Vaucouleurs.
Sky Telescope, Vol. 42, 263 - 264 (1971).

097.066 **Dust storm observations from New Mexico.**
T. B. Kirby, J. C. Robinson.
Sky Telescope, Vol. 42, 264 - 265 (1971).

097.067 **A map of Mars' south pole.**   R. N. Watts, Jr.
Sky Telescope, Vol. 42, 271 (1971).

097.068 **Mapping Martian clouds from 1969 photographs.**
L. J. Martin, J. L. Smith.
Publ. Astron. Soc. Pacific, Vol. 83, 606 (1971). – Abstr.
Astron. Soc. Pacific.

097.069 **Mars 1969 – The north polar region – ALPO report**
II.   C. F. Capen, T. R. Cave.
Strolling Astronomer, Vol. 23, 79 - 85 (1971).

097.070 **The great Martian yellow cloud of 1971 – Preliminary report.**   C. F. Capen.
Strolling Astronomer, Vol. 23, 110 - 112 (1971).

097.071 **The areo-equatorial system of coordinates.**
S. V. Serova (Shilova).
Vestn. Leningr. un-ta, 1971, No. 7, p. 135 - 141. In Russian.
Abstr. in Referativ. Zhurn. 51. Astron., 12.51.272 (1971).

097.072 **Radio observations of Mars at 8.57 millimeters.**
P. M. Kalaghan, L. E. Telford.
Astrophys. Journ., (*Letters*), Vol. 170, L77 - L79 (1971).
    Radio emission of Mars at 8.57 mm has been observed during the planet's close approach in 1971. The ratio of the Martian brightness temperature to that of Jupiter was found to be 1.23 ± 0.03. If the brightness temperature of Jupiter is taken to be $142° \pm 12°$ K, then the resultant brightness temperature of Mars was $175° \pm 15°$ K.

097.073 **Astronomical infrared spectroscopy with a Connes-type interferometer: II – Mars, 2500 - 3500 cm^{-1}.**
R. Beer, R. H. Norton, J. V. Martonchik.
Icarus, Vol. 15, 1 - 10 (1971).
    New spectra of Mars in the 3 - 4 micron region at significantly higher resolution than previously available were obtained near the 1969 opposition. No features positively identifiable as being due to the Martian atmosphere could be detected. The existence of an albedo drop, probably due to surface water of hydration, is confirmed.

097.074 **The heat balance of the Martian polar caps.**
C. A. Cross.
Icarus, Vol. 15, 110 - 114 (1971).
    Calculations based on a model of the polar caps which has been simplified by neglecting the thermal conductivity of the ground confirm the main results obtained by Leighton and Murray (1966).

097.075 **A prediction concerning the cloud activity on Mars.**
R. A. Wells.
IAU Circ., No. 2372 (1971).

097.076 **La próxima oposición de Marte.**   C. Brean.
El Universo, Vol. 25, 87 - 89 (1971).

097.077 **¡Marte!**   F. Diego Q.
El Universo, Vol. 25, 138 - 141 (1971).

097.078 **Martian blue clearing in June 1969.**
V. V. Prokofjeva, N. A. Uschakova.
Astron. Tsirk., No. 648, p. 3 - 5 (1971). In Russian.

097.079 Mars: Has nitrogen escaped? R. T. Brinkmann.
Science, Vol. 174, 944 - 945 (1971).

If eddy mixing is about as effective on Mars as it is on earth, then there seems to be less nitrogen present on Mars than we would expect if terrestrial-type outgassing were the source. However, in this event a nonthermal escape mechanism involving the predissociation of exospheric nitrogen can be invoked to explain the low nitrogen concentration.

097.080 Martian craters and a scarp as seen by radar.
G. H. Pettengill, A. E. E. Rogers, I. I. Shapiro.
Science, Vol. 174, 1321 - 1324 (1971).

Radar observations of Mars with a surface resolution of 1.3° in latitude and 0.8° in longitude have been carried out during the opposition of 1971. It has been possible to measure the detailed characteristics of a number of craters. Many of these can be identified with craters shown in Mariner photographs of Mars. In addition, a scarp has been seen at 41° west, 14° south with an average slope of about 6° extending over about 40 kilometers.

097.081 Mars radar observations, a preliminary report.
G. S. Downs, R. M. Goldstein, R. R. Green, G. A. Morris.
Science, Vol. 174, 1324 - 1327 (1971).

097.082 Aeolian transport on Mars and the nature of Hellas.
R. J. Fryer.
Earth Planet. Sci. Letters, Vol. 13, 6 - 10 (1971).

The Mariner 1969 photographic observations of Hellas are briefly reviewed and limits on the possible scale of roughness of the surface deduced. Previously suggested mechanisms for the creation of this surface are examined. It is hypothesised that aeolian transport and deposition within the Hellas basin adequately accounts for all observations simply and without ad hoc assumptions.

097.083 On the surface layer covering continents and seas of Mars. M. P. Barabashov, D. F. Lupishko.
Dopovidi AN URSR, 1971, B, No. 8, p. 703 - 705. In Ukrainian. – Abstr. in Referativ. Zhurn. 51. Astron., 1.51.287 (1972).

097.084 The strengths of H_2O lines in the 8200 Å region and their application to high dispersion spectra of Mars.
C. B. Farmer.
Icarus, Vol. 15, 190 - 196 (1971).

The improvement in the quality of spectroscopic plates taken in recent years in the search for water vapor in the atmosphere of Mars has dictated the need for improved laboratory data with which to interpret the spectra. This paper presents the results of measurements of the strengths of 41 lines of the 8200 Å water vapor band. The measured values show evidence of vibration-rotation interactions on the line intensities, beyond the principal stretching effect.

097.085 High dispersion spectroscopic studies of Mars. V. A search for oxygen in the atmosphere of Mars.
J. S. Margolis, R. A. J. Schorn, L. D. G. Young.
Icarus, Vol. 15, 197 - 203 (1971).

In order to set a new upper limit on the amount of O_2 in the atmosphere of Mars, we have reduced a number of high-dispersion spectra of the 7620 Å band of O_2 obtained during the 1969 apparition of Mars. Our new upper limit is $w = 15$ cm atm_{stp} for the Martian abundance in a single vertical path. This result confirms and lowers the 1963 upper limit by Kaplan, Münch, and Spinrad. The features reported by Hunten and Belton do not appear on our spectra.

097.086 The distribution of CO_2 on Mars: A spectroscopic determination of surface topography.

M. J. S. Belton, D. M. Hunten.
Icarus, Vol. 15, 204 - 232 (1971).

This paper is a full report of an attempt to map the surface distribution of the partial pressure of CO_2 and surface topography at the 1969 opposition. Emphasis is placed on the fundamental principles underlying the experiment and the experimental techniques used. The results of this experiment are compared with other available data. Comparisons with the results of two experiments on the Mariner 6 and 7 missions are found to be satisfactory. Less satisfactory is the comparison with the Haystack Radar data. The cause of this latter discrepancy is not resolved in this paper, although some possibilities are discussed.

097.087 The origins of Martian nomenclature.
T. L. MacDonald.
Icarus, Vol. 15, 233 - 240 (1971).

The classical origins of the contemporary nomenclature for Mars' surface features are outlined.

097.088 Mariner photography of Mars and aerial photography of earth: Some analogies.
D. Belcher, J. Veverka, C. Sagan.
Icarus, Vol. 15, 241 - 252 (1971).

Tentative characterizations of several Mariner 6 and 7 Martian surface features, made by the senior author in the absence of previous knowledge about Mars, are presented. The ridges in 7N17 are interpreted as a glacial moraine; barchane or parabolic sand dunes are identified in 6N5; and thermokarst collapse features, possibly produced in permafrost by Martian geothermal activity, are proposed in 6N8 and 6N14, in agreement with the suggestion of Sharp et al. (1971).

097.089 Observational consequences of Martian wind regimes. C. Sagan, J. Veverka, P. Gierasch.
Icarus, Vol. 15, 253 - 278 (1971).

The connection with past and future observations of Mars of the high velocity relief winds deduced by Gierasch and Sagan (1971) is examined with the assistance of a large topographic map of Mars. A unimodal hypsometric curve is derived. Seasons and locales of wind velocities at the half surface pressure level > 80 m sec^{-1}, sufficient to lift dust at the surface, are identified.

097.090 History of Martian volatiles: Implications for organic synthesis. F. P. Fanale.
Icarus, Vol. 15, 279 - 303 (1971).

This study reconstructs theoretically the chemical evolution of Martian surface volatiles and considers whether the Martian surface environment was ever chemically suited for the abiotic origination of life. The assumption is made that life must be indigenous to a planet. In this context, the chemical evolution of Martian volatiles emerges as the link between Martian planetology and possible Martian biology.

097.091 Erratum: "Mars and Jupiter: Radio emission at 2.3 mm and 8.15 mm" [Icarus, Vol. 14, 198 - 203 (1971)].
V. A. Efanov, I. G. Moiseev, A. G. Kislyakov, A. I. Naumov.
Icarus, Vol. 15, 361 (1971).

097.092 The contrast level of some Martian details in July 1971. V. V. Avramchuk, A. R. Gajduk, N. B. Ibragimov, I. K. Koval, V. D. Krugov.
Astron. Tsirk., No. 656, p. 1 - 3 (1971). In Russian.

097.093 The developing stages of the Martian yellow storm of 1971. C. F. Capen, L. J. Martin.
Lovell Obs. Bull., *Flagstaff, Arizona*, No. 157, Vol. 7, 211 - 216 (1971).

097.094   Internal constitution of Mars.
          A. E. Ringwood, S. P. Clark.
Nature, Vol. 234, 89 - 92 (1971).
     We investigated the hypothesis which maintains that
Mars is composed of chondritic material in a higher state of
oxidation than is displayed by the earth or ordinary chondrites. Our work shows that the properties of Mars are consistent with an overall oxidized chondritic composition, in
which the relative abundances of metals are believed to be
similar to those in the sun.

097.095   Atmosphere and surface of Mars.   I. K. Koval'.
          Astron. vestn., Vol. 5, 129 - 143 (1971). In Russian.
Abstr. in Referativ. Zhurn. 51. Astron., 2.51.261 (1972).

097.096   Brightness distribution along the equatorial zone of
          Mars.   N. P. Barabashov, V. I. Garazha.
Astron. vestn., Vol. 5, 150 - 152 (1971). In Russian. – Abstr.
in Referativ. Zhurn. 51. Astron., 2.51.263 (1972).

097.097   TV observations of Martian cloud formations dur-
          ing the period August 4 – September 5 in 1971.
A. N. Abramenko, A. K. Dabachov, M. N. Naugolnaja, V. V.
Prokofjeva.
Astron. Tsirk., No. 666, p. 1 - 3 (1971). In Russian.

097.098   The violet clouds of Mars in July–August 1971.
          A. R. Gajduk.
Astron. Tsirk., No. 666, p. 3 - 5 (1971). In Russian.

097.099   Mie scattering calculations of the contribution of
          atmospheric aerosols to the Martian opposition ef-
fect.   J. M. Mead.
Thesis, Georgetown Univ., Washington, D.C. [Available from
Univ. Microfilms, Ann Arbor, Mich., U.S.A. Order No.
70–26677], 84 pp. (1970).
     The Mie theory is used to compute the integrated scatter-
ing intensities for spherical submicron and aerosol particles
with various indices of refraction and several size distributions
in an effort to determine if the presence of atmospheric aero-
sols can account for the Martian opposition effect as observed
in 1967. This nonlinear surge in brightness, as the planet approaches a phase angle of 0°, is reported to be much more pronounced in the ultraviolet than in the infrared.

097.100   Metastable O^+ ions in Martian atmosphere.
          D. C. Agarwal.
Indian Journ. Meteorol. Geophys., Vol. 21, 391 - 397 (1970).
     Production rates of atomic oxygen ions in the metastable
states (^2P and ^2D) by photo-ionization have been calculated.
Equilibrium distributions of these species have been calculated
using recently measured rate-coefficients of ion-atom interchange and other ionic reactions. Finally, the zenith intensity

of the $(7318.6-7330 \text{Å})$ multiplet of O^+ in the dayglow has
been calculated.

097.101   Emission of $\lambda 6300$ Å in Martian atmosphere.
          D. C. Agarwal.
Indian Journ. Meteorol. Geophys., Vol. 21, 398 - 400 (1970).
     Deals with the emission of $\lambda 6300$ Å of OI in the Martian
atmosphere. Using the recently measured rate of coefficients
of ion-atom interchange and other ionic reactions, the volume
emission of this radiation is calculated.

097.102   Dissipation in atmospheres: The thermal structure
          of the Martian lower atmosphere with and without
viscous dissipation.   P. J. Gierasch.
Journ. Atmosph. Sci., Vol. 28, 315 - 324 (1971).

          Dust cloud activity on Mars.
British Astron. Ass., Circ. No. 537 (1971).

          Mariner 9: Into orbit around Mars.
Nature, Vol. 234, 67 (1971).

          Mariner 9: First results may aid Russian lander.
Nature, Vol. 234, 168 (1971).

          Mars: Continuing progress.
Nature, Vol. 234, 500 (1971).

          A portfolio of amateurs' Mars photographs.
Sky Telescope, Vol. 42, 310 - 314 (1971).

          2.5-km low-temperature multiple-reflection cell.
See Abstr. 034.039.

          Kepler's laws and the Mars orbiters.
See Abstr. 042.028.

          Results of daytime observations of Venus and Mars
obtained with the Wanschaff vertical circle in 1969.
See Abstr. 093.026.

          Determination of astrodynamic constants and a test
of the general relativistic time delay with S-band range and
Doppler data from Mariners 6 and 7.   See Abstr. 094.189.

          On reasons for the different brightness distribution
on the disks of the moon and Mars during phases near zero.
See Abstr. 094.205.

          Spectrophotometry of the moon, Mars, and Uranus.
See Abstr. 094.320.

          Microwave radiation of Uranus and Neptune.
See Abstr. 101.006.

098   Minor Planets

098.001   Minor planets (1967).
N. S. Samojlova-Yakhontova.
Byull. Inst. Teoret. Astron., *Leningrad,* Vol. 12, 641 - 648 (1971). In Russian.

098.002   Structure of the asteroid belt.
G. A. Chebotarev, M. J. Shmakova.
Byull. Inst. Teoret. Astron., *Leningrad,* Vol. 12, 649 - 684 (1971). In Russian.

The system of minor planets is investigated by statistical methods. All conclusions are in good agreement with Kuiper's hypothesis about the origin of the asteroid belt. The protoplanets seemed to move in the zone $700''-900''$. Their orbits probably had eccentricities of the order $0.1-0.2$ and inclinations less than $5°$. All known asteroids seem to be fragments of the protoplanets.

098.003   Observations of minor planets made at the Crimean Astrophysical Observatory, (14th report).
L. I. Chernykh.
Byull. Inst. Teoret. Astron., *Leningrad,* Vol. 12, 742 - 752 (1971). In Russian.

098.004   La petite planète Amor.   J. Meeus.
L'Astronomie, 85ᵉ année, p. 352 - 357 (1971).

098.005   Minor planets and related objects. VII. Asteroid 1971 FA.   T. Gehrels, E. Roemer, B. G. Marsden.
Astron. Journ., Vol. 76, 607 - 608 (1971).

1971 FA is a new Apollo-type asteroid with high orbital eccentricity and inclination; the perihelion distance is 0.56 a.u. The shape is found to be rough and elongated, about twice as long as it is wide, and the period of rotation is $8^h 34^m$.

098.006   Ephemerides of minor planets for 1972.
Editor: Institut Teoreticheskoj Astronomii Akademii Nauk SSSR, under the editorship of G. A. Chebotarev. Izdatel'stvo "Nauka", Leningradskoe Otdelenie, Leningrad. 178 pp. Price 2 Rbl. 26 Kop. (1971). In Russian and English. Contents: Introduction, p. 3 - 9; Information on new elements, p. 10 - 11; Elements, p. 12 - 44; Opposition dates, p. 45 - 55; Ephemerides, p. 56 - 170; Ephemerides of some unusual planets, p. 171 - 175; Critical list, p. 176.

098.007   On a new method of calculating disturbed ephemerides of minor planets.
E. A. Trebenikov, G. F. Sultanov, V. A. Cheprasov.
Soobshch. Shemakhinsk. Astrofiz. Obs., vyp. (No.) 5, p. 13 - 18 (1971). In Russian.

098.008   Distribution of the asteroids.
M. Lecar, F. A. Franklin.
Bull. American Astron. Soc., Vol. 3, 369 (1971). — Abstr. AAS.

098.009   The fragmentation of the asteroids—II. Numerical calculations.   B. Hellyer.
Monthly Notices Roy. Astron. Soc., Vol. 154, 279 - 291 (1971).

The equations describing the evolution in time of the asteroidal mass distribution under collisional fragmentation have been solved numerically. Recently published observational data on the smaller asteroids are compared with the theory, and are shown to be in good agreement. Some further discussion is given to the special problem of the mass distribution of the largest asteroids.

098.010   Conditions for magnetic interaction of asteroids with the solar wind.   E. W. Greenstadt.
Icarus, Vol. 14, 374 - 381 (1971).

Conditions are presented for maintenance of asteroid magnetospheres by dipole moments and for propagation of whistler mode noise in the solar wind at asteroid distances. Surface field intensities less than one thousandth that of the earth are found adequate for supporting magnetospheres in the quiet solar wind surrounding the larger asteroids.

098.011   1948 EA.   B. G. Marsden.
IAU Circ., No. 2345 (1971).

098.012   1932 HA (Apollo).
IAU Circ., No. 2349 (1971).

098.013   1971 FA.   E. Roemer, R. Shuart.
IAU Circ., No. 2351 (1971).

098.014   1556 Wingolfia.   J. A. Bruwer.
IAU Circ., No. 2366, 2368 (1971).

098.015   Fast-moving object Kohoutek.   L. Kohoutek.
IAU Circ., No. 2367 (1971).

098.016   The origin of asteroids.   V. A. Bronshtehn.
Zemlya i Vselennaya, No. 5, p. 53 - 59 (1971). In Russian.

098.017   The dust population in the asteroid belt.
W. Kokott.
Space Research XI, (see 012.004), 215 - 224 (1971).

098.018   Photographic observations of the minor planet (433) Eros.   D. A. Pierce.
Astron. Journ., Vol. 76, 943 - 945 (1971).

Photographic observations of the minor planet (433) Eros were taken from 21 August 1935 to 28 June 1947, using the long-focus refractors at the Allegheny and the Yale-Johannesburg Observatories and the Ross cameras at the Yale-Johannesburg and Yale-New Haven Observatories.

098.019   Heldere kleine planeten in 1972.   J. Meeus.
Hemel en Dampkring, Vol. 69, 331 - 333 (1971).

098.020   Étude sur les magnitudes absolues des astéroïdes.
M.-A. Combes.
L'Astronomie, 85ᵉ année, p. 413 - 433 (1971).

098.021   Photographic observations of minor planets, observed at the Republic Observatory Annexe, Hartbeespoort, with the Franklin-Adams star camera.
J. A. Bruwer, M. Klerk.
Republic Obs. Johannesburg, Circ. No. 131, Vol. 8, 2 - 4 (1971)

098.022   Observations des petites planètes à l'Observatoire Astronomique de Belgrade en 1967, 1968 et 1969.
D. Olević.
Bull. Obs. Astron. Beograd, Vol. 28, (No. 123), 53 - 62 (1970).

On donne ici les positions précises des petites planètes plus brillants au cours de 1967 (28), 1968 (52) et 1969 (55).

098.023   Observations photographiques des petites planètes et des comètes à Beograd.   M. B. Protitch.

Bull. Obs. Astron. Beograd, Vol. 28, (No. 124), 171 - 172 (1970). – Concerning the planetoids Nos. 1, 2, 3, 4, 6, 15, 116, and 451, and the comets 1969g and 1969i.

098.024 **The polarization curve and the absolute diameter of Vesta.** J. Veverka.
Icarus, Vol. 15, 11 - 17 (1971).

A new photoelectric polarization curve for Vesta is presented. It is incompatible with Lyot's photographic curve which appears to be inaccurate. The new polarization curve indicates that the reflectivity of Vesta is higher than that of the moon, and a preliminary value of 0.25 ± 0.07 for the reflectivity in the V is suggested. This implies an absolute diameter of about 510 km, a value consistent with the diameter of Vesta calculated from its mass, assuming an achondritic composition.

098.025 **Object Kohoutek.** L. Kohoutek, B. G. Marsden.
IAU Circ., No. 2369 (1971).

098.026 **Object Kohoutek.** L. Kohoutek, K. Aksnes.
IAU Circ., No. 2373 (1971).

098.027 **1971 UA (Object Kohoutek).** L. Kohoutek.
IAU Circ., No. 2374 (1971).

098.028 **1971 UA.** J. B. Gibson.
IAU Circ., No. 2377 (1971).

098.029 **2008 P–L.** C. T. Kowal, J. G. Williams, E. Roemer.
IAU Circ., No. 2377 (1971).

098.030 **Object Piksaev.** V. A. Piksaev.
IAU Circ., No. 2379 (1971).

098.031 **Notes on two minor planets.** M. Dirikis.
Astron. Tsirk., No. 648, p. 5 - 6 (1971). In Russian.

098.032 **Results of observations of Icarus by a television system.** S. G. Braunfeld, Z. N. Grigoryeva, E. K. Denisyuk, E. S. Yeroshevich, V. F. Kartashov, L. N. Kondratyeva, V. S. Matyagin, L. P. Sorokina, L. A. Usoltseva, A. A. Schipenstein.
Byull. Inst. Teoret. Astron., *Leningrad*, Vol. 13, 59 - 60 (1971). In Russian.

098.033 **Observations of 1566 Icarus in Abastumani.** A. Sh. Khatisov.
Byull. Inst. Teoret. Astron., *Leningrad*, Vol. 13, 61 - 62 (1971). In Russian.

098.034 **Results of photographic observations of minor planets at Zvenigorod.** T. A. Guseva.
Byull. Inst. Teoret. Astron., *Leningrad*, Vol. 13, 63 (1971). In Russian.

098.035 **Photographic positions of minor planets observed in Tartu.** H. K. Raudsaar.
Byull. Inst. Teoret. Astron., *Leningrad*, Vol. 13, 64 (1971). In Russian.

098.036 **Observations of minor planets made at the Crimean Astrophysical Observatory (16th report).**
L. I. Chernykh.
Byull. Inst. Teoret. Astron., *Leningrad*, Vol. 13, 65 - 72 (1971). In Russian.

098.037 **Positions of minor planets from observations with the 400-mm astrograph of the Main Astronomical Observatory of the Ukrainian Academy of Sciences.**
I. M. Demenko.
Astrometriya i Astrofiz., *Kiev*, No. 13, (see 003.039), p. 108 - 110 (1971). In Russian.

098.038 **Observations of minor planets at Nikolayev in 1967.** V. I. Voronenko, G. K. Gorel, F. F. Kalichevich, R. T. Fedorova.
Byull. Inst. Teoret. Astron., *Leningrad*, Vol. 12, 910 - 922 (1971). In Russian.

098.039 **Observations of minor planets made at the Crimean Astrophysical Observatory (15th report).**
L. I. Chernykh.
Byull. Inst. Teoret. Astron., *Leningrad*, Vol. 12, 934 - 941 (1971). In Russian.

098.040 **Minor Planet Circulars, (MPC), Nos. 3165 - 3292** (1971).
Edited by Cincinnati Observatory, under the supervision of P. Herget.

A repository of nearly all new data for numbered and unnumbered minor planets: Observations, elements and ephemerides, identifications, newly assigned numbers and names, occultations.

098.041 **Posizioni de pianetini nel 1969.** M. A. Vogliotti, V. Zappalà.
Mem. Soc. Astron. Italiana, Nuova Ser., Vol. 42, 475 - 479 (1971).

098.042 **Fast moving asteroid.** L. Kohoutek.
Yamamoto Circ., No. 1743 (1971). In Japanese.

098.043 **Fast moving object Kohoutek.**
Yamamoto Circ., Nos. 1744, 1745, 1746 (1971). In Japanese.

098.044 **2008 P–L.** J. G. Williams, C. T. Koval, E. Roemer.
Yamamoto Circ., No. 1746 (1971). In Japanese.

Observations of Pluto, minor planets 10 Hygiea and 433 Eros, Saturn's satellites VII, VIII, IX made at the Crimean Astrophysical Observatory.
See Abstr. 041.001.

Meridian observations of major planets and some minor planets 1963 – 1967. See Abstr. 041.003.

Meridiankreis-Beobachtungen von Planeten 1950 bis 1955. See Abstr. 041.032.

Radio location, minor planets and astrometry. See Abstr. 041.044.

The planetary masses and the orbits of the first four minor planets. See Abstr. 043.012.

General relativity and the orbit of Icarus. See Abstr. 066.016.

Derivation of the mass of Jupiter and the orbit of 33 Polyhymnia. See Abstr. 099.012.

The mass of Jupiter from the motion of (76) Freia. See Abstr. 099.047.

Photographic positions of comets and minor planets observed during 1968–1970. See Abstr. 103.006.

099 Jupiter

099.001 Studies of methane absorption in the Jovian atmosphere. III. The reflecting-layer model.
J. S. Margolis.
Astrophys. Journ., Vol. 167, 553 - 558 (1971).

The spectrum of the $3\nu_3$ band of CH_4 in the Jovian spectrum is examined in terms of a model of the atmosphere similar to Trafton's. It is shown that a reflecting-layer model is consistent with the observed temperature, half-width of the absorption lines, and mixing ratio CH_4/H_2 if observations are restricted to the center of the Jovian disk. The failure of the reflecting-layer model of the atmosphere near the edge of the disk is tentatively explained on the basis of a two-layer cloud distribution.

099.002 Thermal inertia of Ganymede from 20-micron eclipse radiometry. D. Morrison, D. P. Cruikshank, R. E. Murphy, T. Z. Martin, J. G. Beery, J. P. Shipley.
Astrophys. Journ., (*Letters*), Vol. 167, L107 - L111 (1971).

We present simultaneous visual photometry and $20\text{-}\mu$ infrared radiometry of the eclipse of Ganymede (Jupiter III) 1971 March 17. The infrared flux was measured throughout the eclipse.

099.003 Jupiter and Io occult Beta Scorpii.
S. Larson, C. Papadopoulos, J. Dragesco, D. Uckotter, R. J. Poole.
Sky Telescope, Vol. 42, 112 - 116 (1971).

099.004 Hydrogen and helium at high temperatures.
V. P. Trubitsyn.
Astron. Zhurn. Akad. Nauk SSSR, Vol. 48, 810 - 814 (1971). In Russian. English translation in Soviet Astron. AJ, Vol. 15, No. 4.

Thermal dissociation and ionization of the hydrogen and helium gas under conditions which are near the adiabatic models of Jupiter and Saturn are investigated. The possibility for the existence of dissociation and partial ionization of the matter in Jupiter and Saturn is noted.

099.005 Een heldere ster nabij Jupiter. J. Meeus.
Hemel en Dampkring, Vol. 69, 226 - 227 (1971).

099.006 Jupiter Section: Report of apparitions 1969 - 1970. W. E. Fox.
Journ. British Astron. Ass., Vol. 81, 396 - 404 (1971).

099.007 Jupiter: Présentation 1970. S. Cortesi.
Orion, 29. Jahrgang, p. 75 - 79 (1971). – Rapport No. 21 du «Groupement planétaire SAS».

099.008 Observations des occultations de β Scorpii les 13 et 14 mai 1971. J. Dragesco.
Orion, 29. Jahrgang, p. 108 - 109 (1971).

099.009 Observation du passage de Io devant l'étoile double β Scorpii, le 14 mai 1971. P. Couteau.
L'Astronomie, 85e année, p. 347 - 351 (1971).

099.010 Determination of the mass of Jupiter from modern observations of the minor planet 10 Hygiea.
N. S. Chernykh.
Astron. Tsirk., No. 617, p. 6 - 8 (1971). In Russian.

099.011 Observations of the occultation of β Sco by Jupiter on May 13, 1971. V. P. Dzapiashvili.
Astron. Tsirk., No. 632, p. 6 (1971). In Russian.

099.012 Derivation of the mass of Jupiter and the orbit of 33 Polyhymnia. P. M. Janiczek.
Thesis, Georgetown Univ., Washington, D. C. [Available from Univ. Microfilms, Ann Arbor, Mich., U.S.A. Order No. 70– 21286], 112 pp. (1970).

In this investigation, accurate observations of Polyhymnia, extending from 1854 to 1969 were systematically adjusted to the system of the Fourth Fundamental Catalog. A provisional ephemeris and partial derivatives were integrated by a modified Cowell method, including perturbations by all the major planets. Comparison with the observations yielded 940 conditional equations for correction of the provisional elements of Polyhymnia and the mass of Jupiter.

099.013 Jupiter: Its captured satellites. J. M. Bailey.
Science, Vol. 173, 812 - 813 (1971).

The conditions whereby Jupiter can capture satellites have been examined. Relationships derived on the basis of the three-body problem for planets in elliptical orbits enable the dimensions of the capture orbits around Jupiter to be calculated. It is found that Jupiter may capture satellites through the inner Lagrangian point when at perihelion or at aphelion.

099.014 Experimental Jovian photochemistry: Initial results.
C. Sagan, B. N. Khare.
Astrophys. Journ., Vol. 168, 563 - 569 (1971).

Experimental simulations of the ultraviolet photochemistry of the lower Jovian NH_4SH and NH_4OH clouds have been performed.

099.015 Theory of Io's effect on Jupiter's decametric emissions. P. M. McCulloch.
Planet. Space Sci., Vol. 19, 1297 - 1312 (1971).

The effect of Io on Jupiter's decametric emissions is explained by a disturbance generated at Io, and coupled to streams of moderately relativistic electrons close to Jupiter which emit cyclotron radiation. The coupling occurs through whistler mode electromagnetic waves which interact with the electron streams by means of the gyroresonant interaction. Predicted distributions of emission probability and power are given as function of Io's departure from superior geocentric conjunction and Io's magnetic longitude.

099.016 Decameter radio radiation of Jupiter as an indicator of high-speed streams and shock waves in the solar wind. V. A. Kovalenko, V. N. Malyshkin.
Geomagn. Aeronom., Vol. 11, 888 - 890 (1971). In Russian. Brief information.

099.017 The occultation of Beta Scorpii C by Io.
B. O'Leary.
Bull. American Astron. Soc., Vol. 3, 373 (1971). – Abstr. AAS.

099.018 Occultation of β^2 Scorpii by Io on 1971 May 14.
G. E. Taylor.
Bull. American Astron. Soc., Vol. 3, 373 (1971). – Abstr. AAS.

099.019 Observation of an occultation of Beta Sco C by Io on 14 May 1971.
P. Bartholdi, F. Owen.
Bull. American Astron. Soc., Vol. 3, 373 (1971). – Abstr. AAS.

099.020 Upper limits for an atmosphere on Io.

B. A. Smith, S. A. Smith.
Bull. American Astron. Soc., Vol. 3, 373 (1971). – Abstr. AAS.

099.021 Observation of the occultation of β Sco C by Io.
F. W. Fallon, E. J. Devinney.
Bull. American Astron. Soc., Vol. 3, 373 (1971). – Abstr. AAS.

099.022 Astrometric results from the Beta Scorpii occulta-tion. T. C. Van Flandern.
Bull. American Astron. Soc., Vol. 3, 373 - 374 (1971). Abstr. AAS.

099.023 Three-channel observations of the occultation of Beta Scorpii by Jupiter. J. Veverka, J. L. El-liot, W. Liller, C. Sagan, L. Wasserman.
Bull. American Astron. Soc., Vol. 3, 374 (1971). – Abstr. AAS.

099.024 The occultation of Beta Scorpii by Jupiter; an ultra-violet light curve and its interpretation.
W. A. Baum.
Bull. American Astron. Soc., Vol. 3, 374 (1971). Abstr. AAS.

099.025 Observations of Jupiter occultation phenomena on 13 and 14 May 1971. D. S. Evans.
Bull. American Astron. Soc., Vol. 3, 374 (1971). – Abstr. AAS.

099.026 Interpretation of the Jupiter occultation data.
W. B. Hubbard.
Bull. American Astron. Soc., Vol. 3, 374 (1971). – Abstr. AAS.

099.027 Computer instrumentation for the Jupiter occulta-tion expedition. D. C. Wells.
Bull. American Astron. Soc., Vol. 3, 374 (1971). – Abstr. AAS.

099.028 The mass of Jupiter from the motion of (76) Freia. W. J. Klepczynski, P. M. Janiczek,
A. D. Fiala.
Bull. American Astron. Soc., Vol. 3, 415 (1971). – Abstr. AAS.

099.029 On the interpretation of polarimetric observations of Jupiter. V. M. Loskutov.
Astron. Zhurn. Akad. Nauk SSSR, Vol. 48, 1046 - 1050 (1971). In Russian. English translation in Soviet Astron. AJ, Vol. 15, No. 5.
The polarization degree of light from the center of the Jovian disk is computed as a function of the planet's phase. With the assumptions made, the theoretical curves agree with the observational data of B. Lyot.

099.030 On the anomalous brightening of Io after eclipse.
B. O'Leary, J. Veverka.
Icarus, Vol. 14, 265 - 268 (1971).
Photometric observations of Io during eclipse reappear-ances, made by various observers, are in conflict and cannot be easily reconciled. We present new observations, some of which show a small brightness anomaly similar to that first reported by Binder and Cruikshank. However, we feel that the standard photometric technique is not sufficiently sensitive to give an unambiguous answer.

099.031 Ultraviolet reflectivity of Jupiter observed from a rocket. Y. Kondo.
Icarus, Vol. 14, 269 - 272 (1971).

The ultraviolet spectral energy distribution of Jupiter was observed with a rocket-borne 13-in. Cassegrain telescope equipped with a scanning spectrometer. The geometric reflec-tivity of the planet has been computed for the wavelength range 3600–2100 Å, and the results are discussed. There was no evidence of the previously reported strong absorption fea-ture at about 2600 Å. The reflectivity decreases rapidly below 2200 Å.

099.032 The upper atmosphere of Jupiter.
M. Shimizu.
Icarus, Vol. 14, 273 - 281 (1971).
Diurnal variations of the Jovian exospheric tempera-tures are computed by solving time dependent heat balance equations. The departure of the daytime exospheric tempera-ture from the nighttime value is found to be slight, due to the rapid rotation of this planet and the weak solar ultraviolet radiation at its orbit. However, the exospheric temperature at the polar region is concluded to be lower than at the equator. The profiles of atomic hydrogen and its ions are obtained for various solar activities, eddy diffusion coefficients, and meso-pause temperatures.

099.033 On the structure and motions of Jupiter's Red Spot.
W. B. Streett, H. I. Ringermacher, G. Veronis.
Icarus, Vol. 14, 319 - 342 (1971).
A new hypothesis – called the Cartesian diver hypothesis – is proposed to explain the physical nature and observed variations in longitude, size and intensity of Jupiter's Great Red Spot.

099.034 Jupiter: Its Red Spot and other features in 1969 - 1970. E. J. Reese.
Icarus, Vol. 14, 343 - 354 (1971).
Photographic observations of Jupiter and its Red Spot between 13 November 1969 and 21 September 1970 are reported. The Red Spot continues its 90-day oscillation in longitude with considerable regularity. An outstanding event of the apparition was the appearance of a new disturbance in the South Tropical Zone. A bright spot at zenographic latitude $23°8$ N displayed the shortest rotation period ever recorded on Jupiter, $9^h 47^m 3^s$.

099.035 Polarization measurements of the Galilean satellites of Jupiter. J. Veverka.
Icarus, Vol. 14, 355 - 359 (1971).
Polarization curves for the four Galilean satellites are presented. They indicate that the surfaces of Io, Ganymede, and Europa are covered mostly by a bright, transparent mate-rial, possibly frost. The surface of Callisto is different. It is more similar to that of the moon, but some frost patches may also be present.

099.036 Plasma densities in the Jovian magnetosphere: Plasma slingshot or Maxwell demon?
G. Ioannidis, N. Brice.
Icarus, Vol. 14, 360 - 373 (1971).
The plasma density distribution in the earth's magneto-sphere is described and the principal factors producing this distribution are discussed. These are diffusion of the iono-spheric thermal plasma inside the plasmapause, and ejection of terrestrial plasma and injection of solar wind plasma beyond the plasmapause. For Jupiter, the ionospheric thermal plasma cannot escape into the magnetosphere and the density distri-bution is dominated by diffusion of photoelectrons at low L values, and interchange instabilities at larger distances (beyond about $8R_J$).

099.037 The Faraday rotation of Jovian decametric radio noise in the earth's ionosphere. A. D. M. Walker.
Monthly Notes Astron. Soc. Southern Africa, Vol. 30, 124 -

126 (1971). – Originally presented at the 16th annual conference of the South African Institute of Physics, University of Cape Town, 15th July 1971.

099.038 Jupiter. E. J. Reese.
IAU Circ., No. 2338 (1971).

099.039 Occultation of β Sco C by Jupiter I.
W. W. Weiss.
IAU Circ., No. 2341 (1971).

099.040 Jupiter. E. J. Reese.
IAU Circ., No. 2344 (1971).

099.041 Jupiter. V. E. Bell.
IAU Circ., No. 2351 (1971).

099.042 First results of the occultation of β Sco by Jupiter.
M. Combes, J. Lecacheux, L. Vapillon.
Astron. Astrophys., Vol. 15, 235 - 238 (1971).

099.043 The main problems of study of the planet Jupiter.
V. G. Teifel.
Space Research XI, (see 012.004), 191 - 201 (1971).

099.044 Nature and topography of Jupiter's Galilean moons.
G. N. Katterfel'd, E. I. Nesterovich.
Vestn. Leningr. un-ta, 1971, No. 12, p. 132 - 141. In Russian.
Abstr. in Referativ. Zhurn. 51. Astron., 11.51.338 (1971).

099.045 The spectrum of Jupiter at millimeter wavelengths.
G. T. Wrixon, W. J. Welch, D. D. Thornton.
Astrophys. Journ., Vol. 169, 171 - 183 (1971).

Measurements of the disk temperature of Jupiter at eight frequencies spanning the range 20.5–35.5 GHz are presented. The measurements are compared with model calculations in which saturated ammonia is the source of emission in the Jovian atmosphere. Agreement is quite good both in terms of the suggested dip in the observed data corresponding to the center of the 1.25-cm ammonia band and in terms of the absolute intensities.

099.046 Errata: 'Studies of Jupiter's equatorial thermal limb darkening during the 1965 apparition'.
[Astrophys. Journ., Suppl. Ser., Vol. 23, 1 - 34 (1971)].
R. L. Wildey, L. M. Trafton.
Astrophys. Journ., Vol. 169, 447 (1971).

099.047 The mass of Jupiter from the motion of (76) Freia.
W. J. Klepczynski, P. M. Janiczek, A. D. Fiala.
Astron. Journ., Vol. 76, 939 - 942 (1971).

Using the visual and photographic observations of the minor planet during three observational periods (1864–1893), (1912–1933) and (1947–1971) a new orbit was determined. The reciprocal mass of Jupiter, simultaneously found, is 1047.366 ± 0.007 (m.e.).

099.048 Ultraviolet absorption in the continuous spectrum of Jupiter. V. D. Krugov.
Astrometriya i Astrofiz., *Kiev,* No. 14, (see 003.014), p. 9 - 23 (1971). In Russian.

Results are presented of a spectrophotometric investigation of Jupiter (3300 - 4800 Å) for the purpose of determining ultraviolet absorption for various disk regions of Jupiter.

099.049 Spectrophotometric peculiarities of some satellites of planets. Yu. D. Davudov, I. K. Koval.
Astrometriya i Astrofiz., *Kiev,* No. 14, (see 003.014), p. 49 - 54 (1971). In Russian.

Stebbins' and Jacobsen's photoelectric measurements of the solar phase-brightness dependence for Jupiter's Galilean satellites are used to show that the surface layers of the satellites are very porous. Results of spectrophotometric observations of satellites are discussed briefly from the point of view of the possible nature of their surfaces.

099.050 Correction to the mass of Jupiter derived from the motion of (153) Hilda, (279) Thule, and (334) Chicago. H. Scholl.
Celestial Mechanics, Vol. 4, 250 - 252 (1971). – Presented at IAU Colloquium No. 9 (see 012.006).

099.051 Isodensitometry of Jupiter's Red Spot and Jupiter.
C. J. Banos, C. E. Alissandrakis.
Astron. Astrophys., Vol. 15, 424 - 432 (1971).

Isodensity tracings for Jupiter in 1968 and for the Red Spot in 1968–1969–1970 are given. The photometric profiles of both derived from the isophotes and for λ = 5500 Å are given in the east-west and north-south directions.

099.052 On the vertical structure of Jupiter's atmosphere.
M. F. Khodyachikh.
Vestn. Khar'kov. Univ., No. 65, (Ser. Astron., No. 6), p. 52 - 59 (1971). In Russian.

099.053 Photoelectric observations of Jupiter.
M. F. Khodyachikh.
Vestn. Khar'kov. Univ., No. 65, (Ser. Astron., No. 6), p. 60 - 63 (1971). In Russian.

099.054 The highest frequency of Jupiter's decametric radiation. E. E. Baart, J. G. Greener, A. Wulff.
Astrophys. Letters, Vol. 9, 211 - 213 (1971).

The position of the satellite Io relative to Jupiter is shown to be correlated with the highest frequency in Jupiter's decametric radiation which is received on earth. The variation in the highest frequency with sub-Io longitude provides a test for theories of the generation of the radiation.

099.055 Spectral geometric albedo of the Galilean satellites, 0.3 to 2.5 microns.
T. V. Johnson, T. B. McCord.
Astrophys. Journ., Vol. 169, 589 - 594 = Contr. Planet. Astron. Lab., Dep. Earth Planet. Sci., Mass. Inst. Technology, No. 33 (1971).

The spectral geometric albedos for the four Galilean satellites of Jupiter were determined over the spectral range 0.9–2.5 μ with a spectral resolution of 0.05 μ. The spectral albedo curves confirm the decrease in albedo beyond 1.0 μ for J II and J III. Jupiter I and J IV have relatively constant albedos from 1.0 to 2.5 μ. The curve for J III appears to show a relative minimum near 1.6 μ. No spectral absorption feature is identifiable in the curve of J III near 2.0 μ.

099.056 Origin of the outer satellites of Jupiter.
J. M. Bailey.
Journ. Geophys. Res., Vol. 76, 7827 - 7832 (1971).

The conditions whereby Jupiter can capture satellites have been examined. It was found that Lagrangian-point capture of satellites can take place when Jupiter is at perihelion or aphelion. Captures at perihelion should give rise to satellites in direct orbits with semimajor axes of about 11.48×10^6 km, whereas capture at aphelion produces satellites in retrograde orbits with semimajor axes of about 21.7×10^6 km. The correspondence between these values and those observed for the seven outer satellites of Jupiter is discussed.

099.057 Jupiter and Beta Scorpii.
D. S. Evans, W. B. Hubbard.
Sky Telescope, Vol. 42, 337 - 341 (1971).

099.058 Erratum: 'Studies of Jupiter's equatorial thermal

limb darkening during the 1965 apparition'.
[Astrophys. Journ., Suppl. Ser., No. 194, Vol. 23, 1 - 34
(1971)].   R. L. Wildey, L. M. Trafton.
Astrophys. Journ., Suppl. Ser., No. 199, Vol. 23, 321 (1971).

**099.059 Thermal inertia of Ganymede from 20-micron eclipse
radiometry.**   D. Morrison, D. P. Cruikshank,
R. E. Murphy, T. Z. Martin, J. G. Beery, J. P. Shipley.
Publ. Astron. Soc. Pacific, Vol. 83, 607 (1971). – Abstr.
Astron. Soc. Pacific.

**099.060 Organic synthesis in a simulated Jovian atmosphere –
II.**   M. S. Chadha, J. J. Flores, J. G. Lawless,
C. Ponnamperuma.
Icarus, Vol. 15, 39 - 44 (1971).
      Reactions which may occur in the Jovian atmosphere
were simulated by passing a semicorona discharge through a
mixture of methane and ammonia. The nonvolatile fraction
consisted of a reddish product which on acid hydrolysis gave
a number of amino and imino acids.

**099.061 The Jovian ionosphere: Composition and tempera-
tures.**   S. S. Prasad, L. A. Capone.
Icarus, Vol. 15, 45 - 55 (1971).
      The problem of ionosphere formation in the Jovian upper
atmosphere has been reexamined with a view to estimating the
electron and ion densities and their temperature. There is a
complete thermal equilibrium between the neutrals, electrons,
and ions. This is in sharp contrast with the terrestrial case, but
is easily understood in terms of reduced heating of the ambient electrons, and more efficient cooling mechanisms in the
Jovian ionosphere.

**099.062 Activity in Jupiter's atmospheric belts between 1904
and 1963.**   J. H. Focas, A. Dollfus.
Icarus, Vol. 15, 56 - 57 (1971).
      Microphotometer tracings of Lowell Observatory photo-
graphs of Jupiter from 1904 to 1963 reveal periodic variations
in the amount of dark material present in the atmospheric
belts. These variations do not appear to be correlated with
solar activity.

099.063 Contribution to the study of Jupiter's atmosphere.
C. J. Banos.
Icarus, Vol. 15, 58 - 67 (1971).
      In Part I, using the method suggested by Focas and Banos
in 1964, we determine the photometric coefficient of activity
during the period 1963–1967 in three wavelengths, from
photographic plates taken at New Mexico State University
Observatory. We also give coefficients of distribution of activity for the Jovian latitudes between ± 45°. We establish for the
first time a coefficient of asymmetry of the activity and we
study the asymmetry by latitude during 1963 - 1967. A direct
correlation between activity and the rotation period is observed. In Part II, we give the relative intensities of some belts
and zones for the period 1964 - 1967, and the evolution of the
relative intensity of the Red Spot.

**099.064 The atmospheric activity of the planet Jupiter.
Part I: From 1964 to 1968 in yellow light.**
R. Prinz.
Icarus, Vol. 15, 68 - 73 (1971).
      The photometric study of the photographic observations
of Jupiter at the Public Observatory in Munich during the
years 1964 - 1968 is discussed with particular regard to an
earlier publication by Focas and Banos (1964). Data on the
intensity of the total activity and the peculiar activity in the
equatorial area of the planet in the period 1964 - 1968 are
given. Additional graphs show the evolution of the activity
during the years 1952 - 1968 and 1904 - 1968.

**099.065 The atmospheric activity of the planet Jupiter.
Part II: Short-term variations in five spectral ranges.**
R. Prinz.
Icarus, Vol. 15, 74 - 79 (1971).
      Photographic observations of Jupiter, obtained over a
nine month period in 1965–66 at the New Mexico State University Observatory are analyzed in terms of the coefficient
of cloud activity.

099.066 Jupiter.   W. E. Fox, E. J. Reese.
British Astron. Ass., Circ. No. 535 (1971).

**099.067 Occultation of Beta Scorpii C by Io on May 14,
1971.**   G. E. Taylor, B. O'Leary, T. C. Van Flan-
dern, P. Bartholdi, F. Owen, W. B. Hubbard, B. A. Smith, S. A.
Smith, F. W. Fallon, E. J. Devinney, J. Oliver.
Nature, Vol. 234, 405 - 406 (1971).
      The occultation of Beta Scorpii C by Jupiter satellite I
(Io) on 1971 May 14 was observed photo-electrically from
Jamaica, the Virgin Islands and two stations in Florida. Preliminary analysis of the observations indicates that the diameter
of Io is 3659 ± 5 km, leading to a density of 2.82 ± 0.23 g
cm^{-3}. It is deduced that the surface pressure of any atmosphere is less than 10^{-4} mbar. Additional events on the light
curves are consistent with the hypothesis that the occulted
star is double with the companion star in position angle
$308°.2 \pm 1°.2$, separation $0''.097 \pm 0''.002$.

099.068 Jupiter.   J. Rubi.
El Universo, Vol. 25, 144 (1971).

**099.069 Observation of the occultation of β Sco by Jupiter
at the Odessa Observatory.**
Yu. A. Medvedev, S. S. Vychrestjuk.
Astron. Tsirk., No. 649, p. 8 (1971). In Russian.

**099.070 Jupiter: An unidentified feature in the 5-micron
spectrum of the north equatorial belt.**
G. Münch, G. Neugebauer.
Science, Vol. 174, 940 - 941 (1971).
      Grating spectra of the north equatorial belt of Jupiter
between 4.5 and 5.1 microns, obtained with a nominal resolving power of 180, are presented. An absorption feature centered at 4.73 microns and not due to a known constituent has
been found. Its possible identification is discussed.

**099.071 On methane absorption changes in the equatorial
belt of Jupiter.**   Yu. A. Egorov, V. G. Teifel, G.
A. Kharitonova.
Astron. Tsirk., No. 656, p. 3 - 5 (1971). In Russian.

**099.072 Adjustment of Fabry-Perot interferometers for
Doppler-shift compensation of uniformly rotating
sources.**   J. Trauger, F. L. Roesler.
Journ. Optical Soc. America, Vol. 61, 1560 (1971). – Abstr.
Optical Soc. America.

**099.073 On the formation of the outer satellite groups of
Jupiter.**   G. Colombo, F. A. Franklin.
Icarus, Vol. 15, 186 - 189 (1971).
      This paper presents evidence suggesting that Jupiter's
seven outer satellites, which exist in two distinct groups, were
formed by a single collision of an asteroid and a larger satellite.

099.074 Photometric radii of Io and Europa.   M. J. Price,
J. S. Hall, P. B. Boyce, R. Albrecht.
Lovell Obs. Bull., *Flagstaff, Arizona,* No. 156, Vol. 7, 207 -
210 (1971).
      During a research program, observations of the ingress
light curves of Io and Europa were also made. These data have
been analyzed to obtain photometric radii for both satellites.

This paper briefly reports the observations and their interpretation. The results are compared with satellite radii obtained using other observational techniques.

099.075   Rotation of giant planets arising from gas accretion. V. S. Safronov.
Astron. vestn., Vol. 5, 167 - 173 (1971). In Russian. – Abstr. in Referativ. Zhurn. 51. Astron., 2.51.251 (1972).

099.076   On the interpretation of photometric observations of Jupiter. V. M. Loskutov.
Astron. vestn., Vol. 5, 153 - 158 (1971). In Russian. – Abstr. in Referativ. Zhurn. 51. Astron., 2.51.275 (1972).

099.077   Albedo and spectral reflectivity of the Galilean satellites of Jupiter. T. V. Johnson.
Thesis, California Inst. Technology, Pasadena. [Available from Univ. Microfilms, Ann Arbor, Mich., U.S.A. Order No. 70–18039], 101 pp. (1970). – See Phys. Abstr., Vol. 74, No. 62834 (1971).

099.078   Polarization of the decametric radiation from Jupiter. D. J. Kennedy.
Thesis, Univ. Florida, Gainesville. [Available from Univ. Microfilms, Ann Arbor, Mich., U.S.A. Order No. 70–14889], 141 pp. (1969). – See Phys. Abstr., Vol. 74, No. 62835 (1971).

099.079   An analysis of the intensity fluctuations produced by the interaction of the decametre radio emission from Jupiter with the interplanetary medium. D. L. Thompson.
Thesis, Florida State Univ., Tallahassee. [Available from Univ. Microfilms, Ann Arbor, Mich., U.S.A. Order No. 70–16351], 196 pp. (1969).
Interplanetary scintillation of the Jovian decametric radio emission by electron density inhomogeneities in the solar wind is considered for solar elongations of Jupiter greater than 90°.

099.080   Io, an Alfvén-wave generator. E. J. Schmahl.
Thesis, Univ. Colorado, Boulder. [Available from Univ. Microfilms, Ann Arbor, Mich., U.S.A. Order No. 71–5927], 142 pp. (1970). – See Phys. Abstr., Vol. 74, No. 62837 (1971).

099.081   Electron energy and density distribution of the Jovian magnetosphere. J. L. Luthey.
Thesis, Univ. Kansas, Lawrence. [Available from Univ. Microfilms, Ann Arbor, Mich., U.S.A. Order No. 71–13333], 175 pp. (1970).
The energy and numerical density of the electrons causing the Jovian synchrotron emission has been determined as a function of equatorial distance from Jupiter from an analysis of the observed radiation at two different frequencies.

099.082   A search for X-rays from the planet Jupiter. K. C. Hurley.
Thesis, Univ. California, Berkeley. [Available from Univ. Microfilms, Ann Arbor, Mich., U.S.A. Order No. 71–15798], 77 pp. (1970).
Three separate calculations are made to estimate the flux of Jovian X-rays at the earth.

099.083   A study of the spectral and statistical properties of Jupiter in the decameter region. F. A. Bozyan.
Thesis, Yale Univ., New Haven, Conn. [Available from Univ. Microfilms, Ann Arbor, Mich., U.S.A. Order No. 71–13794], 334 pp. (1969).
Flux measurements were made at four frequencies, using an absolute scale established by observations with a calibration dipole, and by observing discrete sources whose signal-to-noise ratio is enhanced by a synthesis technique. Periodogram calculations are reported that indicate the average decameter rotation period is consistent with the microwave period, although possibly significant differences are seen in the periods for radiation coming from Io and non-Io related regions.

099.084   Jupiter u 1971. godini. (Jupiter in the year 1971). P. Krešimir.
Vasiona, Vol. 19, 66 - 68 (1971).

099.085   More data on the occultation of Beta Scorpii by Jupiter, May 1971.
Journ. Astron. Soc. Western Australia, Vol. 29, August, p. 4 - 5 (1971).

Observations of new major disturbances on Jupiter.
Sky Telescope, Vol. 42, 176 - 180 (1971).

Collision-broadened half-widths and shapes of methane lines.   See Abstr. 022.052.

Measurement of hydromagnetic waves as a method of estimating a photon's rest mass.
See Abstr. 062.035.

A semiempirical model for the mean transmission of a molecular band an application to the $10\,\mu$ and $16\,\mu$ bands of NH_3.   See Abstr. 091.025.

Mars and Jupiter: Radio emission at 2.3 mm and 8.15 mm.   See Abstr. 097.032.

The motion of comet Schwassmann-Wachmann I and the mass of Saturn.   See Abstr. 100.003.

Microwave radiation of Uranus and Neptune.
See Abstr. 101.006.

100 Saturn

100.001 **The Mimas-Tethys resonance formation problem.**
S. F. Dermott.
Monthly Notices Roy. Astron. Soc., Vol. 153, 83 - 96 (1971).
It is improbable that the Mimas-Tethys resonance was formed by a process analogous to Goldreich's process for e-type resonance. Random forces, therefore, probably had a part in the formation. Tidal forces alone could not have been responsible for the formation of the resonance and, as the resonance is stable, boundary layer turbulence is probably not at present a substantial source of energy dissipation.

100.002 **Photoelectric observations of Saturn satellites Rhea and Titan.** C. Blanco, S. Catalano.
Astron. Astrophys., Vol. 14, 43 - 47 (1971).
Photoelectric observations in the U, B, V system of Saturn's satellites Rhea and Titan are given. Rhea shows a light variation of 0^m23 with the maximum occurring at $\vartheta=0°$. For Titan a quite clear light dependence from the solar phase angle, not yet detected, is established; no variation depending on the orbital phase is found.

100.003 **The motion of comet Schwassmann-Wachmann I and the mass of Saturn.** H. J. Carr.
Astron. Journ., Vol. 76, 507 (1971).
The reciprocal mass of Saturn is determined to be 3497.48 ± 0.15 (p.e.) from observations of the comet. A simultaneous determination gave 3497.64 ± 0.16 for Saturn and 1047.387 ± 0.014 for Jupiter.

100.004 **On the critical density of Saturn's rings.**
I. F. Ginsburg, V. L. Polyachenko, A. M. Fridman.
Astron. Zhurn. Akad. Nauk SSSR, Vol. 48, 815 - 818 (1971). In Russian. English translation in Soviet Astron. AJ, Vol. 15, No. 4.
The paper is concerned with the much discussed problem of possible instability of Saturn's rings resulting in the fall of their matter to the planet. It is shown that the finite thickness and small density of the rings can ensure their stability to arbitrary disturbances. The criterion of stability is obtained.

100.005 **The 1967–68 and 1968–69 apparitions of Saturn.**
J. L. Benton, Jr.
Strolling Astronomer, Vol. 23, 44 - 51 (1971).

100.006 **The measurement of Saturn's radio emission at 8.2 mm and evaluation of the optical thickness of its rings.** A. D. Kuz'min, B. Ya. Losovskij.
Astron. vestn., Vol. 5, 78 - 81 (1971). In Russian. – Abstr. in Referativ. Zhurn. 51. Astron., 10.51.236 (1971).

100.007 **White spots on Saturn.**
T. J. C. A. Moseley, A. Appleyard, M. Wardley.
IAU Circ., No. 2357 (1971).

100.008 **A review of new data on Saturn's system.**
M. S. Bobrov.
Space Research XI, (see 012.004), 203 - 211 (1971).

100.009 **Distribution of energy in the short-wave region of the spectrum for various parts of Saturn's disk.**
V. D. Krugov.
Astrometriya i Astrofiz., *Kiev*, No. 14 (see 003.014), p. 23 - 33 (1971). In Russian.

Results are presented of measurements of the ultraviolet absorption on Saturn's disk in the region of 3300 - 4800 Å obtained in 1968 - 1969.

100.010 **Weisse Flecke auf Saturn.** R. A. Naef.
Orion Schaffhausen, 29. Jahrgang, p. 185 - 186 (1971).

100.011 **Monochromatic albedos for the disk of Saturn.**
W. M. Irvine, A. P. Lane.
Icarus, Vol. 15, 18 - 26 = Contr. Five College Obs., Univ. Mass., *Amherst,* No. 79 (1971).
The Saturn system was observed as part of a program of multicolor photoelectric photometry at wavelengths $0.315 \leqslant \lambda \leqslant 1.06 \mu$ from 1963 to 1965. The brightness of the disk of Saturn has been separated from that due to the rings by correcting all data to zero phase angle and then determining the correction to zero inclination of the rings. The disk spectrum appears generally similar to that of Jupiter, but is redder for $\lambda \leqslant 6250$ Å; in addition, Saturn is considerably brighter than Jupiter at 1.06μ, probably due to the relative absence of ammonia on Saturn.

100.012 **A dynamical model for the radial structure of Saturn's rings. II.**
F. A. Franklin, G. Colombo, A. F. Cook.
Icarus, Vol. 15, 80 - 92 (1971).
We examine first the possibility that material exists outside ring A. This discussion leads to a revision of the period of Janus to 19^d565. In the second part we investigate the influence of the perturbations of Titan on the inner ring structure and conclude that a resonance between the mean motion of Titan and the apsidal motion of the orbit of a ring particle, owing to the oblateness of Saturn, can explain the division internal to ring C recently found by Guerin. Finally we show that if the space density of particles in ring B is $\lesssim 0.1$ g/cm^3, then the location of the resonance associated with the Cassini division is displaced by the amount necessary to reconcile its measured and predicted positions.

100.013 **Saturn.**
T. J. C. A. Moseley, A. Appleyard, M. Wardley.
British Astron. Ass., Circ. No. 536 (1971).

100.014 **Erratum: 'On the phase curves of the B-ring of Saturn'.** [Astron. Tsirk., No. 612, p. 1 - 2 (1971)].
A. M. Gretsky.
Astron. Tsirk., No. 652, p. 8 (1971). In Russian.

100.015 **White spots on Saturn.**
T. J. C. A. Moseley, A. Appleyard, M. Wardley.
Yamamoto Circ., No. 1741 (1971). In Japanese.

Observations of Pluto, minor planets 10 Hygiea and 433 Eros, Saturn's satellites VII, VIII, IX made at the Crimean Astrophysical Observatory.
See Abstr. 041.001.

Hydrogen and helium at high temperatures.
See Abstr. 099.004.

101 Uranus, Neptune, Pluto, Transplutonian Planet

101.001 **Multicolor photoelectric photometry of Uranus.**
J. F. Appleby, W. M. Irvine.
Astron. Journ., Vol. 76, 617 - 619 = Contr. Five College Obs.,
Univ. Massachusetts, Amherst, No. 108 (1971).

Narrow-band and *UBV* photoelectric measurements of
the magnitude at unit distance and geometric albedo of Uranus are presented for wavelengths $0.315 \leq \lambda \leq 1.06 \mu$. Comparison with *UBVRI* observations of Harris shows differences
at long-wavelengths ($\lambda \geq 6000$ Å) which can be explained in
terms of methane absorption.

101.002 **Visuelle Beobachtungsmöglichkeiten von Pluto mit**
Amateurinstrumenten. F. Zehnder.
Orion, 29. Jahrgang, p. 150 - 151 (1971).

101.003 **Multicolor photoelectric photometry of Uranus.**
J. F. Appleby.
Bull. American Astron. Soc., Vol. 3, 385 (1971). – Abstr.
AAS.

101.004 **Motions of the satellites of Uranus.**
D. Dunham.
Bull. American Astron. Soc., Vol. 3, 415 (1971). – Abstr.
AAS.

101.005 **A dynamical search for a transplutonian planet.**
P. K. Seidelmann.
Astron. Journ., Vol. 76, 740 - 742 (1971).

A dynamical search for three hypothetical planets having
representative characteristics of transplutonian planets, but
within certain magnitude limits, was made by numerical integration of the orbits of the outer planets with, and without, a
hypothetical planet. Careful comparison of these orbits with
observations indicates that, although there may be undiscovered planets beyond Neptune, their dynamical effect on the
known planets is so small that their presence, or absence,
cannot be clearly discerned from the observation residuals.

101.006 **Microwave radiation of Uranus and Neptune.**
C. H. Mayer, T. P. McCullough.

Icarus, Vol. 14, 187 - 191 (1971).

Observations of Uranus and Neptune at wavelengths of
1.65, 2.7, and 6 cm further clarify their radio emission spectra
and atmospheric temperatures. Measurements made at the
same time of Jupiter and Mars at the 1.65- and 2.7-cm wavelengths are presented.

101.007 **Measurements of Uranus radio emission at 8.22 mm.**
A. D. Kuzmin, B. Y. Losovsky.
Icarus, Vol. 14, 196 - 197 (1971).

Measurements of radio emission from the planet Uranus
at 8.22-mm wavelength were made using the 22-meter Lebedev
radio telescope with a low-noise maser amplifier. The brightness temperature ratio of Uranus to Jupiter was found to be
0.91 ± 0.10. Adopting a Jupiter brightness temperature of
$144°$K, we obtain a brightness temperature of Uranus equal
to $131° \pm 15°$K.

101.008 **The estimate of SHF absorption in hydrogen under**
high pressures according to radio astronomical
measurements of Uranus. A. D. Kuzmin, A. G. Soloviev.
Dokl. Akad. Nauk SSSR, Ser. Mat. Fiz., Vol. 201, 1313 -
1315 (1971). In Russian.

101.009 **The mass of Pluto.**
Southern Stars, Vol. 24, 63 - 64 (1971).

Observations of Pluto, minor planets 10 Hygiea
and 433 Eros, Saturn's satellites VII, VIII, IX made at the
Crimean Astrophysical Observatory.
See Abstr. 041.001.

Spectrophotometry of the moon, Mars, and Uranus.
See Abstr. 094.320.

Comet families and transneptunian planets.
See Abstr. 102.025.

Photographic positions of comets and minor planets
observed during 1968–1970. See Abstr. 103.006.

102 Comets

102.001 **Space distribution of the splitting and outbursts of comets.** E. M. Pittich.
Bull. Astron. Inst. Czechoslovakia, Vol. 22, 143 - 153 (1971).
A simple method for the determination of the moment of splitting of a cometary nucleus, based on the recession rate of the separated components, is applied to the observed data, and confronted with results obtained by other methods. The number of ten known moments of disruption is extended by ten more cases. The spatial distribution of the points in which splitting and cometary outbursts occurred is given.

102.002 **Comet brightness variations and outer space conditions.**
D. A. Andrienko, A. A. Demenko, I. M. Demenko, I. D. Zosimovich.
Astron. Zhurn. Akad. Nauk SSSR, Vol. 48, 843 - 849 (1971). In Russian. English translation in Soviet Astron. AJ, Vol. 15, No. 4.
Comet brightness variations study may serve as a good method to determine conditions in the interplanetary space. In the work the brightness curves of comets are compared with the curves of changes of the geomagnetic field. Photometric observations of thirty comets in the years from 1874 to 1937 have been used.

102.003 **The rate of disintegration of comets on different heliocentric orbits.**
O. V. Dobrovolsky, P. Egibekov.
Astrometriya i Astrofiz., *Kiev*, Vyp. (No.) 12, (see 003.003), p. 73 - 78 (1971). In Russian.
The disintegration rate of a cometary nucleus is considered as a function of the orbital elements. The life time of the nucleus is determined.

102.004 **A digest of comets.** M. J. Hendrie.
Spaceflight, Vol. 13, 140 - 145 (1971).

102.005 **Integral brightnesses of comets.** V. Riives.
Publ. Tartu Astrofiz. Obs., Vol. 39, 351 - 361 (1971). In Russian.

102.006 **Physico-chemical phenomena in comets—III. The continuum of comet Burnham (1960 II).**
A. H. Delsemme, D. C. Miller.
Planet. Space Sci., Vol. 19, 1229 - 1257 (1971).
The new model of the cometary head proposed in two previous papers (Planet. Space Sci., Vol. 18, 709 - 715 and 717 - 730 (1970)) is developed and applied to comet Burnham. This paper establishes the photometric shape of the continuum as reflected by the icy grains, and compares it to the observed continuum of comet Burnham.

102.007 **Physico-chemical phenomena in comets—IV. The C_2 emission of comet Burnham (1960 II).**
A. H. Delsemme, D. C. Miller.
Planet. Space Sci., Vol. 19, 1259 - 1274 (1971).
The new model of the cometary head proposed in three previous papers is developed to explain the photometric profiles of the molecular emission bands observed in comets. The model, used to explain the photometric profile of C_2 in comet Burnham (1960 II) points to the existence of a halo of icy grains of the same approximate size as independently deduced from the shape of comet Burnham's continuum.

102.008 **Erratum: 'Correction of cometary orbits including the perturbations in differential coefficients'. [Acta Astron., Vol. 21, 87 - 102 (1971)].** G. Sitarski.

Acta Astron., Vol. 21, 415 (1971).

102.009 **The effect of interstellar clouds on the comet cloud.** S. P. Wyatt, M. B. Faintich.
Bull. American Astron. Soc., Vol. 3, 368 (1971). – Abstr. AAS.

102.010 **To the problem of the origin of comets.** J. M. Witkowski.
Astron. vestn., Vol. 5, 82 - 88 (1971). In Russian. – Abstr. in Referativ. Zhurn. 51. Astron., 10.51.303 (1971).

102.011 **The role of major planets in the discovery of short-period comets and the evolution of their orbits.**
E. I. Kazimirchak-Polonskaya.
Byull. Inst. Teoret. Astron., *Leningrad,* Vol. 12, 796 - 812 (1971). In Russian.

102.012 **Comets and their interaction with the solar wind.** L. Biermann.
Quarterly Journ. Roy. Astron. Soc., Vol. 12, No. 4, (see 012.012), 417 - 431 (1971).

102.013 **On the stability of Oort's cloud.** E. M. Nezhinsky.
Byull. Inst. Teoret. Astron., *Leningrad*, Vol. 13, 31 - 35 (1971). In Russian.
An attempt has been made to estimate the rate of destruction of Oort's cloud by stars passing through it. Different mechanisms for the cloud's dispersing have been investigated (sweeping and cumulative ones). The numerical estimates show that the cumulative dispersing mechanism plays the leading role. The lower boundary for the half-period of the cloud's decay is equal to 1.1×10^9 years.

102.014 **Comets and nongravitational forces. IV.** B. G. Marsden, Z. Sekanina.
Astron. Journ., Vol. 76, 1135 - 1151 (1971).
Orbital elements and nongravitational parameters are derived from observations at every apparition of the periodic comets Honda-Mrkos-Pajdušáková, Faye, Tempel 2, Biela, Brorsen, and Tempel-Swift. For all except the first comet, the observations go back a century and more. It is found that, while most of the reliable determinations indicate that the cometary nongravitational effects decrease with time, there are a few cases where the effects increase slightly. The former situation is discussed in terms of a nuclear core-mantle model, implying that these comets will eventually evolve into inert, asteroidal objects, while the nuclei of the other comets are interpreted as coreless, eventually to disappear completely. A list is given of seven "erratic" comets known or suspected to have experienced large and relatively sudden anomalies in their motions. There is no conflict with observed phenomena when one interprets these anomalies in terms of collisions with other objects, specifically the interplanetary boulders.

102.015 **On irregular forces in cometary motion.** I. Zalkalne.
Latv. ordena trud. krasn. znameni gos. univ. im. P. Stuchki, Uch. zap., Vol. 148, vyp. (No.) 6, p. 91 - 97 (1971). In Russian.

102.016 **A comparison between the distributions of hypothetical and real comets on the basis of computations with an electronic computer.** M. A. Mamedov.
Izv. AN AzSSR. Ser. fiz.-tekhn. i mat. n., 1971, No. 1, p. 60 66. In Russian. – Abstr. in Referativ. Zhurn. 51. Astron.,

1.51.134 (1972).

102.017 **The distribution of hypothetical and real comets from perihelion distances and orbital inclinations.** M. A. Mamedov.
Izv. AN AzSSR. Ser. fiz.-tekhn. i mat. n., 1971, No. 1, p. 67 - 72. In Russian. – Abstr. in Referativ. Zhurn. 51. Astron., 1.51.135 (1972).

102.018 **Processes of disappearance of CO_2 molecules in comets.** V. I. Cherednichenko.
Problems of cosmic physics. Vyp. (No.) 6, (see 003.040), p. 95 - 100 (1971). In Russian.

102.019 **Calculation of the heliographic coordinates of a comet from its orbital elements.** I. M. Demenko.
Problems of cosmic physics. Vyp. (No.) 6, (see 003.040), p. 111 - 114 (1971). In Russian.

102.020 **Dependence of the integral brightness of a comet's head on the angular diameter and distance from the observer.** O. V. Dobrovolsky, R. S. Osherov, M. Z. Markovich.
Problems of cosmic physics. Vyp. (No.) 6, (see 003.040), p. 123 - 128 (1971). In Russian.

102.021 **On the intensity of dust outburst processes of long-periodic comets.** V. N. Lebedinets.
Astron. vestn., Vol. 5, 174 - 180 (1971). In Russian. – Abstr. in Referativ. Zhurn. 51. Astron., 2.51.323 (1972).

102.022 **Cometary spectra.** Y. Yamashita.
Astron. Herald (*Japan*), Vol. 63, 199 - 201 (1970). In Japanese. – Review article.

102.023 **Dust tail of the comet.** K. Saito.
Astron. Herald (*Japan*), Vol. 63, 202 - 206 (1970). In Japanese. – Review article.

102.024 **Origin of comets.** S. Yabushita.
Astron. Herald, (*Japan*), Vol. 64, 236 - 238 (1971). In Japanese.
The author discusses the problem for comets of long period (>200y). He calculates the probability distribution and the magnitude of energy perturbation at the collision of each member of a comet group with a planet. He criticizes two quan-titative theories proposed by Oort and by Lyttleton.

102.025 **Comet families and transneptunian planets.** E. J. Öpik.
Irish Astron. Journ., Vol. 10, 35 - 92 (1971). – Part of Research Seminar, Department of Physics and Astronomy, University of Maryland, 1971.

102.026 **Statistical analysis of unexplained changes in the eccentricities of nearly parabolic comets.** N. Smiriga.
Publ. Astron. Soc. Pacific, Vol. 83, 836 - 848 (1971).
143 comets are listed in increasing order of eccentricity, as well as the change in eccentricity $\Delta e = e_{2000} - e_{1800}$ for each comet. We show that a "switching point" exists between two comet numbers, such that for comet numbers below that point there is a probability of approximately 0.46 that $\Delta e \geqq 0$. For comet numbers above this switching point the probability that $\Delta e \geqq 0$ is approximately 0.75. The existence of the "switching point" is statistically demonstrated; however, no speculations are made as to why such a switching point exists.

102.027 **The comets in space.** A. Z. Dolginov.
Proc. Sixth Winter School on Space Physics, Part II. Apatity 1969, (see 012.034), p. 3 - 14 (1971).

102.028 **Laboratory simulation of cometary phenomena.** E. A. Kaimakov, V. I. Sharkov.
Proc. Sixth Winter School on Space Physics, Part II. Apatity 1969, (see 012.034), p. 15 - 22 (1971).

102.029 **Absolute photometry of cometary nuclei by focal star images.** K. I. Churyumov.
Proc. Sixth Winter School on Space Physics, Part II. Apatity 1969, (see 012.034), p. 23 (1971).

102.030 **Dynamic processes in comets and their analogy with the dynamics of the earth's magnetosphere.** Z. M. Ioffe.
Proc. Sixth Winter School on Space Physics, Part II. Apatity 1969, (see 012.034), p. 27 (1971). – Abstract.

On the dissolution time of a class of binary systems. See Abstr. 117.021.

103 Comets: Listed Objects

103.001 **Observations of comets.**   I. R. Bejtrishvili.
IAU Circ., No. 2337 (1971). – Concerning the comets 1970g, 1970*l*, 1970m.

103.002 **Comets in 1970.**   B. G. Marsden.
Quarterly Journ. Roy. Astron. Soc., Vol. 12, 244 - 273 (1971). – Progress report.

103.003 **Absolute magnitudes and other parameters of the comets of 1970.**   S. K. Vsekhsvyatskij.
Kometn. Tsirk., *Kiev*, No. 122 (1971). In Russian.

103.004 **Comet notes.**   E. Roemer.
Publ. Astron. Soc. Pacific, Vol. 83, 690 - 692 (1971).

103.005 **Roman numeral designations of comets in 1970.**
IAU Circ., No. 2378 (1971).

103.006 **Photographic positions of comets and minor planets observed during 1968–1970.**
I. Nikoloff, M. P. Candy.
Perth Obs., Western Australia, Commun. No. 2, p. 9 - 25 (1970). – Six observations of Pluto are included.

103.007 **Observations of comets at Tartu.**   H. K. Raudsaar.
Byull. Inst. Teoret. Astron., *Leningrad*, Vol. 12, 923 - 924 (1971). In Russian.

103.008 **Observations of comets made at the Crimean Astrophysical Observatory 1966–1968.**
N. S. Chernykh.
Byull. Inst. Teoret. Astron., *Leningrad*, Vol. 12, 925 - 933 (1971). In Russian.

103.009 **Comets in the year 1971.**   J. Bouška.
Vesmír, Vol. 50, 348 (1971).   In Czech.

103.010 **Roman numeral designation for 1970.**
Yamamoto Circ., No. 1746 (1971). In Japanese.

103.011 **Komētas 1970. gadā.**   A. Alksnis.
Zvaigžņotā debess, 1971. gada vasara, p. 31 - 33.

Observations photographiques des petites planètes et des comètes à Beograd.   See Abstr. 098.023.

103.100 **Comet 1970*l* Encke**

Improvement of the orbit of comet Encke from four apparitions in 1953–1964.   G. R. Kastel.
Byull. Inst. Teoret. Astron., *Leningrad*, Vol. 12, 724 - 731 (1971). In Russian.
The improved system of orbital elements has been derived as a result of linking the four apparitions of comet Encke in 1953–1954, 1957, 1960–1961, 1963–1964. The value of the secular acceleration coefficient has been found to be equal to $27.''9$.

Comet Encke, 1970*l*.
Kometn. Tsirk., *Kiev*, No. 120 (1971). In Russian.

103.101 **Comet 1970 II Bennett**

Polarization measurements of the head of comet

Bennett (1969i).   D. Clarke.
Astron. Astrophys., Vol. 14, 90 - 94 (1971).
Polarimetric observations of the light from around the nucleus of comet Bennett are reported. The presented observations are compared with similar ones of previous comets.

Search for microwave H_2O emission in comet Bennett (1969i).   T. A. Clark, B. Donn, W. M. Jackson, W. T. Sullivan III, N. Vandenberg.
Astron. Journ., Vol. 76, 614 - 616 (1971).
The 85-ft radio telescope of the Naval Research Laboratory was used in an attempt to detect the 22235-MHz transition ($6_{16} \rightarrow 5_{23}$) of H_2O during the recent appearance of comet Bennett (1969i). No H_2O emission of antenna temperature greater than $2.5°K$ was observed. We have derived upper limits to the H_2O column density for various temperatures of the cometary gas. These limits have been compared with H_2O column densities calculated from two different cometary models.

One micron spectrum of comet Bennett (1969 i).
D. D. Meisel.
Bull. American Astron. Soc., Vol. 3, 368 (1971). – Abstr. AAS.

Izofotometrie.   M. Druckmüller.
Contr. Obs. and Planetarium Brno, No. 11, p. 1 - 32 (1970).

Kometa 1969i Bennett.   Z. Okáč.
Contr. Obs. and Planetarium Brno, No. 11, p. 33 - 58 (1970).

Izofotometrické obrazy komety 1969i Bennett.
M. Druckmüller.
Contr. Obs. and Planetarium Brno, No. 11, p. 59 - 71 (1970).

The distribution of polarization in the head and tail of comet Bennett 1969i.   R. S. Osherov.
Dokl. AN TadzhSSR, Vol. 14, No. 3, p. 17 - 20 (1971). In Russian. – Abstr. in Referativ. Zhurn. 51. Astron., 11.51.429 (1971).

Investigation of the outflow of matter from observations of the tail of comet Bennett, 1969i.
O. V. Dobrovol'skii, Kh. Ibadinov.
Dokl. AN TadzhSSR, Vol. 14, No. 4, p. 18 - 21 (1971). In Russian. – Abstr. in Referativ. Zhurn. 51. Astron., 12.51.355 (1971).

Comet Bennett (1969i).
Yu. E. Migach, N. M. Shiper.
IAU Circ., No. 2363 (1971).

Observations of comets.   H. R. Soper, H. B. Ridley, S. W. Milbourn.
IAU Circ., No. 2370 (1971).

Laser radar observations of dust from comet Bennett.
G. S. Kent, M. C. W. Sandford, W. Keenliside.
Journ. Atmosph. Terr. Phys., Vol. 33, 1257 - 1262 (1971).
Laser radar observations are described that have been made during the passage of the earth through the plane of the orbit of the comet Bennett. These observations show an enhancement of the scattered signal received from heights between 40 km and 90 km; this enhancement, which descended during the observation period of a few days, is attributed to scattering from dust particles entering the earth's atmosphere from the comet.

The cyanogen bands of comet Bennett 1969i.
G. C. L. Aikman, W. J. Balfour, J. B. Tatum.
Journ. Roy. Astron. Soc. Canada, Vol. 65, 174 (1971).
Abstr. RAS Canada.

Comet Bennett, 1969i.
Kometn. Tsirk., *Kiev*, No. 122 (1971). In Russian.

Studio geometrico sulla forma e sull'orientamento della coda della cometa Bennett (1969i).
A. Bernasconi, L. Pansecchi.
Mem. Soc. Astron. Italiana, Nuova Ser., Vol. 42, 185 - 200 (1971).
    Six pictures of the comet have been examined and reduced on the comet's orbital plane following a bidimensional model. The results lead to a progressive enlargement of the angular aperture of the real tail.

Erratum: 'Studio geometrico sulla forma e sull' orientamento della coda della cometa Bennett (1969i)' [Mem. Soc. Astron. Italiana, Nuova Ser., Vol. 42, 185 - 200 (1971)].
A. Bernasconi, L. Pansecchi.
Mem. Soc. Astron. Italiana, Nuova Ser., Vol. 42, 659 (1971).

Photographic observations of comets, observed at the Republic Observatory Annexe, Hartbeespoort, with the Franklin-Adams star camera.   J. A. Bruwer, M. Klerk.
Republic Obs. Johannesburg, Circ. No. 131, Vol. 8, 5 (1971).

103.102   Comet 1971a Toba

Comet Toba (1971a).   I. R. Bejtrishvili, T. Seki.
IAU Circ., No. 2337 (1971).

Comet Toba (1971a).   J. A. Bruwer.
IAU Circ., No. 2346 (1971).

Comet Toba (1971a).   D. P. Elias.
IAU Circ., No. 2356 (1971).

Comet Toba (1971a).   J. A. Bruwer.
IAU Circ., No. 2361 (1971).

Comet Toba, 1971a.
Kometn. Tsirk., *Kiev*, No. 120 (1971). In Russian.

Comet Toba, 1971a.
Kometn. Tsirk., *Kiev*, No. 121 (1971). In Russian.

Comet Toba, 1971a.
Kometn. Tsirk., *Kiev*, No. 122 (1971). In Russian.

Nouvelles des comètes: Comète Toba (1971a).
L'Astronomie, 85ᵉ année, p. 338 (1971).

Der Komet Toba, 1971a.   T. Kleine.
Orion Schaffhausen, 29. Jahrgang, p. 175 - 177 (1971).

Komet Toba (1971a).   F. Mitschke.
SuW, Vol. 10, 244 (1971).

Beobachtungen des Kometen Toba (1971a).
B. Flach.
SuW, Vol. 10, 244 (1971).

Comet Toba (1971a).
Yamamoto Circ., No. 1744 (1971). In Japanese.

103.103   Comet 1970 XV Abe

Comet Abe (1970g).   G. R. Kastel', V. L. Afanas'ev.
IAU Circ., No. 2365 (1971).

Observations of comets.   H. R. Soper, H. B. Ridley, S. W. Milbourn.
IAU Circ., No. 2370 (1971).

Comet Abe, 1970g.
Kometn. Tsirk., *Kiev*, No. 120 (1971). In Russian.

Comet Abe, 1970g.
Kometn. Tsirk., *Kiev*, No. 121 (1971). In Russian.

Helligkeitsschätzungen des Kometen Abe (1970g).
C. Kowalec.
SuW, Vol. 10, 244 - 245 (1971).

103.104   Comet 1960 II Burnham

Physico-chemical phenomena in comets– III. The continuum of comet Burnham (1960 II).
See Abstr. 102.006.

Physico-chemical phenomena in comets–IV. The C_2 emission of comet Burnham (1960 II).
See Abstr. 102.007.

103.105   Comet 1910 II Halley

The orbit of Halley's comet and the apparition of 1986.   J. L. Brady, E. Carpenter.
Astron. Journ., Vol. 76, 728 - 739 (1971).
    The motion of Halley's comet from 1682 to 1986 has been investigated and a search ephemeris is given for the 1986 apparition extending 4 years on each side of perihelion. The last four apparitions have been linked by a continuous integration of 230 years. To do so, however, required the addition of a secular term in the equations of motion.

103.106   Comet 1971d Tsuchinshan 2

Periodic comet Tsuchinshan 2 1971d.
E. Roemer.
British Astron. Ass., Circ. No. 537 (1971).

Periodic comet Tsuchinshan 2 (1965 II).
G. Sitarski.
IAU Circ., No. 2337 (1971).

Periodic comet Tsuchinshan 2 (1971d).
E. Roemer, G. Reskin.
IAU Circ., No. 2357 (1971).

Ephemeris of comet Tsuchinshan 2, 1965 II.
Kometn. Tsirk., *Kiev*, No. 121 (1971). In Russian.

Comet P/Tsu-chin-shan 2, 1971d.
Yamamoto Circ., No. 1741 (1971). In Japanese.

103.107   **Comet 1971b Holmes**

Periodic comet Holmes 1971b.   E. Roemer.
British Astron. Ass., Circ. No. 534 (1971).

Periodic comet Holmes (1971b).
E. Roemer, D. C. Ferguson, A. H. Ferguson, J. Q. Latta.
IAU Circ., No. 2338 (1971).

Short-period comet Holmes, 1971b.
Kometn. Tsirk., *Kiev*, No. 121 (1971). In Russian.

Periodic comet Holmes (1971b).
Yamamoto Circ., No. 1739 (1971). In Japanese.

103.108   **Comet 1967 IX Finlay**

Periodic comet Finlay (1967 IX).   K. Tomita.
IAU Circ., No. 2339 (1971).

103.109   **Comet 1970o Wolf-Harrington**

Periodic comet Wolf-Harrington (1970o).
G. Sitarski.
IAU Circ., No. 2341 (1971).

Periodic comet Wolf-Harrington (1970o).
A. Mrkos.
IAU Circ., No. 2355 (1971).

Periodic comet Wolf-Harrington (1970o).
T. Seki.
IAU Circ., No. 2369 (1971).

103.110   **Comet 1970 XIV Whipple**

Periodic comet Whipple (1969c).
IAU Circ., No. 2341 (1971).

103.111   **Comet 1971c Kearns-Kwee**

Periodic comet Kearns–Kwee 1971c.
E. Roemer, L. M. Vaughn.
British Astron. Ass., Circ. No. 535 (1971).

Periodic comet Kearns-Kwee (1971c).
E. Roemer, L. M. Vaughn.
IAU Circ., No. 2344 (1971).

Short-period comet Kearns-Kwee, 1971c.
Kometn. Tsirk., *Kiev*, No. 121 (1971). In Russian.

Comet P/Kearns-Kwee (1971c).
Yamamoto Circ., No. 1740 (1971). In Japanese.

103.112   **Comet 1971f Tsuchinshan 1**

Periodic comet Tsuchinshan 1 (1965 I).
G. Sitarski.
IAU Circ., No. 2346 (1971).

Periodic comet Tsuchinshan 1 (1971f).
E. Roemer, L. M. Vaughn.
IAU Circ., No. 2379 (1971).

Ephemeris of comet Tsuchinshan 1, 1965 I.
Kometn. Tsirk., *Kiev*, No. 121 (1971). In Russian.

103.113   **Comet 1852 III Biela**

Periodic comet Biela.   B. G. Marsden.
IAU Circ., No. 2347 (1971).

Short-period comet Biela.
Kometn. Tsirk., *Kiev*, No. 121 (1971) In Russian.

Comet Biela.
Yamamoto Circ., No. 1743 (1971). In Japanese.

Meteors from periodic comet Biela.
See Abstr. 104.040.

Meteors from comet P/Biela.
See Abstr. 104.093.

103.114   **Comet 1960 I Wild**

Periodic comet Wild (1960 I).   P. Wild.
IAU Circ., No. 2352 (1971).

103.115   **Comet 1969 II Gunn**

Periodic comet Gunn (1969 II).
E. Roemer.
IAU Circ., No. 2359 (1971).

Periodic comet Gunn, 1969 II.
Kometn. Tsirk., *Kiev*, No. 121 (1971). In Russian.

Continuation of the ephemeris of comet Gunn, 1969 II.
Kometn. Tsirk., *Kiev*, No. 124 (1971). In Russian.

103.116   **Comet 1971e Shajn-Schaldach**

Periodic comet Shajn-Schaldach 1971e.
C. T. Kowal.
British Astron. Ass., Circ. No. 537 (1971).

Periodic comet Shajn-Schaldach (1971e).
C. T. Kowal.
IAU Circ., No. 2360 (1971).

Periodic comet Shajn-Schaldach (1971e).
C. T. Kowal.
IAU Circ., No. 2364 (1971).

Periodic comet Shajn-Schaldach (1971e).
R. L. Waterfield, M. J. Hendrie.
IAU Circ., No. 2368 (1971).

Periodic comet Shajn-Schaldach (1971e).
R. L. Waterfield.
IAU Circ., No. 2372 (1971).

Periodic comet Shajn-Schaldach (1971e).
R. L. Waterfield, R. H. South.
IAU Circ., No. 2378 (1971).

Rediscovery of the periodic comet Shajn-Schaldach
1949 VI = 1971 e.
Kometn. Tsirk., *Kiev*, No. 124 (1971). In Russian.

Comet P/Shajn-Schaldach (1971e).
Yamamoto Circ., Nos. 1741, 1742, 1743, 1744 (1971).
In Japanese.

103.117 Comet 1967 I Grigg-Skjellerup

Periodic comet Grigg-Skjellerup.   G. Sitarski.
IAU Circ., No. 2361 (1971).

Comet P/Grigg-Skjellerup.
Yamamoto Circ., Nos. 1743, 1745 (1971). In Japanese.

Possible meteor shower from periodic comet Grigg-
Skjellerup.   See Abstr. 104.090.

103.118 Comet 1879 III Tempel 1

Periodic comet Tempel 1.   G. Schrutka.
IAU Circ., No. 2363 (1971).

Comet P/Tempel 1.
Yamamoto Circ., No. 1743 (1971). In Japanese.

103.119 Comet 1968 I Ikeya-Seki

Comet Ikeya-Seki (1968 I).   Yu. E. Migach,
G. R. Kastel'.
IAU Circ., No. 2365 with corrections to the times of Odessa
observations in IAU Circ., No. 2371 (1971).

Comet Ikeya-Seki (1968 I).   B. G. Marsden.
IAU Circ., No. 2376 (1971).

Comet Ikeya-Seki, 1967 n = 1968 I.   G. Kastel'.
Kometn. Tsirk., *Kiev*, No. 124 (1971). In Russian.

A study of the brightness of comet Ikeya–Seki
1967n during three solar revolutions.
A. A. Demenko, K. I. Churyumov.
Proc. Sixth Winter School on Space Physics, Part II. Apatity
1969, (see 012.034), p. 25 - 26 (1971).

Comet Ikeya-Seki (1968I).
Yamamoto Circ., No. 1746 (1971). In Japanese.

103.120 Comet 1969 VI Faye

Periodic comet Faye (1969 VI).
G. R. Kastel', V. L. Afanas'ev, S. I. Gerasimenko.
IAU Circ., No. 2366 (1971).

Comet Faye, 1969 VI.
Kometn. Tsirk., *Kiev*, No. 124 (1971). In Russian.

103.121 Comet 1970q Väisälä 1

Periodic comet Väisälä 1 (1970q).
IAU Circ., No. 2367 (1971).

103.122 Comet 1970 X Suzuki-Sato-Seki

Comet Suzuki-Sato-Seki (1970m).
S. I. Gerasimenko.
IAU Circ., No. 2370 (1971).

Comet Suzuki-Sato-Seki, 1970m.
Kometn. Tsirk., *Kiev*, No. 120 (1971). In Russian.

Photographic observations of comets, observed at
the Republic Observatory Annexe, Hartbeespoort, with the
Franklin-Adams star camera.   J. A. Bruwer, M. Klerk.
Republic Obs. Johannesburg, Circ. No. 131, Vol. 8, 5 (1971).

103.123 Comet 1957 IV Schwassmann-Wachmann 1

Ephemeris of comet Schwassmann-Wachmann 1.
Kometn. Tsirk., *Kiev*, No. 124 (1971). In Russian.

103.124 Comet 1970e Ashbrook-Jackson

Ephemeris of the comet Ashbrook-Jackson,
obtained from elements corrected according to observations
in 1971.   N. A. Belyaev.
Kometn. Tsirk., *Kiev*, No. 122 (1971). In Russian.

103.125 Comet 1969 IV Churyumov-Gerasimenko

Ephemeris of the comet Churyumov-Gerasimenko
for the years 1962 - 1963.
Kometn. Tsirk., *Kiev*, No. 123 (1971). In Russian.

103.126 Comet 1967 II Rudnicki

Comet Rudnicki 1966e.
J. E. Bortle, D. Milon.
Strolling Astronomer, Vol. 23, 99 - 102 (1971).

103.127 Comet 1967 X Tempel 2

Periodic comet Tempel 2.
IAU Circ., No. 2370 (1971).

103.128 Comet 1966 I Giacobini-Zinner

Periodic comet Giacobini-Zinner.
Yu. V. Evdokimov.
IAU Circ., No. 2372 (1971).

Nongravitational forces affecting the motion of
comet Giacobini-Zinner.   D. K. Yeomans.

Thesis, Univ. Maryland, Catonsville. [Available from Univ. Microfilms, Ann Arbor, Mich., U.S.A. Order No. 71–4533], 67 pp. (1970).

After extending the differential correction scheme to include nongravitational terms in the equations of motion, the observations were successfully represented over the time intervals 1900–1946, 1913–1959 and 1939–1965. There is an indication that the magnitude of the nongravitational forces increases with time and a motion of the comet was discontinuous between 1959 and 1965.

103.129 **Comet 1970 III Kohoutek**

Comet Kohoutek (1969b).   S. I. Gerasimenko.
IAU Circ., No. 2374 (1971).

Comet Kohoutek, 1969 b.
Kometn. Tsirk., *Kiev,* No. 124 (1971). In Russian.

103.130 **Comet 1951 V Neujmin 3**

Periodic comet Neujmin 3.
IAU Circ., No. 2375 (1971).

103.131 **Comet 1948 V Pajdusakova-Mrkos**

Definitive orbit of the comet 1948 V Pajdúšáková-Mrkos.   O. N. Barteneva.
Byull. Inst. Teoret. Astron., *Leningrad*, Vol. 13, 21 - 26 (1971). In Russian.

103.132 **Comet 1954 XII Kresák-Peltier**

Determination of original and future orbits of the comets 1948 V Pajdúšáková-Mrkos and 1954 XII Kresák-Peltier.   O. N. Barteneva.
Byull. Inst. Teoret. Astron.,*Leningrad,* Vol. 13, 27 - 30 (1971). In Russian.

Determination of original and future orbits of the comets 1948 V Pajdúšáková-Mrkos and 1954 XII Kresák-Peltier.   See Abstr. 103.131.

103.133 **Comet 1969 IX Tago-Sato-Kosaka**

Positions of comet 1969g Tago-Sato-Kosaka.
I. M. Demenko.
Astrometriya i Astrofiz., *Kiev,* No. 13, (see 003.039), p. 112 - 113 (1971). In Russian.

Absolute photometry of comet Tago – Sato – Kosaka, 1969g.   K. I. Churyumov, I. R. Beytrishvili.
Problems of cosmic physics. Vyp. (No.) 6, (see 003.040), p. 101 - 110 (1971). In Russian.

103.134 **Comet 1968 VI Honda**

Investigation of comet Honda 1968c from two points of view.   K. I. Churyumov, T. Maizlina.
Proc. Sixth Winter School on Space Physics, Part II. Apatity 1969, (see 012.034), p. 24 (1971).

Polarization observations of the comet Honda, 1968c.   R. S. Osherov.
Problems of cosmic physics. Vyp. (No.) 6, (see 003.040), p. 115 - 122 (1971). In Russian.

104 Meteors, Meteor Streams

104.001 Ablation and breakup of large meteoroids during atmospheric entry. B. Baldwin, Y. Sheaffer.
Journ. Geophys. Res., Vol. 76, 4653 - 4668 (1971).

An ablation model is described that can be used to estimate the effect on a large meteoroid of passage through a planetary atmosphere. The effect on ablation and deceleration of breakup due to aerodynamic pressure is investigated. Results are given from a series of calculations of the ablation and breakup of bronzite and carbonaceous chondrite meteoroids in the earth's atmosphere.

104.002 Light emission measurements of calcium and magnesium at simulated meteor conditions. I. Cross-section measurements. H. F. Savage, C. A. Boitnott.
Astrophys. Journ., Vol. 167, 341 - 348 (1971).

The emission cross-sections of the strongest spectral features of magnesium and calcium were measured in a crossed beam for collisions involving N_2, N_2^+, N^+, O_2, O_2^+, O^+, Ar^+, and Na^+ at energies from 150 to 2000 eV. Also measured were the cross-sections for ionization of magnesium and calcium by N_2 at energies between 400 and 2000 eV.

104.003 Light emission measurements of calcium and magnesium at simulated meteor conditions. II. Spectral luminous efficiencies. C. A. Boitnott, H. F. Savage.
Astrophys. Journ., Vol. 167, 349 - 355 (1971).

The luminous efficiency of meteors has been calculated from emission and ionization cross-sections for the prominent spectral lines of calcium and magnesium. Characteristics of meteor spectra were interpreted based on experimental measurements, and conditions were identified for the enhancement of ion spectra in certain meteors. Luminous efficiencies for meteoric constituents were obtained for use with observing instruments that incorporate both blue-sensitive and panchromatic photographic emulsions and for the dark-adapted eye.

104.004 Berekeningen aan een tweetal simultaan gefotografeerde meteoren. B. Apeldoorn, E. J. Kaptein.
Hemel en Dampkring, Vol. 69, 227 - 230 (1971).

104.005 Heldere vuurbol in Loenen gefotografeerd.
P. A. Koning.
Hemel en Dampkring, Vol. 69, 194 - 195 (1971).

104.006 The dependence of meteor ionization on meteor velocity. V. M. Kolmakov.
Astron. Tsirk., No. 614, p. 7 - 8 (1971).   In Russian.

104.007 On the precision of meteoroid densities.
A. R. Kolomiets.
Astron. Tsirk., No. 620, p. 7 - 8 (1971).   In Russian.

104.008 On the distribution of the random gravitational perturbations in the Perseid meteor stream.
E. N. Kramer, N. G. Pavlenko.
Astron. Tsirk., No. 621, p. 6 - 7 (1971).   In Russian.

104.009 A method for determinating the radiant position and the radiation area of meteor streams.
R. P. Chebotarev.
Astron. Tsirk., No. 624, p. 7 - 8 (1971).   In Russian.

104.010 Study of the radial distribution of electrons in ionized meteor trains.
G. I. Kolomiets, E. I. Fialko, R. I. Mojsya.
Astron. Tsirk., No. 625, p. 5 - 7 (1971).   In Russian.

104.011 Measurements of the initial radii of meteor trails.
G. I. Kolomiets.
Astron. Tsirk., No. 625, p. 7 - 8 (1971).   In Russian.

104.012 Investigation of meteor phenomena in the earth's atmosphere by radio physical methods.
E. I. Fialko, I. V. Bajrachenko, R. T. Mojsya.
Vyisnik Kiyiv. un-tu. Ser. fiz., 1970, No. 11, p. 50 - 57. In Ukrainian. – Abstr. in Referativ. Zhurn. 51. Astron., 9.51.303 (1971).

104.013 Meteor notes – July to December 1970.
K. B. Hindley.
Journ. British Astron. Ass., Vol. 81, 475 - 478 (1971).

104.014 On the influence of the multicomponent composition of meteor bodies on the characteristics of radio reflections from meteor traces. Yu. I. Portnyagin.
Geomagn. Aeronom., Vol. 11, 626 - 629 (1971). In Russian.

104.015 Osservazioni di meteore effettuate da alcuni membri del Gruppo Astrofili Valdinievole.
M. Niccolai, A. Giorgietti, L. Fanucci.
Coelum, Vol. 39, 198 - 199 (1971).

104.016 Meteor showers.   A. T. Blackwell.
Journ. Roy. Astron. Soc. Canada, Vol. 65, L22 - L24 (1971).

104.017 Prospects for abundances of elements in meteors.
A. F. Cook, C. L. Hemenway, P. M. Millman.
Bull. American Astron. Soc., Vol. 3, 369 (1971). – Abstr. AAS.

104.018 Radio echoes from randomly ionized meteor trails.
N. Brown, W. G. Elford.
Journ. Atmosph. Terr. Phys., Vol. 33, 1659 - 1666 (1971).

The average rate of decay of radio singals reflected from under-dense meteor trails has been shown to follow a simple dependence on trail height. However the decay of a radio echo from a single trail can vary widely from the average for its height. The possibility that this could be caused by an irregular ionization line density along the trail is examined, and decays are calculated for a trail which has a random variation in ionization along its length.

104.019 Investigations of meteoric matter.
V. V. Fedynskij.
Astron. vestn., Vol. 5, 57 - 77 (1971). In Russian. – Abstr. in Referativ. Zhurn. 51. Astron., 10.51.310 (1971).

104.020 Meteor showers in March 1969.
T. L. Korovkina, V. V. Martynenko, V. V. Frolov.
Astron. vestn., Vol. 5, 119 - 123 (1971). In Russian. – Abstr. in Referativ. Zhurn. 51. Astron., 10.51.311 (1971).

104.021 Light curves of telemeteors. O. I. Stepashina.
Astron. vestn., Vol. 5, 124 - 127 (1971). In Russian.
Abstr. in Referativ. Zhurn. 51. Astron., 10.51.313 (1971).

104.022 Spectra of meteors obtained in the USSR 1957 - 1967. I. S. Astapovich.
Astron. vestn., Vol. 5, 89 - 97 (1971). In Russian. – Abstr. in Referativ. Zhurn. 51. Astron., 10.51.314 (1971).

104.023 The fragmentation of small meteor bodies according to radar observations. N. V. Novoselova.

Astron. vestn., Vol. 5, 112 - 116 (1971). In Russian. – Abstr. in Referativ. Zhurn. 51. Astron., 10.51.315 (1971).

104.024 Energy spectra of electrons by modelling meteor phenomena. Yu. F. Bydin, V. I. Ogurtsov.
Astron. vestn., Vol. 5, 98 - 106 (1971). In Russian. – Abstr. in Referativ. Zhurn. 51. Astron., 10.51.320 (1971).

104.025 Determination of concentration of excited natrium atoms in a meteor train.
E. V. Sandakova, A. A. Demenko.
Vestn. Kiev. Un-ta, Ser. Astron., No. 12, p. 31 - 35 (1970). In Russian.

The paper presents an analysis of the meteor train obtained 17.-18. November 1966. Seven photographs of the train are used. The concentration of the excited natrium atoms on the 3p level is determined.

104.026 Some questions of hydrodynamics of the fusing surface of a meteor body.
V. G. Kruchinenko, A. N. Shaido.
Vestn. Kiev. Un-ta, Ser. Astron., No. 12, p. 36 - 41 (1970). In Russian.

The behaviour of the liquid film on a meteor body surface is studied. The meteor body is considered to be in the condition of a free-molecular shower for the case of constant ablation. The numerical values of the film thickness, the velocities of its flow, the mass velocities of ablation and evaporation are found for a concrete meteor.

104.027 The numerical integration program of the equations of motion of a meteor body by Cowell's method on the computer M-220. L. M. Sherbaum.
Vestn. Kiev. Un-ta, Ser. Astron., No. 12, p. 42 - 45 (1970). In Russian.

104.028 The influence of the Poynting-Robertson effect on meteor shower structure.
L. M. Sherbaum, A. N. Shaido.
Vestn. Kiev. Un-ta, Ser. Astron., No. 12, p. 46 - 50 (1970). In Russian.

The influence of the Poynting-Robertson effect on a shower formed by emission of particles with velocities $0.001, 0.01, 0.1$ km sec^{-1} from the comet 1866 I is considered. On the base of the method given by Robertson and developed by Wyatt and Whipple the changes of a and of the particle orbits in $5 \times 10^2, 10^3, 10^4, 10^5, 10^6$ years are calculated.

104.029 The problems and tasks of meteor radio electronics.
E. I. Fialko.
Vestn. Kiev. Un-ta, Ser. Astron., No. 12, p. 51 - 56 (1970). In Russian.

104.030 Some results of meteor observations on three wavelengths.
R. I. Moisya, V. I. Melnik, G. I. Kolomiets.
Vestn. Kiev. Un-ta, Ser. Astron., No. 12, p. 57 - 60 (1970). In Russian.

The results of radar observations of meteors at wavelengths 9.59, 6.49 and 8.7 m are given. The initial radii of ionized meteor trails are obtained from analysis of the general amplitude–time characteristics of the radio echo.

104.031 A phase sensitivity rising method for a meteor radar system. G. I. Kolomiets.
Vestn. Kiev. Un-ta, Ser. Astron., No. 12, p. 61 - 63 (1970). In Russian.

The method proposed allows to increase the number of reflections by which one can determine the velocity of a drifting meteor train; it can also be used for phase investigations of ionized meteor trains.

104.032 Photographic observations of meteors in Kiev 1965.
S. S. Trjashin, V. V. Benyuch, A. A. Demenko, V. G. Kruchinenko, L. M. Sherbaum, N. A. Hinkulova.
Vestn. Kiev. Un-ta, Ser. Astron., No. 12, p. 64 - 67 (1970). In Russian.

104.033 Investigation of the motion of a hypothetical meteor stream. L. M. Sherbaum.
Vestn. Kiev. Un-ta, Ser. Astron., No. 13, p. 57 - 64 (1971). In Russian.

The paper deals with the evolution of the orbit of a hypothetical meteor stream under the influence of Jupiter and Saturn. The change of orbital elements for 6 meteor particles chosen along the meteor stream is given.

104.034 The problem of statistic characteristics of the Geminid meteor shower from results of radar observations.
E. I. Fialko, V. F. Romanjuk, Y. V. Bitsenko, V. N. Dony.
Vestn. Kiev. Un-ta, Ser. Astron., No. 13, p. 65 - 68 (1971). In Russian.

Results of radar observations of the Geminid meteor shower 1967 at $\lambda = 8.7$ m are given. The distribution of the meteor bodies of the shower according to the masses in the region of small masses at the beginning and the end of the shower has been studied.

104.035 Investigation of the drift velocity of meteor trains by means of radar with high phase sensitivity.
G. I. Kolomiets, G. I. Solod.
Vestn. Kiev. Un-ta, Ser. Astron., No. 13, p. 69 - 72 (1971). In Russian.

104.036 The choice of radar parameters for measurement of initial radii of meteor trails.
R. I. Moisya, G. I. Kolomiets.
Vestn. Kiev. Un-ta, Ser. Astron., No. 13, p. 73 - 79 (1971). In Russian.

A method which allows to choose the optimum radar parameters for the measurement of initial radii of meteor trails is described. Some errors of measurement are discussed.

104.037 Distribution of meteor bodies according to their kinetic energies (The case of background and stream).
E. I. Fialko, V. F. Romanjuk.
Vestn. Kiev. Un-ta, Ser. Astron., No. 13, p. 80 - 85 (1971). In Russian.

The distribution of the meteor bodies according to the kinetic energies for the whole region of energies in the case of background and stream is examined. The cases of a "new" stream and background, and also of an "old" stream and background are investigated.

104.038 Helle Feuerkugeln im Juli 1971.
R. A. Naef.
Orion, 29. Jahrgang, p. 152 (1971).

104.039 Spectral data on terminal flare and wake of double-station meteor No. 38421 (Ondřejov, April 21, 1963). Z. Ceplecha.
Bull. Astron. Inst. Czechoslovakia, Vol. 22, 219 - 304 (1971).

The paper contains data on a bright meteor terminated by a flare of –12.4 absolute magnitude and photographed by 5 cameras of the double-station program of the Ondřejov Observatory. The heights and distances of the trajectory, the velocity, the orbit, the light curve and the wave lengths, intensities and identifications of 1007 spectral lines measured and traced at the terminal flare are given. 19 elements and 4 compounds were identified in the spectrum with sufficient certainty.

104.040 Meteors from periodic comet Biela.

L. Kresák.
IAU Circ., No. 2362 (1971).

104.041 On the possibility of return of the Draconid meteor stream. Yu. V. Evdokimov.
Kometn. Tsirk., *Kiev,* No. 122 (1971). In Russian.

104.042 Meteor streams. B. A. Lindblad.
Space Research XI, (see 012.004), 287 - 297 (1971).

104.043 Investigations of the meteor incident flux density by radio methods.
N. S. Andrianov, O. I. Belkovich, L. B. Gussakovskaya, K. V. Kostylyov, V. V. Sidorov, D. I. Stepanov, Y. A. Pupyshev.
Space Research XI, (see 012.044), 299 - 305 (1971).

104.044 Radar meteor influx and its comparison with direct cosmic dust measurement data. V. N. Lebedinets.
Space Research XI, (see 012.004), 307 - 317 (1971).

104.045 Temporal variations in the mass distribution of particles in meteor streams. D. W. Hughes.
Space Research XI, (see 012.004), 319 - 328 (1971).

104.046 Meteor particle studies from space vehicles.
T. N. Nazarova, A. K. Rybakov.
Space Research XI, (see 012.004), 357 - 361 (1971).

104.047 Cosmic dust and meteor showers. E. P. Mazets.
Space Research XI, (see 012.004), 363 - 369 (1971).

104.048 On the concentration of meteor particles in the vicinity of the earth's orbit. E. N. Kramer.
Space Research XI, (see 012.004), 371 - 376 (1971).

104.049 Semi-diurnal variations of the electron attachment rate in meteor trails.
P. B. Babadzhanov, R. Sh. Bibarsov.
Astron. Tsirk., No. 647, p. 6 - 7 (1971). In Russian.

104.050 Overzicht Perseïden 1970. E. J. A. Meurs.
Hemel en Dampkring, Vol. 69, 296 (1971).

104.051 Further studies of the effect of the earth's magnetic field on meteor trains. C. D. Watkins, R. Eames, T. F. Nicholson.
Journ. Atmosph. Terr. Phys., Vol. 33, 1907 - 1921 (1971).
The influence of the geomagnetic field on radar echoes from meteor trains has been studied during the occurrence of the Geminid meteor shower by means of a radar equipment in the UHF and VHF bands. Effects due to the magnetic field were apparent for most of the time that the shower was above the horizon, and the results indicate that, in the presence of the magnetic field, a meteor train develops irregularities which are elongated in the direction of the field lines.

104.052 Meteorfotografering. H. Pedersen.
Astron. Tidsskr., Årg. 4, 130 - 135 (1971).

104.053 The influence of deionization processes on meteor echo duration. R. Sh. Bibarsov.
Byull. Inst. Astrofiz., *Dushanbe,* No. 55, p. 3 - 9 (1970).
In Russian.
Formulas are obtained for meteor echo duration considering attachment, photodetachment, recombination and vortical diffusion.

104.054 The errors of measurement of the attachment rate by meteor methods. R. Sh. Bibarsov.
Byull. Inst. Astrofiz., *Dushanbe,* No. 55, p. 10 - 17 (1970).
In Russian.

The errors of measurements are investigated and previously obtained results are corrected.

104.055 On the radio echo method for determining the rate of attachment. R. Sh. Bibarsov.
Byull. Inst. Astrofiz., *Dushanbe,* No. 55, p. 18 - 20 (1970).
In Russian.
A new method for determining the rate of attachment by radio echo observations of meteors at night time is proposed.

104.056 On the determination of the parameter s by radio echo observations of meteors. R. Sh. Bibarsov.
Byull. Inst. Astrofiz., *Dushanbe,* No. 55, p. 21 - 23 (1970).
In Russian.

104.057 Complex of techniques for radio echo studies of meteors in Dushanbe.
R. P. Tshebotaryov, V. N. Sidorin, G. A. Polushkin, R. Sh. Bibarsov, Sh. O. Isamutdinov, V. M. Kolmakov.
Byull. Inst. Astrofiz., *Dushanbe,* No. 55, p. 24 - 28 (1970).
In Russian.
A complex of techniques for the determination of coordinates, heights, radiants and velocities of meteors and for research in physics of meteors and the upper atmosphere is described.

104.058 System for accurate ranging of meteor trails.
R. P. Tshebotaryov, V. N. Sidorin.
Byull. Inst. Astrofiz., *Dushanbe,* No. 55, p. 29 - 33 (1970).
In Russian.

104.059 Multibeam indicator for a meteor radar.
R. P. Tshebotaryov, Sh. O. Isamutdinov.
Byull. Inst. Astrofiz., *Dushanbe,* No. 55, p. 34 - 39 (1970).
In Russian.
Demands to the indicator of a meteor radar and to the photofilm registration system are considered. The block-scheme and the principle of action of a multibeam indicator fitting in a best way that demands are described. Results of work with such an indicator are described.

104.060 The picture of the drift of meteor trails in space.
V. M. Kolmakov.
Byull. Inst. Astrofiz., *Dushanbe,* No. 55, p. 40 - 45 (1970).
In Russian.

104.061 Parameters of the M-zone based on a photographically observed meteor train. U. Shodiev.
Byull. Inst. Astrofiz., *Dushanbe,* No. 55, p. 46 - 52 (1970).
In Russian.
The magnitude and direction of wind and turbulent diffusion in the M-zone, as well as photometric parameters of a train are given.

104.062 The effect of deionization processes on the distribution of meteor echo durations. R. Sh. Bibarsov.
Byull. Inst. Astrofiz., *Dushanbe,* No. 57, p. 3 - 9 (1970).
In Russian.
The effect of attachment, photodetachment and vortical diffusion on the distribution of meteor echo durations is examined. Formulas are obtained which allow to take into account the effect of these processes on the treatment of the observational results.

104.063 Some results of investigations of attachment processes from radar observations of meteors.
R. Sh. Bibarsov.
Byull. Inst. Astrofiz., *Dushanbe,* No. 57, p. 10 - 17 (1970).
In Russian.
The results of measurements of the attachment rate for heights of 78 + 118 km are given.

104.064 De juli-augustusaktie 1971: conventionele waarne-
mingsmethoden, resultaten en ervaringen.
B. Apeldoorn.
Hemel en Dampkring, Vol. 69, 325 - 329 (1971).

104.065 Maxima van meteoorzwermen.   J. Meeus.
Hemel en Dampkring, Vol. 69, 329 (1971).

104.066 Radar measurements of meteor drifts at two spaced
stations.   K. A. Karimov, V. V. Sidorov.
Investigation of the ionosphere and meteors, (see 003.015),
p. 21 - 33 (1971). In Russian.

104.067 On a one-dimensional expansion of a meteor train.
N. A. Arkabaev.
Investigation of the ionosphere and meteors, (see 003.015),
p. 34 - 38 (1971). In Russian.

104.068 On average monthly characteristics of the wind.
K. A. Karimov, V. Ya. Ogurtsov, V. V. Sidorov.
Investigation of the ionosphere and meteors, (see 003.015),
p. 39 - 46 (1971). In Russian.

104.069 Variant of a radar station for meteor-wind patrol.
K. A. Karimov, V. D. Kostromin, V. Ya. Ogurtsov.
Investigation of the ionosphere and meteors, (see 003.015),
p. 47 - 55 (1971). In Russian.

104.070 Sectorial radar measurements of the drifts of
meteor trails with low transmitter frequency of the
radar.   K. A. Karimov, V. Ya. Ogurtsov.
Investigation of the ionosphere and meteors, (see 003.015),
p. 56 - 62 (1971). In Russian.

104.071 Determination of diffusion in the meteor zone.
A. Ryskulov.
Investigation of the ionosphere and meteors, (see 003.015),
p. 63 - 67 (1971). In Russian.

104.072 On the determination of the parameter s.
V. D. Kostromin.
Investigation of the ionosphere and meteors, (see 003.015),
p. 68 - 72 (1971). In Russian.

104.073 Resonance effects during scattering and absorption
of radio waves on models of traces of fast moving
bodies in the ionosphere.   L. A. Zhizhimov.
Investigation of the ionosphere and meteors, (see 003.015),
p. 73 - 80 (1971). In Russian.

104.074 The influence of mountainous obstacles on the way
cf meteor propagation of radio waves.
N. N. Kalinina.
Investigation of the ionosphere and meteors, (see 003.015),
p. 81 - 88 (1971). In Russian.

104.075 Graphical method for determining the active curve
of a point-like meteor radiant on a given path.
N. N. Kalinina.
Investigation of the ionosphere and meteors, (see 003.015),
p. 89 - 97 (1971). In Russian.

104.076 A stream search among 865 precise photographic
meteor orbits.   B.-A. Lindblad.
Smithsonian Contr. Astrophys., Cambridge, Mass., No. 12,
p. 1 - 13 (1971).
    A search for meteor streams was made among 865 pre-
cise photographic meteor orbits collected in the Harvard Meteor Program. An automatic computer program was utilized.
In all, 80 meteor streams were detected. Of these, 21 represent
19 previously known, well-studied photographic meteor showers and 17 represent new meteor streams found by McCrosky
and Posen and by Southworth and Hawkins. Five previously
unstudied photographic streams with four or more members
were discovered. Of these, three were identified with streams
reported by visual observers.

104.077 A computerized stream search among 2401 photographic meteor orbits.   B.-A. Lindblad.
Smithsonian Contr. Astrophys., Cambridge, Mass., No. 12,
p. 14 - 24 (1971).
    A computer stream search has been made among 2401
photographic meteor orbits. The resulting meteor streams are
presented in tabular form. For known photographic streams,
the mean orbital elements, as determined by the search, are
similar to those previously obtained by conventional methods
of stream classification. Many new photographic meteor
streams have been detected by the search. Some have been
identified with visual showers. Several streams are split into a
northern and a southern branch, with their orbital planes
symmetrical with respect to the plane of the ecliptic. Four
streams move in orbits similar to those of well-known comets.

104.078 A meteor spectrum in the infrared region.
K. Nagasawa.
Tokyo Astron. Bull., Second Ser., No. 213, p. 2505 - 2513
(1971).
    Early morning on January 4th, 1968, a bright meteor
appeared in the sky of Northern-Kanto district in Japan. This
meteor was photographed not only by the meteor camera at
Mitaka, but also by the spectrocamera at Dodaira Station of
the Tokyo Astronomical Observatory. This is the first photographic infrared spectrum of a meteor obtained in Japan. The
spectral line identification was carried out. This report contains these results about the meteor.

104.079 Die Beobachtungen von Meteorströmen.
H.-L. Neumann.
SuW, Vol. 10, 337 - 338 (1971).

104.080 Meteor notes – January to June 1971.
K. B. Hindley.
Journ. British Astron. Ass., Vol. 82, 54 - 56 (1971). – Report
of the Meteor Section.

104.081 The Quadrantid meteor stream.   K. B. Hindley.
Journ. British Astron. Ass., Vol. 82, 57 - 64 (1971).
Report of the Meteor Section.

104.082 Radar complex for the study of faint meteors.
B. L. Kashcheev, I. A. Delov, B. S. Dudnik, A. A.
Tkachuk.
Radiotekhnika. Resp. mezhved. nauch.-tekhn. sb., 1971, vyp.
(No.) 16, p. 11 - 18. In Russian. – Abstr. in Referativ. Zhurn.
51. Astron., 12.51.362 (1971).

104.083 The use of a radar complex for measuring meteor
heights.   N. V. Novoselova, A. A. Tkachuk.
Radiotekhnika. Resp. mezhved. nauch.-tekhn. sb., 1971, vyp.
(No.) 16, p. 18 - 25. In Russian. – Abstr. in Referativ. Zhurn.
51. Astron., 12.51.363 (1971).

104.084 Radar technique for measuring meteor heights.
V. V. Zhukov, B. S. Dudnik.
Radiotekhnika. Resp. mezhved. nauch.-tekhn. sb., 1971, vyp.
(No.) 16, p. 25 - 29. In Russian. – Abstr. in Referativ. Zhurn.
51. Astron., 12.51.364 (1971).

104.085 Optimum recorders of meteor radar stations.
V. A. Nechitajlenko.
Radiotekhnika. Resp. mezhved. nauch.-tekhn. sb., 1971, vyp.
(No.) 16, p. 33 - 41. In Russian.

104.086 **Optimum parameters and selectivity of discrete recorders of meteor radar stations.**
V. A. Nechitajlenko.
Radiotekhnika. Resp. mezhved. nauch.-tekhn. sb., 1971, vyp. (No.) 16, p. 41 - 48. In Russian.

104.087 **On a certain approach to the problem of automation of the analysis of radar data in the study of meteor phenomena.** A. A. D'yakov.
Radiotekhnika. Resp. mezhved. nauch.-tekhn. sb., 1971, vyp. (No.) 16, p. 55 - 58. In Russian. – Abstr. in Referativ. Zhurn. 51. Astron., 12.51.367 (1971).

104.088 **A study of the algorithm of computing meteor velocities from data of radio observations.**
A. A. D'yakov.
Radiotekhnika. Resp. mezhved. nauch.-tekhn. sb., 1971, vyp. (No.) 16, p. 58 - 62. In Russian. – Abstr. in Referativ. Zhurn. 51. Astron., 12.51.368 (1971).

104.089 **Radar study of the geocentric velocity distribution of meteors.** B. L. Kashcheev, V. M. Ushakov.
Radiotekhnika. Resp. mezhved. nauch.-tekhn. sb., 1971, vyp. (No.) 16, p. 62 - 66. In Russian. – Abstr. in Referativ. Zhurn. 51. Astron., 12.51.369 (1971).

104.090 **Possible meteor shower from periodic comet Grigg-Skjellerup.** H. B. Ridley.
IAU Circ., No. 2371 (1971).

104.091 **Quadrantid meteors.** K. B. Hindley.
IAU Circ., No. 2376 (1971).

104.092 **Bright fireball.** S. W. Milbourn.
British Astron. Ass., Circ. No. 537 (1971).

104.093 **Meteors from comet P/Biela.** L. Kresak.
British Astron. Ass., Circ. No. 537 (1971).

104.094 **Transformation of a meteor stream in the gravitational field.** V. V. Andrejev, O. I. Belkovich.
Astron. Tsirk., No. 648, p. 6 - 7 (1971). In Russian.

104.095 **Recent advances in meteor spectroscopy.**
P. M. Millman.
Meteoritics, Vol. 6, 293 (1971). – Abstract.

104.096 **The auroral green line in Perseid spectra near sunspot maximum.** J. A. Russell.
Meteoritics, Vol. 6, 308 - 309 (1971). – Abstract.

104.097 **Reflection of radio waves from meteor traces. III. Observableness of radio meteors.** V. N. Lebedinetz, V. N. Korpusov, A. K. Sosnova, V. B. Shushkova.
Geomagn. Aeronom., Vol. 11, 1011 - 1020 (1971). In Russian.

104.098 **Heliocentric velocity distribution of meteors.**
B. L. Kashcheev, N. V. Novoselova, V. M. Ushakov.
Vestn. Khar'kov. politekhn. in-ta, 1971, No. 54, p. 5 - 9. In Russian. – Abstr. in Referativ. Zhurn. 51. Astron., 1.51.348 (1972).

104.099 **On the influence of diffusive broadening of a meteor train on the accuracy of determination of the wind velocity from radar observations.**
Yu. D. Il'ichev, Yu. I. Portnyagin.
Trudy In-t ehksperim. meteorol. Gl. upr. gidrometeorol. sluzhby pri Sov. Min. SSSR, 1971, vyp. (No.) 24, p. 86 - 91. In Russian. – Abstr. in Referativ. Zhurn. 51. Astron., 1.51.355 (1972).

104.100 **Study of meteor streams from radar observations.**
V. N. Lebedinets, V. N. Korpusov, A. K. Sosnova.
Trudy In-t ehksperim. meteorol. Gl. upr. gidrometeorol. sluzhby pri Sov. Min. SSSR, 1971, vyp. (No.) 24, p. 100 - 113. In Russian. – Abstr. in Referativ. Zhurn. 51. Astron., 1.51.356 (1972).

104.101 **Theories of meteor stream formation. II.**
L. A. Katasev, N. V. Kulikova.
Trudy In-t ehksperim. meteorol. Gl. upr. gidrometeorol. sluzhby pri Sov. Min. SSSR, 1971, vyp. (No.) 24, p. 114 - 121. In Russian. – Abstr. in Referativ. Zhurn. 51. Astron., 1.51.357 (1972).

104.102 **Method of calculation of coordinates of a meteor radiant from radar observations.**
V. N. Korpusov, V. N. Lebedinets, A. K. Sosnova.
Trudy In-t ehksperim. meteorol. Gl. upr. gidrometeorol. sluzhby pri Sov. Min. SSSR, 1971, vyp. (No.) 24, p. 122 - 125. In Russian.

104.103 **Radar observations of meteor train drifts over the equator.** B. V. Kal'chenko, B. L. Kashcheev.
Geofiz. byull. Mezhduved. geofiz. kom. pri Prezidiume AN SSSR, 1971, No. 23, p. 52 - 55. In Russian.

104.104 **Secular perturbations of some meteor streams.**
I. V. Galibina.
Byull. Inst. Teoret. Astron., *Leningrad,* Vol. 12, 870 - 881 (1971). In Russian.
  The secular perturbations of 14 meteor streams have been investigated for 4000 years from 50 B.C. to 3950 A.D. The computations are based on the Gauss-Halphen-Goryachev method. The results of computations of the distances between the earth's orbit and the orbits of the meteor streams in the plane of the ecliptic are given. These distances define conditions of the meteor stream's meeting with the earth.

104.105 **Some meteor events according to ancient chronicles of Armenia.** I. S. Astapovich, B. E. Tumanian.
Problems of cosmic physics. Vyp. (No.) 6, (see 003.040), p. 158 - 163 (1971). In Russian.

104.106 **Nomogram for the determination of meteor luminous efficiency.** I. N. Kovshun.
Astron. Tsirk., No. 658, p. 6 - 8 (1971). In Russian.

104.107 **On the theory of formation of meteor streams.**
N. V. Kulikova.
Astron. vestn., Vol. 5, 181 - 184 (1971). In Russian. – Abstr. in Referativ. Zhurn. 51. Astron., 2.51.324 (1972).

104.108 **Scale of photometric meteor masses for photographic observations of meteors.** I. N. Kovshun.
Astron. vestn., Vol. 5, 185 - 189 (1971). In Russian. – Abstr. in Referativ. Zhurn. 51. Astron., 2.51.325 (1972).

104.109 **Ionization distribution along faint meteor trains.**
I. A. Delov, B. L. Kashcheev.
Astron. vestn., Vol. 5, 190 - 195 (1971). In Russian. – Abstr. in Referativ. Zhurn. 51. Astron., 2.51.329 (1972).

104.110 **The influence of the geomagnetic field on the shape of meteor trains.** U. Shodiev.
Astron. vestn., Vol. 5, 196 - 198 (1971). In Russian. – Abstr. in Referativ. Zhurn. 51. Astron., 2.51.330 (1972).

104.111 **Observations of meteors in August 1969.**
O. P. Batylova.
Astron. vestn., Vol. 5, 199 (1971). In Russian. – Abstr. in Referativ. Zhurn. 51. Astron., 2.51.331 (1972).

104.112 **On the correlation of hourly rates of radiometeors recorded at different sensitivity levels.**
V. N. Donij, E. I. Fialko.
Astron. Tsirk., No. 664, p. 5 - 7 (1971). In Russian.

104.113 **Measurements of atmospheric parameters from radar observations of the ambipolar diffusion coefficient of meteor trails.**   G. M. Teptin.
Astron. Tsirk., No. 665, p. 6 - 8 (1971). In Russian.

104.114 **Photometric analysis of spectrograms of two Perseid meteors.**   A. F. Cook, I. Halliday, P. M. Millman.
Canadian Journ. Phys., Vol. 49, 1738 - 1749 (1971).

Calibrated photometry has been carried out on two Perseid meteor spectra photographed near Ottawa in 1957. Some 80 atomic and molecular features were identified in the two spectra, which corresponded with the normal spectral type for bright Perseids. Absolute fluxes for the radiations from atomic lines, molecular nitrogen bands, and meteor train features have been listed.

104.115 **Quadrantid meteors.**   K. B. Hindley.
Yamamoto Circ., No. 1746 (1971). In Japanese.

104.116 **Gamma-ray measurements from Cosmos-135 satellite with a view to possible detection of antimatter meteor streams.**
B. P. Konstantinov, R. L. Aptekar', M. M. Bredov, S. V. Gole-netskii, Yu. A. Gur'yan, V. N. Il'inskii, E. P. Mazets, V. N. Panov.
Proc. Sixth Winter School on Space Physics, Part I. Apatity 1969, (see 012.033), p. 82 - 97 (1971).

Australians observe meteors.
Sky Telescope, Vol. 42, 154 (1971).

Investigation of the ionosphere and meteors.
See Abstr. 003.015.

Spectral measurements of nitrogen continuum radiation behind incident shocks at speeds up to 13 km/sec.
See Abstr. 022.017.

Microphotometer for photometry of meteors.
See Abstr. 034.041.

Meteoritic contamination of the upper atmosphere by the Quadrantid shower.   See Abstr. 082.045.

Canadian national report on geomagnetism and aeronomy.   See Abstr. 084.316.

Meteor effects of cosmic rays.
See Abstr. 143.110.

The meteoric variation of cosmic rays near the new solar-activity peak in 1967—1968.   See Abstr. 143.149.

105 Meteorites, Meteorite Craters

105.001 Ivory Coast microtektites: Corrected values of uranium content. S. A. Durrani, H. A. Khan.
Nature, Phys. Sci., Vol. 232, 175 (1971).
This note is meant to correct the values of uranium content in Ivory Coast microtektites as published in a recent article of the authors in Nature, Vol. 232, 320 - 323 (1971).

105.002 Aromatic hydrocarbons in the Murchison meteorite. K. L. Pering, C. Ponnamperuma.
Science, Vol. 173, 237 - 239 (1971).
Polynuclear aromatic hydrocarbons in the Murchison meteorite have been identified by the combined techniques of gas chromatography and mass spectrometry. The distribution of the aromatic compounds suggests that they are the products of a high-temperature synthesis.

105.003 Amino acids indigenous to the Murray meteorite. J. G. Lawless, K. A. Kvenvolden, E. Peterson, C. Ponnamperuma, C. Moore.
Science, Vol. 173, 626 - 627 (1971).
Analysis of the Murray meteorite, a type II carbonaceous chondrite, has led to the identification of 17 amino acids. The results suggest that these amino acids, like the amino acids of the Murchison meteorite, are extraterrestrial in origin.

105.004 Meteoritic rutile: A niobium bearing mineral. A. El Goresy.
Earth Planet. Sci. Letters, Vol. 11, 359 - 361 (1971).
Quantitative electron microprobe analysis of rutile grains in several meteorites revealed the presence of variable Nb amounts. The present findings establish a Ti/Nb coherency and the strong lithophilic behavior of Nb in meteorites.

105.005 Meteoritenfälle in Deutschland. H. Eisenlohr.
SuW, Vol. 10, 217 - 220 (1971).

105.006 Some modern aspects of cosmic mineralogy. I. E. K. Lazarenko, A. A. Yasinskaya.
Mineral. sb. L'vov. un-ta, 1970, No. 24, vyp. 4, p. 367 - 384. In Russian. – Abstr. in Referativ. Zhurn. 62. Issled. kosm. prostranstva, 8.62.186 (1971).

105.007 Earth's space scars. P. M. Millman.
Journ. Roy. Astron. Soc. Canada, Vol. 65, 165 - 166 (1971).

105.008 Size distribution of magnetic spheroids. C. T. Nagamoto, J. Rosinski.
Journ. Atmosph. Terr. Phys., Vol. 33, 1559 - 1566 (1971).
Concentration peaks of magnetic spherules in the lower troposphere at different latitudes is indicative of the extraterrestrial origin of the majority of these spherules (Rosinski, 1970). An investigation of size distribution of the spherules gives no certain indication of Rosinski's proposal.

105.009 The distribution of total nitrogen in iron meteorites. E. K. Gibson, Jr., C. B. Moore.
Geochim. Cosmochim. Acta, Vol. 35, 877 - 890 (1971).
Total nitrogen abundances in 123 iron meteorites have been determined by inert carrier-gas fusion extraction-gas chromatography. The median value for the iron meteorites was found to be 18 ppm N. A table contains the single results.

105.010 The mystery of the Popigai basin. M. V. Mikhailov, T. V. Selivanovskaya.
Priroda, No. 9.71, p. 78 - 82 (1971). In Russian.

105.011 Mass spectrometric evidence for organic constituents in tektites. D. W. Muenow, S. J. Steck, J. L. Margrave.
Geochim. Cosmochim. Acta, Vol. 35, 1047 - 1058 (1971).
Various tektite specimens were heated to temperatures up to 1500°C and the vaporizing molecular species identified with a mass spectrometer. In this technique, volatile constituents are selectively vaporized from the sample, ionized to electrically charged positive particles by bombardment with high energy electrons, and separated in an electric and magnetic field according to their respective molecular weights. Trace amounts of organic molecules were observed to vaporize in the temperature range 350°–425°C.

105.012 Organische stoffen in meteorieten. F. P. Israel.
Hemel en Dampkring, Vol. 69, 252 - 255 (1971).

105.013 A search for inhomogeneities in the interplanetary micrometeoroid environment. R. G. Roosen, O. E. Berg, N. H. Farlow.
Bull. American Astron. Soc., Vol. 3, 368 (1971). – Abstr. AAS.

105.014 Implied superheavy element decay lifetime from meteorites. D. N. Schramm.
Nature, Vol. 233, 258 - 260 (1971).
It is shown that if the so called carbonaceous chondrite fission xenon is due to the fissioning of a superheavy element, then the following strong limits can be placed on the half life of the longest lived superheavy element in the decay chain leading to the fission. ($1.65 \times 10^7 \, \mathrm{yr} \leq \tau_{SH} \leq 6.80 \times 10^7 \, \mathrm{yr}$).

105.015 On the velocity of propagation of air waves of the Tunguska explosion of 1908. A. V. Zolotov.
Astron. vestn., Vol. 5, 107 - 111 (1971). In Russian. – Abstr. in Referativ. Zhurn. 51. Astron., 10.51.335 (1971).

105.016 Pyroxenes from non-carbonaceous chondrite meteorites. R. A. Binns.
Mineral. Magazine, Vol. 37, 649 - 669 (1970). – Abstr. in Icarus, Vol. 14, 436 - 437 (1971).

105.017 A metallographic and microprobe study of the metal phases in the Weekeroo Station meteorite. H. J. Axon, P. L. Smith.
Mineral. Magazine, Vol. 37, 670 - 673 (1970). – Abstr. in Icarus, Vol. 14, 436 (1971).

105.018 A zoned perovskite-bearing chondrule from the Lance meteorite. M. J. Frost, R. F. Symes.
Mineral. Magazine, Vol. 37, 724 - 725 (1970). – Abstr. in Icarus, Vol. 14, 439 (1971).

105.019 Medanitas and Putinga, two South American meteorites. R. F. Symes, R. Hutchinson.
Mineral. Magazine, Vol. 37, 721 - 723 (1970). – Abstr. in Icarus, Vol. 14, 443 (1971).

105.020 Isotopic composition of ^{244}Pu fission xenon in meteorites: Reevaluation using lunar spallation xenon systematics. F. A. Podosek, J. C. Huneke.
Earth Planet. Sci. Letters, Vol. 12, 73 - 82 (1971).
The isotopic composition of fission xenon in four meteorites (Pasamonte, Kapoeta, Angra dos Reis, and St. Severin) has been reevaluated. Spallation corrections have been made

by means of spallation systematics based entirely on lunar data, with explicit allowance made for variations in the spallation xenon spectrum at different temperatures in thermal degassing experiments.

105.021 **Assam: A gas rich hypersthene chondrite.**
L. Schultz, P. Signer, P. Pellas, G. Poupeau.
Earth Planet. Sci. Letters, Vol. 12, 119 - 123 (1971).

Concentrations and isotopic compositions of He, Ne and Ar as well as track densities due to galactic and solar flare irradiations were determined in light and dark fractions of the L-chondrite Assam. The investigation showed Assam to be a gas rich dark-light structured hypersthene chondrite.

105.022 **Hibonite $[Ca_2(Al, Ti)_{24}O_{38}]$ from the Leoville and Allende chondritic meteorites.**
K. Keil, L. H. Fuchs.
Earth Planet. Sci. Letters, Vol. 12, 184 - 190 (1971).

In the present paper, the first occurrence of hibonite in meteorites, namely in Ca–Al–Ti-rich xenoliths from polymict-brecciated HL-group chondrites is described, and compositional and X-ray data are presented for the mineral.

105.023 **The age and the origin of Köfels structure, Austria.**
D. Storzer, P. Horn, B. Kleinmann.
Earth Planet. Sci. Letters, Vol. 12, 238 - 244 (1971).

105.024 **Relationship between siderophilic-element content and oxidation state of ordinary chondrites.**
O. Müller, Ph. A. Baedecker, J. T. Wasson.
Geochim. Cosmochim. Acta, Vol. 35, 1121 - 1137 (1971).

The concentrations of Ni and Ir have been determined by neutron activation in a suite of ordinary chondrites for which accurate ferromagnesian-mineral compositional data were available. A significant negative correlation is observed between the abundance of Ni or Ir and the Fe content of the ferromagnesian minerals in the H and L groups, and for Ir in the LL group, as expected if the metal-silicate fractionation and the variation in oxidation states were produced by the same or related processes. The Ir/Ni ratio decreases by a factor of 1.2 between the H and LL groups. This fractionation must have occurred at an early stage in the condensation phase of the solar nebula.

105.025 **Oxygen isotope ratios in the crust of iron meteorites.** K. Heinzinger, C. Junge, M. Schidlowski.
Zeitschr. Naturforschung, Vol. 26a, 1485 - 1490 (1971).

The separation factor, $(^{18}O/^{16}O)$ magnetite$/(^{18}O/^{16}O)$ atmospheric oxygen, between the magnetite crust of iron meteorites and atmospheric oxygen has been determined to be 0.9946 ± 0.0005. It is concluded that this fractionation of the oxygen isotopes is the consequence of an equilibrium isotope effect at high temperatures. It can be assumed that this is also valid for cosmic spherules, which are mainly ablation products of iron meteorites. The difference of the oxygen isotope ratios between magnetite from the lithosphere and airborne magnetite can be used to distinguish between terrestrial and extraterrestrial material.

105.026 **Tracing a cosmic catastrophe.** V. L. Masajtis.
Zemlya i Vselennaya, No. 5, p. 31 - 36 (1971).
In Russian.

105.027 **Fall of meteorites observed in the Far East.**
Kometn. Tsirk., *Kiev,* No. 122 (1971). In Russian.

105.028 **New studies of the Sikhote-Alin iron meteorite shower.** E. L. Krinov.
Meteoritics, Vol. 6, 127 - 138 (1971).

105.029 **An equation for the determination of iron-meteorite**

cooling rates. J. T. Wasson.
Meteoritics, Vol. 6, 139 - 147 (1971).

105.030 **On the possible number and mass of fragments from Pułtusk meteorite shower, 1868.**
B. Lang, M. Kowalski.
Meteoritics, Vol. 6, 149 - 158 (1971).

105.031 **Potassium-argon age of the Raco meteorite (Argentina).** R. R. González, M. A. Cabrera.
Meteoritics, Vol. 6, 159 - 160 (1971).

The Raco meteorite fell in Raco, a small village in the Province of Tucumán, Argentina, on November 17, 1957. Its K-Ar age was determined to 4.400 ± 28 m.y.

105.032 **The Iron River iron meteorite.** V. D. Chamberlain.
Meteoritics, Vol. 6, 161 - 171 (1971).

105.033 **Chemical analyses of thirty-eight iron meteorites.**
C. F. Lewis, C. B. Moore.
Meteoritics, Vol. 6, 195 - 205 (1971).

105.034 **Lost City meteorite: Determination of the temperature gradient induced by atmospheric friction using thermoluminescence.** J. E. Vaz.
Meteoritics, Vol. 6, 207 - 216 (1971).

105.035 **Deposition of extraterrestrial nickel in marine sediments.** K. Yamakoshi, Y. Tazawa.
Nature, Vol. 233, 542 - 543 (1971).

The excess part of nickel in deep-sea sediments is thought to be supplied from extraterrestrial matter. A simple model is proposed; Nickel is supplied through grain settling and ionic (or colloidal) sedimentation. The data of nickel contents in dated core samples revealed that nickel fraction supplied through grain settling is negligible small.

105.036 **Australasian tektite geographic pattern, crater and ray of origin, and theory of tektite events.**
D. R. Chapman.
Journ. Geophys. Res., Vol. 76, 6309 - 6338 (1971).

105.037 **Physical chemistry of the Aouelloul glass.**
J. A. O'Keefe.
Journ. Geophys. Res., Vol. 76, 6428 - 6439 (1971).

105.038 **Liverpool and Strangways craters, Northern Territory: Two structures of probable impact origin.**
D. J. Guppy, R. Brett, D. J. Milton.
Journ. Geophys. Res., Vol. 76, 5387 - 5393 (1971).

105.039 **The Rochechouart meteorite impact structure, France: Preliminary geological results.**
F. Kraut, B. M. French.
Journ. Geophys. Res., Vol. 76, 5407 - 5413 (1971).

105.040 **Impactite of the Charlevoix structure, Quebec, Canada.** J. Rondot.
Journ. Geophys. Res., Vol. 76, 5414 - 5423 (1971).

105.041 **Potassium-argon dating of shock-metamorphosed rocks from the Brent impact crater, Ontario, Canada.**
J. B. Hartung, M. R. Dence, J. A. S. Adams.
Journ. Geophys. Res., Vol. 76, 5437 - 5448 (1971).

105.042 **Shock metamorphism of the Coconino sandstone at meteor crater, Arizona.** S. W. Kieffer.
Journ. Geophys. Res., Vol. 76, 5449 - 5473 (1971).

105.043 **Coesite and stishovite in shocked crystalline rocks.**
D. Stöffler.

Journ. Geophys. Res., Vol. 76, 5474 - 5488 (1971).

105.044 Shock metamorphism of silicate glasses.
R. V. Gibbons, T. J. Ahrens.
Journ. Geophys. Res., Vol. 76, 5489 - 5498 (1971).

105.045 Dynamic compression of enstatite.
T. J. Ahrens, E. S. Gaffney.
Journ. Geophys. Res., Vol. 76, 5504 - 5513 (1971).

105.046 Progressive metamorphism and classification of shocked and brecciated crystalline rocks at impact craters. D. Stöffler.
Journ. Geophys. Res., Vol. 76, 5541 - 5551 (1971).

105.047 Impact melts. M. R. Dence.
Journ. Geophys. Res., Vol. 76, 5552 - 5565 (1971).

105.048 Detrital impact formations. W. von Engelhardt.
Journ. Geophys. Res., Vol. 76, 5566 - 5574 (1971).

105.049 Origin of igneous rocks associated with shock metamorphism as suggested by geochemical investigations of Canadian craters. K. L. Currie.
Journ. Geophys. Res., Vol. 76, 5575 - 5585 (1971).

105.050 Crustal thickness and the forms of impact craters. C. S. Beals.
Journ. Geophys. Res., Vol. 76, 5586 - 5595 (1971).

105.051 Pueblito de Allende penetration craters and experimental craters formed by free fall.
D. P. Elston, G. R. Scott.
Journ. Geophys. Res., Vol. 76, 5756 - 5764 (1971).

105.052 More than two years of micrometeorite data from two Pioneer satellites. O. E. Berg, U. Gerloff.
Space Research XI, (see 012.004), 225 - 235 (1971).

105.053 Configuration particulière des rapports d'abondances des éléments dans les sphérules.
T. Grjebine, Y. Yokoyama, P. Bristeau.
Space Research XI, (see 012.004), 261 - 273 (1971).

105.054 The fall of cosmic material in Italy.
F. di Benedetto.
Space Research XI, (see 012.004), 329 - 333 (1971).

105.055 A simultaneous collection and detection experiment for cosmic dust.
H. Fechtig, M. Feuerstein, P. Rauser.
Space Research XI, (see 012.004), 335 - 346 (1971).

105.056 Sounding rocket samplings of cosmic dust.
D. S. Hallgren, C. L. Hemenway.
Space Research XI, (see 012.004), 377 - 381 (1971).

105.057 Electron microprobe studies of cosmic dust impact craters.
O. K. Griffith, T. S. Renzema, D. S. Hallgren, C. L. Hemenway.
Space Research XI, (see 012.004), 383 - 392 (1971).

105.058 A new high altitude balloon-top cosmic dust collection technique.
C. L. Hemenway, D. S. Hallgren, A. T. Laudate, H. Patashnick, T. S. Renzema, O. K. Griffith.
Space Research XI, (see 012.004), 393 - 395 (1971).

105.059 A model for predicting the results of in situ meteoroid experiments: Pioneer 8 and 9 results and phenomenological evidence. U. Gerloff, O. E. Berg.

Space Research XI, (see 012.004), 397 - 413 (1971).

105.060 Review of in situ measurements of cosmic dust particles in space. J. A. M. McDonnell.
Space Research XI, (see 012.004), 415 - 435 (1971).

105.061 First nitride (CrN) in iron meteorites.
V. F. Buchwald, E. R. D. Scott.
Nature, Phys. Sci., Vol. 233, 113 - 114 (1971).
The first nitride discovered in iron meteorites is a chromium nitride, stoichiometric CrN, with very small amounts of Fe, Mn and Ni. Microprobe analyses gave Cr 76.3, Fe 1.9, Ni 0.3, Mn 0.04 and N 21.4% (by difference) on typical particles extracted from a number of iron meteorites. The nitride, named Carlsbergite for the Carlsberg Foundation, Copenhagen, has a cubic unit cell, a = 4.16 Å, and space group Fm3m. It occurs as a solid state precipitate in the kamacite phase of at least 70 iron meteorites of group III A, II A and I, and reaches typical dimensions of $30 \times 5 \times 2\mu$.

105.062 Tungusic meteorites are falling every year.
I. T. Zotkin.
Priroda, No. 11.71, p. 83 - 84 (1971). In Russian.

105.063 On the isotopic composition of trapped helium and neon in carbonaceous chondrites.
B. Srinivasan, O. K. Manuel.
Earth Planet. Sci. Letters, Vol. 12, 282 - 286 (1971).
The covariance observed in the isotopic composition of primordial He, Ne and Ar in carbonaceous chondrites can be explained on the basis of simple mass-dependent fractionation.

105.064 Xe and Kr analyses of silicate inclusions from iron meteorites. D. D. Bogard, J. C. Huneke, D. S. Burnett, G. J. Wasserburg.
Geochim. Cosmochim. Acta, Vol. 35, 1231 - 1254 (1971).
The purpose of this study was to measure the amounts and isotopic composition of Xe and Kr in silicate inclusions from a comparatively large number of iron meteorites (Copiapo, Four Corners, Linwood, Pine River, Weekeroo Station and Woodbine). The results given in two tables are discussed in terms of correlation diagrams of the measured isotopic ratios, originally used by Hohenberg et al. (1967).

105.065 The production rate of Al^{26} from target elements in the Bruderheim chondrite. P. J. Cressy, Jr.
Geochim. Cosmochim. Acta, Vol. 35, 1283 - 1296 (1971).
An 840-g specimen of the Bruderheim chondrite was subjected to magnetic and heavy-liquid mineral separation procedures, resulting in a number of chemically distinct samples. These samples were analyzed for cosmogenic Al^{26} by nondestructive gamma-gamma coincidence counting. The observed Al^{26} specific activities were correlated with the chemical composition of potential target elements by a weighted least-squares fitting technique.

105.066 Tektite debate continues. I. Halliday.
Journ. Roy. Astron. Soc. Canada, Vol. 65, 296 - 298 (1971).

105.067 Entry trajectory and orbital calculations for the Crater 9 meteorite, Campo del Cielo, Argentina.
M. L. Renard, W. A. Cassidy.
Journ. Geophys. Res., Vol. 76, 7916 - 7923 (1971).
The possible masses and velocities at entry of the Crater 9 meteorite, Campo del Cielo, Argentina, were determined from the dynamic conditions at impact. The assumption of a hyperbolic orbit relative to the earth allows one to put an upper limit on the mass at impact. The dependence of the heliocentric orbital parameters on the hour of entry is also studied.

105.068 Neue Formen meteoritischen Graphits und mögliche Beziehungen zum Cliftonit. P. Ramdohr.
Naturwissenschaften, 58. Jahrgang, p. 613 - 615 (1971).

The iron of the Mundrabilla meteorite contains graphite in very exceptional needlelike forms, obviously in many different polytypes. They are similar to graphite in the iron of the Khairpur meteorite.

105.069 Fossil tracks in the meteorite Angra dos Reis: A predominantly fission origin.
N. Bhandari, S. Bhat, D. Lal, G. Rajagopalan, A. S. Tamhane, V. S. Venkatavaradan.
Nature, Vol. 234, 540 - 543 (1971).

We have studied a few meteorites having relatively high uranium concentrations for fossil tracks as well as for neutron induced tracks in previously annealed samples. In particular we studied the meteorite Angra dos Reis (ADR), an augite achondrite, which is known to have a high uranium concentration (> 170 p.p.b. by weight) and for which the concentrations of rare gases, both fissiogenic and cosmogenic, are well known. ADR is rich in pyroxene occurring as dark brown but fairly large thin transparent crystals having dimensions of $300 - 500 \, \mu m$.

105.070 Ries structure, southern Germany, a review. J. G. Dennis.
Journ. Geophys. Res., Vol. 76, 5394 - 5406 (1971).

105.071 On diffusive losses of radiogenic argon by meteorites. S. B. Brandt, N. V. Volkova.
Geokhimiya, 1971, No. 8, p. 1012 - 1015. In Russian. Abstr. in Referativ. Zhurn. 51. Astron., 12.51.378 (1971).

105.072 Cosmogonic isotopes in 42 fragments of the Sikhote-Alin meteorite. L. K. Levskij.
Geokhimiya, 1971, No. 8, p. 932 - 937. In Russian. – Abstr. in Referativ. Zhurn. 51. Astron., 12.51.379 (1971).

105.073 Synthesis of majorite and other high pressure garnets and perovskites. A. E. Ringwood, A. Major.
Earth Planet. Sci. Letters, Vol. 12, 411 - 418 (1971).

The synthesis is described at a pressure of 250 - 300 kb of the garnet, majorite, previously found to occur as a shock produced phase in a chondritic meteorite. A series of new high pressure garnets containing sodium and/or titanium is also described.

105.074 The origin of nickel component in marine sediments. K. Yamakoshi.
Mem. Fac. Sci., Kyoto Univ., Ser. Phys., Astrophys., Geophys., Chemistry, Vol. 33, 311 - 323 (1971).

The origin of nickel component of the deep sea sediments is discussed. The extraterrestrial nickel fraction is thought to be supplied through grain sedimentation and terrestrial nickel fraction is through ionic (or colloidal) sedimentation into deep sea sediments.

105.075 Más meteoritos. E. Lastra.
El Universo, Vol. 25, 91 - 92 (1971).

105.076 La texture des chondrites observée en plaques minces par fluorescence ultra-violette.
M. C. Michel-Levy, J. P. Ragot.
Meteoritics, Vol. 6, 217 - 224 (1971).

A technique of impregnation by fluorescent resins followed by observations of thin sections in UV light has been applied to the study of porosity, brecciation and shock effects in chondritic meteorites.

105.077 Petrological features of shock metamorphism in chondrites: Alfianello. G. R. Levi-Donati.
Meteoritics, Vol. 6, 225 - 235 (1971).

The Alfianello meteorite was inspected by optical microscopy, both by transmitted and reflected light, in order to look for evidence of shock metamorphism.

105.078 On the constancy of cosmic spherule influx during the quaternary. K. Utech.
Meteoritics, Vol. 6, 237 - 239 (1971).

The influx rate of cosmic spherules can be measured on sediments of which the rate of deposition is known. It proved to be constant over long geological periods.

105.079 The composition of the Johnstown meteorite. B. Mason, E. Jarosewich.
Meteoritics, Vol. 6, 241 - 245 (1971).

A new analysis of the Johnstown meteorite, a hypersthene achondrite, is presented.

105.080 Effects of ultra-high static and dynamic pressures on silicate glasses. J. Arndt.
Meteoritics, Vol. 6, 248 (1971). – Abstract.

105.081 Low temperature inclusions in C3–4 carbonaceous chondrites. G. Arrhenius.
Meteoritics, Vol. 6, 248 - 249 (1971). – Abstract.

105.082 Distribution of elements in tektites and comparable materials. N. A. Askouri.
Meteoritics, Vol. 6, 249 (1971). – Abstract.

105.083 Description and origin of 12.8-kg layered tektite from Thailand. V. E. Barnes.
Meteoritics, Vol. 6, 249 (1971). – Abstract.

105.084 Diffusion of solar wind He and Ne in lunar and meteoritic matter.
H. Baur, U. Frick, H. Funk, L. Schultz, P. Signer.
Meteoritics, Vol. 6, 250 (1971). – Abstract.

105.085 Experiments for iron-meteorite simulation. M. R. Bloch, O. Müller, H. Wirth.
Meteoritics, Vol. 6, 250 - 251 (1971). – Abstract.

105.086 The Cape York shower, a typical group III A iron meteorite, formed by directional solidification in a gravity field. V. F. Buchwald.
Meteoritics, Vol. 6, 252 - 253 (1971). – Abstract.

105.087 The Landes silicate-bearing iron meteorite. T. E. Bunch, K. Keil, G. Huss.
Meteoritics, Vol. 6, 253 - 254 (1971). – Abstract.

105.088 The Piancaldoli meteorite. M. Carapezza, M. Nuccio.
Meteoritics, Vol. 6, 255 (1971). – Abstract.

105.089 On drusy chondrites. M. C. Michel-Levy.
Meteoritics, Vol. 6, 256 (1971). – Abstract.

105.090 The production rates of ^{21}Ne and ^{38}Ar from target elements in the Bruderheim chondrite.
P. J. Cressy, Jr., D. D. Bogard.
Meteoritics, Vol. 6, 257 (1971). – Abstract.

105.091 The process of tektite formation. E. David.
Meteoritics, Vol. 6, 258 (1971). – Abstract.

105.092 Shatter cones (shock fractures) in astroblemes. R. S. Dietz.
Meteoritics, Vol. 6, 258 - 259 (1971). – Abstract.

105.093 Sudbury astrobleme: A review. R. S. Dietz.
Meteoritics, Vol. 6, 259 - 260 (1971). – Abstract.

105.094 Calcium and aluminum as thermal tracers in chondritic olivine. R. T. Dodd.
Meteoritics, Vol. 6, 261 (1971). – Abstract.

105.095 Brecciation processes in extraterrestrial matter.
J. C. Dran, J. P. Duraud.
Meteoritics, Vol. 6, 262 (1971). – Abstract.

105.096 Ivory Coast microtektites: Fission track age and geomagnetic reversal. S. A. Durrani, H. A. Khan.
Meteoritics, Vol. 6, 262 - 263 (1971). – Abstract.

105.097 Al-Hadida and Um-Hadid meteorites, Saudi Arabia.
F. El-Baz, A. El Goresy.
Meteoritics, Vol. 6, 265 - 266 (1971). – Abstract.

105.098 Uranium content and radiogenic ages of hypersthene, bronzite, amphoterite and carbonaceous chondrites. D. E. Fisher.
Meteoritics, Vol. 6, 270 (1971). – Abstract.

105.099 Rare gases in the meteorite Weston.
U. Frick, E. H. Hebeda, L. Schultz, P. Signer.
Meteoritics, Vol. 6, 271 (1971). – Abstract.

105.100 New uranium and lead measurements in stony meteorites. N. H. Gale, J. Arden, R. Hutchison.
Meteoritics, Vol. 6, 272 - 273 (1971). – Abstract.

105.101 Some limits on the micrometeoroid flux at 1 au imposed by lunar erosion rates. D. E. Gault.
Meteoritics, Vol. 6, 274 (1971). – Abstract.

105.102 Cogenesis of the Ries crater and moldavites and the origin of tektites. W. Gentner.
Meteoritics, Vol. 6, 274 - 275 (1971). – Abstract.

105.103 Metallic spherules – their formation on the earth and on the moon. J. I. Goldstein, P. J. Blau.
Meteoritics, Vol. 6, 275 - 276 (1971). – Abstract.

105.104 New ^{26}Al and ^{60}Co data of chondrites with respect to saturation, exceptionally short exposure ages and self-shielding. M. Heimann, U. Herpers, W. Herr.
Meteoritics, Vol. 6, 277 (1971). – Abstract.

105.105 Ca variation in olivines of the Murchison and Vigarano meteorites. R. Hutchison, R. F. Symes.
Meteoritics, Vol. 6, 278 (1971). – Abstract.

105.106 The Allende meteorite – a cooperative study of chemical analysis. E. Jarosewich, R. S. Clarke,Jr.
Meteoritics, Vol. 6, 279 (1971). – Abstract.

105.107 A study of the Juvinas achondrite by means of the ion microanalyzer. D. Y. Jérome.
Meteoritics, Vol. 6, 279 - 280 (1971). – Abstract.

105.108 The Malvern meteorite – a comparison with other achondrite breccias. D. Y. Jérome.
Meteoritics, Vol. 6, 280 (1971). – Abstract.

105.109 Petrography and chemistry of the Faucett meteorite, Buchanan County, Missouri.
E. A. King, Jr., E. Jarosewich, D. G. Brookins.
Meteoritics, Vol. 6, 280 - 281 (1971). – Abstract.

105.110 Cosmic-ray exposure ages of the different Sikhote-Alin meteorite fall fragments. E. M. Kolesnikov, A. K. Lavrukhina, A. V. Fisenko, L. K. Levskii.
Meteoritics, Vol. 6, 282 (1971). – Abstract.

105.111 Quantitative petrographical and chemical data on moldavites and their mutual relations. J. Konta.
Meteoritics, Vol. 6, 283 (1971). – Abstract.

105.112 Crystallogenesis of some meteorite minerals.
I. Kostov.
Meteoritics, Vol. 6, 283 (1971). – Abstract.

105.113 Hedjaz, an L-3, L-4, L-5 and L-6 chondrite.
F. Kraut, K. Fredriksson.
Meteoritics, Vol. 6, 284 (1971). – Abstract.

105.114 New studies of the Sikhote-Alin iron meteorite shower. E. L. Krinov.
Meteoritics, Vol. 6, 284 - 285 (1971). – Abstract.

105.115 Solubilities of noble gases in magnetite: Implications for planetary gases in meteorites.
M. S. Lancet, E. Anders.
Meteoritics, Vol. 6, 286 - 287 (1971). – Abstract.

105.116 On the size distribution of fragments from the Lowicz meteorite shower 1935. B. Lang.
Meteoritics, Vol. 6, 287 (1971). – Abstract.

105.117 On elemental differentiation in the matter of meteorites. A. K. Lavrukhina.
Meteoritics, Vol. 6, 288 (1971). – Abstract.

105.118 Preatmospheric sizes of iron meteorites.
A. K. Lavrukhina, T. A. Ibraev.
Meteoritics, Vol. 6, 288 - 289 (1971). – Abstract.

105.119 Cosmogenic radionuclides in stones and meteorite orbits. A. K. Lavrukhina, G. K. Ustinova.
Meteoritics, Vol. 6, 289 (1971). – Abstract.

105.120 Petrological features of shock metamorphism in chondrites: Alfianello. G. R. Levi-Donati.
Meteoritics, Vol. 6, 290 - 291 (1971). – Abstract.

105.121 Spatial distribution of uranium in meteorites, tektites, and other geological materials by spark counter. S. R. Mallik, S. A. Durrani.
Meteoritics, Vol. 6, 291–292 (1971). – Abstract.

105.122 A search for "gas-rich" meteorites.
G. H. Megrue, F. Steinbrunn.
Meteoritics, Vol. 6, 292 - 293 (1971). – Abstract.

105.123 Potassium-sodium relations in carbonaceous chondrites. W. Nichiporuk, C. B. Moore.
Meteoritics, Vol. 6, 295 - 296 (1971). – Abstract.

105.124 Style and sequence of deformation at the Decaturville, Missouri impact structure.
T. W. Offield, H. A. Pohn.
Meteoritics, Vol. 6, 296 - 297 (1971). – Abstract.

105.125 Gas chromatographic-mass spectrometric identification of organic compounds in the Murchison and other carbonaceous chondrites.
J. Oró, J. M. Gibert, H. Lichtenstein, S. A. Wikstrom.
Meteoritics, Vol. 6, 297 (1971). – Abstract.

105.126 Elemental abundances in meteoritic chondrules.
T. W. Osborn, R. A. Schmitt.

Meteoritics, Vol. 6, 297 - 298 (1971). – Abstract.

105.127 The breccia structure of gas-rich meteorites.
P. Pellas, L. L. Wilkening.
Meteoritics, Vol. 6, 298 - 299 (1971). – Abstract.

105.128 Paleomagnetic results from the Rochechouart (France) impact site. J. Pohl, H. Soffel.
Meteoritics, Vol. 6, 299 (1971). – Abstract.

105.129 Correlation between heavy ion tracks and implanted rare gases in gas rich meteorites.
G. Poupeau, T. Kirsten, F. Steinbrunn, D. Storzer, S. Thio.
Meteoritics, Vol. 6, 299 - 300 (1971). – Abstract.

105.130 Thermoluminescence of the Lost City meteorite.
W. Prachyabrued, S. A. Durrani, J. H. Fremlin.
Meteoritics, Vol. 6, 300 - 301 (1971). – Abstract.

105.131 On the mineralogy of the Mundrabilla meteorite: Nullbor Plain, West Australia.
P. Ramdohr, A. El Goresy.
Meteoritics, Vol. 6, 302 - 303 (1971). – Abstract.

105.132 Evidence for a triplet cratering event in the Ries area formed by fission of a single meteoroid under the earth's tidal forces.
P. Rauser, F. Steinbrunn, D. Storzer.
Meteoritics, Vol. 6, 304 (1971). – Abstract.

105.133 Recent drilling studies at meteor crater, Arizona.
D. J. Roddy, J. M. Boyce, G. W. Colton, A. L. Dial.
Meteoritics, Vol. 6, 306 - 307 (1971). – Abstract.

105.134 Les brèches d'impact de Charlevoix. J. Rondot.
Meteoritics, Vol. 6, 307 - 308 (1971). – Abstract.

105.135 Cosmic-ray produced K-40 and V-50 in metal phase of chondrites. K. Sato, M. Shima.
Meteoritics, Vol. 6, 309 (1971). – Abstract.

105.136 Shock-induced planar deformation structures in olivine from the Chassigny meteorite.
C. B. Sclar, S. P. Morzenti.
Meteoritics, Vol. 6, 310 - 311 (1971). – Abstract.

105.137 The distribution of elements in iron meteorites and its bearing on their origin. E. R. D. Scott.
Meteoritics, Vol. 6, 311 - 312 (1971). – Abstract.

105.138 The occurrence of carbides in iron meteorites.
E. R. D. Scott, S. O. Agrell.
Meteoritics, Vol. 6, 312 - 313 (1971). – Abstract.

105.139 Cosmic-ray produced stable nuclides and potassium-40 in the Trenton meteorite.
M. Shima, N. Takaoka, L. Schultz, H. Hintenberger.
Meteoritics, Vol. 6, 314 - 315 (1971). – Abstract.

105.140 A comparison between a Java tektite (J2) and lunar rock 12013. D. L. Showalter, H. Wakita, R. H. Smith, R. A. Schmitt, D. E. Gillum, W. D. Ehmann.
Meteoritics, Vol. 6, 315 - 316 (1971). – Abstract.

105.141 Cosmic-ray gradient from the meteoritic $^{37}Ar/^{39}Ar$ ratio. G. Spannagel, E. L. Fireman.
Meteoritics, Vol. 6, 317 (1971). – Abstract.

105.142 Classification of shocked quartzofeldspathic crystalline rocks: A review. D. Stöffler.
Meteoritics, Vol. 6, 317 - 318 (1971). – Abstract.

105.143 Properties of feldspar and silica glasses produced by natural and experimental shock.
D. Stöffler, U. Hornemann.
Meteoritics, Vol. 6, 318 (1971). – Abstract.

105.144 Fission track dating of some impact craters in the age range between 6.000 y and 300 m.y.
D. Storzer.
Meteoritics, Vol. 6, 319 (1971). – Abstract.

105.145 The age and the origin of the fused rock from the Köfels structure, Austria.
D. Storzer, P. Horn, B. Kleinmann.
Meteoritics, Vol. 6, 319 - 320 (1971). – Abstract.

105.146 Chemical composition of 61 individual Allende chondrules.
R. G. Warren, T. W. Osborn, R. A. Schmitt.
Meteoritics, Vol. 6, 321 - 322 (1971). – Abstract.

105.147 Formation of ordinary chondrites. J. T. Wasson.
Meteoritics, Vol. 6, 322 - 323 (1971). – Abstract.

105.148 Stony iron meteorites: Rare gases and exposure ages.
H. W. Weber, E. Vilcsek, F. Begemann, H. Hintenberger.
Meteoritics, Vol. 6, 323 - 324 (1971). – Abstract.

105.149 Source of chondritic meteorites. G. W. Wetherill.
Meteoritics, Vol. 6, 324 (1971). – Abstract.

105.150 Chondrules: First occurrence in an iron meteorite.
E. Olsen, E. Jarosewich.
Science, Vol. 174, 583 - 585 (1971).

Complete chondrules and fragments of chondrules have been found within silicate inclusions from the octahedrite iron meteorite Netschaevo. The bulk chemical composition, mineralogy, and mineral chemistry indicate that this chondritic material has properties intermediate between those of the H-group chondrites and those of the enstatite chondrites.

105.151 Stopfenheim Kuppel, Ries Kessel and Steinheim Basin: A triplet cratering event.
D. Storzer, W. Gentner, F. Steinbrunn.
Earth Planet. Sci. Letters, Vol. 13, 76 - 78 (1971).

105.152 Noble gases in the Haverö ureilite.
H. W. Weber, H. Hintenberger, F. Begemann.
Earth Planet. Sci. Letters, Vol. 13, 205 - 209 (1971).

The rare gases He, Ne, Ar, Kr and Xe were measured in two bulk samples and one of dark inclusions from the Haverö ureilite. It is shown that they consist of at least two components, one of which has been found to reside in the dark inclusions. In these graphite-diamond-kamacite intergrowths primordial Ar is more abundant by a factor of 29, Kr and Xe about 20 fold. The ^{3}He- and ^{21}Ne-exposure model ages are found to be 18 my and 13.5 my, respectively.

105.153 The chemical composition of the basaltic achondrites. L. H. Ahrens, R. V. Danchin.
Physics and chemistry of the earth, Vol. 8, (see 003.031), 265 - 303 (1971).

105.154 Estelas en la nube micrometeoroidea causadas por las erupciones cromosféricas solares.
C. Sánchez-Magro, F. Sánchez.
Urania Barcelona, Año 56, No. 273, p. 3 - 9 (1971).

A simplified dynamical study of the interaction of plasma shells originating in chromospheric solar flares with micrometeoroids yields the model of scars, proposed by F. Sánchez.

In this case the scar or wake is a region deficient in the smallest particles.

105.155 Mass fractionation and the isotopic anomalies of xenon and krypton in ordinary chondrites.
E. W. Hennecke, O. K. Manuel.
Zeitschr. Naturforschung, Vol. 26a, 1980 - 1986 (1971).

The abundance and isotopic composition of all noble gases are reported in the Wellman chondrite, and the abundance and isotopic composition of xenon and krypton are reported in the gases released by stepwise heating of the Tell and Scurry chondrites.

105.156 An alternative model for the formation of iron meteorites. M. R. Bloch, O. Müller.
Earth Planet. Sci. Letters, Vol. 12, 134 - 136 (1971).

As an alternative method for iron meteorite formation, a mechanism working primarily at low temperatures could have great advantages. We propose as a mechanism the thermal decomposition of metal carbonyls ($Me_x(CO)_y$).

105.157 The meteorite of April 25th 1969.
T. W. Rackham.
Irish Astron. Journ., Vol. 9, 297 - 307 (1970).

105.158 The sonic boom of the Boveedy meteorite.
E. Öpik.
Irish Astron. Journ., Vol. 9, 308 - 310 (1970).

105.159 La magnétite de la météorite d'Orgueil vue au microscope électronique à balayage. J. Jedwab.
Icarus, Vol. 15, 319 - 340 (1971).

Observations under the scanning electron microscope have made it possible to confirm the fact that abnormal magnetite forms do exist in the Orgueil meteorite: platelets, stackings of platelets, and framboids. The presence of spiral morphologies has been proved positively. A new typical form was identified as nodules with concavities. The morphological details of these various forms are described.

105.160 Yanites – very large meteor spherules of lower Yana (Yakutia). I. S. Astapovich, V. P. Pereyaslov.
Problems of cosmic physics. Vyp. (No.) 6, (see 003.040), p. 148 - 157 (1971). In Russian.

105.161 Examination of cratering formulas and scaling methods. J. W. White.
Journ. Geophys. Res., Vol. 76, 8599 - 8603 (1971).

Techniques that have been used for understanding the cratering process and predicting crater size are reviewed. The cratering process is shown to be unscalable, although many investigators have considered it to be scalable. Results of some specific work are presented, and their distinguishing features are discussed.

105.162 Electron microscope photographs of extraterrestrial particles.
E. K. Bigg, Z. Kviz, W. J. Thompson.
Tellus, Vol. 23, 247 - 260 = Separate print Division Radiophys., C.S.I.R.O., Sydney, Australia (1971).

105.163 Tektites and the moon. R. Taylor.
Comments Earth Sci. Geophys., Vol. 1, 111 - 116 (1971).

Reviews the current state of knowledge on tektites in the light of work on the first lunar samples.

105.164 Analytical possibilities of reactor neutron activation method in nondestructive analysis of meteorites.
M. Vobecky, J. Frana, Z. Randa, J. Benada, J. Kuncir.
Radiochem. Radioanalyt. Letters (*Switzerland*), Vol. 6, 237 -

247 (1971).
The analytical possibilities of the nondestructive reactor activation analysis of meteorites are dealt with.

105.165 On the study of cosmic dust. VII. The accretion rate of cosmic dust. S. Yabuki, M. Shima.
Rep. Inst. Phys. Chem. Res. (*Japan*), Vol. 47, No. 2, p. 33 - 39 (1971). In Japanese.

Influx measurements of extraterrestrial materials are summarized on the following four classifications: (1) Direct collection of cosmic spherules from air, polar ice and sea-sediment, (2) Trace elements or cosmogenic nuclides in various objects, (3) Spatial distribution of dust particles detected by apparatus mounted on spacecrafts and (4) Astronomical observation and calculation.

105.166 Implications of the hypothesis of extinct fissioning isotope in the primitive meteorites. M. Darkowski.
Phys. Letters B, Vol. 35B, 557 - 559 (1971). – See Phys. Abstr., Vol. 74, No. 62848 (1971).

105.167 Trapped helium, neon and argon in meteorites: Boundary conditions on the formation and evolution of the solar system. D. C. Black.
Thesis, Univ. Minnesota, Minneapolis. [Available from Univ. Microfilms, Ann Arbor, Mich., U.S.A. Order No. 70–20178], 112 pp. (1970).

105.168 Thermoluminescence in meteorites and tektites.
S. A. Durrani.
Modern Geol., (*GB*), Vol. 2, 247 - 262 (1971).

The main interest in studying the thermoluminescence of meteorites and tektites lies in using this method for determining their ages as well as in drawing useful inferences as to their radiation and thermal histories. The author reviews the work done in this field and the success attained so far.

105.169 Chemical links between stony meteorites and the lunar surface. L. H. Ahrens.
Comments Earth Sci. Geophys., Vol. 2, 22 - 27 (1971).

Considers the chemical composition of meteorites and compares them with the chemical composition of the rocks on the lunar surface as determined from the Apollo program.

105.170 Geophysical effects of the Tunguska blast of 1908.
A. V. Zolotov.
Proc. Sixth Winter School on Space Physics, Part II. Apatity 1969, (see 012.034), p. 29 - 30 (1971).

Handbook of elemental abundances in meteorites.
See Abstr. 003.027.

Cross-section for ^{10}Be production of high energy fragmentation of oxygen. See Abstr. 061.015.

Über die Empfindlichkeit astronomisch-geodätischer Lotabweichungen gegenüber Dichte-Anomalien des Untergrundes, nebst einer Anwendung auf die Bestimmung der Tiefenstrukturen im Nördlinger Ries. See Abstr. 045.016.

The low abundance of sulfur in the earth's core. See Abstr. 081.041.

Spherulitic textures in glassy and crystalline rocks. See Abstr. 094.165.

Chondrules of lunar origin. See Abstr. 094.269.

Ablation and breakup of large meteoroids during atmospheric entry. See Abstr. 104.001.

106 Interplanetary Matter, Interplanetary Magnetic Field, Zodiacal Light

106.001 **Interplanetary hydrogen and helium from cosmic dust and the solar wind.** P. M. Banks.
Journ. Geophys. Res., Vol. 76, 4341 - 4348 (1971).
In this report it is suggested that the deionizing effect of interplanetary dust on the solar wind may lead to the production of interplanetary ^{1}H and ^{4}He in sufficient quantities to explain recent photometric observations of interplanetary H Ly β and the presence of ^{4}He$^+$ in the solar wind.

106.002 **Single spacecraft method of estimating shock normals.** R. P. Lepping, P. D. Argentiero.
Journ. Geophys. Res., Vol. 76, 4349 - 4359 (1971).
By assuming the validity of the Rankine-Hugoniot conservation relations for interplanetary shocks in an isotropic medium it is demonstrated that improved shock normals can be calculated by employing a least squares technique to combined magnetic field and plasma data from a single spacecraft. As an example of the method a corrected normal and improved shock parameters are obtained for a real case: the August 29, 1966, shock observed by the Pioneer 7 spacecraft.

106.003 **The interplanetary medium and its interaction with the earth's magnetosphere.** J. V. Kovalevsky.
Space Sci. Rev., Vol. 12, 187 - 257 (1971).
This paper reviews the principal results of direct measurements of the plasma and magnetic field by spacecraft close to the earth (within the heliocentric distance range 0.7–1.5 AU). The paper gives an interpretation of the results for periods of decrease, minimum and increase of the solar activity.

106.004 **An investigation of the lunar ejecta theory of the cause of the Gegenschein.** A. J. Jeffries.
Planet. Space Sci., Vol. 19, 841 - 850 (1971).
The ejecta theory proposes that the Gegenschein, the enhancement in the brightness of the zodiacal light observed in the antisolar region, is caused by the ejection of dust particles from the moon by the impact of meteorites, the particles being pushed in the antisolar direction by solar radiation pressure to form a dust-tail seen visually as the Gegenschein. An IBM 360 computer has been used to calculate the trajectories of particles ejected from various positions on the moon and for different positions of the moon in its orbit, the particles being subject to the gravitational forces of the earth, sun and moon, solar radiation pressure and the Coriolis force of the rotating heliocentric coordinate system.

106.005 **The interplanetary hydrogen cone and its solar cycle variations.** H. J. Fahr.
Astron. Astrophys., Vol. 14, 263 - 274 (1971).
The paper investigates the motion of interstellar hydrogen particles within the solar system taking into account the thermal motions of the particles.

106.006 **Errata: 'Some effects of finite electrical conductivity on solar flare-induced interplanetary shock waves'** [Cosmic Electrodynamics, Vol. 1, 348 - 370 (1970)].
M. Dryer.
Cosmic Electrodynamics, Vol. 2, 246 - 248 (1971).

106.007 **Dependence of the scintillation index and the velocity of interplanetary irregularities on solar activity.** V. I. Vlasov.
Astron. Tsirk., No. 628, p. 1 - 2 (1971). In Russian.

106.008 **Characteristics of the interplanetary medium and changes of the kinetic energy of the long waves in the troposphere.** R. V. Smirnov.
Astron. Tsirk., No. 630, p. 3 - 5 (1971). In Russian.

106.009 **Rocket-infrared observations of the interplanetary medium.** B. T. Soifer, J. R. Houck, M. Harwit.
Astrophys. Journ., *(Letters)*, Vol. 168, L73 – L78 (1971).
Upper limits on the diffuse background radiation in the intermediate infrared ($5 \mu \leq \lambda \leq 23 \mu$), as measured from a sounding rocket, are presented. Evidence is given for the detection of thermal emission from the interplanetary medium.

106.010 **Signature in the interplanetary medium for substorms.** R. L. Arnoldy.
Journ. Geophys. Res., Vol. 76, 5189 - 5201 (1971).
A detailed signature for individual substorms is sought in the interplanetary medium. Hourly values of interplanetary field and plasma parameters are correlated with hourly averages of the AE index. An interplanetary variable involving the southward component of the interplanetary field in the solar magnetospheric coordinate system is shown to be singularly important for the generation of substorms.

106.011 **Interpretation of interplanetary scintillations.** A. T. Young.
Astrophys. Journ., Vol. 168, 543 - 562 (1971).
Radio observations of interplanetary scintillations are interpreted by means of a theory previously used for the detailed interpretation of optical observations of atmospheric scintillations. The theory allows a number of restrictive assumptions to be removed, and thus gives a more realistic picture of the interplanetary medium and a more accurate representation of the observations. Observers are requested to report power spectra of the logarithm of the intensity, to avoid artifacts due to "spikes".

106.012 **Interplanetarer Staub.** R. H. Giese.
SuW, Vol. 10, 261 - 268 (1971).

106.013 **Orientation of interplanetary shock waves (sound measurements) and position of chromospheric flares.** K. G. Ivanov.
Astron. Zhurn. Akad. Nauk SSSR, Vol. 48, 998 - 1003 (1971). In Russian. English translation in Soviet Astron. AJ, Vol. 15, No. 5.
A relation between interplanetary shock waves (according to measurements obtained on space sounds near the earth) and the position of powerful chromospheric flares is studied. It is confirmed that there are essential deviations from spherical and axial symmetry in interplanetary shock propagation.

106.014 **Large-scale properties of the interplanetary magnetic field.** K. H. Schatten.
Rev. Geophys. Space Phys., Vol. 9, 773 - 812 (1971).
Our knowledge of the large-scale properties of the interplanetary magnetic field is reviewed. The early theoretical work of Parker is presented, along with the observational evidence supporting his Archimedes spiral model. The variations present in the interplanetary magnetic field from the spiral angle are related to structures in the solar wind. The coronal magnetic models are related to the connection between the solar magnetic field and the interplanetary magnetic field. The direct extension of solar field-magnetic nozzle controversy is discussed, along with the coronal magnetic mo-

dels. The effect of active regions on the interplanetary magnetic field is discussed with particular reference to the evolution of interplanetary sectors. The suggested influence of the sun's polar field on the interplanetary field and alternative views of the magnetic field structure out of the ecliptic plane are presented. In addition, a variety of significantly different interplanetary field structures are discussed.

106.015 Observations of the interplanetary medium: Vela 3 and Imp 3, 1965 - 1967.
N. F. Ness, A. J. Hundhausen, S. J. Bame.
Journ. Geophys. Res., Vol. 76, 6643 - 6660 (1971).

Simultaneous observations were made of the interplanetary plasma by the Vela 3 satellites and of the interplanetary magnetic field by the Imp 3 satellite from July 1965 to July 1967. Certain derived plasma properties are computed and statistically summarized.

106.016 North-south component of interplanetary magnetic field: Explorer 33 and 35 data.
R. L. Rosenberg, P. J. Coleman, Jr., D. S. Colburn.
Journ. Geophys. Res., Vol. 76, 6661 - 6665 (1971).

Measurements of the interplanetary magnetic field taken with Explorer 33 and 35 in 1967 and 1968 indicate that for a given polarity the north-south component in a spherical polar coordinate system has, in general, a 27-day mean value that is significantly different from zero.

106.017 Relationship of interplanetary parameters and occurrence of magnetospheric substorms.
J. C. Foster, D. H. Fairfield, K. W. Ogilvie, T. J. Rosenberg.
Journ. Geophys. Res., Vol. 76, 6971 - 6975 (1971). – Letter.

106.018 Cosmic ray effects on interplanetary dust and dust detectors. K. Sitte.
Space Research XI, (see 012.004), 236 - 247 (1971).

106.019 The zodiacal light lines in the particle flux diagram.
C. Leinert.
Space Research XI, (see 012.004), 248 - 253 (1971).

106.020 Model computations concerning zodiacal light measurements by space missions. R. H. Giese.
Space Research XI, (see 012.004), 255 - 260 (1971).

106.021 Observational evidence for interplanetary atomic hydrogen. H. J. Fahr, P. W. Blum.
Space Research XI, (see 012.004), 1239 - 1245 (1971).

106.022 Correction: 'Twenty-seven day deviations of the interplanetary magnetic field and plasmas from the Parker spiral model' [Journ. Geophys. Res., Vol. 75, 5310 - 5318 (1970)]. R. L. Rosenberg.
Journ. Geophys. Res., Vol. 76, 4708 (1971).

106.023 Interplanetary magnetic sector polarity inferred from polar geomagnetic field observations.
E. Friis-Christensen, K. Lassen, J. M. Wilcox, W. Gonzalez, D. S. Colburn.
Nature, Phys. Sci., Vol. 233, 48 - 50 (1971).

With the use of a prediction technique it is shown that the polarity (toward or away from the sun) of the interplanetary magnetic field can be reliably inferred from observations of the polar geomagnetic field.

106.024 Radio scintillations due to plasma irregularities with power law spectra: The interplanetary medium.
D. N. Matheson, L. T. Little.
Planet. Space Sci., Vol. 19, 1615 - 1624 (1971).

The scintillation effects which are produced when radiation is incident on a medium containing phase changing irregularities which have power law spectra are discussed. The results are then applied to the interplanetary medium.

106.025 Early interplanetary magnetic fields and the remanent magnetization in meteorites. A. Brecher.
Publ. Astron. Soc. Pacific, Vol. 83, 602 - 603 (1971). – Abstr. Astron. Soc. Pacific.

106.026 Theoretical constraints on the microscale fluctuations in the interplanetary medium. A. Barnes.
Journ. Geophys. Res., Vol. 76, 7522 - 7526 (1971).

Theoretical constraints on the character of microscale fluctuations in the solar wind are examined. The microfluctuations could be either 'discontinuities' or more smoothly varying structures. If the fluctuations are waves, they are probably mostly Alfvén waves (or rotational discontinuities), and if they are stationary structures, they are tangential pressure balances (or tangential discontinuities).

106.027 A rotating solar magnetic "dipole" observed from 1926 to 1968. J. M. Wilcox, W. Gonzalez.
Science, Vol. 174, 820 - 821 (1971).

A recurring pattern with a period to $26^7/_8$ days observed in the polar geomagnetic field during the interval from 1926 to 1941 appears to persist in the interplanetary magnetic field polarity observed with spacecraft during the interval from 1963 to 1968. This observation suggests the existence of a rotating solar magnetic "dipole" with a period of $26^7/_8$ ± 0.003 days.

106.028 Reconnecting of force lines of the interplanetary magnetic field and the geomagnetic tail.
A. P. Kropotkin.
Geomagn. Aeronom., Vol. 11, 1075 - 1077 (1971). In Russian. Brief information.

106.029 Planung und Vorbereitung von extraterrestrischen Experimenten zur Messung des Zodiakallichts.
C. Leinert.
Bundesministerium für Bildung und Wissenschaft, Forschungsber. BMwF – FB W 69-18, [Available from Zentralstelle für Luftfahrtdokumentation und -information (ZLDI), Deutsche Forschungs- und Versuchsanstalt für Luft- und Raumfahrt, München. Price DM 18.27], 87 pp. (1969).

For further preparation a rocket experiment is proposed to observe the zodiacal light round the sun on circles with radii between 15° - 30°. Measurements on a prototype have shown that the required suppression of straylight by a factor of about 10^{13} is feasable.

106.030 Structure of the interplanetary magnetic field.
J. O. Stenflo.
Cosmic Electrodynamics, Vol. 2, 309 - 325 (1971).

A method to calculate the three-dimensional structure of the interplanetary magnetic field from the observed line-of-sight component of the field in the solar photosphere is described. Computer-drawn pictures of the coronal and interplanetary fields out to the orbit of earth are presented. The interplanetary magnetic field around the time of the 12 November 1966 total solar eclipse shows a clean dipole-type structure, while the field around the time of the 7 March 1970 solar eclipse is more complicated and has four sectors. The field calculated in the ecliptic plane at 1 AU shows good agreement with spacecraft observations.

106.031 The magnetic structure of interplanetary space.
N. F. Ness.
Proc. 11th international conference on cosmic rays, Budapest 1969, (see 012.025), p. 41 - 83 (1970).

106.032 Interplanetary plasma and interplanetary magnetic

fields. I. V. Kovalevsky.
Problems of cosmic physics. Vyp. (No.) 6, (see 003.040),
p. 19 - 44 (1971). In Russian.

106.033 Angular distribution of radio waves scattered by the interplanetary medium.
D. N. Matheson, L. T. Little.
Nature, Phys. Sci., Vol. 234, 29 - 31 (1971).

It has been suggested, that the exponential appearance of the fluctuation spectrum is due to high random velocities of the irregularities near the sun, which are greater than the systematic drift, rather than to the intrinsic shape of the scattered angular spectrum. To test this idea we have studied the auto-correlation function of complex amplitude $\rho_A(\xi)$ in more detail, using the near occultation of the Crab nebula by the sun.

106.034 The Gegenschein. R. G. Roosen.
Thesis, Univ. Texas, Austin. [Available from Univ. Microfilms, Ann Arbor, Mich., U.S.A. Order No. 70–18283], 177 pp. (1970).

The optical properties of the interplanetary medium in the direction of the Gegenschein are investigated in detail both theoretically and observationally.

105.035 Interplanetary debris in a new light. K. Hindley.
New Scient. and Sci. Journ., Vol. 51, 153 - 156 (1971). – See Phys. Abstr., Vol. 74, No. 62850 (1971).

106.036 Interplanetary dust streams: Observation by satellites and lidar. E. C. Silverberg.
Thesis, Univ. Maryland, Baltimore. [Available from Univ. Microfilms, Ann Arbor, Mich., U.S.A. Order No. 70–23319], 182 pp. (1970).

The interplanetary dust data from the orbiting satellites carrying 'microphone type' detectors are studied in detail. It is shown that the dust showers seen by these satellites are related to low inclination, periodic comets. The orbital parameters of the dust leaving each of the indicated comets have been calculated.

106.037 Theory and analysis of interplanetary scintillations. R. V. E. Lovelace.
Thesis, Cornell Univ., Ithaca, N.Y. [Available from Univ. Microfilms, Ann Arbor, Mich., U.S.A. Order No. 71–14651], 420 pp. (1970).

The thesis investigates several problems of the theory of interplanetary scintillations and the analysis of observations. The propagation of a wave deep into a medium having random spatial variations of index of refraction is discussed with geometrical optics and a wave optics model. The theory of scattering and scintillations for power law spectra is worked out. The power law theory is compared with observations of interplanetary scattering and weak and strong scintillations. A theory connecting density and magnetic field fluctuations is proposed for the interplanetary medium.

106.038 Structure of the interplanetary space from observations of low-energy cosmic rays in 1965–1967.
S. N. Vernov, G. P. Lyubimov, N. V. Pereslegina.
Proc. Sixth Winter School on Space Physics, Part I. Apatity 1969, (see 012.033), p. 201 - 210 (1971).

106.039 Dynamic alignment of atomic and molecular spins as a new method of determination of the interplanetary magnetic field. D. A. Varshalovich.
Proc. Sixth Winter School on Space Physics, Part I. Apatity 1969, (see 012.033), p. 220 - 227 (1971).

106.040 A contribution to the theory of particle acceleration in interplanetary plasma.

M. E. Kats, A. K. Yukhimuk.
Proc. Sixth Winter School on Space Physics, Part I. Apatity 1969, (see 012.033), p. 253 (1971).

106.041 The effect of solar wind velocity inhomogeneities on the structure of the interplanetary field.
I. I. Alekseev, A. P. Kropotkin, A. R. Shister.
Proc. Sixth Winter School on Space Physics, Part I. Apatity 1969, (see 012.033), p. 327 (1971).

Sector structure of the solar magnetic field.
See Abstr. 071.041.

Nature and origin of directional discontinuities in the solar wind. See Abstr. 074.002.

Active solar radio regions at metric frequencies and the interplanetary sector structures.
See Abstr. 077.006.

Interaction between solar burst and planetary atmopshere. See Abstr. 077.055.

Detailed directional and temporal properties of solar energetic particles associated with propagating interplanetary shock waves. See Abstr. 078.013.

Low-energy proton increases associated with interplanetary shock waves. See Abstr. 078.030.

North-south asymmetry of solar-particle fluxes in polar-cap regions. See Abstr. 078.044.

Energy changes of cosmic rays in interplanetary space. See Abstr. 078.045.

Solar magnetic fields—extended.
See Abstr. 080.048.

Polar substorm and interplanetary magnetic field.
See Abstr. 084.210.

Spectral characteristics of the earth's and the interplanetary magnetic field in the frequency band $1 - 10^{-4}$ Hz.
See Abstr. 084.294.

Correlated changes between the daily variation of geomagnetic field and interplanetary plasma parameters.
See Abstr. 084.303.

A connection between some parameters of the outer radiation belt and the interplanetary medium.
See Abstr. 084.405.

Peculiarities of propagation of radio waves in the atmosphere of Venus and in interplanetary plasma from data of Venera 7. See Abstr. 093.015.

The solar diurnal variation of cosmic rays at the 1954 solar minimum. See Abstr. 143.011.

Non-linear interaction of galactic cosmic rays with interplanetary magnetic fields and a possible geometry of the solar wind. See Abstr. 143.017.

Interaction of galactic cosmic rays with the interplanetary magnetic field. See Abstr. 143.027.

Cosmic-ray variations and the interplanetary sector structures. See Abstr. 143.047.

Radial gradient of galactic protons in the inner solar system. See Abstr. 143.090.

Characteristics of cosmic ray fluctuations in the

frequency range of 10^{-6} to 4.15×10^{-3} cycles per second. See Abstr. 143.108.

Investigation of anisotropy in the surrounding of a shock wave front. See Abstr. 143.123.

107 Cosmogony of the Planetary System

107.001 **The floccule theory and planetary formation.**
C. Aust, M. M. Woolfson.
Monthly Notices Roy. Astron. Soc., Vol. 153, 21P - 25P (1971).

A three-dimensional model of the formation of planets by the floccule theory has been examined with the aid of a computer. It is found to lead to a non-planar system with many proto-planets in retrograde orbits.

107.002 **Meteorites and the early solar system.**
E. Anders.
Annual Rev. Astron. Astrophys., Vol. 9, (see 003.001), 1 - 34 (1971).

The principal conclusion of this paper is that substantial chemical fractionations took place in the inner solar system. Theses processes were independent of each other, and thus the known meteorite types represent only a limited sampling of the possible range of planetary materials. The proper building blocks for the construction of planetary models are not the few known classes of meteorites, but the four components that behaved independently during chemical fractionations: early condensate, metal, remelted dust, and unremelted dust. Thus at least eight degrees of freedom are available for the composition of a planet: one each for the amounts and formation temperatures of the four components.

107.003 **Planetary formation from charged bodies.**
I. P. Williams.
Astrophys. Space Sci., Vol. 12, 165 - 171 (1971).

The suggestion of Sarvajna that a charged body which has been ejected from the sun can be captured in orbit because of electromagnetic effects is reinvestigated. It is concluded that the charge assumed by Sarvajna is too high by many orders of magnitude. An alternative scheme is proposed in which the charge requirement is much more realistic and it is shown that this scheme is feasible.

107.004 **Origin of organic matter in early solar system − III. Amino acids: Catalytic synthesis.**
D. Yoshino, R. Hayatsu, E. Anders.
Geochim. Cosmochim. Acta, Vol. 35, 927 - 938 (1971).

In the first two papers (Geochim. Cosmochim. Acta, Vol. 32, 151 - 174 and 175 - 190 (1968)) of this series, we showed that many of the organic compounds found in meteorites are produced spontaneously from CO, H_2 and NH_3 upon heating with nickel-iron or other simple catalysts. We therefore suggested that a major part of prebiotic organic matter in the solar system was made by such processes, either in the solar nebula or on planetary surfaces. The present paper and its sequel (Hayatsu et al., 1971) report results on amino acids.

107.005 **Origin of organic matter in early solar system−IV. Amino acids: Confirmation of catalytic synthesis**

by mass spectrometry.
R. Hayatsu, M. H. Studier, E. Anders.
Geochim. Cosmochim. Acta, Vol. 35, 939 - 951 (1971).

107.006 **The mass distribution of protoplanetary bodies.**
E. V. Zvjagina, V. S. Safronov.
Astron. Zhurn. Akad. Nauk SSSR, Vol. 48, 1023 - 1032 (1971). In Russian. English translation in Soviet Astron. AJ, Vol. 15, No. 5.

The mass distribution of protoplanetary bodies has been investigated with the aid of general methods of the coagulation theory.

107.007 **On the formation of the solar system.**
S. S. Kumar.
Nature, Vol. 233, 473 - 474 (1971).

The problem of the stability of the solar system is discussed, and the relevance of the orbital eccentricity and the mass of the most massive planet (Jupiter) to this problem is pointed out. Had Jupiter been much more massive in the past our planetary system would have become unstable by now. It would have also become unstable if the orbital eccentricity of Jupiter were as small as 0.5. Since double stars are known to occur with high frequency, it is concluded that the occurrence of planetary systems can not be a universal phenomenon.

107.008 **Mass distribution in the solar system.** V. Mitra.
Astrophys. Space Sci., Vol. 12, 471 - 483 (1971).

The present-day observed mass distribution in the solar system including the sun is shown to be compatible with the idea of the splitting of a number of ring-shaped rotating clouds of particles in the equatorial plane of a single contracting nebula. The formation of such a nebula is discussed and it is inferred that during the course of contraction this nebula has remained a sphere of uniform density spinning with the Keplerian velocity of its surface layer.

107.009 **The origin of the system earth − moon.**
G. P. Tamrazyan.
Bol. Acad. Cie. Fis., Mat., Nat., Venezuela, Vol. 30, No. 88, 210 pp. (1971). In Spanish and English.

107.010 **Nucleosynthesis of ^{26}Al in the early solar system and in cosmic rays.** D. N. Schramm.
Astrophys. Space Sci., Vol. 13, 249 - 266 (1971).

The nucleosynthetic yields of ^{26}Al by silicon burning, carbon burning, and spallation are discussed. It is shown that ^{26}Al can be synthesized in carbon and/or silicon- burning supernovae. However, time scales in the early solar system make it more likely that ^{26}Al, if present in planets, was synthesized by a proton irradiation in the early solar system. An integrated proton flux $> 4 \times 10^{18}\,cm^{-2}$ is shown to be necessary in order for ^{26}Al to be a significant heat source. No conclusive evidence has been observed for an irradiation of this magnitude. Therefore, unless such evidence is found, it should be assumed that ^{26}Al was not involved in the formation of the solar system. In addition, the production of ^{26}Al in cosmic rays is discussed.

107.011 Evolutionary processes in the solar system.
G. Colombo.
Mem. Soc. Astron. Italiana, Nuova Ser., Vol. 42, 279 - 291 (1971).

Evolutionary processes like tidal dissipation, close approach and collision are discussed.

107.012 On the equivalence of the planet-satellite formation processes. D. C. Black.
Icarus, Vol. 15, 115 - 119 (1971).

This communication is an effort to obtain a model which permits a quantitative description of features found in both the planetary system and the satellite systems of Jupiter and Uranus. Application of the model to the Saturnian satellite system suggests the presence of a hitherto undiscovered "satellite" located between Rhea and Titan.

107.013 A solar system formation model based on supernova shell fragmentation. W. K. Brown.
Icarus, Vol. 15, 120 - 134 (1971).

A new model of solar system formation is presented which rests on the postulated breakup of an ejected supernova shell into massive fragments. Each fragment subsequently evolves into a separate and complete solar system. The model produces a powerful description of our solar system in both its overall regularity and its many anomalies.

107.014 Origin of the solar nebula.
F. Hoyle (Invited speaker).
Highlights of Astronomy, Vol. 2, (see 012.014), 195 - 203 (1971).

107.015 Lunar composition as a clue to the early history of the solar system. S. F. Singer.
Meteoritics, Vol. 6, 316 (1971). — Abstract.

107.016 Indications of eruptive evolution of planets.
S. K. Vsekhsvyatsky.
Problems of cosmic physics. Vyp. (No.) 6, (see 003.040), p. 73 - 94 (1971). In Russian.

107.017 High-temperature protoplanetary processes (On the formation of metallic cores of the planets).
A. P. Vinogradov.
Geokhimiya, 1971, No. 11, p. 1283 - 1296. In Russian.
Abstr. in Referativ. Zhurn. 51. Astron., 2.51.252 (1972).

107.018 The origin of planetary systems. M. M. Woolfson.
Phys. Bull. (*GB*), Vol. 22, 266 - 272 (1971).

107.019 Extinct ^{129}I, ^{244}Pu and superheavy elements in the early history of the solar system. D. York.
Comments Earth Sci. Geophys., Vol. 2, 14 - 21 (1971).

Plasma physics, space research, and the origin of the solar system. See Abstr. 062.057.

Il problema dell'elio. See Abstr. 065.114.

Of time and the moon.
See Abstr. 094.014.

Stars

111 Stellar Parallaxes

111.001 **First results from the new Yerkes Observatory parallax program.** W. F. Van Altena.
Bull. American Astron. Soc., Vol. 3, 372 (1971). – Abstr. AAS.

111.002 **The application of multivariate analysis to parallax solutions. I. Choice of reference frames.**
A. R. Upgren, S. J. Kerridge.
Astron. Journ., Vol. 76, 655 - 664 (1971).

The statistical method of principal component analysis is investigated as a technique for the formulation of parameters which measure effects of the choice of reference stars upon the parallax and its formal error. This multivariate analysis method is applied to 15 star fields measured in the course of the continuing parallax program of the Van Vleck Observatory. Between 65 and 348 solutions were made for each star field; each solution used a different subset among the total comparison stars measured. Two situations are discussed; the first employs five geometrical parameters in each coordinate, but a second using eight variables in both coordinates was found to provide further evidence on the dependence of the mean error on the parameters.

111.003 **On the completeness of the catalogue of nearby stars.** Z. F. Seidov.
Tsirkulyar Shemakin. astrofiz. observ., 1971, No. 4 (10), p. 14 - 16. In Russian. – Abstr. in Referativ. Zhurn. 51. Astron., 11.51.801 (1971).

111.004 **Trigonometric parallaxes determined with the Yerkes Observatory 40-inch refractor. I. Methods of observation, measurement, and reduction, and the first results.**
W. F. van Altena.
Astron. Journ., Vol. 76, 932 - 939 (1971).

The first results of the new Yerkes Observatory 40-inch refractor parallax program are presented. The program has implemented many of the suggestions made by Vasilevskis to increase the accuracy of trigonometric parallaxes. The 60 stars selected for this program have an average visual magnitude of $V \sim 14.4$ and are divided between white dwarfs and large proper-motion late-type stars.

111.005 **Parallaxes and proper motions. VI.**
W. S. Mesrobian, A. R. Upgren.
Astron. Journ., Vol. 76, 1133 - 1134 (1971).

Relative parallaxes and proper motions are given for 30 stars, 12 of which have no previous trigonometric parallax determination.

Parallaxes of faint stars.
Sky Telescope, Vol. 42, 212 (1971).

Parallax and mass ratio of Eta Cassiopeiae.
See Abstr. 118.027.

112 Proper Motions, Radial Velocities, Space Motions

112.001 **Radial velocities of southern B stars determined at the Radcliffe Observatory – V.** P. W. Hill.
Mem. Roy. Astron. Soc., Vol. 75, 1 - 20 (1971).

Radial velocities are given for 85 O and B type stars at intermediate and high galactic latitudes. These include some brighter than $m_v = 6$ as well as some very distant stars fainter than $m_v = 10$. A number of high velocity stars are noted, of which at least one is probably subluminous.

112.002 **Barnards Pfeilstern – Fixstern mit der größten Eigenbewegung.** B. Wedel.
SuW, Vol. 10, 199 (1971).

112.003 **Selection of 9th magnitude stars to serve as radial-velocity standards.** J. F. Heard.
Journ. Roy. Astron. Soc. Canada, Vol. 65, 172 (1971). Abstr. RAS Canada.

112.004 **BD + 6°2461 : une étoile B à grande vitesse dans la région du Pôle Nord galactique.**
J. Berger, A.-M. Fringant, E. Rebeirot.
Comptes Rendus Acad. Sci. Paris, Sér. B, Vol. 273, 217 - 220 (1971).

L'étoile BD + 6°2461 est probablement une étoile éjectée du disque galactique dans le halo, avec une grande vitesse initiale. Sa grande vitesse résiduelle et sa distance au plan galactique supérieure à 5 kiloparsecs sont compatibles avec son âge estimé à 2×10^7 années. Toutefois, il n'est pas exclu que cette étoile soit une étoile sous-lumineuse de la population du halo.

112.005 **Catalogue of proper motions of 8,790 stars with reference to galaxies.**
A. R. Klemola, S. Vasilevskis, C. D. Shane, C. A. Wirtanen.
Publ. Lick Obs., Univ. California, Santa Cruz, Vol. 22, Part 2, 76 pp. (1971).

The results of a pilot program giving the absolute proper motions of 8790 stars measured with reference to the system of galaxies is presented here. The proper motions are based on two series of plates taken with the 20-inch Carnegie Astrograph of the Lick Observatory at epochs separated on the average by 19.2 years. There are 83 fields in the program with an average of 56 galaxies per field. Methods of measurement and reduction and studies of various errors are also presented. The mean error of the galaxy reference frame for a single field affects the proper motion in each coordinate by an average amount of 0.″16 per century. The mean error of an individual proper motion is about 0.″7 per century in each coordinate, and that of a blue magnitude is about $0^m 2$.

112.006 Stellar-statistical formulation of the problem of setting up an astronomical radial velocity system.
H. Eelsalu.
Publ. Tartu Astrofiz. Obs., Vol. 39, 163 - 170 (1971).

112.007 On proper motions of stars of the latitude program.
I. M. Kalinina.
Soobshch. Gos. Astron. Inst. Shternberga, No. 170, p. 33 - 36 (1971). In Russian.

112.008 Reduced-proper-motion diagrams.
E. M. Jones.
Bull. American Astron. Soc., Vol. 3, 371 (1971). – Abstr. AAS.

112.009 Space motions of nearby K3–M2 dwarfs.
A. R. Upgren.
Bull. American Astron. Soc., Vol. 3, 371 (1971). – Abstr. AAS.

112.010 Space velocities of G and K giants.
K. M. Yoss, T. E. Lutz.
Mem. Roy. Astron. Soc., Vol. 75, 21 - 50 (1971).

Machine reduction procedures have been developed for processing digitized microphotometer data for the derivation of absolute magnitudes and CN anomalies. Absolute magnitudes and CN anomalies have been derived for 631 programme stars. The absolute magnitudes have then been combined with additional observational data to determine space velocities. The degree of correlation between CN anomaly and the dispersions in the U, V and W components of space velocity has been investigated.

112.011 Photoelectric radial velocities, Paper IV. 528 7^m – 10^m stars in the +15° Selected Areas.
R. F. Griffin.
Monthly Notices Roy. Astron. Soc., Vol. 155, 1 - 49 (1971).

A total of nearly 1500 radial-velocity observations of more than 500 $7^m - 10^m$ stars has been made with a photoelectric spectrometer. The standard errors of individual observations are mostly in the range 1.0 to 1.5 km/s. Eighty-five per cent of the mean velocities have standard errors of 1.0 km/s or better. A comparison of the photoelectric results with the 41 b-quality velocities available for the same stars in the Radial Velocity Catalogue shows the standard error of the catalogue to be 3.2 ± 0.4 km/s.

112.012 The radial velocities of 129 stars in the years 1906 to 1917. W. R. Beardsley, assisted by M. Matthew, E. M. Erskine, T. L. Gandet, Jr., E. N. Hubbell, M. Jacobson, G. Jones, K. Kobus, S. Levy, P. Lowrey, D. Namisnak.
Publ. Allegheny Obs. Univ. Pittsburgh, Vol. 8, No. 7, 171 pp. (1969).

In the years 1906–1917, an extensive program of radial velocity determination was undertaken at the Allegheny Observatory. During this interval, 7144 spectra of primarily B and A stars were obtained using a dispersion of 40 Å/mm. A total of 34 orbits of spectroscopic binaries was published in Volumes 1, 2, and 3 of these publications, as well as radial velocities for a number of miscellaneous stars. A list of 2935 radial velocities is presented for 129 stars. The velocities of each star have been analyzed for binary motion. Preliminary

elements are presented for two stars, ϕ Dra and 6 Lac, not previously known to be binaries. A GCRV – Allegheny comparison has been made for 47 stars which had no evidence of binary motion and mean velocity differences tabulated as a function of spectral type.

112.013 Accuracy of radial velocities determined with a fiber optic electrostatic image tube. W. R. Beardsley, J. K. de Jonge, D. J. Haring, J. R. Hansen.
Publ. Allegheny Obs. Univ. Pittsburgh, Vol. 11, No. 1, 26 pp. (1969).

The astrometric fidelity at the output face of a Westinghouse WL–30677 image tube has been investigated. Stringent fidelity is required in order to provide accurate radial velocities of late-type IAU Standard Velocity Stars. These Standard Velocities have been achieved within the probable errors. A Baum Test Pattern has also been used to investigate the fidelity. A performance gain of 16 has also been derived. Tables are presented of performance parameters of the tube as well as individual radial velocities.

112.014 Radial velocities and spectral classification of A-type stars near the south galactic pole.
H. E. Bond, C. L. Perry, W. P. Bidelman.
Publ. Astron. Soc. Pacific, Vol. 83, 643 - 647 = Contr. Louisiana State Univ. Obs., *Baton Rouge,* No. 52 (1971).

Radial velocities and spectral types are listed for over 100 A-type stars in the region of the south galactic pole (SGP). Comparison of published photometry with intrinsic color indices derived from the spectral types yields a small but significant interstellar reddening at the SGP of E(B – V) = $0^m 03$.

112.015 Nomogram for the reduction of radial velocities to the sun. E. A. Vitrichenko.
Peremennye Zvezdy, Vol. 17, 680 - 681 (1971). In Russian.

The nomogram for the reduction of radial velocities to the sun is given. Accuracy is about 1 km/sec.

112.016 Radial velocities of 69 G, K and M-type stars.
M. B. K. Sarma.
Nizamiah Obs., Osmania Univ., *Hyderabad,* Contr. No. 3, 10pp. (1967).

The present list of radial velocities for 69 stars obtained from 107 Mount Wilson and Palomar spectrograms is a continuation of an earlier list (K. D. Abhyankar, Nizamiah Obs. Contr. No. 2, 1964). The standard lines chosen for the radial veloctiy measurements and the method of measurement are the same as those given in the earlier list.

112.017 Proper Motion Survey with the forty-eight inch Schmidt telescope: The zones +66 and +60, 6^h to 20^h. W. J. Luyten.
Separate print Univ. Minnesota, Minneapolis, Minnesota. 195 pp. (1971).

The catalogue contains all the stars found in the +66 and +60 zones to which have been added all the stars from the +72 zone whose declinations are south of +70, and are between the limits of right ascension of the present zones. Altogether the catalogue contains 17418 entries which are divided as follows: 888 B.D. stars, 334 stars first announced by other authors, and 16196 stars found in the Bruce and Palomar Surveys.

112.018 Stars with large proper motions in the astrographic zones +34° and +35°. III.
R. S. Khandelwal, A. N. Goyal.
Indian Journ. Pure Applied Math., Vol. 2, 155 - 229 (1971).

Twenty-seven regions common to Potsdam Photographische Himmelskarte and Paris catalogues of declination zones +34° and +35° have been compared. Annual relative proper motions of 1300 stars with probable error ±0.″010 in either coordinate are listed. The corrections for the reduction to absolute proper motions have also been given for each region.

112.019 **Fundamental reference system and stellar motion.**
H. Yasuda.
Astron. Herald, (*Japan*), Vol. 64, 95 (1971). In Japanese.
See Phys. Abstr., Vol. 74, No. 77456 (1971).

112.020 **Internal probable errors of radial velocity determina-
tions.** E. E. Fenimore, D. N. McGrath, P. M.
Nachman, C. R. O'Dell, F. C. Sanner.
Publ. Astron. Soc. Pacific, Vol. 83, 780 - 782 (1971).

The origin and goals of the automated stellar proper motion survey. See Abstr. 031.068.

The automation of the stellar proper motion survey. See Abstr. 031.069.

Preliminary catalogue of the positions and proper motions of stars between declinations −70° and −90°, reduced to the equinox of 1950 without applying proper motions. See Abstr. 041.030.

Stars with large proper motions in the astrographic zones +32° and +33°. (Final list). See Abstr. 041.037.

General principles of studying differences in positions and proper motions of stars as a random field. See Abstr. 041.038.

An analysis of the AGK3 comparison star positions. See Abstr. 041.048.

On determination of correction to precession from stellar proper motions. See Abstr. 043.001.

Parallaxes and proper motions. VI. See Abstr. 111.005.

The distances of two faint OB star groups in Monoceros. See Abstr. 113.049.

New wavelengths' table for the determination of radial velocities of late type stars. See Abstr. 114.012.

The radial velocity of 60 Serpentis. See. Abstr. 119.013.

The proper motions of RR Lyrae variables, II. See Abstr. 122.059.

Radial velocity observations of the Delta Scuti star 20 Canum Venaticorum. See Abstr. 122.089.

Proper motions of 84 cepheids. See Abstr. 122.113.

The Lick Observatory program on proper motions of RR Lyrae stars. See Abstr. 122.127.

Proper motions of 35 δ Scuti type stars. See Abstr. 122.150.

The Lick Observatory program on proper motions of RR Lyrae stars. See Abstr. 122.151.

Catálogo de nebulosas extragalácticas de la zona −5°/−25° de declinación, seleccionadas para la determinación de un sistema absoluto de movimientos propios estelares. II − Coordenadas rectilíneas de nebulosas, estrellas de referencia y estrellas de control. See Abstr. 158.124.

113 Stellar Magnitudes, Colors, Photometry

113.001 Erratum: "Optical studies of Cassiopeia A. II. *UBV* **photometry of field stars"** [Astrophys. Journ., Vol. 165, 259 - 263 (1971)]. S. van den Bergh.
Astrophys. Journ., Vol. 167, 559 (1971).

113.002 An observational evidence of non-LTE effects for O-type stars. A. Maeder.
Astron. Astrophys., Vol. 13, 444 - 446 (1971).

An UV excess of $0\overset{m}{.}01-0\overset{m}{.}02$ exists for Of, O9 I and O9.5 I stars in comparison with absorption-line O stars of same spectral type. The Of stars have a higher luminosity relative to absorption-line O stars of the same spectral type. If so, the observed effect for Of, O9 I and O9.5 I stars is in strong desagreement with LTE-models for O-type stars. On the other hand, the agreement with non-LTE models by Auer and Mihalas (1971) is excellent.

113.003 Three-color photometry of a field in the large Sagittarius cloud. H. Fünfschilling.
Astron. Astrophys., Vol. 13, 454 - 470 (1971).

In a field of 0.26 square degrees which is part of the well known galactic window, 1400 stars down to a limiting magnitude of $G = 16.5$ mag have been measured photometrically in the *RGU* system on 23 plates taken with the 48-inch Palomar-Schmidt. The interstellar reddening in the field has been studied by means of the two-color diagrams for stars in given intervals of apparent magnitude. Because of the complicated behavior of the reddening function, the interpretation of the two-color diagrams presents some difficulties. Two different kinds of density function have been calculated for main-sequence stars with absolute magnitudes between +2 and +7 — a minimum function which includes only stars with a single-valued absolute magnitude, and a maximum function for all stars.

113.004 Five-colour photometry of 12 magnetic variable stars. A. M. van Genderen.
Astron. Astrophys., Vol. 14, 48 - 65 (1971).

A discussion is presented of photo-electric five-colour observations of 12 magnetic variable stars made in 1966. The periods of these variables range from $0\overset{d}{.}58$ to $314\overset{d}{.}$ The light- and colour-curves are given and their ranges are schematically represented in three colour-colour diagrams.

113.005 Three UBV sequences in Centaurus. C. H. McGruder, G. Schnur.
Astron. Astrophys., Vol. 14, 164 - 166 (1971).

Photoelectric UBV measurements of 67 stars in Centaurus are described.

113.006 A note concerning the determination of the Geneva Observatory photometric system's pass-bands.
F. Rufener, A. Maeder.
Astron. Astrophys., Suppl. Ser., Vol. 4, 43 - 49 (1971).

A computational method leading to a better knowledge of a photometric system's pass-bands is proposed. It enables the determination of pass-band profiles in agreement with the best spectrophotometric calibrations published, without the need of carrying out an absolute spectrophotometric calibration. A table is presented, containing the pass-bands of the Geneva Observatory photometric system obtained on the one hand by Code's (1960) calibrations and on the other hand by those of Hayes (1970).

113.007 Hβ photometry of A-type stars near the North Galactic Pole. A. G. D. Philip, L. E. Tifft.
Astron. Journ., Vol. 76, 567 - 570 (1971).

Hβ measures have been obtained for 52 B- and A-type stars in an area near the North Galactic Pole, for which Strömgren four-color photometry had been obtained by Philip. The color excesses of the population I stars in the sample indicate a color excess of $E_{B-V} = 0.00$ in areas close to the pole, and a color excess of $E_{B-V} = 0.05$ in an area just north (in declination) of the pole. The intrinsic color indices of 14 field horizontal-branch stars found in the area are calculated.

113.008 An extraordinary red object in Sagittarius. J. W. Warner, R. F. Wing.
Astrophys. Journ., (*Letters*), Vol. 167, L53 - L54 (1971).

The infrared source IRC−20385 has been found to coincide with an extended object, at least 2 arc min in diameter, which is conspicuous on red plates but invisible in the blue.

113.009 Origin of the reported millimeter continuum in IRC + 10216.
N. J. Woolf, P. M. Solomon, A. A. Penzias.
Astrophys. Journ., (*Letters*), Vol. 167, L65 (1971).

It is suggested that the reported 3.5-mm continuum in IRC + 10216 is due to one or more emission lines within the receiver bandpass.

113.010 Infrared sources of radiation. G. Neugebauer, E. Becklin, A. R. Hyland.
Annual Rev. Astron. Astrophys., Vol. 9, (see 003.001), 67 - 102 (1971).
Infrared surveys; Characteristics of infrared excesses in stars; Infrared excesses in late-type stars; Infrared excesses in intermediate and early-type stars; Infrared associated with selected galactic objects; Infrared emission from the galactic nucleus; Extragalactic sources.

113.011 Intermediate band photometry of V Puppis. E. E. Mendoza V.
Bol. Obs. Tonantzintla y Tacubaya, No. 36, Vol. 6, 89 - 94 (1971).

Some 150 observations were made on V Puppis on the (33, 35, 37, 40, 45, 52, 58, 63)−photometric system. Absolute visual magnitudes are derived at primary minimum and maximum light.

113.012 Temperaturabhängigkeit des photographischen Farbsystems der Tautenburger Schmidtkamera.
R. Ziener.
Astron. Nachr., Vol. 292, 231 - 233 (1971).

The coefficients of the colour equation for the blue magnitudes determined with the Schmidt telescope at Tautenburg show a correlation with the temperature in the open air. Possible causes of this effect are discussed.

113.013 Schwache blaue Objekte in der Nähe des galaktischen Nordpols (Umgebung von M 3).
W. Bronkalla.
Astron. Nachr., Vol. 292, 263 - 270 (1971).

The results of a complete UBV photometry in a field of 3.1 square degrees to the limiting magnitude $B = 20.0$ are given. The number of blue objects (with $U - B \leqq - 0.4$) per square degree brighter than magnitude B is given by the relation $\log N = (0.66 \pm 0.08) (B - 18) - 0.04 \pm 0.07$. The percentage of the blue objects is 2% of the total number of stars brighter than magnitude $V = 19.5$.

113.014 Photométrie photoélectrique et classification spectrale. B. Hauck.
Orion, 29. Jahrgang, p. 69 - 75 (1971).

113.015 The effect of line and band absorption on B, V magnitudes of K–M stars. M. H. Rodriguez.
Astrometriya i Astrofiz., *Kiev*, Vyp. (No.) 12, (see 003.003), p. 24 - 26 (1971). In Russian.

113.016 A method for increasing the dynamic response of photometric systems. P. Lee, C. L. Perry.
Astron. Journ., Vol. 76, 619 - 620 = Contr. Louisiana State Univ. Obs., *Baton Rouge*, No. 51 (1971).

The effects of a neutral-density attenuator on the measured magnitudes and colors of stars are discussed. Such a scheme offers a way of measuring bright stars with a pulse-counting system.

113.017 Four-color and Hβ photometry for bright B-type stars in the southern hemisphere.
D. L. Crawford, J. V. Barnes, J. C. Golson.
Astron. Journ., Vol. 76, 621 - 630 (1971).

Photoelectric photometry is presented for 325 B-type stars south of Dec. $-10°$ and with $5^m.0 < m_V < 6^m.5$.

113.018 Photometric systems. V. L. Straižys.
Methods of investigation of variable stars, (see 003.008), p. 225 - 278 (1971).

113.019 Photometric standards. A. S. Sharov.
Methods of investigation of variable stars, (see 003.008), p. 279 - 306 (1971).

113.020 Very blue stellar objects appearing near galaxies.
D. W. Weedman.
Astrophys. Letters, Vol. 9, 49 - 51 = Arthur J. Dyer Obs., Vanderbilt Univ., Nashville, Tennessee, Ser. 2, Repr. No. 7 (1971).

Because of Arp's discovery that the quasar Markarian 205 is connected to the spiral galaxy NGC 4319, the entire Palomar Observatory Sky Survey has been searched for very blue stellar objects appearing in or near galaxies. Nine such objects were found, and finding charts are given for these.

113.021 Values of U–B for some bright southern stars.
B. S. Carter, P. M. Corben, G. M. Harvey.
Monthly Notes Astron. Soc. Southern Africa, Vol. 30, 109 - 111 (1971).

113.022 Three colour photometry at a city site.
B. F. Marino.
Southern Stars, Vol. 24, 47 - 58 (1971).

113.023 L'absorption de la lumière dans la région de la Voie Lactée sur les données de la photométrie photographique. III. Les magnitudes stellaires et les classes spectrales des étoiles dans deux parcelles centrées (19^h15^m, $+9°45'$) et (18^h24^m, $+22°30'$) (1900.0). T. A. Uranova.
Soobshch. Gos. Astron. Inst. Shternberga, No. 171, p. 10 - 14 (1971). In Russian.

On donne les résultats de la détermination des classes spectraux et des magnitudes B et V pour les étoiles O–F5, dans les parcelles n° 3 et n° 9 de la région n° 1 de P. P. Parenago (1956), situées dans les constellations Aquila et Hercules.

113.024 Faint O–B2 stars in the Vela, Carina, and Centaurus sections of the Milky Way. E. W. Miller.
Bull. American Astron. Soc., Vol. 3, 371 (1971). – Abstr. AAS.

113.025 Near-ultraviolet surface photometry of the southern Milky Way. · J. Pfleiderer, U. Mayer.
Astron. Journ., Vol. 76, 691 - 700 (1971).

Isophote maps are presented of the Milky Way (except the northern region from Per to Cep/Cyg): color – U of the UBV system; field of view– ~1 deg²; resolution $-1°$ to $2°$. The mean intensity of the inner Milky Way is about $100\,S_{10}(U)$ [intensity at poles 15 to 20 $S_{10}(U)$]. The brightest areas, with intensities up to several hundreds $S_{10}(U)$, are in Sgr/Sco, Car, Vel/Pup, and Cyg. Most bright areas coincide with OB associations and extended Hα regions, and many other such objects are clearly visible, probably due to Balmer continuum emission. The reduction procedure is described in detail.

113.026 High speed photometry.
E. J. Kibblewhite, R. V. Willstrop.
Monthly Notices Roy. Astron. Soc., Vol. 154, 301 - 319 (1971).

Equipment for recording the light of rapidly variable objects, and the methods of analysis of the data are described. The equipment has been used in searches for regular fluctuations in the light received from the directions of pulsars and supernovae remnants, and in the light of white dwarf stars, a flare star, the nuclei of normal and Seyfert galaxies, some radio sources, the remnant of Nova (DQ) Herculis 1934 and the central star of M 57. The known variability of DQ Her and NP 0532 has been confirmed, and upper limits are placed on any other variations.

113.027 Theoretical colours for F and G dwarf stars.
R. A. Bell.
Monthly Notices Roy. Astron. Soc., Vol. 154, 343 - 383 (1971).

Synthetic spectra have been computed for F and G dwarf stars, using a number of values of chemical abundance, Doppler broadening velocity and damping constant. Filter transmission functions have been convolved with these spectra to yield the $U–B$, $B–V$, $R–I$, m_1, c_1, $b–y$, β, 33–52, 35–52, 37–52, 40–52, 45–52, 52–58 and 52–63 colours of the models. The metal abundances of a number of stars have been obtained using computed and observed m_1 and 40–52 colours. The c_1 colours of stars with accurately known trigonometric parallaxes have been used in order to determine how accurately absolute magnitudes can be predicted from the colours. It is found that the predicted variation of ultra-violet excess, $\delta(U–B)$, is only half that which is observed. The effects of interstellar reddening on the colours of the models have been examined.

113.028 Optical identification of infrared sources.
M. Cohen.
Astrophys. Letters, Vol. 9, 95 - 100 (1971).

Optical identifications are presented for eight infrared sources from the Two Micron Sky Survey which have unusual photographic appearance or location. One source may belong to an open cluster, a second lies in an obscured region of an emission nebula, and a third may be associated with a globular cluster.

113.029 An investigation of the consistency of the calibrations of the $uvby\beta$ and $gnkmf$ photometric systems.
E. H. Olsen.
Astron. Astrophys., Vol. 15, 161 - 172 (1971).

Wide physical binary stars with G or K giant stars as primary components and A or F main-sequence stars as secondary components have been observed photoelectrically; the primary components in the narrow-band $gnkmf$ photometric system and the secondary components in the $uvby\beta$ photometric system. Data in the $gnkmf$ system are given for 32 double star components and data in the $uvby\beta$ system for 34 double star components. The two photoelectric photometries have both been calibrated in terms of absolute visual magnitude. Both photometric systems have also been calibrated in terms of the logarithmic iron-to-hydrogen ratio [Fe/H]. The effects of a systematic error in the present $uvby\beta$ photometry, of duplicity, of a possible dependence of $M_v(K)$ on the metal abundance, and of interstellar extinction are discussed.

113.030   Six-color photometry of 13 F−G supergiants in the Large Magellanic Cloud.
J. P. Brunet, P. Mianes, M. N. Perrin, L. Prévot, J. Rousseau.
Astron. Astrophys., Vol. 15, 320 - 324 (1971).

The application of six-color photometry to the study of 13 F−G supergiants among the most luminous stars of the LMC shows that their colors are not obviously different from galactic stars of the same luminosity. It is shown that six-color photometry does not indicate any distinction between Ia and Ib supergiants slightly reddened. Color indices of equatorial standard stars are also given.

113.031   The temperatures, abundances and gravities of F dwarf stars.   R. A. Bell.
Monthly Notices Roy. Astron. Soc., Vol. 155, 65 - 83 (1971).

Theoretical colours, computed using laboratory line data and model stellar atmospheres, have been used to interpret the colours of about 150 F and early G dwarfs. Effective temperatures have been derived from the $H\beta$ index and from $R-I$, abundances have been obtained from m_1 and $b-y$ and gravities have been obtained from c_1 and $b-y$. Absolute magnitudes have been obtained from the effective temperatures and the gravities, the latter being used with assumed stellar masses to yield radii.

113.032   Some red giants of the old disk population.
O. J. Eggen.
Publ. Astron. Soc. Pacific, Vol. 83, 423 - 432 (1971).

($UBVRI$) photometry and space motions of several old disk population red stars are presented. These stars include 14 new variables.

113.033   Photometry of three peculiar A-type stars.
N. D. Morrison, S. C. Wolff.
Publ. Astron. Soc. Pacific, Vol. 83, 474 - 477 (1971).

Four-color ($uvby$) photoelectric observations have yielded provisional periods for three peculiar A-type stars. For 108 Aquarii, the period is $3^{d}73$. For HD 184905, the period is either $1^{d}84$ or $2^{d}17$, and for θ^1 Microscopii the most probable periods are $0^{d}941$ and $1^{d}062$.

113.034   Revised indices for some G and K dwarf stars, on the six-color system of the Lick Observatory.
J. Rousseau.
Astron. Astrophys., Vol. 15, 468 - 470 (1971).

For most of the G and K dwarf stars measured by Stebbins and Whitford (1945) the $B-G$ index is found to be too small by a mean value of 0.08.

113.035   Six photoelectric standard sequences in the Southern Milky Way.   W. Seggewiss.
Veröff. Astron. Inst. Bonn, No. 82, 17 pp. (1971).

UBV magnitudes and colors of stars in six standard sequences near galactic longitudes 212°, 223°, 243°, 265°, 279°, and 300° are presented. The number of stars in each sequence is 21 to 24, the limiting V magnitudes are $14.^{m}8$ to $15.^{m}6$. Identification charts, two color diagrams and estimates of spectral types derived from these diagrams are given. The observations were carried out in February and March 1969 with the photoelectric equipment of the ESO 1m-telescope at La Silla, Chile.

113.036   Photoelectric observations of stars in the southern open clusters NGC 2335, NGC 2343, NGC 2453, NGC 4439 and H 5.   W. Seggewiss.
Veröff. Astron. Inst. Bonn, No. 83, 17 pp. (1971).

Photoelectric UBV measurements were obtained for small numbers of stars in the southern open clusters NGC 2335, 2343, 2453, 4439 and H 5. In all cases the data are sufficient for the determination of distances and color excesses of the clusters from the two-color and color-magnitude diagrams. The measurements were obtained in connection with the observations of six standard sequences in the Southern Milky Way (Seggewiss 1971) with the ESO photometric telescope at La Silla, Chile, in February and March 1969.

113.037   Stellar photometry in the region 1300−2000 Å, part II.   J. W. Campbell.
Astrophys. Space Sci., Vol. 13, 189 - 202 (1971).

Observations have been made at 1450 Å of 94 early type stars. Comparison is made with other observations, and in general, good agreement has been found. Errors in absolute calibration of earlier observations have led to some disagreement. The presented data agree well with theoretical stellar models which include the effects of line blanketing.

113.038   The nature of Becklin's star.
M. V. Penston, D. A. Allen, A. R. Hyland.
Astrophys. Journ., (*Letters*), Vol. 170, L33 - L37 (1971).

It is suggested that Becklin's star is an extremely reddened star.

113.039   Calibration of the $uvby\beta$ system in terms of intrinsic color indices and absolute magnitude for B-type stars.   D. L. Crawford.
Publ. Astron. Soc. Pacific, Vol. 83, 604 (1971). − Abstr. Astron. Soc. Pacific.

113.040   Recent photometric observations of IRC stars.
G. W. Lockwood.
Publ. Astron. Soc. Pacific, Vol. 83, 606 (1971). − Abstr. Astron. Soc. Pacific.

113.041   Photometry of peculiar A stars with long periods.
S. C. Wolff, N. D. Morrison.
Publ. Astron. Soc. Pacific, Vol. 83, 609 - 610 (1971). − Abstr. Astron. Soc. Pacific.

113.042   A UBV equatorial-extinction star network.
D. L. Crawford, J. C. Golson, A. U. Landolt.
Publ. Astron. Soc. Pacific, Vol. 83, 652 - 655 (1971).

Well-observed data are presented for 18 pairs of stars around the celestial equator suitable for use as 'extinction stars' in UBV photometry.

113.043 $uvby\beta$ photometry of stars in the direction of the association Perseus OB 2.   A. E. Rydgren.
Publ. Astron. Soc. Pacific, Vol. 83, 656 - 662 (1971).

The interstellar absorption in the direction of the association Perseus OB 2 is studied by $uvby\beta$ photometry of 35 early-type stars in four fields. Most of the obscuration seems to be concentrated between 100 and 200 parsecs from the sun, rather than between 200 and 300 parsecs as previously suggested by Lynds (1969).

113.044   On the variability of the "blue objects".
L. Richter, N. B. Richter, W. Wenzel.
Astron. Nachr., Vol. 293 119 - 123 (1971).

The reality of the high percentage of variability among blue objects which was found in previous investigations was examined on 170 Tautenburg Schmidt plates of two test fields near M 31. Out of 37 blue objects examined, only two (van den Bergh 5 and 12) are distinctly variable.

113.045   Observations of OJ 287 between 0.36 and 3.4 μm.
H. M. Dyck, T. D. Kinman, G. W. Lockwood, A. U. Landolt.
Nature, Phys. Sci., Vol. 234, 71 - 72 (1971).

Rapid variations in the optical flux and linear polarization of OJ 287 are confirmed. In September and October 1971, observations in the wavelengths range 0.36 to 3.4 microns were consistent with OJ 287 having a power-law energy distribution $F_\nu = \nu^{-1.0}$.

113.046 Motion of A0 stars perpendicular to the galactic plane. V. (U, B, V) photometry of A stars in the south galactic cap.   J. B. Alexander, B. S. Carter.
Roy. Obs. Bull., Greenwich – Cape, No. 169, p. 353 - 361 (1971).

(U, B, V) photometry is presented for 165 A stars in the south galactic cap. For the majority of these stars, radial velocities and MK spectral types have been published in Paper IV (Roy. Obs. Bull., No. 165 (1971)). It appears that the mean interstellar reddening in the direction of the south galactic cap is small.

113.047 Four-color, Hβ, and UBV photometry for bright B-type stars in the northern hemisphere.
D. L. Crawford, J. V. Barnes, J. C. Golson.
Astron. Journ., Vol. 76, 1058 - 1071 (1971).

Photoelectric photometry is presented for 491 B-type stars north of Dec. $-10°$ and with $m_V \leqslant 6^m.5$.

113.048 UBV photometry of early-type stars in two regions at high galactic latitudes.   J. S. Drilling.
Astron. Journ., Vol. 76, 1072 - 1078 = Contr. Louisiana State Univ. Obs., *Baton Rouge*, No. 58 (1971).

Magnitudes and color indices on the UBV system have been determined for 125 stars near $l = 0°$, $b = -45°$ and for 85 stars near $l = 180°$, $b = -45°$ which are of spectral class A7 or earlier and brighter than $V = 14.6$ according to the finding lists of Philip and Drilling. The run of space density with z has been determined for stars of spectral classes A2-A7 in the two regions and is shown to be similar to that of A2-A7 stars in the vicinity of the north galactic pole.

113.049 The distances of two faint OB star groups in Monoceros.   J. A. Graham.
Astron. Journ., Vol. 76, 1079 - 1081 (1971).

UBV and Hβ photometry are presented for 11 stars in a field in Monoceros ($l = 218°$, $b = -0°.5$). Radial velocities have been measured for nine of the stars. The data suggest that there are two groups of OB stars in this direction, one at a distance of 2.4 kpc and one less well established at 3.8 kpc.

113.050 UBV photometry in the nuclear bulge of the Galaxy.   S. van den Bergh.
Astron. Journ., Vol. 76, 1082 - 1098, 1167 - 1171 (1971).

This paper reports new photoelectric and photographic photometry to $V \simeq 17.7$, $B \simeq 19.0$ and $U \simeq 18.3$ in Baade's low-absorption window that is centered on NGC 6522. The results are reported and discussed. The main thrust of the paper is directed towards study of the following problems: (i) What is the stellar population mix in the nuclear bulge of the Galaxy and (ii) what is the distance to the center of the Galaxy?

113.051 A finding list of faint blue stars in the anticenter region of the Galaxy.   V. C. Rubin, J. M. Losee.
Astron. Journ., Vol. 76, 1099 - 1101, 1173 - 1176 (1971).

Finding charts, coordinates, color, and magnitude estimates are given for 63 faint blue stars in the anticenter region of the Galaxy. The stars have been detected on a three-color plate taken with the Palomar 48-inch Schmidt telescope, following the method developed by Haro and Luyten. It is likely that some of these stars are OB main-sequence stars at great distances from the center of the Galaxy.

113.052 K-line photometry of southern A stars.
R. C. Henry, J. E. Hesser.
Astrophys. Journ., Suppl. Ser., No. 202, Vol. 23, 421 - 451

(1971).

Extension of the photoelectric measurements of the strength of the calcium K-line to 223 stars of predominantly southern or equatorial declinations and well distributed in right ascension has expanded the existent list to 369 field stars for which a k-index is available, including many more Am stars. All available k-index data for field stars are presented. A number of stars that probably exhibit spectral peculiarities are identified by means of K-line and intermediate-band photometry.

113.053 K-line photometry of stars in population I clusters.
J. E. Hesser, R. C. Henry.
Astrophys. Journ., Suppl. Ser., No. 202, Vol. 23, 453 - 476 (1971).

Photoelectric photometry of the K-line of calcium has been performed for the A stars of five open clusters (Hyades, Pleiades, IC 2391, IC 2602, and NGC 6475) and one association (Orion). From discussion of the observations in these two papers, evidence supporting three constraints on the metallicity anomaly has emerged.

113.054 The photometric classification of stars by the method of comparison of color-indices.   G. Kakaras.
Bull. Vilnius Astron. Obs., No. 32, p. 3 - 17 (1971). In Russian.

The mean values of normal color-indices of 430 standard stars with M_V from -8 to $+11$ and spectral classes from O to M6 were determined on the basis of observations of about 630 normal single stars in the Vilnius Observatory system UPXYZVS. Linear interpolation of color-indices between the neighbouring standard stars provides the minimum step of classification $0^m.5$ in M_V and 0.5 in spectral subclass. A photometric classification of about 630 stars used in the determination of normal color-indices was made.

113.055 Comparison of observed and computed color-indices of the Vilnius system and the system UBV.
G. Kavaliauskaité, V. Straižys, A. Ažusienis.
Bull. Vilnius Astron. Obs., Vol. 32, p. 18 - 33 (1971).
In Russian.

Computed color-indices are compared with observed ones for two photometric systems – the intermediate band photometric system UPXYZVS and the system UBV. The energy curves of about 350 stars determined by different authors were used.

113.056 Calculation of physical parameters according to the data of multi-colour photometric observations by the method of maximum likelihood.   R. Bartkus.
Bull. Vilnius Astron. Obs., No. 32, p. 34 - 41 (1971).
In Russian.

Calculation of the most probable values of physical parameters of a star with measured magnitudes in various colours is studied. In the case of linear dependence and Gaussian distribution of errors the method of maximum likelihood gives the system of linear equations which just permits to determine the parameters sought for.

113.057 L'extinction interstellaire et ses propriétés dans la photométrie en 7 couleurs. Système de l'Observatoire de Genève.   G. Goy.
Archiv. Sci. Genève, Vol. 23, 559 - 607 (1970) = Publ. Obs. Genève, Ser. A, Fasc. 78/I (1971).

Using the known stellar continua, the available extinction laws and the response curves, it is possible, with the aid of electronic computers, to simulate the color excesses by exact integration. Such a study makes it possible to avoid systematic errors as those which the UBV system, for example, is subjected. In the first part of this paper we study the conditions needed for a good simulation. The results are then compared with the seven-colors stars catalogue of the Geneva Observatory. In the second part, the question of the known parameters d, g, $m2$, Δ is reconsidered, but with respect to the interstellar reddening problem. Interesting properties of certain diagrams are shown which allow to discriminate individual pecularities in the different extinction laws.

113.058 Note sur la détermination de séquences de magnitudes apparentes. F. Rufener, A. Maeder.
Archiv. Sci. Genève, Vol. 23, 609 - 624 (1970) = Publ. Obs. Genève, Ser. A, Fasc. 78/II (1971).

This paper shows a method to get a good sequence of apparent magnitude [V] from semi-differential measurements.

113.059 Photographic photometry with calcite filter calibration.(a) Inherent errors in the method.
P. W. J. L. Brand.
Monthly Notices Roy. Astron. Soc., Vol. 153, 523 = Commun. Roy. Obs. Edinburgh, No. 116 (1971). — The full text of this paper has appeared in Publ. Roy. Obs. Edinburgh, Vol. 7, No. 3, 1971 (see 05.113.039).

113.060 Neutral oxygen observations in stars.
E. E. Mendoza V.
Bol. Obs. Tonantzintla y Tacubaya, No. 37, Vol. 6, 137 - 141 (1971).

We have performed photoelectric photometry for 31 stars in a narrow-band system, which allows the measurements of total absorption of neutral oxygen at $\lambda 7774$ Å. The results showed, that supergiant stars can be clearly separated from other luminosity classes, and that the total absorptions depend strongly on the stellar luminosity.

113.061 Three colour photometry in Auckland.
W. S. G. Walker.
Southern Stars, Vol. 24, 67 - 74 (1971).

113.062 On the determination of magnitudes with plates of the 40-cm astrograph. G. A. Starikova.
Peremennye Zvezdy, Vol. 18, 223 - 230 (1971). In Russian.

Application of Hardorp's method to the plates of the Sternberg Institute 40-cm astrograph is described. It is shown that this method permits to determine stellar magnitudes more precisely than the usual method of comparison.

113.063 Zonal spectrophotometric standards.
V. M. Tereshchenko.
Astron. Tsirk., No. 664, p. 1 - 2 (1971). In Russian.

113.064 Stars observed photoelectrically near quasars and related objects.
M. J. Penston, M. V. Penston, A. Sandage.
Publ. Astron. Soc. Pacific, Vol. 83, 783 - 799 (1971).

UBV photometry is listed for 59 stars near 18 quasars and two unidentified radio sources and for 54 stars near ten Seyfert and four N-type galaxies. Some of the stars are suitable for use as secondary photoelectric standards. In some fields, the sequences are extensive enough to be used for calibration of photographic monitoring programs. Finding charts are given for all listed stars.

113.065 The reddening of stars in the direction of RR Lyrae.
D. H. McNamara, S. K. Croft.
Publ. Astron. Soc. Pacific, Vol. 83, 828 - 829 (1971).

BV photoelectric observations of twelve field stars near RR Lyrae are combined with earlier (khg) photometry to derive the color excess in the direction of the variable. The observational data yield $E(B - V) = 0\overset{m}{.}02$.

113.066 Faint O—B2 stars in the Vela, Carina, Centaurus, and Crux sections of the Milky Way. E. W. Miller.
Separate print Dep. Astron. Steward Obs., Univ. Arizona,

Tucson. 14 + 15 pp. (1971). — Presented at the 135th meeting of the American Astronomical Society Amherst, Massachusetts.

113.067 Photometric standards for the southern hemisphere. II. B. J. Bok, P. F. Bok, E. W. Miller.
Separate print Steward Obs., Univ. Arizona, Tucson. 32 + 15 pp. (1971).

UBV photoelectric observations by Miller (1970) and by Bok and Bok (1971) at the Cerro Tololo Inter-American Observatory have resulted in seven new standard sequences for the southern Milky Way from Vela to Norma. Sequence stars have been added to the previously published (Paper I, 1969) standard sequences in Selected Area 193, Centaurus (now Centaurus I), Norma I, Norma II, and Norma III in order to enlarge and to extend the sequences to fainter magnitudes. Sequence stars in two of Loden's fields in Vela and one field in Carina have been observed, and the three sequences have been extended to fainter magnitudes. Identification charts for 15 sequence fields are included in the present paper.

Impulstaellingsteknik anvendt i fotoelektrisk fotometri. See Abstr. 031.019.

On the neutrality of terrestrial cloud extinction. See Abstr. 034.086.

Determination of atmospheric parameters for G and K giants by means of photoelectric indices. See Abstr. 064.021.

Stellar compositions from narrow-band photometry—I. Iron abundances in 180 G and K giants. See Abstr. 114.004.

Investigation of hot stars with excessive reddening in NGC 6913. See Abstr. 114.022.

V 1057 Cygni. See Abstr. 114.042.

Spectrophotometry of cool stars in the near infrared. II. Results for a region in the direction of the galactic anticentre. See Abstr. 114.073.

Two young bright infrared objects. See Abstr. 114.098.

Abundances in open clusters: F dwarfs in Coma. See Abstr. 114.104.

Carbon stars with strong C^{13} and lithium spectral features. See Abstr. 114.110.

On the red giants normal colour line on the (U—B)—(B—V) diagram. See Abstr. 115.003.

Luminosities and motions of the F-type stars. I. Luminosity and metal abundance indices for disk population stars. See Abstr. 115.017.

Intermediate-band photometry of RR Lyrae variables—I. Definition and characteristics of the system. See Abstr. 122.038.

UBV photometry of the β Cephei type variable stars. III. KP Persei (HD 21803). See Abstr. 122.119.

The color excess of Polaris. See Abstr. 122.134.

The nature of the blue stragglers in the old disk population. See Abstr. 122.158.

Wavelength dependence of polarization. XXIV. Infrared objects.   See Abstr. 131.008.

Étude de la radiosource 3C 173 et de certains astres du même champ.   See Abstr. 141.159.

Am stars as possible members of two subgroups of the Scorpius-Centaurus association.   See Abstr. 152.006.

uvby and Hβ observations of B-type stars in the Scorpio-Centaurus association.   See Abstr. 152.008.

Photometry of main-sequence stars in the Hyades and the field on the BVr system.   See Abstr. 153.017.

The masses of the 'horizontal-branch' stars in M67.   See Abstr. 153.022.

The blue stars above the turn-off in M67: Horizontal branch or blue stragglers?   See Abstr. 153.029.

114 Stellar Spectra, Temperatures, Spectroscopy

114.001 The colours and chemical composition of the
G dwarf HR 72.   D. Branch, R. A. Bell.
Monthly Notices, Roy. Astron. Soc., Vol. 153, 57 - 72 (1971).
One purpose of the present paper is to provide a detailed
analysis of the element abundances in HR 72. By using model
atmosphere techniques, by deriving separately abundances
based on neutral and ionized lines, and by including a greater
number of Fe I lines in our analysis, we provide a check on
the iron abundance given by the curve of growth analysis of
Spinrad and Luebke (1970). We have also derived relative
abundances of fourteen other elements. A second purpose
is to investigate the narrow- and broad-band colours of HR
72, 20 LMi and other stars with an ultra-violet deficiency
relative to the Hyades, such as δ Pav. We have used model
atmospheres and laboratory data on atomic and molecular
lines to compute theoretical broad- and narrow-band colours
and have compared these colours with the observed colours.

114.002 Analysis of the spectrum of the A0 III halo star
HD 106304.   A. Przybylski.
Monthly Notices, Roy. Astron. Soc., Vol. 153, 111 - 118
(1971).
In a coarse analysis the population II horizontal branch
star HD 106304 has been compared with α Lyrae. It has
been found that: 1. Helium is normal or nearly normal in
HD 106304; 2. Metals are deficient by a factor of 7.

114.003 Spectrum of Welch's red variable in Crux.
P. C. Keenan.
Monthly Notices, Roy. Astron. Soc., Vol. 153, 1P - 2P
(1971).
Spectrograms of the red variable discovered by Welch
in Crux in 1969 suggest O/C ≈ 1 and over-abundance of
heavy metals. It may be a member of the SC class recently
defined by Catchpole and Feast.

114.004 Stellar compositions from narrow-band photometry—I. Iron abundances in 180 G and K giants.
P. M. Williams.
Monthly Notices, Roy. Astron. Soc., Vol. 153, 171 - 193
(1971).
Using narrow-band indices observed with the Cambridge
spectrophotometer, red–infra-red colours, and independent
luminosity estimates, the iron abundances of 180 G and K
type giant stars have been determined. The indices were
analysed using synthetic spectra computed from model atmospheres. The abundances of a number of the possible members
of the Hyades, Pleiades, Wolf 630, ζ Herculis, σ Puppis, 61
Cygni, η Cephei and γ Leonis moving stellar groups are discussed.

114.005 Ultra-violet continuum brightnesses of stars
measured by a rocket-borne photoelectric spectrophotometer.   G. C. Sudbury.
Monthly Notices, Roy. Astron. Soc., Vol. 153, 241 - 249
(1971).
Photoelectric spectrophotometric measures of 41 early-
type stars in the range 1800-2800 Å and with 200 Å resolution have been obtained by precessing scans of an objective
dispersion monochromator. The present paper discusses these
data, comparing with other recent observations of spectra.
Additional information is also presented on the origin and
density of the interstellar flux in the neighbourhood of the
sun around 2400 Å.

114.006 Neutral-helium line strengths. IV. Fourteen

"normal" stars of population I.   J. Norris.
Astrophys. Journ., Suppl. Ser., No. 197, Vol. 23, 193 - 212
(1971).
For fourteen sharp-lined stars in the spectral range B0–
B8 the following observations have been obtained: continuum
measures over the wavelength interval 3400–5600 Å, profiles
of the hydrogen lines Hγ and Hδ, and line-strength measurements for twelve neutral-helium lines. These data are analyzed
by the use of model-atmosphere techniques and the assumption of LTE to determine the atmospheric parameters—
effective temperature, surface gravity, and helium abundance.
The helium line-broadening theories employed are those of
Griem, Baranger, Kolb, and Oertel; Griem; and Gieske and
Griem.

114.007 Neutral-helium line strengths. V. The weak-helium-
line stars of population I.   J. Norris.
Astrophys. Journ., Suppl. Ser., No. 197, Vol. 23, 213 - 233
(1971).
Effective temperatures, gravities and apparent helium
deficiencies have been determined for twelve weak-helium-
line stars. The first aim of this work is to provide a quantitative description of the helium weakness of these stars and the
behavior of other elements compared with normal stars. The
second is to relate them to the remainder of the Ap stars.

114.008 Neutral-helium line strengths. VI. The variations
of the helium spectrum variable α Centauri.
J. Norris.
Astrophys. Journ., Suppl. Ser., No. 197, Vol. 23, 235 - 255
(1971).
Spectra (λλ 3800–6700) and continuum measures
(λλ 3400–5600) have been obtained of α Cen during the
years 1967-1969. These have been used to determine the
variations of several observational parameters: hydrogen line
profiles, helium and metallic line strengths, continuum magnitudes and colors, and radial velocities.

114.009 The lithium content of Capella.
A. M. Boesgaard.
Astrophys. Journ., Vol. 167, 511 - 519 (1971).
The lithium content of the two components of the
spectroscopic binary α Aur is determined from seven spectrograms of 2 Å mm⁻¹ dispersion widened to 2 mm. A Li I line
of 22 mÅ was detected in the G star in addition to the Li I
line previously found in the F star by Wallerstein. An abundance analysis gives the result that the Li content of the F
star is 15 times greater than in the G star. Measurements made
to determine the Li isotope ratio indicate that there is little or
no ⁶Li in either star.

114.010 Some intrinsic properties of carbon stars.
H. B. Richer.
Astrophys. Journ., Vol. 167, 521 - 535 (1971).
Near-infrared spectrograms of moderate dispersion have
been employed to establish a new two-dimensional classification scheme for carbon stars which involves temperature and
luminosity. Intrinsic colors of the spectral subtypes were
determined, and these were employed to construct a color-
color diagram which clearly separated out the early- and late-
type stars. A galactic-rotation solution, using radial velocities
and intrinsic colors, yielded a mean absolute magnitude of
–2.7 for the middle carbon stars. This value was employed
to construct the (M_{bol}, log T_e)-diagram for these stars which
were found to form an extension of the normal giant branch
to cooler temperature.

114.011   Star S22 of the Large Magellanic Cloud showing emission lines of Fe II and [Fe II].
C. Fehrenbach.
Astron. Astrophys., Vol. 13, 437 - 443 (1971). In French.

We again find the emission lines of the following elements: H, Fe II and [Fe II] but we add the two lines of the multiplet 1 F of [S II] that are certain. Identification of [Ni II] is confirmed; we must add the following elements: Cr II, Si II, Ti II which we find with weak or rather weak lines. We think [Fe III] might also be present (line 4658). We find He I lines but we confront a problem because of certain weak lines that can be identified as He II lines, but their presence is difficult to understand.

114.012   New wavelengths' table for the determination of radial velocities of late type stars.
M. F. Chériguene.
Astron. Astrophys., Vol. 13, 447 - 453 (1971). In French.

We have established a new system of stellar lines in the spectral region 4800–5500 Å which permits us to obtain the radial velocities of the late type stars (F6–M2). Radial velocities of 23 standard stars, 7 stars in the Hyades cluster and 13 stars in Coma Berenice.

114.013   A catalogue of carbon stars in the Southern Milky Way.   B. E. Westerlund.
Astron. Astrophys., Suppl. Ser., Vol. 4, 51 - 74 (1971).

An objective-prism survey in infrared light has been carried out in the Southern Milky Way from l^{II} = 235° to l^{II} = 7° and between b^{II} = ±5°. The present catalogue gives positions and estimated visual and infrared magnitudes for 1124 carbon stars identified in the survey.

114.014   Effective temperatures and gravities for A- and F-type stars in the Delta Scuti region.
R. J. Dickens, A. J. Penny.
Monthly Notices Roy. Astron. Soc., Vol. 153, 287 - 302 (1971).

Effective temperatures and surface gravities are derived from spectrophotometry between 3500 Å and 8000 Å for 31 A- and F-type stars in or near the Delta Scuti region of the HR diagram. The stars include four Delta Scuti variables and ten stars known to be non-variable. Excellent agreement is obtained with temperatures derived from VRI photometry. A brief discussion of the properties of all known variable and non-variable stars in this region for which temperature, gravity and an independent measurement of the absolute magnitude are available is given.

114.015   Wolf-Rayet stars in H II regions.
D. Crampton.
Monthly Notices Roy. Astron. Soc., Vol. 153, 303 - 314 (1971).

A catalogue of WR stars coincident with H II regions is presented. Approximately 44 per cent of the WR stars surveyed are found to be 'possibly' associated with nebulosity while the association is considered to be 'probable' for 25 per cent of the cases. A list of WR stars in open clusters is also given.

114.016   Interstellar D-lines in the spectra of carbon stars, HD 182040, T Lyrae, and V Aquilae.
K. Utsumi, Y. Yamashita.
Publ. Astron. Soc. Japan, Vol. 23, 437 - 441 (1971).

Interstellar D-lines are found in the spectra of the carbon stars HD 182040, T Lyr, and V Aql. The distances of the stars, and then the absolute magnitudes are estimated from the equivalent widths of interstellar D-lines.

114.017   The possible identification of promethium in S stars.
D. N. Davis.

Astrophys. Journ., Vol. 167, 327 - 330 (1971).

Evidence is presented for Pm I and Pm II in V Cnc and T Sgr.

114.018   Computed and observed cyanide-radical spectra of three N stars in the infrared.   T. D. Faÿ, Jr.
Astrophys. Journ., Vol. 168, 99 - 107 = Publ. Goethe Link Obs., Indiana Univ., *Bloomington,* No. 128 (1971).

Synthetic spectra of the cyanide radical, CN, have been computed for the (1,0) and (0,2) band sequences at wavelengths near 1.0 and 2.3 μ, by means of a *Milne–Eddington* model. Vibrational and rotational temperatures of 3000°K were used, and five values of the $^{12}C/^{13}C$ ratio (1, 2, 5, 20, and 1000) have been considered. These spectra were then broadened with a running Gaussian profile to simulate the instrumental profiles of observed infrared spectra of RY Dra, Y CVn, and 19 Psc. The $^{12}C/^{13}C$ ratios that give the best agreement with infrared observations range from 2 to 5 for RY Dra and Y CVn, but the ratio is greater than 20 for 19 Psc.

114.019   The old disk metal-rich subgiant 31 Aquilae.
J. B. Hearnshaw.
Astrophys. Journ., Vol. 168, 109 - 114 (1971).

A differential coarse curve-of-growth analysis has been carried out with respect to the sun for the high-velocity field subgiant 31 Aql. The results show most metals to be 4 times as abundant as in the sun. The results for 31 Aql are contrary to the hypothesis that the oldest stars in the Galaxy should be metal poor.

114.020   μ Cassiopeiae and the primordial helium abundance – A critique.   J. Faulkner.
Phys. Rev. Letters, Vol. 27, 206 - 208 (1971).

An error in the application of Kepler's law vitiates the claim of Hegyi and Curott to have measured a helium abundance in μ Cas *A* significantly low in terms of cosmological theories. The accuracy required for a significant result is probably unattainable at this time.

114.021   New emission objects in the region α = 17^h56^m, δ = $-22°40'$.
V. P. Arhipova, O. D. Dokuchaeva.
Astron. Zhurn. Akad. Nauk SSSR, Vol. 48, 752 - 754 (1971). In Russian. English translation in Soviet Astron. AJ, Vol. 15, No. 4.

17 new objects having Hα in emission are found in the region with diameter 4°.5 near α = 17^h56^m, δ = $-22°40'$. Positions and identification charts of the stars are given. For 19 emission stars, previously known in the same region, the character of the Hα line is described.

114.022   Investigation of hot stars with excessive reddening in NGC 6913.   R. M. Raznik.
Astron. Zhurn. Akad. Nauk SSSR, Vol. 48, 755 - 759 (1971). In Russian. English translation in Soviet Astron. AJ, Vol. 15, No. 4.

Spectral and photoelectric observations of $8^m - 9^m$ stars of O9–B1 type of NGC 6913 were carried out.

114.023   A finding list of stars of spectral type A 5 and earlier in regions at high galactic latitudes. V. 1 HLF 3.   A. G. D. Philip, L. J. Relyea.
Bol. Obs. Tonantzintla y Tacubaya, No. 36, Vol. 6, 69 - 72 (1971).

An objective prism survey has been made in a 19.3 square degree region in the 1 HLF 3 area (l^{II} = 76°, b^{II} = 45°) continuing an investigation of galactic structure perpendicular to the galactic plane. A finding list containing positions and spectral types for 16 stars is presented with finding charts for the stars too faint to be included in the Bonner Durchmusterung.

114.024 **Distances to stars in the Perseus arm.**
D. Crampton.
Journ. Roy. Astron. Soc. Canada, Vol. 65, 175 (1971).
Abstr. RAS Canada.

114.025 **Some intrinsic properties of carbon stars.**
H. B. Richer.
Journ. Roy. Astron. Soc. Canada, Vol. 65, 182 - 183 (1971).
Abstr. RAS Canada.

114.026 **Determination of the temperature of M-type atmospheres from the molecular spectrum of TiO.**
A. V. Shavrina.
Astrometriya i Astrofiz., *Kiev*, Vyp. (No.) 12, (see 003.003), p. 26 - 32 (1971). In Russian.
General methods for determining the temperature of M-type stars from the spectrum of TiO molecule are discussed. The method of vibrational band intensities was applied to four M-stars (α Sco, δ Sge, R Lyr, α Her) and gave the values 2900°, 2800°, 2900°, 2700°K, respectively. The method of temperature determination from an analysis of the rotational structure of the (0,0) band of TiO γ-system is shown to be uncertain.

114.027 **Spectral classification of unbroadened low-dispersion stellar spectra. I. Catalogue of spectra of faint stars around NGC 6913.** V. I. Kuznetsov.
Astrometriya i Astrofiz., *Kiev*, Vyp. (No.) 12, (see 003.003), p. 32 - 40 (1971). In Russian.

114.028 **Spectral classification of unbroadened low-dispersion stellar spectra. II. New data on the structure of the Galaxy in the direction of NGC 6913 based on the additional list of faint stellar spectra.** V. I. Kuznetsov.
Astrometriya i Astrofiz., *Kiev*, Vyp. (No.) 12, (see 003.003), p. 40 - 44 (1971). In Russian.
New results are presented concerning the distribution of absorbing matter and stars in space at great distances from the sun towards the spiral branch of Car-Cyg.

114.029 **Observations of interstellar Ca-II, H and K, absorption lines of bright O- and B-type stars in the Orion region.** S. Isobe.
Tokyo Astron. Bull., Second Ser., No. 210, p. 2473 - 2479 (1971).
The radial velocities and the equivalent widths of interstellar Ca-II, H and K, absorption lines are obtained for nine bright O- and B-type stars in the Orion region. The dispersion of each spectrum is 1.32 Å/mm. The separation of the lines depends on the conditions of exposure.

114.030 **Two new carbon stars.** M. V. Dolidze.
Astron. Tsirk., No. 616, p. 7 - 8 (1971). In Russian.

114.031 **Spectra of symbiotic stars AG Peg and BF Cyg in the near infrared.** E. A. Kolotilov.
Astron. Tsirk., No. 624, p. 2 - 5 (1971). In Russian.

114.032 **Time variations in the emission lines of Of stars.**
R. J. Brucato.
Monthly Notices Roy. Astron. Soc., Vol. 153, 435 - 452 (1971).
Spectrograms covering the region between Hβ and Hγ at a reciprocal dispersion of 28 Å mm^{-1} have been obtained for the Of stars HD 108, HD 188001, HD 190429N, HD 192639, HD 210839, and, at 42 Å mm^{-1}, HD 66811. Equivalent widths have been measured for several spectral features, particularly N III $\lambda\lambda$4634–40–41 and He II λ4686. The strengths of these lines are shown to vary in several of these stars with time scales on the order of 10 min, with the most striking changes occurring at λ4686 in the earliest spectral types.

114.033 **On the energy distributions of main sequence stars.**
D. J. Stickland.
Monthly Notices Roy. Astron. Soc., Vol. 153, 501 - 520 (1971).
A wide range of MK standard main sequence stars have been observed with a spectrum scanner and spectral energy distributions derived. These have been corrected, where necessary, for the effects of interstellar reddening and line blanketing. The flux curves so obtained have been compared with the predictions of several recent sets of model stellar atmospheres through a scheme of parametrization in which the main characteristics of the curves are represented by a Balmer discontinuity parameter and three spectral gradients.

114.034 **Emission objects in Cygnus.** M. V. Dolidze.
Astron. Tsirk., No. 629, p. 6 - 7 (1971). In Russian.

114.035 **New C stars.** M. V. Dolidze.
Astron. Tsirk., No. 629, p. 8 (1971). In Russian.

114.036 **New MS stars.** M. V. Dolidze.
Astron. Tsirk., No. 632, p. 7 (1971). In Russian.

114.037 **New S stars.** M. V. Dolidze.
Astron. Tsirk., No. 632, p. 7 - 8 (1971). In Russian.

114.038 **An approximate Stark broadening formula for use in spectrum synthesis.** C. R. Cowley.
Observatory, Vol. 91, 139 - 140 (1971).
A simple relationship is obtained for the damping constant due to Stark broadening.

114.039 **Determinación de las abundancias de los elementos en la atmósfera de la estrella de alta velocidad 31 Aql.** M. E. Rego.
Urania Barcelona, Año 55, Nos. 271–272, p. 3 - 84 (1970).

114.040 **A study of carbon stars on the basis of R-N and C-system classification.** M. V. Dolidze.
Byull. Abastumansk. Astrofiz. Obs., No. 40, p. 29 - 38 (1971). In Russian.
On the basis of intercomparison of classification data of two spectral systems (R-N and C) for the carbon stars the properties of red, blue, and such stars with anomalous ratios of colours in the spectrum are considered.

114.041 **Measurement of the spectrophotometric temperature of stars of the spectral classes B and A with transition from the center of their disk to the limb.**
S. M. Azimov.
Soobshch. Shemakhinsk. Astrofiz. Obs. vyp. (No.) 5, p. 3 - 12 (1971). In Russian.

114.042 **V 1057 Cygni.** A. Alksnis.
Astron. Tsirk., No. 635, p. 4 - 5 (1971). In Russian.

114.043 **Spectral observations of M stars in the photographic infrared region.** T. E. Derviz.
Trudy Astron. Obs., *Leningrad*, Vol. 28 (= Uchenye Zapiski Leningr. Un-ta, No. 359 = Seriya Matem. Nauk, vyp. (No.) 47), p. 52 - 57 (1971). In Russian.
The grating coudé spectrograph of the 18″ reflector AZT-3 is described. Spectrograms in the photographic infrared region of a number of M-type stars are obtained. Line wavelengths and identifications are given.

114.044 **Spectroscopic observations of SC stars.**
R. M. Catchpole, M. W. Feast.
Monthly Notices Roy. Astron. Soc., Vol. 154, 197 - 208 (1971).
A search has been carried out for stars similar to the SC

type star UY Cen, and twelve are listed in a table. The six stars newly assigned to the group were selected for observation on the basis of an objective prism survey by Henize. The spectroscopic characteristics defining the group are discussed and the differences from CS stars like R CMi noted. Radial velocities are listed for eight SC stars (making a total of ten with known velocity). A rough estimate of the absolute magnitude ($M_v \sim -2$) and a velocity dispersion of 31 ± 7 km s^{-1} are derived.

114.045 On the estimate of metal abundances in F dwarfs from low dispersion spectra. V. D. Malyuto.
Publ. Tartu Astrofiz. Obs., Vol. 39, 69 - 81 (1971). In Russian.

The index δ (the ratio of mean equivalent width of two narrow bands on either side of H_δ to equivalent width of the narrow band centered on H_δ) is measured at H_γ for some F and G dwarfs. The correlations between d (vertical deviations from the relation between δ and $V - I$ for solar composition stars) and [Fe/H] and different indices of metal abundances are found.

114.046 Spectrophotometry of Wolf-Rayet stars HD 191765, HD 192163 and HD 192103.
T. Nugis, L. Luud.
Publ. Tartu Astrofiz. Obs., Vol. 39, 116 - 136 (1971).
In Russian.

By medium dispersion spectrograms the equivalent widths, intensities and half-widths of spectral lines in the interval 3600 - 4900 Å are determined. For the red region equivalent widths are determined using Underhill's published registrograms. The methods of determining electron and Zanstra temperatures are discussed, and temperatures found.

114.047 Spectrum of P Cygni in 1968 - 1969.
L. Luud, Ü. Ibrus, I. Kolka.
Publ. Tartu Astrofiz. Obs., Vol. 39, 307 - 320 (1971).

114.048 Abundances of heavy elements in late-type stars.
G. A. Bakos.
Journ. Roy. Astron. Soc. Canada, Vol. 65, 222 - 238 = Contr. Univ. Waterloo Obs., No. 8 (1971).

Twenty-one late-type stars of luminosity classes I, III and V have been analysed for abundances of selected elements. By the curve of growth method, an overabundance of heavy elements in both barium and supergiant stars has been found.

114.049 Photoelectric measurements of Hγ line strengths in early-type stars. P. A. Blanchard.
Bull. American Astron. Soc., Vol. 3, 368 (1971). – Abstr. AAS.

114.050 Measurements of cyanide and oxide band strengths in cool stars with a rapid scanner.
T. D. Fay, Jr., R. K. Honeycutt.
Bull. American Astron. Soc., Vol. 3, 379 (1971). – Abstr. AAS.

114.051 The calibration of stellar classification schemes.
E. B. Newell.
Bull. American Astron. Soc., Vol. 3, 381 (1971). – Abstr. AAS.

114.052 Rocket ultraviolet spectra of Delta Scorpii and Zeta Ophiuchi. T. A. Matilsky.
Bull. American Astron. Soc., Vol. 3, 400 (1971). – Abstr. AAS.

114.053 The unusual star, HD 45166.
L. H. Aller, S. Heap.
Bull. American Astron. Soc., Vol. 3, 400 - 401 (1971). Abstr. AAS.

114.054 Four southern A-type supergiants.
W. Buscombe.
Bull. American Astron. Soc., Vol. 3, 401 (1971). – Abstr. AAS.

114.055 Two-dimensional spectral classification of the southern HD stars. N. Houk.
Bull. American Astron. Soc., Vol. 3, 401 (1971). – Abstr. AAS.

114.056 Anomalous distribution of helium on the surface of the silicon Ap star CU Vir. V. L. Khokhlova.
Astron. Zhurn. Akad. Nauk SSSR, Vol. 48, 939 - 941 (1971). In Russian. English translation in Soviet Astron. AJ, Vol. 15, No. 5.

114.057 On the temperature scale of peculiar and metalline stars. Y. V. Glagolevsky, V. V. Leushin,
K. I. Kozlova, N. M. Chunakova.
Astron. Zhurn. Akad. Nauk SSSR, Vol. 48, 942 - 950 (1971). In Russian. English translation in Soviet Astron. AJ, Vol. 15, No. 5.

114.058 Observations of O and Of stars.
S. R. Heap.
Astron. Astrophys., Vol. 15, 77 - 89 (1971).

Spectrograms with a dispersion of 16.2 Å/mm covering the spectral range λ3300 to λ4900 were obtained for three early O stars and three early Of stars. Detailed line profiles of selected hydrogen and helium lines were obtained, and equivalent widths of nearly all visible lines were measured. Comparison of the spectroscopic properties between Of stars and O stars suggest that Of stars have lower surface gravities than do O stars. Comparison of their photometric properties confirms this suggestion.

114.059 Search for interstellar silicate absorption in spectrum of VI Cyg No. 12. W. A. Stein, F. C. Gillett.
Nature, Phys. Sci., Vol. 233, 72 - 73 (1971).

A search for absorption at $11\,\mu m$ in the energy distribution from the heavily reddened B star VI Cyg No. 12 has been made. No absorption at this wavelength that could be attributed to interstellar silicates has been found. It is concluded that a significant fraction of the grain material in interstellar space must be in the form of other materials.

114.060 The emission object AS 299. E. Hu.
IAU Circ., No. 2354 (1971).

114.061 MHα 208-92. F. Ciatti.
IAU Circ., No. 2359 (1971).

114.062 Galactic infrared astronomy. A. R. Hyland.
Proc. Astron. Soc. Australia, Vol. 2, 14 - 20 (1971).

This short review has covered some aspects of infrared techniques and astronomical observations of current importance. The following subjects are discussed: Common types of infrared observations; Infrared observations of late-type stars; Infrared excess in early type stars: (a) Emission line B stars; (b) Stars of high luminosity; (c) T Tauri stars; Infrared emission from H II regions and planetary nebulae; Observations of the galactic nucleus; Infrared observations of extragalactic objects.

114.063 Helium abundance determinations in main sequence B stars. R. W. Simpson.
Proc. Astron. Soc. Australia, Vol. 2, 27 - 30 (1971).

114.064 On the nature of Eta Carinae. K. Davidson.
Monthly Notices Roy. Astron. Soc., Vol. 154, 415 - 427 (1971).

Various details of the spectrum of η Carinae are found to be consistent with a model in which a compact H II region is photoionized by radiation from a very massive star whose surface temperature is about 30000°K. The infra-red spectrum is presumed to be thermal re-emission from dust surrounding the ionized region.

114.065 On the excitation mechanism of N III emission in the Of stars. R. J. Brucato, D. Mihalas.
Monthly Notices Roy. Astron. Soc., Vol. 154, 491 - 503 (1971).

A simplified statistical equilibrium calculation has been carried out to examine the mechanism of the N III $\lambda\lambda$ 4634-41 emission in Of stars. It is shown that the *Swings mechanism*, based on pumping in the ultra-violet $\lambda\lambda$ 374 and 452 transitions is probably correct. The possible importance of dielectronic recombinations to the $3d$ state is pointed out.

114.066 A lithium rich star in Sco—Cen. R. M. Catchpole.
Monthly Notices Roy. Astron. Soc., Vol. 154, 15P - 17P (1971).

The Sco—Cen star HD 113703 ft (K0Ve) is shown to have a high lithium abundance (log $N_{Li} \simeq$ + 2.8).

114.067 The role of magnetic fields in Ap stars.
P. A. Strittmatter, J. Norris.
Astron. Astrophys., Vol. 15, 239 - 250 (1971).

Evidence is presented in favor of the hypothesis that the abundance peculiarities in Ap stars are confined to the surface layers. The effects of rotational circulation, convection, accretion and mass loss on surface peculiarities are considered. The existence of a zone in the gravity-effective temperature plane within which surface peculiarities can occur is demonstrated and its boundaries shown to agree satisfactorily with the observational evidence. In general, however, a substantial surface magnetic field is required if surface peculiarities are to be preserved even in this region. It is suggested that evolution to the Ap state depends on the initial ratio of magnetic to rotational energy.

114.068 Opacity probability distribution functions for electronic systems of CN and C_2 molecules including their stellar isotopic forms. F. Querci, M. Querci, V. G. Kunde.
Astron. Astrophys., Vol. 15, 256 - 274 (1971).

The basis and techniques are presented for generating opacity probability distribution functions for the CN molecule (red and violet systems) and the C_2 molecule (Swan, Phillips, Ballik-Ramsay systems), two of the more important diatomic molecules in the spectra of carbon stars, with a view to including these distribution functions in equilibrium model atmosphere calculations. Comparisons to the CO molecule are also shown.

114.069 The spectrum of ζ Puppis (O5f).
B. Baschek, M. Scholz.
Astron. Astrophys., Vol. 15, 285 - 291 (1971).

We present new observations of the line spectrum of ζ Pup (O5f) from λ 3150 to λ 8600.

114.070 A group of peculiar stars in Centaurus.
L. O. Lodén.
Astron. Astrophys., Vol. 15, 332 - 333 (1971).
Research note.

114.071 Comments on the broadening of He I lines with forbidden components.
B. J. O'Mara, R. W. Simpson.
Astron. Astrophys., Vol. 15, 334 - 336 (1971).

A comparison between the observed and theoretical absorption profiles of the neutral helium lines λ 4471 and λ 4026 is carried out for the B0V star τ Sco for those parts of the line where deviations from local thermodynamic equilibrium are unimportant. There is considerable discrepancy between the theoretical and observed values of the flux near the peak of the forbidden component λ 4471. The agreement between theory and observation is better for the λ 4026 line.

114.072 A search for metal-deficient stars in the southern hemisphere. J. B. Alexander.
Monthly Notes Astron. Soc. Southern Africa, Vol. 30, 139 - 145 (1971).

114.073 Spectrophotometry of cool stars in the near infrared. II. Results for a region in the direction of the galactic anticentre. K. Nandy, F. Smriglio.
Publ. Roy. Obs. Edinburgh, Vol. 7, (No. 6), 73 - 83 (1971).

Spectral types and infrared magnitudes of M and C stars brighter than V = 15^{m}0 have been obtained for a region of about 10 square degrees in the direction of the galactic anticentre. The spectral classification is based on the relative strengths of molecular bands visible in the infrared in low dispersion objective prism spectra. It is found that the number of M stars fainter than infrared magnitude 10^{m}0 decreases sharply in this direction. The total visual extinction at the distance of 2.5 kpc is estimated to be over 5^{m}0 in this direction. From the study of the space distribution of M stars it is concluded that the Perseus arm may extend to ~2 kpc from the sun in the anticentre direction.

114.074 A closer look at interstellar Lyman-alpha absorption. E. B. Jenkins.
Astrophys. Journ., Vol. 169, 25 - 32 (1971).

Previous measurements of the equivalent widths of the interstellar Lα absorption in the spectra of hot stars observed on rocket flights in most cases have yielded considerably lower column densities of atomic hydrogen than those derived from low-resolution observations by the Wisconsin far-ultraviolet spectrometer aboard the OAO-A2 satellite. This disagreement has prompted a reexamination of some of the rocket data and the development of a maximum-likelihood analysis for the detailed fitting of theoretical absorption profiles to the observations. The rocket spectra of δ, ϵ, and ζ Ori and ζ Pup have been chosen for the more detailed reanalysis, and column densities were derived for the respective stars. The present results for the four stars are still about a factor of 10 less than the OAO measurements. Hence the blending of strong stellar lines into the Lα profile observed by the OAO appears for these stars to be more severe than previously anticipated.

114.075 Millimeter observations of CO, CN, and CS emission from IRC+10216. R. W. Wilson, P. M. Solomon, A. A. Penzias, K. B. Jefferts.
Astrophys. Journ., *(Letters)*, Vol. 169, L35 - L37 (1971).

We report the discovery of three additional molecular lines from an object in which we have previously detected carbon monoxide. The data presented tend to support our model of a small expanding shell with high brightness temperature.

114.076 Technetium in the Mira variable RZ Pegasi.
B. F. Peery, Jr., P. C. Keenan, I. R. Marenin.
Publ. Astron. Soc. Pacific, Vol. 83, 496 = Publ. Goethe Link Obs., Indiana Univ., *Bloomington*, No. 129 (1971).

Absorption lines of the unstable element technetium appear in considerable strength in the spectrum of RZ Pegasi (P = 439^d) as photographed with the 200-inch coudé spectrograph at Mount Palomar on 18 September 1970, three weeks after maximum.

114.077 The incidence of metallicism among mid A-type

stars. M. A. Smith.
Astron. Journ., Vol. 76, 896 - 900 (1971).

Quantitative measures of the Ca II K-line strength and of a Sc II/Sr II line ratio are taken from a random sample of 70 bright mid-to-late A stars to determine the incidence of Am stars for this spectral region. It is found that some 31 % are Am by our criteria, 10 % are Ap, and at most only 59 % are "normal". This incidence for metallicism peaks in the center of the Am spectral domain, at A6–A7. The merging of our indices between the Am and normal A groups implies that a transitional rather than a discrete boundary exists between them.

114.078 **Calculation of C III lines for the vacuum ultraviolet region of B stars.** T. Feklistova.
Izv. Akad. Nauk Ehstonskoj SSR, Vol. 20, (Fiz., Mat., 1971, No. 3), 289 - 294 = Tartu Astron. Obs. Teated, No. 34. In Russian.

Probabilities of spontaneous electric dipole transitions and oscillator strengths of C III ions for the wavelength region 1425 - 1620 Å are calculated for the lines established by D. C. Morton and L. Spitzer from rocket observations of δ and π Scorpii. The equivalent widths of these lines in the spectra of B stars are estimated.

114.079 **Experiments with the digital reduction of stellar spectrograms.** W. K. Bonsack.
Astron. Astrophys., Vol. 15, 374 - 382 (1971).

Experiments have been made with techniques of computer processing of data resulting from the measurement of stellar spectrograms with a digital microphotometer. Fairly simple prescriptions have been developed for grain noise smoothing and the removal of instrumental broadening. The summing of underexposed spectrograms is investigated as a possible substitute for multinight exposures on faint objects.

114.080 **On some structural characteristics of WR star spectra. I. A study of the infrared spectra of C II, C III, C IV.** A. A. Nikitin.
Vestn. Leningr. un-ta, 1971, No. 1, p. 134 - 146. In Russian. Abstr. in Referativ. Zhurn. 51. Astron., 9.51.470 (1971).

114.081 **Spectrophotometric studies of non-stable stars. II. On the spectrum of RW Aurigae in the region 3080 –6100 Å.** D. Chalonge, L. Divan, L. V. Mirzoyan.
Astrofizika, Vol. 7, 345 - 362 (1971). – Reprinted in Astrophysics, Vol. 7, No. 3.

The results of a spectrophotometric study of RW Aur based on five short dispersion spectra covering the region 3080–6100 Å are presented.

114.082 **On the motion of matter in the envelope of P Cygni.** E. R. Astafev.
Astrofizika, Vol. 7, 377 - 387 (1971). In Russian. – English translation in Astrophysics, Vol. 7, No. 3.

The displacement velocities of external sections of absorption lines of P Cygni from five spectrograms of high dispersion have been measured. The velocities, normalized according to the intensity of lines, show direct dependence on the ionization potential.

114.083 **On the role of the collisionally broadened wings of strong lines in the blanketing problem.**
A. Natta, A. Preite-Martinez.
Astrophys. Space Sci., Vol. 13, 148 - 153 (1971).

The importance in the blanketing effect of the wings of strong lines broadened by collisions is evaluated. It is found that in stars of spectral types F5–K5 the cumulative absorption by a group of lines is due half to the center of the lines and half to the wings. For the same stars it is evaluated the influence of the collisional broadening on the computed color indices (U–B), (B–V).

114.084 **Nuclear and non-nuclear processes in the production of peculiar A stars.** B. N. G. Guthrie.
Astrophys. Space Sci., Vol. 13, 168 - 179 (1971).

The aim of the present paper is to find out which processes are most important for each kind of Ap star. The results of previous discussions of nuclear processes in the context of the supernova theory are summarized. Magnetic accretion and diffusion processes are considered.

114.085 **On criteria to detect new binaries among Wolf-Rayet stars.** V. N. de Monteagudo, J. Sahade.
Observatory, Vol. 91, 220 - 221 (1971). – Letter.

114.086 **Quantitative Analyse des Spektrums des Metalllinien-sterns 63 Tauri.** E. Hundt.
Diss. Ruprecht-Karl-Univ., Heidelberg. 3 + 57 + 38 pp. (1971).

114.087 **Curve-of-growth analysis of the spectrum of Procyon.** R. Griffin.
Monthly Notices Roy. Astron. Soc., Vol. 155, 139 - 152 (1971).

The spectrum of Procyon (F 5 IV–V) is investigated in the region λλ 4000–7500 Å from spectrograms having reciprocal dispersions of 1 to 1.5 Å mm^{-1}. A differential curve-of-growth analysis with respect to the sun confirms that most elements in the atmosphere of Procyon have abundances, relative to hydrogen, similar to those in the sun, although heavy elements ($Z \gtrsim 50$) appear to be slightly deficient. Particular problems which arise in the curve-of-growth comparison of these two stars are discussed.

114.088 **Stellar compositions from narrow-band photometry – II. Sodium and manganese in the Hyades and field giants.** P. M. Williams.
Monthly Notices Roy. Astron. Soc., Vol. 155, 215 - 229 (1971).

Using narrow-band indices observed with the Cambridge spectrophotometer, and the techniques described in Paper I, the sodium abundances of about 160 and manganese abundances of about 120 G and K giants have been determined. The r.m.s. external errors in [Na/H] and [Mn/H] are found to be 0.20 and 0.27 respectively. The compositions of a number of super-metal-rich stars are discussed. The sodium-richness, relative to iron, of ε Vir and the Hyades giants is confirmed, while the Hyades dwarfs appear to have normal sodium abundances.

114.089 **Departures from local thermodynamic equilibrium in the neutral helium lines of early type stars.**
A. G. Hearn.
Monthly Notices Roy. Astron. Soc., Vol. 155, 3P - 5P (1971).

Recent, apparently conflicting, theoretical papers are discussed together with recent observations. It appears that there is in fact agreement in general terms that in B type stars some strong lines of neutral helium are affected by departures from local thermodynamic equilibrium which cause serious errors in the determination of the helium abundance, while in some other lines these effects are negligible.

114.090 **A possible identification of Nd II in emission in Arcturus.** R. A. E. Fosbury.
Monthly Notices Roy. Astron. Soc., Vol. 155, 7P - 10P (1971).

A feature in the wing of the K-line in Arcturus is identified with an emission line of the singly ionized Lanthanide rare-earth Nd II. The occurrence of this line is briefly discussed in terms of Canfield's analysis of solar rare-earth lines.

114.091 **The Hg II line in HR 465.**
C. R. Cowley, M. F. Aller.
Astrophys. Letters, Vol. 9, 159 - 160 (1971).

The shape of the Hg II line λ3984 in HR 465 changed

markedly from 1961 to 1969. The interpretation of this observation is not unique, but an explanation in terms of surface nuclear activity is a distinct possibility.

114.092 Abundances of the elements in Sirius and Merak.
D. W. Latham.
SAO, *Cambridge, Mass.*, Special Report, No. 321, 7 + 37 + 43 + 9 + 19 pp. (1970). – Presented as a thesis to the Department of Astronomy, Harvard University, August 1969.

New instrumentation and techniques for measuring spectral-energy distributions and line strengths were developed and then applied to obtain the necessary observations for a model-atmosphere analysis of the abundances in Merak (β UMa) relative to Sirius (α CMa).

114.093 What is peculiar about A stars?   W. Buscombe.
Astron. Soc. Pacific, Leaflet No. 508, 8 pp. (1971).

114.094 Análisis cualitativo del espectro de la estrella peculiar HD 18474.   M. J. Fernández-Figueroa.
Rev. Real Acad. Cie. Exactas, Fis., Nat., Madrid, Vol. 65, 113 - 164 = Univ. Madrid, Fac. Cie., Seminario Astron. Geod., Publ. No. 68 (1971).

The $\lambda\lambda$ 400–500 Å region has been identified in the spectrum of the G4, III giant peculiar HD 18474, spectrograms of 2.9 Å/mm have been used. Relative intensities have been assigned to each line. An important weakness has been observed in the intensities of the $B^2 \Sigma \rightarrow X^2 \Sigma$ system of CN band at $\lambda\lambda$ 4216 Å, of the G-band of CH at $\lambda\lambda$ 4300 Å and also of the Swan system corresponding to $v' - v'' = +1$, $\lambda\lambda$ 4737 (1, 0). The BaII lines are stronger than in normal stars of same spectral type.

114.095 Erratum: 'Abundances of heavy elements in late-type stars' [Journ. Roy. Astron. Soc. Canada, Vol. 65, 222 - 238 (1971)].   G. A. Bakos.
Journ. Roy. Astron. Soc. Canada, Vol. 65, 303 (1971).

114.096 Energy distributions and spectra of Orion B stars.
R. E. Schild, F. Chaffee.
Astrophys. Journ., Vol. 169, 529 - 536 (1971).

New MK spectral types and energy distributions are presented for B stars in Orion for which far-ultraviolet flux excesses have recently been discovered. Significant differences between HD spectral types and the new MK types are found. Moreover, the energy distributions show the Orion late B stars to have smaller Balmer discontinuities than do field stars of the same spectral types. For the late B stars, these effects cause the 1500 Å fluxes to be underestimated by approximately 0.5 mag. No comparable systematic effects were found for the early B stars.

114.097 The spectrum of LkHα-101 in the near-infrared.
G. H. Herbig.
Astrophys. Journ., Vol. 169, 537 - 541 = Contr. Lick Obs., No. 337 (1971).

LkHα-101 is the m_{pg} = 17 star that appears to be the source of illumination of NGC 1579. In the near-infrared, the spectrum contains many narrow emission lines of H, O I, [O II], [Fe II], [Cr II], ... , together with a number of unidentified features that were found in η Car by Thackeray. The emission-line spectrum is quite unlike that of the T Tauri stars, or of VY CMa. The distance, inferred from *UBV* data for two B-type stars lying in the same dark cloud, is at least 800 pc.

114.098 Two young bright infrared objects.
M. Cohen, N. J. Woolf.
Astrophys. Journ., Vol. 169, 543 - 547 (1971).
Infrared photometric observations between 2 μ and 22 μ are reported for LkHα-101 and LkHα-190 = V 1057 Cygni.

LkHα-101 is currently the brightest infrared source known at the center of a cometary nebula. The spectral distribution is a broad peak resembling a blackbody of ~750° K. Lick Hα-190, resembling FU Ori, is about 4 mag fainter than LkHα-101 at 3–5 μ, but is equally bright at 20 μ. The energy distribution from 2 to 5 μ is interpreted as a reddened continuum. A double peaked structure at 10 and 20 μ is interpreted as a moderately thick silicate shell at ~110° K.

114.099 Position measurements of main-line OH/infrared stars.   E. G. Hardebeck, W. J. Wilson.
Astrophys. Journ., (*Letters*), Vol. 169, L123 - L126 (1971).

Interferometer position measurements are reported for five main-line OH emission sources associated with infrared stars. The results of the measurements show that in all cases the 1667-MHz OH emission originates from the stellar position, with typical errors of ± 5 to ± 15" of arc. In addition to the observations of the OH/infrared stars, the position of the main-line OH source OH 2019+ 37 (ON−2) was measured. This source was found not to coincide with the red star BC Cyg as previous authors have suggested, but instead to be associated with an H II region.

114.100 G61−29, a helium emission-line star.
E. M. Burbidge, P. A. Strittmatter.
Astrophys. Journ., (*Letters*), Vol. 170, L39 - L42 (1971).
Spectrograms of G61−29 reveal that it is a helium emission-line star, probably a white dwarf. The spectrum is described, and possible interpretations are briefly discussed.

114.101 Some spectroscopic characteristics of the OB stars: An investigation of the space distribution of certain OB stars and the reference frame of the classification.
N. R. Walborn.
Astrophys. Journ., Suppl. Ser., No. 198, Vol. 23, 257 - 282 (1971).

The distances of stars from the Victoria list of revised Hγ spectrophotometric absolute magnitudes have been reinvestigated, by means of MK spectral classification at 63 Å mm^{-1}. The two systems are compared with respect to spectral types, absolute magnitudes, and space distribution of the individual stars; marked systematic and accidental differences are found to exist for the O9−B2 stars. In terms of the present classifications, it is found that the OB stars investigated delineate the Local and Perseus spiral arms. Because the work was done with a dispersion approximately twice that of the MK atlas, the spectral classification has been investigated in detail. With a reference frame based upon ratios of the lines of helium and silicon, an increased classification resolution in the range O9−B1 is obtained. Also, a luminosity classification for the earlier O stars is proposed.

114.102 Infrared observations of some early-type shell stars.
D. A. Allen.
Publ. Astron. Soc. Pacific, Vol. 83, 602 (1971). – Abstr.
Astron. Soc. Pacific.

114.103 N III and C III emission in Of stars.
H. Nussbaumer.
Astrophys. Journ., Vol. 170, 93 - 108 (1971).

It has been shown that absorption of continuous stellar radiation in Of stars, rather than the Bowen mechanism, is responsible for the N III emission. The apparent contradiction of emission in N III $3d \rightarrow 3p$ but absorption in $3s \rightarrow 3p$ is due to the atomic configuration. The approach that was successful in interpreting the N III features also explains, though by a more involved pattern, the occurrence of C III λ5696 in emission with a simultaneous absence of absorption in C III λ4649.

114.104 Abundances in open clusters: F dwarfs in Coma.

F. H. Chaffee, Jr.
Publ. Astron. Soc. Pacific, Vol. 83, 603 - 604 (1971). – Abstr. Astron. Soc. Pacific.

114.105 **Determination of the abundances of the rare earths in the Ap star β Coronae Borealis.**
J. Hardorp, S. S. Shore.
Publ. Astron. Soc. Pacific, Vol. 83, 605 (1971). – Abstr. Astron. Soc. Pacific.

114.106 **The relative isotopic abundances of mercury in mercury stars.**
G. W. Preston, A. H. Vaughan, R. E. White, J.-P. Swings.
Publ. Astron. Soc. Pacific, Vol. 83, 607 (1971). – Abstr. Astron. Soc. Pacific.

114.107 **The chromosphere of Arcturus.** T. Simon.
Publ. Astron. Soc. Pacific, Vol. 83, 607 - 608 (1971).
Abstr. Astron. Soc. Pacific.

114.108 **The periodic variations of the peculiar A-type star HD 111133.** S. C. Wolff, R. J. Wolff.
Publ. Astron. Soc. Pacific, Vol. 83, 610 (1971). – Abstr. Astron. Soc. Pacific.

114.109 **On the presence of He^3 in the photosphere of Rho Leonis.** G. Wallerstein.
Publ. Astron. Soc. Pacific, Vol. 83, 664 - 666 (1971).

From measurement of 17 spectrograms we find no evidence for the He^3 that was previously reported to be present in ρ Leonis. The profile of $H\alpha$ was found to show emission on one occasion and to be extremely broad and shallow at another time.

114.110 **Carbon stars with strong C^{13} and lithium spectral features.** C. P. Gordon.
Publ. Astron. Soc. Pacific, Vol. 83, 667 - 673 = Contr. Five College Obs., *Amherst, Mass.*, No. 101 (1971).

This paper describes the spectral characteristics of the type J subclass of carbon stars, in particular noting the positive correlation between strong C^{13} and lithium features. We list four new members of this class and give a new definition of this class suitable for spectra of moderate dispersion.

114.111 **The half-widths of stellar $H\alpha$ profiles deduced from spectrograms obtained with the McMath solar telescope.** J. C. LoPresto.
Publ. Astron. Soc. Pacific, Vol. 83, 674 - 676 (1971).

Profiles of $H\alpha$ were obtained photoelectrically for some late-type stars. The relationship between the absolute magnitude and the half-width of the Doppler core of $H\alpha$ is better established than that with the half-width of the total profile found by Kraft, Preston, and Wolff in 1964.

114.112 **γ Corvi and the rotation of the Hg-Mn stars.**
A. Cowley, C. Cowley.
Publ. Astron. Soc. Pacific, Vol. 83, 689 (1971).

γ Corvi is shown to be a Hg-Mn star with moderate rotation making the detection of the peculiar features difficult.

114.113 **Interstellar gas in the direction of the Vela pulsar.**
G. Wallerstein, J. Silk.
Astrophys. Journ., Vol. 170, 289 - 296 (1971).

High-velocity interstellar Ca II components have been found in two stars within three degrees of the Vela pulsar, 0833−45. One component, at −172 km s^{-1} relative to the local standard of rest, is believed to be the highest velocity hitherto observed in an interstellar absorption line. These lines appear to be associated with the Vela X, Y, Z, supernova-remnant complex. The measured velocity enables us to derive

significant new limits on the age of the remnant and the initial kinetic energy associated with the supernova outburst, together with a self-consistent interpretation of the associated soft X-ray emission.

114.114 **Lower limit to the interstellar $^{12}C/^{13}C$ ratio in the direction of 20 Tauri.**
P. Vanden Bout, P. Thaddeus.
Astrophys. Journ., Vol. 170, 297 - 298 (1971).

The spectrum of 20 Tau has been photoelectrically scanned in the vicinity of λ4232 of interstellar CH$^+$. From the observed absence of the isotopically-shifted line of ^{13}CH$^+$ it is deduced that the $^{12}C/^{13}C$ isotopic abundance ratio in this direction is greater than 33, and hence consistent with the terrestrial value.

114.115 **Spectroscopic studies of O-type stars. I. Classification and absolute magnitudes.**
P. S. Conti, W. R. Alschuler.
Astrophys. Journ., Vol. 170, 325 - 344 = Contr. Lick Obs., No. 340 (1971).

Our intention in this series of papers is to offer a homogeneous set of spectra of many O and Of stars to describe the classification, derive the absolute-magnitude scale and discuss the absorption and emission line strengths. Particular longstanding problems we would like to solve include the relation between the O and Of stars, the definition of a normal O star, and the isolation of those stars with anomalous spectra. This paper will discuss the spectral classification and absolute magnitude scale. Later papers in this series will consider other topics, including questions of the temperature scale, and measures of the radial velocities of O stars.

114.116 **Ultrarapid activity at $H\alpha$ in the spectra of Be stars.**
J. B. Hutchings, J. R. Auman, A. C. Gower, G. A. H. Walker.
Astrophys. Journ., (*Letters*), Vol. 170, L73 - L76 (1971).

High-resolution spectra of some Be stars, taken with an Isocon television camera, show changes in the $H\alpha$ line profile in times of the order of a minute or less.

114.117 **Observations of interstellar Ca I lines.**
L. M. Hobbs.
Astrophys. Journ., (*Letters*), Vol. 170, L85 - L88 (1971).

Interferometric, photoelectric scans of the interstellar Ca I line at λ4226 are reported for four stars having strong interstellar Ca II K-lines. The Ca I line is detected in three, and possibly all four, cases, of which one was known previously. Negative Ca I results are also obtained for six stars with weaker K-lines. The significance of the resulting electron densities deduced from the Ca I and Ca II lines is discussed.

114.118 **A comment on the interpretation of the broad component of N III λλ4634−4640 emission in Of stars.**
D. Mihalas.
Astrophys. Journ., Vol. 170, 541 - 545 (1971).

The stabilizing transition from the N III autoionizing term $2s2p(^1P^0)3d\ ^2F$ to the bound double-excitation term $2s2p(^1P^0)3p\ ^2D$ gives rise to a very broadened doublet at the wavelengths 4623 and 4630 Å. A study of the rate at which the process occurs suggests that this stabilizing transition may appreciably contribute to the *broad* emission ("band") component near λλ4634−4640 as observed by Wilson and Underhill.

114.119 **HCN in IRC + 10216.** M. Morris, B. Zuckerman, P. Palmer, B. E. Turner.
Astrophys. Journ., (*Letters*), Vol. 170, L109 - L112 (1971).

$H^{12}C^{14}N$ and $H^{13}C^{14}N$ were observed and $H^{12}C^{15}N$ was searched for in the infrared object IRC + 10216. The [^{13}C] [^{14}N] / [^{12}C] [^{15}N] ratio is different from the terrestrial one. If we assume that the HCN lines originate in the same region as CO lines from this source, then the $^{12}C/^{13}C$ abundance ratio is ~15.

114.120 OAO observations of magnesium II emission in late-type stars. L. R. Doherty.
Phil. Trans. Roy. Soc. London A, Vol. 270, No. 1202, (see 012.009), 189 - 195 (1971).

114.121 The interpretation of space observations of stars and interstellar matter. D. C. Morton.
Highlights of Astronomy, Vol. 2, (see 012.017), 466 - 475 (1971).

114.122 The scale of effective temperatures and bolometric corrections. M. Kubiak.
Postępy Astron., Vol. 19, 285 - 298 (1971). In Polish.
The article gives a short historical review of the earlier attempts to establish the scales of effective temperatures and bolometric corrections for the stars. It contains also the effective temperatures obtained by the author for B-type stars from the spectrophotometric observations and the revised effective temperatures scale based on the interferometric observations by Hanbury Brown et al.

114.123 The variability of emission lines in Wolf-Rayet stars. A. A. Gusejn-zade, M. B. Babaev.
Izv. AN AzSSR. Ser. fiz.-tekhn. i mat. n., 1971, No. 1, p. 79 - 81. In Russian. – Abstr. in Referativ. Zhurn. 51. Astron., 1.51.567 (1972).

114.124 Variations of the Hβ line profile in the spectrum of γ Cas. I. D. Kupo.
Trudy Astrofiz. Inst., *Alma-Ata*, Vol. 16, 65 - 69 (1971). In Russian.
Hβ emission profiles and equivalent widths in the spectrum of γ Cas obtained during 1965 - 1968 are reported. Rapid changes during some ten minutes in both profiles and equivalent widths, similar to the variations found by Hutchings for the line H$_\gamma$, were detected. Rapid variations were observed side by side with a systematical increase of the equivalent widths of Hβ.

114.125 Zonal spectrophotometric standards. I. Selection of stars and method of their study.
V. M. Tereshchenko, A. V. Kharitonov.
Trudy Astrofiz. Inst., *Alma-Ata*, Vol. 17, 40 - 50 (1971). In Russian.
A list of 109 selected zonal star standards, which covers the celestial sphere from −15° to +90° in declination, and methods of study of the extra-atmospheric energy distribution in their spectra are presented.

114.126 Levé spectrophotométrique du ciel dans l' ultraviolet. Détermination de la trajectoire de l' axe optique. Méthode statistique. E. Blondelot.
Centre Univ. Mons, Fac. Sci., Dép. d' Astrophys., Commun. No. 20, 9 pp. (1971).

114.127 Levé spectrophotométrique du ciel dans l' ultraviolet. Counting phase determination from the photometric channel. E. Blondelot.
Centre Univ. Mons, Fac. Sci., Dép. d' Astrophys., Commun. No. 21, 9 pp. (1971).

114.128 New variable carbon stars near the cluster NGC 7419. I. Daube.
Astron. Tsirk., No. 661, p. 6 - 7 (1971). In Russian.

114.129 Catalogue and bibliography of B type emission line stars. C. Jaschek, L. Ferrer, M. Jaschek.

Obs. Astron. Univ. Nacional La Plata, Ser. Astron., Vol. 37, 4 + 69 pp. (1971).
The catalogue is divided in three parts. Section I is the main catalogue; section II gives Be stars in clusters, associations or special regions and section III contains the bibliographic references.

114.130 Absolute distribution of energy in the spectra of 8 bright stars.
A. V. Kharitonov, V. M. Tereshchenko.
Astron. Tsirk., No. 664, p. 2 - 3 (1971). In Russian.

114.131 Spectrophotoelectric study of the emission lines λλ 4686, 4861 and 5411 of the Wolf-Rayet star HD 192163. T. Nugis, T. Jevsejenko.
Astron. Tsirk., No. 664, p. 3 - 5 (1971). In Russian.

114.132 Forbidden lines λ6548 [N II] and λ6583 [N II] in the spectrum of T Tau? I. R. Salmanov.
Astron. Tsirk., No. 666, p. 5 - 6 (1971). In Russian.

114.133 A high dispersion spectral analysis of the Ba II star HD 204075 (ζ Capricorni). J. L. Tech.
National Bureau of Standards, Washington, D.C. NBS Monograph 119, 170 pp. (1971).
A double differential curve of growth analysis using both the sun and ε Virginis (G9 II-III) as comparison stars, has been performed for the Ba II star ζ Capricorni. Atmospheric abundances have been derived for 37 elements. The results obtained with respect to the two comparison stars are in good agreement.

114.134 The ultraviolet and visual spectra of Gamma Cassiopeiae. R. C. Bohlin.
Thesis, Princeton Univ., N.J. [Available from Univ. Microfilms, Ann Arbor, Mich., U.S.A. Order No. 70–23602], 47 pp. (1970). – See Phys. Abstr., Vol. 74, No. 62791 (1971).

114.135 High resolution observation of stellar and interstellar lithium. H. E. Utiger.
Thesis, Univ. Wisconsin, Madison. [Available from Univ. Microfilms, Ann Arbor, Mich., U.S.A. Order No. 71–5667], 137 pp. (1971). – See Phys. Abstr., Vol. 74, No. 62792 (1971).

114.136 Diffusion of elements in Ap and Am stars. Y. Osaki.
Astron. Herald, (*Japan*), Vol. 64, 69 - 73 (1971). In Japanese.
A review of the recent theories including results obtained by the author.

114.137 Peculiar and metallic-like A-type stars. K. Nariai.
Astron. Herald, (*Japan*), Vol. 64, 63 - 64 (1971). In Japanese.
Describes the classification of the Ap's and the Am's by their spectra.

114.138 Early-type peculiar stars and blue stragglers in star clusters. F. Imagawa.
Astron. Herald, (*Japan*), Vol. 64, 64 - 66 (1971). In Japanese.
The abnormal position of the blue stragglers on the H-R diagram is explained by several theories on its origin. In relation to the blue stragglers the author presents a summary of the Ap and Am stars in star clusters, star groups and associations.

114.139 Observations of ultraviolet stars from a stratospheric balloon. M. Lehmann.
Sci. Industries Spatiales, (*France*), Vol. 7, No. 3, p. 7 - 22 (1971). In German and French.
Describes the observations of stellar spectra in the range

2-5000Å and in an area of sky 80° × 12° in the Orion-Gemini region. The measurements were taken from a stabilized platform attached to a balloon at a height of 39.6 kilometers.

114.140 Stellar spectra (spectres stellaires).
G. Cayrel de Strobel.
Trans. IAU, Vol. XIVB, (see 05.003.010), p. 189 - 190 = Radcliffe Obs. Repr. No. 99 (1971). − Report of meetings, Brighton, 19 and 25 August 1970, of Commission 29.

·114.141 Time scales for Ca II emission decay, rotational braking and lithium depletion.   A. Skumanich.
Preprint High Altitude Observatory, National Center Atmosph. Res., Boulder, Colorado. 8 pp. (1971). − To be published in Astrophys. Journ., Vol. 171 (1972).

Centre de Données Stellaires. Inform. Bull. No. 2. See Abstr. 002.026.

Carbon stars.   See Abstr. 003.004.

Pressure broadening of UV lines. See Abstr. 022.102

Phase modulation in far infrared (submillimetre-wave) interferometers. I−Mathematical formulation. See Abstr. 031.062.

Quantitative analysis of the atmospheres of two F stars.   See Abstr. 064.010.

The violet opacity in S, C-S and N stars and circumstellar silicon carbide grains.   See Abstr. 064.016.

Microturbulence in main sequence A and F stars. See Abstr. 064.018.

Spontaneous fission of heavy transuranium elements in the surface layers of the peculiar A star HR 465 explains promethium abundance.   See Abstr. 064.022.

Curves of growth and line profiles for neutral helium lines in early type stars.   See Abstr. 064.041.

Turbulence velocities in the atmosphere of Alpha Orionis.   See Abstr. 064.044.

Interstellar molecules from cool stars. See Abstr. 064.053.

Ages and masses for nine "super-metal-rich"field stars.   See Abstr. 065.021.

Nuclear processes associated with peculiar A-type stars.   See Abstr. 065.095.

Iron in the sun and the stars. See Abstr. 071.073.

Space velocities of G and K giants. See Abstr. 112.010.

Radial velocities and spectral classification of A-type stars near the south galactic pole.   See Abstr. 112.014.

Photométrie photoélectrique et classification spectrale.   See Abstr. 113.014.

L'absorption de la lumière dans la région de la Voie Lactée sur les données de la photométrie photographique. III. Les magnitudes stellaires et les classes spectraux des étoiles dans deux parcelles centrées (19^{h}15^m, +9°45') et (18^{h}24^m, +22°30') (1900.0).   See Abstr. 113.023.

Faint O−B2 stars in the Vela, Carina, and Centaurus sections of the Milky Way.   See Abstr. 113.024.

The temperatures, abundances and gravities of F dwarf stars.   See Abstr. 113.031.

Photometry of three peculiar A-type stars. See Abstr. 113.033.

The nature of Becklin's star. See Abstr. 113.038.

Observations of OJ 287 between 0.36 and 3.4 μm. See Abstr. 113.045.

Neutral oxygen observations in stars. See Abstr. 113.060.

Magnetic Ap stars.   See Abstr. 116.009.

The photometric variability of Ap stars. See Abstr. 116.011.

A search for southern OH−IR objects. See Abstr. 122.041.

An analysis of the iron spectrum of four cepheids. See Abstr. 122.102.

Effective temperature, radius and gravitational redshift of Sirius B.   See Abstr. 126.016.

Monitoring of time variations in the microwave and infrared continuum flux from OH/IR emission sources. See Abstr. 131.056.

On the masses, luminosities, and compositions of horizontal-branch stars.   See Abstr. 154.013.

115 Stellar Luminosities, Masses, Diameters, HR-Diagrams and Others

115.001 **Spectral synthesis of low-dispersion luminosity criteria in A and F type stars.** C. T. Bolton.
Astron. Astrophys., Vol. 14, 233 - 242 (1971).

Synthetic spectra have been computed for three regions, $\lambda\lambda$4172-78 Å, λ4417 Å, and λ4481 Å, used for the luminosity classification of A and F stars on low-dispersion spectra. The ratio λ4178: λ4172 is found to be well correlated with surface gravity for stars with solar abundances. In the LTE approximation used here the ratio λ4417: λ4481 is found to be at least as sensitive to microturbulence as it is to surface gravity. The NTE interpretation of this result is also discussed. The consequences for spectral classification of possible variations of abundances and microturbulence at a given effective temperature and surface gravity are discussed.

115.002 **Comments on the instability strip for halo population variables.** I. Iben, Jr., J. Huchra.
Astron. Astrophys., Vol. 14, 293 - 305 (1971).

Results of pulsation calculations in the linear, non-adiabatic approximation are presented. Major emphasis is placed on the location in the H-R diagram of the blue edges of the theoretical instability strip for pulsation in the first harmonic mode and in the fundamental mode. Pulsation results are compared with the observations and with the results of stellar evolution calculations.

115.003 **On the red giants normal colour line on the (U−B) − (B−V) diagram.** A. E. Vasilevskij.
Astron. Tsirk., No. 631, p. 1 - 4 (1971). In Russian.

115.004 **The absolute magnitudes of the BaII stars.**
A. R. Upgren, D. J. MacConnell, R. L. Frye.
Bull. American Astron. Soc., Vol. 3, 372 (1971). − Abstr. AAS.

115.005 **Wie groß sind Fixsterne? Neue Methoden zur Messung der Durchmesser.** F. Schmeidler.
Umschau, 71. Jahrgang, p. 803 - 806 (1971).

115.006 **The luminosity function of old-disk red giants compared with theoretical rates of evolution.**
B. M. Tinsley.
Astrophys. Letters, Vol. 9, 105 - 108 (1971).

In this paper, the members of six very similar old-disk groups of stars, with apparently normal population I composition, are combined. The luminosity function of their giants and subgiants is compared with the rate of evolution on Iben's (1967, 1968) theoretical track for a star of 1.25 solar masses ($M_\odot$), up to core helium ignition, and with the nuclear fuel available for later stages of evolution.

115.007 **Lines of constant periods of RR Lyrae stars in the color−magnitude diagram.** R. M. Russev.
Astron. Tsirk., No. 642, p. 3 - 5 (1971). In Russian.

115.008 **Erratum: 'Luminosities, temperatures, and kinematics of K-type dwarfs' [Astrophys. Journ., Suppl. Ser., No. 191, Vol. 22, 389 - 417 (1971)].** O. J. Eggen.
Astrophys. Journ., Suppl. Ser., No. 200, Vol. 23, 369 (1971).

115.009 **The empirical mass-luminosity relation and Hertzsprung-Russell diagram.**
G. M. Popović, T. D. Angelov.
Bull. Obs. Astron. Beograd, Vol. 28, (No. 124), 147 - 157 (1970).

For the three sequences of HR-diagram, new empiric mass–luminosity relations are given. The results are basing on the known magnitudes of the visual orbital binary systems.

115.010 **Masses of red giants on the asymptotic branch in globular clusters.** R. M. Russev.
Astron. Tsirk., No. 649, p. 4 - 5 (1971). In Russian.

115.011 **Angular diameters of the red giants 46 Leo and ϕ Aqr and parameters of some binary systems from occultation observations.** H. L. Poss.
Highlights of Astronomy, Vol. 2, (see 012.018), 692 - 699 (1971).

115.012 **The determination of angular diameters of stars.** J. Davis.
Highlights of Astronomy, Vol. 2, (see 012.018), 713 - 720 (1971).

115.013 **On the scale of bolometric corrections.**
N. S. Komarov, L. F. Nosova.
Astron.Tsirk., No. 659, p. 3 - 6 (1971). In Russian.

115.014 **Stellar diameters from occultations.** R. A. Berg.
Thesis, Univ. Virginia, Charlottesville. [Available from Univ. Microfilms, Ann Arbor, Mich., U.S.A. Order No. 70−26603], 90 pp. (1970).

In search of angular structure in stellar sources, the Virginia dual-channel photometer was used to record occultations of stars by the moon. In terms of gross angular structure, occultations of stars provide excellent opportunities for discovery of very close binaries, especially those with low inclination orbits where spectroscopic methods fail.

115.015 **Measuring the angular diameters of stars.**
R. Hanbury Brown.
Contemporary Phys., Vol. 12, 357 - 377 (1971). − See Phys. Abstr., Vol. 74, No. 62871 (1971).

115.016 **Luminosity of high temperature stars.**
M. Cattani, N. C. Fernandes, J. Osada.
An. Acad. Brasil. Cienc., Vol. 42, 677 - 678 (1970).

The luminosity is studied by dividing the stellar envelope into two regions: the region where the free-free transition is predominant and the region where Compton scattering is the main effect.

115.017 **Luminosities and motions of the F-type stars. I. Luminosity and metal abundance indices for disk population stars.** O. J. Eggen.
Publ. Astron. Soc. Pacific, Vol. 83, 741 - 761 (1971).

The present paper examines the luminosity and metal abundance parameters for Hyades group stars, other young disk stars including the Sirius group, and old disk stars, including the Wolf 630 and ζ Herculis group members. A procedure is developed for computing luminosities from the intermediate-band indices for both young and old disk population objects. Ultrashort-period cepheids in the young disk population and a comparison between the metal abundance parameters are also briefly discussed.

Determining the angular diameter of a small luminous coherent disk through the diffraction amplitude on axis of imaging lens. See Abstr. 031.058.

Space velocities of G and K giants.
See Abstr. 112.010.

The temperatures, abundances and gravities of F dwarf stars.   See Abstr. 113.031.

Spectral classification of unbroadened low-dispersion stellar spectra. II. New data on the structure of the Galaxy in the direction of NGC 6913 based on the additional list of faint stellar spectra.   See Abstr. 114.028.

On the energy distributions of main sequence stars. See Abstr. 114.033.

Spectrophotometry of cool stars in the near infrared. II. Results for a region in the direction of the galactic anticentre.   See Abstr. 114.073.

Some spectroscopic characteristics of the OB stars: An investigation of the space distribution of certain OB stars and the reference frame of the classification. See Abstr. 114.101.

Spectroscopic studies of O-type stars. I. Classification and absolute magnitudes.   See Abstr. 114.115.

The interpretation of space observations of stars and interstellar matter.   See Abstr. 114.121.

The scale of effective temperatures and bolometric corrections.   See Abstr. 114.122.

On the existence of stable stars in the cepheid instability strip.   See Abstr. 122.046.

The absolute magnitudes of the RR Lyrae stars. See Abstr. 122.123.

Absolute magnitudes of RR Lyrae stars. See Abstr. 122.124.

On the masses, luminosities, and compositions of horizontal-branch stars.   See Abstr. 154.013.

The nearby stars.   See Abstr. 155.008.

116 Stellar Magnetic Field, Figure, Rotation

116.001 High effective magnetic field strengths of the magnetic star 53 Camelopardalis. G. Scholz.
Astron. Nachr., Vol. 292, 279 - 280 = Mitt. Astrophys. Obs. Potsdam, No. 135 (1971).
New measurements of the effective magnetic field strengths of 53 Cam yield essentially higher values than formerly published measurements.

116.002 Reversal of the effective magnetic field of Gamma Equulei. G. Scholz.
Astron. Nachr., Vol. 292, 281 = Mitt. Astrophys. Obs. Potsdam, No. 136 (1971).
The line displacements from a Zeeman-spectrogram of the magnetic star γ Equ yield an effective magnetic field strength of −1100 Gauß. Hitherto the star only showed positive polarity.

116.003 An attempt to measure weak magnetic fields for some bright stars. J. R. Auman, G. A. H. Walker, V. L. Buchholz, B. A. Goldberg, A. C. Gower, B. C. Isherwood.
Journ. Roy. Astron. Soc. Canada, Vol. 65, 171 (1971).
Abstr. RAS Canada.

116.004 Photoelectric observations of magnetic stars. III. HD 124224, HD 140160, and HD 224801.
C. Blanco, F. A. Catalano.
Astron. Journ., Vol. 76, 630 - 633 (1971).
Photoelectric observations in three colors of the magnetic stars HD 124224, HD 140160, and HD 224801 are reported. An improved period is given for HD 224801. For all three stars the oblique rotator model seems to be reliable.

116.005 New southern magnetic stars.
H. J. Wood.
Bull. American Astron. Soc., Vol. 3, 401 (1971). − Abstr. AAS.

116.006 On the variation of specific angular momentum among main sequence stars.
S. P. Tarafdar, M. S. Vardya.
Astrophys. Space Sci., Vol. 13, 234 - 248 (1971).
The specific angular momentum is found to vary with mass for earlier and later-type main sequence stars. The non-rigid rotation may account for this difference in specific angular momentum as well as its gradient, if faster angular velocity in the interior for later-type and/or slower angular velocity for earlier-type stars than the surface value is allowed. A few other possibilities have also been briefly considered to understand this difference.

116.007 Surface characteristics of the magnetic stars.
G. W. Preston.
Publ. Astron. Soc. Pacific, Vol. 83, 571 - 584 (1971). − Invited symposium paper presented at the Hawaii meeting of the Astronomical Society of the Pacific, 22–25 June 1971.

116.008 Theoretical aspects of magnetic stars.
L. Woltjer.
Publ. Astron. Soc. Pacific, Vol. 83, 592 - 593 (1971). − Invited symposium paper presented at the Hawaii meeting of the Astronomical Society of the Pacific, 22-25 June 1971.

116.009 Magnetic Ap stars.
S. B. Pikeľ'ner, V. L. Khokhlova.
Comments Astrophys. Space Phys., Vol. 3, 190 - 195 (1971).
Quantitative fine and coarse analyses of chemical composition were made up to now for about 35 Ap stars. The variability was not taken into account in most cases, although He and Si abundance in silicon stars depends strongly on phase. The results of these analyses permit us to compile a list of anomalies as compared to the sun.

116.010 The theory of magnetic stars. ·L. Mestel.
Quarterly Journ. Roy. Astron. Soc., Vol. 12, No. 4, (see 012.012), 402 - 416 (1971).

116.011 The photometric variability of Ap stars.
R. Radkov.
Acta Astron., Vol. 21, 533 - 539 (1971).
Using a model with a small region of two temperatures different from the temperature of the rest of the star, a qualitative explanation of some of the phase relations between the amplitudes of the observed light curves in the *U, B, V* system of Ap stars is given.

116.012 Magnetic stars with an external non-linear force-free field. M. A. Raadu.
Astrophys. Space Sci., Vol. 14, 464 - 472 (1971).
The possible existence of strong magnetic fields in stars is discussed and a method of constructing highly distorted models of magnetic, rotating stars developed. The force-free equations and the structure equations for a white dwarf are solved simultaneously by a finite difference method.

116.013 Field generation in magnetic stars.
E. M. Drobyshevskii.
Proc. Sixth Winter School on Space Physics, Part I. Apatity 1969, (see 012.033), p. 62 - 64 (1971).

The effect of rapid rotation on radiation from stars. III. Strong helium I lines. See Abstr. 065.006.

Magnetic accretion processes in peculiar A stars. See Abstr. 065.009.

The Schönberg-Chandrasekhar limit and rotation. See Abstr. 065.041.

Axisymmetric multipole magnetic fields in polytropic stars. See Abstr. 065.086.

Rotating stars with very large magnetic fields. See Abstr. 065.087.

Origin of the magnetic fields in early-type peculiar stars. See Abstr. 065.149.

Five-colour photometry of 12 magnetic variable stars. See Abstr. 113.004.

The role of magnetic fields in Ap stars. See Abstr. 114.067.

γ Corvi and the rotation of the Hg-Mn stars. See Abstr. 114.112.

Rotation of stars in binary systems. See Abstr. 117.013.

117 Binary and Multiple Stars, Theory

117.001 **The evolution of contact binary systems of moderate mass.** D. L. Moss.
Monthly Notices, Roy. Astron. Soc., Vol. 153, 41 - 55 (1971).

The evolution during the core hydrogen burning stage of closely binary systems with total masses in the range 1.37 to 2.25 $M_\odot$ is investigated. The initial separation is chosen so that after the mass loss on a thermal time scale a system with a common convective envelope is obtained. The subsequent evolution of the contact system is followed until contact is broken or hydrogen is exhausted at the centre of the primary. Such systems would be identified as W Ursae Majoris systems, but it is pointed out that all W UMa systems cannot arise in this manner.

117.002 **A new double star survey. Its motivation.** P. Couteau.
Astron. Astrophys., Vol. 13, 345 - 347 (1971). In French.

We prove that surveys of close binaries are not finished. Preliminary results of the survey at the Nice Observatory are given. We have set up a relation giving, as a function of the observed elements, the probable period of a binary at the time of discovery. This relation is verified with the observed periods, then is used for the calculation of the probable period and parallax of the newly discovered binaries.

117.003 **On the possibility of existence of distant satellites of stars.** V. A. Antonov, I. N. Latyshev.
Astron. Zhurn. Akad. Nauk SSSR, Vol. 48, 854 - 861 (1971). In Russian. English translation in Soviet Astron. AJ, Vol. 15, No. 4.

The present paper deals with an estimate of maximum distances on which the motion of a material particle with respect to the sun or another star is stable according to Hill in the galactic field of regular force. The equation of the corresponding zero-velocity surface has been obtained, the difference between the galactic potential and that of the mass-point being taken into account.

117.004 **Evolutionary processes in close binary systems.** B. Paczyński.
Annual Rev. Astron. Astrophys., Vol. 9, (see 003.001), 183 - 208 (1971).

The aim of this review is to present current ideas about the evolution of close binaries. Model computations with mass exchange are emphasized.

117.005 **The relative frequencies of binary stars in population I and II.** A. H. Batten.
Journ. Roy. Astron. Soc. Canada, Vol. 65, 173 (1971). Abstr. RAS Canada.

117.006 **The period of the light variation of the peculiar Wolf-Rayet binary CV Ser.** A. M. Cherepashchuk.
Astron. Tsirk., No. 620, p. 4 - 7 (1971). In Russian.

117.007 **Is δ Gem a component of a collapsed star?** O. H. Guseynov, H. J. Novrusova.
Astron. Tsirk., No. 628, p. 7 - 8 (1971). In Russian.

117.008 **Late evolution of close binaries.** S. C. Vila.
Astrophys. Journ., Vol. 168, 217 - 223 (1971).

The evolution of binaries whose primary is a white dwarf and whose secondary is a star with mass less than 0.1 $M_\odot$ that fills its critical lobe has been calculated, with account taken of the loss of angular momentum by gravitational radiation given by Einstein's gravitational theory. The stars are assumed to possess no nuclear-energy sources and simply to be radiating their internal energies. The loss of angular momentum results in a transfer of mass from the secondary to the primary that causes the period and the separation of the stars to increase with time. The mass of the secondary tends asymptotically to zero with increasing time.

117.009 **Shock waves in gaseous streams in close binary systems of dwarf stars.** V. I. Taranov.
Astrofizika, Vol. 7, 295 - 302 (1971). In Russian. – English translation in Astrophysics, Vol. 7, No. 2.

The behaviour of a shock wave front in gaseous streams is investigated. The dependence of the period and the amplitude of the generated oscillations on the system parameters is found.

117.010 **Mass distribution in close binaries.** L. C. Green, E. K. Kolchin.
Bull. American Astron. Soc., Vol. 3, 394 (1971). – Abstr. AAS.

117.011 **Double and multiple stars in four regions of Taurus.** N. M. Bronnikova, L. W. Sazonova.
Astron. Zhurn. Akad. Nauk SSSR, Vol. 48, 1089 - 1094 (1971). In Russian. English translation in Soviet Astron. AJ, Vol. 15, No. 5.

Double and multiple stars in four regions of the Taurus constellation were found using proper motions obtained by Bronnikova. Nine of the stars appear to be really double and multiple. Equatorial coordinates, positional angle and distance between the components were determined.

117.012 **Evidence for black holes in binary star systems.** G. W. Gibbons, S. W. Hawking.
Nature, Vol. 232, 465 - 466 (1971).

The sudden loss of mass associated with gravitational collapse would cause the orbits of a close binary system to become eccentric. We have compared the eccentricities of spectroscopic binaries where the lines of both stars are seen with those binaries where only one set of lines are observed. The anomalously high eccentricities of some of the latter systems suggest that the unseen companions may be black holes.

117.013 **Rotation of stars in binary systems.** K. Nariai.
Publ. Astron. Soc. Japan, Vol. 23, 529 - 538 (1971).

The period–rotational velocity relation of binary systems is discussed by using published rotational velocities for double-line spectroscopic binaries and Olson's (1968) data for eclipsing binaries.

117.014 **New companion of θ Coronae Borealis.** P. Couteau.
IAU Circ., No. 2339 (1971).

117.015 **θ Coronae Borealis.** C. E. Worley, R. K. Honeycutt, D. P. Hube.
IAU Circ., No. 2340 (1971).

117.016 **θ Coronae Borealis.** K. Locher.
IAU Circ., No. 2342 (1971).

117.017 **On a Bernoulli's integral pertaining to gas flow in close binary systems.** Y. Sobouti.
Astrophys. Space Sci., Vol. 12, 408 - 410 (1971). – Research note.

117.018 **Statistics of double star measures.** J. M. Luck.

Proc. Astron. Soc. Australia, Vol. 2, 31 - 34 (1971).
Some statistics from 368 plates of 111 double stars observed between July 1965 and January 1969 with the Yale-Columbia 26-inch refractor at Mount Stromlo Observatory, are presented.

117.019 **The perturbation of G 24-16.** R. S. Harrington.
Astron. Journ., Vol. 76, 930 - 931 (1971).
The star G 24-16 has a Keplerian perturbation with a semimajor axis of $0\overset{''}{.}029$ and a period of 1.5 yr from its proper and parallactic motions. The unseen companion is probably an underluminous dwarf with a mass in the range 0.07 to 0.11 $M_\odot$.

117.020 **Etude du spectre de l'enveloppe de AX Monocerotis.** A. Peton.
Comptes Rendus Acad. Sci. Paris, Sér. B, Vol. 273, 1062 - 1065 (1971).
Les modifications de structure du spectre de l'enveloppe de *AX Mon* peuvent être caractérisées par les variations d'intensité et de profil de la raie Fe II λ 4233. L'intensité de cette raie décroît en moyenne pendant les 10 années couvertes par nos observations, mais reste liée quant à ses maximums à la période orbitale.

117.021 **On the dissolution time of a class of binary systems.** C. Cruz-González, A. Poveda.
Astrophys. Space Sci., Vol. 13, 335 - 349 (1971).
To test the various theories of the dissolution time of binary systems, we have performed a series of numerical experiments. Various masses and velocities of the field stars were used to exhibit their effects on the dissolution time. Because of simplicity, the pairs considered consisted of a primary of one solar mass plus a secondary of negligible mass. A class of binary systems of particular interest is the cloud of bound comets which surrounds the sun. The computations gave, among other things, the evolution in time of the energies and excentricities of the secondary components. The times of dissolution found in the present calculations are between a factor of two and a factor of fifteen longer than predicted by existing theories.

117.022 **Evolution in close binary systems.** Z.Kopal.
Publ. Astron. Soc. Pacific, Vol. 83, 521 - 538 (1971).
Review article. – A comparison of the consequences of current theories of stellar evolution with known observational aspects of close binary systems is presented. We consider the evolution of close binary systems from their origin and through their sojourn on the main sequence, up to the time by which their more massive component begins to evolve away from it. Section III will be concerned with the evolutionary problems in the post-main-sequence stage and in section IV we elaborate some of its more hydrodynamical aspects; while in the concluding section V we turn our attention to dwarf close binaries of the W Ursae Majoris type, in an attempt to discern their position in the general framework of stellar evolution.

117.023 **Ultrashort-period binaries, gravitational radiation, and mass transfer. I. The standard model, with applications to WZ Sagittae and Z Camelopardalis.**
J. Faulkner.
Astrophys. Journ., *(Letters)*, Vol. 170, L99 - L104 = Contr. Lick Obs., No. 344 (1971).
Gravitational-radiation losses, by reducing the scale size of the system, induce mass transfer in the U Geminorum cataclysmic variables at the observed rate. The systems evolve from those having periods $\sim 10^h$ to the shortest period observed, $\sim 1^h 21^m 6$, in times typically of a few billion years or less.

117.024 **Absorption in ring envelopes.** Z. Šima.
Bull. Astron. Inst. Czechoslovakia, Vol. 22, 334 -
342 (1971).
Gaseous rings which rotate round the primary star in some close binaries systems were proved on the basis of their emission lines. Here the profiles of the spectral lines due to the absorption in the rings are discussed. The computed profiles of lines are very wide but so very shallow that it is nearly impossible to observe them. The phenomenon is caused by great Doppler effect. The profiles are also computed for half eclipse case. They are deeper in some parts which causes the illusory shift of spectral lines.

117.025 **Black holes and binary stars.**
A. H. Batten, R. P. Olowin.
Nature, Vol. 234, 341 - 342, with some remarks by S. W. Hawking and G. W. Gibbons, 342 (1971).
The authors discuss critically the claim of Gibbons and Hawking that some single-line binaries of high orbital eccentricity may contain black holes. They point out that observational errors and circumstellar matter within binary systems may lead to erroneous values being deduced for the orbital eccentricity. In addition, the probability of discovery of a spectroscopic binary is dependent on the eccentricity, and selection effects may vitiate statistical inferences made from the observed distribution of orbital eccentricities.

117.026 **Further evidence for collapsed objects in binary star systems.** J. R. Gott III.
Nature, Vol. 234, 342 - 343 (1971).
The list of short period single-line binaries (unseen components > 1.4 $M_\odot$) contains an unusual number of high eccentricity systems. In the present study a strong correlation is found between high velocities and high eccentricities. This evidence supports the suggestion that one or both of the high eccentricity systems has suffered a sudden mass loss and may harbor a neutron star or a black hole. The two high eccentricity systems, HD 176318 and HD 194495, (unseen components ~ 2.2 $M_\odot$ and ~ 2.7 $M_\odot$ respectively) merit further investigation.

117.027 **The photometric proximity effects in close binary systems. V. The average brightness weighted values of gravity and effective temperature for stars filling the Roche Lobes.** S. M. Ruciński.
Acta Astron., Vol. 21, 455 - 466 (1971).
The average values of gravity and effective temperature for the outer parts of the star (opposite to the companion) are related in a similar way as the local values of these quantities. Variations of these averages with phase depend on the mass ratio, the inclination and the effective temperature.

117.028 **The binary system AH Virginis.** K. Kalchayev.
Trudy Astrofiz. Inst., *Alma-Ata*, Vol. 17, 18 - 25 (1971). In Russian.
Two-colour light curves of AH Virginis are presented. The relative and absolute elements of AH Vir are determined. The system of AH Vir is shown to be unstable. The density of the gas stream directed from the more massive component of the system towards the less massive one appears to be somewhat higher than the density of the sun's chromosphere.

117.029 **The distribution of the double stars in the Galaxy.** D. Mihăilesu, O. Gherega.
An. Univ. Timişoara Şt. mat., Vol. 8, No. 1, p. 51 - 57 (1970). In Romanian.

117.030 **Photographic observations of Trapezium type multiple stars.** G. N. Salukvadze.
Soobshch. AN GruzSSR, Vol. 63, 589 - 592 (1971). In Russian. – Abstr. in Referativ. Zhurn. 51. Astron., 2.51.637 (1972).

117.031 **Outflow of matter from a companion in close binary systems in the case of asynchronous rotation.**
Yu. P. Korovyakovskij.
Astron. Tsirk., No. 662, p. 5 - 7 (1971). In Russian.

117.032 **Sphere-ellipsoid model for close binary systems.**
M. I. Lavrov, N. V. Lavrova.
Astron. Tsirk., No. 663, p. 3 - 5 (1971). In Russian.

117.033 **The effects of mass exchange and ejection on the orbit of an evolving binary system.**
E. L. Van Dessel.
Meded. Koninkl. Vlaamse Acad. Wet., Letteren, Schone Kunsten van België; Kl. Wet., Jaargang 30, No. 14, 22 pp. (1968).
It is assumed that in a binary system with an evolving primary mass exchange takes place first, later followed by rapid (explosive) mass ejection during the final transition of the primary into a white dwarf. The results are applied to peculiar A stars.

117.034 **The light variations of close binaries conforming to the Roche model.** G. V. Cochran.
Thesis, Univ. Virginia, Charlottesville. [Available from Univ. Microfilms, Ann Arbor, Mich., U.S.A. Order No. 71–6702], 109 pp. (1970).
Close binary models, based on the Roche stellar model, have been computed and monochromatic light curves deduced.

Catalogue of orbital elements, masses and luminosities of close binaries. See Abstr. 003.002.

Close binary systems with spherical components. See Abstr. 003.006.

The Roche problem in an eccentric orbit. See Abstr. 042.034.

Photoelectric measurements of lunar occultations. V. Observational results. See Abstr. 096.040.

On criteria to detect new binaries among Wolf-Rayet stars. See Abstr. 114.085.

Eruptive binaries. IV. On the light variations of VV Puppis. See Abstr. 119.014.

The eclipsing binary KZ Pavonis as a member of the visual multiple h5231. See Abstr. 121.031.

Monochromatic reflection effect of close binary stars. See Abstr. 121.053.

On the production of white dwarfs in binary systems of small mass. See Abstr. 126.001.

Importance of pulsar observations for planetary physics. See Abstr. 141.160.

Numerical experiments on the escape from non-isolated clusters and the formation of multiple stars. See Abstr. 151.034.

Binary evolution in stellar systems. See Abstr. 151.035.

Recent developments of integrating the gravitational problem of n-bodies. See Abstr. 151.038.

118 Visual Binaries

118.001 **Orbita de la estrella doble visual A.D.S. 1990 – 400 – Hu 1216.** J. M. Costa.
Urania Barcelona, Año 55, Nos. 271–272, p. 127 - 128 (1970).

118.002 **Results of double stars astrometric measurements obtained with a Lallemand's electronic camera.**
P. Laques, A. Bücher, R. Despiau.
Astron. Astrophys., Vol. 15, 179 - 192 (1971). In French.
We give the results of 106 astrometric measurements of double stars obtained with a Lallemand's electronic camera. The use of very short exposure-times, has enabled us to improve the photographic resolution noticeably. The electronographic method brings the well known precision of photographic measurements to some category of stars which were hardly or not accessible with the traditional photographic method.

118.003 *UBV* **observations of visual double stars.**
T. E. Lutz.
Publ. Astron. Soc. Pacific, Vol. 83, 488 - 490 (1971).
The data from *UBV* observations for 28 visual binaries were examined to see if there was any evidence that observations of individual components of the binary stars were contaminated by light from the companion star. No such effect was observed, but the amount of data available is not sufficient to make a very positive statement.

118.004 **Micrometer measures of double stars.**
J. L. Newburg.
Republic Obs. Johannesburg, Circ. No. 131, Vol. 8, 5 (1971).

118.005 **UBV observations of selected double systems, III.**
A. U. Landolt.
Publ. Astron. Soc. Pacific, Vol. 83, 650 - 651 = Contr. Louisiana State Univ. Obs., *Baton Rouge*, No. 55 (1971).
UBV magnitudes and color indices have been obtained for 17 double systems listed in the Catalogue of Bright Stars for which these data have been lacking.

118.006 **Sur la mesure des photographies d'étoiles doubles par interférences.** J. Rösch.
Comptes Rendus Acad. Sci. Paris, Sér. B, Vol. 273, 1131 - 1132 (1971).

118.007 **The new double stars discovered in Belgrade with the Zeiss refractor 65/1055 cm - Supplement II.**
G. M. Popović.
Bull. Obs. Astron. Beograd, Vol. 28, (No. 124), 69 - 74 (1970).
The 66 measurements of 28 new-discovered systems of stars with the Zeiss refractor 65/1055 cm in Belgrade are

shown. The coordinates of systems for epochs 1900, 1950 and 2000 and their differential positions related on BD stars are also given.

118.008 Orbite du système ADS 3021 = Ho 326. D. Olević.
Bull. Obs. Astron. Beograd, Vol. 28, (No. 124),
75 - 76 (1970).

118.009 Orbite de deux étoiles doubles visuelles.
D. J. Zulević.
Bull. Obs. Astron. Beograd, Vol. 28, (No. 124), 77 - 80 (1970).
On donne pour la première fois les éléments préliminaires des couples ADS 896 = AG 14, et 7341 = A 2477.

118.010 Mesures micrométriques des étoiles doubles sur le réfracteur Zeiss 65/1055 cm au cours de 1969 à 1970 (serie 20).
P. M. Djurković, G. M. Popović, D. J. Zulević, D. M. Olević.
Bull. Obs. Astron. Beograd, Vol. 28, (No. 124), 81 - 90 (1970).
Dans cette série il y a 377 des mesures sur les 234 couples visuelles.

118.011 Les orbites de deux étoiles doubles. V. Erceg.
Bull. Obs. Astron. Beograd, Vol. 28, (No. 124),
137 - 139 (1970).
On a donné pour la premièr fois les éléments des orbites pour les étoiles doubles ADS 1990=Hu 1216 et ADS 10542 = Hu 922.

118.012 Zwei visuelle Doppelsternbahnen. G. M. Popović.
Bull. Obs. Astron. Beograd, Vol. 28, (No. 124),
141 - 145 (1970).
Für die zwei Systeme: Hu 1168=ADS 9730 und β 1127 = ADS 11010 werden zum ersten Male Bahnelemente mitgeteilt.

118.013 Orbites nouvelles. P. Muller.
Circ. Inform. (U.A.I. Commission des Etoiles Doubles), Obs. Meudon, No. 54 (1971).

118.014 Etoiles doubles découvertes à Nice, lunette de 50 cm.
P. Couteau, P. Muller.
Circ. Inform. (U.A.I. Commission des Etoiles Doubles), Obs. Meudon, No. 54 (1971).

118.015 Errata: 'Untersuchungen an 12 visuellen Doppelsternen. II'. [Astron. Mitt. Wien, No. 5 (1970) by J. Hopmann]. W. S. Finsen.
Circ. Inform. (U.A.I. Commission des Etoiles Doubles), Obs. Meudon, No. 54 (1971).

118.016 Etoiles doubles découvertes à Nice, lunette de 50 cm.
P. Couteau, P. Muller.
Circ. Inform. (U.A.I. Commission des Etoiles Doubles), Obs. Meudon, No. 54/II (1971).

118.017 Compagnon proche de β Cep.
D. Gezari, A. Labeyrie, R. Stachnik.
Circ. Inform. (U.A.I. Commission des Etoiles Doubles), Obs. Meudon, No. 54/II (1971).

118.018 Orbites nouvelles. P. Muller.
Circ. Inform. (U.A.I. Commission des Etoiles Doubles), Obs. Meudon, No. 55 (1971).

118.019 Etoiles doubles découvertes à Beograd, lunette de 65 cm. G. M. Popović.
Circ. Inform. (U.A.I. Commission des Etoiles Doubles), Obs. Meudon, No. 55 (1971).

118.020 Etoiles doubles découvertes à Nice, lunette de 50 cm.
P. Couteau, P. Muller.

Circ. Inform. (U.A.I. Commission des Etoiles Doubles), Obs. Meudon, No. 55 (1971).

118.021 Errata: 'Third catalogue of orbits of visual binary stars'. [Republic Obs. Johannesburg Circ., No. 129, Vol. 7, 203 - 254 (1970)]. W. S. Finsen, C. E. Worley.
Circ. Inform. (U.A.I. Commission des Etoiles Doubles), Obs. Meudon, No. 55 (1971).

118.022 Theta CrB. S. W. Milbourn.
British Astron. Ass., Circ. No. 534 (1971).

118.023 Orbite de l'étoile double visuelle φ 309 et les paramètres physiques correspondants.
A. Simões da Silva, M. Coelho Balça.
Comun. Obs. Astron. Univ. Coimbra, No. 10, 21 pp. (1971).
The orbital elements of the visual binary φ 309 are determined by using the method of Thiele-Innes. The dynamical parallax, the main physical constants, and an ephemeris are derived.

118.024 Orbits of the visual binaries ADS 1158 and Rst 321.
G. A. Starikova.
Astron. Tsirk., No. 649, p. 5 - 7 (1971). In Russian.

118.025 Orbits of the visual binaries β^2 Tuc and ADS 9094.
G. A. Starikova.
Astron. Tsirk., No. 650, p. 2 - 4 (1971). In Russian.

118.026 On the apsidal motion in the system of HS Herculis.
D. Ya. Martynov.
Astron. Tsirk., No. 651, p. 1 - 3 (1971). In Russian.

118.027 Parallax and mass ratio of Eta Cassiopeiae.
P. van de Kamp, M. D. Worth.
Astron. Journ., Vol. 76, 1129 - 1130 (1971).
Measures on 881 plates with 3234 exposures, taken on 251 nights over the interval 1912 to 1970, with a total weight of 652, give a value of $+0\overset{''}{.}162\pm0\overset{''}{.}002$ (p.e.) for the relative parallax and a value of $4\overset{''}{.}57\pm0\overset{''}{.}06$ for the semi-axis major of the orbit of the primary component. The adopted value of $0\overset{''}{.}170\pm0\overset{''}{.}002$ for the absolute parallax yields masses of 0.93 $\odot$ and 0.59 $\odot$ for the primary and secondary components, respectively.

118.028 Correction of the orbital elements of the visual binary ADS 9626 (BC). G. A. Starikova.
Soobshch. Gos. Astron. Inst. Shternberga, No. 172, p. 27 - 29 (1971). In Russian.

118.029 Untersuchungen an 12 visuellen Doppelsternen. III.
J. Hopmann.
Sitzungsber. Österreich. Akad. Wiss., Math.-naturwiss. Kl., Abt. II, Vol. 179, 253 - 279 (1970) = Astron. Mitt. Wien, No. 7 (1971).
This is a continuation of earlier work (1969). The aim is the determination of parallaxes, masses, absolute magnitudes and so on of longperiod-binaries. If possible, also orbits are computed.

Sur une méthode d'exploitation des images d'étoiles doubles obtenues au moyen de la caméra électronique.
See Abstr. 031.031.

Lunar occultations of two multiple systems.
See Abstr. 096.041.

Astrometric results from the Beta Scorpii occultation. See Abstr. 099.022.

Occultation of Beta Scorpii C by Io on May 14, 1971.
See Abstr. 099.067.

An investigation of the consistency of the calibrations of the *uvbyβ* and *gnkmf* photometric systems.
See Abstr. 113.029.

A new double star survey. Its motivation.
See Abstr. 117.002.

Statistics of double star measures.
See Abstr. 117.018.

HR 6773 — A possible eclipsing binary.
See Abstr. 121.093.

Three somewhat overlooked facts of VY Canis
Majoris.   See Abstr. 122.112.

119 Spectroscopic Binaries

119.001 **Spectroscopic binaries with circular orbits.**
L. B. Lucy, M. A. Sweeney.
Astron. Journ., Vol. 76, 544 - 556 (1971).

Because of observational errors, a spectroscopic binary with a truly circular orbit will be found to have an elliptical orbit of small, but nonzero, eccentricity. This effect is analyzed and shown to lead to spurious eccentricities comparable with those assigned to a great many binaries. We then argue, following Luyten, that such elliptical orbits should be rejected in favor of circular orbits. In order to eliminate these spurious eccentricities, an extensive program of orbit recomputation has been carried out for single-lined systems.

119.002 **Rapid variations in the spectrum of Zeta Tauri.**
J. D. R. Bahng.
Astrophys. Journ., (*Letters*), Vol. 167, L75 - L77 (1971).

Photoelectric spectrum scans of ζ Tau show an indication of extremely rapid variations in the equivalent widths of Hβ and Hγ, in the time scale of about 10 minutes.

119.003 **Orbital elements of the spectroscopic binaries HD 24733 and HD 861.**   A. Acker.
Astron. Astrophys., Vol. 14, 189 - 197 (1971). In French.

119.004 **The spectroscopic binary frequency of the mercury-manganese stars.**   G. C. L. Aikman.
Journ. Roy. Astron. Soc. Canada, Vol. 65, 173 (1971).
Abstr. RAS Canada.

119.005 **Two new chapters in the story of U Cephei — I.**
A. H. Batten, M. Plavec.
Sky Telescope, Vol. 42, 147 - 150 (1971).

119.006 **Discovery of flare activity on YY Geminorum.**
T. J. Moffett, B. W. Bopp.
Astrophys. Journ., (*Letters*), Vol. 168, L117 - L120 (1971).

Flares have been detected both spectroscopically and photometrically on the eclipsing, double-lined spectroscopic binary YY Gem. Previous spectroscopic observations may be explained by intermittent flare activity.

119.007 **Star spots or grey veils? : CC Eri and others.**
D. S. Evans.
Monthly Notices Roy. Astron. Soc., Vol. 154, 329 - 338 (1971).

A discussion of the photometric phenomena exhibited by CC Eri in the light of a theory of star spots produces an explanation in which the 'spots' are not on the surface but are neutral absorbing clouds at two of the collinear Lagran-

gian points. The theory of spots on the star surface accounts for the light curves of the stars described by Krzeminski but some unexplained differences between these stars and CC Eri remain.

119.008 **Spectroscopic binaries—11th complementary catalogue.**   A. Pedoussaut, N. Ginestet.
Astron. Astrophys., Suppl. Ser., Vol. 4, 253 - 264 (1971). In French.

This catalogue is a continuation of the ten catalogues published in Ann. Obs. Astron. météorol. Toulouse. The orbital elements of the known or not yet known spectroscopic binaries are given.

119.009 **Orbital elements of the spectroscopic binary HD 201359 — AGK2 + 47°1705.**   A. Acker.
Astron. Astrophys., Vol. 15, 304 - 305 (1971). In French.

119.010 **Four suspected spectroscopic binaries in the Pleiades.**   J. A. Pearce, G. Hill.
Publ. Astron. Soc. Pacific, Vol. 83, 493 - 495 (1971).

Radial velocities are presented for four suspected spectroscopic binaries in the Pleiades. Published elements are confirmed for one object, yet another may be variable, and two are probably constant velocity objects.

119.011 **HD 21242, a spectroscopic binary with H and K emission.**   R. C. Carlos, D. M. Popper.
Publ. Astron. Soc. Pacific, Vol. 83, 504 - 507 (1971).

Orbital elements are given for the spectroscopic binary HD 21242, (6^m5, G5 V). The period is 6^d4. The cooler component has strong Ca II emission with the same velocities as its absorption lines and is slightly more massive than the G5 star. HD 21242 is a noneclipsing counterpart of a sizable group of eclipsing binaries.

119.012 **The spectroscopic orbits of the binary systems H.D. 91948 and H.D. 176318.**   W. L. Gorza.
Journ. Roy. Astron. Soc. Canada, Vol. 65, 277 - 283 = Commun. David Dunlap Obs., *Richmond Hill, Ontario*, No. 301 (1971).

From spectroscopic observations, orbital elements have been derived for the binary systems H.D. 91948 and H.D. 176318. The large systemic velocity of H.D. 91948, γ = −69.0 km/sec., places the system among the high-velocity binaries. Two previous solutions exist for H.D. 176318. The present solution seems to indicate a variation in two of the orbital elements.

119.013 The radial velocity of 60 Serpentis.
H. A. Abt, S. G. Levy.
Publ. Astron. Soc. Pacific, Vol. 83, 687 - 688 (1971).

Coudé measures of this K0 III star do not confirm the published evidence that it is a single-lined spectroscopic binary with a period of three days.

119.014 Eruptive binaries. IV. On the light variations of VV Puppis. J. Smak.
Acta Astron., Vol. 21, 467 - 478 (1971).

Slow variations of VV Pup are re-discussed and a loose correlation is established between the brightness of the extra light seen at the shoulder of the 100-minute light curve and the system's brightness at minimum. This overall trend can be interpreted within a simple model.

119.015 The mass ratio in spectroscopic binaries.
C. Jaschek.
Archiv. Sci. Genève, Vol. 24, 167 - 175 = Publ. Obs. Genève, Ser. A. Fasc. 78/III (1971).

The mass ratio of spectroscopic binaries showing a single spectrum is derived anew, correcting methods formerly used. An average mass ratio of 0.35 is found, which corresponds to an average magnitude difference of about 4^m for main sequence binaries. A lower limit for the mass ratios is calculated and found to be 0.2, in agreement with theory.

119.016 On δ Geminorum. S. I. Blinnikov, D. A. Ptitsyn.
Astron. Tsirk., No. 663, p. 5 - 6 (1971). In Russian.

119.017 Analysis of a variable spectroscopic double star [computer program].
D. Herbison-Evans, N. R. Lomb.
Computer Phys. Commun., (Netherlands), Vol. 2, 368 - 380 (1971).

The program 'bispec' written in Algol derives the orbital parameters of the stars from the velocities. The program will also find the best fitting sine wave to the velocities, measured at various times, of an oscillating star. It can solve both problems simultaneously if one of the members of a double star is oscillating. This has been done for the star α Virginis.

119.018 A revised orbit for 4 Ursae Minoris.
C. D. Scarfe.
Publ. Astron. Soc. Pacific, Vol. 83, 807 - 809 (1971).

Radial velocities from 28 coudé spectrograms have been used to redetermine the elements of this single-line K giant binary.

The radial velocities of 129 stars in the years 1906 to 1917. See Abstr. 112.012.

The lithium content of Capella.
See Abstr. 114.009.

Stellar diameters from occultations.
See Abstr. 115.014.

Black holes and binary stars.
See Abstr. 117.025.

Two new chapters in the story of U Cephei – II.
See Abstr. 121.033.

Spectroscopic orbits of the binary systems H.D. 128661, AR Cas, β Ari and H.D. 209813.
See Abstr. 121.056.

120 Variable Stars: Catalogues, Ephemerides, Miscellanea

120.001 Visual estimates of variable stars by means of television pictures.
Yu. V. Voroshilov, A. V. Mironov.
Peremennye Zvezdy, Vol. 18, 97 - 103 (1971). In Russian.

120.002 Discovery of variable stars. N. E. Kurochkin.
Methods of investigation of variable stars, (see 003.008), p. 11 - 48 (1971).

120.003 Visual estimates of brightness and reduction of observations of variable stars. V. P. Tsesevich.
Methods of investigation of variable stars, (see 003.008), p. 49 - 90 (1971).

120.004 Photographic photometry of variable stars.
P. N. Kholopov.
Methods of investigation of variable stars, (see 003.008), p. 91 - 116 (1971).

120.005 Methods of photoelectric observations.
P. F. Chugajnov.
Methods of investigation of variable stars, (see 003.008), p. 117 - 166 (1971).

120.006 Application of instruments of photoelectric representation (electron photography, image-converter, television). V. V. Prokof'eva.
Methods of investigation of variable stars, (see 003.008), p. 167 - 198 (1971).

120.007 Methods of investigation of the polarization of variable stars' radiation. N. M. Shakhovskoj.
Methods of investigation of variable stars, (see 003.008), p. 199 - 224 (1971).

120.008 Determination of the periods of brightness variation of variable stars by means of electronic computers.
P. N. Kholopov.
Methods of investigation of variable stars, (see 003.008), p. 307 - 329 (1971).

120.009 An estimate of the accuracy of photoelectric measurements of variable stars by means of a one-channel photometer. Yu. S. Efimov, A. G. Thotochava.
Byull. Abastumansk. Astrofiz. Obs., No. 40, p. 171 - 184 (1971). In Russian.
An expression is given to estimate the minimum r.m.s. error of the U, B, V – magnitudes. The results are applied to compare the observations obtained with telescopes of different size and different durations of observations. A dependence of the r.m.s. errors of the colour-index measurements of the flare star UV Cet on the telescope size is given.

120.010 Variable stars in a field centred at $l^{II} = 0°$, $b^{II} = -10°$ (field 3 of the Palomar–Groningen variable-star survey). L. Plaut.
Astron. Astrophys., Suppl. Ser., Vol. 4, 75 - 230 (1971).
The paper gives a description of the observations of faint variable stars in field 3 of the Palomar–Groningen variable-star survey and of the reduction methods used. A series of tables and figures shows the detailed data as obtained for various kinds of variables.

120.011 Sequences for binocular variables. D. J. Mullan.
Journ. British Astron. Ass., Vol. 81, 454 - 459 (1971).

120.012 Progressi nell'interpretazione e studio delle stelle variabili. L. Rosino.
Coelum, Vol. 39, 165 - 171 (1971).

120.013 A research program on southern red dwarfs.
S. Ferraz Mello, C. A. O. Torres.
Inform. Bull. Variable Stars, (IAU Commission 27), Konkoly Obs., Budapest, No. 577 (1971).
The present observing season is being devoted, at the I. T. A. Astronomical Observatory, São José dos Campos, to a survey on some red dwarfs, in order to make a search for variable stars. In a first survey these stars have been studied in the colors B and G of the 6-color system of Stebbins and Whitford, and only one comparison star has been taken for each program star. The amount of stars which had shown some variability was far greater than expected.

120.014 Tests of two suspected variables. M. A. Seeds.
Inform. Bull. Variable Stars, (IAU Commission 27), Konkoly Obs., Budapest, No. 591 (1971).

120.015 Information on photoelectric observations of variable stars deposited at the Odessa Astronomical Observatory. V. P. Tsesevich, E. N. Makarenko.
Astron. Tsirk., No. 646, p. 1 - 2 (1971). In Russian.

120.016 Chronique des observateurs d'étoiles variables.
A. Brun.
L'Astronomie, 85e année, p. 412 (1971).

120.017 Notes on programmes for visual observers.
T. A. Cragg, F. M. Bateson.
Roy. Astron. Soc. New Zealand, Variable Star Section, Circ. No. 179, 4 pp. (1971).
The attention of visual observers is directed to notes on three programmes: (1) Classical cepheid programme, T. A. Cragg; (2) Visual observation of flare stars, F. M. Bateson; (3) Visual observation of U Gem type variables, F. M. Bateson.

120.018 Programme of cooperative flare star observations for 1972. P. F. Chugainov.
Inform. Bull. Variable Stars, (IAU Commission 27), Konkoly Obs., Budapest, No. 605 (1971).

120.019 Rocznik Astronomiczny Obserwatorium Krakowskiego 1972. International Supplement No. 43.
Prepared under the supervision of K. Kozieł.
Komitet Astronomii, Polskiej Akademii Nauk, Kraków, 5 + 125 pp. Price zł 72.00 (1971). – Contents: Eclipsing binaries (K. Kordylewski, J. Kordylewska); Eclipsing binaries in other galaxies (J. M. Kreiner); RR Lyrae-type variables (W. Zessewitsch, J. M. Kreiner); Auxiliary tables; Occultation of stars by the moon 1972 (L. Orkisz); Geocentric ephemeris of the libration points L_4 and L_5 in the earth-moon system for the year 1972 (A. Szczepanowska).

120.020 Ephemeriden für Kurzperiodische. W. Braune.
BAV Rundbrief, 20. Jahrgang, p. 8 - 10 (1971).

120.021 Bedeckungsveränderliche ohne Ephemeriden.
W. Braune.
BAV Rundbrief, 20. Jahrgang, p. 23 - 26 (1971).

120.022 TV observations of faint variables.
A. N. Abramenko, V. V. Prokofjeva.

Peremennye Zvezdy, Vol. 18, 157 - 170 (1971). In Russian.

The TV method is successfully used for observations of variable stars. The accuracy of the stellar brightness measurements is the same as by ordinary photography, but the limiting magnitude is by 3^m - 4^m fainter.

120.023 **Observation of variable stars.** K. Menzel.
Journ. Astron. Soc. Western Australia, Vol. 28, July, p. 10 - 14 (1971).

Three colour photometry at a city site.
See Abstr. 113.022.

121 Eclipsing Variables

121.001 The light curve and orbital elements of V 539 Arae.
G. F. G. Knipe.
Astron. Astrophys., Vol. 14, 70 - 77 (1971).

The light curve of the eclipsing binary V 539 Arae has been observed on the *B, V,* system, and several times of minimum obtained. The light curve shows orbital eccentricity: after removing this, orbital elements were obtained which combined with the data of Sahade and Dessy give individual masses, absolute magnitudes, and dimensions of the system.

121.002 Light curve and apsidal motion of AR Cas.
S. Catalano, M. Rodonò.
Astron. Journ., Vol. 76, 557 - 561 (1971).

The photoelectric observations of AR Cas carried out at the Catania Astrophysical Observatory during the 1968 international campaign organized by T. Herczeg, are presented. A rediscussion of the apsidal motion, which is based upon the ω advancing obtained from photometric data alone, leads to an apsidal motion period of about 10^3 yr. The value of the orbital eccentricity $e = 0.21$, deduced from the relative durations of photometric eclipses, is utilized.

121.003 The close binary system R Canis Majoris.
K. Sato.
Publ. Astron. Soc. Japan, Vol. 23, 335 - 362 (1971).

Photoelectric studies in UBV of the eclipsing variable R CMa were carried out with the 36-inch reflector of Dodaira Station of Tokyo Astronomical Observatory. Photoelectric observations for Hβ-indices were also made with the 36-inch reflector of Okayama Astrophysical Observatory, which confirmed that the primary component of R CMa is an F1V-type star. Photometric elements of the eclipsing system are discussed with the use of the light curve in yellow (V). Combining with the spectrographic observations, we obtain the absolute values of masses and radii of the system.

121.004 Eclipsing variable IU Aurigae. P. Mayer.
Bull. Astron. Inst. Czechoslovakia, Vol. 22, 168 - 187 (1971).

UBV observations of IU Aur made in the years 1964 to 1971, numbering nearly 2500, are presented. Several new minima times are given and a periodic term in light elements is suspected.

121.005 UBV and Hβ photometry of the eclipsing binary S Velorum. R. F. Sisteró.
Bull. Astron. Inst. Czechoslovakia, Vol. 22, 188 - 196 (1971).

Photoelectric observations of 750 $U, B,$ and V magnitudes and 39 Hβ strengths of the binary system S Vel are presented. The period of the system ($5^d9336663$) and orbital elements are studied. In combining photoelectric and spectroscopic elements the absolute dimensions are discussed.

121.006 Elements of the orbit of β Lyr. M. Ju. Skulsky.
Astron. Zhurn. Akad. Nauk SSSR, Vol. 48, 766 - 776 (1971). In Russian. English translation in Soviet Astron. AJ, Vol. 15, No. 4.

The line Ca II $\lambda3933.7$, which belongs to the second component of β Lyr, is discovered. The elements of the orbit, masses and radii of both components are obtained. The mass of the third body in the triple system β Lyr is equal to $2.5 M_\odot$.

121.007 Zur Deutung des Lichtwechsels von SV Cephei durch zirkumstellare Phänomene.
W. Wenzel, J. Dorschner, C. Friedemann.
Astron. Nachr., Vol. 292, 221 - 224 (1971).

Photoelectric UBV observations of the variable star SV Cephei show quasi-periodical minima of changing depth and form not accompanied by variations of the colour indices. Circumstellar clouds consisting of centimeter-sized meteoritic bodies, which diminish the brightness of the star in a statistical manner, are proposed in order to explain the character of the light curve.

121.008 Determination of non-linear effects of limb-darkening in eclipsing variables from observations of their light minima. Z. Kopal.
Astrophys. Space Sci., Vol. 12, 147 - 150 (1971).

The aim of the present note is to point out that observations of eclipsing variables within minima do not, in general, allow a separation of the quadratic terms of limb-darkening from the first-order effects of the gravity-darkening of distorted components undergoing eclipse. Only a difference of the two can be deduced from the observations, but − especially in close binaries − the net effect will be dominated by gravity-darkening.

121.009 V 1010 Ophiuchi, ein einfacher Bedeckungsveränderlicher für den Feldstecher. K. Locher.
Orion, 29. Jahrgang, p. 90 - 91 (1971).

121.010 Ergebnisse der Beobachtungen von Bedeckungsveränderlichen. R. Diethelm, K. Locher.
Orion, 29. Jahrgang, p. 91 - 92 (1971).

121.011 Résultats des observations d'étoiles variables à éclipse. R. Diethelm, K. Locher.
Orion, 29. Jahrgang, p. 111 - 112 (1971).

121.012 The Be component of VV Cephei.
J. B. Hutchings, K. O. Wright.
Journ. Roy. Astron. Soc. Canada, Vol. 65, 172 (1971).
Abstr. RAS Canada.

121.013 The dependence of the Barr effect for eclipsing variables on the geometrical probability of their discovery.
V. V. Radzievsky, L. P. Surkova, V. P. Tolstijch.
Peremennye Zvezdy, Vol. 18, 31 - 51 (1971). In Russian.

The dependence of the discovery probability of eclipsing binaries on the longitude of the periastron ω for the eclipsed component is investigated. The formulae are deduced and a table is computed for the determination of the geometrical depth of the eclipse and the loss of brightness depending on ω, inclination i and eccentricity e of the orbit, and also on the ratio of radii and of luminosities.

121.014 Trois cents ans d'observations de l'étoile Algol.
I. Todoran.
L'Astronomie, 85e année, p. 339 - 345 (1971).

121.015 Photoelectric observations of the close binary system TX Cancri. M. Kitamura, A. Yamasaki.
Tokyo Astron. Bull., Second Ser., No. 209, p. 2451 - 2472 (1971).

The present work contains more than two thousand photoelectric observations of this variable in different colours. The observations were made with the 36-inch reflectors at the Okayama Astrophysical Observatory and the Dodaira Station of Tokyo Astronomical Observatory on sixteen nights during the winters of 1962−63, 1964−65, 1965−66, 1967−68, and 1970−71.

121.016 Photoelectric observations of the eclipsing variable

U Pegasi. K. Saito.
Tokyo Astron. Bull., Second Ser., No. 211, p. 2481 - 2489 (1971).

Photoelectric observations during December 1961 are presented and discussed.

121.017 On solutions of the light curves of eclipsing binaries by a generalized least squares method.
V. M. Tabachnik.
Astron. Tsirk., No. 618, p. 6 - 7 (1971). In Russian.

121.018 Dependence of the dimensions of the binary system components of W UMa-type on mass ratio.
L. F. Istomin, M. A. Svechnikov.
Astron. Tsirk., No. 625, p. 1 - 3 (1971). In Russian.

121.019 Two-colour photoelectric observations of AR Lac.
M. B. Babaev.
Astron. Tsirk., No. 628, 5 - 7, with a correction, No. 656, p. 8 (1971). In Russian.

121.020 A new method of computing the function $^D\alpha$ for eclipsing binary systems. M. I. Lavrov.
Astron. Tsirk., No. 631, p. 4 - 6 (1971). In Russian.

121.021 A spectroscopic reconnaissance of a new β Lyrae system HD 72754. A. D. Thackeray.
Monthly Notices Roy. Astron. Soc., Vol. 154, 103 - 123 (1971).

This paper presents the spectroscopic observations and what must be regarded as a preliminary orbit and discussion based on medium dispersion. The object is clearly an important one with similarities to the eclipsing systems β Lyr and W Cru.

121.022 Photoelectric light variation elements for the eclipsing variable EQ Tauri.
N. L. Magalashvili, J. I. Kumsishvili.
Byull. Abastumansk. Astrofiz. Obs., No. 40, p. 3 - 12 (1971). In Russian.

121.023 On the period of TY Puppis.
C. J. van Houten.
Astron. Astrophys., Vol. 14, 487 - 488 (1971).

The period derived by Campbell for the eclipsing variable TY Puppis is incorrect; the correct period is shown to be $0^d.819235$.

121.024 Mean parallaxes and absolute magnitudes of eclipsing stars. M. Yu. Volyanskaya.
Astron. Tsirk., No. 633, p. 1 - 3 (1971). In Russian.

121.025 A spectroscopic study of Algol.
G. Hill, J. V. Barnes, J. B. Hutchings, J. A. Pearce.
Astrophys. Journ., Vol. 168, 443 - 460 = Contr. Dominion Astrophys. Obs., Victoria, No. 164 (1971).

The results of a radial-velocity study of Algol A, AB, and C are presented and revised elements for both the short-period (2.8673 ... days) and long-period (1.862 years) orbits derived. A mass ratio $m_{AB}/m_C = 2.63 \pm 0.20$ is deduced. Reliable observational data on the orbit of Algol A are used to show that the line of apsides rotates with a period of 32 years. A new treatment of the rotation effect provides an explanation of the discrepancy between results based on broad and narrow lines, and a rotational velocity for Algol A of 55 ± 4 km s^{-1} has been derived. This and other considerations lead to a mass ratio $m_A/m_B = 4.6$. A complete physical model for the system is presented, and its evolutionary state is discussed.

121.026 The period changes of Epsilon Aurigae.
H. Albo.
Publ. Tartu Astrofiz. Obs., Vol. 39, 82 - 88 (1971). In Russian.

The main aim of the present paper is to investigate the period changes of ϵ Aurigae. As observational material the photometric data of years 1848 - 1957 from the literature are used.

121.027 Investigation of changes in periods of eclipsing variables. J. M. Kreiner.
Acta Astron., Vol. 21, 365 - 390 (1971).

This paper contains $O-C$ diagrams for 137 eclipsing variables based on data up to the year 1970. The collected material served for a new examination of the dependence of the changes in periods of these stars on the spherical coordinates.

121.028 Analysis of the photoelectric light curve of 31 Cygni.
A. Galatola.
Bull. American Astron. Soc., Vol. 3, 402 (1971). – Abstr. AAS.

121.029 UV observations of the eclipsing binary CW Cephei.
S. Sobieski.
Bull. American Astron. Soc., Vol. 3, 403 (1971). – Abstr. AAS.

121.030 The puzzle of the eclipses in CV Serpentis.
A. P. Cowley.
Bull. American Astron. Soc., Vol. 3, 403 (1971). – Abstr. AAS.

121.031 The eclipsing binary KZ Pavonis as a member of the visual multiple h5231. J. S. Shaw.
Bull. American Astron. Soc., Vol. 3, 403 (1971). – Abstr. AAS.

121.032 Determination of the elements of photometric orbits of eclipsing binary systems by a direct method. M. I. Lavrov.
Astron. Zhurn. Akad. Nauk SSSR, Vol. 48, 951 - 956 (1971). In Russian. English translation in Soviet Astron. AJ, Vol. 15, No. 5.

Mathematical foundation and the program of determination of elements of a photometric orbit by a direct method are described. The method has been tested successfully on the light curves of AT Peg and W Del.

121.033 Two new chapters in the story of U Cephei – II.
A. H. Batten, M. Plavec.
Sky Telescope, Vol. 42, 213 - 215 (1971).

121.034 An analytic model of eclipsing binary star systems.
D. B. Wood.
Astron. Journ., Vol. 76, 701 - 710 (1971).

A new approach to the modeling of eclipsing binary star systems is described. The model takes advantage of a digital computer and thus allows a reduction in the number of simplifying assumptions and eliminates rectification altogether. The validity of this model is tested through analysis of numerical integration errors, comparison with the spherical model, parametric studies, and application to observational data. It is concluded that the model is a valid representation of eclipsing systems, and that it is a useful tool for the analysis of such systems.

121.035 Spectrophotometric study of the continuous spectrum of the eclipsing variable AW Peg.
M. B. Babaev.
Izv. AN ASSR. Ser. fiz.-tekhn. i mat. n., 1970, No. 4, p. 109 - 115. In Russian. – Abstr. in Referativ. Zhurn. 51. Astron., 10.51.596 (1971).

121.036 Ergebnisse der Beobachtungen von Bedeckungsveränderlichen. R. Diethelm, J. Isles, K. Locher.
Orion, 29. Jahrgang, p. 142 - 144 (1971).

121.037 Photoelectric observations of TX Cancri.
M. Kitamura, A. Yamasaki.
Inform. Bull. Variable Stars, (IAU Commission 27), Konkoly Obs., Budapest, No. 567 (1971).
Observations at the Okayama Astrophysical Observatory and the Dodaira Station are reported.

121.038 About the variability of the eclipsing variable star NQ Her. C. Blanco.
Inform. Bull. Variable Stars, (IAU Commission 27), Konkoly Obs., Budapest, No. 571 (1971).

121.039 Minima of eclipsing variables. Z. Klimek.
Inform. Bull. Variable Stars, (IAU Commission 27), Konkoly Obs., Budapest, No. 573 (1971). − Concerning U CrB, UW Cyg, Z Dra, TU Her, Y Leo, UZ Lyr, Beta Per, TX UMa, BE Vul.

121.040 The period of MW Pavonis. R. M. Williamon.
Inform. Bull. Variable Stars, (IAU Commission 27), Konkoly Obs., Budapest, No. 574 = Rosemary Hill Obs., Dep. Phys. Astron., Univ. Florida, Gainesville, Florida, Contr. No. 24 (1971).

121.041 The period of AQ Tucanae. T. F. Collins.
Inform. Bull. Variable Stars, (IAU Commission 27), Konkoly Obs., Budapest, No. 575 = Rosemary Hill Obs., Dep. Phys. Astron., Univ. Florida, Gainesville, Florida, Contr. No. 25 (1971).

121.042 HD 101799, a completely eclipsing W UMa system.
R. F. Sisteró, M. E. Castore de Sisteró.
Inform. Bull. Variable Stars, (IAU Commission 27), Konkoly Obs., Budapest, No. 576 (1971).
We report in this note the first results of 1350 U, B, and V observations secured at Cerro Tololo Inter-American Observatory. The variable star was observed differentially in relation to the comparison star HD 101834. We derived eleven times of minimum for each light curve in U, B, and V. They are listed in a table.

121.043 Photoelectric and spectrographic observations of the eclipsing variable SZ Camelopardalis.
M. Kitamura, A. Yamasaki.
Inform. Bull. Variable Stars, (IAU Commission 27), Konkoly Obs., Budapest, No. 582 (1971).

121.044 Minima of eclipsing variables. P. Flin.
Inform. Bull. Variable Stars, (IAU Commission 27), Konkoly Obs., Budapest, No. 584 (1971). − Concerning CX Aqr, OO Aql, ZZ Cas, WZ Cyg, V 401 Cyg, TT Del, SZ Her, TZ Lyr, DI Peg, RS Sct, BE Vul.

121.045 Possible black hole in Beta Lyrae. E. J. Devinney, Jr.
Nature, Vol. 233, 110 - 112 (1971).
By combining β Lyrae's out-of-eclipse light variation with that of a computed model, it is found that the luminous (B8) component possesses less than about 35% of the system's total mass. It is further noted that the companion is remarkably underluminous for its mass. The limiting solutions for the masses are discussed and it is concluded that the peculiarities of the system are best understood if the secondary is a black hole, surrounded by a ring.

121.046 HD 105507, an eclipsing variable.
A. W. J. Cousins.
Monthly Notes Astron. Soc. Southern Africa, Vol. 30, 150 (1971).

121.047 Relation luminosité−rayon aux binaires à eclipse.
C. Popovici, A. Dumitrescu.
Stud. Cerc. Astron., Vol. 16, 123 - 129 (1971). In Roumanian.
On utilise les étoiles doubles à eclipse pour trouver la relation magnitude absolue−rayon dans le cas des étoiles de la séquence principale des couples détachés et semi-détachés et des étoiles sous-géantes des couples semi-détachés. On trouve une différence notable dans les deux cas qui peut être interprétée comme un effet d'évolution.

121.048 Considerations on the determination of long periodic apsidal motion. Application to the eclipsing binary system RU Monocerotis. I. Todoran.
Stud. Cerc. Astron., Vol. 16, 177 - 189 (1971).
The aim of the present paper is to investigate the way of determination of long periodic apsidal motion when we cannot dispose of observed minima during a long interval of time. Also the ephemeris formulae for primary and secondary minima are considered. A practical application for RU Mon is made and it is pointed out the need of an apart consideration of the two series of observations: before and after the jump of the orbital period.

121.049 Eclipsing variable star AB Andromedae.
H. Minţi, M. Ganea.
Stud. Cerc. Astron., Vol. 16, 197 - 205 (1971). In Roumanian.
A new estimate of the elements of the eclipsing system AB And is given. The AB And system is analysed using the photoelectric observations in the V-filter obtained by Binnendijk in 1959.

121.050 The equivalent widths of the absorption lines in the spectrum of β Lyrae.
M. Yu. Skul'skij, E. B. Vovchik.
Tsirk. L'vov. Astron. Obs., No. 45, p. 25 - 31 (1971). In Russian.

121.051 A remarkable eclipsing star in Lyncis − SVS 1740.
G. A. Lange, O. E. Mandel.
Astron. Tsirk., No. 637, p. 7 - 8 (1971). In Russian.

121.052 De eclips van Zeta Aurigae. G. W. E. Beekman.
Hemel en Dampkring, Vol. 69, 289 - 290 (1971).

121.053 Monochromatic reflection effect of close binary stars. K.-Y. Chen, W. J. Rhein.
Publ. Astron. Soc. Pacific, Vol. 83, 449 - 458 = Contr. Rosemary Hill Obs., *Gainesville*, No. 18 (1971).
The stars of a binary system are assumed to exchange radiation like spherical blackbodies with limb darkening. The temperatures of the facing surfaces are calculated as functions of the radii, the intrinsic temperatures, and the coefficients of limb darkening of the two stars. The light curves, neglecting eclipses, are then calculated as functions of the wavelength of the radiation, the inclination of the system and its phase, using numerical integration on a digital computer. Results for several typical close binary systems are shown and compared with the light curves deduced from observations.

121.054 A *UBV* photometric study of HS Herculis.
D. S. Hall, G. S. Hubbard.
Publ. Astron. Soc. Pacific, Vol. 83, 459 - 470 = Arthur J. Dyer Obs., Vanderbilt Univ., *Nashville, Tenn.*, Repr. Ser. 2, No. 5 (1971).
About 250 new *UBV* observations are presented. $(B-V)$ and $(U-B)$, corrected for reflection, implied B4 ± 1 and A4 ± 1 for the primary and secondary, with $E(B-V) = 0^{\text{m}} 15$. With the mass function of Cesco and Sahade and the mass-luminosity relation of Iben applied to the hot star, we derived the abso-

lute dimensions $4.7M_\odot$, $2.8 R_\odot$ and $1.6M_\odot$, $1.6 R_\odot$ for the primary and secondary, respectively. The orbital period is variable and only apsidal motion, with a period of $15 \, ^1/_2$ years, can explain the available times of minimum and the spectroscopic elements.

121.055 **The light variations of XY Bootis and LS Herculis.**
L. Binnendijk.
Astron. Journ., Vol. 76, 923 - 929 (1971).

A total of 441 observations in yellow light and 440 observations in blue light are presented for the system XY Bootis. The period is variable and the light curve of this W UMa-type system shows partial eclipses which after rectification become very shallow. LS Herculis was observed during six consecutive nights. The period and the shape of the light curve change during this relative short time interval. Reasons are given to indicate that LS Her is probably an RR Lyrae-type variable.

121.056 **Spectroscopic orbits of the binary systems H.D. 128661, AR Cas, β Ari and H.D. 209813.**
W. L. Gorza, J. F. Heard.
Publ. David Dunlap Obs., Univ. Toronto, *Richmond Hill*, Vol. 3, 99 - 111 (1971).

From spectroscopic observations there have been obtained the orbital elements of two eclipsing binary systems (H.D. 128661, AR Cas) and two spectroscopic binaries (β Ari and H.D. 209813). For one system (H.D. 128661), no solution was previously available. The other systems are well known and were investigated for possible changes in their orbital elements.

121.057 **Rotationally extended stellar envelopes – III. The Be component of VV Cephei.**
J. B. Hutchings, K. O. Wright.
Monthly Notices Roy. Astron. Soc., Vol. 155, 203 - 214 (1971).

A new orbit is derived for the system of VV Cephei from recent observations. Emission line profiles of Hα in the spectrum obtained during the 1956–58 eclipse are studied and the variation of width, height, equivalent width and velocity distortion presented as a function of phase. Various simplified calculations of the velocity and equivalent width are made and their inadequacy shown. A program is described for computing the Hα profile at any orbital phase, and a model for the system and the Be envelope that gives the best fit to the observational quantities is derived. Finally, out of eclipse light variations of $0^m_.1$ or more are predicted for this system and 32 Cygni, by mutual distortion effects.

121.058 **Some double-lined eclipsing binaries with metallic-line spectra.** D. M. Popper.
Astrophys. Journ., Vol. 169, 549 - 562 (1971).

Spectroscopic orbits are obtained for the metallic-line eclipsing binaries XY Cet, RR Lyn, and MY Cyg. Provisional orbits are also given for the A-type giant SZ Cen. These systems are shown in the $(\log m, \log R)$- and $[(B-V), \log g]$-planes. The regions of metallicism in these planes are roughly delineated.

121.059 **Addendum: "The synthesis of close-binary light curves. II. Double distortion and the systems AS Eridani, Lambda Tauri, and RS Vulpeculae." [Astrophys. Journ., Vol. 166, 373 - 386 (1971)].**
J. B. Hutchings, G. Hill.
Astrophys. Journ., Vol. 169, 635 (1971).

121.060 **Minima of U Ophiuchi.** G. F. G. Knipe.
Republic Obs. Johannesburg, Circ. No. 131, Vol. 8, 6 (1971).

121.061 **The light curve and minima of BU Velorum.**
G. F. G. Knipe.
Republic Obs. Johannesburg, Circ. No. 131, Vol. 8, 7 (1971).

121.062 **The light curve and minima of TU Muscae.**
G. F. G. Knipe.
Republic Obs. Johannesburg, Circ. No. 131, Vol. 8, 8 - 12 (1971).

121.063 **Minima and light elements of V 453 Cygni.**
H. L. Cohen.
Publ. Astron. Soc. Pacific, Vol. 83, 677 - 679 = Contr. Rosemary Hill Obs., Univ. Florida, *Gainesville*, No. 16 (1971).

Old unpublished times of minima light of V 453 Cygni and new photoelectric times of minima light are presented. The observations show that the light elements published in *A Finding List for Observers of Eclipsing Variables* appear to be slightly in error and that secondary minimum is displaced from one-half period.

121.064 **Résultats des observations d'étoiles variables à éclipse.** R. Diethelm, J. Isles, K. Locher.
Orion Schaffhausen, 29. Jahrgang, p. 182 - 183 (1971).

121.065 **A model of Epsilon Aurigae.** R. E. Wilson.
Astrophys. Journ., Vol. 170, 529 - 539 (1971).

A more conventional dust ring, with a semitransparent central opening, is found to represent the eclipse observations of ε Aur satisfactorily and to account for the general features of the secondary spectrum. There is good evidence that an eclipse of the primary by the secondary "star" was actually observed in 1956 and that the secondary's dimensions, for the purpose of producing this eclipse, were those of a giant star. It is also shown that ε Aur may be somewhat closer and less luminous than it has been regarded in most other theoretical studies.

121.066 **Photoelectric observations of LY Aurigae.**
P. Mayer, T. B. Horák.
Bull. Astron. Inst. Czechoslovakia, Vol. 22, 327 - 333 (1971).

417 *UBV* measurements of the eclipsing variable LY Aur are published. The eclipses are complete. New light elements are given, and a tentative solution is presented.

121.067 **Analysis of periods of some eclipsing variables.**
A. Kizilirmak.
Sci. Rep. Fac. Sci. Ege Univ., *Izmir*, No. 120 (Astron. No.12), 14 pp. (1971).

Using the (O–C) values of eclipsing variables a practical method for the analysis of their periods is presented.

121.068 **The eclipsing binary AB Cassiopeiae as a Delta Scuti star.** P. Tempesti.
Inform. Bull. Variable Stars, (I. A. U. Commission 27), Konkoly Obs., *Budapest*, No. 596, 4 pp. (1971).

The eclipsing binary AB Cas has been observed in several nights from 1967 to 1971 with the 40-cm refractor of the Teramo Observatory. A $0^m_.10$ deep secondary minimum is detectable, but the most interesting feature is the presence of brightness fluctuations of amplitude $0^m_.05$ and period $1^h 24^m$.

121.069 **W UMa type variables in the cluster NGC 188.**
P. N. Kholopov, A. S. Sharov.
Peremennye Zvezdy, Prilozhenie, Vol. 1, 77 - 88 (1971). In Russian.

Photographic magnitudes in the B, V system and elements of brightness variation based on observations in 1964–1970 are given for four W UMa type variables in the cluster NGC 188 (EP, EQ, ER and ES Cep). The period of ES Cep brightness variations is variable.

121.070 TV observations of the variables EQ, ER and ES Cephei in the cluster NGC 188.
V. V. Prokofieva, S. N. Klotchkov.
Peremennye Zvezdy, Prilozhenie, Vol. 1, 89 - 100 (1971). In Russian.

TV observations of EQ Cep, ER Cep, ES Cep were made at the Crimean Astrophysical Observatory in autumn 1969. In blue and yellow regions about 350 photos were obtained in four nights with the half-metre telescope. The light curves of the variables are given.

121.071 Investigation of variable stars in the open cluster NGC 188.   O. P. Vasilyanovskaya.
Peremennye Zvezdy, Prilozhenie, Vol. 1, 101 - 103 (1971). In Russian.

Results of a photometric study for ER and ES Cep are given.

121.072 Photographic observations of the eclipsing binary BC Herculis and of the new variable star BD +
$12°3680$.   M. E. Kiperman.
Peremennye Zvezdy, Vol. 17, 673 - 679 (1971). In Russian.

121.073 Photoelectric observations of the eclipsing variable AI Draconis.   M. Winiarski.
Acta Astron., Vol. 21, 517 - 527 (1971).

This paper contains results of the B and V photoelectric observations obtained during 42 nights from March 28, 1968 to October 13, 1969. The light curve based on the normal points is given along with the improved photometric elements and the $(O - C)$ diagram.

121.074 A narrow-band photoelectric photometry of the peculiar WR-type eclipsing binary CV Ser.
A. M. Cherepashchuk.
Astron. Zhurn. Akad. Nauk SSSR, Vol. 48, 1201 - 1211 (1971). In Russian. English translation in Soviet Astron. AJ, Vol. 15, No. 6.

It is shown that CV Ser is an eclipsing variable in the emission line λ 4653 Å with depth of the primary minimum $\sim 0^m19$ and the secondary one $\sim 0^m04$. This star shows only small amplitude of variability in the continuum near λ 4795 Å and through U-, B-, V-filters. Light elements are obtained.

121.075 Variation of the gradient and Balmer discontinuity during the eclipse of the variable RW Tauri: Comparison between observations and theory.
P. L. Battistini, M. Fracassini, L. E. Pasinetti.
Astrophys. Space Sci., Vol. 14, 438 - 445 (1971).

The variations of φ_b and D during the eclipse of RW Tau have been observed. The observations have been performed at the Observatories of Jungfraujoch and Haute Provence with the photoelectric system in seven colors of Geneva. The comparison between the observed curves of D and the theoretical ones, provides the limb darkening coefficients; the same comparison between the observed φ_b (reddened) and theoretical ones (not reddened) together with that of D, points out some anomalies of the variable.

121.076 Three-color photoelectric observations of five short period eclipsing variables.   R. B. Carr.
Publ. Goodsell Obs., No. 16, 5 + 35 pp. (1971).

Instrumental differential photoelectric observations in the three UBV wavelength bands are reported for the five short period eclipsing variables TY Boo, TZ Boo, RW CrB, WY Hya, and BF Vir. The observations fully define all phases of the light curves for each of the variables.

121.077 A minimum of ζ Phoenicis.   G. F. G. Knipe.
Monthly Notices Astron. Soc. Southern Africa, Vol. 30, 156 (1971).

121.078 The eclipsing binary ST Aquarii.   G. F. G. Knipe.
Monthly Notices Astron. Soc. Southern Africa, Vol. 30, 157 - 162 (1971).

121.079 Introduction to "Instationary stars and methods of their investigation. Eclipsing variables".
V. P. Tsesevich.
Eclipsing variables, (see 003.037), p. 9 - 20 (1971). In Russian.

121.080 Photometric phases of eclipses.   V. P. Tsesevich.
Eclipsing variables, (see 003.037), p. 21 - 44 (1971). In Russian.

121.081 Eclipses of spherical stars moving in circular orbits.   A. M. Shul'berg.
Eclipsing variables, (see 003.037), p. 45 - 88 (1971). In Russian.

121.082 Darkening of spherical stars to the disc limb.   A. M. Shul'berg.
Eclipsing variables, (see 003.037), p. 89 - 112 (1971). In Russian.

121.083 Determination of elements with computers.   V. M. Tabachnik.
Eclipsing variables, (see 003.037), p. 113 - 153 (1971). In Russian.

121.084 Eclipsing systems with deformed components. Fine effects.   D. Ya. Martynov.
Eclipsing variables, (see 003.037), p. 155 - 208 (1971). In Russian.

121.085 Unique systems.   V. P. Tsesevich.
Eclipsing variables, (see 003.037), p. 209 - 260 (1971). In Russian.

121.086 Eclipses of spherical stars with an arbitrary darkening law.   A. M. Cherepashchuk.
Eclipsing variables, (see 003.037), p. 261 - 312 (1971). In Russian.

121.087 Elliptic orbits. Motion of the line of apsides. The influence of a third body on the epochs of minima.
D. Ya. Martynov.
Eclipsing variables, (see 003.037), p. 313 - 347 (1971). In Russian.

121.088 Machine solution of light curves of eclipsing binary systems with eccentric orbits.   M. I. Lavrov.
Astron. Tsirk., No. 656, p. 5 - 7 (1971). In Russian.

121.089 Further evidence for a black hole in β Lyrae.   R. E. Wilson.
Nature, Vol. 234, 406 - 407 (1971).

OAO ultraviolet observations support Devinney's proposal that the secondary component of β Lyrae is a black hole. The depth ratio of primary and secondary eclipses requires the primary star to be the hotter component if both are Planckian sources, but the wavelength dependence of this ratio requires the secondary to be the hotter. Therefore they cannot both be Planckian and the secondary (disk) must shine by dilute high-temperature radiation. Applied restrictions lead to a highly condensed secondary mass, which probably generates thermal bremsstrahlung by accretion, and whose mass of 20 $M_\odot$ suggests that it is a black hole.

121.090 Spectrophotometry of the eclipsing variable AW Peg.   M. B. Babaev.
Izv. AN AzSSR. Ser. fiz.-tekhn. i mat. n.; 1971, No. 1, p. 82 - 88. In Russian. −Abstr. in Referativ. Zhurn. 51. Astron.,

2.51.632 (1972).

121.091 **The application of linear and non-linear limb darkening laws in the determination of orbital elements of select eclipsing binary systems by the computer oriented best-fit iterative method.**   M. L. Cooper.
Thesis, Georgetown Univ., Washington, D.C. [Available from Univ. Microfilms, Ann Arbor, Mich., U.S.A. Order No. 70–21299], 541 pp. (1970).

Under the assumption of the spherical model, non-linear laws of limb darkening based upon non-gray model atmosphere calculations in addition to the usual linear laws of limb darkening are utilized in the determination of the photometric orbital elements of six select detached and semi-detached systems (e.g. YZ Cas, SZ Cam, VV Ori, S Cnc, TW Dra and X Tri).

121.092 **Companion of ε Aurigae.**   M. Kitamura.
Astron. Herald, (*Japan*), Vol. 64, 239 - 241 (1971).
In Japanese.

A review of the study around the Black Hole theory on ε Aurigae.

121.093 **HR 6773 – A possible eclipsing binary.**
D. P. Hube.
Publ. Astron. Soc. Pacific, Vol. 83, 805 - 806 (1971).

On the basis of the spectroscopic observations presented here, we suggest that the relatively bright, B8-type star HR 6773 = HD 165814, $\alpha = 18^h05^m8$, $\delta = -25°29'$ (1950) is an eclipsing binary with a period of a few days. It is noted that the star is also a visual binary.

121.094 *UBVRI* photometry of V642 Orionis.

B. B. Bookmyer.
Publ. Astron. Soc. Pacific, Vol. 83, 824 - 827 (1971).

121.095 **Observations of eclipsing binaries in 1969—70.**
Compiled by J. Šilhán, O. Obůrka.
Contr. Obs. and Planetarium Brno, No. 12, p. 3 - 18 (1971). In Czech.

The present paper contains the times of minima derived from visual and several photographic and photoelectric observations that have been performed at Czechoslovak observatories and in Astronomical Clubs mostly in 1970.

121.096 **CH Cygni.**   M. Mattei.
AAVSO Abstr.,October 1971, p. 16.

121.097 **PEP light curve of DV Aquarii.**   L. Kalish.
AAVSO Abstr., October 1971, p. 18 - 19.

The Isaac Newton Telescope Cassegrain spectrograph: Description and results of radial-velocity observations of stars near δ^2 Lyrae.   See Abstr. 034.004.

Two new chapters in the story of U Cephei – I. See Abstr. 119.005.

RW Arietis – an RR Lyrae variable star in an eclipsing system.   See Abstr. 122.050.

Measurement of limb darkening on the white dwarf BD + 16° 516 B.   See Abstr. 126.008.

Eclipsing binary model of Cygnus XR-1.   See Abstr. 142.040.

122 Physical Variables, Flare Stars, Pulsation Theory

122.001 An abundance analysis of the Delta Scuti variable 20 CVn. R. J. Dickens, V. A. French, P. W. Owst, A. J. Penny, A. L. T. Powell.
Monthly Notices, Roy. Astron. Soc., Vol. 153, 1 - 7 (1971).

The equivalent widths measured from two blue high dispersion spectrograms and the temperature deduced from spectral scans have been used for a differential curve-of-growth analysis. This shows the star to have a metal abundance greater than the sun's and similar to the Hyades. No distinct abundance anomalies are apparent. The present analysis has been compared with previous analyses of Delta Scuti stars.

122.002 On the X-ray emission of flare stars. G. A. Gurzadyan.
Astron. Astrophys., Vol. 13, 348 - 352 (1971).

An analysis is made of the possibility of generating X-ray emission during stellar flares, within the framework of the hypothesis of "fast electrons". It is shown that the non-thermal bremsstrahlung of fast electrons may be the basic mechanism for generating X-ray photons in flare stars. The theoretical spectrum of the X-ray emission has been derived. It has a maximum at 0.05 Å, does not depend on the physical parameters of the star, and is the same in all flares.

122.003 Short-period variables. VIII. Evolution and pulsation of δ Scuti stars. C. Chevalier.
Astron. Astrophys., Vol. 14, 24 - 31 (1971).

Two evolutionary sequences of model stars of 1.8 and $2 M_\odot$ are constructed in the post main sequence phase. The data on δ Scuti variables are analyzed in order to place the stars in a theoretical HR diagram. The periods of the fundamental mode and the first overtone of the models are compared with the observed periods of the variables.

122.004 BCD classification for cepheids. II. A quasistatic interpretation of the results. M. H. Schneider.
Astron. Astrophys., Vol. 14, 128 - 142 (1971). In French.

This paper presents the interpretation of the individual observations of cepheids with the BCD parameters. As for the other classifications based on atmospheric parameters, the proper physical parameters for calibration are not the luminosity but the effective temperature and the gravity. For a cepheid the gravity is replaced by the effective gravity in the quasistatic approximation. After a discussion on the validity of this approximation in most recent theoretical papers from Hillendahl, Castor and Keller we conclude that we may apply this interpretation to our results.

122.005 Temperature and gravity variations of the classical cepheid Beta Doradus. S. B. Parsons.
Astron. Journ., Vol. 76, 562 - 566 (1971).

Large differences exist in reported values of the range in effective temperature and of the mean radius for β Dor. Comparing these with results from analysis of six-color photometry in the Stebbins and Kron system, we conclude that the range in T_e is probably about 950 ± 70°K, the mean radius 77 ± 10 $R_\odot$, and the mean visual absolute magnitude –4.7 ± 0.4. This star may be pulsating in the first-overtone mode. Evidence is now strong, from the variation in surface gravity as obtained from the Balmer jump and from the phase lag of the Hα core radial velocity, that the outer envelope of β Dor is characterized by running-wave pulsation.

122.006 Atmospheric structure, mass loss, and chemical composition in R Andromedae and R Cygni.

T. Tsuji.
Publ. Astron. Soc., Japan, Vol. 23, 275 - 312 (1971).

High-dispersion spectra of the Se-type, Mira variables, R Andromedae and R Cygni, show certain spectroscopic features in the near infrared which are discussed in detail. On analysis, an empirical model of several layers is proposed. The physical condition in the circumstellar envelope of R Andromedae is inferred from a curve-of-growth analysis of the circumstellar lines. The lower limit of the number density is of the order of $10^9 \, cm^{-3}$, and the rate of mass loss is estimated to be at least $7 \times 10^{-6} \, M_\odot$/year. Finally, with the atmospheric structure discussed, an analysis of the chemical composition is given.

122.007 Spectral analyses of some Mira-type long-period variable stars. H. Maehara.
Publ. Astron. Soc. Japan, Vol. 23, 313 - 333 (1971).

Spectral analyses were made for the spectrograms obtained near the light maximum of seven Mira-type long-period variable stars: χ Cygni, o Ceti, R Leonis, R Hydrae, R Trianguli, T Ursae Majoris, R Cassiopeiae. For each spectrogram, identifications of spectral lines were made, and equivalent widths were measured.

122.008 The dissociation equilibrium of some diatomic molecules in Delta Cephei.
M. C. Pande, B. M. Tripathi, C. S. Murthy, V. P. Gaur.
Bull. Astron. Inst. Czechoslovakia, Vol. 22, 196 - 199 (1971).

The dissociation equilibrium of CO, CN, C_2, CH, NH and OH is considered for the maximum and minimum phases of the light curve of δ Cep.

122.009 Absolute magnitudes of cepheids. III. Amplitude as a function of position in the instability strip: A period-luminosity-amplitude relation.
A. Sandage, G. A. Tammann.
Astrophys. Journ., Vol. 167, 293 - 310 (1971).

The amplitude of cepheid variation is maximum at the blue edge of the instability strip, and decreases monotonically toward the red in the period range $0.40 < \log P < 0.86$. Amplitude as a function of strip position permits formulation of a period-luminosity-amplitude relation which is equivalent to, and nearly as accurate as, the usual period-luminosity-color relation. Its advantage is that no colors are necessary. What were previously considered fundamental differences between cepheids in our Galaxy and those in the SMC have largely disappeared in the present formulation when the amplitude effect is considered.

122.010 On the origin of outbursts of U Gem stars.
V. G. Gorbatsky.
Astron. Zhurn. Akad. Nauk SSSR, Vol. 48, 676 - 683 (1971). In Russian. English translation in Soviet Astron. AJ, Vol. 15, No. 4.

A hypothesis is put forward to explain the origin of energy emerged during outbursts of U Gem stars. It is supposed that the output of energy in the central region of a U Gem star is constant. This energy is stored periodically in the star envelope. The outburst is observed when stored energy emerges to the surface of the star.

122.011 A spectroscopic search for southern δ Scuti stars.
D. H. P. Jones.
Roy. Obs. Bull., Greenwich–Cape, No. 163, p. 241 - 248 (1971).

The radial velocities of eight stars have been intensively

observed in search of short-period variation. Two variables were discovered. Their pulsational and other characteristics are compared with the acknowledged δ Scuti variables.

122.012 The dependence of physical properties of flare-ups on the absolute luminosity of UV Cet stars.
G. A. Gurzadyan.
Bol. Obs. Tonantzintla y Tacubaya, No. 36, Vol. 6, 39 - 46 (1971).
  It is important to analyze the dependence of the basic physical properties of the flare-ups (frequency, mean amplitude, equivalent duration and relative and absolute energy) on the absolute luminosity of flare stars. The present paper is aimed at proving that such a dependence exists for objects of the UV Cet type. To this effect the data of more than 900 recorded flare-ups of about twenty flare stars in the vicinity of the sun have been used.

122.013 Beobachtete Maxima von Mirasternen 1970.
R. Lukas.
SuW, Vol. 10, 243 (1971).

122.014 Atmospheres of pulsating stars.   K. Stępień.
  Postępy Astron., Vol. 19, 225 - 245 (1971).
In Polish.
  Obervational data on pulsating stars are given. They are mainly spectrophotometric observations by the Oke's group and photometric and spectroscopic observations by the Preston's group. Evidences are presented that shock waves develop in the atmospheres of W Virginis stars and some RR Lyrae stars. A numerical kinematic model of W Virginis developed by Whitney is presented. This model is based on the assumption that the flux of strong shock waves is applied to the atmosphere of a W Virginis cepheid. Recent results on models of a δ Cephei type star are also discussed.

122.015 Photometry of R Coronae Borealis.   J. D. Fernie.
  Journ. Roy. Astron. Soc. Canada, Vol. 65, 172 - 173 (1971). — Abstr. RAS Canada.

122.016 In (a) — variable stars in extremely young clusters.
V. I. Kardopolov.
Peremennye Zvezdy, Vol. 18, 3 - 30 (1971).   In Russian.
  Five-years series of photoelectric UBV observations of seven Orion variable stars of spectral classes O7-B9—typical representatives of In(a) stars in extremely young clusters — were obtained. Photoelectric UBV measurements by some other authors allowed to discuss the light curves of the examined stars for an interval up to 18 years. The photoelectric data show brightness constancy of the In(a) variable stars within the observational errors. On the basis of spectral observations within the region λλ3120 - 7000 Å, an excessive intensity of about 0^m3-0^m4 was discovered in the ultraviolet and the near infrared regions for the majority of investigated stars, as compared with the main sequence O-, B-stars reddening according to the normal extinction law.

122.017 On period instability of RV Tauri stars.
G. E. Erleksova.
Peremennye Zvezdy, Vol. 18, 53 - 70 (1971). In Russian.
  The period changes of 21 RV Tauri stars were studied. It was established that the instability of the periods of the RV Tauri stars belonging to the spectroscopic group B (according to the identification by G. W. Preston et al.) was smaller than for the group A with the same periods.

122.018 RR Lyncis.   R. A. Botsula.
  Peremennye Zvezdy, Vol. 18, 71 - 83 (1971).
In Russian.
  Orbital elements and absolute dimensions of RR Lyn have been obtained from Linnell's (1966) observations in UBV.

122.019 Seven variable stars.   N. E. Kurochkin.
  Peremennye Zvezdy, Vol. 18, 85 - 90 (1971).
In Russian.
  New or improved elements are derived for the stars LS Per, PZ Per, BB UMa, BD UMa, BE UMa, BF UMa, and SVS 1666; the mean light curves are also given.

122.020 Proper motion of HR Aurigae.   N. M. Artiukhina.
  Peremennye Zvezdy, Vol. 18, 91 - 92 (1971).
In Russian.

122.021 HR Aurigae — a W Virginis type variable.
P. N. Kholopov.
Peremennye Zvezdy, Vol. 18, 93 - 96 (1971).   In Russian.

122.022 On the period of DX Delphini.
R. K. Kanishcheva.
Peremennye Zvezdy, Vol. 18, 105 - 113 (1971).   In Russian.

122.023 Investigation of RR Lyrae type stars in the globular cluster M5 (NGC 5904).
B. V. Kukarkin, N. P. Kukarkina.
Peremennye Zvezdy, Prilozhenie, Vol. 1, No. 1, 76 pp. (1971).   In Russian.
  The period variability of 51 RR Lyr type stars in the globular cluster M5 was investigated. For the reduction were used observations by Bailey (1895 - 1912), Shapley (1917) and Oosterhoff (1934), as also estimates from 180 plates of the Moscow Observatory and the Crimean Station of the Sternberg Astronomical Institute (1952, 1958 - 1968). Elements of light variation, amplitudes and O—C diagrams are given for each star.

122.024 Sporadic outbursts of red dwarf stars.   B. Lovell.
  Quarterly Journ., Roy. Astron. Soc., Vol. 12, 98 - 131 (1971). — Presidential address delivered at the anniversary meeting of the Royal Astron. Soc. on 1971 February 12.

122.025 Brightness, colour and spectrum of GN Her.
A. F. Pugach.
Astrometriya i Astrofiz., *Kiev*, Vyp. (No.) 12, (see 003.003), p. 3 - 8 (1971). In Russian.
  Results are given of more than 70 photoelectric UBV observations made in 1966—1968. Colour temperature determined from low dispersion spectrograms agrees with that obtained from $B-V$.

122.026 An analysis of the brightness of AG Dra.
F. I. Lukatskaya.
Astrometriya i Astrofiz., *Kiev*, Vyp. (No.) 12, (see 003.003), p. 10 - 13 (1971). In Russian.
  The distribution and autocorrelative functions of brightness of AG Dra were obtained for observations made by L. Robinson after JD 2426100.

122.027 Investigation of pulsating variables. I. Analysis of the diagram of gradients for cepheids.
I. G. Kolesnik.
Astrometriya i Astrofiz., *Kiev*, Vyp. (No.) 12, (see 003.003), p. 13 - 23 (1971). In Russian.
  Interpretation of the cepheid sequence on gradient diagrams is given on the basis of photometric data for model atmospheres. It is shown that the gradients dU/dB and dV/dB of a star permit to determine relations between parameters of effective levels in the atmosphere that correspond to the UBV magnitudes.

122.028 Odd behaviour of polarization in the radiation of BL Lacertae = VRO 42.22.01.   V. A. Dombrovsky.
Astron. Tsirk., No. 614, p. 1 - 3 (1971).   In Russian.

122.029 Method of differential determination of pulsating stars' radii using the diagram $\Delta R - \Delta m$.
M. S. Frolov.
Astron. Tsirk., No. 616, p. 3 - 4 (1971).   In Russian.

122.030 New variable star of RR Lyrae type SVS 1732.
A. S. Sharov, A. K. Alksnis.
Astron. Tsirk., No. 616, p. 6 - 7 (1971).   In Russian.

122.031 New activity cycle of RU Cam.
G. V. Zaitseva, V. M. Lyutyj.
Astron. Tsirk., No. 617, p. 1 - 3 (1971).   In Russian.

122.032 On the evolutionary phase of cepheids in globular clusters.   B. V. Kukarkin, Yu. V. Voroshilov.
Astron. Tsirk., No. 617, p. 3 (1971).   In Russian.

122.033 Period−luminosity and period−radius relations for dwarf cepheids.   M. S. Frolov.
Astron. Tsirk., No. 619, p. 4 - 5 (1971).   In Russian.

122.034 Photoelectric observations of rapid irregular variables.   G. V. Zajtseva.
Astron. Tsirk., No. 628, p. 3 - 4 (1971).   In Russian.

122.035 Photometric variability of the Be star θ Coronae Borealis.   T. P. Roark.
Astron. Journ., Vol. 76, 634 - 638 (1971).

Four-color $uvby$ and $H\alpha$ filter observations were made of the Be star θ CrB in June and September 1970. The variability is interpreted in terms of particulate matter in the atmosphere of the star. The size distribution of the particles must change rapidly with time to explain the detailed structure of the light curves in a single night and from night to night.

122.036 The spectrum of SZ Mon.   T. Lloyd Evans.
Observatory, Vol. 91, 159 - 160 (1971).

Spectroscopic and photometric observations of SZ Mon suggest it is a type II cepheid with unusually pronounced alternation of cycles rather than an RV Tauri star of unprecedentedly short period.

122.037 Visual companions of two classical cepheids.
T. Lloyd Evans, R. S. Stobie.
Observatory, Vol. 91, 160 - 162 (1971).

Plaut has listed cepheids which have visual companions. A physical companion can, in favourable cases, enable one to estimate the luminosity of the cepheid. We report here the results of observations of the visual companions of BB Sgr and RY Sco.

122.038 Intermediate-band photometry of RR Lyrae variables−I. Definition and characteristics of the system.
D. H. P. Jones.
Monthly Notices Roy. Astron. Soc., Vol. 154, 79 - 101 (1971).

An intermediate-band photometric system is described together with the manner of setting it up. Magnitudes and colours of 125 stars are presented. Their relation to each other and to comparable photometric systems is discussed.

122.039 A conjecture concerning the transition period for variable stars.   I. Iben, Jr.
Astrophys. Journ., Vol. 168, 225 - 230 (1971).

It is conjectured that a helium-sensitive relationship between period and luminosity that is obtained with linear theory is closely related to a transition curve obtained with nonlinear theory. It follows that the luminosity inferred for RR Lyrae stars in populous clusters by using the transition curve may be a function of the assumed helium abundance.

122.040 Data on three new VV Cephei stars: BD + 63°3,
BD + 54°2698, BD + 61°219.   M. Barbier.
Astron. Astrophys., Vol. 14, 396 - 400 (1971). In French.

122.041 A search for southern OH−IR objects.
J. L. Caswell, B. J. Robinson, H. R. Dickel.
Astrophys. Letters, Vol. 9, 61 - 64 (1971).

We have searched for 1612-MHz OH emission from 94 long-period variable stars in the southern sky. V Microscopii gave the only positive detection. In addition we discovered by chance a new 1612-MHz emission source (OH 338.5 + 0.1) with the characteristics of an OH−IR object.

122.042 Fuors.   V. A. Ambartsumian.
Byurakan Astrophys. Obs., Preprint No. 3, 23 pp. (1971).   In Russian.

The FU Orionis stars (fuors) have the peculiarity that during comparatively short time they strongly increase their luminosity in the observable part of the spectrum. An explanation of this phenomenon is given, based on the assumption that before the increase of brightness directly around the star there are some energy sources which radiate mostly the high energy particles.

122.043 Line weakening in the infrared spectrum of χ Cyg near the minimum of light.   T. E. Dervis.
Astron. Tsirk., No. 634, p. 3 - 5 (1971). In Russian.

122.044 On the period of TV Lyn.
G. A. Lange, R. I. Chuprina.
Astron. Tsirk., No. 634, p. 7 - 8 (1971). In Russian.

122.045 Changes of polarization of high-luminosity red variables. II.
V. A. Dombrovsky, T. A. Polyakova, V. A. Jakovleva.
Trudy Astron. Obs., *Leningrad*, Vol. 28 (= Uchenye Zapiski Leningr. Un-ta, No. 359 = Seriya Matem. Nauk, vyp. (No.) 47), p. 25 - 32 (1971).   In Russian.

The results of polarimetric and photometric observations of high-luminosity red variables μ Cep, R Sct, V CVn, X Her, AK Peg and S UMa are given. The behaviour of polarization of the stars is different. For some of them a correlation exists between the changes of polarization, brightness and colour.

122.046 On the existence of stable stars in the cepheid instability strip.   J. D. Fernie, J. O. Hube.
Astrophys. Journ., Vol. 168, 437 - 442 (1971).

Forty-eight stars with spectral types near G0 Ib have been examined for light variability. Five stars are suspected of being variable by about 0.1 mag. The remaining forty-three stars appear to be stable to within a few hundredths of a magnitude. Intrinsic colors for many of the nonvariable stars were assembled from published $BVRI$ photometry. Most of these stars appear to lie within the cepheid instability strip on the H-R diagram.

122.047 UBV observations of CH Cygni.
L. Luud, M. Ruusalepp, T. Kuusk.
Publ. Tartu Astrofiz. Obs., Vol. 39, 106 - 110 (1971).

122.048 Observations of flare and suspected flare stars.
T. Kuusk, M. Ruusalepp, L. Luud.
Publ. Tartu Astrofiz. Obs., Vol. 39, 111 - 115 (1971).

122.049 A photoelectric minimum of β Persei.
H. Albo.
Publ. Tartu Astrofiz. Obs., Vol. 39, 301 - 306 (1971).
In Russian.

122.050 RW Arietis − an RR Lyrae variable star in an eclipsing system.   W. Z. Wiśniewski.
Acta Astron., Vol. 21, 307 - 310 (1971).

RW Ari was observed photoelectrically in the *UBV* system on 19 nights. Peculiar variations of the usually stable light curve observed on three nights can be interpreted as a superposition of RR Lyrae type variability and eclipsing type variability.

122.051 On the polarization of radiation of flare stars.
V. P. Grinin, H. Domke.
Astrofizika, Vol. 7, 211 - 222 (1971). In Russian. – English translation in Astrophysics, Vol. 7, No. 2.

The problem of diffuse reflection of light of a point source (a flare) by a semi-infinite plane-parallel atmosphere of a cold dwarf is considered. Rayleigh scattering and pure absorption in the atmosphere are supposed. The radiation of the source is supposed to be non-polarized. The degree of polarization of the reflected radiation and of the total radiation are calculated as functions of the cosine of the reflection angle and of the particle albedo.

122.052 On some characteristics of the flare activity of UV Ceti type stars. II.
V. S. Oskanian, V. Yu. Terebizh.
Astrofizika, Vol. 7, 281 - 294 (1971). In Russian. – English translation in Astrophysics, Vol. 7, No. 2.

The results of B-colour photoelectric observations of AD Leo, EV Lac, YZ CMi and UV Cet during 1967 - 1970 are discussed. The distributions of amplitudes and amounts of radiated energies of flares as well as their frequency functions show that the properties of the first three stars are very similar. The influence of the observational selection on the detection of small-amplitude flares as well as the problem of determination of the absolute flare activity of these stars are discussed in detail.

122.053 Polarimetric and photometric observations of EV Lac during flares.
K. A. Grigorian, M. A. Eritsian.
Astrofizika, Vol. 7, 303 - 306 (1971). In Russian. – English translation in Astrophysics, Vol. 7, No. 2. – Note.

122.054 The rate of change of the period of BW Vulpeculae.
J. R. Percy.
Journ. Roy. Astron. Soc. Canada, Vol. 65, 217 - 221 = Commun. David Dunlap Obs., Richmond Hill, Ontario, No. 300 (1971).

New photometric observations of the β Cephei star BW Vulpeculae are presented, and are combined with previous observations to show that the rate of change of period of this star has (i) remained near +3.7 sec/century from 1924 to 1970, on the average and (ii) shown fluctuations, including a decrease in 1953–54.

122.055 A 7% reduction in the factor 24/17 and in Wesselink radii of cepheids. S. B. Parsons.
Bull. American Astron. Soc., Vol. 3, 402 (1971). – Abstr. AAS.

122.056 The spectra of the strong flare of AD Leo on March 2, 1970. R. E. Gershberg, N. I. Shakhovskaya.
Astron. Zhurn. Akad. Nauk SSSR, Vol. 48, 934 - 938 (1971). In Russian. English translation in Soviet Astron. AJ, Vol. 15, No. 5.

122.057 On the evolutionary phase of cepheids in globular clusters. B. V. Kukarkin, Yu. V. Voroshilov.
Astron. Zhurn. Akad. Nauk SSSR, Vol. 48, 1087 - 1089 (1971). In Russian. English translation in Soviet Astron. AJ, Vol. 15, No. 5.

A comparison of the surface density gradients of cepheids and stars of RR Lyrae type in globular clusters undoubtedly indicates a considerably greater mass of cepheids.

This is in contradiction with M. Schwarzschild and R. Härm's ideas on the evolutionary phase of cepheids in globular clusters.

122.058 The activity phase of CH Cygni during the period 1967 to 1970. R. Faraggiana, M. Hack.
Astron. Astrophys., Vol. 15, 55 - 76 (1971).

The spectrum of CH Cygni during the period July 1967 – December 1970, following the explosion of June 1967, has been studied using high dispersion spectrograms. In section II we give a qualitative description of the characteristics of each spectrogram. In sections III, IV, and V we describe the information which can be derived from the quantitative measurements of the blue continuum, emission lines and ultraviolet absorption lines. In section VI we discuss the radial velocity behavior and the stratification effects. In section VII the various hypotheses proposed to explain the symbiotic stars are examined in order to see which can best explain our observations.

122.059 The proper motions of RR Lyrae variables, II.
S. V. M. Clube, Z. Aslan, T. W. Russo, E. D. Clements.
Roy. Obs. Bull., Greenwich–Cape, No. 161, p. 175 - 211 (1971).

The proper motions of RR Lyraes presented in this paper result from a continuation of the programme outlined and developed in Paper I of this series (Clube 1968). Most of the present second-epoch material involved has been secured with the Cape 13-inch astrographic telescope. The first-epoch positions have been taken from the Astrographic Catalogue throughout, save in a few cases where plate material has been made available to us. A description of the reduction programme is followed by a presentation of the proper motions. These results are discussed very briefly.

122.060 The light curve parameters of photoelectrically observed galactic cepheids.
R. Schaltenbrand, G. A. Tammann.
Astron. Astrophys., Suppl. Ser., Vol. 4, 265 - 314 (1971).

The *UBV* light curve parameters of 323 galactic cepheids are derived from nearly 12000 photoelectric observations by Fourier analysis.

122.061 A study on differential radial velocities in the spectra of Chi Cygni and Omicron Ceti.
H. Maehara.
Publ. Astron. Soc. Japan, Vol. 23, 503 - 528 (1971).

The spectra of χ Cygni and o Ceti were studied in order to investigate the dynamical structure of their atmospheres. A linear relationship between the radial velocity and the line strength was obtained for χ Cygni. For o Ceti the relationship is not linear. By means of the velocity-intensity relationship the relative concentration of atoms was determined. The results of abundance determinations for χ Cygni are briefly mentioned. A dynamical model of the atmosphere of χ Cygni was constructed from the observational point of view. Several atmospheric layers are introduced to represent several spectral features. From the data published by Joy (1954) for o Ceti, a strong cycle variation was found in the radial velocities of both emission and absorption lines.

122.062 WX Hydri. W. S. G. Walker, A. F. Jones.
IAU Circ., No. 2348 (1971).

122.063 CH Cygni. W. Liller, C. Y. Shao, M. Mattei.
IAU Circ., No. 2359 (1971).

122.064 V 1057 Cygni. L. J. Robinson, M. Harwood.
Inform. Bull. Variable Stars, (IAU Commission 27), Konkoly Obs., Budapest, No. 568 (1971).

122.065 **Orion flare stars in February 1971.**
A. N. Kulapova.
Inform. Bull. Variable Stars, (IAU Commission 27), Konkoly Obs., Budapest, No. 579 (1971).

Observations have been carried out at the 40 cm astrograph of the Sternberg Institute South Station; ZU-2 plates have been used. Multiple exposure method has been used to search for flares of stars.

122.066 **V 1057 Cygni.** G. Welin.
Inform. Bull. Variable Stars, (IAU Commission 27), Konkoly Obs., Budapest, No. 581 (1971).

In a further search for early photographs of this star V 1057 Cygni seems to have brightened just slightly between 1896 and 1921, when its magnitude was 16.3. It then remained as a low-amplitude T Tauri variable (16.0 ± 0.3) up to the brightening of about 6 magnitudes and spectral change in 1969.

122.067 **Observations of BV 1041.** L. Plaut.
Inform. Bull. Variable Stars, (IAU Commission 27), Konkoly Obs., Budapest, No. 589 (1971).

During August and September, 1970, some observations of the RR Lyrae-type star BV 1041 (Strohmeier, 1967) were made with the 1 meter photometric telescope of the European Southern Observatory on La Silla (Chile). The observations are presented.

122.068 **Power spectrum analysis of the light curve of RR Tauri.**
S. M. Silverman, F. W. Ward, Jr., R. Shapiro.
Astrophys. Space Sci., Vol. 12, 319 - 324 (1971).

Power spectrum analysis is applied to the light curve of RR Tauri. Periodicities are found at 80, 200 and 533 days with some variation in the peak positions and power for different decades. Factors involved in the classification of the variable are discussed.

122.069 **Secondary bumps in cepheid variables.**
S. C. B. Gascoigne.
Proc. Astron. Soc. Australia, Vol. 2, 34 (1971).

The lag of the secondary bump was calculated for a number of Robertson's models, chosen at various points in the instability strip, from the time taken for a sound wave to travel from a layer near the surface, specified by its temperature, to the centre and back. Periods were found from the homology relations as discussed by Cogan. The data are communicated.

122.070 **Light curves and period variations of DT Cyg and T Vul.** K. T. Johansen.
Astron. Astrophys., Vol. 15, 311 - 319 (1971).

Photoelectric observations of the cepheid variables DT Cyg and T Vul have been obtained in B and V. The period variation of DT Cyg is studied using all available observations of the light curve, and the variability of the period is confirmed. Using observations later than J. D. 2410500, T Vul was investigated. No variation of the period could be detected.

122.071 **35 Crucis (BV 476).**
A. W. J. Cousins, H. C. Lagerwey.
Monthly Notes Astron. Soc. Southern Africa, Vol. 30, 146 - 148 (1971).

122.072 **Comparison stars for RR Scorpii.**
A. W. J. Cousins, H. C. Lagerwey.
Monthly Notes Astron. Soc. Southern Africa, Vol. 30, 149 (1971).

122.073 **Etude de la céphéide à courte période SW Piscium.**
V. Ureche.

Stud. Cerc. Astron., Vol. 16, 141 - 149 (1971). In Roumanian.

L'auteur présente 270 observations photographiques de la céphéide à courte période SW Piscium, effectuées à l'Observatoire Astronomique de Cluj entre le 4 septembre 1964 et le 11 octobre 1966.

122.074 **Interpretation of some spectral peculiarities of long-period variables.** I. A. Klimishin, A. F. Novak.
Tsirk. L'vov. Astron. Obs., No. 45, p. 3 - 10 (1971). In Russian.

122.075 **On four long-period cepheids.** Yu. V. Fridel'.
Tsirk. L'vov. Astron. Obs., No. 45, p. 32 - 36 (1971). In Russian.

122.076 **Light curves of five RR Lyrae variables in the globular cluster M 15 obtained by means of television techniques.** A. V. Mironov.
Astron. Tsirk., No. 637, p. 1 - 3 (1971). In Russian.

122.077 **On the variability of 37 stars in the globular cluster NGC 3201 and in its vicinity.** B. V. Kukarkin.
Astron. Tsirk., No. 637, p. 4 - 5 (1971). In Russian.

122.078 **Light flare observations of V 1057 Cyg.**
O. E. Mandel.
Astron. Tsirk., No. 637, p. 5 - 7 (1971). In Russian.

122.079 **On the possible existence of associations of cepheids.** Yu. N. Efremov.
Astron. Tsirk., No. 639, p. 3 - 5 (1971). In Russian.

122.080 **Var 156 in the globular cluster M3—an RR Lyrae type variable.** P. N. Kholopov.
Astron. Tsirk., No. 640, p. 3 - 4 (1971). In Russian.

122.081 **Cepheid FW Lup.** T. G. Nikulina.
Astron. Tsirk., No. 640, p. 7 - 8 (1971). In Russian.

122.082 **Model of a U Gem type variable.** F. I. Lukatskaja.
Astron. Tsirk., No. 642, p. 5 - 7 (1971). In Russian.

122.083 **Light curves of Mira variables at 1.04 microns.**
G. W. Lockwood, R. F. Wing.
Astrophys. Journ., Vol. 169, 63 - 86 (1971).

Near-infrared light curves have been compiled for twenty-five Mira variables of types M and S from photoelectric measurements of the radiation contained in a narrow bandpass near 1.04 μ. An average of twenty-seven observations per star was collected over a 5-year interval. Spectral types have been derived from measurements of TiO and VO bands made simultaneously with the measurements of I (104). We give 595 spectral types for the twenty-two M and mild S stars and a judgment as to each star's most representative type at minimum light.

122.084 **Spectroscopic observations of VY Canis Majoris during 1969—1971.** G. Wallerstein.
Astrophys. Journ., Vol. 169, 195 - 197 (1971).

In the interval between December 1969 and March 1971 five measurable spectrograms of VY CMa at high dispersion have been obtained at the Hale Observatories in the visual-red and near-infrared. Our material includes one infrared plate of considerably higher dispersion (13.5 Å mm^{-1}) than heretofore obtained. We can therefore report the wavelengths and identifications of a number of emission lines not previously recorded.

122.085 **Photoelectric observations of variable stars. III. Long-period variable stars and period changes.**
O. P. Vasiljanovskaja, N. N. Kiselev, T. K. Kiseleva.

Byull. Inst. Astrofiz., *Dushanbe,* No. 56, p. 3 - 22 (1970). In Russian.

The results obtained from observations 1964 - 1966 in blue and yellow light of 12 long-period variable stars are given. Period changes of these variables were investigated.

122.086 Visuelle Beobachtungen von 8 Mirasternen und R Scuti. E. Scheller.
MVS, *Sonneberg,* Vol. 6, 8 (1971).

122.087 Atmospheric abundances in the Beta Cephei stars. R. D. Watson.
Astrophys. Journ., Vol. 169, 343 - 356 (1971).

Extensive data on helium and metal line strengths are given for β Cephei stars. These data are used to determine atmospheric helium abundances for the β Cephei stars and relative metal abundances, the latter being obtained from a straightforward curve-of-growth procedure. A sample of other early B stars is used for comparison. The normality of the atmospheric composition of the β Cephei stars is the significant fact which emerges. No encouragement is given to Stothers and Simon's μ-mechanism hypothesis.

122.088 The red RR Lyrae variable DK Velorum. D. H. P. Jones.
Publ. Astron. Soc. Pacific, Vol. 83, 471 - 473 (1971).

DK Vel appears to be a short-period, metal-rich variable, cooler than any other RR Lyrae star.

122.089 Radial velocity observations of the Delta Scuti star 20 Canum Venaticorum. J. E. Penfold.
Publ. Astron. Soc. Pacific, Vol. 83, 497 - 501 (1971).

Radial-velocity observations ot the Delta Scuti star 20 CVn taken over four nights indicate a probable variable radial velocity with an amplitude of about 1.5 km/sec. Photometric observations in the V band obtained on one of the nights show that maximum light apparently occurs on the rising branch of the velocity curve near minimum velocity.

122.090 Investigation of long-period cepheids of population I. III. O. P. Vasiljanovskaja, G. E. Erleksova.
Byull. Inst. Astrofiz., *Dushanbe,* No. 57, p. 18 - 30 (1970). In Russian.

Statistical relations between the logarithm of the period and the asymmetry of the light curve, the amplitude, the normal color index, the form of the light curve, as well as the relation between period change and amplitude, and the positions of the cepheids in the HR–diagram are investigated. The division of long-period cepheids belonging to population I into two groups was confirmed. The characteristic peculiarity of sudden period changes for these groups was found.

122.091 Investigation of periods for 12 cepheids. T. G. Nikulina.
Byull. Inst. Astrofiz., *Dushanbe,* No. 57, p. 31 - 35 (1970). In Russian.

122.092 Studies of the variables in the globular cluster NGC 6171. C. M. Coutts, H. Sawyer Hogg.
Publ. David Dunlap Obs., Univ. Toronto, *Richmond Hill,* Vol. 3, 61 - 77 (1971).

The purpose of this investigation is to study periods of the variables in NGC 6171 over a long time interval and to look for period changes in the RR Lyrae stars. The study is based on a collection of 47 photographs taken at the David Dunlap Observatory between 1946 and 1969 and 24 photographs at Cerro Tololo in 1970 combined with published observations of other investigators dating back to 1935. Twenty-three variable stars have been studied. Twenty-two of these are RR Lyrae stars, 10 of which show period changes. One of the variables is a long period variable with a period of 332

days. All the variables inside the cluster radius are RR Lyraes.

122.093 Variables in Messier 5: A study of Mount Wilson 1917 observations. C. M. Coutts.
Publ. David Dunlap Obs., Univ. Toronto, *Richmond Hill,* Vol. 3, 81 - 96 (1971).

This paper portrays the light curves and gives the epochs of maximum light for 62 variables from Shapley's 1917 collection of photographs of M5. This completes the publication of their magnitudes.

122.094 Flare stars in the Pleiades. II.
V. A. Ambartsumian, L. V. Mirzoyan, E. S. Parsamian, H. S. Chavushian, L. K. Erastova.
Astrofizika, Vol. 7, 319 - 331 (1971). In Russian. – English translation in Astrophysics, Vol. 7, No. 3.

The results of observations of flare stars of the Pleiades during the winter 1969–1970 and partly during the second part of 1970 are given. They are supplemented by observations carried out in Asiago and Budapest. 44 new flare stars were found. Evidence in favour of strong, but slow variations of activity of some flare stars in the Pleiades is given.

122.095 Spectrophotometric study of some T Tauri stars and rapid irregular variables. G. V. Zajtseva.
Astrofizika, Vol. 7, 333 - 344 (1971). In Russian. – English translation in Astrophysics, Vol. 7, No. 3.

The absolute energy distribution in the continuous spectrum of four Isa and two T Tau stars has been investigated in the region 4000–6000 Å. The spectrophotometric temperatures, corrected for interstellar extinction, and line intensities are determined.

122.096 The results of spectral observations of CH Cygni for 1967–1969. G. N. Jimsheleishvili.
Astrofizika, Vol. 7, 363 - 375 (1971). In Russian. – English translation in Astrophysics, Vol. 7, No. 3.

The results of spectral observations of the semi-regular variable CH Cyg carried out in 1967–1969 with the Abastumani Observatory meniscus prismatic camera are discussed. The variations in the continuous spectrum have been studied and compared with those of relative monochromatic brightness of the star. The relation between the continuous radiation in the violet and the variation of emission in hydrogen lines is presented.

122.097 The frequency of flare-ups in the star Haro 18 in the Pleiades. E. S. Parsamian.
Astrofizika, Vol. 7, 507 - 509 (1971). In Russian. – English translation in Astrophysics, Vol. 7, No. 3.

11 flare-ups in the star Haro 18 were found. The frequency of flares in this star is equal to one flare-up per 13 hours.

122.098 Spectrum of AP Lib ($\equiv$ PKS 1514–24).
A. W. Rodgers.
Nature, Phys. Sci., Vol. 233, 75 (1971).

Optical spectra of AP Lib show a pure continuum between $\lambda 6600$Å and $\lambda 3600$Å and show AP Lib to be very similar to BL Lac.

122.099 A note on models for the envelopes of long-period variable stars. G. E. Langer.
Monthly Notices Roy. Astron. Soc., Vol. 155, 199 - 202 (1971).

Keeley has made an extensive investigation of the envelopes of long-period variable stars; he has constructed a grid of static envelope models and made a non-linear analysis of their oscillations. The purpose of this note is to present a simple linear stability analysis of four similar envelope models. Several of the results are compared to the results of Keeley's calculations.

122.100 **Photometric observations of V 1057 Cygni.**
E. E. Mendoza V.
Astrophys. Journ., (Letters), Vol. 169, L117 - L118 (1971).

UBVRI photometry and total absorptions of neutral-oxygen lines at λ7774 were obtained for V 1057 Cyg. These observations show a high luminosity for this object and give more evidence to affirm that the brightening of V 1057 Cyg is a manifestation of the same event that caused the brightening of FU Ori. The infrared excess can be interpreted as the presence of a circumstellar dust cloud; however, the change in spectral type from K to A indicates that an intrinsic change took place, in addition to a partial dissipation of the circumstellar dust cloud.

122.101 **Variability of radiation from circumstellar grains surrounding R Coronae Borealis.**
W. J. Forrest, F. C. Gillett, W. A. Stein.
Astrophys. Journ., (Letters), Vol. 170, L29 - L31 (1971).

Radiation at infrared wavelengths from circumstellar grains surrounding R CrB has been observed for approximately 3 years. Variations of flux with time have been larger at $\lambda = 3.5\ \mu$ than at $\lambda = 11\ \mu$. The change in the shape of the spectrum is discussed.

122.102 **An analysis of the iron spectrum of four cepheids.**
E. G. Schmidt.
Astrophys. Journ., Vol. 170, 109 - 129 (1971).

The iron spectra of four classical cepheids, η Aql, U Sgr, S Nor and Y Oph, are analyzed using both the curve of growth method and model atmospheres. The coarse curve of growth analysis is carried out with three sets of oscillator strengths, and systematic differences which depend on temperature are found. A comparison of the fine analysis with the curve of growth shows consistency when correct oscillator strengths are used.

122.103 **Addendum: 'Absolute magnitudes of cepheids. III'.**
[Astrophys. Journ., Vol. 167, 293 - 310 (1971)].
A. Sandage, G. A. Tammann.
Astrophys. Journ., Vol. 170, 191 (1971).

122.104 **Further observations of HD 193516.** P. Lee.
Publ. Astron. Soc. Pacific, Vol. 83, 648 - 649 = Contr. Louisiana State Univ. Obs., Baton Rouge, No. 53 (1971).

Additional observations of the suspected β Cephei star HD 193516 are presented. No statistically significant variations were found.

122.105 **On the distance to BL Lacertae.**
J. C. Pigg, M. H. Cohen.
Publ. Astron. Soc. Pacific, Vol. 83, 680 - 682 (1971).

The 21-cm absorption spectrum of BL Lacertae indicates that it is more than 200 pc, and possibly less than 2 kpc $(z = -350\ pc)$ distant. Arguments are presented for the position that the data are consistent with BL Lac's being extragalactic.

122.106 **Ausgewählte Maxima von Mirasternen 1972.**
R. Lukas.
Orion Schaffhausen, 29. Jahrgang, p. 195 (1971).

122.107 **The variations of Beta Cephei stars.** R. D. Watson.
Astrophys. Journ., Vol. 170, 345 - 351 (1971).

Three-color light-curve observations of a number of β Cephei stars are reported, together with some observations using a Hγ filter. The large observed $2K/\Delta m$ ratio of the β Cephei stars is discussed.

122.108 **An attempt to analyse the flare polarization of some UV Ceti stars.**

A. Kubičela, J. Arsenijević.
Bull. Obs. Astron. Beograd, Vol. 28, (No. 123), 3 - 10 (1970).

1965 - 1966 at Belgrade Observatory regular patrol observations of UV Ceti stars were done with the aim to establish eventual changes of polarization parameters during flares. The variation of polarization parameters has in all three observed cases been bigger during and after the flare than immediately before it. In the most thoroughly analysed case (EQ Peg, 24.10.65) the mean values indicated the possibility that the changed state in polarization of radiation lasted longer than the optical flare.

122.109 **Polarimetric observation of RR Lyr.**
J. Arsenijević.
Bull. Obs. Astron. Beograd, Vol. 28, (No. 123), 11 - 14 (1970).

The mean values of polarization for the photometric regions B and V are $p = 0^{m}0051 \pm 0^{m}0007$, $\theta = 59° \pm 15°$ and $p = 0^{m}0055 \pm 0^{m}0015$, $\theta = 81° \pm 7°$.

122.110 **V 605 Aquilae: A nova-like variable in an old planetary nebula.** H. C. Ford.
Astrophys. Journ., Vol. 170, 547 - 549 (1971).

The outburst of V 605 Aql is shown to have occurred near the center of the low-surface-brightness nebula Abell 58. Hα photographs show no perceptible expansion since 1951. A spectrum of the nebula shows Hα and [N II] λ6584 in emission, with [N II] λ6584 stronger than Hα. The new observations associate the outburst of V 605 Aql with the central star of the old planetary nebula A 58.

122.111 **Flare monitoring of AD Leo.** R. B. Herr.
Inform. Bull. Variable Stars, (I. A. U. Commission 27), Konkoly Obs., Budapest, No. 597, 4 pp. (1971).

122.112 **Three somewhat overlooked facts of VY Canis Majoris.** L. J. Robinson.
Inform. Bull. Variable Stars, (I. A. U. Commission 27), Konkoly Obs., Budapest, No. 599, 10 pp. (1971).

122.113 **Proper motions of 84 cepheids.**
D. K. Karimova, E. D. Pavlovskaya.
Peremennye Zvezdy, Vol. 17, 591 - 598 (1971). In Russian.

122.114 **On some characteristics of pulsating stars with periods from one to two days.** O. E. Mandel.
Peremennye Zvezdy, Vol. 17, 599 - 609 (1971). In Russian.

On the basis of a semi-empirical period-luminosity-colour relation calibrated with the help of globular cluster variables the colour indices $(B-V)_0$ at light minimum and the mean absolute magnitudes for 32 cepheids with periods from one to two days were obtained. The interstellar absorption was taken into account. By their position in the Galaxy the majority of the investigated stars at distances up to 3 kpc from the sun form a specific system, the plane of symmetry of which coincides with the basic plane of the Local System.

122.115 **Observations of variable stars in the globular cluster NGC 3201.** B. V. Kukarkin.
Peremennye Zvezdy, Vol. 17, 610 - 619 (1971). In Russian.

During my short stay at the Astronomical Observatory of the National University of Chile 23 plates of the globular cluster NGC 3201 were obtained. Light curves, epochs and periods of some not investigated stars were derived.

122.116 **On the period of RU Piscium.** R. K. Kanishcheva.
Peremennye Zvezdy, Vol. 17, 651 - 660 (1971). In Russian.

122.117 **On the period of BH Pegasi.**
G. A. Lange, Yu. E. Migach.
Peremennye Zvezdy, Vol. 17, 660 - 673 (1971). In Russian.

122.118 **Analysis of light change of four RW Aur type variables.** I. G. Sdanchuk.
Peremennye Zvezdy, Vol. 17, 681 - 687 (1971). In Russian.

122.119 **UBV photometry of the β Cephei type variable stars. III. KP Persei (HD 21803).** M. Jerzykiewicz.
Acta Astron., Vol. 21, 501 - 515 (1971).

Eight B and five UBV light curves of KP Persei are presented. The mean period of a single cycle of the light variation is equal to $0^d201753 \pm 0^d000032$, and the range in $O - C$ from the linear elements is found to be 0^d042, or 0.21 of the mean period. The star is most "ultraviolet" at maximum light and the range in $U - B$ increases with the light-range.

122.120 **RR Lyrae type variables SVS 1365, 1367 and 1371 in the nucleus of the globular cluster M3.**
P. N. Kholopov.
Astron. Tsirk., No. 651, p. 7 - 8 (1971). In Russian.

122.121 **On the emission variability of δ Sct.**
V. T. Doroshenko, I. N. Glushneva.
Astron. Tsirk., No. 652, p. 3 - 6 (1971). In Russian.

122.122 **RR Lyrae type variable Var 201 in the centre of the globular cluster M3.** P. N. Kholopov.
Astron. Tsirk., No. 652, p. 7 - 8 (1971). In Russian.

122.123 **The absolute magnitudes of the RR Lyrae stars.**
R. Woolley.
Highlights of Astronomy, Vol. 2, (see 012.020), 771 - 776 (1971).

122.124 **Absolute magnitudes of RR Lyrae stars.**
R. F. Christy.
Highlights of Astronomy, Vol. 2, (see 012.020), 777 - 780 (1971).

122.125 **Review of observational data on RR Lyrae stars.**
G. van Herk.
Highlights of Astronomy, Vol. 2, (see 012.020), 781 - 787 (1971).

122.126 **Absolute magnitudes of RR Lyrae variables.**
S. V. M. Clube.
Highlights of Astronomy, Vol. 2, (see 012.020), 788 - 789 (1971).

122.127 **Photoelectric observations of the flare star EV Lac during the 1971, September 11 - 27 International Patrol.** S. Cristaldi, M. Rodonò.
Inform. Bull. Variable Stars, (IAU Commission 27), Konkoly Obs., Budapest, No. 600, 6 pp. (1971).

122.128 **Photoelectric observations of the flare star UV Cet during the 1971 October 11 - 27 International Patrol.** S. Cristaldi, M. Rodonò.
Inform. Bull. Variable Stars, (IAU Commission 27), Konkoly Obs., Budapest, No. 601, 6 pp. (1971).

122.129 **Photoelectric observations of V 1216 Sgr during the 1971, June 16 - 30 International Patrol.**
S. Cristaldi, M. Rodonò.
Inform. Bull. Variable Stars, (IAU Commission 27), Konkoly Obs., Budapest, No. 602, 2 pp. (1971).

122.130 **On the space distribution of flare stars in Pleiades.**
L. V. Mirzoyan, M. A. Mnatsakanian.
Inform. Bull. Variable Stars, (IAU Commission 27), Konkoly Obs., Budapest, No. 604, 3 pp. (1971).

122.131 **The changing period of DL Cassiopeiae.**
D. Hoffleit.
Inform. Bull. Variable Stars, (IAU Commission 27), Konkoly Obs., Budapest, No. 607, 4 pp. (1971).

122.132 **Some peculiarities of classical cepheids of the Magellanic Clouds.** N. N. Yakimova.
Astron. Zhurn. Akad. Nauk SSSR, Vol. 48, 1265 - 1268 (1971). In Russian. English translation in Soviet Astron. AJ, Vol. 15, No. 6.

For classical cepheids of the Galaxy and Magellanic Clouds the interaction (along lines of constant period) of amplitudes of brightness and colour in maximum, minimum of brightness, and in the mean phase turns out to be qualitatively identical but with considerable quantitative differences due, probably, to a dissimilarity of the chemical composition in the centre and periphery of stellar systems. A general tendency of amplitude growth of variations of brightness and colour to the low-temperature border of the instability strip, more developed, apparently, at the periphery, is confirmed theoretically.

122.133 **Further studies of variable stars of the Large Magellanic Cloud.** F. W. Wright, P. W. Hodge.
Astron. Journ., Vol. 76, 1003 - 1016, 1163 - 1165 (1971).

This paper reports the completion of our study of the variable stars in the section of the Large Magellanic Cloud centered at $\alpha = 5^h12^m$, $\delta = -67°.5$. Measures in two colors, B and V, were obtained on a series of 80 ADH Schmidt plates, and measures in B were made on approximately 400 Harvard Bruce plates for each variable star. Fifteen of the variables were found to be classical cepheids with periods from 1.97 to 10.5 days. All but two fit well on the period—luminosity relationship previously published by us.

122.134 **The color excess of Polaris.**
E. G. Schmidt.
Astron. Journ., Vol. 76, 1102 - 1104 (1971).

The color excess of Polaris is found from measurements of the Hα profile and the G-band strength. The Hα profile indicates that this star is unreddened while the G band indicates a color excess of $E_{B-V} = 0^m07$. Comparison with another cepheid of similar period, SZ Tau, indicates that the G band in Polaris is slightly weak for its temperature. It is therefore concluded that Polaris is unreddened and that its amplitude is reasonably consistent with its location in the instability strip.

122.135 **The low-luminosity boundary of the β Cephei instability strip.** J. R. Percy.
Astron. Journ., Vol. 76, 1105 - 1107 = Commun. David Dunlap Obs., Univ. Toronto, Richmond Hill, No. 302 (1971).

Several suspected β Cephei stars, less luminous than known β Cephei stars, have been tested for short-period light variations and were found to be constant, the upper limit to their variations being between 0.005 to 0.010 mag. The β Cephei instability strip appears to terminate at its low-luminosity end near $M_V = -3.0$.

122.136 **An interpretation of the changes observed in V 1057 Cygni.** P. Pişmiş.
Bol. Obs. Tonantzintla y Tacubaya, No. 37, Vol. 6, 131 - 133 (1971).

It is argued that the brightening of the T-Tauri star V 1057 Cygni from photographic magnitude 15.5 to 10.0 and its change of spectral type from about K0 to A1, at which the star appears to remain, may be explained by a fast readjustment of mass distribution in the star and the consequent release of potential energy as radiation.

122.137 **Infrared photometry of V 1057 Cygni.**
E. E. Mendoza V.
Bol. Obs. Tonantzintla y Tacubaya, No. 37, Vol. 6, 135 - 136 (1971).

122.138 Infrared photometry of UV Ceti stars.
B. Iriarte Erro.
Bol. Obs. Tonantzintla y Tacubaya, No. 37, Vol. 6, 143 - 147 (1971).

Infrared photometry is performed on 27 M-type dwarf stars, of which nine are stars of the UV Ceti type. These stars do not show infrared excesses.

122.139 Two-colour photoelectric observations of RW Coronae Borealis. K. K. Kalchaev.
Trudy Astrofiz. Inst., Alma-Ata, Vol. 16, 77 - 85 (1971). In Russian.

Elements of the RW CrB system from two-colour photoelectric observations are determined. It is shown that RW CrB is a non-stationary system.

122.140 Flare stars in the Coal Sack.
E. M. Lindsay.
Irish Astron. Journ., Vol. 9, 315 - 318 (1970).

122.141 Report of the committee on variable stars in clusters. H. B. Sawyer Hogg.
Trans. IAU, Vol. 14A, (see 03.003.028), 291 - 297 = Commun. David Dunlap Obs. Univ. Toronto, *Richmond Hill*, Ontario No. 282 (1970). – Appendix II from the report of Commission 27 (see 03.120.011).

122.142 Some physical properties of the shock waves in the atmospheres of RR Lyrae stars. V. E. Panchuk.
Problems of cosmic physics. Vyp. (No.) 6, (see 003.040), p. 199 - 204 (1971). In Russian.

122.143 Eta Carinae. R. H. Garstang.
Southern Stars, Vol. 24, 65 - 67 (1971).

122.144 Analysis of light variations of variable stars in the Ursa Minor dwarf galaxy based on van Agt's observations. P. N. Kholopov.
Peremennye Zvezdy, Vol. 18, 117 - 130 (1971). In Russian.

Using van Agt's observations (1968) the periods of twelve variables in the Ursa Minor dwarf galaxy were derived. The period-luminosity relation for variable stars in spheroidal dwarf galaxies is defined more precisely. The presence in these systems of W Vir type stars having periods less than 1^d is confirmed. A unique eclipsing Algol type system with one component being probably an RR Lyr variable (V 80) is discovered.

122.145 On the polarization of some variable stars in the Orion nebula. V. S. Shevchenko, V. I. Kardopolov.
Peremennye Zvezdy, Vol. 18, 131 - 139 (1971). In Russian.

The polarization of light of θ^1 Ori A, θ^1 Ori C, θ^2 Ori A, θ^2 Ori B, NU Ori, BM Ori, KX Ori, V 359 Ori, V 361 Ori, Π 1938, Π 2271, S Mon has been observed at the 122-cm reflector with an electropolarimeter. The results are discussed in detail.

122.146 The z-coordinate distribution of RR Lyrae variables. II. Results. G. G. Borzov.
Peremennye Zvezdy, Vol. 18, 141 - 156 (1971). In Russian.

The density function of the RR Lyrae variables in Harvard fields was obtained by means of the corrected version of the method published earlier (1970). The results of this work are in good accordance with those of Keenman (1966). It is shown that the logarithmic gradient of the density function for the RR Lyrae stars is a function of the z-coordinate.

122.147 On the variability of red giants in the globular cluster M3. R. M. Russev.
Peremennye Zvezdy, Vol. 18, 171 - 181 (1971). In Russian.

The results of statistical investigations on instability of four red giants in M3: von Zeipel No. 238, 297, 837 and 1397 are given. The mean light curve of V95 and the amplitude-magnitude relation of the known red variables in M3 have been obtained.

122.148 Flare stars on the colour-luminosity diagram.
A. N. Sedyakina.
Peremennye Zvezdy, Vol. 18, 213 - 217 (1971). In Russian.

M. Hayashi's evolutionary tracks of small mass stars and the lines of constant age were transformed into M_V, (V−R) coordinates. From the available UBVR-photometry the $V° - (V−R)°$ diagram for the flare stars in Orion was drawn and discussed in detail.

122.149 New flare stars. A. N. Sedyakina.
Peremennye Zvezdy, Vol. 18, 218 - 222 (1971). In Russian.

A number of new flare stars was discovered in the regions of Coma Berenices, Hyades and α Persei clusters, and in the control region in Cygnus. Identification charts are given.

122.150 Proper motions of 35 δ Scuti type stars.
L. A. Khrushcheva (Kolesova).
Peremennye Zvezdy, Vol. 18, 231 - 234 (1971). In Russian.

122.151 The Lick Observatory program on proper motions of RR Lyrae stars. A. R. Klemola.
Highlights of Astronomy, Vol. 2, (see 012.020), 790 - 791 (1971).

122.152 Elements and light curve of CSV 6956.
N. E. Kurochkin.
Astron. Tsirk., No. 662, p. 7 - 8 (1971). In Russian.

122.153 Spectrophotometry of CH Cyg. E. B. Gusev.
Astron. Tsirk., No. 667, p. 6 - 8 (1971). In Russian.

122.154 Observed and predicted intensities of absorption in Mira variables. S. V. Morris.
Thesis, Ohio State Univ., Columbus. [Available from Univ. Microfilms, Ann Arbor, Mich., U.S.A. Order No. 70−26334], 90 pp. (1970).

The purpose of this thesis was to obtain a quantitative description of the spectrum peculiarities, and to see whether the measured peculiarities could be understood in terms of the atmospheric structure of these stars. Thirty-five high dispersion spectra of M-type stars were employed for the study, both normal giant stars of type M and Mira stars of corresponding spectral types.

122.155 Observational aspects of cepheid evolution.
R. J. Havlen.
Thesis, Univ. Arizona, Tucson. [Available from Univ. Microfilms, Ann Arbor, Mich., U.S.A. Order No. 70−13733], 204 pp. (1970).

Three galactic classical cepheids are studied with respect to their relationship to the surrounding stars and interstellar material. The observations are interpreted within the existing framework of present theories of cepheid evolution and of the ·origin in groups.

122.156 Beta Cephei stars: A linear non-adiabatic analysis of radial oscillations. W. R. Davey.
Thesis, Univ. Colorado, Boulder. [Available from Univ. Microfilms, Ann Arbor, Mich., U.S.A. Order No. 71−5883], 92 pp. (1970).

Attempts to provide a theoretical interpretation of this class of variable stars, by investigating the pulsational stability toward infinitesimal radial oscillations of stellar evolutionary models which lie near the observed region of the H-R diagram occupied by Beta Cephei stars.

122.157   The Beta Canis Majoris stars: A binary hypothesis.
M. A. Seeds.
Thesis, Indiana Univ., Bloomington. [Available from Univ. Microfilms, Ann Arbor, Mich., U.S.A. Order No. 71−13558], 272 pp. (1970).

The subject of this thesis is a re-examination of our knowledge of the βCMa stars in light of the suspicion that they are binary systems, and a report of an observational test of the binary hypothesis. Rough models of mass exchange between close binaries have been calculated to demonstrate that such a process could have deposited an envelope of heavy material on the secondary of the system.

122.158   The nature of the blue stragglers in the old disk population.   O. J. Eggen.
Publ. Astron. Soc. Pacific, Vol. 83, 762 - 767 (1971).

It is suggested, on the basis of four new variables, that there is more evidence for the presence of ultrashort-period cepheids than of close binaries among the blue stragglers of the old disk population. The possibility that one of the newly discovered variables, HD 100366, and the known, large-amplitude ultrashort-period SX Phoenicis are blue stragglers in the halo population is discussed.

122.159   Light variation of HR 5329.
K. Desikachary, M. Parthasarathy, N. Kameswara Rao.
Publ. Astron. Soc. Pacific, Vol. 83, 832 - 833 (1971).

From the photoelectric observations of HR 5329 on four nights, a probable fundamental period of $0.^{d}07306$ (13.6875 c/d) is derived and a beat period around 16 days is suggested.

122.160   New elements of the variable star CY Aqr.
V. Znojil.
Contr. Obs. and Planetarium Brno, No. 12, p. 19 - 36 (1971). In Czech.

122.161   Uzliesmojošãs zvaigznes.   J. Francmanis.
Zvaigžņotā debess, 1970./71. gada ziema, p. 10 - 14. Concerning UV Ceti stars.

122.162   Problems on DL Cassiopeiae.   D. Hoffleit.
AAVSO Abstr., October 1971, p. 14 - 15.

122.163   Blazko effect on an RR Lyrae star.
M. P. Bonnell.
AAVSO Abstr., October 1971, p. 15.

A periodic variation in the shape and amplitude of the light curve of TU Comae Berenices on the order of tens of days is reported.

Long period variable stars.
See Abstr. 003.005.

Calculations of shock waves in an RR Lyrae atmosphere.   See Abstr. 064.020.

The atmosphere of Delta Cephei. I. A coarse analysis.   See Abstr. 064.051.

Sternentwicklung auf dem Horizontalast und Zustandsgrößen von RR-Lyrae-Sternen in Kugelhaufen verschiedenen Alters.   See Abstr. 065.023.

Pulsations of massive main-sequence stars.
See Abstr. 065.030.

Quantitative results of stellar evolution and pulsation theories.   See Abstr. 065.037.

Non-radial oscillations and the Beta Canis Majoris phenomenon.   See Abstr. 065.061.

A radiative-transfer model of a cepheid.
See Abstr. 065.079.

Pulsational characteristics of a low mass cepheid model including convection effects.   See Abstr. 065.141.

Three colour photometry in Auckland.
See Abstr. 113.061.

Effective temperatures and gravities for A- and F-type stars in the Delta Scuti region.   See Abstr. 114.014.

Measurement of the spectrophotometric temperature of stars of the spectral classes B and A with transition from the center of their disk to the limb.
See Abstr. 114.041.

Spectroscopic observations of SC stars.
See Abstr. 114.044.

Helium abundance determinations in main sequence B stars.   See Abstr. 114.063.

Technetium in the Mira variable RZ Pegasi.
See Abstr. 114.076.

Comments on the instability strip for halo population variables.   See Abstr. 115.002.

Lines of constant periods of RR Lyrae stars in the color−magnitude diagram.   See Abstr. 115.007.

Ultrashort-period binaries, gravitational radiation, and mass transfer. I. The standard model, with applications to WZ Sagittae and Z Camelopardalis.   See Abstr. 117.023.

The light variations of XY Bootis and LS Herculis.
See Abstr. 121.055.

The eclipsing binary AB Cassiopeiae as a Delta Scuti star.   See Abstr. 121.068.

OH observations at the positions of T Tauri stars in the Taurus-Auriga region.   See Abstr. 131.005.

OH emission sources associated with long-period variable-infrared stars.   See Abstr. 131.006.

On the explanation of the intrinsic polarization of the light of some red long period variables.
See Abstr. 131.037.

H_2O line emission from Orion A, VY CMa, and W49.   See Abstr. 132.028.

Photographic V and R magnitudes of T Tauri stars and related objects in Orion.   See Abstr. 132.034.

Photometry of variables in globular clusters. III. M 14 and the period−luminosity relation of population II cepheids.   See Abstr. 154.014.

Variables in globular clusters.   See Abstr. 154.022.

Galactic X-rays from unresolved flare stars.
See Abstr. 155.038.

The variable stars of the Large Magellanic Cloud. See Abstr. 159.010.

Comparison of the cepheid variables in the Magellanic Clouds and the Galaxy.   See Abstr. 159.018.

The short-period variable stars in the Magellanic Clouds.   See Abstr. 159.019.

Preliminary results of a photometric study of the NGC 371 region in the SMC.   See Abstr. 159.023.

Cepheid variables and the Magellanic Clouds. See Abstr. 159.029.

123 Variable Stars: Lists of Observations, Individual Observations

123.001 **De veranderlijke van de maand: SS Cygni.**
H. Feijth.
Hemel en Dampkring, Vol. 69, 220 - 222 (1971).

123.002 **Veränderlichenbeobachtung 1970.**
E. Heiser.
SuW, Vol. 10, 202 - 203 (1971).

123.003 **Further observations of V 1216 Sagittarii.**
A. H. Jarrett, J. P. Eksteen.
Monthly Notes Astron. Soc. Southern Africa, Vol. 30, 93 - 94 (1971).

123.004 **Beobachtungsergebnisse über veränderliche Sterne.**
R. Lukas.
Sterne, 47. Jahrgang, p. 151 - 154 (1971).

123.005 **Variable star notes.** M. W. Mayall.
Journ. Roy. Astron. Soc. Canada, Vol. 65, 191 - 194 (1971).

123.006 **Veränderliche in einem Feld um M31 auf Tautenburger Schmidt-Aufnahmen.** L. Meinunger.
MVS, *Sonneberg*, Vol. 5, 177 - 195 (1971).
On plates of M31 taken with the Tautenburg Schmidt telescope 34 new variable objects (S 10724 – S 10757) have been found. For these objects and a few other variable stars types of brightness variation, light curves, coordinates, and charts are given.

123.007 **Sternverzeichnis – MVS Band 5.**
MVS, *Sonneberg*, Vol. 5, 1 - 3 (1971).

123.008 **CSV 2550.** V. Satyvaldyev.
Astron. Tsirk., No. 618, p. 7 - 8 (1971).
In Russian.

123.009 **Note on some constant variables.**
P. N. J. Wisse, M. Wisse.
Monthly Notes Astron. Soc. Southern Africa, Vol. 30, 112 (1971).

123.010 **On seventeen variable stars.** V. Satyvaldiev.
Astron. Tsirk., No. 633, p. 7 - 8 (1971).
In Russian.

123.011 **Confirmation of light variations of CSV 6956.**
G. A. Starikova.
Astron. Tsirk., No. 634, p. 5 - 6 (1971). In Russian.

123.012 **New long-period variable SVS 1739 in Scutum.**
V. P. Tsessevich.
Astron. Tsirk., No. 634, p. 6 (1971). In Russian.

123.013 **De veranderlijke van de maand: X Camelopardalis.**
H. Feijth.
Hemel en Dampkring, Vol. 69, 268 - 269 (1971).

123.014 **Variable star notes.** M. W. Mayall.
Journ. Roy. Astron. Soc. Canada, Vol. 65, 247 - 250 (1971).

123.015 **HD 34409 - a suspected new variable star.**
R. W. Hilditch.
Inform. Bull. Variable Stars, (IAU Commission 27), Konkoly Obs., Budapest, No. 569 (1971).

123.016 **Identification and spectrum of the new variable in Serpens.** D. J. MacConnell.
Inform. Bull. Variable Stars, (IAU Commission 27), Konkoly Obs., Budapest, No. 570 (1971).

123.017 **Concerning variable 14 in M 5.** C. Coutts.
Inform. Bull. Variable Stars, (IAU Commission 27), Konkoly Obs., Budapest, No. 572 (1971).

123.018 **Observations of variables on Sonneberg plates.**
E. Splittgerber.
Inform. Bull. Variable Stars, (IAU Commission 27), Konkoly Obs., Budapest, No. 572 (1971). – Concerning HP Aur, AE Boo, HI Gem.

123.019 **UBV photometry of the suspected variable BD +18°4586.** R. C. Tate, E. W. Burke, Jr.
Inform. Bull. Variable Stars, (IAU Commission 27), Konkoly Obs., Budapest, No. 578 (1971).

123.020 **HDE 302013 - a new short-period variable star.**
R. S. Cannon, O. J. Eggen.
Inform. Bull. Variable Stars, (IAU Commission 27), Konkoly Obs., Budapest, No. 580 (1971).

123.021 **Elements and curves of the two infrared variables MO and MP Cassiopeiae.**
L. Rosino, D. di Martino.
Inform. Bull. Variable Stars, (IAU Commission 27), Konkoly Obs., Budapest, No. 585 (1971).

123.022 **New bright southern variable stars.** R. Bloomer.
Inform. Bull. Variable Stars, (IAU Commission 27), Konkoly Obs., Budapest, No. 586 = Veröff. Remeis-Sternw. Bamberg, Astron. Inst. Univ. Erlangen-Nürnberg, Vol. 8, No. 97 = Rosemary Hill Obs., Dep. Phys. Astron., Univ. Florida, Gainesville, Florida, Contr. No. 27 (1971).

123.023 **Elements for two Bamberg variables.** R. Bloomer.
Inform. Bull. Variable Stars, (IAU Commission 27), Konkoly Obs., Budapest, No. 587 = Veröff. Remeis-Sternw. Bamberg, Astron. Inst. Univ. Erlangen-Nürnberg, Vol. 8, No. 98 = Rosemary Hill Obs., Dep. Phys. Astron., Univ. Florida, Gainesville, Florida, Contr. No. 28 (1971).

123.024 **HD 184077: Another red variable.** M. Wisse.
Inform. Bull. Variable Stars, (IAU Commission 27), Konkoly Obs., Budapest, No. 588 (1971).

123.025 **Der veränderliche Stern T Ursae Maioris.**
R. Lukas.
SuW, Vol. 10, 314 (1971).

123.026 **On five variables in Cygnus.** M. B. Girnyak.
Tsirk. L'vov. Astron. Obs., No. 45, p. 17 - 24 (1971). In Russian.

123.027 **Elements of three variable stars.**
V. P. Tsesevich.
Astron. Tsirk., No. 641, p. 7 - 8 (1971). In Russian.

123.028 **Notes on variables.** E. Splittgerber.
Inform. Bull. Variable Stars, (IAU Commission 27), Konkoly Obs., Budapest, No. 578 (1971).

123.029 **Ten new variable stars in the Cygnus cloud, VV 271 - 280.** W. J. Miller.

Ric. Astron., Specola Vaticana, *Castel Gandolfo*, Vol. 8, (No. 10), 167 - 188 (1971).

Ten new variable stars have been studied on a total of 5661 plates from the Vatican, Hamburg, Heidelberg, Harvard and Hale Observatories. Five of the stars are irregular red variables, two are cepheids, two are long period variables, and one is an SS Cygni variable. The results are summarized in five tables of data and nine pages of either fragmentary or mean light curves.

123.030 **Variable V 14 in the globular cluster M 5.**
B. V. Kukarkin.
Astron. Tsirk., No. 646, p. 2 - 4 (1971). In Russian.

123.031 **Observations of variable stars. January - June 1971. Report No. 20.** L. Plaut, H. Feijth.
Kapteyn Astron. Lab., Groningen–Netherlands. 3 + 8 pp. (1971).

This report gives 3582 visual observations of 161 variable stars, 1971 January - June.

123.032 **SS Cygni i 1970.** O. Klinting.
Astron. Tidsskr., Årg. 4, 138 (1971).

123.033 **SV Cassiopeiae.** O. Klinting.
Astron. Tidsskr., Årg. 4, 139 (1971).

123.034 **Beobachtungen Veränderlicher Sterne.**
P. Ahnert.
MVS, *Sonneberg*, Vol. 6, 9 - 10 (1971).

Results of photographic or visual observations are given for the following 32 variables: R Aql, R Ari, R Boo, V Boo, RS Boo, V Cnc, V Cas, T Cep, S Com, R CrB, S CrB, R Cyg, RT Cyg, SS Cyg, ZZ Cyg, R Dra, S Her, T Her, TW Her, S Lac, VX Lac, R Leo, W Lyr, X Oph, Z Oph, DI Peg, R Sct, X Tri, R UMa, S UMa, T UMa, R Vir.

123.035 **Bearbeitung von 48 Veränderlichen am Südhimmel. (Feld η Ara, Teil II).** I. Meinunger.
MVS, *Sonneberg*, Vol. 6, 11 - 12 (1971). – S 5783 - 5830.

123.036 **Bearbeitung von 50 Veränderlichen am Südhimmel. (Feld β Apodis, Teil II).** H. Geßner.
MVS, *Sonneberg*, Vol. 6, 12 - 13 (1971).

123.037 **VRO 20 08 01 Cancri.** W. Wenzel.
MVS, *Sonneberg*, Vol. 6, 13 - 14 (1971).

Data of 11 maxima and of 5 minima are given. The time scale of a maximum is roughly 2 months, the amplitude up to 3 mag.

123.038 **Photographische Beobachtungen von Veränderlichen auf Platten der Sonneberger Himmelsüberwachung.**
E. Splittgerber.
MVS, *Sonneberg*, Vol. 6, 15 - 16 (1971).

The results of photographic estimations are given for DD Aquarii (6 minima), WX Hydri (17 observations in the course of 3 maxima), and Giacconi's star no. 9 near Cyg X–2 (21 observations).

123.039 **The R Coronae Borealis stars.** M. G. Connors.
Journ. Roy. Astron. Soc. Canada, Vol. 65, L27 - L28 (1971).

123.040 **A short period variable with changing period.**
M. W. Mayall, M. E. Baldwin.
Journ. Roy. Astron. Soc. Canada, Vol. 65, 307 - 310 (1971).
The RR Lyrae type variable XZ Cygni has been discussed.

123.041 **Étoiles variables observées par les membres de la Société en 1970.** M. Dumont.

L'Astronomie, 85ᵉ année, p. 434 - 437 (1971). – Concerning R Boo, χ Cyg, R Sct.

123.042 **A minimum of ER Orionis.** G. F. G. Knipe.
Republic Obs. Johannesburg, Circ. No. 131, Vol. 8, .8 (1971).

123.043 **New variable in Musca.** B. F. Marino, B. Ward.
Roy. Astron. Soc. New Zealand, Variable Star Section, Circ. No. 163, 3 pp. (1970).

The discovery of a new variable in Musca is reported. A chart is reproduced showing its location. V and B observations are listed and photographic results are given.

123.044 **FN Sagittarii.** F. M. Bateson.
Roy. Astron. Soc. New Zealand, Variable Star Section, Circ. No. 164, 5 pp. (1971).

Visual estimates, in steps, are listed for FN Sgr for interval J.D. 2,435,638 to 2,440,918. This star shows a continuous small variation superimposed on larger semi-regular variations, for which there appears to be periods of around 400 and 1,000 days. A plot of the larger variations based on 50 day means is reproduced.

123.045 **Sequences for southern variables.**
F. M. Bateson, P. J. Gordon, B. Menzies.
Roy. Astron. Soc. New Zealand, Variable Star Section, Circ. No. 177, 6pp. (1971).

Sequences, determined photo-electrically at the Auckland Observatory, are published for the fields of: ST PsA; AA, UU Tuc; SU Dor; RV Pup; W Cha; AS Pup; CU Vel; UW Cen; TT Cen; S Hor; RZ Ind; T Vol; CH Pup; Z Cru; V Cha; TU Pup; V Pyx; CM Vel and a suspected variable in Centaurus.

123.046 **RZ Indi.** F. M. Bateson, A. F. Jones.
Roy. Astron. Soc. New Zealand, Variable Star Section, Circ. No. 178, 3 pp. (1971).

Visual observations of RZ Indi are published for the interval 1957 to 1970. Eighteen observed maxima are tabulated and elements are derived.

123.047 **CH Puppis.** F. M. Bateson, A. F. Jones.
Roy. Astron. Soc. New Zealand, Variable Star Section, Circ. No. 180, 2 pp. (1971).

From visual observations of CH Pup eight maxima were determined in the interval J.D. 2,435,864 to 2,440,160. A period of 498.66 days is determined with a mean visual range of 9.56 to < 14.0.

123.048 **Z Crucis.** F. M. Bateson, A. F. Jones.
Roy. Astron. Soc. New Zealand, Variable Star Section, Circ. No. 181, 2 pp. (1971).

Visual observations during the interval 2,438,299 to 2,440,108 are published together with a list of observed maxima and minima. These give a faint indication of a period of about 353 days although Z Cru in general oscillates very slightly around a mean visual magnitude of 9.82.

123.049 **TU Puppis.** F. M. Bateson, A. F. Jones.
Roy. Astron. Soc. New Zealand, Variable Star Section, Circ. No. 182, 3 pp. (1971).

From visual observations during the interval 2,435,836 to 2,440,160, TU Pup is shown to be a Mira type variable. The individual observations, dates of observed maxima and minima and details of the mean light curve are tabulated.

123.050 **W Chamaeleonis.** F. M. Bateson.
Roy. Astron. Soc. New Zealand, Variable Star Section, Circ. No. 183, 2 pp. (1971).

Ten day means, from visual observations, are listed together with details of seven maxima determined.

123.051 **One new and four up-dated variable stars.**
D. Hoffleit.
Inform. Bull. Variable Stars, (I. A. U. Commission 27),
Konkoly Obs., *Budapest*, No. 592 (1971). – Concerning
V734 Cyg, MM Sgr, NW Sgr, S Com and a new RR Lyr-star.

123.052 **7 new variable stars in NGC 1261.**
C. Bartolini, F. Grilli, J. W. Robertson.
Inform. Bull. Variable Stars, (I. A. U. Commission 27), Konkoly Obs., *Budapest*, No. 594 (1971).

123.053 **Further remarks concerning variable 14 in M 5.**
W. Osborn.
Inform. Bull. Variable Stars, (I. A. U. Commission 27), Konkoly Obs., *Budapest*, No. 598 (1971).

123.054 **Suspected variable.** G. E. D. Alcock.
British Astron. Ass., Circ. No. 536 (1971).

123.055 **30 new variable stars in the region M 56.**
N. E. Kurochkin.
Peremennye Zvezdy, Vol. 17, 620 - 637 (1971). In Russian.
The results of an investigation of 30 new variable stars
SVS 1636–1665 are given. The variables were discovered by
the author by means of the colour contrast method in the region $10° \times 10°$ around the globular cluster M 56.

123.056 **Six new variable stars in Virgo, Serpens and Libra.**
R. M. Russev.
Peremennye Zvezdy, Vol. 17, 638 - 646 (1971). In Russian.
The results of the investigation of six new variable stars
SVS 1611 - 1616 discovered by the author in the region
$10° \times 10°$ with the centre $\alpha = 15^h 00^m$, $\delta = +2°$ are presented.

123.057 **Photoelectric observations of RW Coronae Borealis.**
K. Kalchaev, Yu. L. Trutse, V. G. Hamidulina.
Peremennye Zvezdy, Vol. 17, 647 - 651 (1971). In Russian.

123.058 **List of stars.**
Peremennye Zvezdy, Vol. 17, 688 - 696 (1971).
In Russian.
This list contains all variables in Peremennye Zvezdy, Vol. 17
(1969 - 1971) ordered with respect to their constellation.

123.059 **A new variable star in Ophiuchus.** M. Beyer.
Inform. Bull. Variable Stars, (IAU Commission 27),
Konkoly Obs., Budapest, No. 603 (1971).

123.060 **A late type variable in NGC 6819.** U. Lindoff.
Inform. Bull. Variable Stars, (IAU Commission 27),
Konkoly Obs., Budapest, No. 606, 3 pp. (1971).

123.061 **SV Sagittae and BH Lacertae.** E. B. Vovchik.
Tsirk. L'vov. Astron. Obs., No. 46, p. 32 - 33 (1971).
In Russian.

123.062 **AC Geminorum.** O. S. Yatsyk.
Tsirk. L'vov. Astron. Obs., No. 46, p. 34 - 35 (1971).
In Russian.

123.063 **Observations visuelles de la variable irrégulière V348 Sgr en 1970.** M. Duruy.
Centre Univ. Mons, Fac. Sci., Dép. d' Astrophys., Commun.
No. 22, 2 pp. (1971).

123.064 **Mitteilungen über Veränderliche der Bamberger Liste.** R. Knigge.
Veröff. Remeis-Sternw. Bamberg, Astron. Inst. Univ. Erlangen–Nürnberg, Vol. 8, No. 93, 14 pp. (1971).
The positions and the maximal magnitudes of 43 new
and 6 suspected faint variables BV 1310–BV 1358 are given.

The variables were found on plates taken at Bloemfontein
(South Africa) in 1969–1970.

123.065 **Mitteilungen über Veränderliche der Bamberger Liste.** D. Friedrich, E. Schöffel.
Veröff. Remeis-Sternw. Bamberg, Astron. Inst. Univ. Erlangen–Nürnberg, Vol. 8, No. 95, 19 pp. (1971).
The positions and magnitudes of 32 new and 33 suspected faint variables of the southern sky (BV 1405–BV 1469)
are given.

123.066 **Für den Amateur beobachtbare Eruptivveränderliche.** B. Wybranski.
BAV Rundbrief, 20. Jahrgang, p. 1 - 2 (1971).

123.067 **Beobachtungen von Langperiodischen im Jahr 1970.**
E. Heiser.
BAV Rundbrief, 20. Jahrgang, p. 3 - 5 (1971).

123.068 **Beobachtungsdifferenzen bei visuellen Schätzungen.**
E. Heiser.
BAV Rundbrief, 20. Jahrgang, p. 17 - 21 (1971).

123.069 **New variable in Ursa Majoris.** V. P. Goranskij.
Astron. Tsirk., No. 656, p. 7 - 8 (1971). In Russian.

123.070 **Observations of FG Sagittae in 1967–70.**
V. P. Arhipova.
Peremennye Zvezdy, Vol. 18, 183 - 194 (1971). In Russian.
Photoelectric UBV observations of FG Sge were made
during 1967–1970 and are discussed in detail.

123.071 **Variable stars in the Orion nebula.**
V. N. Sincheskul.
Peremennye Zvezdy, Vol. 18, 201 - 211 (1971). In Russian.
In the course of compiling a photometric UBV catalogue
some observations of 103 variable stars in the Orion nebula
were obtained. It was shown that the stars of late spectral
type deviate from the main sequence. They have a surplus
luminosity and a strong ultraviolet colour excess. A list of 28
stars suspected variable is also given.

123.072 **Seven new variable stars.** G. Romano.
Mem. Soc. Astron. Italiana, Nuova Ser., Vol. 42,
639 - 640 (1971).

123.073 **Photoelectric observations of Y Cygni near minima.**
G. V. Zaitseva, V. M. Lyutyj, D. Ya. Martynov.
Astron. Tsirk., No. 662, p. 1 - 5 (1971). In Russian.

123.074 **Observation of AK Cygni.** H. Honda.
Yamamoto Circ., No. 1746 (1971). In Japanese.

123.075 **Peculiar changes of SU Lacertae.**
W. M. Lowder.
AAVSO Abstr., October 1971, p. 13 - 14.
SU Lacertae (1900 coordinates: $22^h 19^m 10^s +55° 00'$)
is described as a Mira-type variable with a mean period of
294.4 days with a note to the effect that this period is probably increasing rapidly. Recent observations reported below tend
to confirm this suggestion.

123.076 **V734 Cygni and MM Sagittarii.** K. B. Kwitter.
AAVSO Abstr., October 1971, p. 15 - 16.

123.077 **Three variables.** E. Hu.
AAVSO Abstr., October 1971, p. 16.

Photographic observations of the eclipsing binary BC Herculis and of the new variable star BD + 12° 3680.
See Abstr. 121.072.

124 Novae

124.001 Wavelength dependence of polarization. XXII. Observations of novae.
B. Zellner, N. D. Morrison.
Astron. Journ., Vol. 76, 645 - 650 (1971).

Extensive observations of HR Delphini (nova Delphini 1967) showed the polarization to be variable by at least 1.2% at intermediate wavelengths but almost constant in the ultraviolet and infrared. Nova Serpentis 1970 also fluctuated in polarization by more than 1%, but no polarization variability was detected for nova Vulpeculae No. 1 1968.

124.002 Wavelength dependence of polarization. XXIII. Dust grains in novae. B. Zellner.
Astron. Journ., Vol. 76, 651 - 654 (1971).

The Mie theory for spherical dust grains was used to compute the polarizations produced by circumstellar dust clouds of low optical depth. Polarization observations during the early stages of nova HR Delphini 1967 can be explained by grains of iron or graphite, but not by pure silicates.

124.003 Catalogue of precise positions of 42 novae.
A. Sh. Khatisov.
Byull. Abastumansk. Astrofiz. Obs., No. 40, p. 13 - 28 (1971). In Russian.

124.004 The principal spectrum of novae. M. Friedjung.
Astron. Astrophys., Vol. 14, 440 - 450 (1971).

The excitation and ionization of the parts of a nova envelope producing the principal absorption system are discussed. Deductions concerning the importance of the emission produced by these regions have been made.

124.005 Some accurate positions of early Korean 'guest stars'. F. R. Stephenson.
Astrophys. Letters, Vol. 9, 81 - 84 (1971).

Accurate positions of 'guest stars' observed in Korea in A D 1163, 1356 and 1399 are deduced. The close agreement between the positions of the stars of 1163 and 1356 and the radio sources 3C 358 and 4C 28.17, although possibly accidental, is commented upon.

124.006 Nova in the Large Magellanic Cloud.
J. A. Graham.
IAU Circ., No. 2353 (1971).

124.007 Novae in the Magellanic Clouds during the 1970− 1971 observing season. J. A. Graham, G. Araya.
Astron. Journ., Vol. 76, 768 - 774 (1971).

A new survey for Magellanic Cloud novae has been initiated at Cerro Tololo Inter-American Observatory. During the period 7 September 1970 to 30 March 1971, two novae were discovered in the Large Magellanic Cloud. The methods used for the survey are discussed. Two photographs identifying the two novae in the surrounding star fields are published. Some photometric observations of the decline of the two novae are also presented.

124.008 Identification of four novae and a supernova in Palomar Sky Atlas. H. Kosai.
Tokyo Astron. Bull., Second Ser., No. 214, p. 2515 - 2518 (1971).

Identifications of pre-nova stars were attempted for four novae and a supernova in Palomar Sky Atlas. They are nova Serpentis 1970, nova Aquilae 1970, nova Scuti 1970, nova Cephei 1971, and supernova in M 63 1971.

124.009 On the cause of the nova outburst − II. Evolution

at $1.00\,M_\odot$.   S. Starrfield.
Monthly Notices Roy. Astron. Soc., Vol. 155, 129 - 137 (1971).

The recent investigation of thermonuclear runaways in the hydrogen-rich envelopes of helium rich white dwarfs as an explanation for the nova outburst has been extended to a mass of $1.00\,M_\odot$. Two models are considered; one is constructed as a more evolved version of the other. Both models are more violent than the models of lower mass. They produced over 10^{46} erg from nuclear reactions, their peak temperatures were greater than $10^8\,^\circ$K, and the maximum rate of energy generation surpassed 10^{14} erg^{-1} g^{-1} s^{-1}.

124.010 Observations of recurrent novae. F. M. Bateson.
Roy. Astron. Soc. New Zealand, Variable Star Section, Circ. No. 165, 2 pp. (1971).

Observations of recurrent novae are summarized as well as observations of stars suspected to belong to this class.

124.011 Aktuelle Beobachtungsobjekte. J. Hübscher.
BAV Rundbrief, 20. Jahrgang, p. 30 - 32 (1971).

124.012 Search for novae in the Andromeda nebula on Abastumani plates. A. S. Sharov.
Astron. Tsirk., No. 655, p. 6 - 8 (1971). In Russian.

The structure of the subsystem of novae in the Andromeda nebula. See Abstr. 158.112.

124.100 Nova Herculis 1963

A study of the Balmer absorption lines of nova Herculis 1963. M. Friedjung.
Astron. Astrophys., Vol. 14, 246 - 251 (1971).

A study of line profiles of some spectra of nova Herculis 1963 has led to demonstrations of the occurrence of inhomogeneities, and the importance of the ionization produced by the absorption of Balmer continuum quanta.

On some characteristics of N Her 1963.
Sh. G. Gordeladze, V. A. Antonyuk.
Problems of cosmic physics. Vyp. (No.) 6, (see 003.040), p. 193 - 198 (1971). In Russian.

124.101 Nova Vulpeculae 1968

Photoelectric photometry of N Vul 1968.
L. N. Kolesnik, A. F. Pugach.
Astrometriya i Astrofiz., *Kiev*, Vyp. (No.) 12, (see 003.003), p. 8 - 10 (1971). In Russian.

124.102 Nova Cephei 1971

Discovery of a nova.
Astron. Tsirk., No. 642, p. 8 (1971). In Russian.

Photovisual light curve of nova Cephei.
V. P. Tsesevich.
Astron. Tsirk., No. 664, p. 7 - 8 (1971). In Russian.

Nova Cephei 1971. Y. Kuwano, M. Matteu.

British Astron. Ass., Circ. No. 534 (1971).

Observation de la nova Cephei 1971.
C. Fehrenbach, Y. Andrillat.
Comptes Rendus Acad. Sci. Paris, Sér. B, Vol. 273, 572 - 576 (1971).

Les spectres de la nova, obtenus à partir du 14 juillet, couvrent tout le domaine spectral de 3400 à 9600 Å. Dans l'ultraviolet et le bleu, ils ont été obtenus au télescope de 152 cm de l'Observatoire de Haute-Provence avec une dispersion de 12 à 20 Å mm⁻¹ ; dans le rouge et l'infrarouge, les observations ont été faites au télescope de 120 cm avec une dispersion de 230 Å mm⁻¹. Dans le bleu, les spectres permettent une analyse très détaillée. Par l'étude des raies H et K du calcium interstellaire qui ont une structure double, les auteurs avons essayé d'éstimer la distance de la nova.

Nova Cephei 1971. Y. Kuwano, K. Ishida, K. Ichimura, M. Mattei, J. Ashbrook, M. Seslar, J. Bortle, R. K. Honeycutt, R. Chaldu.
IAU Circ., No. 2340 (1971).

Nova Cephei 1971. K. Tomita, H. Kosai, W. P. Bidelman, A. J. Weitenbeck, K. Locher, E. H. Mayer, C. E. Scovil, K. Simmons, R. R. Bailey.
IAU Circ., No. 2342 (1971).

Nova Cephei 1971. K. Tomita, H. Kosai, C. Bertaud, T. Seki, G. Comello.
IAU Circ., No. 2343 (1971).

Nova Cephei 1971. P. Wild, E. H. Mayer.
IAU Circ., No. 2345 (1971).

Nova Cephei 1971. G. Comello.
IAU Circ., No. 2350 (1971).

Nova Cephei 1971. R. Burchi.
IAU Circ., No. 2351 (1971).

Nova Cephei 1971. P. Wellmann.
IAU Circ., No. 2353 (1971).

Nova Cephei 1971. A. Kunert, R. Lukas.
IAU Circ., No. 2354 (1971).

Nova Cephei 1971. D. P. Elias.
IAU Circ., No. 2365 (1971).

Accurate position and pre-nova identification for nova Cephei 1971.
G. K. Walker, M. J. Keyes.
Inform. Bull. Variable Stars, (IAU Commission 27), Konkoly Obs., Budapest, No. 583 (1971).

La nova Cephei 1971.
C. Fehrenbach, Y. Andrillat.
L'Astronomie, 85ᵉ année, p. 409 - 411 (1971).

Nova Cephei 1971. R. Lukas.
SuW, Vol. 10, 282 (1971).

Nova Cephei 1971.
Yamamoto Circ., Nos. 1739, 1740, 1741 (1971). In Japanese.

124.103 Nova Delphini 1967

Positions of nova Delphini 1967. I. M. Demenko.
Astrometriya i Astrofiz., *Kiev*, No. 13, (see 003.039), p. 110 - 111 (1971). In Russian.

Nova Delphini 1967. Observations du spectre nebulaire entre 5600 et 8800 Å en 1970.
Y. Andrillat, L. Houziaux.
Astrophys. Space Sci., Vol. 13, 100 - 105 (1971).

Five 230 Å/mm infrared spectra of nova Delphini 1967 have been obtained from May to September 1970 on Kodak hypersensitized IN plates. The continuous spectrum remains strong and a mean value of 2.35 is found for the gradient in the region 6000–8000 Å, indicating a slight increase in temperature since 1969. Permitted lines of H I, O I, He I, Fe II are weakening, while intensities of forbidden transitions due [N II], [Fe VII], [O I], [O II], [A III], [A V], [Fe VII], [Fe X], [Fe XI], [Ni XV] are much more stable.

The spectrum of nova Delphini 1967.
F. M. Stienon.
Bull. American Astron. Soc., Vol. 3, 400 (1971). – Abstr. AAS.

The spectrum of N Del 67 and some remarks on the chemical composition of nova envelopes.
M. Ruusalepp, L. Luud.
Publ. Tartu Astrofiz. Obs., Vol. 39, 89 - 105 (1971).

Observations of nova Delphini (HR Del).
V. Novotný.
Říše hvězd, Vol. 52, 237 (1971). In Czech.

48 visual observations between 9 September 1967 and 24 November 1968.

124.104 Nova Serpentis 1970

Observations of N Ser 1970. N. N. Kiselev.
Astron. Tsirk., No. 638, p. 7 - 8 (1971). In Russian.

Observations of the recovery and final decline of nova Serpentis 1970.
E. F. Borra.
Publ. Astron. Soc. Pacific, Vol. 83, 447 - 448 (1971).

Ten nights of photoelectric observations of nova Serpentis 1970 are presented. Three spectrograms in the nebular stage have been taken. After a small brightening, nova Serpentis 1970 appears to be declining slowly towards minimum.

Some spectroscopic observations of the early decline of nova Serpentis 1970.
N. R. Walborn.
Publ. Astron. Soc. Pacific, Vol. 83, 813 - 816 = Commun. David Dunlap Obs., Univ. Toronto, *Richmond Hill*, No. 303 (1971).

Classification-dispersion spectrograms obtained on the night after maximum light and throughout the early decline of nova Serpentis 1970 are illustrated. The behavior of the diffuse enhanced absorption system is particularly interesting. In addition, one spectrogram of nova Aquilae 1970 is illustrated.

124.105 Nova Sagittarius SVS 1728

Study of the brightness of nova SVS 1728.
V. P. Arhipova, O. D. Dokuchaeva, T. G. Nikulina.
Peremennye Zvezdy, Vol. 18, 195 - 200 (1971). In Russian.

The magnitudes of nova Sgr SVS 1728 were estimated during the pre-maximum rise and in several points after the maximum brightness (J.D. 2440470 - 2440771). The light curve of the nova was constructed.

124.106   Nova Sagittae 1783

Photometry of WY Sge (nova Sagittae 1783).
B. Warner.
Publ. Astron. Soc. Pacific, Vol. 83, 817 - 818 (1971).

124.107   Nova V605 Aquilae

The strange case of V605 Aquilae.
S. van den Bergh.
Publ. Astron. Soc. Pacific, Vol. 83, 819 - 821 (1971).
A 200-inch plate shows that the peculiar slow nova V605 Aql occurred within a few seconds of arc of the center of planetary nebula 37 −5° 1.

125   Supernovae, Supernova Remnants

125.001   A blast-wave model for the Vela X supernova remnant and the origin of the Gum nebula.
W. H. Tucker.
Astrophys. Journ., (Letters), Vol. 167, L85 - L87 (1971).
It is discussed how the various radiations from the Vela X supernova can be understood in terms of the transformation ultimately into radiation of the kinetic energy of the matter ejected in the explosion and the rotational energy of a relict neutron star. In addition, it is shown that for an appropriate choice of parameters, the ionization of the Gum nebula can be explained by the transformation of the kinetic energy of the ejecta into ultraviolet photons over a period $\sim 10^4$ years.

125.002   Radio observations of the supernova remnants IC 443 and Puppis A.   D. K. Milne.
Australian Journ. Phys., Vol. 24, 429 - 440 (1971).
High-resolution 5000 and 2700 MHz brightness distributions and the 2700 MHz polarization distribution are presented for the supernova remnants IC 443 and Puppis A. Directions of the magnetic field are deduced for IC 443. The distribution of spectral index obtained between 1400 and 5000 MHz for IC 443 reflects the annular brightness distribution and shows that there is considerable thermal radio emission from the nebula.

125.003   Expansion of the optical remnant of B Cassiopeiae = 3C 10.   S. van den Bergh.
Astrophys. Journ., Vol. 168, 37 - 39 (1971).
Intercomparison of four 200-inch plates taken between 1949 and 1970 shows that the filamentary shell of Tycho's supernova has suffered considerable deceleration. The presently observed expansion rate is consistent with a Sedov similarity solution. This result suggests that the expanding shell of Tycho's supernova has swept up a mass of interstellar gas that is several times greater than its own mass. Individual filaments exhibit significant changes on a time scale of 10 years.

125.004   On the physical characteristics of the envelopes of type I supernovae during the first period of their expansion. I. The temperature characteristics of the envelopes and an estimate of the number of absorbing atoms.
E. R. Mustel.
Astron. Zhurn. Akad. Nauk SSSR, Vol. 48, 665 - 675 (1971).
In Russian. English translation in Soviet Astron. AJ, Vol. 15, No. 4.
The problem of temperature of the envelopes of type I supernovae around light maximum is discussed in more detail than previously (Astron. Zhurn. Akad. Nauk SSSR, Vol. 48, 3 - 13, 1971). Additional spectrophotometric data and values $B − V$, $U − B$ are taken into account. It is concluded that for

this period of evolution the emission from the envelopes of type I supernovae is essentially thermal. Some important differences between type I supernovae and novae are discussed. A blend of absorption lines $D_{1,2}$ Na I is used for the calculation of the number of the Na atoms in the envelope of the supernova NGC 4496, 1960.

125.005   Supernovae.   B. Baschek.
SuW, Vol. 10, 189 - 193 (1971).

125.006   Observations of six supernovae.
F. Ciatti, R. Barbon.
Mem. Soc. Astron. Italiana, Nuova Ser., Vol. 42, 145 - 161 (1971).
This paper reports observations, carried out at Asiago during the last years, of the following six supernovae: 1965-i, 1965-l in NGC 4753 and in NGC 3631; SN 1966-a, 1966-n in anonymous galaxies at $9^h 12^m + 47°$ and $4^h 34^m −3°$; SN 1968-e in NGC 2713 and SN 1969-b in NGC 3556. From the light curves and spectroscopic data the membership to type I or II is discussed, and other characteristic parameters are deduced.

125.007   Laboratory simulation of absorption bands in type I supernova spectra.
W. W. Duley, W. R. M. Graham.
Journ. Roy. Astron. Soc. Canada, Vol. 65, 181 (1971).
Abstr. RAS Canada.

125.008   Supernova remnants.
P. Gorenstein, W. Tucker.
Sci. American, Vol. 225, No. 1, p. 74 - 85 (1971).

125.009   Absorption lines of Fe II whose lower atomic states are metastable in the spectra of type I supernovae.
E. R. Mustel.
Astron. Tsirk., No. 621, p. 1 - 4 (1971).   In Russian.

125.010   Fossil Strömgren spheres from supernova explosions.
M. C. Kafatos, P. Morrison.
Astrophys. Journ., Vol. 168, 195 - 201 (1971).
Brandt et al. have shown that consistency in the combined observations of the Gum nebula requires a giant H II region, presumably formed by the Vela X supernova explosion. Morrison and Sartori had concluded on the basis of their He II fluorescence theory of type I supernovae that a giant H II region would be formed as result of the ultraviolet burst. (Bottcher et al., by integrating over the light curve, expect a smaller H II region.) We present in brief some consequences of the fluorescence model as illustrated by the Vela X and the

Tycho supernovae. We conclude that such giant H II regions might not in general be as easily detectable as the Vela X region. The Tycho region may just be detectable in the O II, O III forbidden optical lines or as a "hole" in the 21-cm emission-line profiles.

125.011 **The effect of beta processes on the dynamic evolution of carbon-detonation supernovae.**
S. W. Bruenn.
Astrophys. Journ., Vol. 168, 203 - 215 (1971).

The effect of β-processes on the detonation and post-detonation evolution of a degenerate carbon-oxygen core which explosively ignites $^{12}C + {}^{12}C$ at $\rho = 2 \times 10^9$ g cm^{-3} is studied by numerical hydrodynamic calculations. It is found that β-processes, even when increased by a factor of 3, have no qualitative and only minor quantitative effects on the postdetonation evolution. Some preliminary results for nucleosynthesis are discussed.

125.012 **Magneto-rotational explosion of a supernova.**
P. R. Amnuel, O. H. Guseinov, F. K. Kasumov.
Astron. Tsirk., No. 634, p. 1 - 3 (1971). In Russian.

125.013 **Laboratory simulation of absorption bands in type I supernova spectra.**
W. R. M. Graham, W. W. Duley.
Bull. American Astron. Soc., Vol. 3, 389 (1971). – Abstr. AAS.

125.014 **Supernova in edge-on spiral galaxy.**
M. Schmidt.
IAU Circ., Nos. 2352, 2353 (1971).

125.015 **Supernovae.**   M. Schmidt.
IAU Circ., No. 2353 (1971).

125.016 **Supernovae.**   M. Schmidt, J. Kormendy.
IAU Circ., No. 2356 (1971).

125.017 **Observations at 408 MHz of supernova remnants in the LMC.**   J. N. Clarke.
Proc. Astron. Soc. Australia, Vol. 2, 44 - 45 (1971).

125.018 **Soft X-rays from nonthermal galactic radio sources: Implications concerning the galactic background and the interstellar medium.**   S. A. Ilovaisky, C. Ryter.
Astron. Astrophys., Vol. 15, 224 - 234 (1971).

Using presently available data on soft X-ray emission from supernova remnants (SNR) with known distances, empirical correlations have been found between the intrinsic luminosity (in the 0.2 to 1 keV band and at 0.27 keV) and the linear source diameter, which is an age parameter in the theory of SNR. Assuming all SNR discovered in radio surveys follow the correlations, their 0.27 keV flux densities at earth have been computed allowing for absorption in a clumpy interstellar medium. An evaluation of the power input into the interstellar medium due to X-rays between 0.2 and 1 keV emitted from SNR has been made using the corrected number-diameter function, proper galactic distribution parameters, and the luminosity-diameter correlation.

125.019 **A southern supernova search.**   A. P. Fairall.
Monthly Notes Astron. Soc. Southern Africa, Vol. 30, 135 - 138 (1971).

125.020 **Wavelength shifts of intensity minima in type I supernova spectra.**   D. Branch, B. Patchett.
Nature, Phys. Sci., Vol. 233, 29 - 30 (1971).

Wavelength shifts of intensity minima in spectra of type I supernova do not support recent interpretations of the minima as solid state absorption bands.

125.021 **Discovery of a supernova.**
Astron. Tsirk., No. 646, p. 1 (1971). In Russian.

125.022 **A faint supernova in an anonymous southern galaxy.**
S. van Agt, C. Coutts.
Publ. Astron. Soc. Pacific, Vol. 83, 478 (1971).

On a series of eleven Curtis Schmidt plates of the Sculptor dwarf galaxy obtained between 4 August and 11 August 1970 by Coutts, van Agt has found a supernova at $\alpha = 1^h 1^m 8$, $\delta = -35°22'.8$ (1950). It is located in a spiral arm of a faint anonymous galaxy of type Sc II, 17" north and 0" west of the nucleus.

125.023 **Supernovae discovered since 1885.**
C. T. Kowal, W. L. W. Sargent.
Astron. Journ., Vol. 76, 756 - 764 (1971).

All confirmed supernovae discovered between 1885 and May 1971 are listed in two tables.

125.024 **Upper limits to the X-ray luminosity from five supernovae.**
M. Ulmer, V. Grace, H. Hudson, D. Schwartz.
Publ. Astron. Soc. Pacific, Vol. 83, 608 (1971). – Abstr. Astron. Soc. Pacific.

125.025 **The supernova of 1006 A.D.**
D. K. Milne.
Australian Journ. Phys., Vol. 24, 757 - 767 (1971).

A radio investigation is presented of the region in Lupus believed to contain the supernova of 1006 A.D. These observations, which were made with the Parkes 64 m telescope, show a spur or plateau extending from the galactic plane. There are two probable supernova remnants near the position of SN 1006: a large diffuse shell source, the "Lupus Loop", and a more compact object, MSH 14−4*15*, believed to be the remnant of the 1006 A.D. event. The radio emission from both of these sources is strongly polarized.

125.026 **Radio emission from the supernova remnant Puppis A.**   A. J. Green.
Australian Journ. Phys., Vol. 24, 773 - 774 (1971).

This note presents a radio map of the supernova remnant Puppis A at 408 MHz.

125.027 **Supernova in faint ScI galaxy.**   W. K. Ford, V. C. Rubin, T. Zinter.
IAU Circ., No. 2378 (1971).

125.028 **On the abundance of heavy elements in the envelopes of type I supernovae.**   E. R. Mustel.
Astron. Tsirk., No. 649, p. 1 - 3 (1971). In Russian.

125.029 **Observations of IC 443 at 408 MHz.**
G. Colla, C. Fanti, R. Fanti, A. Ficarra, L. Formiggini, E. Gandolfi, C. Lari, B. Marano, L. Padrielli, C. J. Salter, G. Setti, P. Tomasi.
Astron. Journ., Vol. 76, 953 - 956, 1155 - 1156 (1971).

High-resolution observations at 408 MHz (3 by 10 arc min) of the supernova remnant IC 443, made using the Northern Cross radiotelescope are presented. An exceptional coincidence between the radio and optical structures of the nebula is shown to exist. The spectral-index distribution is studied from 408 MHz to 5000 and 6600 MHz. The spectral index is found to be constant over the object, excepting the zones of two small-diameter sources.

125.030 **Relativistic hydrodynamics in supernovae.**
E. Teller.
Physics of high energy density. Course 48, Italian Phys. Soc., 1969, (see 012.023), p. 402 - 418 (1971).

125.031 Supernovae and neutron stars.
S. Calamai, G. Righini.
Scientia, (*Italy*), Vol. 106, 92 - 101 (1971).

125.032 The evolution of supernova remnants.
J. W. Erkes.
Thesis, Graduate College, Univ. Illinois (1968). [Available from Univ. Microfilms, Ann. Arbor, Mich., U.S.A.], 127 pp. (1970).

Relativistic hydrodynamics in one dimension.
See Abstr. 066.107.

Nuclear and non-nuclear processes in the production of peculiar A stars.   See Abstr. 114.084.

Interstellar gas in the direction of the Vela pulsar.
See Abstr. 114.113.

Some accurate positions of early Korean 'guest stars'.   See Abstr. 124.005.

Identification of four novae and a supernova in Palomar Sky Atlas.   See Abstr. 124.008.

Photometry of WY Sge (nova Sagittae 1783).
See Abstr. 124.106.

OH absorption in the direction of W 44.
See Abstr. 131.032.

Considerations on the origin of the Platt particles.
See Abstr. 131.041.

H I absorption measurements on two galactic regions containing supernova remnants and H II regions.
See Abstr. 131.089.

The Cygnus loop at 1420 MHz.
See Abstr. 132.011.

Ionization and heating of the Gum nebula by energetic particles from the Vela X supernova.
See Abstr. 132.036.

Observations of a thin filamentary nebula – The supernova remnant in Monoceros.   See Abstr. 132.041.

Observations of Cygnus Loop at 408 MHz.
See Abstr. 132.042.

Optical studies of Cassiopeia A. IV. Physical conditions in the gaseous remnant.   See Abstr. 141.021.

3C391– a galactic supernova remnant with an unusual radio spectrum.   See Abstr. 141.039.

On the absorption of gamma rays by photons in pulsars, quasi-stellar objects, and other source objects.
See Abstr. 141.186.

The nature of 3C 391.   See Abstr. 141.223.

The polarization of extended radio sources at 6 cm wavelength. II. Galactic sources.   See Abstr. 141.244.

X-ray sources and final stages of stellar evolution.
See Abstr. 142.064.

Origin of cosmic electrons from about 10^2 to 10^6 GeV.   See Abstr. 143.039.

Are the galactic loops supernova remnants?
See Abstr. 155.009.

The radio continuum of galaxies. II. The origin of the continuum emission in spiral galaxies.
See Abstr. 158.064.

125.100 Supernova in NGC 6384

Supernova in NGC 6384.   K. Locher.
IAU Circ., No. 2339 (1971).

Supernova in NGC 6384.   K. Locher.
IAU Circ., No. 2343 (1971).

Supernova in NGC 6384.   F. Seiler.
IAU Circ., No. 2350 (1971).

Supernova in NGC 6384.
Yamamoto Circ., Nos. 1739, 1740 (1971). In Japanese.

125.101 Supernova in NGC 1058

The type II supernova 1969 *l* in NGC 1058.
F. Ciatti, L. Rosino, F. Bertola.
Mem. Soc. Astron. Italiana, Nuova Ser., Vol. 42, 163 - 184 (1971).
   The results of spectroscopic and photographic UBV observations of the type II supernova 1969 *l*, found by Rosino on Dec 2, 1969 at a distance of 227'' from the centre of the Sc galaxy NGC 1058, are given.

125.102 Supernova in NGC 4165

Supernova in NGC 4165.
Astron. Tsirk., No. 618, p. 1 (1971).   In Russian.

Supernova in NGC 4165.   G. N. Kimeridze.
Astron. Tsirk., No. 629, p. 1 (1971).   In Russian.

On the supernova in NGC 4165.
P. G. Kulikovskij, N. A. Gorynya.
Astron. Tsirk., No. 631, p. 1 (1971).   In Russian.

On a supernova in NGC 4165.   G. N. Kimeridze.
Astron. Tsirk., No. 650, p. 7 - 8 (1971).   In Russian.

Photometric observations of supernova 1971 in NGC 4165.
G. de Vaucouleurs, A. de Vaucouleurs, G. S. Brown.
Astrophys. Letters, Vol. 9, 77 - 78 (1971).
   UBV photometry of the supernova in NGC 4165 on April 23, 24, 26 and 27, 1971, gives results consistent with a Type I supernova that reached a maximum $m = 13.2 \pm 0.2$ (B) on April 11 ± 2. The total magnitudes and colors of the galaxies measured through an aperture of 2.50 arc min are $V_t = 13.57$, $(B{-}V)_t = +0.74$, $(U{-}B)_t = +0.21$.

Supernova in NGC 4165.
Zemlya i Vselennaya, No. 5, p. 81 (1971). In Russian.

125.103 **Supernova in NGC 5055**

Supernova in NGC 5055.
Astron. Tsirk., No. 630, p. 1 (1971).   In Russian.

Observations of a supernova in NGC 5055.
G. N. Kimeridze.
Astron. Tsirk., No. 648, p. 8 (1971).   In Russian.

Supernova in NGC 5055.
C. E. Scovil, K. Locher.
IAU Circ., No. 2338 (1971).

Supernova in NGC 5055.
J. D. Wiseman, K. Locher, E. H. Mayer.
IAU Circ., No. 2341 (1971).

Supernova in NGC 5055.
C. Bertaud, C. Pollas, E. H. Mayer, F. Seiler.
IAU Circ., No. 2347 (1971).

Supernova 11. Grösse in Messier 63.   K. Locher.
Orion, 29. Jahrgang, p. 110 - 111 (1971).

Spectrum of supernova in NGC 5055.
S. Kikuchi.
Publ. Astron. Soc. Japan, Vol. 23, 593 - 596 (1971).
    Two spectrograms of the supernova in NGC 5055 were
obtained on May 25.57 and 25.66, 1971 UT. The spectrum
shows the features of type I supernova.

Observations of the supernova in M63.
Sky Telescope, Vol. 42, 85 (1971).

Supernova in NGC 5055.
Yamamoto Circ., Nos. 1739, 1740 (1971). In Japanese.

125.104 **Supernova in NGC 3811**

Supernova in the galaxy NGC 3811.
Astron. Tsirk., No. 630, p. 1 (1971).   In Russian.

Supernova in NGC 3811.
Yamamoto Circ., No. 1739 (1971). In Japanese.

125.105 **Supernova in NGC 7319**

Possible supernova in NGC 7319.   L. Rosino.
IAU Circ., No. 2355 (1971).

Supernovae.   M. Schmidt.
IAU Circ., No. 2356 (1971).

125.106 **Supernova in IC 4798**

Supernova in IC 4798.   A. P. Fairall.
IAU Circ., No. 2359 (1971).

Supernova in IC 4798.   W. J. H. Fisher,
A. C. Gilmore, R. E. Millington.
IAU Circ., No. 2363 (1971).

125.107 **Supernova in NGC 493**

Supernova in NGC 493.
Astron. Tsirk., No. 666, p. 1 (1971). In Russian.

Supernova in NGC 493.   Pigatto.
IAU Circ., No. 2371 (1971).

Supernova in NGC 493. L. Rosino, Pigatto,
Yamamoto Circ., No. 1744 (1971). In Japanese.

125.108 **Supernova in NGC 1090**

Possible supernova in NGC 1090.
Astron. Tsirk., No. 666, p. 1 (1971). In Russian.

Probable supernova in NGC 1090.   C. T. Kowal.
IAU Circ., No. 2376 (1971).

Probable supernova in NGC 1090.
Yamamoto Circ., No. 1745 (1971). In Japanese.

126 Low-luminosity Stars, Subdwarfs, White Dwarfs

126.001 On the production of white dwarfs in binary systems of small mass. S. Refsdal, A. Weigert.
Astron. Astrophys., Vol. 13, 367 - 373 (1971).

For most binary systems of small mass which end up with a He-white dwarf, the evolution through mass transfer is very simple. Therefore, simple formulae can be derived which predict the properties of the final system, e.g. final masses, or orbital velocities. These formulae are discussed, for instance, in view of applications to blue stragglers.

126.002 Quantitave analysis of the O-subdwarf HD 49798. D. Richter.
Atron. Astrophys., Vol. 14, 415 - 427 (1971).

HD 49798, the brightest O-subdwarf known, is analyzed by evaluation of a blue plate, 25 image-tube spectra, scans and broad- and narrow-band colors. We derive the atmospheric parameters by fitting the outer wings of typical lines to profiles computed with LTE-models. First flux-constant unblanketed models are used for computing line-widths. From the interstellar lines Ca II K and Na D an absolute visual magnitude $M_H \approx -0^m2$ and a mass of about $2\,M_\odot$ are estimated.

126.003 The cooling of white dwarfs. Z. F. Seydov, T. A. Eminzade.
Astrofizika, Vol. 7, 306 - 310 (1971). In Russian. – English translation in Astrophysics, Vol. 7, No. 2.

The time of cooling of a white dwarf has been found on the basis of more accurate values of opacity, thermal energy and chemical composition.

126.004 On accretion and convective envelopes in DB white dwarfs. E. M. Sion.
Bull. American Astron. Soc., Vol. 3, 395 (1971). – Abstr. AAS.

126.005 Masses and radii of white dwarfs. H. L. Shipman.
Bull. American Astron. Soc., Vol. 3, 395 (1971). Abstr. AAS.

126.006 Rotating magnetic white dwarf stars as X-ray sources. K. M. V. Apparao.
Nature, Phys. Sci., Vol. 232, 153 - 154 (1971).

Rotating magnetic white dwarf stars with their magnetic axes inclined to their rotation axes (oblique rotators) emit low frequency radiation which can be absorbed by surrounding gas to heat it. This heated gas is shown to be at sufficiently high temperature to emit X-rays. This model is applied to the X-ray source Sco X-1.

126.007 Element abundances in white dwarfs of spectral type DF. G. Wegner.
Proc. Astron. Soc. Australia, Vol. 2, 30 - 31 (1971).

Model atmospheres were computed consistent with T_{eff}, g, and assumed chemical composition. In all the models He^- is the dominant opacity source. Radiative models were first constructed using Lucy's temperature correction method.

126.008 Measurement of limb darkening on the white dwarf BD + 16° 516 B.
B. Warner, E. L. Robinson, R. E. Nather.
Monthly Notices Roy. Astron. Soc., Vol. 154, 455 - 465 (1971).

We have observed primary eclipse in the U band with half second time resolution. A least squares fit of theoretical eclipse curves to two ingresses and one egress leads to an estimate of $u = 0.366 \pm 0.037$ for the limb darkening coefficient. This is in excellent agreement with that for a grey atmosphere.

126.009 The energy of cold white dwarfs. T. A. Éminzade.
Tsirkulyar Shemakhin. astrofiz. observ., 1971, No. 3 (9), p. 10 - 13. In Russian. – Abstr. in Referativ. Zhurn. 51. Astron., 11.51.557 (1971).

126.010 HD 149 382, une sous-naine OB de 9ᵉ magnitude. J. Berger, A.-M. Fringant, E. Rebeirot.
Comptes Rendus Acad. Sci. Paris, Sér. B, Vol. 273, 880 - 883 (1971).

L'étoile HD 149382 classée jusqu'ici B 5 est en fait un des membres les plus brillants actuellement connus de la classe des sous-naines chaudes.

126.011 Spectra of white dwarfs with circular polarization. J. L. Greenstein, J. E. Gunn, J. Kristian.
Astrophys. Journ., *(Letters)*, Vol. 169, L63 - L69 (1971).

Spectra of circularly polarized stars, EG 248, GR 289, and EG 250 (Giclas G99–37, G99–47, and G195–19) were examined, together with spectrophotometric scans. It is suggested that circularly polarized degenerate stars are convective, diluting surface hydrogen and bringing unusual elements to the surface from what were cores of old red giants.

126.012 The subluminous B-type star CD −42° 14462. H. E. Bond, A. U. Landolt.
Publ. Astron. Soc. Pacific, Vol. 83, 485 - 487 = Contr. Louisiana State Univ. Obs., *Baton Rouge,* No. 49 (1971).

Observations of the tenth-magnitude, newly discovered peculiar B-type star CD −42° 14462 are described. Photometrically this object shows a large ultraviolet excess, and its spectrum contains only broad and shallow absorption lines of H and He I. It may be related to the white-dwarf stars, and certainly it is one of the brightest of the subluminous stars now known.

126.013 Rotating white dwarfs in general relativity. G. G. Arutyunian, D. M. Sedrakian, E. V. Chubarian.
Astrofizika, Vol. 7, 467 - 479 (1971). In Russian. – English translation in Astrophysics, Vol. 7, No. 3.

The main integral parameters of rotating white dwarfs are obtained in the Ω^2 (Ω – angular velocity) approximation with and without regard of the neutronization effect. A comparison with the results of the analogical calculations in Newtonian theory is done.

126.014 On helium-rich white dwarfs and cooling sequences. I. Bues.
Observatory, Vol. 91, 221 - 222 (1971). – Letter.

126.015 Vibrational stability of DA white dwarfs. G. Vauclair.
Astrophys. Letters, Vol. 9, 161 - 164 (1971).

The vibrational stability of twenty-five models of DA white dwarfs is studied. Evidence is shown that the efficiency of κ-mechanism can drive the radial pulsations on both the fundamental and the first harmonic modes. The influence of an external shell burning is discussed. The instability region is delimited in the H-R diagram.

126.016 Effective temperature, radius, and gravitational redshift of Sirius B. J. L. Greenstein, J. B. Oke, H. L. Shipman.
Astrophys. Journ., Vol. 169, 563 - 566 (1971).

Analysis of the $H\alpha$ and $H\gamma$ line profiles in Sirius B, along with the apparent magnitude and parallax, yields $T_{eff} = 32000° \pm 1000°$ K, $\log g = 8.65$, and $R/R_\odot = 0.0078 \pm 0.0002$. These values put Sirius B on the mass-radius relation for degenerate configurations of helium. The measured gravitational redshift is 89 ± 16 km s^{-1}, which is in excellent agreement with that predicted by the radius and gravity.

126.017 Measurement of limb darkening on the white dwarf BD + 16° 516 B.
B. Warner, E. L. Robinson, R. E. Nather.
Publ. Astron. Soc. Pacific, Vol. 83, 608 (1971). – Abstr. Astron. Soc. Pacific.

126.018 Radio observations of magnetic white dwarfs.
V. A. Hughes, P. A. Feldman, A. Woodsworth.
Astrophys. Journ., *(Letters)*, Vol. 170, L125 - L126 (1971).
Radio observations at 10.63 GHz (2.8 cm) have been made of a number of magnetic white dwarfs. Upper limits of a few times 10^{-2} flux units are obtained.

126.019 Helium emission white dwarfs. B. Warner.
IAU Circ., No 2374 (1971).

126.020 White dwarf envelope structure and its effect on the thermal cooling rate. A. A. Lacis.
Thesis, Univ. Iowa, Iowa City. [Available from Univ. Microfilms, Ann Arbor, Mich., U.S.A. Order No. 71–5778], 244 pp. (1970).
White dwarf envelope models are computed for a pure helium composition as well as for a composition containing a 0.1 percent metal abundance. The envelope equation is developed in terms of a temperature-degeneracy formulation. Asymptotic relations are derived for the temperature distribution in the nondegenerate surface layers and the highly degenerate interior. The effect of different opacities on the interior core temperature are analyzed.

126.021 A spectroscopic investigation of four O-type subdwarfs. A. V. Peterson.
Thesis, California Inst. Technology, Pasadena. [Available from Univ. Microfilms, Ann Arbor, Mich., U.S.A. Order No. 70–18044], 156 pp. (1970).
The spectra of four O-type subdwarfs, of the class with

strong nitrogen lines and very weak carbon and oxygen lines, have been studied in some detail. Model atmospheres have been constructed for the stars HZ 44, +25°4655, and HD 127493, and the computed profiles of selected hydrogen and helium lines have been compared with the observed profiles.

126.022 Red shifts, white dwarfs, and the Stark effect.
National Bureau of Standards, Techn. News Bull., Vol. 55, No. 9, p. 211 (1971).
A significant portion of the red shift of light coming from hydrogen in white dwarfs is related to interatomic Stark broadening.

126.023 Mass-luminosity relation for white-dwarf stars.
J. Osada, H. V. Capelato, J. R. M. Bonilha.
Rev. Brasil. Fis., Vol. 1, 123 - 128 (1971).
It was found that, though the masses are the same, the luminosity of the white dwarfs which contain light elements in their interior is greater than that of those which contain heavier elements.

126.024 Possible identification of molecular helium, He$_2$ in a white dwarf. D. J. Mullan.
Irish Astron. Journ., Vol. 10, 25 - 28 (1971), with a note added in proof.

Thomson scattering in a strong magnetic field.
See Abstr. 062.046.

Evolution of a 0.6 M$_\odot$ white dwarf.
See Abstr. 065.096.

The ground state of matter at high densities: Equation of state and stellar models. See Abstr. 065.098.

On the origin of magnetic fields in white dwarfs and meson stars. See Abstr. 065.139.

G61–29, a helium emission-line star.
See Abstr. 114.100.

Interstellar Matter, Gaseous Nebulae, Planetary Nebulae

131 Interstellar Space, Interstellar Matter, Polarization of Starlight

131.001 Optical properties of graphite-iron-silicate grain mixtures. N. C. Wickramasinghe, K. Nandy.
Monthly Notices, Roy. Astron. Soc., Vol. 153, 205 - 227 (1971).

We present here a detailed account of the optical properties of graphite-iron-silicate mixtures. We shall discuss our calculations in relation to the observed data on the extinction curve, diffuse galactic light and the backscattering function derived from studies of reflection nebulae. We confine our attention to homogeneous, isotropic spherical particles so that the question of interstellar polarization falls outside the scope of our discussion.

131.002 Detection of H 137α recombination-line emission from an H I region in the direction of NGC 2024. D. A. Cesarsky.
Astrophys. Journ., (Letters), Vol. 167, L89 - L92 (1971).

The H 137α recombination-line emission from an H I region has been observed in the spectrum of NGC 2024 (Orion B). This observation confirms a previous detection of the same feature at the frequency of H 157α. The fraction of ionized hydrogen in the H I region is in good agreement with the earlier determination.

131.003 Observations of 1.35-centimeter H_2O emission in the southern hemisphere. K. J. Johnston, S. H. Knowles, W. T. Sullivan III.
Astrophys. Journ., (Letters), Vol. 167, L93 - L96 (1971).

Results of a search for 1.35-cm H_2O sources with $\delta < -38°$ are presented.

131.004 Discovery of interstellar silicon monoxide. R. W. Wilson, A. A. Penzias, K. B. Jefferts, M. Kutner, P. Thaddeus.
Astrophys. Journ., (Letters), Vol. 167, L97 - L100 (1971).

Line emission attributed to the $J = 3 \rightarrow 2$ rotational transition of SiO has been detected in Sgr B2, at a radial velocity comparable to that of other molecules. The line has not been found in Ori A, IRC+10216, Sgr A, W 51, DR 21, or NML Cyg.

131.005 OH observations at the positions of T Tauri stars in the Taurus-Auriga region. G. F. Gahm, A. Winnberg.
Astron. Astrophys., Vol. 13, 489 - 492 (1971).

The positions of some of the T Tauri stars in Taurus and Auriga were searched for OH 18 cm radiation with the 25.6 m radio telescope of Onsala Space Observatory. Positive results were obtained for SU Aur and RY Tau at 1667 MHz. Radial velocities of T Tauri stars and dust clouds are discussed.

131.006 OH emission sources associated with long-period variable-infrared stars. Nguyen-Quang-Rieu, R. Fillit, M. Gheudin.
Astron. Astrophys., Vol. 14, 154 - 159 (1971).

Four new OH emission sources associated with W Hya, U Her, S CrB and R Cas have been detected during a search for OH emission of the long-period variable-IR stars which show 1.35 cm H_2O emission and/or strong 1.9 μ H_2O absorption. These new sources emit mainly in the 1665−1667 MHz lines. The water-vapor laser line pumping mechanism may contribute to the OH maser process operating in these stars.

131.007 Comparative study of radio and optical photometry of several H II regions. E. F. Schmitter.
Astron. Journ., Vol. 76, 571 - 575 (1971).

Values of optical depth for NGC 896, NGC 1976, NGC 2024, NGC 2237-46, and IC 1795 were calculated at different points of each object. The H II regions were observed at Hα with the 26−31-inch Schmidt telescope at the Observatory of Tonantzintla in Mexico. Expected values of the Hα intensity were calculated from radio data obtained by other observers. The differences between the observed and calculated Hα intensity values were assumed to be due to interstellar extinction produced by material far away from the nebulae.

131.008 Wavelength dependence of polarization. XXIV. Infrared objects. A. Kruszewski.
Astron. Journ., Vol. 76, 576 - 580 (1971).

Polarimetric observations of 11 infrared objects are presented. All observed stars show intrinsic polarization or large interstellar polarization. The interstellar origin is evident for CIT 11. For NML Cygnus, in addition to a large interstellar polarization, there is also a variable intrinsic polarization component. NML Taurus and CIT 6 both show large amplitude variations of the polarization with maximum polarization at light minimum.

131.009 On interstellar grain alignment by a magnetic field. P. G. Martin.
Monthly Notices Roy. Astron. Soc., Vol. 153, 279 - 285 (1971).

The role of the galactic magnetic field in the alignment of interstellar grains believed to cause optical polarization is discussed. The different predictions of Davis-Greenstein alignment and alignment combined with precession are presented. Recent direct observational evidence concerning magnetic alignment raises both positive support and contradictions.

131.010 Physical characteristics of W49A as determined from radio recombination lines. M. A. Gordon, D. C. Wallace.
Astrophys. Journ., Vol. 167, 235 - 243 (1971).

Observations of the 109α and 137β lines of hydrogen at thirteen positions over the thermal component of W49 show the nebula to have a velocity gradient, to have central regions well out of thermodynamic equilibrium, and to have electron temperatures ranging from 8000° K to greater than 10000° K. The nebula is highly clumped, being composed of dense condensations embedded in a low-density gas. The small-scale density structure of W49A, one of the largest H II regions in the Galaxy, appears to be similar to that of the compact H II region M42.

131.011 Non-LTE analysis of data on radio recombination lines for five H II regions. M. H. Andrews, R. M. Hjellming, E. Churchwell.
Astrophys. Journ., Vol. 167, 245 - 248 (1971).

Data on hydrogen recombination lines for M8, W43, W49, W3, and W51 are analyzed to obtain average electron temperatures, emission measures, and electron concentrations.

131.012 Detection of an unidentified emission feature in the microwave spectrum of W3. E. J. Chaisson.

Astrophys. Journ., (*Letters*), Vol. 167, L61 - L64 (1971).

An anomalous emission line reported in the 10.5-GHz microwave spectrum of W3A has also been detected at 7.8 GHz. It has been determined to be a recombination line, and is probably an additional carbon emission.

131.013   A search for near-infrared emission of interstellar molecular hydrogen. T. R. Gull, M. O. Harwit.
Astrophys. Journ., Vol. 168, 15 - 27 (1971).

A search for interstellar molecular hydrogen (H_2) by detection of emission lines within the 3—0 vibrational band has been unsuccessful. No emission lines attributed to H_2 were detected in spectra of regions associated with the Orion nebula, IC 5146, NGC 2264, or IC 1499, or in the close vicinity of ξ Persei.

131.014   Detection of radio recombination-line emission associated with distributed ionized hydrogen.
P. D. Jackson, F. J. Kerr.
Astrophys. Journ., Vol. 168, 29 - 35 (1971).

We have detected (at $l = 31°.2$, $b = 0°.0$) H109α–H110α line emission which is believed to be associated with a distributed component of the galactic ionized gas. We base this association on a comparison of the velocity profile of the line with the corresponding 21-cm-line profile of neutral hydrogen at the same position, and on an estimate of the electron temperature of the emitting gas.

131.015   Influence of cosmic plasma inhomogeneities upon the radio wave absorption in interstellar medium.
Yu. V. Tokarev.
Astron. Zhurn. Akad. Nauk SSSR, Vol. 48, 710 - 715 (1971).
In Russian. English translation in Soviet Astron. AJ, Vol. 15, No. 4.

The radio wave absorption in clouds of interstellar ionized hydrogen is considered. At finite beam width, the equivalent optical depth of a layer of these clouds increases with the decrease of frequency more slowly than for the homogeneous absorbing layer. The results of measuring the cosmic radio emission spectrum at frequencies below 10 MHz are discussed.

131.016   Physical conditions and chemical constitution of dark clouds. C. Heiles.
Annual Rev. Astron. Astrophys., Vol. 9, (see 003.001), 293 - 322 (1971).

This article is primarily concerned with dust clouds, which have gained interest in recent years because they contain molecules in amounts observable by radio astronomy. Clarification of their environment will hopefully result in increased understanding of molecular formation and pumping processes. However, attention should also be directed towards less dense unobscured regions, which have relative molecular abundances comparable to those in the dust clouds.

131.017   Comparative study of radio and optical photometry of several H II regions. II. Calibration procedure, and tables of Hα intensity and optical depth.
E. F. Schmitter, E. Recillas-Cruz.
Bol. Obs. Tonantzintla y Tacubaya, No. 36, Vol. 6, 47 - 67 (1971).

This paper presents tables and contour maps of the Hα intensity and optical depth of the gaseous nebulae NGC 896, NGC 1976, NGC 1982, NGC 2024 and IC 1795. The calibration procedure is discussed in detail.

131.018   Untersuchungen der inneren Struktur interstellarer Wolken. I. Photographisch-photometrische Methode.
S. Marx.
Astron. Nachr., Vol. 292, 235 - 241 = Mitt. Univ.-Sternw. Jena, No. 102 (1971).

The possibilities for investigating the inner structure of interstellar clouds with aid of the photographic photometry are discussed.

131.019   Untersuchungen zur inneren Struktur interstellarer Wolken. II. Ergebnisse für drei Wolken in Cassiopeia.
S. Marx.
Astron. Nachr., Vol. 292, 243 - 250 = Mitt. Univ.-Sternw. Jena, No. 103 (1971).

In a region at a (1900) = $00^h 08^m$ and δ (1900) = $60°.6$ the dependence of the interstellar extinction on the distance was investigated with aid of photographic photometry in the UBV-system. Three clouds of interstellar dust with distances of 800 pc, 1000 pc, and 1100 pc were found and investigated with respect to their inner structure.

131.020   Interstellare Spektren. I. Entdeckungen mehratomiger organischer Moleküle im interstellaren Gas.
H. Lambrecht.
Sterne, 47. Jahrgang, p. 133 - 136 (1971).

131.021   He$^+$/H$^+$ concentration ratios in H I regions.
M. Jura, A. Dalgarno.
Astron. Astrophys., Vol. 14, 243 - 245 (1971).

The observation of helium recombination lines in the hot intercloud medium may allow a choice between the cosmic ray and soft X-ray steady-state models of the interstellar medium.

131.022   Intrinsic variations of interstellar extinction laws.
K. Nandy, W. M. Napier, G. I. Thompson.
Astrophys. Space Sci., Vol. 12, 151 - 157 (1971).

The intrinsic ultraviolet colours of early type stars have been derived from the rocket data of J. W. Campbell and A. M. Smith. There exists a considerable spread in these colours. One possible source is the predicted rms variation due to effect of rotation. Predicted rms errors due to different sources (rotation, uncertainty of MK spectral type, photometric errors) have been compared with the dispersion of the observed extinction data and it is found that part of the large scatter could be due to intrinsic differences of the properties of interstellar grains.

131.023   A study of the wavelength dependence of interstellar polarization using a scanning spectropolarimeter.
R. D. Wolstencroft, K. Nandy.
Astrophys. Space Sci., Vol. 12, 158 - 164 (1971).

A scanning spectropolarimeter has been constructed and used in a preliminary search for conspicuous features of the interstellar polarization curve between $\lambda^{-1} = 1.58 \mu^{-1}$ and $2.50 \mu^{-1}$. Scans were made on HD 2905, HD 21389, α And and β Tau with slitwidths of 50 Å and 100 Å.

131.024   A model of interstellar grains with surface roughness. G. A. Shah, M. S. Vardya.
Astrophys. Space Sci., Vol. 12, 250 - 255 (1971).

A simple model of surface roughness on the interstellar spherical grains has been considered. A test case with dirty ice spherical grains reveals that the effect of surface roughness is to significantly enhance the extinction in the far ultraviolet compared to the equivalent smooth spheres.

131.025   Studies of the 21-cm line in dense dust clouds.
M. J. Mahoney, W. L. H. Shuter.
Journ. Roy. Astron. Soc. Canada, Vol. 65, 175 - 176 (1971).
Abstr. RAS Canada.

131.026   Laboratory studies of interstellar molecules at microwave frequencies. G. Winnewisser.
Journ. Roy. Astron. Soc. Canada, Vol. 65, 182 (1971).
Abstr. RAS Canada.

131.027 **On the interstellar extinction in the direction to the galactic center from observations of planetary nebulae.** V. P. Arhipova, O. D. Dokuchaeva.
Astron. Tsirk., No. 623, p. 3 - 5 (1971). In Russian.

131.028 **Measurement of the one-dimensional distribution function of the signals from some galactic OH radio sources.**
M. I. Paschenko, G. M. Rudnitskij, V. I. Slysh, R. Fillit.
Astron. Tsirk., No. 626, p. 1 - 4 (1971). In Russian.

131.029 **On the polarization of some O-B stars within nebulae.** V. S. Shevchenko, V. I. Kardopolov.
Astron. Tsirk., No. 630, p. 5 - 8 (1971). In Russian.

131.030 **Interstellar carbon monosulfide.**
A. A. Penzias, P. M. Solomon, R. W. Wilson, K. B. Jefferts.
Astrophys. Journ., *(Letters)*, Vol. 168, L53 - L58 (1971).

We have observed line emission from the 146969.16-MHz $J = 3$ to $J = 2$ transition in CS in four sources (Orion A, W51, IRC+10216, and DR21). Typical column densities are near 10^{14} molecules cm^{-2}. The excitation rates required to produce the observed line intensities are used to derive densities for the central regions of the sources.

131.031 **The bright condensation 3 C 153.1 inside the H II region NGC 2175.**
R. Garnier, M. C. Lortet-Zuckermann.
Astron. Astrophys., Vol. 14, 408 - 414 (1971).

Physical parameters are derived for the bright knot inside NGC 2175 using optical dimensions $1' \times 2'$.

131.032 **OH absorption in the direction of W 44.**
W. M. Goss, J. L. Caswell, B. J. Robinson.
Astron. Astrophys., Vol. 14, 481 - 486 (1971).

Four 1667 MHz OH absorption components are seen in the direction of the supernova remnant W 44 at velocities of 12.5, 22, 30 and 42 km s^{-1}. The distribution of the opacity of each component has been determined by observations on a 24-point grid with basic spacing 6 arc min (half the half-power beamwidth). The +42 km s^{-1} feature has a velocity similar to that of the 1720 MHz OH emission lines, and maximum opacity near the position of 1720 MHz emission.

131.033 **A distance determination program for some low-latitude high-velocity neutral hydrogen clouds around longitude 125°.** A. N. M. Hulsbosch.
Astron. Astrophys., Vol. 14, 489 - 492 (1971).

A list is given of 31 early-type stars down to magnitude 11.4 which are seen in projection against some low-latitude high-velocity H I clouds. These stars are very suitable to investigate cloud distances by studying interstellar absorption lines. The expected line strengths are also estimated.

131.034 **Review of research on interstellar dust.**
C. Schalén.
Astron. Nachr., Vol. 293, 1 - 3 (1971). – Introductory remarks to the IAU Colloquium "Interstellar Dust" Jena, August 1969.

131.035 **The orientation parameter of dust grains in three special cases.** P. Cugnon.
Astron. Nachr., Vol. 293, 5 - 8 (1971).

The orientation parameter $F = \langle \cos^2 \vartheta - 1/3 \rangle$ has been obtained by solving a Fokker-Planck equation taking into account a two parameter model of collision and Jones and Spitzer's correction to Davis and Greenstein's interpretation of grain alignment by magnetic relaxation. Three cases have been considered.

131.036 **On the possible existence of dust grains at large distances from the galactic plane.** K.-H. Schmidt.
Astron. Nachr., Vol. 293, 11 - 16 = Mitt.Univ.-Sternw. Jena, No. 104 (1971).

The effect of the radiation pressure of the interstellar radiation field on interstellar grains high above the galactic plane is estimated. It is shown that interstellar dust particles reaching large distances from the galactic plane are concentrated between 140 and 250 pc from the galactic plane.

131.037 **On the explanation of the intrinsic polarization of the light of some red long period variables.**
C. Friedemann.
Astron. Nachr., Vol. 293, 17 - 23 = Mitt. Univ.-Sternw. Jena, No. 105 (1971).

The time dependent polarization of the light of some red long period variables can be explained using the simple model of the oblique rotator. For the stars V CVn, L_2 Pup, R Leo, and R Peg the spatial orientation of the rotational axis, the angle between the magnetic poles and the rotational axis and the period of the revolution of the magnetic poles around the rotational axis were derived from the observational material. The proposed hypothesis gives an explanation of the essential observational facts concerning the intrinsic polarization of some of the red variables.

131.038 **The diffuse interstellar band at 4430 Å and the spectrum of atomic calcium trapped in solid hydrocarbons.** W. W. Duley, W. R. M. Graham.
Astron. Nachr., Vol. 293, 33 - 36 (1971).

Stoeckly and Dressler's suggestion that the diffuse band at 4430 Å might be due to absorption by Ca atoms trapped in interstellar grains has been examined in a series of laboratory experiments. It has been assumed that Ca atoms are trapped in the dielectric mantles of interstellar grains.

131.039 **A diffuse interstellar band in the far ultra-violet?** K. Nandy, H. Seddon.
Astron. Nachr., Vol. 293, 37 - 38 (1971).

A peak of the interstellar extinction curve near $1/\lambda = 4.5\,\mu m^{-1}$ is tentatively interpreted as caused by a diffuse interstellar band.

131.040 **Absorbing ice model of interstellar grains.**
K. S. K. Swamy, W. M. Jackson, B. D. Donn.
Astron. Nachr., Vol. 293, 43 - 48 (1971).

Extinction curves have been calculated for Mie scattering by grains with a refractive index $m = n - ik$ where n was between 1.3 and 2.0 and k varied from 0 to 0.5. A good fit for the observed interstellar extinction in the wavelength region of 0.1 to 2.0 μm was obtained for particles with radius 0.1 μm and $m = 1.3 - 0.2$ i. Theoretical extinction curves for the composite medium of H_2O, NH_3, CH_4 and graphite show a general agreement with the observed interstellar reddening curve over the whole spectral range.

131.041 **Considerations on the origin of the Platt particles.** S. Codina.
Astron. Nachr., Vol. 293, 49 - 51 (1971).

The possibility of the formation of polycyclic hydrocarbon molecules in the atmospheres of very cool stars and in the gases ejected by supernovae and Seyfert galaxies is analyzed. Donn has suggested that such molecules may be Platt particles. It is found out that these can only be originated in some of those sources if their original carbon abundance is very high in relation to the one usually attributed to our Galaxy.

131.042 **Infrared spectra of silicate grains.** J. Dorschner.
Astron. Nachr., Vol. 293, 53 - 55 = Mitt. Univ.-Sternw. Jena, No. 106 (1971).

The observed $10 \mu m$ emission is interpreted as due to Si–O-stretching vibrations of the SiO_4-tetrahedrons of circumstellar silicate grains.

131.043 Evidence for the circumstellar formation of dust.
K.-H. Schmidt.
Astron. Nachr., Vol. 293, 57 - 63 = Mitt. Univ.-Sternw. Jena, No. 107 (1971).

From an examination of the catalogue of early-type stars by Neckel it follows that O stars generally have a mean circumstellar extinction $A_V = 0.89 \pm 0.55$ mag. On the basis of this result the consequences of the hypothesis are considered that all interstellar dust was formed in circumstellar regions in the process of star formation. If dust formation occurs in dense clouds with radii of about 0.1 pc in the neighbourhood of massive hot stars the observed relative abundances of Ca and Na in the interstellar gas can be explained.

131.044 Dust production in circumstellar space.
J. Dorschner.
Astron. Nachr., Vol. 293, 65 - 70 = Mitt. Univ.-Sternw. Jena, No. 108 (1971).

Observational indications which point out that the formation of dust particles in the circumstellar space usually accompanies the formation of most stars are discussed. An analytic treatment of circumstellar crushing processes leading to dust grains is presented. A necessary and sufficient condition for the escape of dust grains produced in circumstellar space by fragmentation is given, and dynamical effects on the grains are estimated.

131.045 The dynamical separation of dust particles during cloud collisions. H. Zimmermann.
Astron. Nachr., Vol. 293, 71 - 74 = Mitt. Univ.-Sternw. Jena, No. 109 (1971).

The dynamical behaviour of dust particles during cloud collisions is calculated.

131.046 Polarization in Be-type stars. M. T. Martel.
Astron. Nachr., Vol. 293, 9 - 10 (1971).

At $1/\lambda = 2.79 \mu m^{-1}$, the comparison of a Be star with a nearby normal B star shows that the interstellar polarization of the Be star is diminished by a factor 0.7.

131.047 Interstellar bands.
G. E. Bromage, M. T. Brück, K. Nandy.
Astron. Nachr., Vol. 293, 39 - 41 (1971).

The presence of an apparent emission wing in the interstellar absorption band at $\lambda = 4430$ Å is confirmed. The observed profile of the bands is in qualitative agreement with recent theoretical predictions.

131.048 Graph of scattering for spherical particles and application to interstellar extinction. S. Isobe.
Ann. Tokyo Astron. Obs., Second Ser., Vol. 12, 263 - 285 (1971).

Extinction coefficients, absorption coefficients, scattering coefficients, radiation pressure coefficients and albedo of ice particles, graphite particles, and graphite core with ice mantle particles, are calculated. The calculations are performed by assuming that the grains are homogeneous spheres and by using Mie's formulae for the simple particle and Güttler's formulae for the composite particle.

131.049 Temperature and thermal evaporation of the interstellar grains surrounding early type stars.
S. Isobe.
Ann. Tokyo Astron. Obs., Second Ser., Vol. 12, 286 - 302 (1971).

Equilibrium temperatures of interstellar grains, which are ice grains, graphite grains and graphite core with ice mantle grains, were calculated in the regions surrounding early type stars. The temperatures sensitively depend on both the grain size and the distance from the central stars. The smaller the grain size and the distance are, the higher the equilibrium temperature.

131.050 Electron density and temperature in the diffuse interstellar medium determined from recombination lines. M. A. Gordon, S. T. Gottesman.
Astrophys. Journ., Vol. 168, 361 - 371 (1971).

An analysis is presented of the H157α and H197β radio recombination lines from regions near the galactic plane believed to be free of discrete radio sources. The results are consistent with a distribution of H II having a mean temperature of approximately 10^3 °K and a mean emission measure of 280 pc cm^{-6} over a path length of 14 kpc, and thus a lower limit of approximately 0.15 cm^{-3} to the rms electron density.

131.051 Ionization equilibrium of interstellar nitrogen: A probe for the intercloud medium?
G. Steigman, M. W. Werner, F. M. Geldon.
Astrophys. Journ., Vol. 168, 373 - 380 (1971).

Using the orbiting-collision approach devised by Field and Steigman, we have calculated the rate coefficients for the charge-exchange reactions $N^+ + H \rightleftarrows N + H^+$ for temperatures between $10°$ and 10^4 °K. It is found that the ionization equilibrium of nitrogen in interstellar space is strongly influenced by this charge-exchange process. The N II resonance line at 1084 Å will be formed only in the intercloud medium, and its strength is shown to be a sensitive indicator of the temperature of this region.

131.052 Detection of millimeter emission lines from interstellar methyl cyanide.
P. M. Solomon, K. B. Jefferts, A. A. Penzias, R. W. Wilson.
Astrophys. Journ., (*Letters*), Vol. 168, L107 - L110 (1971).

The 2.7-mm $J = 6 \rightarrow 5$ transition of methyl cyanide has been detected in the Sgr B and Sgr A molecular clouds.

131.053 Detection of interstellar carbonyl sulfide.
K. B. Jefferts, A. A. Penzias, R. W. Wilson, P. M. Solomon.
Astrophys. Journ., (*Letters*), Vol. 168, L111 - L113 (1971).

The 109.5-GHz, $J = 9 \rightarrow 8$ transition of carbonyl sulfide (OCS) has been detected in the Sgr B molecular cloud.

131.054 Limiting ^{3}He abundances in four H II regions.
C. R. Predmore, H. C. Goldwire, Jr., G. K. Walters.
Astrophys. Journ., (*Letters*), Vol. 168, L125 - L129 (1971).

A search for the 3.46-cm hyperfine line of singly ionized ^{3}He in galactic H II regions has given for M17 a 4 σ upper limit of ^{3}He/H = 5.3 × 10^{-5}. Similar limits were obtained for W3, the Orion nebula, and W51. In conjunction with this measurement, the H171,7 line was detected in the four H II regions observed.

131.055 The collapse of a rotating cloud. R. B. Larson.
Bull. American Astron. Soc., Vol. 3, 366 (1971).
Abstr. AAS.

131.056 Monitoring of time variations in the microwave and infrared continuum flux from OH/IR emission sources. K. P. Bechis, P. Harvey, W. Wilson, E. Becklin, G. Neugebauer.
Bull. American Astron. Soc., Vol. 3, 380 (1971). – Abstr. AAS.

131.057 Damping of density inhomogeneities by drift motions in the interstellar medium.
T. E. Tascione, E. R. Harrison.
Bull. American Astron. Soc., Vol. 3, 381 (1971). – Abstr.

AAS.

131.058 **The effect of spiral structure on the interstellar medium.** W. J. Quirk.
Bull. American Astron. Soc., Vol. 3, 381 - 382 (1971). Abstr. AAS.

131.059 **Properties of H I clouds as determined from radio-frequency recombination lines.** A. K. Dupree.
Bull. American Astron. Soc., Vol. 3, 382 (1971). − Abstr. AAS.

131.060 **Cloud distributions in the Perseus arm.** J. Perry, H. L. Helfer.
Bull. American Astron. Soc., Vol. 3, 382 (1971). − Abstr. AAS.

131.061 **Time variations in interstellar OH and H_2O masers.** W. T. Sullivan III.
Bull. American Astron. Soc., Vol. 3, 388 (1971). − Abstr. AAS.

131.062 **Detection of interstellar isocyanic acid, methyl-acetylene and hydrogen isocyanide.** L. E. Snyder, D. Buhl.
Bull. American Astron. Soc., Vol. 3, 388 (1971). − Abstr. AAS.

131.063 **Simulated interstellar organic chemistry.** B. N. Khare, C. Sagan.
Bull. American Astron. Soc., Vol. 3, 389 (1971). − Abstr. AAS.

131.064 **Broadband structure in the interstellar extinction curve.** D. S. Hayes, J. M. Greenberg, G. E. Mavko, K. H. Rex.
Bull. American Astron. Soc., Vol. 3, 389 - 390 (1971). Abstr. AAS.

131.065 **A general law of interstellar polarization.** G. V. Coyne, T. Gehrels, K. Serkowski.
Bull. American Astron. Soc., Vol. 3, 390 (1971). − Abstr. AAS.

131.066 **H I self-absorption: The galactic center cold cloud and properties of the interstellar medium.** K. W. Riegel, R. M. Crutcher.
Bull. American Astron. Soc., Vol. 3, 395 - 396 (1971). Abstr. AAS.

131.067 **Radio fine structure in H II regions.** B. Balick.
Bull. American Astron. Soc., Vol. 3, 396 (1971). − Abstr. AAS.

131.068 **Similarity models of interstellar loop structures.** W. W. Zuzak.
Astron. Astrophys., Vol. 15, 95 - 109 (1971).

It is hypothesized that the interstellar loop structures are continuously driven by means of cosmic rays emitted by some source near the centre. Spherical similarity solutions, in which the shock position varies with time according to $R = At^\alpha$, are obtained in the approximation that all fluid motion is radial, the magnetic field is tangential, and the diffusion coefficient η is zero. Any desired shell thicknesses may be obtained by considering the radiation losses to occur at the shock and thereafter treating the shocked gas adiabatically. Information available on Loops I and II (North Polar Spur and Cetus Arc) indicate that these have undergone relatively large cooling and are driven by a relatively small energy input.

131.069 **A 1665 MHz OH survey of the southern Milky Way.** B. J. Robinson, J. L. Caswell, W. M. Goss.
Astrophys. Letters, Vol. 9, 5 - 8 (1971).

A search at 1665 MHz of 218 galactic continuum sources with $256° < l < 352°$ has detected 18 new OH emission sources. Many of the new sources have been observed on all four 18 cm OH transitions with both hands of circular polarization. One of the new sources (OH 330.9−0.4) is the strongest main-line emitter yet discovered, with a peak flux density of 498 flux units at 1665 MHz. In addition, our 1665 MHz search detected prominent OH absorption against 13 of the more intense continuum sources; subsequent measurements on the other 18-cm transitions show that half of these features observed in absorption at 1665 and 1667 MHz display emission on the satellite lines.

131.070 **Sputtering of ice grains in H II regions.** M. J. Barlow.
Nature, Phys. Sci., Vol. 232, 152 - 153 (1971).

It is shown that previous treatments of the problem of the thermal sputtering of ice grains in H II regions (e.g. W. G. Mathews, Astrophys. Journ., Vol. 157, 583 - 599, 1969) have used incorrect values for the sputtering coefficients of ice bombarded by hydrogen and helium ions. New estimates for the appropriate sputtering yield coefficients increase the lifetime of ice grains against thermal sputtering by a factor of several thousand over the previous estimates, leading to the conclusion that ice grains in H II regions will not be significantly affected by thermal sputtering.

131.071 **Irradiated quartz particles as interstellar grains.** N. C. Wickramasinghe.
Nature, Phys. Sci., Vol. 234, 7 - 10 (1971).

Quartz particles irradiated by cosmic rays or X-rays may be able to account for the observed properties of interstellar grains over the wavelength range ∼ 1600 Å to 20 μm.

131.072 **The profiles of radio recombination lines.** M. Brocklehurst, M. J. Seaton.
Astrophys. Letters, Vol. 9, 139 - 142 (1971).

Using the theory of impact broadening and the observed profiles of radio recombination lines, Brocklehurst and Leeman (1971) have shown that the high lines must be formed in regions of low density. In order to interpret all of the observed line profiles and intensities it is necessary to consider models of variable density. A convenient approximate solution of the transfer equation is obtained and it is shown that, using plausible density distributions, one can obtain calculated results in agreement with observations.

131.073 **Amino-acid synthesis from gases detected in interstellar space.** G. Wollin, D. B. Ericson.
Nature, Vol. 233, 615 - 616 (1971).

Compounds which play an important role in the origin of life may have been formed without water in interstellar space. Experiments show that reaction products from different combinations of ammonia, methanol, formic acid, and formaldehyde gases when subjected to ultraviolet light react to produce amino-acids and peptides. The absence of water in the experiments suggests that perhaps such compounds as amino-acids can be formed on the waterless moon.

131.074 **Radio detection of interstellar acetaldehyde.** J. A. Ball, C. A. Gottlieb, A. E. Lilley, H. E. Radford.
IAU Circ., No. 2350 (1971).

131.075 **Detection of the O^{18} isotope of formaldehyde.** F. F. Gardner, J. C. Ribes, B. F. C. Cooper.
IAU Circ., No. 2354 (1971).

131.076 Radio detection of interstellar thioformaldehyde.
M. W. Sinclair, J.-C. Ribes, N. Fourikis, R. D. Brown, P. D. Godfrey.
IAU Circ., No. 2362 (1971).

131.077 Transient OH source in W75.   J. Elldér.
IAU Circ., No. 2364 (1971).

131.078 New OH emission sources.
B. J. Robinson, J. L. Caswell, W. M. Goss.
Proc. Astron. Soc. Australia, Vol. 2, 36 - 38 (1971).

We have searched for OH in 218 galactic sources in the longitude range 256° to 352°. The sensitivity was a factor of six better than in the original McGee et al. survey, and the radial velocity coverage was increased by a factor of four. The new search was at 1665 MHz and was intended primarily to detect new emission sources. Eighteen OH emitters were found; all of these have since been observed at 1667 MHz and 11 have been observed on the satellite transitions.

131.079 Effect of shock wave dissipation on the stability of the interstellar medium.   J. A. Burke.
Monthly Notices Roy. Astron. Soc., Vol. 154, 385 - 391 (1971).

Heating of the interstellar medium by dissipation of weak shock waves adds somewhat to its stability against gravitational collapse or thermal changes. The range of gravitationally unstable long wavelengths is generally diminished, and the tendency toward thermal instability at short wavelengths is lessened.

131.080 Galactic absorption.   T. W. Noonan.
Astron. Soc. Pacific, Leaflet No. 506, 8 pp. (1971).

131,081 Interaction between interstellar helium and the solar wind.   T. E. Holzer, W. I. Axford.
Journ. Geophys. Res., Vol. 76, 6965 - 6970 (1971). − Letter.

131.082 On the interstellar absorption in the direction of the Crab nebula.   V. V. Golovatyj.
Tsirk. L'vov. Astron. Obs., No. 45, p. 11 - 16 (1971). In Russian.

131.083 Microwave detection of interstellar formamide.
R. H. Rubin, G. W. Swenson, Jr., R. C. Benson, H. L. Tigelaar, W. H. Flygare.
Astrophys. Journ., *(Letters)*, Vol. 169, L39 - L44 (1971).

Formamide was detected by its microwave emission from the $2_{11} \rightarrow 2_{12}$ rotational transition at ~4620 MHz in the direction of Sgr B2 and possibly Sgr A. There is evidence that all three of the $\Delta F = 0$ hyperfine components are present.

131.084 Eine Untersuchung des Entfernungsverlaufes der interstellaren Extinktion in der Umgebung von SV Cephei.   H.-E. Fröhlich, S. Rößiger.
MVS, *Sonneberg*, Vol. 6, 1 - 8 (1971).

For a small region around the variable star SV Cep the relation between interstellar extinction and distance from the earth is derived from partly new photometric and spectrographic data of surrounding stars.

131.085 Interpretation of recombination-line emission from the interstellar medium.   C. J. Cesarsky, D. A. Cesarsky.
Astrophys. Journ., Vol. 169, 293 - 298 (1971).

The recombination-line emission detected by Gottesman and Gordon, which they attribute to a diffuse ionized component of the interstellar medium, can also be explained as resulting from a discrete distribution of cold and dense clouds.

131.086 Polarimetry of red and infrared stars at 1 to 4 microns.   H. M. Dyck, F. F. Forbes, S. J. Shawl.
Astron. Journ., Vol. 76, 901 - 915 (1971).

Polarimetric data in the $1-4-\mu$ spectral range are presented and discussed for 64 stars, mostly of late spectral type. It is shown that large polarization in the near infrared is usually associated with extreme circumstellar shell characteristics among the cool stars. Particular attention has been paid to VY CMa which has previously been shown to be peculiar polarimetrically by Forbes. We develop a qualitative model in which the circumstellar shell contains *two* discrete particle sizes, one of the order of 1 μ (radius) and the other of the order of 0.1 μ. For a hypothetical optically thin analog, it is demonstrated that all of the polarimetric features observed in VY CMa can be reproduced in an asymmetric circumstellar envelope containing spherical grains of (Mg, Fe) SiO_3.

131.087 Chemical constituents of interstellar clouds.   D. Buhl.
Nature, Vol. 234, 332 - 334 (1971).

The application of organic chemistry to the interstellar medium has considerably heightened the appreciation of the complexity and variety of interstellar clouds. Results obtained during the past three years suggest that only a start has been made in the identification of the molecular constituents of these clouds.

131.088 From radio astronomy towards astrochemistry.   D. Buhl, L. E. Snyder.
Technology Rev., Vol. 73, No. 6, p. 1 - 10 = National Radio Astron. Obs., Green Bank, Repr. Ser. A, No. 207 (1971).

131.089 H I absorption measurements on two galactic regions containing supernova remnants and H II regions.   I. Kazès.
Astron. Astrophys., Vol. 15, 460 - 467 (1971).

Through H I absorption measurements, distances have been estimated for three supernova remnants and their nearby H II regions. No evident physical association exists between these two kinds of galactic objects. However, G348.8−0.6, an H II region, and G350.0−0.3, presumably a thermal radio source, have been found to be related.

131.090 Erratum: 'A distance determination program for some low-latitude high-velocity-neutral hydrogen clouds around longitude 125°' [Astron. Astrophys., Vol. 14, 489 - 492 (1971)].   A. N. M. Hulsbosch.
Astron. Astrophys., Vol. 15, 473 (1971).

131.091 Observations of high velocities in H II regions with a two-etalon Fabry-Perot spectrometer.
J. Meaburn.
Astrophys. Space Sci., Vol. 13, 110 - 127 (1971).

High internal motions of the ionized material in the H II regions M8, M16, M17 and the Orion nebula were searched for with a two-etalon Fabry-Perot monochromator. The profiles of the [O III], 5007 Å and in one case the 4959 Å line were obtained at many positions from these nebulae.

131.092 Polarization at 4430 Å.   A. Kelly.
Astrophys. Space Sci., Vol. 13, 211 - 218 (1971).

The wavelength dependence of polarization in the region of the interstellar absorption band at 4430 Å is investigated theoretically by variation of the relevant parameters over a wide range. Comparison is made with observations, and the suggestion that the 4430 Å band is produced by impurities in a silicate matrix is found to be not inconsistent.

131.093 Mechanisms of molecule formation.   D. A. Williams.
Observatory, Vol. 91, 225 - 227 (1971). − Letter.

131.094 **Observations of the excited lines of OH near 4700 MHz.** F. F. Gardner, J. C. Ribes.
Astrophys. Letters, Vol. 9, 175 - 179 (1971).

Profiles of the narrow-band emission at 4765 ($1 \rightarrow 0$) and 4660 ($0 \rightarrow 1$) MHz with 1-kHz resolution are presented together with the first astronomical detection of the $F = 1 \rightarrow 1$ transition at 4750 MHz. Upper limits to broad-band or narrow-band emission from four other southern OH sources are given.

131.095 **Detection of the O^{18} isotope of formaldehyde at 4388 MHz.**
F. F. Gardner, J. C. Ribes, B. F. Cooper.
Astrophys. Letters, Vol. 9, 181 - 183 (1971).

In this paper we compare absorption profiles for the three isotopic species of formaldehyde. The optical depths are now low for both the C^{13} and O^{18} species, and this permits an accurate estimate of the C^{13}/O^{18} abundance ratio on the assumption that the excitation temperatures are the same.

131.096 **The mean and the mean squared electron density in interstellar space.** M. Walmsley, M. Grewing.
Astrophys. Letters, Vol. 9, 185 - 188 (1971).

We suggest that the late O and early B type stars situated in low density regions could account for much of the current observational data relating to the ionisation in the interstellar medium. The recently observed recombination line radiation, however, requires a separate explanation.

131.097 **Chemicals in the sky.** S. Mitton.
Astron. Soc. Pacific, Leaflet No. 507, 8 pp. (1971).

131.098 **Radiofrequency detection of an anomalous interstellar recombination line.**
E. J. Chaisson, J. A. Ball.
Astrophys. Journ., Vol. 169, 495 - 501 (1971).

An anomalous microwave emission line discovered in the 94α spectrum toward W 49 A and independently observed at 85α is shown to be the result of electronic recombination. This paper describes the observed physical characteristics of the spectral line and considers several possible origins.

131.099 **Excitation temperatures of the 18-centimeter OH transitions in an absorbing cloud.**
R. N. Manchester, M. A. Gordon.
Astrophys. Journ., Vol. 169, 507 - 514 (1971).

Observations at the 18-cm OH transition frequencies have been made on, and adjacent to, the strong absorption source W 12. The on-source observations show that the line ratios do not vary significantly across the main 9.5 km s^{-1} feature, and they thus imply that the excitation temperatures are approximately constant. In the weaker absorption between 10 and 13 km s^{-1} the excitation temperatures depart greatly from thermodynamic equilibrium and are a strong function of radial velocity. The results show that for the OH in the strong absorption feature, all four excitation temperatures are close to the background-radiation temperature. The implications of this result are discussed, and we conclude that the OH is most probably imbedded in a dense H I region having a kinetic temperature of 5° K or less. Optical depths, projected density, and filling factor of the OH cloud are also calculated.

131.100 **New ammonia lines and sources in the Galaxy.**
B. Zuckerman, M. Morris, B. E. Turner, P. Palmer.
Astrophys. Journ., (*Letters*), Vol. 169, L105 - L108 (1971).

Microwave emission from two nonmetastable rotational levels of ammonia (NH$_3$) has been observed in the direction of the continuum source Sgr B2. Microwave lines from NH$_3$ were also observed in W3-OH, Orion A, W43, W51, DR 21-OH, and possibly in Cloud 4, a dense dust cloud.

131.101 **High-speed interstellar gas dynamics: Shocks moder-** ated by cosmic rays. D. G. Wentzel.
Astrophys. Journ., Vol. 170, 53 - 63 (1971).

Cosmic rays interact with the thermal interstellar gas through a 'cosmic-ray sound speed' of the order of (cosmic-ray energy density/thermal gas density)$^{1/2}$, typically $10 - 10^2$ km s^{-1}. Gas motions faster than the thermal sound speed may be moderated by the cosmic rays. The cosmic-ray sound speed and the jump conditions across cosmic-ray-dominated shocks are derived here on the assumption that the cosmic rays and the interstellar gas interact via resonant hydromagnetic waves.

131.102 **Formation of OH through inverse predissociation.**
P. S. Julienne, M. Krauss, B. Donn.
Astrophys. Journ., Vol. 170, 65 - 70 (1971).

Formation of OH can occur by inverse predissociation from continuum levels to the $\nu = 1$, $k = 1$ level of the $A^2\Sigma^+$ state. A rate constant of $1 - 3 \times 10^{-20}$ cm^3 s^{-1} is calculated for temperatures greater than 20° K. Predicted OH densities are consistent with observations in dense, heavily obscured clouds but appear to be somewhat low for H I clouds.

131.103 **The meaning of the OH-H$_2$O maser maps.**
M. M. Litvak.
Astrophys. Journ., Vol. 170, 71 - 80 (1971).

The approximate conditions of density, diameter, and temperature of the OH and H$_2$O maser regions are deduced from the very long baseline interferometry, from the presence of microwave saturation, from the energy requirements, from the absence of excessive pressure or thermal broadening, from the microwave optical depths, and from the selection of hyperfine components. The amplifier surrounding each emission point has a diameter comparable to the distance between neighboring points and a mass close to 1 $M_\odot$. Calculations indicate that the nonlinear effect of self-focusing is probably not occurring, that stimulated Raman scattering is yielding only the lowest hyperfine component in H$_2$O with a lineshape asymmetry toward lower frequency, and that for both OH and H$_2$O the interferometers are observing "hot spots" of unsaturated amplification.

131.104 **Etat actuel de la connaissance du milieu interstellaire.** P. Cugnon.
Ciel et Terre, Vol. 87, 596 - 605 (1971).

131.105 **Kondensierte Materie im Kosmos. IV. Interstellarer Staub.** J. Dorschner.
Sterne, 47. Jahrgang, p. 161 - 173 (1971).

131.106 **The absence of formaldehyde radiation toward cold regions of the galactic plane.**
M. A. Gordon, M. S. Roberts.
Astrophys. Journ., Vol. 170, 277 - 279 (1971).

A search for H$_2$CO along the galactic equator in directions free of discrete continuum sources failed to detect the 6-cm line in either absorption or emission. By using observations of Cas A and regions near it, we suggest that the excitation temperature of H$_2$CO in these directions lies between 1.3 and 4.3° K.

131.107 **Accurate position measurements in the 1720-MHz line of OH.** E. G. Hardebeck.
Astrophys. Journ., Vol. 170, 281 - 288 (1971).

Positions accurate to ±10″ were measured with the Owens Valley interferometer for seven sources in the 1720-MHz line of OH. These included three Class I sources (W3, W49, and W51) and four Class IIa sources (W28 A$_1$, W44, OH 1959+33 [ON-3], and G111.5+0.8). OH 1959+33 was found to lie close to component C of the complex radio source near NGC 6857, instead of near component A (the unusual nebula K 3-50).

131.108 Nonthermal OH emission in interstellar dust clouds. B. E. Turner, C. Heiles.
Astrophys. Journ., Vol. 170, 453 - 462 (1971).

Satellite-line emission from OH molecules has been detected for the first time in dust clouds. Although the main lines have strengths characteristic of LTE, the satellite lines do not. The 1720-MHz transition is anomalously strong compared with the main-line emission, while the 1612-MHz transition is anomalously weak. The ratio of main lines is used to derive column densities and excitation temperatures.

131.109 New galactic H_2O sources associated with H II regions. B. E. Turner, R. H. Rubin.
Astrophys. Journ., *(Letters)*, Vol. 170, L113 - L118 (1971).

Eight new galactic H_2O emission sources have been found in H II regions at positions all of which appear to coincide with type I OH emission. No H_2O emission was detected in 20 continuum sources which show OH in absorption, even though these sources all had apparent OH opacities of at least 0.15. These facts are discussed in terms of two possible pumping mechanisms for the H_2O and type I OH emission.

131.110 Radiofrequency recombination lines as diagnostics. of the cool interstellar medium. A. K. Dupree.
Astrophys. Journ., *(Letters)*, Vol. 170, L119 - L123 (1971).

Intensity ratios of α and β transitions of recombination lines can be used to determine the temperature and electron density in H I clouds. The few available observations of the carbon line in Orion A and NGC 2024 suggest preliminary values of $N_e \gtrsim 1$ cm^{-3} for $T_e \lesssim 100°$K.

131.111 Radio detection of $2_{12} - 2_{11}$ transition of interstellar acetaldehyde. N. Fourikis, M. W. Sinclair, R. D. Brown, P. D. Godfrey, G. Lackman.
IAU Circ., No. 2379 (1971).

131.112 Formaldehyde and ammonia as precursors to prebiotic amino acids. H. R. Hulett, Y. Wolman, S. L. Miller, J. Ibanez, J. Oró, S. W. Fox, C. R. Windsor.
Science, Vol. 174, 1038 - 1041 (1971).

131.113 Cosmic rays and interstellar matter. F. D. Kahn.
Quarterly Journ. Roy. Astron. Soc., Vol. 12, No. 4, (see 012.012), 384 - 401 (1971).

131.114 Interstellar molecules — the optical region. D. McNally.
Highlights of Astronomy, Vol. 2, (see 012.016), 339 - 349 (1971).

131.115 Infrared observations and interstellar molecules. N. J. Woolf.
Highlights of Astronomy, Vol. 2, (see 012.016), 350 - 358 (1971).

131.116 Microwave evidence for interstellar molecules. C. H. Townes.
Highlights of Astronomy, Vol. 2, (see 012.016), 359 - 365 (1971).

131.117 Molecules in dense clouds and protostars. P. G. Mezger.
Highlights of Astronomy, Vol. 2, (see 012.016), 366 - 377 (1971).

131.118 OH as a constituent of the interstellar medium. B. E. Turner.
Highlights of Astronomy, Vol. 2, (see 012.016), 378 - 390 (1971).

131.119 Λ doublet radiation from OH excited rotational states. B. Zuckerman.
Highlights of Astronomy, Vol. 2, (see 012.016), 391 - 393 (1971).

131.120 Interstellar formaldehyde. P. Palmer.
Highlights of Astronomy, Vol. 2, (see 012.016), 394 - 401 (1971).

131.121 The relative density of H, OH and H_2CO in interstellar clouds. R. D. Davies.
Highlights of Astronomy, Vol. 2, (see 012.016), 402 - 403 (1971).

131.122 NH_3 and H_2O emission in our Galaxy. D. M. Rank.
Highlights of Astronomy, Vol. 2, (see 012.016), 404 - 406 (1971).

131.123 Radio emission from interstellar hydrogen cyanide and X-ogen. L. E. Snyder, D. Buhl.
Highlights of Astronomy, Vol. 2, (see 012.016), 407 - 412 (1971).

131.124 Laboratory studies of the spectra of interstellar molecules. G. Herzberg.
Highlights of Astronomy, Vol. 2, (see 012.016), 415 - 420 (1971).

131.125 Interstellar molecule formation; radiative association and exchange reactions. W. Klemperer.
Highlights of Astronomy, Vol. 2, (see 012.016), 421 - 428 (1971).

131.126 Molecule formation on grain surfaces. E. E. Salpeter.
Highlights of Astronomy, Vol. 2, (see 012.016), 429 - 431 (1971).

131.127 Photochemistry of atoms and molecules in the adsorbed state. H. D. Breuer, H. Moesta.
Highlights of Astronomy, Vol. 2, (see 012.016), 432 - 437 (1971).

131.128 Interstellar molecules: Final remarks. J. L. Greenstein.
Highlights of Astronomy, Vol. 2, (see 012.016), 460 - 462 (1971).

131.129 The beginnings of organic cosmochemistry. B. Kuchowicz.
Postępy Astron., Vol. 19, 299 - 312 (1971). In Polish.

Since three years the application of radio telescopes to a search for molecular spectral lines from space started what may be now called the organic cosmochemistry. Sixteen kinds of complex molecules are known to exist in the gas clouds of interstellar space (in April 1971). Sophisticated methods of laboratory investigations on the moon dust and meteorite matter provide additional tools for studying the extraterrestrial chemical evolution.

131.130 A search for OH in nine high-latitude Selected Areas. F. J. Kerr, G. R. Knapp.
Astron. Journ., Vol. 76, 993 - 994 (1971).

A search has been made for OH emission at nine points at high galactic latitudes. A positive result was obtained for one region, and is thought to originate from a dust cloud at that position.

131.131 Some OH and formaldehyde properties of W 3 and

W 51 regions.   J. Elldér.
Res. Lab. Electronics, Chalmers Univ. Technology, Gothenburg, Sweden, Res. Report No. 96, 28 pp. = Preprint Onsala Space Obs., Onsala (1970).

The purpose of this investigation was to compare the continuum radiation and the spectral line radiation in the directions of W 3 and W 51. Both sources have comparatively strong OH and H_2CO features.

131.132 On the evolution of Strömgren spheres.
     R. Bartkus.
Bull. Vilnius Astron. Obs., No. 32, p. 42 - 50 (1971). In Russian.

Evolution of a typical massive cloud (R = 6.5 pc, N_H = 20 cm^{-3}) immersed in a hot inter-cloud medium of unionized hydrogen under action of a massive star (11 $M_\odot$ – 30 $M_\odot$) born in the centre of the cloud is studied semi-quantitatively.

131.133 Condensation of solid hydrogen in contracting interstellar clouds.   T. Nakano.
Progr. Theor. Phys., Japan, Vol. 45, 1737 - 1746 = National Radio Astron. Obs., *Green Bank,* Repr. Ser. A, No. 216 (1971).

We investigate the condensation of solid hydrogen in contracting clouds in order to clarify how much hydrogen can condense and how much the contraction of the cloud is affected by the condensation.

131.134 Planetary nebulae. IV. Predicted chemical composition and interstellar enrichment.
S. Torres-Peimbert, M. Peimbert.
Bol. Obs. Tonantzintla y Tacubaya, No. 37, Vol. 6, 101 - 111 (1971).

Based on stellar evolution models and on observations of planetary nebulae we have studied the change of chemical abundances in the interstellar medium. We analyze the chemical abundances of the nuclei of M 51 and M 81 from stellar evolution results and the observed emission lines.

131.135 The luminescence of isolated absorbing clouds affected by the general radiation field of galactic stars.
D. A. Rozkovsky.
Trudy Astrofiz. Inst., *Alma-Ata,* Vol. 16, 86 - 97 (1971). In Russian.

An approximate computation of the luminescence of a spherical cloud of interstellar grains scattering the integral radiation of stars of the Galaxy is made. The application of the theory to the interpretation of observed brightnesses of two obscuring regions of the Milky Way showed a comparatively low value of the albedo of interstellar grains.

131.136 A study of dark nebulae.
B. J. Bok, C. S. Cordwell.
Separate print Steward Obs., Univ. Arizona, Tucson, Arizona. 2 + 54 pp. (1971).

We present first a brief summary of different types of dark nebulae together with representative properties for each variety we discuss. Next, we discuss the methods for finding approximate absorptions and distances of dark clouds through the use of star count data and from results obtained with modern color techniques. A summary of available catalogs and photographic atlases of dark nebulae is given. In the concluding sections of the paper we present three tables with lists of positions and properties of some known dark nebulae.

131.137 Interstellar molecules and dense clouds.
     D. M. Rank, C. H. Townes, W. J. Welch.
Science, Vol. 174, 1083 - 1101 (1971). – Review article: Optical identification of interstellar molecules; Radio molecular lines; Molecules as probes of interstellar clouds; Probes of isotopic abundances; Probes of the radiation field; Formation and disappearance of molecules; Conclusions and future prospects.

131.138 OH spectral line measurements of radiation from the galactic H II regions W 3 and W 49 at 1665 MHz.
J. Elldér.
Res. Lab. Electronics, Chalmers Univ. Technology, Gothenburg, Sweden, Res. Report No. 87, 3 + 94 pp. (1968).

Spectral line measurements of radiation from the 18 cm Λ-type multiplet of OH in the ground state have been made at Onsala Space Research Observatory. During the years 1966 and 1967, two regions in the sky, viz. W 3 and W 49, were studied more closely at a frequency of 1665.401 MHz. Special attention was devoted to long term frequency and intensity variations of the spectral line features. No frequency variations exceeding the limits set by the frequency stability of the receiver (10^{-9} per day) have been detected. The intensity variations were less than ±20 % during the same period.

131.139 OH excited state emissions from W 75 B and W 3, OH.
O. E. H. Rydbeck, E. Kollberg, J. Elldér.
Res. Lab. Electronics, Chalmers Univ. Technology, Gothenburg, Sweden, Res. Report No. 97, 1 + 20 pp. (1970).

Strongly varying 6035 MHz emission from the F = 3 → 3 transition of the $^2\Pi_{3/2}$, J = $^5/_2$ state of OH has been detected in W 75 B with the Onsala 84-foot, maser equipped telescope. Detailed Onsala observations of the F = 3 → 3, and 2 → 2 lines from W 3 – OH, have shown that they are also polarized much like the W 75 B emissions, but lack detectable temporal variations.

131.140 Very long baseline interferometry of galactic OH sources.
B. O. Rönnäng, O. E. H. Rydbeck, J. M. Moran.
Res. Lab. Electronics, Chalmers Univ. Technology, Gothenburg, Sweden, Res. Report No. 100, 1 + 16 pp. (1970).

Several hydroxyl radical (OH) microwave emission sources were studied in July 1969 using very long baseline interferometry. Previous spectral line interferometric observations have given detailed information about the angular sizes and spatial separations of the 1665 MHz OH sources in W 3. The measurements reported here confirm these results and give additional information about the complex structure of the different features. Four other galactic OH emission sources with unknown angular sizes were also investigated. Only one, the 5 km/s component in the 1667 MHz spectrum of W 49 gave reliable fringes.

131.141 Between the stars.   I. R. Gordon.
     Southern Stars, Vol. 24, 62 - 63 (1971).

131.142 Results of observations of faint H II regions using a Fabry-Perot interferometer.
V. F. Zhidkov, G. V. Novikova.
Astron. Tsirk., No. 655, p. 4 - 6 (1971). In Russian.

131.143 Chemistry of interstellar medium.   J. Svatoš.
     Říše hvězd, Vol. 52, 161 - 164 (1971). In Czech.

131.144 Radioastronomy and microwave spectroscopy.
     R. D. Davies.
Phys. Bull. (*GB*), Vol. 22, 141 - 144 (1971).

Discusses the relative abundance of the elements and the types of molecule likely to be most abundant in interstellar space. The main features of the molecules detected so far are presented.

131.145 Interstellar molecular spectroscopy.   D. Buhl.
     IEEE International Convention Digest 1971 (IEEE, New York), p. 158 - 159. – See Phys. Abstr., Vol. 74, No. 50066 (1971).

131.146 Interstellar warm gas.   S. Souffrin.

Sci. Progress Découverte, No. 3432, p. 9 - 16 (1971). In French.

131.147 Interferometry of galactic H II regions.
W. J. Webster, Jr.

Thesis, Case Western Reserve Univ., Cleveland, Ohio. [Available from Univ. Microfilms, Ann Arbor, Mich., U.S.A. Order No. 70–25928], 100 pp. (1969).

The N.R.A.O. interferometer has been employed to prepare supersynthesis maps of the galactic H II regions W3(IC 1795), DR 21, Orion A (M 42) and M 17 at 11.1 cm.

131.148 High-velocity cloud collisions. T. L. Chow.
Thesis, Univ. Rochester, Rochester, N.Y. [Available from Univ. Microfilms, Ann Arbor, Mich., U.S.A. Order No. 71–1433], 115 pp. (1970).

Attempt a critical examination of the consequences when high momentum inflowing extragalactic gas clouds collide with galactic gases, the extragalactic gases having initial velocities of perhaps −500 km/sec with respect to the local standard of rest. We are interested in the cooling time, velocity, and other properties of the gas which recombines after ionization.

131.149 Observations of radio recombination lines of hydrogen, helium, and carbon. E. B. Churchwell.
Thesis, Indiana Univ., Bloomington. [Available from Univ. Microfilms, Ann Arbor, Mich., U.S.A. Order No. 71–6831], 210 pp. (1970).

A review is given of the basic formulae which describe the emission of radio recombination lines. Using these formulae an analysis of observed hydrogen, helium, and carbon radio recombination lines is carried out. From the helium and hydrogen radio recombination line intensities the relative abundance of ionized helium in twenty-two galactic H II regions is inferred. Carbon line emission has been observed in seven galactic H II regions, the line parameters for which are reported. The nature of the line formation process for the carbon radio recombination lines is considered.

131.150 Astrophysical interests in r.f. molecular spectroscopy. J. Lequeux.
Revue Phys. Appliquée, Vol. 6, 255 - 258 (1971). In French.

Gives a bibliography for the molecular spectroscopy, both optical and radiofrequency of the OH radical, NH_3, H_2O and H_2CO.

131.151 Formaldehyde absorption in three dark galactic clouds. P. C. Myers, A. H. Barrett.
Quarterly Progr. Report, (USA), No. 102, p. 21 - 24 (1971).

A study of formaldehyde absorption at 4.83 GHz in three dark galactic clouds was made in April 1971 with the 140 ft telescope of the National Radio Astronomy Observatory in Green Bank. Each cloud was mapped extensively in order to gain information about its formaldehyde distribution, its spatial structure, its group and internal motions.

131.152 Photoionization of gas clouds by non-thermal radiation spectra. K. D. Davidson.
Thesis, Cornell Univ., Ithaca, N.Y. [Available from Univ. Microfilms, Ann Arbor, Mich., U.S.A. Order No. 71–12126], 218 pp. (1970).

In an effort to account for the major emission-line intensities in the spectra of certain astronomical objects such as quasi-stellar objects and the Crab nebula, a programme has been developed for calculating the ionization equilibrium in a gas cloud excited by a source of ionizing photons, which may extend to X-ray frequencies. The observed emission-line intensities of quasistellar objects are discussed. An attempt is made to reproduce a 'typical' QSO line spectrum by calculating models of ionized emitting regions. Finally, a limited investigation is made of a radiative excitation process for producing emission lines in QSO spectra.

131.153 From radio astronomy towards astrochemistry.
D. Buhl, L. E. Snyder.

Technol. Rev., Vol. 73, No. 6, p. 1 - 10 (1971). – Popular review. – *RXM*

131.154 Microwave radiation of singly charged helium 3 from H II regions. H. C. Goldwire, Jr.
Thesis, Rice Univ., Houston, Texas (1967). [Available from Univ. Microfilms, Ann Arbor, Mich., U.S.A.], 121 pp. (1970).

131.155 An absorption-line study of the galactic neutral-hydrogen at 21 cm.
M. P. Hughes, A. R. Thompson, R. S. Colvin.

Obs. Owens Valley Radio Obs., 1971, No. 2, 43 pp.

Galactic H I absorption measured on 64 extragalactic sources. The mean temperature of the cool absorbing gas was found to be 71 ± 9 °K. The "highest lower-limit" for the temperature of the hotter gas was about 600°K. – *RXM*

131.156 Galactic 21-cm observations in the direction of 35 extragalactic sources.
V. Radhakrishnan, J. D. Murray, P. Lockhart, R. P. J. Whittle.

Australian C.S.I.R.O. Division Radiophys., Rep. No. RPP 1429, 24 pp. (1971).

By comparison of emission and absorption spectra in the direction of 35 extragalactic sources the authors obtain an unambigious separation of contributions from a diffuse high-temperature optically thin component of spin temperature >750°K and from colder, denser local concentrations of spin temperatures 60 to 80°K. – *RXM*

131.157 The space between the stars. D. J. Mullan.
Irish Astron. Joun., Vol. 10, 1 - 12 (1971). – Contents of a lecture given to the Armagh Centre of the Irish Astronomical Society, April 14, 1970.

Dark nebulae, globules, and protostars. See Abstr. 003.091.

Laboratory measurement of the 6-centimeter formaldehyde transitions. See Abstr. 022.072.

The condensation and evaporation of hydrogen on liquid-helium-cooled surfaces. See Abstr. 022.081.

Precise laboratory determination of rotational transition frequencies in cyanoacetylene. See Abstr. 022.088.

On the dispersion of electromagnetic waves in interstellar space. See Abstr. 062.056.

Interstellar matter and the location of the shock front. See Abstr. 074.013.

uvbyβ **photometry of stars in the direction of the association Perseus OB 2.** See Abstr. 113.043.

L'extinction interstellaire et ses propriétés dans la photométrie en 7 couleurs. Système de l'Observatoire de Genève. See Abstr. 113.057.

Wolf-Rayet stars in H II regions. See Abstr. 114.015.

Search for interstellar silicate absorption in spectrum of VI Cyg No. 12. See Abstr. 114.059.

A closer look at interstellar Lyman-alpha absorp-

tion.   See Abstr. 114.074.

Position measurements of main-line OH/infrared stars.   See Abstr. 114.099.

Interstellar gas in the direction of the Vela pulsar. See Abstr. 114.113.

Lower limit to the interstellar $^{12}C/^{13}C$ ratio in the direction of 20 Tauri.   See Abstr. 114.114.

Observations of interstellar Ca I lines. See Abstr. 114.117.

The interpretation of space observations of stars and interstellar matter.   See Abstr. 114.121.

High resolution observation of stellar and interstellar lithium.   See Abstr. 114.135.

Zur Deutung des Lichtwechsels von SV Cephei durch zirkumstellare Phänomene.   See Abstr. 121.007.

Variability of radiation from circumstellar grains surrounding R Coronae Borealis.   See Abstr. 122.101.

Wavelength dependence of polarization. XXII. Observations of novae.   See Abstr. 124.001.

Wavelength dependence of polarization. XXIII. Dust grains in novae.   See Abstr. 124.002.

Soft X-rays from nonthermal galactic radio sources: Implications concerning the galactic background and the interstellar medium.   See Abstr. 125.018.

Infrared radiation from grains in Orion. See Abstr. 132.017.

Radio recombination lines.   See Abstr. 132.023.

Aperture-synthesis observations of M17 and W49A at 2.695 GHz.   See Abstr. 132.033.

High-frequency confirmation of a radio recombination line from an H I region.   See Abstr. 132.037.

Internal motions in diffuse nebulae. Origin of comet-like nebulae, globulae and stars connected with them. See Abstr. 132.040.

Pulsar JP 1933: 21 cm line absorption profile and interstellar scintillation.   See Abstr. 141.061.

The nature of the galactic radio source G 45.5 + 0.1. See Abstr. 141.120.

Scattering of pulsar radiation in the interstellar medium.   See Abstr. 141.139.

Microsecond intensity variations in the radio emissions from CP 0950.   See Abstr. 141.181.

An absorption-line study of the galactic neutral hydrogen at 21 centimeters wavelength. See Abstr. 141.192.

Role of soft galactic X-rays in the alignment of interstellar grains.   See Abstr. 142.039.

Galactic antiproton cosmic radiation. See Abstr. 143.026.

The production of the elements Li, Be, B by galactic cosmic rays in space and its relation with stellar observations.   See Abstr. 143.040.

Isolated low mass clusters in the interstellar medium. See Abstr. 153.010.

Outward drift of the interstellar medium in the disk of the Galaxy.   See Abstr. 155.005.

The interpretation of the absolute intensity of the diffuse galactic light.   See Abstr. 155.028.

Correlation of neutral hydrogen and radiocontinuum loop IV.   See Abstr. 157.005.

Detection of interstellar OH in two external galaxies. See Abstr. 158.004.

132 Emission Nebulae, Reflection Nebulae

132.001 **The Gum nebula: Further evidence from spacecraft and ground-based instruments.**
J. K. Alexander, J. C. Brandt, S. P. Maran, T. P. Stecher.
Astrophys. Journ., Vol. 167, 487 - 490 (1971).
　　Measurements by the RAE-1 and OGO-5 satellites are combined with data from ground-based telescopes to yield more accurate parameters for the Gum nebula, including a somewhat smaller size and an estimate of the electron temperature. The supernova that ionized the nebula must have been observed at the earth, and it is possible that records exist.

132.002 **Radiofrequency observations of symmetric nebulae around Wolf-Rayet stars and an O7f star.**
H. M. Johnson.
Astrophys. Journ., Vol. 167, 491 - 498 (1971).
　　New observations have been made of the symmetric nebulae around the population I Wolf-Rayet stars HD 50896 and HD 192163 and around the O7f star BD +60°2522. Nine more Wolf-Rayet stars not centered in symmetric nebulae were scanned at 7795 MHz, and one of them, HD 211853, coincides with a source in the telescope beam. Existing optical data have been used to estimate some physical characteristics of the objects. The peculiar nebula NGC 7635 and the star observed in it, BD +60°2522, are discussed in some detail. NGC 7635 is probably a planetary nebula, and not a diffuse nebula as it has been classified lately.

132.003 **On the expansion of the Orion-nebula cluster.**
S. Vasilevskis.
Astrophys. Journ., Vol. 167, 537 - 539 = Contr. Lick Obs., No. 326 (1971).
　　Expansion of the Orion nebula cluster was derived by Strand under an assumption that the scale of the Yerkes 40-inch refractor did not change during a 50-year interval. It is shown that the scale did actually change due to the use of various filters, to readjustments, and to the variable thermal condition of the lens. Consequently, there is now no observational evidence for expansion of the cluster.

132.004 **Temperature and density in gaseous nebulae. II.**
M. Perinotto.
Astron. Astrophys., Vol. 14, 78 - 89 (1971).
　　The electron temperature and density conditions have been extensively investigated in the bright planetary nebulae of quite different degree of excitation NGC 7027, NGC 6543, IC 4997 and IC 418 by using the more recent level populations computations.

132.005 **Capture-cascade intensities of the helium singlets in nebulae.** R. R. Robbins, E. L. Robinson.
Astrophys. Journ., Vol. 167, 249 - 256 (1971).
　　Effective recombination coefficients and recombination intensities of the helium singlets have been calculated for a range of temperatures and densities appropriate to diffuse and planetary nebulae. The equilibrium equations for the 210 lowest singlet levels have been solved simultaneously and the effect of cascades from levels with $n > 20$ has been approximately allowed for, as has the effect of collisional ($n, l \to n, l \pm 1$) interactions.

132.006 **Internal dust in gaseous nebulae. II. Absorption of Lyman-continuum radiation by dust.**
J. S. Mathis.
Astrophys. Journ., Vol. 167, 261 - 271 (1971).
　　The ionization structure of model nebulae containing hydrogen, helium, and isotropically scattering dust is discussed. Definite results for real nebulae must wait on better determi-

nations of optical properties of dust. For purposes of illustrating the magnitude of possible effects, it was assumed that the ultraviolet optical depth for absorption equals the total optical depth (absorption plus scattering) for Hβ. The τ (Hβ) was determined for NGC 6514 and NGC 6523 from the Hβ photometry of O'Dell, Hubbard, and Peimbert.

132.007 **The Gum nebula.** B. J. Bok.
Sky Telescope, Vol. 42, 64 - 69, 94 - 95 (1971).

132.008 **Radiative transfer in spherically symmetric dust nebulae.** J. Dorschner.
Astron. Nachr., Vol. 292, 225 - 229 = Mitt. Univ.-Sternw. Jena, No. 101 (1971).
　　The mathematical tools for future calculations of the diffuse radiation field inside a homogeneous spherically symmetric dust nebula are made available.The method developed here is more general as well as more rigorous than previous work in this field.

132.009 **Der Gum–Nebel – ein Fossil.** H. Rohr.
Orion, 29. Jahrgang, p. 88 (1971).

132.010 **The emission line spectrum of the Orion nebula in the wavelength range 4959 to 8665 Å.**
L. A. Morgan.
Monthly Notices Roy. Astron. Soc., Vol. 153, 393 - 399 (1971).
　　Measurements of the emission line spectrum of the Orion nebula in the region λλ4959−8665 are presented. Among the new lines identified are eight O I lines whose intensities cannot be accounted for by recombination. The [Cl IV] line λ7531 is seen for the first time. From its intensity we find that the abundance of Cl^{+++} is small compared to that of Cl^{++}. Using this result and the intensities of the [Cl III] lines given by Aller and Liller we find that $N(Cl)/N(O^{++}) \simeq 0.008$.

132.011 **The Cygnus loop at 1420 MHz.** P. H. Moffat.
Monthly Notices Roy. Astron. Soc., Vol. 153, 401 - 418 (1971).
　　Part of the Cygnus loop has been mapped at 1420 MHz, using the Cambridge Half-Mile telescope with a resolution of 140″ × 280″. The source has a complex structure which is closely associated with that of the optical nebula. The implications of the observations are discussed with particular reference to van der Laan's theory of the radio emission and the possibility that the optical 'filaments' are excited by the shock wave at the boundary of the supernova remnant.

132.012 **Calculations of the level populations for the low levels of hydrogenic ions in gaseous nebulae.**
M. Brocklehurst.
Monthly Notices Roy. Astron. Soc., Vol. 153, 471 - 490 (1971).
　　The level populations, b_{nl} of hydrogen and singly ionized helium are calculated making full allowance for collisional redistribution of angular momentum and energy. Intensities of the most important line series are presented for $n \leq 40$ and for a wide range of electron temperatures and electron densities. Calculated relative intensities in the spectra of H I are compared with observed intensities in the planetary nebula NGC 7662.

132.013 **Spectrophotometric investigations of diffuse nebulae. I. Absolute intensities of the Hα line in the spectrum of the nebulae NGC 2068 and S-57.**
S. V. Karyagina, Yu. I. Glushkov.

Astron. Tsirk., No. 632, p. 1 - 3 (1971).   In Russian.

132.014   Spectrophotometric investigations of diffuse
nebulae. II. Mi I-19 − a compact H II region.
Yu. I. Glushkov, S. V. Karyagina.
Astron. Tsirk., No. 632, p. 3 - 6 (1971).   In Russian.

132.015   Discovery of para-formaldehyde and the 2-millimeter
formaldehyde distribution in the Orion infrared ne-
bula.   P. Thaddeus, R. W. Wilson, M. Kutner, A. A. Penzias,
K. B. Jefferts.
Astrophys. Journ., *(Letters)*, Vol. 168, L59 - L65 (1971).

The 150.5-GHz line of ortho- and the 145.6-GHz line of
para-H_2CO have been discovered in Ori A, and a detailed map
of the previously detected 140.8-GHz ortho line has been obtained; H_2CO emission extends over a region whose dimensions are ~3′ × 5′, whose neutral-particle density is calculated
to be ~2 × 10^5 cm^{-3}, and whose total mass is ~200 $M_\odot$; it is
argued that the Kleinmann-Low infrared nebula is the central
condensation of this cloud. The 140.8-GHz line has also now
been found in Sgr A, W3(OH), and W51.

132.016   6-centimeter formaldehyde absorption and emission
in the Orion nebula.   M. Kutner, P. Thaddeus.
Astrophys. Journ., *(Letters)*, Vol. 168, L67 - L71 (1971).

Weak 6-cm H_2CO absorption has been found over much
of the Orion nebula, and is attributed to "anomalous" absorption of the universal microwave radiation by H_2CO behind the
H II continuum source. In the Kleinmann-Low infrared nebula the line is self-reversed, with a core of about 0.1° K that
is probably emission from the H_2CO observed there at 2 mm.
These observations are interpreted as indicating that neutral-
particle collisions suppress the pumping mechanism responsible for anomalous absorption.

132.017   Infrared radiation from grains in Orion.
K. S. Krishna Swamy.
Astron. Astrophys., Vol. 14, 405 - 407 (1971).

The calculated thermal emission from model grains are
compared with some of the available infrared measurements
for Orion.

132.018   The dust continuum in the Orion nebula.
M. Perinotto, K. Wurm.
Astron. Nachr., Vol. 293, 25 - 31 (1971).

Spectra of the Orion nebula with a long slit and with exposures to the appearance of the visual continuum have been
obtained for several positions in the field including, in particular, areas of low surface brightness. The variation of the ratio
q = intensity of Balmer emission/intensity of continuous emission is studied.

132.019   On the genetic relation between reflection nebulae
and their illuminating stars.   K.-H. Schmidt.
Astron. Nachr., Vol. 293, 75 - 77 = Mitt.Univ.-Sternw. Jena, No.
110 (1971).

From a discussion of catalogues of reflection nebulae
based on the Palomar Observatory Sky Survey it is shown that
a genetic relation exists between the nebulae and the illuminating stars.

132.020   Sur le spectre de la nébuleuse A 21.
M. Chopinet, M.-C. Lortet-Zuckermann.
Comptes Rendus Acad. Sci. Paris, Sér. B, Vol. 273, 513 -
516 (1971).

La nébuleuse A 21, classée «particulière» par certains
auteurs, a été observée récemment à l'aide d'un tube-image.
Le spectre obtenu ne confirme pas les anomalies qui lui ont
été attribuées.

132.021   Dust in the Orion nebula.

K. Nandy, N. C. Wickramasinghe.
Monthly Notices Roy. Astron. Soc., Vol. 154, 255 - 264
(1971).

Observations relating to extinction and scattering by dust
grains in the Orion nebula are compared with theoretical predictions for graphite-iron-silicate grain mixtures.

132.022   Detection of methyl alcohol in Orion at a wave-
length of ~1 centimeter.
A. H. Barrett, P. R. Schwartz, J. W. Waters.
Astrophys. Journ., *(Letters)*, Vol. 168, L101 - L106 (1971).

Five transitions of CH_3OH, corresponding to the J = 4,
5, 6, 7, and 8 rotational levels, at frequencies of approximately 25 GHz, have been detected in Orion A. The source of
emission is less than 1 arc minute in angular size and appears
to be coincident with the infrared nebula.

132.023   Radio recombination lines.   L. Goldberg.
National Bureau Standards Special Publ. 353, (see
012.001), p. 169 - 181 (1971).

132.024   Detection of methyl alcohol (CH_3OH) in Orion at
$\lambda \sim 1$ cm.   A. H. Barrett, P. R. Schwartz,
J. W. Waters.
Bull. American Astron. Soc., Vol. 3, 388 (1971). − Abstr.
AAS.

132.025   Spectrophotometric studies of reflection nebulae.
A. N. Witt, W. F. Rush.
Bull. American Astron. Soc., Vol. 3, 389 (1971). − Abstr.
AAS.

132.026   A discussion of the distance to the Hα filamentary
nebulae in Cygnus and their excitation.
T. A. Matthews, S. C. Simonson III.
Bull. American Astron. Soc., Vol. 3, 396 (1971). − Abstr.
AAS.

132.027   The Gum nebula reappraised.
J. C. Brandt, S. P. Maran.
Bull. American Astron. Soc., Vol. 3, 396 (1971). − Abstr.
AAS.

132.028 H_2O line emission from Orion A, VY CMa, and
W49.   K. J. Johnston, S. H. Knowles, W. T.
Sullivan III, J. M. Moran, B. F. Burke, K. Y. Lo, D. C. Papa,
G. D. Papadopoulos, P. R. Schwartz, C. A. Knight, I. I. Sha-
piro, W. J. Welch.
Bull. American Astron. Soc., Vol. 3, 416 (1971). − Abstr.
AAS.

132.029   Balmer line intensities near the series limit in
gaseous nebulae.   L. E. Goad, L. Goldberg,
J. L. Greenstein.
Bull. American Astron. Soc., Vol. 3, 417 (1971). − Abstr.
AAS.

132.030   The formation of nebulae by Wolf-Rayet stars.
V. S. Avedisova.
Astron. Zhurn. Akad. Nauk SSSR, Vol. 48, 894 - 901 (1971).
In Russian. English translation in Soviet Astron. AJ, Vol. 15,
No. 5.

The motion of the interstellar gas surrounding WR stars,
swept with the stellar wind, can be represented by two model
solutions corresponding to the adiabatic motion of a shock
wave and the shell state, characterized by intensive radiation.
The behaviour of the physical parameters inside the gas flow
for both models is computed. It is shown that the observed
ellipsoidal form of the shells cannot be obtained with typical
values of magnetic fields of the interstellar gas of the Galaxy.
The age of the nebula NGC 6888 and the power of the matter

ejection from the central star is estimated.

132.031 Observational evidence for Stark broadening in radio recombination lines. E. Churchwell.
Astron. Astrophys., Vol. 15, 90 - 94 (1971).

Observations of the H109α and adjacent higher order lines (H137β, H157γ, and H172δ) from Orion A and M17 indicate successively increasing line widths with increasing order. Since the telescope beamwidth, optical depth, and Doppler width are essentially the same for each line the excess broadening with increasing order is interpreted in terms of Stark broadening.

132.032 Polarization in reflection nebulae. B. H. Zellner, III.
Thesis, Univ. Arizona, Tucson. [Available from Univ. Microfilms, Ann Arbor, Mich., U.S.A. Order No. 70–22018], 201 pp. (1970).

Photoelectric measurements of color and polarization in seven reflection nebulae were made between November 1967 and December 1969. Detailed observations were made in NGC 2068, IC 5076, and NGC 7023 with the 154-cm Catalina reflector and the 229-cm Steward reflector.

132.033 Aperture-synthesis observations of M17 and W49A at 2.695 GHz.
W. J. Webster, Jr., W. J. Altenhoff, J. E. Wink.
Astron. Journ., Vol. 76, 677 - 682 (1971).

The Omega nebula (M17) and the thermal component of W49 (3C398) have been mapped with high resolution at 2.695 GHz. W49A has been resolved into a clustering of components near the radio peak position, a component 2 arc min southeast of the peak position and a pair of components 1 arc min southwest of the peak position, while M17 consists of a broad component with a pair of small components superimposed. We show that the components we detect in W49A are of high apparent excitation and contain a large mass of ionized hydrogen while the components of M17 are comparable with components of more normal H II regions like W3. It may be that W49A surrounds a young OB association.

132.034 Photographic V and R magnitudes of T Tauri stars and related objects in Orion. M. T. Brück.
Publ. Roy. Obs. Edinburgh, Vol. 7, (No. 5), 63 - 72 (1971).

Photographic V and R magnitudes to R = 13$\overset{m}{.}$2 are given for 67 T Tauri, flare and Hα emission stars in the Orion nebula region. Three additional possible T Tauri stars are identified on the basis of colour excesses. Details of variability are given for 15 stars. It is suggested that T Tauri and related stars are intrinsically more reddened than the generality of stars in the association.

132.035 On the Hβ profiles in the central region of NGC 1976. N. I. Grachev.
Astron. Tsirk., No. 646, p. 4 - 6 (1971). In Russian.

132.036 Ionization and heating of the Gum nebula by energetic particles from the Vela X supernova.
R. Ramaty, E. A. Boldt, S. A. Colgate, J. Silk.
Astrophys. Journ., Vol. 169, 87 - 96 (1971).

We investigate a model in which the Gum nebula is ionized and heated by energetic particles from the supernova associated with the Vela X remnant. We investigate the consequences of this model for the ionization and heating of the interstellar medium, the generation of the light elements, X-ray production, and observable cosmic rays.

132.037 High-frequency confirmation of a radio recombination line from an H I region. E. J. Chaisson.
Astrophys. Journ., Vol. 170, 81 - 84 (1971).

The existence of a hydrogen recombination line arising

from an H I region in the microwave spectrum of NGC 2024 has been confirmed at 94α. Frequency dependence of the line radiation from the H I cloud differs significantly from that of the associated H II region.

132.038 Half-Angstrom filtergrams of NGC 2392. R. R. Fisher, S. D. Cain.
Publ. Astron. Soc. Pacific, Vol. 83, 604 - 605 (1971). – Abstr. Astron. Soc. Pacific.

132.039 Abundances of helium in gaseous nebulae. M. J. Seaton.
Highlights of Astronomy, Vol. 2, (see 012.015), 288 - 295 (1971).

132.040 Internal motions in diffuse nebulae. Origin of comet-like nebulae, globulae and stars connected with them.
E. A. Dibay.
Astron. Zhurn. Akad. Nauk SSSR, Vol. 48, 1134 - 1144 (1971). In Russian. English translation in Soviet Astron. AJ, Vol. 15, No. 6.

The internal motions in diffuse nebulae responsible for the origin of comet-like nebulae and globulae are considered. The elongated "proboscis" structures are suggested to be formed within H I zones when small inhomogeneities (globulae) move to the gravity centre of a spacious nebula with a frozen magnetic field. Various geometry of the magnetic and gravitational fields produces other forms of the nebulae. Within H II zones the globulae are compressed by converging shock waves, which arise at the boundary between neutral and ionized hydrogen.

132.041 Observations of a thin filamentary nebula – The supernova remnant in Monoceros.
T. A. Lozinskaya.
Astron. Zhurn. Akad. Nauk SSSR, Vol. 48, 1145 - 1149 (1971). In Russian. English translation in Soviet Astron. AJ, Vol. 15, No. 6.

A large series of observations of the faint filamentary nebula in Monoceros with a Fabry–Perot etalon and a contact image converter was made. The mean radial velocity of the object (reduced to LSR) is determined equal to +12 km/sec ± 5 km/sec. The effect of the nebula expansion is considered; the velocity of expansion is found to be equal to 45 km/sec ± 10 km/sec. Three independent estimates of the nebula radius give the value of 20 pc. The age of the nebula, equal to 150000 years, and the kinetic energy of the ejected shell, equal to 7×10^{49} erg, are determined. The nebula is shown to be a supernova remnant of type II, like Cygnus Loop and IC443.

132.042 Observations of Cygnus Loop at 408 MHz.
G. Colla, C. Fanti, R. Fanti, A. Ficarra, L. Formiggini, E. Gandolfi, C. Lari, B. Marano, L. Padrielli, C. J. Salter, G. Setti, P. Tomasi.
Astron. Journ., Vol. 76, 956 - 957, 1157 (1971).

High-resolution observations of the Cygnus Loop at 408 MHz are presented. An exceptional coincidence between radio and optical filamentary structure is shown to exist.

132.043 The Gum nebula. S. P. Maran.
Sci. American, Vol. 225, No. 6, p. 20 - 29 (1971).

132.044 The Gum nebula – A new kind of astronomical object. S. P. Maran, J. C. Brandt, T. P. Stecher.
Phys. Today, Vol. 24, No. 9, p. 42 - 47 (1971).

Did radiation from a supernova explosion ionize this huge mass of hydrogen? Four theories propose ways that the nebula could have been created by energy from the supernova.

132.045 Nomograms for calculating the ionization of hydro-

gen and helium in gaseous nebulae.
V. V. Golovatyj.
Tsirk. L'vov. Astron. Obs., No. 46, p. 23 - 27 (1971).
In Russian.

132.046 Polarization of the radiation of the Orion nebula in the visual spectral region. K. G. Dzhakusheva.
Trudy Astrofiz. Inst., *Alma-Ata,* Vol. 16, 3 - 12 (1971).
In Russian.

132.047 Photometric features and structures of the nebulae NGC 6914a, IC 5076 and Ced 201.
Ju. I. Glushkov, E. S. Eroshevich.
Trudy Astrofiz. Inst., *Alma-Ata,* Vol. 16, 13 - 21 (1971).
In Russian.

Data for fluxes and brightness distribution over the disks of nebulae NGC 6914a, IC 5076 and Ced 201 for spectral regions with λ_{eff} = 3750, 4060, 4740 Å are given. Isophotes for every nebula are obtained. An analysis of the structure of nebulae is given. The depression of the ultraviolet radiation of the nebulae relative to the violet radiation is observed.

132.048 An elementary method of evaluating the role of multiple scattering in reflecting nebulae.
D. A. Rozkovsky.
Trudy Astrofiz. Inst., *Alma-Ata,* Vol. 16, 128 - 130 (1971).
In Russian.

Formulae allowing to evaluate the role of multiple scattering for two models of reflecting nebulae illuminated by distant stars are given.

132.049 On the polarization of radiation in the nebulae IC 4592 and IC 4601. E. S. Yeroshevich.
Trudy Astrofiz. Inst., *Alma-Ata,* Vol. 17, 51 - 55 (1971). In Russian.

132.050 On the polarization of diffuse nebulae in the Orion constellation. I. NGC 2024. K. G. Dzhakusheva.
Astron. Tsirk., No. 660, p. 6 - 8 (1971). In Russian.

132.051 Radio astronomy Explorer 1 observations of the Gum nebula. J. K. Alexander.
U. S. Goddard Space Flight Center, Greenbelt, Rep. X-683-71-375, p. 34 - 38 (1971).

132.052 Some calculations of cooling rates and spectral emission of gaseous nebulae. D. P. Cox.
Thesis, Univ. California, San Diego. [Available from Univ. Microfilms, Ann Arbor, Mich., U.S.A. Order No. 70–14347], 189 pp. (1970).

A general method of computing hydrogenic recombination decrements in gaseous nebulae is described. Presents the results of calculations of the ionization equilibrium and radiative cooling rate of a high temperature low density plasma. The elements H, He, C, N, O, Ne, Mg, Si and S are considered, and the temperature range is taken to be 10^3-10^8 °K. The dynamic and radiative properties are calculated for a steady state shock wave traveling at about 100 km/sec into the interstellar medium.

132.053 An interstellar calcium feature in the region of Eta Carinae. N. R. Walborn.
Publ. Astron. Soc. Pacific, Vol. 83, 811 - 812 = Commun. David Dunlap Obs., Univ. Toronto, *Richmond Hill,* No. 304 (1971).

The stars in the immediate vicinity of η Carinae show considerably stronger interstellar calcium lines than the surrounding stars on all sides. There is some evidence that the enhanced interstellar absorption is due to material associated with the stars and further observations would be of interest.

Optical line spectrum. See Abstr. 061.014

On the abundance of chlorine in the sun. See Abstr. 071.012.

The nature of Becklin's star. See Abstr. 113.038.

Energy distributions and spectra of Orion B stars. See Abstr. 114.096.

The spectrum of LkHα-101 in the near-infrared. See Abstr. 114.097.

Two young bright infrared objects. See Abstr. 114.098.

Three somewhat overlooked facts of VY Canis Majoris. See Abstr. 122.112.

On the polarization of some variable stars in the Orion nebula. See Abstr. 122.145.

Variable stars in the Orion nebula. See Abstr. 123.071.

A blast-wave model for the Vela X supernova remnant and the origin of the Gum nebula. See Abstr. 125.001.

Fossil Strömgren spheres from supernova explosions. See Abstr. 125.010.

Interstellar carbon monosulfide. See Abstr. 131.030.

Observations of high velocities in H II regions with a two-etalon Fabry-Perot spectrometer. See Abstr. 131.091.

Radiofrequency recombination lines as diagnostics of the cool interstellar medium. See Abstr. 131.110.

Spectrophotometric studies of gaseous nebulae. XIX. The moderate-excitation planetary NGC 6826. See Abstr. 133.009.

Reduced helium abundances in nebulae. See Abstr. 133.024.

On the distances of the open clusters Tr 14, Tr 15, Tr 16 and the η Carinae nebula. See Abstr. 153.001.

Some unusual southern hemisphere objects. See Abstr. 158.090.

On the [S II] and [O I] line intensities in gaseous nebulae and nuclei of galaxies. See Abstr. 158.119.

Emission nebulae in the Magellanic Clouds at 408 MHz. See Abstr. 159.002.

133 Planetary Nebulae

133.001 The profiles of the [O III], 5007 Å line from the Dumb-bell nebula. J. Meaburn.
Astron. Astrophys., Vol. 13, 478 - 486 (1971).

The profiles of the [O III] line have been observed at 23 positions across the Dumb-bell nebula (NGC 6853) with a two-etalon scanning Fabry-Perot. It is shown that the [O III] region is not simply an expanding shell. Very involved nebular motions are indicated. The application of the Jones and Misell (1970) method of deconvolution to this type of observation is examined in some detail.

133.002 New planetary nebulae. L. Kohoutek.
Astron. Astrophys., Vol. 13, 493 - 495 (1971).

New planetary nebulae of low surface brightness were discovered on the Palomar Sky Atlas, especially in the declination zones −36° and −42°. The co-ordinates, short description and identification charts of the individual objects are given.

133.003 A survey of microwave radiation from planetary nebulae. L. A. Higgs.
Monthly Notices Roy. Astron. Soc., Vol. 153, 315 - 336 (1971).

Radio observations of 121 planetary nebulae have been made at wavelengths of 9.3, 4.5 and 2.8 cm, using the 46-metre telescope at the Algonquin Radio Observatory. Detectable radio emission was found for approximately one half of these nebulae. Flux densities (or upper limits to flux densities) have been determined with an accuracy of the order of 0.04 flux units.

133.004 Central stars of planetary nebulae. E. E. Salpeter.
Annual Rev. Astron. Astrophys., Vol. 9, (see 003.001), 127 - 146 (1971).

This review treats mainly a restricted aspect of the central stars of planetary nebulae, the present-day theoretical ideas on the evolution of these objects.

133.005 Planetary nebulae. I. Photoelectric photometry. M. Peimbert, S. Torres-Peimbert.
Bol. Obs. Tonantzintla y Tacubaya, No. 36, Vol. 6, 21 - 28 (1971).

Photoelectric observations of emission lines in thirteen planetary nebulae and of the continuum near the Balmer discontinuity for three of these are presented. The absolute flux at Hβ for nine of these objects is determined. From the observed Balmer decrement and the normal extinction law the total absorption at Hβ is obtained for all the objects. An independent determination of the reddening is obtained by comparing the Hβ flux with the radio emission in five of the brightest planetary nebulae. No deviations from the normal reddening law are detected in the direction of these objects.

133.006 Planetary nebulae. II. Electron temperatures and electron densities. M. Peimbert.
Bol. Obs. Tonantzintla y Tacubaya, No. 36, Vol. 6, 29 - 37 (1971).

The electron temperature is obtained for thirteen planetary nebulae by means of forbidden lines of oxygen doubly ionized and nitrogen singly ionized. From the ratio of the Balmer continuum to the Balmer emission line intensities, temperatures in the 7000 to 8000°K range are derived for NGC 6572, IC 418, and NGC 7009. Electron densities for these objects are derived by means of auroral to nebular line intensity ratios of O II and from transauroral to nebular line intensity ratios of S II.

133.007 The absolute energy distribution in the Balmer continuum spectral region of 16 planetary nebulae. E. B. Kostyakova.
Astron. Tsirk., No. 623, p. 5 - 7 (1971). In Russian.

133.008 Preliminary results of a spectral investigation of some planetary nebulae. L. N. Kondratieva.
Astron. Tsirk., No. 629, p. 4 - 6 (1971). In Russian.

133.009 Spectrophotometric studies of gaseous nebulae. XIX. The moderate-excitation planetary NGC 6826.
S. J. Czyzak, L. H. Aller, J. B. Kaler.
Astrophys. Journ., Vol. 168, 405 - 411 (1971).

Photoelectric and photographic spectrophotometric observations of the relatively bright, moderate-excitation planetary NGC 6826 obtained at Lick, Mount Wilson, and Kitt Peak Observatories are combined to yield estimates of its density, temperature, and ionic concentrations. Data for ions of helium, carbon, nitrogen, oxygen, neon, sulfur, chlorine, and argon are presented.

133.010 Planetary nebulae. III. Chemical abundances. M. Peimbert, S. Torres-Peimbert.
Astrophys. Journ., Vol. 168, 413 - 421 (1971).

Chemical abundances of thirteen planetary nebulae have been obtained from new photoelectric observations. The helium/hydrogen abundance ratio of planetary nebulae is very similar to that of H II regions. It is found that the nitrogen/oxygen abundance ratio is a factor of 3–5 times higher in planetary nebulae than in H II regions.

133.011 The dynamics and infrared radiation of young, dust-filled planetary nebulae. W. S. Kovach.
Astrophys. Journ., Vol. 168, 423 - 436 (1971).

Five models of very young, dust-filled planetary nebulae are constructed. We calculate a density gradient in the neutral shell and discuss the resultant dynamics. The problem of Lα transfer in the neutral shell is solved and the lifetime of the grain in the H II region is analyzed. The grain temperature and the infrared radiation are found. The stability of the neutral shell is analyzed, and the models are discussed.

133.012 Chemical composition of typical planetary nebulae. L. H. Aller.
National Bureau Standards Special Publ. 353, (see 012.001), p. 161 - 168 (1971).

133.013 Internal motions and kinematics of planetary nebulae. W. Liller.
National Bureau Standards Special Publ. 353, (see 012.001), p. 182 - 189 (1971).

133.014 Filamentary structure of planetary nebulae. D. H. Menzel.
National Bureau Standards Special Publ. 353, (see 012.001), p. 190 - 203 (1971).

133.015 The Bowen fluorescence mechanism in planetary nebulae. J. P. Harrington.
Bull. American Astron. Soc., Vol. 3, 397 (1971). – Abstr. AAS.

133.016 Origin of filamentary structure in planetary nebulae. D. J. Van Blerkom.
Bull. American Astron. Soc., Vol. 3, 397 (1971). – Abstr. AAS.

133.017 Interpretation of the neutral helium triplet spectrum in planetary nebulae. S. E. Persson.
Bull. American Astron. Soc., Vol. 3, 397 (1971). – Abstr. AAS.

133.018 Hydrogen Paschen and HeI λ10830 emission in the spectrum of IC 418. R. P. Kovar, A. E. Potter, N. S. Kovar, L. Trafton, B. Ulrich.
Bull. American Astron. Soc., Vol. 3, 417 (1971). – Abstr. AAS.

133.019 The formation of planetary nebulae. B. Paczyński.
Astrophys. Letters, Vol. 9, 33 - 34 (1971).

Recent estimates of the rates of mass loss from Mira variables are sufficient to explain the origin of planetary nebulae.

133.020 The dynamics and thermal stability of planetary nebulae. J. H. Hunter, S. Sofia.
Monthly Notices Roy. Astron. Soc., Vol. 154, 393 - 413 (1971).

Dynamical models of planetary nebulae are constructed utilizing a semi-analytical approach. The influence of the various physical parameters upon the thermal histories of the expanding nebular shells is investigated. Also, the thermal stability of the nebulae is rediscussed from a fundamentally new viewpoint.

133.021 The Hβ profile in NGC 6853. N. I. Grachev.
Astron. Tsirk., No. 644, p. 7 - 8 (1971). In Russian.

133.022 Absolute intensities of continua of planetary nebulae in the spectral region λ 9000 Å.
R. I. Noskova.
Astron. Tsirk., No. 647, p. 7 - 8 (1971). In Russian.

133.023 Line intensities in NGC 7027.
J. B. Kaler, S. J. Czyzak, L. H. Aller.
Astrophys. Journ., Vol. 169, 199 - 201 (1971).

New photographic observations of NGC 7027 confirm Miller's finding that earlier spectrophotometry of NGC 7027 is afflicted with strong systematic errors.

133.024 Reduced helium abundances in nebulae.
R. R. Robbins, E. Daltabuit, D. P. Cox.
Astrophys. Journ., *(Letters)*, Vol. 169, L77 - L81 (1971).

Recent capture-cascade calculations by Robbins and Robinson for the helium singlets, and also calculations by Cox and Daltabuit concerning collisional excitation in the helium triplets, imply a reduction in nebular helium abundances. Some consequences of these reductions are discussed.

133.025 Monochromatic photographs and isophotic contours of planetary nebulae, III: NGC 2392, 6210, 6826, 6720 and 6853. W. A. Feibelman.
Journ. Roy. Astron. Soc. Canada, Vol. 65, 251 - 262 (1971).

A brief description for each of five planetary nebulae is given. Monochromatic photographs for three nebulae and isophotic contours for four nebulae are presented. The measured diameters are derived from λ5007 observations.

133.026 Central stars of planetary nebulae showing O or Of-type spectra. S. R. Heap.
Thesis, Univ. California, Los Angeles. [Available from Univ. Microfilms, Ann Arbor, Mich., U.S.A. Order No. 71–16324], 296 pp. (1970).

Attempts to resolve the discrepancy between the temperatures obtained from analysis of the line spectrum of O and Of-type central stars and the Zanstra temperatures, obtained from indirect estimates of the far-UV stellar flux.

133.027 The infrared spectrum of IC 418.
R. P. Kovar, A. E. Potter, N. S. Kovar, L. Trafton.
Astrophys. Journ., Vol. 170, 449 - 452 (1971).

Observations of the intensities of the Paschen lines in the spectrum of IC 418 together with published measurements of the Balmer-line intensities are utilized to derive a reddening constant, $c = 0.31$. The corrected flux value for Paschen β is $\log [I(P\beta)/I(H\beta)] = -0.76$. Two additional emission lines, one at 9202 cm^{-1} and the second at 5890 cm^{-1}, are briefly discussed.

133.028 Evolution of single stars. VI. Model nuclei of planetary nebulae. B. Paczyński.
Acta Astron., Vol. 21, 417 - 435 (1971).

Model evolutionary computations are presented for population I ($X_0 = 0.7$, $Z = 0.03$) stars of 0.6, 0.8, and 1.2 $M_\odot$. Each model consists of a degenerate carbon-oxygen core, helium and hydrogen burning shell sources and a low mass hydrogen rich envelope. Neutrino energy losses are included in the computations and the evolution of these models in the H–R diagram is considered.

133.029 Infrared emission from planetary nebulae. I. Observations of some planetary nebulae in the 1.0 - 2.5 micron region. G. S. Khromov, V. I. Moroz.
Astron. Zhurn. Akad. Nauk SSSR, Vol. 48, 1122 - 1133 (1971). In Russian. English translation in Soviet Astron. AJ, Vol. 15, No. 6.

16 planetary nebulae were observed in the spectral region $1.0 - 2.5 \mu$. Together with the relevant observational data by other authors these results were used to construct the total electromagnetic continua of 9 planetary nebulae. The empirical data were compared with the common theory of thermal continuous emission. It was found that the observational intensities in the radio, near infrared and near ultraviolet regions could be satisfactorily linked together with the aid of the theoretical spectrum.

133.030 Observations of the planetary nebula NGC 6853 at [N II] 6584 Å. A. C. Danks.
Astrophys. Space Sci., Vol. 14, 480 - 484 (1971).

New observations of the [N II] 6584 Å line have been made over the surface of the Dumbbell nebula (NGC 6853). The observed lines at the centre of the nebula disc exhibited line splitting of $\simeq 54.3$ km s^{-1}. The lines appeared double at the centre of the nebula and became single at the boundary. These observations are discussed and compared with those obtained by previous workers.

133.031 The effects of dust and Lyman alpha radiation on the dynamical evolution of planetary nebulae.
W. S. Kovach.
Thesis, Ohio State Univ., Columbus. [Available from Univ. Microfilms, Ann Arbor, Mich., U.S.A. Order No. 71–7495], 110 pp. (1970).

The dynamics of spherically expanding envelopes illuminated by a very hot central star such as observed in some planetary nebula is investigated. It is assumed that the expanding envelope consists of an H II region surrounded by an optically thick H I region containing a mixture of hydrogen and dust grains. The resulting reradiation of the Lyman alpha energy into the infra-red region is evaluated for different dust densities as a function of time.

133.032 Catalog of radio observations of planetary nebulae and related optical data. L. A. Higgs.
National Res. Council Canada, NRC 12129. Publ. Astrophys. Branch, Ottawa, PAB Vol. 1, No. 1, 3 + 454 pp. (1971).

All radio observations of the total flux density of planetary nebulae, in the literature up to 1971, are presented in catalog form. For each nebula included in the catalog, morphological parameters, observations of Balmer line (Hα and Hβ) fluxes, and physical parameters derived from the radio and optical data are listed. The latter quantities are obtained by fitting model nebulae to the observed radio spectra and by assuming that all nebulae have the same ionized mass (i.e., using the Shklovskii method).

V605 Aquilae: A nova-like variable in an old planetary nebula.　See Abstr. 122.110.

The strange case of V605 Aquilae. See Abstr. 124.107.

Planetary nebulae. IV. Predicted chemical composition and interstellar enrichment.　See Abstr. 131.134.

Temperature and density in gaseous nebulae. II. See Abstr. 132.004.

Capture-cascade intensities of the helium singlets in nebulae.　See Abstr. 132.005.

Calculations of the level populations for the low levels of hydrogenic ions in gaseous nebulae. See Abstr. 132.012.

Sur le spectre de la nébuleuse A 21. See Abstr. 132.020.

Non-stellar objects used in preparing the SAO Star Atlas, coordinates at equinox 1950.0.　See Abstr. 158.125.

134 Crab Nebula

134.001 Search for polarized X-rays from the Crab nebula using a focusing graphite crystal polarimeter.
G. Epstein, R. Novick, M. C. Weisskopf.
Bull. American Astron. Soc., Vol. 3, 392 (1971). − Abstr. AAS.

134.002 Search for polarized X-rays from the Crab nebula using an incoherent scattering polarimeter.
R. Linke, R. Novick, R. S. Wolff.
Bull. American Astron. Soc., Vol. 3, 392 (1971). − Abstr. AAS.

134.003 Detection of pulsed gamma radiation from the Crab nebula. R. Browning, D. Ramsden, P. J. Wright.
Nature, Phys. Sci., Vol. 232, 99 - 101 (1971).
Data recorded from a high altitude balloon flight from Palestine, Texas on 9 January 1971 to study gamma rays greater than 70 MeV from the Crab nebula has been analysed for periodicity. The results indicate that the pulsed emission from NP0532 extends into the high energy gamma ray region. The estimated source strength is consistent with an extrapolation from measurements at lower energies.

134.004 Measurement of the $10\mu m$ flux from the Crab nebula. D. K. Aitken, P. G. Polden.
Nature, Phys. Sci., Vol. 233, 45 - 46 (1971).
Observations of the 10 micron flux from the Crab nebula are reported, using a 4 arc min beam on the University of London Observatory's 24″ telescope. The measured flux is significantly in excess of the extrapolation from measurements at shorter wavelengths and from the radio data for $\lambda \geqslant 3$ mm.

134.005 The synchrotron spectrum of the Crab nebula and ionization of gas in filaments.
V. V. Golovatyj, V. I. Pronik.
Astron. Tsirk., No. 640, p. 1 - 3 (1971). In Russian.

134.006 Dust in the Crab nebula.
D. K. Aitken, P. G. Polden.
Nature, Phys. Sci., Vol. 234, 72 - 73 (1971).
The implications of an excess flux from the Crab at 10μ is discussed in terms both of synchrotron processes and a dust model. It is shown that small graphite particles in the ambient synchrotron radiation field will have a temperature ~ 100°K and that about $0.03M_{\odot}$ of such material can account for the observed excess flux.

134.007 A radio map of the Crab nebula at 3.5 mm.
L. I. Matveyenko.
Astron. Zhurn. Akad. Nauk SSSR, Vol. 48, 1154 - 1159 (1971). In Russian. English translation in Soviet Astron. AJ, Vol. 15, No. 6.
Observations of the Crab nebula at 3.5 mm are carried out with the 11-m radiotelescope of NRAO at Kitt Peak Observatory. A radio map of the Crab nebula is obtained.

134.008 On the ionization of gas in the filaments of the Crab nebula. V. V. Golovatyj.
Tsirk. L'vov. Astron. Obs., No. 46, p. 16 - 22 (1971). In Russian.

134.009 Story of Crab nebula. H. Yoko-o.
Astron. Herald, *(Japan)*, Vol. 64, No. 11, 1 pp. (1971). In Japanese.
A historical survey of the nebula is given.

The bending of the synchrotron spectrum at high energies. See Abstr. 022.055.

On electron acceleration in an alternating magnetic field under astrophysical conditions. See Abstr. 061.004.

Short-term stability of the Crab pulsar. See Abstr. 141.007.

On the mechanism of the glitches in the Crab nebula pulsar. See Abstr. 141.012.

Observations of the sporadic bright pulses from the Crab nebula pulsar. See Abstr. 141.044.

Observation of low-energy gamma radiation from NP 0532. See Abstr. 141.059.

Disappearance of an interstellar-scattering phenomenon shown by the Crab nebula pulsar. See Abstr. 141.094.

Crab nebula pulsar radio pulse arrival times at Arecibo Observatory. See Abstr. 141.108.

Observation of the Crab nebula at 30−100 keV. See Abstr. 141.111.

The optical emission of the Crab nebula pulsar. See Abstr. 141.114.

A possible interpretation of the precursor pulse in NP 0532. See Abstr. 141.154.

A positional determination of NP 0532. See Abstr. 141.156.

The precursor pulse from NP 0532: A result of scattering? See Abstr. 141.175.

On the absorption of gamma rays by photons in pulsars, quasi-stellar objects, and other source objects. See Abstr. 141.186.

Upper limit to circular polarization of optical pulsar NP 0532. See Abstr. 141.187.

Mechanism for the delay of polarization minima in the optical pulsar NP 0532. See Abstr. 141.212.

Erratum: 'Crab nebula pulsar radio pulse arrival times at Arecibo Observatory'. See Abstr. 141.226.

Possible evidence for pulsed ~ 10^{12} eV gamma rays from NP 0532. See Abstr. 141.239.

Crab pulsar radiation mechanism. See Abstr. 141.241.

Production of γ-rays in the Crab nebula pulsar. See Abstr. 141.256.

Extensive air shower studies of cosmic gamma rays and cosmic ray composition. See Abstr. 142.075.

Radio Sources, Quasars, Pulsars, X Ray-, Gamma Ray-Sources, Cosmic Radiation

141 Radio Sources, Quasars, Pulsars

141.001 **The rate of change of period of the pulsars.**
 G. C. Hunt.
Monthly Notices, Roy. Astron. Soc., Vol. 153, 119 - 131 (1971).

The periods, and the rates of change of period, of 16 pulsars have been determined. There is no clear functional relation between these parameters, although there is some support for theories suggesting that radio luminosity decays with age, and that the braking mechanism is a magnetic field which decays with age.

141.002 **Observations of the distribution of polarized emission of Cygnus A at 6-cm wavelength.**
S. Mitton.
Monthly Notices, Roy. Astron. Soc., Vol. 153, 133 - 143 (1971).

High resolution observations, at a wavelength of 6 cm, of the distribution of polarized emission from Cygnus A are presented. It has been found that there are significant variations of the magnitude and position angle of the polarization across each source component. A comparison of the integrated polarization from each component over the range 1.55 cm < λ < 6 cm confirms the earlier conclusion that the rotation measure of the east component is much greater than that of the west component. The observations are related to a number of specific source models but none of them provide an entirely satisfactory explanation for the distribution of polarized emission.

141.003 **Circular polarization of quasars.**
 A. G. Pacholczyk, T. L. Swihart.
Monthly Notices, Roy. Astron. Soc., Vol. 153, 3P - 5P (1971).

A synchrotron optical depth effect, which might be responsible for the observed sign reversal of the circular polarization in certain compact radio sources, is suggested and briefly discussed.

141.004 **The dynamics of extended extragalactic radio sources.** D. S. De Young.
Astrophys. Journ., Vol. 167, 541 - 551 (1971).

The propagation of extended radio sources into an intergalactic medium is examined through a series of time-dependent, axisymmetric calculations. The sources considered range in mass from $10^6 - 10^8 M_\odot$ and in energy from 10^{57} to 10^{59} ergs. Sources are ejected in a narrow cone at $\nu \sim 0.1c$ into an intergalactic medium of various densities and temperatures. It is found that confinement by the medium produces results in agreement with observations only for ambient densities $\gtrsim 10^{-29}$ g cm^{-3}. Implications of this result are discussed, together with the dominant dynamical processes and time scales revealed by the calculations.

141.005 **Properties of PSR 0525+21.** R. N. Manchester.
 Astrophys. Journ., (*Letters*), Vol. 167, L101 - L105 (1971).

Observations at frequencies between 250 and 450 MHz have shown that the pulsar PSR 0525+21 is highly linearly polarized with position angle varying continuously through the pulse by almost 180°. The pulse profile and its variation with frequency are discussed. Dispersion and rotation measures for the pulsar have been determined from measurements of pulse arrival times and position angle at six frequencies.

141.006 **Possible discretization of quasar redshifts.**
 K. G. Karlsson.
Astron. Astrophys., Vol. 13, 333 - 335 (1971).

A number of new peaks in the distribution of redshifts of quasi-stellar objects have been found. These, together with the well known peaks at $z = 1.956$ and $z = 0.061$, form a geometrical series.

141.007 **Short-term stability of the Crab pulsar.**
 J. Pfleiderer.
Astron. Astrophys., Vol. 13, 496 - 497 (1971). -- Research note.

141.008 **Two-dimensional structures of 76 extragalactic radio sources at 1425 MHz.** E. B. Fomalont.
Astron. Journ., Vol. 76, 513 - 524 (1971).

The two-dimensional structures of 76 extragalactic radio sources at a frequency of 1425 MHz are given. All of the observations were made using the Owens Valley Radio Observatory twin-element interferometer with a resolution of 45 arc sec in the east—west directions and 45" sec ($\delta - 37°$) in the north—south direction. The inversion of the interferometric data was obtained using a model-fitting technique. The structure for each source and the optical identification are given in tabular and graphical form.

141.009 **Observations of extragalactic radio sources at centimeter wavelengths.** M. B. Bell, E. R. Seaquist, L. D. Braun.
Astron. Journ., Vol. 76, 524 - 529 (1971).

Flux-density data at 6630 and 10700 MHz are presented for 101 presumably extragalactic radio sources. The results of this investigation provide data at high frequencies for approximately 30 sources whose spectra are very flat or show concave spectral curvature. The remaining sources were found to have straight spectra and for each of these a spectral index is given. Three sources were found to be variable.

141.010 **Flux density measurements of optically identified B2 radio sources at 5 GHz.** G. Grueff.
Astron. Journ., Vol. 76, 530 - 536 (1971).

Observations of 224 optically identified B2 radio sources at 5 GHz are reported. The mean spectral index between 408 and 5000 MHz has been computed; according to the source's optical identification, an indication of a correlation between the spectrum shape and the absolute power (or cosmological epoch) has been found for radio galaxies.

141.011 **Radio sources: 3.3-mm flux and variability measurements.** W. G. Fogarty, E. E. Epstein, J. W. Montgomery, M. M. Dworetsky.
Astron. Journ., Vol. 76, 537 - 543 (1971).

We present here 3.3-mm flux-density measurements of several types of discrete galactic and extragalactic sources and some unidentified sources. We have selected for observation some objects of particular interest and those sources for which the longer-wavelength data indicated the possibility of detection at 3 mm. Also, since April 1965 we have been monitoring the 3-mm variability of quasars and Seyfert galaxies.

141.012 On the mechanism of the glitches in the Crab nebula pulsar. J. D. Scargle, F. Pacini.
Nature, Phys. Sci., Vol. 232, 144 - 149 (1971).

New observations of the wisps in the Crab nebula show further examples of activity connected with sudden changes in the rotation rate of the pulsar. The evidence leads to a model of the "glitches" in which plasma is released explosively from the closed magnetosphere of the neutron star.

141.013 Precise positions of twenty-eight radio sources. J. W. Smith.
Nature, Phys. Sci., Vol.232, 150 - 152 (1971).

This communication presents a series of observations using the Cambridge 1 mile telescope, in which an absolute declination system is obtained with an accuracy believed to be $0''.2 \operatorname{cosec} \delta$ arc. Because of surveying difficulties in the instrument, an absolute zero for the right ascension scale has not been established but an accuracy of about $0''.2$ arc is obtained by reference to the optical positions given by Murray, Tucker and Clements(Nature, Vol. 221, 1229, 1969) and by Argue and Kenworthy (Nature, Vol. 228, 1076 - 1077, 1970).

141.014 The radio source counts and their implications for cosmic evolution.
W. Davidson, M. Davies, B. G. Cox.
Australian Journ. Phys., Vol. 24, 403 - 428 (1971).

A new evolutionary scheme is devised to reach a more accurate assessment of the evolutionary behaviour underlying the radio source counts. Certain empirical linear relations are found to connect any pair of the five model parameters giving a satisfactory fit to the count data. A single independent parameter (taken to be the cutoff redshift z^*) finally determines a satisfactory fit in a given cosmology. Comparison of the predictions of well-fitting models with assembled information on identified sources reveals a very hopeful means of determining the parameter z^* in the near future.

141.015 Pulsar energy fluxes at 80 MHz.
O. B. Slee, E. R. Hill.
Australian Journ. Phys., Vol. 24, 441 - 443 (1971).

During 1969 and 1970 the radioheliograph operated by the Division of Radiophysics at Culgoora has been used to measure the average pulse energies of most of the known pulsars situated within the declination range of the instrument (+34° to −45°). The averaged nightly values of pulse energy are listed in a table, which shows that 10 of the 21 observed pulsars were detected with certainty; upper limits to the average pulse energy are given for the remaining 11 pulsars. The results are compared with values obtained by Hamilton, Ables and Komesaroff, who recorded simultaneously at five frequencies between 150 and 1410 MHz.

141.016 Errata: Observations at 408 MHz of radio sources from the 4C catalogue. I. Declination range −7° to −3°.[Australian Journ. Phys., Vol. 24, 263 - 291 (1971)].
R. E. B. Munro.
Australian Journ. Phys., Vol. 24, 449 (1971).

141.017 The Parkes 2700 MHz Survey. Catalogues for the ±4° declination zone and for the selected regions.
J. V. Wall, A. J. Shimmins, J. K. Merkelijn.
Australian Journ. Phys., Astrophys. Suppl. No. 19, 68 pp. = Separate print Division Radiophys. C. S. I. R. O., Sydney (1971).

Two catalogues of extragalactic radio sources obtained from sky surveys at 2700 MHz are presented. The first catalogue comprises 500 radio sources in the declination zone +4° to −4°, all of which are at least 10° from the galactic plane. The catalogue is complete to a limiting flux density of 0.35 f.u. at 2700 MHz over an area of 0.73 sr. The second catalogue is of 300 sources, obtained from relatively deep surveys of six selected areas each approximately 6.°5 square, and is complete to a limiting flux density of 0.10 f.u. at 2700 MHz. The results of optical identifications from the Palomar Sky Survey prints and additional Schmidt plates taken by J. G. Bolton are given, together with source counts from the two catalogues.

141.018 Characteristics of the radio pulses from the pulsars. A. G. Lyne, F. G. Smith, D. A. Graham.
Monthly Notices Roy. Astron. Soc., Vol. 153, 337 - 382 (1971).

The shape and structure of radio pulses from the majority of the pulsars have been studied at frequencies of 151, 240, 408 and 610 MHz, using polarimeter receivers. Individual pulses from the stronger pulsars have been recorded photographically. The integrated pulse profiles have been obtained from on-line integration of the four Stokes parameters. Many of the results are presented graphically; the main characteristics are collected in a table. A discussion of the pulse widths and the change of polarization angle within the pulses supports the view that the integrated pulse profile represents a longitude distribution of emission, while the individual pulses represent individual beams of radiation. New measurements of the rotation measures of five pulsars are presented together with a compilation of nine previous results. Some new measurements of period and position are also presented.

141.019 Evolutionary origin of the magnetic field on pulsars and its relation to other types of stars.
M. Imoto, M. Kanai.
Publ. Astron. Soc. Japan, Vol. 23, 363 - 370 (1971).

We propose a possible origin of the magnetic field of pulsars, and generalize it as a schematic picture of the magnetic field in all stages of stellar evolution. The origins of the magnetic fields on DC white dwarfs and on peculiar A stars are also discussed.

141.020 Properties of gas that produces absorption lines in some quasi-stellar objects.
Y. W. T. Chan, E. M. Burbidge.
Astrophys. Journ., Vol. 167, 213 - 222 (1971).

Equivalent widths of the Fe II and Mg II absorption lines which have a redshift of 0.6127, in the spectrum of the QSO PHL 938 ($z_{em} = 1.955$), are used to derive velocity dispersions, column densities, and relative abundances of iron and magnesium in the absorbing region by the Strömgren method of line pairs. The same analysis is applied to the C IV $\lambda\lambda 1548, 1551$ doublet in two QSOs with multiple absorption-line redshifts, PKS 0237−23 and Ton 1530. The physical conditions in the absorbing regions are discussed, as is the origin of the very different redshifts. New evidence is presented for the reality of the redshift system at $z = 2.055$ in Ton 1530.

141.021 Optical studies of Cassiopeia A. IV. Physical conditions in the gaseous remnant.
M. Peimbert, S. van den Bergh.
Astrophys. Journ., Vol. 167, 223 - 234 (1971).

Emission-line intensity ratios have been used to study physical conditions in the remnant Cas A.

141.022 Observations of rapid fluctuations of intensity and phase in pulsar emissions.
J. H. Taylor, G. R. Huguenin.
Astrophys. Journ., Vol. 167, 273 - 291 = Contr. Five College Obs., *Amherst, Mass.,* No. 102 (1971).

We present an analysis of the pulse-to-pulse fluctuations of the intensity and phase of twenty pulsars. Observations were made with a new type of receiver which removes most of the effects of dispersion from the pulsar signals, and enhances sensitivity by approximately a factor of 7 over that obtainable with conventional receivers. The results suggest that the pulsars studied can be divided into three classes according to their pulse shapes and the nature of their rapid fluctuations of intensity.

141.023 The optically variable radio source PKS 1514–24 = AP Librae. H. E. Bond.
Astrophys. Journ., *(Letters)*, Vol. 167, L79 = Contr. Louisiana State Univ. Obs., Baton Rouge, No. 50 (1971).

The optically variable object AP Librae is identified with the radio source PKS 1514–24, an N-type galaxy whose variations were previously unknown.

141.024 The reddening of Cassiopeia A. L. Searle.
Astrophys. Journ., Vol. 168, 41 - 43 (1971).

Photoelectric measurements of emission-line strengths from a condensation in the nebulosity associated with the radio source Cas A are reported, and the interstellar reddening is found from the ratio of the auroral to the transauroral lines of [S II].

141.025 Extragalactic sources of cosmic radio emission. III. Double radio sources. V. N. Kurilchik.
Astron. Zhurn. Akad. Nauk SSSR, Vol. 48, 684 - 696 (1971). In Russian. English translation in Soviet Astron. AJ, Vol. 15, No. 4.

The spectral characteristics of the radio emission of double structures are considered. The important feature of the spectrum of the components is the presence of break points. The break points, as a rule, are shifted in the frequency scale. A model of the double radio structure with anisotropic streams of relativistic particles in a quasi-dipole configuration of the magnetic field, which may result from the evolution of the dipole field of the nucleus is qualitatively considered. Some observational aspects of this model are discussed.

141.026 Optical positions of radio sources.
C. A. Murray, R. H. Tucker, E. D. Clements.
Roy. Obs. Bull., Greenwich–Cape, No. 162, p. 215 - 238 (1971).

New optical positions, on the FK4 system, of sixteen radio sources are presented. These have been derived from observations made with the Cooke transit circle and plates taken with the 13-inch and 26-inch refractors, and at the prime focus of the Isaac Newton telescope. The standard errors do not exceed ±0$\overset{s}{.}$013 sec δ in R. A. and ±0$\overset{''}{.}$15 in Dec. The transit circle positions of reference stars are compared with positions taken from AGK3, and the errors associated with both sets of positions are discussed.

141.027 Quasars revisited: Rapid time variations observed via very-long-baseline interferometry.
A. R. Whitney, I. I. Shapiro, A. E. E. Rogers, D. S. Robertson, C. A. Knight, T. A. Clark, R. M. Goldstein, G. E. Marandino, N. R. Vandenberg.
Science, Vol. 173, 225 - 230 (1971).

Recent Goldstone-Haystack radio interferometric observations of the quasars 3C 279 and 3C 273 reveal rapid variations in their fine structure. Most notably, the data for 3C 279, interpreted in terms of a symmetric double-source model and the accepted red-shift distance, indicate differential proper motion corresponding to an apparent speed about ten times that of light. A number of possible mechanisms that might give rise to such an apparent speed are considered.

141.028 Rapidly changing radio images.
A. Cavaliere, P. Morrison, L. Sartori.
Science, Vol. 173, 525 - 528 (1971).

Differences in total transit time can give rise to images that expand at arbitrarily high speed. Two versions of a model based on this idea can account for the varying microwave structure reported for the quasar 3C 279. Other possible examples are suggested.

141.029 Helligkeitsschätzungen von 12 quasistellaren Objekten auf photographischen Himmelsaufnahmen.
G. Jackisch.
Astron. Nachr., Vol. 292, 271 - 274 (1971).

Twelve QSO's have been investigated for variability on plates of the "Sonneberger Himmelsüberwachung". Besides for 3C 273 and Ton 616 no variability exceeding the mean error (±0.08 mag for $m < 17.0$) was found.

141.030 A Green Bank sky survey in search of radio sources at 1400 MHz. I. A spectral analysis of the 5 C1 sources. J. Maslowski.
Astron. Astrophys., Vol. 14, 215 - 222 (1971).

The NRAO 300-foot telescope has been used to make a survey of the sky at 1400 MHz in a 6° strip of declination between +45$\overset{\circ}{.}$8 and +51$\overset{\circ}{.}$7 and between right ascension $7^h 17^m$ and $16^h 23^m$. This paper presents a map of the full survey and gives a brief description of the observations and the reduction procedure. The region of the survey that overlapped the 5C1 survey is also briefly discussed.

141.031 Polarization and the magnetic field scale in radio sources. G. C. Perola.
Astron. Astrophys., Vol. 14, 337 - 339 (1971).

A statistical analysis of the intensity of the polarized radiation may help to determine about the scale of the magnetic field in the extragalactic extended radio sources.

141.032 Pulsar radio emission from expanding charge sheets.
E. Tademaru.
Astrophys. Space Sci., Vol. 12, 193 - 203 (1971).

We semi-quantitatively calculate the distribution of energy in frequency and angle emitted from a sheet of charges that are moving out relativistically along dipolar magnetic field lines originating near the magnetic polar caps of a rotating neutron star. The derived features are in good general agreement with the observed characteristics of the intensity, pulse shape, and frequency spectrum of the radio pulses from pulsars.

141.033 Estimates of radio emission intensity of pulsars CP 0808 and CP 1133 at 25 MHz frequency.
Yu. M. Bruk.
Izv. vyssh. uchebn. zavedenij. Radiofizika, Vol. 13, 1814 - 1817 (1970). In Russian. – Abstr. in Referativ. Zhurn. 51. Astron.,8.51.469 (1971).

141.034 Time variations of maxima and the shape of pulses of the pulsar CP 1133.
V. N. Brezgunov, V. A. Udal'tsov.
Ivz. vyssh. uchebn. zavedenij. Radiofizika, Vol. 13, 1827 - 1832 (1971). In Russian. – Abstr. in Referativ. Zhurn. 51. Astron., 8.51.470 (1971).

141.035 Observations of γ-quanta with energies more than 100 MeV from the radio source 3C 120.
S. A. Volobuev, A. M. Gal'per, V. G. Kirillov-Ugryumov, B. I. Luchkov, Yu. V. Ozerov.
Pis'ma v ZhurnEhTF, Vol. 13, 43 - 46 (1971). In Russian.

141.036 Le teorie sulle pulsars. G. Calamai.
Mem. Soc. Astron. Italiana, Nuova Ser., Vol. 42, 201 - 220 (1971).

The theories many authors have derived to explain the pulsars are given in their fundamental ideas. They are divided in three groups: I. Binary theories, II. Oscillatory theories and III. Rotatory theories.

141.037 **PKS 1127—14, a quasar with an unusually strong magnetic field?**   E. R. Seaquist.
Journ. Roy. Astron. Soc. Canada, Vol. 65, 176 (1971).
Abstr. RAS Canada.

141.038 **Observations of structure in quasi-stellar radio sources.**   T. H. Legg, N. W. Broten, J. A. Galt, S. J. Goldstein, J. L. Locke, J. L. Yen.
Journ. Roy. Astron. Soc. Canada, Vol. 65, 176 (1971).
Abstr. RAS Canada.

141.039 **3C 391 — a galactic supernova remnant with an unusual radio spectrum.**
M. J. L. Kesteven, A. H. Bridle.
Journ. Roy. Astron. Soc. Canada, Vol. 65, 177 (1971).
Abstr. RAS Canada.

141.040 **A study of 3C 452 and 3C 270 (NGC 4261).**
P. P. Kronberg.
Journ. Roy. Astron. Soc. Canada, Vol. 65, 177 (1971).
Abstr. RAS Canada.

141.041 **Number counts of discrete radio sources at 1400 MHz.**   A. H. Bridle.
Journ. Roy. Astron. Soc. Canada, Vol. 65, 177 - 178 (1971).
Abstr. RAS Canada.

141.042 **Model pulsar envelopes.**
M. H. L. Pryce, I. Easson.
Journ. Roy. Astron. Soc. Canada, Vol. 65, 180 (1971).
Abstr. RAS Canada.

141.043 **Pulsar pulse amplitude analysis.**
W. H. McCutcheon, W. L. H. Shuter.
Journ. Roy. Astron. Soc. Canada, Vol. 65, 180 (1971).
Abstr. RAS Canada.

141.044 **Observations of the sporadic bright pulses from the Crab nebula pulsar.**   J. F. R. Gower.
Journ. Roy. Astron. Soc. Canada, Vol. 65, 180 (1971).
Abstr. RAS Canada.

141.045 **Polarization of radio sources at $\lambda = 73$ cm.**
R. G. Strom, R. G. Conway, P. P. Kronberg.
Journ. Roy. Astron. Soc. Canada, Vol. 65, 183 (1971).
Abstr. RAS Canada.

141.046 **On the spectra of radio emission similar to the spectrum of the quasar 3C 119.**
V. N. Kurilchik.
Astron. Tsirk., No. 613, p. 1 - 3 (1971).   In Russian.

141.047 **On the nature of the jet of 3C 273.**
V. N. Kurilchik, V. M. Charugin.
Astron. Tsirk., No. 626, p. 4 - 5 (1971).   In Russian.

141.048 **The UBV photometry of the variable radio sources NGC 1275 and 3C 120.**   V. M. Lyutyj.
Astron. Tsirk., No. 626, p. 5 - 8, with a correction in No. 631, p. 8 (1971).   In Russian.

141.049 **Optical monitoring of radio sources—III. Further observations of quasars.**
K. P. Tritton, R. A. Selmes.
Monthly Notices Roy. Astron. Soc., Vol. 153, 453 - 469 (1971).
Results for 20 quasars are reported from the Herstmonceux programme of optical monitoring of radio sources. New

variables are 3C 175 and 3C 232, and variations suspected in earlier work have been confirmed for 3C 249.1, 3C 263 and 3C 380. All quasars in this sample have now been found variable either by us or by other workers. A new major outburst is recorded for 3C 454.3. No well-defined periodicities are apparent for any quasar in the sample.

141.050 **A possible model of radio-variable structures.**
V. N. Kurilchik.
Astron. Tsirk., No. 629, p. 2 - 3 (1971).   In Russian.

141.051 **Analysis of flux density variations of quasars PKS 0736+01 and PKS 1510—08.**
A. G. Gorshkov, N. G. Rogov.
Astron. Tsirk., No. 630, p. 1 - 3 (1971).   In Russian.

141.052 **Occultation studies of 3C 245, MSH 14-1*21* and MSH 19-*21*.**   C. Hazard, J. Sutton.
Astron. Journ., Vol. 76, 609 - 614 (1971).
This paper describes the structure of the radio sources 3C 245, MSH 12-1*21*, and MSH 19-*21* derived from occultation observations. Each source is double with the majority of the radio emission arising in structure ≤ 1 arc sec.

141.053 **Counts of sources and theories.**
K. Brecher, G. Burbidge, P. A. Strittmatter.
Comments Astrophys. Space Phys., Vol. 3, 99 - 110 (1971).
From different surveys which have been extended to fainter flux levels it has been found that three parts of the log N - log S curve must be explained: the bright end where the slope is steeper than −1.5, the intermediate region, where the slope is close to the Euclidian value, and the faint end, where it becomes significantly flatter than −1.5. The possible ways of interpreting the observations within the framework of different cosmological models are summarized.

141.054 **Mode changing in pulsar radiation.**   A. G. Lyne.
Monthly Notices Roy. Astron. Soc., Vol. 153, P27 - P32 (1971).
The integrated profiles of the radio pulses from PSR 1237 + 25 and PSR 0329 + 54 change abruptly at intervals of some thousands of pulses. The changes are observed mainly in the intensity profiles; the polarization characteristics are only slightly affected.

141.055 **Optical identification of southern radio sources.**
R. W. Hunstead, B. M. Lasker, B. Mintz, M. G. Smith.
Australian Journ. Phys., Vol. 24, 601 - 607 (1971).
The fields of 12 southern radio sources have been photographed using a fibre-optics image tube at the 60 in. Cerro Tololo reflector. Identifications are suggested for 10 of the 12 sources on the basis of positional agreement with accurate 408 MHz radio positions. The remaining two fields contain faint galaxies within 10″ of the radio position.

141.056 **Observations at 408 MHz of radio sources from the 4C catalogue. II. Declination range −3° to 0°.**
R. E. B. Munro.
Australian Journ. Phys., Vol. 24, 617 - 630 (1971).
Radio positions and flux densities measured at 408 MHz with the Molonglo radio telescope are given for 235 sources from the Fourth Cambridge catalogue in the declination range −3° to 0°. Optical identifications are suggested for 70 of the sources.

141.057 **On the V/V_m test applied to quasi-stellar radio sources.**   M. J. Rees, M. Schmidt.
Monthly Notices Roy. Astron. Soc., Vol. 154, 1 - 7 (1971).
The V/V_m test for a sample of quasi-stellar radio sources complete to given radio and optical flux densities has been recently discussed critically by Longair and Scheuer. It seems to

us, however, that Longair and Scheuer have under-rated the important role of optical selection effects—and of the optical data generally—in the V/V_m procedure. In this note we shall consider these effects in the hope of further clarifying this question.

141.058 Relativistic beaming of pulsars.—The effect of the emission spectrum. F. G. Smith.
Monthly Notices Roy. Astron. Soc., Vol. 154, 5P - 6P (1971).
The width of the pulses formed by the relativistic beaming effect is shown to depend on the spectrum of the emission process. The analysis is applied to the question of the existence of detectable radiation between the pulses.

141.059 Observation of low-energy gamma radiation from NP 0532. J. D. Kurfess.
Astrophys. Journ., *(Letters)*, Vol. 168, L39 – L42 (1971).
Gamma-ray emission from the Crab pulsar has been observed up to energies above 1 MeV. The 100–400-keV light curve exhibits a strong flux in the region between the primary and the secondary peaks. The flux ratio of the secondary peak to primary peak is 2.3 ± 0.2. Comparison of the pulsed flux measured in this experiment with previous results for the total Crab emission indicates that the pulsar contributes nearly half the total emission in the 100–400-keV region.

141.060 Evolution of a stabilized oblique rotator: Behavior over short and long time scales.
W. Y. Chau, R. N. Henriksen, D. R. Rayburn.
Astrophys. Journ., *(Letters)*, Vol. 168, L79 - L85 (1971).
We have deduced properties of the Crab pulsar from the observations by treating it as a nonaligning or "stabilized" oblique rotator that, on the radiation time scale, evolves as a fluid object. We have given a description of the behavior over a short time scale (several months) that should be expected on such a model, and we have fitted this short-term behavior to observations centered on the "starquake" of September 1969. We discuss the conventional starquake theory for the present model, and we also suggest an alternative interpretation based on the instability of a stressed magnetic field.

141.061 Pulsar JP 1933: 21 cm line absorption profile and interstellar scintillation.
M. Guélin, P. Encrenaz, S. Bonazzola.
Astron. Astrophys., Vol. 14, 387 - 389 (1971).
A lower limit of 6 kpc to the distance of JP 1933 is derived from the 21 cm line absorption profile, using the galactic rotation model of Schmidt. The scintillation parameters observed at 1420 MHz for this source seem consistent with the mean parameters derived from low frequency pulsar observations in the thin-screen model.

141.062 Lunar occultation observations of 25 radio sources made with the Ooty radio telescope: List I.
G. Swarup, V. K. Kapahi, N. V. G. Sarma, G. Krishna, M. N. Joshi, A. P. Rao.
Astrophys. Letters, Vol. 9, 53 - 59 (1971).
A survey of radio sources by the method of lunar occultation is being carried out at Ootacamund in India with a large steerable radio telescope whose collecting area is 8700 m². Results for 25 weak radio sources with flux densities lying in the range 0.4 to 4 fu at 327 MHz are presented here. Optical identifications have been suggested for 8 of the 25 sources, 5 of which are possibly QSOs. The unidentified radio sources seem to be galaxies with high redshifts.

141.063 Interstellar scattering and the low-frequency spectrum of PSR 0833-45.
C. S. Higgins, M. M. Komesaroff, O. B. Slee.
Astrophys. Letters, Vol. 9, 75 - 76 (1971).
A small diameter radio source has been detected at

80 MHz in the direction of PSR 0833-45. The measured flux of 12 ± 2 fu is consistent with the source being the 'smeared out' low frequency radiation from the pulsar. This supports the view of Ables *et al.* (1970) that the spectral turnover in the pulsed component of the emission at higher frequencies is the result of radiation scattering.

141.064 Pulsars. (Theoretical conceptions). V. L. Ginzburg.
Uspekhi fiz. nauk, Vol. 103, 393 - 429 (1971). In Russian.
Abstr. in Referativ. Zhurn. 51. Astron., 9.51.501 (1971).

141.065 UBV photometry of the quasar 3C 273. K. P. Tritton.
Monthly Notes Astron. Soc. Southern Africa, Vol. 30, 113 (1971).

141.066 Fragmentation of supermassive disks. E. E. Salpeter.
Nature, Phys. Sci., Vol. 233, 5 - 7 (1971).
Supermassive rotating disks, invoked by some models for quasars and active nuclei of galaxies, are unstable against fragmentation. A hierarchy of successive fragmentation into smaller and smaller pieces is described. Questions are raised about pulsar-like and gravitational radiation, which become important at the later stages of fragmentation.

141.067 Cuasares. G. Iturbe.
El Universo, Vol. 25, 46 - 48 (1971).

141.068 Accurate radio and optical positions of 3C 273 B.
C. Hazard, J. Sutton, A. N. Argue, C. M. Kenworthy, L. V. Morrison, C. A. Murray.
Nature, Phys. Sci., Vol. 233, 89 - 91 (1971).
New measurements have confirmed that 3C 273 B and the associated quasistellar object are coincident.

141.069 On peculiarities of quasar redshifts and hypotheses to explain them. A. Kipper.
Tartu Astron. Obs. Teated, No. 30, p. 3 - 45 (1971).
A hypothesis is presented according to which redshift of the spectral lines of distant cosmic objects can be explained as a continuous flow of the energy of photons to the cosmological vacuum.

141.070 Preliminary results of observations of radio sources at short millimeter wavelengths.
V. F. Zabolotnyj, I. G. Moiseiev, A. V. Pavlov, V. I. Slysh, V. A. Soglasnova, G. B. Sholomitskij, M. B. Shcherbina-Samojlova.
Astron. Tsirk., No. 635, p. 1 - 2 (1971). In Russian.

141.071 On the existence of ultra-short flares from the quasar 3C 273.
L. M. Ozernoy, V. E. Chertoprud.
Astron. Tsirk., No. 635, p. 2 - 4 (1971). In Russian.

141.072 Method and equipment for the study of QSS in millimeter-wavelengths.
G. P. Apushkinsky, V. V. Vitkovsky.
Trudy Astron. Obs., *Leningrad*, Vol. 28 (= Uchenye Zapiski Leningr. Un-ta, No. 359 = Seriya Matem. Nauk, vyp. (No.) 47), p. 46 - 51 (1971). In Russian.
Harmonic analysis of the flux variations of QSS is discussed. The possibilities of the method are illustrated. The receiver for observations at millimeter wavelengths is described.

141.073 Observations of the structure of radio sources in the 3C catalogue – V. The properties of sources in a complete sample. C. D. Mackay.

Monthly Notices Roy. Astron. Soc., Vol. 154, 209 - 227 (1971).

High resolution maps of every source in a statistically complete sample of 200 sources have been examined for common features, in the hope that these may provide clues to the nature of the sources and thereby indicate clearly what must be explained by a successful model of the structure and evolution of sources.

141.074 Radio emission from Antares B.
R. M. Hjellming, C. M. Wade.
Astrophys. Journ., (*Letters*), Vol. 168, L115 - L116 (1971).
New interferometric data on the Antares radio source show that it is variable, detectable at both 3.7 and 11.1 cm, and associated with the B3 V companion to the red supergiant.

141.075 Flux des pulsars à 1420 MHz.
P. Encrenaz, É. Falgarone, O. Franquelin, M. Guélin.
Comptes Rendus Acad. Sci. Paris, Sér. B, Vol. 273, 686 - 688 (1971).
Dans une note précédente, nous avons présenté les observations des pulsars à 1420 MHz à l'aide du radiotélescope de Nançay et nous avons donné quelques résultats préliminaires. Nous rapportons ici des observations précises du flux moyen de 27 pulsars en polarisation verticale à 21 cm de longueur d'onde, et la détermination de l'écart quadratique moyen de ce flux. Les mesures de flux ont été effectuées avec une bande de 5 MHz.

141.076 Electromagnetic radiation of a rotating magnetic multipole. S. A. Kaplan, V. Ya. Eidman.
Astrofizika, Vol. 7, 310 - 313 (1971). In Russian. — English translation in Astrophysics, Vol. 7, No. 2.
It is shown that the complex multipole magnetic field on the surface of a pulsar or in its vicinity can radiate electromagnetic waves on high harmonics of the fundamental frequency (the angular velocity of rotation).

141.077 Optical spectra of quasi-stellar objects.
E. M. Burbidge.
Nuclei of galaxies. Conference 1970, (see 012.002), p. 121 - 146 (1971).

141.078 Absorption redshifts in QSOs.
W. H. McCrea.
Nuclei of galaxies. Conference 1970, (see 012.002), p. 189 - 190 (1971).

141.079 Compact radio sources in the nuclei of galaxies.
K. I. Kellermann.
Nuclei of galaxies. Conference 1970, (see 012.002), p. 217 - 233 (1971).

141.080 Expansion models of eruptions in quasars and radio galaxies. H. van der Laan.
Nuclei of galaxies. Conference 1970, (see 012.002), p. 245 - 266 (1971).

141.081 Space distribution and luminosity functions of quasi-stellar objects. M. Schmidt.
Nuclei of galaxies. Conference 1970, (see 012.002), p. 387 - 394 (1971).

141.082 Quasar statistics for Lemaître cosmologies.
E. E. Salpeter.
Nuclei of galaxies. Conference 1970, (see 012.002), p. 399 - 401 (1971).

141.083 Rotation and pulsation periods for pulsar models of quasars. W. A. Fowler.

Nuclei of galaxies. Conference 1970, (see 012.002), p. 511 - 514 (1971).

141.084 Mechanisms for jets. J. A. Wheeler.
Nuclei of galaxies. Conference 1970, (see 012.002), p. 539 - 567 (1971).

141.085 The evolution of radio sources. M. J. Rees.
Nuclei of galaxies. Conference 1970, (see 012.002), p. 633 - 651 (1971).

141.086 The curious mystery of log N – log S. F. Hoyle.
Nuclei of galaxies. Conference 1970, (see 012.002), p. 655 - 660 (1971).

141.087 Radio source counts, cosmology, and evolution.
G. F. Mitchell.
Journ. Roy. Astron. Soc. Canada, Vol. 65, 195 - 205 (1971).
A method of calculating number counts in uniform model universes is described and recent attempts to interpret the radio source counts are discussed. It is shown that source evolution must occur and is probably restricted to powerful sources only. There is little possibility of choosing a cosmological model using the N(S) test without a prior knowledge of radio source evolution rates.

141.088 Observations of Sgr A at 22.2 GHz with a beamwidth of 1.5 arcminutes. B. G. Leslie, M. L. Meeks, S. Rogers.
Bull. American Astron. Soc., Vol. 3, 364 (1971). – Abstr. AAS.

141.089 On cosmic ray production by pulsars.
L. C. Rosen, A. G. W. Cameron.
Bull. American Astron. Soc., Vol. 3, 365 (1971). – Abstr. AAS.

141.090 Generalization of the drifting subpulse phenomenon in pulsars. D. C. Backer.
Bull. American Astron. Soc., Vol. 3, 365 (1971). – Abstr. AAS.

141.091 Polarization of drifting subpulses in pulsar 0809+74. J. H. Taylor, G. R. Huguenin, R. N. Manchester.
Bull. American Astron. Soc., Vol. 3, 365 - 366 (1971). Abstr. AAS.

141.092 Observations of long-term intensity variations in pulsars. G. R. Huguenin, J. H. Taylor.
Bull. American Astron. Soc., Vol. 3, 366 (1971). – Abstr. AAS.

141.093 Pulsar rotation measures. R. N. Manchester.
Bull. American Astron. Soc., Vol. 3, 382 (1971). Abstr. AAS.

141.094 Disappearance of an interstellar-scattering phenomenon shown by the Crab nebula pulsar.
C. C. Counselman, J. M. Rankin.
Bull. American Astron. Soc., Vol. 3, 383 (1971). – Abstr. AAS.

141.095 The spectra of sources found in an 8000 MHz survey. G. W. Brandie.
Bull. American Astron. Soc., Vol. 3, 383 (1971). – Abstr. AAS.

141.096 The polarization of 3C279 at 8 GHz during the 1966 – 1970 outburst. H. D. Aller, E. T. Olsen.
Bull. American Astron. Soc., Vol. 3, 383 (1971). – Abstr.AAS.

141.097 Variations in the fine structure of the quasars 3C279 and 3C273. T. A. Clark, R. M. Goldstein, H. F. Hinteregger, C. A. Knight, G. E. Marandino, D. S. Robertson, A. E. E. Rogers, I. I. Shapiro, D. J. Spitzmesser, N. R. Vandenberg, A. R. Whitney.
Bull. American Astron. Soc., Vol. 3, 383 (1971). – Abstr. AAS.

141.098 The radio structure of quasars.
J. F. C. Wardle, G. K. Miley.
Bull. American Astron. Soc., Vol. 3, 384 (1971). – Abstr. AAS.

141.099 Optical monitoring of quasi-stellar objects.
P. K. Lü.
Bull. American Astron. Soc., Vol. 3, 384 (1971). – Abstr. AAS.

141.100 Characteristic behavior of four optically variable quasars. G. H. Folsom, A. G. Smith, R. L. Hackney, K. R. Hackney.
Bull. American Astron. Soc., Vol. 3, 384 (1971). – Abstr. AAS.

141.101 Tidally interacting galaxies as radio sources.
C. R. Purton, A. E. Wright.
Bull. American Astron. Soc., Vol. 3, 390 (1971). – Abstr. AAS.

141.102 Radio emission from Antares B.
R. M. Hjellming, C. M. Wade.
Bull. American Astron. Soc., Vol. 3, 399 - 400 (1971). Abstr..AAS.

141.103 Detection of double-source structure in the nucleus of the quasi-stellar radio source 3C279.
C. A. Knight, A. E. E. Rogers, I. I. Shapiro, A. R. Whitney, T. A. Clark, R. M. Goldstein, G. E. Marandino, N. R. Vandenberg.
Bull. American Astron. Soc., Vol. 3, 416 (1971). – Abstr. AAS.

141.104 Long baseline interferometry at a decametric wavelength. W. M. Cronyn, W. K. Klemperer, C. L. Rufenach, T. A. Clark, W. C. Erickson.
Bull. American Astron. Soc., Vol. 3, 416 (1971). – Abstr. AAS.

141.105 Alignment of particle spins in quasar envelopes.
D. A. Varshalovitch, B. V. Komberg.
Astron. Zhurn. Akad. Nauk SSSR, Vol. 48, 1085 - 1087 (1971). In Russian. English translation in Soviet Astron. AJ, Vol. 15, No. 5. – Short note.

141.106 The evolution at 8 GHz of the linear polarization of 3C 279. H. D. Aller, E. T. Olsen.
Astron. Journ., Vol. 76, 671 - 676 (1971).
The linear polarization of 3C 279 has been monitored at 8 GHz since 1965. During the single extended outburst in flux density since that time, the plane of polarization of the variable components has rotated by about 100 deg and is now nearly parallel to the direction of the line through the centers of the two source components found by recent very long-baseline-interferometer measurements. The apparent evolution in the magnetic field structure of the emitting regions indicates that the outward moving variable components are colliding with an ambient medium. The irregular effects of this medium may account for the short-term fluctuations in the flux density since 1968.

141.107 Identification of radio sources from the Ohio survey.
M. M. Radivich, J. D. Kraus.
Astron. Journ., Vol. 76, 683 - 685, 743 - 746 (1971).
Tentative identifications are made for 50 Ohio survey sources, many of which have flat or unusual radio spectra. About 40% of the identifications are with galaxies, 50% with stellar objects, and the remaining 10% are blank fields.

141.108 Crab nebula pulsar radio pulse arrival times at Arecibo Observatory.
J. M. Rankin, C. C. Counselman III, D. W. Richards.
Astron. Journ., Vol. 76, 686 - 690 (1971).
U. T. C. times of reception at Arecibo Observatory of 430-MHz radio pulses from the Crab nebula pulsar are tabulated at approximately 3-day intervals from 10 May 1969 to 6 April 1971. Some arrival times recorded at various frequencies prior to May 1969 are also given.

141.109 Radio stars. R. M. Hjellming, C. M. Wade.
Science, Vol. 173, 1087 - 1092 (1971).
Within the past 10 years radio emissions have been detected from six distinct classes of stars: red dwarf flare stars, red supergiants, a blue dwarf companion to a red supergiant, novas or exploding stars, pulsars, and X-ray stars. In this article we shall first review briefly the radio emission from the sun, and then we shall discuss what is known and not known about the radio stars. We shall also try to assess some of the prospects for this new area of radio astronomy.

141.110 Extragalactic Faraday rotation. H. Arp.
Nature, Vol. 232, 463 - 465 (1971).
Quasars and compact radio galaxies which show Faraday rotation are discussed. It is argued that the behavior of the Faraday rotation as a function of position on the sky, and as a function of redshift, furnishes evidence that these objects are distributed at distances within the local super cluster of galaxies instead of at the large distances indicated by their redshifts.

141.111 Observation of the Crab nebula at 30–100 keV.
H. W. Smathers, T. A. Chubb, D. Sadeh.
Nature, Phys. Sci., Vol. 232, 120 -121 (1971).
On October 17, 1970, a measurement of the pulsed emission of hard X-rays from the Crab nebula pulsar NP 0532 was made with a balloon-borne X-ray telescope from Palestine, Texas. The data fit the general pattern which shows that the energy spectrum of the main pulse falls off more steeply than that of the interpulse.

141.112 The local luminosity function and the secular evolution of extragalactic radio sources. W. Davidson.
Monthly Notices Roy. Astron. Soc., Vol. 154, 339 - 342 (1971).
It is argued that recent criticisms of an earlier paper (Davidson) by Longair & Scheuer are unfounded. The count models of Doroshkevich, Longair & Zeldovich are shown to be based on the wrong luminosity function and an incorrect evolutionary theory for the radio source population.

141.113 On calculation of quasar scintillations.
S. M. Rytov.
Izv. vyssh. uchebn. zavedenij. Radiofizika, Vol. 14, 645 - 658 (1971). In Russian. – Abstr. in Referativ. Zhurn. 51. Astron., 10.51.363 (1971).

141.114 The optical emission of the Crab nebula pulsar.
M. J. Disney.
Astrophys. Letters, Vol. 9, 9 - 12 (1971).
A model already proposed to explain the Vela pulsar is shown to give an excellent fit to the much more detailed observations of NP 0532 and its surrounding nebula.

141.115 Comprehensive observations of the rapidly varying radio source VRO 42.22.01 (BL Lac).
J. M. MacLeod, B. H. Andrew, W. J. Medd, E. T. Olsen.
Astrophys. Letters, Vol. 9, 19 - 26 (1971).

This paper presents the results of three years of radio observations of the flux density and linear polarization of VRO 42.22.01 at wavelengths of 2.8, 3.75 and 4.5 cm. During this period, nine radio outbursts have occurred. The data for the first eight outbursts are compared with several models of variable radio sources. The behavior of the source is compatible with the expanding source theory, although some difficulties remain. However, there are features which indicate that the outbursts do not occur randomly, suggesting possible rotation of the source. The ninth outburst represents a striking departure from the other eight, because of the presence of short-lived radio 'spikes'. This outburst is discussed in detail.

141.116 The structure of compact extragalactic radio sources. D. S. De Young.
Astrophys. Letters, Vol. 9, 43 - 46 (1971).

The radio spectra of 46 compact extragalactic objects are examined, and it is shown that the data are reproduced by sources which have successive uncorrelated outbursts. Models of compact sources which employ energetic particle injection into radial or multipole magnetic fields are eliminated, and it is shown that models which result in successive outbursts originating from a single object have difficulty reproducing the observed time scale between outbursts. It is noted that models, which employ multiple events occurring in physically separate regions are consistent with the spectral data.

141.117 Erratum: 'The multiplicity of secondary periodicities in pulsar CP 0834' [Astrophys. Letters, Vol. 8, 7 - 9 (1971)]. O. B. Slee, P. S. Mulhall.
Astrophys. Letters, Vol. 9, 47 (1971).

141.118 Pulsar magnetosphere. L. Mestel.
Nature, Phys. Sci., Vol. 233, 149 - 152 (1971).
The arguments of Goldreich and Julian (Astrophys. Journ., Vol. 157, 869 - 880, 1969) for a dense pulsar magnetosphere are extended to apply to the oblique rotator model.

141.119 Linear polarization measurements of PSR 0833 − 45 and their relevance to pulsar models.
M. M. Komesaroff, J. G. Ables, P. A. Hamilton.
Astrophys. Letters, Vol. 9, 101 - 104 (1971).
Measurements between 300 and 1420 MHz show that for PSR 0833 − 45 the pulse shape, as well as the degree and position angle of the linear polarization, are frequency-invariant when allowance has been made for interstellar scattering and Faraday rotation. The polarization is thus not measurably affected by propagation through the dense plasma and intense magnetic field believed to be associated with the pulsar magnetosphere. An explanation is suggested in terms of a pulsar model due to Sturrock, which involves relativistic streaming of charged particles along polar magnetic field lines.

141.120 The nature of the galactic radio source G 45.5 + 0.1.
C. G. Wynn-Williams, D. Downes, T. L. Wilson.
Astrophys. Letters, Vol. 9, 113 - 116 (1971).
The galactic radio source G 45.5 + 0.1, recently suggested to be a supernova remnant with a flat spectrum, has been mapped in the continuum with the Cambridge One Mile Telescope at 2.7 and 5 GHz and resolved into three components with angular sizes of the order of 20 arc sec. A new search at NRAO for the H 109α recombination line has been successful and the source is found to have a normal ratio of line-to-continuum temperatures, in contradiction to previous published results. It is concluded that G 45.5 + 0.1 is an H II region located in the Sagittarius spiral arm.

141.121 The low-frequency cutoff in the radio spectrum of Sagittarius A.
V. N. Brezgunov, R. D. Dagkesamansky, V. A. Udal'tsov.
Astrophys. Letters, Vol. 9, 117 - 119 (1971).

The upper limits of flux densities of the radio source Sgr A were obtained at several frequencies in the range 100-120 MHz by observations with the east-west arm of the cross radiotelescope of the Lebedev Physical Institute. It is pointed out that in the case of the uniform model of a source only the thermal absorption of radiation outside the source can explain the observational data, the optical depth of the absorbing layer τ being equal to 1 at 200 MHz. The observed spectrum can also be explained by assuming the presence of two components in the source, the nucleus and the halo, the nucleus contribution to the total emission of Sgr A dominating at frequencies higher than 5 GHz and abruptly falling at lower frequencies.

141.122 Optical variability of a bright new quasi-stellar source. S. C. Lucchetti, P. D. Usher.
Nature, Vol. 232, 622 - 623 (1971).

Using plate material from the Harvard plate stacks, the quasi-stellar source ON 325, i. e. B 2 1215 + 30, has been found to vary by approximately 2 magnitudes.

141.123 Possible selection effect in the measurement of quasar redshifts. R. C. Roeder.
Nature, Phys. Sci., Vol. 233, 74 - 75 (1971).

It is shown that there is a statistically significant correlation between the numbers of quasars in intervals of 0.1 in redshift, and the average number of emission lines measured in the spectra of those quasars. This is interpreted as meaning that at some redshifts it is easier to measure redshifts and that at these redshifts more quasar redshifts have been measured.

141.124 Expanding-source model of radio outbursts in QSS and in nuclei of Seyfert galaxies. T. Kogure.
Publ. Astron. Soc. Japan, Vol. 23, 449 - 465 (1971).

The radio outbursts which occurred in 3C120, 3C273, and 3C279 in 1966−1967 are considered under the revised expanding-source model of van der Laan (1966) type, in which the energy supply in the form of amplification of magnetic field and acceleration of electrons is taken into account. The effect of ionized gas is also examined with respect to two points for the nuclei of Seyfert galaxies.

141.125 Circular polarization of 3C 273 and NGC 4151.
A. B. Severnyj, N. S. Nikulin, V. M. Kuvshinov.
IAU Circ., No. 2343 (1971).

141.126 Possible identification of a red star with a radio source. S. P. Maran, A. A. Hoag, J. B. De-Veny.
IAU Circ., No. 2344 (1971).

141.127 Two new pulsars. G. Swarup, D. K. Mohanty, V. Balasubramanian.
IAU Circ., No. 2356 (1971).

141.128 OJ 287 = VRO 20.08.01. N. E. Kurochkin.
IAU Circ., No. 2365 (1971).

141.129 OJ 287 = VRO 20.08.01
H. Kosai, T. Seki, M. S. Burkhead.
IAU Circ., No. 2366 (1971).

141.130 NP 0532. E. Lohsen, C. Papaliolios, N. P. Carleton.
IAU Circ., No. 2368 (1971).

141.131 Time variations in high energy cosmic rays.

T. C. Weekes.
Nature, Phys. Sci., Vol. 233, 129 - 130 (1971).

Conditions under which a periodic flux of 10^{14} eV cosmic rays from nearby pulsars would be detectable are considered; while these conditions are stringent, it is concluded that a detailed statistical analysis of cosmic ray air shower arrival times on time scales of milliseconds to seconds is warranted.

141.132 The interpretation of non-linear radio spectra of discrete radio sources by a general mechanism.
S. Ya. Braude, B. P. Ryabov, I. N. Zhouk.
Astrophys. Space Sci., Vol. 12, 349 - 365 (1971).

The analysis of discrete radio sources spectra in the range 10–5000 MHz reveals that deviations from a power law in the low-frequency region may be due to distortion of differential energy spectra of relativistic electrons at low energies. An empirical expression for an energy-spectrum law was found to be in a good agreement with most of the radio spectra measured. The main physical parameters of 92 sources are evaluated.

141.133 Doppler period variations of high-velocity pulsars.
R. Gallino, G. Silvestro.
Astrophys. Space Sci., Vol. 12, 415 - 423 (1971).

The effect of the source motion on the period variation of pulsars is investigated.

141.134 Errata: 'Pulsar radio emission from expanding charge sheets' [Astrophys. Space Sci., Vol. 12, 193 - 203 (1971)].   E. Tademaru.
Astrophys. Space Sci., Vol. 12, 501 (1971).

141.135 Measurement of optical positions for identified radio sources.   R. W. Hunstead.
Proc. Astron. Soc. Australia, Vol. 2, 43 - 44 (1971).

141.136 The flux density scale for radio sources at 81.5 MHz.
P. F. Scott, J. R. Shakeshaft.
Monthly Notices Roy. Astron. Soc., Vol. 154, 19P - 23P (1971).

The flux densities at 81.5 MHz of 13 radio sources have been carefully measured with respect to those of Cas A and Cyg A.

141.137 Pulsed gamma emission above 50 MeV from NP 0532.   J. P. Leray, B. Parlier, J. Vasseur, J. Paul, M. Forichon, B. Agrinier, R. Buccheri, L. Scarsi, G. Boella, L. Maraschi, A. Treves.
Space Research XI, (see 012.004), 1385 - 1390 (1971).

141.138 Radio sources similar to BL Lac.
G. D. Nicolson.
Nature, Phys. Sci., Vol. 233, 155 (1971).

It is shown that the source PKS 1514–24, identified with the rapid variable star AP Lib, has not varied significantly at 2295 MHz over the past 4 years. However, PKS 0727–11, a compact object at low galactic latitude, has exhibited rapid variations over the same period, and these are briefly discussed in terms of results of intercontinental interferometry.

141.139 Scattering of pulsar radiation in the interstellar medium.   J. M. Sutton.
Monthly Notices Roy. Astron. Soc., Vol. 155, 51 - 64 (1971).

Electron scattering in the interstellar medium is responsible for some of the observed properties of pulsar radiation. This paper is concerned with two of these properties, the pulse broadening Δt and the range of frequency over which intensity fluctuations are correlated Δf. It is the purpose of this paper to compare the measurements of Δf made by different observers, to establish the relationship between Δf and Δt more precisely, and finally to use these results to investi-

gate the relationship between Δf and the integrated electron content DM.

141.140 A method of allowing for known observational selection in small samples applied to 3CR quasars.
D. Lynden-Bell.
Monthly Notices Roy. Astron. Soc., Vol. 155, 95 - 118 (1971).

A useful method is developed which minimizes numerical fluctuations when deriving luminosity functions and density evolution from data subject to observational selection. The distribution of ratios of radio power to optical power for quasars is derived from the 3CR quasars. The density evolution of the quasi-stellar sources is derived direct from the data and a possible interpretation is an exponential decay with cosmic time. The density evolution is then used to determine the optical luminosity function of quasars.

141.141 Spectroscopy and photography of southern hemisphere quasar identifications.   K. P. Tritton.
Monthly Notices Roy. Astron. Soc., Vol. 155, 1P - 2P (1971). Short communication.

141.142 Hard X-ray spectrum of NP 0532.
C. Cavani, F. Frontera, F. Fuligni, D. Brini.
Nature, Phys. Sci., Vol. 233, 153 - 155 (1971).

Results of a balloon borne experiment intended to measure the spectrum of the pulsating emission from NP 0532 in the energy range 20–200 KeV are reported. Useful data, referring to an observational time of 10^4 sec, have been obtained by using an azimuth stabilized gondola. Four pulse profiles, corresponding to four different energy intervals were obtained. A power spectrum was then fitted giving I (E) = (1.4 ± 0.3) E $^{-2.1 \pm 0.2}$ photons (cm² s KeV)⁻¹ for the pulsating emission. The slope of this spectrum, which fits satisfactorily all data available at higher energies, appears remarkably similar to that of the steady emission of the Crab nebula.

141.143 Quasi-stellar objects and gravitational lenses.
N. Sanitt.
Nature, Vol. 234, 199 - 203 (1971).

The space density of gravitational lens images as a function of redshift is calculated for various cosmological models. One cannot interpret all QSOs as gravitational lens images, unless there is a substantial amount of evolution with z in the sources of the lens images, and also a large fraction of the mass of the dark universe in the form of compact dark objects. A conservative estimate, assuming no evolution of the sources, and a smoothed-out density of 7×10^{-31} g/cm^{-3}, for the objects which act as lenses, results in a discrepancy by a factor $\sim 10^4$ between the comoving density of lens images and QSOs, out to a redshift of $z \simeq 2.8$.

141.144 Pulsar distances and energy.
B. Basu, R. Bandyopadhaya.
Stud. Cerc. Astron., Vol. 16, 169 - 176 (1971).

The distances of pulsars have been calculated on the assumption $<n_e> = 0.01$ cm^{-3}. This value of $<n_e>$ has been taken as the most plausible value if partial ionization of the interstellar hydrogen by a high flux of low-energy cosmic rays is assumed. It has been shown that the pulsars so far discovered are all galactic and also that they are, in general, weak emitters. Although the pulsars are believed to be physically associated with supernovae, the fact that no known galactic supernova except the Crab nebula has been identified with a pulsar, casts doubt on such an association. The association of NP 0532 with the Crab nebula may be only a positional coincidence.

141.145 Structure of non-thermal radio source G 55.7 + 3.4 in the direction of CP 1919.

M. R. Kundu, T. Velusamy.
Nature, Phys. Sci., Vol. 234, 54 (1971).

Intensity contour maps of G 55.7 + 3.4 at 11 and 21 cm wavelengths, with resolution of 5 and 10 min of arc, respectively, are presented. The 11 cm map suggests that the source has a shell structure typical of supernova remnants. It has a non-thermal spectrum with a spectral index of 0.54. Possible association of the source G 55.7 + 3.4 with the pulsar CP 1919 is discussed.

141.146 **On radio structures with synchrotron selfabsorption.**
V. N. Kurilchik, T. G. Sitnik.
Astron. Tsirk., No. 636, p. 1 - 3 (1971). In Russian.

141.147 **Pulsar P 1749 - 28 a possible XR source.**
G. S. Tsarevskij.
Astron. Tsirk., No. 636, p. 4 - 6 (1971). In Russian.

141.148 **The nature of the apparent expansion of radio variable structures with super-light velocities.**
V. N. Kurilchik.
Astron. Tsirk., No. 639, p. 1 - 3 (1971). In Russian.

141.149 **Photographic observations of radio sources.**
N. E. Kurochkin.
Astron. Tsirk., No. 644, p. 1 - 4 (1971). In Russian.

141.150 **An attempt of discovering ultrashort optical flares of 3C 273.** V. M. Lyutyj, A. M. Cherepashchuk.
Astron. Tsirk., No. 647, p. 1 - 4 (1971). In Russian.

141.151 **High-resolution observations of compact radio sources at 6 and 18 centimeters.**
K. I. Kellermann, D. L. Jauncey, M. H. Cohen, B. B. Shaffer, B. G. Clark, J. Broderick, B. Rönnäng, O. E. H. Rydbeck, L. Matveyenko, I. Moiseyev, V. V. Vitkevitch, B. F. C. Cooper, R. Batchelor.
Astrophys. Journ., Vol. 169, 1 - 24 (1971).

This paper presents data obtained in two series of observations made in 1967 April at 18 cm between Lincoln Laboratory in Massachusetts and Green Bank, West Virginia, and during 1969 at 18 and 6 cm using stations in the United States, in Sweden, Australia, and the U.S.S.R. The new data have been combined with the previously published material to determine in more detail the small-scale radio structure over a wide range of wavelength, and also to estimate, where possible, the spectra of the individual components within the compact sources. The sources studied include identified QSOs and radio galaxies as well as some unidentified sources.

141.152 **Properties of pulsars.**
G. R. Huguenin, R. N. Manchester, J. H. Taylor.
Astrophys. Journ., Vol. 169, 97 - 104 (1971).

The correlation of various observed properties of pulsars is discussed with reference to the subdivision of pulsars into three groups on the basis of their mean pulse profile and sub-pulse phase modulation.

141.153 **The period-age distribution of pulsars.**
P. A. Sturrock.
Astrophys. Journ., *(Letters)*, Vol. 169, L7 - L10 (1971).

A recently proposed model of pulsars leads to a constraint on the values of period and "age" of radio-emitting neutron stars and a similar constraint for optically emitting neutron stars. These constraints compare satisfactorily with available data.

141.154 **A possible interpretation of the precursor pulse in NP 0532.** F. Pacini.
Astrophys. Journ., *(Letters)*, Vol. 169, L11 - L12 (1971).

The properties of the precursor pulse of NP 0532 are consistent with the idea that it is emitted along the proton field lines. The sharp spectral cutoff above 430 MHz implies a cutoff in the proton energy spectrum around 2×10^{11} eV.

141.155 **Daily and hourly variations in flux density of radio sources.** B. J. Wills.
Astrophys. Journ., Vol. 169, 221 - 233 (1971).

Flux-density measurements with rms accuracy of ± 1.6 percent have been made in an attempt to detect day-to-day and hour-to-hour variations in flux density at a frequency of 2700 MHz, for radio sources selected from the Parkes Catalog. Day-to-day variations of about 2–4 percent detected for the sources PKS 0106 + 01, 0336 – 01 (CTA 26), 0440–00 (NRAO 190), and 1510–08 are probably significant, and their flux densities have been compared with data at higher frequencies. Upper limits to variability are given for the other sources.

141.156 **A positional determination of NP 0532.**
B. J. McNamara.
Publ. Astron. Soc. Pacific, Vol. 83, 491 - 492 = Lick Obs. Bull., No. 616 (1971).

A newly determined position, based on recent observational data, is given for the pulsar NP 0532 (Crab nebula).

141.157 **The Ohio Survey between declinations of 40° and 63° north.**
R. K. Brundage, R. S. Dixon, J. R. Ehman, J. D. Kraus.
Astron. Journ., Vol. 76, 777 - 889 (1971).

A 1415-MHz continuum survey with the Ohio State University (OSU) 110-m by 21-m radio telescope has been made between declinations of 40° and 63° north covering 4407 deg² of sky. Results are presented by 67 maps of the regions surveyed and by a list of 3475 sources at or above 0.18 f.u. Of these sources, 2388 are previously uncatalogued. This is the fifth installment of the Ohio Survey. The five installments include a total of 11808 sources in 5.87 steradians of sky.

141.158 **Evolution of quasar radio and optical luminosity.**
L. M. Golden.
Nature, Phys. Sci., Vol. 234, 103 - 106 (1971).

Arakelian's method of distinguishing between evolution of the optical and radio luminosities of quasars has been applied to samples freed from the selection effects that he did not allow for. It is found, contrary to his conclusion, that the optical and radio luminosities of quasars evolve similarly with time.

141.159 **Étude de la radiosource 3C 173 et de certains astres du même champ.**
G. Wlérick, G. Lelièvre, avec la collaboration technique de A. Sellier, J. P. Lemonnier, D. Michet.
Comptes Rendus Acad. Sci. Paris, Sér. B, Vol. 273, 989 - 992 (1971).

La mesure de clichés électronographiques pris dans le système *UBV* permet d'identifier la radiosource 3C 173 à un quasar ayant les propriétés particulières suivantes: magnitude élevée, forte variation du flux lumineux, variation très importante de l'indice de couleur B – V. La magnitude limite des clichés est la plus élevée atteinte à ce jour dans chacune des trois couleurs; l'électronographie permet de mesurer des astres à excès d'ultraviolet du type naine blanche ou quasar plus faibles que tous ceux qui ont été mesurés jusqu'ici avec les grands télescopes.

141.160 **Importance of pulsar observations for planetary physics.** A. Treves.
Astron. Astrophys., Vol. 15, 471 - 473 (1971).

The possibility of observing planetary systems of pulsars is examined. The expected perturbation in the time of arrival of pulses is computed and the detectability is discussed.

141.161 Quasi-stellar radio sources and optical quasi-stellar objects. M. A. Arakelian.
Astrofizika, Vol. 7, 457 - 465 (1971). In Russian. – English translation in Astrophysics, Vol. 7, No. 3.

The distributions of redshifts and luminosities of quasi-stellar radio sources and optical quasi-stellar objects are compared. The differences between distributions of these two types of objects make it possible to conclude that their luminosity functions are different.

141.162 Faraday rotation and signal dispersion: The geometrical optics: approximation, an exact solution, and first order smoothing theory. I. Lerche.
Astrophys. Space Sci., Vol. 13, 48 - 52 (1971).

We critically discuss the three approximations which have been employed to estimate the influence of interstellar fluctuations in both electron density and magnetic field on Faraday rotation measure and signal dispersion measure in the radio band.

141.163 Comments on 'The Faraday rotation and signal dispersion' by I. Lerche.
V. L. Ginzburg, L. M. Erukhimov.
Astrophys. Space Sci., Vol. 13, 53 - 55 (1971). Concerning the paper by I. Lerche cited in Abstr. 141.162.

141.164 On the formation of pulsar radiation diagrams.
V. V. Zheleznyakov.
Astrophys. Space Sci., Vol. 13, 74 - 86, 87 - 99 (1971). In Russian and English.

A model of pulsars is discussed in which formation of a polar diagram of the radiation is influenced by the motion of the source around a neutron star with a velocity close to that of light. For a power-law frequency-spectrum of the radiation and isotropy of the diagram in a system of coordinates rotating with the source, the width of the observed pulse is shown to be independent of frequency. The proposed explanation of the second period characteristics of type CP 1919 pulsars is based on the effect of relativistic motion of the radiation source. The positions are established (relative to the axis of rotation of the star) of the local sources of radiation in the optical and in the radio ranges for the pulsar NP 0532.

141.165 Search for pulsed gamma ray emission above 50 MeV from NP 0532.
J. Vasseur, J. Paul, B. Parlier, J. P. Leray, M. Forichon, B. Agrinier, G. Boella, L. Maraschi, A. Treves, L. Buccheri, A. Cuccia, L. Scarsi.
Nature, Phys. Sci., Vol. 233, 46 - 48 (1971).

A search for pulsed gamma ray emission above 50 MeV from the pulsar NP 0532 has been conducted with a balloon borne spark chamber. The analysis of the data of two of the flights performed in 1969 to search for a periodic pulsar profile in phase with radio and optical measurements shows a two standard deviation excess of counts at the place of the secondary peak observed at lower wavelengths. This result is not considered to be a statistically positive effect.

141.166 Compact radio sources in the galactic nucleus.
D. Downes, A. H. M. Martin.
Nature, Vol. 233, 112 - 114 (1971).

This letter describes new observations of Sgr A with the Cambridge one-mile-telescope at frequencies of 2.7 and 5 GHz. The beamwidths in right ascension were respectively 11 and 6 arc sec, the latter being the highest resolution yet applied to the galactic centre. The observations suggest a three-component model for Sgr A, and confirm the existence of radio structure on a scale of 10 arc sec. The relation of these compact radio sources to the infrared emission is also examined.

141.167 Second decrease in the period of the Vela pulsar.

P. E. Reichley, G. S. Downs.
Nature, Phys. Sci., Vol. 234, 48 (1971).

The period of the Vela pulsar (PSR 0833–45) decreased 179 ns between August 21 and September 4, 1971. A preliminary analysis of the data shows that the rate of change of period increased. This discontinuity in period is very similar to the discontinuity in period that occurred between February 24 and March 3, 1969. The time span between discontinuities of 2.5 yr. would seem to rule out most starquake theories. The age estimate will have to be increased and could very well be as old as the Vela X supernova remnant.

141.168 Non-thermal shell source close to the direction of CP 1919. W. M. Goss, U. J. Schwarz.
Nature, Phys. Sci., Vol. 234, 52 - 53 (1971).

Using the Westerbork radio synthesis telescope it is shown that the extended source G 55.7 + 3.4 (Caswell and Goss, Astrophys. Letters, Vol. 7, 141 (1970)) has a ring structure, suggestive for a SNR; the pulsar is at the edge of the ring. Both the estimates of distances and ages of the two objects are mostly different, therefore the reality of the association can be questioned. However, if the association is real, then the extended source must be a new type of galactic object.

141.169 Period-luminosity function for pulsars.
V. R. Venugopal.
Nature, Phys. Sci., Vol. 234, 55 (1971).

The validity of the period-luminosity relation as given by Cavallo (Nature, Phys. Sci., Vol. 231, 35 (1971)) appears doubtful since the distances derived from this relation are not in agreement with more directly determined distances to some pulsars.

141.170 Quasars and cosmology. M. Schmidt.
Observatory, Vol. 91, 209 - 214 (1971). – The Halley Lecture for 1971, delivered in Oxford on May 6.

141.171 Further data on the optical variable PKS 1514–24.
R. W. Hunstead.
Nature, Vol. 233, 401 - 402 (1971).

Accurate positions have been measured for the radio source PKS 1514–24 and its optical counterpart, the irregular variable 'star' AP Lib. A radio-optical displacement of about 5" arc may indicate the presence of small scale radio structure. There is no definite evidence at present for radio variability. The optical appearance of AP Lib on the plate copies of the Palomar Sky Survey is quite unusual and warrants further investigation.

141.172 Interplanetary scintillation of radio sources at metre wavelengths – II. Theory. A. C. S. Readhead.
Monthly Notices Roy. Astron. Soc., Vol. 155, 185 - 197 (1971).

Interplanetary scintillation is being used, in a survey of radio sources, to study angular structure in the range 0".1–1" at a frequency of 81.5 MHz. The application of diffraction theory to scintillation at this frequency is discussed, and the diffracting parameters of the interplanetary medium are derived. The dependence of scintillation index on angular structure and receiver bandwidth is determined.

141.173 Observations of OJ 287 at optical and millimeter wavelengths. T. D. Kinman, E. K. Conklin.
Astrophys. Letters, Vol. 9, 147 - 149 (1971).

The radio source OJ 287 varies with a time scale of a few days at both optical and millimeter wavelengths. At optical wavelengths it has a blue continuous spectrum and has a high variable polarization.

141.174 OJ 287: An exceptionally active variable source.

B. H. Andrew, G. A. Harvey, W. J. Medd.
Astrophys. Letters, Vol. 9, 151 - 154 (1971).

Observations of the radio source OJ 287 at centimeter wavelengths have shown that it has both rapid variations with time scales of a few days and very strong long term variations. The rapid variations resembled optical variations which were observed at the same time.

141.175 The precursor pulse from NP 0532: A result of scattering? M. M. Komesaroff.
Astrophys. Letters, Vol. 9, 195 - 200 (1971).

It is shown that most of the observed properties of the precursor pulse from NP 0532 can be explained by assuming that the precursor results from radiation scattered out of the main pulse beam by a ring of material centered on the pulsar. If the radius of the ring is just greater than that of the velocity-of-light cylinder, the approximate shape, the arrival time, and the high degree and invariant position angle of the linear polarization are explained in simple geometric terms.

141.176 An investigation of pulse height fluctuations of four pulsars. W. H. McCutcheon, W. L. H. Shuter.
Astrophys. Letters, Vol. 9, 201 - 204 (1971).

The pulsars CP0329, CP0950, CP1133, and CP1919 have been investigated for periodic fluctuations in their pulse amplitudes. Power spectral analysis of integrated pulse amplitudes at 408 MHz shows no periodicities that could be classed as intrinsic to the source. However, the same analysis over narrow time windows of the pulse profiles of CP1133 and CP1919 exhibits the periodicities that have been seen over the integrated pulse at lower radio frequencies. CP0329 and CP0950 exhibit only pulse-to-pulse correlations.

141.177 Polarization of the drifting subpulses of pulsar 0809 + 74.
J. H. Taylor, G. R. Huguenin, R. M. Hirsch, R. N. Manchester.
Astrophys. Letters, Vol. 9, 205 - 208 (1971).

Individual pulses from PSR 0809 + 74, which exhibits well-organized drifting subpulses, are found to be highly linearly polarized. Variations of the polarization parameters are accurately synchronized with the drifting subpulse pattern. It is suggested that the symmetry axis determining the polarization variation is associated with an emission region which moves systematically with respect to the neutron star.

141.178 Bursts in the radioemission of pulsar PP0943.
V. V. Vitkevich, Yu. I. Alekseev, V. F. Zhuravlev, Yu. P. Shitov.
Astrophys. Letters, Vol. 9, 209 - 210 (1971).

Occasional strong increases in intensity occur in the radioemission from pulsar PP0943.

141.179 Lunar occultation observations of PKS 1514–24 (=AP Lib). V. K. Kapahi.
Nature, Phys. Sci., Vol. 234, 49 - 50 (1971).

From an occultation observed at 327 MHz with the Ooty radio telescope, accurate radio position and brightness distribution over the optically variable source AP Lib have been determined. Whereas Hunstead's (1971) observations at 408 MHz indicated a significant difference between the radio and optical positions, we find excellent agreement between the two.

141.180 Two variable radio sources.
D. G. MacDonell, A. H. Bridle.
Nature, Phys. Sci., Vol. 234, 88 - 89 (1971).

Two radio sources (DW 0224+67 and DW 0727−11) with opaque microwave spectra have shown large variations at 3.24, 6.63 and 10.6 GHz since November 1968. The variation of DW 0224+67 may be consistent with van der Laan's model; that of DW 0727−11 shows a long period of apparent stability during two years in which the overall increase in flux density was ~ 180 per cent. The opaque source DA 193 has not varied by more than 10 per cent since August 1969.

141.181 Microsecond intensity variations in the radio emissions from CP 0950. T. H. Hankins.
Astrophys. Journ., Vol. 169, 487 - 494 (1971).

A technique for the removal of the effect of dispersion by the interstellar medium from the radio emission from pulsars has been developed. Observations of the pulsar CP 0950 over a bandwidth of 125 kHz at 111.5 MHz have revealed time structure as short as the reciprocal bandwidth limit of 8 microseconds. Occasional isolated subpulses of 40000 flux units have been observed.

141.182 Radio sorgenti e oggetti quasi stellari.
C. Barbieri.
Mem. Soc. Astron. Italiana, Nuova Ser., Vol. 42, 363 - 371 (1971).

In this paper several problems of structure and evolution of the extragalactic radio sources (galaxies and quasi-stellar objects) are reviewed, among these the source of energy, the emission mechanism, the life time of the radio activity, the variability.

141.183 Chronology of the peculiar objects. A. Cavaliere.
Mem. Soc. Astron. Italiana, Nuova Ser., Vol. 42, 407 - 418 (1971).

The collective evolution of quasars and the information they can provide about the chronology of the condensed matter are discussed. A simple model for the energy source and its individual evolution is used.

141.184 Ohio radio-source spectra: List II.
B. J. Wills, J. D. Kraus, B. H. Andrew.
Astrophys. Journ., (Letters), Vol. 169, L87 - L91 (1971).

Spectra and improved positions are presented for 30 selected Ohio radio sources. The selection has been limited to those sources whose spectra are flat or show an increase in flux density with increasing frequency, or are otherwise complex. Flux densities have been measured between frequencies of 0.6 and 10.6 GHz, with some additional measurements at 13.5 GHz. The new position measurements have typical rms errors of ± 15″.

141.185 Observations of the $^2\Pi_{1/2}$, $J = 1/2$ line of OH at 4660 MHz in Sagittarius B2.
F. F. Gardner, J. C. Ribes, M. W. Sinclair.
Astrophys. Journ., (Letters), Vol. 169, L109 - L112 (1971).

The emission of the 4660-MHz $^2\Pi_{1/2}$, $J = 1/2$ OH line from Sgr B2 consists of one narrow feature, possibly two, superposed on a broad emission spectrum some 300 kHz (20 km s^{-1}) wide. The emission is centered 0.5 north of the position of the 6-cm continuum peak.

141.186 On the absorption of gamma rays by photons in pulsars, quasi-stellar objects, and other source objects. J. B. Pollack, P. D. Guthrie, B. S. P. Shen.
Astrophys. Journ., (Letters), Vol. 169, L113 - L116 (1971).

Because of pair production, cosmic γ-rays above a certain critical energy are destroyed by collisions with low-energy photons before they can escape from their source objects. The critical energy is a sensitive function of the size of the photon-production zone of pulsars and QSOs and can be used for obtaining their sizes. Critical energies for supernova explosions and the Crab nebula are estimated.

141.187 Upper limit to circular polarization of optical pulsar NP 0532.
W. J. Cocke, G. W. Muncaster, T. Gehrels.
Astrophys. Journ., (Letters), Vol. 169, L119 - L121 (1971).

We have measured the circular polarization of optical pulsar NP 0532 at an effective wavelength of 0.5 μ. The measurements were made throughout the light curve with time-resolution windows 2.65 ms wide. No circular polarization was found; the precision about the peak of the main pulse was ±0.07 percent (rms).

141.188 **Optical changes in eleven Ohio radio sources with unusual spectra.** G. H. Folsom, A. G. Smith, R. L. Hackney, K. R. Hackney, R. J. Leacock.
Astrophys. Journ., (*Letters*), Vol. 169, L131 - L135 = Rosemary Hill Obs., Univ. Florida, *Gainesville*, Contr. No. 26 (1971).

Two and a half years of photographic monitoring of 17 Ohio radio sources with flat or peaked spectra has resulted in the detection of short-term optical variations in eight of the sources and possible changes in three of the remaining sources.

141.189 **Counts of radio sources at 6-centimeter wavelength.** K. I. Kellermann, M. M. Davis, I. I. K. Pauliny-Toth.
Astrophys. Journ., (*Letters*), Vol. 170, L1 - L5 (1971).

Counts of radio sources at 6 cm down to a source density of 5×10^3 sources per steradian show no compelling evidence that the space density of powerful radio sources depends strongly on epoch.

141.190 **Angular distribution of quasistellar objects.** D. Wills.
Nature, Phys. Sci., Vol. 234, 168 - 172 (1971).

An examination of the distribution on the sky of quasistellar objects identified with 4C radio sources shows no significant departures from a uniform distribution of redshift, apparent magnitude or total number as a function of position on the sky.

141.191 **Observations of pulsar polarization at 410 and 1665 MHz.** R. N. Manchester.
Astrophys. Journ., Suppl. Ser., No. 199, Vol. 23, 283 - 319 (1971).

Linear-polarization parameters and mean pulse profiles with good time resolution have been determined for twenty-one pulsars, with comparable data at two frequencies, 410 and 1665 MHz, for most of these. In addition, circular polarization has been measured at 410 MHz for about two-thirds of the pulsars discussed. The observations show that the fractional linear polarization varies over a wide range and that the amount of polarization is generally less at 1665 MHz than at 410 MHz. Circular polarization is generally weak. However, when it does occur, it frequently reaches a peak at the center of the pulse. The observations give strong support to the pulsar model in which the radiation originates close to the surface of the star. Depolarization effects occurring in the pulsar magnetosphere are proposed to account for the variation in fractional polarization.

141.192 **An absorption-line study of the galactic neutral hydrogen at 21 centimeters wavelength.**
M. P. Hughes, A. R. Thompson, R. S. Colvin.
Astrophys. Journ., Suppl. Ser., No. 200, Vol. 23, 323 - 367 (1971).

The interferometer at the Owens Valley Radio Observatory, together with a multichannel receiving system with filters of bandwidth 4 or 6 kHz, has been used to observe ninety-seven sources for absorption at 21-cm wavelength. All but four of the sources are believed to be extragalactic. Measurable absorption was detected in sixty-four cases, and for these the profiles are presented. Optical depths greater than 0.5 are found only within 20° of the galactic plane, except in the region of the Cetus Arc where values up to 1.4 occur near $b = -40°$. The mean temperature of the cool absorbing gas is found to be 71°± 9°K, and this figure is corrected for the

effects of the surrounding hot gas. A comparison of 21-cm absorption with published measurements of the free-free absorption for three of the sources results in a mean value for the ionization rate of $2.0 \times 10^{-15}\,s^{-1}$. This figure refers to gas in the cool state and within 2 kpc of the sun. The distance of the excess absorbing gas in the direction of the Cetus Arc is found to be ≤ 120 pc from a comparison of velocities of 21-cm absorption features with those of absorption lines in nearby stars. For the galactic source 3C 58, a lower limit of 8.5 kpc is obtained for the distance.

141.193 **Electromagnetic pulsar models.** J. E. Gunn.
Publ. Astron. Soc. Pacific, Vol. 83, 594 - 598 (1971).
Invited symposium paper presented at the Hawaii meeting of the Astronomical Society of the Pacific, 22–25 June 1971.

141.194 **On the absorption-line spectrum of 4C 05.34.** J. N. Bahcall, S. Goldsmith.
Astrophys. Journ., Vol. 170, 17 - 24 (1971).

Eight acceptable absorption redshifts are found in the spectrum of 4C 05.34 reported by Lynds. The absorption redshifts range from $z = 2.875$ to $z = 1.776$. An average nonsense spectrum has 1.4 acceptable redshifts. The absence of absorption from excited fine-structure states implies that the absorbing region has an electron density $\lesssim 10^2\,cm^{-3}$ and is at a distance $\gtrsim 1$ kpc from the continuum source of the QSO.

141.195 **On the possibility of pulsar action in quasars.** P. A. Sturrock.
Astrophys. Journ., Vol. 170, 85 - 92 (1971).

A recently proposed theory of pulsars is taken as a basis for evaluating Morrison's proposal that quasars may be giant pulsars. It is found that an object with mass $10^{43.4}$g, radius $10^{16.5}$ cm, and magnetic field strength 10^4 gauss (Morrison's parameters) would not exhibit pulsar activity but would be an intense source of γ-rays with luminosity exceeding 10^{48} ergs s^{-1}. Pulsar activity would occur if the field strength were $10^{5.7}$ gauss or more, but the lifetime would be only $10^{2.6}$ years.

141.196 **Charged particle motion near pulsars.** P. Goldreich.
Publ. Astron. Soc. Pacific, Vol. 83, 599 - 601 (1971). – Invited symposium paper presented at the Hawaii meeting of the Astronomical Society of the Pacific, 22 - 25 June 1971.

141.197 **The radio luminosity function and the luminosity diameter function of extragalactic radio sources.**
G. M Richter.
Astron. Nachr., Vol. 293, 111 - 117 (1971).

On the basis of a radio index-surface brightness diagram recently published, the luminosity function and the luminosity diameter function are obtained. The distinction of two populations is supported. The density of the weak population ($P < 10^{25}$ W Hz^{-1} ster^{-1} at 1400 MHz) follows nearly a power law in P. The density of the strong population has a maximum between 10^{25} and 10^{26} W Hz^{-1} ster^{-1} and around 100 kpc. A strong evolution effect is clearly present and is in a good agreement with the models obtained from the log N-log S counts.

141.198 **Abschätzung von Quasarmassen.** K.-H. Schmidt.
Astron. Nachr., Vol. 293, 125 - 126 (1971).

The masses of quasi-stellar objects having absorption lines with redshifts larger than the redshifts of the corresponding emission lines have an average value of $5 \times 10^{12}\,M_\odot$, if the emission lines are attributed to gas clouds which are moving radially against the quasar.

141.199 **A catalogue of quasars.** J. B. De Veny, W. H. Osborn, K. Janes.
Publ. Astron. Soc. Pacific, Vol. 83, 611 - 625 (1971).

We have catalogued important optical data and collected a bibliography for all quasars with redshifts published prior to May or June 1971, including objects with only one spectral line. References to optical identifications, alternate designations, photometry, redshifts, proper motions, variability, spectrophotometry, and polarization measurements are given.

141.200 **What the pulsars are?** K. Ziołkowski.
Urania Kraków, Vol. 42, 274 - 277 (1971).
In Polish.

141.201 **Observations at 408 MHz of radio sources from the 4C catalogue. III. Studies of identified sources.**
R. E. B. Munro.
Australian Journ. Phys., Vol. 24, 743 - 755 (1971).

This paper considers a sample of 4C sources selected from the data of two earlier papers of this series, which is essentially complete down to 2 f.u. at 178 MHz and contains 416 sources of which 152 are identified. It has been used to study a number of properties of radio sources, including the relation between the emission from associated radio and optical objects, the ratio of galaxies to quasi-stellar sources among radio sources, the angular extent of the sources, the distribution of the radio spectral indices, and the source-count relationship.

141.202 **The small-scale structure of radio galaxies and quasistellar sources at 3.8 centimeters.**
M. H. Cohen, W. Cannon, G. H. Purcell, D. B. Shaffer, J. J. Broderick, K. I. Kellermann, D. L. Jauncey.
Astrophys. Journ., Vol. 170, 207 - 217 (1971).

We have observed fringes from 31 compact radio sources, including eight known or suspected galaxies and 20 known or suspected QSSs, by using the Goldstack interferometer at $\lambda = 3.8$ cm $(d/\lambda = 10^8)$. Fringe visibility curves were obtained for nine sources showing structure on a scale of 10^{-3} sec of arc, and simple models are fitted to the data. Results for 3C 273 and 3C 279 are compared with data taken by Knight et. al. at an earlier epoch. The apparent changes in brightness distribution of 3C 273 and 3C 279 are difficult to explain.

141.203 **A search for redshifted neutral hydrogen in 3C 48 and other compact objects.**
G. A. Seielstad, B. Höglund, E. Kollberg.
Astrophys. Journ., Vol. 170, 219 - 222 (1971).

Upper limits to the possible redshifted H I absorption or emission in 3C 48, PKS 1217+02, PKS 1229−02, 3C 277.1, and 3C 171 are reported.

141.204 **On quasar evolution.**
A. Cavaliere, P. Morrison, K. Wood.
Astrophys. Journ., Vol. 170, 223 - 231 (1971).

We examine the consequences for quasar statistics of a class of models describing the evolution of individual strong sources. The continuity equation for the change of density and luminosity with cosmological epoch determines the population, once a model for the evolution of an individual object is chosen. A genetic relationship between quasars and radio galaxies which qualitatively fits the observations is suggested by the model.

141.205 **On the mass and chemical composition of Cassiopeia A.** M. Peimbert.
Astrophys. Journ., Vol. 170, 261 - 263 (1971).

The mass and chemical composition of the moving knots in Cassiopeia A are estimated from new photoelectric observations by Searle.

141.206 **Identification of the Ryle-Neville radio sources.**
M. V. Penston.
Astrophys. Journ., Vol. 170, 395 - 399 (1971).

A complete search for optical identifications of the Ryle-Neville (1962) radio survey has been performed. A particularly interesting identification is that of RN 8 = 3C 61.1 with a star with an ultraviolet excess in a small cluster of galaxies. The complete identification of this catalog gives constraints to the cosmological model and the optical and radio luminosity functions of QSOs and radio galaxies.

141.207 **Polarization of radio sources. III. Absorption effects on circular polarization in a synchrotron source.** A. G. Pacholczyk, T. L. Swihart.
Astrophys. Journ., Vol. 170, 405 - 408 (1971).

The transfer problem for polarized radiation is solved for a uniform synchrotron source. The absolute value of the degree of circular polarization V/I has the same dependence on frequency and magnetic field in the two extremes of very large and very small optical thickness of the source. It is remarkable that V reverses sign at an intermediate optical thickness of the source.

141.208 **Coherent synchrotron radiation.**
P. Goldreich, D. A. Keeley.
Astrophys. Journ., Vol. 170, 463 - 477 = Contr. Lick Obs., No. 338 (1971).

A simple model consisting of a distribution of charges constrained to move on a ring is the basis of an investigation of coherent synchrotron radiation. The radiation produced as a result of a nonrandom particle distribution on the ring is examined from the viewpoint of the interaction of individual particles with the total electric field of the system. A linear stability analysis shows that, under reasonable conditions, a uniform distribution of particles is unstable to clumping. The model is applied to pulsars.

141.209 **Abundance ratios from the absorption spectrum of the quasar PKS 0237−23.** F. D. A. Hartwick.
Astrophys. Journ., *(Letters)*, Vol. 170, L127 - L129 (1971).

Using the measurements of absorption features in the spectrum of the quasar PKS 0237−23 by Bahcall, Greenstein, and Sargent, we find $N(C) / N(Si) \sim 2.6$ and $N(Al) / N(Si) \sim 0.04$ to within factors of 3. On the basis of recent explosive-nucleosynthesis calculations and observations of objects within the Galaxy, we conclude that PKS 0237−23 is chemically old.

141.210 **Variable radio object 20.08.01.** W. Wenzel.
Inform. Bull. Variable Stars, (I. A. U. Commission 27), Konkoly Obs., *Budapest*, No. 593 (1971).

141.211 **Two new pulsars.** G. Colla, C. Salter, J. Sutton.
IAU Circ., No. 2374 (1971).

141.212 **Mechanism for the delay of polarization minima in the optical pulsar NP 0532.** D. C. Ferguson.
Nature, Phys. Sci., Vol. 234, 86 - 87 (1971).

The observed delays in the polarization minima of the optical radiation after the pulse peaks in the Crab nebula pulsar NP 0532 may arise from the relativistic beaming of radiation from localized emitting regions orbiting relativistically around a neutron star, with the emission maxima in the direction of the local magnetic field lines in the emitting frames.

141.213 **Pulsare und Neutronensterne.**
L. Biermann.
Physik und Kosmologie, (see 003.024), p. 80 - 89 (1971).

141.214 **Pulsar theory I. Dynamics and electrodynamics.**
P. Goldreich, F. Pacini, M. J. Rees.
Comments Astrophys. Space Phys., Vol. 3, 185 - 189 (1971).

In this comment we wish to examine critically some currently popular theories of pulsar electrodynamics. We adopt the hypothesis that pulsars are spinning neutron stars deriving their power from rotational energy, since there seems – within the framework of conventional physics – to be no feasible alternative to this view.

141.215 The influence of selection on the redshifts distribution of quasars. I. Semeniuk, A. Kruszewski.
Acta Astron., Vol. 21, 437 - 447 (1971).

It is shown that the observational selection strongly affects the redshift distribution of quasars. The apparent clustering of redshifts around $z = 2.0$ is most likely spurious.

141.216 Pulsars. A. Hewish.
Highlights of Astronomy, Vol. 2, (see 003.026), 3 - 19 (1971). – Invited discourse.

141.217 Pulsars (theoretical considerations). V. L. Ginzburg.
Highlights of Astronomy, Vol. 2, (see 003.026), 20 - 62 (1971). – Invited discourse.

141.218 The calibration of Cas A and Cyg A fluxes in the range of 300 - 9375 MHz.
V. S. Troitsky, K. S. Stankevich, N. M. Tseitlin, V. D. Krotikov, L. N. Bondar, K. M. Strezhneva, V. L. Rakhlin, V. P. Ivanov, S. A. Peljushenko, M. M. Zubov, R. A. Samoilov, G. K. Titov, V. A. Porfirjev, S. P. Chekalev.
Astron. Zhurn. Akad. Nauk SSSR, Vol. 48, 1150 - 1153 (1971). In Russian. English translation in Soviet Astron. AJ, Vol. 15, No. 6.

The present paper is a continuation of precise absolute measurements of flux densities of discrete sources by the method of "artificial moon". The purpose of this program of measurements is obtaining detailed spectra of radioemission of the most powerful discrete sources as well as the observation of its possible changes.

141.219 The mean magnetic field towards the pulsar AP 1237 + 25 and its linear polarization.
V. V. Vitkevich, V. M. Malofeev, Yu. P. Shitov.
Astron. Zhurn. Akad. Nauk SSSR, Vol. 48, 1333 - 1335 (1971). In Russian. English translation in Soviet Astron. AJ, Vol. 15, No. 6.

The observations which have been carried out in Arecibo and Pushchino at frequencies 111 and 85.5 MHz, have shown, on the average, 45 percentage of the linear polarization of radioemission of pulsar AP 1237 + 25. The rotation measure, as it has been defined, is equal to 16 ± 4 rad/m² and the mean longitudinal component of the interstellar magnetic field towards this pulsar has been determined to be equal to $2.1 \pm 0.5 \times 10^{-6}$ gauss.

141.220 Some astrophysical aspects of pulsars. L. Woltjer.
Highlights of Astronomy, Vol. 2, (see 012.019), 725 - 726 (1971).

141.221 Pulsars and the origin of cosmic rays. T. Gold.
Highlights of Astronomy, Vol. 2, (see 012.019), 727 - 730 (1971).

141.222 Remarks on the role of pulsars in cosmic ray production. V. L. Ginzburg.
Highlights of Astronomy, Vol. 2, (see 012.019), 737 - 739 (1971).

141.223 The nature of 3C 391. A. H. Bridle, M. J. L. Kesteven.
Astron. Journ., Vol. 76, 958 - 964 (1971).

Observations of the neutral-hydrogen absorption profile and microwave continuum emission of 3C 391 are presented.

The classification of the source as a supernova remnant is supported, and the nature of its unusual radio continuum spectrum is discussed. The spectrum is most simply explained by postulating that the radiation from the source passes through an H II region situated between 3C 391 and the sun. Alternative mechanisms, intrinsic to the source, are also discussed. We suggest five experiments which may provide further understanding of the spectrum.

141.224 The structure of P 1934–63.
J. S. Gubbay, A. J. Legg, D. S. Robertson, N. Craske, G. D. Nicolson.
Astron. Journ., Vol. 76, 965 - 969, 1159 (1971).

Radio telescopes of the NASA-JPL Deep Space Network near Woomera and near Canberra were operated as an interferometer in the latter part of 1969 and in May 1970 to study the dependence of correlated flux from P 1934–63 at 2.3 GHz on the orientation of the baseline. It was found that two discrete components accounted for 60 % of the flux received.

141.225 The NRAO 5-GHz radio source survey. I. A survey of faint sources. M. M. Davis.
Astron. Journ., Vol. 76, 980 - 992 (1971).

Positions and flux densities are given for 254 sources detected in a 190 deg² survey at 6-cm wavelength, using the newly re-surfaced NRAO 300-ft telescope. In addition, the spectral index distribution of the sources can be used to determine a characteristic redshift of the flat-spectrum population, as discussed by Pauliny-Toth, Kellermann, and Davis (1970). The data found in this survey are listed in a table.

141.226 Erratum: 'Crab nebula pulsar radio pulse arrival times at Arecibo Observatory' [Astron. Journ., Vol. 76, 686 - 690 (1971)].
J. M. Rankin, C. C. Counselman III, D. W. Richards.
Astron. Journ., Vol. 76, 1154 (1971).

141.227 Quasi-stellar objects: Their importance for cosmology and general relativity.
G. R. Burbidge, E. M. Burbidge.
General relativity and cosmology. Course 47, Italian Phys. Soc., 1969, (see 012.022), p. 284 - 305 (1971).

141.228 Pulsars and the origin of high energy particles. T. Gold.
Proc. 11th international conference on cosmic rays, Budapest 1969, (see 012.025), p. 163 - 175 (1970).

141.229 Relaxation of electron velocity in a rotating neutron superfluid: Application to the relaxation of a pulsar's slowdown rate. P. J. Feibelman.
Phys. Rev. D, Particles and Fields, Third Ser., Vol. 4, 1589 - 1597 (1971).

We estimate the relaxation time τ of an average electron velocity relative to a dilute array of vortex cores in a rotating, s-wave-paired neutron superfluid. For reasonable choices of Δ and ϵ_F, we find values of τ which include the values of a year and of several days observed, respectively, in the post-speedup relaxation of the Vela and Crab pulsar's slowdown rates.

141.230 Observations of intensity modulation of starlight at discrete radio frequencies. G. J. Morris.
Phys. Rev. Letters, Vol. 27, 1600 - 1604 (1971).

Components modulated in intensity at discrete radio frequencies have been detected in the light from several stars. The effect is believed to be due to time-dependent, very small-angle scattering of the starlight by enhanced electron density fluctuations in the ionosphere.

141.231 Neutron starquakes and pulsar speedup.

G. Baym, D. Pines.
Ann. Physics, Vo. 66, 816 - 835 (1971).

We give here a simple model of pulsar speedup due to starquakes that enables one to predict the time to the next starquake from the magnitude of the prior one. The parameters of the theory are estimated for recent models of neutron stars. The starquake explanation of speedup leads one to conclude that the Crab pulsar is a fairly light neutron star and the pulsar in Vela lighter still. The energy release in starquakes is discussed, and it is estimated that the Crab pulsar released at best 7×10^{40} ergs in its quake of September, 1969. The geometry of starquakes is considered for the simple model of a self-gravitating elastic incompressible sphere, and we conclude with a brief discussion of plastic flow in neutron star crusts.

141.232 Rapidly rotating pulsars and Jacobi ellipsoids.
W. Y. Chau, P. Srulovicz.
Phys. Rev. D, Particles and Fields, Third Ser., Vol. 3, 1999 - 2000 (1971).

In connection with the evolution of a rotating Jacobi ellipsoid through the emission of gravitational radiation, we discuss the possibility that rapidly rotating pulsars can assume such triaxial, nonaxisymmetrical configurations.

141.233 News of optical variations of the radio source OJ 287. N. E. Kurochkin.
Astron. Tsirk., No. 654, p. 7 - 8 (1971). In Russian.

141.234 Results of measuring the polarization of subimpulses in the radio emission of pulsar CP 1133 at 3.5 m. Yu. I. Aleksejev.
Astron. Tsirk., No. 655, p. 1 - 2 (1971). In Russian.

141.235 Note on the nature of the fast optical variability of quasars. V. N. Kurilchik.
Astron. Tsirk., No. 658, p. 2 - 4 (1971). In Russian.

141.236 The theories of the pulsars. Chapter III. Rotational theories. G. Calamai.
Mem. Soc. Astron. Italiana, Nuova Ser., Vol. 42, 617 - 634 (1971). − Review article.

141.237 Magnetic field effects on the outermost crusts of pulsars. R. O. Mueller, A. R. P. Rau, L. Spruch.
Nature, Phys. Sci., Vol. 234, 31 - 32 (1971).

The intense magnetic fields generally presumed to be associated with pulsars affect atoms in the outermost layers of the crust. The resulting pressure-density equation of state changes the extent of these layers (with $\rho \lesssim 10^4$ gcm^{-3}) from the no-field situation and, further, results in a much higher value for the density that marks the surface of the star. These results may be of interest for calculations of opacity coefficients and other properties associated with these outer reaches of the star.

141.238 Parameters of twelve weak pulsars at 327 MHz.
S. K. Mohan, V. Balasubramanian, G. Swarup.
Nature, Phys. Sci., Vol. 234, 151 - 153 (1971).

Declination, average pulse shape and energy have been measured for 12 pulsars. Barycentric period is reported for 4 pulsars. Three low period pulsars have complex pulse shape which indicates that the dependence of pulse shape on period is not as marked as suggested earlier.

141.239 Possible evidence for pulsed ~ 10^{12} eV gamma rays from NP 0532. J. E. Grindlay.
Nature, Phys. Sci., Vol. 234, 153 - 155 (1971).

Using new techniques for the Čerenkov detection of extensive air showers (EAS) allowing arrival direction resolution ~ 6×10^{-5} str and > 70% rejection of cosmic ray primaries, a

$\gtrsim 3\sigma$ (confidence level) detection of pulsed γ-ray initiated EAS from NP-0532 is reported. The data (42 scans) were recorded with absolute arrival times, analyzed at the correct period of the Crab pulsar and added in phase yielding a ~ 4σ excess of (only) EAS selected as γ-ray initiated at the expected phase of the interpulse. The flux implied is F ($\geqslant 6 \times 10^{11}$ eV) ~ 1.5×10^{-11} photons/cm^{2}sec, consistent with an extrapolation of the X-ray pulsed spectrum but requiring confirmation.

141.240 Interferometric observations of pulsars at 2.7 and 8.1 GHz.
G. R. Huguenin, J. H. Taylor, R. M. Hjellming, C. M. Wade.
Nature, Phys. Sci., Vol. 234, 50 - 51 (1971).

Twenty pulsars were observed during the period May 20 to 25, 1971. The principal aims of the observations were to obtain accurate position measurements independent of pulse timing considerations; to determine the flux densities of pulsars at 2.7 and 8.1 GHz; and to detect (or place upper limits on) any continuous, non-pulsed radio emission from regions of small angular size near the pulsars.

141.241 Crab pulsar radiation mechanism.
R. N. Manchester, E. Tademaru.
Nature, Phys. Sci., Vol. 232, 164 - 165, with a reply by F. G. Smith, p. 165 (1971).

Manchester and Tademaru present some remarks to the oblique rotator model for the Crab pulsar proposed by Smith (Nature, Phys. Sci., Vol. 231, 191 - 193 (1971)). They wish to show that Smith's model cannot explain the observations. In his reply Smith writes: Manchester and Tademaru (Nature, Phys. Sci., Vol. 232, 164 - 165 (1971)), have misunderstood the physical mechanism proposed for pulse formation. This reply emphasises that relativistic compression applies only to individual pulses; the superposition of many individual pulses generated over a range of longitude gives rise to the integrated profile. After referring to the basic evidence for this interpretation, the reply re-asserts the validity of the model proposed for the Crab nebula pulsar.

141.242 Some identifications for weak sources in the Parkes catalogue for declinations +20° to −20°.
J. G. Bolton, J. V. Wall, A. J. Shimmins.
Australian Journ. Phys., Vol. 24, 889 - 898 = Separate print Division Radiophys., C.S.I.R.O., Sydney (1971).

Accurate positions and flux densities at 2700 MHz have been measured for 156 weak sources in the Parkes catalogue for declinations 0° to +20° and 0° to −20°. Identifications are suggested for 22 of the sources, 6 with galaxies and 16 with possible quasi-stellar objects.

141.243 The polarization of extended radio sources at 6 cm wavelength. I. Extragalactic sources.
F. F. Gardner, J. B. Whiteoak.
Australian Journ. Phys., Vol. 24, 899 - 911 = Separate print Division Radiophys., C.S.I.R.O., Sydney (1971).

Maps are presented of the linear polarization distributions over eight extragalactic radio sources. They were obtained at 6 cm wavelength with a 4′ arc resolution. Low brightness extensions show polarization of up to 70% with magnetic fields aligned perpendicular to their elongations.

141.244 The polarization of extended radio sources at 6 cm wavelength. II. Galactic sources.
J. B. Whiteoak, F. F. Gardner.
Australian Journ. Phys., Vol. 24, 913 - 924 = Separate print Division Radiophys., C.S.I.R.O., Sydney (1971).

Maps are presented of the polarization and total intensity distributions over the supernova remnants 13S6A, MSH 15−56, RCW 103, and W44. They were obtained at 6 cm wavelength with a 4′ arc resolution. It has not proved possible to interpret the polarization characteristics in terms of a simple expansion against the interstellar magnetoionic medium.

141.245 **2700 MHz observations of 4C radio sources in the declination zone +4° to −4°.** J. V. Wall.
Australian Journ. Phys., Astrophys. Suppl. No. 20, 30 pp. = Separate print Division Radiophys., C.S.I.R.O., Sydney (1971).

Most sources in the 4C catalogue with declinations between +4° and −4° have been observed at 2700 MHz. Accurate positions, flux densities, and the results of a search for optical identifications are presented. A brief comparison is made with results from other investigations.

141.246 **The Parkes 2700 MHz survey. Catalogue for 03^h, 11^h, 19^h, and 23^h zone, declinations −33° to −75°.** A. J. Shimmins.
Australian Journ. Phys., Astrophys. Suppl. No. 21, 34 pp. = Separate print Division Radiophys., C.S.I.R.O., Sydney (1971).

A catalogue of 618 extragalactic radio sources obtained from a sky survey at 2700 MHz is presented. The catalogue is complete to a limiting flux density of 0.32 f.u. at 2700 MHz and is thought to be 90 % complete at a flux density of 0.20 f.u. The positions are accurate to 15″ arc or slightly better in both coordinates for sources stronger than 0.32 f.u.; the flux densities of the weaker sources are accurate to 0.02 f.u., and for sources stronger than 1 f.u. the accuracy is 3 %.

141.247 **Method of determining the parameters of extended cosmic radio sources.** L. G. Sodin.
Izv. vyssh. uchebn. zavedenij. Radiofizika, Vol. 14, 1143 - 1148 (1971). In Russian. − Abstr. in Referativ. Zhurn. 51. Astron., 2.51.592 (1972).

141.248 **Rapid survey of the sky at the frequency 8550 MHz in the region of declinations +0° − +30°.**
M. G. Larionov, A. G. Gorshkov, M. V. Popov, I. G. Moiseev.
Astron. Tsirk., No. 665, p. 1 - 6 (1971). In Russian.

141.249 **Radio variability of OK 290.**
A. G. Gorshkov, M. G. Larionov, M. V. Popov.
Astron. Tsirk., No. 667, p. 1 - 2 (1971). In Russian.

141.250 **Period and declination of pulsar MP 0450.**
D. K. Mohanty, V. Balasubramanian, G. Swarup.
Proc. Indian Acad. Sci., Section A, Vol. 72, 246 - 248 (1970). − See Phys. Abstr., Vol. 74, No. 46230 (1971).

141.251 **Red shift without reason.** P. Stubbs.
New Scient. and Sci. Journ., Vol. 50, 254 - 256 (1971).

A long baseline experiment, involving the radio dishes at Goldstone, California, and Haystack, Massachusetts, has produced the startling result that a quasar − if it is as far away as its red shift implies − must be flying apart at ten times the speed of light. Since this is improbable it questions the entire basis of the red shift/distance relationship.

141.252 **An observational study of the optical variability of quasi-stellar objects.** R. J. Angione.
Thesis, Univ. Texas, Austin. [Available from Univ. Microfilms, Ann Arbor, Mich., U.S.A. Order No. 71−11509], 305 pp. (1970). − See Phys. Abstr., Vol. 74, No. 66558 (1971).

141.253 **Observations of quasi-stellar sources and radio galaxies at millimeter wavelengths.** J. D. G. Rather.
Thesis, Univ. California, Berkeley. [Available from Univ. Microfilms, Ann Arbor, Mich., U.S.A. Order No. 71−835], 161 pp. (1970).

Fluxes of 49 extragalactic objects have been measured at wavelengths near 8 millimeters and 1.2 centimeters.

141.254 **Radio observations of the pulse profiles and dispersion measures of twelve pulsars.** H. D. Craft, Jr.
Thesis, Cornell Univ., Ithaca, N.Y. [Available from Univ. Microfilms, Ann Arbor, Mich., U.S.A. Order No. 71−4972], 354 pp. (1970).

Discusses some of the techniques for and results of measurements of pulse profiles and dispersion measures of the first twelve pulsars known to be within view of the Arecibo Observatory. All measurements were made at that observatory utilizing frequencies between 430 MHz and 40.12 MHz.

141.255 **The brightness distribution of core-halo radio sources.** R. A. Sramek.
Thesis, California Inst. Technology, Pasadena. [Available from Univ. Microfilms, Ann Arbor, Mich., U.S.A. Order No. 70−24313], 179 pp. (1970).

The east-west visibility of 28 extragalactic radio sources with both large and very small components was obtained at 605 MHz using the radio interferometer at the Owens Valley Radio Observatory. A 2′ HPBW fan beam was synthesized from observations made at twelve spacings between 62 and 977 wavelengths. The availability of such low spatial frequencies permitted the calculation of the brightness distribution of components as large as 56′.

141.256 **Production of γ-rays in the Crab nebula pulsar.** A. Treves.
Nuovo Cimento B, Ser. 11, Vol. 4B, 88 - 96 (1971).

The γ-ray flux at energies above 50 MeV from NP 0532 is evaluated for a pulsar model in which the optical and X radiations are produced by the synchrotron effect. Synchrotron radiation and inverse Compton scattering are considered as production mechanisms of the γ-rays. The theoretical estimates are compared with the experimental values.

141.257 **Radiation mechanisms of pulsars.** S. Ichimaru.
Astron. Herald, (*Japan*), Vol. 64, 105 - 108 (1971). In Japanese.

141.258 **Optical timing of pulsars shows accurate results.** L. D. Shergalis.
Electronics Australia, Vol. 33, No. 5, p. 15 - 17 (1971).

The article describes how light from the pulsar in the Crab nebula is monitored. It also states that light observations can be much more accurate than radio observations.

141.259 **Theory of quasars.** A. Aizu, H. Tawara.
Kagaku, (*Japan*), Vol. 41, 375 - 383 (1971). In Japanese.

Reviews the author's theory proposed in 1964 and extends it.

141.260 **Pulsating stars.** S. Ichimaru.
Butsuri, (*Japan*), Vol. 26, 826 - 835 (1971). In Japanese.

Analyses the pulsar's radiation intensity, polarization properties, dependence of the pulse width on the period, distribution of the pulse period, spectrum distribution and fine structures of the pulse. Considers what sort of plasma is expected around the rotating neutron star and also examines the radiation near the magnetic pole and that near the light-velocity cylinder. The variable X-ray star discovered lately appears to comply with this pulsar model.

141.261 **Observation of radiostars.**
Metalurgia y Electricidad, (*Spain*), Vol. 35, 230 - 231 (1971). In Spanish. − See Phys. Abstr., Vol. 75, No. 7476 (1972).

141.262 **Brightness distributions of radio sources at 2695 MHz.** F. N. Bash.
Diss. Graduate Fac. Univ. Virginia, Charlottesville, Virginia

(1967). [Available from Univ. Microfilms, Ann Arbor, Mich., U.S.A.], 86 pp. (1970).

141.263 Extragalactic radio sources. D. M. Mills.
Stanford Univ. Inst. Plasma Res. SUIPR Rep., No. 408, 75 pp. (1971).

Detailed review of extragalactic radio sources with particular interest in the double radio 'clouds' which are the observed characteristics of these sources. Specific results given for Cygnus A and 3C 33. – *GD*

141.264 Extragalactic radio sources in uniform model universes. R. R. Ringenberg.
Thesis, Graduate College, Univ. Illinois (1969). [Available from Univ. Microfilms, Ann Arbor, Mich., U.S.A.], 95 pp. (1970).

141.265 A catalog of the north-south visibility functions of radio sources at 1425 MHz. E. B. Fomalont.
Publ. Owens Valley Radio Obs., Vol. 1, No. 5, p. 1 - 5 (1971).

The author tabulates visibility functions for over 100 sources from observations with the two 90-foot aerials at the Owens Valley Radio Observatory. – *OBS*

141.266 Flux density measurements of faint radio sources at 5 GHz. G. Grueff.
Obs. Owens Valley Radio Obs., 1971, No. 1, 23 pp.

The flux densities of 224 radio sources located in the area between Dec +29 deg and +35 deg and RA 07h 30m and 18h 30m are tabulated. All sources come from the B2 survey made at 408 MHz with the northern cross. Mean spectral indices are tabulated and an optical identification sought for each source. – *OBS*

141.267 Redshift without reason. P. Stubbs.
New Scient. Vol. 50, 254 - 255 (1971).

Popular account of new, unpublished, long-baseline interferometry work which raises doubts on the interpretation of quasar redshifts as "cosmological". – *JLC*

141.268 *UBV* observations of 3C 273. IV.
M. S. Burkhead, W. L. Stein.
Publ. Astron. Soc. Pacific, Vol. 83, 830 - 831 = Publ. Goethe Link Obs., Indiana Univ., *Bloomington*, No. 132 (1971).

Observations of 3C 273 for 1970-71 are presented. A rapid change in *V* magnitude on 15 April 1971 is discussed.

141.269 OJ 287 = VRO 20.08.01.
Yamamoto Circ., No. 1744 (1971). In Japanese.

141.270 Pulsars, eine neue Klasse kosmischer Objekte.
G. Traving.
Phys. Meeting, Hannover (*Germany*), 1970 [B. G. Teubner, Stuttgart], p. 279 - 289 (1970).

141.271 Use of pulsar signals as clocks.
P. Reichley, G. Downs, G. Morris.
Jet Propulsion Lab., Techn. Rev., Vol. 1, No. 1, p. 80 - 86 (1971).

Discussion on the calibration of "pulsar clocks". – *MWS*

141.272 The period-age distribution of pulsars.
P. A. Sturrock.
Stanford Univ. Inst. Plasma Res., SUIPR Rep. No. 427, 8 pp. (1971).

The author discusses limitations on the period and age of a pulsar based on a model that he recently proposed.

Does annihilation power quasars?
Physics Today, Vol. 24, No. 2, p. 32 (1971).

The bending of the synchrotron spectrum at high energies. See Abstr. 022.055.

Terrestrial and extraterrestrial limits on the photon mass. See Abstr. 022.061.

Quantum treatment of electron emission in a strong electromagnetic wave. See Abstr. 022.066.

Atoms in superstrong magnetic fields.
See Abstr. 022.128.

Chambre à étincelles optique pour la recherche de sources de rayons gamma. See Abstr. 034.050.

Neutron star models and pulsars.
See Abstr. 065.007.

Frictional heating in neutron stars.
See Abstr. 065.011.

Matter in superstrong magnetic fields: The surface of a neutron star. See Abstr. 065.083.

A radio scintillation method of estimating the small-scale structure in the ionosphere. See Abstr. 083.053.

Optical and radio occultation analysis.
See Abstr. 096.026.

Stars observed photoelectrically near quasars and related objects. See Abstr. 113.064.

Odd behaviour of polarization in the radiation of BL Lacertae = VRO 42.22.01. See Abstr. 122.028.

Spectrum of AP Lib (≡ PKS 1514−24).
See Abstr. 122.098.

On the distance to BL Lacertae.
See Abstr. 122.105.

Radio observations of the supernova remnants IC 443 and Puppis A. See Abstr. 125.002.

Observations of IC 443 at 408 MHz.
See Abstr. 125.029.

Excitation temperatures of the 18-centimeter OH transitions in an absorbing cloud. See Abstr. 131.099.

Cosmic rays and interstellar matter.
See Abstr. 131.113.

OH excited state emissions from W 75 B and W 3, OH. See Abstr. 131.139.

Observations of Cygnus Loop at 408 MHz.
See Abstr. 132.042.

X-rays from Puppis A and the vicinity of Vela X.
See Abstr. 142.069.

Further radio observations of Scorpius X-1.
See Abstr. 142.077.

Registration of energetic gamma-ray flux from the extragalactic source 3C 120. See Abstr. 142.080.

New point γ-ray source Lib γ-1: Evidence for time variation and possible identification with PKS 1514−24.
See Abstr. 142.088.

Radio spectrum of Cygnus X-1.
See Abstr. 142.090.

Measurements of hard X and gamma radiation from Virgo A and Centaurus A.   See Abstr. 142.093.

Pulsars and cosmic-ray prehistory.
See Abstr. 143.062.

High-resolution observations of the galactic center at 5 GHz.   See Abstr. 155.034.

A low-latitude survey from $l = 355°$ to 5° at 1410 MHz.   See Abstr. 157.008.

Optical line spectrum of a gas heated by hard UV radiation or energetic particles.   See Abstr. 158.009.

Variability of N-galaxies 3C 371, 3C 390.3 and quasar 3C 345.   See Abstr. 158.013.

On the nature of the galaxy identified with 3C 386.
See Abstr. 158.030.

Physical conditions in the active nuclei of galaxies and quasi-stellar objects deduced from line spectra.
See Abstr. 158.039.

Space densities and time scales of Seyfert galaxies, radio galaxies and quasi-stellar objects.
See Abstr. 158.042.

Spinars: A progress report.
See Abstr. 158.045.

Nuclei of galaxies and quasars as sources of infrared emission.   See Abstr. 158.059.

The redshift: Another model.
See Abstr. 158.061.

The radio continuum of galaxies. I. Observations.
See Abstr. 158.063.

Note on N galaxies and mini-quasars.
See Abstr. 158.080.

A polarimetric study of compact extragalactic objects.   See Abstr. 158.094.

Radio emission from the nucleus of NGC 5128.
See Abstr. 158.098.

Apparent associations between bright galaxies and quasi-stellar objects.   See Abstr. 158.105.

Entwicklung der Sternsysteme und Quasare aus radio-astronomischer Sicht.   See Abstr. 158.107.

On the spectra of radio galaxies between 10 and 10 000 MHz.   See Abstr. 158.113.

Observational paradoxes in extragalactic astronomy.
See Abstr. 162.055.

142 X Ray-, Gamma Ray-Sources

142.001 Weak X-ray sources in the southern hemisphere.
P. J. N. Davison, G. Buselli, M. C. Clancy,
R. M. Thomas.
Astrophys. Journ., Vol. 167, 479 - 486 (1971).
The first statistically significant observations of hard X-rays (energy > 27 keV) from the source Nor X-1 is reported herein. Measurements of the flux and energy spectrum of the source Ara X-1 are also given. A statistically marginal observation of the variable X-ray source Cen X-2 is reported and compared with observations made several months earlier by Lewin et al. Upper limits to the flux above 27 keV are reported for Lup X-1, Nor X-2, Sco X-4, and the quiet sun.

142.002 A strong X-ray source in the Coma cluster observed by Uhuru. H. Gursky, E. Kellogg, S. Murray,
C. Leong, H. Tananbaum, R. Giacconi.
Astrophys. Journ., (Letters), Vol. 167, L81 - L84 (1971).
X-rays have been observed from a source in the Coma cluster of galaxies. The source is extended, with a size of about 45'. Its X-ray luminosity is 2.6×10^{44} ergs s^{-1}, and its spectrum is consistent with thermal bremsstrahlung at $7.3 \times 10^{7\,\circ}$ K or a power law. If the source is hot gas, its mass is $3 \times 10^{13} M_{\odot}$, which is about 1 percent of the mass required to stabilize the cluster.

142.003 Radio emission from Scorpius X-1 at 21.2 cm.
L. L. E. Braes, G. K. Miley.
Astron. Astrophys., Vol. 14, 160 - 163 (1971).
Radio emission has been detected from Sco X-1 at 1415 MHz. The brightness distribution is similar to that previously observed at 2695 and 8085 MHz; it shows three components, the central one of which is strongly variable.

142.004 Erratum: 'Gamma rays (250 keV−2.3 MeV) from NP 0532'. [Nature, Vol. 231, 171 (1971)].
L. E. Orwig, E. L. Chupp, D. J. Forrest.
Nature, Vol. 232, 664 (1971).

142.005 Discovery of periodic X-ray pulsations in Centaurus X-3 from UHURU.
R. Giacconi, H. Gursky, E. Kellogg, E. Schreier, H. Tananbaum.
Astrophys. Journ., (Letters), Vol. 167, L67 - L73 (1971).
A search for X-ray sources exhibiting pulsating characteristics similar to the ones recently discovered in Cyg X-1 by Oda et al., and confirmed by Holt et al., has revealed the existence of periodic pulsations in the X-ray emission from Cen X-3.

142.006 A cocoon pulsar model for Scorpius X-1.
K. Davidson, F. Pacini, E. E. Salpeter.
Astrophys. Journ., Vol. 168, 45 - 55 (1971).
A model is proposed for the thermal X-ray emission from Sco X-1 in terms of a hot gaseous region surrounding a rotating neutron star. This gaseous cocoon is heated by the emissions from the pulsar and, in this model, is kept at distances far outside the speed-of-light circle by a Poynting–Robertson effect.

142.007 X-ray observations of Virgo XR-1.
M. Lampton, S. Bowyer, J. E. Mack, B. Margon.
Astrophys. Journ., (Letters), Vol. 168, L1 - L6 (1971).
Virgo XR-1 was observed on 1969 June 14 with two rocket-borne proportional counters. Comparisons with other measurements are made, and the question of possible variability is discussed. We conclude that present evidence for variability is not compelling.

142.008 Measurement of the location of the X-ray source

Cygnus X-1. S. Miyamoto, M. Fujii,
M. Matsuoka, J. Nishimura, M. Oda, Y. Ogawara, S. Ohta,
M. Wada.
Astrophys. Journ., (Letters), Vol. 168, L11 - L14 (1971).
The location of Cyg X-1 was determined with balloon experiments. It is in agreement with that determined by the X-ray satellite Uhuru. The measurements altogether have considerably improved the accuracy of the previous location.

142.009 X-ray source positions for Cygnus X-1, Cygnus X-2, and Cygnus X-3. A. Toor, R. Price, F. Seward,
J. Scudder.
Astrophys. Journ., (Letters), Vol. 168, L15 - L16 (1971).
On 1970 September 24, the Cygnus region was scanned with a large scintillation counter. From the analysis of the data between 5 and 35 keV, we report source locations for Cyg X-1, Cyg X-2, and Cyg X-3.

142.010 On the location of Cygnus X-1.
S. Rappaport, W. Zaumen, R. Doxsey.
Astrophys. Journ., (Letters), Vol. 168, L17 - L20 (1971).
The position of Cyg X-1 is determined with a precision of ∼30″ with a rocket-borne rotating modulation collimator.

142.011 Radio emission from X-ray sources.
R. M. Hjellming, C. M. Wade.
Astrophys. Journ., (Letters), Vol. 168, L21 - L24 (1971).
Variable radio sources probably associated with the X-ray sources GX 17+2 and Cyg X-1 have been found. Efforts to detect radio emission near the positions of Cyg X-2, GX 5−1, GX 9+1, and GX 3+1 have been unsuccessful.

142.012 X-ray observations of GX 17+2 from Uhuru.
H. Tananbaum, H. Gursky, E. Kellogg, R. Giacconi.
Astrophys. Journ., (Letters), Vol. 168, L25 - L28 (1971).
We have determined an improved location for GX 17+2 which supports the identification of the radio candidate found by Hjellming and Wade. Our data show significant variations in the intensity and temperature of GX 17+2 over a few thousand seconds. The spectrum is exponential, characteristic of thermal bremsstrahlung. The striking resemblance between GX 17+2 and Sco X-1 in both the X-ray and radio strongly suggests that they belong to the same class of object.

142.013 Observation of cosmic soft X-rays.
S. Hayakawa, T. Kato, F. Makino, H. Ogawa,
Y. Tanaka, K. Yamashita, M. Matsuoka, S. Miyamoto, M. Oda,
Y. Ogawara.
Astrophys. Space Sci., Vol. 12, 104 - 117 (1971).
Cosmic soft X-rays in the energy range between 0.14 and 7 keV were observed with thin polypropylene window proportional counters on board a sounding rocket. The field of view crossed the galactic plane in the Cygnus-Cassiopeia region at a large angle and reached the galactic latitudes of –55° and +30°. Referring also to the result with Be window counters, we obtained the energy spectrum of Cyg XR-2, the flux from the Cas A region and the distribution of the intensity of diffuse X-rays over the scanned region.

142.014 Kosmische Röntgen- und Gammastrahlung.
K. Pinkau.
SuW, Vol. 10, 221 - 227 (1971).

142.015 Recent development in X-ray astronomy.
B. G. Wilson.
Journ. Roy. Astron. Soc. Canada, Vol. 65, 178 (1971).
Abstr. RAS Canada.

142.016 **Errata: 'Low-energy cosmic X-rays'** [Astrophys.
Journ., Vol. 164, 265 - 273 (1971)].
P. G. Shukla, B. G. Wilson.
Astrophys. Journ., Vol. 168, 319 (1971).

142.017 **A search for absorption of the soft X-ray diffuse
flux by the Small Magellanic Cloud.**
D. McCammon, A. N. Bunner, P. L. Coleman, W. L. Kraushaar.
Astrophys. Journ., *(Letters)*, Vol. 168, L33 - L37 (1971).

An upper limit of 25 percent is placed on the fraction of
the observed 120–284 eV X-ray flux that originates beyond
the Small Magellanic Cloud if 21-cm measurements of total
columnar hydrogen density and the effective absorption
cross-sections of Brown and Gould are assumed. This result
is still consistent with the extragalactic origin of an $E^{-1.4}$
extrapolation of the isotropic X-ray background observed
above 2 keV.

142.018 **A search for X-ray pulsations from Cygnus X-1.**
S. Rappaport, R. Doxsey, W. Zaumen.
Astrophys. Journ., *(Letters)*, Vol. 168, L43 - L47 (1971).

X-ray data from Cyg X-1 with a timing resolution of 1 ms
have been obtained during a sounding-rocket flight. We find
flaring activity on time scales down to 50 ms in which the
X-ray intensity changes by a factor of 2. There are no regular
X-ray pulsations which comprise more than 5 percent of the
total X-ray flux in the range 0.010–1.0 s. About 30 percent
of the X-ray intensity is modulated with periodicities in the
range 1.3–5 s.

142.019 **X-ray intensity fluctuations in Cygnus XR-1.**
S. Shulman, G. Fritz, J. F. Meekins, H. Friedman,
M. Meidav.
Astrophys. Journ., *(Letters)*, Vol. 168, L49 - L51 (1971).

A reanalysis of 1967 September data shows apparent
fluctuations in the X-ray intensity of Cyg XR-1 on a time
scale of several seconds. There is no evidence, however, for
the periodicities that have recently been reported by other
observers.

142.020 **Cosmic gamma-ray measurements in the range
0.3–3.7 MeV.**
S. V. Golenetskii, E. P. Mazets, V. N. Il'inskii, R. L. Aptekar', M. M. Bredov, Yu. A. Gur'yan, V. N. Panov.
Astrophys. Letters, Vol. 9, 69 - 74 (1971).

A study of the spectrum, latitude dependence and time
variations of the low-energy gamma-radiation intensity has
been carried out in the vicinity of the earth on the Cosmos
135 and Cosmos 163 satellites by means of a multichannel
scintillation spectrometer. An upper limit for the flux of
primary gamma-radiation has been obtained in some energy
intervals. The estimates obtained lie considerably below the
values derived from measurements in the interplanetary space.
Possible reasons for these discrepancies are discussed.

142.021 **Erratum: "Results of gamma-ray balloon astronomy".** [Astrophys. Journ., Vol. 158, 193 - 206
(1969)].
C. E. Fichtel, D. A. Kniffen, H. B. Ögelman.
Astrophys. Journ., Vol. 168, 581 (1971).

142.022 **On the optical identification of Cygnus X-1.**
J. Kristian, R. Brucato, N. Visvanathan, H. Lanning,
A. Sandage.
Astrophys. Journ., *(Letters)*, Vol. 168, L91 - L93 (1971).

We believe that the bright star at the location of the radio
source near Cyg X-1 may not be associated with the X-ray
source. Analysis of presently incomplete data suggests possible
peculiarity of a very red faint star, which is the only other object in the radio error box brighter than $V \sim 19$.

142.023 **A search for an optical source at the position of
Centaurus XR-3.**
J. L. Elliot, P. Horowitz, W. Liller, C. Papaliolios, J. Veverka.
Astrophys. Journ., *(Letters)*, Vol. 168, L95 - L96 (1971).

On plates reaching to the sixteenth magnitude (photographic), we found no pulsating optical source identifiable
with the pulsating X-ray source Cen XR-3 within 3 standard
deviations of the position given by Giacconi *et al.* No unusually blue stars exist down to nineteenth B-magnitude in the
same area of the sky.

142.024 **X-ray background radiation.**   H. Friedman.
Nuclei of galaxies. Conference 1970, (see 012.002),
p. 669 - 691 (1971).

142.025 **X-ray sources near the center of our Galaxy.**
E. Kellogg, H. Gursky, H. Tananbaum, S. Murray,
R. Giacconi.
Bull. American Astron. Soc., Vol. 3, 364 (1971). – Abstr.
AAS.

142.026 **Soft X-rays from Vela-X and Puppis A.**
F. D. Seward, G. A. Burginyon, R. J. Grader,
R. W. Hill, T. M. Palmieri.
Bull. American Astron. Soc., Vol. 3, 393 (1971). – Abstr.
AAS.

142.027 **A search for pulsed radio emission from Sco X-1.**
R. M. Hirsch, G. R. Huguenin, J. H. Taylor.
Bull. American Astron. Soc., Vol. 3, 393 (1971). – Abstr.
AAS.

142.028 **Further observations of the pulsating X-ray source
Cygnus X-1 from UHURU.**
E. Schreier, R. Giacconi, H. Gursky, E. Kellogg, H. Tanan-
baum.
Bull. American Astron. Soc., Vol. 3, 393 (1971). – Abstr.
AAS.

142.029 **A model for pulsating X-ray stars.**
W. Tucker, G. Blumenthal, A. Cavaliere, W. Rose.
Bull. American Astron. Soc., Vol. 3, 393 - 394 (1971).
Abstr. AAS.

142.030 **An X-ray source near M82.**
E. Kellogg, H. Gursky, R. Giacconi, H. Tananbaum,
A. Cavaliere, W. Forman.
Bull. American Astron. Soc., Vol. 3, 399 (1971). – Abstr.
AAS.

142.031 **Extraterrestrial γ ray contribution between 0.7 MeV
and 4.5 MeV at balloon altitude.**
G. Vedrenne, F. Albernhe, I. Martin, R. Talon.
Astron. Astrophys., Vol. 15, 50 - 54 (1971).

The comparison of γ ray results obtained through balloon
flights at different latitudes enables an extraterrestrial contribution for γ rays to be detected in the 1–5 MeV energy range.
The method for obtaining this γ ray flux and its spectrum is
described for a flight at low latitude (Guiana 10 °N). Our results seem to confirm the ERS 18 measurements in a similar
energy range and to prove that the γ ray cosmic component
can be detected through balloon experiments.

142.032 **Upper limits on pulsations from Cyg X-1 in hard
X-rays.**   R. K. Manchanda, V. S. Iyengar,
P. C. Agrawal, G. S. Gokhale, P. K. Kunte, B. V. Sreekantan.
Nature, Phys. Sci., Vol. 232, 190 - 191 (1971).

A balloon flight was carried out from Hyderabad on 5th
April, 1971 using an oriented X-ray telescope and an exposure
of 3 hours was obtained on Cyg X-1. Using a fast folding analysis the data has been examined for the presence of pulsations

in hard X-ray intensity in the energy range 18-88 keV over the period range 50 ms – 8 sec. Upper limits on pulsations are 3.5% of steady flux for the period range 1 – 8 sec, 9% for 160 ms – 1280 ms and 13% for 50 ms – 320 ms. It is concluded that either there are no pulsations in hard X-rays or the phase or frequency of pulsation does not remain the same even for periods of the order of a few minutes.

142.033 Search for pulsed radio emission from Cyg X-1 at 327 MHz.
D. K. Mohanty, V. Balasubramanian, G. Swarup.
Nature, Phys. Sci., Vol. 232, 191 - 192 (1971).

A search was made for a pulsar in the region of X-ray source Cygnus X-1 at 327 MHz. An upper limit of $1 \times 10^{-28} \, \text{JM}^{-2} \, \text{Hz}^{-1}$ for the average energy per pulse was found for any probable pulsar with a period lying in the range of 0.064 to 1.32 seconds.

142.034 On the nature of Centaurus X-2. S. Sofia.
Monthly Notices Roy. Astron. Soc., Vol. 154, 9P - 13P (1971).

New astrometric evidence is presented which indicates that the variable star WX Cen, which has been proposed as the optical object associated with the X-ray source Cen X-2, is probably not a member of the Scorpio-Centaurus association. Instead, it may belong to one of two possible groups of stars at a distance of approximately 200 or 2000 pc, respectively.

142.035 Upper luminosity limits of some extragalactic objects in the hard γ-ray region.
L. S. Bratolyubova-Tsulukidze, L. F. Kalinkin, A. S. Melioranskij, O. F. Prilutskij, E. A. Pryakhin, I. A. Savenko, V. Ya. Yufarkin.
Pis'ma v ZhEhTF, Vol. 13, 566 - 569 (1971). In Russian.
Abstr. in Referativ. Zhurn. 51. Astron., 10.51.498 (1971).

142.036 Model for Centaurus X-3. J. Gribbin.
Nature, Phys. Sci., Vol. 233, 18 - 19 (1971).

The discovery of periodic variations in the X-ray flux from Cen X-3 prompts the author to consider a model based on white dwarf pulsations. All the properties of the source can be explained plausibly by the pulsations of a rotating white dwarf of some 1 M⊙.

142.037 Optical identification of Cygnus X-1.
P. Murdin, B. L. Webster.
Nature, Vol. 233, 110 (1971).

The star HD 226868, coincident within the errors with Cygnus X-1, has the spectrum and colours of a normal B0 Ib supergiant at a distance of 2.0 kpc.

142.038 Short term variability of pulsations in the X-ray flux from Cygnus X-1.
A. M. Cruise, A. C. Newton, C. E. Chapman.
Nature, Vol. 233, 468 - 469 (1971).

Two observations of the X-ray source Cygnus X-1 were made from the same sounding rocket flight. In the first observation several periods of pulsation were detected. There was no trace of any pulsation in the X-ray flux observed 55 seconds later. It is concluded that the pulsed fraction varies on a time scale of minutes.

142.039 Role of soft galactic X-rays in the alignment of interstellar grains. N. C. Wickramasinghe.
Nature, Phys. Sci., Vol. 232, 110 - 111, with a reply by J. Mack, p. 111 (1971).

If the effects of interstellar absorption are included, the existing data relating to galactic soft X-ray sources are consistent with the occurrence of a flux of soft X-rays which is adequate to produce grain alignment over a substantial fraction of the galactic disk.

142.040 Eclipsing binary model of Cygnus XR-1.
J. F. Dolan.
Nature, Vol. 233, 109 - 110 (1971).

Recent observations of Cygnus XR-1 at energies above 20 kev are compared with the author's previously outlined model of the variable component of the source derived from the X-ray observations alone (a binary system in which a black body source is eclipsed every 2.9850 days by a non X-radiating companion). In particular, one set of observations may refer to the ingress of the black body source into eclipse.

142.041 Corrigendum: 'Rocket observations and the cosmic X-ray background' [Nature, Phys. Sci., Vol. 231, 52 - 53 (1971)]. A. C. Fabian, P. W. Sandford.
Nature, Phys. Sci., Vol. 234, 20 (1971).

142.042 Transition radiation as a cosmic X-ray source.
S. A. E. Johansson.
Astrophys. Letters, Vol. 9, 143 - 146 (1971).

Transition radiation is emitted when a charged particle traverses a boundary separating two media. Such radiation must therefore be emitted as X-rays and ultraviolet radiation in interstellar space, supernova remnants and some other objects where cosmic ray electrons pass through dust grains. Calculations show that the transition radiation emitted in this way might account for the soft X-rays emitted from various cosmic sources. The possibility of observing transition radiation emitted in the ultraviolet region is also discussed.

142.043 "X 1" im Sternbild Schwan – ein neuartiger Pulsar?
H. Rohr.
Orion, 29. Jahrgang, p. 139 - 140 (1971).

142.044 Possible optical identification of Cyg X-1.
W. P. Bidelman, A. J. Weitenbeck.
IAU Circ., No. 2345 (1971).

142.045 Strong X-ray source. E. T. Byram.
IAU Circ., Nos. 2348, 2350 (1971).

142.046 Transient X-ray source in Norma.
H. Tananbaum, H. Gursky, E. Kellogg, T. Matilsky, S. Murray, R. Giacconi.
IAU Circ., No. 2355 (1971).

142.047 Possible optical identification of X-ray source.
N. Sanduleak, W. P. Bidelman.
IAU Circ., No. 2356 (1971).

142.048 Time variation of the X-ray spectrum and optical luminosity of SCO X-1.
T. Kitamura, M. Matsuoka, S. Miyamoto, M. Nakagawa, M. Oda, Y. Ogawara, K. Takagishi, U. R. Rao, E. V. Chitnis, U. B. Jayanthi, A. S. Prakasa-Rao, S. M. Bhandari.
Astrophys. Space Sci., Vol. 12, 378 - 393 (1971).

Results of rocket observations of SCO X-1 over the spectral range of 2 ~ 20 keV are presented and compared with results of similar observations carried out by LRL (Lawrence Radiation Laboratory) group. Some of these X-ray observations were accompanied by simultaneous optical observations. Relationships between the hardness of the X-ray spectrum and the X-ray intensity and between the hardness and the optical luminosity are compiled. The relationships among the parameters (temperature, density and size) which characterize the postulated isothermal cloud model of SCO X-1 are given.

142.049 Observation of the diffuse component of cosmic soft X-rays. S. Hayakawa, T. Kato, F. Makino, H. Ogawa, Y. Tanaka, K. Yamashita.

Space Research XI, (see 012.004), 1359 - 1365 (1971).

142.050 Observation of Sco X-1.
T. Kitamura, M. Matsuoka, S. Miyamoto, M. Naka-
gawa, M. Oda, Y. Ogawara, K. Takagishi.
Space Research XI, (see 012.004), 1367 - 1371 (1971).

**142.051 A balloon observation of diffuse background
Gamma rays in the energy range from 100 keV to
1 MeV.** A. Danjo, S. Hayakawa, M. Ikeda, F. Makino,
Y. Tanaka, P. C. Agrawal, G. S. Gokhale, B. V. Sreekantan.
Space Research XI, (see 012.004), 1373 - 1378 (1971).

142.052 Progress in the observational X-ray astronomy.
M. Oda.
IAU Symposium No. 41, (see 012.005), p. 89 - 103 (1971).

**142.053 Measurement of the polarisation, spectra and accu-
rate locations of cosmic X-ray sources.**
K. A. Pounds.
IAU Symposium No. 41, (see 012.005), p. 165 - 167 (1971).

**142.054 Experiment to measure hard solar and celestial
X-rays from the fifth Orbiting Solar Observatory.**
K. J. Frost, B. R. Dennis, R. J. Lencho.
IAU Symposium No. 41, (see 012.005), p. 185 - 191 (1971).

**142.055 Upper luminosity boundaries for some extragalactic
objects in the region of hard γ-rays.** L. S. Bratolyubova-Tsulukidze, L. F. Kalinkin, A. S. Melioranskij, O. F.
Prilutskij, E. A. Pryakhin, I. A. Savenko, V. Ya. Yufarkin.
Pis'ma v ZhEhTF, Vol. 13, 566 - 569 (1971). In Russian.
Abstr. in Referativ. Zhurn. 62. Issled. kosm. prostranstva,
11.62.181 (1971).

142.056 Diffuse cosmic X-ray flux from 0.2 to 2 keV.
T. M. Palmieri, G. A. Burginyon, R. J. Grader, R. W.
Hill, F. D. Seward, J. P. Stoering.
Astrophys. Journ., Vol. 169, 33 - 39 (1971).

Observations of the low-energy diffuse X-ray background
have been made with a new detector system sensitive in the
range from 0.2 to 2 keV. The spectrum above 2 keV agrees
with that reported by other authors. Between 0.2 and 1.5 keV
a good fit to the spectrum is given by $dN/dE \propto E^{-1} \exp(-E/
0.45)$ photons (cm s keV sterad)$^{-1}$. The spatial distribution of
the flux below 1 keV is complicated. Possible interpretations
are discussed. The flux at 0.25 keV is basically in agreement
with that obtained by other authors.

142.057 X-rays from a new variable source GX 1+4.
W. H. G. Lewin, G. R. Ricker, J. E. McClintock.
Astrophys. Journ., (Letters), Vol. 169, L17 - L21 (1971).

On 1970 October 15 - 16 we carried out balloon X-ray
observations from Australia (energies above 15 keV). We detected a variable flux from a source at $l^{II} = 1°.4 \pm 0°.7$, $b^{II} =
3°.9 \pm 0°.8$ (GX 1+4). The location of this source does not
coincide with any of the accurately known locations of X-ray
sources reported so far from rocket and satellite observations.
Our data suggest that during our observations the flux changes
may have been periodic with an approximate frequency of 1
cycle per 2.3 minutes. A high-energy X-ray flux was also
detected from at least two more sources near the galactic center (quite likely GX 3+1 and GX 5–1).

142.058 A pulsing X-ray source in Circinus.
B. Margon, M. Lampton, S. Bowyer, R. Cruddace.
Astrophys. Journ., (Letters), Vol. 169, L23 - L25 (1971).

A pulsing X-ray source with period 685 ± 30 ms has been
observed in Circinus. A spectrum of the source is presented
and may be fitted by a thermal bremsstrahlung emission mechanism with $T = 3.5 \times 10^7$ °K, or by a blackbody at $T =$

1.1×10^7 °K, both subject to heavy absorption.

142.059 Røntgenobservationer fra satelliten UHURU.
S. Frandsen.
Astron. Tidsskr., Årg. 4, 123 - 129 (1971).

142.060 Evidence for a highly compact X-ray source.
B. Margon, S. Bowyer, M. Lampton, R. Cruddace.
Astrophys. Journ., (Letters), Vol. 169, L45 - L48 (1971).

A detailed spectrum of the X-ray source GX 340 + 0 is
presented. The spectrum may be fitted from 1 to 10 keV by
a blackbody model at $T = 1.5 \times 10^7$ °K, subject to heavy
photoelectric absorption. Unlike X-ray sources previously reported, the spectrum is not also compatible with power-law
or thermal-bremsstrahlung models. If the source is a blackbody
at the distance indicated by its X-ray absorption, its radius is
8 km and luminosity 3×10^{37} ergs s^{-1}. This source may provide direct evidence for the existence of a neutron star.

**142.061 Correlated transient short-period oscillation in the
optical and X-ray flux from Scorpius X-1.**
H. Kestenbaum, J. R. P. Angel, R. Novick, W. J. Cocke.
Astrophys. Journ., (Letters), Vol. 169, L49 - L55 (1971).

Correlated oscillations with a 20-s period have been detected in a simultaneous observation of the X-ray and optical
flux from Sco X-1. The oscillations persist for about 2 minutes
and have amplitudes of (0.81 ± 0.17) and (0.56 ± 0.17) percent in the X-ray and optical bands, respectively.

**142.062 Evidence of high-frequency oscillations in the X-ray
flux from Scorpius X-1.** J. R. P. Angel, H. Kestenbaum, R. Novick.
Astrophys. Journ., (Letters), Vol. 169, L57 - L61 (1971).

An analysis of the X-ray flux from Sco X-1 taken over
a 4-minute period shows evidence for oscillations in the frequency range 1–10 Hz which persist for typically 1 minute.

142.063 Some remarks on the universal X-ray background.
P. Raychaudhuri, P. Bandyopadhyay.
Astrophys. Space Sci., Vol. 13, 185 - 188 (1971).

It is shown that a universal steady X-ray background with
the energy flux $\simeq 10^{-7}$ erg cm^{-2} s^{-1} sr^{-1} can arise as a superposition of radiation from pulsars (neutron stars) in various gal-
axies when it is taken into account that supernova outburst
occurs in a galaxy at the rate of 10^{-2}/year.

142.064 X-ray sources and final stages of stellar evolution.
V. Weidemann.
Astrophys. Letters, Vol. 9, 155 - 157 (1971).

Stellar statistical arguments favor the existence of stars
which neither become white dwarfs nor undergo supernova
explosion, and could be identified with X-ray sources of the
Cyg X-1 type. It is suggested that not the parent mass but
the angular momentum history as described by the rotation
parameter GM^2/Jc, decides whether a star at the end of its
nuclear evolution approaches a spherical symmetrical situation with supernova explosion and pulsar formation, or
stores its gravitational energy in the form of rotating disks.

**142.065 Photometric observations of Sco X-1 in 1970 and
1971.** K. Osawa, K. Ichimura, K. Tomita.
Tokyo Astron. Bull., Second Ser., No. 215, p. 2519 - 2523
(1971).

The results of photographic and photoelectric photometry of Sco X-1 performed at the stations of Tokyo Astronomical Observatory in 1970 and 1971 are presented.

142.066 Soft X-ray emission from galactic radio spurs.
S. A. Ilovaisky, S. Bowyer.
Nature, Vol. 233, 469 - 471 (1971).

The possibility that galactic radio spurs are sources of

soft X-ray emission is explored. Theories of X-ray emission from supernovae shells are summarized and the Cetus Arc is discussed in detail in regards to these theories. It is shown that this object is not likely to be a soft X-ray source. Experimental results at soft X-ray energies are reviewed and show no obvious enhancements coincident with galactic spurs.

142.067 **What flux limits can be set for X-ray pulses accompanying Weber's pulses?** J. V. Jelley.
Nature, Vol. 234, 142 - 143 (1971).

The paper discusses the flux limits which can be set for X-rays which might accompany Weber's pulses of gravitational radiation (GR), using the upper air X-ray fluorescence technique. It is shown that over the band 5 keV – 100 keV, X-ray fluxes as small as 1.8×10^{-8} to 1.5×10^{-7} of the GR-fluxes could be detected with a simple ground-based phototube, for pulse durations of 0.1 s to 7 s respectively. Photoelectric absorption in the interstellar gas is shown to be negligible, though scattering by dust, particularly toward the galactic centre, may cause time dispersion.

142.068 **80 MHz observations of the Scorpius X-1 source.**
O. B. Slee, C. S. Higgins.
Nature, Vol. 234, 210 (1971).

The 80 MHz heliograph at Culgoora N.S.W. has been used to survey an area of $15' \times 16'$ arc centred on the Sco X–1 source. The area was kept under observation for a total of 14 hours distributed among 4 nights. Hour-long integrations failed to detect any flare emission from the variable radio source identified with Sco X–1. The flux density of the source was < 1.5 flux units at 80 MHz, suggesting that the spectral indices of up to 1.5 noted in the centimetric radio flares are not continued into the metre waveband. The upper limit to the spectral index between 80 MHz and 2695 MHz is about 1.2.

142.069 **X-rays from Puppis A and the vicinity of Vela X.**
F. D. Seward, G. A. Burginyon, R. J. Grader, R. W. Hill, T. M. Palmieri, J. P. Stoering.
Astrophys. Journ., Vol. 169, 515 - 524 (1971).

A new rocket observation of the Vela region confirms that Vel X and Pup A are strong sources of soft X-rays. X-rays from Vel X come from a broad region of sky with a diameter of $\approx 5°$. This region is centered neither on the Vel X radio source nor on the pulsar. Optical filaments lie along one edge of the X-ray emitting region. X-ray spectra are derived for Vel X and Pup A. Neither source was detectable above ~2 keV.

142.070 **GX 349+2 and GX 340+0: Locations and X-ray pulsation limits.**
S. Rappaport, W. Zaumen, R. Doxsey, W. Mayer.
Astrophys. Journ., (Letters), L93 - L97 (1971).

The celestial positions of GX 349+2 and GX 340+0 are determined with a precision of better than 1 arc min with a rocket-borne rotating modulation collimator. Limits on periodic X-ray pulsations from these sources are set.

142.071 **X-ray sources near the galactic center observed by**
Uhuru. E. Kellogg, H. Gursky, S. Murray, H. Tananbaum, R. Giacconi.
Astrophys. Journ., (Letters), L99 - L103 (1971).

X-rays emitted from the direction of the galactic nucleus have been detected for the first time. The emitting region is of the order of one-tenth as strong as the strongest X-ray emitters in the Sagittarius complex, and is extended by about 2° in galactic longitude. Its spectrum indicates considerable absorption at low energies. A highly variable discrete source close to the galactic center has also been observed; it lies about 2° off the galactic plane.

142.072 **Further observations of the pulsating X-ray source**

Cygnus X-1 from *Uhuru.*
E. Schreier, H. Gursky, E. Kellogg, H. Tananbaum, R. Giacconi.
Astrophys. Journ., (Letters), Vol. 170, L21 - L27 (1971).

The temporal behavior of the source are: (1) Large fluctuations of intensity exist on all observed time scales ranging from 50 ms to 10 s containing up to 50 percent of the power. (2) Periodic pulse trains with periods from 0.3 s to over 10 s exist containing 10–25 percent of the power. (3) No single period is consistently present.

142.073 **X-ray observation of a new soft source in Cygnus.**
P. L. Coleman, A. N. Bunner, W. L. Kraushaar, D. McCammon.
Astrophys. Journ., (Letters), Vol. 170, L47 - L49 (1971).

We have detected a new point X-ray source in Cygnus whose incident flux is confined to the energy range 0.5–1.3 keV. Its spectrum is consistent with line emission at ~ 1 keV or thermal bremsstrahlung at ~ 10^6 °K with significant interstellar absorption.

142.074 **X-ray emission from degenerate stars.**
K. Brecher, G. Burbidge, P. Strittmatter.
Publ. Astron. Soc. Pacific, Vol. 83, 603 (1971). – Abstr.
Astron. Soc. Pacific.

142.075 **Extensive air shower studies of cosmic gamma rays and cosmic ray composition.** J. E. Grindlay.
SAO, *Cambridge, Mass.*, Special Report, No. 334, 9 + 147 pp. (1971). – Presented as a thesis to the Department of Astronomy, Harvard University, May 1971.

A relatively high yield of gamma-rays near ~ 10^{12} eV was found. The muon component and longitudinal development were found to be most important for initiating an extensive air shower (EAS). A unique component of the Čerenkov radiation from the penetrating shower cores seems to exist, tentatively identified as the muon component. Observations of EAS from the direction of the Crab nebula are carried out. From these a new upper limit is estimated for the detection (> 3 σ) of continuous gamma-rays with $E_\gamma > 6 \times 10^{11}$ eV to be $F(>E_\gamma) < 3.5 \times 10^{-11}$ photons/cm^2-sec. Finally, the EAS identification methods were applied to a study of the relative strength of the cores in some 300 EAS.

142.076 **On the circular polarization of some peculiar objects.** N. S. Nikulin, V. M. Kuvshinov, A. B. Severny.
Astrophys. Journ., (Letters), Vol. 170, L53 - L58 (1971).

Circular polarization of several peculiar objects has been measured with the stellar magnetograph of the 104-inch Crimean reflecting telescope. Rapid variations of the amount and sense of circular polarization are observed in the X-ray source Sco X-1 with time scales of the order of 30 minutes. An appreciable effect is also detected in the nucleus of the planetary nebula +30°3639.

142.077 **Further radio observations of Scorpius X-1.**
C. M. Wade, R. M. Hjellming.
Astrophys. Journ., Vol. 170, 523 - 528 (1971).

Continuing interferometric observations of Sco X-1 at 2695 and 8085 MHz have provided more complete data on its structure and variability. To date, nearly 300 hours of observation have shown that major flaring activity is in progress about one-sixth of the time. The triple structure of the source is a persistent feature, but the position of one component may have changed somewhat during the past year.

142.078 **Variable X-ray sources in Norma and Cetus.**
J. R. Harries, I. R. Tuohy, A. J. Broderick, K. B. Fenton, A. P. J. Luyendyk.
Nature, Phys. Sci., Vol. 234, 149 - 151 (1971).

Two wall-less proportional counters flown on a Skylark

rocket from Woomera, Australia, on July 10, 1970, observed X-rays from several celestial sources. The proportional counters were surrounded by guard counters for charged particle discrimination and had continuous gain calibration throughout the flight. Norma X-2 was found to be more intense in the energy range 2.4 to 10.5 keV than previously measured. An upper limit of 0.24 photons $cm^{-2} sec^{-1}$ was established for the previously intense source Cetus X-2.

142.079 **High energy astronomy: Observations of gamma radiation.** W. D. Metz.
Science, Vol. 174, 1314 (1971).

142.080 **Registration of energetic gamma-ray flux from the extragalactic source 3C 120.**
S. A. Volobuev, A. M. Galper, V. G. Kirillov-Ugryumov, B. I. Luchkov, Yu. V. Ozerov.
Astron. Zhurn. Akad. Nauk SSSR, Vol. 48, 1105 - 1113 (1971). In Russian. English translation in Soviet Astron. AJ, Vol. 15, No. 6.

A comparison of data obtained by means of a gamma-telescope on board Cosmos 251 and Cosmos 264 allowed to conclude on the existence of a discrete gamma-source at energies greater than 100 MeV. The source is likely to be in the region of the following coordinates: right ascension $(3.6 - 5.0)^h$, declination $4° - 9°$. The region also includes 3C 120, an extragalactic source of irregular variable radiation at $2 - 6$ cm wavelengths.

142.081 **On the nature of sources of Sco XR-1 type.**
I. S. Shklovsky.
Astron. Zhurn. Akad. Nauk SSSR, Vol. 48, 1114 - 1121 (1971) In Russian. English translation in Soviet Astron. AJ, Vol. 15, No. 6.

The interpretation of absorption in the soft X-ray region of the spectrum of Sco XR-1 leads to the conclusion that there is a comparatively cool plasma shell round the X-ray source. The main parameters of this shell are derived. A new interpretation of the "mirror-like" variations of radial velocities of He II and He I lines in the spectra of Sco XR-1 and Cyg XR-2 is presented. An advanced hypothesis that "nova" X-ray sources are objects similar to Sco XR-1 is stated.

142.082 **Cosmic flux of low energy gamma rays.**
S. V. Damle, R. R. Daniel, G. Joseph, P. J. Lavakare.
Astrophys. Space Sci., Vol. 14, 473 - 479 (1971).

Measurements have been made on the cosmic gamma rays of energy between 0.25 and 4.2 MeV from a balloon experiment made near the geomagnetic equator. The depth-intensity curves obtained were used to estimate the contribution due to the diffuse cosmic gamma rays in the above energy interval; an unfolding of the counting rates was then performed to obtain the energy spectrum. A critical examination is made of all the observational data between 1 keV and 100 MeV to deduce information on the spectral shape in this energy region. Upper limits on low energy gamma ray fluxes from Sco X-1 and the galactic centre region are also reported.

142.083 **Investigation of cosmic γ-radiation.**
A. M. Gal'per, V. G. Kirillov-Ugryumov, B. I. Luchkov, O. F. Prilutskij.
Uspekhi fiz. nauk, Vol. 105, 209 - 250 (1971). In Russian. – Abstr. in Referativ. Zhurn. 62. Issled. kosm. prostranstva, 1.62.162 (1972).

142.084 **Consideration of the observed isotropic γ-ray background.** G. Cavallo.
Phys. Rev. D, Particles and Fields, Third Ser., Vol. 3, 299 - 305 (1971).

The experimental data on the isotropic component of the diffuse γ-ray background at energies $E_\gamma > 100$ MeV are shown to be inconsistent with the explanation of the isotropic X-ray background as due to the inverse Compton effect of secondary electrons on the $3°K$ (or $8°K$) blackbody radiation. The possibility of the coexistence of a dense intergalactic medium and a universal cosmic-ray flux is also examined, in the context of an evolutionary and of a steady-state cosmology, and restrictions are found.

142.085 **On the nature of the X-ray source in the Coma cluster.** V. N. Kurilchik.
Astron. Tsirk., No. 655, p. 2 - 4 (1971). In Russian.

142.086 **Observations of circular polarization in the optical radiation of the X-ray star Sco X-1.**
Yu. N. Gnedin, O. S. Shulov.
Astron. Tsirk., No. 658, p. 1 - 2 (1971). In Russian.

142.087 **X-ray source in the galactic center and origin of cosmic rays.** I. S. Shklovskij.
Astron. Tsirk., No. 661, p. 1 - 4 (1971). In Russian.

142.088 **New point γ-ray source Lib γ-1: Evidence for time variation and possible identification with PKS 1514–24.** G. M. Frye, Jr., P. A. Albats, A. D. Zych, J. A. Staib, V. D. Hopper, W. R. Rawlinson, J. A. Thomas.
Nature, Vol. 233, 466 - 468 (1971).

In this article we report a new point γ-ray source Libra γ-1 detected on the balloon flight of November 26, 1969. The signal-to-noise ratio is 6σ for $E_\gamma > 100$ MeV, better than for any of the previous sources. For the first time we have direct experimental evidence for the time variation of a γ-ray source. The angular resolution circle in the direction of Lib γ-1 includes PKS 1514–24 which is also an optical variable.

142.089 **Limits on the small scale structure of the diffuse cosmic X-rays.** D. A. Schwartz, E. A. Boldt, S. S. Holt, P. J. Serlemitsos, R. D. Bleach.
Nature, Phys. Sci., Vol. 233, 110 - 112 (1971).

In this communication, we compare the experimentally measured autocorrelation function with the calculations of Wolfe and Burbidge (1970) and confirm their conclusions that the X-ray isotropy is not consistent with the observed superclustering, and probably not with clustering, of galaxies.

142.090 **Radio spectrum of Cygnus X-1.**
R. M. Hjellming, C. M. Wade, V. A. Hughes, A. Woodsworth.
Nature, Vol. 234, 138 (1971).

Recent observations at 2,795 and 1,415 MHz have shown that a point radio source appeared in the field of the X-ray source Cyg X-1 some time between March 22 and April 28, 1971. On the basis of strong variability and close proximity, this radio source is identified with Cyg X-1. Observations with the NRAO interferometer at 2,695 MHz have shown that the source appeared constant in flux from May 13 to July 10, 1971.

142.091 **Gamma rays and the distribution of cosmic rays in the Galaxy.** M. P. Ulmer.
Thesis, Univ. Wisconsin, Madison. [Available from Univ. Microfilms, Ann Arbor, Mich., U.S.A. Order No. 70–22077], 168 pp. (1970).

This thesis provides an interpretation of the OSO III gamma ray results of Clark, Garmire, and Kraushaar, 1968.

142.092 **A study of Sco X-1.** D. E. Mook, II.
Thesis, Univ. Michigan, Ann Arbor. [Available from Univ. Microfilms, Ann Arbor, Mich., U.S.A. Order No. 70–21736], 151 pp. (1970).

Photometric monitoring of Sco X-1 simultaneously in the V and B bands of the UBV system is used to establish a mean color magnitude relation for the object. As an example of how this relation can be used in evaluating models for the Sco X-1 system, the self-absorption bremsstrahlung mode is discussed in some detail. Spectra, polarization, and UBV measurements of field stars around Sco X-1 are used to estimate both the distance to the object, and the amount of interstellar reddening suffered by its radiation.

142.093 Measurements of hard X and gamma radiation from Virgo A and Centaurus A.   G. J. Fishman.
Thesis, Rice Univ., Houston, Texas. [Available from Univ. Microfilms, Ann Arbor, Mich., U.S.A. Order No. 70–23507], 176 pp. (1970).
Balloon-borne measurements were made of two nearby radio galaxies, Virgo A and Centaurus A, in the hard X and gamma ray regions in an effort to relate the radiation to lower energy regions of the electromagnetic spectrum. Measurements of this type make it possible to evaluate various models of the emission mechanism and energy source of these unusual objects.

142.094 A search for high energy gamma rays from the Cygnus region.   J. V. Valdez.
Thesis, Univ. Minnesota, Minneapolis. [Available from Univ. Microfilms, Ann Arbor, Mich., U.S.A. Order No. 70–15829], 113 pp. (1969).
A nuclear emulsion-spark chamber detector was flown on July 7, 1967 to search for gamma rays (E>100 MeV) from the Cygnus region. This part of the sky includes the strongest radio galaxy Cygnus A, the X-ray source Cygnus XR-I and that portion of the galactic plane between longitude $60° < l^{II} < 90°$.

142.095 Transient gamma-ray sources.
K. M. V. Apparao. T. N. Rengarajan.
Proc. Indian Acad. Sci., Section A, Vol. 73, 257 - 260 (1971).
Gamma-ray production by particles escaping from a pulsar into the surrounding nebula is considered. The gamma-ray emission decreases with time and such pulsar-nebula complexes will be observed as transient sources.

142.096 A measurement of the primary X-ray diffuse component in the range from 25 to 200 keV.
E. Horstman-Moretti, F. Fuligni, D. Brini.
Nuovo Cimento B, Ser. 11, Vol. 6B, 68 - 82 (1971).
An apparatus for the measurement of X-rays from 25 to 200 keV was flown aboard a rocket. A wide scan of the sky was performed and a spectrum of the primary diffuse component was obtained which is best fitted by a power law. The detectors also looked toward the earth and a rough spectrum of the earth albedo is shown.

142.097 Cosmic gamma and X rays.   S. I. Syrovatskii.
Proc. Sixth Winter School on Space Physics, Part I.
Apatity 1969, (see 012.033), p. 111 - 127 (1971).

142.098 Low-energy gamma rays (0.1–10 MeV) in the atmosphere and in outer space.   A. M. Romanov.
Proc. Sixth Winter School on Space Physics, Part I. Apatity 1969, (see 012.033), p. 128 - 142 (1971).

142.099 X-ray and gamma-ray observations from artificial earth satellites.
M. M. Anisimov, L. S. Bratolyubova-Tsulukidze, N. L. Grigorov, L. F. Kalinkin, A. S. Melioranskii, E. A. Pryakhin, I. A. Savenko, V. Ya. Yufarkin.
Proc. Sixth Winter School on Space Physics, Part I. Apatity 1969, (see 012.033), p. 143 - 151 (1971).

142.100 Variable X-ray stars.   L. J. Boss.
AAVSO Abstr., October 1971, p. 12 - 13.

A gas-Čerenkov telescope experiment to observe cosmic gamma rays.   See Abstr. 034.001.

X-ray spectrometry of galactic sources in the energy range 30–200 keV.   See Abstr. 034.054.

Contribution to the background rate of a satellite X-ray detector by spallation products in a caesium iodide crystal.   See Abstr. 034.096.

Survey on new techniques for X-ray astronomy. See Abstr. 051.020.

The present state of gamma-ray astronomy. See Abstr. 061.020.

Structure and evolution of supermassive rotating magnetic polytropes.   See Abstr. 065.085.

Cen XR-3: A neutron star younger than the Crab? See Abstr. 065.088.

On the X-ray emission of flare stars. See Abstr. 122.002.

Soft X-rays from nonthermal galactic radio sources: Implications concerning the galactic background and the interstellar medium.   See Abstr. 125.018.

Rotating magnetic white dwarf stars as X-ray sources.   See Abstr. 126.006.

Search for polarized X-rays from the Crab nebula using a focusing graphite crystal polarimeter. See Abstr. 134.001.

Search for polarized X-rays from the Crab nebula using an incoherent scattering polarimeter. See Abstr. 134.002.

Detection of pulsed gamma radiation from the Crab nebula.   See Abstr. 134.003.

Observations of γ-quanta with energies more than 100 MeV from the radio source 3C 120. See Abstr. 141.035.

Pulsed gamma emission above 50 MeV from NP 0532.   See Abstr. 141.137.

Hard X-ray spectrum of NP 0532. See Abstr. 141.142.

Pulsar P 1749 - 28 a possible XR source. See Abstr. 141.147.

Search for pulsed gamma ray emission above 50 MeV from NP 0532.   See Abstr. 141.165.

Possible evidence for pulsed ~ 10^{12} eV gamma rays from NP 0532.   See Abstr. 141.239.

Some remarks on the universal X-ray background. See Abstr. 142.063.

Cosmic electron acceleration and the spectrum of metagalactic X-rays.   See Abstr. 143.013.

Measurements of high-energy Gamma-ray intensity in the primary cosmic rays aboard Cosmos 208 satellite. See Abstr. 143.029.

The physics of cosmic X-ray, γ-ray, and particle sources. See Abstr. 143.032.

Electron-photon component of cosmic rays. See Abstr. 143.066.

Spectrum and galactic isotropy of diffuse cosmic X-rays. See Abstr. 155.026.

Galactic X-rays from unresolved flare stars. See Abstr. 155.038.

Optical aspects of X-ray sources in the Large Magellanic Cloud. See Abstr. 159.026.

143 Cosmic Radiation

143.001 Solar modulation origin of 'sidereal' cosmic ray anisotropies. D. B. Swinson.
Journ. Geophys. Res., Vol. 76, 4217 - 4223 (1971).

On the basis of earlier work that suggested that the sidereal diurnal variation in cosmic ray intensity could be due to the cooperative effects of the interplanetary magnetic field and a radial heliocentric cosmic ray density gradient, a more comprehensive data analysis has been made to test the model. In the present paper a longer series of data from both the northern and the southern hemispheres is analyzed in sidereal time for phase changes in the diurnal variation of the cosmic ray flux underground in relation to the periodic reversal of the interplanetary magnetic field.

143.002 Charged-particle observations from OSO 3.
G. D. Badhwar, M. F. Kaplon, D. A. Valentine.
Journ. Geophys. Res., Vol. 76, 4224 - 4229 (1971).

The results on the nucleonic component of the primary cosmic rays obtained from the analysis of 5800 orbits of data from the University of Rochester telescope on OSO 3 satellite are presented. They are in good agreement with values measured on balloon-borne detectors.

143.003 The modulation of galactic protons by the solar wind: A Monte Carlo approach.
T. A. Moss, R. T. Giuli.
Astrophys. Journ., Vol. 167, 331 - 340 (1971).

We have made a Monte Carlo analysis of a radial-magnetic-field model for the solar wind. By comparison of the results with observations made during solar minimum, we are able to reach some conclusions, appropriate to the time of solar minimum, concerning the real values of the solar-wind parameters discussed above, an approximate shape of the spectrum of unmodulated galactic protons, and the directional energy flow of the cosmic rays in the solar wind.

143.004 A comparative study of the rigidity dependence of Forbush decreases and 11-year variation in cosmic ray intensity. T. Mathews, P. H. Stoker, B. G. Wilson.
Planet. Space Sci., Vol. 19, 981 - 991 (1971).

Results of a comparative study of Forbush decreases and the 11-year variation observed at Calgary, Sulphur Mountain and other neutron monitor stations during the current cycle of solar activity are presented. No definite difference between modulation functions applicable to Forbush decreases and 11-year variation from a comparative analysis of data on daily average counting rates from different latitude stations was found.

143.005 The relative abundances of the isotopes of lithium, beryllium and boron in the primary cosmic radiation. N. Durgaprasad.
Astrophys. Space Sci., Vol. 12, 98 - 103 (1971).

Studies have been made to determine the relative abundances of the isotopes of lithium, beryllium and boron in primary cosmic rays in the low energy interval 180–400 MeV per nucleon recorded in the emulsion stack flown from Fort Churchill. Two independent measurements of mass, whereever possible, were made on each track. Out of nine boron tracks, 6 particle tracks are consistent with B^{11} and 3 with B^{10}. Amongst 2 Li tracks, one is consistent with Li^6 and the other with Li^7.

143.006 On the direct observation of cosmic-ray energy losses. L. J. Gleeson, I. D. Palmer.
Astrophys. Space Sci., Vol. 12, 137 - 146 (1971).

During the decay of solar cosmic-ray events cosmic rays with kinetic energies of about 1 MeV are convected outward with the solar wind. It is shown that, with currently available observations, it should be possible to demonstrate directly the energy losses which are occurring. In this paper observations from Venera-4 and Imp-F have been used.

143.007 The relative abundances of the isotopes of light nuclei (Li, Be and B) in the primary cosmic radiation. N. Durgaprasad.
Astrophys. Space Sci., Vol. 12, 243 - 249 (1971).

Studies have been made to determine the relative abundances of the isotopes of lithium and boron in the primary cosmic rays in the low energy interval 120 - 400 MeV per nucleon recorded in the emulsion stack flown from Fort Churchill. Two independent measurements of mass, wherever possible, were made on each track by the range vs ionisation method. The results obtained on nine stopping boron nuclei and two lithium nuclei are presented.

143.008 Radio signals from cosmic ray air showers.
J. H. Hough, R. W. Clay, J. R. Prescott.
Journ. Roy. Astron. Soc. Canada, Vol. 65, 178 - 179 (1971).
Abstr. RAS Canada.

143.009 Correlation analysis of the intensity of galactic cosmic rays in the universe and on the earth.
R. M. Golynskaya, N. V. Pereslegina.
Kosmich. Issled., Vol. 9, 565 - 569 (1971). In Russian.

143.010 Periodic variations of the cosmic radiation—II. The Sidereal Time Period. W. Messerschmidt.
Planet. Space Sci., Vol. 19, 1025 - 1040 (1971).

The present paper makes an approach to an evidence of the Sidereal Time Period of the cosmic radiation. In order to observe the properties of the Sidereal Time Period for a long time, the data were classified into time intervals according to the different solar activity, or they were summed up over eight to ten years. The result of the paper confirms the evidence of the Sidereal Time Period. Its amplitude amounts to 0.03–0.10 per cent. In the southern hemisphere the maximum lies between 05 and 06 Local Sidereal Time, and in the northern hemisphere between 20 and 21 LST, in good agreement with other authors.

143.011 The solar diurnal variation of cosmic rays at the 1954 solar minimum. D. M. Thomson.
Planet. Space Sci., Vol. 19, 1169 - 1183 (1971).

A re-examination of cosmic ray data at the 1954 solar minimum shows that the anomalous diurnal variation during July, August and September 1954 was not a sidereal time effect but was associated with a streaming of cosmic rays in a direction perpendicular to the interplanetary field lines and towards the sun. The phenomenon is discussed in terms of the theory of diffusion and scattering of cosmic rays in the interplanetary medium.

143.012 Median primary energy of response of a cosmic ray telescope underground. H. S. Ahluwalia.
Journ. Geophys. Res., Vol. 76, 5358 - 5360 (1971). – Letter.

143.013 Cosmic electron acceleration and the spectrum of metagalactic X-rays.
A. Z. Dolginov, Yu. N. Gnedin.
Astrophys. Letters, Vol. 9, 91 - 94 (1971).

The power spectrum of cosmic relativistic electrons is explained as due to the acceleration and scattering of electrons in turbulent plasma with a frozen-in magnetic field. A possibility is discussed of explaining the bend in the metagalactic X-ray spectrum as a result of the bend in the electron spectrum.

143.014 **Measurements of high-energy γ-quanta streams in the primary cosmic radiation on Cosmos 208.**
L. S. Bratolyubova-Tzulukidze, N. L. Grigorov, L. F. Kalinkin, A. S. Melioransky, E. A. Pryakhin, I. A. Savenko, V. Ya. Yufarkin.
Geomagn. Aeronom., Vol. 11, 585 - 589 (1971). In Russian.

143.015 **Angular distributions of the intensity of energetic charged particles above the atmosphere in the region of the cosmic ray equator.**
R. N. Basilova, I. A. Savenko.
Geomagn. Aeronom., Vol. 11, 590 - 594 (1971). In Russian.

143.016 **Correction: 'Solar modulation origin of 'sidereal' cosmic ray anisotropies'. [Journ. Geophys. Res., Vol. 76, 4217 - 4223 (1971)].** D. B. Swinson.
Journ. Geophys. Res., Vol. 76, 6211 (1971).

143.017 **Non-linear interaction of galactic cosmic rays with interplanetary magnetic fields and a possible geometry of the solar wind.** I. V. Dorman, L. I. Dorman.
Geomagn. Aeronom., Vol. 11, 776 - 779 (1971). In Russian.

143.018 **Foundation of cosmic ray transport theory in random magnetic fields.** A. J. Klimas, G. Sandri.
Bull. American Astron. Soc., Vol. 3, 364 - 365 (1971). Abstr. AAS.

143.019 **Model for cosmic ray transport.**
G. Sandri, A. Klimas.
Bull. American Astron. Soc., Vol. 3, 365 (1971). – Abstr. AAS.

143.020 **Transurane in der kosmischen Strahlung?**
H. Pilkuhn.
Phys. Blätter, 27. Jahrgang, p. 448 - 451 (1971).

143.021 **Determination of the energy spectrum of cosmic particles.**
A. M. Gal'per, A. V. Kurochkin, B. I. Luchkov, Yu. T. Yurkin.
Pribory i tekhn. ehksperimenta, 1971, No. 3, p. 60 - 62. In Russian. – Abstr. in Referativ. Zhurn. 62. Issled. kosm. prostranstva, 10.62.202 (1971).

143.022 **Measurements of the charge and isotope composition of cosmic ray Li, Be and B nuclei.**
W. R. Webber, S. V. Damle, J. M. Kish.
Astrophys. Letters, Vol. 9, 125 - 129 (1971).
We have measured cosmic ray Li, Be, and B nuclei using two new detectors, a large-area dE/dx-E-Range telescope, and a dE/dx-Cerenkov-Range telescope. The L/M ratio is found to be 0.23 ± 0.01, essentially constant with energy between 100 MeV/nuc to > 2 BeV/nuc. The observed ratio Be/(Li+B) is in generally good agreement with that predicted on the basis that Be^{10} has decayed. The average lifetime of the cosmic rays producing Be must therefore be $\gtrsim 1 \times 10^7$ yr. This value taken together with the value of $X_0 \sim 5$ g/cm² obtained from L/M ratio leads to an average matter density of $\lesssim 0.2$ g/cm² along the interstellar trajectories of the cosmic ray particles producing L nuclei.

143.023 **Extragalactic cosmic rays – A reappraisal.**
G. Burbidge, K. Brecher.
Comments Astrophys. Space Phys., Vol. 3, 140 - 148 (1971).
Modern theories concerned with the origin of cosmic rays fall into three main categories. There are the theories in which it has been argued that cosmic rays all originate and are confined to the solar system, those in which it is argued that the cosmic rays are largely confined to the Galaxy and originate in supernova remnants and other galactic objects, and theories in which it is argued that the cosmic rays are an extragalactic phenomenon. In modern times the extragalactic theory has been continuously attacked by Ginzburg and Syrovatskii, and their arguments against this hypothesis have been . repeated quite uncritically by others. Many of the objections to this theory are unjustified, and in this paper we discuss the present situation.

143.024 **Cosmic rays of ultra-high energy.** S. I. Syrovatskii.
Comments Astrophys. Space Phys., Vol. 3, 155 - 162 (1971).
The observed cosmic ray spectrum in the whole relativistic energy range can be due to the galactic sources with constant slope of the injection spectrum up to the maximum observed energy. The kinks in the spectrum are interpreted in this model by the changes of the mode of propagation of cosmic rays in the galaxy: from diffusion to drift motion at first kink at $E \simeq 3 \times 10^{15}$ eV and from drift to quasirectilinear propagation at the second kink at $E \simeq 4 \times 10^{17}$ eV. The crucial point of the model as also for other galactic models would be the test of the predicted anisotropy in our case with the minimum in the directions of galactic poles.

143.025 **The direct mode of propagation of cosmic rays to geostationary satellites.** R. Gall, D. F. Smart, M. A. Shea.
Planet. Space Sci., Vol. 19, 1419 - 1430 (1971).
Using a mathematical model of the magnetosphere, we have computed trajectories of low energy cosmic rays arriving at the position of the ATS 1 synchronous orbit satellite to determine the propagation characteristics. Analysis of these calculations leads to the determination of the geomagnetic cutoffs, the specific regions of penetration at the magnetospheric boundary and the directions of approach from the interplanetary medium.

143.026 **Galactic antiproton cosmic radiation.**
P. K. Suh.
Astron. Astrophys., Vol. 15, 206 - 215 (1971).
The energy spectrum for the cosmic antiproton flux in the galactic interstellar space is examined. The antiproton flux exhibits a peak around one BeV of antiproton energy, where the maximum flux lies in the proximity of 0.1 to 0.2 percent of the cosmic proton flux at the same energy.

143.027 **Interaction of galactic cosmic rays with the interplanetary magnetic field.**
S. N. Vernov, E. V. Gorchakov, P. P. Ignatyev.
Space Research XI, (see 012.004), 1247 - 1251 (1971).

143.028 **The daily variation of cosmic-ray cutoff at the altitude of a geostationary satellite.**
R. Gall, D. F. Smart, M. A. Shea.
Space Research XI, (see 012.004), 1259 - 1264 (1971).

143.029 **Measurements of high-energy Gamma-ray intensity in the primary cosmic rays aboard Cosmos 208 satellite.** L. S. Bratolubova-Tzulukidze, N. L. Grigorov, E. A. Pryakhin, I. A. Savenko, V. Ya. Yufarkin, L. F. Kalinkin, A. S. Melioransky.
Space Research XI, (see 012.004), 1379 - 1383 (1971).

143.030 **High-energy cosmic rays on the Proton-4 cosmic scientific station.** N. L. Grigorov, I. D. Rapoport, I. A. Savenko, V. E. Nesterov.
Space Research XI, (see 012.004), 1391 - 1395 (1971).

143.031 **Charge composition of relativistic primary cosmic**

rays between beryllium and iron.
M. Casse, O. Corydon-Petersen, B. Dayton, L. Koch, N. Lund, K. Melgaard, P. Mestreau, J. P. Meyer, K. Omø, T. Risbo, D. Roussel.
Space Research XI, (see 012.004), 1397 - 1408 (1971).

143.032 The physics of cosmic X-ray, γ-ray, and particle sources. K. Greisen.
Astrophysics and general relativity. Brandeis Univ. Summer Inst. 1968, Vol. 2, (see 003.011), 37 - 150 (1971).

143.033 Olivines: Revelation of tracks of charged particles.
S. Krishnaswami, D. Lal, N. Prabhu, A. S. Tamhane.
Science, Vol. 174, 287 - 291 (1971).

In this report we describe an etchant for satisfactorily revealing tracks in olivine, regardless of the crystallographic orientation.

143.034 Origin and composition of ultrahigh energy cosmic rays. F. W. Stecker.
Nature, Phys. Sci., Vol. 234, 28 - 29 (1971).

Five alternative hypotheses for the origin and composition of ultrahigh energy cosmic rays are critically examined for consistency with recent data on the characteristics of the cosmic-ray energy spectrum and upper limits on the flux of isotropic gamma-radiation. The gamma-ray upper limits rule out cosmological origin hypotheses including the neutrino hypothesis. Other extragalactic origin models are ruled out by examining spectral characteristics. The galactic origin hypothesis alone seems to be consistent with present data and it seems most likely that the highest energy cosmic-rays observed are heavy nuclei.

143.035 Foundation of the theory of cosmic-ray transport in random magnetic fields.
A. J. Klimas, G. Sandri.
Astrophys. Journ., Vol. 169, 41 - 56 (1971).

A new closed pair of coupled integro-differential equations for the cosmic-ray omnidirectional intensity and flux in a random magnetic field with uniform mean part are derived from first principles. The only assumptions made to obtain this result are: (1) the cosmic-ray pressure tensor is isotropic, (2) higher than two-point correlations in the random field are neglected, (3) the two-point correlation tensor corresponds to isotropic, homogeneous turbulence, and (4) spatial inhomogeneities in the omnidirectional intensity and flux are negligible over the scale of the correlation length. These equations are valid for all energies compatible with assumption (2).

143.036 The origin of fluorine, sodium, and aluminum in the galactic cosmic radiation. B. G. Cartwright.
Astrophys. Journ., Vol. 169, 299 - 310 (1971).

The abundances of F, Ne, Na, Mg, Al, and Si relative to oxygen in the galactic cosmic radiation in the 106 to 207 MeV kinetic energy per nucleon interval measured by the University of Chicago charged particle telescope on board the IMP-5 satellite are reported, as well as the abundance ratio Γ (F/O) above 1 GeV per nucleon kinetic energy measured by the same experiment. The abundances of the odd-Z elements F, Na, and Al are then investigated in the light of three models for their production.

143.037 Steplike changes in the long-term modulation of cosmic rays. P. H. Stoker, H. Carmichael.
Astrophys. Journ., Vol. 169, 357 - 368 (1971).

An experimental investigation is reported of the rigidity dependence of the solar modulation of relativistic cosmic-ray particles at the orbit of the earth by comparison of the monthly averages of the neutron monitor at Deep River at geomagnetic cutoff rigidity 1 GV with those of the neutron monitor at Kula at cutoff 13 GV during the period of in-

creasing solar activity which began in 1964–1965. The steplike long-lasting changes between the successive segments of the regression are believed to be demonstrated here for the first time. They are probably associated with the state of the interplanetary medium far beyond the orbit of earth.

143.038 Cross section for $^{11}B(p, 2p)^{10}$Be at 150 and 600 MeV: Implications for cosmic-ray studies.
G. M. Raisbeck, F. Yiou.
Phys. Rev. Letters, Vol. 27, 875 - 877 (1971).

The cross section for $^{11}B(p, 2p)^{10}$Be has been measured to be 19 ± 6 and 25 ± 8 mb at 150 and 600 MeV, respectively. These values show that the tertiary production of ^{10}Be in cosmic rays is very important. However, it still appears likely that an actual isotope abundance measurement in cosmic rays will be necessary to utilize this nuclide as a cosmic-ray "clock".

143.039 Origin of cosmic electrons from about 10^2 to 10^6 GeV. R. Ramaty, R. E. Lingenfelter.
Phys. Rev. Letters, Vol. 27, 1309 - 1312 (1971).

The origin of high-energy cosmic electrons is considered. It is found that electrons of energies $\lesssim 10^3$ GeV could have been produced by local supernovae associated with known radio remnants. At higher energies, observations of muon-poor air showers indicate the existence of electrons at 10^6 GeV which may have originated entirely from the supernova Vela X.

143.040 The production of the elements Li, Be, B by galactic cosmic rays in space and its relation with stellar observations.
M. Meneguzzi, J. Audouze, H. Reeves.
Astron. Astrophys., Vol. 15, 337 - 359 (1971).

The L-element (Li, Be, B) contamination rate of the interstellar gas by nuclear reactions induced by the galactic cosmic rays (G.C.R.) is calculated using a diffusion model of fast moving particles in the Galaxy. The presence of helium in the G.C.R. flux and in the interstellar gas is taken into account.

143.041 The energy spectrum of primary cosmic ray electrons from 20 MeV to 20 GeV in 1968 and 1969.
D. Hovestadt, P. Meyer, P. J. Schmidt.
Astrophys. Letters, Vol. 9, 165 - 168 (1971).

The energy spectrum of primary cosmic ray electrons between 20 MeV and 20 GeV was measured with a balloon-borne instrument in 1968 and 1969 from Ft. Churchill, Manitoba. The instrument consists of an absorption spectrometer, designed to discriminate effectively against nuclear cosmic ray components and to achieve high energy resolution. The measured electron spectrum is presented and compared with results by other investigators.

143.042 Anisotropy of high energy cosmic-ray electrons in the discrete source model. C. S. Shen, C. Y. Mao.
Astrophys. Letters, Vol. 9, 169 - 174 (1971).

Because of the rapid loss of energy due to radiation, high-energy cosmic-ray electrons observed at earth probably all come from a few young and nearby sources. Diffusion of electron flux from these nearby sources gives rise to an anisotropy which becomes unidirectional at high energy. A survey of possible cosmic ray sources indicates anisotropy of the order of 10 per cent for cosmic ray electrons in the energy range above 100 GeV. Implications of the detection of this large anisotropy in astrophysics are discussed.

143.043 Antiprotons in the primary cosmic radiation near the geomagnetic equator.
N. Durgaprasad, P. K. Kunte.
Nature, Phys. Sci., Vol. 234, 74 - 75 (1971).

An estimate on the abundance of the antiprotons in the cosmic radiation has been made from the measured asymmetric fluxes of singly charged particles at the zenith angle $\Theta = 40°$ in the East, West, North and South directions and the flux in the vertical direction. These fluxes were measured previously by a Cerenkov-scintillation counter telescope flown near the geomagnetic equator, Hyderabad, India. The flux of anti-protons of energy $E \geqslant 16$ GeV at the top of the atmosphere is determined as 2.0 ± 7.9 particles/m² sec sr. The ratio of anti-protons to all singly charged particles in the primary radiation of energy $E \geqslant 16$ GeV in the vertical direction is obtained as $(1.7 \pm 6.5) \times 10^{-2}$. The reentrant albedo flux is determined as 13.9 ± 10.5 particles/m² sec sr.

143.044 **Effects of active solar regions on the galactic cosmic ray intensity.**
E. Antonucci, G. C. Castagnoli, M. A. Dodero.
Solar Physics, Vol. 20, 497 - 506 (1971).

The solar modulation of galactic cosmic ray intensity during 1969 was dominated by effects resulting from the activity in the two zones. In fact all the decreases can be related to the passage at the central meridian of the active centres. Persistence of the effects connected to solar regions is found also during rotations in which they do not produce flares in front of the earth.

143.045 **Balloon measurements of cosmic ray protons and helium over half a solar cycle 1965–1969.**
T. A. Rygg, J. A. Earl.
Journ. Geophys. Res., Vol. 76, 7445 - 7469 (1971).

Differential energy spectra for protons and helium covering the energy range 100–260 MeV/nucleon were obtained from balloon flights made each summer (1965–1969) at Churchill. The observed proton spectra are characterized over a wide range of energy (30–300 MeV) by a simple relationship between cosmic-ray intensity J and kinetic energy T: $J = AT$. The helium spectra also follow this law at solar maximum but rise less steeply near solar minimum.

143.046 **The energy dependence of the cosmic-ray neutron leakage flux in the range 0.01–10 MeV.**
R. W. Jenkins, S. O. Ifedili, J. A. Lockwood, H. Razdan.
Journ. Geophys. Res., Vol. 76, 7470 - 7478 (1971).

We report the results of four months of simultaneous measurements of leakage neutrons in the ranges 1–10 MeV and 0.001–1 MeV made with the University of New Hampshire neutron detector on the polar-orbiting OGO 6 satellite. From these data, information is deduced about the energy spectrum in the range 0.001–10 MeV and its variation with latitude.

143.047 **Cosmic-ray variations and the interplanetary sector structures.**
S. Yoshida, N. Ogita, S.-I. Akasofu.
Journ. Geophys. Res., Vol. 76, 7801 - 7803 (1971).

143.048 **Correlation of galactic cosmic-ray intensity with $\lambda 5303$ coronal intensity.** P. N. Pathak.
Journ. Geophys. Res., Vol. 76, 7804 - 7807 (1971). – Letter.

143.049 **Acceleration and formation of the electron spectrum of galactic cosmic rays.**
Yu. N. Gnedin, A. Z. Dolginov.
"Trudy Mezhdunar. seminara po probl. "Generatsiya kosmich. luchej na Solntse", 1970". Moskva, 1971, p. 442 - 453. In Russian. – Abstr. in Referativ. Zhurn. 51. Astron., 12.51.656 (1971).

143.050 **Cosmic rays in the Galaxy: Convection or diffusion?**
J. Skilling.
Astrophys. Journ., Vol. 170, 265 - 273 (1971).

The theory of the self-trapping of cosmic rays in our Galaxy shows that cosmic-ray particles stream steadily away from their sources at a speed given by that of hydromagnetic waves. Random diffusion along magnetic field lines is a comparatively negligible process. A crucial role in confining high-energy cosmic rays is played by the less dense material on the edges of the galactic disk or in the halo.

143.051 **Ursprung der kosmischen Strahlung.**
K. Pinkau.
Physik und Kosmologie, (see 003.024), p. 129 - 136 (1971).

143.052 **Origins of cosmic rays.**
R. Cowsik, P. B. Price.
Physics Today, Vol. 24, No. 10, p. 30 - 38 (1971).

We shall discuss some of the properties of the three classes of cosmic rays – the nuclei, the electrons (negatrons and positrons) and the electromagnetic component – with an eye to explaining the mystery of their origins.

143.053 **On the galactic electrons produced by cosmic rays.**
K. Arai.
Sci. Rep. Tôhoku Univ., First Ser., Vol. 54, 1 - 12 = Sendai Astron. Rap., No. 118 (1971).

The energy spectra of secondary electrons (negatrons and positrons) produced by cosmic-ray nuclei through the interaction of the interstellar medium (90% hydrogen and 10% helium) are recalculated. The equilibrium spectra are obtained in the frame work of the galactic disk model.

143.054 **Rayos cósmicos.** J. Gómez.
El Universo, Vol. 25, 115 - 118 (1971).

143.055 **Effective modulation layer in the interplanetary space.**
A. K. Lavrukhina, G. K. Ustinova, A. N. Simonenko.
Meteoritics, Vol. 6, 290 (1971). – Abstract.

143.056 **Regression analysis of atmospheric effects of the neutron component of cosmic rays.**
N. Akhababian, L. Alexandrov.
Geomagn. Aeronom., Vol. 11, 949 - 952 (1971). In Russian.

143.057 **Composition and galactic confinement of cosmic rays.** M. M. Shapiro.
Highlights of Astronomy, Vol. 2, (see 012.019), 740 - 756 (1971).

143.058 **The energy spectrum of primary cosmic ray electrons from 2 GeV to 200 GeV.**
J. L. Fanselow, R. C. Hartman, P. Meyer, P. J. Schmidt.
Astrophys. Space Sci., Vol. 14, 301 - 313 (1971).

A balloon borne counter telescope with a gas Cerenkov counter is used to measure the energy of primary cosmic ray electrons between 2 and 200 GeV. The results from six balloon flights are combined to obtain the electron energy spectrum. Up to about 30 GeV the spectrum measured in this experiment can be directly checked with calibrations and agrees well with results from other experiments. Above this energy the flux reported here is somewhat higher than the determinations reported by most other authors.

143.059 **On the modulation and energy spectrum of highly charged cosmic ray nuclei.**
T. F. Cleghorn, P. S. Freier, C. J. Waddington.
Astrophys. Space Sci., Vol. 14, 422 - 430 (1971).

This paper reports the intensities of VH-nuclei observed at a time when appreciable additional solar modulation was occurring, and compares the amount of modulation experienced by these high charged nuclei with that of helium nuclei.

143.060 **A possible mechanism for the injection of particles into a cosmic ray acceleration region.**

K. Kristiansson.
Astrophys. Space Sci., Vol. 14, 485 - 500 (1971).

The heavy element over-abundance is discussed and new experimental data are added to the comparison. In the succeeding sections there is a discussion of a hitherto unconsidered mechanism which may be responsible for the selection of cosmic ray nuclei. A source model in which the plasma clouds originate in type II supernova explosions is discussed.

143.061 Forbush decreases in the cosmic radiation.
J. A. Lockwood.
Space Sci. Rev., Vol. 12, 658 - 715 (1971).

The experimental observations of Forbush decreases in recent years are reviewed and related to different theoretical models which have been proposed. The observational data from both ground-based and spacecraft experiments were selected to illustrate the important characteristics of Forbush decreases.

143.062 Pulsars and cosmic-ray prehistory.
R. E. Lingenfelter.
Proc. 11th international conference on cosmic rays, Budapest 1969, (see 05.012.018), p. 557 - 563 (1970).

143.063 The nature of the primary cosmic radiation.
C. F. Powell.
Proc. 11th international conference on cosmic rays, Budapest 1969, (see 012.025), p. 3 - 16 (1970).

143.064 High-energy interactions at accelerator energies.
O. Czyzewski.
Proc. 11th international conference on cosmic rays, Budapest 1969, (see 012.025), p. 177 - 230 (1970).

143.065 Origin and acceleration mechanisms.
S. I. Syrovatskii.
Proc. 11th international conference on cosmic rays, Budapest 1969, (see 012.025), p. 233 - 239 (1970).

143.066 Electron-photon component of cosmic rays.
Y. Pal.
Proc. 11th international conference on cosmic rays, Budapest 1969, (see 012.025), p. 241 - 274 (1970).

143.067 Protons and nuclei in the galactic radiation. Charge composition and energy spectra. W. R. Webber.
Proc. 11th international conference on cosmic rays, Budapest 1969, (see 012.025), p. 275 - 311 (1970).

143.068 Extensive air showers (experiment). J. Trümper.
Proc. 11th international conference on cosmic rays, Budapest 1969, (see 012.025), p. 497 - 518 (1970).

143.069 Search for cosmic magnetic monopoles.
R. L. Fleischer, H. R. Hart, Jr., I. S. Jacobs, P. B. Price, W. M. Schwarz, R. T. Woods, F. Aumento, H. G. Goodell.
Proc. 11th international conference on cosmic rays, Vol. 3, (see 012.027), 27 - 30 (1970).

143.070 Search for quarks beyond the atmosphere on the Proton-3 satellite.
N. L. Grigorov, G. P. Kakhidze, I. D. Rapoport, I. A. Savenko.
Proc. 11th international conference on cosmic rays, Vol. 3, (see 012.027), 37 - 39 (1970).

143.071 High energy nuclear interactions observed at 5200 m above sea level. K. Kamata, T. Maeda, P. K. MacKeown, C. Agüirre, A. Trepp, G. R. Mejia, Y. Toyoda, K. Suga, K. Uchino, K. Murakami, M. La Pointe.
Proc. 11th international conference on cosmic rays, Vol. 3, (see 012.027), 49 - 56 (1970).

143.072 Nuclear interactions in light elements above 10^{12} eV energies. R. E. Gibbs, J. J. Lord.
Proc. 11th international conference on cosmic rays, Vol. 3, (see 012.027), 113 - 118 (1970).

143.073 The energy spectrum of cosmic radiation from 10^{10} to 2×10^{20} eV. R. G. Brownlee, G. J. Chapman, S. A. David, A. J. Fisher, L. Horton, L. Goorevich, P. C. Kohn, C. B. A. McCusker, A. Outhred, A. F. Parkinson, L. S. Peak, M. H. Rathgeber, M. J. Ryan, M. M. Winn.
Proc. 11th international conference on cosmic rays, Vol. 3, (see 012.027), 377 - 382 (1970).

143.074 The arrival directions of high energy cosmic ray primaries. R. G. Brownlee, S. A. David, A. J. Fisher, L. Horton, L. Goorevich, P. C. Kohn, C. B. A. McCusker, A. Outhred, D. E. Page, A. F. Parkinson, L. S. Peak, M. H. Rathgeber, R. J. O. Reid, M. J. Ryan, M. M. Winn.
Proc. 11th international conference on cosmic rays, Vol. 3, (see 012.027), 383 - 388 (1970).

143.075 Possible change in the chemical composition of cosmic particles at energies about 10^{16} eV.
P. Catz, R. Maze, J. Gawin.
Proc. 11th international conference on cosmic rays, Vol. 3, (see 012.027), 395 - 397 (1970).

143.076 The chemical composition of cosmic radiation at 10^{16} eV. J. Trümper, E. Böhm, R. Fritze, M. Samorski, R. Staubert.
Proc. 11th international conference on cosmic rays, Vol. 3, (see 012.027), 447 - 450 (1970).

143.077 Mass composition of primary cosmic rays above 10^{13} eV. M. G. Thompson, M. J. L. Turner, A. W. Wolfendale, J. Wdowczyk.
Proc. 11th international conference on cosmic rays, Vol. 3, (see 012.027), 615 - 620 (1970).

143.078 Changes of rigidity dependence and daily variation during cosmic ray decreases.
B. Östman, E. Awadalla.
Proc. 11th international conference on cosmic rays, Vol. 4, (see 012.028), 7 - 11 (1970).

143.079 High energy primary cosmic ray program of Goddard Space Flight Center.
J. F. Ormes, V. K. Balasubrahmanyan.
Proc. 11th international conference on cosmic rays, Vol. 4, (see 012.028), 397 - 402 (1970).

143.080 Satellite borne semiconductor telescope for identification of relativistic heavy primary cosmic rays.
S. Nakagawa, M. Tsukuda, M. Yoshimori, H. Murakami, K. Nagata, A. Nakamoto, A. Sasaki, T. Doke.
Proc. 11th international conference on cosmic rays, Vol. 4, (see 012.028), 449 - 455 (1970).

143.081 The method of "global survey" for investigating cosmic ray modulation.
A. M. Altukhov, G. F. Krymsky, A. I. Kuzmin.
Proc. 11th international conference on cosmic rays, Vol. 4, (see 012.028), 457 - 460 (1970).

143.082 On a method of finding Dirac's magnetic monopoles.
V. L. Dadykin.
Proc. 11th international conference on cosmic rays, Vol. 4, (see 012.028), 461 - 463 (1970).

143.083 Energy losses which are not measured by ionization spectrometers. W. V. Jones.

Proc. 11th international conference on cosmic rays, Vol. 4, (see 012.028), 505 - 512 (1970).

143.084 A magnet spectrograph for primary cosmic ray studies. R. Cowsik, S. V. Damle, Y. Pal, T. N. Rengarajan, S. N. Tandon, R. P. Verma, P. A. Vidwans.
Proc. 11th international conference on cosmic rays, Vol. 4, (see 012.028), 535 - 540 (1970).

143.085 Sea-level search for cosmic magnetic monopoles. R. L. Fleischer, H. R. Hart, Jr., G. E. Nichols, P. B. Price.
Phys. Rev. D, Particles and Fields, Third Ser., Vol. 4, 24 - 27 (1971).

On the hypothesis that the highest-energy cosmic rays consist at least in part of magnetic monopoles, heavily ionizing particles have been sought at ground level using an 18-m^2 detector array exposed for 630 days.

143.086 Cosmic-ray study of properties of nuclear interactions in the 10–300-GeV energy range.
E. R. Goza, R. W. Huggett, W. V. Jones, E. G. Stafford.
Phys. Rev. D, Particles and Fields, Third Ser., Vol. 4, 30 - 36 (1971).

143.087 Evidence for a primary cosmic-ray particle with energy 4 × 10^{21} eV.
K. Suga, H. Sakuyama, S. Kawaguchi, T. Hara.
Phys. Rev. Letters, Vol. 27, 1604 - 1607 (1971).

We describe the analysis of an extremely energetic air shower produced by a primary cosmic-ray particle of energy 4 × 10^{21} eV. The arrival direction of this cosmic ray is right ascension 20 h 14.5 min and declination 24°. The directions of 3C409 (radio source) and AP2015 + 28 (pulsar) are inside the uncertainty of the arrival direction.

143.088 Study of the charge spectrum of extremely heavy cosmic rays using combined plastic detectors and nuclear emulsions. P. B. Price, P. H. Fowler, J. M. Kidd, E. J. Kobetich, R. L. Fleischer, G. E. Nichols.
Phys. Rev. D, Particles and Fields, Third Ser., Vol. 3, 815 - 823 (1971).

Tracks of 99 nuclei with $Z > 36$ have been analyzed in a stack of Lexan polycarbonate, cellulose triacetate, and nuclear emulsion launched from Palestine, Texas, with an area-time factor of 900 m^2 h. One nucleus with charge estimated to be $Z \approx 96$ ($\beta \approx 0.95$) was identified in all three detectors. Our results favor supernova explosions within the Galaxy as the major source of cosmic rays and suggest that their average lifetime is less than 10^7 yr.

143.089 Transition radiation from magnetic monopoles. J. Dooher.
Phys. Rev. D, Particles and Fields, Third Ser., Vol. 3, 2652 - 2660 (1971).

143.090 Radial gradient of galactic protons in the inner solar system. R. C. Englade.
Journ. Geophys. Res., Vol. 76, 8394 - 8400 (1971).

Numerical solutions of a realistic particle propagation model for the inner solar system are used to investigate the radial gradient of galactic protons with 50 Mev $< E <$ 2000 Mev. The model assumes anisotropic, energy-dependent diffusion in a stochastic interplanetary magnetic field whose average configuration is the Archimedian spiral pattern. The results are compared with the Mariner 4 measurements.

143.091 Cosmic-ray decreases and the occurrence of solar flares.
J. R. Ballif, D. E. Jones, E. N. Skousen, D. T. Smith.
Journ. Geophys. Res., Vol. 76, 8401 - 8408 (1971).

We report the results of a more extensive study of the relationship between the occurrence of cosmic-ray decreases and solar flares. In particular, we consider (a) whether it is possible to relate all large cosmic-ray decreases to flare-producing regions on the sun and (b) whether there is any compelling evidence relating the time of a large flare and the occurrence of a decrease to substantiate the long-standing hypothesis that at least some large cosmic-ray decreases are due to the effects of a single energetic flare.

143.092 The spectrum of the cosmic ray solar diurnal modulation.
R. M. Jacklyn, S. P. Duggal, M. A. Pomerantz.
Proc. 11th international conference on cosmic rays, Vol. 2, (see 012.030), 47 - 54 (1970).

143.093 Variations of the diurnal anisotropy with periods of one and two solar cycles.
S. P. Duggal, S. E. Forbush, M. A. Pomerantz.
Proc. 11th international conference on cosmic rays, Vol. 2, (see 012.030), 55 - 59 (1970).

143.094 The first and second harmonics of the daily variation recorded by NM−64 neutron monitors during 1965 - 68.
A. C. Willets, W. K. Griffiths, C. J. Hatton, P. L. Marsden.
Proc. 11th international conference on cosmic rays, Vol. 2, (see 012.030), 61 - 67 (1970).

143.095 Measurements of the cosmic ray solar diurnal variation.
D. J. Sumner, D. M. Thomson, A. Hashim, T. Thambyahpillai.
Proc. 11th international conference on cosmic rays, Vol. 2, (see 012.030), 69 - 75 (1970).

143.096 Semi-diurnal anisotropy of cosmic radiation in the energy range 1–200 GeV.
U. R. Rao, S. P. Agrawal.
Proc. 11th international conference on cosmic rays, Vol. 2, (see 012.030), 77 - 82 (1970).

143.097 Semi-diurnal anisotropy of cosmic radiation.
Z. Fujii, K. Fujimoto, H. Ueno, I. Kondo, K. Nagashima.
Proc. 11th international conference on cosmic rays, Vol. 2, (see 012.030), 83 - 88 (1970).

143.098 The effective angles of acceptance and the energy spectrum of the diurnal variation.
P. Chaloupka, J. Dubinský, S. Fischer, T. Kowalski.
Proc. 11th international conference on cosmic rays, Vol. 2, (see 012.030), 89 - 94 (1970).

143.099 Statistical reliability tests of cosmic ray intensity data. E. Dyring, B. Sporre.
Proc. 11th international conference on cosmic rays, Vol. 2, (see 012.030), 95 - 98 (1970).

143.100 The latitude effect on the solar diurnal variations of cosmic ray intensities. M. Kitamura.
Proc. 11th international conference on cosmic rays, Vol. 2, (see 012.030), 99 - 104 (1970).

143.101 Approximations in the theory of solar-cycle modulation. L. A. Fisk, L. J. Gleeson, W. I. Axford.
Proc. 11th international conference on cosmic rays, Vol. 2, (see 012.030), 105 - 110 (1970).

143.102 The solar diurnal variation of cosmic rays underground since 1958. V. H. Regener, D. B. Swinson, J. H. Ericksen, H. S. Ahluwalia.

Proc. 11th international conference on cosmic rays, Vol. 2, (see 012.030), 133 - 137 (1970).

143.103   Solar diurnal variation of cosmic-ray intensity underground during solar activity cycle–20.
H. S. Ahluwalia, J. H. Ericksen.
Proc. 11th international conference on cosmic rays, Vol. 2, (see 012.030), 139 - 146 (1970).

143.104   On the diurnal variations of the cosmic ray intensity observed 70 m w.e. underground.
E. Antonucci, G. C. Castagnoli, M. A. Dodero.
Proc. 11th international conference on cosmic rays, Vol. 2, (see 012.030), 157 - 161 (1970).

143.105   Analysis of annual and semi-annual periodicities in neutron monitor data.
S. M. Schneider, S. A. Korff.
Proc. 11th international conference on cosmic rays, Vol. 2, (see 012.030), 163 - 168 (1970).

143.106   The yearly variation of the cosmic ray intensity from 1963 - 1967, its relation to the observed
variation of solar activity with heliolatitude and time.
M. C. Barker, C. J. Hatton.
Proc. 11th international conference on cosmic rays, Vol. 2, (see 012.030), 177 - 181 (1970).

143.107   Underground response functions and the upper limiting rigidity to solar modulation.
D. S. Peacock.
Proc. 11th international conference on cosmic rays, Vol. 2, (see 012.030), 189 - 194 (1970).

143.108   Characteristics of cosmic ray fluctuations in the frequency range of 10^{-6} to 4.15×10^{-3} cycles per
second.   M. S. Dhanju, V. Sarabhai.
Proc. 11th international conference on cosmic rays, Vol. 2, (see 012.030), 237 - 240 (1970).

143.109   A search for rapid periodic variations in the galactic cosmic ray intensity.   S. Ruthberg, E. Dyring,
S. Lindgren, B. Sporre, B. Östman, P. Tanskanen.
Proc. 11th international conference on cosmic rays, Vol. 2, (see 012.030), 241 - 245 (1970).

143.110   Meteor effects of cosmic rays.   S. A. Belsky.
Proc. 11th international conference on cosmic rays, Vol. 2, (see 012.030), 253 - 256 (1970).

143.111   Microvariations of cosmic ray intensity.
N. P. Chirkov, V. I. Ipatjev.
Proc. 11th international conference on cosmic rays, Vol. 2, (see 012.030), 257 - 260 (1970).

143.112   The effects of solar modulation on the energy spectrum of heavy cosmic ray nuclei.
T. F. Cleghorn, P. S. Freier, C. J. Waddington.
Proc. 11th international conference on cosmic rays, Vol. 2, (see 012.030), 275 - 278 (1970).

143.113   Synoptic analysis of the cosmic ray neutron monitor daily data for the study of modulation problems.
F. Bachelet, M. C. Fazzini, N. Iucci, G. Villoresi, B. Sporre.
Proc. 11th international conference on cosmic rays, Vol. 2, (see 012.030), 295 - 300 (1970).

143.114   On the role of the heliolatitudes of sunspots in the 11-year galactic cosmic ray modulation.
Yu. I. Stozhkov, T. N. Charakhchyan.
Proc. 11th international conference on cosmic rays, Vol. 2,

(see 012.030), 301 - 304 (1970).

143.115   Solar activity and dimension of modulation region.
E. V. Kolomeets, Yu. A. Shakhova, A. G. Zusmanovich.
Proc. 11th international conference on cosmic rays, Vol. 2, (see 012.030), 305 - 310 (1970).

143.116   Variation of the secondary cosmic ray spectra due to solar modulation.   O. C. Allkofer, W. D. Dau.
Proc. 11th international conference on cosmic rays, Vol. 2, (see 012.030), 311 - 317 (1970).

143.117   Cosmic ray modulation during Forbush decreases in 1968 - 1969.   J. A. Lockwood, P. Singh.
Proc. 11th international conference on cosmic rays, Vol. 2, (see 012.030), 319 - 325 (1970).

143.118   Spectral variations in short term (Forbush) decreases and in long term changes in cosmic ray intensity.
V. K. Balasubrahmanyan, D. Venkatesan.
Proc. 11th international conference on cosmic rays, Vol. 2, (see 012.030), 327 - 336 (1970).

143.119   A new approach to the study of cosmic ray modulation anisotropy.
R. L. Chasson, R. E. Gold, S. Lal, D. S. Peacock.
Proc. 11th international conference on cosmic rays, Vol. 2, (see 012.030), 337 - 344 (1970).

143.120   Anomalous streaming of cosmic rays.
S. Lindgren.
Proc. 11th international conference on cosmic rays, Vol. 2, (see 012.030), 345 - 349 (1970).

143.121   Transient north-south asymmetries of cosmic radiation.   S. P. Duggal, M. A. Pomerantz.
Proc. 11th international conference on cosmic rays, Vol. 2, (see 012.030), 351 - 358 (1970).

143.122   Cosmic ray intensity distribution over the territory of the USSR, the latitude effect and coupling coefficients of the cosmic ray muon component at sea level.
L. I. Dorman, V. A. Kovalenko, N. P. Milovidova, S. B. Chernov.
Proc. 11th international conference on cosmic rays, Vol. 2, (see 012.030), 359 - 363 (1970).

143.123   Investigation of anisotropy in the surrounding of a shock wave front.
A. M. Altukhov, V. P. Mamrukova, G. V. Shafer.
Proc. 11th international conference on cosmic rays, Vol. 2, (see 012.030), 365 - 367 (1970).

143.124   The geomagnetic cut-off at Ft. Churchill and the primary cosmic ray electron spectrum from 10 MeV
to 12 GeV in 1968.   D. Hovestadt, P. Meyer.
Proc. 11th international conference on cosmic rays, Vol. 2, (see 012.030), 525 - 531 (1970).

143.125   On the application of trajectory-derived cutoff rigidities to cosmic-ray intensity variations.
M. A. Shea, D. F. Smart.
Proc. 11th international conference on cosmic rays, Vol. 2, (see 012.030), 533 - 537 (1970).

143.126   Time dependence of the position of the cosmic ray equator.   B. Sporre, M. A. Pomerantz.
Proc. 11th international conference on cosmic rays, Vol. 2, (see 012.030), 545 - 551 (1970).

143.127 **The flux of fast neutrons in the atmosphere.**
1. The effect of solar modulation of galactic cosmic
rays. M. Merker, E. S. Light, R. B. Mendell, S. A. Korff.
Proc. 11th international conference on cosmic rays, Vol. 2,
(see 012.030), 739 - 744 (1970).

143.128 **Search for a systematic change in the phase of the**
second harmonic of the daily variation of cosmic
rays in the course of a year. H. J. Müller.
Zeitschr. Geophys., Vol. 37, 39 - 45 (1971). In German.
It is pointed out that an influence of a galactic magnetic
field on the cosmic ray particles should be seen as systematic
phase variation in the course of the year in the second harmonic of the daily variation of cosmic ray intensity. The data
of the super neutron monitor at Deep-River are harmonically
analysed for the time period May 1962 – December 1968. No
statistical significance is found for a systematic phase variation.

143.129 **Transport of high energy cosmic rays in the inter-**
stellar medium. R. E. Turner.
Thesis, Washington Univ., St. Louis, Mo. [Available from Univ.
Microfilms, Ann Arbor, Mich., U.S.A. Order No. 70–18935],
458 pp. (1970). – See Phys. Abstr., Vol. 74, No. 62747
(1971).

143.130 **A study of low energy cosmic rays at 1 A.U.**
J. H. Kinsey.
Thesis, Univ. Maryland, Catonsville. [Available from Univ. Microfilms, Ann Arbor, Mich., U.S.A. Order No. 70–22571],
160 pp. (1970).
The results from the two scintillator ΔE versus-ΔE telescopes on IMP-III and IMP-IV and the solid state telescope on
IMP-IV are analyzed and the resulting proton and alpha particle fluxes presented. The low energy flux time histories and
energy spectra are shown for the energy interval 18.7 to
81.7 MeV/nucleon from June 1965 to April 1967, and in the
interval from 5.2 to 81.7 MeV/nucleon from May 1967 to
August 1968.

143.131 **A balloon borne gas Čerenkov scintillator telescope**
for the measurement of the charge spectrum of cos-
mic ray nuclei at high energies. T. T. von Rosenvinge.
Thesis, Univ. Minnesota, Minneapolis. [Available from Univ.
Microfilms, Ann Arbor, Mich., U.S.A. Order No. 70–20243],
125 pp. (1970). – See Phys. Abstr., Vol. 74, No. 62749
(1971).

143.132 **Cosmic rays in the solar system space.** I. Kondo.
Kagaku, (*Japan*), Vol. 41, 315 - 322 (1971).
In Japanese.
Review of recent studies on the solar system space using
cosmic rays as a probe.

143.133 **Determination of the antiproton flux in primary**
cosmic rays.
E. A. Bogomolov, V. K. Karakad'ko, N. D. Lubyanaya, V. A.
Romanov, M. G. Totubalina, M. A. Yamshchikov.
Proc. Sixth Winter School on Space Physics, Part I. Apatity
1969, (see 012.033), p. 98 - 101 (1971).

143.134 **Variations of the electron cosmic-ray component**
with energies between 100 and 1500 MeV in the
upper atmosphere. A. M. Gal'per, B. I. Luchkov.
Proc. Sixth Winter School on Space Physics, Part I. Apatity
1969, (see 012.033), p. 152 - 157 (1971).

143.135 **Investigation of proton fluxes in the 1.50–50 MeV**
range with Zond-4 and Zond-5 automatic interplane-
tary stations.
M. M. Bredov, A. A. Kolchin, V. V. Lebedev, G. P. Skrebtsov.

Proc. Sixth Winter School on Space Physics, Part I. Apatity
1969, (see 012.033), p. 216 - 219 (1971).

143.136 **Radiation transfer in space physics.**
Yu. N. Gnedin, A. Z. Dolginov.
Proc. Sixth Winter School on Space Physics, Part I. Apatity
1969, (see 012.033), p. 228 - 239 (1971).

143.137 **The energy spectrum of cosmic rays in interstellar**
space. A. N. Charakhch'yan, T. N. Charakhch'yan.
Proc. Sixth Winter School on Space Physics, Part II. Apatity
1969, (see 012.034), p. 38 - 43 (1971).

143.138 **The diffusion coefficients of cosmic rays.**
A. N. Charakhch'yan, T. N. Charakhch'yan.
Proc. Sixth Winter School on Space Physics, Part II. Apatity
1969, (see 012.034), p. 44 - 52 (1971).

143.139 **The diffusion of relativistic electrons in the Galaxy**
and their energy spectrum.
V. A. Dogel', S. I. Syrovatskii.
Proc. Sixth Winter School on Space Physics, Part II. Apatity
1969, (see 012.034), p. 53 - 55 (1971).

143.140 **Ultrahigh-energy cosmic rays.** G. B. Khristiansen.
Proc. Sixth Winter School on Space Physics, Part II.
Apatity 1969, (see 012.034), p. 56 - 64 (1971).

143.141 **27-day cosmic-radiation variation from ground and**
stratospheric data (1966–1967).
L. Kh. Shatashvili.
Proc. Sixth Winter School on Space Physics, Part II. Apatity
1969, (see 012.034), p. 65 - 71 (1971).

143.142 **Measurement of the cosmic-ray anisotropy during a**
solar cycle. K. K. Fedchenko.
Proc. Sixth Winter School on Space Physics, Part II. Apatity
1969, (see 012.034), p. 72 - 76 (1971).

143.143 **The 11-year intensity modulation of galactic cosmic**
rays and the heliographic distribution of sunspots.
Yu. I. Stozhkov.
Proc. Sixth Winter School on Space Physics, Part II. Apatity
1969, (see 012.034), p. 77 - 81 (1971).

143.144 **The modulation of galactic cosmic rays.**
A. G. Zusmanovich, E. V. Kolomeets, Yu. A. Shakhova.
Proc. Sixth Winter School on Space Physics, Part II. Apatity
1969, (see 012.034), p. 82 - 86 (1971).

143.145 **The 27-day variations of cosmic rays and solar ac-**
tivity during 1966–1967. V. P. Okhlopkov.
Proc. Sixth Winter School on Space Physics, Part II. Apatity
1969, (see 012.034), p. 87 (1971). – Abstract.

143.146 **Galactic cosmic-ray fluctuations of interplanetary**
origin. V. I. Shishov.
Proc. Sixth Winter School on Space Physics, Part II. Apatity
1969, (see 012.034), p. 88 (1971). – Abstract.

143.147 **The barometric coefficient of the neutron cosmic-**
ray component with allowance for proton and muon
absorption. M. V. Alaniya, O. G. Rogava.
Proc. Sixth Winter School on Space Physics, Part II. Apatity
1969, (see 012.034), p. 89 (1971). – Abstract.

143.148 **Estimate of the partial barometric coefficient of**
integral multiplicity for the neutron component of
secondary cosmic rays. L. I. Dorman, A. V. Sergeev.
Proc. Sixth Winter School on Space Physics, Part II. Apatity

1969, (see 012.034), p. 90 - 91 (1971).

143.149 **The meteoric variation of cosmic rays near the new solar-activity peak in 1967−1968.** S. A. Bel'skii. Proc. Sixth Winter School on Space Physics, Part II. Apatity 1969, (see 012.034), p. 93 (1971). − Abstract.

Attempt to determine the elastic proton-nucleon cross section at 83 GeV. See Abstr. 022.130.

The connection of solar activity indices with cosmic ray intensity. See Abstr. 072.020.

On the propagation, acceleration and deceleration of cosmic rays in different regions of the interplanetary space. See Abstr. 078.028.

Report on quasi-stationary modulation papers. See Abstr. 078.037.

Modulation dynamics. See Abstr. 078.038.

Solar flare increases in cosmic ray intensity on November 18, 1968; February 25, 1969, and March 30, 1969. See Abstr. 078.053.

Nucleosynthesis of ^{26}Al in the early solar system and in cosmic rays. See Abstr. 107.010.

High-speed interstellar gas dynamics: Shocks moderated by cosmic rays. See Abstr. 131.101.

Cosmic rays and interstellar matter. See Abstr. 131.113.

On cosmic ray production by pulsars. See Abstr. 141.089.

Time variations in high energy cosmic rays. See Abstr. 141.131.

Pulsars and the origin of cosmic rays. See Abstr. 141.221.

Remarks on the role of pulsars in cosmic ray production. See Abstr. 141.222.

Pulsars and the origin of high energy particles. See Abstr. 141.228.

Extensive air shower studies of cosmic gamma rays and cosmic ray composition. See Abstr. 142.075.

Consideration of the observed isotropic γ-ray background. See Abstr. 142.084.

X-ray source in the galactic center and origin of cosmic rays. See Abstr. 142.087.

Stellar Systems

151 Kinematics and Dynamics of Stellar Systems

151.001   A partially thermalized model of a stellar system.
E. Infeld, A. Skorupski.
Astron. Astrophys., Vol. 14, 12 - 14 (1971).
A spherically symmetric model for a stellar system that includes effects of the first stage of thermalization is presented.

151.002   Jeans-type criterion in an isothermally contracting turbulent gas sphere.   T. Sasao.
Publ. Astron. Soc. Japan, Vol. 23, 433 - 436 (1971).
A Jeans-type criterion in an isothermally contracting turbulent medium is obtained according to Chandrasekhar's (1951) procedure.

151.003   On some general properties of gravitational systems.
L. E. Gurevich.
Astron. Zhurn. Akad. Nauk SSSR, Vol. 48, 716 - 721 (1971). In Russian. English translation in Soviet Astron. AJ, Vol. 15, No. 4.
In the evolution process of gravitational systems the tendency of approaching statistical equilibrium must reveal, although such an equilibrium cannot be reached. It is shown that such a tendency must show itself in two manifestations.

151.004   Numerical investigations in stellar statistical mechanics.   F. C. House, K. A. Innanen.
Journ. Roy. Astron. Soc. Canada, Vol. 65, 175 (1971).
Abstr. RAS Canada.

151.005   On the problem of phase mixing in stellar systems.
L. P. Osipkov.
Astron. Tsirk., No. 623, p. 1 - 2 (1971).   In Russian.

151.006   An investigation of the relative orbits of cluster stars. I.   R. M. Dzigvashvili.
Byull. Abastumansk. Astrofiz. Obs., No. 40, p. 101 - 122 (1971). In Russian.
The relative orbits in five galactic clusters NGC 5460, 5617, 6067, 6405, 6494 have been studied. The minimum and maximum distances from the cluster center were estimated and some kinematic and dynamic parameters determined.

151.007   An investigation of plane galactic orbits of stars for different expressions of the gravitational potential.
G. A. Malasidze.
Byull. Abastumansk. Astrofiz. Obs., No. 40, p. 123 - 170 (1971). In Russian.
The Galaxy is considered to be a stationary stellar system with axial symmetry having such a form of the gravitational potential in the plane of symmetry, for which the solution of the plane problem in elliptic integrals is obtainable. The task of the paper is to investigate plane galactic orbits for individual stars and to search for orbital element dependence on the structural parameters of the potential. Stellar orbits were studied in the gravitational field conforming to four different values of the structural parameters, and some conclusions are drawn concerning the dependence of general orbital elements upon the rate of approximation to the real galactic potential.

151.008   Tidal actions on small star clusters.

P. Bouvier.
Astron. Astrophys., Vol. 14, 341 - 350 (1971).
The major aim of this paper is to combine the tidal action of the galactic field with that of passing interstellar clouds on a small star containing, say, from one to three dozens of stars.

151.009   The exact determination of the oscillation spectrum of stellar systems as represented by the model of a plane homogeneous layer.   V. A. Antonov.
Trudy Astron. Obs., *Leningrad,* Vol. 28 (= Uchenye Zapiski Leningr. Un-ta, No. 359 = Seriya Matem. Nauk, vyp. (No.) 47), p. 64 - 85 (1971).   In Russian.
A one-dimensional finite stationary model with uniform stellar density and anomalous gradient of the phase density is considered. The co-operative phenomena do not lead to unstable modes.

151.010   Construction of a model of a rotating stellar system by a numerical experiment.
T. A. Agekian, S. P. Yakimov.
Trudy Astron. Obs., *Leningrad,* Vol. 28 (= Uchenye Zapiski Leningr. Un-ta, No. 359 = Seriya Matem. Nauk, vyp. (No.) 47), p. 85 - 96 (1971).   In Russian.
A model of a quasistationary rotating stellar system is constructed by numerical integration of the equations of motion of the problem of five bodies. The distributions of density, centroid velocity, dispersions of peculiar velocities are obtained. The ratio of rotational to total kinetic energy is 0.08.

151.011   On the origin of galactic rotation.   E. R. Harrison.
Monthly Notices Roy. Astron. Soc., Vol. 154, 167 - 186 (1971).
Current theories on the origin of galactic rotation are discussed and it is shown that they encounter difficulties. A more general cosmogony is then discussed, which takes a less extreme view of the initial state, and starts from initial conditions consisting of both density and spin inhomogeneities.

151.012   Energy and angular momentum exchanges between a density wave and stars at Lindblad resonances in a disk-like stellar system.   S. Kato.
Publ. Astron. Soc. Japan, Vol. 23, 467 - 483 (1971).
The energy and angular momentum exchanges between a wave and stars due to resonant interaction at Lindblad resonances are examined for mildly wound density waves, the form of the perturbed gravitational potential being assumed. The possibility of the maintenance of spiral waves in disk-like galaxies by this process of energy and angular momentum exchanges with stars is discussed.

151.013   Numerical experiments with a disk of stars.
F. Hohl.
Astrophys. Journ., Vol. 168, 343 - 359 (1971).
The evolution of an initially balanced rotating disk of stars with an initial velocity dispersion given by Toomre's local criterion is investigated by means of a computer model for isolated disks of stars. The final mass distribution for the disk gives a high-density central core and a disk population of stars that is closely approximated by an exponential variation.

151.014 Optical appearance of a collisionless gas of stars surrounding a black hole. U. H. Gerlach.
Astrophys. Journ., Vol. 168, 481 - 493 (1971).

We consider a thin-shelled ensemble of noncolliding luminous particles, such as stars, surrounding a collapsed configuration, a black hole, and determine (a) the spectral flux density and (b) the brightness across the observed disk as seen by a distant observer.

151.015 On the spiral structure of barred galaxies. O. V. Chumak.
Astrofizika, Vol. 7, 197 - 201 (1971). In Russian. – English translation in Astrophysics, Vol. 7, No. 2.

A possible formation mechanism of the spiral structure in barred galaxies is discussed. A bar is regarded as a mechanical source which causes two density waves of spiral form in the near-by medium.

151.016 The stability of gravitating systems of point masses. II. The cylinder with mono-energy streams and the sphere with circular orbits. G. S. Bisnovaty-Kogan.
Astrofizika, Vol. 7, 223 - 236 (1971). In Russian. – English translation in Astrophysics, Vol. 7, No. 2.

The consideration of stability of a uniform, rotating, infinitely long, cylinder of finite radius is proceeded. Distribution functions are considered which contain mono-energy streams along the z-axis. It is shown that in this case the Jeans instability is always present. The stability of the sphere with circular orbits is proved.

151.017 Dynamical evolution of dense spherical star systems. L. Spitzer, Jr.
Nuclei of galaxies. Conference 1970, (see 012.002), p. 443 - 471 (1971).

151.018 Shock waves in barred spiral galaxies. W. W. Roberts, Jr.
Bull. American Astron. Soc., Vol. 3, 369 - 370 (1971). Abstr. AAS.

151.019 The stability of an inhomogeneous spherically symmetric system of rotating masses. A. M. Fridman.
Astron. Zhurn. Akad. Nauk SSSR, Vol. 48, 910 - 921 (1971). In Russian. English translation in Soviet Astron. AJ, Vol. 15, No. 5.

A model, proposed by A. Einstein, of a globular star cluster as a spherically symmetric system of masses rotating on circular trajectories is considered. In Newtonian approximation the stability of such a system relative to arbitrary perturbations is shown.

151.020 The velocity distribution function for stars in star clusters. V. S. Kaliberda.
Astron. Zhurn. Akad. Nauk SSSR, Vol. 48, 969 - 975 (1971). In Russian. English translation in Soviet Astron. AJ, Vol. 15, No. 5.

The distribution function of stellar velocities in non-rotating systems has been found. Two cases are considered: a) the stellar system consists of stars with masses m and $m/2$; b) it consists of stars with masses m and $2m$. The effect of multiplicity of stellar encounters has been taken into account.

151.021 The persistence of the wave spiral pattern in galaxies. L. S. Marochnik, A. A. Suchkov.
Astrophys. Letters, Vol. 9, 37 - 38 (1971).

It is shown that the Landau instability of spiral density waves in a galaxy is an absolute one. Therefore the obliteration of the wave spiral pattern would not take place in spite of the dispersion of the wavepacket. It is found that the distortion of the wave spiral pattern by the differential rotation

in our Galaxy takes place in 20 revolutions of the Galaxy.

151.022 The third integral of motion in stellar systems and in an axisymmetric potential field. T. A. Agekian.
Dokl. Akad. Nauk SSSR, Ser. Mat. Fiz., Vol. 200, 1310 - 1312 (1971). In Russian.

151.023 Density wave theory of spiral structure. C. C. Lin, F. H.-S. Shu.
Astrophysics and general relativity. Brandeis Univ. Summer Inst. 1968, Vol. 2, (see 003.011), 235 - 329 (1971).

151.024 On the quasilinear approximation in the dynamics of rotating stellar systems. S. G. Pomagaev.
Byull. Inst. Astrofiz., *Dushanbe,* No. 56, p. 23 - 36 (1970). In Russian.

A rotating stellar system of stars and interstellar gas in the presence of a magnetic field is considered. The dispersion relation of the small perturbations and the quasilinear equations are found. It is shown that instable fluctuations of the gravitational field lead to the diffusion of stars in the velocity space.

151.025 On the wave nature of the spiral structure of galaxies. I. The linear theory. (Part 1). L. S. Marochnik, A. A. Suchkov.
Byull. Inst. Astrofiz., *Dushanbe,* No. 58, p. 3 - 11 (1971). In Russian.

A selfgravitating thin disc consisting of differentially rotating interstellar gas and population I stars and of non-rotating population II stars is considered. In the linear approximation the dispersion relation of spiral waves is found.

151.026 On the wave nature of the spiral structure of galaxies. II. The linear theory. (Part 2). L. S. Marochnik, A. A. Suchkov.
Byull. Inst. Astrofiz., *Dushanbe,* No. 58, p. 12 - 22 (1971). In Russian.

Due to the Landau effect a selfgravitating thin disc consisting of differentially rotating population I and non-rotating population II stars is unstable against spiral waves of small amplitude. In addition to Landau instability, Jeans instability is possible. However, it cannot, apparently, lead to formation of spiral arms.

151.027 On the wave nature of the spiral structure of galaxies. III. The nonlinear (quasi-linear) theory. L. S. Marochnik, A. A. Suchkov.
Byull. Inst. Astrofiz., *Dushanbe,* No. 58, p. 23 - 34 (1971). In Russian.

The effect of Landau–unstable waves (in a thin selfgravitating system of differentially rotating population I and non-rotating population II stars) upon the "background" and the reverse of slowly changing "background" on the density waves is considered.

151.028 On the wave nature of the spiral structure of galaxies. IV. Physical interpretation. Astrophysical conclusions. L. S. Marochnik, A. A. Suchkov.
Byull. Inst. Astrofiz., *Dushanbe,* No. 58, p. 35 - 44 (1971). In Russian.

Obvious physical considerations illustrating the mathematical theory developed in parts I–III are given. Some numerical estimates for the solar neighbourhood are made.

151.029 Determination of the potential energy of a gravitational system with spheroidal symmetry. V. L. Afanasjev.
Astrofizika, Vol. 7, 481 - 487 (1971). In Russian. – English translation in Astrophysics, Vol. 7, No. 3.

Ambartsumian "strips" method has been generalized on

systems with spheroidal symmetry. Relationships obtained were applied to the determination of mass-luminosity ratio for two elliptical galaxies NGC 4406 ($f/f_\odot$ = 46), NGC 4486 ($f/f_\odot$ = 48). The calculated results accord satisfactorily with the values of other authors.

151.030 Collective instabilities and waves for inhomogeneous stellar systems. II. The normal modes problem of the self-consistent plane-parallel slab. J. W -K. Mark. Astrophys. Journ., Vol. 169, 455 - 475 (1971).

Studies have been made of the normal modes of instabilities and waves for models of inhomogeneous, nonrotating stellar systems whose basic equilibrium has one-dimensional planar symmetry (slabs) while the perturbations are not restricted to this symmetry. Comparisons are also made with gaseous slabs, the energy principle, and the Hartree-Fock operator principle.

151.031 Polarization clouds and dynamical friction. A. J. Kalnajs. Astrophys. Space Sci., Vol. 13, 279 - 283 (1971).

We argue that dynamical friction can be viewed as the drag exerted on a 'test' star by the wake it induces in the field stars. We compute the wakes for the uniform infinite medium and a flat rotating sheet. In the first case we obtain a result which differs by a factor of two from the classical result. In the second case the drag vanishes.

151.032 Monte Carlo models of star clusters. M. Hénon. Astrophys. Space Sci., Vol. 13, 284 - 299 (1971).

The dynamical evolution of spherical star clusters under the effect of internal encounters is followed numerically using a Monte Carlo procedure. Some provisional results are presented. Once more it is found that N-body systems develop a very high central density peak. The velocity distribution becomes isotropic in the central parts, radially elongated in the halo. Models started with widely different initial conditions tend to become similar after a few relaxation times. The presence of a tidal field, or a distribution of masses, accelerate the evolution of the system. A companion paper gives a detailed technical description of the method.

151.033 On the lifetimes of galactic clusters. R. Wielen. Astrophys. Space Sci., Vol. 13, 300 - 308 (1971).

From the observed age distribution of galactic clusters within 1 kpc we deduce that the typical total lifetime of a galactic cluster is about 2×10^8 yr. The individual lifetimes vary between 10^8 and 10^{10} yr. The observed lifetimes are compared with the evaporation times which are found from numerical experiments with star cluster models. These models contain up to 250 stars with a realistic mass spectrum. The effect of the galactic tidal field is taken into account and enhances the rate of escape significantly.

151.034 Numerical experiments on the escape from non-isolated clusters and the formation of multiple stars. A. Hayli. Astrophys. Space Sci., Vol. 13, 309 - 323 (1971).

Isolated and non-isolated clusters with a mass distribution have been studied by numerical techniques. The rates of escape of stars and of kinetic energy are compared with Hénon's theoretical expressions. Multiple encounters play a very important role in the escape phenomenon, at least for clusters with a small number of stars. For non-isolated clusters, the tidal field of the Galaxy is responsible for one half of the rate of escape of the stars. Very stable subsystems are formed which are not destroyed under the influence of the galactic tide. Separation between stars can be as low as 1000 AU.

151.035 Binary evolution in stellar systems. S. J. Aarseth. Astrophys. Space Sci., Vol. 13, 324 - 334 (1971).

Three new star cluster models containing 250 members and one case with 500 particles have been studied by numerical methods of direct integration. The evolution is dominated by one central binary in all systems with a realistic mass spectrum and more than 50 % of the total energy is absorbed by one heavy pair after only 6–18 mean crossing times. General conditions for binary formation and disruption are discussed and a qualitative explanation is given for the energy sink behavior.

151.036 On the reproducibility of run-away stars formed in collapsing clusters. C. Allen, A. Poveda. Astrophys. Space Sci., Vol. 13, 350 - 359 (1971).

To test the stability of the trajectories of run-away stars we present the results of a comparative study of 26 star clusters involving very strong encounters. Each one of these clusters was computed with different time steps, different techniques of integration and on different machines.

151.037 A numerical experiment on relaxation times in stellar dynamics. M. Lecar, C. Cruz-González. Astrophys. Space Sci., Vol. 13, 360 - 364 (1971).

The deflection of the velocity vector of a massless test star in the field of 100 stars was determined by numerical integration. The deflection due to each field star independently (with the other field stars removed) was also determined. The square of the deflection caused by the combined action of the field stars agreed quantitatively with the sum of the squares of the individual deflections and also with the theoretical estimate of Williamson and Chandrasekhar.

151.038 Recent developments of integrating the gravitational problem of n-bodies. V. Szebehely, D. G. Bettis. Astrophys. Space Sci., Vol. 13, 365 - 376 (1971).

This paper discusses the formulation and the numerical integration of large systems of differential equations occurring in the gravitational problem of n-bodies. Different forms of the pertinent differential equations of motion are presented, and various regularizing and smoothing transformations are compared. A method is described in which some of the phase variables are treated in the regularized system and others in the ordinary system. This mixed method of numerical regularization offers some advantages. Numerical results are described with 5, 25 and 500 bodies participating. These examples compare the various integration techniques, several regularization methods and different logics in treating binaries.

151.039 Collisionless stellar dynamics. G. Contopoulos. Astrophys. Space Sci., Vol. 13, 377 - 386 (1971). – Invited paper presented at the IAU Colloquium No. 10, in Cambridge, England.

151.040 On the stability of an encounterless self-gravitating constant density system. S. Goldstein. Astrophys. Space Sci., Vol. 13, 387 - 396 (1971).

The stability of a constant density, self-gravitating system is investigated. The system considered is one-dimensional, collisionless and described by the sheet model. The equilibrium distribution function $F(E)$, E being the energy, is such that the system has constant density in real space over a finite region. An analytical treatment as well as computer experiment show stability for symmetric disturbances.

151.041 Numerical experiments on Lynden-Bell's statistics. M. Lecar, L. Cohen. Astrophys. Space Sci., Vol. 13, 397 - 410 (1971).

We performed computer experiments on 13 different initial configurations of one-dimensional self-gravitating systems. The three most and the three least violently relaxed systems were compared with the predictions of Lynden-Bell's statistical mechanics. The agreement between the experimental results and the theoretical predictions became worse as the relaxation became more violent.

151.042 **A phase-space boundary integration of the Vlasov equation for collisionless one-dimensional stellar systems.** S. Cuperman, A. Harten, M. Lecar.
Astrophys. Space Sci., Vol. 13, 411 - 424 (1971).

The evolution of one dimensional (stratified) self-gravitating systems of stars with constant phase-space density ('water bag' model) is investigated by following the motion of the boundary curves defining the systems. The results are compared with those obtained by sheet-model computer experiments and good agreement is found. New aspects of the evolution, revealed by the present method, are discussed.

151.043 **The collective relaxation of two-phase-space-density collisionless one-dimensional selfgravitating systems.**
S. Cuperman, A. Harten, M. Lecar.
Astrophys. Space Sci., Vol. 13, 425 - 445 (1971).

The evolution of one-dimensional two-phase-space-density selfgravitating systems of stars is investigated by following the motion of the boundary curves of the systems in phase space. A qualitative agreement with Lynden-Bell's theory predicting, for the most probably state, velocity dispersions inversely proportional to the phase space density of the component at the star formation, is found.

151.044 **Stability properties for encounterless self-gravitational stellar gas and plasma.**
M. R. Feix, J. P. Doremus, G. Baumann.
Astrophys. Space Sci., Vol. 13, 478 - 495 (1971).

The stability properties of the collisionless plasma and encounterless stellar gas described by the Vlasov equations are studied. The introduction of the 'multiple Water Bag' model allows, for one-dimensional plane geometry, a treatment of the general case. We will consider in this paper the following points. Introduction of the Water Bag Calculation of δW and connection to the continuum case. Stability property of a decreasing function of the energy in the plasma and gravitational case. For plane geometry we show that the limiting case must be studied through the multiple Water-Bag concept. Introduction of the marginal stability mode. Computation of the stability property for the plasma case through a virtual modification of the property of the medium.

151.045 **On necessary and sufficient global stability criteria for axisymmetric perturbations of the stellar dynamic disk.** J. W-K. Mark.
Studies Applied Math., Vol. 50, 1 - 12 (1971).

Necessary and sufficient global stability criteria are obtained which govern the stellar-dynamic stability to axisymmetric perturbations for disk models of galaxies. They are more complicated and difficult to apply than Toomre's criterion because they are global and allow general orbits of disk equilibria. If local and epicyclic assumptions are added, they then reduce to Toomre's criterion. But unlike the latter, in these new criteria, stability is determined by some average velocity dispersion rather than its local values, thus local variations in Toomre's Q parameter appears possible.

151.046 **Galactic winds.** W. G. Mathews, J. C. Baker.
Astrophys. Journ., Vol. 170, 241 - 259 = Contr. Lick Obs., No. 349 (1971).

Gas ejected from stars in elliptical galaxies is heated by supernova explosions and produces outward-flowing galactic winds. Except possibly for the dust component, steady-state galactic winds are impossible to observe. However, some galactic winds have thermally unsteady cores which can be observed in optical emission lines. If the gas which is thermally unsteady remains ionized as it goes into free fall at the galactic center, objects more massive than stars tend to form. It is likely that nonthermal radio emission and optical line emission can occur only in those ellipticals with thermally unsteady galactic winds.

151.047 **Relaxation time in clusters with pair correlations.** E. M. Butterworth, R. H. Miller.
Astrophys. Journ., Vol. 170, 275 - 276 (1971).

The results are described of numerical experiments on relaxation times in stellar clusters whose members have a deliberately introduced pair correlation. Pairs behaved on the average as single stars, even though members of correlated pairs were separated at least as far as the mean distance between nearest neighbors as measured in clusters with no pair correlation.

151.048 **The evolution of galaxies. I. Formulation and mathematical behavior of the one-zone model.**
R. J. Talbot, Jr., W. D. Arnett.
Astrophys. Journ., Vol. 170, 409 - 422 (1971).

A procedure is discussed for calculating the evolution of a closed system of gas and stars. Numerical and analytical solutions are given for a simple set of prescriptions for the stellar birth rate and the evolutionary end state of stars. A fundamental result, which previous investigators have not stressed, is that the metal content Z of the gas in a galaxy need not be a monotonically increasing function of time even if the system is homogeneous in space.

151.049 **On equipartition in galactic nuclei and gravitating systems.** W. C. Saslaw, D. S. De Young.
Astrophys. Journ., Vol. 170, 423 - 429 (1971).

We show that under very general conditions, self-gravitating systems with particles of different mass cannot be in equipartition. We also derive a prescription for determining whether equipartition can exist for a given distribution.

151.050 **Collisional processes in stellar systems.**
I. H. Gilbert.
Astrophys. Space Sci., Vol. 14, 3 - 10 (1971).

The theory of collisional relaxation in stellar systems is discussed in terms of an expansion in powers of $1/N$, the inverse of the total number of stars. The results are expressed in terms of the concept of gravitational polarization.

151.051 **A certain discontinuous Markov process in stellar dynamics.** W. Tscharnuter.
Astrophys. Space Sci., Vol. 14, 11 - 14 (1971).

The aim of these considerations is to show that Chandrasekhar's basic assumption (Chandrasekhar, 1942) is more general than the choice of a diffusion process.

151.052 **Relaxation times in strictly disk systems.**
G. B. Rybicki.
Astrophys. Space Sci., Vol. 14, 15 - 19 (1971).

It is shown that the time of relaxation by particle encounters of self-gravitating systems in the plane interacting by $1/r^2$ forces is of the same order of magnitude as the mean orbit time. Therefore such a system does not have a Vlasov limit for large numbers of particles, unless appeal is made to some non-zero thickness of the disk. The relevance of this result to numerical experiments on galactic structure is discussed.

151.053 **The hose-pipe instability in stellar systems.**
R. M. Kulsrud, J. W. K. Mark, A. Caruso.
Astrophys. Space Sci., Vol. 14, 52 - 55 (1971).

In this paper an instability in the non-rotating slab equilibrium is discussed.

151.054 **Exact statistical mechanics of a one-dimensional self-gravitating system.** G. B. Rybicki.

Astrophys. Space Sci., Vol. 14, 56 - 72 (1971).

The statistical mechanics of an isolated self-gravitating system consisting of N uniform mass sheets is considered using both canonical and microcanonical ensembles. The one-particle distribution function is found in closed form. The limit for large numbers of sheets with fixed total mass and energy is taken and is shown to yield the isothermal solution of the Vlasov equation. The order of magnitude of the approach to Vlasov theory is found to be $O(1/N)$. Numerical results for spatial density and velocity distributions are given.

151.055　Numerical experiments in collisionless systems.
R. H. Miller.
Astrophys. Space Sci., Vol. 14, 73 - 90 (1971).

Some difficulties with the gravitational n-body calculations are considered.

151.056　Dynamics of plane stellar systems.　F. Hohl.
Astrophys. Space Sci., Vol. 14, 91 - 109 (1971).

The evolution of initially balanced rotating disks of stars is investigated with a computer model for isolated disks of stars. An isolated, initially cold balanced disk is found to be violently unstable and after about two rotations most disks tend to assume a bar-shaped structure. It is found that the final mass distribution over most of the disk can be closely approximated by an exponential variation, irrespective of the initial mass distribution.

151.057　On the number of isolating integrals in systems with three degrees of freedom.　C. Froeschle.
Astrophys. Space Sci., Vol. 14, 110 - 117 (1971).

We have found that even for a very weak coupling a dynamical system with three degrees of freedom has in general either two or zero isolating integrals (besides the usual energy integral).

151.058　The Monte Carlo method.　M. Hénon.
Astrophys. Space Sci., Vol. 14, 151 - 167 (1971).

We give here a detailed technical description of a Monte Carlo scheme for the dynamical evolution of spherical stellar systems.

151.059　Integration methods where force is obtained from the smoothed gravitational field.　F. Hohl.
Astrophys. Space Sci., Vol. 14, 168 - 178 (1971).

The dynamics of collisionless stellar systems can be studied by representing the system by large numbers of representative stars. The numerical methods that are used to integrate the motion of the system in time are presented in some detail.

151.060　Theory of spiral structure.
C. C. Lin.
Highlights of Astronomy, Vol. 2, (see 003.026), 88 - 121 (1971). – Invited discourse.

151.061　Le problème des N corps unidimensionnel et le mouvement rectiligne des gaz.　F. Nahon.
Ann. Inst. Henri Poincaré, Section A, Vol. 14, 249 - 284 (1971).

Appelons 'système autogravitant' le système de deux fonctions: f, densité dans l'espace des phases, φ fonction de forces, vérifiant les deux équations de Liouville et de Poisson. A tout problème des N corps, particularisé par ses conditions initiales, correspond un système autogravitant qui en est la formulation statistique. D'autre part à tout système autogravitant correspond par la méthode des équations hydrodynamiques, le mouvement d'un gaz. Le but de cet article est de préciser les rapports entre ces trois problèmes dans le cas unidimensionnel.

151.062　Dynamical parameters of clusters of galaxies as a consequence of cosmological turbulence.

L. M. Ozernoy.
Astron. Zhurn. Akad. Nauk SSSR, Vol. 48, 1160 - 1173 (1971). In Russian. English translation in Soviet Astron. AJ, Vol. 15, No. 6.

In the framework of the idea about the vortex nature of pregalactic structure a quantitative model of formation of groups and clusters of galaxies is given. The main dynamical characteristics of clusters (mean density or radius, velocity dispersion) are expressed through the ratio of a cluster mass to the maximal mass of a galaxy with weak dependence on the parameter of cosmological deceleration. In spite of a number of simplifications of the quantitative scheme considered the calculated characteristics of groups and clusters of galaxies are in good agreement with observational data. Qualitative consideration of some questions concerning space correlation of velocities in metagalactic turbulence allows to explain the dependence between the mean density of a cluster of galaxies and its morphological type. Estimates are made of velocity dispersion for galaxies outside of clusters as well as the relative contribution of rotation and chaotic velocities to the kinetic energy of a cluster.

151.063　On the connection between morphological forms and main parameters of spiral galaxies.
V. L. Polyachenko, V. S. Synakh, A. M. Fridman.
Astron. Zhurn. Akad. Nauk SSSR, Vol. 48, 1174 - 1182 (1971). In Russian. English translation in Soviet Astron. AJ, Vol. 15, No. 6.

The heterogeneous system, consisting of two rotating disks, differing in angular velocities, densities and thermal velocity dispersions, is considered as a model of the spiral galaxy. It is shown that such a system is unstable. Moreover, there is a strong hydrodynamic instability, leading to the growth of spiral density disturbances. The formation conditions of different spiral galaxies are investigated. The classification of galaxies by their specific angular momenta and the connected hypothesis on the nature of barred galaxies are proposed.

151.064　A variation principle in the theory of equilibrium and stability of rotating bodies.
G. S. Bisnovaty-Kogan.
Astron. Zhurn. Akad. Nauk SSSR, Vol. 48, 1183 - 1189 (1971). In Russian. English translation in Soviet Astron. AJ, Vol. 15, No. 6.

The equation of equilibrium and the condition of stability relative to radial perturbations for a disk and a cylinder, taking into account rotation, magnetic field and thermal energy, are obtained with the help of a direct variation of energy, being an analogy of the potential energy of a conservative system in mechanics.

151.065　Dynamic evolution of rich galactic star clusters. I.
S. W. Prata.
Astron. Journ., Vol. 76, 1017 - 1028 (1971).

Results are presented of a theoretical study of the dynamic evolution of rich galactic star clusters. The model uses a distribution of stellar masses and takes into account tidal limitation, star loss, gas loss, and tidal shocks. The evolution of the mass function depends primarily on the mass function itself and only secondarily on the cluster structure. M67's original mass function was strongly deficient in low-mass stars. M67's continued existence is the result of a fortunate combination of initial conditions. Effects of a mass mixture on star loss are discussed. General comments on cluster evolution are made.

151.066　Dynamic evolution of rich galactic star clusters. II.
S. W. Prata.
Astron. Journ., Vol. 76, 1029 - 1040 (1971).

A method is presented for calculating star cluster dynamic evolution using a series of equilibrium models of the King type. Effects of star loss, gas loss, tidal shocks, tidal limitation, and of a mass function are included. Existing theories for the first three effects are carried another step further.

151.067   On the dynamical evolution of the Trapezium of
Orion.   G. N. Duboshin, A. I. Rybakov, E. P. Kalinina, P. N. Kholopov.
Soobshch. Gos. Astron. Inst. Shternberga, No. 175, 53 pp. (1971). In Russian.

A numerical solution of the equations of motion for the Trapezium stars observed in the centre of the Trapezium cluster of the Orion nebula was carried out. Accepting advisable suggestions on the distribution and number of faint members of the cluster we can conclude that the observable velocity dispersion for the Trapezium stars is probably insufficient to allow them to escape from the cluster where they have been formed. The Trapezium components should be considered as the most massive and brightest members of the stationary star cluster, their velocity dispersion being determined by the mass of this cluster.

151.068   The stability of relativistic star clusters.
J. R. Ipser.
General relativity and cosmology. Course 47, Italian Phys. Soc., 1969, (see 012.022), p. 356 - 358 (1971).

151.069   Markoffsche Prozesse und Stellardynamik. II.
W. Tscharnuter.
Sitzungsber. Österreich. Akad. Wiss., Math.-naturwiss. Kl., Abt. II, Vol. 179, 385 - 398 (1970) = Astron. Mitt. Wien, No. 8 (1971).

In this part of a series of papers dedicated to the connections between Markov processes and collisional stellar dynamics some numerical results concerning relaxation times, evaporation times and quasi-steady distributions of the star velocities are presented mainly with respect to galactic cluster models.

151.070   On the stages of evolution of stellar systems.
I. L. Genkin.
Trudy Astrofiz. Inst., *Alma-Ata*, Vol. 16, 98 - 102 (1971). In Russian.

Some features of the evolution of stellar systems in different stages were considered. The form of kinetic equations in different evolutionary stages was qualitatively analysed.

151.071   Regular and irregular forces in stellar systems.
I. L. Genkin.
Trudy Astrofiz. Inst., *Alma-Ata*, Vol. 16, 103 - 110 (1971). In Russian.

Principles of apportionment of regular and irregular components of the gravitational field of a stellar system are discussed. Interrelations of collective and individual phenomena in gravitating systems and other problems dealing with relaxation are considered. Various forms of kinetic equations are discussed.

151.072   Diffusion of star streams in the field of Coriolis
forces.   I. L. Genkin.
Trudy Astrofiz. Inst., *Alma-Ata*, Vol. 17, 68 - 72 (1971). In Russian.

An exact solution of the equation of the star stream dif-

fusion in the phase space under the influence of irregular and Coriolis forces is presented.

151.073   Transfer equations for stellar systems.
I. L. Genkin.
Trudy Astrofiz. Inst., *Alma-Ata*, Vol. 17, 73 - 81 (1971). In Russian.

151.074   Computer simulations of processes in galactic dy-
namics.   W. J. Quirk.
Thesis, Columbia Univ., New York. [Available from Univ. Microfilms, Ann Arbor, Mich., U.S.A. Order No. 71−6242], 108 pp. (1970).

Computer n-body calculations have been carried out to simulate galactic evolution. Three types of systems are simulated: pure gas, pure star, and pure gas evolving into mixed star and gas.

The possibility of capture in the restricted problem of three bodies and formation of bridges between galaxies. See Abstr. 042.001.

Numerical experiments on the N-body problem. See Abstr. 042.045.

The use of integrals in numerical integrations of the N-body problem.   See Abstr. 042.047.

Direct integration methods of the N-body problem. See Abstr. 042.048.

Computer simulation of plasmas. See Abstr. 062.026.

Enhancement of relaxation processes by collective effects.   See Abstr. 062.027.

The structure of rapidly rotating relaxed globular clusters.   See Abstr. 154.005.

The density computation for the galactic model of L. Perek.   See Abstr. 155.006.

On the kinematics of nearby stars. I. The vertex deviation of the velocity ellipsoids.   See Abstr. 155.015.

Observational evidence for galactic spiral structure. See Abstr. 155.039.

Stellar movement.   See Abstr. 155.046.

The spiral wave of our Galaxy near inner Lindblad resonance.   See Abstr. 155.048.

Formation and evolution of bright black holes. See Abstr. 158.046.

On numerical models for galaxies with intermediate ellipticity.   See Abstr. 158. 121.

Mass of a galaxy and dissipative process in the hot universe.   See Abstr. 162.067.

152 Stellar Associations

152.001 Spectral types of stars in the Orion and Hydra stellar rings. R. E. Schild, A. P. Cowley.
Astron. Astrophys., Vol. 14, 66 - 69 (1971).

New MK spectral types support the identity of the Orion ring as a coeval group of stars at common distance. New MK spectral types for the Hydra ring stars, however, are not consistent with their sharing common distance or age.

152.002 Evaporation of dirty ice particles surrounding early type stars. II. The Orion association.
S. Isobe.
Publ. Astron. Soc. Japan, Vol. 23, 371 - 386 (1971).

The evaporation processes and the radiative segregation of ice particles surrounding early type stars are calculated. The ice particle model for dust grains is justified by observational evidence found in the Orion association.

152.003 Intermediate band photometry of the Scorpio-Centaurus association and southern bright stars.
E. E. Mendoza V.
Bol. Obs. Tonantzintla y Tacubaya, No. 36, Vol. 6, 73 - 88 (1971).

We have made photometric observations in the (33, 35, 37, 40, 45, 52, 58, 63)—system of 333 stars of the southern hemisphere. Approximately one third of them are located in the Scorpio-Centaurus association region. Possibly most of them can be considered physical members of this stellar aggregate. The observational data have been used to derive provisional reddening-free indices. An absolute magnitude calibration is also given based on the parallaxes of 70 B-type stars in the Scorpio-Centaurus region. Satisfactory results are obtained under simple assumptions.

152.004 Formation stellaire dans l'association Sco OB$_1$.
A. Laval.
Comptes Rendus Acad. Sci. Paris, Sér. B, Vol. 273, 642 - 644 (1971).

La découverte de trois étoiles de grande luminosité intrinsèque au centre d'une importante nébulosité, permet de retracer les étapes successives de la formation stellaire dans cette association.

152.005 Photometry of the members of stellar rings from measurements on Palomar Sky Survey chart
(18^h, -18°). T. A. Uranova.
Soobshch. Gos. Astron. Inst. Shternberga, No. 171, p. 15 - 22 (1971). In Russian.

A list of B and V magnitudes for members of some stellar rings, noted by Isserstedt (1968) and the author (1969), is published. Spectral classes are given.

152.006 Am stars as possible members of two subgroups of the Scorpius-Centaurus association. J. W. Glaspey.
Astrophys. Journ., Vol. 169, 525 - 527 (1971).

Photometric evidence suggests that the Am star HD 148321 is not a member of the Upper Scorpius subgroup of the Scorpius-Centaurus association. An Am star in the Upper Centaurus subgroup, HD 119674, is a possible member.

152.007 The A stars in the UMa stream. I. N. Latyshev.
Astron. Zhurn. Akad. Nauk SSSR, Vol. 48, 1269 - 1279 (1971). In Russian. English translation in Soviet Astron. AJ, Vol. 15, No. 6.

The stars brighter than 6^{m}00 belonging to the spectral class A have been investigated using their radial velocities and proper motions by a method proposed by the author. The aim of this investigation is to prove the existence of the UMa stream. About 80 of 1036 stars considered here have been found to belong to the stream.

152.008 *uvby* and Hβ observations of B-type stars in the Scorpio-Centaurus association. J. W. Glaspey.
Astron. Journ., Vol. 76, 1041 - 1047 (1971).

We present *uvby* and Hβ observations of B-type stars in the upper Scorpius and upper Centaurus subgroups, and we intercompare these two subgroups. The data include: (i) the bright B-type stars considered by Bertiau (1958) to be members of the association based on the similarity of their proper motions and radial velocities, and (ii) a large number of the later B-type stars in the upper Scorpius suggested as members by Garrison (1967).

152.009 Catalogue of stellar magnitudes, color indices and spectral classes of stars in the region of the I Orion association. V. N. Sincheskul.
"Naukova dumka", Kiev. 38 pp. Price 17 Kop. (1971). In Russian.

152.010 On the star streams. N. Owaki.
Astron. Herald, (*Japan*), Vol. 64, 103 - 105 (1971). In Japanese. – See Phys. Abstr., Vol. 74, No. 77487 (1971).

L'età delle stelle di popolazione I.
See Abstr. 065.091.

The spectrum of LkHα-101 in the near-infrared.
See Abstr. 114.097.

Molecules in dense clouds and protostars.
See Abstr. 131.117.

Four-color and Hβ photometry for open clusters. VII. NGC 6231 and the I Sco association.
See Abstr. 153.025.

Weitere OB-Assoziationen in den äußeren Randzonen von M31. See Abstr. 158.007.

153   Galactic Clusters

153.001   On the distances of the open clusters Tr 14, Tr 15, Tr 16 and the η Carinae nebula.
P. S. Thé, G. Vleeming.
Astron. Astrophys., Vol. 14, 120 - 127 (1971).

Using Loden's (1968) photoelectric sequence near the η Carinae nebula, the distances of the open clusters Tr 14 and Tr 15 are determined by photographic photometry. We have redetermined the distance of Tr 16 using Feinstein's (1969) photoelectric and our photographic data. Starting from newly determined data on Tr 14 and Tr 16, and the assumption that they are embedded in the η Carinae nebulae, the distance of this nebula is discussed, and compared with the mean distance obtained by several astronomers from measurements of its exciting stars.

153.002   On the ages of the galactic clusters NGC 188, M67 and NGC 6791.   S. Torres-Peimbert.
Bol. Obs. Tonantzintla y Tacubaya, No. 36, Vol. 6, 3 - 14 (1971).

Theoretical isochrones for three chemical compositions are presented and compared to the color-magnitude diagrams of NGC 188, M67 and NGC 6791.

153.003   Untersuchungen über den Aufbau offener Sternhaufen.   W. Lohmann.
Astron. Nachr., Vol. 292, 193 - 204 = Astron. Rechen-Inst. Heidelberg, Mitt. Ser. B (1971).

From strip counts in 20 open star clusters the characteristic parameters of their structure are derived.

153.004   An improved method for computing membership probabilities in open clusters.   W. L. Sanders.
Astron. Astrophys., Vol. 14, 226 - 232 (1971).

Relative proper motions in the region of open clusters are fitted to a model by a maximum likelihood procedure for determining the frequency distribution function parameters for field and cluster stars. The clusters NGC 2168, 2281, 2420, 6633, 6823 and 7062 are reexamined for membership.

153.005   The stellar groups Ba 8, Ba 9, and Be 68.   R. Wagner.
Astron. Astrophys., Vol. 14, 283 - 292 (1971).

Three-color photometry of these three stellar groups in the RGU system on plates taken with the 48-inch Palomar-Schmidt leads to the conclusion that they are physical groups at distances of 1320 pc, 5520 pc, and 3160 pc respectively. The earliest color types are as follows: b 2 in Ba 9, a 0 in Ba 8, and b 8 in Be 68. Ba 8 and Be 68 include a red-giant branch.

153.006   Luminosity functions of stars in the Pleiades cluster.   P. N. Kholopov, N. M. Artyukhina.
Astron. Tsirk., No. 614, p. 6 - 7 (1971).   In Russian.

153.007   Very old open star cluster Collinder 110.
G. S. Tsarevskij, I. E. Abakumov.
Astron. Tsirk., No. 631, p. 6 - 8 (1971).   In Russian.

153.008   Photometric study of the galactic cluster NGC 5460.
J. J. Clariá.
Astron. Journ., Vol. 76, 639 - 644 (1971).

Three-color photoelectric photometry has been carried out for 64 stars in the vicinity of the southern galactic cluster NGC 5460. Hβ photometry for 16 stars is also presented. The stars range in V magnitude from $6\overset{m}{.}49$ to $13\overset{m}{.}07$.

153.009   Photometry of M67 to M_V = +12.   R. Racine.
Astrophys. Journ., Vol. 168, 393 - 404 (1971).

The present study is an attempt to (a) refine the definition of the M67 color-magnitude diagram in the vicinity of the turnoff and of the gap, (b) ascertain the location of the unevolved main-sequence stars in the color-magnitude and color-color plots, (c) investigate the luminosity function of the cluster to the limits of currently available observational facilities, and (d) detect the white dwarfs which should be found close to the photographic limit of the 200-inch.

153.010   Isolated low mass clusters in the interstellar medium.
A. F. Aveni, J. H. Hunter.
Bull. American Astron. Soc., Vol. 3, 367 (1971). – Abstr. AAS.

153.011   Photometry of the galactic cluster H 20.
B. Hidajat.
Bull. American Astron. Soc., Vol. 3, 367 (1971). – Abstr. AAS.

153.012   Photometry of the A and F stars in the new open cluster in line of sight with the Large Magellanic Cloud.   A. G. D. Philip.
Bull. American Astron. Soc., Vol. 3, 367 (1971). – Abstr. AAS.

153.013   Motions and membership of super metal rich stars in NGC 188.   W. S. Mesrobian, S. J. Kerridge, A. R. Upgren.
Bull. American Astron. Soc., Vol. 3, 371 - 372 (1971). Abstr. AAS.

153.014   Luminosity functions of stars in the Pleiades cluster.   P. N. Kholopov, N. M. Artiukhina.
Astron. Zhurn. Akad. Nauk SSSR, Vol. 48, 962 - 968 (1971). In Russian. English translation in Soviet Astron. AJ, Vol. 15, No. 5.

Luminosity functions of stars brighter than $18\overset{m}{.}0$ pg (M_{pg} = +12.3) for the nucleus and corona regions of the Pleiades cluster are derived. Possible members of the Pleiades brighter than 15^m were selected according to their proper motions.

153.015   A catalogue of galactic star clusters observed in three colours.   W. Becker, R. Fenkart.
Astron. Astrophys., Suppl. Ser., Vol. 4, 241 - 252 (1971).

A catalogue of 216 galactic star clusters is given for which observations in three colours are available. The distances have been calculated or recalculated using the method A, applied at the Basel Observatory.

153.016   On the distribution of flare stars in the Pleiades cluster.   P. N. Kholopov.
Inform. Bull. Variable Stars, (IAU Commission 27), Konkoly Obs., Budapest, No. 566 (1971).

In the paper entitled 'Unusual Distribution of flare stars in Pleiades' Mirzoyan and Mnatsakanian affirmed, that there are no flare stars in the central volume of the Pleiades cluster within the radius 1.4 pc. They suggested that the existence of such a 'cavity' might be explained 'in the frames of the idea of expansion of stellar associations'. However, there is no need to such an idea as the phenomenon treated by the above mentioned authors is only an apparent one.

153.017   Photometry of main-sequence stars in the Hyades and the field on the BVr system.
E. J. Mannery, G. Wallerstein.
Astron. Journ., Vol. 76, 890 - 895 (1971).

Observations of stars in the Hyades and stars earlier than type M0 with trigonometric parallaxes greater than 0''.080 have been made on the *BVr* system. The color index *V−r* is shown to correlate well with MK spectral type and effective temperature. Color-magnitude diagrams have been drawn using M_v and *V−r* for the field stars and *V* and *V−r* for the Hyades stars. By comparing color-magnitude diagrams we find a distance modulus for the Hyades of 3.24 mag in agreement with the modulus derived from dynamical parallaxes.

153.018 Membership of the open cluster NGC 6811.
W. L. Sanders.
Astron. Astrophys., Vol. 15, 368 - 373 (1971).

Probabilities of membership, based on relative proper motions, for 296 stars in the field of NGC 6811 are given. The cluster proper motion dispersion (m.e.) of 0''.0007 yields 97 probable members.

153.019 The open clusters NGC 6613 (M18) and NGC 6716.
U. Lindoff.
Astron. Astrophys., Vol. 15, 439 - 449 (1971).

UBV-magnitudes have been determined for stars in the open clusters NGC 6613 and NGC 6716, using a combination of photoelectric and photographic photometry. Spectral classes for the brighter stars have been determined from slit-spectra and objective prism-plates. The absorption has been determined to $1^m.4$ in front of NGC 6613 and to $0^m.4$ in front of NGC 6716. Their respective distances are 1250 pc and 600 pc.

153.020 Die absolute Eigenbewegung offener Sternhaufen.
H. van Schewick.
Veröff. Astron. Inst. Bonn, No. 84, 36 pp. (1971).

Absolute proper motions in the FK4 system have been obtained for 61 open star clusters. Own measurements and re-reduced data from other authors were used separately and in combination. A discussion is given for every cluster. The velocity of the sun relative to open clusters turns out to be smaller than the 'standard solar motion'. However, it nearly equals the 'basic solar motion'. The solar apex for open clusters is similar to that obtained for objects with small space velocities and OB stars from their proper motions and radial velocities. Two values of Oort's constant B have been obtained.

153.021 Four-color and Hβ photometry for the open cluster NGC 6231.
D. L. Crawford, J. V. Barnes, G. Hill, C. L. Perry.
Publ. Astron. Soc. Pacific, Vol. 83, 604 (1971). − Abstr. Astron. Soc. Pacific.

153.022 The masses of the 'horizontal-branch' stars in M 67.
H. E. Bond, C. L. Perry.
Publ. Astron. Soc. Pacific, Vol. 83, 638 - 642 = Contr. Louisiana State Univ. Obs., *Baton Rouge*, No. 54 (1971).

Four-color Strömgren photometry shows that the 'blue-straggler' and 'horizontal-branch' stars in the old galactic cluster M 67 do not have low surface gravities and that they are more massive than the cluster stars at the main-sequence turnoff. Therefore they are not highly evolved single stars, but most probably close-binary systems in various stages of evolution with mass transfer, as suggested by McCrea.

153.023 Untersuchungen über die Struktur junger Sternhaufen und die Entwicklungsphasen ihrer Mitglieder.
I. NGC 2264. W. Götz.
Astron. Nachr., Vol. 293, 81 - 104 (1971).

It could be shown that the structure of the cluster is closely combined with the continuous formation and the evolution of its members. A time-scale was fixed to state the age of determined regions and the phases of stellar evolution. As to the behaviour of T-Tauri-stars and other objects with small masses new results could be obtained.

153.024 The proper motions of two open star clusters.
L. S. Koroleva.
Astron. Zhurn. Akad. Nauk SSSR, Vol. 48, 1280 - 1288 (1971). In Russian. English translation in Soviet Astron. AJ, Vol. 15, No. 6.

From the data of our catalogues, relative and absolute proper motions of the two open clusters NGC 6866 and NGC 7789 have been determined. The cluster members were selected according to a photometric criterion and proper motions of stars. The tangential velocities of the investigated clusters were determined using distances known from the literature.

153.025 Four-color and Hβ photometry for open clusters. VII. NGC 6231 and the I Sco association.
D. L. Crawford, J. V. Barnes, G. Hill, C. L. Perry.
Astron. Journ., Vol. 76, 1048 - 1057 = Contr. Louisiana State Univ. Obs., *Baton Rouge*, No. 60 (1971).

Photoelectric photometry has been obtained at the Cerro Tololo Inter-American Observatory for 154 stars in the neighborhood of the open cluster NGC 6231. Analysis of the data leads to a calculated distance modulus for the cluster and the association of $V_0 - M_V = 11^m.5$, determined from data for 110 probable member stars. An average interstellar absorption of $1^m.4$ is derived, mostly caused by foreground absorbing matter. Several of the stars are brighter than $M_V = -7^m.0$.

153.026 Radio emission of galactic clusters at 1.4 GHz and 2.7 GHz. R. Schwartz.
Astrophys. Space Sci., Vol. 14, 286 - 300 (1971).

41 galactic clusters containing stars with spectral types from O5 to B9 have been observed at frequencies of 1.4 GHz and 2.7 GHz. Only clusters with spectral types earlier than B1 show thermal radio emission. Emission measure, mean electron density, and mass of HII gas have been computed from the observed data, and a comparison between the radio flux and Hα- and Hβ-flux densities has been made. The ratio of hydrogen mass and total stellar mass of the clusters and the gas to dust ratio are given.

153.027 Faint members of the Praesepe cluster.
N. M. Artiukhina.
Soobshch. Gos. Astron. Inst. Shternberga, No. 172, p. 3 - 26 (1971). In Russian.

Photographic proper motions and magnitudes of 784 stars brighter than $16^m.6$ pg in a circle with radius of 45' around the center of the cluster have been measured. The mean square error of one component of the proper motion is $\pm 0''.004 - 0''.005$. The catalogue contains all stars up to $16^m.0 - 16^m.2$ pg. Twenty-three new possible members of the cluster fainter than 14^m are indicated. The luminosity function for the nucleus of the cluster is derived.

153.028 A study on the galactic cluster Stock 2.
A. Martini.
Mem. Soc. Astron. Italiana, Nuova Ser., Vol. 42, 523 - 545 (1971).

U, B, V values for the stars of the galactic cluster Stock 2 have been derived by photographic photometry on plates taken with the 60/90 Schmidt telescope at the Asiago Observatory. The ratio of total to selective absorption distance modulus and age of the cluster have been derived. It is shown that the method developed by Krzeminski and Serkowski for deriving the $(U-B)_o$ values is physically and mathematically wrong.

153.029 The blue stars above the turn-off in M67: Horizontal branch or blue stragglers?
S. E. Strom, K. M. Strom, J. N. Bregman.
Publ. Astron. Soc. Pacific, Vol. 83, 768 - 779 (1971).

The nature of the pseudohorizontal branch in the galactic cluster M67 is investigated. The stars are found to have masses in the range $1.3 < M/M_\odot < 2.5$; a mass lower than or equal to the turnoff-point mass seems to be ruled out. This result suggests that these stars are analogs of the blue stragglers found in other galactic clusters rather than Population I analogs of Population II blue horizontal-branch stars.

153.030   The open cluster NGC 6025.
            A. Feinstein.
Publ. Astron. Soc. Pacific, Vol. 83, 800 - 804 (1971).
     Photoelectric *UBV* measures are presented for 78 stars in the open cluster NGC 6025. Distance and age estimates derived are 760 parsecs and 1×10^8 years. The brightest star, HD 143448, is a known emission star and it is very probably a member of the cluster.

153.031   Erratum: "Spectroscopy of stars in the galactic cluster NGC 1039 (M34)" [Publ. Astron. Soc. Pacific, Vol. 82, 825 - 829 (1970)].   P. A. Ianna.
Publ. Astron. Soc. Pacific, Vol. 83, 871 (1971).

     L'età delle stelle di popolazione I.
See Abstr. 065.091.

     Photoelectric observations of stars in the southern open clusters NGC 2335, NGC 2343, NGC 2453, NGC 4439 and H 5.   See Abstr. 113.036.

     The distances of two faint OB star groups in Monoceros.   See Abstr. 113.049.

     K-line photometry of stars in population I clusters.
See Abstr. 113.053.

     Stellar compositions from narrow-band photometry—I. Iron abundances in 180 G and K giants.
See Abstr. 114.004.

     Wolf-Rayet stars in H II regions.
See Abstr. 114.015.

     Stellar compositions from narrow-band photometry — II. Sodium and manganese in the Hyades and field giants.   See Abstr. 114.088.

     Abundances in open clusters: F dwarfs in Coma.
See Abstr. 114.104.

     Early-type peculiar stars and blue stragglers in star clusters.   See Abstr. 114.138.

     Time scales for Ca II emission decay, rotational braking and lithium depletion.   See Abstr. 114.141.

     Luminosities and motions of the F-type stars. I. Luminosity and metal abundance indices for disk population stars.   See Abstr. 115.017.

     Four suspected spectroscopic binaries in the Pleiades.   See Abstr. 119.010.

     W UMa type variables in the cluster NGC 188.
See Abstr. 121.069.

     TV observations of the variables EQ, ER and ES Cephei in the cluster NGC 188.   See Abstr. 121.070.

     Investigation of variable stars in the open cluster NGC 188.   See Abstr. 121.071.

     Flare stars in the Pleiades. II.   See Abstr. 122.094.

     On the space distribution of flare stars in Pleiades.
See Abstr. 122.130.

     The changing period of DL Cassiopeiae.
See Abstr. 122.131.

     The nature of the blue stragglers in the old disk population.   See Abstr. 122.158.

     Problems on DL Cassiopeiae.   See Abstr. 122.162.

     A late type variable in NGC 6819.
See Abstr. 123.060.

     On the lifetimes of galactic clusters.
See Abstr. 151.033.

     Numerical experiments on the escape from non-isolated clusters and the formation of multiple stars.
See Abstr. 151.034.

     Binary evolution in stellar systems.
See Abstr. 151.035.

     On the reproducibility of run-away stars formed in collapsing clusters.   See Abstr. 151.036.

     Dynamic evolution of rich galactic star clusters. I.
See Abstr. 151.065.

     Dynamic evolution of rich galactic star clusters. II.
See Abstr. 151.066.

     On the dynamical evolution of the Trapezium of Orion.   See Abstr. 151.067.

     A spectroscopic study of Magellanic Cloud globular-type clusters.   See Abstr. 159.024.

154 Globular Clusters

154.001 A photometric study of the metal-rich globular cluster M71.   H. C. Arp, F. D. A. Hartwick.
Astrophys. Journ., Vol. 167, 499 - 509 (1971).

From an analysis of photoelectric and photographic UBV observations of stars in M71 we have found (a) E_{B-V} = 0.31 ± 0.02, (b) $(m - M)_0$ = 13.07 ± 0.21, (c) [Fe/H] = –0.3 ± 0.2, and (d) age = 7.6 (+3.1, –2.3) × 10^9 years, assuming Y_{M71} = 0.29.

154.002 The globular cluster NGC 6541.   G. Alcaino.
Astron. Astrophys., Vol. 13, 399 - 404 (1971).

A UBV photometric investigation of the globular cluster NGC 6541 was carried out at the European Southern Observatory using the 1-meter reflector for the photoelectric work and at the Cerro Tololo Inter-American Observatory using the 1.5 meter reflector for the photographic work. Fifteen stars were observed photoelectrically to apparent magnitude V = 15.67. With the sequence, 283 stars were calibrated photographically.

154.003 Asymptotic giant and red giant branch stars in globular clusters.   S. E. Strom, K. M. Strom.
Astron. Astrophys., Vol. 14, 111 - 119 (1971).

Stars of unusual composition were found in the course of a spectroscopic survey of the red-giant and asymptotic-giant branch stars of the metal-weak globular clusters M92 and M22. These stars have unusually strong CN and CH band strengths and apparently strong metal (Fe, Ca, Mn) lines as well. They may be mixed stars in which the atmospheric hydrogen is depleted and the C, N and O abundances enhanced.

154.004 The bright red giants of the globular cluster M92.   Z. I. Kadla.
Astron. Zhurn. Akad. Nauk SSSR, Vol. 48, 760 - 765 (1971). In Russian. English translation in Soviet Astron. AJ, Vol. 15, No. 4.

67 members of M 92 brighter than V = 13.40 have been found on the basis of V magnitudes and proper motions. All these bright red giants are at a distance r < 5.'9 from the cluster center, 34 of them being at r ≤ 38."5.

154.005 The structure of rapidly rotating relaxed globular clusters.   J. Kormendy, S. P. S. Anand.
Astrophys. Space Sci., Vol. 12, 47 - 57 (1971).

A model of rapidly rotating globular clusters is constructed assuming uniform angular velocity and a truncated Maxwellian distribution in the stellar velocities. Since the first-order theory developed by Woolley and Dickens becomes inaccurate for rapid rotation, a small-mass envelope is fitted to their models by the method of Monaghan and Roxburgh. A comparison is made of the critical values derived by the two methods.

154.006 Space distribution of evolving stars in M13.   A. Blaghikh, V. Castellani.
Astrophys. Space Sci., Vol. 12, 208 - 218 (1971).

Information about space distribution is collected for selected classes of evolving stars in the globular cluster M13. After a rigorous elimination of field stars, three samples are examined, corresponding to the red giant stage (G), the blue (B) and the yellow (YG) parts of the horizontal branch. Theoretical interpretations are briefly discussed with reference to the reliability and usefulness of this type of investigation.

154.007 Hoe licht is het in de kern van een bolvormige sterhoop?   A. G. Jansen.
Hemel en Dampkring, Vol. 69, 186 (1971).

154.008 UBV photometry of globular clusters in M87.   D. A. Hanes, R. Racine.
Journ. Roy. Astron. Soc. Canada, Vol. 65, 183 (1971). Abstr. RAS Canada.

154.009 On the variability of horizontal branch stars in the globular clusters M10, M2, M5.   Yu. V. Voroshilov.
Astron. Tsirk., No. 623, p. 7 - 8 (1971).   In Russian.

154.010 On the difference between the Oosterhoff types I and II globular clusters.   R. S. Stobie.
Astrophys. Journ., Vol. 168, 381 - 391 (1971).

The globular clusters M3 and ω Cen are the most populous representatives of Oosterhoff types I and II, respectively. By studying their period-frequency distributions in detail we conclude that the variables in ω Cen are either more massive by $\Delta \log M$ = 0.10 or more helium rich by ΔY = 0.20 than the variables in M3. Reasons are given for preferring the former alternative. It is shown that this difference is a general property of the Oosterhoff types I and II.

154.011 Highly reddened objects in Sagittarius.   J. D. Wray.
Astrophys. Journ., (*Letters*), Vol. 168, L97 - L99 (1971).

Photographs taken in the near-infrared of IRC–20385 reveal the object to be a highly reddened globular cluster. Two other objects discovered in a search of the National Geographic Society-Palomar Observatory Sky Survey prints for morphologically similar objects in the Sagittarius area have also been observed.

154.012 Bright membership of the globular cluster NGC 6838.   W. L. Sanders.
Astron. Astrophys., Vol. 15, 173 - 178 (1971).

Probabilities of membership, based on relative proper motions, for 401 stars in the field of NGC 6838 are given. To an estimated plate limiting magnitude of about 15 visual, 81 probable members are found over a field of 38' × 31'.

154.013 On the masses, luminosities, and compositions of horizontal-branch stars.
T. S. van Albada, N. Baker.
Astrophys. Journ., Vol. 169, 311 - 326 (1971).

For RR Lyrae stars in globular clusters, three different relations between mass and luminosity can be obtained from a combination of available observations and theoretical models. By requiring compatibility among these relations, we infer the masses and luminosities of these stars, as functions of the composition parameters assumed for the models. This requirement also restricts the composition parameters. The ratio of horizontal-branch luminosity to that at the main-sequence turnoff is a sensitive function of Z; values of this function are given. The problem of determining Y in RR Lyrae stellar envelopes from the blue edge of the instability region is reexamined with the help of linear nonadiabatic models of pulsating stars. The theoretical uncertainties, and the consequent uncertainties in the inferred values of M, L, Y, and Z are emphasized.

154.014 Photometry of variables in globular clusters. III. M 14 and the period–luminosity relation of population II cepheids.   S. Demers, A. Wehlau.
Astron. Journ., Vol. 76, 916 - 922, 951 (1971).

B and V photographic photometry of five population II cepheids in M 14 is presented. The mean magnitudes and colors of the M 14 variables are combined with population II cepheids in ω Cen, M2, and M13. These 17 variables define a period-luminosity relation given as $\langle M_V \rangle = -0.16 - 1.65 \log P$. No period-color relation is evident for all stars but the M 14 and ω Cen variables do follow a $P-C$ relation with almost identical slope.

154.015   The globular clusters NGC 1851 and NGC 2808.
G. Alcaino.
Astron. Astrophys., Vol. 15, 360 - 367 (1971).

A BV photographic investigation on the globular clusters NGC 1851 and NGC 2808 was carried out at the Cerro Tololo Inter-American Observatory (CTIO) with the 1.5 meter reflector and the plate material was reduced with the iris-photometer of the European Southern Observatory. For NGC 1851, 117 stars were calibrated photographically from a photoelectric sequence of 32 stars previously obtained by the author at CTIO. For NGC 2808, 255 stars were calibrated photographically from 57 stars of two photoelectric sequences obtained at CTIO by different observers.

154.016   Two new CN-strong globular cluster stars.
W. Osborn.
Observatory, Vol. 91, 223 - 224 (1971). – Letter.

154.017   Stability of a model of a globular star cluster with non-zero rotational moment.
V. S. Synakh, A. M. Fridman, I. G. Shukhman.
Dokl. Akad. Nauk SSSR, Ser. Mat. Fiz., Vol. 201, 827 - 830 (1971). In Russian.

154.018   Sulle funzioni di luminosità degli ammassi globulari.
V. Castellani, F. A. D'Antona.
Mem. Soc. Astron. Italiana, Nuova Ser., Vol. 42, 441 - 473 (1971).

The meaning of the observed luminosity functions for globular clusters is shortly recalled, and the questions regarding their theoretical derivations are defined. The available theoretical results for the giant branch are investigated in order to test the possibility of determining, by observational results, the main evolutionary parameters: age and initial chemical composition. The experimental functions (given in the appendix) are compared to the theoretical ones.

154.019   On the dependence of the form of the horizontal branch on the chemical abundance of globular clusters.   A. V. Mironov.
Astron. Tsirk., No. 667, p. 2 - 4 (1971). In Russian.

154.020   The distribution functions of star clusters.
J. Broderick.
Thesis, Brandeis Univ., Waltham, Mass. [Available from Univ. Microfilms, Ann Arbor, Mich., U.S.A. Order No. 70–17115], 96 pp. (1970).

A model globular cluster is constructed and studied from a point of view different than that usually used. The model is defined by an analytical surface density law found by King (1962) to be in good agreement with observed surface densities of star clusters. From this surface density law the phase space distribution function is sought.

154.021   Globular-cluster stars: Results of theoretical evolution and pulsation studies compared with the observations.   I. Iben, Jr.
Publ. Astron. Soc. Pacific, Vol. 83, 697 - 740 (1971). – Review article.

154.022   Variables in globular clusters.   H. Sawyer Hogg.
AAVSO Abstr., October 1971, p. 9 - 11.

Sternentwicklung auf dem Horizontalast und Zustandsgrößen von RR-Lyrae-Sternen in Kugelhaufen verschiedenen Alters.   See Abstr. 065.023.

La determinazione dell'età delle stelle di popolazione II.   See Abstr. 065.092.

On the dependence of Q metallicity index from basic parameters of stellar evolution.   See Abstr. 065.116.

Masses of red giants on the asymptotic branch in globular clusters.   See Abstr. 115.010.

Investigation of RR Lyrae type stars in the globular cluster M5 (NGC 5904).   See Abstr. 122.023.

On the evolutionary phase of cepheids in globular clusters.   See Abstr. 122.032.

On the evolutionary phase of cepheids in globular clusters.   See Abstr. 122.057.

Light curves of five RR Lyrae variables in the globular cluster M 15 obtained by means of television techniques.   See Abstr. 122.076.

On the variability of 37 stars in the globular cluster NGC 3201 and in its vicinity.   See Abstr. 122.077.

Var 156 in the globular cluster M3—an RR Lyrae type variable.   See Abstr. 122.080.

Studies of the variables in the globular cluster NGC 6171.   See Abstr. 122.092.

Variables in Messier 5: A study of Mount Wilson 1917 observations.   See Abstr. 122.093.

Observations of variable stars in the globular cluster NGC 3201.   See Abstr. 122.115.

RR Lyrae type variables SVS 1365, 1367 and 1371 in the nucleus of the globular cluster M3. See Abstr. 122.120.

RR Lyrae type variable Var 201 in the centre of the globular cluster M3.   See Abstr. 122.122.

Report of the committee on variable stars in clusters.   See Abstr. 122.141.

On the variability of red giants in the globular cluster M3.   See Abstr. 122.147.

7 new variable stars in NGC 1261. See Abstr. 123.052.

Further remarks concerning variable 14 in M5. See Abstr. 123.053.

30 new variable stars in the region M 56. See Abstr. 123.055.

Monte-Carlo models of star clusters. See Abstr. 151.032.

Non-stellar objects used in preparing the SAO Star Atlas, coordinates at equinox 1950.0.   See Abstr. 158.125.

155 Structure and Evolution of the Galaxy

155.001 The ratio of total to selective extinction and the distance to the galactic centre. P. G. Martin.
Monthly Notices, Roy. Astron. Soc., Vol. 153, 251 - 260 (1971).

A method for determining the ratio of total to selective extinction is discussed. It is shown that reliable distances, expressed as a fraction of the distance to the galactic centre, can be calculated from radial velocities and a model for the differential rotation of the Galaxy when care is taken to select suitable stars. A measure of the total extinction can be derived when the apparent distance modulus is known.

155.002 Attempt to explain the motions of the gas in the central region of the Galaxy by explosive events in its nucleus. P. C. van der Kruit.
Astron. Astrophys., Vol. 13, 405 - 425 (1971).

The possibility is investigated that the well known expanding features in the galactic plane, such as the three-kpc arm, would be due to an expulsion. It appears that these can be explained as the result of an expulsion of matter from the nucleus some 12 - 13 million years ago in two opposite directions making an angle of 25° to 30° with the plane. Orbits of gas clouds were calculated on the basis of a rough new model for the mass distribution in the central region, and on the assumption that at distances larger than 0.1 kpc only gravitational forces and interaction with the quiescent gas layer around the plane are important. A possible model for the entire velocity structure in the central region has been constructed by computing velocity-longitude diagrams on the basis of a specified field and density distribution.

155.003 Moment equations in the study of the total mass density in the neighbourhood of the sun and of the galactic force law K_z. C. Turon Lacarrieu.
Astron. Astrophys., Vol. 14, 95 - 102 (1971).

A study of stellar motions perpendicular to the galactic plane, using King's method of the "pseudomoments" has been made. The estimation of the "pseudomoments" from observational data allows an evaluation of the total mass density in the neighbourhood of the sun and of the galactic force law $K_z(z)$.

155.004 Galactic orbits and integrals of motion for "high-velocity" stars (I). L. Martinet, A. Hayli.
Astron. Astrophys., Vol. 14, 103 - 110 (1971).

We study properties of stellar motions in the non-separable galactic Schmidt potential for many different values of the energy and angular momentum corresponding to the occupation of Lindblad's diagram by different samples of nearby stars.

155.005 Outward drift of the interstellar medium in the disk of the Galaxy. E. R. Harrison.
Monthly Notices Roy. Astron. Soc., Vol. 153, 12P - 16P (1971).

It is proposed that the gaseous component in the disk of the Galaxy has an outward radial drift velocity relative to the stellar component. The drift is the results of non-gravitational forces acting on the gas, and the dynamic coupling between gas and stars. Some comments are made on the possible importance of the motion concerning the origin of spiral structure.

155.006 The density computation for the galactic model of L. Perek. V. I. Rodionov.
Bull. Astron. Inst. Czechoslovakia, Vol. 22, 162 - 168 (1971).

The phase-space model of the Galaxy proposed by L. Pe-

rek (1966) is examined. The behaviour of the density function outside the plane of symmetry of the model is studied.

155.007 The statistics of the nearby stars.
R. Woolley, S. B. Pocock, E. A. Epps, R. Flinn.
Roy. Obs. Bull., Greenwich—Cape, No. 166, p. 275 - 299 (1971).

The data, presented in Roy. Obs. Ann. No. 5 (1970), are used to construct a relation between spectral type and absolute magnitude. The luminosity function is calculated and it is suggested that the data imply an uneven rate of star formation in the sense that the number of stars formed recently (i.e. in the last 1.5×10^8 years) is greater than the average rate of star formation relating to stars now found close to the sun. The proportion of double stars in the catalogue is calculated; it is found that 25 per cent of stars are visual doubles and that this percentage is practically independent of spectral type. On the other hand, the proportion of spectroscopic binaries does appear to vary significantly from 25 per cent in A stars to about 10 per cent for late-type main-sequence stars.

155.008 The nearby stars. P. van de Kamp.
Annual Rev. Astron. Astrophys., Vol. 9, (see 003.001), 103 - 126 (1971).
Long-focus photographic astrometry; Parallax determinations; General survey; Density function; Velocity distribution; Luminosity function; Spectrum-luminosity relation; Double stars; Mass ratio and masses; Perturbations; Individual objects.

155.009 Are the galactic loops supernova remnants?
E. M. Berkhuijsen, C. G. T. Haslam, C. J. Salter.
Astron. Astrophys., Vol. 14, 252 - 262 (1971).

Observational evidence is presented on the nature of the galactic continuum loops. Their small circle geometry is discussed. A relationship between neutral hydrogen features at intermediate and high latitudes and the continuum loops has been found. The available evidence would seem to favour the supernova remnant hypothesis for the origin of the galactic loops.

155.010 On the diffuse radiation of the Galaxy in far UV according to the investigations by S. Hayakawa, K. Yamashita and S. Yoshioka.
D. A. Rožkovskij, V. S. Matjagin.
Astrophys. Space Sci., Vol. 12, 204 - 207 (1971).

It has been found that certain optical properties of interstellar grains obtained by Hayakawa et al. (1969) from their study of the diffuse radiation of the Galaxy in far UV are not in reality consistent with the more accurate theoretical calculations of diffuse radiation for the model of the Galaxy used by Hayakawa et al. in their work.

155.011 Hydrogen clouds in the halo and the origin of the Galaxy. A. Żytkow.
Postępy Astron., Vol. 19, 263 - 268 (1971). In Polish.

155.012 Die Entfernung des Milchstraßenzentrums.
K.-H. Schmidt.
Sterne, 47. Jahrgang, p. 129 - 133 (1971).

155.013 A study of solar motion and galactic rotation.
A. R. Klemola, S. Vasilevskis.
Publ. Lick Obs., Univ. California, Santa Cruz, Vol. 22, Part 3, 14 pp. (1971).

Preliminary results for the constants of solar motion and galactic rotation are derived from the analysis of absolute stellar proper motions measured with respect to galaxies in 83 fields north of declination −23°. Solutions based on nearly 9000 stars of mag. 9 − 17, separated into three ranges of magnitude, are made for all fields together and for fields separated into three zones of latitude. The position of the solar apex changes with stellar magnitude; the right ascension increases from about 272° to 292° and the declination from +44° to +54° for common stars with blue magnitudes 9 to 16, respectively. In the same range of magnitudes the mean centennial secular parallax for the whole sky decreases from 2″6 to 1″2; as expected, it increases with galactic latitude. No reliable values for the constants of galactic rotation are obtained at this stage due to the availability of motions for only a small number of fields at low galactic latitudes.

155.014 Galactic spurs as possible sources of soft X-ray radiation. I. S. Shklovsky, E. K. Sheffer.
Astron. Tsirk., No. 613, p. 3 - 6 (1971).   In Russian.

155.015 On the kinematics of the nearby stars. I. The vertex deviation of the velocity ellipsoids.
C. Yuan.
Astron. Journ., Vol. 76, 664 - 669 (1971).
      A statistical approach has been adopted to construct theoretical space velocity distributions for the nearby stars using the density-wave theory of galactic spirals. The idea that the observed vertex deviation of nearby young stars is due to the places of origin is studied in great detail. It is shown that the synthesized space velocity distributions are in good agreement with observations. The theoretical spiral pattern as well as the pattern speed and the strength of the spiral gravitational field determined in previous studies is reaffirmed in the present analysis.

155.016 The structure of the stellar field in the direction to NGC 2129. N. B. Kalandadze, L. N. Kolesnik, V. I. Kuznetsov.
Byull. Abastumansk. Astrofiz. Obs., No. 40, p. 39 - 54 (1971). In Russian.
      A stellar area of 1 square degree in the Milky Way, located in the direction to the galactic anticenter, with its center coinciding with NGC 2129, has been examined. On the basis of B, V magnitudes and spectral classes for 525 stars the interstellar absorption and spatial distribution of stars have been studied.

155.017 Composition and activity of the nucleus of our Galaxy, and comparison with M 31.
J. H. Oort.
Nuclei of galaxies. Conference 1970, (see 012.002), p. 321 - 344 (1971).

155.018 High-velocity neutral hydrogen in the inner region of the Galaxy. G. L. Mader.
Bull. American Astron. Soc., Vol. 3, 363 (1971). − Abstr. AAS.

155.019 Further evidence of explosive events in the galactic nucleus. R. Sanders, G. Wrixon, A. Penzias.
Bull. American Astron. Soc., Vol. 3, 363 (1971). − Abstr. AAS.

155.020 Results of lunar occultation studies of the galactic center region in HI, OH and CH₂O lines.
Aa. Sandqvist, F. J. Kerr.
Bull. American Astron. Soc., Vol. 3, 363 - 364 (1971). Abstr. AAS.

155.021 100 micron survey of the galactic plane.
W. F. Hoffmann, C. L. Frederick, R. J. Emery.
Bull. American Astron. Soc., Vol. 3, 364 (1971). − Abstr.

AAS.

155.022 The spiral wave of our Galaxy near inner Lindblad resonance. J. W.-K. Mark.
Bull. American Astron. Soc., Vol. 3, 370 (1971). − Abstr. AAS.

155.023 Galactic structure from binebulous objects.
W. E. Greig.
Bull. American Astron. Soc., Vol. 3, 370 (1971). − Abstr. AAS.

155.024 OB star distribution in Puppis.
R. J. Havlen.
Bull. American Astron. Soc., Vol. 3, 370 - 371 (1971). Abstr. AAS.

155.025 Observations of the diffuse galactic light.
F. E. Roach, L. L. Smith.
Bull. American Astron. Soc., Vol. 3, 396 - 397 (1971). Abstr. AAS.

155.026 Spectrum and galactic isotropy of diffuse cosmic X-rays. P. J. N. Davison, R. M. Thomas.
Nature, Phys. Sci., Vol. 233, 27 - 29 (1971).
      The spectrum of diffuse cosmic X-rays from the vicinity of the galactic centre has been measured using an actively shielded and collimated detector covering the energy range 27−167 keV. The results are consistent with other measurements. Upper limits are also placed on the isotropy of diffuse X-rays relative to the galactic plane, in the longitude range $320° < l < 40°$.

155.027 Motion of A0 stars perpendicular to the galactic plane. IV. Radial velocities in the south galactic cap. G. A. Harding, F. Fahim, C. M. Haslam.
Roy. Obs. Bull., Greenwich−Cape, No. 165, p. 259 - 271 (1971).
      Radial velocity observations and MK spectral types are presented for 125 stars classified as A0 in the Henry Draper Catalogue and appearing in the south galactic cap. The new radial velocities and spectral types are used to produce velocities (w) and distances (z) perpendicular to the galactic plane. The new data are then combined with those contained in Paper III and analysed to produce mean w velocities and velocity dispersions in bands of z.

155.028 The interpretation of the absolute intensity of the diffuse galactic light. K. Mattila.
Astron. Astrophys., Vol. 15, 292 - 298 (1971).
      An interpretation of the absolute intensity of the diffuse galactic light at the galactic equator is presented. The discrete cloud structure of interstellar dust is taken into account, and it is shown to have a considerable effect on the determination of the grain properties. The calculated intensity of the diffuse light is compared with the observations of Witt (1968) in Cygnus and Taurus-Auriga.

155.029 Erratum: 'Four new star clusters in the direction of the central area of the Galaxy' [Astron. Astrophys., Vol. 12, 477 - 481 (1971)].   A. Terzan.
Astron. Astrophys., Vol. 15, 336 (1971).

155.030 Ultraviolet radiation in space and in the Venus atmosphere. V. G. Kurt, A. S. Smirnov.
Space Research XI, (see 012.004), 1345 - 1349 (1971).

155.031 Map of the galactic nucleus at 10 μm.
G. H. Rieke, F. J. Low.
Nature, Vol. 233, 53 - 54 (1971).
      A contour map of the center of our Galaxy at a wave-

length of 10 μm and with a beam 5.5 arcsec in diameter is presented. Four sources can be distinguished; an additional broad zone of emission extends to the north. The relation of these sources to those detected at 2.2 μm, 5 μm, 22 μm, and in the radio region is discussed.

155.032 **Rocket-infrared four–color photometry of the Galaxy's central regions.** J. R. Houck, B. T. Soifer, J. L. Pipher, M. Harwit.
Astrophys. Journ., *(Letters)*, Vol. 169, L31 - L34 (1971).

The central portion of the Galaxy was observed in the bandwidths 5–6, 12–14, 16–23, and 85–115 μ during an Aerobee 170 rocket flight launched on 1971 July 16. We report on measurements made during a 100-second time interval around 21:56 MST. In addition to the galactic center, we also observed four new sources.

155.033 **Interferometric study of the Milky Way between Carina and Aquila.**
H. Dottori, G. Carranza.
Astrophys. Space Sci., Vol. 13, 180 - 184 (1971).

Large field Hα observations of the Milky Way between Carina and Aquila were made through a narrow interference filter 15 Å wide. Characteristic large-scale features of the observed region are extended emission areas in Carina, Norma-Scorpius and Scutum-Sagittarius and some weak isolated nebulosities near the Coal Sac, α Centauri and γ Normae. Hα photographs, a chart mapping the emission, and a list of identified emission regions are given.

155.034 **High-resolution observations of the galactic center at 5 GHz.** R. D. Ekers, D. Lynden-Bell.
Astrophys. Letters, Vol. 9, 189 - 193 (1971).

Interferometric observations at 5 GHz with a resolution of 6 × 18 arc sec indicate the presence of compact components in Sagittarius A and in some of the other radio sources in the galactic center region.

155.035 **Progress report on studies of southern spiral structure.** B. J. Bok, E. W. Miller.
Separate print Steward Obs., Univ. Arizona, Tucson. 4 pp. (1971).

155.036 **100-micron survey of the galactic plane.**
W. F. Hoffmann, C. L. Frederick, R. J. Emery.
Astrophys. Journ., *(Letters)*, Vol. 170, L89 - L97 (1971).

A survey of a portion of the galactic plane has been carried out at a wavelength of 100 μ with a beamwidth of 12' and a sensitivity of 10^{-22} W m^{-2} Hz^{-1}. The survey covers 750 square degrees of the sky including most of the galactic plane between $l^{II} = 335°$ and $l^{II} = 88°$ and a number of other selected areas of interest. Seventy-two sources have been detected, 60 of which are identified with continuum radio sources, bright nebulae, dark nebulae, and infrared stars.

155.037 **Evolutionary models of nucleosynthesis in the Galaxy.** J. W. Truran, A. G. W. Cameron.
Astrophys. Space Sci., Vol. 14, 179 - 222 (1971).

A model of the Galaxy is constructed and evolved in which the integrated influence of stellar and supernova nucleosynthesis on the composition of the interstellar gas is traced numerically. The implications of our model for other features of the Galaxy, including supernova nucleosynthesis, the cosmic ray production of the light elements, and cosmochronology, are discussed in detail.

155.038 **Galactic X-rays from unresolved flare stars.**
P. J. Edwards.
Nature, Phys. Sci., Vol. 234, 75 - 76 (1971).

The X-ray brightness of the galactic disc due to unresolved red dwarf flare stars is estimated from recent optical observations of nearby flare stars together with solar optical and X-ray data. The calculated brightness is close to the present observational upper limits to a disc component. Owing to a type-setting error, the space density of flare stars used in the calculation (0.03 pc^{-3}) was incorrectly published as 0.3 pc^{-3}.

155.039 **Observational evidence for galactic spiral structure.** B. J. Bok.
Highlights of Astronomy, Vol. 2, (see 003.026), 63 - 87 (1971). – Invited discourse.

155.040 **Distribution of B8–A3 stars near the galactic plane. I. Galactic longitudes 50° to 150°.**
S. W. McCuskey, N. Houk.
Astron. Journ., Vol. 76, 1117 - 1128 (1971).

Counts of B8–A3 stars to a limiting magnitude $V=13$ along the galactic plane from $l=50°$ to 150° and, in general, between $b=-5°$ and $+5°$ have yielded: (i) contour maps showing the surface distribution, and indicating several high-density areas; (ii) approximate space densities (corrected for interstellar absorption) to a distance of 1 kpc from the sun. The detailed results of the survey are displayed in histogram form.

155.041 **Evaluation of the diffuse galactic radiation in the region $\alpha = 18^h 50^m$, $\delta = 1°.5$.** L. N. Kondratjeva.
Trudy Astrofiz. Inst., *Alma-Ata*, Vol. 16, 22 - 31 (1971). In Russian.

For the evaluation of the diffuse galactic radiation in the region $\alpha = 18^h 50^m$, $\delta = 1°.5$ 12 photographs of this and a comparison region (b = 30°) were obtained. The diffuse galactic radiation was determined with consideration of stellar and zodiacal components of night airglow and the tropospheric component.

155.042 **On the calculation of stellar and diffuse radiation for a plane-parallel semi-infinite model of the Galaxy.** D. A. Rozhkovsky.
Trudy Astrofiz. Inst., *Alma-Ata*, Vol. 17, 3 - 17 (1971). In Russian.

Formulae for the calculation of the integral radiation of stars in the inner points of a plane-parallel model of the Galaxy with uniform distribution of stars and absorbing matter are presented. A simple and sufficiently precise calculation of the diffuse galactic radiation intensity by isotropically scattering interstellar dust particles is given. The derived equations are compared with the results obtained by other authors. The intensity of the radiation scattered by a spherical absorbing cloud located in the stellar and diffuse radiation field of the Galaxy is determined.

155.043 **Distribution of stars of different spectral classes in the z-coordinate.** G. G. Borzov.
Astron. Tsirk., No. 659, p. 6 - 8 (1971). In Russian.

155.044 **A study of galactic structure in a region of Cassiopeia with the help of the M and C type stars.**
C. Poulakos.
Mem. Soc. Astron. Italiana, Nuova Ser., Vol. 42, 421 - 439 (1971).

170 M2-M7 stars in an area of 20.9 sq deg centered at R.A. $0^h 27^m 30^s$, Dec. $+59°41'$; $l^{II} = 120°.3$, $b^{II} = -2°.8$ are studied by means of photographic photometry. Twenty stars have been characterized as suspected of variability.

155.045 **Some implications of a new value for the primordial solar deuterium-hydrogen ratio.** D. C. Black.
Nature, Phys. Sci., Vol. 234, 148 - 149 (1971).

The interesting aspects of the D/H ratio in galactic evolution and cosmology are discussed.

155.046 **Stellar movement.** R. Woolley.

Contemporary Phys., Vol. 12, 395 - 409 (1971).

The article sketches the technique by which velocities relative to the sun are found and gives a brief account of the mathematics by means of which these velocities are shown to correspond to galactic orbits, and the article concludes with a sketch of the interpretation of the statistics of galactic orbits in terms of the history of the Galaxy.

155.047 Nucleus of our Galaxy.
H. Okuda, T. Maihara.
Astron. Herald, (*Japan*), Vol. 64, 184 - 188 (1971). In Japanese.

A review of the infrared observations of the nucleus.

155.048 The spiral wave of our Galaxy near inner Lindblad resonance. J. W-K. Mark.
Proc. National Acad. Sci. USA, Vol. 68, 2095 - 2098 (1971).

The dispersion relationship for short-wavelength spiral density waves in our Galaxy has been refined to remove the divergences that occurred in wave number and in amplitude as inner Lindblad resonance is approached.

155.049 Stellar distributions at high galactic latitudes.
J. J. Schreur.
Thesis, Univ. Arizona, Tucson. [Available from Univ. Microfilms, Ann Arbor, Mich., U.S.A. Order No. 71–14507], 200 pp. (1971).

A system of four-colour photographic photometry is presented which is very similar to the Strömgren photoelectric four-colour (uvby) system. The photographic four-colour system is applied to a study of the stellar distributions in Selected Areas 56 and 27. The four-colour indices are used to separate the stars in each of the two survey fields into four groups; the halo giants and dwarfs, and the disk giants and dwarfs. Space densities at various distances from the sun are computed for each group.

155.050 Lunar occultations of the galactic center region in H I, OH and CH$_2$O lines.   A. Sandqvist.
Diss. Fac. Graduate School Univ. Maryland, College Park, Maryland (Univ. Maryland Astronomy Program), 196 pp. (1971). − *RXM*

Selected exercises in galactic astronomy.
See Abstr. 003.025.

Luminous stars in the Southern Milky Way.
See Abstr. 041.023.

The present state of gamma-ray astronomy.
See Abstr. 061.020.

Gamma-ray spectrometry in the energy range 0.5−5 MeV.   See Abstr. 061.022.

Contribution to the helium content of the Galaxy from pulsationally unstable stars evolving inhomogeneously.
See Abstr. 065.122.

Focusing of gravitational radiation by the galactic core.   See Abstr. 066.115.

Catalogue of proper motions of 8,790 stars with reference to galaxies.   See Abstr. 112.005.

Three-color photometry of a field in the large Sagittarius cloud.   See Abstr. 113.003.

Infrared sources of radiation.
See Abstr. 113.010.

Near-ultraviolet surface photometry of the southern Milky Way.   See Abstr. 113.025.

UBV photometry in the nuclear bulge of the Galaxy.   See Abstr. 113.050.

A finding list of faint blue stars in the anticenter region of the Galaxy.   See Abstr. 113.051.

Faint O−B2 stars in the Vela, Carina, Centaurus, and Crux sections of the Milky Way.   See Abstr. 113.066.

Some spectroscopic characteristics of the OB stars: An investigation of the space distribution of certain OB stars and the reference frame of the classification.
See Abstr. 114.101.

The distribution of the double stars in the Galaxy.
See Abstr. 117.029.

On some characteristics of pulsating stars with periods from one to two days.   See Abstr. 122.114.

Detection of radio recombination-line emission associated with distributed ionized hydrogen.
See Abstr. 131.014.

Cloud distributions in the Perseus arm.
See Abstr. 131.060.

H I self-absorption: The galactic center cold cloud and properties of the interstellar medium.
See Abstr. 131.066.

Similarity models of interstellar loop structures.
See Abstr. 131.068.

New ammonia lines and sources in the Galaxy.
See Abstr. 131.100.

The absence of formaldehyde radiation toward cold regions of the galactic plane.   See Abstr. 131.106.

NH$_3$ and H$_2$O emission in our Galaxy.
See Abstr. 131.122.

The Ohio Survey between declinations of 40° and 63° north.   See Abstr. 141.157.

Compact radio sources in the galactic nucleus.
See Abstr. 141.166.

An absorption-line study of the galactic neutral hydrogen at 21 centimeters wavelength.
See Abstr. 141.192.

X-ray sources near the center of our Galaxy.
See Abstr. 142.025.

Diffuse cosmic X-ray flux from 0.2 to 2 keV.
See Abstr. 142.056.

Soft X-ray emission from galactic radio spurs.
See Abstr. 142.066.

X-ray sources near the galactic center observed by *Uhuru*.   See Abstr. 142.071.

Cosmic rays in the Galaxy: Convection or diffusion?
See Abstr. 143.050.

Relaxation times in strictly disk systems.
See Abstr. 151.052.

An investigation of plane galactic orbits of stars for different expressions of the gravitational potential.
See Abstr. 151.007.

The persistence of the wave spiral pattern in galaxies. See Abstr. 151.021.

Theory of spiral structure. See Abstr. 151.060.

Die absolute Eigenbewegung offener Sternhaufen.
See Abstr. 153.020.

A survey of the continuum radiation at 820 MHz between declinations −7° and + 85°. II. A study of the galactic radiation and the degree of polarization with special reference to the loops and spurs.
See Abstr. 157.002.

Correlation of neutral hydrogen and radiocontinuum loop IV. See Abstr. 157.005.

Neutral hydrogen in an interior region of the Galaxy. The longitude interval 22° to 42°. II. Analysis of the observations. See Abstr. 157.006.

Radio confusion in the galactic plane.
See Abstr. 157.007.

The radio continuum of galaxies. II. The origin of the continuum emission in spiral galaxies.
See Abstr. 158.064.

The separation of galaxies into the halo and the disk subsystem. See Abstr. 158.140.

Comparison of the cepheid variables in the Magellanic Clouds and the Galaxy. See Abstr. 159.018.

Nucleosynthesis in the Magellanic Clouds and the Galaxy. See Abstr. 159.030.

A fluid dynamical study of the accretion process.
See Abstr. 161.001.

156   Galactic Magnetic Field

156.001   Photon mass and the galactic magnetic field.
E. Williams, D. Park.
Phys. Rev. Letters, Vol. 26, 1651 - 1652 (1971).

If the photon has a finite rest mass m, a filament of magnetic flux sustained by a partially ionized gas decays exponentially at a rate proportional to m^2. Arguing from assumptions regarding the galactic magnetic field that are plausible though not yet rigorously established, we find an upper limit $m \leqslant 3 \times 10^{-56}$ gm, corresponding to a Compton wavelength of 6 lt yr.

156.002   The generation of magnetic fields in astrophysical bodies. VII. The internal small-scale fields.
I. Lerche, E. N. Parker.
Astrophys. Journ., Vol. 168, 231 - 237 (1971).

The paper investigates the small-scale magnetic fields δB generated, along with the large-scale magnetic fields B, by cyclonic turbulence and large-scale shear. The results are directly applicable to the problem of the generation of magnetic fields in the Galaxy, for which the observations show only the complexity of the small- and large-scale fields together, making it difficult to discover the actual configuration of the large-scale galactic field. The calculations show that the local fluctuations in the generation of field in the Galaxy contribute a $\langle(\delta B)^2\rangle$ comparable to $\langle B^2\rangle$.

156.003   The generation of magnetic fields in astrophysical bodies. VIII. Dynamical considerations.
E. N. Parker.
Astrophys. Journ., Vol. 168, 239 - 249 (1971).

This paper examines some of the dynamical questions associated with the hydromagnetic dynamos of the sun and the Galaxy. The outstanding new point which emerges is that the galactic cosmic rays are a major driving force in generating the poloidal magnetic field of the Galaxy.

156.004   The generation of the large-scale magnetic field of the Galaxy. S. I. Vainshtein, A. A. Ruzmaikin.
Astron. Zhurn. Akad. Nauk SSSR, Vol. 48, 902 - 909 (1971). In Russian. English translation in Soviet Astron. AJ, Vol. 15, No. 5.

Poloidal field components due to differential rotation yield a toroidal component. A feed back — the generation of poloidal components by the toroidal field — is produced due to the turbulence mechanism suggested by Steenbeck.

156.005   A dynamic equation for stochastic magnetic field lines in the Galaxy. F. C. Jones.
Astrophys. Journ., Vol. 169, 477 - 485 (1971).

We derive here an equation that describes the development of the probability density of a given field line as a function of distance along the average field. We explain why a Fokker-Planck type of equation is quite unsuitable in this case and show that the answer must be sought in the theory of stationary Gaussian processes. We also derive the distribution function for the "endpoints" of a field line that starts on the galactic central plane.

Terrestrial and extraterrestrial limits on the photon mass. See Abstr. 022.061.

On interstellar grain alignment by a magnetic field.
See Abstr. 131.009.

157 Galactic Radio Radiation

157.001 **Neutral hydrogen in an interior region of the Galaxy, the longitude interval 22° to 42° . I. Observations.** W. W. Shane.
Astron. Astrophys., Suppl. Ser., Vol. 4, 1 - 42 (1971).

Measurements of the 21-cm hydrogen emission near the galactic plane have been made using the 25-m telescope in Dwingeloo. The region $l = 22°.3$ to $42°.3$ has been surveyed at 1° intervals in longitude, with an extent to between 6° and 9° on either side of the galactic equator. The observing and reduction procedures are described briefly. The observations consist in line profiles and latitude scans. The reduced measurements are reproduced graphically as Part VIII of the "Dwingeloo Atlas of 21-cm Profiles".

157.002 **A survey of the continuum radiation at 820 MHz between declinations −7° and + 85°. II. A study of the galactic radiation and the degree of polarization with special reference to the loops and spurs.** E. M. Berkhuijsen.
Astron. Astrophys., Vol. 14, 359 - 386 (1971).

A study of the distribution of the continuum radiation at 820 MHz was made, based on observations between $\delta = -7°$ and $+ 85°$ with a resolution of $1°.2$ (paper I). The distribution along the galactic plane was compared with that of the integrated brightness of the H ı radiation. In the areas of loops I, II and III ridges were collected from different surveys. The temperature spectral index of parts of the loops and spurs is derived. A combination of the 820 MHz continuum survey with a polarization survey at 820 MHz also made with the Dwingeloo telescope (Brouw and Spoelstra, 1971) yielded the distribution of polarization percentages over the sky. After subtraction of the black body radiation and the contribution of extragalactic sources from the minimum temperature a residual temperature of about 2°K at 820 MHz remains.

157.003 **A general survey of 21-cm line radiation at high galactic latitudes ($|b| \geqslant 15°$).** C. R. Tolbert.
Astron. Astrophys., Suppl. Ser., Vol. 3, 349 - 454, with a correction, p. 455 (1971).

A general survey of 21-cm line radiation has been made on a grid of 5° or less for galactic latitudes $|b| \geqslant 15°$. The observations were made with the 25-m telescope at Dwingeloo with a half-power beamwidth of 0°.6 and using a bandwidth of 16 kHz. This material is presented in the form of a profile atlas (Part VII of the "Dwingeloo Atlas of 21-cm Profiles") and as contour diagrams. The hydrogen columnar density has been determined as a function of velocity at each observed position and the general distribution determined.

157.004 **Rocket measurements of cosmic radio noise between 1.16 MHz and 2.40 MHz at 1600 km altitude.** S. Hoang.
Astron. Astrophys., Vol. 15, 383 - 402 (1971).

A rocket cosmic radio noise experiment is described in which two crossed 36-m dipole antennas were each connected to receivers through separate input circuits which were tuned by servo loops to obtain a high sensitivity for the overall system. The measured antenna impedance and the calculated electromagnetic radiation resistance were used to interpret the cosmic noise data obtained near apogee at 1636 km altitude in the direction $l = 172°$, $b = 36°$. The following intensities were deduced at 2.40, 1.65, 1.61 and 1.16 MHz respectively:

$4.2 \pm 16\%$, $3.3 \pm 24\%$, $3.2 \pm 24\%$, and $3.9 \pm 24\% \times 10^{-20}$ W $m^{-2} Hz^{-1} sterad^{-1}$.

157.005 **Correlation of neutral hydrogen and radiocontinuum loop IV.** I. Fejes.
Astron. Astrophys., Vol. 15, 419 - 423 (1971).

A low-velocity neutral-hydrogen spur is found to coincide with the radiocontinuum loop IV at galactic latitudes above $b = +40°$. The neutral-hydrogen spur follows loop IV within $2°.5$ along an arc of 180°.

157.006 **Neutral hydrogen in an interior region of the Galaxy. The longitude interval 22° to 42°. II. Analysis of the observations.** W. W. Shane.
Astron. Astrophys., Suppl. Ser., Vol. 4, 315 - 393 (1971).

Twenty-one centimeter observations reported in paper I (Shane 1971) are presented here in the form of contour maps at constant longitude. Further analysis of the line profiles is described. These are resolved into Gaussian components, whose variation as a function of latitude is studied in order to derive a representation of the data at each longitude in terms of a sum of bivariate normal distributions. The results of the one- and two-dimensional analyses are presented. Certain corrections are applied in the course of the analysis, the most significant being the conversion of observed brightness temperatures to optical depths and the removal of a neutral-hydrogen background component. The influence of these and other corrections is discussed.

157.007 **Radio confusion in the galactic plane.** L. A. Higgs.
Journ. Roy. Astron. Soc. Canada, Vol. 65, 263 - 272 (1971).

A method has been developed to estimate r.m.s. confusion levels in the galactic plane, on the basis of a model distribution of thermal radio sources (HII regions) in the Galaxy. This method has been used to calculate the confusion levels for the frequencies and beamwidths used at the Algonquin Radio Observatory. Comparison with observations indicates that at frequencies above 3 GHz, thermal emission can account for most of the galactic background.

157.008 **A low-latitude survey from l = 355° to 5° at 1410 MHz.** M. W. Sinclair, F. J. Kerr.
Australian Journ. Phys., Vol. 24, 769 - 772 (1971).

This paper presents a contour map of the region of the Milky Way between longitudes 355° and 5°, latitudes $\pm 2°$, at 1410 MHz. A list of sources is also given with values of peak antenna temperature and flux density.

157.009 **Linienspektroskopische Beobachtungen in der Radio-Astronomie.** O. Hachenberg.
Physik und Kosmologie, (see 003.024), p. 110 - 119 (1971).

Rutile traveling wave masers for the frequency range 1300−3400 MHz and their application in the Onsala 84 foot radio telescope to galactic spectral line emission studies. See Abstr. 033.026.

The Ohio Survey between declinations of 40° and 63° north. See Abstr. 141.157.

Map of the galactic nucleus at 10 μm. See Abstr. 155.031.

158 Single and Multiple Galaxies

158.001 Neutral hydrogen in M33.
G. de Jager, R. D. Davies.
Monthly Notices, Roy. Astron. Soc., Vol. 153, 9 - 27 (1971).

The neutral hydrogen survey of M33 taken with an angular resolution of $14' \times 18'$ arc and a velocity resolution of 8.4 km s^{-1} has been analysed and compared with optical data. These resolutions were adequate to give the density distribution within M33, to obtain a useful rotation curve and to give interesting new information about the local kinematics within the galaxy.

158.002 Observations of NGC 4151 during 1970 in the optical and infra-red. M. V. Penston,
M. J. Penston, G. Neugebauer, K. P. Tritton, E. E. Becklin, N. Visvanathan.
Monthly Notices, Roy. Astron. Soc., Vol. 153, 29 - 40 (1971).

Observations of NGC 4151 at seven wavelengths from 0.3 to 3.4 microns made during the 1970 season are presented. Variations are found at all observed wavelengths but the optical and infra-red light curves are different. The energy distributions of the point source and the background galaxy have been separated and that of the point source closely resembles that of the quasar 3C273. The general form of the light curves can possibly be attributed to a dust model for the infra-red emission but this would be ruled out if suspected rapid infra-red variations are confirmed.

158.003 Compact galaxies. A. P. Fairall.
Monthly Notices Roy. Astron. Soc., Vol. 153, 383 - 392 (1971).

An observational investigation has been carried out on compact galaxies which have been identified on the Palomar Sky Survey. Their colours and spatial distributions are discussed. Results of image tube spectroscopy, made at the McDonald Observatory, of 44 compact galaxies are reported.

158.004 Detection of interstellar OH in two external galaxies.
L. Weliachew.
Astrophys. Journ., (Letters), Vol. 167, L47 - L52 (1971).

Molecular OH lines at 1665 and 1667 MHz have been detected in absorption in the spiral galaxy NGC 253 and the irregular galaxy M82. This is the first time that interstellar molecular lines, either radio or optical, have been observed outside the Galaxy. Observations were made with the Owens Valley interferometer; thus the amplitude and phase of the absorption features could both be measured.

158.005 A supermassive double galaxy in the cluster Abell 1775.
G. Chincarini, H. J. Rood, G. N. Sastry, G. A. Welch.
Astrophys. Journ.. Vol. 168, 11 - 14 (1971).

Spectrographic and photometric data are presented for the giant double galaxy in the cluster Abell 1775. Several independent pieces of evidence are offered to support our belief that the system is a stable physical entity.

158.006 Dwarf galaxies. P. W. Hodge.
Annual Rev. Astron. Astrophys., Vol. 9, (see 003.001), 35 - 66 (1971).

Dwarf galaxies make up a large percentage of the galaxies in the local neighborhood and probably in the universe, though their total mass is insignificant in comparison with that of the rarer giant galaxies. Local group dwarfs have in the past been most useful because of their proximity and the ease with which it has been possible to study their stellar and interstellar content. Their intrinsic interest centers around the question of their origin and its possible dependence upon the

mode of formation of our Galaxy.

158.007 Weitere OB-Assoziationen in den äußersten Randzonen von M31. G. A. Richter.
Astron. Nachr., Vol. 292, 275 - 278 (1971).

In comparing a U- and a B-plate of the Tautenburg 2 m Schmidt telescope 7 scattered groupings of blue objects have been discovered in the outer regions of the large Andromeda nebula. A closer investigation showed these objects to be apparently so far not known OB associations, well fitting to a spiral structure.

158.008 Evolution of the M31 disk population.
B. M. Tinsley, H. Spinrad.
Astrophys. Space Sci., Vol. 12, 118 - 136 (1971).

An evolutionary model for the M31 inner disk population is described, which at age 12 billion years agrees closely with the narrow-band colors and line indices recently measured by Spinrad et al. (1971), and with the broad-band colors from 0.36 to 3.4 μ. Assuming that gE galaxies have the same stellar population as the M31 inner disk, this model is used to derive evolutionary effects in cosmology.

158.009 Optical line spectrum of a gas heated by hard UV radiation or energetic particles.
J. Bergeron, S. Souffrin.
Astron. Astrophys., Vol. 14, 167 - 188 (1971).

The line emission spectrum of moderately ionized regions is calculated for cases of homogeneous heating. The heating sources considered are either hard UV radiation (energy of order of 10^2 eV) or energetic particles (MeV protons). The physical conditions in the diffuse gas are calculated for various primary ionization rates per total hydrogen number. The ionization equilibria of hydrogen, helium, nitrogen, oxygen and neon are determined and forbidden line as well as hydrogen and helium line emissivities are computed. The general characteristics of both heating mechanisms are given and the range of application of homogeneous heating is discussed. Observations of line intensities of nuclei of galaxies are compared with our models, and fair agreement is found for radiation heating. Furthermore the theoretical total power radiated in the nucleus of Seyfert galaxies, mainly in the form of 200 eV photons, is computed. Good agreement with recent soft X-ray observations is found.

158.010 Nieuwe buren in het heelal. F. P. Israel.
Hemel en Dampkring, Vol. 69, 210 - 213 (1971).

158.011 The post-eruptive galaxy M82.
S. van den Bergh.
Journ. Roy. Astron. Soc. Canada, Vol. 65, 176 (1971).
Abstr. RAS Canada.

158.012 Étude qualitative de quelques galaxies de Markarian.
N. Carozzi, M. Chopinet, R. Duflot.
Comptes Rendus Acad. Sci. Paris, Sér. B, Vol. 273, 151 - 153 (1971).

Les galaxies de Markarian présentent des spectres d'aspect très différents. Dans Ma 2, Ma 100 et Ma 146, on remarque un «renversement» des raies de l'hydrogène.

158.013 Variability of N-galaxies 3C 371, 3C 390.3 and quasar 3C 345. M. K. Babadzhanjanz.
Astron. Tsirk., No. 614, p. 3 - 6 (1971). In Russian.

158.014 Spectral observations of Markarian galaxies. I.
E. K. Denisjuk.

Astron. Tsirk., No. 615, p. 4 - 6 (1971).   In Russian.

158.015 On the radio emission of some Markarian galaxies.
V. N. Kurilchik, V. N. Semenov.
Astron. Tsirk., No. 615, p. 6 - 7 (1971).   In Russian.

158.016 Spectra of two diffuse objects in the region of the Andromeda nebula.
V. F. Esipov, A. S. Sharov.
Astron. Tsirk., No. 615, p. 7 - 8 (1971).   In Russian.

158.017 Comparison stars for observations of the variability of the nuclei of Seyfert galaxies.   V. M. Lyutyj.
Astron. Tsirk., No. 619, p. 1 - 3 (1971).   In Russian.

158.018 Photoelectric observations of the variability of the nuclei of Seyfert galaxies.   V. M. Lyutyj.
Astron. Tsirk., No. 620, p. 1 - 4 (1971).   In Russian.

158.019 Spectral observations of Markarian galaxies. II.
E. K. Denisjuk.
Astron. Tsirk., No. 621, p. 7 - 8 (1971).   In Russian.

158.020 Spectral observations of Markarian galaxies. III.
E. K. Denisjuk.
Astron. Tsirk., No. 624, p. 1 - 2 (1971).   In Russian.

158.021 How to find young galaxies.
W. L. W. Sargent, L. Searle.
Comments Astrophys. Space Phys., Vol. 3, 111 - 114 (1971).
The primary purpose of this article is to discuss the criteria which might be used to identify young galaxies, supposing they exist.

158.022 A compact radio component in M82.
P. N. Wilkinson.
Monthly Notices Roy. Astron. Soc., Vol. 154, 1P - 4P (1971).
The nearby galaxy M82 contains a small radio component ($< 0.''7$), emitting ≈ 7 per cent of the total radio flux at 1423 MHz, whose spectrum exhibits a low frequency cut off. Several explanations for the cut off in the spectrum are considered. M82 appears to be another example of a galaxy having an 'active' core in Heeschen's definition.

158.023 Radius of maximum rotational velocity in spiral galaxies.   T. W. Noonan.
Astron. Astrophys., Vol. 14, 437 - 439 (1971).
An attempt to find a correlation between the radius R of maximum rotational velocity of spiral galaxies and other galaxian parameters gives negative results for the sparse data available, except for a possible increase in R with galaxy type a, b, c.

158.024 Galaxias vecinas.   A. L. de la Barra.
El Universo, Vol. 25, 44 - 45 (1971).

158.025 Search for the variability of the Hα-line intensity in the nucleus of the Seyfert galaxy NGC 4151.
V. M. Lyutyj, A. M. Cherepashchuk.
Astron. Tsirk., No. 633, p. 3 - 6 (1971).   In Russian.

158.026 An attempt to observe Hα-emission in M31, M33 and M51.   V. F. Zhidkov.
Astron. Tsirk., No. 633, p. 6 - 7 (1971).   In Russian.

158.027 Detailed UBV photometry of the peculiar galaxy NGC 1023.   V. G. Derevjanko.
Trudy Astron. Obs., *Leningrad,* Vol. 28 (= Uchenye Zapiski Leningr. Un-ta, No. 359 = Seriya Matem. Nauk, vyp. (No.) 47), p. 32 - 39 (1971).   In Russian.
The UBV intensity distribution and the distribution of

the colours U−B and B−V along the image of NGC 1023 are found. The colour indices and the spectrum of the peculiar region I show that this region is not an ejection from the galaxy.

158.028 Interacting radio galaxies.
J. M. Hill, M. S. Longair.
Monthly Notices Roy. Astron. Soc., Vol. 154, 125 - 139 (1971).
Optical identifications have been found for the radio sources 3C 129 and 3C 129.1 which suggest that they are associated. Similiar associations of radio sources have previously been found. 3C 129, the sources 3C 83.1B, IC 310 and 5C4.81 form a new class of radio galaxy in which the source of energy lies outside the associated galaxy. The energy originates within a nearby galaxy which is also a radio source. The energy appears to be transmitted in the form of plasma streams with velocities which must be greater than or equal to about 0.01 c. The existence of these streams supports the hypothesis of an intergalactic medium in clusters.

158.029 Evidence for composition gradients across the disks of spiral galaxies.   L. Searle.
Astrophys. Journ., Vol. 168, 327 - 341 (1971).
The integrated spectra of H II regions located in the inner spiral arms of Sc galaxies are systematically different from those of H II regions in the outer arms. This is, in part at least, an abundance effect. The N/O ratio (and probably also the abundance ratios O/H and N/H) decreases from the inner to the outer arms.

158.030 On the nature of the galaxy identified with 3C 386.
R. Lynds.
Astrophys. Journ., (*Letters*), Vol. 168, L87 - L89 (1971).
New spectroscopic and photometric observations indicate that Griffin's optical identification for 3C 386 is a galaxy with a foreground galactic star superposed. The galaxy has a redshift of 0.0177 and an absolute visual magnitude of at least -20.

158.031 Optical polarization of two Seyfert galaxies.
A. Kruszewski.
Acta Astron., Vol. 21, 311 - 327 (1971).
Multicolor polarimetric and photometric observations of NGC 1068 and NGC 4151 are presented. In NGC 1068 the polarization is confined to the nuclear region smaller than $10''$. There is no convincing evidence for time variability. In NGC 4151 the degree of polarization changes with time and these variations are positively correlated with the brightness variations. In both objects the position angles do not change with time or wavelength and the flux density of the polarized radiation per unit frequency is only slightly dependent on the frequency.

158.032 UBV photometry of spiral galaxies Markarian 10 and 79 with Seyfert type nuclei.
E. A. Dibay, V. M. Lyutyj.
Astrofizika, Vol. 7, 169 - 175 (1971).   In Russian. – English translation in Astrophysics, Vol. 7, No. 2.
UBV data for the two Markarian galaxies 10 and 79 with Seyfert type nuclei are given. The distributions of surface brightness and colours along the radius of the galaxy have been calculated. We used two-colour diagrams $(U-B)_0 - (B-V)_0$ to compare the Markarian galaxies 10 and 79 with the galaxies NGC 1275 and 7469. The distribution of colour along radius of Markarian 10 is similar to that of NGC 1275, that of Markarian 79 to that of NGC 7469. The colour of the outer regions of Markarian 10 and NGC 1275 corresponds to G-stars, that of Markarian 79 and NGC 7469 to redder stars of K type.

158.033 The spectra of Markarian galaxies. III.

M. A. Arakelian, E. A. Dibay, V. F. Yesipov, B. E. Markarian.
Astrofizika, Vol. 7, 177 - 187 (1971). In Russian. — English translation in Astrophysics, Vol. 7, No. 2.

The results of spectral observations of fifty objects from Markarian's third list of galaxies with ultraviolet continuum are presented. Emission lines are detected in the spectra of thirty-eight objects. The spectra of objects 231, 268, 270, 273, 279 and 290 contain wide emission lines, which are typical for the nuclei of Seyfert galaxies. Less prominent Seyfert-type features exist in the spectra of 291 and 298.

158.034 On the surface brightness of the bars of spiral galaxies.   A. T. Kalloghlian.
Astrofizika, Vol. 7, 189 - 195 (1971). In Russian. — English translation in Astrophysics, Vol. 7, No. 2.

A direct photometric study of 30 barred spirals shows that the mean surface brightness of the majority of the bars has a small dispersion regardless of the subtypes of galaxies. Some differences in physical properties of nuclei of barred and normal spirals are noted.

158.035 Nuclei of galaxies: Introduction.
     V. A. Ambartsumian.
Nuclei of galaxies. Conference 1970, (see 012.002), p. 9 - 24 (1971).

158.036 An optical form morphology of Seyfert galaxies.
     W. W. Morgan, N. R. Walborn, J. W. Tapscott.
Nuclei of galaxies. Conference 1970, (see 012.002), p. 27 - 40 (1971).

158.037 The stellar content and evolution of galaxy nuclei.
     H. Spinrad.
Nuclei of galaxies. Conference 1970, (see 012.002), p. 45 - 72 (1971).

158.038 The optical line and continuous spectra of radio galaxies, compact galaxies, and Seyfert galaxies.
W. L. W. Sargent.
Nuclei of galaxies. Conference 1970, (see 012.002), p. 81 - 109 (1971).

158.039 Physical conditions in the active nuclei of galaxies and quasi-stellar objects deduced from line spectra.
D. E. Osterbrock.
Nuclei of galaxies. Conference 1970, (see 012.002), p. 151 - 185 (1971).

158.040 Infrared emission of galaxies.   F. J. Low.
     Nuclei of galaxies. Conference 1970, (see 012.002), p. 195 - 208 (1971).

158.041 Optical properties of nuclei.   A. Sandage.
     Nuclei of galaxies. Conference 1970, (see 012.002), p. 271 - 309 (1971).

158.042 Space densities and time scales of Seyfert galaxies, radio galaxies and quasi-stellar objects.
M. Schmidt.
Nuclei of galaxies. Conference 1970, (see 012.002), p. 395 - 397 (1971).

158.043 Theoretical considerations regarding non-thermal emission and ejection of matter from galactic nuclei.
G. Burbidge.
Nuclei of galaxies. Conference 1970, (see 012.002), p. 411 - 433 (1971).

158.044 Massive rotators in galactic nuclei.
     L. Woltjer.

Nuclei of galaxies. Conference 1970, (see 012.002), p. 477 - 483 (1971).

158.045 Spinars: A progress report.
     P. Morrison, A. Cavaliere.
Nuclei of galaxies. Conference 1970, (see 012.002), p. 485 - 509 (1971).

158.046 Formation and evolution of bright black holes.
     D. Lynden-Bell.
Nuclei of galaxies. Conference 1970, (see 012.002), p. 527 - 538 (1971).

158.047 Of the nature of compact objects.   F. Hoyle.
     Nuclei of galaxies. Conference 1970, (see 012.002), p. 583 - 591 (1971).

158.048 The age of the galaxies and globular clusters: Problems of finding the Hubble constant and deceleration parameter.   A. Sandage.
Nuclei of galaxies. Conference 1970, (see 012.002), p. 601 - 622 (1971).

158.049 Nuclei of galaxies: Summary of observational results.   E. M. Burbidge.
Nuclei of galaxies. Conference 1970, (see 012.002), p. 713 - 731 (1971).

158.050 Nuclei of galaxies: Summary from the theoretical point of view.   L. Woltjer.
Nuclei of galaxies. Conference 1970, (see 012.002), p. 741 - 753 (1971).

158.051 Gli oggetti Maffei.   G. Parmeggiani.
     Coelum, Vol. 39, 172 - 183 (1971).

158.052 A high resolution radio continuum survey of M51 and NGC 5195 at 1415 MHz.
D. S. Mathewson, P. C. van der Kruit, W. N. Brouw.
Bull. American Astron. Soc., Vol. 3, 369 (1971). — Abstr. AAS.

158.053 The dynamical mass of supergiant galaxies.
     G. A. Welch.
Bull. American Astron. Soc., Vol. 3, 390 (1971). — Abstr. AAS.

158.054 Theoretical model of NGC 4038/39.
     A. Toomre, J. Toomre.
Bull. American Astron. Soc., Vol. 3, 390 - 391 (1971). Abstr. AAS.

158.055 Probably most precise modulus of the great Andromeda nebula.   S. I. Gaposhkin.
Bull. American Astron. Soc., Vol. 3, 398 (1971). — Abstr. AAS.

158.056 X-ray emission from Seyfert galaxies.
     A. Cavaliere, G. Blumenthal, W. Tucker.
Bull. American Astron. Soc., Vol. 3, 399 (1971). — Abstr. AAS.

158.057 Photometry of galaxies with integrating television.
     P. Crane.
Bull. American Astron. Soc., Vol. 3, 399 (1971). — Abstr. AAS.

158.058 Line profiles of the E7 galaxy NGC 7332.
     R. A. Chevalier, D. C. Morton.
Bull. American Astron. Soc., Vol. 3, 399 (1971). — Abstr. AAS.

158.059 Nuclei of galaxies and quasars as sources of infrared emission. G. S. Bisnovaty-Kogan, R. A. Sunyaev.
Astron. Zhurn. Akad. Nauk SSSR, Vol. 48, 881 - 893 (1971).
In Russian. English translation in Soviet Astron. AJ, Vol. 15, No. 5.

A model of the nuclei of galaxies and of quasars is discussed, representing a disk consisting of stars, with often supernova explosions and a large number of neutron stars. The possibility of energy release is considered during accretion of gas on a collapsed star.

158.060 Diameter function and luminosity function of galaxies.
I. D. Karachentsev, V. E. Karachentseva, A. I. Shapovalova, T. Jaakkola.
Astron. Zhurn. Akad. Nauk SSSR, Vol. 48, 922 - 933 (1971).
In Russian. English translation in Soviet Astron. AJ, Vol. 15, No. 5.

For galaxies with known radial velocities, the luminosity function φ (M) and the diameter function ψ (lg D) are obtained. The influence of the minimum radial velocity restriction on the form of φ (M) is considered. Using the lists of dwarf galaxies, the diameter function for the 88 members of the local group and the M 81, M 101 groups is obtained. Results of counts of galaxies on the Palomar prints for clusters of galaxies are presented.

158.061 The redshift: Another model. S. I. Urbanovich.
Astron. Zhurn. Akad. Nauk SSSR, Vol. 48, 957 - 961 (1971). In Russian. English translation in Soviet Astron. AJ, Vol. 15, No. 5.

A possible model for describing the redshift or the excess (lack) of the redshift and, as a rule, the wide spectral lines is proposed. The model is based on the assumption that emitting (absorbing) atoms are subjected to an external influence.

158.062 A determination of the luminosity function of radiogalaxies at 400 and 2700 MHz.
J. K. Merkelijn.
Astron. Astrophys., Vol. 15, 11 - 29 (1971).

This paper presents the results of a statistical treatment of the radiogalaxy identifications from the Parkes catalogue. Identification statistics are also given for quasi-stellar objects. A special effort was made to minimize the errors in apparent optical magnitude and flux density. The effects of galactic absorption, limiting apparent optical magnitude and limiting flux densities on the completeness of the samples are investigated. The resulting complete samples are used in the construction of luminosity distributions and a luminosity function at frequencies of 400 and 2700 MHz.

158.063 The radio continuum of galaxies. I. Observations
J. Lequeux.
Astron. Astrophys., Vol. 15, 30 - 41 (1971).

Positions, flux densities in the continuum near 1420 MHz and east-west brightness distributions have been measured for 54 spiral and irregular galaxies and 20 radio-weak elliptical galaxies. Upper limits for the flux density are given for 16 spiral and irregular galaxies and 7 elliptical galaxies. The observations have been made with the two-element interferometer at the Owens Valley Radio Observatory. We confirm that the radio emission of spiral and irregular galaxies comes from a flat radio disk and/or radio nucleus, the properties of which are discussed. NGC 5253 is an exceptional object with purely thermal radio emission. Elliptical galaxies fall into two distinct categories from their radio properties.

158.064 The radio continuum of galaxies. II. The origin of the continuum emission in spiral galaxies.
J. Lequeux.

Astron. Astrophys., Vol. 15, 42 - 49 (1971).

We compare the distribution of young population I and the distribution and rate of appearance of supernovae in our Galaxy and external spiral galaxies with the diameter and distribution of the disk continuum radio observed by Lequeux (see the preceding paper) and others in these galaxies. Marked correlations are found, and it is suggested that the relativistic electrons which produce the disk radio emission are accelerated in supernovae and propagate by diffusion in the galaxy, leaving it at about 500 pc from their source, in agreement with the model of Jokipii and Parker (1969). Radio emission in the nuclei of spiral galaxies may also be due to electrons ejected by supernovae, at least for a fraction of them.

158.065 Observations of core sources in Seyfert and normal galaxies with the Westerbork synthesis radio telescope at 1415 MHz. P. C. van der Kruit.
Astron. Astrophys., Vol. 15, 110 - 122 (1971).

Eight Seyfert and ten normal galaxies have been observed with the Westerbork synthesis radio telescope at a frequency of 1415 MHz. Unresolved sources ($<7''$) were detected in the nuclei of all galaxies previously classified as Seyferts. In only three of the ten "normal" galaxies radio emission was detected. The observations show that complex structures observed in radio galaxies exist also within normal galaxies. The 1415 MHz power of the other 11th to 12th magnitude galaxies is shown to be less than about 10^{19} W Hz^{-1} sterad^{-1}. The analysis shows that there is a relation between the 1415 MHz and 10μ power for the nuclei of Seyfert galaxies. Similar relations may exist between the powers at 1415 MHz and those in the optical continuum and H$_\beta$. This is briefly discussed in terms of the recent model given by Cavaliere et al. (1970), where the electrons radiating at radio wavelengths originate from the small core, where the infrared radiation is emitted.

158.066 Sculptor-type dwarfs in nearby groups of galaxies. V. E. Karachentseva.
Vestn. Kiev. Un-ta, Ser. Astron., No. 12, p. 98 - 102 (1970). In Russian.

In four nearby groups (M 81, Eridanus, NGC 1068, NGC 1084) the distribution of galaxies along the radii of the groups is examined. The functions of linear diameters of Sculptor-type galaxies are drawn. A positive correlation between the luminosities of the brightest normal members of groups and the linear dimensions of Sculptor-type dwarfs is obtained.

158.067 On the statistics of double galaxies.
I. D. Karachentsev.
Vestn. Kiev. Un-ta, Ser. Astron., No. 12, p. 103 - 115 (1970). In Russian.

A list of 96 pairs of galaxies with known radial velocities of components is given. The list contains some data about the members of pairs: angular and linear distances between the members, the difference of radial velocities, apparent and absolute magnitudes, the angular and linear diameters, morphological types, and the calculated values of virial mass to luminosity ratios of double galaxies. The distribution of members of pairs on different parameters is discussed. The reasons indicating disintegration of double galaxies are presented.

158.068 A connection between the spiral galaxy NGC 4319 and the quasi-stellar object Markarian 205.
H. Arp.
Astrophys. Letters, Vol. 9, 1 - 4 (1971).

A luminous filament connects the quasi-stellar object Markarian 205 whose redshift is 0.070 with the nucleus of the spiral galaxy NGC 4319 whose redshift is 0.006.

158.069 Hubble, Lundmark and the classification of non-galactic nebulae. R. Hart, R. Berendzen.

Journ. History Astron., Vol. 2, 200 (1971). – Remarks to an earlier paper (Journ. History Astron., Vol. 2, 109 - 119 (1971)).

158.070 Redshift effect of galaxies in systems of galaxies.
T. Jaakkola.
Vestn. Kiev. Un-ta, Ser. Astron., No. 13, p. 97 - 103 (1971). In Russian.

6 clusters, 47 groups and 54 pairs of galaxies (altogether 614 objects) are studied regarding systematic differences between redshifts of different galaxy type. A systematic effect is found in every kind of systems. The results suggest a gravitational reason to the effect. This is in favour of the gravitational origin of the redshifts of quasars, too. A consequence of the effect is an increase of the calculated expansion time scales of systems of galaxies.

158.071 UBV photometry of irregular galaxies. I. NGC 4449.
A. I. Shapovalova.
Vestn. Kiev. Un-ta, Ser. Astron., No. 13, p. 104 - 115 (1971). In Russian.

The results of a detailed three-colour photometry of the irregular galaxy NGC 4449 are presented. Integral magnitudes, U - B and B - V colors, and the distribution of brightness and colour along the axes of the galaxy are given, as well as U - B, B - V and M_B values of some agglomerations. The results allow to point out the main structural peculiarities of the galaxy.

158.072 Observational consequences of inverse Compton models for Seyfert galaxies.
J. Bergeron, E. E. Salpeter.
Astrophys. Letters, Vol. 9, 121 - 124 (1971).

A simple model is reviewed for relativistic electrons in a compact region, where infrared emission is produced by synchrotron radiation and X-ray emission by the inverse Compton scattering of the infrared photons. The model parameters are given for the nuclei of three Seyfert galaxies and one quasar, which suggest rapid intensity variations (days or weeks). Production, by further Compton-scattering of the X-rays, of a γ-ray flux is predicted, which should be observable at least for NGC 1275.

158.073 Variations in the μm flux from NGC 1068.
F. J. Low, G. H. Rieke.
Nature, Vol. 233, 256 - 257 (1971).

The flux detected at 10 μm from the nucleus of NGC 1068 has decreased by 30 % over a period of about a year. If this flux is generated by thermal radiation from dust grains, the observed variations place constraints on the spatial distribution or on the wavelength dependence of the emissivity of the grains.

158.074 Spectra of seven compact galaxies.
K. Kodaira.
Publ. Astron. Soc. Japan, Vol. 23, 589 - 591 (1971).

Spectroscopic observations of seven compact galaxies are reported. Two of these show sharp emission lines superposed on F-type spectra, one shows a featureless continuum, and one has an A-type spectrum without emission lines, while the other three show G-type spectra.

158.075 Chain of galaxies. B. A. Vorontsov-Vel'yaminov.
IAU Circ., No. 2357 (1971).

158.076 Possible new companion to M31.
S. van den Bergh.
IAU Circ., No. 2366 (1971).

158.077 The color-redshift relation for giant elliptical galaxies. B. M. Tinsley.
Astrophys. Space Sci., Vol. 12, 394 - 407 (1971).

The colors of giant elliptical (gE) galaxies in clusters out to redshift z = 0.2, observed by Oke and Sandage (1968), are studied for systematic color-redshift effects. The color-redshift trends, interpreted as evolutionary changes, are related to evolution in the magnitude-redshift relation by means of models of stellar evolution in a gE galaxy.

158.078 Mapping of neutral hydrogen in galaxies by aperture synthesis techniques.
J. E. Baldwin, C. Field, P. J. Warner, M. C. H. Wright.
Monthly Notices Roy. Astron. Soc., Vol. 154, 445 - 454 (1971).

The problems of observing neutral hydrogen in external galaxies with high angular resolution are discussed. The operation of an aperture synthesis telescope with a cross correlation spectrometer giving an angular resolution better than 1' arc with a velocity resolution of 39 km s⁻¹ is described.

158.079 Photometric measurements at $\lambda = 11 \mu$m of NGC 4151. W. A. Stein, F. C. Gillett.
Nature, Phys. Sci., Vol. 233, 16 - 17 (1971).

The nucleus of NGC 4151 has been observed at $\lambda = 11$ μm over a period of about two years. No significant changes in flux at this wavelength have been detected.

158.080 Note on N galaxies and mini-quasars.
D. Lynden-Bell.
Monthly Notices Roy. Astron. Soc., Vol. 155, 119 - 127 (1971).

A comparison of the properties of quasars and N galaxies suggests that they form a continuous distribution. The observed numbers of N galaxies then provide further information on the luminosity function of the quasars.

158.081 The period (age) gradient of cepheids across the spiral branch S4 in the Andromeda nebula.
Yu. N. Efremov.
Astron. Tsirk., No. 639, p. 5 - 8 (1971). In Russian.

158.082 Distribution of Markarian galaxies with ultraviolet continua according to diameters and absolute magnitudes. I. M. Yankulova.
Astron. Tsirk., No. 647, p. 4 - 6 (1971). In Russian.

158.083 Search for extragalactic H_2O.
D. F. Dickinson, E. J. Chaisson.
Astrophys. Journ., Vol. 169, 207 - 208 (1971).

Three giant H II regions in M 33 examined for 22-GHz H_2O emission showed no maser as intense as the one in W 49.

158.084 Energy distributions and K-corrections for the total light from giant elliptical galaxies.
R. Schild, J. B. Oke.
Astrophys. Journ., Vol. 169, 209 - 214 (1971).

New absolute spectral-energy distributions of giant elliptical galaxies in the Virgo cluster are presented. New K-corrections are calculated for B, V, and R_s magnitudes, and predicted color changes in $B - V$ and $V - R_s$ are given to z = 0.28 and 0.60, respectively.

158.085 Absolute energy curves and K-corrections for giant elliptical galaxies. A. E. Whitford.
Astrophys. Journ., Vol. 169, 215 - 220 = Contr. Lick Obs., No. 330 (1971).

Scanner comparison of five typical giant elliptical galaxies with standard stars calibrated on the Hayes system has been used to derive the absolute energy curve over the range $3400 < \lambda < 11000$ Å. The energy curve shows the galaxies to be bluer than found by Oke and Sandage. The K-corrections calculated from the curve are smaller than the Oke-Sandage values.

158.086 A 21-centimeter study of the spiral galaxy Messier 33. K. J. Gordon.
Astrophys. Journ., Vol. 169, 235 - 270 = Contr. Five College Obs., *Amherst, Mass.,* No. 110 (1971).

The distribution of neutral atomic hydrogen and the kinematic properties of the late-type spiral galaxy M33 are derived from observations obtained at 21-cm wavelength with a multichannel radiometer. By considering these atoms as test particles in the gravitational field of the galaxy, we draw conclusions about the galaxy's total mass and mass distribution. We summarize the results and relate them to findings for other galaxies of similar type. The Appendix contains details of the method of model fitting used in the analysis of the kinematics of the galaxy.

158.087 Hydrogen line and continuum study of Maffei 2 by radio interferometry. G. S. Shostak, L. Weliachew.
Astrophys. Journ., *(Letters),* Vol. 169, L71 - L76 (1971).

The suspected spiral galaxy Maffei 2 has been interferometrically observed in the 21-cm line and adjacent continuum using a north-south baseline having a 6.5 fringe spacing. Velocity resolution was 21 km s^{-1}. Positions of the emission centroids at each velocity allowed determination of the orientation and dynamic parameters of the object and a lower limit to the total mass. The fractional hydrogen content of Maffei 2 is less than 0.005D, where D is the distance in Mpc. The continuum source associated with it is shown to be extended.

158.088 Stellar populations in galaxies. I. R. King.
Publ. Astron. Soc. Pacific, Vol. 83, 377 - 400 (1971).

This article reviews present-day knowledge of stellar populations and the relation of populations to the forms of galaxies. A historical outline of the concept of populations is followed by a summary of the present picture. After a discussion of the distribution of populations in the various types of galaxies, it is shown that the whole picture can be understood in terms of a simple scheme of the development of galaxies. Problems of heavy-element enrichment through nucleosynthesis are briefly discussed.

158.089 The mean ratio of mass to the three-halves power of luminosity for elliptical and lenticular galaxies. T. W. Noonan.
Publ. Astron. Soc. Pacific, Vol. 83, 479 - 484 (1971).

The available data on masses of ellipticals and lenticulars are summarized with emphasis on the uncertainties in the ratio R of mass to $10^{-0.6M}$, where M is the absolute magnitude. With correction for galactic absorption the result is $R = 0.30^{+0.27}_{-0.14}$ $M_\odot$ based on the magnitude scale in the de Vaucouleurs catalog. Implications for the mean cosmic density are discussed.

158.090 Some unusual southern hemisphere objects. P. K. Lü.
Astron. Journ., Vol. 76, 775 - 776, 947 - 949 = Western Connecticut State Coll. Obs., *Danbury,* Contr. No. 15 (1971).

Positions, finding charts, and descriptions are given for 40 unusual objects in the southern hemisphere. These unusual objects, which appear to be mostly peculiar galaxies or galactic nebulosities, were found from an examination of photographic plates taken with the 20-inch double astrograph of the Yale-Columbia Southern Observatory.

158.091 Possibility that the far ultraviolet excess in M 31 is due to main-sequence stars. B. M. Tinsley.
Astron. Astrophys., Vol. 15, 403 - 405 (1971).

158.092 A spectroscopic study of luminous galactic nuclei. E. Ye. Khachikian, D. W. Weedman.

Astrofizika, Vol. 7, 389 - 406 = Repr. Arthur J. Dyer Obs., Vanderbilt Univ., *Nashville, Tennessee,* No. 58 (1971). – Reprinted in Astrophysics, Vol. 7, No. 3.

Spectroscopic observations of a sample of galactic nuclei from the lists of Markarian and Seyfert are presented in order to consider the nature of the nuclei with broad emission lines. It is found from photographic spectrophotometry and emission line profiles that such galaxies can be consistently classified into two classes: objects like NGC 5548 that are found to have small, dense nuclei containing low velocity gas and objects like NGC 1068 with larger nuclei containing gas of lower density but higher velocity.

158.093 Large-scale structural characteristics and photometry of NGC 3031 determined from equidensity curves. W. Hoegner, Z. Kadla, N. Richter, A. Strugatskaya.
Astrofizika, Vol. 7, 407 - 415 (1971). – Reprinted in Astrophysics, Vol. 7, No. 3.

The method of integral equidensity curves was applied for determining the large-scale structure of the galaxy NGC 3031. Altogether 49 equidensity curves were obtained from four UBVR plates taken with the two-meter telescope of the Tautenburg Observatory. The axial ratio and position angle of the major axis in dependence on the apparent distance to the center of the galaxy were derived. The relative surface brightness in UBV in dependence on the semi-major axis was determined by means of a photographic wedge printed on the original photographs. A comparison with photoelectric measurements in V and B enabled a determination of the zero-point and to obtain the apparent surface brightness.

158.094 A polarimetric study of compact extragalactic objects. V. A. Dombrovsky, M. K. Babadzhanianz, V. A. Hagen-Thorn, S. M. Houtkevich.
Astrofizika, Vol. 7, 417 - 434 (1971). In Russian. – English translation in Astrophysics, Vol. 7, No. 3.

The present paper gives the results of polarimetric and in some cases photometric observations of compact extragalactic objects – QSS and the nuclei of different galaxies.

158.095 On the interpretation of the Hα profile of the Seyfert galaxy NGC 5548. K. S. Anderson.
Astrophys. Journ., Vol. 169, 449 - 453 (1971).

Spectroscopic observations of the Hα profile in the Seyfert galaxy NGC 5548 suggest that electron scattering is not the dominant factor in producing the extensive emission wings. The profile shows indications of discrete cloud structures. The predictions of a simple model based upon radial expansion about an optically thick central region are compared with the observations.

158.096 Sequenze evolutive di galassie. F. Bertola.
Mem. Soc. Astron. Italiana, Nuova Ser., Vol. 42, 373 - 374 (1971).

158.097 Identification of the nucleus of NGC 5128. W. E. Kunkel, H. V. Bradt.
Astrophys. Journ., *(Letters),* Vol. 170, L7 - L10 (1971).

A 'hot spot', 3″ by 5″ in size at the center of NGC 5128 (Cen A), has been detected photographically at 8000 Å and identified with a relatively inconspicuous feature in blue photographs. From photometry in *UBVRI*, its luminosity is 2.4×10^{41} ergs s^{-1} if the distance is 5 Mpc and absorption by foreground dust is A$_v$= 1.9 magnitudes. This leads to a mass of 1.5×10^9 $M_\odot$.

158.098 Radio emission from the nucleus of NGC 5128. C. M. Wade, R. M. Hjellming, K. I. Kellermann, J. F. C. Wardle.
Astrophys. Journ., *(Letters),* Vol. 170, L11 - L13 (1971).

A compact nonthermal radio source has been found within the infrared object suggested by Kunkel and Bradt as the nucleus of NGC 5128. The measured angular size is less than $0\overset{''}{.}5$, corresponding to linear dimensions of 13 pc or less. The increase in flux density toward shorter wavelengths indicates that the opacity is appreciable at centimeter wavelengths; this implies that the true angular diameter is of the order of $0\overset{''}{.}01$ or less.

158.099 **Infrared observations of the core of Centaurus A, NGC 5128.**   E. E. Becklin, J. A. Frogel, D. E. Kleinmann, G. Neugebauer, E. P. Ney, D. W. Strecker. Astrophys. Journ., (*Letters*), Vol. 170, L15 - L19 (1971).

Infrared radiation from the nucleus of NGC 5128 has been measured from 1 to 10μ. Scans and aperture-size data show that there is only one bright infrared region within the central 1' of the galaxy and that it is coincident to within 5'' with the 'hot spot' of Kunkel and Bradt.

158.100 **Radial velocities and line strengths of emission lines across the nuclear disk of M31.** V. C. Rubin, W. K. Ford, Jr. Astrophys. Journ., Vol. 170, 25 - 52 (1971).

Image-tube spectra have been obtained at sixteen position angles across the nuclear bulge of M31, at a dispersion of 28 Å mm^{-1}. Emission lines of Hα, [N II] $\lambda\lambda$6548 and 6583, [S II] $\lambda\lambda$6717 and 6731, He I λ5876, and [O III] λ5007 are observed. From the measured line-of-sight velocities of the excited gas across the inner 400 pc, we obtain a model of the nuclear disk and describe it in detail.

158.101 **Neutral hydrogen motions in the central regions of M 82.**   L. Weliachew. Publ. Astron. Soc. Pacific, Vol. 83, 609 (1971). – Abstr. Astron. Soc. Pacific.

158.102 **The intrinsic color index of Maffei No. 2.** S. van den Bergh. Publ. Astron. Soc. Pacific, Vol. 83, 663 (1971).

The maximum rotational velocities in late-type galaxies are strongly correlated with their integrated intrinsic color indices. This correlation is used to obtain an intrinsic color index $(B-V)_0 = 0.60 \pm 0.08$ m.e. for Maffei No. 2.

158.103 **Größe und Spiralstruktur des Adromedanebels.** G. A. Richter. Sterne, 47. Jahrgang, p. 173 - 184 (1971).

158.104 **Maffei 1, ein neues massereiches Sternsystem der Lokalen Gruppe?**   K.-H. Schmidt. Sterne, 47. Jahrgang, p. 184 - 188 (1971).

158.105 **Apparent associations between bright galaxies and quasi-stellar objects.**   E. M. Burbidge, G. R. Burbidge, P. M. Solomon, P. A. Strittmatter. Astrophys. Journ., Vol. 170, 233 - 240 (1971).

A comparison of the spatial distribution of the 47 identified QSOs in the 3C and 3CR catalogs with the small-redshift galaxies contained in the *Reference Catalog of Bright Galaxies* shows that four QSOs with redshifts in the range 0.5−1.4 are much closer to bright galaxies than would be expected if the 47 QSOs were distributed randomly. An extensive analysis of the distributions shows that the probability that this is a change occurrence is less than 5×10^{-3}.

158.106 **Photographic photometry of compact galaxies around the globular cluster M3.**   Ş. Bozkurt. Sci. Rep. Fac. Sci. Ege Univ., *Izmir*, No. 118 (Astron. No. 11), 27 pp. (1971).

This work deals with a photographic photometry of compact galaxies in the vicinity of globular cluster M3 on photographic plates which have been taken in 1964 with the 2m tel-

escope of Karl Schwarzschild Observatory at Tautenburg. The photometry of the galaxies is based on standard stars, using the photoelectric magnitudes of 18 stars in the globular cluster M3. We tried to find out whether there exists a correlation between the classification of compact galaxies by Richter and any quantitative and diffuse parameter which could be derived by photographic photometry.

158.107 **Entwicklung der Sternsysteme und Quasare aus radio-astronomischer Sicht.**   O. Hachenberg. Physik und Kosmologie, (see 003,024), p. 100 - 109 (1971).

158.108 **Satellites of stellar systems in general.** E. B. Holmberg. The Magellanic Clouds. Astrophys. Space Sci. Library, Vol. 23, (see 012.010), 109 - 113 (1971).

158.109 **On the relative number and space distribution of double and multiple galaxies.** W. Zonn, B. Juchniewicz. Acta Astron., Vol. 21, 487 - 499 (1971).

The analysis of two independent counts of single, double and multiple galaxies on the National Geographic Society Palomar Observatory Sky Survey maps shows, that the mean number of double systems is not a monotonous function of the number N of single galaxies in the same region, but the mean number of multiple systems seems to be an increasing function of N.

158.110 **On the Balmer jump in the continuous spectrum of the nucleus of NGC 4303.** L. S. Nasarova, A. G. Shcherbakov. Astron. Tsirk., No. 648, p. 1 - 3 (1971).   In Russian.

158.111 **Preliminary data on the frequency function of angular momenta of galaxies.** I. L. Genkin, L. M. Genkina. Astron. Tsirk., No. 651, p. 4 - 5 (1971).   In Russian.

158.112 **The structure of the subsystem of novae in the Andromeda nebula.**   A. S. Sharov. Astron. Zhurn. Akad. Nauk SSSR, Vol. 48, 1258 - 1264 (1971). In Russian. English translation in Soviet Astron. AJ, Vol. 15, No. 6.

The structure of the subsystem of novae in the Andromeda nebula is considered up to a distance above 20 kpc. The subsystem has a nucleus with a radius about 2.4 kpc. At the distance about 1−2.4 kpc the space density gradient is 0.81. Further up to the distance of 17 kpc the subsystem forms a disc with a space density gradient 0.16 similar to the value of a gradient of the subsystem of novae in our Galaxy in the neighbourhood of the sun.

158.113 **On the spectra of radio galaxies between 10 and 10 000 MHz.**   M. A. Stull. Astron. Journ., Vol. 76, 970 - 979 (1971).

Measurements have been made of the 8000-MHz flux densities of 60 Parkes Catalogue radio galaxies with the University of Michigan 85-ft radio telescope. The radio spectra of these galaxies were formed by combining the 8000-MHz data with as many flux-density measurements at frequencies between 10 and 10 000 MHz as could be found for them in the literature. The results are reported in detail.

158.114 **A unitary classification for N galaxies.** W. W. Morgan. Astron. Journ., Vol. 76, 1000 - 1002, 1161 - 1162 (1971).

A general N-type classification for galaxies having unstable nuclei is described; it derives out of morphologies by L. M. Ozernoy, and by the writer. The N characteristic is represented by three subgroups (N+, N, N −), which are situated between the normal spirals on one side, and the QSS on the other. This arrangement formalizes the conclusions of a number of investigators. The classification is applied to a group of direct photographs of Zwicky compact objects obtained by W. L. W. Sargent with the 200-inch Hale reflector.

158.115 The jet in M 87.   P. Stewart.
Astrophys. Space Sci., Vol. 14, 261 - 264 (1971).
The jet is assumed to be ejected from the nucleus supersonically into a gas which is at rest; the knots that result are related to those which are observed and an estimate of the jet temperature follows. The magnetic field and particle density inside the jet are discussed and observations are suggested.

158.116 On the evolutionary stage of stars in the Draco dwarf galaxy.   F. Caputo, V. Castellani.
Astrophys. Space Sci., Vol. 14, 323 - 331 (1971).
An analysis of the colour-magnitude diagram of stars in Draco dwarf galaxy is performed on the basis of Baade and Swope observational results. Some correlations among lifetimes of evolutionary stages are derived.

158.117 Infra-red emission from galactic nuclei.
K. M. V. Apparao, T. N. Rengarajan.
Astrophys. Space Sci., Vol. 14, 460 - 463 (1971).
We have examined here the coherent and incoherent synchrotron mechanisms from compact objects and find that these are unlikely to explain the observed infra-red emission.

158.118 Optical observations relevant to cosmology: Hubble diagram.   E. M. Burbidge.
General relativity and cosmology. Course 47, Italian Phys. Soc., 1969, (see 012.022), p. 306 - 314 (1971).

158.119 On the [S II] and [O I] line intensities in gaseous nebulae and nuclei of galaxies.   M. Peimbert.
Bol. Obs. Tonantzintla y Tacubaya, No. 37, Vol. 6, 97 - 100 (1971).
We present new photoelectric observations of the [S II] line intensities in the nuclei of M 51 and M 81. These objects comprise normal H II regions, nuclei of galaxies and a supernova remnant.

158.120 On the frequency function of the absolute magnitudes of field galaxies.   E. K. Denisjuk.
Trudy Astrofiz. Inst., *Alma-Ata*, Vol. 16, 119 - 122 (1971). In Russian.
A method to construct the luminosity function of galaxies is suggested. This function is not distorted by selection of observations. The forms of this function are given both for all types of galaxies in common, and individually for *S-*, *S0-* and *E-*galaxies.

158.121 On numerical models for galaxies with intermediate ellipticity.   L. M. Genkina.
Trudy Astrofiz. Inst., *Alma-Ata*, Vol. 17, 56 - 62 (1971). In Russian.
The possibility of constructing models of systems with intermediate ellipticity systems on the basis of rotation curve observations is considered. For constructing such models a system of two equations is to be solved: the mass distribution equation and the equation relating the circular velocity to the observed rotation velocity. The inapplicability of the usual method of successive approximations is demonstrated. A model for NGC 3115 consisting of three subsystems is calculated.

158.122 Angular diameters of galaxies as distance indicators.
N. N. Pavlova.
Trudy Astrofiz. Inst., *Alma-Ata*, Vol. 17, 63 - 67 (1971). In Russian.
A description of a method of independent determination of the distance to galaxies on the basis of angular dimensions is presented. The system of these angular diameters is related to Holmberg's and de Vaucouleurs' photometric systems.

158.123 Explosiones en galaxias elípticas.   J. L. Sérsic.
Ciencia e Investigación, Vol. 25, 338 - 442 = Obs. Córdoba, Tirada Aparte No. 182 (1969).

158.124 Catálogo de nebulosas extragalácticas de la zona −5°/−25° de declinación, seleccionadas para la determinación de un sistema absoluto de movimientos propios estelares. II − Coordenadas rectilíneas de nebulosas, estrellas de referencia y estrellas de control.   M. López Palacios.
Separate print Inst. y Obs. de Marina, San Fernando, 3 + 144 pp. (1971).
This volume contains the measured coordinates of nebulae, reference stars and control stars on 144 plates, 3 plates for each of the 48 assigned areas, obtained with the Carte-du-Ciel astrograph, in the −5° through −25° zone of the Pulkovo program for the determination of an absolute proper motion system.

158.125 Non-stellar objects used in preparing the SAO Star Atlas, coordinates at equinox 1950.0.
K. Haramundanis.
Separate print Smithsonian Astrophys. Obs., Cambridge, Mass., 166 pp. (1970).

158.126 A new outstanding chain of galaxies.
B. A. Vorontsov-Velyaminov.
Astron. Tsirk., No. 654, p. 1 - 2 (1971). In Russian.

158.127 Photometric peculiarities of Seyfert galaxies.
A. V. Zasov, V. M. Lyutyj.
Astron. Tsirk., No. 658, p. 4 - 6 (1971). In Russian.

158.128 Note to the problem of Seyfert type nuclei of galaxies. I.   E. A. Dibaj.
Astron. Tsirk., No. 660, p. 1 - 3 (1971). In Russian.

158.129 Note to the problem of Seyfert type nuclei of galaxies. II.   E. A. Dibaj.
Astron. Tsirk., No. 660, p. 3 - 6 (1971). In Russian.

158.130 Morphology and related properties of compact galaxies.   F. Bertola, F. Lucchin, E. Nasi.
Mem. Soc. Astron. Italiana, Nuova Ser., Vol. 42, 517 - 521 (1971).
The first 600 compact galaxies contained in Zwicky's lists were photographed from the blue and red prints of the Palomar Sky Survey. A morphological description is given and an attempt to correlate morphological properties with color and apparent magnitude has been made. A high percentage of double and multiple galaxies was found.

158.131 On the redshifts of galaxies.
T. Jaakkola.
Nature, Vol. 234, 534 - 535 (1971).
E, S0 and Sa galaxies in clusters, groups and pairs have excessive negative and Sb and Sc ones excessive positive residual redshifts. For the latter there are correlations between the redshift on one hand and colour index, inclination and possibly magnitude on the other. Part of the redshifts, even of normal galaxies, cannot be explained by systemic velocity.

158.132 On the variability of the emission-line spectrum of the Seyfert galaxy NGC 3227.   I. I. Pronik.
Astron. Tsirk., No. 663, p. 1 - 3 (1971). In Russian.

158.133 Strong infrared galaxies.   N. Wickramasinghe.
New Scient. and Sci. Journ., Vol. 50, 694 - 695

(1971).

The available data on infrared emission from galactic nuclei are consistent with a model wherein a strong source of cosmic rays of X-rays is surrounded by a shell of dust grains. The dust effectively absorbs the primary particle or photon energy and re-emits this in the infrared.

158.134   Compact galaxies.   A. P. Fairall.
Thesis, Univ. Texas, Austin. [Available from Univ. Microfilms, Ann Arbor, Mich., U.S.A. Order No. 71−122], 199 pp. (1970).

The 44 compact galaxies observed at McDonald are reported on individually in this paper. Three multiple systems were involved in the spectroscopic investigation.

158.135   Supergiant galaxies with multiple nuclei.
D. C. Jenner.
Thesis, Univ. Wisconsin, Madison. [Available from Univ. Microfilms, Ann Arbor, Mich., U.S.A. Order No. 70−24753], 51 pp. (1970). − See Phys. Abstr., Vol. 74, No. 62760 (1971).

158.136   The stellar content of M82.   R. W. O'Connell.
Thesis, California Inst. Technology, Pasadena. [Available from Univ. Microfilms, Ann Arbor, Mich., U.S.A. Order No. 70−24308], 191 pp. (1970).

The stellar contents of two regions on the peculiar galaxy M82 are investigated by spectral synthesis. One of the regions is identified as the nucleus of the galaxy on the basis of its surface brightness and inferred mass density; the other is spectrally representative of the disk of M82.

158.137   Nuclei of galaxies.   K. Wakamatsu, K. Sakka.
Astron. Herald, (Japan), Vol. 64, 180 - 183 (1971).
In Japanese.
A review of the author's study on NGC 2782.

158.138   Nebulous objects near Maffei 1.
S. van den Bergh.
Publ. Astron. Soc. Pacific, Vol. 83, 822 - 823 (1971).

Three nebulous objects have been found within 350″ of the center of Maffei 1. The discovery of a prominent dust lane in Maffei 1 by Ford and Jenner is confirmed.

158.139   Galaxies & the universe.   S. E. Williams.
Journ. Astron. Soc. Western Australia, Vol. 32, November, p. 2 - 5; Vol. 33, December, p. 2 - 5 (1971).

158.140   The separation of galaxies into the halo and the disk subsystem.
L. E. Gurevich, A. D. Chernin.
Proc. Sixth Winter School on Space Physics, Part I. Apatity 1969, (see 012.033), p. 34 - 36 (1971).

Brücke zwischen zwei Galaxien.
Umschau, 71. Jahrgang, p. 948 (1971).

Compton scattering of plasma and magnetohydro-dynamic waves on relativistic electrons as a source of radio emission from metagalactic objects.
See Abstr. 062.016.

Final stages of evolution of a magnetoid and observations.   See Abstr. 062.049.

Infrared sources of radiation.
See Abstr. 113.010.

On the variability of the "blue objects".
See Abstr. 113.044.

Stars observed photoelectrically near quasars and related objects.   See Abstr. 113.064.

Analysis of light variations of variable stars in the Ursa Minor dwarf galaxy based on van Agt's observations. See Abstr. 122.144.

Search for novae in the Andromeda nebula on Abastumani plates.   See Abstr. 124.012.

Considerations on the origin of the Platt particles.
See Abstr. 131.041.

Planetary nebulae. IV. Predicted chemical composition and interstellar enrichment.   See Abstr. 131.134.

The optically variable radio source PKS 1514−24 = AP Librae.   See Abstr. 141.023.

Fragmentation of supermassive disks.
See Abstr. 141.066.

Compact radio sources in the nuclei of galaxies.
See Abstr. 141.079.

Expansion models of eruptions in quasars and radio galaxies.   See Abstr. 141.080.

Mechanisms for jets.   See Abstr. 141.084.

The spectra of sources found in an 8000 MHz survey.   See Abstr. 141.095.

Extragalactic Faraday rotation.
See Abstr. 141.110.

Expanding-source model of radio outbursts in QSS and in nuclei of Seyfert galaxies.   See Abstr. 141.124.

Circular polarization of 3C 273 and NGC 4151.
See Abstr. 141.125.

The small-scale structure of radio galaxies and quasi-stellar sources at 3.8 centimeters.   See Abstr. 141.202.

Determination of the potential energy of a gravitational system with spheroidal symmetry.   See Abstr. 151.029.

UBV photometry of globular clusters in M87.
See Abstr. 154.008.

Composition and activity of the nucleus of our Galaxy, and comparison with M 31.   See Abstr. 155.017.

Redshifts and absolute spectral energy distributions of galaxies in distant clusters.   See Abstr. 160.012.

Cosmological evolution in radio galaxies.
See Abstr. 162.036.

Observational paradoxes in extragalactic astronomy.
See Abstr. 162.055.

159 Magellanic Clouds

159.001 X-rays from the Magellanic Clouds.
R. E. Price, D. J. Groves, R. M. Rodrigues, F. D. Seward, C. D. Swift, A. Toor.
Astrophys. Journ., *(Letters)*, Vol. 168, L7 - L9 (1971).

X-rays from the vicinities of both Magellanic Clouds were observed with a spatial resolution of ~ 1° on 1970 September 24. Each Cloud appears as an extended source. The center of X-ray emission is displaced from the region of maximum optical emission. The X-ray spectrum of the Large Cloud is much softer than that of the Small Cloud. The data are more easily interpreted as due to a few strong sources rather than many unresolved weak sources. It is possible that the 30 Doradus nebula is one of these strong sources.

159.002 Emission nebulae in the Magellanic Clouds at 408 MHz. B. Y. Mills, L. H. Aller.
Australian Journ. Phys., Vol. 24, 609 - 615 (1971).

Eleven emission complexes in the LMC and six in the SMC have been observed at 408 MHz with a resolution of about 3' arc. Electron densities and total masses have been calculated, assuming uniform spherical models, and the values are compared with equivalent optical data. Although comparable, the optically derived values are significantly smaller than the radio results.

159.003 Kinematic properties of 71 H II regions in the Large Magellanic Cloud. M. G. Smith, D. W. Weedman.
Bull. American Astron. Soc., Vol. 3, 398 (1971). – Abstr. AAS.

159.004 Red stars in the 30 Doradus nebula region.
E. E. Mendoza V, T. Gomez.
Bull. American Astron. Soc., Vol. 3, 398 (1971). – Abstr. AAS.

159.005 X-ray observations of the Magellanic Clouds.
C. Leong, E. Kellogg, H. Gursky, R. Giacconi, H. Tananbaum, J. Bishop.
Bull. American Astron. Soc., Vol. 3, 399 (1971). – Abstr. AAS.

159.006 Large radial velocity stars in the wing of the Small Magellanic Cloud. N. Carozzi, Y. Peyrin, A. Robin.
Astron. Astrophys., Suppl. Ser., Vol. 4, 231 - 240 (1971). In French.

49 early stars with high radial velocity have been measured on objective prism plates in the wing of the Small Magellanic Cloud. Twenty of them can be found in Sanduleak's lists but 29 are unknown stars. We have determined for 5 of them accurate radial velocities and spectral types on RV Cas spectrograms (73 Å/mm).

159.007 Note on LMC variable stars.
P. W. Hodge, F. W. Wright.
Inform. Bull. Variable Stars, (IAU Commission 27), Konkoly Obs., Budapest, No. 590, 9 pp. (1971).

Of the 53 LMC variables studied by us in Region 35, (Wright and Hodge 1971) 36 have also appeared in the massive study published by Payne-Gaposchkin (1971). It is unusual to have so many stars common to different studies in the Magellanic Clouds, and it is particularly important, therefore, that a comparison be made so that previous suspicions of systematic errors – those due, for example, to selection effects – be checked. This note compares results and draws what conclusions seem reasonable.

159.008 Kinematic properties of seventy-one H II regions in the Large Magellanic Cloud.
M. G. Smith, D. W. Weedman.
Astrophys. Journ., Vol. 169, 271 - 280 = Arthur J. Dyer Obs., Vanderbilt Univ., *Nashville,* Repr. No. 60 (1971).

Using a pressure-scanned interferometer, we have obtained 314 emission-line profiles in Hα and [O III] λ5007 emission from H II regions scattered throughout the Large Magellanic Cloud (LMC). Of these line profiles, 219 were obtained in the region of the 30 Doradus nebula, and 95 were of seventy other H II regions in the LMC. These observations show that the 30 Doradus nebula is unique, characterized by much more rapid, disordered, and complex motions than any other region observed. We present the kinematic properties of the seventy other H II regions observed in the LMC.

159.009 A search for light variations in some supergiants in the Large Magellanic Cloud.
J. D. Rosendhal, M. S. Snowden.
Astrophys. Journ., Vol. 169, 281 - 288 = Rosemary Hill Obs., Dep. Phys. Astron., Univ. Florida, *Gainesville,* Contr. No. 23 (1971).

A careful photometric study has been made of five of the most luminous supergiants in the Large Magellanic Cloud. The results are (1) the maximum amplitude of variability on short time scales (minutes to hours) is less than ± 0.015 mag in both the blue and visual wavelength ranges for all five program stars, and (2) all five stars appear to be variable when observed over a time interval of a month or more. Implications of these results for recent theoretical calculations of the stability of massive stars are briefly discussed.

159.010 The variable stars of the Large Magellanic Cloud.
C. H. Payne-Gaposchkin.
Smithsonian Contr. Astrophys., *Cambridge, Mass.,* No. 13, 41 pp. (1971).

The variable stars in the Large Magellanic Cloud have been studied on the basis of estimates made on all available plates at the Harvard College Observatory. Of the 2184 stars measured, 1830 have been judged to be variable. About 800000 estimates were made. The numerical data derived from a discussion of the material are listed in tabular form.

159.011 Color-magnitude diagrams of OB associations in the Large Magellanic Cloud. P. B. Lucke.
Publ. Astron. Soc. Pacific, Vol. 83, 606 (1971). – Abstr. Astron. Soc. Pacific.

159.012 X-ray emission from the Magellanic Clouds observed by *Uhuru*. C. Leong, E. Kellogg, H. Gursky, H. Tananbaum, R. Giacconi.
Astrophys. Journ., *(Letters),* Vol. 170, L67 - L71 (1971).

Three sources in the Large Magellanic Cloud and one in the Small Cloud have been discovered by *Uhuru*. These sources each emit about 10^{38} ergs s^{-1} in the 2–7-keV band, and they account for more than 90 percent of the total emission from the Clouds. The source in the Small Cloud shows time variability on time scales of hours.

159.013 Survey of principal characteristics of the Magellanic Clouds. A. D. Thackeray.
The Magellanic Clouds. Astrophys. Space Sci. Library, Vol. 23, (see 012.010), 3 - 8 (1971).

159.014 Colour-magnitude arrays of the brightest stars.
A. D. Thackeray.

The Magellanic Clouds. Astrophys. Space Sci. Library, Vol. 23, (see 012.010), 9 - 18 (1971).

159.015 **Colour-magnitude diagrams of faint stars in the associations and in the field of the Magellanic Clouds.** B. E. Westerlund.
The Magellanic Clouds. Astrophys. Space Sci. Library, Vol. 23, (see 012.010), 19 - 24 (1971).

159.016 **Clusters in the Magellanic Clouds.** S. C. B. Gascoigne.
The Magellanic Clouds. Astrophys. Space Sci. Library, Vol. 23, (see 012.010), 25 - 30 (1971).

159.017 **Supernova remnants, planetary nebulae and red stars in the Magellanic Clouds.** B. E. Westerlund.
The Magellanic Clouds. Astrophys. Space Sci. Library, Vol. 23, (see 012.010), 31 - 33 (1971).

159.018 **Comparison of the cepheid variables in the Magellanic Clouds and the Galaxy.** C. Payne-Gaposchkin.
The Magellanic Clouds. Astrophys. Space Sci. Library, Vol. 23, (see 012.010), 34 - 46 (1971).

159.019 **The short-period variable stars in the Magellanic Clouds.** J. Landi Dessy, J. R. Laborde.
The Magellanic Clouds. Astrophys. Space Sci. Library, Vol. 23, (see 012.010), 47 - 49 (1971).

159.020 **Properties of the neutral hydrogen in the Magellanic Clouds.** F. J. Kerr.
The Magellanic Clouds. Astrophys. Space Sci. Library, Vol. 23, (see 012.010), 50 - 65 (1971).

159.021 **Intermediate band photometry of the brightest stars in the Magellanic Clouds.** E. E. Mendoza V.
The Magellanic Clouds. Astrophys. Space Sci. Library, Vol. 23, (see 012.010), 69 - 73 (1971).

159.022 **A search for red variable stars in the Magellanic Clouds.** T. L. Evans.
The Magellanic Clouds. Astrophys. Space Sci. Library, Vol. 23, (see 012.010), 74 - 78 (1971).

159.023 **Preliminary results of a photometric study of the NGC 371 region in the SMC.** P. J. Andrews.
The Magellanic Clouds. Astrophys. Space Sci. Library, Vol. 23, (see 012.010), 79 - 87 (1971).

159.024 **A spectroscopic study of Magellanic Cloud globular-type clusters.** P. J. Andrews, T. L. Evans.
The Magellanic Clouds. Astrophys. Space Sci. Library, Vol. 23, (see 012.010), 88 - 91 (1971).

159.025 **L'émission Hα dans le GrandNuage de Magellan. Programmes en développement à Córdoba.** G. J. Carranza.
The Magellanic Clouds. Astrophys. Space Sci. Library, Vol. 23, (see 012.010), 92 - 94 (1971).

159.026 **Optical aspects of X-ray sources in the Large Magellanic Cloud.** H. M. Johnson.
The Magellanic Clouds. Astrophys. Space Sci. Library, Vol. 23, (see 012.010), 95 - 97 (1971).

159.027 **Radio continuum observations of the Magellanic Clouds.** D. S. Mathewson.
The Magellanic Clouds. Astrophys. Space Sci. Library, Vol. 23, (see 012.010), 98 - 108 (1971).

159.028 **Physical parameters of supergiants in the Magellanic Clouds.** T. Walraven, J. Walraven.
The Magellanic Clouds. Astrophys. Space Sci. Library, Vol. 23, (see 012.010), 117 - 135 (1971).

159.029 **Cepheid variables and the Magellanic Clouds.** R. F. Christy.
The Magellanic Clouds. Astrophys. Space Sci. Library, Vol. 23, (see 012.010), 136 - 143 (1971).

159.030 **Nucleosynthesis in the Magellanic Clouds and the Galaxy.** G. Burbidge.
The Magellanic Clouds. Astrophys. Space Sci. Library, Vol. 23, (see 012.010), 156 - 162 (1971).

159.031 **Importance of Magellanic Clouds studies for extragalactic work.** E. M. Burbidge.
The Magellanic Clouds. Astrophys. Space Sci. Library, Vol. 23, (see 012.010), 163 - 168 (1971).

159.032 **Possibilities of narrow-band photometry, especially for main-sequence problems.** B. Strömgren.
The Magellanic Clouds. Astrophys. Space Sci. Library, Vol. 23, (see 012.010), 171 - 173 (1971).

159.033 **Image tube work on the Magellanic Clouds.** M. F. Walker.
The Magellanic Clouds. Astrophys. Space Sci. Library, Vol. 23, (see 012.010), 174 - 180 (1971).

159.034 **Application of objective prism techniques in the Magellanic Clouds.** J. Stock.
The Magellanic Clouds. Astrophys. Space Sci. Library, Vol. 23, (see 012.010), 181 - 183 (1971).

159.035 **The Magellanic Clouds: Summary and desiderata.** J. H. Oort.
The Magellanic Clouds. Astrophys. Space Sci. Library, Vol. 23, (see 012.010), 184 - 189 (1971).

159.036 **The kinematics and stellar content of populous clusters in the Magellanic Clouds.** H. C. Ford.
Thesis, Univ. Wisconsin, Madison. [Available from Univ. Microfilms, Ann Arbor, Mich., U.S.A. Order No. 71–289], 157 pp. (1970).
Image-tube spectra have been obtained of twenty-six blue populous clusters and ten red globular clusters in the LMC and three globular clusters in the SMC. The B-type spectra of the blue clusters are classified in a sequence from Type I to Type VI. A synthetic spectral type of NGC 2100 (Type I) close to the observed type is derived by combining its observed luminosity function with the galactic initial luminosity function.

Six-color photometry of 13 F—G supergiants in the Large Magellanic Cloud. See Abstr. 113.030.

Star S22 of the Large Magellanic Cloud showing emission lines of Fe II and [Fe II]. See Abstr. 114.011.

Wolf-Rayet stars in H II regions. See Abstr. 114.015.

Some peculiarities of classical cepheids of the Magellanic Clouds. See Abstr. 122.132.

Further studies of variable stars of the Large Magellanic Cloud. See Abstr. 122.133.

Novae in the Magellanic Clouds during the 1970– 1971 observing season. See Abstr. 124.007.

Observations at 408 MHz of supernova remnants in the LMC.   See Abstr. 125.017.

A search for absorption of the soft X-ray diffuse flux by the Small Magellanic Cloud.   See Abstr. 142.017.

160   Clusters of Galaxies

160.001   On the mass discrepancy in clusters of galaxies.
K.-H. Schmidt, H. Oleak.
Astron. Nachr., Vol. 292, 207 - 210 (1971).
The change of the slope in the luminosity function of the members of the Coma cluster near $M \approx -21$ is interpreted as an evolutionary effect.

160.002   Dynamics of the Perseus cluster of galaxies.
G. Chincarini, H. J. Rood.
Astrophys. Journ., Vol. 168, 321 - 325 (1971).
Carnegie image-tube spectra obtained in November 1970 with the 84-inch telescope of the Kitt Peak National Observatory have yielded radial velocities for forty-two more galaxies in a central region of the Perseus cluster. This brings the total number of known velocities to forty-nine. The average velocity (relative to the Local Group of galaxies) is $(V_0) = 5460 \pm 200$ km s^{-1}. NGC 1265, a prominent galaxy whose membership is indicated by radio observations, has a radial velocity $V_0 = 7660$ km s^{-1}.

160.003   On the non-existence of clusters of clusters of galaxies. VI.   M. Karpowicz.
Acta Astron., Vol. 21, 391 - 394 (1971).
The distribution of clusters of galaxies is analysed on the basis of volume III of Catalogue of Galaxies and of Clusters of Galaxies by Zwicky and Herzog (1966). It is found that the distribution of clusters is random and almost uniform.

160.004   Is cluster Abell 2199 a supergalaxy?
H. J. Rood, G. N. Sastry.
Bull. American Astron. Soc., Vol. 3, 391 (1971). − Abstr. AAS.

160.005   The correlation of redshift with magnitude in the Coma cluster.   W. G. Tifft.
Bull. American Astron. Soc., Vol. 3, 391 (1971). − Abstr. AAS.

160.006   Small clusters of galaxies and the mass-to-light ratio.
L. G. Taff, U. DeAngelis.
Astron. Astrophys., Vol. 15, 1 - 10 (1971).
We have investigated the stability of 19 clusters of galaxies in an attempt to combine the virial theorem with the current mass-to-light ratios to obtain the mass distribution function for galaxies. A detailed numerical analysis of possible masses for the galaxies leads us to three conclusions, 1) most of these systems are probably not stable, 2) in those that we do consider to be gravitationally bound the usual problem of high mass-to-light ratios is not encountered and 3) current mass-to-light ratios are not statistically significant. Due to the first and third conclusions we could not define the mass distribution function.

160.007   The Hubble constant and X-rays from galaxy clusters.   P. D. Noerdlinger.
Nature, Vol. 232, 393 (1971).
It is shown that recent X-ray observations of the Coma cluster of galaxies are consistent with the virial mass being in intergalactic gas provided that the Hubble constant is about 40 km/sec-Mpc.

160.008   Velocity dispersions and discrepant redshifts in groups of galaxies.   E. M. Burbidge, W. L. W. Sargent.
Nuclei of galaxies. Conference 1970, (see 012.002), p. 351 - 378 (1971).

160.009   The Coma cluster as an X-ray source: Some cosmological implications.   J. R. Gott III, J. E. Gunn.
Astrophys. Journ., *(Letters)*, Vol. 169, L13 - L15 (1971).
Recent positive X-ray observations of the Coma cluster make it possible to set firm upper limits on the mass of hot gas within the cluster. A theory of infall from the intergalactic medium relates this gas to the gas in intergalactic space and sets severe limits on the total gas density in the universe; if most of the matter in the universe is gaseous, the Friedmann deceleration parameter q_0 must be less than about 0.1, corresponding to an open ("hyperbolic") model.

160.010   The effect of the Einstein light deflection on observed properties of clusters of galaxies.
T. W. Noonan.
Astron. Journ., Vol. 76, 765 - 767 (1971).
Inhomogeneities in the background of field galaxies due to the deflection of light by the mass of a cluster of galaxies introduces an overestimate of the cluster population no larger than the order of a percent, depending on the nature of the unseen cluster mass. Focusing by the cluster mass causes a brightening of background galaxies seen through the cluster.

160.011   An angular diameter − redshift relation for rich clusters of galaxies.   G. Paal.
Astrofizika, Vol. 7, 435 - 456 (1971). − Reprinted in Astrophysics, Vol. 7, No. 3.
A simple method of determining a characteristic size of clusters of galaxies has been developed and applied to all rich clusters with known redshifts to establish a new observational relation for cosmology: an apparent-size − redshift relation. Empirical evidence has been obtained suggesting the presence of large-scale inhomogeneities in the distribution of clusters of galaxies, the evolution of the clusters, the reality of the expansion of the universe and the possibility of correcting the value of the Hubble "constant" as well as the deceleration parameter of the universal expansion.

160.012   Redshifts and absolute spectral energy distributions of galaxies in distant clusters.   J. B. Oke.
Astrophys. Journ., Vol. 170, 193 - 198 (1971).
Absolute spectral energy distributions have been obtained for galaxies in three distant clusters. Comparison with the energy distribution of the integrated light from nearby giant elliptical galaxies yields redshifts and information concerning evolu-

tionary effects. The Hydra cluster (0855+0321) gives the known result that $z = 0.20$. A galaxy in the cluster (0024 + 1654) yields $z = 0.38$, and the absorption spectrum of 3C 295 gives $z = 0.46$ which agrees with Minkowski's redshift. All have errors of ±0.01. The energy distributions for these galaxies show no evidence within the accuracy of the data for any evolution during the last 3 to 6 billion years.

160.013 **Groups of galaxies: Hidden mass or quick disintegration?**   G. B. Field, W. C. Saslaw.
Astrophys. Journ., Vol. 170, 199 - 206 (1971).

In small groups of galaxies there are correlations between the value of the mass discrepancy (defined as the ratio of virial mass to luminous mass) and other characteristics of the group, including the average morphological type of galaxies in the group, and the time for galaxies to cross the group. The significance of these correlations for possible explanations of the mass discrepancy is discussed. It is concluded that likely explanations include binding by large amounts of ionized gas, or quick disintegration, perhaps because of gravitational mass loss or ejection of intergalactic gas.

160.014 **Rectangular coordinates of rich clusters of galaxies on the *Palomar Sky Survey* charts.**
G. N. Sastry, H. J. Rood.
Astrophys. Journ., Suppl. Ser., No. 201, Vol. 23, 371 - 419 (1971).

Rectangular coordinates of the 2712 rich clusters listed in Abell's catalog have been calculated that allow quick identifications to be made on the charts of the *National Geographic-Palomar Observatory Sky Survey*. Right ascensions and declinations are given for epoch 1950 to facilitate the identification of Abell clusters with radio sources and clusters listed in other references.

160.015 **The problem of super-clustering of galaxies.**
M. Karpowicz.
Postępy Astron., Vol. 19, 313 - 324 (1971). In Polish.

This article describes two essential views on the distribution of matter in the universe. The methods and conclusions of the basic papers concerning the existence or the non-existence of super-clusters of galaxies are presented. Attention is paid to the difficulties connected with inevitable systematic errors of the data derived from the catalogues and to the uncertainty of applied statistical tests.

160.016 **The distribution of galaxies in the Ursa Major II cluster.**   N. A. Bahcall.
Astron. Journ., Vol. 76, 995 - 999 (1971).

Galaxy counts for the region of the Ursa Major II cluster of galaxies ($Z = 0.134$) are presented and the static properties of the cluster are derived. The cluster center is located by the symmetry of the galaxy distribution around a high-density point. We find that the Ursa Major II cluster has a diameter of about 52 min of arc and contains approximately 140 members to a photo-red magnitude of about 18^m.

160.017 **On the form of clusters of galaxies.**
I. P. Kostyuk.
Problems of cosmic physics. Vyp. (No.) 6, (see 003.040), p. 205 - 209 (1971). In Russian.

160.018 **Star clustering near the NGC 68-72 group of galaxies.**   W. G. Tifft, S. A. Gregory.
Publ. Astron. Soc. Pacific, Vol. 83, 810 (1971).

A study of this star clustering indicates it is apparently only a fluctuation in foreground star density.

160.019 **Clusters of galaxies and the cosmological expansion.**
A. D. Chernin.
Proc. Sixth Winter School on Space Physics, Part I. Apatity 1969, (see 012.033), p. 37 - 43 (1971).

Extragalactic Faraday rotation.
See Abstr. 141.110.

A strong X-ray source in the Coma cluster observed by Uhuru.   See Abstr. 142.002.

Interacting radio galaxies.   See Abstr. 158.028.

Photographic detection of "intergalactic" matter in the Coma cluster.   See Abstr. 161.004.

Possible selection effects and their consequences on the determination of the model of the universe from the redshift-magnitude relation of the brightest cluster galaxies.
See Abstr. 162.031.

161 Intergalactic Matter

161.001 A fluid dynamical study of the accretion process.
R. Hunt.
Monthly Notices Roy. Astron. Soc., Vol. 154, 141 - 165 (1971).

A numerical solution is found for a fluid dynamical treatment of the accretion process in which a gravitating point source is moving through an adiabatic gas (having specific heat ratio 5/3); the calculations being performed for Mach speeds 0.6, 1.4 and 2.4.

161.002 On possible absorbing matter in systems of galaxies.
V. A. Lipovetsky.
Vestn. Kiev. Un-ta, Ser. Astron., No. 12, p. 116 - 119 (1970). In Russian.

The light absorption in groups, pairs and around bright galaxies has been studied.

161.003 The density of intergalactic dust: An upper limit.
B. G. Nickerson, R. B. Partridge.
Astrophys. Journ., Vol. 169, 203 - 205 (1971).

Recently reported values of the magnitude and redshift of distant galaxies are used to set a limit on the total absorption by intergalactic dust. We then show that the density of such dust is likely to be well below the cosmologically critical density of 1.8×10^{-29} g cm^{-3}.

161.004 Photographic detection of "intergalactic" matter in the Coma cluster. G. A. Welch, G. N. Sastry.
Astrophys. Journ., *(Letters)*, Vol. 169, L3 - L5 (1971).

Isodensitometry of deeply exposed plates of the Coma cluster of galaxies reveals the presence of diffuse luminous material at the center of the cluster. The material not only envelopes the two supergiant galaxies NGC 4874 and NGC 4889 but extends outward to the southwest for at least 16 arc minutes. This extension may represent matter flowing out from NGC 4874.

161.005 Interaction of fast particles with intergalactic matter. J. Arons, R. McCray, J. Silk.
Astrophys. Journ., Vol. 170, 431 - 447 (1971).

We explore the intergalactic model first proposed by Silk and McCray (1969) in greater detail, in which the X-rays are produced by the bremsstrahlung of suprathermal, nonrelativistic electrons moving in a fully ionized intergalactic gas. The motivation for working out this theory with some care is that, if the X-ray background can be reasonably explained by this source mechanism, the "break" in the approximately power-law spectrum of observed X-rays at $\epsilon \sim 40$ keV provides information on the density of the intergalactic gas.

161.006 Average density of the metagalactic matter.
L. M. Genkina.
Astron. Tsirk., No. 650, p. 1 - 2 (1971). In Russian.

161.007 Statistical search for the intergalactic matter.
T. Kwast.
Postępy Astron., Vol. 19, 335 - 342 (1971). In Polish.

Colour excess of about 0^m3 was estimated from counts of galaxies in some regions of the sky suspected to be obscured by intergalactic matter near clusters of galaxies Zw. 156−5 and Zw. 156−14.

Interacting radio galaxies. See Abstr. 158.028.

162 Structure and Evolution of the Universe, Cosmology

162.001 Observational approximative dependences in the cosmology of a homogeneous isotropic universe.
O. A. Kalinin, A. M. Finkelshtein.
Byull. Inst. Teoret. Astron., *Leningrad*, Vol. 12, 714 - 723 (1971). In Russian.

Within the framework of the universe model specified by the Robertson−Walker fundamental form some observational dependences have been obtained with an accuracy not less than z^3 using a series expansion in powers of a small parameter $R - R_\odot \sim z$, $R_\odot$ being the value of the scale factor at the present epoch. Consideration has been given to Einstein's field theory with the cosmological term, to the first version of Hoyle's theory with the tensor of matter creation, and to the particular version of Jordan's theory.

162.002 Time scale of the expansion for cosmological models with radiation. D. Edwards.
Monthly Notices Roy. Astron. Soc., Vol. 153, 17P - 20P (1971).

Exact expressions are derived in terms of the basic parameters for Friedmann models containing non-interacting matter and radiation.

162.003 The effect of cosmological expansion on self-gravitating ensembles of particles.
P. D. Noerdlinger, V. Petrosian.
Astrophys. Journ., Vol. 168, 1 - 9 (1971).

We consider the possible expansion of clusters or superclusters of galaxies, of mean rest-mass density ρ_c, immersed in a universe containing a gas of particles having zero rest mass and having energy density ρc^2. The universe may also contain dust.

162.004 Gravitational waves in closed universes.
R. H. Gowdy.
Phys. Rev. Letters, Vol. 27, 826 - 829 (1971).

New boundary conditions on the Einstein-Rosen-Bondi gravitational-wave metrics yield closed inhomogeneous universes which solve Einstein's vacuum field equations exactly.

162.005 The conception "universe" in modern scientific literature. E. G. Aniskin.
Vestn. AN KazSSR, 1971, No. 2 (310), p. 55 - 60. In Russian. Abstr. in Referativ. Zhurn. 51. Astron., 8.51.664 (1971).

162.006 Impossibility of mixing in a cosmological model of the Bianchi IX type.
A. G. Doroshkevich, V. N. Lukash, I. D. Novikov.
Zhurn. ehksperim. i teor. fiz., Vol. 60, 1201 - 1205 (1971).
In Russian. − Abstr. in Referativ. Zhurn. 51. Astron., 8.51.672 (1971).

162.007 The cosmical constant. W. H. McCrea.
Quarterly Journ., Roy. Astron. Soc., Vol. 12, 140 - 153 (1971).

Like that of many other features of relativistic cosmology, the history of the cosmical constant in Einstein's equations abounds in peculiarities and paradoxes. The status of this constant in general relativity is still as much a question for debate as when it was first introduced. The question is of fundamental significance in cosmology and its discussion raises fundamental issues in the interpretation of general relativity itself. The purpose of this article is to review various possible ways of thinking about the problem.

162.008 Qualitative cosmology.
C. B. Collins, J. M. Stewart.
Monthly Notices Roy. Astron. Soc., Vol. 153, 419 - 434 (1971).

This paper discusses the role of dissipative processes in the early universe, in accordance with the 'chaotic cosmology programme' outlined by Misner. The conclusions drawn are that (I) for arbitrary initial conditions the shear anisotropy could be arbitrarily large now, (II) the universe need not have been in thermal equilibrium during the early stages, and (III) matter effects are unimportant in most, but not all, of the cosmological models considered.

162.009 Uniform model universes containing matter and blackbody radiation−II.
T. L. May, G. C. McVittie.
Monthly Notices Roy. Astron. Soc., Vol. 153, 491 - 500 (1971).

A method previously developed for this problem (Monthly Notices Roy. Astron. Soc., Vol. 148, 407 - 416 (1970)) is applied to the case of models in which space is spherical. It is also shown how to determine the constants of a model from observational data. The dependence of the temperature of the radiation on the scale-factor is found for any model. The temperature is not always inversely proportional to the scale-factor but can vary in a more complicated fashion which depends on the law of energy exchange between matter and radiation inherent in the definition of the model universe.

162.010 A generalized redshift−magnitude formula.
S. E. Kaufman, E. L. Schucking.
Astron. Journ., Vol. 76, 583 - 587 (1971).

A closed formula for the relation between luminosity and redshift in an expanding Friedman universe is derived for the case of a positive cosmological constant, positive space curvature, and vanishing pressure.

162.011 On the physical nature of cosmic electromagnetic absorption. III. The Einstein−de Sitter cosmology with adiabatic plasma. R. Burman.
Observatory, Vol. 91, 141 - 146 (1971).

This paper deals with a stage of an evolving universe during which the cosmic plasma is fully ionized and undergoing adiabatic expansion. An approximation is obtained for the refractive index of electromagnetic waves. The effects of collisions and radiation reaction on both retarded and advanced waves are investigated using the Einstein−de Sitter cosmology.

162.012 On the physical nature of cosmic electromagnetic absorption. IV: Effects of electron thermal motions. R. Burman.
Observatory, Vol. 91, 147 - 154 (1971).

Dispersion relationships are obtained for wave propagation in a warm plasma with tensor pressure perturbations; losses due to radiation damping, electron−heavy-particle collisions and electron−electron collisions are included. Approximate results are obtained for electromagnetic waves in constant-density and evolving universes. Absorption of retarded and advanced waves in the steady-state and Einstein−de Sitter cosmologies is discussed.

162.013 Entropy generation and the survival of protogalaxies in an expanding universe.
S. Weinberg.
Astrophys. Journ., Vol. 168, 175 - 194 (1971).

General formulae are derived for the bulk viscosity, shear viscosity, and heat transport due to radiation, and for the damping rate of sound waves, in an imperfect relativistic fluid.

These results are used to evaluate the cosmological entropy production associated with a nonvanishing mean free time of photons, neutrinos, or gravitons, and to calculate the rate of damping of protogalactic fluctuations in the period immediately prior to the recombination of hydrogen.

162.014 **Was there really a big bang?** G. Burbidge.
Nature, Vol. 233, 36 - 40 (1971).
The evidence in favour of a big bang cosmology is much less definite than is widely realized, and it is not impossible that we are living in a steady state universe.

162.015 **On the nature of mass.** F. Hoyle, J. V. Narlikar.
Nature, Vol. 233, 41 - 44 (1971).
The increasing number of observations of discrepant redshifts means that no longer can these be passed off as chance juxtapositions. A possible explanation of the data is given here in terms of a theory that incorporates a gravitational "constant" that is decreasing with time.

162.016 **Pressure in inhomogeneous world models.**
E. Saar.
Tartu Astron. Obs. Teated, No. 30, p. 47 - 58 (1971).
To make the theory of inhomogeneous cosmological models developed by the author physically more realistic, pressure terms are included. An approximate equation of state for the present stage of the universe is proposed and some possibilities of its refinement are discussed.

162.017 **Effects of very long wavelength primordial gravitational radiation.** M. J. Rees.
Monthly Notices Roy. Astron. Soc., Vol. 154, 187 - 195 (1971).
The universe may contain gravitational waves with wavelength 1 - 10 Mpc, of primordial origin, associated with the initial irregularities that give rise to galaxies and clusters. If their energy density were comparable with the 'critical density', these waves would have interesting effects on the apparent dynamics of groups of galaxies.

162.018 **On determining the parameters of cosmological models.** E. Saar.
Publ. Tartu Astrofiz. Obs., Vol. 39, 171 - 182 (1971). In Russian.
The application of the least-squares method to determine the parameters of cosmological models is considered. As the accuracy of the estimates is small, it is important to find their confidence levels.

162.019 **Cosmological models with pressure, and observations.** E. Saar, I. Saar.
Publ. Tartu Astrofiz. Obs., Vol. 39, 183 - 205 (1971).
Using the redshift-magnitude relation the parameters of Friedmann models with pressure are determined. A method of finding estimates if the confidence ellipsoids are very elongated is described.

162.020 **Inhomogeneous model universes I. Basic equations.**
E. Saar.
Publ. Tartu Astrofiz. Obs., Vol. 39, 206 - 233 (1971).
Model universes, inhomogeneous on a small scale, are studied. It is shown that the Einstein equations split up into two coupled sets. One of them can be regarded as governing the evolution of inhomogeneities. The other, when averaged, can be considered as an effective energy-momentum tensor, by means of which the feedback of inhomogeneities to the overall dynamics of the universe is realized.

162.021 **Inhomogeneous model universes II. Gauge invariance.** E. Saar.
Publ. Tartu Astrofiz. Obs., Vol. 39, 234 - 248 (1971).

The invariance of the equations describing inhomogeneous model universes, derived in the preceding paper under coordinate transformations is studied.

162.022 **Inhomogeneous model universes III. Lorentz gauge.**
E. Saar.
Publ. Tartu Astrofiz. Obs., Vol. 39, 249 - 272 (1971).
The equations describing inhomogeneous model universes are studied by using a special gauge and supposing the fluctuating fields to be stochastic.

162.023 **Barions and antibarions in an anisotropic universe.**
I. N. Mishustin.
Astrofizika, Vol. 7, 271 - 279 (1971). In Russian. – English translation in Astrophysics, Vol. 7, No. 2.
The process of barions and antibarions freezing in the stage of anisotropic expansion of the universe is investigated. Consideration is given for symmetric and nonsymmetric charge of the universe. The analytical dependence of residual barion and antibarion concentration on the anisotropy parameter is constructed.

162.024 **Energy in the universe.** F. J. Dyson.
Sci. American, Vol. 225, No. 3, p. 50 - 59 (1971).
Review article.

162.025 **On the observable redshift of gamma rays in an expanding universe.** G. R. Blumenthal.
Bull. American Astron. Soc., Vol. 3, 391 (1971). – Abstr. AAS.

162.026 **Night sky darkness in the Eddington–Lemaître universe.** D. T. Pegg.
Monthly Notices Roy. Astron. Soc., Vol. 154, 321 - 327 (1971).
It is the purpose of this paper to find how dark the night sky should be if the universe corresponds to the expanding Eddington–Lemaître model. This will be done in the spirit of the original Olbers paradox in which stars are considered as the basic unit light emitters.

162.027 **Mode of oscillations by approaching the singularity in homogeneous cosmological models with rotating axes.** V. A. Belinskij, E. M. Lifshits, I. M. Khalatnikov.
Zhurn. ehksperim. i teor. fiz., Vol. 60, 1969 - 1979 (1971).
In Russian. – Abstr. in Referativ. Zhurn. 51. Astron., 10.51.657 (1971).

162.028 **Thermodynamics and cosmology.** A. Kovetz.
Astrophys. Letters, Vol. 9, 17 - 18 (1971).
The equations of cosmology are reformulated to account for the existence of a bulk viscosity, and some of the consequences are discussed.

162.029 **Dynamics of the cosmic fireball in the scalar-tensor theory of gravitation.** A. V. Manjos.
Vestn. Kiev. Un-ta, Ser. Astron., No. 13, p. 91 - 96 (1971).
In Russian.
Results of cosmological researches on the Jordan-Dicke field equations are given.

162.030 **Charge symmetry and the universe.**
G. L. Vardenga, E. O. Okonov.
Priroda, No. 10.71, p. 56 - 63 (1971). In Russian.

162.031 **Possible selection effects and their consequences on the determination of the model of the universe from the redshift-magnitude relation of the brightest cluster galaxies.** A. Vignato, R. Marcucci.
Astrophys. Space Sci., Vol. 12, 456 - 470 (1971).
The purpose of the present paper is, first, to attempt

taking into account quantitatively the richness selection effects in the m-z relation, in order to see how much they affect the choice of the present value of the deceleration parameter of the universe; the second purpose is a preliminary attempt of taking into account the overall selection effect by eliminating the large difference in the density of data points between the lower and the upper part of the Hubble plot. A discussion of the results obtained by the two methods as well as a comparison with the results given by other tests, i. e. log N-log$S$ and N-z relations, is also given.

162.032 **Relativistic cosmological models with pressure.**
R. F. Sisteró.
Astrophys. Space Sci., Vol. 12, 484 - 492 (1971).

Cosmological models with pressure are considered. A general method of solving Einstein's equations is presented in terms of the scale factor $R(t)$. Its general application to models containing interacting radiation and matter is explicitly given. A particular class of models is studied which are 'radiation-like' in their early history and 'matter-like' at present. Their analytic solutions are given for all three values of the space curvature constant.

162.033 **Matter-antimatter hydrodynamics: the coalescence effect.** R. Omnès.
Astron. Astrophys., Vol. 15, 275 - 284 (1971).

The hydrodynamical motion of a system consisting of matter and antimatter embedded in thermal radiation is investigated. These motions are generated by the pressure which is due to the annihilation products and by thermal convection. Thermal convection is dominant and its velocity can be simply written. The coalescence of matter systems into regions having a size of the order of the mean free path for high-energy photons is established for simple cases.

162.034 **Cosmogonic processes.** D. Layzer.
Astrophysics and general relativity. Brandeis Univ.
Summer Inst. 1968, Vol. 2, (see 003.011), 151 - 233 (1971).

162.035 **Kinetic theory and cosmology.**
R. K. Sachs, J. Ehlers.
Astrophysics and general relativity. Brandeis Univ. Summer
Inst. 1968, Vol. 2, (see 003.011), 331 - 383 (1971).

162.036 **Cosmological evolution in radio galaxies.**
C. D. Mackay.
Nature, Vol. 233, 402 - 403, with a reply by M. Rowan-Robinson, p. 403 - 404 (1971).

Mackay criticised an earlier analysis by Rowan-Robinson (Nature, Vol. 229, 388, 1971) because a) much of Rowan-Robinson's discussion did not take into account the discontinuous nature of the data he used and b) because the significance of the effect found by Rowan-Robinson depended strongly on an extrapolation of available data on the colours of radio galaxies. Rowan-Robinson replied by denying point a) made by Mackay and saying that b) was not important, as the effect found by Rowan-Robinson was essentially independent of the optical data on the radio galaxies.

162.037 **A complete redshift-magnitude formula.**
S. E. Kaufman.
Astron. Journ., Vol. 76, 751 - 755 (1971).

A closed formula for the relation between luminosity and redshift in an expanding Friedman universe with vanishing pressure is found which is valid for all values of the cosmological constant and the three possible space curvatures. The formula is tested with the well-known special cases.

162.038 **Errata: 'Gravitational waves in closed universes**
[Phys. Rev. Letters, Vol. 27, 826 - 827 (1971)].
R. H. Gowdy.

Phys. Rev. Letters, Vol. 27, 1102 (1971).

162.039 **Possible evidence for the existence of antimatter on a cosmological scale in the universe.**
F. W. Stecker, D. L. Morgan, Jr., J. Bredekamp.
Phys. Rev. Letters, Vol. 27, 1469 - 1472 (1971).

We present some initial results of a detailed calculation of the cosmological γ-ray spectrum from matter-antimatter annihilation in the universe. The similarity of the calculated spectrum with the present observations of the γ-ray background spectrum above 1 MeV suggests that such observations may be evidence of the existence of antimatter on a large scale in the universe.

162.040 **L'età delle galassie.** L. Gratton.
Mem. Soc. Astron. Italiana, Nuova Ser., Vol. 42,
373 - 396 (1971).

The age of galaxies is discussed in the general frame of the "big bang" cosmological theory; the possibility of the formation of a galaxy through gravitational instability and through a fluctuation of the velocity distribution of the primeval gas is considered. In an appendix some cosmological implications of the thermodynamics of strong interactions (Hagedorn) and of their meaning in connection with the problem of galactic formation are reviewed.

162.041 **Cosmologia: Aspetti evolutivi.** G. Setti.
Mem. Soc. Astron. Italiana, Nuova Ser., Vol. 42,
397 - 406 (1971).

In the first part of this paper we briefly review the present status of our knowledge of the values of the cosmological parameters H_0 and q_0. In the second part we discuss the evidence of strong cosmological evolutionary effects as derived from radio source counts and spatial distribution of quasi-stellar objects. Possible interpretations of these effects are suggested.

162.042 **Antimaterie.** B. Gonsior.
Phys. Blätter, 27. Jahrgang, p. 552 - 558 (1971).

162.043 **Expansion anisotropy and the spectrum of the cosmic background radiation.** S. N. Rasband.
Astrophys. Journ., Vol. 170, 1 - 15 (1971).

The spectrum of primeval-fireball radiation is considered in a homogeneous axisymmetric anisotropic world model which includes the effects of electron neutrinos on expansion dynamics. Characteristic deviations from a thermal spectrum are found, but current measurements of the background intensity do not further restrict expansion anisotropy beyond the limits set by isotropy measurements of the fireball.

162.044 **Perturbations dans un univers de Gödel.**
M. Teboul.
Comptes Rendus Acad. Sci. Paris, Sér. A, Vol. 273, 1335 -
1338 (1971).

Nous montrons que les perturbations qui J. Silk a considérées sont en grande partie fictives, ce qui nous a conduit à reprendre cette étude à l'aide d'un modèle de perturbation simplifié respectant les symétries de l'univers de Gödel.

162.045 **Heißes Weltall.**
J. B. Seldowitsch.
Astron. in der Schule, 8. Jahrgang, p. 98 - 102 (1971).

162.046 **Analogues of homogeneous anisotropic models of general relativity in Newtonian cosmology.**
I. S. Shikin.
Zhurn. ehksperim. i teor. fiz., Vol. 61, 445 - 453 (1971). In Russian. – Abstr. in Referativ. Zhurn. 51. Astron., 12.51.809 (1971).

162.047 The hadron barrier in cosmology and gravitational collapse. J. N. Bahcall, S. Frautschi.
Astrophys. Journ., (Letters), Vol. 170, L81 - L84 (1971).

It is shown that the early stages of a big-bang cosmology are hadron-dominated if the density of hadron states per unit mass interval, $\rho(m)$, increases at least as fast as m^{-1} and if the initial gravitational anisotropy is not large. It is further shown that if these conditions are satisfied, then quantum effects must modify the classical equations of general relativity at times $t \sim 10^{-23}$ s before the attainment of the singularities predicted in classical relativistic cosmologies and stellar collapse.

162.048 The electron-neutrino cross-section and its effect on the cosmological helium abundance. H. F. Hecht.
Astrophys. Journ., Vol. 170, 401 - 404 (1971).

This paper contains the result of a calculation of the ^{4}He abundance produced in the early stages of expansion of the universe. It is assumed in the calculation that the electron-neutrino scattering cross-section is sufficiently large to maintain thermal equilibrium between the neutrinos and the photons and electrons in the primordial gas.

162.049 Are the astrophysical and statistical schools of irreversibility compatible? B. Gal-Or.
Nature, Vol. 234, 217 - 218 (1971).

The new Hogarth-Gold-Narlikar-Gal-Or's Astrophysical School of Thermodynamics which claims that the origin of irreversibility and all time asymmetries observed in nature can be traced back to initial conditions, which had given rise to the present expansion of the universe as a whole, is shown to be compatible with the statistical school of Thermodynamics only in non uniform expanding world. However, at the earliest period of the world, the statistical irreversibility can not be properly defined.

162.050 Fortschritte und Ziele der empirischen Kosmologie. L. Biermann.
Physik und Kosmologie, (see 003.024), p. 90 - 99 (1971).

162.051 Classification of uniform cosmological models containing both matter and radiation. M. Kubo.
Sci. Rep. Tôhoku Univ., First Ser., Vol. 53, 103 - 109 = Sendai Astron. Rap., No. 117 (1970).

A classification of uniform cosmological models containing both matter and radiation without cosmological constant is presented. Models are specified by two parameters: the matter and the radiation energy density parameters. The corresponding relation between magnitude and redshift is briefly discussed.

162.052 Improving the Hubble diagram for cosmology in the 1970s. H. Spinrad.
Comments Astrophys. Space Phys., Vol. 3, 168 - 172 (1971).

The extension and improvement of the Hubble diagram, galaxy magnitude m, versus redshift z, for distant objects seems an important − perhaps the most important −possible advance in observational cosmology in the next few years. After having outlined the history and the present status of the problem some remarks on prospective results are presented.

162.053 Two old cosmological tests. P. J. E. Peebles.
Comments Astrophys. Space Phys., Vol. 3, 173 - 178 (1971).

In 1935 Hubble and Tolman discussed two possible tests of the new expanding cosmology of Friedmann and Lemaitre. The first test is the count of galaxies as a function of apparent magnitude (measured energy flux f from the galaxy). The second test deals with the question: Is the universe expanding? In the present paper a reconsideration of these two tests in relation to the large scale structure of the universe is given.

162.054 Helium production in the different cosmological models. A. G. Doroshkevich, I. D. Novikov, R. A. Sunyaev, Ya. B. Zeldovich.
Highlights of Astronomy, Vol. 2, (see 012.015), 318 - 327 (1971).

162.055 Observational paradoxes in extragalactic astronomy. H. Arp.
Science, Vol. 174, 1189 - 1200 (1971).

For some extragalactic objects evidence contradicts the usual assumptions about red shifts, ages, and origins.

162.056 Space-time structure from a global viewpoint. R. Geroch.
General relativity and cosmology. Course 47, Italian Phys. Soc., 1969 (see 012.022), p. 71 - 103 (1971).

162.057 Relativistic cosmology. G. F. R. Ellis.
General relativity and cosmology. Course 47, Italian Phys. Soc., 1969 (see 012.022), p. 104 - 182 (1971).

162.058 Astrophysical cosmology. D. W. Sciama.
General relativity and cosmology. Course 47, Italian Phys. Soc., 1969, (see 012.022), p. 183 - 236 (1971).

162.059 Some current ideas on galaxy formation. M. J. Rees.
General relativity and cosmology. Course 47, Italian Phys. Soc., 1969, (see 012.022), p. 315 - 346 (1971).

162.060 Elementary particles in De Sitter space. G. Börner.
General relativity and cosmology. Course 47, Italian Phys. Soc., 1969, (see 012.022), p. 362 - 364 (1971).

162.061 Antimatter and cosmology. G. Steigman.
General relativity and cosmology. Course 47, Italian Phys. Soc., 1969, (see 012.022), p. 373 - 382 (1971).

162.062 Electromagnetic radiation in the universe. M. S. Longair, R. A. Syunyaev.
Uspekhi fiz. nauk, Vol. 105, 41 - 96 (1971). In Russian. Abstr. in Referativ. Zhurn. 51. Astron., 1.51.833 (1972).

162.063 Formation of "priming" magnetic fields during the formation of protogalaxies. I. N. Mishustin, A. A. Ruzmajkin.
Zhurn. ehksperim. i teor. fiz., Vol. 61, 441 - 444 (1971). In Russian. − Abstr. in Referativ. Zhurn. 51. Astron., 1.51.836 (1972).

162.064 Brans-Dicke cosmologies in arbitrary units: Solutions in flat Friedmann universes. R. E. Morganstern.
Phys. Rev. D, Particles and Fields, Third Ser., Vol. 4, 278 - 282 (1971).

162.065 Exact solutions to radiation-filled Brans-Dicke cosmologies. R. E. Morganstern.
Phys. Rev. D, Particles and Fields, Third Ser., Vol. 4, 282 - 286 (1971).

Using the Robertson-Walker metric, exact general solutions to the Brans-Dicke cosmologies for $p = \frac{1}{3}\rho$ are found.

162.066 Exact solutions to Brans-Dicke cosmologies in flat Friedmann universes. R. E. Morganstern.
Phys. Rev. D, Particles and Fields, Third Ser., Vol. 4, 946 - 954 (1971).

The Brans-Dicke cosmological equations for flat Friedmann-type expanding universes with $p = \epsilon\rho$ are solved parametrically for time $t(z)$, density $\rho(z)$, expansion parameter $a(z)$, and scalar field $\lambda(z)$. These results reduce to a previously obtained exact solution to the radiation cosmology when $\epsilon = {}^1/_3$.

162.067 Mass of a galaxy and dissipative process in the hot universe. H. Sato.
Progr. Theor. Phys., Japan, Vol. 45, 370 - 385 (1971).

Decay of the acoustic motions in an early stage of the hot universe is studied in detail. Considering a growth of density perturbation by gravitation and its decay by dissipative processes, i.e. viscosity and thermal conductivity, we can get a gross feature of the size spectrum of density inhomogeneity at the stage of recombination of the cosmic plasma.

162.068 Dissipation of primordial turbulence and thermal history of the universe.
T. Matsuda, H. Sato, H. Takeda.
Progr. Theor. Phys., Japan, Vol. 46, 416 - 432 (1971).

The assumption that in the early stage of the universe there existed turbulence of photons and plasma dragged by them can explain the formation of galaxies plausibly. Using simple expressions to represent the decay law of the primordial turbulence, the thermal history of gas at the pre-galactic stage is followed and the residual ionization degree of hydrogen is computed.

162.069 On the removal of initial singularity in a big-bang universe in terms of a renormalized theory of gravitation. I. Examination of the present status and a new approach. H. Nariai.
Progr. Theor. Phys., Japan, Vol. 46, 433 - 438 (1971).

The present status on the problem of initial singularity characteristic in any model-universe compatible with the cosmological interpretation of the $3°$K black-body radiation is examined from the standpoints of both classical and quantum cosmologies. A new approach is proposed, the essence of which lies in modifying the Einstein field equations in a physically reasonable manner on the basis of a renormalized theory of gravitation due to Utiyama and DeWitt.

162.070 On the removal of initial singularity in a big-bang universe in terms of a renormalized theory of gravitation. II. Criteria for obtaining a physically reasonable model.
H. Nariai, K. Tomita.
Progr. Theor. Phys., Japan, Vol. 46, 776 - 786 (1971).

On the basis of the renormalized theory of gravitation proposed in a previous paper, an attempt is made to construct a homogeneous and isotropic expanding model-universe which is free from the initial singularity of infinite density and tends asymptotically to the usual Friedmann universe.

162.071 Closed rotating cosmologies containing matter described by the kinetic theory. A: Formalism.
R. A. Matzner.
Ann. Physics, Vol. 65, 438 - 481 (1971).

In this and the accompanying paper, we investigate the kinetic theory of collisionless systems in closed, anisotropic (Bianchi type IX) cosmologies including the effects of rotation. We discuss the behavior of the distribution function for two special cases (subsets of the full rotating situation). We present a Lagrangian for the behavior of the anisotropy in these models. We discuss the meaning of the angular momentum constraints embodied in the $T_{0i} = G_{0i}$ equations. We analyze the behavior of the models in the large anisotropy regime in terms of "wall collisions". We discuss the geodesic equation, and show that it can be reduced to a one-dimensional time dependent Hamiltonian system.

162.072 Closed rotating cosmologies containing matter described by the kinetic theory. B: Small anisotropy calculations; Application to observations. R. A. Matzner.
Ann. Physics, Vol. 65, 482 - 505 (1971).

We continue the discussion of anisotropy in closed, rotating cosmologies begun in the accompanying paper. We present an explicit small anisotropy model, taking parameters suggested by the observed universe. We give an explicit discussion of the geodesic equation in the small anisotropy regime, with applications to the microwave temperature anisotropy.

162.073 Qualitative cosmology: Diagrammatic solutions for Bianchi type IX universes with expansion, rotation, and shear. I. The symmetric case. M. P. Ryan, Jr.
Ann. Physics, Vol. 65, 506 - 537 (1971).

By means of mathematical techniques due to Arnowitt, Deser, and Misner we investigate the problem of universes which exhibit simultaneously expansion, rotation, and shear. We are able to obtain a solution of the Einstein equations for such a universe whose three-space sections are spaces of Bianchi type IX in the sense of a pictorial or diagrammatic solution.

162.074 Qualitative cosmology: Diagrammatic solutions for Bianchi type IX universes with expansion, rotation, and shear. II. The general case. M. P. Ryan, Jr.
Ann. Physics, Vol. 68, 541 - 555 (1971).

In this paper we extend our investigation of expanding, rotating, shearing Bianchi type IX universes to the most general possible case. As in a previous paper, we use the techniques of Arnowitt, Deser, and Misner. We are able to show that the conclusion made in that paper is true in general, i.e. rotation changes the singularity of type IX universes very little.

162.075 Quantized fields and particle creation in expanding universes. II. L. Parker.
Phys. Rev. D, Particles and Fields, Third Ser., Vol. 3, 346 - 356, with a correction, p. 2546 (1971).

We consider the quantized spin-${}^1/_2$ field which satisfies the fully covariant generalization of the Dirac equation. The metric, which is not quantized, is that of an expanding universe with Euclidean 3-space. In general, there will be production of spin-${}^1/_2$ particles as a result of the expansion of the universe. However, we show that in the limits of zero and infinite mass there is no spin-${}^1/_2$ particle production. We also consider the Friedmann expansion of a radiation-filled universe, emphasizing the effect of the initial stage of the expansion. We obtain the asymptotic form of the created particle density for large momenta, and thus show that the particle density, integrated over all momenta, is finite, in contrast to the previous case.

162.076 Bemerkungen zu Kosmologie. H.-U. Keller.
Separate print Sternw. Bochum, 14 pp. (1971).

At first the fundamental theories of cosmology are shown under a critical point of view. Then the author points out the necessity to consider thermodynamics about all cosmological researches and theories.

162.077 Photon propagation in a perturbed Einstein-de Sitter universe containing partially ionized gas.
P. C. White.
Thesis, Univ. Texas, Austin. [Available from Univ. Microfilms, Ann Arbor, Mich., U.S.A. Order No. 70–18310], 108 pp. (1970).

Photon propagation is studied in a locally inhomogeneous cosmological model containing partially ionized gas. Starting from the Boltzmann equation, detailed derivation of the general relativistic photon transport equation is presented. Some of the usual approximations used in the transport equation are discussed, and an approximate solution for the photon distribution function is constructed in the limit of small optical depth. The approximate solution of the transport equation is used to investigate the effects of photon scattering on two cosmological observables. The combined effects of local inhomogeneities and photon scattering on the luminosity-redshift relation are discussed.

162.078 Scale factor in cosmological homogeneous models.
A. G. Agnese, A. Wataghin.
Nuovo Cimento Lettere, Ser. 2, Vol. 1, 857 - 860 (1971).

The authors discuss an alternative cosmological time scale which has the properties that the speed of light, unit of time and the gravitational constant vary with the age of the universe, and in which the value for the age of the universe tends to infinity.

162.079 An investigation of the gravitational interaction of electromagnetic radiation and matter as a cause of the cosmological red shift. J. C. Hegarty.
Thesis, Univ. Boston, Mass. [Available from Univ. Microfilms, Ann Arbor, Mich., U.S.A. Order No. 70−22367], 127 pp. (1970).

The author connects the work of Tolman and Westervelt in which has is shown that the gravitational field produced by a pulse of electromagnetic radiation transmits an impulse to a test particle in the direction of propagation of the radiation which is equal in magnitude and opposite in direction to the momentum change of the photon which occurs in the gravitational red shift. The cosmological implications of this effect are noted. A class of solutions to the Einstein field equations for null electromagnetic fields are studied. The gravitational interaction between matter and null electromagnetic fields is studied in more detail to ascertain whether Zwicky's suggested mechanism for the red shift is plausible.

162.080 A cosmological model in the projective theory of relativity. G. Macheleidt.
Acta Phys. Polonica B, Vol. B2, 145 - 150 (1971).

A simple system of field-equations of the projective theory of relativity is solved for a homogeneous, isotropic cosmological model in the case of incoherent matter. The consequences of this projective cosmology are discussed and compared with those of the classical Einsteinian and projective Jordanian theories.

162.081 Rotation does not enhance mixing in the Mixmaster universe. R. A. Matzner, D. M. Chitre.
Commun. Math. Phys., Vol. 22, 273 - 289 (1971).

The authors investigate closed rotating cosmologies to determine if rotation leads to enhancement of causal mixing proposed by Misner to guarantee the homogeneity of such models. It is concluded that rotation cannot lead to significantly more efficient mixing than occurs in non-rotating models.

162.082 On the stability of the Taub universe.
S. Bonanos.
Commun. Math. Phys., Vol. 22, 190 - 222 (1971).

An analysis of the stability of the Taub universe for arbitrary, initially small perturbations is carried out. It is found that the perturbation decrease during the expansion and increase during the contraction of the unperturbed space.

162.083 Spatially homogeneous rotating world models.
I. Ozsvath.
Journ. Math. Physics, (USA), Vol. 12, 1078 - 1082 (1971).

The Lagrangian function for four different rotating universes is derived simultaneously. These models correspond in a certain sense to Gödel's 'symmetric case'.

162.084 Hamiltonian cosmology.
F. B. Estabrook, H. D. Wahlquist.
Phys. Letters A, Vol. 35A, 453 - 454 (1971).

A simple polynomial Hamiltonian is given for type VIII and IX vacuum cosmologies. This Hamiltonian also describes solutions with time-like homogeneous 3-surfaces, and so suggests quantized versions which involve fluctuations of 3-space signature and topology.

162.085 Perturbations in anisotropic Euclidean-homogeneous cosmologies.
R. A. Matzner, T. E. Perko, L. C. Shepley.
Phys. Letters A, Vol. 35A, 467 - 468 (1971).

The role of anisotropy in the Galaxy formation process is considered. It is found that the growth rate of a density perturbation does not appreciably differ from the isotropic growth rate.

162.086 On the pulsating universe of Sengupta.
M. A. H. Maccallum.
Phys. Letters A, Vol. 35A, 474 (1971).

It is pointed out that the recently announced pulsating universe of Sengupta does not satisfy Einstein's field equations. A discussion of the possibility of pulsating solutions is given.

162.087 Newtonian hierarchical cosmology. J. R. Wertz.
Thesis, Univ. Texas, Austin. [Available from Univ. Microfilms, Ann Arbor, Mich., U.S.A. Order No. 71−209], 148 pp. (1970). − See Phys. Abstr., Vol. 74, No. 62746 (1971).

162.088 Qualitative cosmology. Diagrammatic solutions for Bianchi type IX universes with expansion, rotation and shear. M. P. Ryan, Jr.
Thesis, Univ. Maryland, Catonsville. [Available from Univ. Microfilms, Ann Arbor, Mich., U.S.A. Order No. 71−4093], 141 pp. (1971). − See Phys. Abstr., Vol. 74, No. 70418 (1971).

162.089 A simple magnetic universe.
L. K. Patel, P. C. Vaidya.
Current Sci., (India), Vol. 40, 288 (1971).

The authors call a distribution of magnetic field together with its associated geometry 'a magnetic universe'. The object of the present note is to report on a simple magnetic universe which is different from that of Melvin.

162.090 More qualitative cosmology. C. B. Collins.
Commun. Math. Phys., Vol. 23, 137 - 158 (1971).

Standard geometric techniques of differential equation theory are employed to determine the qualitative behaviour of a set of non-rotating perfect-fluid cosmologies, whose spatially homogeneous hypersurfaces admit a 3-parameter group of isometries of Bianchi types I, II, III, V, or VI. In this way some new exact solutions of the field equations are obtained.

162.091 Formation of galaxies due to thermal instability.
M. Kondo, Y. Sofue, W. Unno.
Astron. Herald, (Japan), Vol. 64, 123 - 127 (1971). In Japanese.

A review of the recent theory developed by the authors.

162.092 Topology and cosmology. G. F. R. Ellis.
General Relativity and Gravitation, Vol. 2, 7 - 21 (1971).

Discusses the topological properties one might expect in any reasonable model of the universe; the properties of some exact solutions which might serve as reasonable simple cosmological models; and some comments on properties one might expect in more realistic universe models.

162.093 Velocity-dominated singularities in irrotational dust cosmologies.
D. Eardley, E. Liang, R. Sachs.
Journ. Math. Phys., (USA), Vol. 13, 99 - 107 (1971). − See

Phys. Abstr., Vol. 75, No. 2091 (1972).

162.094 Charged particle creation in cosmology.
    P. C. W. Davies.
Nuovo Cimento B, Ser. 11, Vol. 6B, 164 - 178 (1971).
    Three modifications of Maxwell's equations are discussed which allow for the possibility of nonconservation of electric charge.

162.095 Uniform model universes containing matter and black-body radiation.   T. L. May.
Thesis, Univ. Illinois, Urbana. [Available from Univ. Microfilms, Ann Arbor, Mich., U.S.A. Order No. 71—14861], 60 pp. (1970).
    The recently discovered 3°K background radiation is assumed to be a remnant of the primeval fireball. The Robertson-Walker metric is adopted, and solutions to Einstein's equations for the cosmological problem are sought. The universe is assumed to consist of matter and black-body radiation.

162.096 The fluctuating electromagnetic field of the universe.   M. Surdin.
Ann. Inst. Henri Poincaré, Section A, Vol. 15, 203 - 241 (1971). In French.
    Based on the concept of the random electromagnetic field of the universe a stochastic electrodynamics is developed.

162.097 Isotropization of inhomogeneous centrally symmetric cosmological models.
V. A. Ruban, A. D. Chernin.
Proc. Sixth Winter School on Space Physics, Part I. Apatity 1969, (see 012.033), p. 13 - 25 (1971).

162.098 Formation of galaxies in an expanding universe.
A. G. Doroshkevich.
Proc. Sixth Winter School on Space Physics, Part I. Apatity 1969, (see 012.033), p. 26 - 33 (1971).

162.099 The spectrum of the extragalactic background radiation.   R. A. Syunyaev.
Proc. Sixth Winter School on Space Physics, Part I. Apatity 1969, (see 012.033), p. 65 - 72 (1971).

162.100 Charge symmetry of the universe.   N. A. Vlasov.
    Proc. Sixth Winter School on Space Physics, Part I.
Apatity 1969, (see 012.033), p. 77 - 81 (1971).

162.101 Perturbations of a cosmological model and angular variations of the microwave background.
A. M. Wolfe.
Diss. Fac. Graduate School, Univ. Texas, Austin, Texas (1967). [Available from Univ. Microfilms, Ann Arbor, Mich., U.S.A.], 86 pp. (1970).

162.102 Unser Wissen vom Ursprung der Welt.
    W. Kundt, M. Reinhardt.
Phys. unserer Zeit, [Verlag Chemie, Weinheim (Germany)], Vol. 2, 24 - 28 (1971).

162.103 Survey of cosmology. Is "our world" implied by thermal equilibrium in the hadron era?
W. Kundt.
Springer-Tracts Modern Phys. (Ergebnisse exakt. Naturwiss.), Vol. 58, 1 - 47 (1971). — Review article.

    Plasma physics applied to cosmology.
See Abstr. 062.031.

        Neutrino emission, population II stars and big-bang cosmology.   See Abstr. 065.064.

    Production of helium in massive objects.
See Abstr. 065.112.

    Observable effects of primordial gravitational waves.   See Abstr. 066.023.

    Tetrads, anholonomic coordinates, and space-time geometry.   See Abstr. 066.055.

    Exact cosmological solutions in Brans and Dicke's scalar-tensor theory, I.   See Abstr. 066.071.

    Cosmological density fluctuations during hadron stage.   See Abstr. 066.080.

    Creation of particles by gravitational fields.
See Abstr. 066.081.

    Experimental verification of the gravitation theory in cosmology.   See Abstr. 066.147.

    Galactic absorption.   See Abstr. 131.080.

    The radio source counts and their implications for cosmic evolution.   See Abstr. 141.014.

    On peculiarities of quasar redshifts and hypotheses to explain them.   See Abstr. 141.069.

    Quasar statistics for Lemaître cosmologies.
See Abstr. 141.082.

    The curious mystery of log N – log S.
See Abstr. 141.086.

    Radio source counts, cosmology, and evolution.
See Abstr. 141.087.

    Quasars and cosmology.   See Abstr. 141.170.

    X-ray background radiation.
See Abstr. 142.024.

    The evolution of galaxies. I. Formulation and mathematical behavior of the one-zone model.
See Abstr. 151.048.

    Some implications of a new value for the primordial solar deuterium-hydrogen ratio.   See Abstr. 155.045.

    Radius of maximum rotational velocity in spiral galaxies.   See Abstr. 158.023.

    The age of the galaxies and globular clusters: Problems of finding the Hubble constant and deceleration parameter.   See Abstr. 158.048.

    The mean ratio of mass to the three-halves power of luminosity for elliptical and lenticular galaxies.
See Abstr. 158.089.

    Optical observations relevant to cosmology; Hubble diagram.   See Abstr. 158.118.

    The Coma cluster as an X-ray source: Some cosmological implications.   See Abstr. 160.009.

    An angular diameter – redshift relation for rich clusters of galaxies.   See Abstr. 160.011.

Author Index

BARABASHOV, N. P.
091.021
094.205
097.096
BARAKAT, R.
063.025
BARAN, P. I.
046.035
BARANOV, A. V.
072.049
BARANOV, B. A.
044.017
BARANOV, V. B.
074.010
BARANOV, V. K.
094.193
BARANOV, V. N.
045.002
BARATTA, G. B.
034.008
BARBARO, G.
065.091
BARBER, D. J.
073.005
094.056
BARBIER, M.
122.040
BARBIERI, C.
031.066
141.182
BARBON, R.
125.006
BARCZA, S.
064.038
BARDEEN, J. M.
065.002
066.001
BARKAT, Z.
065.005 .017
BARKER, E. S.
093.009
BARKER, F. C.
065.135
BARKER, M. C.
143.106
BARKER, W. A.
091.004
097.005
BARLIER, F.
082.077
BARLOW, B. V.
034.132
BARLOW, M. J.
131.070
BARNES, A.
062.060
106.026
BARNES, C. A.
065.081
BARNES, J. V.
113.017 .047
121.025
153.021 .025
BARNES, V. E.
105.083
BAROCAS, V.
003.107
005.002
080.050
BAROUCH, E.
078.005

BARR, E. S.
005.016
BARRA, A. L. DE LA
158.024
BARRAR, R. B.
042.068
BARRETT, A. H.
061.054
131.151
132.022 .024
BARRETT, E. W.
082.007
BARRETTE, L.
022.143
BARSUKOV, V. M.
084.217
BARTENEVA, O. N.
103.131
BARTH, C. A.
097.009
BARTHOLDI, P.
099.019 .067
BARTKUS, R.
113.056
131.132
BARTLEY, W. C.
078.003
BARTOLINI, C.
123.052
BARTOLINI, U.
075.017
BARTON, D.
066.014
BARTON, D. K.
033.036
BASCHEK, B.
114.069
125.005
BASH, F. N.
141.262
BASHKIN, S.
022.112 .137
BASHKIRTSEV, V. S.
073.019
BASILOVA, R. N.
143.015
BASS, A. M.
022.048
BASS, M. N.
094.061 .132
BASSETT, A. B.
003.104
BASTIDAS, A.
015.015
BASU, B.
141.144
BASU, D.
077.001
BASU, S.
083.020
BASU, SUNANDA
083.020
BATALLI-COSMOVICI, C.
051.012
BATCHELOR, R.
141.151
BATE, R. R.
003.166
BATES, B.
034.064
071.061

BATESON, F. M.
010.024
120.017
123.044 .045 .046 .047
     .048 .049 .050
124.010
BATISHKO, C. R.
084.016
BATRAKOV, YU. V.
044.003
054.014 .015
BATTEN, A. H.
117.005 .025
119.005
121.033
BATTEN, R. L.
003.062
BATTEY, M. H.
094.325
BATTISTINI, P. L.
121.075
BATTISTON, L.
065.137
BATUEVA, N. B.
052.029
BATYLOVA, O. P.
104.111
BAUER, E.
082.042
BAUER, P.
082.078
BAUER, S. J.
082.148
BAUM, W. A.
097.064
099.024
BAUMAN, EH. I.
035.002
BAUMANN, G.
151.044
BAUR, H.
105.084
BAUSSUS-VON LUETZOW, H.
081.004
BAVASSANO, B.
084.236
BAXTER, W. M.
010.012
BAYM, G.
065.098 .151
141.231
BAZER, J.
062.040 .041
BAZILEVSKAYA, G. A.
078.047
BAZILEVSKY, A. T.
094.196
BEALS, C. S.
105.050
BEAMS, J. W.
043.016
BEARDSLEY, W. R.
112.012 .013
BEATY, E. C.
082.001
BEAVERS, W. I.
096.041
BEC, A.
043.015
BECHIS, K. P.
131.056

CLANCY, M. C.
142.001
CLARIA, J. J.
153.008
CLARK, B. G.
141.151
CLARK, D. D.
034.066
CLARK, G.
061.020
CLARK, G. A.
033.042
CLARK, I. D.
091.026
CLARK, K. C.
084.016 .027
CLARK, R. S.
094.061
CLARK, S. P.
097.094
CLARK, T. A.
071.062
103.101
141.027 .097 .103 .104
CLARKE, D.
032.017
103.101
CLARKE, J. N.
125.017
CLARKE JR., R. S.
105.106
CLARRICOATS, P. J. B.
033.049 .051 .069 .070
CLASSEN, J.
004.002 .032
CLAY, R. W.
143.008
CLAYDON, B.
033.047
CLAYTON, D. D.
061.044
065.018 .030 .124 .129
CLAYTON, R. N.
094.097
CLEAVER, A. V.
011.008
CLEGHORN, T. F.
143.059 .112
CLEMENT, M. J.
065.013
CLEMENTS, E. D.
122.059
141.026
CLIFFORD, S. F.
082.110 .111
CLINE, T. L.
078.041
CLOUET, B.
010.028
CLOUTIER, P. A.
084.031
CLUBE, S. V. M.
041.053
122.059 .126
COBLEIGH, T.
094.061
COCHRAN, G. V.
117.034
COCKE, C. L.
022.068 .144

COCKE, W. J.
141.187
142.061
CODE, A. D.
013.009
064.016
CODINA, S.
131.041
COELHO BALCA, M.
118.023
COFFARO, P.
034.046
COGGER, L. L.
083.039
COHEN, H.
053.017
063.005
COHEN, H. L.
121.063
COHEN, L.
151.041
COHEN, M.
022.079
073.095
113.028
114.098
COHEN, M. H.
122.105
141.151 .202
COLBURN, D. S.
094.064 .090 .244 .307
106.016 .023
COLE, G. H. A.
094.008
COLE, K. D.
066.117
COLEMAN, C. J.
066.130
COLEMAN, P. L.
142.017 .073
COLEMAN JR., P. J.
106.016
COLGATE, S. A.
132.036
COLLA, G.
125.029
132.042
141.211
COLLINS, B. S.
034.006
COLLINS, C. B.
162.008 .090
COLLINS, J.
064.017
COLLINS, T. F.
121.041
COLLINSON, D. W.
094.325
COLOMBO, G.
099.073
100.012
107.011
COLTHARP, R. N.
082.040
COLTON, G. W.
105.133
COLVIN, R. S.
033.002
131.155
141.192

COMBES, M.
034.076
099.042
COMBES, M.-A.
098.020
COMELLA, P. A.
034.105
COMELLO, G.
124.102
COMPSTON, W.
094.133
COMSTOCK, G. M.
094.107
CONDOS, T.
004.052
CONEL, J. E.
094.066 .080
CONGEDUTI, F.
082.104 .120
CONKLIN, E. K.
141.173
CONNERADE, J. P.
022.095
062.054
CONNORS, M. G.
123.039
CONTI, P. S.
065.009
114.115
CONTOPOULOS, G.
015.010
151.039
CONTRO, W. S.
004.016
CONWAY, R. G.
141.045
COOK, A. F.
100.012
104.017 .114
COOK, A. H.
003.056
081.015
COOMBS, A. E.
032.019
COOPER, B. F.
131.095
COOPER, B. F. C.
131.075
141.151
COOPER, D. N.
033.027 .028
COOPER, G.
062.047
COOPER, J.
022.015 .038 .039
COOPER, M. L.
121.091
COPPI, B.
073.062 .071
CORBEN, P. M.
113.021
CORDELL, B. M.
097.057
CORDES, E.
094.105
CORDWELL, C. S.
131.136
CORNWALL, J. M.
084.001
CORONITI, F. V.
084.001

GAWIN, J.
143.075
GAY, P.
094.251
GEBBIE, H. A.
031.063
082.006 .130
GEBBIE, K. B.
012.001
064.006
073.034
GEHRELS, T.
098.005
131.065
141.187
GEISS, J.
094.139 .199 .252 .254
.255
GELDON, F. M.
131.051
GENDEREN, A. M. VAN
113.004
GENDRIN, R.
084.277
GENKIN, I. L.
061.039 .040
151.070 .071 .072 .073
158.111
GENKINA, L. M.
158.111 .121
161.006
GENTNER, W.
105.102 .151
GEORGIEV, N.
046.031
GEORGOBIANI, G. G.
004.005
GERARDI, G.
034.046
GERARDO, J. B.
022.075 .118
GERASIMENKO, S. I.
103.120 .122 .129
GERLACH, U. H.
151.014
GERLACH, W.
005.008 .014
GERLOFF, U.
105.052 .059
GEROCH, R.
066.140
162.056
GERSHBERG, R. E.
003.151
122.056
GERSHMAN, B. N.
083.003
GERSTBACH, G.
045.018
GESSNER, H.
123.036
GETMANTSEV, G. G.
062.016
GEYLING, F. T.
003.068
GEZARI, D.
118.017
GHAFFARI, A.
042.018
GHEREGA, O.
117.029

GHEUDIN, M.
131.006
GHOBRIAL, S.
033.052
GHOSH, S. N.
076.040
GHOZEIL, I.
031.059
GIACAGLIA, G. E. O.
042.043
052.020
GIACCONI, R.
034.058
051.020
142.002 .005 .012 .025
.028 .030 .046 .071
.072
159.005 .012
GIANNONE, P.
065.036 .114 .115 .117
GIBBONS, G. W.
066.093
117.012
GIBBONS, R. V.
105.044
GIBBS, R. E.
143.072
GIBERT, J. M.
105.125
GIBSON, J. B.
098.028
GIBSON JR., E. K.
094.061
105.009
GIELINGH, W. F.
097.056
GIERASCH, P.
097.039 .089
GIERASCH, P. J.
011.016
097.027 .102
GIESE, R. H.
106.012 .020
GILBERT, F.
081.040
GILBERT, I. H.
151.050
GILBODY, H. B.
003.148
GILLE, J. C.
097.027
GILLESPIE, B.
071.057
GILLETT, F. C.
114.059
122.101
158.079
GILLETT, H. R.
034.024
GILLUM, D. E.
105.140
GILMAN, P. A.
080.006
GILMARTIN, D. E.
094.308
GILMORE, A. C.
031.014
125.106
GILRA, D. P.
064.016

GINDILIS, L. M.
051.019
GINESTET, N.
119.008
GINGERICH, O.
004.011 .024
073.001
GINSBURG, I. F.
100.004
GINTSBURG, M. A.
062.035 .075 .076
GINZBURG, V.
013.002
GINZBURG, V. L.
141.064 .163 .217 .222
GIORDANO, A.
085.003
GIORGIETTI, A.
104.015
GIOVANELLI, R. G.
034.024
072.033
GIRAUD, A.
083.070
GIRICHEV, V. P.
042.024
GIRNYAK, M. B.
082.132
123.026
GIROTTI, H. O.
066.129
GIULI, R. T.
143.003
GLAGOLEVSKY, Y. V.
114.057
GLASBY, J. S.
003.069
GLASPEY, J. W.
152.006 .008
GLASS, B. P.
094.166
GLASS, E. N.
066.143
GLAZMAN, V. N.
033.012
GLEBOVA, N. I.
097.015
GLEESON, L. J.
078.022 .045
143.006 .101
GLEISSBERG, W.
072.068
GLENCROSS, W. M.
034.136
073.097
076.015
GLUSHKO, V. P.
051.003
GLUSHKOV, JU. I.
132.047
GLUSHKOV, YU. I.
132.013 .014
GLUSHNEVA, I. N.
122.121
GNEDIN, YU. N.
142.086
143.013 .049 .136
GNEVYSHEV, M. N.
085.006
GNEVYSHEVA, R. S.
075.001

HARA, T.
143.087
HARADA, K.
094.015
HARAMUNDANIS, K.
158.125
HARDEBECK, E. G.
114.099
131.107
HARDING, G. A.
008.021 .041
034.004
155.027
HARDING, P. J. R.
082.024
HARDORP, J.
-065.006
074.006
114.105
HARDY, D. A.
015.008
HARE, P. E.
094.015
HARGREAVE, D.
004.014
HARING, D. J.
112.013
HARPER, R. I.
004.045
HARRIES, J. E.
082.129 .130
HARRIES, J. R.
142.078
HARRINGTON, J. P.
133.015
HARRINGTON, R. S.
072.053
117.019
HARRIS, B.
034.058
HARRIS, B. J.
091.040
HARRIS, K. K.
084.033 .285
HARRIS, R. D.
083.009
HARRIS, W. E.
065.013
HARRISON, E. R.
066.022
131.057
151.011
155.005
HART, A. M.
084.040
HART, M. R.
094.107
HART, R.
158.069
HART, W. R.
094.061
HARTEN, A.
074.053
151.042 .043
HART JR., H. R.
143.069 .085
HARTLE, R. E.
091.039
HARTMAN, R. C.
143.058

HARTMANN, R.
004.016
072.069 .085
HARTMANN, W. K.
011.045
094.003
HARTNETT, J. P.
003.081
HARTUNG, J. B.
094.116 .177 .266
105.041
HARTWICK, F. D. A.
141.209
154.001
HARTZ, T. R.
084.042 .316
HARVEY, G. A.
141.174
HARVEY, G. M.
113.021
HARVEY, J.
034.027 .123
073.041
080.046
HARVEY, J. W.
072.036
HARVEY, K. L.
072.036
HARVEY, P.
131.056
HARWIT, M.
061.049
106.009
155.032
HARWIT, M. O.
131.013
HARWOOD, M.
122.064
HASEGAWA, A.
084.055 .251
HASER, L.
034.073
HASHEMI-TAFRESHI, J.
074.097
HASHIM, A.
143.095
HASLAM, C. G. T.
155.009
HASLAM, C. M.
155.027
HASTIE, R. J.
084.403
HATANAKA, Y.
094.323
HATTON, C. J.
143.094 .106
HAUCK, B.
002.026
113.014
HAUG, E.
076.027 .038
HAUG, H.
079.101
HAUGE, OE.
071.070
HAUS, H. A.
033.078
HAUSKA, H.
078.048
HAVLEN, R. J.
122.155

HAVLEN, R. J.
155.024
HAVNES, O.
065.009
HAWKING, S. W.
066.093
117.012
HAYAKAWA, S.
065.132
142.013 .049 .051
HAYATSU, R.
107.004 .005
HAYES, D. S.
131.064
HAYLI, A.
151.034
155.004
HAYMES, R. C.
003.072
HAYS, P. B.
084.018 .022
HAYSHAM, H.
003.073
HAZARD, C.
096.026
141.052 .068
HEAP, S.
114.053
HEAP, S. R.
114.058
133.026
HEARD, J. F.
112.003
121.056
HEARN, A. G.
022.104
114.089
HEARN, D.
034.048
HEARNSHAW, J. B.
114.019
HEASLEY, J. N.
065.073
HEBEDA, E. H.
105.099
HECHT, H. F.
162.048
HEDGECOCK, P. C.
078.021
084.253
HEFFRON, W. G.
094.088
HEGARTY, J. C.
162.079
HEGGESTAD, H. M.
082.112
HEGGIE, D. C.
042.046
HEICKLEN, J.
091.022
HEIER, K. S.
094.018
HEIKEN, G.
094.132
HEIKEN, G. H.
094.061
HEILES, C.
131.016 .108
HEIMANN, M.
105.104

KANAI, M.
141.019
KANAL, M.
063.019
KANDAUROVA, K. A.
072.016 .077
KANDEL, R. S.
071.023
073.036 .089
KANE, J. A.
083.071
KANE, R. P.
084.220 .303
KANE, S. R.
073.082 .111
078.051
KANE, T. R.
042.020
KANGEDAL, B.
078.048
KANISHCHEVA, R. K.
122.022 .116
KANITSCHNEIDER, B.
003.082
KANONIDI, KH. D.
084.014
KAPAHI, V. K.
141.062 .179
KAPITSA, S. P.
005.009
022.060
KAPITZKY, J. E.
066.022
KAPLAN, I. R.
003.027
KAPLAN, S. A.
061.025
062.007
141.076
KAPLON, M. F.
143.002
KAPTEIN, E. J.
104.004
KAPUSTIN, I. N.
034.018
KAPUSTIN, P. A.
079.102
KAPUSTKIN, A. A.
033.022 .023
KARACHENTSEV, I. D.
158.060 .067
KARACHENTSEVA, V. E.
158.060 .066
KARA JR., C.
094.101
KARAKAD'KO, V. K.
143.133
KARAL, F.
062.040
KARANJAI, S.
063.034
KARAS, R. H.
084.258
KARASTOYANOV, A.
066.119 .120
KARDOPOLOV, V. I.
082.062
122.016 .145
131.029
KARIMOV, K. A.
082.072

KARIMOV, K. A.
104.066 .068 .069 .070
KARIMOVA, D. K.
041.016
122.113
KARLSSON, K. G.
141.006
KARPINSKY, V. N.
071.042
KARPMAN, V. I.
084.298
KARPOWICZ, M.
160.003 .015
KARTASHOV, V. F.
003.007
098.032
KARYAGINA, S. V.
132.013 .014
KASHCHEEV, B. L.
035.005
104.082 .089 .098 .103
.109
KASINSKY, V. V.
073.045
KASPER, U.
066.062 .063
KASPER JR., J. F.
081.034
KASSINSKY, V. V.
073.021
KASTEL', G.
103.119
KASTEL, G. R.
103.100 .103 .119 .120
KASUMOV, F. K.
065.074
066.010
125.012
KATASEV, L. A.
082.097
104.101
KATO, S.
065.149
082.074
151.012
KATO, T.
142.013 .049
KATO, Y.
062.067
KATOH, M.
033.053
KATS, M. E.
106.040
KATTAWAR, G. W.
082.086
093.022
KATTERFELD, G. N.
091.024
099.044
KATYUSHINA, V. V.
082.037
KATZ, J. M.
063.035
072.022
KATZ, L.
084.269
KAUFMAN, A. S.
074.056
KAUFMAN, M.
032.004

KAUFMAN, S. E.
162.010 .037
KAUFMAN, V.
022.113
KAUFMANN, R. L.
084.302
KAULA, W. M.
094.188
KAUTZLEBEN, H.
046.004
KAVALIAUSKAITE, G.
113.055
KAVERIN, A. A.
014.014
KAWAGUCHI, S.
143.087
KAWASAKI, K.
084.240 .259 .284
KAYE, M.
094.037
KAZANTSEV, A. N.
097.050
KAZES, I.
131.089
KAZIMIRCHAK-POLONSKAYA,
E. I.
102.011
KAZIMIROVSKII, E. S.
082.008
KAZYUTINSKIJ, V.
004.033
KAZYUTINSKIJ, V. V.
013.005
KEATH, E. P.
078.002
KEBABIAN, P. L.
034.133
KEBULADZE, T. V.
078.049
KEELEY, D. A.
141.208
KEENAN, P. C.
114.003 .076
KEENLISIDE, W.
103.101
KEGEL, W. H.
062.021
063.030
KEIL, K.
094.042 .269 .272
105.022 .087
KEIL, S. L.
071.023
KEILHACKER, M.
062.008
KEIRLE, P.
061.021
KEITH, J. E.
094.061
KELLAWAY, G. A.
004.008
KELLER, H.-U.
044.009
162.076
KELLERMANN, K. I.
033.009
141.079 .151 .189 .202
158.098
KELLEY, M. C.
084.013

MANDRYKIN, S. S.
080.064
MANFROID, J.
061.051
MANGUS, J. D.
031.061
MANIE, H. J.
085.004
MANJOS, A. V.
162.029
MANKA, R. H.
094.055
MANNERY, E. J.
082.059
153.017
MANN PATERSON, A.
003.102
MANSINHA, L.
044.014
045.028
081.012 .014
MANSON, A. J.
094.325
MANSUROV, S. M.
084.230
MANSUROVA, L. G.
084.230
MANUEL, O. K.
061.006
082.094
105.063 .155
MAO, C. Y.
143.042
MARAN, S. P.
132.001 .027 .043 .044
141.126
MARANDINO, G. E.
141.027 .097 .103
MARANO, B.
125.029
132.042
MARASCHI, L.
034.050
141.137 .165
MARCHAL, C.
042.041
MARCHAL, J.
008.072
MARCINKOWSKI, C. J.
066.091
MARCUCCI, R.
162.031
MARCUS, W. DE
091.028
MARDER, L.
003.092
MAREK, K.-H.
046.031
MARENIN, I.
022.011
064.015
MARENIN, I. R.
114.076
MARGOLIS, J. S.
022.016
091.016
097.085
099.001
MARGON, B.
142.007 .058 .060

MARGRAVE, J. L.
105.011
MARIANI, F.
084.236
MARIIN, B. V.
034.017
MARIN, M.
034.022
MARINO, B. F.
113.022
123.043
MARIS, D.
072.070
MARIS, G.
073.064
075.031
MARK, J. W. K.
151.053
155.022
MARK, J. W-K.
151.030 .045
155.048
MARKACHEV, V. V.
094.021 .024
MARKARIAN, B. E.
158.033
MARKEEV, A. K.
077.049
MARKEEV, A. P.
042.007
MARKOV, M. S.
094.079
MARKOVICH, M. Z.
102.020
MARLBOROUGH, J. M.
064.031 .032
MAROCHNIK, L. S.
151.021 .025 .026 .027
.028
MAROV, M. YA.
082.053
093.003 .030
MAROVICH, E.
084.026
MARQUES DOS SANTOS, P.
077.001
MARSDEN, B. G.
098.005 .011 .025
102.014
103.002 .113 .119
MARSDEN, P. L.
143.094
MARSH, E. L.
042.020
MARSH, J. G.
052.018
MARTEL, M. T.
131.046
MARTI, K.
094.290
MARTIN, A. H. M.
141.166
MARTIN, C. N.
003.106
MARTIN, D. H.
033.076
MARTIN, I.
142.031
MARTIN, I. M.
061.022

MARTIN, J. R.
094.061
MARTIN, L.
097.064
MARTIN, L. J.
097.068 .093
MARTIN, P. G.
131.009
155.001
MARTIN, T. Z.
099.002 .059
MARTIN, W.
094.189
MARTIN, W. C.
022.111
MARTINET, L.
155.004
MARTINEZ, J.
034.022
MARTINEZ-GARCIA, M.
022.154
MARTINI, A.
153.028
MARTINO, D. DI
123.021
MARTINSON, I.
012.032
022.112
MARTJANOVA, G. N.
085.011
MARTONCHIK, J. V.
093.008
097.073
MARTRES, M. J.
071.037
073.087
074.007
075.014
MARTSVALADZE, N. M.
082.025
MARTYNENKO, V. V.
104.020
MARTYNOV, D. YA.
003.093
005.006
013.004
118.026
121.084 .087
123.073
MARUSSI, A.
011.041
MARUYAMA, K.
074.099
MARVIN, U. B.
094.158 .159 .277 .281
MARX, G.
061.037
MARX, S.
031.045
131.018 .019
MASAJTIS, V. L.
105.026
MASAKI, I.
063.002
MASANI, A.
009.004
065.090 .154
MASLENNIKOVA, L. B.
062.017
MASLOWSKI, J.
141.030

MOBLEY, R. E.
032.027
MOELLENSTEDT, G.
076.019
MOELLER PEDERSEN, B.
074.055
MOERGELI, M.
094.254 .255
MOESTA, H.
131.127
MOFFAT, P. H.
132.011
MOFFETT, T. J.
119.006
MOGILEVSKY, E. I.
072.039
077.031
080.060 .061
MOHAN, S. K.
141.238
MOHANTY, D. K.
141.127 .250
142.033
MOHNEN, V. A.
083.069
MOHR, J. M.
004.051
MOISEEV, I. G.
097.032 .091
141.248
MOISEEV, YU. N.
085.008
MOISEJEV, I. G.
141.070
MOISEYEV, I.
141.151
MOISYA, R. I.
104.030 .036
MOJSYA, R. I.
104.010
MOJSYA, R. T.
104.012
MOLER, R. B.
094.100
MOLLWO, L.
077.002
MOLODENSKY, M. M.
073.104
MOLTON, P. M.
051.008
MONAGHAN, J. J.
065.001 .038 .086 .087
MONIN, I. F.
081.042 .043
MONTAG, H.
046.031
MONTBRIAND, L. E.
084.053
MONTEAGUDO, V. N. DE
114.085
MONTGOMERY, J. W.
141.011
MONTGOMERY, M. D.
074.018
078.013
084.238 .257 .283
MOOK II, D. E.
142.092
MOORE, C.
105.003

MOORE, C. B.
003.027
094.048 .061
105.009 .033 .123
MOORE, C. E.
022.093
MOORE, E. O.
034.093
MOORE, H. J.
094.063 .176
MOORE, J. G.
084.011
MOORE, P.
003.154 .165 .168
005.003
094.027 .241
MOORWOOD, A. F. M.
082.087
MOOS, H. W.
082.135
093.020
MORAN, J. M.
131.140
132.028
MORANDO, B.
003.097
041.043
MORBEY, C. L.
096.038
MOREFIELD, F.
009.014
MOREIRINHAS PINHEIRO, M.
079.101
MORENO, G.
084.253
MORFILL, G.
078.020
084.201
MORFILL, G. E.
034.096
078.012
084.214
MORFORD, J. M.
054.002
MORGAN, F. J.
076.006 .025
MORGAN, J. W.
003.027
MORGAN, L. A.
132.010
MORGAN, T. A.
066.050 .106
MORGAN, W. W.
158.036 .114
MORGAN JR., D. L.
162.039
MORGANSTERN, R. E.
066.105 .114
162.064 .065 .066
MORGANTE, O.
073.105
MORGENTHALER, G. W.
011.025
MORITZ, H.
081.038
MOROZ, V. I.
093.019
097.006 .028
133.029
MOROZHENKO, A. V.
097.007 .059

MOROZHENKO, N. N.
073.033 .090 .099
MOROZOV, D. KH.
084.415
MORRIS, G.
141.271
MORRIS, G. A.
097.081
MORRIS, G. J.
141.230
MORRIS, M.
114.119
131.100
MORRIS, S. V.
122.154
MORRISBY, A. G. F.
010.007
MORRISON, D.
011.036
099.002 .059
MORRISON, D. A.
094.061 .167
MORRISON, F.
081.023
MORRISON, J. A.
094.160
MORRISON, L. V.
096.023
141.068
MORRISON, N. D.
113.033 .041
124.001
MORRISON, P.
066.021
125.010
141.028 .204
158.045
MORSE, F. A.
083.027
MORTON, A. E.
034.091
MORTON, D. C.
114.121
158.058
MORZENTI, S. P.
105.136
MOSALOV, I. V.
033.012
MOSELEY, T.
094.240
MOSELEY, T. J. C. A.
100.007 .013 .015
MOSER, E.
073.018
MOSER, J. K.
003.112
MOSKALEVA, L. P.
097.051
MOSS, D. L.
117.001
MOSS, G. E.
033.019
MOSS, T. A.
143.003
MOTOVILOV, E. A.
094.022
MOTOVILOV, EH. A.
094.028
MOUTHAAN, K.
033.080

MOUTSOULAS, M.
002.047
094.110 .301
MOZER, F. S.
084.013 .409
MOZER, M.
076.019
MOZHERIN, V. M.
055.007
MRKOS, A.
103.109
MUCKE, H.
047.020
053.024
MUDGETT, P. S.
063.017
MUELLER, D. D.
003.166
MUELLER, G.
094.271
MUELLER, H.
046.001
MUELLER, H. J.
143.128
MUELLER, I. I.
031.073
MUELLER, O.
094.263
105.024 .085 .156
MUELLER, R.
075.012
MUELLER, R. O.
141.237
MUELLER, W. F.
094.258
MUENCH, G.
097.029
099.070
MUENCH, J. W.
084.271
MUENOW, D. W.
105.011
MUFF, E.
044.025
MUGGLESTONE, D.
063.022 .023
MUGGLETON, L. M.
083.011 .012
MUHLEMAN, D. O.
094.189
MUIR, P.
094.037
MUKHERJEE, N. R.
094.093
MULDER, F. G.
084.248
MULHALL, P. S.
141.117
MULHOLLAND, J. D.
094.188
MULLAN, D. J.
003.151
062.001 .055
064.028 .058
071.072
120.011
126.024
131.157
MULLER, A. B.
012.010

MULLER, G. J.
096.005
MULLER, P.
094.161
118.013 .014 .016 .018
.020
MUNCASTER, G. W.
141.187
MUNGALL, A. G.
044.031
MUNRO, E. W.
033.073
MUNRO, R. E. B.
141.016 .056 .201
MUNRO, R. H.
073.022
MURAKAMI, H.
143.080
MURAKAMI, K.
143.071
MURATOV, M. V.
081.031
MURCRAY, D. G.
080.040
082.006
MURDIN, P.
096.010
142.037
MURGATROYD, R. J.
082.117
MURPHY, J. K.
041.034
MURPHY, J. O.
061.008
080.035 .036
MURPHY, J. P.
052.013
MURPHY, R. E.
084.040
099.002 .059
MURRAY, B. C.
003.058
092.004
MURRAY, C. A.
041.022
141.026 .068
MURRAY, J. B.
094.248
MURRAY, J. D.
131.156
MURRAY, S.
142.002 .025 .046 .071
MURTHY, C. S.
122.008
MURTHY, V. R.
094.145 .313
MURTY, S. S. R.
062.015
MUSEN, P.
042.039
MUSGROVE, R. G.
003.138
MUSMAN, S.
071.005 .032 .049
MUSORIN, M. I.
032.030
MUSTEL, E. R.
125.004 .009 .028
MUZZIO, J. C.
065.082

MYERS, P. C.
131.151
NACHMAN, P. M.
112.020
NACOZY, P. E.
042.047
NAEBAUER, M.
046.022 .024
NAEF, R. A.
047.021
075.004
100.010
104.038
NAGAMOTO, C. T.
105.008
NAGARAJAN, S.
062.013
NAGASAWA, K.
104.078
NAGASHIMA, K.
143.097
NAGATA, K.
143.080
NAGATA, S.
065.130
NAGATA, T.
084.272
094.185
NAGEL, R. H.
093.002
NAGY, A. F.
083.017
NAHON, F.
151.061
NAIDENOV, V. O.
061.059
NAKAGAWA, M.
142.048 .050
NAKAGAWA, S.
143.080
NAKAGAWA, Y.
062.028
072.041 .046
073.013
NAKAMOTO, A.
143.080
NAKAMURA, Y.
094.232 .243
NAKANO, T.
131.133
NAKAZAWA, K.
065.130
NAMISNAK, D.
112.012
NANCE, R. L.
094.233
NANCE, W. B.
094.061
NANDY, K.
114.073
131.001 .022 .023 .039
.047
132.021
NAPIER, W. M.
131.022
NAPIER, W. MCD.
093.006
NARDI, V.
062.014
NARIAI, H.
162.069 .070

RUTGERS, G. A. W.
034.104
RUTHBERG, S.
143.109
RUUSALEPP, M.
122.047 .048
124.103
RUZMAIKIN, A. A.
156.004
RUZMAJKIN, A. A.
162.063
RYABOV, B. P.
141.132
RYABOV, YU. A.
003.012 .045
RYABOVA, T. YA.
084.419
RYAN, J. A.
094.093
RYAN, M. J.
143.073 .074
RYAN JR., M. P.
162.073 .074 .088
RYBAKOV, A. I.
151.067
RYBAKOV, A. K.
104.046
RYBAKOV, A. V.
094.148
RYBANSKY, M.
073.083
074.045 .068
RYBICKI, G.
063.004
RYBICKI, G. B.
022.021
151.052 .054
RYCROFT, M. J.
012.004
RYDBECK, O. E. H.
022.135
033.026
131.139 .140
141.151
RYDGREN, A. E.
113.043
RYGG, T. A.
143.045
RYKHLOVA, L. V.
045.004
RYLE, M.
033.006
RYNIN, N. A.
003.109
RYSKULOV, A.
104.071
RYTER, C.
125.018
RYTOV, S. M.
141.113
RZHEVSKII, V. V.
094.023
RZHIGA, O. N.
053.033
SAAR, E.
162.016 .018 .019 .020
.021 .022
SAAR, I.
162.019
SABAUD, L.
034.054

SABITOV, SH. N.
032.029
066.084
SACCHETTI, F.
034.038
SACHDEV, P. L.
065.146
SACHS, A.
004.042
SACHS, M.
066.066
SACHS, R.
162.093
SACHS, R. K.
012.022
162.035
SACOTTE, D.
076.037
SADEH, D.
079.100
141.111
SADZAKOV, S.
032.025
SAFRONOV, V. S.
099.075
107.006
SAGAN, C.
012.004
015.002
097.035 .039 .088 .089
099.014 .023
131.063
SAGDEEV, R. Z.
084.298
SAGGION, A.
061.042
SAGITOV, M. U.
081.010
SAHA, P. K.
033.049 .069 .070
SAHA, S. K.
061.018
SAHADE, J.
114.085
SAHAL-BRECHOT, S.
022.103
SAID-UZ-ZAFAR CHAGHTAI,
M.
022.057
SAINT-MARC, L.
084.014
SAITO, K.
102.023
121.016
SAKAI, K.
066.073
SAKASHITA, S.
061.035
065.122
SAKIBAYEV, O.
066.086
SAKKA, K.
158.137
SAKURAI, K.
072.019
073.061
077.006 .038
SAKUYAMA, H.
143.087
SALEMA, C. E. R. C.
033.051

SALETIC, D.
032.025
SALIE, H.
003.118
SALISBURY, J. W.
002.047
SALISBURY, W. W.
053.031
SALMANOV, I. R.
114.132
SALPETER, E. E.
131.126
133.004
141.066 .082
142.006
158.072
SALTER, C.
141.211
SALTER, C. J.
125.029
132.042
155.009
SALUKVADZE, G. N.
074.016 .043
117.030
SAMAIN, D.
071.053
SAMOILOV, R. A.
141.218
SAMOJLOVA-YAKHONTOVA,
N. S.
098.001
SAMONENKO, YU. A.
034.017
SAMORSKI, M.
143.076
SAMPSON, D. H.
022.089 .090
SAMSONOV, A. V.
083.003
SANCHEZ, F.
008.103
034.112
082.029
105.154
SANCHEZ-MAGRO, C.
034.112
105.154
SANDAGE, A.
113.064
122.009 .103
142.022
158.041 .048
SANDAKOVA, E. V.
104.025
SANDERS, R.
155.019
SANDERS, W. L.
153.004 .018
154.012
SANDFORD, B. P.
084.040
SANDFORD, M. C. W.
103.101
SANDFORD, P. W.
142.041
SANDIG, H. U.
032.028
SANDLIN, G. D.
076.023

SILK, J.
114.113
132.036
161.005
SILVA, A. SIMOES DA
008.028
118.023
SILVER, L. T.
094.051
SILVERBERG, E. C.
106.036
SILVERMAN, S. M.
122.068
SILVESTRO, G.
065.154
141.133
SIMA, Z.
117.024
SIMIC, M.
096.017
SIMMONS, J. E.
031.059
SIMMONS, J. W.
003.111
SIMMONS, K.
124.102
SIMNETT, G. M.
078.025
SIMO, C.
081.045
SIMOES DA SILVA, A.
008.028
118.023
SIMON, G. W.
074.040
SIMON, M.
073.060
077.016
SIMON, P.
074.007
082.028
SIMON, T.
114.107
SIMONAITIS, R.
091.022
SIMONEIT, B. R.
094.204
SIMONENKO, A. N.
143.055
SIMONSON III, S. C.
132.026
SIMPSON, R. W.
064.029 .041
114.063 .071
SIMS, J. S.
097.005
SINANOGLU, O.
022.150
SINCHESKUL, B. F.
031.007
SINCHESKUL, V. N.
031.007
123.071
152.009
SINCLAIR, M. W.
131.076 .111
141.185
157.008
SINCLAIR, W. S.
043.011

SINELNIKOV, V. M.
083.023
SINGER, S.
078.013
084.058 .238 .266
SINGER, S. F.
093.007
094.210
107.015
SINGH, P.
143.117
SINTON, W. M.
094.120
SINYAEV, V. A.
031.064
041.041
SINZI, A.
041.055
SINZI, A. M.
043.014
096.030
SION, E. M.
126.004
SIPPEL, R. F.
094.095
SIROKY, J.
011.051
SISCOE, G. L.
084.219
SISTERO, M. E.
CASTORE DE
121.042
SISTERO, R. F.
121.005 .042
162.032
SITARSKI, G.
102.008
103.106 .109 .112 .117
SITNIK, G. F.
034.111
SITNIK, T. G.
141.146
SITTE, K.
106.018
SIVJEE, G. G.
082.048
SIVTSEVA, L. D.
083.042
SJOGREN, W.
094.161
SJOGREN, W. L.
094.220
SKALAFURIS, A. J.
071.024
SKILLING, J.
062.073
143.050
SKILLMAN, T. L.
084.282
SKOBELEVA, T. P.
094.032 .149
SKORUPSKI, A.
151.001
SKOUSEN, E. N.
143.091
SKREBTSOV, G. P.
143.135
SKRIPIN, G. V.
078.050
SKUL'SKIJ, M. YU.
121.050

SKULSKY, M. JU.
121.006
SKUMANICH, A.
064.045
073.002
114.141
SKURIDIN, G.
013.017
SLABINSKI, V. J.
055.003
SLANGER, T. G.
082.137
SLATER, R. H.
033.042
SLATTERY, W. L.
022.004
SLAUCITAJS, S. J.
046.038
SLAUGHTER, C. D.
071.003
072.036
SLEE, O. B.
141.015 .063 .117
142.068
SLEPTSOV-SHEVLEVICH,
B. A.
085.009
SLOCUM, G.
082.007
SLONIMSKAYA, M. V.
094.079
SLOTTJE, C.
077.064
SLOVOKHOTOVA, N. P.
094.245
SLUTSKY, L. F.
097.059
SLUTSKY, V. E.
082.062
SLYSH, V. I.
131.028
141.070
SMAK, J.
119.014
SMART, D. F.
078.056
143.025 .028 .125
SMART, N. C.
065.001
SMATHERS, H. W.
141.111
SMERD, S. F.
077.024
SMERNOFF, B. J.
065.140
SMEYERS, P.
065.024
SMIRIGA, N.
102.026
SMIRNOV, A. N.
035.005
SMIRNOV, A. S.
155.030
SMIRNOV, A. V.
034.017
SMIRNOV, N. P.
072.094
SMIRNOV, R. V.
106.008
SMIRNOV, V. V.
054.032

SWAMY, K. S. K.
131.040
132.017
SWANENBURG, B. N.
061.021
SWANN, G. A.
094.058 .068
SWANSON, P. N.
077.079
SWARUP, G.
141.062 .127 .238 .250
142.033
SWEENEY, M. A.
119.001
SWEET, P. A.
072.037
SWEIGART, A. V.
065.016
SWENSON, C.
022.144
SWENSON, J. W.
003.110
SWENSON JR., G. W.
131.083
SWIFT, C. D.
159.001
SWIHART, T. L.
141.003 .207
SWINGS, J. P.
071.025
114.106
SWINGS, P.
061.062
SWINSON, D. B.
143.001 .016 .102
SWITZER, P.
073.110
SY, W.
077.025
SYKORA, J.
032.036
080.021
SYMES, R. F.
105.018 .019 .105
SYNAKH, V. S.
151.063
154.017
SYNITSYN, V. M.
082.054
SYROVATSKII, S. I.
142.097
143.024 .065 .139
SYROVATSKIJ, S. I.
062.052
078.039
084.289
SYUNYAEV, R. A.
162.062 .099
SZCZEPANOWSKA, A.
120.019
SZEBEHELY, V.
042.023 .049
151.038
SZEKERES, P.
066.101
SZUMIEJKO, E.
073.103
TABACHNIK, V. M.
121.017 .083
TADEMARU, E.
141.032 .134 .241

TAEUSCH, D. R.
083.073
TAFF, L. G.
160.006
TAJIRI, M.
062.067
TAKAGISHI, K.
142.048 .050
TAKAKURA, T.
072.035
073.015
TAKAOKA, N.
105.139
TAKATSUKA, T.
065.131
TAKEDA, H.
094.132 .273
162.068
TAKEUTI, M.
064.051
TALBOT JR., R. J.
151.048
TALON, R.
061.022
142.031
TAM, C. K. W.
062.058
TAM, W. G.
022.158
TAMAGAKI, R.
065.131
TAMHANE, A. S.
105.069
143.033
TAMMANN, G. A.
122.009 .060 .103
TAMRAZYAN, G. P.
107.009
TANAKA, H.
074.033
TANAKA, K.
071.002
TANAKA, Y.
065.122
142.013 .049 .051
TANANBAUM, H.
142.002 .005 .012 .025
    .028 .030 .046 .071
    .072
159.005 .012
TANDBERG-HANSSEN, E.
073.002 .004 .032 .039
    .049
TANDON, J. N.
064.025
073.053
TANDON, S. N.
143.084
TANENBAUM, A. S.
071.036
TANIUTI, T.
062.067
TANNER, R. W.
032.038
TANSKANEN, P.
143.109
TANTASHEV, M. V.
063.006
TAN TUNG ARJUN
066.137

TANZI, E. G.
078.018
TAPPERE, E. J.
073.010
TAPSCOTT, J. W.
158.036
TARADY, V. K.
093.026
TARAFDAR, S. P.
116.006
TARANOV, V. I.
117.009
TARASOV, A. V.
051.018
TARCSAI, G.
066.031
TARLING, D. H.
003.122
TARLING, M. P.
003.122
TARNSTRÖM, G. L.
077.019 .020 .081
TARTER, C. B.
022.029
TARTER, J.
065.097
TARTOIS, L.
010.028
TASCIONE, T. E.
131.057
TATARSKII, V. I.
082.133
TATE, R. C.
123.019
TATEVJAN, S. K.
013.014
055.006
TATUM, J. B.
103.101
TAUBENHEIM, J.
083.046
TAUBER, G. E.
065.120
TAWAKLEY, V. B.
052.027
TAWARA, H.
141.259
TAYLER, R. J.
061.032
062.032
TAYLOR, D. B.
011.030
TAYLOR, E. G. R.
003.123
TAYLOR, G. E.
099.018 .067
TAYLOR, G. J.
094.158 .159 .277 .281
TAYLOR, H. E.
084.017 .403
TAYLOR, J. H.
141.022 .091 .092 .152
    .177 .240
142.027
TAYLOR, L. A.
094.053 .257 .299
TAYLOR, P. A.
063.029
TAYLOR, R.
105.163

VOGES, W.
034.047
061.021
VOGLIOTTI, M. A.
098.041
VOISKOVSKY, M. I.
082.054
VOLCHKOVA, L. I.
094.187
VOLKOV, M. S.
042.060
VOLKOVA, N. V.
105.071
VOLLAND, H.
082.131
083.066
VOLOBUEV, S. A.
141.035
142.080
VOLOBUYEV, S. A.
078.027
VOLYANSKAYA, M. YU.
121.024
VOLYNOV, B. V.
082.088
VOLYNSKIJ, B. A.
003.020
VONDRAK, R. R.
084.030
VORONENKO, V. I.
098.038
VORONTSOV-VEL'YAMINOV,
B. A.
014.016
158.075 .126
VOROSHILOV, V. I.
003.003
VOROSHILOV, YU. V.
120.001
122.032 .057
154.009
VORPAHL, J.
073.029
VOSHAGE, H.
094.267
VOSKRESENSKIJ, L. L.
031.005
VOVCHIK, E. B.
121.050
123.061
VRABEC, D.
071.035
VRABEL, J.
031.060
VRIES, T. DE
053.005 .006
VSEKHSVYATSKAYA, I. S.
083.032
VSEKHSVYATSKIJ, S. K.
003.040
103.003
VSEKHSVYATSKY, S. K.
074.087
107.016
VUKOVICH, F. M.
082.093
VYCHRESTJUK, S. S.
099.069
WACHI, F. M.
094.308

WADA, M.
142.008
WADDINGTON, C. J.
143.059 .112
WADE, C. M.
141.074 .102 .109 .240
142.011 .077 .090
158.098
WAECHTER, S.
046.013
WAENKE, H.
094.197 .278 .280 .318
WAGNER, L. S.
083.045
WAGNER, R.
062.022
153.005
WAGONER, R. V.
065.112
066.001
WAHL, J. J.
003.129
WAHLEN, M.
094.252
WAHLQUIST, H. D.
162.084
WAKAMATSU, K.
158.137
WAKITA, H.
094.054
105.140
WALBORN, N. R.
114.101
124.104
132.053
158.036
WALD, R. M.
066.008 .030
WALDMEIER, M.
008.120
034.097
072.054 .063
074.009 .069 .070 .071
.072 .073
075.011 .013 .021 .022
.036
079.100
WALES, I. M.
052.002
WALKER, A. D. M.
099.037
WALKER, D. M. C.
082.075
WALKER, G. A. H.
010.023 .037
114.116
116.003
WALKER, G. K.
124.102
WALKER, J. C. G.
083.065
084.009
WALKER, M. F.
034.087
082.071
159.033
WALKER, R. M.
094.322
WALKER, T. R.
084.296

WALKER, W. S. G.
113.061
122.062
WALL, J. V.
141.017 .242 .245
WALL, R. E.
081.035
WALLACE, B. G.
066.135
WALLACE, D. C.
131.010
WALLACE, L.
093.002
WALLENHAUER, A.
046.026
WALLENQUIST, A.
094.151
WALLER, A. J.
033.068
WALLERSTEIN, G.
114.109 .113
122.084
153.017
WALLIS, M.
074.023
WALLIS, R. E.
041.021
WALMSLEY, M.
131.096
WALRAVEN, J.
159.028
WALRAVEN, T.
159.028
WALSH, W. J.
084.260
WALT, M.
084.279 .404 .412
WALTER, H.
031.033
WALTER, H. G.
021.001
052.002
WALTER, K.
005.015
WALTERS, G. K.
131.054
WANG, C. G.
065.062 .101
WANG, L.
034.103
WANG LUIG
003.155
WARBURTON, D.
094.138
WARD, B.
123.043
WARD, S. H.
094.117 .162
WARD JR., F. W.
122.068
WARDLE, J. F. C.
141.098
158.098
WARDLEY, M.
100.007 .013 .015
WARE, N. G.
094.295
WARNER, B.
124.106
126.008 .017 .019

WARNER, J.
082.147
094.061 .132 .226
WARNER, J. W.
113.008
WARNER, P. J.
158.078
WARNOW, J. N.
002.018
WARREN, R. G.
105.146
WASHIMI, H.
074.099
WASSENBERG, W.
077.032
WASSERBURG, G. J.
094.045 .046 .130 .131
.136 .292 .296
105.064
WASSERMAN, L.
099.023
WASSON, J. T.
003.027
105.024 .029 .147
WATAGHIN, A.
162.078
WATAGHIN, G.
022.069
WATANABE, S.
074.094
WATANABE, T.
074.099
WATERFIELD, R. L.
096.012
103.116
WATERS, B. E.
080.032 .033 .034
WATERS, J. W.
132.022 .024
WATKINS, C. D.
104.051
WATKINS, J.
094.163
WATSON, P. A.
033.052
WATSON, R. D.
122.087 .107
WATTENBERG, D.
005.018
007.000
009.022
WATTS JR., R. N.
053.002 .003 .013 .021
.022 .023
054.001 .010 .012 .017
.018
097.067
WAYNE, R. P.
082.124
WDOWCZYK, J.
143.077
WEBBER, J. C.
072.096
WEBBER, W. R.
143.022 .067
WEBER, H. W.
105.148 .152
WEBER, J.
066.125 .128
WEBER, S. E.
074.073

WEBROVA, L.
044.019
WEBSTER, B. L.
142.037
WEBSTER JR., W. J.
131.147
132.033
WEDEL, B.
009.006
031.036
112.002
WEEDMAN, D. W.
113.020
158.092
159.003 .008
WEEKES, T. C.
141.131
WEEKS, L. H.
076.018
WEFER, F. L.
077.080
WEGNER, G.
126.007
WEGNER, M. W.
094.093
WEHLAU, A.
154.014
WEIBLEN, P. W.
094.043
WEICKMANN, H. K.
082.007
WEIDEMANN, V.
065.126
142.064
WEIGERT, A.
003.130
126.001
WEILL, D. F.
094.293
WEINBERG, S.
162.013
WEINER, C.
002.018
WEINSTEIN, D. H.
066.002
WEISBACH, M. F.
062.006
WEISS, A. W.
022.149
WEISS, N. O.
061.031
080.025
WEISS, W. W.
099.039
WEISSKOPF, M. C.
134.001
WEISSMAN, P.
065.053
WEITENBECK, A. J.
124.102
142.044
WELCH, G. A.
158.005 .053
161.004
WELCH, W. J.
099.045
131.137
132.028
WELIACHEW, L.
158.004 .087 .101

WELIN, G.
122.066
WELLER, C. S.
082.076
WELLMANN, P.
124.102
WELLS, D. C.
099.027
WELLS, M. B.
051.014
WELLS, R. A.
004.025
097.075
WELTHER, B. L.
010.001
WENTINK JR., T.
082.049
WENTZEL, D. G.
131.101
WENZEL, W.
113.044
121.007
123.037
141.210
WERNER, M. W.
011.046
066.056
131.051
WERNER, N. E.
080.053
WERNER, W.
034.055 .072
WERTLIEB, A. B.
072.015 .074 .080
WERTZ, J. R.
162.087
WESCOTT, E. M.
084.057
WESTERLUND, B. E.
114.013
159.015 .017
WESTERMAN, H. R.
003.068
WESTERVELT, P. J.
066.113
WESTFALL, J. E.
094.073 .215 .216
WESTON, L. B.
047.018
WESTPHAL, J. A.
097.003 .037
WETHERILL, G. W.
003.027
094.014
105.149
WHALEN, B. A.
084.046
WHALING, W.
022.140 .154
WHANG, Y. C.
074.054 .063
WHEELER, J. A.
066.060
141.084
WHEELER, J. C.
061.023
065.005 .035
WHITAKER, A. J. T.
033.073
WHITE, J. E.
003.166

Subject Index

GALAXIES
MASSES
158.006
GALAXIES
MASS-LUMIN RELATION
151.029
158.089
GALAXIES
NUCLEI
012.002
131.134
141.066 .124
151.049
158.009 .059 .073
.079 .092 .094
.097 .098 .099
.117 .119 .128
.129 .133 .135
.137
GALAXIES
OH EMISSION
158.004
GALAXIES
PECULIAR
158.090
GALAXIES
PHOTOMETRY
113.026
158.005 .027 .032
.034 .071 .093
.106
GALAXIES
POPULATION II STARS
065.092
GALAXIES
POSITIONS
158.063
GALAXIES
RADIO RADIATION
158.022 .063 .064
.065
GALAXIES
REDSHIFTS
158.061 .070 .131
160.012
GALAXIES
ROTATION
151.011
158.023
GALAXIES
SEYFERT GALAXIES
141.124
158.002 .009 .017
.018 .031 .065
.072 .095 .127
.132
GALAXIES
SPECTRA
158.012 .033 .074
GALAXIES
SPIRAL STRUCTURE
151.015 .025 .026
.027 .028 .063
GALAXIES
STELLAR POPULATIONS
158.088
GALAXIES
SUPERMASSIVE
158.005

GALAXIES
VARIABLES
122.144
GALAXIES MULTIPLE
158.000
GALAXIES SINGLE
158.000
GALAXY
21 CM RADIATION
157.006
GALAXY
EVOLUTION
155.000
GALAXY
H ALPHA
155.033
GALAXY
HELIUM ABUNDANCE
065.122
GALAXY
H I REGIONS
131.089
GALAXY
NUCLEOSYNTHESIS
155.037
GALAXY
STRUCTURE
155.000
GALAXY
UV PHOTOMETRY
113.025
GALAXY
UV RADIATION
155.010
GAMMA RAY ASTRONOMY
061.000
GAMMA RAY BACKGROUND
142.031 .084
GAMMA RAYS
CRAB NEBULA
134.003
GAMMA RAYS
GALACTIC CENTER
142.082
GAMMA RAYS
MARS ATMOSPHERE
097.051
GAMMA RAYS
PULSARS
141.059 .165 .186
.239 .256
GAMMA RAYS
QUASI-STELLAR OBJECTS
141.186
GAMMA RAYS
RADIO GALAXIES
142.093
GAMMA RAYS
RADIO SOURCES
141.035
142.080
GAMMA RAYS
SOLAR
076.000
GAMMA RAYS
SOLAR FLARES
073.017
GAMMA RAYS
VARIATIONS
142.020

GAMMA RAYS
X RAY SOURCES
142.082
GAMMA RAY SOURCES
142.000
GAMMA RAY SOURCES
VARIATIONS
142.088
GAS
GALACTIC CENTER
155.002
GAS CLOUDS
HYDROMAGNETIC SHOCKS
062.001
GASEOUS NEBULAE
012.001
GASEOUS NEBULAE
CENTRAL STARS
132.030
GASEOUS NEBULAE
DUST
132.006
GASEOUS NEBULAE
HELIUM ABUNDANCE
132.039
GASEOUS NEBULAE
LINE INTENSITIES
132.012
GASEOUS NEBULAE
SPECTRA
132.012
158.119
GASEOUS NEBULAE
WOLF RAYET STARS
132.030
GAUNT FACTORS
022.053 .092
G DWARFS
COLORS
113.027
114.001
G DWARFS
ELEMENT ABUNDANCES
114.001
GEGENSCHEIN
106.004 .034
GEMINIDS
104.034
GENERAL RELATIVITY
TESTS
066.016
094.189
GEODETIC ASTRONOMY
046.000
GEOMAGNETIC FIELD
084.000
GEOPOTENTIAL
081.005 .022 .023
.036 .042 .043
.044
G GIANTS
ATMOSPHERES
064.021
G GIANTS
ELEMENT ABUNDANCES
114.088
G GIANTS
IRON ABUNDANCE
114.004